ENERGY TRANSFER PARAMETERS OF AROMATIC COMPOUNDS

ENERGY TRANSFER PARAMETERS OF AROMATIC COMPOUNDS

Isadore B. Berlman

MICROWAVE DIVISION
RACAH INSTITUTE OF PHYSICS
HEBREW UNIVERSITY
JERUSALEM, ISRAEL

ACADEMIC PRESS New York and London 1973

A Subsidiary of Harcourt Brace Jovanovich, Publishers

ACADEMIC PRESS, INC.
111 Fifth Avenue, New York, New York 10003

United Kingdom Edition published by
ACADEMIC PRESS, INC. (LONDON) LTD.
24/28 Oval Road, London NW1

Library of Congress Catalog Card Number: 72-12191

PRINTED IN THE UNITED STATES OF AMERICA

Dedicated to my father and mother to whom I owe so much

CONTENTS

PREFACE

Following the appearance of the fluorescence and absorption data contained in both editions of my previous book, ("Handbook of Fluorescence Spectra of Aromatic Molecules," Second Edition, Academic Press, 1971), I received requests to develop the data now contained in this work. The amount of data is so massive it could not be incorporated in the second edition of above handbook. Thus this separate publication was planned.

The volume has been written in an extensive rather than an intensive fashion and is intended to supplement other existing work already available to workers in the field. A short historical sketch is provided to give the reader a proper perspective of some of the concepts. References are copiously given to provide sources for more detailed information. Because of the logjam of publications, some relevant references may have been inadvertently omitted. For this we apologize in advance.

I wish to thank Professors W. Low, and S. G. Cohen, Racah Institute of Physics, Hebrew University and Dr. R. E. Rowland, Director, Radiological and Environmental Research Division, Argonne National Laboratory for their support of this project. I also wish to thank Dr. F. Hirayama for suggesting these calculations, Dr. M. Inokuti for stimulating discussions, Mrs. C. A. Yack for typing assistance, and my wife Saralee for assistance in compiling the author index.

1

INTRODUCTION

Many reactions in radiation and photochemistry, physics, and biology depend on the transfer of electronic energy. That radiationless energy transfer can occur in the gas phase, in liquid and solid solutions, and in crystals further attests to the generality of the phenomenon. Thus, any advance in the understanding of the energy-transfer process will find immediate application in many areas of research.

The number of publications on electronic energy transfer has been increasing dramatically. Some assistance in coping with the ever-increasing supply has been afforded by a concomitant proliferation of review papers; to list a few, Livingston (1957), Ganguly and Chaudhury (1959), Ermolaev (1963), Wilkinson (1964), Jortner *et al.* (1965b), Windsor (1965), Wilkinson (1966), Bennett and Kellogg (1967), Förster (1967), Lamola (1969a), Cundall (1969), and Wagner (1971). A yearly bibliography on energy transfer and related subjects has been assembled by Lipsett (e.g., 1967, 1968). The well known books of Parker (1968) and Birks (1970b) are also recommended.†

The tabulated data in this monograph are restricted to a specific type of nonradiative mechanism, a dipole–dipole interaction. This process may be

† However, this author strongly disagrees with claims made by Birks concerning quantum yield and decay time (Birks, 1970b, p. 104, 109).

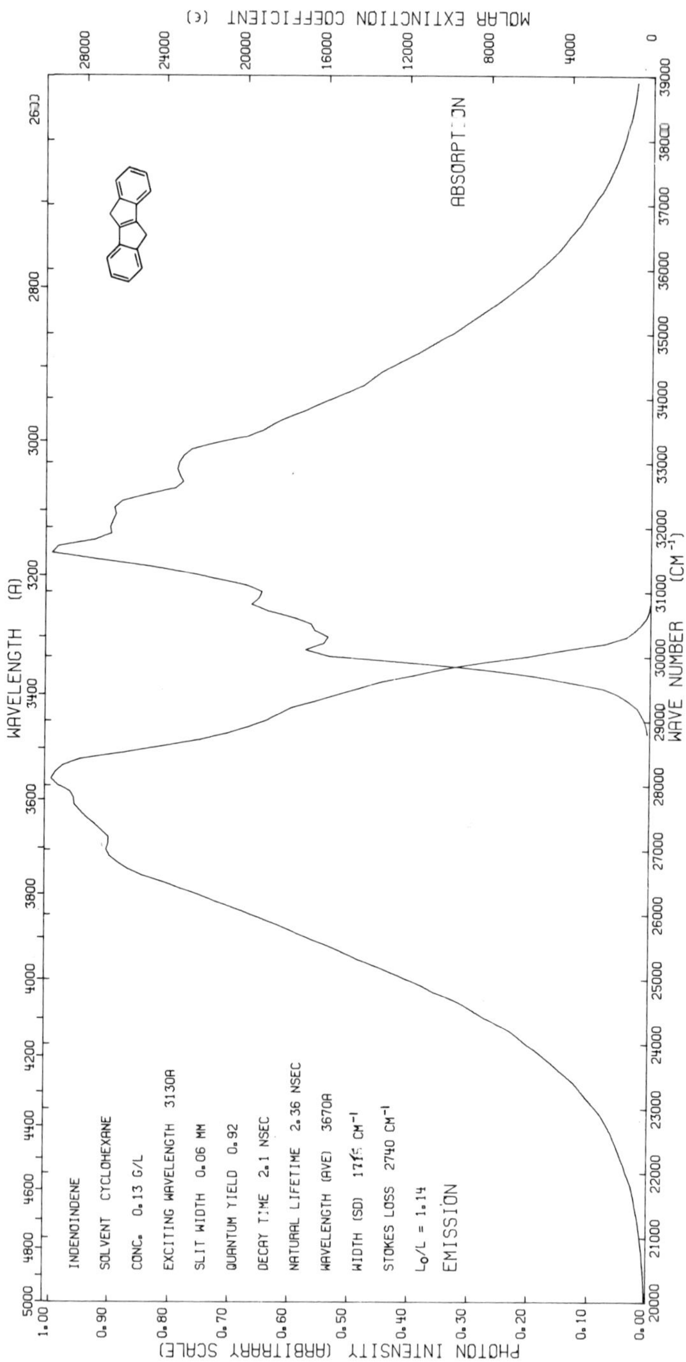

FIG. 1. The absorption and fluorescence characteristics of indenoindene. This compound is number 210.

represented by the equation

$$D^* + A \rightarrow D + A^*,$$

where D and A refer to donor and acceptor molecules, respectively, and an asterisk represents a molecule excited to its first excited singlet state. The donor and acceptor molecules can be like or unlike chemical species. In the initial condition, the donor is electronically excited to its first excited singlet state; in the final condition, the excitation energy has been transferred to the acceptor. For purposes of completeness, other processes and mechanisms are discussed in the text, but no data by the author are presented in their support.

Förster (1949) has suggested that certain standard reference parameters called R_0 and C_0 are useful in evaluating the effectiveness of resonance energy transfer between different compounds in a rigid or viscous medium. The quantity R_0 is defined as a mean distance, in angstroms, between a donor molecule and an acceptor molecule where the probability for resonance energy transfer is equal to the probability for emission. A related parameter C_0 is the concentration of acceptor molecules in moles per liter wherein there is, on the average, one acceptor molecule inside a sphere of radius R_0. The tabulation of these quantities (plus an overlap integral to be defined later) in a meaningful manner is the prime concern of this publication.

The first 209 compounds are listed in the order found in the "Handbook of Fluorescence Spectra of Aromatic Molecules, Second Edition" (Berlman, 1971). Compound 210, indenoindene, was not available at the time of the above publication and is included here as Fig. 1.

Cyclohexane is the standard solvent. In those cases where solubility is a problem, other solvents such as alcohol or benzene are used. The type of solvent employed is indicated by a letter after each graph number: an A for alcohol, a B for benzene, and a C for cyclohexane.

All of the spectroscopic data (Berlman, 1971) were taken on solutes that were soluble in cyclohexane. Solutes that were soluble in other solvents were studied only partially, i.e., their absorption and fluorescence spectra were measured but not their quantum yield. Therefore in the calculations of R_0, only the former compounds were employed as both donors and acceptors; the latter compounds were used only as acceptors—their fluorescence quantum yields were not available.

The short-wavelength portion of the fluorescence spectrum is sometimes distorted by a process called self-absorption. This distortion is most prominent when the spectra overlap appreciably; it is minimized by using very dilute solutions. The fluorescence spectra of compounds in trace concentrations, when available, were therefore used in the calculation of R_0.

2

SPECTROSCOPIC CONCEPTS

2.1 Energy Levels

Aromatic compounds possess electrons, called π electrons, that occupy nonlocalized molecular orbitals. It is the transitions of these electrons from one orbital to another that produce the phenomena of absorption and fluorescence. Although π electrons contribute to the chemical binding, they are not primarily concerned with holding the molecular framework together.

Since the ground state of these compounds, S_0 in Fig. 2, is a singlet state, this means that their lowest molecular orbitals have paired electrons of antiparallel spin. The small arrows in the boxes at the extreme left in the figure represent such electrons. In the absorption of ultraviolet radiation one of the π electrons in a filled orbital is raised to a higher unfilled orbital, as shown by a solid arrow pointing upward in the figure. When the spin direction is retained, the molecule ends up in an excited singlet state, and when reversed, in a triplet state. For each excited electronic singlet state there is a corresponding triplet state of lower energy. When excitation is by uv the probability of a singlet–singlet transition is many orders of magnitude higher than that of a singlet–triplet transition. When excitation is by means of an ionizing particle, the probability of populating a triplet state relative to a singlet state is a function of the energy of the particle. The upper states are termed π, π^* states and the transitions, π–π^* transitions. With each electronic state there are associated vibrational and

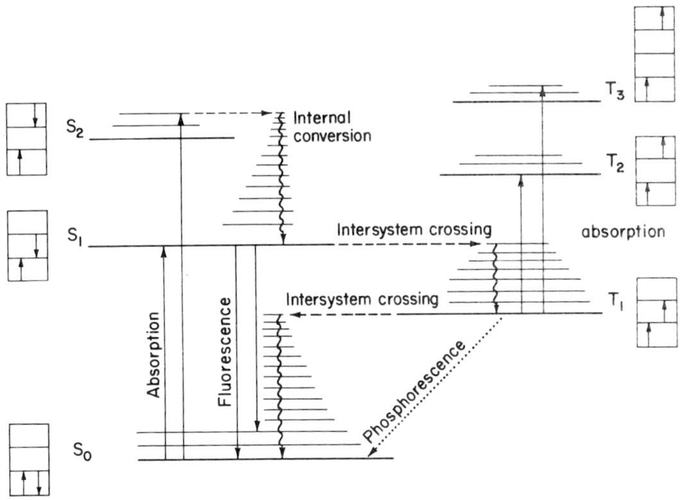

FIG. 2. An energy-level diagram that illustrates the relative spacing of electronic and vibrational levels. The large arrows depict the directions of the various transitions.

rotational states. In Fig. 2, several electronic and vibronic levels but not rotational levels are depicted. It should be noted that the energy interval between the ground level and first excited electronic level is much larger than the interval between successively higher levels.

At room temperature most of the molecules are found in their zero vibrational state, so that in the absorption process an electron is raised from the zero vibrational state of the ground state to one of the higher electronic and vibrational states, depending on the energy of the absorbed photons. The pathways of deexcitation can be by internal and external processes, as explained below. In a condensed medium, excess vibrational energy can be transferred as vibrational quanta to the medium by collisions (external process) and the compound is cooled to its lowest vibrational level (zero) in about 10^{-11} sec. In other words, the vibrationally hot molecule is brought into thermal equilibrium with its environment by collisions. This vibrational relaxation is shown as a wavy line in Fig. 2. If the molecule is in the zeroth vibrational state of an upper electronic state, it can be transformed into a molecule with a lower excited electronic state, but with high vibrational energy by a process called internal conversion (internal process). Here, too, the vibrationally rich molecule is cooled by interacting with the medium. Because internal conversion is adiabatic, i.e., involves no net energy change, it is depicted by a horizontal dashed arrow in the figure.

When a molecule ends up in the zero vibrational state of its first excited state, it can remain in this state for relatively long times, i.e., nanoseconds, because the energy gap between this state and the ground state is relatively large so that the (competing) rate constant for internal conversion is relatively small. The molecule can lose its remaining excitation energy by radiation, called fluorescence, by energy transfer to another molecule, or by converting adiabatically to the triplet manifold, called intersystem crossing. In the latter case, the triplet–state molecule donates its surplus vibrational energy (if any) to the environment and ends up in the zero vibrational state, from which it can either produce radiation, called phosphorescence, or transfer its energy to another molecule, or, under the proper conditions, cross back to the singlet manifold by intersystem crossing, as shown in the figure. Kotsubanov *et al.* (1968) have shown that the triplet manifold forms a relatively isolated system of states and that upon excitation to a high triplet state relaxation proceeds to the lowest triplet state. There is little or no crossing from a higher triplet state to the singlet manifold.

In summary, internal conversion takes place between states of like multiplicity and intersystem crossing between states of different multiplicity. The first step in each is adiabatic, followed by vibrational relaxation.

Whereas absorption takes place from the zero vibrational level of the ground state to one of the vibrational levels of an upper state, fluorescence and phosphorescence both take place from the zero vibrational level of the excited state to one of the vibrational levels of the ground state. In general, fluorescence is defined as a transition between states of like multiplicity, i.e., singlet–singlet or triplet–triplet transitions, and phosphorescence is between states of different multiplicity, i.e., triplet–singlet. In our study, fluorescence always (with the exception of azulene) involves a transition from the first excited singlet state to the ground state.

The natural fluorescence lifetime for organic molecules is usually in the range from 10^{-7} to 10^{-9} sec. As discussed later, the fluorescence quantum yield is unity when the competing processes for the excitation energy are weak, and the measured decay time is then equal to the theoretical natural lifetime. When there are effective competing processes, the measured decay time is less than the theoretical value, and the quantum yield is less than one.

Phosphorescence is a much slower process. Because of the spin-forbidden character of the transition, the natural phosphorescence lifetime is many orders of magnitude longer than that of fluorescence, being sometimes many seconds. Here, too, competing processes reduce the measured decay time. It is of interest to note that intersystem crossing from an *excited* singlet to the triplet manifold is much faster, having a typical rate constant of about 10^7 sec^{-1}. Some interesting theoretical work (Hunt *et al.*, 1962; Robinson and Frosch, 1963; and Siebrand, 1967) has been generated to explain the dependence of the rate constant on the singlet–triplet energy gap.†

Because of the slowness of the phosphorescence transition, even a small amount of impurity may effectively quench the excited state. To minimize collisional deactivation, measurements of spectra and decay time are normally made on compounds supported in a rigid medium.

Atoms of high atomic weight (high Z) increase the probability of an excited singlet molecule ending up in its excited triplet state. By increasing the spin–orbit interaction such atoms increase the rate of intersystem crossing from the singlet to the triplet manifold and reduce the fluorescence quantum yield. Although these heavy atoms are most effective when they are component parts of the molecule, they are also efficacious as components of the solvent or matrix.

Molecules such as benzophenone and pyrazine contain nitrogen and oxygen atoms that have lone pair electrons (designated n or nonbonding). One of these n electrons may be promoted to an unoccupied π orbital producing an n, π^* state that often is the lowest excited singlet state. Since the fluorescence transition is from this state, and the transition is a very weak one and the radiative lifetime is very long. This long lifetime renders the excited molecule susceptible to radiationless processes, with the result that conversion to the ground state or to the lowest triplet takes place at the expense of fluorescence. The fluorescence quantum yield of the forementioned compounds is of the order of 1000 times less than some of the well known and efficient fluors. The phosphorescence yield, on the other hand, is very high.

In the early literature there was some discussion as to the type of radiation involved in the fluorescence process. Was it electric-dipole, electric-quadrupole, or magnetic-dipole radiation? To answer this question Selenyi (1939) studied the angular distribution of the fluorescence radiation from fluorescein by means of a wide-angle interference technique which he developed (Selenyi, 1911). He ably demonstrated that fluorescence radiation was electric-dipole radiation. Weissman and Lipkin (1942), using a similar technique, were able to conclude that phosphorescence radiation from fluorescein can also be classified as electric-dipole radiation. More

† One may explain this large difference in rates by the fact that in one case the excited molecule has to create a phonon of small energy and in the other case it has to create a photon of large energy.

recently, Bücher *et al.* (1967) improved upon the above technique by a very clever scheme employing monomolecular layers of fluorescent material. They recorded the angular distribution of radiation from these layers and again established the electric-dipole character of the radiation.

In weakly bound molecular crystals such as anthracene, the exciton bands correspond closely to the lowest excited states of the constituent molecules in an isolated state. Davydov (1948, 1951, 1962) was the first to apply Frenkel's exciton theory to the calculation of energy states of molecular crystals. He (Davydov) envisaged a weak coupling model of a molecular crystal in which the intermolecular interaction energy is small compared to the intramolecular energy. Under these assumptions, crystallization does not appreciably affect the electronic structure of a molecule and the broad features of the absorption and fluorescence spectra remain essentially unchanged. One of the significant results of his calculations is that for each excited level of an isolated molecule there exists a band of energy levels in the crystal. In addition, corresponding to a transition in the free molecule, several optical transitions may take place in the crystal, even though the level is nondegenerate in the isolated case. This effect is called Davydov splitting.

In crystals, excitation is not confined to a single molecule but may be spread over many molecules. The degree of delocalization of the exciton is dependent on the strength of the mutual interaction between the molecules composing the crystal.

The following effects of the crystalline medium on the manifold of electronic states have been proposed (Jortner *et al.*, 1965b).

A. Spectral shifts. The absorption and fluorescence spectra are shifted to the red.

B. Removal of electronic degeneracy. The crystal field leads to a splitting of degenerate electronic states.

C. Violation of selection rules. Symmetry-forbidden electronic transitions which are vibronically induced in the isolated molecule become formally allowed in the crystal. Typical examples are transitions to the $^1B_{2u}$ and $^1B_{1u}$ states of benzene.

D. The Davydov splitting. (Mentioned above).

E. Crystal field mixing. Configuration interaction between exciton states arising from crystal field perturbation is of importance in determining the intensities and splitting of the Davydov components.

F. Appearance of excited states of the crystal that are not observed in the isolated molecule.

G. Effects of exciton trapping on the emission spectrum. When the rate of intermolecular electronic excitation transfer is less than the rate of formation of an excited deformed region in the crystal, excitation trapping will occur. The formation of a localized exciton may lead to a marked change in the emission spectrum.

2.2 Energy Transfer

The term *energy transfer* is used to describe the transfer of electronic excitation energy from one molecule to another, generally of another species. When transfer is between identical molecules it is called energy migration. Energy transfer or migration is usually between separated molecules, i.e., intermolecular, but similar processes can occur in an intramolecular fashion, e.g., excitation migration between segments of an aromatic polymer.

The mechanisms involved in energy transfer are often divided into radiative and nonradiative processes. A radiative process involves the physical emission of a photon by a donor molecule and its subsequent absorption by an acceptor molecule. All other energy-transfer mechanisms are nonradiative processes.

Inductive resonance transfer is a nonradiative transfer process and is based on classical concepts. Since light emission by an atom or molecule can be attributed to the movement of electric charges, an excited donor is assumed to have a field about itself. If a second oscillator is placed in this varying electric field, it (the oscillator) will be induced to oscillate, i.e., energy will be transferred. Maximum transfer occurs at resonance, when the frequency of the oscillation in the donor is equal to that of the acceptor. In the quantum-mechanical model, it still remains true that for efficient energy transfer, the frequencies of the transferring and receiving oscillators should be matched. In practice, energy transfer is observed when there is some overlap between the fluorescence spectrum of the donor and the absorption spectrum of the acceptor: the greater the overlap, the faster is the transfer rate and the further is the transfer distance. Transfer distances are generally much larger than molecular diameters. More detailed information may be found in Chapter IV.

2.3 Material Diffusion or Collisional Transfer

In gases and liquids, molecules are more mobile than in solids and are able to migrate (diffuse) and encounter and collide with each other. Many collisions may be associated with each encounter. Collisional transfer means an excitation transfer process in which a close approach (van der Waals diameters) of the donor and acceptor molecules is required. If every encounter between an excited donor and an acceptor leads to transfer, the transfer is said to be diffusion controlled. As an example, the quenching of excited singlet states of aromatic compounds by molecular oxygen has been found to be diffusion controlled (Berlman and Walter, 1962). Viscosity is an important parameter in material migration. Diffusion-controlled rate constants for benzene and cyclohexane are approximately 10^{10} liter mole^{-1} sec^{-1}.

2.4 Energy Migration

When the concentration of donor molecules is very high, as in the case where the solvent or crystal is the donor, multistep excitation migration from donor to donor can occur before the energy is transferred to the acceptor. In the case of crystals, the energy can travel thousands of angstroms is this manner (Klöpffer, 1969a).

2.5 Solvent and Host Effects

In a liquid solution, the position and structure of the absorption and fluorescence spectra of a solute are affected by the solvent, the displacement being related to the dielectric constant and the index of refraction. Hydrogen bonding is a common phenomenon in solutions composed of a polar solvent and solute, either of which contains a hydroxyl group (OH) or an amino group (NH_2). The solvent can also perturb the fluorescence decay time, e.g., the decay time of TMPD in ethanol is about twice that in a nonpolar solvent (Berlman, 1971). Any modification of the fluorescence characteristics of a molecule will also affect its transfer efficiency.

Lamola (1969a) has pointed out that the solubility of a donor or acceptor can be an important parameter: a "poor" solvent may lead to an astatistical distribution of molecules and result in transfer rates and efficiencies that are atypical.

It has been known for a long time that an excited solute molecule may interact with an unexcited molecule of the same species in the form of concentration quenching, fluorescence depolarization, or excimer formation. An excimer is a transient excited dimer that emits radiation whose spectrum is structureless and shifted about 6000 cm^{-1} to the red of the monomer emission. The process of excimer formation and decay may be written as

$$S^* + S \rightleftarrows (SS)^* \rightarrow S + S + h\nu.$$

On the other hand, a transient excited dimer composed of *unlike* species is sometimes called an exciplex. The excimer interaction becomes more prominent as the concentration of the solute is increased and the temperature decreased. Here, too, certain characteristics of the solvent such as viscosity and polar strength may play a role in affecting the strength and stability of the interaction.

The formation of excimers from an excited molecule and an unexcited one can be employed as a cogent example of the fact that the attractive forces of a molecule generally become stronger when it becomes electronically excited. The chemical behavior of the electronically excited species indicates that it is electrophilic in nature. In classical terms an excited molecule has a positive core that is less shielded than when in the ground state. Thus, the excited species exhibits a greater attraction (or larger cross section) for other molecules than does the same species in the ground state.

In a crystal, the host material may sometimes participate in the transfer of energy between the donor and the acceptor. Hirota and Hutchison (1965) believe that there are cases where the donor transfers its energy to the host material, and this excitation travels by exciton migration until it meets an acceptor trap and transfers its energy to it. They found that in a biphenyl host containing phenanthrene-d_{10} and naphthalene as guest molecules, the phenanthrene-d_{10} transferred triplet energy to the biphenyl crystal which in turn transferred the energy to the naphthalene. Since this was so, even though the triplet energy of the host was greater than that of the phenanthrene-d_{10}, the authors concluded that the transfer process received assistance from phonons.

2.6 Phenomenological Processes

The following processes (except for V) have been reported in the literature and will be discussed individually.

I.	$S_D^1 + S_A^0 \rightarrow S_D^0 + S_A^1$	(singlet → singlet)
II.	$T_D^1 + S_A^0 \rightarrow S_D^0 + T_A^1$	(triplet → triplet)
III.	$T_D^1 + S_A^0 \rightarrow S_D^0 + S_A^1$	(triplet → singlet)
IV.	$S_D^1 + S_A^0 \rightarrow S_D^0 + T_A^1$	(singlet → triplet)
V.	$S^1 + T^1 \rightarrow S^0 + S^2$	(singlet + triplet → higher singlet)
VI.	$S^1 + T^1 \rightarrow S^0 + T^2$	(singlet + triplet → higher triplet)
VII.	$S^1 + S^1 \rightarrow S^0 + S^2$	(singlet + singlet → higher singlet)
VIII.	$T^1 + T^1 \rightarrow S^0 + T^2$	(triplet + triplet → higher triplet)
IX.	$T^1 + T^1 \rightarrow S^0 + S^1$	(triplet + triplet → singlet)
X.	$S^1 + S^0 \rightarrow T^1 + T^1$	(singlet → triplet + triplet)
XI.	$e + R^+ \rightarrow R^* \rightarrow R + h\nu$	(ion recombination)

The subscripts D and A refer to the donor and acceptor, respectively, 1 and 0 designate the first excited state and the ground state, and R* represents a molecule excited to its singlet or triplet state. Processes V through X do not have subscripts because they can take place with either like or unlike species. Sometimes, several of these processes may occur concurrently and/or successively in regions of high excitation density, as along an ionized track.

Under proper conditions, processes I, III, V, VI, and VII (those processes in which the transition in the acceptor is allowed) involve primarily a nonradiative resonance transfer mechanism. This type of transfer is long range, with transfer distances to about 50 Å. In processes II, IV, VIII, IX, and X, transfer is primarily by means of an exchange mechanism, and is short range, <15 Å.

Process I is called both singlet–singlet† transfer and sensitized fluorescence. This process can take place by means of radiation transfer, by radiationless transfer, or by collision transfer. For emphasis, let us repeat that the data assembled in this handbook pertain exclusively to nonradiative resonance transfer.

Historically, Perrin (1925) observed fluorescence depolarization in viscous solvents and explained his observations by process I. He was the first to invoke classically a long-range singlet energy transfer to explain his results. A quantum-mechanical interpretation of the phenomenon was made by Perrin (1932). More recent experimental work on energy transfer may be found in the following references: Cohen and Weinreb (1956), Basile (1964), Bennett (1964a), and Powell (1971). This process is discussed in more detail in Chapter IV.

Process II is called both triplet–triplet transfer and sensitized phosphorescence. Terenin and Ermolaev (1952) were the first to observe that when a system containing naphthalene and benzophenone was irradiated with 3650-Å radiation, a wavelength absorbed only by benzophenone, phosphorescence radiation from naphthalene was detected. In a subsequent experiment (Terenin and Ermolaev, 1956), conducted at 77°K and in glasses, the donor yield was studied as a function of the acceptor concentration and it was concluded that the results could be explained by a formula developed by Perrin (1924):

$$I_0/I_{\mathrm{A}} = e^{\alpha C_A}$$

† The notation singlet–singlet or triplet–triplet is employed in the text so as to be consistent with the literature. In the listing of the processes at the beginning of this section, the notations singlet → singlet and triplet → triplet are used to differentiate them from singlet + singlet or triplet + triplet processes.

where I_0 is the donor yield in the absence of acceptor, I_A is the yield in the presence of acceptors in a concentration C_A, and α is a constant. This formula is based on the concept of a "sphere of action" around an excited molecule. When an acceptor molecule is within this sphere, transfer is immediate and absolute. Outside this sphere, the probability of transfer is zero. When this equation is used, it is assumed that there is no appreciable change in the relative positions of the molecules during the lifetime of the excited state and that there are no association effects that interfere with the statistical distribution of the molecules. Because the interaction is instantaneous in this formulation, the donor decay time should be independent of acceptor concentration; the early investigators found it so. This constancy of decay time with a change in acceptor concentration may have been fortuitous because the phosphorescence decay time is long, sometimes seconds, and the transfer times may be very short, nanoseconds. As an example, if the time constant of the decay-measuring apparatus is longer than a millisecond, the apparatus will respond to the long component but not to a component that is less than a microsecond.

Triplet–triplet energy transfer has been observed in all possible states of organic compounds: vapor (Ishikawa and Noyes, 1962); liquids (Bäckström and Sandros, 1958); solid solutions and crystals (Ermolaev, 1963); and polymers (Eisenthal and Murashige, 1963). Generally, the experiments have been performed on cooled samples. Recently, the phenomenon has been illustrated in a solution at room temperature by measuring the triplet–triplet absorption of the acceptor after it has been sensitized by a donor (El-Sayed and Bhaumik, 1963).

In the triplet–triplet transfer process an exchange mechanism is operative so that the quenching radii are small, about 11–15 Å. A dipole–dipole interaction is negligible because of the spin-forbidden transition in the acceptor. Transfer by means of higher multipole interactions can exist at short range, but according to Dexter (1953) the probabilities for such processes should be small compared to the exchange mechanism when the latter is allowed by spin selection rules.

Smaller *et al.* (1965) have shown by an ESR experiment that triplet–triplet transfer from phenanthrene to naphthalene-d_8 is by means of an exchange mechanism and is diffusion controlled at low viscosities. Porter and Wilkinson (1961) used a flash photolysis technique to determine the rate of triplet–triplet transfer. Their technique involved the measurement of the rate of quenching of the triplet donors. They found that the quenching constant fell into three categories when the experiment was conducted in fluid solution and at room temperature:

1. When the triplet level of the donor is higher than that of the ac-

ceptor by at least 3 kcal/mole, the transfer rate is diffusion controlled, i.e., every encounter leads to transfer. Nordin and Strong (1968) disagree with these limits.

2. As the difference between triplet levels falls below 3 kcal/mole, the transfer rate drops quickly and becomes two orders of magnitude less than the diffusion-controlled rate as the triplet states become isoenergetic.

3. When the triplet level of the acceptor is above that of the donor by more than 3 kcal/mole, the quenching rate constant falls to a value of more than 10^6 less than that of the diffusion-controlled rate. These differences in transfer rates are related to the differences in spectral overlap between the two species.

When the triplet energy level in the donor is below that in the acceptor, it is still possible to activate the process thermally. Liu (1968) and Lamola (1970) have observed this reversible triplet–triplet energy transfer in two separated chromophores.

El-Sayed *et al.* (1969) found a conservation of spin direction and production of spin alignment in triplet–triplet energy transfer. The results of Eisenthal (1969) concur with those of the above authors.

The feasibility of energy transfer from higher triplet states was proposed by Terenin *et al.* (1966) and by Brocklehurst *et al.* (1966). This transfer process is just an extension of process II and may be written as

$$T_D^2 + S_A^0 \rightarrow S_D^0 + T_A^1.$$

Direct experimental evidence of transfer from higher triplet states was provided by the spectroscopic evidence of Kobyshev *et al.* (1966), by the chemical studies of Liu and Edman (1969), and the spectroscopic studies of Liu and Kellogg (1969) and Alfimov *et al.* (1969). A quantitative study of transfer from higher triplet states has been made by Alfimov *et al.* (1971).

Since phosphorescence radiation from large organic molecules takes place because spin–orbit coupling produces some singlet–triplet mixing of states (Weissman and Lipkin, 1942), it would seem that triplet–triplet transfer might also involve a dipole–dipole interaction. From the published experimental values of R_0 for molecules of low atomic number, this mixing appears to be too weak for this mechanism to be important.

Process III, triplet–singlet transfer, was first predicted by Förster (1948) and observed by Ermolaev and Sveshnikova (1962). In the latter experiment the reduction of the phosphorescence decay time of the donor triphenylamine as a function of the concentration of an acceptor was measured. Certain acceptors such as chrysoidin, chlorophyll, and pheophytin were chosen because they yield little or no phosphorescence and process II was obviated. Pulsed excitation was employed and the compounds were dissolved in rigid organic glasses at 93°K. In a subsequent experiment, the intensity of the "delayed" fluorescence of the acceptor, as well as the donor lifetime, were measured (Ermolaev and Sveshnikova, 1963) and a correspondence between the two was found. Process III is possible even though spin is not conserved because all electronic states are impure, each having admixtures of other types (singlets and triplets).

Lamola (1969a) has pointed out that the efficiency of transfer $\tau_D k_t$, where τ_D is the donor decay time and k_t is the rate constant for transfer, is often more important than the transfer rate k_t alone. If f_D is the oscillator strength of the donor transition, it then follows that $\tau_D \propto f_D^{-1}$ and $k_t \propto f_D$, and the efficiency of transfer $\tau_D k_t$ should be relatively independent of the nature of the donor transition. [See Eq. (2) in Section 4.1]. In other words, in an energy-transfer process involving a symmetry- or spin-forbidden transition in the donor, the weakness of the transition is compensated for by the long lifetime of the excited donor. It then follows that process III should be as efficient as process I, and it usually is.

Quantitative measurements of this process were made by Bennett *et al.* (1964) by measuring the phosphorescence decay time for the donor, phenanthrene-d_{10}, for various concentrations of acceptor, rhodamine B. The compounds were incorporated in an inert plastic film and pulsed excitation was employed. The large experimental value of 47 Å for R_0 from decay-time data is in good agreement with their calculated value of 45 Å from spectral overlap data. Process III was also used as a method of measuring the phosphorescence efficiency of aromatic molecules (Kellogg and Bennett, 1964).

A detailed kinetic analysis of triplet–singlet energy transfer has been carried out using triphenylene-d_{12} as donor and rhodamine B as acceptor (Kellogg, 1967). His results indicated that in his experiment transfer was pure dipole–dipole transfer.

Singlet–triplet energy transfer (process IV) in liquid solution has been reported by Ermolaev and Sveshnikova (1970) and Berlman (1973). The original evidence for this process comes from the work of Terenin and Ermolaev (1956) in which they observed that the fluorescence of anthracene derivatives was quenched by naphthalene. In the work of Ermolaev and Sveshnikova the quenching of the fluorescence and the reduction in the decay time of 9,10-di-X-anthracene (where X is *n*-propyl, chloro, or bromo) in toluene at 293°K by naphthalene, fluorene, diphenyl, and stilbene was measured. The quenchers were chosen as to have their lowest triplet and singlet levels higher than those of the anthracene derivatives. Fluorene proved to be the most effective quencher and 9,10-dibromoanthracene the most easily quenched.

In the experiment of Berlman (1973), a study was made of the quenching of the fluorescence state by bromobenzene and ethylbromide. The high-Z atom produces a breakdown of the spin-selection rules so that when bromobenzene is the quencher, two normally forbidden processes become more allowed: these are singlet–triplet transfer (when energetically favorable) and intramolecular intersystem crossing. When ethylbromide is the quencher, only intramolecular intersystem crossing is the path of deactivation. Since steric hindrance interferes strongly with the efficiency of this quenching process, a close encounter is necessary for quenching to occur.

Vorobev and Shpak (1971) believe that in a three-component frozen solution consisting of α-iodonaphthalene, toluene, and biphenyl they have observed singlet–triplet transfer. The singlet excitation energy from the donor α-iodonaphthalene is transferred to the triplet state of toluene and from there to the triplet state of biphenyl.

Experimental work related to process V, the interaction between an excited singlet and a triplet to produce a more highly excited singlet, does not appear to be reported in the literature. One of the difficulties has been that most compounds do not readily emit from their second excited electronic state. Although azulene does emit from its second excited singlet state, it still may not be a candidate to test process V because the measured lifetime of the first excited singlet state is only a few picoseconds (Rentzepis, 1969).

Process VI, the interaction of excited singlet and triplet state molecules to produce a more highly excited triplet, was studied by Bennett (1964b). He was able to demonstrate that the fluorescence from perylene was quenched by resonance transfer to the lowest triplet state of phenanthrene-d_{10}. The compounds were incorporated in cellulose acetate film and the experiment was performed at 77°K. The very long lifetime of the triplet state of phenanthrene-d_{10} makes it possible to achieve a suitably high concentration of triplets. Transfer is facilitated because there is an appreciable overlap between the fluorescence spectrum of perylene and the triplet–triplet absorption spectrum of phenanthrene-d_{10}. The observed quenching of the perylene fluorescence as a function of the triplet acceptor concentration corresponds to what is predicted by resonance transfer theory (Förster, 1948).

Process VII involves an interaction between two singlet excited molecules to generate a molecule in a higher excited singlet state. This process leads to the quenching of the fluorescence intensity without a comparable reduction of the fluorescence decay time. Process VII may be observed in regions of a high concentration of excited singlet states, as along an alpha track in a scintillation solution or when excitation is by means of a strong laser beam.

Tolstoi and Abramov (1967) report that the fluorescence quantum yield of anthracene decreases at high light intensities. Their explanation is that two singlets interact to produce a single highly excited molecule and a ground-state molecule, with a resultant loss of fluorescence efficiency (process VII). One of the conclusions of the above authors is that the overlap between the fluorescence and absorption spectra of two excited anthracene molecules is greater than between an excited and an unexcited molecule. Because of this, quenching is more efficient than excitation migration.

A very large overlap between the fluorescence spectrum of perylene and the absorption spectrum of excited perylene may have been in part the nemesis that prevented Stockman *et al.* (1964) from achieving laser action with a perylene solution and being the first to build an organic-"dye" laser. Another example of process VII may be the excimer–excimer interaction in pyrene crystals and solutions at high excitation intensities (Goldschmidt *et al.*, 1968; Babenko and Benderskii, 1971) in which the fluorescence quantum efficiency decreases at high excitation intensities.

Process VIII involves an interaction between two triplets with the result that one ends up in a higher triplet state and the second in its ground state. Kellogg (1964) demonstrated that the transfer of triplet excitation from one phenanthrene-d_{10} to another triplet-excited phenanthrene-d_{10} results in the excitation of the latter to a higher triplet state. This transfer takes place with high efficiency and with transfer distances as large as 40 Å, because the triplet–triplet absorption spectrum of phenanthrene-d_{10} and its phosphorescence spectrum overlap significantly. This interaction is so efficient that it produces a depletion and an astatistical distribution of triplet states. In fact, the depletion is so effective that Kellogg did not observe any delayed fluorescence (process IX).

In process IX, delayed fluorescence is generated by triplet + triplet annihilation. It was first observed by two independent sets of investigators: by Sponer *et al.* (1958) in a naphthalene crystal, and by Blake and McClure (1958) in both a naphthalene and a phenanthrene crystal. Since then, delayed fluorescence has been observed in gases by Williams (1958), Stevens *et al.* (1960), and Hutton and Stevens (1962), in liquid solutions by Parker and Hatchard (1961, 1962), in glassy solutions by Azumi and McGlynn (1963a, b), in pure crystals by Avakian *et al.* (1963) and Hall *et al.* (1963), and in mixed crystals by Nieman and Robinson (1962), El-Sayed *et al.* (1962), and Misra and McGlynn (1966).

Singh and Lipsett (1964) have extended the experimental work of Avakian *et al.* (1963) by studying the intensity of delayed fluorescence from anthracene versus temperature over a range from 4 to 300°K. They found that the intensity is a complicated function of temperature with three pronounced maxima at 7.5, 35, and 192°K, respectively. Siebrand (1965) has suggested that these maxima are due to the longer lifetime of exciton triplets at impurities and defects in the crystal.

The intensity of the delayed fluorescence is proportional to both rate constant for triplet + triplet annihilation and the square of the triplet lifetime. Both of these quantities have been found to be virtually independent of temperature from 200 to 300°K (Singh *et al.*, 1965). At lower temperatures it has been assumed that trapping of triplet excitons would lead to a reduction in the rate of bimolecular annihilation and a comcomitant reduction in the delayed fluorescence intensity. Yet, if the trapped excitons have a longer lifetime than free triplets, the reverse may be true in a certain range of temperatures because of the square dependence of the annihilation rate on the triplet lifetime (Siebrand, 1965).

Azumi and McGlynn (1963a) found a second-order dependence of delayed fluorescence on phosphorescence intensity (or on the excitation intensity). In agreement with this finding, Stevens *et al.* (1960) observed that when the first-order triplet decay is dominant, the delayed fluorescence decays exponentially with a decay time that is half that of the phosphorescence decay.

The delayed fluorescence, measured in isotopically mixed benzene crystals (Nieman and Robinson, 1962) and in naphthalene crystals (El-Sayed *et al.*, 1962), has been interpreted by Sternlicht *et al.* (1963) as being due to a trap-to-trap migration of triplet excitons and an eventual annihilation of two such excitons.

The effect of temperature (4.5–10°K) on the steady-state phosphorescence and delayed fluorescence intensities in crystalline benzene was studied by Aurich and Marquard (1971). As the temperature is lower, two shallow traps at 5.3 and 10°K are indicated. At these values, the phosphorescence intensity is increased and the delayed fluorescence decreased. Then the following interesting experiment was made: The temperature was kept at 5°K (the region of one of the traps) and the sample was irradiated with a pulse of red light. Concurrently with this pulse, the delayed fluorescence was observed to increase dramatically and the phosphorescence to decrease abruptly. The traps are depopulated by the red-light pulse, and a pulse of uv radiation is produced by the mutual annihilation of the released excitons.

Parker (1963) proposed that mixed triplet annihilation can occur in a dilute solution of naphthacene (4×10^{-8} mole) and anthracene (5×10^{-5} mole) and produce delayed naphthacene fluorescence. A concentration of 10^{-6} mole of naphthacene alone does not produce delayed fluorescence. A detailed review of delayed fluorescence is given by Parker (1964a), one of the major contributors to this field. For more recent work by Parker and co-workers see Parker (1964b), Parker and Joyce (1966, 1967), and Parker *et al.* (1965).

In process X, a molecule that is excited to its first excited electronic state interacts with a ground-state molecule. For crystals and excitons, two triplet excitons are generated from a singlet exciton. This latter process has been called singlet–exciton fission (Pope *et al.*, 1969) and is the inverse of process IX, which has been called exciton fusion.

Crystalline anthracene has a fluorescence quantum yield of about 0.95 (Wright, 1955), whereas tetracene (also called naphthacene) has a quantum yield of only about 0.002 at 298°K (Bowen *et al.*, 1949). On the other hand, the isolated molecule in solution exhibits a fluorescence quantum yield of about 0.21 (Berlman, 1965) and as a guest molecule in anthracene has a quantum yield of unity (Bowen and Williams, 1939). Swenberg and Stacy (1968) have advanced a theoretical explanation of the low quantum yield in tetracene crystals. They suggested that the singlet states are depleted by process X, i.e., a singlet exciton generates two triplet excitons. Energetically, this mechanism is feasible at room temperature in the larger polyacene compounds such as tetracene and pentacene, but not in anthracene. In tetracene, twice the energy of the triplet level is slightly higher than the singlet energy level, so that vibrational quanta must contribute energy to make this process go. It follows that if this is a bona fide and favorable process, a reduction of temperature should dramatically inhibit this process and concomitantly increase the fluorescence. Geacintov *et al.* (1969) found that the fluorescence quantum yield of a tetracene crystal does indeed increase rapidly as the temperature is lowered.

Whereas Tolstoi and Abramov (1967) find that the fluorescence quantum yield of anthracene decreases at high light intensities, Pope *et al.* (1969) report that the quantum yield in crystalline tetracene increases with increasing light intensity. In tetracene, the triplet excitons formed from the fission process recombine to form singlets when the excitation intensity is high (process IX), but in anthracene under the same conditions the two singlets annihilate to produce a single highly excited molecule and a ground-state molecule, with a resultant loss of fluorescence efficiency (process VII).

The fluorescence efficiency of tetracene can be increased (Donnini and Abetino, 1968) by a factor of 1.4, (Geacintov *et al.*, 1969) in a magnetic field of about 2000 G, substantiating the theoretical prediction of Johnson *et al.* (1967) and Merrifield (1968). At a field strength of 4000 G, the rate of fission is decreased by the magnetic field (Pope *et al.*, 1969) and the intensity of delayed fluorescence is decreased as a result of the reduction in the number of triplets. In similar experiments by Merrifield *et al.* (1969), the effect of a magnetic field on singlet fission was investigated and similar conclusions were drawn.

The importance of an orientation factor in affecting the efficiency of the bimolecular triplet annihilation has been studied by independent groups employing different techniques (Roy and El-Sayed, 1964; Siegel and Goldstein, 1965; Judeikis and Siegel, 1970). Their results are conflicting: In the first experiment, an orientation dependence was found; in the latter two, it was not.

Srinivasan *et al.* (1968) studied the rise time and decay time of delayed fluorescence in various rigid glasses and crystals. They found that the rise time was invariably shorter than the decay time, which they explain as follows: In rigid glasses, triplet–triplet absorption depopulates the number of triplet states and decreases the rise time; in mixed crystals, on the other hand, triplet + triplet annihilation events (process IX) abbreviate the rise time.

Sometimes the singlet state is dissociative. In a theoretical study Krishna (1967) investigated a mechanism for delayed fluorescence, in which the interaction of two triplet-state molecules is immediately followed by nonradiative transitions to a dissociative singlet state. Delayed fluorescence comes afterwards. The work of Babenko and Benderskii (1970) provides some experimental evidence for this process.†

Processes leading to ionization have not been included in our tabulation because they can be considered as extensions of processes V, VI, VII, and VIII, i.e.,

$$S^2 \rightarrow S^+ + e \qquad \text{or} \qquad T^2 \rightarrow S^+ + e,$$

where S^+ is a molecular ion and e is an electron. Delayed ionization from triplet + triplet annihilation in fluid media have been observed by Kawada and Jarnagin (1966), Pilloff and Albrecht (1966), and de Groot *et al.* (1968) and identified for anthracene, naphthalene, phenanthracene, and pyrene in methyltetrahydrofuran, tetrahydrofuran, benzonitrile, and acetonitrile. On the other hand, ionization was not found in solvents such as hexane and cyclohexane in which delayed fluorescence was observed. It was concluded that the ionization potential of the above solutes is

† In atomic spectroscopy such "energy pooling" (Arnold *et al.*, 1966) between metastable argon atoms has been observed (Hurt, 1966). Since the energy of a metastable argon atom is relatively high, the probability that a collision between two excited argon atoms will lead to ionization is large. Likewise the pink afterglow in a discharge through nitrogen has been interpreted as being due to a similar mechanism (Lund and Oskam, 1969; Bromer and Hesse, 1969).

reduced sufficiently by the dielectric constant of certain solvents that ionization can be produced by the mutual annihilation of two triplets.

In an interesting experiment, Needler (1965) found that singlet energy could be transferred from a dye polymer to an aromatic amine, producing sensitized photoconduction. A build-up of the unpaired electrons was detected by and identified from ESR spectra. It is the contention of Needler that the dye–amine interaction is a nearest-neighbor interaction.

Ion recombination followed by molecules in excited states (process XI) has become more amenable to investigation with the advent of nanosecond and subnanosecond pulsed accelerators. In a detailed study (Hunt and Thomas, 1967) of the radiolysis of naphthalene and anthracene in cyclohexane and benzene in which 3- to 11-nsec pulses of 3-MeV electrons were used, the triplet states of the solutes were observed via their characteristic absorption spectra, and the excited singlet states were detected via their fluorescence. The time of formation of most of the triplet states was less than a nanosecond; a small portion, however, did grow in with a half-life of about 100 nsec. Two mechanisms were proposed for the direct production of the excited states: ion recombination and subexcitation electrons. However, in the study by Land and Swallow (1968) of the yield of triplet states in the pulse radiolysis of an aliphatic hydrocarbon glass containing naphthalene, it was deduced that the major source of triplet states (and singlets) was by ion recombination and not by subexcitation electrons.† Other references to process XI may be found in the work of Perkins (1968), Ludwig and Huque (1968), and Spurny (1970).

† The present author believes that some of the triplets formed immediately after the pulse of electrons are produced via process IV. Paramagnetic species such as radicals and anions are responsible for the breakdown of the spin-selection rules.

3

HISTORICAL SKETCH

That electronic energy could be transferred between atoms of different species and that the efficiency of transfer should be increased as the difference between the excited levels is decreased was predicted by Franck (1922) and verified by Cario and Franck (1923). A mixture of mercury† and thallium vapors at low pressure was excited by a mercury resonance line at 2537 Å and lines from mercury and thallium were emitted. Since the exciting radiation was not absorbed by the thallium atoms, these atoms must have been indirectly excited by a transfer of energy from excited mercury atoms. This experiment with mercury and thallium vapors has recently been repeated by Hudson and Curnette (1966). Energy transfer between molecules in the vapor phase was first reported for the cases of indigo sensitized by aniline (Prileshayeva, 1935) and of aniline sensitized by benzene (Prileshayeva and Klimova, 1937).

Early polarization studies have provided some of the convincing evidence that excitation energy is transferred from molecule to molecule. The work of Gaviola and Pringsheim (1924), Weigert and Käppler (1924), and Levshin (1924) are classic. More recently, Knox (1968) and Ore and Eriksen (1971) have critically reexamined this subject.

The first observation of sensitized fluorescence in solution was made by

† Mercury vapor is a very convenient material for experimental study, and a vast literature dealing with its photochemical behavior has accumulated over the years.

Perrin and Choucroun (1927). A solution containing two fluorescent dyes was irradiated with uv radiation of such a wavelength that only one of the dyes was excited and emission from the other was detected. Here, too, the second dye must receive its excitation energy from the first. Förster (1949) was able to demonstrate that in a system containing trypaflavin and rhodamine, the donor (trypaflavin) fluorescence yield was quenched as a result of nonradiative energy transfer to the acceptor. The transfer distance (R_0) was estimated to be about 70 Å. Watson and Livingston (1950) were among the first to investigate the change in fluorescence yield of both the donor and acceptor. On the other hand, the decrease in donor fluorescence decay time as a function of the solute concentration was first measured accurately by Galanin and Levshin (1951) and Galanin (1955). More recent studies of the change in the donor decay time as a function of acceptor concentration have been made by Kuznetsova *et al.* (1957), Berlman (1960), Sveshinikov *et al.* (1960), Ware (1961), and Bennett (1964a).

The first theoretical treatment based on classical physics of excitation transfer between molecules in solution was by Perrin (1927). This theory was later developed on quantum-mechanical principles for atoms by Kallmann and London (1928) and for molecules by Perrin (1932). Transfer distances of several hundred angstroms were predicted. A theoretical treatment for electric dipole–dipole energy transfer based on experimentally accessible quantities such as fluorescence, absorption, and decay time was developed by Förster (1948, 1949). His formulation was for a viscous medium where translational motion was small but where the rotational motion was faster than the transfer rate. The limits of applicability of Förster's theory of weak coupling are discussed in papers by Robinson and Frosch (1963) and Förster (1966).

A formula for dipole–dipole energy transfer similar to that of Förster was developed by Galanin (1955), although a slightly different averaging technique was employed. Extensions to the theory were as follows: multipoles of higher order such as dipole–quadrupole and quadrupole–quadrupole interaction were included by Dexter (1953). Their probability for transfer as a function of distance is as R^{-8} and R^{-10}, respectively. Energy transfer by an exchange mechanism was also developed by Dexter.

In dipole–dipole energy transfer, the theory has been extended to include transfer to other than nearest neighbors (Bojarski, 1962). Another question that has also been considered is what happens to the excitation energy after it is transferred from molecule A to molecule B? In the Förster mechanism, the energy is assumed to have the largest probability of being transferred back to A. Important contributions have been made by Ore (1959, 1960),

Eriksen and Ore (1967), and Knox (1968), who observed that while B might be the closest neighbor to A, the reverse is not necessarily true, and they suggested corrections. Expressions have been derived for the transfer yield averaged over all possible arrangements of the acceptor molecules (Golubov and Konobev, 1972). A many-body treatment has been made by Dow (1968).

The influence of molecular diffusion on energy transfer has been investigated by Kurskii and Selivanienko (1960), Bagdasaryan and Muler (1965), Voltz *et al.* (1966), Feitelson (1966), Yokota and Tanimoto (1967), Lavorel (1967), Elkana *et al.* (1968), Birks and Conte (1968), Steinberg and Katchalski (1968), Birks and Georghiou (1968), and Birks and Leite (1970). Long-range transfer to include excitation migration has been considered by Swenberg and Stacy (1969). In addition, retardation effects have been considered by Kuhn (1970) and the effects of nonuniform broadening of spectral lines on resonance energy transfer have been studied by Agabekyan (1970).

Johnson and Offen (1972) have found a dependence of energy transfer on pressure. They found that the efficiency of singlet–singlet energy transfer from pyrene to perylene in polymethylmethacrylate decreased with increasing pressure, i.e., at 77°K the value of R_0 decreased from 44 Å at 1 atm to 35 Å at 30 kbar. The factors responsible for the reduced efficiency are listed in order of decreasing importance: A decrease in the fluorescence decay time of the donor, an increase in the index of refraction of the medium, and a decrease in the spectral overlap between the pyrene fluorescence and the perylene absorption.

After World War II, the interest in nuclear physics stimulated the development of efficient devices for detecting nuclear radiation and measuring their energy. One of these was the liquid scintillation solution. Independently and almost simultaneously three groups—Ageno *et al.* (1949), Reynolds *et al.* (1950), and Kallmann (1950)—suggested the applicability of such a system to detect nuclear radiations. For about a decade Kallmann and his collaborators were most prolific in the publication of data on the scintillation efficiency of many combinations of solvents and solutes when excitation was by gamma rays and/or alpha particles, e.g., Kallmann and Furst (1950).

When an ionizing particle loses its energy in a scintillation solution, it does so by exciting and ionizing the solvent. Many initial physicochemical processes occur in a time $<10^{-10}$ sec. By measuring the emission yield of the solute as a function of its concentration, Furst and Kallmann (1954) were able to propose that energy is transferred from the solvent to the solute primarily via a nonradiative process and that the rate of energy transfer is proportional to the solute concentration. At about the same time, Cohen and Weinreb (1954) were among the first to excite a benzene-*p*-terphenyl solution with uv radiation of such a wavelength as to excite preferentially the solvent and to observe a large solute fluorescence. When the solvent was replaced with chloroform, there was very little solute fluorescence. These results were explained by assuming that energy transfer occurred from the excited benzene to the solute. Moreover, the fluorescence from a terphenyl and xylene solution was then measured as a function of the cell thickness. As the fluorescence yield was found to be independent of the thickness, it was assumed that radiative transfer did not play an important role.

In spite of the large accumulation of experimental data by 1953, the exact mechanism of energy transfer remained to be proven. Was it radiative or nonradiative? Birks (1958) maintained until a rather late date that transfer is primarily radiative. In a classic experiment devised by Bowen and collaborators (Bowen and Brocklehurst, 1953, 1955; Bowen and Livingston, 1954), the radiative component was purposely eliminated as described below. A series of benzene solutions of 1-chloroanthracene and perylene in various concentrations but in constant molar ratio was irradiated by 3650-Å radiation. Although the chloroanthracene was excited directly, it was the perylene luminescence that was monitored. A crucial point in these experiments was the choice of the donor, chloroanthracene, whose fluorescence quantum yield is much smaller than that of the donor, perylene. As the concentration of the donor and acceptor was increased in constant molar ratio, the total fluorescence yield of the system was enhanced. If transfer were first by means of radiation by the donor, and then followed by absorption by the acceptor, the fluorescence yield would not have been nearly as great. Thus, nonradiative transfer must take place between the two compounds and in a time much shorter than the other nonradiative processes, i.e., intersystem crossing, acting in chloroanthracene.

When the system *p*-xylene and PPO (2,5-diphenyloxazole) was excited by uv radiation of the proper wavelength, the decrease in the fluorescence decay time of the donor with increased solute concentration coincided with both the decrease in the donor yield and to the increase in the acceptor yield (Berlman, 1960). Moreover, in the concentration range investigated, i.e., $<10^{-2}$ mole, energy transfer was shown to take place only from the first electronically excited level of the donor. At much higher concentrations, evidence of possible transfer from higher levels has been reported by Laor and Weinreb (1965).

With the advent of high-powered lasers, these instruments were employed to excite an anthracene crystal to its fluorescent state. Fluorescence from an excited singlet state excited by a ruby laser was first observed by Peticolas *et al.* (1963) and Singh and Stoicheff (1963). Both groups attributed their success to the Goeppert–Mayer (1931) mechanism of double-photon absorption. Subsequently, a delayed fluorescence emitted from the same excited singlet state and attributed to triplet + triplet annihilation was discovered by Kepler *et al.* (1963). This latter group of scientists also used a ruby laser. The interrelationship between these two processes was clarified by Hall *et al.* (1963). At high irradiation power, the bimolecular annihilation process becomes saturated so that the fluorescence due to this effect is a linear function of the laser-beam intensity. The biphotonic process, which is proportional to the square of the laser intensity, then predominates. On the other hand, at low levels of excitation intensity, the bimolecular annihilation process also varies as the square of the excitation intensity, and its magnitude is significantly larger than that of the double-photon absorption process. Thus delayed fluorescence is the predominant process at low irradiation intensities.

4

INTERMOLECULAR ELECTRONIC ENERGY TRANSFER

4.1 Definitions

The total electrostatic interaction between excited states of the donor and acceptor molecules is usually separated into Coulomb and exchange terms (Dexter, 1953). In the Coulomb term the electrostatic potential can be replaced by a multipole expansion, the leading term of which is the dipole–dipole term. The dipole–dipole interaction is a function of distance R as R^{-3}, and the probability of transfer is the square of this quantity or R^{-6} (Eq. 1 below). This interaction term is dominant if the transitions in the donor and acceptor are allowed. On the other hand, if the transition in the acceptor is spin forbidden, the dipole–dipole term is very small and other interaction terms such as the dipole–quadrupole term and/or the exchange mechanism may be the major contributors to the transfer.

A detailed theory of electronic energy transfer has been developed by Förster (1948) in terms of a resonance dipole–dipole interaction mechanism. In the above mechanism, the transitions are coupled to each other as well as to the radiation field. It was shown by Förster that the rate constant for dipole–dipole energy transfer between a donor D and an acceptor A, i.e., $k_{\mathrm{D}\to\mathrm{A}}$ is given by

$$k_{\mathrm{D}\to\mathrm{A}} = \frac{9000K^2 \ln 10}{128\pi^5 n^4 N \tau_0 R^6} \int \frac{f_{\mathrm{D}}(\nu)\, \epsilon_{\mathrm{A}}(\nu)\, d\nu}{\nu^4}, \tag{1}$$

where ν is a wave number, $\epsilon(\nu)$ is the molar decadic extinction coefficient, $f(\nu)$ is the spectral distribution of fluorescence normalized such that $\int f(\nu)\, d\nu = 1$, the limits of integration are over the relevent fluorescence spectrum. N is Avogadro's number, τ_0 is the natural fluorescence lifetime of the donor, n is the refractive index of the solvent, and R is the distance between donor and acceptor.

$$K = \cos \phi_{DA} - 3 \cos \phi_D \cos \phi_A,$$

where ϕ_{DA} is the angle between the transition moment vectors of both molecules, while ϕ_D and ϕ_A are the angles between these respective vectors and the direction of R. Thus, for rigid systems, K^2 may have values from 0 to 4. When the molecules are free to rotate at a rate much faster than the transfer rate, the average value of K^2 is $\frac{2}{3}$ (Förster, 1951, 1959, 1960b, 1966). For the case where the orientation of the donor and acceptor is random but where the rotation is much slower than the transfer time, the average value of K^2 is 0.475 (Maksimov and Rozman, 1962; Steinberg, 1968).

As a measure of the degree of interaction between donor and acceptor molecules, a reference distance R_0 has been defined (Förster, 1949) as a mean distance in angstroms between a donor and an acceptor molecule where the probability for resonance transfer of energy between the molecules is 50%. In other words, R_0 is that distance where the rate constant for energy transfer $k_{D\to A}$ is equal to the rate constant for fluorescence by the donor in the absence of acceptors $1/\tau$. Equation (1) can now be written as

$$k_{D\to A} = \frac{1}{\tau}\left(\frac{R_0}{R}\right)^6, \tag{2}$$

and when $R = R_0$, then $k_{D\to A} = \tau^{-1}$.

In terms of rate constants, $\tau_0^{-1} = k_e$, $\tau^{-1} = k_e + k_n$, and $Q_y = k_e/(k_e + k_n)$, where τ_0 is the natural fluorescence lifetime of the donor, k_e is the rate constant for fluorescence, k_n is the rate constant for all other competing processes, and Q_y is the fluorescence quantum yield of the donor in the absence of acceptor. It follows that $\tau = Q_y\tau_0$.

From Eqs. (1) and (2),

$$R_0^6 = \frac{9000 K^2 Q_y \ln 10}{128\pi^5 n^4 N} \int \frac{f_D(\nu)\, \epsilon_A(\nu)}{\nu^4}\, d\nu. \tag{3}$$

A large value of R_0 is thus dependent on the following factors: (1) a good overlap between the fluorescence spectrum of the donor and the absorption spectrum of the acceptor, (2) a large value of ϵ_A, and (3) a large value of fluorescence quantum yield of the donor, Q_y. Because of a sixth-power dependence of $k_{D\to A}$ on R_0, a change in R_0 by just a few angstroms makes a large change in $k_{D\to A}$. In the above equation, the value of R_0 is affected by the strengths of the transitions of the donor and acceptor through the values of Q_y and ϵ_A respectively. Processes such as diffusion, donor–donor transfer, and fluorescence followed by absorption are not considered in the present definition of R_0.

Several investigators (Förster, 1949; Galanin, 1955; Kurskii and Selivanenko, 1960), using different approaches, have arrived at the same formula for the probability $P_{D\to A}$ of dipole–dipole transfer between donors and acceptors that are statistically distributed. They show that in a viscous medium,

$$P_{D\to A} = 2x \exp(x^2) \int_x^{\infty} \exp(-t^2)\, dt = \pi^{1/2} x \exp(x)^2 [1 - \mathrm{erf}(x)] \quad (4)$$

where $\mathrm{erf}(x)$ is the error function and $x = C/C_0$ and

$$C_0 = \frac{3000}{2\pi^{3/2} N R_0^3} = \left(\frac{7.66 \times 10^{-8}}{R_0}\right)^3. \quad (5)$$

The quantity C_0 may be regarded as a critical transfer concentration† corresponding to a transfer efficiency of 76%. Since

$$P_{D\to A} = 1 - \frac{Q}{Q_0},$$

where Q_0 is the fluorescence yield of the donor in the absence of acceptor and Q is the yield with acceptor, $P_{D\to A}$ will become larger as Q becomes smaller. In the presence of an acceptor, both the donor yield and the donor decay time are diminished. For the case of viscuous and solid solutions, transfer occurs preferentially between close-lying donor–acceptor pairs, with the result that the normal statistical distribution is disturbed, the resultant fluorescence decay is nonexponential, and its value is dependent on the duration of the excitation.

Eisenthal and Siegel (1964) have formulated decay laws for pulsed and steady-state excitation. They verify (see Inokuti and Hirayama, 1965) that Eq. (4) remains the same for steady-state and pulsed excitation.

As stated above and repeated here for completeness, the theory of long-

† The definition of C_0 has been changed over the years. In an earlier publication (Förster, 1948) its value was given as $C_0 = 3000/4\pi N R_0^3$, and this form, which is more general and can be used for cases other than dipole–dipole, was employed in our earlier publication (Berlman, 1971). Since Eq. (5) appears to be more useful in the dipole–dipole case (Förster, 1967), it was adopted for the present publication.

$C_0 = \left(\frac{7{,}34 \cdot 10^{-8}}{R_0}\right)^3$

range dipole–dipole transfer has been extended recently to include Brownian translation in solutions of low viscosity (Kurskii and Selivanenko, 1960; Kilin *et al.*, 1964; Yokota and Tanimoto, 1967; Elkana *et al.*, 1968, to name a few). Long-range transfer to include excitation migration has been considered by Swenberg and Stacy (1969).

The assumptions made in the development of Eqs. (1) and (4) are as follows:

1. The molecules (dipoles) are assumed to be point molecules. This is not a limitation as long as R_0 is much larger than molecular dimensions.
2. The distance between the donor and acceptor is assumed to be larger than molecular dimensions and does not change during the lifetime of the excited donor.
3. Contributions from higher multipoles and exchange interactions are neglected.
4. It is assumed that there is no medium or solvent interaction other than dielectric shielding.
5. Thermal equilibrium is assumed to be established between the vibrational modes of the excited donor before transfer takes place. The possibility of energy transfer from higher vibrational or electronic states is ignored.
6. The transitions in both donor and acceptor are assumed to be allowed.
7. The fluorescence and absorption spectra are assumed to be continuous because of collision broadening.
8. The oscillator strength of the donor transition is assumed to be independent of the distribution and number of the acceptor molecules.
9. The rates of transfer between an excited donor molecule and all the acceptor molecules are assumed to be additive.
10. In the transfer of excitation energy from a donor to an acceptor, the possibility of another acceptor being in the path of the interaction is ignored. This may be an important consideration at high concentrations or in intramolecular energy transfer in proteins.
11. Although the acceptor cannot occupy the space of the donor, no correction is made for this limitation.
12. The number of excited donors or acceptors is assumed to be much smaller than the number of unexcited donors or acceptors.
13. The excited donors are assumed to be randomly distributed with respect to the acceptors.
14. The donor–donor interaction is assumed to be negligible. This places a limitation on the donor concentration and eliminates excimers.

Generally, the presentation of the treatment of energy transfer begins with a rate equation (Förster, 1948; Dexter, 1953; Galanin, 1955). Some questions have been raised as to the validity of such a starting point (Davydov, 1968; Grant, 1971). A rebuttal has been given by Dexter *et al.* (1969). The justification presented is that as long as the vibrational relaxation of the acceptor molecule is faster than energy transfer, their treatment is valid. However, at very close distances where back transfer can compete with vibrational relaxation, the formulae may become incorrect. The latter authors also cite many experimental results that agree with their theory.

Hirayama (1963) and Basile (1964) applied Förster kinetics to the study of energy transfer in plastic scintillators, e.g., a polystyrene–tetraphenylbutadiene system. They showed that energy transfer agrees closely with Eq. (4), using the value of R_0 obtained from spectroscopic data (Eq. 3). In a similar fashion, Powell (1971) studied energy transfer in polyvinyltoluene, Volosyuk *et al.* (1971) studied energy transfer and quenching in halogen-containing scintillators, Rozman (1961) and Terskoi *et al.* (1968) studied electronic energy transfer between two solutes in rigid solutions.

The following experiments were devised specifically to test the Förster theory. The effect of radiationless energy transfer on the donor excited-state decay time in a solid system, where donor–acceptor separations are randomly oriented, has been examined by Bennett (1964a). Pyrene was used as the donor and nonfluorescent dyes as the acceptor. The fluorescence decay time of pyrene was measured as a function of the acceptor concentration, and C_0 was determined for each value of concentration. Within experimental error, all of the computed values of C_0 were constant and independent of acceptor concentration. This constancy of C_0 was considered to be strong evidence that the radiation transfer mechanism can be adequately described by a dipole–dipole interaction. More recently, fluorescence decay curves of solid solutions of pyrene and perylene in polymethylmethacrylate were measured by Mataga *et al.* (1967). From the shape of the decay curves a value of $R_0 = 35 \pm 2.5$ Å was obtained, and from Eq. (3) a value of 38 Å was calculated. Since the values agree, within the experimental error, they maintain that the Förster mechanism has been verified.

When two chromophores are separated by a molecular chain, as in the "spectroscopic ruler" experiments of Stryer and Haugland (1967), the values of R_0 are often in agreement with Förster kinetics. In progressively more spectacular experiments, Kuhn and co-workers devised "molecular sheet" experiments to test the Förster theory. Zwick and Kuhn (1962) found that if a molecular layer of donors is separated from a layer of acceptors by an inert transparent paraffin layer of greater than 100 Å, there is no nonradiative transfer. If the paraffin layer is less than 100 Å, the intensity of the donor is quenched by the presence of the acceptor layer. Drexhage *et al.* (1963) studied the variation of the acceptor yield as a function of the thickness of the paraffin layer and also found agreement with Förster theory.

As stated above, Eq. (1) refers to a system in which vibrational relaxation is much faster than energy transfer and in which the dipole–dipole interaction is predominant. In many cases, vibronic coupling reduces the symmetry of the molecule and the theoretically forbidden transitions are made partially allowed. In other cases, spin–orbit coupling introduces triplet character in the singlet states and singlet character in the triplet states and thus increases the transition moment.

If the optical transition is forbidden in the donor but allowed in the acceptor, large transfer distances are still predicted by Eq. (3), provided the yield Q_y of the donor is high. This is so because the slower transfer rate calculated from Eq. (1) is compensated by the longer lifetime of the triplet state. In other words, if processes competing for the excitation energy are weak, the transfer probability is large. By contrast, however, if the optical transition in the acceptor is symmetry forbidden so that the molar extinction coefficient is very small, the value of R_0 as calculated from Eq. (3) will generally be smaller than molecular dimensions. In this case, the dipole–quadrupole interaction or the exchange mechanism may be predominant. Even for this case, it has been shown by Dexter (1953) that transfer can occur over distances larger than molecular diameters. Finally, if the transition in the acceptor is spin forbidden so that neither dipole–dipole nor high multipole interaction leads to strong coupling, the exchange terms in the interaction operator must be considered. Since some overlap of the electronic clouds of both molecules is needed for the exchange interaction to be effective, transfer can obviously take place only over short distances, i.e., less than 15 Å.

Dexter (1953) has shown that the rate constant for energy transfer by the exchange mechanism is proportional to $\exp(-2R/L)$, where R is the donor–acceptor distance. More recently, Inokuti and Hirayama (1965) have extended the Dexter formulation and developed an expression for calculating the donor yield and decay time as functions of the acceptor concentration for transfer by the exchange mechanism. Their formalization is a generalization of the Förster (1949) treatment for dipole–dipole transfer, and their calculated data show very good agreement with the experimental data of Ermolaev (1959) on triplet–triplet transfer.

4.2 Radiation Transfer—The Trivial Process

This process has been nicknamed "trivial" because it is easy to conceive and explain, not because it is unimportant. It is a two-step process: one molecule emits a photon and another molecule absorbs the radiation, i.e.,

$$D^* + A \rightarrow D + h\nu + A \rightarrow D + A^*.$$

As there is no direct interaction between the donor (D) and the acceptor (A) molecules, the donor decay time is not affected by the process or by the concentration of acceptor molecules. For this case, the efficiency of transfer is governed by the amount of overlap between the fluorescence spectrum of the donor and the absorption spectrum of the acceptor, and by the values of the molar extinction coefficient of the absorption spectrum (acceptor). Thus, as the concentration of the acceptor is increased, the fluorescence yield of the donor is decreased but the fluorescence decay time of the donor remains unaffected. Understandably, viscosity is unimportant. Since the probability for radiative transfer falls off with distance (R) as R^{-2}, the transfer distance can be very large, often being limited only by the physical size of the container.

4.3 Radiationless Transfer

A. General

In contrast to radiative transfer, nonradiative transfer is a one-step process involving the simultaneous deexcitation of the donor and the excitation of the acceptor. Only virtual photons are considered. Since there must be an interaction between the donor and acceptor, the donor yield and decay time are both affected by the presence of acceptors.

The prevalent situation, as stated above, is the one in which vibrational relaxation is faster than energy transfer. Thus, if the temperature is not too high, energy transfer takes place between donor and acceptor molecules that are in their zero vibrational state. Because resonance transfer involves a frequency that is common to the donor emission spectrum and the acceptor absorption spectrum, it follows that the larger the spectral overlap, the greater the number of possible coupled transitions, and the larger the probability of energy transfer.

As mentioned above but repeated here for completeness, the interaction between donor and acceptor is composed principally of a Coulomb and an exchange interaction. The Coulomb term, which is the long-range term, can be represented by a multipole expansion, the first term of which is the dipole–dipole term, the second the dipole–quadrupole, and so on. If the optical transitions are allowed, the dipole–dipole term predominates and the higher terms and the exchange term can be neglected. The classic papers on this subject are by Förster (1948, 1949).

The exchange interaction is a quantum-mechanical effect arising from the symmetry properties required of electronic wave functions with regard to the interchange of the space and spin coordinates of any two electrons in the system (donor plus acceptor). This short-range term is appreciable only in the region where the electron clouds of the donor and acceptor overlap. Because the electron density outside of the boundaries of the molecule falls off exponentially, the range of the interaction is very short. This term predominates in cases where the transition in the acceptor is spin forbidden.

The exchange mechanism differs from dipole–dipole transfer in the following ways:

1. The transfer rate does not obey Eq. (4) but varies roughly as $\exp(-aC)$, where C is the acceptor concentration and a is a positive constant.
2. The change in the donor fluorescence decay time with an increase in acceptor concentration is much smaller than the corresponding change for dipole–dipole transfer.
3. The efficiency of transfer is independent of the oscillator strength of the transition in the acceptor. Ermolaev (1968) claims that this statement is true only for those cases where the oscillator strength is primarily influenced by the spin–orbit coupling.

Items 1 and 2 indicate that the exchange mechanism is of much shorter range than dipole–dipole transfer, and item 3 indicates that the transfer mechanism is not electromagnetic. The classic papers on the exchange mechanism are by Dexter (1953) and Inokuti and Hirayama (1965).

B. Vapor Phase

The transfer of electronic excitation energy between molecules in the vapor phase was first observed by Prileshayeva (1935) for an indigo–aniline system, and Prileshayeva and Klimova (1937) for a benzene–aniline system. In these early experiments the authors were more interested in anti-Stokes fluorescence than in energy transfer.

Neporent (1950) noticed that when he irradiated β-naphthylamine vapor with 2537-Å radiation, the fluorescence was enhanced when benzene was present. He ascribed the phenomenon to a transfer of electronic energy from the excited benzene, because benzene vapor absorbs at 2537 Å and an internal filter would diminish the fluorescence rather than increase it.

Acridine fluorescence can be sensitized by naphthalene in the vapor state (Terenin and Karyakin, 1951). Moreover, electronic excitation can be transferred from benzene vapor to β-naphthylamine (Dubois, 1959) and to anthracene (Stevens, 1959). Delayed fluorescence of biacetyl and its quenching by oxygen has been studied in the vapor phase by Sheats and Noyes (1955).

Parmenter and Ring (1967) have developed a flash-sensitization technique for observing directly triplet–triplet transfer between benzene and biacetyl. They arrive at a triplet decay time of about 2.6×10^{-5} sec at 300°K. The rate constant for triplet transfer is about 5.7×10^{11} cm^3 sec^{-1}, a value about three times that for singlet–singlet transfer and a value many orders of magnitude greater than that for intersystem crossing in a rigid medium at 77°K. For an acetone–biacetyl system, Heicklen and Noyes (1959) found a much slower rate constant and estimated that transfer of energy occurs once every 10–100 collisions. Ishawaka and Noyes (1962) have observed triplet–triplet transfer in a benzene–biacetyl system, and Finger and Zahlon (1969) have detected delayed fluorescence in naphthalene, anthracene, and tetracene.

A thorough treatment of energy transfer between atoms, diatomic molecules, and polyatomic molecules in the vapor phase may be found in a review article by Cundall (1969).

C. Liquid Phase

In liquids the effects of diffusion and possibly excitation migration must be considered in addition to that of resonance transfer (Brown *et al.*, 1959). The effect of diffusion on energy transfer by a dipole–dipole interaction was first treated by Kurskii and Selivanenko (1960). Only the limiting cases of high and low viscosity were treated. The approach of Antonov-Romanovski and Galanin (1957) for rigid solutions was expanded to include liquids by Kilin *et al.* (1964), and then by Agrest *et al.* (1969). The Förster formulation was modified to include Brownian translational motion by Bagdasaryan and Muler (1965). Their mathematical solution is limited to steady-state excitation, so that no variation in donor decay rate with acceptor concentrations could be indicated. In a more recent treatment, a quantitative description of the decrease in donor decay time as a function of the diffusion coefficient D has been presented by Yokota and Tanimoto (1967). When D is very large and transfer is diffusion controlled, the probability of energy transfer is proportional to D. When D is small and the long-range dipole–dipole interaction is dominant, the transfer probability of energy transfer varies as $D^{3/4}$. Similar results are obtained by Steinberg and Katchalski (1968) and Birks and Leite (1970). The above ideas have been expanded to include molecular displacements during the lifetime of the excited state (Mikhelashvili, 1971).

The theoretical developments can generally be divided according to whether they are extensions of Förster theory or whether they follow diffusion theory and extrapolate to Stern–Volmer kinetics. Those that start with Förster's formulations are Kurskii and Selivanenko (1960), Samson (1962), Bagdasaryan and Muler (1965), Feitelson (1966), Yokota and Tanimoto (1967), Nafisi-Movaghar *et al.* (1967), and Birks and Leite (1970). Those that follow diffusion kinetics are Belikova and Galanin (1958), King and Voltz (1966), Voltz *et al.* (1966), and Agrest *et al.* (1969).

Experimentally, Weinreb and Cohen (1956) and Weinreb (1961) studied the efficiency of singlet–singlet transfer in a naphthalene–anthracene system as a function of the diffusion coefficient D. The solvent was a solution of toluene and mineral oil in varying concentrations. These authors observed that as D approached zero, the energy-transfer rate approached the value predicted by Förster's theory. Kurskii and Selivanenko (1960), Samson (1962), and Melhuish (1963) reported similar observations. It is of interest that the value Melhuish calculated for the mean transfer distance between 9-methylanthracene and perylene, extrapolated to zero fluidity, is 36 Å. The value given herein is 36.1 Å.

If the mean diffusion length in a liquid r is defined (Birks and Leite, 1970) as

$$r = (2D\tau_{0D})^{1/2},$$

where τ_{0D} is the fluorescence decay time of the donor in the absence of acceptor, and $D = D_D + D_A$, where D_D and D_A are the diffusion coefficients of donor and acceptor, then the studies of energy transfer can be divided into three regimes:

1. A diffusion-controlled or Stern–Volmer region where $r > R_0$.
2. An intermediate region where $r \sim R_0$.
3. An inductive-resonance or Förster region where $r < R_0$.

Experiments generally fall into region 1 or 3. Examples of studies limited to region 1 are Cohen and Weinreb (1956) for a toluene–paraterphenyl system, Belikova and Galanin (1956) for an anthracene–toluene system, Berlman (1960) for a paraxylene–PPO system, Dubois and Cox (1963) for cyclohexane solutions containing ketones as donors and biacetyl as acceptor, Alfimov *et al.* (1968) for toluene solutions containing carbazole, phenanthrene, and naphthalene-d_8 as solutes, and Birks and Leite (1970) for 9,10-diphenylanthracene in solvents of different viscosity. Examples of studies in region 3 are Basile (1964) for a polystyrene–paraterphenyl system and Birks and Leite (1970) for PPO (donor) and 9,10-diphenylanthracene (acceptor) in cyclohexanol.

Both Birks and Leite (1970) and Feitelson (1966) find that they get a better fit to their data if they assume a slightly larger value for R_0 than they compute by the Förster equation (Eq. 1). They suggest that this implies that higher multipole terms contribute to the transfer. Cyclohexanol was the solvent common to both experiments. The question still remains whether, on a microscopic scale, regions of low viscosity in cyclohexanol contribute to their anomalous results. Generally, Birks and Leite (1970) find good agreement between their other experimental results and the theoretical predictions of Steinberg and Katchalski (1968) and Yokota and Tanimoto (1967).

Tomura *et al.* (1967, 1968) studied the probability of energy transfer from pyrene to perylene as a function of the temperature and viscosity of the solution. Actually, they measured the fluorescence decay of the donor as a function of the variables. The transfer rate varied as D at large values of D and varied as $D^{3/4}$ at low values of D, as predicted by Yokota and Tanimoto (1967).

Singlet quenching of fluorescein by rhodamine B and of PPO by 9,10-dibromoanthracene has been investigated by Kawski *et al.* (1971). Their results are in agreement with Förster theory. The critical transfer radius R_0 between naphthalene and anthranilic acid was measured by Alkana *et al.* (1968). Their experimental value was 22.4 Å and their calculated value was 27 Å. Our computed value of 23.3 Å is much closer to their experimental value.

Triplet–triplet energy transfer in liquid solution was first detected by Bäckström and Sandros (1958) when the phosphorescence decay time of biacetyl was observed to decrease on addition of aromatic hydrocarbons. The decrease was according to the Stern–Volmer formulation. In a liquid deoxygenated benzene solution, the rate constant for triplet–triplet energy transfer was equal to the diffusion rate constant, highlighting the high efficiency of transfer on collision. The transfer of triplet energy between two aromatic hydrocarbons was first studied by Porter and Wilkinson (1961), using a flash-photolysis technique developed by Porter and Windsor (1953) and Porter and Wright (1955). Triplet-state absorption spectra were measured for both species; the mechanism of the transfer process was established, and the rate of energy transfer was confirmed as being diffusion controlled. Complex formation between donor and acceptors was assumed to be unimportant under the conditions of their experiment.

Bäckström and Sandros (1960) showed that benzophenane in liquid deoxygenated benzene or isoproponol at 293°K could sensitize the phosphorescence of biacetyl, benzil, or anisil. They found the lifetime of the triplet state of benzophenane in benzene to be 2×10^{-6} sec, and in isopropanol to be 6×10^{-6} sec.

Sandros (1964) also studied reversible energy transfer. He found that at room temperature, an acceptor molecule with a triplet level slightly higher than that of the donor was still capable of quenching the donor phosphorescence. Since the efficiency of quenching was a function of the donor concentration, he concluded that reversible energy transfer was taking place.

The emission from biacetyl, optically excited to its first absorption band, consists of both triplet and singlet transitions to the ground state with an intensity ratio R_0 (phosphorescence to fluorescence radiation) of about 60 to 1 (Okabe and Noyes, 1958). This emission may also be sensitized by singlet–singlet and triplet–triplet transfer from suitably chosen donors. Evidence for triplet–triplet transfer can be shown to occur if the sensitized phosphorescence-to-fluorescence ratio R of biacetyl is greater than R_0. Values of $R > R_0$ have been observed (Lipsky, 1963) in a degassed cyclohexane solution at 25°C with benzene as the donor and biacetyl as the acceptor. Similar experiments were performed by Lisovskaya *et al.* (1967) in their study of triplet–triplet energy transfer from benzene and toluene to biacetyl.

Bonnier and Jardon (1970, 1971) have also used biacetyl as the acceptor, but they used biphenyl, fluorene, phenanthrene, and heterocyclic compounds as donors. In these experiments they employed triplet–triplet energy transfer to measure the phosphorescence decay time and quantum yield of the above donors. The heterocyclic compounds containing an oxygen or nitrogen atom appear to have high ratios of phosphorescence-to-fluorescence yield.

In the process of searching for compounds that emit phosphorescence in liquid solutions at room temperatures, Richtol and Klappmeier (1966) investigated triplet energy transfer from benzophenone to aliphatic α-diketones. They found that the phosphorescence yield of the α-diketones decreases with increasing length of the side chains.

A relatively simple technique for measuring the integrated number of triplets in solution has been developed by Lamola and Hammond (1965). The compound to be studied is used as a photosensitizer for the cis-trans isomerization of selected olefins. An olefin is so chosen that its excited singlet state is higher and its triplet state lower than that of the sensitizer, so that isomerization can be actuated only by triplet transfer from the sensitizer. From a knowledge of the total number of sensitizer triplets, it is possible to determine the quantum yield for intersystem crossing from singlet to the triplet manifolds in the sensitizer. This technique is an extension of the work of Cundall *et al.* (1964) in the vapor phase.

Ermolaev (1962, 1963) used triplet–triplet transfer to determine the quantum yield of sensitized phosphorescence γ. He accomplished this by measuring the ratio of the number of phosphorescence quanta emitted by the acceptor q_{SPA} to the number of quenched phosphorescence quanta of the donor $q_{0D} - q_D$, where q_{0D} and q_D are the phosphorescence quantum yields of the donor without and with the acceptor, respectively: i.e.,

$$\gamma = \frac{q_{SPA}}{q_{0D} - q_D}$$

This value γ may be determined graphically from the ratio of the areas (assuming the spectra have been calibrated relative to wavelength) under the sensitized emission spectrum and the difference in the areas of the unquenched and quenched areas of the donor emission spectra. It was shown that the value of γ for any given pair of transferring molecules is independent of the concentration of both the donor and the acceptor molecules. The value of γ was found to vary from 0.07 to 1.0, depending on the combination of donor–acceptor molecules: its value always increased in the series of acceptors, naphthalene, 1-chloronaphthalene, 1-bromonaphthalene, and 1-iodonaphthalene.

The triplet states of large molecules can behave as metastable "energy sinks," from which many interesting physical and chemical phenomena can originate. The sensitized state in a triplet–singlet transfer process, in contrast to triplet–triplet transfer, is an emissive state and is more easily detected and measured. Because of the dipole–dipole nature of the triplet–singlet process, large intermolecular transfer distances are measured; moreover, the efficiency of this process may be orders of magnitude larger than a triplet–triplet process in solution (which proceeds via an exchange mechanism). Diffusion is a particularly helpful mechanism by which triplet states, because of their long decay time, can effectively transfer their energy.

Vaudo and Hercules (1970) and Ermolaev and Svesnikova (1964) have studied triplet–singlet transfer in fluid solutions and find that the transfer rate is very fast, 10^7–10^9 mole^{-1} sec^{-1}, the value depending on the relative strength of the dipole–dipole interaction of the components of the system. The efficiency of the transfer process is a function of the decay time of the donor and of its ability to diffuse within the critical transfer distance during its lifetime. In a subsequent work, Vaudo and Hercules (1971) have confirmed the dipole–dipole nature of triplet–singlet transfer process, and found a linear dependence on the probability of triplet quenching as a function of D from 10^{-6}–10^{-5} cm^2 sec^{-1}.

Excitation migration in liquids is a more recent subject of investigation. Lipsky (1964) believes that excitation migration as well as molecular diffusion must be invoked to explain the high transfer rates in benzene solutions. Dillon and Burton (1965) conclude that excitation migration is at least as important as molecular diffusion for excitation transfer from benzene to a solute. Van Loben Sels and Dubois (1968) find that the rate constant for quenching exceeds the rate constant for diffusion. Here, too, excitation migration would explain this difference.

Small regions of order called domains are associated with excitation migration in liquids—especially benzene. In order to account for variations in the behavior of luminescent solutions containing low concentrations of benzene as contrasted with those having high concentrations, Nosworthy *et al.* (1961) support the concept of domains. Burton and Dreeskamp (1957) have shown that the rate of transfer from a domain to a solute is 10^{11} sec^{-1}, a time slightly shorter than that expected from a diffusion-controlled process. In addition, there appears to be some physical evidence that implies the existence of small ordered regions in benzene, e.g., Bell and Davey (1941), Hisatsune and Jayadevappa (1960), Wood (1947), Munster (1950), and Coumou and Mackor (1964). Förster (1960a), however, is critical of the domain theory.

In studies of scintillation solutions, one of the remarkable findings is that a saturated hydrocarbon can transfer energy efficiently to a common scintillator such as anthracene. Burton and Dreeskamp (1957) find that the excitation transfer time from cyclohexane to paraterphenyl is less than 0.3 nsec, which is equivalent to a rate constant of 10^{11}–10^{12} mole^{-1} sec^{-1}.

From some recent studies (Hirayama and Lipsky, 1969, 1971; Hirayama *et al.*, 1970; Rothman *et al.*, 1973), it would appear that the olefins can form transient excited bound states and emit this excitation energy as fluorescence. Energy transfer also takes place from these bound states. In a pulsed-electron experiment, Baxendale *et al.* (1969) find that in a solution of dioxane and anthracene (10^{-2} mole) the efficiency of energy transfer from the solvent to the solute is equivalent to that in a benzene and anthracene (10^{-2} mole) solution. What is surprising is that there has been no definitive study showing that water has excited bound states and can transfer energy. All of the good experiments have produced negative results.

Polarization studies in liquids were some of the earliest studies that provided evidence of energy transfer. When a system is irradiated with polarized light, the fluorescence emission is expected to be polarized as long as the molecular relaxation time is larger than the fluorescence decay time.

Depolarization in dilute viscous solutions as a function of solute concentration was observed by Gaviola and Pringsheim (1924), Weigert and Käppler (1924), and Levshin (1924). When concentration of the solute was increased, the degree of polarization was decreased.

In a very ingenious experiment, Feofilov and Sveshnikov (1940) were able to show that concentration depolarization is a function of the lifetime of the excited state. A quenching substance was added to a fluorescent solution that already had a large solute concentration, and the effect was to decrease both the lifetime of the excited state and the depolarization. Thus, it was concluded that depolarization is lifetime dependent. On the other hand, in the absence of an external quencher, an increase in the solute concentration will often lead to a reduction in its lifetime but to an increase in the depolarization. The depolarization in these latter results is interpreted as evidence of excitation migration. Other publications on concentration depolarization are Vavilov (1943), Förster (1948), Kudryashov *et al.* (1960), Kawski (1961, 1963), Kawski and Krosnicka (1965), Weber (1966), Eriksen and Ore (1967), Knox (1968), Jablonski (1970), and Ore and Eriksen (1971).

D. Solid Phase

One of the fascinating observations has been that an anthracene crystal with a trace of tetracene (naphthacene) has an emission spectrum that resembles completely that of pure tetracene. Analogous results are obtained with a naphthalene crystal containing a trace of anthracene. The emission spectrum is in the blue (anthracene). On the other hand, if the impurity in naphthalene is 0.1% phenazine, a nonfluorescent compound, the resultant solid solution is practically nonfluorescent in any region (Bowen *et al.*, 1949). In these studies, the impurity is in such small concentration that the fraction of the host radiation absorbed by the impurity (the trivial mechanism) is negligible (Bowen and Mikiewicz, 1947). To explain these large transfer distances, energy migration, in addition to long-range transfer, must take place. Zhevandrov *et al.* (1962) were able to show by means of a polarization technique that singlet excitation migrates between naphthacene molecules incorporated in an anthracene crystal lattice over distances as large as 25 lattice constants. These experimental observations all point to the phenomenon of energy migration and transfer in the solid state.

The diffusion length of singlet excitons in crystalline tetracene has been measured by Vaubel and Baessler (1970) and found to be 580 ± 50 Å below 190°K, to decrease in value at higher temperatures, and to approach a value of 120 ± 10 Å at room temperature. The explanation given for such a dependence of diffusion length on temperature is that singlet fission (process X, Section 2.6) increases with increasing temperature, with a resultant decrease in the exciton lifetime and diffusion length.

A descriptive but nonmathematical treatment of the energy-transfer mechanism and the molecular exciton model for molecular aggregates has been written by Kasha (1963). The processes intermolecular resonance transfer, intermolecular nonresonance transfer, and intramolecular electronic relaxation in solid media may be considered as special cases of radiationless transitions between nonstationary states of the entire system (Robinson and Frosch, 1963).

Schmillen (1965) measured the effect of a guest concentration (C) on the decay time τ of the host. His results fit the equation

$$\tau_0/\tau = 1 + aC.$$

With 2,3-dimethylanthracene as the host, he finds values of a equal to 1.05×10^5 and 8.7×10^4 liter mole^{-1} for anthracene and perylene, respectively. Although the singlet energies of these guests differ by about 0.5 eV, the important consideration in the comparison of guests is the relative overlap between the fluorescence spectrum of the host and the absorption spectrum of the guest.

Jortner *et al.* (1963) propose that triplet + triplet annihilation in aromatic crystals proceeds by charge-pair formation, rather than singlet formation. DeLosh and Grant (1970) have developed a model for electronic energy transfer between impurity ions in a solid based on virtual-phonon processes. When the energy being transferred is much greater than the maximum phonon energy in the solid, the transfer reduces to the expected R^{-6} dependence for dipole–dipole transfer, and R^{-10} dependence for quadrupole–quadrupole transfer. Their treatment includes transfer to other than nearest neighbors.

The following experiments illustrate the various processes and characteristics of crystals. Trlifaj (1958) gauged the mean diffusion length of excitons in anthracene crystals as about 460 Å. Klöpffer (1969a) measured the singlet transfer from a host crystal (N-isopropylcarbazole) to a guest (perylene). He estimated that, on the average, 4×10^4 transfers take place, with an average displacement of about 1350 Å. In amorphous polyvinylcarbazole he (Klöpffer, 1969b) found that the average number of transfers is only 10^3.

Ohigashi *et al.* (1964) observed a pressure effect on the fluorescence yield of a mixed crystal composed of anthracene as host and tetracene as guest. They found a linear increase of the probability of singlet energy transfer with a decrease of the lattice constant to the sixth power and concluded that transfer is by a dipole–dipole interaction.

Recently, multistep triplet–triplet transfer processes have been found to play an important role in molecular crystals. The exchange interaction, the mechanism by which triplet states are coupled, is strong because of the small distance between adjacent molecules in the lattice and the overlap of the charge distribution of adjacent molecules. This overlap is significant in an aromatic hydrocarbon, where the excited electron occupies an orbital that is spread over the whole molecule. Experimental and theoretical estimates of the magnitude of vibronic interaction are about 10^{-1} cm (Förster, 1967). Because this value is small, the rate of each individual transfer process is less than that between singlet states; but because of the longer triplet decay time, the migration is much greater. Large migration distances are responsible for the fact that phosphorescence emission from impurities is a common property of molecular crystals. A diffusion length of about 10 μm has been estimated for anthracene crystals (Avakian and Merrifield, 1964)—a rather large value. An extensive review of triplet excitons in anthracene crystals has been written by Avakian and Merrifield (1968).

In organic molecular crystals, the delocalization of triplet excitation is normally so efficient that pure crystals do not exhibit host phosphorescence (Sternlicht *et al.*, 1963). The triplet states, produced from initially excited singlet states, annihilate and repopulate the singlet states.

The transfer mechanisms in pure pyrene crystals and in perylene-doped pyrene crystals have been investigated (Chu *et al.*, 1971). Pyrene crystals are somewhat special in that the molecules are in a sandwich dimer arrangement, and only excimer fluorescence is emitted by pure crystals at room temperature. Evidence was found of excimer excimer migration in the perylene-doped pyrene crystals, with a transfer rate constant of 4×10^6 sec^{-1} at room temperature. Excimer–excimer migration was also observed in pyrene crystals containing coronene at room temperature (Klöpffer and Bauser, 1970).

Energy transfer from pyrene to perylene in polymethylmethacrylate and polystyrene systems were studied by Mataga *et al.* (1969). Their value of 37 Å is slightly larger than the value reported herein for these compounds, 33.3 Å.

The decay time of a triplet state in a solid medium is long. Therefore, many of the processes involving triplet states are often studied in this environment at room temperature. If quenching processes are still operative, then low temperatures are also required.

Energy transfer between like molecules in the solid phase can be inferred from experiments in which the donor concentration is varied and the acceptor concentration is held constant. In an alcohol–ether glass with benzophenone as donor and naphthalene as acceptor, the donor phosphorescence emission is decreased as the concentration of the benzophenone is increased (Ermolaev, 1964). These results are consistent with the interpretation that triplet energy can migrate among the benzophenone molecules before it is transferred.

Chauduri and El-Sayed (1965) have measured the degree of polarization versus concentration and concluded that depolarization is due to triplet–triplet transfer and that this transfer is based on an exchange mechanism. Azumi and McGlynn (1963a) have studied delayed fluorescence in solid solutions of naphthalene, phenanthrene, and hexahelicene. The delayed fluorescence was produced by a triplet + triplet interaction involving the exchange mechanism.

The luminescence characteristics of 1,2-benzanthracene (BA) as guest in a crystal of benzophenone were studied by Hochstrasser (1964). In BA, the lowest excited singlet state is below the lowest benzophenone singlet exciton band and above the lowest crystal triplet state. Furthermore, the lowest triplet state is at lower energy than the triplet state of the crystal. The ratio of fluorescence to phosphorescence of BA as an isolated photoexcited molecule is about 10^3 and is an index of the relative rates of fluorescence and intersystem crossing. It is assumed that this ratio of BA is not severely altered by BA being a guest in a host lattice. On the other hand, when the BA (at a concentration of 10^{-4} mole/mole) luminescence is sensitized by energy transfer from the host, the ratio is about unity. This result indicates that in the above system, triplet exciton migration is more efficient than singlet migration. The rate constants for intersystem crossing and for triplet exciton migration were found to be of the same order of magnitude, 10^{10} sec^{-1} (Hochstrasser, 1964).

Trap-to-trap migration in crystals of C_6D_6 was studied by Nieman and Robinson (1962). The guests were C_6H_5D and C_6H_6, each in a concentration of 0.4%. Their triplet energy levels are 170 cm^{-1} and 200 cm^{-1}, respectively, below the host triplet. When the host was excited, the fluorescence intensities of both guests were approximately equal, but the phosphorescence intensity from C_6H_6 was some tenfold more intense than from C_6H_5D. When the concentration of C_6H_5D was 0.8% and that of C_6H_6 was 0.008%, the fluorescence and phosphorescence yields from C_6H_6 were $\sim$0 and 30%, in comparison to that from C_6H_5D. From the above work and from similar experiments (Robinson, 1961; Sternlicht *et al.*, 1963), it was observed that triplet energy transfer took place not only between molecules of the host and from the host to the guest but also between widely dispersed guests. This process of "long-range tunneling" also occurs at 4.2°K, where thermal activation is excluded. The following deductions have been made:

1. The value of the nearest-neighbor interaction energy is about 1–10 cm^{-1} for benzene.
2. During their respective lifetimes, a triplet exciton makes 10^{11} jumps and a singlet 10^4–10^6 jumps.
3. The efficient triplet migration leads to a fast triplet plus triplet annihilation in pure crystals, resulting in the quenching of the phosphorescence.
4. Efficient trap-to-trap migration in impure crystals will occur, depending on the temperature, trap depth, and impurity concentration.
5. Delayed fluorescence may be produced by triplet + triplet annihilation in impure crystals.

The observation that triplet–triplet transfer occurs over greater distances than singlet–singlet transfer may be understood more easily if it is remembered that in triplet–triplet migration the competition is between a spin-forbidden radiative transition in the donor molecule and an allowed process (energy transfer), while in the transfer of singlet energy the competition is between allowed processes.

In a continuation of the above work, the phosphorescence from a C_6H_6 crystal was measured as a function of temperature from 2 to 15°K (Hatch and Nieman, 1968). Below 8°K, the phosphorescence yield was approximately constant, but above 8°K, thermally activated annihilation processes—e.g., those that produce delayed fluorescence, quenching of fluorescence, and quenching of phosphorescence—were active.

Exciton migration and transfer, and the effects of host exciton traps in pure and mixed crystals are considered by Birks (1970a). Host exciton traps and radiative migration are shown to be important in thick anthracene and mixed anthracene–tetracene crystals.

Hirota and Hutchinson (1965) measured the triplet decay time of phenanthrene and naphthalene as co-guests in biphenyl by optical and paramagnetic-resonance techniques. The triplet levels of phenanthrene and naphthalene are 1440 and 1550 cm^{-1}, respectively, below that of biphenyl. After selectively exciting phenanthrene, they could follow the decay of the phenanthrene donor and the sensitized naphthalene acceptor, using the respective resonance peaks of the excited guests. They varied both the acceptor concentration and the temperature and found that the transfer process was temperature dependent at low concentrations (0.05%) and temperature independent at high concentrations ($>1.5\%$).

In dilute solutions, the donor triplet is therefore assumed to be thermally excited to the crystal triplet level, and the energy is assumed to be propagated by normal exciton motion to the acceptor site. This phonon-assisted energy transfer is supposed to be a necessary prerequisite to triplet + triplet annihilation. The estimated rate constant for escape of a thermally excited triplet from the phenanthrene into the crystal host is 2×10^8 sec^{-1}. At high concentrations, the temperature independence of transfer and a somewhat long transfer distance imply that the direct exchange mechanism is dominant.

Laser-induced delayed fluorescence, produced by triplet + triplet annihilation, has been studied by Kepler *et al.* (1963). The migration of triplet excitons in anthracene crystals is assumed to be determined by a relatively small exchange interaction ($\sim$5 cm^{-1}), so that the interaction of a triplet exciton with phonons will be strong and the mean free path will be of the order of a lattice spacing. Furthermore, it is believed that the triplet exciton must be treated as a localized exciton. Since the rate of excitation transfer is small compared with the lattice relaxation time, the triplet exciton is surrounded by a lattice deformation. As the exciton propagates through the crystal, so does the deformation.

Avakian *et al.* (1963) have generated triplet excitons in anthracene by irradiating a crystal of anthracene with a wavelength corresponding to the $S_0 \rightarrow T'$ absorption band. They detected the delayed fluorescence resulting from triplet + triplet annihilation and were able to map out the $S_0 \rightarrow T'$ absorption spectrum. The triplet + triplet annihilation rate was 2.4×10^{10} mole^{-1} sec^{-1}.

Zander (1964) has discussed the temperature dependence of the delayed fluorescence of five polyacenes in a phenanthrene host lattice. He has shown that the temperature at which delayed fluorescence is a maximum increases as the energy interval T'(host) $-$ T'(guest) increases, where T' denotes the energy of the lowest triplet state. Zander deduced from these data that delayed fluorescence is facilitated by a low value of the T'(host) $-$ T'(guest) energy interval.

Several authors have reported investigation on orientation effects. Eisenthal (1969) studied the orientation dependence in triplet–triplet transfer. Since the exchange interaction is a short-range interaction sensitive to the overlap of the electronic wavefunction of the participating molecules, there might be a preferred transfer geometry. In the systems that he studied, he did not find any dependence.

Carbonyl compounds such as acetophenone, benzophenone, and fluorenone have (n, π^*) states as their lowest triplet state. Spectral evidence for these compounds indicates that the energies of the triplet states are not very dependent on the aromatic part of the molecule, and the best description of the excited configuration is one in which the positive hole on the oxygen is not far removed from the excited electron (Hochstrasser, 1963). Accordingly, in these cases, the triplet–exciton annihilation process is very inefficient, and pure crystals of the above mentioned carbonyls phosphoresce strongly.

Roy and El-Sayed (1964) find that the most probable configuration of the donor (benzophenone) and the acceptor (phenanthrene-d_{10}) for maximum triplet–triplet transfer is that in which the molecular plane of the acceptor is parallel to the carbonyl axis of the donor. This orientation is precisely the one expected on the basis of maximum overlap of the charge distribution of the triplet states of the two compounds.

Environmental symmetry appears to play a role in the rate of energy transfer. Hunter *et al.* (1969) find that triplet energy transfer from benzophenone to naphthalene is two orders of magnitude less in a glass than in a crystal. They suggest that the difference in mobilities is due to the fact that in crystals the excitons are coherent and in glass they are incoherent.

A theory has been developed on phonon-assisted nonresonance energy transfer between ions in solids (Miyakawa and Dexter, 1970). The dependence of the probability of energy transfer W on the energy gap ΔE is given by

$$W = W_0 \exp(-\beta \Delta E),$$

where β is a constant determined by the phonon nature of the host crystal and by the strength of the lattice coupling. Experimentally, the above formula has been verified by Yamada *et al.* (1972), i.e., the reversible transfer rate between trivalent rare-earth ions was found to be governed by the energy gap.

5

INTRAMOLECULAR ELECTRONIC ENERGY TRANSFER

5.1 Conjugated Chromophores

The absorption spectrum of a molecule containing two conjugated chromophores is often very different from the summed spectra of the individual components. A strong mixing of the electronic states of the two chromophores and delocalization of the π electron produces this effect. On the other hand, some molecules containing unlike chromophores exhibit absorption spectra that resemble the combined spectra of the individual chromophores. In these cases there is little electron delocalization and only a very weak interaction between the chromophores. Excitation transfer appears to take place from one unit to the other. Ermolaev and Terenin (1958, 1960) noted that the long-wavelength absorption spectrum of 4-benzoylbiphenyl is very much like that of benzophenone, i.e., the (n, π^*) singlet state is the lowest excited singlet. Absorption of light in this wavelength region results in an excited singlet state in which the excitation is fairly well localized at the carbonyl group. On the other hand, the phosphorescence spectrum from 4-benzoylbiphenyl corresponds to that of biphenyl. Since the lowest triplet state of biphenyl lies lower than that of benzophenone, intramolecular energy transfer takes place from the (n, π^*) triplet state of the carbonyl to the triplet state of the biphenyl system by means of the exchange mechanism.

Certain chelates of the rare earths are other examples of intramolecular excitation transfer between conjugated chromophores. Excitation of the organic ligand leads to a line emission spectrum that is characteristic of the metal ion, provided the lowest triplet state of the ligand lies above the emitting state of the ion. If no such excited state of the ion exists, the emitted radiation is ligand phosphorescence. Bhaumik and El-Sayed (1965a) provided proof for this model. They employed quenching compounds with high-lying singlet states but with triplet states in between those of the ligand and the ion. In solution, these compounds did not quench the emission from the free ions, but did quench the chelated ions. These quenching compounds were intercepting and inhibiting intramolecular triplet energy transfer.

Polymer chains resemble one-dimensional crystals in their transfer of energy. Cozzens and Fox (1969) have studied intramolecular triplet energy transfer in poly(1-vinylnaphthalene) and have observed delayed fluorescence due to triplet + triplet annihilation, the fluorescence being second order in triplet concentration. They suggest that intramolecular energy transfer may be the mechanism leading to photodegradation of polymers. The degradation takes place because the energy becomes dissipated at "weak links," producing scission, etc. In another publication, Fox and Cozzens (1969) report that in aromatic vinyl polymers the absorbed energy does not remain localized at a photoexcited side group, but can be transferred to other side groups. They explain this transfer process by an exciton model that is similar to that used in aromatic molecular crystals. In support of this proposal, they report that triplet energy is more efficiently transferred from polystyrene to 1-vinylnaphthalene when both are copolymerized than when mixed in a solution.

Lashkov and Ermolaev (1967) studied triplet–triplet energy transfer along polymer chains with aromatic rings in the monomer units. They found that if an impurity in a polymer has the lowest triplet level of the system, the energy will migrate to the impurity. David *et al.* (1970) investigated triplet–triplet energy transfer from polyvinybenzophenone to naphthalene at liquid-nitrogen temperatures. They believe that transfer is by an exchange mechanism and that energy is migrating along the polymer because the transfer distances are more than twice those measured for mixtures of benzophenone and naphthalene in organic glasses.

5.2 Nonconjugated Chromophores

Attention has recently been directed to intramolecular excitation energy transfer between different types of chromophores separated by a saturated system. Theoretical formulae can be tested by this system. Moreover, biological macromolecules are composed of a large number of nonconjugated chromophores. Model compounds incorporating two chemically different chromophores in a nonconjugated configuration have been built. In addition to being able to work at high dilution and to avoid unwanted intermolecular association such as complex formation and co-precipitation, the relative spatial spacing of the donor and acceptor is much better defined in model compounds. This is especially true for rigid compounds having no free rotation about any of the σ bonds connecting the two chromophores. In addition, the value of K^2 in Eqs. (1) and (3) can be more rigorously estimated. Methylene bridges were used in earlier experiments, but more recently more rigid structures have been employed.

When two chromophores, A and B, as components of a molecule are separated by a saturated group such as an alkane chain, there is negligible interaction between them in the ground state. The long-wavelength absorption spectrum of the total molecule is essentially equivalent to the sum of the absorption spectra of the component parts. In such compounds, A and B can be excited separately by the choice of appropriate excitation wavelengths, and intermolecular energy transfer can be studied by measuring the luminescence characteristics of A and B.

Schnepp and Levy (1962) found the absorption spectrum of 1-naphthyl-9-anthrylalkanes, compounds shown in Fig. 3a, are similar to that of an equimolar mixture of 1-methylnaphthalene and 9-methylanthracene. They found that when they excited the naphthyl chromophore, the transfer

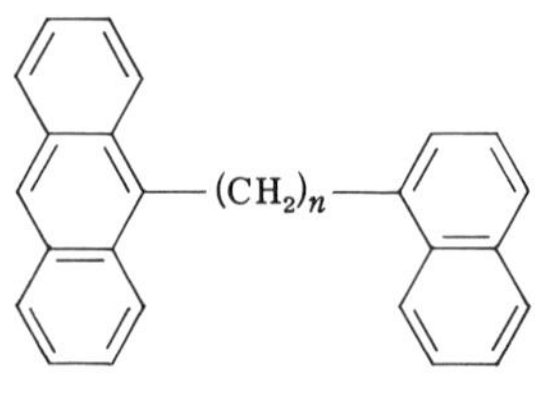

(a)

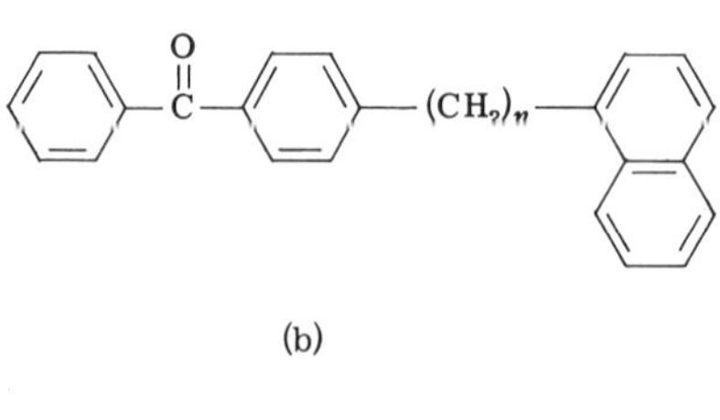

(b)

FIG. 3. (a) Anthracene and naphthalene chromophores are joined by a chain of methylene groups. (b) Benzophenone and naphthalene chromophores are joined by a chain of methylene groups. $n = 1$, 2, or 3.

efficiency to the anthryl component was 30%, and was independent of the value of $n = 1, 2$, or 3, where n is the number of methylene groups in the alkane chain. Breen and Keller (1968) studied the case of energy transfer across a methylene bridge where phthalimide or carbazole was the donor chromophore and naphthalene was the acceptor. In both compounds, triplet energy transfer was 100% efficient.

Cowan and Baum (1970) investigated intramolecular triplet energy transfer from a carbonyl, acetophenone, to an olefin, *trans*-β-methylstyryl, followed by trans–cis isomerization of the olefin. They performed these studies as a function of the length of the methylene chain joining the two chromophores.

Lamola *et al.* (1965) have studied a series of compounds similar to those above but where A is 1-naphthyl and B is *p*-benzophenone, as shown in Fig. 3b. The two chromophores, naphthalene and benzophenone, luminesce independently when excited as independent molecules in a rigid glass at a concentration of 5×10^{-3} mole. However, the composite molecule under the same conditions emits only phosphorescence radiation characteristic of 1-methylnaphthalene.

Triplet sensitized cis $\leftrightarrow$ trans isomerization of piperylene was used as a probe to examine the pathway of the excitation energy. At low piperylene concentration, energy transfer was from the triplet state of naphthalene alone, whereas at high piperylene concentration (>0.2 mole), some transfer from benzophenone triplets was also detected. These results demonstrate that intersystem crossing in benzophenone is very rapid and that intersystem crossing is followed by triplet excitation transfer to the naphthalene. From the concentration of piperylene necessary to compete effectively with this energy transfer, the rate constant for triplet energy transfer in the composite molecule was estimated to be about 10^{10} sec^{-1}. The rate is nearly the same for $n = 1, 2$, or 3, where n is the number of methylene groups in the alkane chain.

When an alkyl chain is used to join two chromophores, rotation about the axis of the chain is still possible. Moreover, the chain may not always be rigid. To overcome these deficiencies, two chromophores were attached to the hydroxyl groups of a "bisteroid" (Latt *et al.*, 1965). In this case, the donor–acceptor distance was a fixed separation of about 20 Å. The value obtained from Förster's equation for dipole–dipole transfer was equal to the separation obtained with a molecular model.

Keller and Dolby (1967, 1969) investigated systems in which the intervening unit was a rigid steroid. They studied intramolecular triplet energy transfer and found that the difference in rate constants between systems paralleled the difference in the triplet decay time of the donors.

Triplet–triplet energy transfer between nonconjugated compounds that had a fixed orientation and separation were studied by Hudson and Hedges (1969) and by Filipescu *et al.* (1969). In a similar study, intramolecular energy transfer between two chromophores held perpendicularly but not rigidly to each other by a spiro linkage was investigated at 77°K by Keller (1968). Complete transfer of both singlet and triplet excitation energy to the chromophore with the lowest excited state of a particular multiplicity was observed.

Intramolecular singlet energy transfer and triplet energy transfer between phenanthrene and cyclopentenone, attached to a rigid spiro-cyclopropane-norbornane molecular frame, has been studied by Bunting and Filipescu (1970). When each chromophore is separately attached to the frame, cyclopentenone emits phosphorescence and spiro-linked phenanthrene emits both phosphorescence and fluorescence. In the model compound, the major emission is phosphorescence from the phenanthrene, and the other emission processes are reduced in intensity. The above authors propose that singlet–singlet energy transfer takes place from phenanthrene to cyclopentenone, followed by back-transfer (triplet–triplet) from cyclopentenone to phenanthrene. DeMember and Filipescu (1968), on the other hand, studied energy transfer between two chromophores held rigidly perpendicular to each other and found that no energy transfer took place. Energy transfer between the same chromophores does take place when they are separate and individual molecules in solution. These authors claim that their results agree with the orientation requirements for electric dipole energy transfer. Lamola (1969b) found entirely different results in his study of a similar system.

Rauh *et al.* (1968) found 100% efficiency in intramolecular energy transfer of singlet excitation between nonconjugated chromophores of the indole alkaloids reserpine, deserpidine, 17-naphthylyohimbol, and rescinnamine. To explain this high efficiency of transfer, they suggested that a combination of exchange and resonance mechanisms were operative. In a subsequent polarization study (Rauh *et al.*, 1969) they surprisingly found that the polarization increased as the solvent viscosity decreased. They conjectured that the energy-transfer process requires the donor and acceptor to explore and find some preferred relative conformation, and this action can occur more easily when the viscosity is low. Bunting and Filipescu (1970) find an experimental rate of energy transfer that is ten times greater than predicted theoretically. They also suggest that an exchange interaction takes place in addition to a dipole–dipole interaction.

By means of magnetic-resonance techniques, DeGroot and van der Waals (1963) studied intramolecular energy transfer in triptycene and tribenzotriptycene, compounds that contain three nonconjugated equivalent aromatic chromophores. Since the long-wavelength absorption bands correspond to the individual subsystems, and since the singlet manifolds are independent, they assumed that the triplet manifolds are also independent. If the triplet exciton in the above compound is localized in one of the chromophores, the ESR signal will not correspond to trigonal symmetry. On the other hand, if the triplet energy migrates between subsystems at a rate that is much faster than the ESR frequency, the spin-resonance signal will correspond to that of a compound with ternary symmetry. They found that at 77°K the triplet exciton in triptycene has a jumping rate $>10^{10}$ sec^{-1}, but at 20°K the speed of the exciton is reduced, and the exciton becomes more localized. This result indicates that vibrational quanta are needed to assist in the propagation of a deformation accompanying the excitation of the subsystem. In tribenzotriptycene, some transfer still takes place at 20°K.

Kottis (1965) has further analyzed the data of DeGroot and van der Waals. In the case of tribenzotriptycene, he computes a triplet exciton exchange frequency of 5×10^7 sec^{-1} at 20°K and 5×10^9 sec^{-1} at 77°K. In addition, he believes that ternary symmetry would be achieved at an exchange frequency of 10^{11} sec^{-1}, but this rate could be attained only at some higher temperature.

5.3 Rare-Earth Chelates

Energy transfer to and between rare-earth atoms is a subject of appreciable current interest for practical as well as theoretical reasons. Rare earths have been used for laser and color-television applications. Care, however, must be exercised in distinguishing between intramolecular energy transfer in chelated materials and intermolecular energy transfer in solution.

In their pioneering studies, Weissman (1942), Yuster and Weissman (1949), and Weissman (1950) demonstrated that rare-earth ions can be excited to their luminescence level through intramolecular energy transfer from excited ligands chelating the ion. While competing processes can decrease the probability of energy reaching the ion, comparatively little quenching occurs once the metal cation becomes excited. Sevchenko and Trofimov (1951), and Sevchenko and Morachevskii (1951) have published similar findings.

Crosby and Kasha (1958), Crosby *et al.* (1961), Crosby *et al.* (1962), and Freeman and Crosby (1963) have shown that intramolecular transfer to the chelated ion occurs via the triplet state, provided the energy of the triplet level of the ligand exceeds that of the (lanthanide) ion. Similar conclusions appear in the work of Ermolaev *et al.* (1967a), and Ermolaev *et al.* (1967b). The latter authors believe that chelated systems can be used to verify the exchange mechanism in triplet–triplet transfer. Bhaumik and El-Sayed (1965a) found that the transfer rate in rare-earth chelates is 10^{10} sec^{-1}. By means of an ingenious technique, i.e., using the competition between intramolecular transfer to an ion and intermolecular triplet–triplet transfer to aromatic molecules in liquid solution, Ermolaev *et al.* (1966) determined the rate constants of intramolecular transfer for a series of chelates of β-diketones. The rate constant varied from 3×10^7 sec^{-1} to about 10^{11} sec^{-1}.

Matovich and Suzuki (1963) were first to observe intermolecular energy transfer from aromatic ketones to europium and terbium, and El-Sayed and Bhaumik (1963) were first to measure intermolecular triplet energy transfer from aromatic ketones to rare-earth chelates. The latter authors maintain that triplet–triplet sensitization in fluid solutions can be observed with rare-earth ions because transfer is diffusion controlled and the sensitized emission involves 4f electrons, electrons that are well protected against strong quenching. Confirming evidence of intermolecular energy transfer from ketones to rare-earth ions has come from Gallagher *et al.* (1964), Ballara and Edward (1965), and Heller and Wasserman (1965). This technique of exciting the rare earths by energy transfer can be used as an analytical tool for detecting trace quantities of rare earths (McCarthy and Winefordner, 1966).

Morina *et al.* (1967) have employed benzophenone, acetophenone, and 2-naphthylmethylketone as sensitizers of chelates and find a rate constant for transfer from benzophenone to Eu^{3+} equal to 6×10^8 $mole^{-1}$ sec^{-1}, a value that agrees with the low rate constants for transfer from a ligand to an ion in rare earth chelates (Ermolaev *et al.*, 1967b).

The rate of transfer of triplet energy from a ketone to a rare-earth ion in acetone at room temperature depends on the type of donor triplet state (Ermolaev and Tachin, 1969; Morina and Sveshnikova, 1971). For a triplet (n, π^*) state, the rate of transfer may exceed by more than an order of magnitude that from a triplet (π, π^*) state. This result confirms the contention that the rate of transfer is a function of the overlap of the electronic clouds of the excited ketone and rare-earth ion. Since the rare-earth ion and the ketone are coordinated through the oxygen atom of the carboxyl group, the overlap of electron clouds of an excited ketone with localized excitation (excitation that is localized at the carboxyl group) and a rare-earth ion is substantially larger than the overlap for ketones whose electronic excitation is spread throughout the molecule.

Lamola and Eisinger (1969) have studied the transfer of singlet and triplet energy from acetophenone to Eu^{3+} and Tb^{3+} in aqueous solutions. They find the rate constant for the quenching of the donor by the lanthanide ions to be 1.3×10^7 mole^{-1} sec^{-1}, which is about 400 times smaller than the diffusion-controlled rate constant. The action of ketones as sensitizers may be the result of the formation of transient ketone–ion complexes through the nonbonding electrons of the carbonyl group.

Aromatic molecules can also activate rare-earth ions. Bhaumik and El-Sayed (1965b) have found that transfer from carbazole and aromatic hydrocarbons (donors) to rare-earth chelates (acceptors) in solutions is diffusion controlled. Briskina *et al.* (1966) and Banfield and Husain (1969) have used dyes as sensitizers of chelates. Ermolaev and Shakhverdov (1971) have observed the "inverse" process of transfer from the rare earth ions to dyes. In a study of the energy transfer from acridine yellow, perylene, and anthracene to rare-earth ions in solution at room temperature, Ermolaev and Shakhverdov (1969) have concluded that transfer from the triplet state of the donor is by an exchange mechanism, whereas transfer from the singlet state is primarily by an induction–resonance mechanism.

The mechanism of energy transfer from rare earth ions and their complexes to organic compounds in liquid solutions at 293°K was studied by Ermolaev and Tachin (1970). From the rate constant for quenching of the ions, energy transfer is assumed to be induced by an exchange–resonance interaction. Triplet–triplet transfer (process II, Section 2.6) is indicated.

In a study of the mechanism of radiationless energy transfer between rare-earth ions in liquid solutions, Antipenko and Ermolaev (1969) observed that transfer was manifested only in those cases that were energetically favorable. The quenching of the donor fluorescence by an acceptor attests to the fact that radiationless transfer processes are active. From a comparison of the rate constants for transfer and the overlap integrals of the donor and acceptor spectra, the above authors conclude that in acetone, transfer is by an exchange–resonance mechanism and, in water, by both exchange– and induction–resonance mechanisms.

The nonradiative energy transfer from one rare-earth ion to another via an intermediate was studied by Tachin and Ermolaev (1970). Energy was transferred from Tb^{3+} ions to Eu^{3+} ions via the triplet state of 2-acetylnaphthalene in acetone at room temperature. This two-step transfer could be achieved with unity quantum yield. Energy transfer between rare-earth ions in tungstates has been investigated by means of self-quenching experiments (Van Uitert, 1971). Compounds were employed that have the steelite structure in which the rare-earth ions occupy dodecahedral sites. Such sites lack a center of symmetry, and the rare-earth ions are amenable to either dipole or quadrupole transitions. Critical interaction distances exceeding 17 Å for dipole–quadrupole quenching of Tm^{3+} have been found. In an earlier report on energy transfer between rare-earth ions, Van Uitert (1967) observed that dipole–dipole interactions are prevalent but that dipole–quadrupole interactions are also present for the higher energy states of the larger rare-earth ions.

6

STUDIES RELATED TO ELECTRONIC ENERGY TRANSFER

6.1 Chemical Studies

These studies are divided into those performed in the vapor phase and those in the condensed phase. A review paper on energy transfer in radiation chemistry has been written by Burton *et al.* (1969).

In a classical experiment, Cario and Franck (1923) showed that in the vapor phase, an Hg atom can transfer sufficient energy to a hydrogen molecule to dissociate it. This is one of the first cases of a photosensitized dissociation reaction. More recently, Dainton and Ivin (1950) found that olefins could quench the photoreaction between SO_2 and paraffins to produce sulfuric acid. Dubois and Noyes (1951) were able to sensitize the dissociation of methyl iodide by excited benzene. Cundall and Palmer (1960) found that electronically excited benzene vapor caused cis–trans isomerization of butene-2 and postulated that the transfer process involved the transfer of triplet excitation. This result suggested the use of the isomerization of butene-2 as an indicator for triplet states and has since yielded some of the most reliable values of G(triplets).

Triplet–triplet energy transfer from acetone to aliphatic aldehydes has been studied in the gas phase by Rebbert and Ausloos (1964). It was found that the electronically excited aldehyde molecule formed by an energy-transfer process shows the same modes of decomposition as by direct photolysis.

In studies involving liquids, West and Paul (1932) showed that the quantum yield of decomposition products of alkyl iodides in benzene solution can be explained in terms of excitation transfer from excited solvent to iodine. Subsequently, West (1941) demonstrated that naphthalene sensitizes the decomposition of ethyl iodide in hexane solutions.

In the irradiation of cyclohexane and cyclohexane plus benzene solutions by electrons, Manion and Burton (1952) explained the decrease in hydrogen yield with increased benzene concentration as being due to energy transfer from the cyclohexane to benzene. In terms of hydrogen yield, i.e., $G(H_2)$, benzene is more radiation resistant to ionic bombardment than cyclohexane.

A similar experiment has been used to demonstrate long-range energy transfer in the condensed phase. The decrease in radiolysis effects was studied by Brodsky *et al.* (1962) as a function of an impurity concentration. The diffusion length and radiation damage were observed to decrease, because the excitation energy was scavenged by the impurity.

In an experiment wherein cyclohexane solutions containing benzene or cyclohexene were irradiated with vacuum uv radiation at 1470 Å, the photolysis yields of hydrogen and cyclohexyl were observed to decrease on addition of the benzene or cyclohexene (Holroyd *et al.*, 1967). "Protection" was provided by energy transfer from the neutral excited cyclohexane.

Alexander and Charlesby (1954) showed that aromatic rings built into polymers lower their sensitivity to radiation effects in terms of chain breakage or cross-linking between chains. All the evidence suggests that the chemical activity is the result of transfer of excitation or ionization. Partridge (1970a, b, c) has written a series of papers on the theoretical and experimental aspects of energy migration in alkanes. His predictions are that scission should occur primarily at CH bonds.

Bäckström and Sandros (1958) were able to show that dianthracene is formed in a solution containing anthracene and biacetyl irradiated by uv radiation. They propose that this photosensitized reaction is initiated via energy transfer from the triplet state of biacetyl.

Dole and Williams (1959) studied the energy transfer in polyethylene and polyethene–polybutadiene mixtures during gamma irradiation. They present evidence to indicate that energy transfer is localized predominantly in spurs along the ionization track.

A study of the effect of a phase change in the host material on radiation-induced chemical changes has been investigated by Collinson *et al.* (1963). The radiation effects were greatly enhanced in those cases where the host became crystalline on cooling because energy-transfer processes became viable.

Hammond *et al.* (1964) have shown that energy transfer plays an important role in the photodimerization of coumarin. This compound has a large fluorescence quantum yield and does not readily form triplet states by intersystem crossing. In the presence of benzophenone, the photodimerization of coumarin takes place via the triplet states. If one excites coumarin molecules to their lowest excited singlet level, these excited molecules quickly transfer their energy to benzophenone molecules, where, by intersystem crossing, the benzophenone molecules first convert to their lowest triplet state and then transfer this energy back to the coumarin compound. A triplet-state coumarin interacts with a ground-state coumarin molecule, and photodimerization takes place.

Energy transfer in liquids can produce isomerization in certain compounds. Hammond *et al.* (1962, 1963) and Krongauz (1964) studied cis–trans isomerization of linear dienes and ethylene derivatives by triplet–triplet transfer. These experiments proved the important role played by triplet states in isomerization. Cundall and Griffiths (1965) were able to estimate the lifetime of the triplet state of benzene ($\sim 10^{-7}$ sec) from the cis–trans isomerization of butene. From the studies of benzene solutions of butene-2, Cundall and Griffiths (1966a) were able to conclude that the triplet rather than the singlet state of the olefin is involved (i.e., a triplet–triplet process) because butene-2 does not appear to quench the benzene fluorescence. Electron capture is unlikely because of the higher electron affinity and much greater concentration of benzene. Golub and Stephens (1966) have reached similar conclusions in their study of photo- and radiation-induced cis–trans isomerization of octene-2 in benzene. Many studies have shown that the ring orientation in stilbene, except under hindered conditions, is very fast and procedes via the triplet state. In benzene solution, benzene triplets were able to generate cis–trans isomerization of stilbene by the mechanism of energy transfer (Hentz *et al.*, 1966). More recently, Stephenson and Hammond (1969) found that energy transfer could produce cis–trans isomerization in 1,3-pentadiene.

Solutions of 10^{-1} mole *trans*-stilbene in benzene in the presence and absence of 3×10^{-5} mole anthracene were studied (Bensasson and Land, 1971). No transient absorption around 3600 Å attributable to triplet–triplet absorption of stilbene was detected, and no sensitization of anthracene triplet absorption occurred. These results are in accord with other findings (Herkstroeter and McClure, 1968; Heinrich *et al.*, 1967) that the triplet–triplet absorption of stilbene is very short lived, except in rigid media where ring rotation is hindered.

Cundall and Griffiths (1966b) have suggested the use of olefins as triplet-state scavengers in irradiated systems. The characteristics that make olefins good scavengers are that they readily undergo triplet–triplet energy transfer from donors with triplet levels above 3 eV, but do not undergo singlet–singlet transfer from donors with singlet levels below 5 eV. Moreover, the lifetime of the olefin triplet level is expected to be less than a nanosecond. Thus, it becomes possible to separate the roles of singlet and triplet states in irradiated systems.

Other reactions have also been investigated. Triplet–triplet transfer is effective in sensitizing: e.g., the addition reaction of maleic anhydride (Hammond and Hardham, 1963), the decomposition of ethylpyruvate and diazomethane (Kopecky *et al.*, 1962), and the photobleaching and photocoloration of spiropyranes (Lashkov *et al.*, 1966). The chemiluminescence of phthalic hydrazide derivatives may have been generated by triplet–singlet transfer (White *et al.*, 1969).

Lamola (1969a) lists four possible pathways for a photochemical reaction, as shown in Fig. 4. On path *a*, the primary photochemical step takes place from the lowest excited singlet state. On path *b*, the chemistry takes place from a high vibrational ground state after internal conversion from the excited singlet state. On path *c*, the primary event takes place from the lowest triplet state; and on path *d*, the reaction takes place from a high vibrational ground state after intersystem crossing from the triplet state. By means of triplet energy transfer, one can preferentially populate the triplet state, precluding paths *a* and *b*. On the other hand, one can selectively quench the triplet state by a suitable quencher, giving priority to paths *a* and *b*. Thus, the course of photochemical reactions can be controlled and determined.

In summary, the mechanisms by which energy may be transferred to produce a chemical change are:

A. Excitation transfer, (e.g., dipole–dipole) between neutral molecules.

B. The localization in a particular component of the system of previously nonlocalized energy (Katsuura and Inokuti, 1964).

C. Transfer of energy via charge transfer.

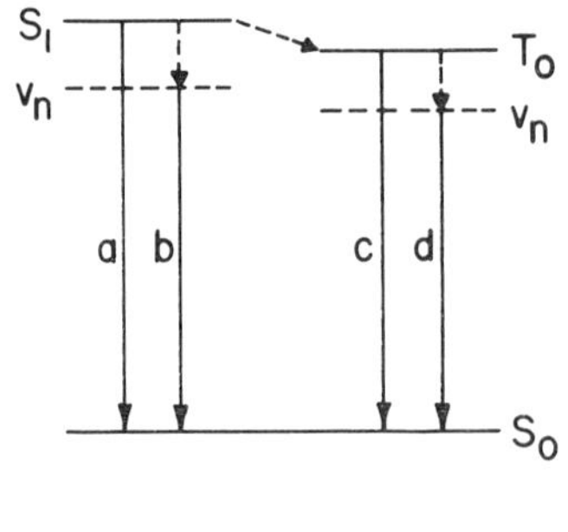

FIG. 4. An energy-level diagram illustrating the various pathways leading to a photochemical reaction (Lamola, 1969a).

6.2 Electron-Spin Resonance

Electron-spin resonance (ESR) is a powerful tool for detecting and measuring the number of unpaired electrons in a system, e.g., radicals, triplet states, and paramagnetic ions. Farmer *et al.* (1961) appear to be among the first to employ ESR to confirm the existence of energy transfer involving triplet states.

Brandon *et al.* (1962) showed by magnetic-resonance techniques that triplet-state energy is transferred from phenanthrene to naphthalene in single crystals of diphenyl at 77°K. In this experiment, they were able to excite the phenanthrene alone because the exciting wavelength was selected by a filter. From an accurate measurement of the zero-field parameters of naphthalene and phenanthrene, they were able to conclude that there is no complex formation between (or juxtaposition of) the above guests and that triplet energy must be transferred through the diphenyl host molecules.

Triplet–triplet transfer has been investigated by monitoring the triplet concentration by ESR and spectroscopic techniques (Siegel and Judeikas, 1964). The rate of triplet transfer was studied as a function of ΔE, the energy difference between the triplet levels of the donor and host. The solvents were varied so that ΔE assumed values from 1200 to 1500 cm^{-1}.

The EPR absorption spectra of the photoexcited triplet states species in four guest–host crystal systems: β-methylnaphthalene-d_{10} in naphthalene-d_8, β-methylnaphthalene-d_{10} in naphthalene-h_8, β-methylnaphthalene-h_{10} in naphthalene-d_8, and β-methylnaphthalene-h_{10} in naphthalene-h_8 have been investigated in the temperature range ~5 to ~25°K (Ashford and Hutchinson, 1971). The triplet-state species that was observed in the β-methylnaphthalene-d_{10} in naphthalene-h_8 system were found to be naphthalene host molecules, whereas in the other three systems the triplet-state species were the β-methylnaphthalene guest molecules.

In a magnetoselection experiment, Siegel and Goldstein (1966) studied the concentration depolarization of naphthalene-d_8 in a glassy matrix at 77°K. Their results on triplet-state energy transfer are consistent with an electron-exchange mechanism.

Brinen and Gerhardt (1970) studied intramolecular energy transfer by means of ESR. They found 100% intersystem crossing from the singlet to the triplet state in benzophenone.

6.3 Biological Studies

Energy transfer can be demonstrated by sensitized fluorescence in biological materials that are composed of a mixture of pigments. Transfer has been found both in solution and in living plants.

Dutton *et al.* (1943) find that the fluorescence quantum yield of chlorophyll in *Nitzschia closterium* is independent of the exciting wavelength, over a wavelength region from 4358 to 6000 Å. They conclude that since the exciting light at the lower wavelengths is absorbed primarily by the carotenoid pigment fucoxanthin, energy transfer must take place from this material to the chlorophyll. Duysen (1952) confirmed the above conclusion, but subsequently became more cautious (Duysen, 1964).

The sensitization of the fluorescence of chlorophyll a by chlorophyll b in fluid solution was demonstrated by Watson and Livingston (1950). The probability of energy transfer in a methanolic solution containing equal concentrations of chlorophyll a and b was found to have a maximum value of 0.4 at a concentration of 10^{-3} mole. These results were confirmed independently by Duysen (1951), employing a different experimental technique. The measured values from both experiments are in agreement with the predictions of Förster (1948).

Chlorophyll a is a molecule responsible for photosynthesis in plants. In brown algae, the carotenoid pigment fucoxanthol transfers excitation energy to chlorophyll a, and in red algae the pigment phycoerythrin does the same with an efficiency close to 100%. Energy migration among the chlorophyll molecules is the means by which energy can be concentrated at particular areas in the chloroplast where the actual photosynthesis occurs. This transfer of energy is responsible for increasing the efficiency of the utilization of solar energy by plants (Rabinowitch, 1956, 1957; Duysen, 1956; Franck, 1958).

The mechanism in chlorophyll itself remains unclear. Livingston and Owen (1956) believe that the triplet state is an essential intermediate in chlorophyll-sensitized photochemical reactions. Livingston and Pugh (1959) suggest a mechanism of photosynthesis in which singlet migration of energy between chlorophyll molecules by inductive resonance leads to trapping of this energy by a pigment molecule. This latter molecule is so sensitized as to initiate the sequence of biochemical events that constitute photosynthesis. Rabinowitch (1959), on the other hand, proposes that the primary photophysical process in photosynthesis is the creation of an exciton in the chloroplast and the intramolecular migration of this exciton to the photoenzymatic center.

Experiments on fluorescence quenching and energy transfer were performed on monomolecular films containing chlorophyll by Tweet *et al.* (1964). For dilute layers, where a single-transfer calculation was adequate, the range of interaction between chlorophyll a and copper pheophytin a was found to be about 40 Å, in good agreement with determinations from spectroscopic data of 38–41 Å. For undiluted monolayers, interactions among the chlorophyll molecules must be considered. A review of the published data on energy migration in chlorophylls may be found in the book by Kamen (1963).

Several peptide hormones as well as a few proteins contain a sufficiently small number of aromatic amino acids that the efficiency of singlet energy transfer between pairs of the aromatic residues can be measured unambiguously. The distance between the donor and acceptor chromophores can be determined with the aid of Förster theory (Eisinger, 1969a). This type of experiment serves as a "molecular ruler" and also detects changes in the polypeptide conformation. The fluorescence of large proteins containing many aromatic amino acids is generally too complex to permit such an analysis. The use of bound fluorescent labels (although they may alter the normal conformation of the protein) leads to information of a similar nature (Stryer, 1960, 1968).

Lamola (1969a) maintains that proteins are made up of twenty different amino acids, but only three have low-lying excited states and support fluorescence. These are the three amino acids phenylalanine, tyrosine, and tryptophan. The lowest excited singlet levels lie in the order of phenylalanine > tyrosine > tryptophan. Förster's equation gives a value of $R_0 = 17$ Å for the transfer of singlet excitation energy from tyrosine to tryptophan. Konev (1967) discusses the merits of published work on energy migration in different donor–acceptor pairs of aromatic amino acids.

Bannister (1954) has reported observing energy transfer from the protein to the chromophore in phycocylanin. He found that the fluorescence quantum yield is the same when the exciting radiation is at 4250 Å, which is absorbed by the chromophore, or at 2750 Å, which is absorbed by the protein component of the molecule. This globulin, with a molecular weight of about 2.8×10^5, has eight or sixteen chromophores attached to it which are molecules of the bile pigment mesobiliviolin. He finds that the intramolecular energy transfer probability is unity.

Many proteins contain several aromatic amino acids that fluoresce and exchange singlet excitation energy. The measurement of the rate of energy transfer among the chromophores offers, in principle, the possibility of determining structure-sensitive parameters (Eisinger, 1969a, b; Edelman and McClure, 1968; Teale, 1960; Weber, 1952; Stryer, 1968).

Energy-transfer experiments to determine intramolecular distances between aromatic amino acids were performed by Eisinger *et al.* (1969), Stryer and Haugland (1967), Perlman *et al.* (1968), and Konev (1967).

Myoglobin contains two heme components as prosthetic groups attached to a protein of molecular weight of about 30,000. Under certain conditions, the iron can combine with CO. When this compound is irradiated with light that is absorbed by the heme, dissociation of CO and heme takes place with unity yield. The same process with the same yield can be achieved by irradiating the system with radiation of such wavelength that is absorbed only by the protein and not by the heme (Bücher and Kaspers, 1947). Energy transfer has obviously taken place from the protein to the heme. The ratio of the fluorescence yield of heme proteins with and without the heme component was found to be 0.01 (Weber and Teale, 1959). Moreover, the probability of transfer of excitation energy from the protein to the heme was determined to be at least 20 times greater than other competing radiationless processes. The values of R_0 for transfer from tryptophan to heme lie in the region 30–40 Å (Stryer, 1960).

Myoglobin has one of the highest degrees of helicity of any protein, and yet it lacks translational symmetry for distances greater than 35 Å. A consequence of this lack of translation symmetry, together with such possibilities as the coupling of side chains and backbone chromophores, is that the mean free path of an exciton is not likely to be very large (Rosen, 1966).

Certain plant seeds and bacteria appear to be more radiation resistant with increasing water content. Rosen (1966) found that water does not transfer energy to macromolecules, but that water may cause swelling of the molecule and may interfere with its hydrogen bonding.

Goedheer (1957) found that the fluorescence from phycobilin macromolecules was depolarized, indicating that migration between like molecules took place. Weber (1960) employed polarization techniques to detect energy transfer in aromatic amino acids.

Many basic dyes bind so strongly to nucleic acids that they are mutagenic (Udenfriend, 1969). The strong binding of acridines and related fluorescent dyes has been employed to investigate the structure of nucleic acids, to strain nucleic acid *in situ*, and to study energy transfer. Lerman (1961, 1963) proposes that acridine and related dyes bind to DNA by intercalation between neighboring base pairs in a plane perpendicular to the helical axis. Space is made available for the dye molecule by extension and twisting of the helix. When an acridine–DNA complex is excited by 2650-Å radiation, the fluorescence is from the dye, indicating that energy transfer has taken place from purine and pyrimidine bases to the dye chromophores (Weill and Calvin, 1963).

Isenberg *et al.* (1965) have examined the luminescence properties of complexes of DNA and certain dyes such as acridine orange. The complexes have a delayed fluorescence identical to that of the dye, and the activation spectrum of the delayed fluorescence is the same as that of the DNA phosphorescence and is different from the absorption spectrum of the dye. The intensity of the delayed fluorescence increases with dye concentration up to a ratio of one dye to 50 DNA phosphates, and the phosphorescence decay time is nonexponential. These results indicate that the transfer process may be a long-range Coulombic interaction between the triplet state of the nucleic acid and the first excited singlet state of the dye. Gueron *et al.* (1967) have discussed the problem associated with singlet energy transfer in polynucleotides.

Because resonance energy transfer is extremely sensitive to distance, Conrad and Brand (1968) explored the potential of energy-transfer measurements as a means of calculating distances between fluorescent residues of polypeptides and proteins in solution. They employed a model system composed of 1-dimethylaminonaphthalene-5-sulfonyl chloride coupled to an amino end of a homologous series of peptides with the structure $NH_2(CH_2)_nC(=O)$-L-tryptophan, where n ranged from 1 to 10. The measurements of intramolecular energy transfer from tryptophan to the 1-dimethylaminonaphthalene-5-sulfonamido group were each made in water, glycerol, ethanol, and *p*-dioxane. Whereas the calculated interchromatic separations in the shorter models are practically independent of solvent, in the larger models they increase significantly with decreasing solvent polarity. They assume that this result may be indicative of a hydrophobic interaction within the methylene chain.

The phosphorescence spectrum of deoxyribonucleohistone was found (Matsuyama and Nagata, 1970) to be greatly different from the spectrum expected from its constituents, histone and DNA. This difference is attributed to triplet–triplet energy transfer from the tyrosine residue of histone to the DNA bases.

Guzzo and Pool (1969) employed triplet–triplet energy transfer to determine the triplet level of all-trans retinal. They correlated the change in the decay time of the donor with the concentration of the acceptor (retinal) and were able to place tentatively the triplet level at 38 kcal.

Electronic energy transfer in oligomers and polymers of L-tyrosine in solution was investigated by Longworth *et al.* (1969). Polymers of tyrosine are suitable for this study, because the singlet and triplet states of the phenoxyl ionized form of tyrosine have lower energies than the corresponding states in unionized tyrosyl residues. The ratio of donor tyrosine to acceptor tyrosinate can be altered by varying the pH of the solution, and the fractional ionization can be determined from absorption spectrophotometry. An added feature of the polymer poly-L-tyrosine is that its molecular conformation in solution has been established. The transfer of electronic energy between two types of chromophores was found, tyrosyl-to-tyrosyl and tyrosyl-to-tyrosinate. Each is in agreement with Förster theory. Energy-transfer rates have been used to establish that hexa-L-tyrosine and poly-L-tyrosine are right-handed α helices (Knopp *et al.*, 1969). Finally, energy transfer in biological systems is discussed extensively by Udenfriend (1969) and Mason (1966).

6.4 Laser Studies

Lasers have been built using rare-earth chelates. Some of the europium chelates, whose luminescence properties were investigated originally by Weissman and later by Crosby and co-workers (See Section 5.3), have recently been used successfully as laser materials (Lempicki and Samelson, 1963; Wolff and Pressley, 1963; Schimitschek, 1963).

Lasing action in liquids was first achieved in organic chelate systems in which the active ion Eu^{3+} was shielded from the solvent by a cage of four benzoylactonate ligands. Initially, operation was achieved only at lower temperatures, i.e., $-160°C$, but the results demonstrated that laser action could take place in a liquid. The emitted beam of red light at 6131 Å had all the properties found in a solid-state laser. A chelate was soon found that would lase at room temperature. It was a complex formed by Eu^{3+}, the ligand benzoyltrifluoracetonate, and the solvent acetonitrile (Lempicki and Samelson, 1966).

More efficient systems are those in which the rare-earth ions are solvated. Moreover, vibrational loss to the solvent can be reduced by employing selenium oxychloride, which contains no light atoms, as the solvent. In this kind of system, energy transfer takes place from Nb^{3+} to Yb^{3+} and from Nd^{3+} to Yb^{3+} and then to Er^{3+} (Heller, 1968).

For other work on the same subject, the reader is referred to Schimitschek *et al.* (1965), who observed laser action in fluorinated europium chelates in acetonitrile, and to El-Sayed and Bhaumik (1963), who studied triplet–triplet energy transfer to chelates.

Energy transfer can take place in solutions composed of two or more component dyes. The intensity and tuning range of lasers employing such mixed dye solutions have been found to be greatly enhanced (Glenn *et al.*, 1968; Peterson and Snavely, 1968).

Lasers, because of their intense beams and rapid pulse, are useful in many photochemical studies. Photodissociation of acceptor molecules has been achieved by a double photon excitation of the donor molecules. The radicals that are formed can be detected from their ESR spectrum. This biphotonic technique has been used by Rylkov *et al.* (1967) to excite a donor which, in turn, excited and dissociated an aliphatic compound (acceptor). With such a mechanism of photosensitization of organic molecules, a valence bond may be broken by photons of lower energy than is necessary for direct photolysis. Their results were interpreted as a triplet–triplet transfer process and not as a triplet–singlet process.

Orientation relaxation phenomena by means of picosecond laser pulses has been investigated by Eisenthal (1970) and Rehm and Eisenthal (1971). Because of these very short excitation pulses, they were able to study the effect of orientation of the donor on the energy-transfer process.

A laser beam can produce a high density of excited singlet states, which, by intersystem crossing, convert to the triplet state. Triplet + triplet annihilation occurs in crystals even at moderate excitation intensities (Nieman and Robinson, 1963). As stated above, this type of annihilation leads to delayed fluorescence.

7

DATA

The tables of data described in this chapter comprise Chapter 8.

7.1 Definitions

The quantity R_0 is defined as

$$R_0^6 = \frac{9000K^2Q_y \ln 10}{128\pi^5 n^4 N} \int \frac{f_D(\nu)\ \epsilon_A(\nu)}{\nu^4}\, d\nu,$$

where ν is the wave number, $\epsilon(\nu)$ is the molar extinction coefficient of the acceptor, $f(\nu)$ is the spectral distribution of donor fluorescence normalized such that $\int f(\nu)\, d\nu = 1$ and the limits of integration are over the relevent fluorescence spectrum, N is Avogadro's number, K^2 is an orientation factor, assumed to be equal to $\frac{2}{3}$, Q_y is the fluorescence quantum yield of the donor, and n is the index of refraction in the overlap region. The units of R_0 are angstroms.

The concentration C_0 is defined as

$$C_0 = \frac{3000}{2\pi^{3/2} N R_0^3} \text{ mole liter}^{-1}.$$

The overlap integral is equal to

$$\int \frac{f_D(\nu)\ \epsilon_A(\nu)}{\nu^4}\, d\nu.$$

Its units are 10^{-15} cm^6 mole^{-1}.

The values of R_0 are tabulated with benzene as the first donor, and all 210 compounds (including benzene) are the acceptors. In the second list, the second compound, deuterated benzene, is the donor, and all 210 compounds are the acceptors. And so on. When there is no overlap between the fluorescence spectrum of the donor and the absorption spectrum of the acceptor, the values of R_0, C_0, and the overlap integral will be zero. For certain donors, such as number 70, there are several pages just of zeroes. In the interest of economy, pages consisting only of zeroes have been omitted. When only one value of R_0 per page is different from zero, that value is listed at the top of the next page of tabulated values.

In the calculation of R_0, the absorption curves have all been extended to about 2600 Å, where possible. Thus the value of R_0 may include transfer to higher electronic states.

7.2 Evaluation of the Data

When R_0 is small, i.e., of molecular dimensions, it represents a lower limit of the transfer distance. Terms other than dipole–dipole terms may be active and may increase the transfer distance. In addition, the values of R_0 in liquids are larger than shown here because of an added diffusion term.

Förster (1959) maintains that transfer distances up to 100 Å have been calculated. The maximum value of R_0 in this collection appears to be 55.6 Å for the case where 9,10-bis(phenyl-ethynyl)anthracene is the donor and rhodamine 6G is the acceptor.

As pointed out in the text, the values are often in good agreement with published experimental and calculated values.

7.3 List of Numbered Compounds

The first table in Chapter 8 is a list of compounds in their numbered sequence, which is the same as that found in Berlman (1971). When the compound is a liquid at room temperature, its concentration in ml/liter has been included after its name. The word *trace* appears in those cases where the calculation was made using the fluorescence spectrum from solutions containing trace quantities of the material. Generally, cyclohexane was the common solvent. When solubility was a problem, other solvents were employed. The letter A indicates that the solvent was ethanol, and B indicates benzene.

REFERENCES

Agabekyan, A. S. (1970). *Opt. Spectrosc.* (*USSR*) **29,** 37.
Ageno, M., Chiozzotto, M., and Querzoli, R. (1949). *Atti Accad. Nazl. Lincei* **6,** 626.
Agrest, M. M., Kilin, S. F., Rikenglaz, M. M., and Rozman, I. M. (1969). *Opt. Spectrosc.* (*USSR*) **27,** 514.
Alexander, P., and Charlesby, A. (1954). *Nature* (*London*) **173,** 578.
Alfimov, M. V., Batekha, I. G., and Sheck, Y. B. (1968). *Bull. Acad. Sci. USSR Phys. Ser.* **32,** 1377.
Alfimov, M. V., Batekha, I. G., and Smirnov, V. A. (1969). *Dokl. Phys. Chem.* **185,** 201.
Alfimov, M. V., Batekha, I. G., Sheck, Y. B., and Gerko, V. I. (1971). *Spectrochim. Acta Part A* **27,** 329.
Antipenko, B. M., and Ermolaev, V. L. (1969). *Opt. Spectrosc.* (*USSR*) **26,** 415.
Antonov–Romanovski, V. V., and Galanin, M. D. (1957). *Opt. Spektrosk.* **3,** 389.
Arnold, S. J., Finlayson, N., and Ogryzlo, E. A. (1966). *J. Chem. Phys.* **44,** 2529.
Ashford, N. A., and Hutchison, Jr., C. A. (1971). *J. Chem. Phys.* **54,** 1165.
Aurich, F., and Marquard, J. (1971). *Chem. Phys. Lett.* **8,** 91.
Avakian, P., and Merrifield, R. E. (1964). *Phys. Rev. Lett.* **13,** 541.
Avakian, P., and Merrifield, R. E. (1968). *Mol. Cryst.* **5,** 37.
Avakian, P., Abramson, E., Kepler, R. G., and Caris, J. C. (1963). *J. Chem. Phys.* **39,** 1127.
Azumi, T., and McGlynn, S. P. (1963a). *J. Chem. Phys.* **38,** 2773.
Azumi, T., and McGlynn, S. P. (1963b). *J. Chem. Phys.* **39,** 1186.
Babenko, S. D., and Benderskii, V. A. (1970). *Opt. Spectrosc.* (*USSR*) **29,** 45.
Babenko, S. D., and Benderskii, V. A. (1971). *Opt. Spectrosc.* (*USSR*) **31,** 485.
Bäckström, H. L. J., and Sandros, K. (1958). *Acta Chem. Scand.* **12,** 823.
Bäckström, H. L. J., and Sandros, K. (1960). *Acta Chem. Scand.* **14,** 48.
Bagdasaryan, K. S., and Muler, A. L. (1965). *Opt. Spectrosc.* (*USSR*) **18,** 558.
Ballera, R. E., and Edward, J. W. (1965). *Spectrochim. Acta* **21,** 1353.
Banfield, T. L., and Husain, D. (1969). *Trans. Faraday Soc.* **65,** 1985.
Bannister, T. (1954). *Arch. Biochem. Biophys.* **49,** 222.
Basile, L. J. (1964). *Trans. Faraday Soc.* **60,** 1702.
Baxendale, J. H., Beaumond, D., and Rodgers, M. A. J. (1969). *Chem. Phys. Lett.* **4,** 3.

Belikova, T. P., and Galanin, M. D. (1956). *Opt. Spectrosk.* **1,** 168.
Belikova, T. P., and Galanin, M. D. (1958). *Bull. Acad. Sci. USSR Phys. Ser.* **22,** 46.
Bell, P. H., and Davey, W. P. (1941). *J. Chem. Phys.* **9,** 441.
Bennett, R. G. (1964a). *J. Chem. Phys.* **41,** 3037.
Bennett, R. G. (1964b). *J. Chem. Phys.* **41,** 3048.
Bennett, R. G., and Kellogg, R. E. (1967). *Progr. React. Kinet.* **4,** 217.
Bennett, R. G., Schwenker, R. P., and Kellogg, R. E. (1964). *J. Chem. Phys.* **41,** 3040.
Bensasson, R., and Land, E. J. (1971). *Trans. Faraday Soc.* **67,** 1904.
Berlman, I. B. (1960). *J. Chem. Phys.* **33,** 1124.
Berlman, I. B. (1965). "Handbook of Fluorescence Spectra of Aromatic Molecules." Academic Press, New York.
Berlman, I. B. (1971). "Handbook of Fluorescence Spectra of Aromatic Molecules," 2nd ed. Academic Press, New York.
Berlman, I. B. (1973). *J. Phys. Chem.* **77,** 562.
Berlman, I. B., and Walter, T. A. (1962). *J. Chem. Phys.* **37,** 1888.
Bhaumik, M. L., and El-Sayed, M. A. (1965a). *J. Chem. Phys.* **42,** 787.
Bhaumik, M. L., and El-Sayed, M. A. (1965b). *J. Phys. Chem.* **69,** 275.
Birks, J. B. (1958). *J. Chem. Phys.* **28,** 730.
Birks, J. B. (1970a). *J. Phys. B* **3,** 1704.
Birks, J. B. (1970b). "Photophysics of Aromatic Molecules." Wiley (Interscience), New York.
Birks, J. B., and Conte, J. C. (1968). *Proc. Roy. Soc. Ser. A* **303,** 85.
Birks, J. B., and Georghiou, S. (1968). *J. Phys. B* **1,** 958.
Birks, J. B., and Leite, M. S. S. C. P. (1970). *J. Phys. B* **3,** 513.
Blake, N. W., and McClure, D. S. (1958). *J. Chem. Phys.* **29,** 722.
Bojarski, C. (1962). *Acta Phys. Pol.* **22,** 211.
Bonnier, J. M., and Jardon, P. (1970). *J. Chim. Phys. Physicochim. Biol.* **67,** 577.
Bonnier, J. M., and Jardon, P. (1971). *J. Chim. Phys. Physicochim. Biol.* **68,** 432.
Bowen, E. J., and Brocklehurst, B. (1953). *Trans. Faraday Soc.* **49,** 1131.
Bowen, E. J., and Brocklehurst, B. (1955). *Trans. Faraday Soc.* **51,** 774.
Bowen, E. J., and Livingston, R. (1954). *J. Amer. Chem. Soc.* **76,** 6300.
Bowen, E. J., and Mikiewicz, E. (1947). *Nature (London)* **159,** 706.
Bowen, E. J., and Williams, A. H. (1939). *Trans. Faraday Soc.* **35,** 765.
Bowen, E. J., Mikiewicz, E., and Smith, F. (1949). *Proc. Phys. Soc. London A* **62,** 26.
Brandon, R. W., Gerkin, R. E., and Hutchison, Jr., C. A. (1962). *J. Chem. Phys.* **37,** 447.
Breen, D. E., and Keller, R. A. (1968). *J. Amer. Chem. Soc.* **90,** 1935.
Brinen, J. S., and Gerhardt, G. E. (1970). *Mol. Photochem.* **2,** 121.
Briskina, C. M., Zolin, V. F., and Samokhina, M. A. (1966). *Opt. Spectrosc. (USSR)* **20,** 599.
Brocklehurst, B., Gibbons, W. A., Lang, F. T., Porter, G., and Savadatti, M. I. (1966). *Trans. Faraday Soc.* **62,** 1793.
Brodsky, A. M., Kolbanovsky, Y. A., and Polak, L. S. (1962). *Int. J. Appl. Radiat. Isotop.* **13,** 143.
Bromer, H. H., and Hesse, J. (1969). *Z. Phys.* **219,** 269.
Brown, F. H., Furst, M., and Kallmann, H. (1959). *Discuss. Faraday Soc.* **27,** 43.
Bücher, T., and Kaspers, J. (1947). *Biochim. Biophys. Acta* **1,** 21.
Bücher, H., Drexhage, K. H., Fleck, M., Kuhn, J., Möbius, D., Schafer, F. P., Sondermann, J., Sperling, W., Tillman, P., and Wiegand, J. (1967). *Mol. Cryst.* **2,** 199.
Bunting, J. R., and Filipescu, N. (1970). *J. Chem. Soc.* p. 1750.
Burton, M., and Dreeskamp, H. (1957). *Trans. Faraday Soc.* **27,** 64.

Burton, M., Funabashi, K., Hentz, R. R., Ludwig, P. K., Magee, J. L., and Mozumder, A. (1969). *In* "Transfer and Storage of Energy in Molecules" (G. M. Burnett and A. M. North, eds.), Vol. 1, p. 161. Wiley, New York.
Cario, C., and Franck, J. (1923). *Z. Phys.* **17,** 202.
Chauduri, N. K., and El-Sayed, M. A. (1965). *J. Chem. Phys.* **42,** 1947.
Chu, N. Y., Kawaoka, K., and Kearns, D. R. (1971). *J. Chem. Phys.* **55,** 3059.
Cohen, S., and Weinreb, A. (1954). *Phys. Rev.* **93,** 1117.
Cohen, S., and Weinreb, A. (1956). *Proc. Phys. Soc. London B* **69,** 593.
Collinson, E., Conlay, J. J., and Dainton, F. S. (1963). *Discuss. Faraday Soc.* **36,** 153.
Conrad, R. H., and Brand, L. (1968). *Biochemistry* **7,** 777.
Coumou, D. J., and Mackor, E. L. (1964). *Trans. Faraday Soc.* **60,** 1726.
Cowan, D. O., and Baum, A. A. (1970). *J. Amer. Chem. Soc.* **92,** 2153.
Cozzens, R. F., and Fox, R. B. (1969). *J. Chem. Phys.* **50,** 1532.
Crosby, G. A., and Kasha, M. (1958). *Spectrochim. Acta* **10,** 377.
Crosby, G. A., Whan, R. E., and Alire, R. M. (1961). *J. Chem. Phys.* **34,** 743.
Crosby, G. A., Whan, R. E., and Freeman, J. J. (1962). *J. Phys. Chem.* **66,** 2493.
Cundall, R. B. (1969). *In* "Transfer and Storage of Energy by Molecules" (G. M. Burnett and A. M. North, eds.), Vol. I, p. 1. Wiley (Interscience), New York.
Cundall, R. B., and Griffiths, P. A. (1965). *Trans. Faraday Soc.* **61,** 1968.
Cundall, R. B., and Griffiths, P. A. (1966a). *In* "Energy Transfer in Radiation Processes" (G. O. Phillips, ed.), p. 60. Elsevier, Amsterdam.
Cundall, R. B., and Griffiths, P. A. (1966b). *Chem. Commun.* p. 195.
Cundall, R. B., and Palmer, T. F. (1960). *Trans. Faraday Soc.* **56,** 1211.
Cundall, R. B., Fletcher, F. J., and Milne, D. G. (1964). *Trans. Faraday Soc.* **60,** 1146.
Dainton, F. S., and Ivin, K. J. (1950). *Trans. Faraday Soc.* **46,** 382.
David, C., Demarteau, W., and Geuskins, G. (1970). *Eur. Polym. J.* **6,** 537.
Davydov, A. S. (1948). *Zh. Eksp. Teor. Fiz.* **18,** 210.
Davydov, A. S. (1951). *Zh. Eksp. Teor. Fiz.* **21,** 673.
Davydov, A. S. (1962). "Theory of Molecular Excitons" (translated by M. Kasha and M. Oppenheimer). McGraw-Hill, New York.
Davydov, A. S. (1968). *Phys. Status Solidi* **30,** 357.
De Groot, K., Gary, L. P., and Jarnagin, R. C. (1968). *J. Chem. Phys.* **48,** 5280.
DeGroot, M. S., and van der Waals, J. H. (1963). *Mol. Phys.* **6,** 545.
DeLosh, R. G., and Grant, W. J. (1970). *Phys. Rev. B* **1,** 1754.
DeMember, J. R., and Filipescu, N. (1968). *J. Amer. Chem. Soc.* **90,** 6425.
Dexter, D. L. (1953). *J. Chem. Phys.* **21,** 836.
Dexter, D. L., Förster, T., and Knox, R. S. (1969). *Phys. Status Solidi* **34,** K159.
Dillon, M. A., and Burton, M. (1965). *In* "Pulse Radiolysis" (M. Ebert, J. P. Keene, A. J. Swallow, and J. H. Baxendale, eds.), p. 259. Academic Press, New York.
Dole, M., and Williams, T. F. (1959). *Discuss. Faraday Soc.* **27,** 74.
Donnini, J. M., and Abetino, F. (1968). *C. R. Acad. Sci. Ser. B* **266,** 1618.
Dow, J. D. (1968). *Phys. Rev.* **174,** 962.
Drexhage, K. H., Zwick, M. M., and Kuhn, H. (1963). *Ber. Bunsenges. Phys. Chem.* **67,** 62.
Dubois, J. T. (1959). *J. Phys. Chem.* **63,** 638.
Dubois, J. T., and Cox, M. (1963). *J. Chem. Phys.* **38,** 2536.
Dubois, J. T., and Noyes, Jr., W. A. (1951). *J. Chem. Phys.* **19,** 1512.
Dutton, H. J., Manning, W. M., and Duggar, B. M. (1943). *J. Phys. Chem.* **47,** 308.
Duysen, L. N. M. (1951). *Nature (London)* **169,** 548.
Duysen, L. N. M. (1952). Thesis, Univ. of Utrecht, Netherlands.

Duysen, L. N. M. (1956). *In* "Photosynthesis" (E. Rabinowitch, ed.), p. 1874. Wiley (Interscience), New York.
Duysen, L. N. M. (1964). *Progr. Biophys. Mol. Biol.* **14,** 1.
Edelman, G. M., and McClure, W. O. (1968). *Accounts Chem. Res.* **1,** 65.
Eisenthal, K. B. (1969). *J. Chem. Phys.* **50,** 3120.
Eisenthal, K. B. (1970). *Chem. Phys. Lett.* **6,** 155.
Eisenthal, K. B., and Murashige, R. (1963). *J. Chem. Phys.* **39,** 2108.
Eisenthal, K. B., and Siegel, S. (1964). *J. Chem. Phys.* **41,** 652.
Eisinger, J. (1969a). *Biochemistry* **8,** 3902.
Eisinger, J. (1969b). *Photochem. Photobiol.* **9,** 247.
Eisinger, J., Feuer, B., and Lamola, A. A. (1969). *Biochemistry* **8,** 3908.
Elkana, Y., Feitelson, J., and Katchalski, E. (1968). *J. Chem. Phys.* **48,** 2399.
El-Sayed, M. A., and Bhaumik, M. L. (1963). *J. Chem. Phys.* **39,** 2391.
El-Sayed, M. A., Wauk, M. T., and Robinson, G. W. (1962). *Mol. Phys.* **5,** 205.
El-Sayed, M. A., Tinti, D. S., and Lee, E. M. (1969). *J. Chem. Phys.* **51,** 5721.
Eriksen, E. L., and Ore, A. (1967). *Phys. Norv.* **2,** 159.
Eriksen, E. L., and Ore, A. (1971). *Phys. Norv.* **5,** 45.
Ermolaev, V. L. (1959). *Opt. Spectrosc.* (*USSR*) **6,** 417.
Ermolaev, V. L. (1962). *Opt. Spectrosc.* (*USSR*) **13,** 49.
Ermolaev, V. L. (1963). *Sov. Phys. Usp.* **6,** 333.
Ermolaev, V. L. (1964). *Opt. Spectrosc.* (*USSR*) **16,** 299.
Ermolaev, V. L. (1968). *Bull. Acad. Sci. USSR Phys. Ser.* **32,** 1193.
Ermolaev, V. L., and Shakhverdov, T. A. (1969). *Opt. Spectrosc.* (*USSR*) **26,** 459.
Ermolaev, V. L., and Shakhverdov, T. A. (1971). *Opt. Spectrosc.* (*USSR*) **30,** 648.
Ermolaev, V. L., and Sveshnikova, E. B. (1962). *Bull. Acad. Sci. USSR Phys. Ser.* **26,** 29.
Ermolaev, V. L., and Sveshnikova, E. B. (1963). *Sov. Phys. Dokl.* **8,** 373.
Ermolaev, V. L., and Sveshnikova, E. B. (1964). *Opt. Spectrosc.* (*USSR*) **16,** 320.
Ermolaev, V. L., and Sveshnikova, E. B. (1970). *Opt. Spectrosc.* (*USSR*) **28,** 324.
Ermolaev, V. L., and Tachin, V. S. (1969). *Opt. Spectrosc.* (*USSR*) **27,** 546.
Ermolaev, V. L., and Tachin, V. S. (1970). *Opt. Spectrosc.* (*USSR*) **29,** 49.
Ermolaev, V. L., and Terenin, A. N. (1958). *J. Chim. Phys. Physico Chim. Biol.* **55,** 698.
Ermolaev, V. L., and Terenin, A. N. (1960). *Sov. Phys. Usp.* **3,** 423.
Ermolaev, V. L., Aleshin, V. G., and Saenko, E. A. (1966). *Sov. Phys. Dokl.* **10,** 1186.
Ermolaev, V. L., Sveshnikova, E. B., and Saenko, E. A. (1967). *Opt. Spectrosc.* (*USSR*) **22,** 86.
Ermolaev, V. L., Saenko, E. A., Domrachev, G. A., Khudenski, Y. K., and Aleshin, V. G. (1967b). *Opt. Spectrosc.* (*USSR*) **22,** 466.
Farmer, J. B., Gardner, C. L., and McDowell, C. A. (1961). *J. Chem. Phys.* **34,** 1058.
Feitelson, J. (1966). *J. Chem. Phys.* **44,** 1497.
Feofilov, P. P., and Sveshnikov, B. Y. (1940). *J. Phys.* (*USSR*) **3,** 493.
Filipescu, N., DeMember, J. R., and Minn, F. L. (1969). *J. Amer. Chem. Soc.* **91,** 4169.
Finger, G., and Zahlon, A. B. (1969). *J. Chem. Phys.* **50,** 25.
Förster, T. (1948). *Ann. Phys.* **2,** 55.
Förster, T. (1949). *Z. Naturforsch. A* **4,** 321.
Förster, T. (1951). "Fluoreszenz Organisher Verbindungen." Vandenhoek and Ruprecht, Göttingen.
Förster, T. (1959). *Discuss. Faraday Soc.* **27,** 7.
Förster, T. (1960a). *In* "Comparative Effects of Radiation" (M. Burton, J. S. Kirby-Smith, and J. L. Magee, eds.), p. 300. Wiley, New York.

Förster, T. (1960b). *Radiat. Res. Suppl.* **2,** 326.
Förster, T. (1966). *In* "Modern Quantum Chemistry" (O. Sinanoglu, ed.), Vol. 3, p. 93. Academic Press, New York.
Förster, T. (1967). *In* "Comprehensive Biochemistry" (M. Florkin and E. H. Stotz, eds.), Vol. 22, p. 61. Elsevier, Amsterdam.
Fox, R. B., and Cozzens, R. F. (1969). *Macromolecules* **2,** 181.
Franck, J. (1922). *Z. Phys.* **9,** 259.
Franck, J. (1958). *Proc. Nat. Acad. Sci. U.S.* **44,** 941.
Freeman, J. J., and Crosby, G. A. (1963). *J. Phys. Chem.* **67,** 2717.
Furst, M., and Kallmann, H. (1954). *Phys. Rev.* **94,** 503.
Galanin, M. D. (1955). *Sov. Phys. JETP* **1,** 317.
Galanin, M. D., and Levshin, V. L. (1951). *Zh. Eksp. Teor. Fiz.* **21,** 121.
Gallagher, P. K., Heller, A., and Wasserman, E. (1964). *J. Chem. Phys.* **41,** 3921.
Ganguly, S. C., and Chaudhury, N. K. (1959). *Rev. Mod. Phys.* **31,** 990.
Gaviola, E., and Pringsheim, P. (1924). *Z. Phys.* **24,** 24.
Geacintov, N. E., Pope, M., and Vogel, F. (1969). *Phys. Rev. Lett.* **22,** 593.
Glenn, W. H., Brienza, M. J., and DeMaria, A. J. (1968). *Appl. Phys. Lett.* **12,** 54.
Goedheer, J. C. (1957). Thesis, Univ. of Utrecht, Netherlands.
Goeppert-Mayer, M. (1931). *Ann. Phys.* **9,** 273.
Goldschmidt, C. R., Tomkiewicz, Y., and Berlman, I. B. (1968). *Chem. Phys. Lett.* **2,** 520.
Golub, M. A., and Stephens, C. L. (1966). *J. Phys. Chem.* **70,** 3576.
Golubov, S. I., and Konobeev, Y. V. (1972). *Sov. Phys. Solid State* **13,** 2679.
Grant, W. J. C. (1971). *Phys. Rev. B* **4,** 648.
Gueron, M., Eisinger, J., and Shulman, R. G. (1967). *J. Chem. Phys.* **47,** 4077.
Guzzo, A. V., and Pool, G. L. (1969). *J. Phys. Chem.* **73,** 2512.
Hall, J. C., Jennings, D. A., and McClintock, R. M. (1963). *Phys. Rev. Lett.* **11,** 364.
Hammond, G. S., and Hardham, W. M. (1963). *Proc. Chem. Soc. London* p. 63.
Hammond, G. S., Turro, N. J., and Leermakers, P. A. (1962). *J. Phys. Chem.* **66,** 1144.
Hammond, G. S., Turro, N. J., and Liu, R. S. H. (1963). *J. Org. Chem.* **28,** 3297.
Hammond, G. S., Stout, C. A., and Lamola, A. A. (1964). *J. Amer. Chem. Soc.* **86,** 3103.
Hatch, G. F., and Nieman, G. C. (1968). *J. Chem. Phys.* **48,** 4116.
Heicklen, J., and Noyes, Jr., W. A. (1959). *J. Amer. Chem. Soc.* **81,** 3858.
Heller, A. (1968). *J. Mol. Spectrosc.* **28,** 208.
Heller, A., and Wasserman, E. (1965). *J. Chem. Phys.* **42,** 949.
Henrich, G., Blume, H., and Schulte-Frohlinde, D. (1967). *Tetrahedron Lett.* p. 4693.
Hentz, R. R., Peterson, D. B., Srivastova, S. B., Barzynski, H. F., and Burton, M. (1966). *J. Phys. Chem.* **70,** 2362.
Herkstroeter, W. G., and McClure, D. S. (1968). *J. Amer. Chem. Soc.* **90,** 4522.
Hirayama, F. (1963). Thesis, Univ. of Michigan, Ann Arbor, Michigan.
Hirayama, F., and Lipsky, S. (1969). *J. Chem. Phys.* **51,** 3616.
Hirayama, F., and Lipsky, S. (1971). *In* "Organic Scintillators and Liquid Scintillation Counting" (D. L. Horrocks and C. T. Peng, eds.), p. 205. Academic Press, New York.
Hirayama, F., Rothman, W., and Lipsky, S. (1970). *Chem. Phys. Lett.* **5,** 296.
Hirota, N., and Hutchison, Jr., C. A. (1965). *J. Chem. Phys.* **42,** 2869.
Hisatsune, I. C., and Jayadevappa, E. S. (1960). *J. Chem. Phys.* **32,** 565.
Hochstrasser, R. M. (1963). *J. Chem. Phys.* **39,** 3153.
Hochstrasser, R. M. (1964). *J. Chem. Phys.* **40,** 1038.

Holroyd, R. A., Yang, J. Y., and Servedio, F. M. (1967). *J. Chem. Phys.* **46,** 4540.
Hudson, B. C., and Curnette, Jr., B. (1966). *Phys. Rev.* **152,** 56.
Hudson, J. A., and Hedges, R. M. (1969). *In* "Molecular Luminescence" (E. C. Lim, ed.), p. 667. Benjamin, New York.
Hunt, G. R., McCoy, E. F., and Ross, I. G. (1962). *Aust. J. Chem.* **15,** 591.
Hunt, J. W., and Thomas, J. K. (1967). *J. Chem. Phys.* **46,** 2954.
Hunter, T. F., McAlpine, R. D., and Hochstrasser, R. M. (1969). *J. Chem. Phys.* **50,** 1140.
Hurt, W. B. (1966). *J. Chem. Phys.* **45,** 2713.
Hutton, E., and Stevens, B. (1962). *Spectrochim. Acta* **18,** 425.
Inokuti, M., and Hirayama, F. (1965). *J. Chem. Phys.* **43,** 1978.
Isenberg, I., Leslie, R. B., Baird, Jr., S. L., Rosenbluth, R., and Bersohn, R. (1965). *Proc. Nat. Acad. Sci. U.S.* **52,** 379.
Ishikawa, H., and Noyes, Jr., A. W. (1962). *J. Chem. Phys.* **37,** 583.
Jablonski, A. (1970). *Acta Phys. Polon. A* **38,** 453.
Johnson, P. C., and Offen, H. W. (1972). *J. Chem. Phys.* **57,** 1473.
Johnson, R. C., Merrifield, R. E., Avakian, P., and Flippen, R. B. (1967). *Phys. Rev. Lett.* **19,** 285.
Jortner, J., Choi, S. I., Katz, J. L., and Rice, S. A. (1963). *Phys. Rev. Lett.* **11,** 323.
Jortner, J., Rice, S. A., Katz, J. L., and Choi, S. I. (1965a). *J. Chem. Phys.* **42,** 309.
Jortner, J., Rice, S. A., and Silbey, R. (1965b). *In* "Modern Quantum Chemistry" (O. Sinanoglu, ed.), Part III, p. 139. Academic Press, New York.
Judeikis, H. S., and Siegel, S. (1970). *J. Chem. Phys.* **53,** 3500.
Kallmann, H. (1950). *Phys. Rev.* **78,** 621.
Kallmann, H., and Furst, M. (1950). *Phys. Rev.* **79,** 857.
Kallmann, H., and London, F. (1928). *Z. Phys. Chem. Abt. B* **2,** 207.
Kamen, M. D. (1963). "Primary Processes in Photosynthesis," p. 117. Academic Press, New York.
Kasha, M. (1963). *Radiat. Res.* **20,** 55.
Katsuura, K., and Inokuti, M. (1964). *J. Chem. Phys.* **41,** 989.
Kawada, A., and Jarnagin, R. C. (1966). *J. Chem. Phys.* **44,** 1919.
Kawski, A. (1961). *Ann. Phys.* **7,** 116.
Kawski, A. (1963). *Z. Naturforsch. A* **18,** 961.
Kawski, A., and Kroonicka, M. (1965). *Z. Naturforsch. A* **20,** 917.
Kawski, A., Kaminski, J., and Kuten, E. (1971). *J. Phys. B* **4,** 609.
Keller, R. A. (1968). *J. Amer. Chem. Soc.* **90,** 1940.
Keller, R. A., and Dolby, L. J. (1967). *J. Amer. Chem. Soc.* **89,** 2768.
Keller, R. A., and Dolby, L. J. (1969). *J. Amer. Chem. Soc.* **91,** 1293.
Kellogg, R. E. (1964). *J. Chem. Phys.* **41,** 3046.
Kellogg, R. E. (1967). *J. Chem. Phys.* **47,** 3403.
Kellogg, R. E., and Bennett, R. G. (1964). *J. Chem. Phys.* **41,** 3042.
Kepler, R. G., Caris, J. C., Avakian, P., and Abramson, E. (1963). *Phys. Rev. Lett.* **10,** 400.
Kilin, S. F., Mikhelashvili, M. S., and Rozman, I. M. (1964). *Opt. Spectrosc.* (*USSR*) **16,** 576.
King, T. A., and Voltz, R. (1966). *Proc. Roy. Soc. Ser. A* **289,** 424.
Klöpffer, W. (1969a). *J. Chem. Phys.* **50,** 1689.
Klöpffer, W. (1969b). *J. Chem. Phys.* **50,** 2337.

Klöpffer, W., and Bauser, H. (1970). *Chem. Phys. Lett.* **6**, 279.
Knopp, J. A., ten Bosch, J. J., and Longworth, J. W. (1969). *Biochem. Biophys. Acta* **188**, 185.
Knox, R. S. (1968). *Physica (Utrecht)* **39**, 361.
Kobyshev, G. I., Lyalin, G. N., and Terenin, A. N. (1966). *Opt. Spectrosc. (USSR)* **21**, 74.
Konev, S. V. (1967). "Fluorescence and Phosphorescence of Proteins and Nucleic Acids," p. 107. Plenum, New York.
Kopecky, K. R., Hammond, G. S., and Leermakers, P. A. (1962). *J. Amer. Chem. Soc.* **84**, 1015.
Kotsubanov, V. D., Naboikin, Y. V., Ogurtsova, L. A., Podgornyi, A. P., and Pokrovskaya, F. S. (1968). *Opt. Spectrosc. (USSR)* **25**, 251.
Kottis, P. (1965). *Phys. Lett.* **17**, 206.
Krishna, V. G. (1967). *J. Chem. Phys.* **46**, 1735.
Krongauz, V. A. (1964). *Dokl. Phys. Chem.* **155**, 333.
Kudryashov, P. I., Sveshnikov, B. Y., and Shirokov, V. I. (1960). *Opt. Spectrosc. (USSR)* **9**, 177.
Kuhn, H. (1970). *J. Chem. Phys.* **53**, 101.
Kurskii, Y. A., and Selivanenko, A. S. (1960). *Opt. Spectrosc. (USSR)* **8**, 340.
Kuznetsova, P. A., Sveshnikov, B. Y., and Tchirokov, V. I. (1957). *Opt. Spektrosk.* **2**, 578.
Lamola, A. A. (1969a). *In* "Energy Transfer and Organic Photochemistry," Vol. XIV, "Techniques of Organic Chemistry" (A. Weissberger, ed.), p. 17. Wiley, New York.
Lamola, A. A. (1969b). *J. Amer. Chem. Soc.* **91**, 4786.
Lamola, A. A. (1970). *J. Amer. Chem. Soc.* **92**, 5045.
Lamola, A. A., and Eisinger, J. (1969). *In* "Molecular Luminescence" (E. C. Lim, ed.), p. 801. Benjamin, New York.
Lamola, A. A., and Hammond, G. S. (1965). *J. Chem. Phys.* **43**, 2129.
Lamola, A. A., Leermakers, P. A., Byers, G. W., and Hammond, G. S. (1965). *J. Amer. Chem. Soc.* **87**, 2322.
Land, E. J., and Swallow, A. J. (1968). *J. Chem. Phys.* **49**, 5552.
Laor, U., and Weinreb, A. (1965). *J. Chem. Phys.* **43**, 1080.
Lashkov, G. I., and Ermolaev, V. L. (1967). *Opt. Spectrosc. (USSR)* **22**, 462.
Lashkov, G. I., Ermolaev, V. L., and Shablya, A. V. (1966). *Opt. Spectrosc. (USSR)* **21**, 305.
Latt, S. A., Cheung, H. T., and Blout, E. R. (1965). *J. Amer. Chem. Soc.* **87**, 995.
Lavorel, J. (1967). *J. Chem. Phys.* **47**, 2235.
Lempicki, A., and Samelson, H. (1963). *Phys. Lett.* **4**, 133.
Lempicki, A., and Samelson, H. (1966). *In* "Lasers" (A. K. Levine, ed.), Vol. 1, p. 181. Dekker, New York.
Lerman, L. S. (1961). *J. Mol. Biol.* **3**, 18.
Lerman, L. S. (1963). *Proc. Nat. Acad. Sci. U.S.* **49**, 94.
Levshin, V. L. (1924). *Z. Phys.* **26**, 274.
Lipsett, F. R. (1967). *Mol. Cryst.* **3**, 1.
Lipsett, F. R. (1968). *Mol. Cryst.* **5**, 9.
Lipsky, S. (1963). *J. Chem. Phys.* **38**, 2786.
Lipsky, S. (1964). *In* "Physical Processes in Radiation Biology" (L. Augenstein, R. Mason, and B. Rosenberg, eds.), p. 215. Academic Press, New York.
Lisovskaya, I. A., Zakharova, M. D., and Kaplan, I. G. (1967). *Opt. Spectrosc. (USSR)* **23**, 348.
Liu, R. S. H. (1968). *J. Amer. Chem. Soc.* **90**, 1899.

Liu, R. S. H., and Edman, J. R. (1969). *J. Amer. Chem. Soc.* **91,** 1492.
Liu, R. S. H., and Kellogg, R. E. (1969). *J. Amer. Chem. Soc.* **91,** 250.
Livingston, R. (1957). *J. Phys. Chem.* **61,** 860.
Livingston, R., and Owen, K. E. (1956). *J. Amer. Chem. Soc.* **78,** 3301.
Livingston, R., and Pugh, A. C. (1959). *Discuss. Faraday Soc.* **27,** 144.
Longworth, J. W., ten Bosch, J. J., Knopp, J. A., and Rahn, R. O. (1969). *In* "Molecular Luminescence" (E. C. Lim, ed.), p. 529. Benjamin, New York.
Ludwig, P. K., and Huque, M. M. (1968). *J. Chem. Phys.* **49,** 805.
Lund, R. E., and Oskam, H. J. (1969). *Z. Phys.* **219,** 131.
McCarthy, W. J., and Wineforder, J. D. (1966). *Anal. Chem.* **38,** 848.
Maksimov, M. Z., and Rozman, I. M. (1962). *Opt. Spectrosc.* (*USSR*) **12,** 337.
Manion, J. P., and Burton, M. (1952). *J. Phys. Chem.* **56,** 560.
Mason, R. (1966). *In* "Energy Transfer in Radiation Processes" (G. O. Phillips, ed.), p. 127. Elsevier, Amsterdam.
Mataga, N., Obashi, H., and Okada, T. (1967). *Chem. Phys. Lett.* **1,** 133.
Mataga, N., Obashi, H., and Okada, T. (1969). *J. Phys. Chem.* **73,** 370.
Matovich, E., and Suzuki, C. K. (1963). *J. Chem. Phys.* **39,** 1442.
Matsuyama, A., and Nagata, C. (1970). *Biochim. Biophys. Acta* **224,** 588.
Melhuish, W. H. (1963). *J. Phys. Chem.* **67,** 1681.
Merrifield, R. E. (1968). *J. Chem. Phys.* **48,** 4318.
Merrifield, R. E., Avakian, P., and Groff, R. P. (1969). *Chem. Phys. Lett.* **3,** 386.
Mikhelashvili, M. S. (1971). *Opt. Spectrosc.* (*USSR*) **30,** 141.
Misra, T. N., and McGlynn, S. P. (1966). *J. Chem. Phys.* **44,** 3816.
Miyakawa, T., and Dexter, D. L. (1970). *Phys. Rev. B* **1,** 2961.
Morina, V. F., and Sveshnikova, E. B. (1971). *Opt. Spectrosc.* (*USSR*) **31,** 316.
Morina, V. F., Ermolaev, V. L., and Khudenski, Y. K. (1967). *Opt. Spectrosc.* (*USSR*) **23,** 349.
Munster, A. (1950). *Trans. Faraday Soc.* **46,** 165.
Nafisi-Movaghar, J., Birks, J. B., and Razi Naqvi, K. (1967). *Proc. Phys. Soc.* (*London*) **91,** 449.
Needler, W. C. (1965). *J. Chem. Phys.* **42,** 2972.
Neporent, B. S. (1950). *Zh. Fiz. Khim.* **24,** 1219.
Nieman, G. C., and Robinson, G. W. (1962). *J. Chem. Phys.* **37,** 2150.
Nieman, G. C., and Robinson, G. W. (1963). *J. Chem. Phys.* **38,** 1326.
Nordin, S., and Strong, R. L. (1968). *Chem. Phys. Lett.* **2,** 429.
Nosworthy, J., Magee, J. L., and Burton, M. (1961). *J. Chem. Phys.* **34,** 83.
Ohigashi, H., Shirotani, I., Inokuchi, H., and Minomura, S. (1964). *J. Phys. Soc. Jap.* **19,** 1996.
Okabe, H., and Noyes, Jr., W. A. (1958). *J. Amer. Chem. Soc.* **79,** 801.
Ore, A. (1959). *J. Chem. Phys.* **31,** 442.
Ore, A. (1960). *J. Chem. Phys.* **33,** 31.
Ore, A., and Eriksen, E. L. (1971). *Phys. Norv.* **5,** 57.
Parker, C. A. (1963). *Proc. Roy. Soc.* (*London*) **A276,** 125.
Parker, C. A. (1964a). *Advan. Photochem.* **2,** 305.
Parker, C. A. (1964b). *Trans. Faraday Soc.* **60,** 1998.
Parker, C. A. (1968). "Photoluminescence of Solutions." Elsevier, Amsterdam.
Parker, C. A., and Hatchard, C. G. (1961). *Trans. Faraday Soc.* **57,** 1894.
Parker, C. A., and Hatchard, C. G. (1962). *Proc. Chem. Soc.* p. 147.
Parker, C. A., and Joyce, T. A. (1966). *Trans. Faraday Soc.* **62,** 2785.
Parker, C. A., and Joyce, T. A. (1967). *Chem. Commun.* p. 744.

Parker, C. A., Hatchard, C. G., and Joyce, T. A. (1965). *Analyst (London)* **90,** 1.
Parmenter, C. S., and Ring, B. L. (1967). *J. Chem. Phys.* **46,** 1998.
Partridge, R. H. (1970a). *J. Chem. Phys.* **52,** 2485.
Partridge, R. H. (1970b). *J. Chem. Phys.* **52,** 2491.
Partridge, R. H. (1970c). *J. Chem. Phys.* **52,** 2501.
Perkins, W. G. (1968). *J. Chem. Phys.* **48,** 931.
Perlman, R. L., von Zyl, A., and Edelhock, H. (1968). *J. Amer. Chem. Soc.* **90,** 2167.
Perrin, F. (1924). *C. R. Acad. Sci.* **178,** 1978.
Perrin, F. (1932). *Ann. Chim. Phys.* **17,** 283.
Perrin, J. (1925). *Cons. Chim. Solvay, 2nd,* p. 322. Gauthier-Villar, Paris.
Perrin, J. (1927). *C. R. Acad. Sci.* **184,** 1097.
Perrin, J., and Choucroun, Mlle. (1927). *C. R. Acad. Sci.* **184,** 1097.
Peterson, O. G., and Snavely, B. B. (1968). *Bull. Amer. Phys. Soc.* **13,** 397.
Peticolas, W. L., Goldsborough, J. R., and Rieckhoff, K. E. (1963). *Phys. Rev. Lett.* **10,** 43.
Pilloff, H. S., and Albrecht, A. C. (1966). *Nature (London)* **212,** 499.
Pope, M., Geacintov, N. E., and Vogel, F. (1969). *Mol. Cryst. Liquid Cryst.* **6,** 83.
Porter, G., and Wilkinson, F. (1961). *Proc. Roy. Soc. Ser. A* **264,** 1.
Porter, G., and Windsor, M. W. (1953). *J. Chem. Phys.* **21,** 2088.
Porter, G., and Wright, F. J. (1955). *Trans. Faraday Soc.* **51,** 1205.
Powell, R. C. (1971). *Phys. Rev. B* **4,** 628.
Prileshayeva, N. (1935). *Acta Physicochim.* **1,** 785.
Prileshayeva, N., and Klimova, A. (1937). *Acta Physicochim.* **7,** 163.
Rabinowitch, E. I. (1956). "Photosynthesis and Related Processes," Vol. 2, p. 812. Wiley (Interscience), New York.
Rabinowitch, E. I. (1957). *J. Phys. Chem.* **61,** 870.
Rabinowitch, E. I. (1959). *Discuss. Faraday Soc.* **27,** 161.
Rauh, R. D., Evans, T. R., and Leermakers, P. A. (1968). *J. Amer. Chem. Soc.* **90,** 6897.
Rauh, R. D., Evans, T. R., and Leermakers, P. A. (1969). *J. Amer. Chem. Soc.* **91,** 1868.
Rebbert, R. E., and Ausloos, P. (1964). *J. Amer. Chem. Soc.* **86,** 4803.
Rehm, D., and Eisenthal, K. B. (1971). *Chem. Phys. Lett.* **9,** 387.
Rentzepis, P. M. (1969). *Chem. Phys. Lett.* **3,** 717.
Reynolds, G. T., Harrison, F. B., and Salvini, G. (1950). *Phys. Rev.* **78,** 488.
Richtol, H. H., and Klappmeier, F. H. (1966). *J. Chem. Phys.* **44,** 1519.
Robinson, G. W. (1961). *J. Mol. Spectrosc.* **6,** 58.
Robinson, G. W., and Frosch, R. P. (1963). *J. Chem. Phys.* **38,** 1187.
Rosen, C. G. (1966). *In* "Energy Transfer in Radiation Processes" (G. O. Phillips, ed.), p. 137. Elsevier, Amsterdam.
Rothman, W., Hirayama, F., and Lipsky, S. (1973). *J. Chem. Phys.* **58,** 1300.
Roy, J. K., and El-Sayed, M. A. (1964). *J. Chem. Phys.* **40,** 3442.
Rozman, I. M. (1961). *Opt. Spectrosc. (USSR)* **10,** 178.
Rylkov, V. V., Kholmogorov, V. E., and Terenin, A. N. (1967). *Opt. Spectrosc. (USSR)* **22,** 129.
Samson, A. M. (1962). *Opt. Spectrosc. (USSR)* **13,** 285.
Sandros, K. (1964). *Acta. Chem. Scand.* **18,** 2355.
Schimitschek, E. J. (1963). *Appl. Phys. Lett.* **3,** 117.
Schimitschek, E. J., Nehrish, R. B., and Trias, J. A. (1965). *J. Chem. Phys.* **42,** 788.
Schmillen, A. (1965). *J. Phys. Chem.* **69,** 751.
Schnepp, O., and Levy, M. (1962). *J. Amer. Chem. Soc.* **84,** 172.
Selenyi, P. (1911). *Ann. Phys.* **35,** 444.

Selenyi, P. (1939). *Phys. Rev.* **56,** 477.
Sevchenko, A. N., and Morachevskii, A. G. (1951). *Izv. Akad. Nauk SSSR Ser. Fiz.* **15,** 628.
Sevchenko, A. N., and Trofimov, A. (1951). *Zh. Eksp. Teor. Fiz.* **21,** 220.
Sheats, G. F., and Noyes, Jr., W. A. (1955). *J. Amer. Chem. Soc.* **77,** 1421.
Siebrand, W. (1965). *J. Chem. Phys.* **42,** 3951.
Siebrand, W. (1967). *J. Chem. Phys.* **47,** 2411.
Siegel, S., and Goldstein, L. (1965). *J. Chem. Phys.* **43,** 4185.
Siegel, S., and Goldstein, L. (1966). *J. Chem. Phys.* **44,** 2780.
Siegel, S., and Judeikis, H. (1964). *J. Chem. Phys.* **41,** 648.
Singh, S., and Lipsett, F. R. (1964). *J. Chem. Phys.* **41,** 1163.
Singh, S., and Stoicheff, B. P. (1963). *J. Chem. Phys.* **38,** 2032.
Singh, S., Jones, W. J., Siebrand, W., Stoicheff, B. P., and Schneider, W. G. (1965) *J. Chem. Phys.* **42,** 330.
Smaller, B., Avery, E. C., and Remko, J. R. (1965). *J. Chem. Phys.* **43,** 922.
Sponer, H., Kanda, Y., and Blackwell, L. A. (1958). *J. Chem. Phys.* **29,** 721.
Spurny, F. (1970). *Collect. Czech. Chem. Commun.* **35,** 565.
Srinivasan, B. N., Kinoshita, M., Rabalais, J. W., and McGlynn, S. P. (1968). *J. Chem. Phys.* **48,** 1924.
Steinberg, I. Z. (1968). *J. Chem. Phys.* **48,** 2411.
Steinberg, I. Z., and Katchalski, E. (1968). *J. Chem. Phys.* **48,** 2404.
Stephenson, L. M., and Hammond, G. S. (1969). *Angew. Chem. Int. Ed. Engl.* **8,** 261.
Sternlicht, H., Nieman, G. C., and Robinson, G. W. (1963). *J. Chem. Phys.* **38,** 1326.
Stevens, B. (1959). *Discuss. Faraday Soc.* **27,** 34.
Stevens, B., Hutton, E., and Porter, G. (1960). *Nature (London)* **185,** 917.
Stockman, D. L., Mallory, W. R., and Tittel, K. F. (1964). *Proc. IEEE* **52,** 318.
Stryer, L. (1960). *Radiat. Res. Suppl.* **2,** 432.
Stryer, L. (1968). *Science* **162,** 526.
Stryer, L., and Haugland, R. P. (1967). *Proc. Nat. Acad. Sci. U.S.* **58,** 719.
Sveshnikov, B. Y., Kudryashov, P. I., and Limareva, L. A. (1960). *Opt. Spectrosc. (USSR)* **9,** 107.
Swenberg, C. E., and Stacy, W. T. (1968). *Chem. Phys. Lett.* **2,** 327.
Swenberg, C. E., and Stacy, W. T. (1969). *Phys. Status Solidi* **36,** 717.
Tachin, V. S., and Ermolaev, V. L. (1970). *Opt. Spectrosc. (USSR)* **29,** 164.
Teale, F. W. J. (1960). *Biochem. J.* **76,** 381.
Terenin, A. N., and Ermolaev, V. L. (1952). *Dokl. Akad. Nauk SSSR Ser. Fiz.* **85,** 547.
Terenin, A. N., and Ermolaev, V. L. (1956). *Trans. Faraday Soc.* **52,** 1042.
Terenin, A. N., and Karyakin, A. V. (1951). *Izv. Akad. Nauk SSSR Ser. Fiz.* **15,** 550.
Terenin, A. N., Rylkov, V. V., and Kholmogorov, V. E. (1966). *Photochem. Photobiol.* **5,** 543.
Terskoi, Y. A., Brudz, V. G., and Korolkova, O. N. (1968). *Opt. Spectrosc. (USSR)* **25,** 287.
Tolstoi, N. A., and Abramov, A. P. (1967). *Sov. Phys. Solid State* **9,** 255.
Tomura, M., Ishiguro, E., and Mataga, N. (1967). *J. Phys. Soc. Jap.* **22,** 1117.
Tomura, M., Ishiguro, E., and Mataga, N. (1968). *J. Phys. Soc. Jap.* **25,** 1439.
Trlifaj, M. (1958). *Czech. J. Phys.* **8,** 510.
Tweet, A. G., Bellamy, W. D., and Gaines, Jr., G. L. (1964). *J. Chem. Phys.* **41,** 2068.
Udenfriend, S. (1969). "Fluorescence Assay in Biology and Medicine," Vol. 2. Academic Press, New York.

Van Loben Sels, J. W., and Dubois, J. T. (1968). *Mol. Cryst.* **4,** 33.
Van Uitert, L. G. (1967). *J. Electrochem. Soc.* **114,** 1048.
Van Uitert, L. G. (1971). *J. Lumin.* **4,** 1.
Vaubel, G., and Baessler, H. (1970). *Mol. Cryst.* **12,** 47.
Vaudo, A. F., and Hercules, D. M. (1970). *J. Amer. Chem. Soc.* **92,** 3573.
Vaudo, A. F., and Hercules, D. M. (1971). *J. Amer. Chem. Soc.* **93,** 2599.
Vavilov, S. I. (1943). *Sov. Phys. JETP* **13,** 13.
Volosyuk, G. P., Petrova, I. B., Prokofeva, L. S., and Gunder, O. A. (1971). *Opt. Spectrosc. (USSR)* **30,** 236.
Voltz, R., Laustriat, G., and Coche, A. (1966). *J. Chim. Phys.* **63,** 1253.
Vorobev, V. P., and Shpak, M. T. (1971). *Opt. Spectrosc. (USSR)* **30,** 253.
Wagner, P. J. (1971). *In* "Creation and Detection of the Excited State" (A. A. Lamola, ed.), p. 173. Dekker, New York.
Ware, W. R. (1961). *J. Phys. Chem.* **83,** 4374.
Watson, W., and Livingston, R. (1950). *J. Chem. Phys.* **18,** 802.
Weber, G. (1952). *Biochem. J.* **51,** 155.
Weber, G. (1960). *Biochem. J.* **75,** 335.
Weber, G. (1966). *In* "Fluorescence and Phosphorescence Analysis" (D. M. Hercules, ed.), p. 217. Wiley (Interscience), New York.
Weber, G., and Teale, F. W. J. (1959). *Discuss. Faraday Soc.* **27,** 134.
Weigert, F., and Käppler, G. (1924). *Z. Phys.* **25,** 99.
Weill, G., and Calvin, M. (1963). *Biopolym.* **1,** 401.
Weinreb, A. (1961). *J. Chem. Phys.* **35,** 91.
Weinreb, A., and Cohen, S. G. (1956). *Proc. Phys. Soc. London B* **69,** 593.
Weissman, S. I. (1942). *J. Chem. Phys.* **10,** 214.
Weissman, S. I. (1950). *J. Chem. Phys.* **18,** 1258.
Weissman, S. I., and Lipkin, D. (1942). *J. Amer. Chem. Soc.* **64,** 1916.
West, W. (1941). *Ann. N. Y. Acad. Sci.* **41,** 203.
West, W., and Paul, B. (1932). *Trans. Faraday Soc.* **28,** 688.
White, E. H., Roberts, D. R., and Roswell, D. F. (1969). *In* "Molecular Luminescence" (E. C. Lim, ed.), p. 479. Benjamin, New York.
Wilkinson, F. (1964). *Advan. Photochem.* **3,** 241.
Wilkinson, F. (1966). *Quart. Rev. Chem. Soc.* **20,** 403.
Williams, R. (1958). *J. Chem. Phys.* **28,** 577.
Windsor, M. W. (1965). *In* "Physics and Chemistry of the Organic Solid State" (D. Fox, ed.), Vol. 2, p. 343. Wiley (Interscience), New York.
Wolff, N. E., and Pressley, R. J. (1963). *Appl. Phys. Lett.* **2,** 152.
Wood, S. E. (1947). *J. Chem. Phys.* **15,** 358.
Wright, G. T. (1955). *Proc. Phys. Soc. London B* **68,** 241.
Yamada, N., Shionoya, S., and Kushida, T. (1972). *J. Phys. Soc. Jap.* **32,** 1577.
Yokota, M., and Tanimoto, O. (1967). *J. Phys. Soc. Jap.* **22,** 779.
Yuster, P., and Weissman, S. I. (1949). *J. Chem. Phys.* **17,** 1182.
Zander, M. (1964). *Ber. Bunsenges. Phys. Chem.* **68,** 301.
Zhevandrov, N. D., Gribkov, V. I., and Khan-Magometova, S. D. (1962). *Opt. Spectrosc. (USSR)* **13,** 52.
Zwick, M. M., and Kuhn, H. (1962). *Z. Naturforsch. A* **17,** 411.

8

TABLES

For explanation of tables, see Chapter 7.

LIST OF COMPOUNDS

CODE NO.	COMPOUND
1	BENZENE (3 ML/L)
2	DEUTERATED BENZENE-C6D6 5ML/L
3	FLUOROBENZENE (TRACE)
4	TOLUENE (TRACE)
5	DEUTERATED TOLUENE-C7D8 5ML/L
6	ETHYLBENZENE (TRACE)
7	PROPYLBENZENE (TRACE)
8	ISOPROPYLBENZENE (TRACE)
9	SEC-BUTYLBENZENE (TRACE)
10	DIPHENYLMETHANE (TRACE)
11	BIBENZYL
12	1,3,5-TRIPHENYLPENTANE 0.05ML/L
13	POLYSTYRENE
14	POLYVINYLTOLUENE
15	PHENYLCYCLOHEXANE (TRACE)
16	BENZYLALCOHOL 3ML/L
17	BENZYLACETATE 5ML/L
18	PHENOL (TRACE)
19	ANISOLE (TRACE)
20	PHENYL ETHER
21	ANILINE 3ML/L
22	N,N-DIMETHYLANILINE 3ML/L
23	TRIPHENYLAMINE
24	P-XYLENE (TRACE)
25	DEUTERATED P-XYLENE C8D10 (TRACE)
26	M-XYLENE (TRACE)
27	O-XYLENE 3ML/L
28	1-METHYL-4-ETHYLBENZENE (TRACE)
29	1,4-BIS(TRIFLUOROMETHYL)BENZENE 5ML/L
30	P-CRESOL
31	P-METHYLANISOLE 3ML/L
32	1,4-DIMETHOXY BENZENE
33	1-METHYL-4-CHLOROBENZENE 5ML/L
34	SODIUM SALICYLATE (A)
35	ANTHRANILIC ACID (A)
36	TMPD
37	1,2,4-TRIMETHYLBENZENE (TRACE)
38	MESITYLENE 5ML/L
39	1,3,5-TRIETHYLBENZENE 5ML/L
40	STYRENE
41	1,1-DIPHENYLETHYLENE 3ML/L
42	BIPHENYL
43	DEUTERATED BIPHENYL-C12D10
44	4-METHYL BIPHENYL
45	MONOISOPROPYLBIPHENYL 5ML/L
46	4-BENZYL BIPHENYL
47	3,3'-DIMETHYL BIPHENYL
48	2,2'-DIMETHYLBIPHENYL 2ML/L
49	OCTAMETHYL BIPHENYL
50	4-METHOXY BIPHENYL
51	4-BIPHENYLYL PHENYL ETHER
52	4,4'-DIHYDROXYBIPHENYL (A)
53	4,4' DIMETHOXY BIPHENYL

CODE NO.	COMPOUND
54	4,4'-DIISOBUTOXYBIPHENYL
55	BENZIDINE
56	4-VINYL BIPHENYL
57	M-TERPHENYL
58	M-QUATERPHENYL
59	1,3,5-TRIPHENYL BENZENE
60	1,3-DIBIPHENYLYLBENZENE (B)
61	1,3,5,-TRIBIPHENYLYLBENZENE (B)
62	FLUORENE (TRACE)
63	1-METHYL FLUORENE
64	1,2-BENZOFLUORENE
65	9,10-DIHYDROPHENANTHRENE
66	CARBAZOLE (TRACE)
67	N-METHYL CARBAZOLE (TRACE)
68	N-VINYLCARBAZOLE
69	N-PHENYL CARBAZOLE
70	N-METHYL-2,3-BENZCARBAZOLE
71	DIBENZOFURAN
72	DIBENZOTHIOPHENE (TRACE)
73	P-TERPHENYL
74	DEUTERATED P-TERPHENYL-C18D14
75	4-METHYL TERPHENYL
76	2-METHYL-P-TERPHENYL
77	3,3''-DIMETHYL-P-TERPHENYL
78	4-(3,3-DIMETHYLBUTOXY)-P-TERPHENYL
79	4,4''-DIHEXAHYDROFARNESOXY-P-TERPHENYL
80	2,2'-METHYLENE-P-TERPHENYL
81	2,2'-ETHYLENE-P-TERPHENYL
82	2,2'-METHYLENE-5',6''-METHYLENE-P-TERPHENYL (TRACE)
83	3-PHENYL DIBENZOFURAN
84	BENZO(1,2-B-4,5-B')BISBENZOFURAN (TRACE) (B)
85	SEXIPHENYL BENZENE
86	P-QUATERPHENYL
87	3,3 -DIMETHYL-P-QUATERPHENYL
88	TETRAMETHYL-P-QUATERPHENYL
89	3,3'''-DIMETHYL-3',2''-METHYLENE-P-QUATERPHENYL
90	3,3'''-DIMETHYL-3',2''-ETHYLENE-P-QUATERPHENYL
91	4-(3-ETHYLBUTOXY)-P-QUATERPHENYL (B)
92	4,4'''-DI(2-BUTYLOCTOXY-1)-P-QUATERPHENYL
93	DIETHYL-P-QUINQUEPHENYL
94	DI(3-ETHYLHEPTYL)-P-QUINQUEPHENYL
95	TETRAMETHYL-P-SEXIPHENYL
96	TETRAMETHYL-P-OCTAPHENYL (B)
97	INDOLE (TRACE)
98	2-PHENYL INDOLE
99	1-METHYL-2-PHENYL INDOLE
100	1-NEOPENTYL-2-PHENYL INDOLE
101	1-N-BUTYL-2(BIPHENYLYL-4) INDOLE
102	1,2-DIPHENYL INDOLE
103	1-BIPHENYLYL-2-PHENYL INDOLE
104	1,3-DIMETHYL-2-PHENYL INDOLE
105	2,3-DIPHENYL INDOLE
106	1-METHYL-2,3-DIPHENYL INDOLE

CODE NO.	COMPOUND
107	2-PHENYL-3,2′-METHYLENE INDOLE (TRACE)
108	2-PHENYL-3,2′-ETHYLENE INDOLE (TRACE)
109	2-PHENYL-3,2′-TRIMETHYLENE INDOLE
110	1-METHYL-3,2′-METHYLENE-2-PHENYL INDOLE (TRACE)
111	1-METHYL-3,2′-ETHYLENE-2-PHENYL INDOLE
112	1-METHYL-3,2′-TRIMETHYLENE-2-PHENYL INDOLE
113	4,5-DIPHENYLIMIDAZOLE (A)
114	PPF (TRACE)
115	PPO (TRACE)
116	2-(4-FLUOROPHENYL)-5-PHENYLOXAZOLE (TRACE)
117	αNPO (TRACE)
118	BBO (B)
119	POPCP
120	DIMETHYL POPOP (TRACE)
121	PPD (TRACE)
122	2,5-DI(2-METHYLPHENYL)OXADIAZOLE
123	PBD
124	BBD (B)
125	αNPD (TRACE)
126	βNPD
127	αNND (TRACE)
128	βNND (TRACE)
129	TRANS-1-(4-BIPHENYLYL)-2-PHENYL-ETHYLENE
130	TRANS-1-(4-BIPHENYLYL)-2-(1-NAPHTHYL)-ETHYLENE
131	DIPHENYLSTILBENE (B)
132	1,1,4,4-TETRAPHENYLBUTADIENE
133	1,6-DIPHENYLHEXATRIENE
134	1,8-DIPHENYL-1,3,5,7-OCTATETRAENE
135	BPSB (TRACE)
136	BIS-MSB
137	ESOULIN (A)
138	AZULENE
139	NAPHTHALENE
140	DEUTERATED NAPHTHALENE-C10D8
141	1-METHYLNAPHTHALENE 5ML/L
142	2-METHYL NAPHTHALENE
143	2,3-DIMETHYL NAPHTHALENE
144	2,6-DIMETHYL NAPHTHALENE (TRACE)
145	1-NAPHTHOL (TRACE)
146	2-NAPHTHOL (TRACE)
147	1-NAPHTHYLAMINE
148	2-CHLORONAPHTHALENE
149	ACENAPHTHENE
150	1-PHENYL NAPHTHALENE
151	2-PHENYL NAPHTHALENE
152	1,4-DIPHENYL NAPHTHALENE
153	1,5-DIPHENYL NAPHTHALENE
154	1,7-DIPHENYL NAPHTHALENE
155	1,4,5,8-TETRAPHENYL NAPHTHALENE
156	1,1-BINAPHTHYL
157	4,4′-DI(N-BUTOXY)-1,1′-BINAPHTHYL
158	2,2′-BINAPHTHYL
159	ACRIDINE (A)

CODE NO.	COMPOUND
160	ANTHRACENE (TRACE)
161	DEUTERATED ANTHRACENE C14D10 (TRACE)
162	9-METHYL ANTHRACENE (TRACE)
163	1-AMINO ANTHRACENE
164	2-AMINO ANTHRACENE
165	9-VINYL ANTHRACENE
166	9-PHENYL ANTHRACENE (TRACE)
167	9,10-DIPHENYL ANTHRACENE
168	9,10-DICHLOROANTHRACENE (TRACE)
169	BIANTHRANYL (B)
170	9,10-DINAPHTHYLANTHRACENE (B)
171	9-(P-VINYLPHENYL)-10-PHENYL ANTHRACENE
172	9,10-BIS(PHENYL-ETHYNYL)ANTHRACENE (TRACE)
173	RUBICENE (B)
174	DIBENZ(A,H)ANTHRACENE (B)
175	TETRACENE (TRACE) (B)
176	RUBRENE (B)
177	TETRAHYDROCANNABINOL ACID METHYL ESTER
178	PHENANTHRENE
179	3,4-BENZOPHENANTHRENE
180	2-PHENYL PHENANTHRENE
181	CHRYSENE (TRACE)
182	TRIPHENYLENE
183	DEUTERATED TRIPHENYLENE-C18D12
184	PYRENE
185	1,3,6,8,-TETRAPHENYL PYRENE
186	BIPYRENYL (B)
187	O-PHENYLENE-PYRENE
188	ANTHANTHRENE (B)
189	FLUORANTHENE
190	3-PHENYL FLUORANTHENE
191	3-AMINO FLUORANTHENE
192	3,3'-BIFLUORANTHENYL (B)
193	BENZO(GHI)FLUORANTHENE
194	9,12-DI(N-OCTADECOXY)-BENZ(E)ACEPHENANTHRYLENE
195	DECACYCLENE (B)
196	10-ETHOXY-BENZO(KL)XANTHENE
197	PERYLENE
198	BENZO(GHI)PERYLENE (B)
199	BBOT (TRACE)
200	QUININE SULFATE (A)
201	ACRIDINE YELLOW (A)
202	ACRIDINE ORANGE (A)
203	ACRIDINE RED (A)
204	URANINE (A)
205	RHODAMINE B (A)
206	RHODAMINE 6G (A)
207	EOSIN B (A)
208	EOSIN Y (A)
209	ROSE BENGAL (A)
210	INDENOINDENE

8.1 Donor Molecule: Benzene (3 ml/L)

ACCEPTOR MOLECULES	RO	CO	OVERLAP
BENZENE (3 ML/L)	3.84	7.913	0.00037
DEUTERATED BENZENE-C6D6 5ML/L	3.25	13.087	0.00013
FLUOROBENZENE (TRACE)	6.72	1.476	0.01054
TOLUENE (TRACE)	6.82	1.413	0.01147
DEUTERATED TOLUENE-C7D8 5ML/L	6.15	1.926	0.00618
ETHYLBENZENE (TRACE)	6.61	1.547	0.00957
PROPYLBENZENE (TRACE)	6.31	1.777	0.00726
ISOPROPYLBENZENE (TRACE)	5.92	2.159	0.00492
SEC-BUTYLBENZENE (TRACE)	5.94	2.132	0.00504
DIPHENYLMETHANE (TRACE)	7.75	0.959	0.02487
BIBENZYL	7.19	1.202	0.01585
1,3,5-TRIPHENYLPENTANE 0.05ML/L	7.38	1.112	0.01850
POLYSTYRENE	6.61	1.548	0.00956
POLYVINYLTOLUENE	8.59	0.707	0.04521
PHENYLCYCLOHEXANE (TRACE)	5.94	2.130	0.00506
BENZYLALCOHOL 3ML/L	5.75	2.358	0.00412
BENZYLACETATE 5ML/L	5.76	2.336	0.00420
PHENOL (TRACE)	12.31	0.239	0.39731
ANISOLE (TRACE)	12.51	0.229	0.43514
PHENYL ETHER	12.86	0.210	0.51392
ANILINE 3ML/L	13.96	0.165	0.83175
N,N-DIMETHYLANILINE 3ML/L	14.81	0.138	1.17676
TRIPHENYLAMINE	21.61	0.044	11.28738
P-XYLENE (TRACE)	9.26	0.564	0.07165
DEUTERATED P-XYLENE C8D10 (TRACE)	9.18	0.578	0.06837
M-XYLENE (TRACE)	7.76	0.956	0.02497
O-XYLENE 3ML/L	7.60	1.020	0.02196
1-METHYL-4-ETHYLBENZENE (TRACE)	8.67	0.686	0.04851
1,4-BIS(TRIFLUOROMETHYL)BENZENE 5ML/L	7.64	1.004	0.02272
P-CRESOL	13.94	0.165	0.83026
P-METHYLANISOLE 3ML/L	13.46	0.183	0.67437
1,4-DIMETHOXY BENZENE	15.02	0.132	1.28965
1-METHYL-4-CHLOROBENZENE 5ML/L	9.76	0.482	0.09820
SODIUM SALICYLATE (A)	15.69	0.116	1.65490
ANTHRANILIC ACID (A)	12.86	0.210	0.50156
TMPD	18.64	0.069	4.65509
1,2,4-TRIMETHYLBENZENE (TRACE)	10.09	0.435	0.12019
MESITYLENE 5ML/L	7.77	0.953	0.02510
1,3,5-TRIETHYLBENZENE 5ML/L	7.57	1.030	0.02149
STYRENE	11.85	0.269	0.31295
1,1-DIPHENYLETHYLENE 3ML/L	14.05	0.161	0.87271
BIPHENYL	13.76	0.172	0.77116
DEUTERATED BIPHENYL-C12D10	13.81	0.170	0.79097
4-METHYL BIPHENYL	15.52	0.120	1.58231
MONOISOPROPYLBIPHENYL 5ML/L	15.07	0.131	1.32974
4-BENZYL BIPHENYL	17.00	0.091	2.73893
3,3'-DIMETHYL BIPHENYL	14.53	0.146	1.06658
2,2'-DIMETHYLBIPHENYL 2ML/L	10.11	0.432	0.12135
OCTAMETHYL BIPHENYL	11.38	0.304	0.24628
4-METHOXY BIPHENYL	18.71	0.068	4.84119
4-BIPHENYLYL PHENYL ETHER	19.26	0.063	5.75666
4,4'-DIHYDROXYBIPHENYL (A)	20.27	0.054	7.77121
4,4'-DIMETHOXY BIPHENYL	20.21	0.054	7.63948
4,4'-DIISOBUTOXYBIPHENYL	20.01	0.056	7.21284
BENZIDINE	22.86	0.037	15.84392
4-VINYL BIPHENYL	22.45	0.040	14.36029
M-TERPHENYL	17.92	0.078	3.71989
M-QUATERPHENYL	18.41	0.072	4.39161
1,3,5-TRIPHENYL BENZENE	19.27	0.063	5.71610
1,3-DIBIPHENYLYLBENZENE (B)	25.32	0.028	29.31607
1,3,5,-TRIBIPHENYLYLBENZENE (B)	27.69	0.021	50.12620
FLUORENE (TRACE)	18.66	0.069	4.74656
1-METHYL FLUORENE	15.77	0.114	1.71533
1,2-BENZOFLUORENE	20.20	0.054	7.52366
9,10-DIHYDROPHENANTHRENE	19.07	0.065	5.39423
CARBAZOLE (TRACE)	19.56	0.060	6.22358
N-METHYL CARBAZOLE (TRACE)	19.05	0.065	5.30347
N-VINYLCARBAZOLE	20.68	0.051	8.66121
N-PHENYL CARBAZOLE	20.67	0.051	8.64989
N-METHYL-2,3-BENZCARBAZOLE	24.61	0.030	24.63207
DIBENZOFURAN	19.91	0.057	7.00703
DIBENZOTHIOPHENE (TRACE)	17.78	0.080	3.51329
P-TERPHENYL	23.08	0.036	16.93131
DEUTERATED P-TERPHENYL-C18D14	22.81	0.038	15.80663
4-METHYL TERPHENYL	23.28	0.035	17.80941
2-METHYL-P-TERPHENYL	20.72	0.050	8.89610
3,3''-DIMETHYL-P-TERPHENYL	22.91	0.037	16.17614
4-(3,3-DIMETHYLBUTOXY)-P-TERPHENYL	23.52	0.034	18.86928
4,4''-DIHEXAHYDROFARNESOXY-P-TERPHENYL	24.23	0.031	22.53458
2,2'-METHYLENE-P-TERPHENYL	23.19	0.036	17.34630
2,2'-ETHYLENE-P-TERPHENYL	23.14	0.036	17.10202
2,2'-METHYLENE-5',6''-METHYLENE-P-TERPHENYL (TRACE)	22.12	0.041	12.98912
3-PHENYL DIBENZOFURAN	22.02	0.042	12.71041
BENZO(1,2-B-4,5-B')BISBENZOFURAN (TRACE) (B)	18.78	0.067	4.85224
SEXIPHENYL BENZENE	18.83	0.067	5.02179
P-QUATERPHENYL	24.04	0.032	21.43883
3,3'''-DIMETHYL-P-QUATERPHENYL	24.24	0.031	22.51834
TETRAMETHYL-P-QUATERPHENYL	22.68	0.038	15.28687
3,3''' -DIMETHYL-3',2''-METHYLENE-P-QUATERPHENYL	22.79	0.038	15.53924
3,3''' -DIMETHYL-3',2''-ETHYLENE-P-QUATERPHENYL	23.95	0.033	20.93592
4-(3-ETHYLBUTOXY)-P-QUATERPHENYL (B)	23.47	0.035	18.53762
4,4'''-DI(2-BUTYLOCTOXY-1)-P-QUATERPHENYL	24.41	0.031	23.47067
DIETHYL-P-QUINQUEPHENYL	25.48	0.027	30.38489

ACCEPTOR MOLECULES	RO	CO	OVERLAP
DI(3-ETHYLHEPTYL)-P-QUINQUEPHENYL	24.80	0.029	25.75674
TETRAMETHYL-P-SEXIPHENYL	25.97	0.026	34.11691
TETRAMETHYL-P-OCTAPHENYL (B)	25.19	0.028	28.29353
INDOLE (TRACE)	15.65	0.117	1.66550
2-PHENYL INDOLE	20.98	0.048	[illegible]
1-METHYL-2-PHENYL INDOLE	[illegible]	0.051	8.57410
1-NEOPENTYL-2-PHENYL INDOLE	20.61	0.051	8.53848
1-N-BUTYL-2(BIPHENYLYL-4) INDOLE	21.49	0.045	10.91507
1,2-DIPHENYL INDOLE	21.42	0.045	10.72731
1-BIPHENYLYL-2-PHENYL INDOLE	23.01	0.037	16.45202
1,3-DIMETHYL-2-PHENYL INDOLE	19.78	0.058	6.65182
2,3-DIPHENYL INDOLE	20.41	0.053	8.01171
1-METHYL-2,3-DIPHENYL INDOLE	20.42	0.052	8.04515
2-PHENYL-3,2'-METHYLENE INDOLE (TRACE)	19.44	0.061	5.98769
2-PHENYL-3,2'-ETHYLENE INDOLE (TRACE)	19.06	0.065	5.31029
2-PHENYL-3,2'-TRIMETHYLENE INDOLE	19.99	0.056	7.07163
1-METHYL-3,2'-METHYLENE-2-PHENYL INDOLE (TRACE)	19.07	0.064	5.33140
1-METHYL-3,2'-ETHYLENE-2-PHENYL INDOLE	18.76	0.068	4.82450
1-METHYL-3,2'-TRIMETHYLENE-2-PHENYL INDOLE	19.44	0.061	5.97647
4,5-DIPHENYLIMIDAZOLE (A)	19.45	0.061	6.02046
PPF (TRACE)	20.43	0.052	8.04984
PPO (TRACE)	22.47	0.039	14.26825
2-(4-FLUOROPHENYL)-5-PHENYLOXAZOLE (TRACE)	21.78	0.043	11.83976
αNPO (TRACE)	20.28	0.054	7.70121
BBO (B)	21.31	0.046	10.39015
POPOP	19.72	0.058	6.51488
DIMETHYL POPOP (TRACE)	20.79	0.050	8.95526
PPD (TRACE)	22.71	0.038	15.39609
2,5-DI(2-METHYLPHENYL)OXADIAZOLE	22.12	0.041	13.10222
PBD	24.09	0.032	21.72001
BBD (B)	23.33	0.035	17.87921
αNPD (TRACE)	20.31	0.053	7.78343
βNPD	23.63	0.034	19.30132
αNND (TRACE)	18.77	0.068	4.84858
βNND (TRACE)	23.74	0.033	19.85663
TRANS-1-(4-BIPHENYLYL)-2-PHENYL-ETHYLENE	22.14	0.041	13.06165
TRANS-1-(4-BIPHENYLYL)-2-(1-NAPHTHYL)-ETHYLENE	21.24	0.047	10.16986
DIPHENYLSTILBENE (B)	19.92	0.057	6.92785
1,1,4,4-TETRAPHENYLBUTADIENE	18.57	0.070	4.53718
1,6-DIPHENYLHEXATRIENE	18.04	0.076	3.81360
1,8-DIPHENYL-1,3,5,7-OCTATETRAENE	17.33	0.086	3.00660
BPSB (TRACE)	19.02	0.065	5.24943
BIS-MSB	19.76	0.058	6.59457
ESCULIN (A)	17.64	0.081	3.34321
AZULENE	22.83	0.038	15.69687
NAPHTHALENE	16.17	0.106	2.00369
DEUTERATED NAPHTHALENE-C10D8	16.29	0.103	2.07914
1-METHYLNAPHTHALENE 5ML/L	17.14	0.089	2.83245
2-METHYL NAPHTHALENE	16.31	0.103	2.10502
2,3-DIMETHYL NAPHTHALENE	16.70	0.096	2.42268
2,6-DIMETHYL NAPHTHALENE (TRACE)	16.46	0.100	2.21853
1-NAPHTHOL (TRACE)	17.16	0.089	2.84453
2-NAPHTHOL (TRACE)	15.92	0.111	1.80698
1-NAPHTHYLAMINE	15.59	0.118	1.59405
2-CHLORONAPHTHALENE	16.33	0.103	2.11632
ACENAPHTHENE	17.74	0.080	3.48577
1-PHENYL NAPHTHALENE	19.29	0.062	5.74773
2-PHENYL NAPHTHALENE	28.68	0.019	61.92044
1,4-DIPHENYL NAPHTHALENE	19.86	0.057	6.79653
1,5-DIPHENYL NAPHTHALENE	19.57	0.060	6.23512
1,7-DIPHENYL NAPHTHALENE	19.75	0.058	6.58781
1,4,5,8-TETRAPHENYL NAPHTHALENE	19.23	0.063	5.60168
1,1'-BINAPHTHYL	20.33	0.053	7.86095
4,4'-DI(N-BUTOXY)-1,1'-BINAPHTHYL	19.66	0.059	6.40015
2,2'-BINAPHTHYL	22.47	0.039	14.31348
ACRIDINE (A)	13.00	0.204	0.53384
ANTHRACENE (TRACE)	11.38	0.304	0.24016
DEUTERATED ANTHRACENE C14D10 (TRACE)	12.07	0.254	0.34281
9-METHYL ANTHRACENE (TRACE)	12.02	0.258	0.33363
1-AMINO ANTHRACENE	20.04	0.056	7.17604
2-AMINO ANTHRACENE	9.71	0.488	0.09125
9-VINYL ANTHRACENE	13.72	0.173	0.74031
9-PHENYL ANTHRACENE (TRACE)	14.50	0.147	1.02887
9,10-DIPHENYL ANTHRACENE	17.09	0.090	2.75823
9,10-DICHLOROANTHRACENE (TRACE)	12.68	0.220	0.45994
BIANTHRANYL (B)	20.31	0.053	7.76844
9,10-DINAPHTHYLANTHRACENE (B)	20.97	0.049	9.41495
9-(P-VINYLPHENYL)-10-PHENYL ANTHRACENE	18.50	0.071	4.44433
9,10-BIS(PHENYL-ETHYNYL)ANTHRACENE (TRACE)	24.09	0.032	21.64264
RUBICENE (B)	23.75	0.033	19.91011
DIBENZ(A,H)ANTHRACENE (B)	25.75	[illegible]	32.34177
TETRACENE (TRACE) (B)	28.98	0.018	65.64200
RUBRENE (B)	25.11	0.028	27.77614
TETRAHYDROCANNABINOL ACID METHYL ESTER	21.15	0.047	10.01448
PHENANTHRENE	19.20	0.063	5.55531
3,4-BENZOPHENANTHRENE	24.55	0.030	24.23678
2-PHENYL PHENANTHRENE	24.76	0.029	25.55186
CHRYSENE (TRACE)	22.60	0.039	14.77608
TRIPHENYLENE	20.66	0.051	8.61266
DEUTERATED TRIPHENYLENE-C18D12	20.74	0.050	8.82233
PYRENE	19.74	0.058	6.55257
1,3,6,8,-TETRAPHENYL PYRENE	23.38	0.035	18.08311
BIPYRENYL (B)	24.18	0.032	22.17258
O-PHENYLENE-PYRENE	22.89	0.037	15.92539
ANTHANTHRENE (B)	22.25	0.041	13.44424

ACCEPTOR MOLECULES	RO	CO	OVERLAP
FLUORANTHENE	20.97	0.048	9.43550
3-PHENYL FLUORANTHENE	21.76	0.043	11.75007
3-AMINO FLUORANTHENE	20.12	0.055	7.35020
3,3'-BIFLUORANTHENYL (B)	23.26	0.036	17.56292
BENZO(GHI)FLUORANTHENE	20.11	0.055	7.33821
9,12-DI(N-OCTADECOXY)-BENZ(E)ACEPHENANTHRYLENE	23.56	0.034	18.96787
DECACYCLENE (B)	22.81	0.038	15.59222
10-ETHOXY-BENZO(KL)XANTHENE	17.02	0.091	2.69565
PERYLENE	15.09	0.130	1.30895
BENZO(GHI)PERYLENE (B)	23.53	0.034	18.83073
BBOT (TRACE)	17.69	0.081	3.39966
QUININE SULFATE (A)	18.24	0.074	4.08188
ACRIDINE YELLOW (A)	21.01	0.048	9.53100
ACRIDINE ORANGE (A)	21.10	0.048	9.78681
ACRIDINE RED (A)	18.14	0.075	3.94203
URANINE (A)	18.19	0.074	4.01561
RHODAMINE B (A)	20.88	0.049	9.19178
RHODAMINE 6G (A)	20.94	0.049	9.34487
EOSIN B (A)	18.63	0.069	4.63762
EOSIN Y (A)	19.17	0.063	5.50464
ROSE BENGAL (A)	19.98	0.056	7.05778
INDENOINDENE	20.08	0.055	7.26698

8.2 Donor Molecule: Deuterated Benzene (C_6D_6) (5 ml/L)

ACCEPTOR MOLECULES	RO	CO	OVERLAP
BENZENE (3 ML/L)	4.57	4.694	0.00092
DEUTERATED BENZENE-C6D6 5ML/L	4.17	6.151	0.00053
FLUOROBENZENE (TRACE)	8.15	0.825	0.02960
TOLUENE (TRACE)	7.39	1.109	0.01635
DEUTERATED TOLUENE-C7D8 5ML/L	7.00	1.304	0.01182
ETHYLBENZENE (TRACE)	7.24	1.178	0.01449
PROPYLBENZENE (TRACE)	7.02	1.290	0.01207
ISOPROPYLBENZENE (TRACE)	6.76	1.448	0.00960
SEC-BUTYLBENZENE (TRACE)	6.74	1.461	0.00943
DIPHENYLMETHANE (TRACE)	8.33	0.775	0.03343
BIBENZYL	7.95	0.889	0.02539
1,3,5-TRIPHENYLPENTANE 0.05ML/L	8.13	0.833	0.02890
POLYSTYRENE	7.21	1.193	0.01411
POLYVINYLTOLUENE	9.19	0.577	0.05954
PHENYLCYCLOHEXANE (TRACE)	6.76	1.451	0.00957
BENZYLALCOHOL 3ML/L	6.47	1.653	0.00736
BENZYLACETATE 5ML/L	6.51	1.621	0.00766
PHENOL (TRACE)	12.66	0.220	0.41191
ANISOLE (TRACE)	12.85	0.211	0.44942
PHENYL ETHER	13.31	0.190	0.55494
ANILINE 3ML/L	14.13	0.159	0.78477
N,N-DIMETHYLANILINE 3ML/L	15.19	0.128	1.20204
TRIPHENYLAMINE	21.75	0.043	10.31376
P-XYLENE (TRACE)	9.95	0.454	0.09712
DEUTERATED P-XYLENE C8D10 (TRACE)	9.92	0.458	0.09545
M-XYLENE (TRACE)	8.33	0.774	0.03348
O-XYLENE 3ML/L	8.13	0.833	0.02886
1-METHYL-4-ETHYLBENZENE (TRACE)	9.40	0.539	0.06893
1,4 BIS(TRIFLUOROMETHYL)BENZENE 5ML/L	8.36	0.766	0.03423
P-CRESOL	14.20	0.156	0.81533
P-METHYLANISOLE 3ML/L	13.71	0.174	0.66060
1,4-DIMETHOXY BENZENE	15.13	0.129	1.18496
1-METHYL-4-CHLOROBENZENE 5ML/L	10.00	0.447	0.10011
SODIUM SALICYLATE (A)	15.77	0.114	1.50172
ANTHRANILIC ACID (A)	13.01	0.203	0.47272
TMPD	19.31	0.062	5.04893
1,2,4-TRIMETHYLBENZENE (TRACE)	10.62	0.374	0.14295
MESITYLENE 5ML/L	8.28	0.789	0.03216
1,3,5-TRIETHYLBENZENE 5ML/L	8.05	0.856	0.02730
STYRENE	12.38	0.236	0.35662
1,1-DIPHENYLETHYLENE 3ML/L	14.82	0.137	1.05297
BIPHENYL	14.63	0.143	0.97765
DEUTERATED BIPHENYL-C12D10	14.70	0.141	1.00853
4-METHYL BIPHENYL	16.34	0.103	1.88976
MONOISOPROPYLBIPHENYL 5ML/L	15.91	0.111	1.61368
4-BENZYL BIPHENYL	17.86	0.079	3.22478
3,3'-DIMETHYL BIPHENYL	15.22	0.127	1.23286
2,2'-DIMETHYLBIPHENYL 2ML/L	10.72	0.363	0.15115
OCTAMETHYL BIPHENYL	11.78	0.274	0.26576
4-METHOXY BIPHENYL	19.40	0.061	5.26668
4-BIPHENYLYL PHENYL ETHER	19.98	0.056	6.30369
4,4'-DIHYDROXYBIPHENYL (A)	20.86	0.049	8.09257
4,4'-DIMETHOXY BIPHENYL	20.82	0.050	8.01490
4,4'-DIISOBUTOXYBIPHENYL	20.59	0.051	7.51200
BENZIDINE	23.20	0.036	15.20366
4-VINYL BIPHENYL	22.95	0.037	14.36581
M-TERPHENYL	18.75	0.068	4.27634
M-QUATERPHENYL	19.37	0.062	5.23292
1,3,5-TRIPHENYL BENZENE	20.26	0.054	6.78202
1,3-DIBIPHENYLYLBENZENE (B)	25.72	0.026	28.25678
1,3,5,-TRIBIPHENYLYLBENZENE (B)	28.06	0.020	47.02915
FLUORENE (TRACE)	19.31	0.062	5.11485
1-METHYL FLUORENE	15.85	0.112	1.54627
1,2-BENZOFLUORENE	20.39	0.053	6.97892

ACCEPTOR MOLECULES	RO	CO	OVERLAP
9,10-DIHYDROPHENANTHRENE	19.69	0.059	5.74164
CARBAZOLE (TRACE)	19.75	0.058	5.79660
N-METHYL CARBAZOLE (TRACE)	19.36	0.062	5.13717
N-VINYLCARBAZOLE	21.15	0.047	8.72593
N-PHENYL CARBAZOLE	20.96	0.049	8.27421
N-METHYL-2,3-BENZCARBAZOLE	[illegible]	0.028	25.34618
DIBENZOFURAN	20.19	0.054	6.68754
DIBENZOTHIOPHENE (TRACE)	18.09	0.076	3.41246
P-TERPHENYL	23.58	0.034	16.87180
DEUTERATED P-TERPHENYL-C18D14	23.31	0.035	15.77795
4-METHYL TERPHENYL	23.73	0.033	17.50055
2-METHYL-P-TERPHENYL	21.37	0.046	9.38728
3,3''-DIMETHYL-P-TERPHENYL	23.37	0.035	15.97268
4-(3,3-DIMETHYLBUTOXY)-P-TERPHENYL	23.87	0.033	18.06816
4,4''-DIHEXAHYDROFARNESOXY-P-TERPHENYL	24.54	0.030	21.33601
2,2'-METHYLENE-P-TERPHENYL	23.51	0.034	16.49418
2,2'-ETHYLENE-P-TERPHENYL	23.44	0.035	16.20719
2,2'-METHYLENE-5',6''-METHYLENE-P-TERPHENYL (TRACE)	22.32	0.040	12.03854
3-PHENYL DIBENZOFURAN	22.27	0.041	11.91343
BENZO(1,2-B-4,5-B')BISBENZOFURAN (TRACE) (B)	18.97	0.066	4.51675
SEXIPHENYL BENZENE	19.45	0.061	5.35486
P-QUATERPHENYL	24.32	0.031	20.15582
3,3'''-DIMETHYL-P-QUATERPHENYL	24.51	0.030	21.14530
TETRAMETHYL-P-QUATERPHENYL	23.35	0.035	15.96659
3,3''' -DIMETHYL-3',2''-METHYLENE-P-QUATERPHENYL	22.99	0.037	14.39120
3,3''' -DIMETHYL-3',2''-ETHYLENE-P-QUATERPHENYL	24.26	0.031	19.89402
4-(3-ETHYLBUTOXY)-P-QUATERPHENYL (B)	23.68	0.034	17.18407
4,4'''-DI(2-BUTYLOCTOXY-1)-P-QUATERPHENYL	24.64	0.030	21.82052
DIETHYL-P-QUINQUEPHENYL	25.77	0.026	28.57474
DI(3-ETHYLHEPTYL)-P-QUINQUEPHENYL	25.01	0.029	23.88373
TETRAMETHYL-P-SEXIPHENYL	26.30	0.025	32.30342
TETRAMETHYL-P-OCTAPHENYL (B)	25.51	0.027	26.89475
INDOLE (TRACE)	16.15	0.106	1.76897
2-PHENYL INDOLE	21.15	0.047	8.72255
1-METHYL-2-PHENYL INDOLE	20.82	0.050	7.95024
1-NEOPENTYL-2-PHENYL INDOLE	20.81	0.050	7.93968
1-N-BUTYL-2(BIPHENYLYL-4) INDOLE	21.86	0.043	10.63180
1,2-DIPHENYL INDOLE	21.72	0.044	10.24810
1-BIPHENYLYL-2-PHENYL INDOLE	23.36	0.035	15.85916
1,3-DIMETHYL-2-PHENYL INDOLE	19.98	0.056	6.21358
2,3-DIPHENYL INDOLE	20.70	0.050	7.66827
1-METHYL-2,3-DIPHENYL INDOLE	20.81	0.050	7.91581
2-PHENYL-3,2'-METHYLENE INDOLE (TRACE)	19.65	0.059	5.60689
2-PHENYL-3,2'-ETHYLENE INDOLE (TRACE)	19.35	0.062	5.11851
2-PHENYL-3,2' -TRIMETHYLENE INDOLE	20.17	0.055	6.55833
1-METHYL-3,2'-METHYLENE-2-PHENYL INDOLE (TRACE)	19.37	0.062	5.15062
1-METHYL-3,2'-ETHYLENE-2-PHENYL INDOLE	19.09	0.064	4.72557
1-METHYL-3,2'-TRIMETHYLENE-2-PHENYL INDOLE	19.63	0.059	5.57899
4,5-DIPHENYLIMIDAZOLE (A)	19.78	0.058	5.83751
PPF (TRACE)	20.63	0.051	7.50907
PPO (TRACE)	22.68	0.038	13.27104
2-(4-FLUOROPHENYL)-5-PHENYLOXAZOLE (TRACE)	22.00	0.042	11.05416
αNPO (TRACE)	20.67	0.051	7.60406
BBO (B)	21.69	0.044	10.14954
POPOP	20.02	0.056	6.26105
DIMETHYL POPOP (TRACE)	21.17	0.047	8.77518
PPD (TRACE)	23.11	0.036	14.98723
2,5-DI(2-METHYLPHENYL)OXADIAZOLE	22.52	0.039	12.81722
PBD	24.32	0.031	20.20047
BBD (B)	23.62	0.034	16.94795
αNPD (TRACE)	20.72	0.050	7.71481
βNPD	24.18	0.032	19.50477
αNND (TRACE)	19.01	0.065	4.59297
βNND (TRACE)	24.39	0.031	20.51244
TRANS-1-(4-BIPHENYLYL)-2-PHENYL-ETHYLENE	22.33	0.040	12.08964
TRANS-1-(4-BIPHENYLYL)-2-(1-NAPHTHYL)-ETHYLENE	21.58	0.045	9.84546
DIPHENYLSTILBENE (B)	20.12	0.055	6.47519
1,1,4,4-TETRAPHENYLBUTADIENE	18.94	0.066	4.49653
1,6-DIPHENYLHEXATRIENE	18.25	0.074	3.60686
1,8-DIPHENYL-1,3,5,7-OCTATETRAENE	17.67	0.081	2.96431
BPSB (TRACE)	19.25	0.063	4.96657
BIS-MSB	19.99	0.056	6.22043
ESCULIN (A)	17.83	0.079	3.13810
AZULENE	23.50	0.034	16.44408
NAPHTHALENE	16.66	0.097	2.09949
DEUTERATED NAPHTHALENE-C10D8	16.80	0.094	2.19284
1-METHYLNAPHTHALENE 5ML/L	17.49	0.084	2.80319
2-METHYL NAPHTHALENE	16.75	0.095	2.16164
2,3-DIMETHYL NAPHTHALENE	17.10	0.089	2.44940
2,6-DIMETHYL NAPHTHALENE (TRACE)	16.91	0.093	2.28499
1-NAPHTHOL (TRACE)	17.38	0.085	2.69696
2-NAPHTHOL (TRACE)	16.30	0.102	1.88629
1-NAPHTHYLAMINE	15.73	0.115	1.47911
2-CHLORONAPHTHALENE	16.75	0.095	2.16144
ACENAPHTHENE	17.99	0.077	3.32449
1-PHENYL NAPHTHALENE	19.56	0.060	5.47983
2-PHENYL NAPHTHALENE	29.23	0.018	60.87752
1,4-DIPHENYL NAPHTHALENE	20.04	0.056	6.31197
1,5-DIPHENYL NAPHTHALENE	19.74	0.058	5.77520
1,7-DIPHENYL NAPHTHALENE	20.25	0.054	6.72386
1,4,5,8-TETRAPHENYL NAPHTHALENE	19.70	0.059	5.69781
1,1'-BINAPHTHYL	20.62	0.051	7.51402
4,4'-DI(N-BUTOXY)-1,1'-BINAPHTHYL	19.86	0.057	5.98086
2,2'-BINAPHTHYL	23.05	0.037	14.62662
ACRIDINE (A)	13.34	0.189	0.54879
ANTHRACENE (TRACE)	11.76	0.275	0.25803

ACCEPTOR MOLECULES	RO	CO	OVERLAP
DEUTERATED ANTHRACENE C14D10 (TRACE)	12.40	0.235	0.35463
9-METHYL ANTHRACENE (TRACE)	12.86	0.210	0.44196
1-AMINO ANTHRACENE	21.12	0.048	8.64582
2-AMINO ANTHRACENE	9.99	0.449	0.09462
9-VINYL ANTHRACENE	14.87	0.136	1.05282
9-PHENYL ANTHRACENE (TRACE)	15.53	0.119	1.36640
9,10-DIPHENYL ANTHRACENE	18.81	0.067	4.32351
9,10-DICHLOROANTHRACENE (TRACE)	14.64	0.142	0.96093
BIANTHRANYL (B)	22.21	0.041	11.70682
9,10-DINAPHTHYLANTHRACENE (B)	21.92	0.042	10.82482
9-(P-VINYLPHENYL)-10-PHENYL ANTHRACENE	19.74	0.058	5.77595
9,10-BIS(PHENYL-ETHYNYL)ANTHRACENE (TRACE)	24.91	0.029	23.28618
RUBICENE (B)	24.15	0.032	19.34213
DIBENZ(A,H)ANTHRACENE (B)	25.93	0.026	29.64299
TETRACENE (TRACE) (B)	29.70	0.017	66.94779
RUBRENE (B)	25.23	0.028	25.15456
TETRAHYDROCANNABINOL ACID METHYL ESTER	21.71	0.044	10.28509
PHENANTHRENE	19.79	0.058	5.85641
3,4-BENZOPHENANTHRENE	25.10	0.028	24.40776
2-PHENYL PHENANTHRENE	25.50	0.027	26.83521
CHRYSENE (TRACE)	23.76	0.033	17.53209
TRIPHENYLENE	21.26	0.047	9.00403
DEUTERATED TRIPHENYLENE-C18D12	21.37	0.046	9.29106
PYRENE	20.59	0.051	7.44078
1,3,6,8,-TETRAPHENYL PYRENE	23.56	0.034	16.68179
BIPYRENYL (B)	24.67	0.030	21.97436
O-PHENYLENE-PYRENE	23.23	0.036	15.33969
ANTHANTHRENE (B)	22.47	0.039	12.55197
FLUORANTHENE	21.27	0.046	9.03877
3-PHENYL FLUORANTHENE	22.03	0.042	11.15359
3-AMINO FLUORANTHENE	20.49	0.052	7.21163
3,3'-BIFLUORANTHENYL (B)	23.60	0.034	16.85805
BENZO(GHI)FLUORANTHENE	20.49	0.052	7.20731
9,12-DI(N-OCTADECOXY)-BENZ(E)ACEPHENANTHRYLENE	24.08	0.032	19.00957
DECACYCLENE (B)	23.32	0.035	15.67096
10-ETHOXY-BENZO(KL)XANTHENE	17.31	0.086	2.62246
PERYLENE	15.81	0.113	1.52216
BENZO(GHI)PERYLENE (B)	23.82	0.033	17.81385
BBOT (TRACE)	18.06	0.076	3.38548
QUININE SULFATE (A)	18.59	0.070	4.02704
ACRIDINE YELLOW (A)	21.78	0.043	10.41346
ACRIDINE ORANGE (A)	21.68	0.044	10.13071
ACRIDINE RED (A)	18.50	0.071	3.90858
URANINE (A)	18.54	0.070	3.96030
RHODAMINE B (A)	21.31	0.046	9.13886
RHODAMINE 6G (A)	21.30	0.046	9.12011
EOSIN B (A)	19.13	0.064	4.77512
EOSIN Y (A)	19.67	0.059	5.64677
ROSE BENGAL (A)	20.60	0.051	7.45192
INDENOINDENE	20.21	0.054	6.65480

8.3 Donor Molecule: Fluorobenzene (trace)

BENZENE (3 ML/L)	4.19	6.080	0.00034
DEUTERATED BENZENE-C6D6 5ML/L	3.80	8.137	0.00019
FLUOROBENZENE (TRACE)	7.90	0.909	0.01497
TOLUENE (TRACE)	7.40	1.102	0.01016
DEUTERATED TOLUENE-C7D8 5ML/L	6.90	1.361	0.00667
ETHYLBENZENE (TRACE)	7.25	1.175	0.00893
PROPYLBENZENE (TRACE)	6.99	1.308	0.00722
ISOPROPYLBENZENE (TRACE)	6.67	1.507	0.00544
SEC-BUTYLBENZENE (TRACE)	6.66	1.514	0.00539
DIPHENYLMETHANE (TRACE)	8.43	0.746	0.02216
BIBENZYL	7.95	0.891	0.01553
1,3,5-TRIPHENYLPENTANE 0.05ML/L	8.13	0.833	0.01776
POLYSTYRENE	7.24	1.178	0.00890
POLYVINYLTOLUENE	9.71	0.488	0.05102
PHENYLCYCLOHEXANE (TRACE)	6.62	1.541	0.00521
BENZYLALCOHOL 3ML/L	6.43	1.681	0.00437
BENZYLACETATE 5ML/L	6.41	1.697	0.00429
PHENOL (TRACE)	13.50	0.182	0.37122
ANISOLE (TRACE)	13.78	0.171	0.41883
PHENYL ETHER	14.31	0.153	0.52560
ANILINE 3ML/L	15.41	0.122	0.81077
N,N-DIMETHYLANILINE 3ML/L	16.36	0.102	1.15590
TRIPHENYLAMINE	23.85	0.033	11.00479
P-XYLENE (TRACE)	10.55	0.381	0.08461
DEUTERATED P-XYLENE C8D10 (TRACE)	10.54	0.382	0.08405
M-XYLENE (TRACE)	8.78	0.661	0.02812
O-XYLENE 3ML/L	8.43	0.746	0.02210
1-METHYL-4-ETHYLBENZENE (TRACE)	9.98	0.451	0.06062
1,4-BIS(TRIFLUOROMETHYL)BENZENE 5ML/L	8.38	0.759	0.02144
P-CRESOL	15.41	0.122	0.81691
P-METHYLANISOLE 3ML/L	14.88	0.136	0.66300
1,4-DIMETHOXY BENZENE	16.55	0.099	1.24534
1-METHYL-4-CHLOROBENZENE 5ML/L	10.66	0.369	0.09013
SODIUM SALICYLATE (A)	17.30	0.086	1.60289
ANTHRANILIC ACID (A)	14.42	0.149	0.53660
TMPD	20.73	0.050	4.74195

ACCEPTOR MOLECULES	RO	CO	OVERLAP
1,2,4-TRIMETHYLBENZENE (TRACE)	11.29	0.311	0.12665
MESITYLENE 5ML/L	8.75	0.668	0.02753
1,3,5-TRIETHYLBENZENE 5ML/L	8.44	0.745	0.02212
STYRENE	13.23	0.193	0.32684
1,1-DIPHENYLETHYLENE 3ML/L	15.65	0.117	0.89705
BIPHENYL	15.35	0.124	0.80200
DEUTERATED BIPHENYL-C12D10	15.42	0.122	0.82315
4-METHYL BIPHENYL	17.29	0.086	1.63378
MONOISOPROPYLBIPHENYL 5ML/L	16.81	0.094	1.37914
4-BENZYL BIPHENYL	18.95	0.066	2.82503
3,3'-DIMETHYL BIPHENYL	16.17	0.106	1.09066
2,2'-DIMETHYLBIPHENYL 2ML/L	11.30	0.310	0.12709
OCTAMETHYL BIPHENYL	12.66	0.221	0.25136
4-METHOXY BIPHENYL	20.79	0.050	4.90185
4-BIPHENYLYL PHENYL ETHER	21.40	0.046	5.83707
4,4'-DIHYDROXYBIPHENYL (A)	22.48	0.039	7.77558
4,4'-DIMETHOXY BIPHENYL	22.42	0.040	7.66486
4,4'-DIISOBUTOXYBIPHENYL	22.19	0.041	7.22456
BENZIDINE	25.28	0.028	15.60371
4-VINYL BIPHENYL	24.86	0.029	14.26421
M-TERPHENYL	19.93	0.057	3.79126
M-QUATERPHENYL	20.50	0.052	4.51032
1,3,5-TRIPHENYL BENZENE	21.47	0.045	5.88686
1,3-DIBIPHENYLYLBENZENE (B)	28.00	0.020	28.83536
1,3,5,-TRIBIPHENYLYLBENZENE (B)	30.59	0.016	49.14160
FLUORENE (TRACE)	20.71	0.050	4.78697
1-METHYL FLUORENE	17.41	0.085	1.66836
1,2-BENZOFLUORENE	22.29	0.040	7.31076
9,10-DIHYDROPHENANTHRENE	21.17	0.047	5.44167
CARBAZOLE (TRACE)	21.61	0.044	6.08416
N-METHYL CARBAZOLE (TRACE)	21.03	0.048	5.17862
N-VINYLCARBAZOLE	22.92	0.037	8.66501
N-PHENYL CARBAZOLE	22.83	0.038	8.47373
N-METHYL-2,3-BENZCARBAZOLE	27.31	0.022	24.80896
DIBENZOFURAN	22.00	0.042	6.87119
DIBENZOTHIOPHENE (TRACE)	19.64	0.059	3.43795
P-TERPHENYL	25.56	0.027	16.79428
DEUTERATED P-TERPHENYL-C18D14	25.26	0.028	15.68182
4-METHYL TERPHENYL	25.76	0.026	17.59812
2-METHYL-P-TERPHENYL	22.99	0.037	8.95126
3,3''-DIMETHYL-P-TERPHENYL	25.36	0.027	16.00194
4-(3,3-DIMETHYLBUTOXY)-P-TERPHENYL	26.00	0.025	18.53693
4,4''-DIHEXAHYDROFARNESOXY-P-TERPHENYL	26.78	0.023	22.11951
2,2'-METHYLENE-P-TERPHENYL	25.65	0.027	17.08646
2,2'-ETHYLENE-P-TERPHENYL	25.58	0.027	16.81109
2,2'-METHYLENE-5',6''-METHYLENE-P-TERPHENYL (TRACE)	24.49	0.030	12.90452
3-PHENYL DIBENZOFURAN	24.31	0.031	12.38773
BENZO(1,2-B-4,5-B')BISBENZOFURAN (TRACE) (B)	21.03	0.048	5.15992
SEXIPHENYL BENZENE	20.92	0.049	5.08952
P-QUATERPHENYL	26.56	0.024	21.03127
3,3'''-DIMETHYL-P-QUATERPHENYL	26.79	0.023	22.10997
TETRAMETHYL-P-QUATERPHENYL	25.15	0.028	15.32484
3,3''' -DIMETHYL-3',2''-METHYLENE-P-QUATERPHENYL	25.26	0.028	15.53866
3,3''' -DIMETHYL-3',2''-ETHYLENE-P-QUATERPHENYL	26.54	0.024	20.90780
4-(3-ETHYLBUTOXY)-P-QUATERPHENYL (B)	25.96	0.026	18.31909
4,4'''-DI(2-BUTYLOCTOXY-1)-P-QUATERPHENYL	27.00	0.023	23.18916
DIETHYL-P-QUINQUEPHENYL	28.16	0.020	29.83535
DI(3-ETHYLHEPTYL)-P-QUINQUEPHENYL	27.43	0.022	25.47072
TETRAMETHYL-P-SEXIPHENYL	28.71	0.019	33.49971
TETRAMETHYL-P-OCTAPHENYL (B)	27.93	0.021	28.38632
INDOLE (TRACE)	17.35	0.086	1.66727
2-PHENYL INDOLE	23.21	0.036	9.36512
1-METHYL-2-PHENYL INDOLE	22.79	0.038	8.38575
1-NEOPENTYL-2-PHENYL INDOLE	22.76	0.038	8.33266
1-N-BUTYL-2(BIPHENYLYL-4) INDOLE	23.82	0.033	10.90899
1,2-DIPHENYL INDOLE	23.69	0.034	10.55927
1-BIPHENYLYL-2-PHENYL INDOLE	25.46	0.027	16.29727
1,3-DIMETHYL-2-PHENYL INDOLE	21.86	0.043	6.53094
2,3-DIPHENYL INDOLE	22.61	0.039	7.98996
1-METHYL-2,3-DIPHENYL INDOLE	22.62	0.039	8.01321
2-PHENYL-3,2'-METHYLENE INDOLE (TRACE)	21.60	0.044	6.07471
2-PHENYL-3,2'-ETHYLENE INDOLE (TRACE)	21.21	0.047	5.44353
2-PHENYL-3,2'-TRIMETHYLENE INDOLE	22.16	0.041	7.08744
1-METHYL-3,2'-METHYLENE-2-PHENYL INDOLE (TRACE)	21.21	0.047	5.43903
1-METHYL-3,2'-ETHYLENE-2-PHENYL INDOLE	20.89	0.049	4.97036
1-METHYL-3,2'-TRIMETHYLENE-2-PHENYL INDOLE	21.53	0.045	5.94862
4,5-DIPHENYLIMIDAZOLE (A)	21.51	0.045	5.93516
PPF (TRACE)	22.71	0.038	8.19560
PPO (TRACE)	24.86	0.029	14.11781
2-(4-FLUOROPHENYL)-5-PHENYLOXAZOLE (TRACE)	24.10	0.032	11.71144
αNPO (TRACE)	22.56	0.039	7.88671
BBO (B)	23.74	0.033	10.70427
POPOP	21.93	0.042	6.65305
DIMETHYL POPOP (TRACE)	23.10	0.036	9.08298
PPD (TRACE)	25.12	0.028	15.16742
2,5-DI(2-METHYLPHENYL)OXADIAZOLE	24.47	0.031	12.93532
PBD	26.63	0.024	21.34994
BBD (B)	25.89	0.026	18.02317
αNPD (TRACE)	22.57	0.039	7.90446
βNPD	26.25	0.025	19.55536
αNND (TRACE)	20.87	0.049	4.94728
βNND (TRACE)	26.36	0.024	20.06083
TRANS-1-(4-BIPHENYLYL)-2-PHENYL-ETHYLENE	24.56	0.030	13.13279
TRANS-1-(4-BIPHENYLYL)-2-(1-NAPHTHYL)-ETHYLENE	23.61	0.034	10.36777
DIPHENYLSTILBENE (B)	22.24	0.041	7.22636
1,1,4,4-TETRAPHENYLBUTADIENE	20.74	0.050	4.75730

ACCEPTOR MOLECULES	RO	CO	OVERLAP
1,6-DIPHENYLHEXATRIENE	20.25	0.054	4.12440
1,8-DIPHENYL-1,3,5,7-OCTATETRAENE	19.37	0.062	3.16183
BPSB (TRACE)	21.27	0.046	5.53871
BIS-MSB	22.05	0.042	6.86427
ESCULIN (A)	19.59	0.059	3.37922
AZULENE	25.40	0.027	16.05348
NAPHTHALENE	17.98	0.077	2.03679
DEUTERATED NAPHTHALENE-C10D8	18.13	0.075	2.12601
1-METHYLNAPHTHALENE 5ML/L	18.97	0.065	2.80566
2-METHYL NAPHTHALENE	18.09	0.076	2.10987
2,3-DIMETHYL NAPHTHALENE	18.49	0.071	2.40344
2,6-DIMETHYL NAPHTHALENE (TRACE)	18.27	0.073	2.23487
1-NAPHTHOL (TRACE)	18.96	0.066	2.78875
2-NAPHTHOL (TRACE)	17.72	0.080	1.85663
1-NAPHTHYLAMINE	17.30	0.086	1.60127
2-CHLORONAPHTHALENE	18.09	0.076	2.10532
ACENAPHTHENE	19.60	0.059	3.41263
1-PHENYL NAPHTHALENE	21.31	0.046	5.62970
2-PHENYL NAPHTHALENE	31.71	0.014	60.98665
1,4-DIPHENYL NAPHTHALENE	21.94	0.042	6.66515
1,5-DIPHENYL NAPHTHALENE	21.61	0.044	6.09613
1,7-DIPHENYL NAPHTHALENE	21.86	0.043	6.52381
1,4,5,8-TETRAPHENYL NAPHTHALENE	21.40	0.046	5.74921
1,1'-BINAPHTHYL	22.46	0.039	7.70978
4,4'-DI(N-BUTOXY)-1,1'-BINAPHTHYL	21.78	0.043	6.37623
2,2'-BINAPHTHYL	24.91	0.029	14.30279
ACRIDINE (A)	14.57	0.145	0.57096
ANTHRACENE (TRACE)	12.85	0.211	0.26883
DEUTERATED ANTHRACENE C14D10 (TRACE)	13.54	0.180	0.36825
9-METHYL ANTHRACENE (TRACE)	13.48	0.183	0.35896
1-AMINO ANTHRACENE	22.32	0.040	7.39167
2-AMINO ANTHRACENE	11.24	0.315	0.11816
9-VINYL ANTHRACENE	15.39	0.123	0.79474
9-PHENYL ANTHRACENE (TRACE)	16.23	0.105	1.09439
9,10-DIPHENYL ANTHRACENE	19.20	0.063	2.99174
9,10-DICHLOROANTHRACENE (TRACE)	14.35	0.152	0.52115
BIANTHRANYL (B)	22.75	0.038	8.28806
9,10-DINAPHTHYLANTHRACENE (B)	23.30	0.035	9.56909
9-(P-VINYLPHENYL)-10-PHENYL ANTHRACENE	20.62	0.051	4.59560
9,10-BIS(PHENYL-ETHYNYL)ANTHRACENE (TRACE)	26.81	0.023	22.22270
RUBICENE (B)	26.25	0.025	19.55966
DIBENZ(A,H)ANTHRACENE (B)	28.43	0.019	31.55445
TETRACENE (TRACE) (B)	32.21	0.013	66.71014
RUBRENE (B)	27.70	0.021	27.00940
TETRAHYDROCANNABINOL ACID METHYL ESTER	23.41	0.035	9.94271
PHENANTHRENE	21.36	0.046	5.67807
3,4-BENZOPHENANTHRENE	27.25	0.022	24.48591
2-PHENYL PHENANTHRENE	27.45	0.022	25.57684
CHRYSENE (TRACE)	24.97	0.029	14.48061
TRIPHENYLENE	22.91	0.037	8.63664
DEUTERATED TRIPHENYLENE-C18D12	23.02	0.037	8.89177
PYRENE	22.41	0.040	7.57275
1,3,6,8,-TETRAPHENYL PYRENE	25.75	0.026	17.43239
BIPYRENYL (B)	26.81	0.023	22.20861
O-PHENYLENE-PYRENE	25.31	0.028	15.72853
ANTHANTHRENE (B)	24.59	0.030	13.22149
FLUORANTHENE	23.16	0.036	9.23515
3-PHENYL FLUORANTHENE	24.01	0.032	11.45177
3-AMINO FLUORANTHENE	22.28	0.040	7.31120
3,3'-BIFLUORANTHENYL (B)	25.71	0.026	17.28348
BENZO(GHI)FLUORANTHENE	22.25	0.041	7.25518
9,12-DI(N-OCTADECOXY)-BENZ(E)ACEPHENANTHRYLENE	26.10	0.023	18.89600
DECACYCLENE (B)	25.27	0.028	15.57225
10-ETHOXY-BENZO(KL)XANTHENE	18.83	0.067	2.66182
PERYLENE	16.77	0.095	1.32940
BENZO(GHI)PERYLENE (B)	25.99	0.025	18.40748
BBOT (TRACE)	19.76	0.058	3.55923
QUININE SULFATE (A)	20.26	0.054	4.13173
ACRIDINE YELLOW (A)	23.32	0.035	9.61049
ACRIDINE ORANGE (A)	23.36	0.035	9.71665
ACRIDINE RED (A)	20.08	0.055	3.92238
URANINE (A)	20.14	0.055	3.99387
RHODAMINE B (A)	23.12	0.036	9.13374
RHODAMINE 6G (A)	23.17	0.036	9.26159
EOSIN B (A)	20.68	0.051	4.67392
EOSIN Y (A)	21.27	0.046	5.54089
ROSE BENGAL (A)	22.20	0.041	7.15435
INDENOINDENE	22.25	0.041	7.25487

8.4 Donor Molecule: Toluene (trace)

BENZENE (3 ML/L)	3.99	7.022	0.00019
DEUTERATED BENZENE-C6D6 5ML/L	3.37	11.657	0.00007
FLUOROBENZENE (TRACE)	6.99	1.310	0.00550
TOLUENE (TRACE)	7.20	1.200	0.00655
DEUTERATED TOLUENE-C7D8 5ML/L	6.51	1.623	0.00358
ETHYLBENZENE (TRACE)	7.00	1.303	0.00556
PROPYLBENZENE (TRACE)	6.70	1.490	0.00425
ISOPROPYLBENZENE (TRACE)	6.28	1.810	0.00288
SEC-BUTYLBENZENE (TRACE)	6.29	1.795	0.00293

ACCEPTOR MOLECULES	RO	CO	OVERLAP
DIPHENYLMETHANE (TRACE)	8.24	0.799	0.01475
BIBENZYL	7.64	1.003	0.00937
1,3,5-TRIPHENYLPENTANE 0.05ML/L	7.84	0.928	0.01093
POLYSTYRENE	7.01	1.298	0.00560
POLYVINYLTOLUENE	9.61	0.504	0.03659
PHENYLCYCLOHEXANE (TRACE)	6.33	1.848	0.00277
BENZYLALCOHOL 3ML/L	6.11	1.958	0.00246
BENZYLACETATE 5ML/L	6.08	1.993	0.00238
PHENOL (TRACE)	13.66	0.176	0.30432
ANISOLE (TRACE)	13.94	0.165	0.34346
PHENYL ETHER	14.53	0.146	0.43959
ANILINE 3ML/L	16.28	0.104	0.86359
N,N-DIMETHYLANILINE 3ML/L	17.26	0.087	1.21763
TRIPHENYLAMINE	25.43	0.027	12.35302
P-XYLENE (TRACE)	10.41	0.396	0.05983
DEUTERATED P-XYLENE C8D10 (TRACE)	10.34	0.404	0.05753
M-XYLENE (TRACE)	8.55	0.715	0.01839
O-XYLENE 3ML/L	8.24	0.801	0.01466
1-METHYL-4-ETHYLBENZENE (TRACE)	9.74	0.483	0.04022
1,4-BIS(TRIFLUOROMETHYL)BENZENE 5ML/L	8.09	0.846	0.01318
P-CRESOL	15.97	0.110	0.77429
P-METHYLANISOLE 3ML/L	15.43	0.122	0.62967
1,4-DIMETHOXY BENZENE	17.62	0.082	1.38334
1-METHYL-4-CHLOROBENZENE 5ML/L	10.82	0.354	0.07502
SODIUM SALICYLATE (A)	18.48	0.071	1.82187
ANTHRANILIC ACID (A)	15.17	0.128	0.55706
TMPD	21.25	0.047	4.21572
1,2,4-TRIMETHYLBENZENE (TRACE)	11.28	0.311	0.09662
MESITYLENE 5ML/L	8.62	0.699	0.01919
1,3,5-TRIETHYLBENZENE 5ML/L	8.32	0.778	0.01550
STYRENE	13.62	0.177	0.29635
1,1-DIPHENYLETHYLENE 3ML/L	15.81	0.113	0.72805
BIPHENYL	15.30	0.125	0.60006
DEUTERATED BIPHENYL-C12D10	15.34	0.124	0.61101
4-METHYL BIPHENYL	17.41	0.085	1.30047
MONOISOPROPYLBIPHENYL 5ML/L	16.87	0.093	1.07923
4-BENZYL BIPHENYL	19.12	0.064	2.28066
3,3'-DIMETHYL BIPHENYL	16.42	0.101	0.91339
2,2'-DIMETHYLBIPHENYL 2ML/L	11.23	0.316	0.09354
OCTAMETHYL BIPHENYL	12.87	0.210	0.21280
4-METHOXY BIPHENYL	21.31	0.046	4.34746
4-BIPHENYLYL PHENYL ETHER	21.90	0.043	5.11599
4,4'-DIHYDROXYBIPHENYL (A)	23.27	0.036	7.31357
4,4'-DIMETHOXY BIPHENYL	23.17	0.036	7.14509
4,4'-DIISOBUTOXYBIPHENYL	22.97	0.037	6.78986
BENZIDINE	26.63	0.024	16.32213
4-VINYL BIPHENYL	25.94	0.026	14.05835
M-TERPHENYL	20.29	0.054	3.22388
M-QUATERPHENYL	20.70	0.050	3.66083
1,3,5-TRIPHENYL BENZENE	21.67	0.044	4.75201
1,3-DIBIPHENYLYLBENZENE (B)	29.48	0.017	30.00785
1,3,5,-TRIBIPHENYLYLBENZENE (B)	32.32	0.013	52.22627
FLUORENE (TRACE)	21.38	0.046	4.42592
1-METHYL FLUORENE	18.78	0.068	2.01142
1,2-BENZOFLUORENE	23.60	0.034	7.88805
9,10-DIHYDROPHENANTHRENE	21.80	0.043	4.95715
CARBAZOLE (TRACE)	22.93	0.037	6.65181
N-METHYL CARBAZOLE (TRACE)	22.30	0.040	5.62427
N-VINYLCARBAZOLE	23.93	0.033	8.58148
N-PHENYL CARBAZOLE	24.11	0.032	8.97613
N-METHYL-2,3-BENZCARBAZOLE	28.38	0.020	23.89521
DIBENZOFURAN	23.11	0.036	7.05726
DIBENZOTHIOPHENE (TRACE)	20.60	0.051	3.49383
P-TERPHENYL	26.69	0.024	16.64626
DEUTERATED P-TERPHENYL-C18D14	26.38	0.024	15.52985
4-METHYL TERPHENYL	26.99	0.023	17.80964
2-METHYL-P-TERPHENYL	23.72	0.034	8.24057
3,3''-DIMETHYL-P-TERPHENYL	26.55	0.024	16.10326
4-(3,3-DIMETHYLBUTOXY)-P-TERPHENYL	27.42	0.022	19.47283
4,4''-DIHEXAHYDROFARNESOXY-P-TERPHENYL	28.32	0.020	23.62308
2,2'-METHYLENE-P-TERPHENYL	27.08	0.023	18.11014
2,2'-ETHYLENE-P-TERPHENYL	27.05	0.023	17.93761
2,2'-METHYLENE-5',6''-METHYLENE-P-TERPHENYL (TRACE)	25.98	0.026	14.06153
3-PHENYL DIBENZOFURAN	25.90	0.026	13.86174
BENZO(1,2-B-4,5-B')BISBENZOFURAN (TRACE) (B)	22.25	0.041	5.52601
SEXIPHENYL BENZENE	21.51	0.045	4.60058
P-QUATERPHENYL	28.14	0.020	22.69425
3,3'''-DIMETHYL-P-QUATERPHENYL	28.38	0.020	23.91239
TETRAMETHYL-P-QUATERPHENYL	26.01	0.025	14.33034
3,3''' -DIMETHYL-3',2''-METHYLENE-P-QUATERPHENYL	26.83	0.023	17.05955
3,3''' -DIMETHYL-3',2''-ETHYLENE-P-QUATERPHENYL	28.01	0.020	22.07522
4-(3-ETHYLBUTOXY)-P-QUATERPHENYL (B)	27.59	0.021	20.16028
4,4'''-DI(2-BUTYLOCTOXY-1)-P-QUATERPHENYL	28.68	0.019	25.43633
DIETHYL-P-QUINQUEPHENYL	29.83	0.017	32.19922
DI(3-ETHYLHEPTYL)-P-QUINQUEPHENYL	29.15	0.018	28.05146
TETRAMETHYL-P-SEXIPHENYL	30.37	0.016	35.84347
TETRAMETHYL-P-OCTAPHENYL (B)	29.52	0.017	30.25512
INDOLE (TRACE)	17.77	0.080	1.47141
2-PHENYL INDOLE	24.72	0.030	10.43593
1-METHYL-2-PHENYL INDOLE	24.24	0.031	9.26639
1-NEOPENTYL-2-PHENYL INDOLE	24.18	0.032	9.16716
1-N-BUTYL-2(BIPHENYLYL-4) INDOLE	25.13	0.028	11.51994
1,2-DIPHENYL INDOLE	25.06	0.028	11.32704
1-BIPHENYLYL-2-PHENYL INDOLE	26.83	0.023	17.05302
1,3-DIMETHYL-2-PHENYL INDOLE	23.22	0.036	7.16375
2,3-DIPHENYL INDOLE	23.89	0.033	8.50373

ACCEPTOR MOLECULES	RO	CO	OVERLAP
1-METHYL-2,3-DIPHENYL INDOLE	23.76	0.033	8.22769
2-PHENYL-3,2'-METHYLENE INDOLE (TRACE)	22.95	0.037	6.67617
2-PHENYL-3,2'-ETHYLENE INDOLE (TRACE)	22.35	0.040	5.69660
2-PHENYL-3,2'-TRIMETHYLENE INDOLE	23.59	0.034	7.87091
1-METHYL-3,2'-METHYLENE-2-PHENYL INDOLE (TRACE)	22.42	0.040	5.80703
1-METHYL-3,2'-ETHYLENE-2-PHENYL INDOLE	21.96	0.042	5.12822
1-METHYL-3,2'-TRIMETHYLENE-2-PHENYL INDOLE	22.89	0.037	6.58078
4,5-DIPHENYLIMIDAZOLE (A)	22.63	0.039	6.14519
PPF (TRACE)	24.06	0.032	8.87512
PPO (TRACE)	26.41	0.024	15.51152
2-(4-FLUOROPHENYL)-5-PHENYLOXAZOLE (TRACE)	25.58	0.027	12.80143
αNPO (TRACE)	23.55	0.034	7.80849
BBO (B)	24.83	0.029	10.71686
POPOP	22.98	0.037	6.73126
DIMETHYL POPOP (TRACE)	24.16	0.032	9.10346
PPD (TRACE)	26.38	0.024	15.55734
2,5-DI(2-METHYLPHENYL)OXADIAZOLE	25.68	0.026	13.22709
PBD	28.29	0.020	23.44923
BBD (B)	27.37	0.022	19.20561
αNPD (TRACE)	23.64	0.034	7.97139
βNPD	27.40	0.022	19.33427
αNND (TRACE)	22.01	0.042	5.19864
βNND (TRACE)	27.43	0.022	19.49650
TRANS-1-(4-BIPHENYLYL)-2-PHENYL-ETHYLENE	26.08	0.025	14.37761
TRANS-1-(4-BIPHENYLYL)-2-(1-NAPHTHYL)-ETHYLENE	24.76	0.029	10.54265
DIPHENYLSTILBENE (B)	23.45	0.035	7.59752
1,1,4,4-TETRAPHENYLBUTADIENE	21.67	0.044	4.73357
1,6-DIPHENYLHEXATRIENE	21.22	0.047	4.17149
1,8-DIPHENYL-1,3,5,7-OCTATETRAENE	20.10	0.055	3.01887
BPSB (TRACE)	22.31	0.040	5.63682
BIS-MSB	23.21	0.036	7.14500
ESCULIN (A)	20.71	0.050	3.60603
AZULENE	26.07	0.025	14.34864
NAPHTHALENE	18.54	0.070	1.86764
DEUTERATED NAPHTHALENE-C10D8	18.67	0.069	1.94031
1-METHYLNAPHTHALENE 5ML/L	19.83	0.057	2.79630
2-METHYL NAPHTHALENE	18.73	0.068	1.98455
2,3-DIMETHYL NAPHTHALENE	19.21	0.063	2.31127
2,6-DIMETHYL NAPHTHALENE (TRACE)	18.91	0.066	2.10225
1-NAPHTHOL (TRACE)	20.04	0.056	2.96841
2-NAPHTHOL (TRACE)	18.29	0.073	1.71342
1-NAPHTHYLAMINE	18.37	0.072	1.75772
2-CHLORONAPHTHALENE	18.74	0.068	1.98769
ACENAPHTHENE	20.67	0.051	3.58451
1-PHENYL NAPHTHALENE	22.49	0.039	5.94052
2-PHENYL NAPHTHALENE	33.35	0.012	63.01592
1,4-DIPHENYL NAPHTHALENE	23.32	0.035	7.35838
1,5-DIPHENYL NAPHTHALENE	22.98	0.037	6.74210
1,7-DIPHENYL NAPHTHALENE	23.00	0.037	6.77462
1,4,5,8-TETRAPHENYL NAPHTHALENE	22.32	0.040	5.65891
1,1'-BINAPHTHYL	23.71	0.034	8.14294
4,4'-DI(N-BUTOXY)-1,1'-BINAPHTHYL	23.11	0.036	6.96343
2,2'-BINAPHTHYL	26.03	0.025	14.22463
ACRIDINE (A)	15.05	0.131	0.53029
ANTHRACENE (TRACE)	13.22	0.194	0.24393
DEUTERATED ANTHRACENE C14D10 (TRACE)	14.04	0.161	0.35097
9-METHYL ANTHRACENE (TRACE)	13.60	0.178	0.28898
1-AMINO ANTHRACENE	22.51	0.039	5.94625
2-AMINO ANTHRACENE	11.39	0.303	0.09770
9-VINYL ANTHRACENE	15.45	0.121	0.62248
9-PHENYL ANTHRACENE (TRACE)	16.29	0.104	0.85287
9,10-DIPHENYL ANTHRACENE	18.93	0.066	2.10633
9,10-DICHLOROANTHRACENE (TRACE)	14.19	0.157	0.37345
BIANTHRANYL (B)	22.32	0.040	5.65152
9,10-DINAPHTHYLANTHRACENE (B)	24.09	0.032	8.94274
9-(P-VINYLPHENYL)-10-PHENYL ANTHRACENE	20.91	0.049	3.81750
9,10-BIS(PHENYL-ETHYNYL)ANTHRACENE (TRACE)	27.55	0.021	19.99313
RUBICENE (B)	27.70	0.021	20.63528
DIBENZ(A,H)ANTHRACENE (B)	30.33	0.016	35.60624
TETRACENE (TRACE) (B)	33.07	0.012	59.83182
RUBRENE (B)	29.68	0.017	31.26043
TETRAHYDROCANNABINOL ACID METHYL ESTER	24.39	0.031	9.69580
PHENANTHRENE	22.14	0.041	5.38371
3,4-BENZOPHENANTHRENE	28.32	0.020	23.59432
2-PHENYL PHENANTHRENE	28.38	0.020	23.88727
CHRYSENE (TRACE)	25.45	0.027	12.41023
TRIPHENYLENE	23.75	0.033	8.20827
DEUTERATED TRIPHENYLENE-C18D12	23.83	0.033	8.37414
PYRENE	22.55	0.039	6.01442
1,3,6,8,-TETRAPHENYL PYRENE	27.45	0.022	19.58212
BIPYRENYL (B)	27.90	0.021	21.57062
O-PHENYLENE-PYRENE	26.70	0.023	16.57556
ANTHANTHRENE (B)	26.17	0.025	14.69206
FLUORANTHENE	24.24	0.031	9.28018
3-PHENYL FLUORANTHENE	25.43	0.027	12.37540
3-AMINO FLUORANTHENE	23.48	0.035	7.67037
3,3'-BIFLUORANTHENYL (B)	27.15	0.022	18.32079
BENZO(GHI)FLUORANTHENE	23.41	0.035	7.52072
9,12-DI(N-OCTADECOXY)-BENZ(E)ACEPHENANTHRYLENE	27.22	0.022	18.58479
DECACYCLENE (B)	26.45	0.024	15.65080
10-ETHOXY-BENZO(KL)XANTHENE	19.88	0.057	2.82599
PERYLENE	17.16	0.088	1.16908
BENZO(GHI)PERYLENE (B)	27.54	0.021	19.94424
BBOT (TRACE)	20.63	0.051	3.52102
QUININE SULFATE (A)	21.18	0.047	4.12622
ACRIDINE YELLOW (A)	23.95	0.033	8.63558

ACCEPTOR MOLECULES	RO	CO	OVERLAP
ACRIDINE ORANGE (A)	24.28	0.031	9.37219
ACRIDINE RED (A)	21.07	0.048	4.00373
URANINE (A)	21.15	0.047	4.08797
RHODAMINE B (A)	24.26	0.031	9.33303
RHODAMINE 6G (A)	24.31	0.031	9.43160
EOSIN B (A)	21.82	0.044	4.66370
EOSIN Y (A)	22.26	0.041	5.55875
ROSE BENGAL (A)	22.97	0.037	6.71284
INDENOINDENE	23.71	0.034	8.13218

8.5 Donor Molecule: Deuterated Toluene (C_7D_8) (5 ml/L)

	RO	CO	OVERLAP
BENZENE (3 ML/L)	4.15	6.246	0.00020
DEUTERATED BENZENE-C6D6 5ML/L	3.64	9.237	0.00009
FLUOROBENZENE (TRACE)	7.61	1.016	0.00740
TOLUENE (TRACE)	7.51	1.055	0.00685
DEUTERATED TOLUENE-C7D8 5ML/L	6.90	1.364	0.00410
ETHYLBENZENE (TRACE)	7.33	1.134	0.00593
PROPYLBENZENE (TRACE)	7.04	1.281	0.00465
ISOPROPYLBENZENE (TRACE)	6.66	1.514	0.00333
SEC-BUTYLBENZENE (TRACE)	6.66	1.516	0.00332
DIPHENYLMETHANE (TRACE)	8.63	0.695	0.01580
BIBENZYL	8.03	0.863	0.01023
1,3,5-TRIPHENYLPENTANE 0.05ML/L	8.23	0.803	0.01182
POLYSTYRENE	7.35	1.124	0.00604
POLYVINYLTOLUENE	10.16	0.427	0.04132
PHENYLCYCLOHEXANE (TRACE)	6.58	1.570	0.00310
BENZYLALCOHOL 3ML/L	6.46	1.657	0.00278
BENZYLACETATE 5ML/L	6.41	1.700	0.00265
PHENOL (TRACE)	14.33	0.152	0.32924
ANISOLE (TRACE)	14.66	0.142	0.37642
PHENYL ETHER	15.27	0.126	0.48049
ANILINE 3ML/L	16.84	0.094	0.85612
N,N-DIMETHYLANILINE 3ML/L	17.81	0.079	1.18769
TRIPHENYLAMINE	26.14	0.025	11.85128
P-XYLENE (TRACE)	11.01	0.335	0.06778
DEUTERATED P-XYLENE C8D10 (TRACE)	10.98	0.338	0.06642
M-XYLENE (TRACE)	9.10	0.594	0.02155
O-XYLENE 3ML/L	8.72	0.675	0.01673
1-METHYL-4-ETHYLBENZENE (TRACE)	10.37	0.401	0.04722
1,4-BIS(TRIFLUOROMETHYL)BENZENE 5ML/L	8.50	0.727	0.01443
P-CRESOL	16.66	0.097	0.80765
P-METHYLANISOLE 3ML/L	16.10	0.107	0.65692
1,4-DIMETHOXY BENZENE	18.17	0.075	1.34619
1-METHYL-4-CHLOROBENZENE 5ML/L	11.34	0.307	0.08065
SODIUM SALICYLATE (A)	18.99	0.065	1.74232
ANTHRANILIC ACID (A)	15.53	0.119	0.52129
TMPD	22.18	0.041	4.41643
1,2,4-TRIMETHYLBENZENE (TRACE)	11.89	0.266	0.10717
MESITYLENE 5ML/L	9.13	0.588	0.02201
1,3,5-TRIETHYLBENZENE 5ML/L	8.79	0.659	0.01749
STYRENE	14.21	0.156	0.30946
1,1-DIPHENYLETHYLENE 3ML/L	16.56	0.098	0.78455
BIPHENYL	16.12	0.107	0.66478
DEUTERATED BIPHENYL-C12D10	16.17	0.106	0.67767
4-METHYL BIPHENYL	18.30	0.073	1.41702
MONOISOPROPYLBIPHENYL 5ML/L	17.75	0.080	1.18251
4-BENZYL BIPHENYL	20.08	0.055	2.47625
3,3'-DIMETHYL BIPHENYL	17.21	0.088	0.97889
2,2'-DIMETHYLBIPHENYL 2ML/L	11.85	0.269	0.10452
OCTAMETHYL BIPHENYL	13.52	0.181	0.23083
4-METHOXY BIPHENYL	22.27	0.040	4.58523
4-BIPHENYLYL PHENYL ETHER	22.90	0.037	5.41848
4,4'-DIHYDROXYBIPHENYL (A)	24.23	0.031	7.55249
4,4'-DIMETHOXY BIPHENYL	24.14	0.032	7.40414
4,4'-DIISOBUTOXYBIPHENYL	23.92	0.033	7.01764
BENZIDINE	27.53	0.021	16.17374
4-VINYL BIPHENYL	26.94	0.023	14.29136
M-TERPHENYL	21.23	0.047	3.42444
M-QUATERPHENYL	21.72	0.044	3.95087
1,3,5-TRIPHENYL BENZENE	22.74	0.038	5.14670
1,3-DIBIPHENYLYLBENZENE (B)	30.49	0.016	29.82662
1,3,5,-TRIBIPHENYLYLBENZENE (B)	33.39	0.012	51.45623
FLUORENE (TRACE)	22.25	0.041	4.55396
1-METHYL FLUORENE	19.27	0.063	1.89928
1,2-BENZOFLUORENE	24.31	0.031	7.65827
9,10-DIHYDROPHENANTHRENE	22.73	0.038	5.16385
CARBAZOLE (TRACE)	23.66	0.034	6.51869
N-METHYL CARBAZOLE (TRACE)	22.99	0.037	5.47874
N-VINYLCARBAZOLE	24.82	0.029	8.68020
N-PHENYL CARBAZOLE	24.91	0.029	8.87059
N-METHYL-2,3-BENZCARBAZOLE	29.48	0.017	24.37853
DIBENZOFURAN	23.97	0.032	7.11423
DIBENZOTHIOPHENE (TRACE)	21.35	0.046	3.52184
P-TERPHENYL	27.71	0.021	16.88474
DEUTERATED P-TERPHENYL-C18D14	27.39	0.022	15.75434
4-METHYL TERPHENYL	27.99	0.020	17.91132
2-METHYL-P-TERPHENYL	24.74	0.030	8.58429
3,3''-DIMETHYL-P-TERPHENYL	27.53	0.021	16.22647

ACCEPTOR MOLECULES	RO	CO	OVERLAP
4-(3,3-DIMETHYLBUTOXY)-P-TERPHENYL	28.35	0.020	19.27177
4,4''-DIHEXAHYDROFARNESOXY-P-TERPHENYL	29.23	0.018	23.18575
2,2'-METHYLENE-P-TERPHENYL	27.98	0.020	17.81909
2,2'-ETHYLENE-P-TERPHENYL	27.92	0.021	17.59847
2,2'-METHYLENE-5',6''-METHYLENE-P-TERPHENYL (TRACE)	26.73	0.023	13.54906
3-PHENYL DIBENZOFURAN	26.63	0.024	13.23471
BENZO(1,2-B-4,5-B')BISBENZOFURAN (TRACE) (B)	22.77	0.038	5.16606
SEXIPHENYL BENZENE	22.47	0.039	4.84079
P-QUATERPHENYL	29.02	0.018	22.16124
3,3'''-DIMETHYL-P-QUATERPHENYL	29.26	0.018	23.30300
TETRAMETHYL-P-QUATERPHENYL	27.10	0.022	14.83430
3,3''' -DIMETHYL-3',2''-METHYLENE-P-QUATERPHENYL	27.57	0.021	16.31404
3,3''' -DIMETHYL-3',2''-ETHYLENE-P-QUATERPHENYL	28.88	0.019	21.56616
4-(3-ETHYLBUTOXY)-P-QUATERPHENYL (B)	28.37	0.020	19.38112
4,4'''-DI(2-BUTYLOCTOXY-1)-P-QUATERPHENYL	29.51	0.017	24.50349
DIETHYL-P-QUINQUEPHENYL	30.75	0.015	31.41707
DI(3-ETHYLHEPTYL)-P-QUINQUEPHENYL	29.98	0.017	26.94069
TETRAMETHYL-P-SEXIPHENYL	31.33	0.015	35.12158
TETRAMETHYL-P-OCTAPHENYL (B)	30.39	0.016	29.25395
INDOLE (TRACE)	18.61	0.069	1.57227
2-PHENYL INDOLE	25.40	0.027	9.96140
1-METHYL-2-PHENYL INDOLE	24.94	0.029	8.94660
1-NEOPENTYL-2-PHENYL INDOLE	24.91	0.029	8.88283
1-N-BUTYL-2(BIPHENYLYL-4) INDOLE	25.88	0.026	11.16505
1,2-DIPHENYL INDOLE	25.83	0.026	11.02384
1-BIPHENYLYL-2-PHENYL INDOLE	27.71	0.021	16.80781
1,3-DIMETHYL-2-PHENYL INDOLE	23.90	0.033	6.92142
2,3-DIPHENYL INDOLE	24.61	0.030	8.25325
1-METHYL-2,3-DIPHENYL INDOLE	24.55	0.030	8.13097
2-PHENYL-3,2'-METHYLENE INDOLE (TRACE)	23.54	0.034	6.31213
2-PHENYL-3,2'-ETHYLENE INDOLE (TRACE)	22.97	0.037	5.46070
2-PHENYL-3,2'-TRIMETHYLENE INDOLE	24.20	0.032	7.46566
1-METHYL-3,2'-METHYLENE-2-PHENYL INDOLE (TRACE)	23.00	0.037	5.49561
1-METHYL-3,2'-ETHYLENE-2-PHENYL INDOLE	22.59	0.039	4.92982
1-METHYL-3,2'-TRIMETHYLENE-2-PHENYL INDOLE	23.50	0.034	6.25543
4,5-DIPHENYLIMIDAZOLE (A)	23.41	0.035	6.11291
PPF (TRACE)	24.70	0.030	8.43858
PPO (TRACE)	27.16	0.022	14.90697
2-(4-FLUOROPHENYL)-5-PHENYLOXAZOLE (TRACE)	26.32	0.025	12.34099
αNPO (TRACE)	24.35	0.031	7.75010
BBO (B)	25.62	0.027	10.50083
POPOP	23.70	0.034	6.58464
DIMETHYL POPOP (TRACE)	24.97	0.029	8.99773
PPD (TRACE)	27.33	0.022	15.56118
2,5-DI(2-METHYLPHENYL)OXADIAZOLE	26.60	0.024	13.22585
PBD	29.12	0.018	22.64837
BBD (B)	28.16	0.020	18.52597
αNPD (TRACE)	24.41	0.031	7.84778
βNPD	28.38	0.020	19.39710
αNND (TRACE)	22.64	0.039	4.99758
βNND (TRACE)	28.41	0.019	19.54263
TRANS-1-(4-BIPHENYLYL)-2-PHENYL-ETHYLENE	26.78	0.023	13.70537
TRANS-1-(4-BIPHENYLYL)-2-(1-NAPHTHYL)-ETHYLENE	25.56	0.027	10.35543
DIPHENYLSTILBENE (B)	24.04	0.032	7.17545
1,1,4,4-TETRAPHENYLBUTADIENE	22.30	0.040	4.57113
1,6-DIPHENYLHEXATRIENE	21.71	0.044	3.89283
1,8-DIPHENYL-1,3,5,7-OCTATETRAENE	20.75	0.050	2.96620
BPSB (TRACE)	22.90	0.037	5.35586
BIS-MSB	23.82	0.033	6.78675
ESCULIN (A)	21.30	0.046	3.46659
AZULENE	27.26	0.022	15.24154
NAPHTHALENE	19.35	0.062	1.93391
DEUTERATED NAPHTHALENE-C10D8	19.49	0.060	2.03357
1-METHYLNAPHTHALENE 5ML/L	20.60	0.051	2.83973
2-METHYL NAPHTHALENE	19.52	0.060	2.05827
2,3-DIMETHYL NAPHTHALENE	20.00	0.056	2.37844
2,6-DIMETHYL NAPHTHALENE (TRACE)	19.71	0.058	2.17656
1-NAPHTHOL (TRACE)	20.70	0.050	2.92265
2-NAPHTHOL (TRACE)	19.05	0.065	1.77734
1-NAPHTHYLAMINE	18.86	0.067	1.67385
2-CHLORONAPHTHALENE	19.53	0.060	2.05947
ACENAPHTHENE	21.38	0.046	3.55941
1-PHENYL NAPHTHALENE	23.26	0.036	5.88153
2-PHENYL NAPHTHALENE	34.51	0.011	62.75249
1,4-DIPHENYL NAPHTHALENE	24.00	0.032	7.09185
1,5-DIPHENYL NAPHTHALENE	23.65	0.034	6.50559
1,7-DIPHENYL NAPHTHALENE	23.73	0.033	6.63942
1,4,5,8-TETRAPHENYL NAPHTHALENE	23.06	0.036	5.58573
1,1'-BINAPHTHYL	24.51	0.030	8.04987
4,4'-DI(N-BUTOXY)-1,1'-BINAPHTHYL	23.76	0.033	6.68383
2,2'-BINAPHTHYL	26.95	0.023	14.21659
ACRIDINE (A)	15.52	0.120	0.51933
ANTHRACENE (TRACE)	13.57	0.179	0.23189
DEUTERATED ANTHRACENE C14D10 (TRACE)	14.45	0.148	0.33780
9-METHYL ANTHRACENE (TRACE)	14.15	0.158	0.29848
1-AMINO ANTHRACENE	23.60	0.034	6.40922
2-AMINO ANTHRACENE	11.63	0.285	0.08998
9-VINYL ANTHRACENE	16.14	0.106	0.65564
9-PHENYL ANTHRACENE (TRACE)	17.03	0.091	0.90510
9,10-DIPHENYL ANTHRACENE	19.92	0.057	2.32283
9,10-DICHLOROANTHRACENE (TRACE)	14.85	0.137	0.39797
BIANTHRANYL (B)	23.53	0.034	6.30419
9,10-DINAPHTHYLANTHRACENE (B)	25.03	0.029	9.12345
9-(P-VINYLPHENYL)-10-PHENYL ANTHRACENE	21.82	0.043	4.00934
9,10-BIS(PHENYL-ETHYNYL)ANTHRACENE (TRACE)	28.69	0.019	20.71618
RUBICENE (B)	28.57	0.019	20.21666

ACCEPTOR MOLECULES	RO	CO	OVERLAP
DIBENZ(A,H)ANTHRACENE (B)	31.18	0.015	34.09608
TETRACENE (TRACE) (B)	34.59	0.011	63.57153
RUBRENE (B)	30.41	0.016	29.35930
TETRAHYDROCANNABINOL ACID METHYL ESTER	25.33	0.028	9.85397
PHENANTHRENE	23.03	0.037	5.54179
3,4-BENZOPHENANTHRENE	29.47	0.017	24.33449
2-PHENYL PHENANTHRENE	29.52	0.017	24.59911
CHRYSENE (TRACE)	26.44	[illegible]	[illegible]
TRIPHENYLENE	24.70	0.030	8.43926
DEUTERATED TRIPHENYLENE-C18D12	24.80	0.029	8.63510
PYRENE	23.65	0.034	6.49315
1,3,6,8,-TETRAPHENYL PYRENE	28.24	0.020	18.85006
BIPYRENYL (B)	29.00	0.018	22.07967
O-PHENYLENE-PYRENE	27.55	0.021	16.22182
ANTHANTHRENE (B)	26.90	0.023	14.05820
FLUORANTHENE	25.17	0.028	9.44767
3-PHENYL FLUORANTHENE	26.22	0.025	12.06762
3-AMINO FLUORANTHENE	24.21	0.032	7.47065
3,3'-BIFLUORANTHENYL (B)	28.02	0.020	17.96827
BENZO(GHI)FLUORANTHENE	24.21	0.032	7.47662
9,12-DI(N-OCTADECOXY)-BENZ(E)ACEPHENANTHRYLENE	28.26	0.020	18.90542
DECACYCLENE (B)	27.33	0.022	15.49210
10-ETHOXY-BENZO(KL)XANTHENE	20.50	0.052	2.75348
PERYLENE	17.88	0.078	1.21284
BENZO(GHI)PERYLENE (B)	28.38	0.020	19.38644
BBOT (TRACE)	21.24	0.047	3.41257
QUININE SULFATE (A)	21.92	0.042	4.11497
ACRIDINE YELLOW (A)	24.97	0.029	9.00086
ACRIDINE ORANGE (A)	25.19	0.028	9.49723
ACRIDINE RED (A)	21.79	0.043	3.97674
URANINE (A)	21.86	0.043	4.05371
RHODAMINE B (A)	25.09	0.028	9.27381
RHODAMINE 6G (A)	25.16	0.028	9.43009
EOSIN B (A)	22.34	0.040	4.61857
EOSIN Y (A)	23.00	0.037	5.49655
ROSE BENGAL (A)	23.85	0.033	6.82883
INDENOINDENE	24.33	0.031	7.69491

8.6 Donor Molecule: Ethylbenzene (trace)

	RO	CO	OVERLAP
BENZENE (3 ML/L)	4.20	6.046	0.00024
DEUTERATED BENZENE-C6D6 5ML/L	3.74	8.563	0.00012
FLUOROBENZENE (TRACE)	7.75	0.962	0.00964
TOLUENE (TRACE)	7.44	1.085	0.00756
DEUTERATED TOLUENE-C7D8 5ML/L	6.87	1.379	0.00469
ETHYLBENZENE (TRACE)	7.27	1.165	0.00656
PROPYLBENZENE (TRACE)	6.99	1.308	0.00521
ISOPROPYLBENZENE (TRACE)	6.63	1.533	0.00380
SEC-BUTYLBENZENE (TRACE)	6.63	1.532	0.00380
DIPHENYLMETHANE (TRACE)	8.47	0.735	0.01648
BIBENZYL	7.95	0.889	0.01127
1,3,5-TRIPHENYLPENTANE 0.05ML/L	8.15	0.827	0.01300
POLYSTYRENE	7.26	1.169	0.00653
POLYVINYLTOLUENE	9.77	0.480	0.03817
PHENYLCYCLOHEXANE (TRACE)	6.60	1.555	0.00369
BENZYLALCOHOL 3ML/L	6.41	1.700	0.00309
BENZYLACETATE 5ML/L	6.39	1.713	0.00304
PHENOL (TRACE)	13.72	0.173	0.29622
ANISOLE (TRACE)	14.01	0.163	0.33458
PHENYL ETHER	14.61	0.143	0.42943
ANILINE 3ML/L	16.31	0.103	0.82388
N,N-DIMETHYLANILINE 3ML/L	17.43	0.084	1.21814
TRIPHENYLAMINE	25.66	0.026	12.31332
P-XYLENE (TRACE)	10.57	0.379	0.06190
DEUTERATED P-XYLENE C8D10 (TRACE)	10.53	0.384	0.06034
M-XYLENE (TRACE)	8.75	0.667	0.01996
O-XYLENE 3ML/L	8.44	0.745	0.01602
1-METHYL-4-ETHYLBENZENE (TRACE)	9.94	0.456	0.04283
1,4-BIS(TRIFLUOROMETHYL)BENZENE 5ML/L	8.40	0.756	0.01561
P-CRESOL	15.97	0.110	0.73253
P-METHYLANISOLE 3ML/L	15.43	0.122	0.59575
1,4-DIMETHOXY BENZENE	17.66	0.081	1.32643
1-METHYL-4-CHLOROBENZENE 5ML/L	10.86	0.349	0.07274
SODIUM SALICYLATE (A)	18.63	0.069	1.80303
ANTHRANILIC ACID (A)	15.73	0.115	0.65305
TMPD	21.45	0.045	4.20382
1,2,4-TRIMETHYLBENZENE (TRACE)	11.40	0.303	0.09702
MESITYLENE 5ML/L	8.77	0.663	0.02019
1,3,5-TRIETHYLBENZENE 5ML/L	8.47	0.736	0.01637
STYRENE	13.70	0.174	0.29102
1,1-DIPHENYLETHYLENE 3ML/L	15.98	0.110	0.73403
BIPHENYL	15.52	0.120	0.61952
DEUTERATED BIPHENYL-C12D10	15.58	0.118	0.63256
4-METHYL BIPHENYL	17.60	0.082	1.30999
MONOISOPROPYLBIPHENYL 5ML/L	17.07	0.090	1.09297
4-BENZYL BIPHENYL	19.31	0.062	2.28454
3,3'-DIMETHYL BIPHENYL	16.56	0.098	0.90846
2,2'-DIMETHYLBIPHENYL 2ML/L	11.40	0.302	0.09674
OCTAMETHYL BIPHENYL	12.94	0.206	0.20731

ACCEPTOR MOLECULES	RO	CO	OVERLAP
4-METHOXY BIPHENYL	21.43	0.045	4.25145
4-BIPHENYLYL PHENYL ETHER	22.03	0.042	5.01235
4,4'-DIHYDROXYBIPHENYL (A)	23.37	0.035	7.09967
4,4'-DIMETHOXY BIPHENYL	23.27	0.035	6.93426
4,4'-DIISOBUTOXYBIPHENYL	23.06	0.036	6.58094
BENZIDINE	26.79	0.023	15.95001
4-VINYL BIPHENYL	26.03	0.025	13.58471
M-TERPHENYL	20.48	0.052	3.22047
M-QUATERPHENYL	20.92	0.049	3.68687
1,3,5-TRIPHENYL BENZENE	21.89	0.043	4.78431
1,3-DIBIPHENYLYLBENZENE (B)	29.61	0.017	29.17363
1,3,5,-TRIBIPHENYLYLBENZENE (B)	32.49	0.013	50.94582
FLUORENE (TRACE)	21.56	0.045	4.40045
1-METHYL FLUORENE	18.84	0.067	1.93774
1,2-BENZOFLUORENE	23.81	0.033	7.84795
9,10-DIHYDROPHENANTHRENE	21.94	0.042	4.87393
CARBAZOLE (TRACE)	23.01	0.037	6.40215
N-METHYL CARBAZOLE (TRACE)	22.40	0.040	5.45124
N-VINYLCARBAZOLE	24.04	0.032	8.33270
N-PHENYL CARBAZOLE	24.18	0.032	8.62984
N-METHYL-2,3-BENZCARBAZOLE	28.52	0.019	23.22537
DIBENZOFURAN	23.14	0.036	6.72034
DIBENZOTHIOPHENE (TRACE)	20.65	0.051	3.35559
P-TERPHENYL	26.79	0.023	16.08839
DEUTERATED P-TERPHENYL-C18D14	26.47	0.024	15.01227
4-METHYL TERPHENYL	27.10	0.022	17.23853
2-METHYL-P-TERPHENYL	23.83	0.033	8.00652
3,3''-DIMETHYL-P-TERPHENYL	26.66	0.024	15.59337
4-(3,3-DIMETHYLBUTOXY)-P-TERPHENYL	27.56	0.021	18.98519
4,4''-DIHEXAHYDROFARNESOXY-P-TERPHENYL	28.50	0.019	23.21177
2,2'-METHYLENE-P-TERPHENYL	27.28	0.022	17.85528
2,2'-ETHYLENE-P-TERPHENYL	27.24	0.022	17.68770
2,2'-METHYLENE-5',6''-METHYLENE-P-TERPHENYL (TRACE)	26.30	0.025	14.29226
3-PHENYL DIBENZOFURAN	26.19	0.025	13.99009
BENZO(1,2-B-4,5-B')BISBENZOFURAN (TRACE) (B)	23.02	0.037	6.39143
SEXIPHENYL BENZENE	21.60	0.044	4.45559
P-QUATERPHENYL	28.34	0.020	22.39928
3,3'''-DIMETHYL-P-QUATERPHENYL	28.62	0.019	23.69792
TETRAMETHYL-P-QUATERPHENYL	26.13	0.025	13.91395
3,3''' -DIMETHYL-3',2''-METHYLENE-P-QUATERPHENYL	27.26	0.022	17.71182
3,3''' -DIMETHYL-3',2''-ETHYLENE-P-QUATERPHENYL	28.37	0.020	22.49080
4-(3-ETHYLBUTOXY)-P-QUATERPHENYL (B)	27.93	0.021	20.49454
4,4'''-DI(2-BUTYLOCTOXY-1)-P-QUATERPHENYL	29.02	0.018	25.78722
DIETHYL-P-QUINQUEPHENYL	30.06	0.016	31.86358
DI(3-ETHYLHEPTYL)-P-QUINQUEPHENYL	29.52	0.017	28.58139
TETRAMETHYL-P-SEXIPHENYL	30.58	0.016	35.32369
TETRAMETHYL-P-OCTAPHENYL (B)	29.99	0.017	31.37436
INDOLE (TRACE)	17.86	0.079	1.43244
2-PHENYL INDOLE	25.06	0.028	10.71335
1-METHYL-2-PHENYL INDOLE	24.45	0.031	9.21914
1-NEOPENTYL-2-PHENYL INDOLE	24.36	0.031	9.05249
1-N-BUTYL-2(BIPHENYLYL-4) INDOLE	25.51	0.027	11.90482
1,2-DIPHENYL INDOLE	25.31	0.028	11.34324
1-BIPHENYLYL-2-PHENYL INDOLE	27.07	0.023	16.97930
1,3-DIMETHYL-2-PHENYL INDOLE	23.44	0.035	7.17265
2,3-DIPHENYL INDOLE	24.22	0.031	8.70616
1-METHYL-2,3-DIPHENYL INDOLE	24.01	0.032	8.27367
2-PHENYL-3,2'-METHYLENE INDOLE (TRACE)	23.45	0.035	7.16776
2-PHENYL-3,2'-ETHYLENE INDOLE (TRACE)	22.88	0.037	6.19225
2-PHENYL-3,2'-TRIMETHYLENE INDOLE	24.02	0.032	8.28595
1-METHYL-3,2'-METHYLENE-2-PHENYL INDOLE (TRACE)	22.99	0.037	6.37312
1-METHYL-3,2'-ETHYLENE-2-PHENYL INDOLE	22.50	0.039	5.60409
1-METHYL-3,2'-TRIMETHYLENE-2-PHENYL INDOLE	23.28	0.035	6.86773
4,5-DIPHENYLIMIDAZOLE (A)	22.75	0.038	5.99492
PPF (TRACE)	24.59	0.030	9.54736
PPO (TRACE)	26.75	0.023	15.80739
2-(4-FLUOROPHENYL)-5-PHENYLOXAZOLE (TRACE)	25.89	0.026	12.99654
αNPO (TRACE)	23.96	0.033	8.17151
BBO (B)	25.35	0.027	11.46324
POPOP	23.42	0.035	7.11163
DIMETHYL POPOP (TRACE)	24.54	0.030	9.41990
PPD (TRACE)	26.49	0.024	15.07214
2,5-DI(2-METHYLPHENYL)OXADIAZOLE	25.81	0.026	12.87006
PBD	28.57	0.019	23.52262
BBD (B)	27.84	0.021	20.09506
αNPD (TRACE)	24.04	0.032	8.32640
βNPD	27.70	0.021	19.48414
αNND (TRACE)	22.48	0.039	5.56340
βNND (TRACE)	27.83	0.021	20.02820
TRANS-1-(4-BIPHENYLYL)-2-PHENYL-ETHYLENE	26.55	0.024	15.12348
TRANS-1-(4-BIPHENYLYL)-2-(1-NAPHTHYL)-ETHYLENE	25.19	0.028	11.02411
DIPHENYLSTILBENE (B)	24.14	0.032	8.54489
1,1,4,4-TETRAPHENYLBUTADIENE	22.29	0.040	5.29437
1,6-DIPHENYLHEXATRIENE	22.08	0.042	5.00039
1,8-DIPHENYL-1,3,5,7-OCTATETRAENE	20.66	0.051	3.35266
BPSB (TRACE)	23.03	0.037	6.43379
BIS-MSB	23.87	0.033	7.98265
ESCULIN (A)	21.08	0.048	3.78733
AZULENE	26.19	0.025	13.91199
NAPHTHALENE	18.60	0.070	1.79983
DEUTERATED NAPHTHALENE-C10D8	18.74	0.068	1.87527
1-METHYLNAPHTHALENE 5ML/L	19.88	0.057	2.68159
2-METHYL NAPHTHALENE	18.80	0.067	1.91483
2,3-DIMETHYL NAPHTHALENE	19.28	0.062	2.22773
2,6-DIMETHYL NAPHTHALENE (TRACE)	19.00	0.065	2.03963

ACCEPTOR MOLECULES	RO	CO	OVERLAP
1-NAPHTHOL (TRACE)	20.15	0.055	2.90206
2-NAPHTHOL (TRACE)	18.43	0.071	1.69491
1-NAPHTHYLAMINE	18.71	0.068	1.84896
2-CHLORONAPHTHALENE	18.80	0.067	1.91651
ACENAPHTHENE	20.75	0.050	3.46846
1-PHENYL NAPHTHALENE	22.58	0.039	5.75284
2-PHENYL NAPHTHALENE	33.51	0.012	61.25920
1,4-DIPHENYL NAPHTHALENE	23.56	0.034	7.37299
1,5-DIPHENYL NAPHTHALENE	23.18	0.036	6.71232
1,7-DIPHENYL NAPHTHALENE	23.26	0.036	6.83267
1,4,5,8-TETRAPHENYL NAPHTHALENE	22.77	0.038	6.00891
1,1'-BINAPHTHYL	23.81	0.033	7.90089
4,4'-DI(N-BUTOXY)-1,1'-BINAPHTHYL	23.46	0.035	7.18658
2,2'-BINAPHTHYL	26.30	0.025	14.29683
ACRIDINE (A)	15.56	0.119	0.61133
ANTHRACENE (TRACE)	13.87	0.168	0.30732
DEUTERATED ANTHRACENE C14D10 (TRACE)	14.55	0.145	0.40868
9-METHYL ANTHRACENE (TRACE)	14.09	0.160	0.33714
1-AMINO ANTHRACENE	22.80	0.038	6.06233
2-AMINO ANTHRACENE	12.36	0.237	0.15088
9-VINYL ANTHRACENE	15.90	0.111	0.69616
9-PHENYL ANTHRACENE (TRACE)	16.73	0.096	0.94498
9,10-DIPHENYL ANTHRACENE	19.46	0.061	2.34116
9,10-DICHLOROANTHRACENE (TRACE)	14.73	0.140	0.44142
BIANTHRANYL (B)	22.93	0.037	6.27023
9,10-DINAPHTHYLANTHRACENE (B)	24.33	0.031	8.95771
9-(P-VINYLPHENYL)-10-PHENYL ANTHRACENE	21.26	0.047	3.98190
9,10-BIS(PHENYL-ETHYNYL)ANTHRACENE (TRACE)	27.88	0.021	20.26026
RUBICENE (B)	27.93	0.021	20.49526
DIBENZ(A,H)ANTHRACENE (B)	30.59	0.016	35.35606
TETRACENE (TRACE) (B)	33.13	0.012	57.04476
RUBRENE (B)	30.08	0.016	31.95956
TETRAHYDROCANNABINOL ACID METHYL ESTER	24.50	0.030	9.42842
PHENANTHRENE	22.25	0.041	5.23123
3,4-BENZOPHENANTHRENE	28.43	0.019	22.78689
2-PHENYL PHENANTHRENE	28.58	0.019	23.50578
CHRYSENE (TRACE)	25.91	0.026	13.06175
TRIPHENYLENE	23.85	0.033	7.93646
DEUTERATED TRIPHENYLENE-C18D12	23.93	0.033	8.10455
PYRENE	23.10	0.036	6.55914
1,3,6,8,-TETRAPHENYL PYRENE	27.60	0.021	19.05765
BIPYRENYL (B)	28.10	0.020	21.25200
O-PHENYLENE-PYRENE	26.97	0.023	16.60177
ANTHANTHRENE (B)	26.51	0.024	14.97910
FLUORANTHENE	24.32	0.031	8.93000
3-PHENYL FLUORANTHENE	25.59	0.027	12.11026
3-AMINO FLUORANTHENE	23.79	0.033	7.81677
3,3'-BIFLUORANTHENYL (B)	27.36	0.022	18.09032
BENZO(GHI)FLUORANTHENE	23.55	0.034	7.35269
9,12-DI(N-OCTADECOXY)-BENZ(E)ACEPHENANTHRYLENE	27.35	0.022	18.07602
DECACYCLENE (B)	26.78	0.023	15.92907
10-ETHOXY-BENZO(KL)XANTHENE	20.08	0.055	2.83218
PERYLENE	17.39	0.085	1.19360
BENZO(GHI)PERYLENE (B)	27.75	0.021	19.70186
BBOT (TRACE)	21.19	0.047	3.90119
QUININE SULFATE (A)	21.44	0.045	4.19279
ACRIDINE YELLOW (A)	24.14	0.032	8.54501
ACRIDINE ORANGE (A)	24.47	0.031	9.25840
ACRIDINE RED (A)	21.27	0.047	3.99122
URANINE (A)	21.34	0.046	4.07881
RHODAMINE B (A)	24.47	0.031	9.27235
RHODAMINE 6G (A)	24.46	0.031	9.23440
EOSIN B (A)	21.94	0.042	4.80683
EOSIN Y (A)	22.58	0.039	5.72122
ROSE BENGAL (A)	23.27	0.036	6.84835
INDENOINDENE	24.14	0.032	8.53683

8.7 Donor Molecule: Propylbenzene (trace)

BENZENE (3 ML/L)	3.93	7.394	0.00021
DEUTERATED BENZENE-C6D6 5ML/L	3.44	10.976	0.00010
FLUOROBENZENE (TRACE)	7.13	1.234	0.00753
TOLUENE (TRACE)	7.03	1.285	0.00694
DEUTERATED TOLUENE-C7D8 5ML/L	6.45	1.664	0.00414
ETHYLBENZENE (TRACE)	6.86	1.383	0.00599
PROPYLBENZENE (TRACE)	6.60	1.559	0.00471
ISOPROPYLBENZENE (TRACE)	6.23	1.852	0.00334
SEC-BUTYLBENZENE (TRACE)	6.23	1.848	0.00337
DIPHENYLMETHANE (TRACE)	8.03	0.863	0.01538
BIBENZYL	7.51	1.057	0.01025
1,3,5-TRIPHENYLPENTANE 0.05ML/L	7.70	0.981	0.01188
POLYSTYRENE	6.87	1.382	0.00599
POLYVINYLTOLUENE	9.36	0.546	0.03794
PHENYLCYCLOHEXANE (TRACE)	6.19	1.888	0.00322
BENZYLALCOHOL 3ML/L	6.03	2.039	0.00276
BENZYLACETATE 5ML/L	6.00	2.068	0.00268
PHENOL (TRACE)	13.19	0.195	0.30039
ANISOLE (TRACE)	13.48	0.183	0.34086
PHENYL ETHER	14.07	0.161	0.43968
ANILINE 3ML/L	15.68	0.116	0.83649

ACCEPTOR MOLECULES	RO	CO	OVERLAP
N,N-DIMETHYLANILINE 3ML/L	16.69	0.096	1.20879
TRIPHENYLAMINE	24.58	0.030	12.27075
P-XYLENE (TRACE)	10.14	0.428	0.06210
DEUTERATED P-XYLENE C8D10 (TRACE)	10.10	0.435	0.06042
M-XYLENE (TRACE)	8.36	0.765	0.01952
O-XYLENE 3ML/L	8.04	0.860	0.01542
1-METHYL-4-ETHYLBENZENE (TRACE)	9.53	0.517	0.04267
1,4-BIS(TRIFLUOROMETHYL)BENZENE 5ML/L	7.93	0.896	0.01428
P-CRESOL	15.39	0.123	0.75271
P-METHYLANISOLE 3ML/L	14.87	0.136	0.61203
1,4-DIMETHOXY BENZENE	16.96	0.092	1.33968
1-METHYL-4-CHLOROBENZENE 5ML/L	10.44	0.393	0.07381
SODIUM SALICYLATE (A)	17.85	0.079	1.79876
ANTHRANILIC ACID (A)	14.94	0.134	0.61752
TMPD	20.58	0.051	4.22912
1,2,4-TRIMETHYLBENZENE (TRACE)	10.95	0.341	0.09793
MESITYLENE 5ML/L	8.40	0.755	0.02001
1,3,5-TRIETHYLBENZENE 5ML/L	8.10	0.842	0.01606
STYRENE	13.16	0.196	0.29419
1,1-DIPHENYLETHYLENE 3ML/L	15.32	0.124	0.73387
BIPHENYL	14.87	0.136	0.61467
DEUTERATED BIPHENYL-C12D10	14.92	0.135	0.62670
4-METHYL BIPHENYL	16.89	0.093	1.31414
MONOISOPROPYLBIPHENYL 5ML/L	16.37	0.102	1.09439
4-BENZYL BIPHENYL	18.53	0.070	2.29729
3,3'-DIMETHYL BIPHENYL	15.90	0.111	0.91458
2,2'-DIMETHYLBIPHENYL 2ML/L	10.92	0.344	0.09604
OCTAMETHYL BIPHENYL	12.46	0.231	0.21238
4-METHOXY BIPHENYL	20.60	0.051	4.30879
4-BIPHENYLYL PHENYL ETHER	21.17	0.047	5.07840
4,4'-DIHYDROXYBIPHENYL (A)	22.47	0.039	7.19721
4,4'-DIMETHOXY BIPHENYL	22.37	0.040	7.03380
4,4'-DIISOBUTOXYBIPHENYL	22.17	0.041	6.67708
BENZIDINE	25.71	0.026	16.06812
4-VINYL BIPHENYL	25.03	0.029	13.77906
M-TERPHENYL	19.65	0.059	3.23102
M-QUATERPHENYL	20.07	0.055	3.69059
1,3,5-TRIPHENYL BENZENE	21.01	0.048	4.79631
1,3-DIBIPHENYLYLBENZENE (B)	28.44	0.019	29.44127
1,3,5,-TRIBIPHENYLYLBENZENE (B)	31.20	0.015	51.29846
FLUORENE (TRACE)	20.68	0.051	4.40412
1-METHYL FLUORENE	18.08	0.076	1.94751
1:2-BENZOFLUORENE	22.82	0.038	7.83979
9,10-DIHYDROPHENANTHRENE	21.08	0.048	4.92173
CARBAZOLE (TRACE)	22.10	0.041	6.48033
N-METHYL CARBAZULE (TRACE)	21.49	0.045	5.47603
N-VINYLCARBAZOLE	23.09	0.036	8.43400
N-PHENYL CARBAZOLE	23.23	0.036	8.73048
N-METHYL-2,3-BENZCARBAZOLE	27.40	0.022	23.51810
DIBENZOFURAN	22.26	0.041	6.85123
DIBENZOTHIOPHENE (TRACE)	19.86	0.057	3.40598
P-TERPHENYL	25.75	0.026	16.31197
DEUTERATED P-TERPHENYL-C18D14	25.45	0.027	15.21869
4-METHYL TERPHENYL	26.04	0.025	17.44821
2-METHYL-P-TERPHENYL	22.91	0.037	8.12556
3,3''-DIMETHYL-P-TERPHENYL	25.62	0.027	15.78281
4-(3,3-DIMETHYLBUTOXY)-P-TERPHENYL	26.46	0.024	19.13132
4,4''-DIHEXAHYDROFARNESOXY-P-TERPHENYL	27.35	0.022	23.31021
2,2'-METHYLENE-P-TERPHENYL	26.18	0.025	17.92699
2,2'-ETHYLENE-P-TERPHENYL	26.14	0.025	17.74598
2,2'-METHYLENE-5',6''-METHYLENE-P-TERPHENYL (TRACE)	25.19	0.028	14.18638
3-PHENYL DIBENZOFURAN	25.08	0.028	13.85558
BENZO(1,2-B-4,5-B')BISBENZOFURAN (TRACE) (B)	21.89	0.043	6.10397
SEXIPHENYL BENZENE	20.78	0.050	4.54005
P-QUATERPHENYL	27.19	0.022	22.44860
3,3'''-DIMETHYL-P-QUATERPHENYL	27.44	0.022	23.71403
TETRAMETHYL-P-QUATERPHENYL	25.12	0.028	14.11082
3,3''' -DIMETHYL-3',2''-METHYLENE-P-QUATERPHENYL	26.07	0.025	17.45557
3,3''' -DIMETHYL-3',2''-ETHYLENE-P-QUATERPHENYL	27.17	0.022	22.36064
4-(3-ETHYLBUTOXY)-P-QUATERPHENYL (B)	26.74	0.023	20.32449
4,4'''-DI(2-BUTYLOCTOXY-1)-P-QUATERPHENYL	27.79	0.021	25.60108
DIETHYL-P-QUINQUEPHENYL	28.83	0.019	31.90550
DI(3-ETHYLHEPTYL)-P-QUINQUEPHENYL	28.26	0.020	28.32040
TETRAMETHYL-P-SEXIPHENYL	29.34	0.018	35.42932
TETRAMETHYL-P-OCTAPHENYL (B)	28.68	0.019	30.95068
INDOLE (TRACE)	17.18	0.088	1.45860
2-PHENYL INDOLE	23.99	0.032	10.58852
1-METHYL-2-PHENYL INDOLE	23.43	0.035	9.20111
1-NEOPENTYL-2-PHENYL INDOLE	23.36	0.035	9.05742
1-N-BUTYL-2(BIPHENYLYL-4) INDOLE	24.40	0.031	11.73801
1,2-DIPHENYL INDOLE	24.25	0.031	11.30088
1-BIPHENYLYL-2-PHENYL INDOLE	25.96	0.026	16.99693
1,3-DIMETHYL-2-PHENYL INDOLE	22.46	0.039	7.14544
2,3-DIPHENYL INDOLE	23.18	0.036	8.62005
1-METHYL-2,3-DIPHENYL INDOLE	23.01	0.037	8.24910
2-PHENYL-3,2'-METHYLENE INDOLE (TRACE)	22.38	0.040	6.99239
2-PHENYL-3,2'-ETHYLENE INDOLE (TRACE)	21.83	0.043	6.00972
2-PHENYL-3,2'-TRIMETHYLENE INDOLE	22.95	0.037	8.12767
1-METHYL-3,2'-METHYLENE-2-PHENYL INDOLE (TRACE)	21.91	0.043	6.15355
1-METHYL-3,2'-ETHYLENE-2-PHENYL INDOLE	21.46	0.045	5.43005
1-METHYL-3,2'-TRIMETHYLENE-2-PHENYL INDOLE	22.25	0.041	6.74520
4,5-DIPHENYLIMIDAZOLE (A)	21.85	0.043	6.04338
PPF (TRACE)	23.47	0.035	9.29844
PPO (TRACE)	25.61	0.027	15.66690
2-(4-FLUOROPHENYL)-5-PHENYLOXAZOLE (TRACE)	24.79	0.029	12.90064

ACCEPTOR MOLECULES	RO	CO	OVERLAP
αNPO (TRACE)	22.92	0.037	8.06189
BBO (B)	24.21	0.032	11.19205
POPOP	22.36	0.040	6.95312
DIMETHYL POPOP (TRACE)	23.47	0.035	9.28226
PPD (TRACE)	25.45	0.027	15.23742
2,5-DI(2-METHYLPHENYL)OXADIAZOLE	24.79	0.029	12.99145
PBD	27.38	0.022	23.43138
BBD (B)	[illegible]	[illegible]	[illegible]
αNPD (TRACE)	22.99	0.037	8.20402
βNPD	26.55	0.024	19.46333
αNND (TRACE)	21.46	0.045	5.42266
βNND (TRACE)	26.64	0.024	19.85976
TRANS-1-(4-BIPHENYLYL)-2-PHENYL-ETHYLENE	25.37	0.027	14.82992
TRANS-1-(4-BIPHENYLYL)-2-(1-NAPHTHYL)-ETHYLENE	24.09	0.032	10.86053
DIPHENYLSTILBENE (B)	22.97	0.037	8.17343
1,1,4,4-TETRAPHENYLBUTADIENE	21.22	0.047	5.07686
1,6-DIPHENYLHEXATRIENE	20.93	0.049	4.67273
1,8-DIPHENYL-1,3,5,7-OCTATETRAENE	19.67	0.059	3.21550
BPSB (TRACE)	21.89	0.043	6.11084
BIS-MSB	22.72	0.038	7.64801
ESCULIN (A)	20.14	0.055	3.71097
AZULENE	25.20	0.028	14.24160
NAPHTHALENE	17.89	0.078	1.83430
DEUTERATED NAPHTHALENE-C10D8	18.03	0.076	1.91079
1-METHYLNAPHTHALENE 5ML/L	19.12	0.064	2.73063
2-METHYL NAPHTHALENE	18.06	0.076	1.94942
2,3-DIMETHYL NAPHTHALENE	18.54	0.070	2.26715
2,6-DIMETHYL NAPHTHALENE (TRACE)	18.27	0.073	2.07273
1-NAPHTHOL (TRACE)	19.35	0.062	2.92091
2-NAPHTHOL (TRACE)	17.71	0.081	1.71324
1-NAPHTHYLAMINE	17.88	0.078	1.81418
2-CHLORONAPHTHALENE	18.08	0.076	1.95043
ACENAPHTHENE	19.94	0.056	3.50545
1-PHENYL NAPHTHALENE	21.69	0.044	5.81024
2-PHENYL NAPHTHALENE	32.18	0.013	61.78088
1,4-DIPHENYL NAPHTHALENE	22.57	0.039	7.34123
1,5-DIPHENYL NAPHTHALENE	22.22	0.041	6.69701
1,7-DIPHENYL NAPHTHALENE	22.27	0.041	6.77864
1,4,5,8-TETRAPHENYL NAPHTHALENE	21.75	0.043	5.88007
1,1′-BINAPHTHYL	22.87	0.037	7.97297
4,4′-DI(N-BUTOXY)-1,1′-BINAPHTHYL	22.44	0.040	7.09489
2,2′-BINAPHTHYL	25.21	0.028	14.26067
ACRIDINE (A)	14.77	0.139	0.57820
ANTHRACENE (TRACE)	13.11	0.198	0.28249
DEUTERATED ANTHRACENE C14D10 (TRACE)	13.82	0.170	0.38667
9-METHYL ANTHRACENE (TRACE)	13.36	0.187	0.31641
1-AMINO ANTHRACENE	21.84	0.043	6.03129
2-AMINO ANTHRACENE	11.62	0.285	0.13417
9-VINYL ANTHRACENE	15.11	0.130	0.66261
9-PHENYL ANTHRACENE (TRACE)	15.92	0.111	0.90412
9,10-DIPHENYL ANTHRACENE	18.50	0.071	2.22757
9,10-DICHLOROANTHRACENE (TRACE)	13.93	0.165	0.40674
BIANTHRANYL (B)	21.81	0.043	5.98981
9,10-DINAPHTHYLANTHRACENE (B)	23.31	0.035	8.91183
9-(P-VINYLPHENYL)-10-PHENYL ANTHRACENE	20.30	0.053	3.89375
9,10-BIS(PHENYL-ETHYNYL)ANTHRACENE (TRACE)	26.72	0.023	20.22081
RUBICENE (B)	26.77	0.023	20.46391
DIBENZ(A,H)ANTHRACENE (B)	29.31	0.018	35.20925
TETRACENE (TRACE) (B)	31.92	0.014	58.75878
RUBRENE (B)	28.76	0.019	31.46287
TETRAHYDROCANNABINOL ACID METHYL ESTER	23.54	0.034	9.52114
PHENANTHRENE	21.38	0.046	5.30532
3,4-BENZOPHENANTHRENE	27.36	0.022	23.30145
2-PHENYL PHENANTHRENE	27.43	0.022	23.67838
CHRYSENE (TRACE)	24.73	0.030	12.71700
TRIPHENYLENE	22.91	0.037	8.04502
DEUTERATED TRIPHENYLENE-C18D12	23.00	0.037	8.22045
PYRENE	22.10	0.041	6.47071
1,3,6,8,-TETRAPHENYL PYRENE	26.46	0.024	19.08664
BIPYRENYL (B)	27.00	0.023	21.52531
O-PHENYLENE-PYRENE	25.84	0.026	16.53641
ANTHANTHRENE (B)	25.37	0.027	14.82444
FLUORANTHENE	23.37	0.035	9.06533
3-PHENYL FLUORANTHENE	24.54	0.030	12.13140
3-AMINO FLUORANTHENE	22.76	0.038	7.72945
3,3′-BIFLUORANTHENYL (B)	26.23	0.025	18.10400
BENZO(GHI)FLUORANTHENE	22.60	0.039	7.40229
9,12-DI(N-OCTADECOXY)-BENZ(E)ACEPHENANTHRYLENE	26.28	0.025	18.31796
DECACYCLENE (B)	25.64	0.027	15.78209
10-ETHOXY-BENZO(KL)XANTHENE	19.23	0.063	2.81189
PERYLENE	16.64	0.097	1.18072
BENZO(GHI)PERYLENE (B)	26.60	0.024	19.68210
BBOT (TRACE)	20.16	0.055	3.73580
QUININE SULFATE (A)	20.54	0.052	4.18020
ACRIDINE YELLOW (A)	23.17	0.036	8.59116
ACRIDINE ORANGE (A)	23.47	0.035	9.28160
ACRIDINE RED (A)	20.39	0.053	3.99072
URANINE (A)	20.46	0.052	4.07791
RHODAMINE B (A)	23.47	0.035	9.28087
RHODAMINE 6G (A)	23.47	0.035	9.29828
EOSIN B (A)	20.99	0.048	4.74812
EOSIN Y (A)	21.60	0.044	5.65339
ROSE BENGAL (A)	22.29	0.040	6.81004
INDENOINDENE	23.07	0.036	8.37971

8.8 Donor Molecule: Isopropylbenzene (trace)

ACCEPTOR MOLECULES	RO	CO	OVERLAP
BENZENE (3 ML/L)	4.11	6.447	0.00032
DEUTERATED BENZENE-C6D6 5ML/L	3.71	8.735	0.00018
FLUOROBENZENE (TRACE)	7.70	0.979	0.01398
TOLUENE (TRACE)	7.21	1.195	0.00936
DEUTERATED TOLUENE-C7D8 5ML/L	6.71	1.484	0.00607
ETHYLBENZENE (TRACE)	7.04	1.280	0.00816
PROPYLBENZENE (TRACE)	6.79	1.428	0.00656
ISOPROPYLBENZENE (TRACE)	6.47	1.649	0.00492
SEC-BUTYLBENZENE (TRACE)	6.47	1.654	0.00489
DIPHENYLMETHANE (TRACE)	8.17	0.820	0.01986
BIBENZYL	7.71	0.975	0.01405
1,3,5-TRIPHENYLPENTANE 0.05ML/L	7.89	0.909	0.01615
POLYSTYRENE	7.03	1.289	0.00804
POLYVINYLTOLUENE	9.29	0.558	0.04232
PHENYLCYCLOHEXANE (TRACE)	6.45	1.665	0.00485
BENZYLALCOHOL 3ML/L	6.23	1.849	0.00391
BENZYLACETATE 5ML/L	6.23	1.850	0.00391
PHENOL (TRACE)	12.97	0.205	0.31709
ANISOLE (TRACE)	13.24	0.193	0.35780
PHENYL ETHER	13.80	0.170	0.45794
ANILINE 3ML/L	15.25	0.126	0.82483
N,N-DIMETHYLANILINE 3ML/L	16.29	0.104	1.21685
TRIPHENYLAMINE	23.86	0.033	11.94249
P-XYLENE (TRACE)	10.05	0.440	0.06867
DEUTERATED P-XYLENE C8D10 (TRACE)	10.02	0.445	0.06728
M-XYLENE (TRACE)	8.36	0.767	0.02267
O-XYLENE 3ML/L	8.07	0.850	0.01846
1-METHYL-4-ETHYLBENZENE (TRACE)	9.47	0.527	0.04802
1,4-BIS(TRIFLUOROMETHYL)BENZENE 5ML/L	8.13	0.832	0.01934
P-CRESOL	15.01	0.132	0.75586
P-METHYLANISOLE 3ML/L	14.50	0.147	0.61453
1,4-DIMETHOXY BENZENE	16.47	0.100	1.31156
1-METHYL-4-CHLOROBENZENE 5ML/L	10.27	0.413	0.07809
SODIUM SALICYLATE (A)	17.32	0.086	1.75021
ANTHRANILIC ACID (A)	14.45	0.148	0.58879
TMPD	20.18	0.054	4.37751
1,2,4-TRIMETHYLBENZENE (TRACE)	10.80	0.355	0.10544
MESITYLENE 5ML/L	8.35	0.769	0.02252
1,3,5-TRIETHYLBENZENE 5ML/L	8.08	0.849	0.01844
STYRENE	12.92	0.207	0.30661
1,1-DIPHENYLETHYLENE 3ML/L	15.15	0.129	0.79853
BIPHENYL	14.77	0.139	0.68865
DEUTERATED BIPHENYL-C12D10	14.82	0.137	0.70489
4-METHYL BIPHENYL	16.68	0.096	1.42658
MONOISOPROPYLBIPHENYL 5ML/L	16.20	0.105	1.19644
4-BENZYL BIPHENYL	18.29	0.073	2.47571
3,3'-DIMETHYL BIPHENYL	15.67	0.116	0.97596
2,2'-DIMETHYLBIPHENYL 2ML/L	10.84	0.351	0.10719
OCTAMETHYL BIPHENYL	12.22	0.245	0.22060
4-METHOXY BIPHENYL	20.21	0.054	4.47964
4-BIPHENYLYL PHENYL ETHER	20.78	0.050	5.29869
4,4'-DIHYDROXYBIPHENYL (A)	21.97	0.042	7.35011
4,4'-DIMETHOXY BIPHENYL	21.89	0.043	7.20051
4,4'-DIISOBUTOXYBIPHENYL	21.68	0.044	6.81653
BENZIDINE	25.02	0.029	15.89675
4-VINYL BIPHENYL	24.41	0.031	13.85169
M-TERPHENYL	19.35	0.062	3.43908
M-QUATERPHENYL	19.82	0.057	3.99888
1,3,5-TRIPHENYL BENZENE	20.74	0.050	5.10643
1,3-DIBIPHENYLYLBENZENE (B)	27.68	0.021	29.19997
1,3,5,-TRIBIPHENYLYLBENZENE (B)	30.34	0.016	50.61817
FLUORENE (TRACE)	20.27	0.054	4.55715
1-METHYL FLUORENE	17.53	0.083	1.88726
1,2-BENZOFLUORENE	22.18	0.041	7.69337
9,10-DIHYDROPHENANTHRENE	20.64	0.051	5.07074
CARBAZOLE (TRACE)	21.47	0.045	6.34133
N-METHYL CARBAZOLE (TRACE)	20.91	0.049	5.42062
N-VINYLCARBAZOLE	22.53	0.039	8.46651
N-PHENYL CARBAZOLE	22.60	0.039	8.62854
N-METHYL-2,3-BENZCARBAZOLE	26.76	0.023	23.77916
DIBENZOFURAN	21.66	0.044	6.78744
DIBENZOTHIOPHENE (TRACE)	19.33	0.062	3.38019
P-TERPHENYL	25.11	0.028	16.37569
DEUTERATED P-TERPHENYL-C18D14	24.82	0.029	15.28725
4-METHYL TERPHENYL	25.38	0.027	17.42749
2-METHYL-P-TERPHENYL	22.42	0.040	8.33906
3,3''-DIMETHYL-P-TERPHENYL	24.97	0.029	15.79183
4-(3,3-DIMETHYLBUTOXY)-P-TERPHENYL	25.75	0.026	18.92717
4,4''-DIHEXAHYDROFARNESOXY-P-TERPHENYL	26.59	0.024	22.95427
2,2'-METHYLENE-P-TERPHENYL	25.45	0.027	17.66094
2,2'-ETHYLENE-P-TERPHENYL	25.41	0.027	17.46902
2,2'-METHYLENE-5',6''-METHYLENE-P-TERPHENYL (TRACE)	24.43	0.031	13.76844
3-PHENYL DIBENZOFURAN	24.36	0.031	13.59152
BENZO(1,2-B-4,5-B')BISBENZOFURAN (TRACE) (B)	21.16	0.047	5.79777
SEXIPHENYL BENZENE	20.34	0.053	4.66615
P-QUATERPHENYL	26.42	0.024	22.04415
3,3'''-DIMETHYL-P-QUATERPHENYL	26.66	0.024	23.26499
TETRAMETHYL-P-QUATERPHENYL	24.57	0.030	14.42721
3,3'''-DIMETHYL-3',2''-METHYLENE-P-QUATERPHENYL	25.28	0.028	16.89332
3,3'''-DIMETHYL-3',2''-ETHYLENE-P-QUATERPHENYL	26.39	0.024	21.88502
4-(3-ETHYLBUTOXY)-P-QUATERPHENYL (B)	25.95	0.026	19.77786
4,4'''-DI(2-BUTYLOCTOXY-1)-P-QUATERPHENYL	26.97	0.023	24.94278

ACCEPTOR MOLECULES	RO	CO	OVERLAP
DIETHYL-P-QUINQUEPHENYL	28.02	0.020	31.31844
DI(3-ETHYLHEPTYL)-P-QUINQUEPHENYL	27.42	0.022	27.55300
TETRAMETHYL-P-SEXIPHENYL	28.52	0.019	34.87751
TETRAMETHYL-P-OCTAPHENYL (B)	27.85	0.021	30.22780
INDOLE (TRACE)	16.85	0.094	1.51377
2-PHENYL INDOLE	23.26	0.036	10.26956
1-METHYL-2-PHENYL INDOLE	22.75	0.038	8.99183
1-NEOPENTYL-2-PHENYL INDOLE	22.69	0.038	8.86948
1-N-BUTYL-2(BIPHENYLYL-4) INDOLE	23.73	0.033	11.57562
1,2-DIPHENYL INDOLE	23.58	0.034	11.14364
1-BIPHENYLYL-2-PHENYL INDOLE	25.25	0.028	16.79221
1,3-DIMETHYL-2-PHENYL INDOLE	21.82	0.043	6.98986
2,3-DIPHENYL INDOLE	22.52	0.039	8.45800
1-METHYL-2,3-DIPHENYL INDOLE	22.41	0.040	8.19925
2-PHENYL-3,2'-METHYLENE INDOLE (TRACE)	21.69	0.044	6.74331
2-PHENYL-3,2'-ETHYLENE INDOLE (TRACE)	21.18	0.047	5.84642
2-PHENYL-3,2'-TRIMETHYLENE INDOLE	22.25	0.041	7.85417
1-METHYL-3,2'-METHYLENE-2-PHENYL INDOLE (TRACE)	21.27	0.047	5.99350
1-METHYL-3,2'-ETHYLENE-2-PHENYL INDOLE	20.84	0.049	5.30975
1-METHYL-3,2'-TRIMETHYLENE-2-PHENYL INDOLE	21.59	0.044	6.55429
4,5-DIPHENYLIMIDAZOLE (A)	21.27	0.046	6.00158
PPF (TRACE)	22.75	0.038	8.97222
PPO (TRACE)	24.85	0.029	15.25285
2-(4-FLUOROPHENYL)-5-PHENYLOXAZOLE (TRACE)	24.06	0.032	12.57967
αNPO (TRACE)	22.29	0.040	7.95044
BBO (B)	23.52	0.034	10.96727
POPOP	21.73	0.044	6.81560
DIMETHYL POPOP (TRACE)	22.83	0.038	9.17226
PPD (TRACE)	24.79	0.029	15.17269
2,5-DI(2-METHYLPHENYL)OXADIAZOLE	24.15	0.032	12.95137
PBD	26.58	0.024	22.87698
BBD (B)	25.83	0.026	19.25606
αNPD (TRACE)	22.37	0.040	8.11640
βNPD	25.88	0.026	19.46300
αNND (TRACE)	20.82	0.050	5.27100
βNND (TRACE)	26.01	0.025	20.05642
TRANS-1-(4-BIPHENYLYL)-2-PHENYL-ETHYLENE	24.59	0.030	14.33200
TRANS-1-(4-BIPHENYLYL)-2-(1-NAPHTHYL)-ETHYLENE	23.40	0.035	10.63893
DIPHENYLSTILBENE (B)	22.24	0.041	7.84868
1,1,4,4-TETRAPHENYLBUTADIENE	20.61	0.051	4.96168
1,6-DIPHENYLHEXATRIENE	20.24	0.054	4.45947
1,8-DIPHENYL-1,3,5,7-OCTATETRAENE	19.11	0.064	3.15460
BPSB (TRACE)	21.20	0.047	5.88564
BIS-MSB	22.02	0.042	7.37656
ESCULIN (A)	19.54	0.060	3.59993
AZULENE	24.65	0.030	14.52758
NAPHTHALENE	17.49	0.084	1.86856
DEUTERATED NAPHTHALENE-C10D8	17.63	0.082	1.94828
1-METHYLNAPHTHALENE 5ML/L	18.64	0.069	2.73449
2-METHYL NAPHTHALENE	17.66	0.081	1.97813
2,3-DIMETHYL NAPHTHALENE	18.10	0.075	2.29310
2,6-DIMETHYL NAPHTHALENE (TRACE)	17.84	0.079	2.10077
1-NAPHTHOL (TRACE)	18.81	0.067	2.87895
2-NAPHTHOL (TRACE)	17.29	0.087	1.73333
1-NAPHTHYLAMINE	17.33	0.086	1.75389
2-CHLORONAPHTHALENE	17.67	0.081	1.97908
ACENAPHTHENE	19.39	0.061	3.46777
1-PHENYL NAPHTHALENE	21.10	0.048	5.74459
2-PHENYL NAPHTHALENE	31.34	0.015	61.55926
1,4-DIPHENYL NAPHTHALENE	21.91	0.043	7.16937
1,5-DIPHENYL NAPHTHALENE	21.58	0.045	6.54249
1,7-DIPHENYL NAPHTHALENE	21.72	0.044	6.80967
1,4,5,8-TETRAPHENYL NAPHTHALENE	21.18	0.047	5.84666
1,1'-BINAPHTHYL	22.25	0.041	7.88453
4,4'-DI(N-BUTOXY)-1,1'-BINAPHTHYL	21.77	0.043	6.90063
2,2'-BINAPHTHYL	24.60	0.030	14.37038
ACRIDINE (A)	14.37	0.151	0.57132
ANTHRACENE (TRACE)	12.74	0.216	0.27710
DEUTERATED ANTHRACENE C14D10 (TRACE)	13.42	0.185	0.37893
9-METHYL ANTHRACENE (TRACE)	13.16	0.196	0.33610
1-AMINO ANTHRACENE	21.60	0.044	6.57043
2-AMINO ANTHRACENE	11.17	0.321	0.12345
9-VINYL ANTHRACENE	14.99	0.133	0.73529
9-PHENYL ANTHRACENE (TRACE)	15.78	0.114	1.00094
9,10-DIPHENYL ANTHRACENE	18.57	0.070	2.65918
9,10-DICHLOROANTHRACENE (TRACE)	13.98	0.164	0.48393
BIANTHRANYL (B)	21.92	0.042	7.18303
9,10-DINAPHTHYLANTHRACENE (B)	22.90	0.037	9.33637
9-(P-VINYLPHENYL)-10-PHENYL ANTHRACENE	20.12	0.055	4.29890
9,10-BIS(PHENYL-ETHYNYL)ANTHRACENE (TRACE)	26.18	0.025	20.86741
RUBICENE (B)	26.07	0.025	20.32569
DIBENZ(A,H)ANTHRACENE (B)	28.45	0.019	34.33207
TETRACENE (TRACE) (B)	31.12	0.015	58.86781
RUBRENE (B)	27.90	0.021	30.55519
TETRAHYDROCANNABINOL ACID METHYL ESTER	23.00	0.037	9.66738
PHENANTHRENE	20.90	0.049	5.39557
3,4-BENZOPHENANTHRENE	26.68	0.024	23.35845
2-PHENYL PHENANTHRENE	26.86	0.023	24.30492
CHRYSENE (TRACE)	24.51	0.030	14.04504
TRIPHENYLENE	22.40	0.040	8.18966
DEUTERATED TRIPHENYLENE-C18D12	22.49	0.039	8.39021
PYRENE	21.61	0.044	6.59363
1,3,6,8,-TETRAPHENYL PYRENE	25.71	0.026	18.69795
BIPYRENYL (B)	26.31	0.025	21.48659
O-PHENYLENE-PYRENE	25.14	0.028	16.36017

ACCEPTOR MOLECULES	RO	CO	OVERLAP
ANTHANTHRENE (B)	24.64	0.030	14.49285
FLUORANTHENE	22.75	0.038	8.97980
3-PHENYL FLUORANTHENE	23.87	0.033	11.98006
3-AMINO FLUORANTHENE	22.16	0.041	7.67444
3,3'-BIFLUORANTHENYL (B)	25.52	0.027	17.89952
BENZO(GHI)FLUORANTHENE	22.01	0.042	7.35820
9,12-DI(N-OCTADECOXY)-BENZ(E)ACEPHENANTHRYLENE	25.64	0.027	18.38412
DECACYCLENE (B)	25.00	0.029	15.81091
10-ETHOXY-BENZO(KL)XANTHENE	18.72	0.068	2.78599
PERYLENE	16.39	0.102	1.25784
BENZO(GHI)PERYLENE (B)	25.87	0.026	19.41105
BBOT (TRACE)	19.59	0.059	3.66475
QUININE SULFATE (A)	19.99	0.056	4.13444
ACRIDINE YELLOW (A)	22.74	0.038	8.96017
ACRIDINE ORANGE (A)	22.96	0.037	9.48379
ACRIDINE RED (A)	19.86	0.057	3.97937
URANINE (A)	19.93	0.057	4.05861
RHODAMINE B (A)	22.87	0.037	9.26396
RHODAMINE 6G (A)	22.86	0.037	9.24866
EOSIN B (A)	20.48	0.052	4.77341
EOSIN Y (A)	21.08	0.048	5.67752
ROSE BENGAL (A)	21.80	0.043	6.94479
INDENOINDENE	22.35	0.040	8.07979

8.9 Donor Molecule: *sec*-Butylbenzene (trace)

BENZENE (3 ML/L)	4.08	6.582	0.00027
DEUTERATED BENZENE-C6D6 5ML/L	3.67	9.013	0.00014
FLUOROBENZENE (TRACE)	7.60	1.020	0.01104
TOLUENE (TRACE)	7.18	1.209	0.00784
DEUTERATED TOLUENE-C7D8 5ML/L	6.67	1.508	0.00504
ETHYLBENZENE (TRACE)	7.02	1.294	0.00685
PROPYLBENZENE (TRACE)	6.77	1.444	0.00550
ISOPROPYLBENZENE (TRACE)	6.44	1.672	0.00410
SEC-BUTYLBENZENE (TRACE)	6.44	1.678	0.00407
DIPHENYLMETHANE (TRACE)	8.16	0.822	0.01696
BIBENZYL	7.69	0.983	0.01186
1,3,5-TRIPHENYLPENTANE 0.05ML/L	7.87	0.916	0.01364
POLYSTYRENE	7.01	1.299	0.00680
POLYVINYLTOLUENE	9.43	0.533	0.03977
PHENYLCYCLOHEXANE (TRACE)	6.40	1.704	0.00396
BENZYLALCOHOL 3ML/L	6.22	1.861	0.00331
BENZYLACETATE 5ML/L	6.20	1.875	0.00327
PHENOL (TRACE)	13.20	0.194	0.30219
ANISOLE (TRACE)	13.49	0.182	0.34379
PHENYL ETHER	14.09	0.160	0.44406
ANILINE 3ML/L	15.66	0.116	0.83006
N,N-DIMETHYLANILINE 3ML/L	16.70	0.096	1.21516
TRIPHENYLAMINE	24.55	0.030	12.14628
P-XYLENE (TRACE)	10.21	0.420	0.06463
DEUTERATED P-XYLENE C8D10 (TRACE)	10.18	0.425	0.06336
M-XYLENE (TRACE)	8.46	0.740	0.02090
O-XYLENE 3ML/L	8.14	0.830	0.01659
1-METHYL-4-ETHYLBENZENE (TRACE)	9.61	0.503	0.04514
1,4-BIS(TRIFLUOROMETHYL)BENZENE 5ML/L	8.11	0.840	0.01627
P-CRESOL	15.37	0.123	0.74917
P-METHYLANISOLE 3ML/L	14.85	0.136	0.60927
1,4-DIMETHOXY BENZENE	16.93	0.092	1.32702
1-METHYL-4-CHLOROBENZENE 5ML/L	10.45	0.392	0.07428
SODIUM SALICYLATE (A)	17.82	0.079	1.78026
ANTHRANILIC ACID (A)	14.94	0.134	0.61668
TMPD	20.63	0.051	4.27732
1,2,4-TRIMETHYLBENZENE (TRACE)	10.99	0.337	0.10009
MESITYLENE 5ML/L	8.47	0.737	0.02102
1,3,5-TRIETHYLBENZENE 5ML/L	8.17	0.821	0.01694
STYRENE	13.21	0.194	0.30017
1,1-DIPHENYLETHYLENE 3ML/L	15.41	0.122	0.75906
BIPHENYL	14.98	0.133	0.64376
DEUTERATED BIPHENYL-C12D10	15.03	0.132	0.65744
4-METHYL BIPHENYL	16.97	0.092	1.35525
MONOISOPROPYLBIPHENYL 5ML/L	16.47	0.100	1.13258
4-BENZYL BIPHENYL	18.62	0.069	2.36110
3,3'-DIMETHYL BIPHENYL	15.96	0.110	0.93672
2,2'-DIMETHYLBIPHENYL 2ML/L	10.99	0.337	0.10027
OCTAMETHYL BIPHENYL	12.47	0.231	0.21395
4-METHOXY BIPHENYL	20.64	0.051	4.35981
4-BIPHENYLYL PHENYL ETHER	21.21	0.047	5.14503
4,4'-DIHYDROXYBIPHENYL (A)	22.48	0.039	7.23338
4,4'-DIMETHOXY BIPHENYL	22.39	0.040	7.07534
4,4'-DIISOBUTOXYBIPHENYL	22.19	0.041	6.70906
BENZIDINE	25.69	0.026	15.97280
4-VINYL BIPHENYL	25.02	0.029	13.76354
M-TERPHENYL	19.72	0.058	3.30722
M-QUATERPHENYL	20.17	0.054	3.80669
1,3,5-TRIPHENYL BENZENE	21.11	0.048	4.93958
1,3-DIBIPHENYLYLBENZENE (B)	28.42	0.019	29.28693
1,3,5,-TRIBIPHENYLYLBENZENE (B)	31.16	0.015	50.95463
FLUORENE (TRACE)	20.72	0.050	4.46111
1-METHYL FLUORENE	18.05	0.076	1.92740

ACCEPTOR MOLECULES	RO	CO	OVERLAP
1,2-BENZOFLUORENE	22.79	0.038	7.76422
9,10-DIHYDROPHENANTHRENE	21.10	0.048	4.96295
CARBAZOLE (TRACE)	22.07	0.042	6.42085
N-METHYL CARBAZOLE (TRACE)	21.49	0.045	5.46544
N-VINYLCARBAZOLE	23.10	0.036	8.43252
N-PHENYL CARBAZOLE	23.21	0.036	8.68162
N-METHYL-2,3-BENZCARBAZOLE	27.41	0.022	[illegible]
DIBENZOFURAN	22.23	0.041	6.80358
DIBENZOTHIOPHENE (TRACE)	19.84	0.057	3.39083
P-TERPHENYL	25.74	0.026	16.28675
DEUTERATED P-TERPHENYL-C18D14	25.44	0.027	15.19967
4-METHYL TERPHENYL	26.03	0.025	17.39368
2-METHYL-P-TERPHENYL	22.93	0.037	8.17952
3,3''-DIMETHYL-P-TERPHENYL	25.60	0.027	15.74408
4-(3,3-DIMETHYLBUTOXY)-P-TERPHENYL	26.43	0.024	19.01739
4,4''-DIHEXAHYDROFARNESOXY-P-TERPHENYL	27.32	0.022	23.14312
2,2' -METHYLENE-P-TERPHENYL	26.14	0.025	17.79873
2,2' -ETHYLENE-P-TERPHENYL	26.10	0.025	17.61856
2,2'-METHYLENE-5',6'-METHYLENE-P-TERPHENYL (TRACE)	25.15	0.028	14.05918
3-PHENYL DIBENZOFURAN	25.05	0.028	13.77072
BENZO(1,2-B-4,5-B')BISBENZOFURAN (TRACE) (B)	21.87	0.043	6.05741
SEXIPHENYL BENZENE	20.80	0.050	4.57112
P-QUATERPHENYL	27.15	0.022	22.27285
3,3'''-DIMETHYL-P-QUATERPHENYL	27.40	0.022	23.52616
TETRAMETHYL-P-QUATERPHENYL	25.14	0.028	14.18577
3,3''' -DIMETHYL-3',2''-METHYLENE-P-QUATERPHENYL	26.03	0.025	17.29442
3,3''' -DIMETHYL-3',2''-ETHYLENE-P-QUATERPHENYL	27.14	0.022	22.19849
4-(3-ETHYLBUTOXY)-P-QUATERPHENYL (B)	26.70	0.023	20.13820
4,4'''-DI(2-BUTYLOCTOXY-1)-P-QUATERPHENYL	27.75	0.021	25.37246
DIETHYL-P-QUINQUEPHENYL	28.79	0.019	31.65658
DI(3-ETHYLHEPTYL)-P-QUINQUEPHENYL	28.22	0.020	28.06357
TETRAMETHYL-P-SEXIPHENYL	29.30	0.018	35.17973
TETRAMETHYL-P-OCTAPHENYL (B)	28.66	0.019	30.75219
INDOLE (TRACE)	17.21	0.088	1.47369
2-PHENYL INDOLE	23.95	0.033	10.48205
1-METHYL-2-PHENYL INDOLE	23.40	0.035	9.11905
1-NEOPENTYL-2-PHENYL INDOLE	23.33	0.035	8.97987
1-N-BUTYL-2(BIPHENYLYL-4) INDOLE	24.40	0.031	11.70820
1,2-DIPHENYL INDOLE	24.23	0.031	11.24000
1-BIPHENYLYL-2-PHENYL INDOLE	25.94	0.026	16.90316
1,3-DIMETHYL-2-PHENYL INDOLE	22.44	0.040	7.08756
2,3-DIPHENYL INDOLE	23.16	0.036	8.57224
1-METHYL-2,3-DIPHENYL INDOLE	23.00	0.037	8.23090
2-PHENYL-3,2'-METHYLENE INDOLE (TRACE)	22.35	0.040	6.93173
2-PHENYL-3,2'-ETHYLENE INDOLE (TRACE)	21.82	0.043	5.98988
2-PHENYL-3,2' -TRIMETHYLENE INDOLE	22.92	0.037	8.05052
1-METHYL-3,2'-METHYLENE-2-PHENYL INDOLE (TRACE)	21.91	0.043	6.14155
1-METHYL-3,2'-ETHYLENE-2-PHENYL INDOLE	21.46	0.045	5.42571
1-METHYL-3,2'-TRIMETHYLENE-2-PHENYL INDOLE	22.22	0.041	6.69312
4,5-DIPHENYLIMIDAZOLE (A)	21.83	0.043	6.01600
PPF (TRACE)	23.45	0.035	9.23009
PPO (TRACE)	25.57	0.027	15.52582
2-(4-FLUOROPHENYL)-5-PHENYLOXAZOLE (TRACE)	24.76	0.029	12.78846
αNPO (TRACE)	22.92	0.037	8.05152
BBO (B)	24.21	0.032	11.18825
POPOP	22.37	0.040	6.95317
DIMETHYL POPOP (TRACE)	23.47	0.035	9.29095
PPD (TRACE)	25.43	0.027	15.17515
2,5-DI(2-METHYLPHENYL)OXADIAZOLE	24.77	0.029	12.94628
PBD	27.34	0.022	23.22212
BBD (B)	26.59	0.024	19.63825
αNPD (TRACE)	22.99	0.037	8.19460
βNPD	26.55	0.024	19.45140
αNND (TRACE)	21.45	0.045	5.40200
βNND (TRACE)	26.66	0.024	19.96526
TRANS-1-(4-BIPHENYLYL)-2-PHENYL-ETHYLENE	25.34	0.027	14.70528
TRANS-1-(4-BIPHENYLYL)-2-(1-NAPHTHYL)-ETHYLENE	24.08	0.032	10.82538
DIPHENYLSTILBENE (B)	22.97	0.037	8.15572
1,1,4,4-TETRAPHENYLBUTADIENE	21.24	0.047	5.09987
1,6-DIPHENYLHEXATRIENE	20.95	0.049	4.69385
1,8-DIPHENYL-1,3,5,7-OCTATETRAENE	19.69	0.059	3.24071
BPSB (TRACE)	21.90	0.043	6.12620
BIS-MSB	22.72	0.038	7.64291
ESCULIN (A)	20.13	0.055	3.69251
AZULENE	25.21	0.028	14.27072
NAPHTHALENE	17.90	0.078	1.84093
DEUTERATED NAPHTHALENE-C10D8	18.04	0.076	1.91819
1-METHYLNAPHTHALENE 5ML/L	19.11	0.064	2.72177
2-METHYL NAPHTHALENE	18.08	0.076	1.95264
2,3-DIMETHYL NAPHTHALENE	18.54	0.070	2.26785
2,6-DIMETHYL NAPHTHALENE (TRACE)	18.27	0.073	2.07626
1-NAPHTHOL (TRACE)	19.32	0.062	2.90036
2-NAPHTHOL (TRACE)	17.71	0.081	1.71710
1-NAPHTHYLAMINE	17.85	0.079	1.79819
2-CHLORONAPHTHALENE	18.09	0.076	1.95327
ACENAPHTHENE	19.91	0.057	3.48374
1-PHENYL NAPHTHALENE	21.67	0.044	5.77307
2-PHENYL NAPHTHALENE	32.16	0.013	61.61696
1,4-DIPHENYL NAPHTHALENE	22.54	0.039	7.27720
1,5-DIPHENYL NAPHTHALENE	22.19	0.041	6.63690
1,7-DIPHENYL NAPHTHALENE	22.29	0.040	6.80673
1,4,5,8-TETRAPHENYL NAPHTHALENE	21.77	0.043	5.90514
1,1'-BINAPHTHYL	22.84	0.038	7.92402
4,4'-DI(N-BUTOXY)-1,1'-BINAPHTHYL	22.41	0.040	7.03555
2,2'-BINAPHTHYL	25.22	0.028	14.29716
ACRIDINE (A)	14.81	0.138	0.58686

ACCEPTOR MOLECULES	RO	CO	OVERLAP
ANTHRACENE (TRACE)	13.16	0.196	0.28887
DEUTERATED ANTHRACENE C14D10 (TRACE)	13.84	0.169	0.39098
9-METHYL ANTHRACENE (TRACE)	13.47	0.183	0.33147
1-AMINO ANTHRACENE	21.97	0.042	6.24282
2-AMINO ANTHRACENE	11.64	0.284	0.13502
9-VINYL ANTHRACENE	15.28	0.125	0.70649
9-PHENYL ANTHRACENE (TRACE)	16.07	0.108	0.95783
9,10-DIPHENYL ANTHRACENE	18.80	0.067	2.45392
9,10-DICHLOROANTHRACENE (TRACE)	14.22	0.155	0.45970
BIANTHRANYL (B)	22.14	0.041	6.55066
9,10-DINAPHTHYLANTHRACENE (B)	23.41	0.035	9.13788
9-(P-VINYLPHENYL)-10-PHENYL ANTHRACENE	20.48	0.052	4.09491
9,10-BIS(PHENYL-ETHYNYL)ANTHRACENE (TRACE)	26.76	0.023	20.41410
RUBICENE (B)	26.76	0.023	20.40848
DIBENZ(A,H)ANTHRACENE (B)	29.27	0.018	34.90934
TETRACENE (TRACE) (B)	31.88	0.014	58.28197
RUBRENE (B)	28.72	0.019	31.16554
TETRAHYDROCANNABINOL ACID METHYL ESTER	23.55	0.034	9.55969
PHENANTHRENE	21.40	0.046	5.33081
3,4-BENZOPHENANTHRENE	27.34	0.022	23.21379
2-PHENYL PHENANTHRENE	27.47	0.022	23.85077
CHRYSENE (TRACE)	24.89	0.029	13.22055
TRIPHENYLENE	22.93	0.037	8.08242
DEUTERATED TRIPHENYLENE-C18D12	23.02	0.037	8.26775
PYRENE	22.15	0.041	6.55483
1,3,6,8,-TETRAPHENYL PYRENE	26.43	0.024	18.94724
BIPYRENYL (B)	26.99	0.023	21.48379
O-PHENYLENE-PYRENE	25.82	0.026	16.46792
ANTHANTHRENE (B)	25.34	0.027	14.70889
FLUORANTHENE	23.35	0.035	9.01179
3-PHENYL FLUORANTHENE	24.52	0.030	12.07914
3-AMINO FLUORANTHENE	22.76	0.038	7.72140
3,3'-BIFLUORANTHENYL (B)	26.22	0.025	18.03255
BENZO(GHI)FLUORANTHENE	22.59	0.039	7.38842
9,12-DI(N-OCTADECOXY)-BENZ(E)ACEPHENANTHRYLENE	26.28	0.025	18.30388
DECACYCLENE (B)	25.65	0.026	15.83042
10-ETHOXY-BENZO(KL)XANTHENE	19.23	0.063	2.80967
PERYLENE	16.72	0.096	1.21567
BENZO(GHI)PERYLENE (B)	26.58	0.024	19.57525
BBOT (TRACE)	20.20	0.054	3.76974
QUININE SULFATE (A)	20.54	0.052	4.16984
ACRIDINE YELLOW (A)	23.22	0.036	8.71602
ACRIDINE ORANGE (A)	23.49	0.034	9.34170
ACRIDINE RED (A)	20.39	0.053	3.98687
URANINE (A)	20.46	0.052	4.07092
RHODAMINE B (A)	23.47	0.035	9.27563
RHODAMINE 6G (A)	23.46	0.035	9.26818
EOSIN B (A)	21.00	0.048	4.77035
EOSIN Y (A)	21.62	0.044	5.67689
ROSE BENGAL (A)	22.32	0.040	6.86455
INDENOINDENE	23.03	0.037	8.28886

8.10 Donor Molecule: Diphenylmethane (trace)

BENZENE (3 ML/L)	4.02	6.863	0.00021
DEUTERATED BENZENE-C6D6 5ML/L	3.34	12.032	0.00007
FLUOROBENZENE (TRACE)	6.97	1.323	0.00573
TOLUENE (TRACE)	7.25	1.175	0.00726
DEUTERATED TOLUENE-C7D8 5ML/L	6.49	1.634	0.00375
ETHYLBENZENE (TRACE)	7.03	1.286	0.00605
PROPYLBENZENE (TRACE)	6.70	1.488	0.00453
ISOPROPYLBENZENE (TRACE)	6.26	1.821	0.00302
SEC-BUTYLBENZENE (TRACE)	6.28	1.803	0.00308
DIPHENYLMETHANE (TRACE)	8.29	0.784	0.01629
BIBENZYL	7.65	0.997	0.01006
1,3,5-TRIPHENYLPENTANE 0.05ML/L	7.86	0.922	0.01177
POLYSTYRENE	7.05	1.278	0.00613
POLYVINYLTOLUENE	9.54	0.515	0.03721
PHENYLCYCLOHEXANE (TRACE)	6.23	1.848	0.00294
BENZYLALCOHOL 3ML/L	6.12	1.951	0.00263
BENZYLACETATE 5ML/L	6.10	1.974	0.00258
PHENOL (TRACE)	13.56	0.179	0.31028
ANISOLE (TRACE)	13.85	0.168	0.35071
PHENYL ETHER	14.40	0.150	0.44217
ANILINE 3ML/L	16.06	0.108	0.84535
N,N-DIMETHYLANILINE 3ML/L	17.08	0.090	1.21335
TRIPHENYLAMINE	25.15	0.028	12.31338
P-XYLENE (TRACE)	10.31	0.409	0.05971
DEUTERATED P-XYLENE C8D10 (TRACE)	10.25	0.416	0.05771
M-XYLENE (TRACE)	8.55	0.716	0.01950
O-XYLENE 3ML/L	8.26	0.794	0.01585
1-METHYL-4-ETHYLBENZENE (TRACE)	9.67	0.495	0.04076
1,4-BIS(TRIFLUOROMETHYL)BENZENE 5ML/L	8.12	0.834	0.01440
P-CRESOL	15.77	0.114	0.76251
P-METHYLANISOLE 3ML/L	15.24	0.126	0.62019
1,4-DIMETHOXY BENZENE	17.38	0.085	1.35509
1-METHYL-4-CHLOROBENZENE 5ML/L	10.74	0.361	0.07659
SODIUM SALICYLATE (A)	18.27	0.073	1.80940
ANTHRANILIC ACID (A)	15.15	0.129	0.58975

ACCEPTOR MOLECULES	RO	CO	OVERLAP
TMPD	21.05	0.048	4.23066
1,2,4-TRIMETHYLBENZENE (TRACE)	11.18	0.320	0.09701
MESITYLENE 5ML/L	8.58	0.709	0.01987
1,3,5-TRIETHYLBENZENE 5ML/L	8.30	0.782	0.01629
STYRENE	13.45	0.184	0.29205
1,1-DIPHENYLETHYLENE 3ML/L	15.67	0.116	0.73351
BIPHENYL	15.20	0.127	0.61347
DEUTERATED BIPHENYL-C12D10	15.25	0.126	0.63550
4-METHYL BIPHENYL	17.27	0.087	1.31465
MONOISOPROPYLBIPHENYL 5ML/L	16.74	0.095	1.09318
4-BENZYL BIPHENYL	18.95	0.066	2.29925
3,3′-DIMETHYL BIPHENYL	16.27	0.104	0.91637
2,2′-DIMETHYLBIPHENYL 2ML/L	11.17	0.321	0.09635
OCTAMETHYL BIPHENYL	12.75	0.216	0.21297
4-METHOXY BIPHENYL	21.08	0.048	4.32851
4-BIPHENYLYL PHENYL ETHER	21.67	0.044	5.10236
4,4′-DIHYDROXYBIPHENYL (A)	23.00	0.037	7.24535
4,4′-DIMETHOXY BIPHENYL	22.90	0.037	7.07755
4,4′-DIISOBUTOXYBIPHENYL	22.70	0.038	6.72092
BENZIDINE	26.31	0.025	16.16492
4-VINYL BIPHENYL	25.62	0.027	13.88256
M-TERPHENYL	20.10	0.055	3.23869
M-QUATERPHENYL	20.53	0.052	3.69335
1,3,5-TRIPHENYL BENZENE	21.48	0.045	4.79965
1,3-DIBIPHENYLYLBENZENE (B)	29.12	0.018	29.65635
1,3,5,-TRIBIPHENYLYLBENZENE (B)	31.94	0.014	51.65141
FLUORENE (TRACE)	21.17	0.047	4.42608
1-METHYL FLUORENE	18.51	0.071	1.95746
1,2-BENZOFLUORENE	23.35	0.035	7.89052
9,10-DIHYDROPHENANTHRENE	21.57	0.045	4.94645
CARBAZOLE (TRACE)	22.62	0.039	6.52096
N-METHYL CARBAZOLE (TRACE)	22.00	0.042	5.51818
N-VINYLCARBAZOLE	23.63	0.034	8.46525
N-PHENYL CARBAZOLE	23.78	0.033	8.80462
N-METHYL-2,3-BENZCARBAZOLE	28.03	0.020	23.62545
DIBENZOFURAN	22.80	0.038	6.92619
DIBENZOTHIOPHENE (TRACE)	20.33	0.053	3.43692
P-TERPHENYL	26.37	0.024	16.43652
DEUTERATED P-TERPHENYL-C18D14	26.05	0.025	15.33404
4-METHYL TERPHENYL	26.67	0.024	17.58317
2-METHYL-P-TERPHENYL	23.45	0.035	8.17324
3,3″-DIMETHYL-P-TERPHENYL	26.23	0.025	15.90416
4-(3,3-DIMETHYLBUTOXY)-P-TERPHENYL	27.09	0.022	19.26357
4,4″DIHEXAHYDROFARNESOXY-P-TERPHENYL	28.00	0.020	23.43753
2,2′-METHYLENE-P-TERPHENYL	26.78	0.023	17.99956
2,2′-ETHYLENE-P-TERPHENYL	26.75	0.023	17.82377
2,2′-METHYLENE-5,6′-METHYLENE-P-TERPHENYL (TRACE)	25.73	0.026	14.11986
3-PHENYL DIBENZOFURAN	25.65	0.027	13.88370
BENZO(1,2-B-4,5-B′)BISBENZOFURAN (TRACE) (B)	22.23	0.041	5.86007
SEXIPHENYL BENZENE	21.26	0.047	4.55805
P-QUATERPHENYL	27.82	0.021	22.55145
3,3‴-DIMETHYL-P-QUATERPHENYL	28.07	0.020	23.80148
TETRAMETHYL-P-QUATERPHENYL	25.71	0.026	14.20179
3,3‴-DIMETHYL-3′,2″-METHYLENE-P-QUATERPHENYL	26.61	0.024	17.29228
3,3‴-DIMETHYL-3′,2″-ETHYLENE-P-QUATERPHENYL	27.75	0.021	22.24005
4-(3-ETHYLBUTOXY)-P-QUATERPHENYL (B)	27.33	0.022	20.27114
4,4‴-DI(2-BUTYLOCTOXY-1)-P-QUATERPHENYL	28.40	0.020	25.54802
DIETHYL-P-QUINQUEPHENYL	29.49	0.017	32.03045
DI(3-ETHYLHEPTYL)-P-QUINQUEPHENYL	28.88	0.019	28.22998
TETRAMETHYL-P-SEXIPHENYL	30.02	0.017	35.60243
TETRAMETHYL-P-OCTAPHENYL (B)	29.28	0.018	30.68135
INDOLE (TRACE)	17.59	0.082	1.46850
2-PHENYL INDOLE	24.50	0.030	10.54100
1-METHYL-2-PHENYL INDOLE	23.97	0.032	9.23284
1-NEOPENTYL-2-PHENYL INDOLE	23.91	0.033	9.10589
1-N-BUTYL-2(BIPHENYLYL-4) INDOLE	24.92	0.029	11.66271
1,2-DIPHENYL INDOLE	24.80	0.029	11.32779
1-BIPHENYLYL-2-PHENYL INDOLE	26.54	0.024	17.02574
1,3-DIMETHYL-2-PHENYL INDOLE	22.97	0.037	7.15746
2,3-DIPHENYL INDOLE	23.68	0.034	8.57946
1-METHYL-2,3-DIPHENYL INDOLE	23.52	0.034	8.24861
2-PHENYL-3,2′-METHYLENE INDOLE (TRACE)	22.82	0.038	6.86635
2-PHENYL-3,2′-ETHYLENE INDOLE (TRACE)	22.23	0.041	5.87975
2-PHENYL-3,2′-TRIMETHYLENE INDOLE	23.42	0.035	8.03130
1-METHYL-3,2′-METHYLENE-2-PHENYL INDOLE (TRACE)	22.32	0.040	6.01744
1-METHYL-3,2′-ETHYLENE-2-PHENYL INDOLE	21.86	0.043	5.30704
1-METHYL-3,2′-TRIMETHYLENE-2-PHENYL INDOLE	22.71	0.038	6.68719
4,5-DIPHENYLIMIDAZOLE (A)	22.36	0.040	6.08207
PPF (TRACE)	23.92	0.033	9.12327
PPO (TRACE)	26.16	0.025	15.61592
2-(4-FLUOROPHENYL)-5-PHENYLOXAZOLE (TRACE)	25.33	0.028	12.86967
αNPO (TRACE)	23.37	0.035	7.94104
BBO (B)	24.67	0.030	10.96321
POPOP	22.80	0.038	6.83287
DIMETHYL POPOP (TRACE)	23.95	0.033	9.18002
PPD (TRACE)	26.06	0.025	15.36077
2,5-DI(2-METHYLPHENYL)OXADIAZOLE	25.38	0.027	13.08402
PBD	28.00	0.020	23.45446
BBD (B)	27.16	0.022	19.54357
αNPD (TRACE)	23.46	0.035	8.11430
βNPD	27.13	0.022	19.41152
αNND (TRACE)	21.87	0.043	5.32102
βNND (TRACE)	27.20	0.022	19.71432
TRANS-1-(4-BIPHENYLYL)-2-PHENYL-ETHYLENE	25.88	0.026	14.64028
TRANS-1-(4-BIPHENYLYL)-2-(1-NAPHTHYL)-ETHYLENE	24.57	0.030	10.70397

ACCEPTOR MOLECULES	RO	CO	OVERLAP
DIPHENYLSTILBENE (B)	23.36	0.035	7.90575
1,1,4,4-TETRAPHENYLBUTADIENE	21.58	0.045	4.91418
1,6-DIPHENYLHEXATRIENE	21.20	0.047	4.42547
1,8-DIPHENYL-1,3,5,7-OCTATETRAENE	20.00	0.056	3.11118
BPSB (TRACE)	22.23	0.041	5.87722
BIS-MSB	23.11	0.036	7.40989
ESCULIN (A)	20.54	0.052	3.65742
AZULENE	25.77	0.026	14.26071
NAPHTHALENE	18.31	0.073	1.84101
DEUTERATED NAPHTHALENE-C10D8	18.45	0.071	1.91766
1-METHYLNAPHTHALENE 5ML/L	19.58	0.060	2.75368
2-METHYL NAPHTHALENE	18.50	0.071	1.96046
2,3-DIMETHYL NAPHTHALENE	18.98	0.065	2.28326
2,6-DIMETHYL NAPHTHALENE (TRACE)	18.69	0.068	2.08199
1-NAPHTHOL (TRACE)	19.80	0.058	2.93905
2-NAPHTHOL (TRACE)	18.10	0.075	1.71330
1-NAPHTHYLAMINE	18.24	0.074	1.79174
2-CHLORONAPHTHALENE	18.51	0.071	1.96303
ACENAPHTHENE	20.41	0.053	3.53300
1-PHENYL NAPHTHALENE	22.21	0.041	5.85842
2-PHENYL NAPHTHALENE	32.94	0.013	62.22754
1,4-DIPHENYL NAPHTHALENE	23.08	0.036	7.35409
1,5-DIPHENYL NAPHTHALENE	22.73	0.038	6.71972
1,7-DIPHENYL NAPHTHALENE	22.77	0.038	6.78123
1,4,5,8-TETRAPHENYL NAPHTHALENE	22.17	0.041	5.78126
1,1'-BINAPHTHYL	23.42	0.035	8.03720
4,4'-DI(N-BUTOXY)-1,1'-BINAPHTHYL	22.91	0.037	7.04819
2,2'-BINAPHTHYL	25.78	0.026	14.28331
ACRIDINE (A)	14.99	0.133	0.55330
ANTHRACENE (TRACE)	13.24	0.193	0.26279
DEUTERATED ANTHRACENE C14D10 (TRACE)	14.01	0.163	0.36801
9-METHYL ANTHRACENE (TRACE)	13.57	0.179	0.30455
1-AMINO ANTHRACENE	22.33	0.040	6.02990
2-AMINO ANTHRACENE	11.61	0.286	0.11659
9-VINYL ANTHRACENE	15.37	0.123	0.64238
9-PHENYL ANTHRACENE (TRACE)	16.20	0.105	0.88127
9,10-DIPHENYL ANTHRACENE	18.83	0.067	2.16876
9,10-DICHLOROANTHRACENE (TRACE)	14.12	0.159	0.38524
BIANTHRANYL (B)	22.24	0.041	5.88384
9,10-DINAPHTHYLANTHRACENE (B)	23.81	0.033	8.87774
9-(P-VINYLPHENYL)-10-PHENYL ANTHRACENE	20.72	0.050	3.85461
9,10-BIS(PHENYL-ETHYNYL)ANTHRACENE (TRACE)	27.32	0.022	20.24917
RUBICENE (B)	27.39	0.022	20.56640
DIBENZ(A,H)ANTHRACENE (B)	29.98	0.017	35.32730
TETRACENE (TRACE) (B)	32.62	0.013	58.69636
RUBRENE (B)	29.41	0.018	31.49530
TETRAHYDROCANNABINOL ACID METHYL ESTER	24.10	0.032	9.59199
PHENANTHRENE	21.86	0.043	5.31689
3,4-BENZOPHENANTHRENE	27.98	0.020	23.35157
2-PHENYL PHENANTHRENE	28.07	0.020	23.81564
CHRYSENE (TRACE)	25.28	0.028	12.70722
TRIPHENYLENE	23.45	0.035	8.10136
DEUTERATED TRIPHENYLENE-C18D12	23.53	0.034	8.27278
PYRENE	22.51	0.039	6.32703
1,3,6,8,-TETRAPHENYL PYRENE	27.10	0.022	19.28072
BIPYRENYL (B)	27.59	0.021	21.46482
O-PHENYLENE-PYRENE	26.42	0.024	16.56581
ANTHANTHRENE (B)	25.93	0.026	14.80971
FLUORANTHENE	23.92	0.033	9.12124
3-PHENYL FLUORANTHENE	25.12	0.028	12.22095
3-AMINO FLUORANTHENE	23.27	0.036	7.72628
3,3'-BIFLUORANTHENYL (B)	26.84	0.023	18.17992
BENZO(GHI)FLUORANTHENE	23.11	0.036	7.42508
9,12-DI(N-OCTADECOXY)-BENZ(E)ACEPHENANTHRYLENE	26.89	0.023	18.38951
DECACYCLENE (B)	26.20	0.025	15.73882
10-ETHOXY-BENZO(KL)XANTHENE	19.66	0.059	2.81514
PERYLENE	17.00	0.091	1.17646
BENZO(GHI)PERYLENE (B)	27.22	0.022	19.80239
BBOT (TRACE)	20.51	0.052	3.61890
QUININE SULFATE (A)	20.98	0.048	4.14809
ACRIDINE YELLOW (A)	23.70	0.034	8.63377
ACRIDINE ORANGE (A)	24.02	0.032	9.35403
ACRIDINE RED (A)	20.85	0.049	3.99509
URANINE (A)	20.92	0.049	4.08256
RHODAMINE B (A)	24.00	0.032	9.29562
RHODAMINE 6G (A)	24.02	0.032	9.34380
EOSIN B (A)	21.43	0.045	4.71899
EOSIN Y (A)	22.07	0.042	5.62169
ROSE BENGAL (A)	22.77	0.038	6.78758
INDENOINDENE	23.54	0.034	8.29111

8.11 Donor Molecule: Bibenzyl

POLYVINYLTOLUENE	3.96	7.184	0.00043
PHENYL ETHER	2.55	27.034	0.00003
ANILINE 3ML/L	10.10	0.434	0.11787
N,N-DIMETHYLANILINE 3ML/L	13.50	0.182	0.66812
TRIPHENYLAMINE	21.60	0.044	11.08891

ACCEPTOR MOLECULES	RO	CO	OVERLAP
BENZENE (3 ML/L)	3.63	9.328	0.00014
DEUTERATED BENZENE-C6D6 5ML/L	3.14	14.442	0.00006
FLUOROBENZENE (TRACE)	6.60	1.553	0.00512
TOLUENE (TRACE)	6.58	1.568	0.00501
DEUTERATED TOLUENE-C7D8 5ML/L	6.00	2.075	0.00286
ETHYLBENZENE (TRACE)	6.42	1.688	0.00433
PROPYLBENZENE (TRACE)	6.15	1.919	0.00335
ISOPROPYLBENZENE (TRACE)	5.80	[illegible]	[illegible]
SEC-BUTYLBENZENE (TRACE)	5.80	2.294	0.00235
DIPHENYLMETHANE (TRACE)	7.58	1.027	0.01169
BIBENZYL	7.03	1.286	0.00745
1,3,5-TRIPHENYLPENTANE 0.05ML/L	7.21	1.194	0.00864
POLYSTYRENE	6.44	1.672	0.00441
POLYVINYLTOLUENE	9.04	0.605	0.03318
PHENYLCYCLOHEXANE (TRACE)	5.71	2.399	0.00215
BENZYLALCOHOL 3ML/L	5.66	2.465	0.00203
BENZYLACETATE 5ML/L	5.60	2.552	0.00190
PHENOL (TRACE)	12.80	0.214	0.26918
ANISOLE (TRACE)	13.09	0.199	0.30889
PHENYL ETHER	13.70	0.174	0.40410
ANILINE 3ML/L	15.49	0.120	0.83554
N,N-DIMETHYLANILINE 3ML/L	16.51	0.099	1.21898
TRIPHENYLAMINE	24.42	0.031	12.72744
P-XYLENE (TRACE)	9.79	0.477	0.05397
DEUTERATED P-XYLENE C8D10 (TRACE)	9.74	0.484	0.05241
M-XYLENE (TRACE)	8.03	0.864	0.01646
O-XYLENE 3ML/L	7.68	0.989	0.01258
1-METHYL-4-ETHYLBENZENE (TRACE)	9.18	0.578	0.03681
1,4-BIS(TRIFLUOROMETHYL)BENZENE 5ML/L	7.43	1.091	0.01037
P-CRESOL	15.10	0.130	0.72216
P-METHYLANISOLE 3ML/L	14.59	0.144	0.58775
1,4-DIMETHOXY BENZENE	16.79	0.094	1.35803
1-METHYL-4-CHLOROBENZENE 5ML/L	10.13	0.431	0.06617
SODIUM SALICYLATE (A)	17.73	0.080	1.86359
ANTHRANILIC ACID (A)	15.02	0.132	0.68926
TMPD	20.15	0.055	4.01918
1,2,4-TRIMETHYLBENZENE (TRACE)	10.60	0.376	0.08676
MESITYLENE 5ML/L	8.09	0.844	0.01724
1,3,5-TRIETHYLBENZENE 5ML/L	7.78	0.951	0.01357
STYRENE	12.89	0.209	0.27922
1,1-DIPHENYLETHYLENE 3ML/L	14.88	0.136	0.66306
BIPHENYL	14.36	0.151	0.53717
DEUTERATED BIPHENYL-C12D10	14.40	0.150	0.54570
4-METHYL BIPHENYL	16.38	0.102	1.17973
MONOISOPROPYLBIPHENYL 5ML/L	15.87	0.112	0.97606
4-BENZYL BIPHENYL	18.00	0.077	2.07521
3,3'-DIMETHYL BIPHENYL	15.48	0.121	0.83775
2,2'-DIMETHYLBIPHENYL 2ML/L	10.54	0.382	0.08366
OCTAMETHYL BIPHENYL	12.14	0.250	0.19542
4-METHOXY BIPHENYL	20.13	0.055	4.03726
4-BIPHENYLYL PHENYL ETHER	20.67	0.051	4.73629
4,4'-DIHYDROXYBIPHENYL (A)	22.03	0.042	6.89826
4,4'-DIMETHOXY BIPHENYL	21.93	0.042	6.71626
4,4'-DIISOBUTOXYBIPHENYL	21.74	0.044	6.39505
BENZIDINE	25.41	0.027	16.13039
4-VINYL BIPHENYL	24.62	0.030	13.44785
M-TERPHENYL	19.15	0.064	2.98303
M-QUATERPHENYL	19.50	0.060	3.33862
1,3,5-TRIPHENYL BENZENE	20.40	0.053	4.33486
1,3-DIBIPHENYLYLBENZENE (B)	28.08	0.020	29.40410
1,3,5,-TRIBIPHENYLYLBENZENE (B)	30.85	0.015	51.70233
FLUORENE (TRACE)	20.28	0.054	4.22275
1-METHYL FLUORENE	17.97	0.077	2.01962
1,2-BENZOFLUORENE	22.61	0.039	7.99917
9,10-DIHYDROPHENANTHRENE	20.64	0.051	4.67885
CARBAZOLE (TRACE)	21.88	0.043	6.57327
N-METHYL CARBAZOLE (TRACE)	21.26	0.047	5.54124
N-VINYLCARBAZOLE	22.73	0.038	8.27572
N-PHENYL CARBAZOLE	22.94	0.037	8.74155
N-METHYL-2,3-BENZCARBAZOLE	26.93	0.023	22.87630
DIBENZOFURAN	21.94	0.042	6.76615
DIBENZOTHIOPHENE (TRACE)	19.59	0.059	3.38975
P-TERPHENYL	25.34	0.027	15.95614
DEUTERATED P-TERPHENYL-C18D14	25.04	0.028	14.88063
4-METHYL TERPHENYL	25.67	0.026	17.21233
2-METHYL-P-TERPHENYL	22.44	0.040	7.72523
3,3''-DIMETHYL-P-TERPHENYL	25.23	0.028	15.53737
4-(3,3-DIMETHYLBUTOXY)-P-TERPHENYL	26.16	0.025	19.20023
4,4''-DIHEXAHYDROFARNESOXY-P-TERPHENYL	27.08	0.023	23.62713
2,2'-METHYLENE-P-TERPHENYL	25.91	0.026	18.16337
2,2'-ETHYLENE-P-TERPHENYL	25.88	0.026	18.01938
2,2'-METHYLENE-5',6''-METHYLENE-P-TERPHENYL (TRACE)	25.06	0.028	14.85561
3-PHENYL DIBENZOFURAN	24.92	0.029	14.37867
BENZO(1,2-B-4,5-B')BISBENZOFURAN (TRACE) (B)	[illegible]	[illegible]	[illegible]
SEXIPHENYL BENZENE	20.33	0.053	4.28763
P-QUATERPHENYL	26.93	0.023	22.88881
3,3'''-DIMETHYL-P-QUATERPHENYL	27.19	0.022	24.25231
TETRAMETHYL-P-QUATERPHENYL	24.63	0.030	13.49399
3,3''' -DIMETHYL-3',2''-METHYLENE-P-QUATERPHENYL	25.99	0.025	18.49496
3,3''' -DIMETHYL-3',2''-ETHYLENE-P-QUATERPHENYL	26.98	0.023	23.12467
4-(3-ETHYLBUTOXY)-P-QUATERPHENYL (B)	26.60	0.024	21.22257
4,4'''-DI(2-BUTYLOCTOXY-1)-P-QUATERPHENYL	27.63	0.021	26.66174
DIETHYL-P-QUINQUEPHENYL	28.57	0.019	32.58426
DI(3-ETHYLHEPTYL)-P-QUINQUEPHENYL	28.11	0.020	29.61172
TETRAMETHYL-P-SEXIPHENYL	29.04	0.018	35.98456

ACCEPTOR MOLECULES	RO	CO	OVERLAP
TETRAMETHYL-P-OCTAPHENYL (B)	28.54	0.019	32.38503
INDOLE (TRACE)	16.78	0.095	1.36344
2-PHENYL INDOLE	23.89	0.033	11.15808
1-METHYL-2-PHENYL INDOLE	23.26	0.036	9.48680
1-NEOPENTYL-2-PHENYL INDOLE	23.17	0.036	9.28376
1-N-BUTYL-2(BIPHENYLYL-4) INDOLE	24.24	0.031	12.17244
1,2-DIPHENYL INDOLE	24.03	0.032	11.55926
1-BIPHENYLYL-2-PHENYL INDOLE	25.69	0.026	17.23814
1,3-DIMETHYL-2-PHENYL INDOLE	22.30	0.040	7.37865
2,3-DIPHENYL INDOLE	23.02	0.037	8.93405
1-METHYL-2,3-DIPHENYL INDOLE	22.77	0.038	8.34883
2-PHENYL-3,2'-METHYLENE INDOLE (TRACE)	22.38	0.040	7.53788
2-PHENYL-3,2'-ETHYLENE INDOLE (TRACE)	21.80	0.043	6.43317
2-PHENYL-3,2'-TRIMETHYLENE INDOLE	22.92	0.037	8.68864
1-METHYL-3,2'-METHYLENE-2-PHENYL INDOLE (TRACE)	21.91	0.043	6.63308
1-METHYL-3,2'-ETHYLENE-2-PHENYL INDOLE	21.41	0.046	5.78143
1-METHYL-3,2'-TRIMETHYLENE-2-PHENYL INDOLE	22.16	0.041	7.14389
4,5-DIPHENYLIMIDAZOLE (A)	21.57	0.045	6.03602
PPF (TRACE)	23.46	0.035	10.00320
PPO (TRACE)	25.47	0.027	16.37017
2-(4-FLUOROPHENYL)-5-PHENYLOXAZOLE (TRACE)	24.64	0.030	13.43355
αNPO (TRACE)	22.74	0.038	8.29717
BBO (B)	24.09	0.032	11.70996
POPOP	22.23	0.041	7.23274
DIMETHYL POPOP (TRACE)	23.26	0.036	9.50218
PPD (TRACE)	25.10	0.028	15.10847
2,5-DI(2-METHYLPHENYL)OXADIAZOLE	24.45	0.031	12.88869
PBD	27.19	0.022	24.25127
BBD (B)	26.51	0.024	20.81060
αNPD (TRACE)	22.82	0.038	8.45745
βNPD	26.22	0.025	19.49205
αNND (TRACE)	21.40	0.046	5.75788
βNND (TRACE)	26.31	0.025	19.90364
TRANS-1-(4-BIPHENYLYL)-2-PHENYL-ETHYLENE	25.32	0.028	15.79920
TRANS-1-(4-BIPHENYLYL)-2-(1-NAPHTHYL)-ETHYLENE	23.94	0.033	11.28486
DIPHENYLSTILBENE (B)	23.02	0.037	8.92908
1,1,4,4-TETRAPHENYLBUTADIENE	21.19	0.047	5.42742
1,6-DIPHENYLHEXATRIENE	21.05	0.048	5.21490
1,8-DIPHENYL-1,3,5,7-OCTATETRAENE	19.57	0.060	3.37031
BPSB (TRACE)	21.92	0.042	6.65380
BIS-MSB	22.74	0.038	8.28902
ESCULIN (A)	20.07	0.055	3.92257
AZULENE	24.65	0.030	13.44639
NAPHTHALENE	17.53	0.083	1.74767
DEUTERATED NAPHTHALENE-C10D8	17.66	0.081	1.81886
1-METHYLNAPHTHALENE 5ML/L	18.81	0.067	2.66548
2-METHYL NAPHTHALENE	17.74	0.080	1.87081
2,3-DIMETHYL NAPHTHALENE	18.20	0.074	2.18759
2,6-DIMETHYL NAPHTHALENE (TRACE)	17.93	0.078	1.99774
1-NAPHTHOL (TRACE)	19.14	0.064	2.95033
2-NAPHTHOL (TRACE)	17.41	0.085	1.66754
1-NAPHTHYLAMINE	17.84	0.079	1.93432
2-CHLORONAPHTHALENE	17.74	0.080	1.87262
ACENAPHTHENE	19.69	0.059	3.50666
1-PHENYL NAPHTHALENE	21.43	0.045	5.82003
2-PHENYL NAPHTHALENE	31.76	0.014	61.49930
1,4-DIPHENYL NAPHTHALENE	22.41	0.040	7.59855
1,5-DIPHENYL NAPHTHALENE	22.06	0.042	6.91020
1,7-DIPHENYL NAPHTHALENE	22.03	0.042	6.85384
1,4,5,8-TETRAPHENYL NAPHTHALENE	21.59	0.044	6.06646
1,1'-BINAPHTHYL	22.60	0.039	7.99467
4,4'-DI(N-BUTOXY)-1,1'-BINAPHTHYL	22.34	0.040	7.46063
2,2'-BINAPHTHYL	24.87	0.029	14.19940
ACRIDINE (A)	14.72	0.140	0.61119
ANTHRACENE (TRACE)	13.15	0.197	0.30943
DEUTERATED ANTHRACENE C14D10 (TRACE)	13.80	0.170	0.41378
9-METHYL ANTHRACENE (TRACE)	13.19	0.195	0.31630
1-AMINO ANTHRACENE	21.21	0.047	5.45872
2-AMINO ANTHRACENE	11.85	0.269	0.16244
9-VINYL ANTHRACENE	14.79	0.138	0.62802
9-PHENYL ANTHRACENE (TRACE)	15.55	0.119	0.84934
9,10-DIPHENYL ANTHRACENE	17.87	0.078	1.95124
9,10-DICHLOROANTHRACENE (TRACE)	13.60	0.178	0.37906
BIANTHRANYL (B)	20.97	0.049	5.09308
9,10-DINAPHTHYLANTHRACENE (B)	22.88	0.037	8.61010
9-(P-VINYLPHENYL)-10-PHENYL ANTHRACENE	19.79	0.058	3.60335
9,10-BIS(PHENYL-ETHYNYL)ANTHRACENE (TRACE)	26.20	0.025	19.39902
RUBICENE (B)	26.48	0.024	20.69473
DIBENZ(A,H)ANTHRACENE (B)	29.10	0.018	36.44536
TETRACENE (TRACE) (B)	31.24	0.015	55.75767
RUBRENE (B)	28.65	0.019	33.18321
TETRAHYDROCANNABINOL ACID METHYL ESTER	23.14	0.036	9.25138
PHENANTHRENE	20.98	0.048	5.11267
3,4-BENZOPHENANTHRENE	26.88	0.023	22.61320
2-PHENYL PHENANTHRENE	26.90	0.023	22.70633
CHRYSENE (TRACE)	24.08	0.032	11.69405
TRIPHENYLENE	22.48	0.039	7.74664
DEUTERATED TRIPHENYLENE-C18D12	22.55	0.039	7.88832
PYRENE	21.74	0.044	6.33959
1,3,6,8,-TETRAPHENYL PYRENE	26.22	0.025	19.50481
BIPYRENYL (B)	26.60	0.024	21.24630
O-PHENYLENE-PYRENE	25.58	0.027	16.80397
ANTHANTHRENE (B)	25.22	0.028	15.42504
FLUORANTHENE	23.04	0.037	8.97834

ACCEPTOR MOLECULES	RO	CO	OVERLAP
3-PHENYL FLUORANTHENE	24.28	0.031	12.29049
3-AMINO FLUORANTHENE	22.56	0.039	7.90811
3,3'-BIFLUORANTHENYL (B)	25.95	0.026	18.33282
BENZO(GHI)FLUORANTHENE	22.32	0.040	7.40871
9,12-DI(N-OCTADECOXY)-BENZ(E)ACEPHENANTHRYLENE	25.86	0.026	17.95424
DECACYCLENE (B)	25.35	0.027	15.90502
10-ETHOXY-BENZO(KL)XANTHENE	19.05	0.065	2.86361
PERYLENE	18.28	0.104	1.11551
BENZO(GHI)PERYLENE (B)	26.34	0.024	20.02412
BBOT (TRACE)	20.10	0.055	3.95895
QUININE SULFATE (A)	20.34	0.053	4.24784
ACRIDINE YELLOW (A)	22.66	0.038	8.10891
ACRIDINE ORANGE (A)	23.06	0.036	9.01523
ACRIDINE RED (A)	20.14	0.055	4.00607
URANINE (A)	20.23	0.054	4.10923
RHODAMINE B (A)	23.18	0.036	9.30047
RHODAMINE 6G (A)	23.16	0.036	9.25725
EOSIN B (A)	20.76	0.050	4.80422
EOSIN Y (A)	21.38	0.046	5.72513
ROSE BENGAL (A)	21.95	0.042	6.70195
INDENOINDENE	23.04	0.037	8.97126

8.13 Donor Molecule: Polystyrene

BENZENE (3 ML/L)	1.55	******	0.00000
DEUTERATED BENZENE-C6D6 5ML/L	0.00	0.000	0.00000
FLUOROBENZENE (TRACE)	2.81	20.145	0.00013
TOLUENE (TRACE)	3.23	13.327	0.00030
DEUTERATED TOLUENE-C7D8 5ML/L	2.87	18.919	0.00015
ETHYLBENZENE (TRACE)	3.17	14.064	0.00027
PROPYLBENZENE (TRACE)	3.02	16.288	0.00020
ISOPROPYLBENZENE (TRACE)	2.81	20.151	0.00013
SEC-BUTYLBENZENE (TRACE)	2.79	20.548	0.00013
DIPHENYLMETHANE (TRACE)	3.89	7.591	0.00092
BIBENZYL	3.51	10.378	0.00049
1,3,5-TRIPHENYLPENTANE 0.05ML/L	3.59	9.634	0.00057
POLYSTYRENE	3.23	13.228	0.00030
POLYVINYLTOLUENE	5.05	3.477	0.00435
PHENYLCYCLOHEXANE (TRACE)	2.59	25.766	0.00008
BENZYLALCOHOL 3ML/L	2.84	19.522	0.00014
BENZYLACETATE 5ML/L	2.70	22.769	0.00010
PHENOL (TRACE)	7.20	1.196	0.03705
ANISOLE (TRACE)	7.40	1.103	0.04350
PHENYL ETHER	7.76	0.956	0.05782
ANILINE 3ML/L	9.15	0.583	0.15402
N,N-DIMETHYLANILINE 3ML/L	10.70	0.366	0.38909
TRIPHENYLAMINE	17.29	0.086	6.88516
P-XYLENE (TRACE)	5.44	2.780	0.00686
DEUTERATED P-XYLENE C8D10 (TRACE)	5.41	2.819	0.00668
M-XYLENE (TRACE)	4.42	5.193	0.00197
O-XYLENE 3ML/L	4.15	6.239	0.00136
1-METHYL-4-ETHYLBENZENE (TRACE)	5.10	3.374	0.00467
1,4-BIS(TRIFLUOROMETHYL)BENZENE 5ML/L	3.70	8.797	0.00069
P-CRESOL	8.65	0.691	0.11036
P-METHYLANISOLE 3ML/L	8.36	0.766	0.08995
1,4-DIMETHOXY BENZENE	10.10	0.434	0.27810
1-METHYL-4-CHLOROBENZENE 5ML/L	5.70	2.415	0.00909
SODIUM SALICYLATE (A)	12.16	0.249	0.82957
ANTHRANILIC ACID (A)	15.72	0.115	3.85564
TMPD	14.39	0.150	2.27438
1,2,4-TRIMETHYLBENZENE (TRACE)	5.94	2.134	0.01163
MESITYLENE 5ML/L	4.49	4.939	0.00218
1,3,5-TRIETHYLBENZENE 5ML/L	4.28	5.723	0.00162
STYRENE	7.37	1.118	0.04206
1,1-DIPHENYLETHYLENE 3ML/L	8.37	0.762	0.09092
BIPHENYL	7.99	0.876	0.06894
DEUTERATED BIPHENYL-C12D10	8.00	0.873	0.06947
4-METHYL BIPHENYL	9.20	0.574	0.16027
MONOISOPROPYLBIPHENYL 5ML/L	8.89	0.636	0.13070
4-BENZYL BIPHENYL	10.13	0.430	0.28524
3,3'-DIMETHYL BIPHENYL	8.76	0.665	0.11893
2,2'-DIMETHYLBIPHENYL 2ML/L	5.86	2.218	0.01073
OCTAMETHYL BIPHENYL	6.88	1.375	0.02793
4-METHOXY BIPHENYL	11.51	0.293	0.61075
4-BIPHENYLYL PHENYL ETHER	11.79	0.273	0.70438
4,4'-DIHYDROXYBIPHENYL (A)	12.89	0.209	1.19683
4,4'-DIMETHOXY BIPHENYL	12.72	0.217	1.10691
4,4'-DIISOBUTOXYBIPHENYL	12.83	0.222	1.06022
BENZIDINE	16.59	0.098	5.38368
4-VINYL BIPHENYL	14.64	0.143	2.56372
M-TERPHENYL	11.06	0.331	0.47858
M-QUATERPHENYL	11.01	0.335	0.46759
1,3,5-TRIPHENYL BENZENE	11.63	0.285	0.64135
1,3-DIBIPHENYLYLBENZENE (B)	17.70	0.081	7.94719
1,3,5,-TRIBIPHENYLYLBENZENE (B)	19.88	0.057	15.99432
FLUORENE (TRACE)	12.04	0.256	0.79764
1-METHYL FLUORENE	10.94	0.342	0.44524
1,2-BENZOFLUORENE	16.01	0.109	4.33825
9,10-DIHYDROPHENANTHRENE	12.14	0.250	0.83718

ACCEPTOR MOLECULES	RO	CO	OVERLAP
CARBAZOLE (TRACE)	14.73	0.140	2.63767
N-METHYL CARBAZOLE (TRACE)	15.18	0.128	3.14532
N-VINYLCARBAZOLE	15.57	0.118	3.67097
N-PHENYL CARBAZOLE	15.56	0.119	3.66161
N-METHYL-2,3-BENZCARBAZOLE	17.81	0.079	8.09997
DIBENZOFURAN	12.86	0.210	1.18681
DIBENZOTHIOPHENE (TRACE)	13.20	0.194	1.37132
P-TERPHENYL	15.16	0.128	3.15892
DEUTERATED P-TERPHENYL-C18D14	14.96	0.134	2.91750
4-METHYL TERPHENYL	15.61	0.118	3.76466
2-METHYL-P-TERPHENYL	12.97	0.205	1.24237
3,3″-DIMETHYL-P-TERPHENYL	15.31	0.125	3.34979
4-(3,3-DIMETHYLBUTOXY)-P-TERPHENYL	16.78	0.095	5.77956
4,4″-DIHEXAHYDROFARNESOXY-P-TERPHENYL	18.07	0.076	9.02407
2,2′-METHYLENE-P-TERPHENYL	17.46	0.084	7.34907
2,2′-ETHYLENE-P-TERPHENYL	17.57	0.082	7.61436
2,2′-METHYLENE-5′,6″-METHYLENE-P-TERPHENYL (TRACE)	20.49	0.052	19.11097
3-PHENYL DIBENZOFURAN	17.46	0.084	7.34271
BENZO(1,2-B-4,5-B′)BISBENZOFURAN (TRACE) (B)	21.12	0.047	22.85035
SEXIPHENYL BENZENE	11.64	0.284	0.65283
P-QUATERPHENYL	18.40	0.072	10.03843
3,3‴-DIMETHYL-P-QUATERPHENYL	18.94	0.066	11.90660
TETRAMETHYL-P-QUATERPHENYL	14.32	0.152	2.25561
3,3‴-DIMETHYL-3′,2″-METHYLENE-P-QUATERPHENYL	21.60	0.044	26.13502
3,3‴-DIMETHYL-3′,2″-ETHYLENE-P-QUATERPHENYL	21.51	0.045	25.47219
4-(3-ETHYLBUTOXY)-P-QUATERPHENYL (B)	20.50	0.052	19.09909
4,4‴-DI(2-BUTYLOCTOXY-1)-P-QUATERPHENYL	21.10	0.048	22.70155
DIETHYL-P-QUINQUEPHENYL	19.81	0.058	15.58888
DI(3-ETHYLHEPTYL)-P-QUINQUEPHENYL	21.81	0.043	27.59585
TETRAMETHYL-P-SEXIPHENYL	19.65	0.059	14.89204
TETRAMETHYL-P-OCTAPHENYL (B)	23.58	0.034	44.07944
INDOLE (TRACE)	9.55	0.514	0.20004
2-PHENYL INDOLE	18.45	0.071	10.18582
1-METHYL-2-PHENYL INDOLE	16.24	0.104	4.73761
1-NEOPENTYL-2-PHENYL INDOLE	15.57	0.118	3.69306
1-N-BUTYL-2(BIPHENYLYL-4) INDOLE	19.52	0.060	14.17409
1,2-DIPHENYL INDOLE	16.84	0.094	5.89352
1-BIPHENYLYL-2-PHENYL INDOLE	18.11	0.075	9.06475
1,3-DIMETHYL-2-PHENYL INDOLE	16.07	0.108	4.44907
2,3-DIPHENYL INDOLE	18.21	0.074	9.34597
1-METHYL-2,3-DIPHENYL INDOLE	16.92	0.092	6.01410
2-PHENYL-3,2′-METHYLENE INDOLE (TRACE)	19.27	0.063	13.20413
2-PHENYL-3,2′-ETHYLENE INDOLE (TRACE)	19.90	0.057	15.96922
2-PHENYL-3,2′-TRIMETHYLENE INDOLE	18.87	0.067	11.64984
1-METHYL-3,2′-METHYLENE-2-PHENYL INDOLE (TRACE)	19.85	0.057	15.71315
1-METHYL-3,2′-ETHYLENE-2-PHENYL INDOLE	19.90	0.057	15.94185
1-METHYL-3,2′-TRIMETHYLENE-2-PHENYL INDOLE	18.01	0.077	8.78827
4,5-DIPHENYLIMIDAZOLE (A)	13.90	0.167	1.86199
PPF (TRACE)	20.89	0.049	21.39680
PPO (TRACE)	19.56	0.060	14.46346
2-(4-FLUOROPHENYL)-5-PHENYLOXAZOLE (TRACE)	18.74	0.068	11.16947
αNPO (TRACE)	20.90	0.049	21.25560
BBO (B)	23.58	0.034	43.81876
POPOP	22.55	0.039	33.44819
DIMETHYL POPOP (TRACE)	22.96	0.037	37.17483
PPD (TRACE)	15.31	0.125	3.36567
2,5-DI(2-METHYLPHENYL)OXADIAZOLE	15.13	0.129	3.11931
PBD	19.73	0.058	15.26719
BBD (B)	22.43	0.040	32.71025
αNPD (TRACE)	19.25	0.063	13.05269
βNPD	18.78	0.068	11.32289
αNND (TRACE)	19.92	0.057	16.02093
βNND (TRACE)	20.79	0.050	20.79584
TRANS-1-(4-BIPHENYLYL)-2-PHENYL-ETHYLENE	21.94	0.042	28.61356
TRANS-1-(4-BIPHENYLYL)-2-(1-NAPHTHYL)-ETHYLENE	22.09	0.042	29.57363
DIPHENYLSTILBENE (B)	23.79	0.033	46.14203
1,1,4,4-TETRAPHENYLBUTADIENE	22.16	0.041	30.05934
1,6-DIPHENYLHEXATRIENE	24.19	0.032	50.98268
1,8-DIPHENYL-1,3,5,7-OCTATETRAENE	23.27	0.036	40.06417
BPSB (TRACE)	24.30	0.031	52.35087
BIS-MSB	23.87	0.033	47.04240
ESCULIN (A)	18.74	0.068	11.06404
AZULENE	15.83	0.113	3.97798
NAPHTHALENE	10.12	0.431	0.28027
DEUTERATED NAPHTHALENE-C10D8	10.33	0.406	0.31410
1-METHYLNAPHTHALENE 5ML/L	11.06	0.330	0.47582
2-METHYL NAPHTHALENE	10.52	0.384	0.35296
2,3-DIMETHYL NAPHTHALENE	10.78	0.357	0.40744
2,6-DIMETHYL NAPHTHALENE (TRACE)	11.23	0.316	0.52053
1-NAPHTHOL (TRACE)	12.62	0.223	1.04713
2-NAPHTHOL (TRACE)	12.43	0.233	0.95232
1-NAPHTHYLAMINE	15.11	0.130	3.06398
2-CHLORONAPHTHALENE	10.55	0.380	0.35883
ACENAPHTHENE	12.31	0.240	0.90294
1-PHENYL NAPHTHALENE	13.36	0.188	1.47269
2-PHENYL NAPHTHALENE	19.72	0.058	15.23634
1,4-DIPHENYL NAPHTHALENE	16.13	0.107	4.54218
1,5-DIPHENYL NAPHTHALENE	15.48	0.121	3.55690
1,7-DIPHENYL NAPHTHALENE	15.32	0.124	3.34111
1,4,5,8-TETRAPHENYL NAPHTHALENE	20.10	0.055	16.77317
1,1′-BINAPHTHYL	14.30	0.153	2.21192
4,4′-DI(N-BUTOXY)-1,1′-BINAPHTHYL	17.91	0.078	8.48751
2,2′-BINAPHTHYL	17.08	0.090	6.42762
ACRIDINE (A)	16.84	0.094	5.80146
ANTHRACENE (TRACE)	15.79	0.114	3.94653
DEUTERATED ANTHRACENE C14D10 (TRACE)	15.68	0.116	3.78411

ACCEPTOR MOLECULES	RO	CO	OVERLAP
9-METHYL ANTHRACENE (TRACE)	15.39	0.123	3.38211
1-AMINO ANTHRACENE	15.10	0.130	2.98902
2-AMINO ANTHRACENE	14.90	0.135	2.72106
9-VINYL ANTHRACENE	15.89	0.111	4.08409
9-PHENYL ANTHRACENE (TRACE)	16.24	0.104	4.66481
9,10-DIPHENYL ANTHRACENE	16.38	0.102	4.90001
9,10-DICHLOROANTHRACENE (TRACE)	15.49	0.120	3.49986
BIANTHRANYL (B)	17.[illegible]	[illegible]	[illegible]
9,10-DINAPHTHYLANTHRACENE (B)	16.95	0.092	6.00513
9-(P-VINYLPHENYL)-10-PHENYL ANTHRACENE	16.62	0.097	5.33557
9,10-BIS(PHENYL-ETHYNYL)ANTHRACENE (TRACE)	18.00	0.077	8.57369
RUBICENE (B)	19.10	0.064	12.24500
DIBENZ(A,H)ANTHRACENE (B)	20.83	0.049	20.74232
TETRACENE (TRACE) (B)	18.39	0.072	9.77882
RUBRENE (B)	20.69	0.050	19.81095
TETRAHYDROCANNABINOL ACID METHYL ESTER	13.73	0.173	1.74608
PHENANTHRENE	12.42	0.233	0.94594
3,4-BENZOPHENANTHRENE	17.04	0.090	6.22897
2-PHENYL PHENANTHRENE	16.27	0.104	4.76321
CHRYSENE (TRACE)	16.40	0.101	4.97653
TRIPHENYLENE	13.35	0.188	1.45627
DEUTERATED TRIPHENYLENE-C18D12	13.35	0.188	1.45667
PYRENE	19.40	0.061	13.62176
1,3,6,8,-TETRAPHENYL PYRENE	20.85	0.049	20.74432
BIPYRENYL (B)	22.69	0.038	34.56493
O-PHENYLENE-PYRENE	19.86	0.057	15.48302
ANTHANTHRENE (B)	19.35	0.062	13.23184
FLUORANTHENE	17.49	0.084	7.26365
3-PHENYL FLUORANTHENE	18.81	0.067	11.21030
3-AMINO FLUORANTHENE	17.83	0.079	8.09796
3,3'-BIFLUORANTHENYL (B)	20.81	0.050	20.48813
BENZO(GHI)FLUORANTHENE	17.01	0.091	6.11327
9,12-DI(N-OCTADECOXY)-BENZ(E)ACEPHENANTHRYLENE	18.94	0.066	11.64531
DECACYCLENE (B)	22.32	0.040	31.17455
10-ETHOXY-BENZO(KL)XANTHENE	16.94	0.092	5.98882
PERYLENE	15.24	0.126	3.16365
BENZO(GHI)PERYLENE (B)	19.33	0.062	13.22301
BBOT (TRACE)	22.82	0.038	35.81815
QUININE SULFATE (A)	17.02	0.091	6.26181
ACRIDINE YELLOW (A)	14.97	0.133	2.84338
ACRIDINE ORANGE (A)	14.88	0.136	2.73998
ACRIDINE RED (A)	15.38	0.123	3.33725
URANINE (A)	15.53	0.119	3.54802
RHODAMINE B (A)	17.49	0.084	7.22046
RHODAMINE 6G (A)	17.46	0.084	7.15522
EOSIN B (A)	16.56	0.099	5.19700
EOSIN Y (A)	16.93	0.092	5.93811
ROSE BENGAL (A)	17.36	0.085	6.91887
INDENOINDENE	18.81	0.067	11.44191

8.14 Donor Molecule: Polyvinyltoluene

P-CRESOL	4.69	4.338	0.00119
P-METHYLANISOLE 3ML/L	4.41	5.232	0.00082
1,4-DIMETHOXY BENZENE	11.87	0.267	0.31068
1-METHYL-4-CHLOROBENZENE 5ML/L	0.00	0.000	0.00000
SODIUM SALICYLATE (A)	15.23	0.127	1.35965
ANTHRANILIC ACID (A)	18.05	0.076	3.75054
TMPD	16.39	0.102	2.10875
1,2,4-TRIMETHYLBENZENE (TRACE)	0.00	0.000	0.00000
MESITYLENE 5ML/L	0.00	0.000	0.00000
1,3,5-TRIETHYLBENZENE 5ML/L	0.00	0.000	0.00000
STYRENE	5.64	2.500	0.00357
1,1-DIPHENYLETHYLENE 3ML/L	5.10	3.369	0.00197
BIPHENYL	0.00	0.000	0.00000
DEUTERATED BIPHENYL-C12D10	0.00	0.000	0.00000
4-METHYL BIPHENYL	5.11	3.349	0.00199
MONOISOPROPYLBIPHENYL 5ML/L	2.75	21.616	0.00005
4-BENZYL BIPHENYL	5.36	2.908	0.00264
3,3'-DIMETHYL BIPHENYL	6.59	1.562	0.00915
2,2'-DIMETHYLBIPHENYL 2ML/L	2.76	21.389	0.00005
OCTAMETHYL BIPHENYL	1.94	61.683	0.00001
4-METHOXY BIPHENYL	10.19	0.423	0.12429
4-BIPHENYLYL PHENYL ETHER	9.98	0.449	0.11016
4,4'-DIHYDROXYBIPHENYL (A)	13.43	0.185	0.64790
4,4'-DIMETHOXY BIPHENYL	12.72	0.217	0.46969
4,4'-DIISOBUTOXYBIPHENYL	12.76	0.215	0.47867
BENZIDINE	20.10	0.054	[illegible].40803
4-VINYL BIPHENYL	16.32	0.103	2.09215
M-TERPHENYL	11.33	0.308	0.23372
M-QUATERPHENYL	9.08	0.598	0.06225
1,3,5-TRIPHENYL BENZENE	10.61	0.374	0.15725
1,3-DIBIPHENYLYLBENZENE (B)	21.31	0.046	10.26453
1,3,5,-TRIBIPHENYLYLBENZENE (B)	24.31	0.031	22.63687
FLUORENE (TRACE)	14.02	0.162	0.84319
1-METHYL FLUORENE	13.33	0.189	0.62213
1,2-BENZOFLUORENE	19.85	0.057	6.68320
9,10-DIHYDROPHENANTHRENE	13.53	0.181	0.67825
CARBAZOLE (TRACE)	17.05	0.090	2.68854
N-METHYL CARBAZOLE (TRACE)	17.17	0.088	2.80108

ACCEPTOR MOLECULES	RO	CO	OVERLAP
N-VINYLCARBAZOLE	17.37	0.085	3.00657
N-PHENYL CARBAZOLE	17.47	0.084	3.10395
N-METHYL-2,3-BENZCARBAZOLE	19.94	0.056	6.78585
DIBENZOFURAN	13.73	0.173	0.74361
DIBENZOTHIOPHENE (TRACE)	15.33	0.124	1.42155
P-TERPHENYL	17.06	0.090	2.74313
DEUTERATED P-TERPHENYL-C18D14	16.82	0.094	2.50520
4-METHYL TERPHENYL	18.16	0.075	3.95304
2-METHYL-P-TERPHENYL	12.68	0.219	0.46088
3,3″-DIMETHYL-P-TERPHENYL	17.73	0.080	3.42335
4-(3,3-DIMETHYLBUTOXY)-P-TERPHENYL	20.40	0.053	7.93148
4,4″-DIHEXAHYDROFARNESOXY-P-TERPHENYL	22.30	0.040	13.50102
2,2′-METHYLENE-P-TERPHENYL	21.68	0.044	11.41208
2,2′-ETHYLENE-P-TERPHENYL	21.73	0.044	11.57508
2,2′-METHYLENE-5′,6″-METHYLENE-P-TERPHENYL (TRACE)	24.37	0.031	22.94240
3-PHENYL DIBENZOFURAN	22.25	0.041	13.33320
BENZO(1,2-B-4,5-B′)BISBENZOFURAN (TRACE) (B)	25.25	0.028	28.35286
SEXIPHENYL BENZENE	9.89	0.462	0.10437
P-QUATERPHENYL	22.77	0.038	15.29596
3,3‴-DIMETHYL-P-QUATERPHENYL	23.44	0.035	18.14806
TETRAMETHYL-P-QUATERPHENYL	14.66	0.142	1.09769
3,3‴-DIMETHYL-3′,2″-METHYLENE-P-QUATERPHENYL	26.10	0.025	34.52647
3,3‴-DIMETHYL-3′,2″-ETHYLENE-P-QUATERPHENYL	25.95	0.026	33.30015
4-(3-ETHYLBUTOXY)-P-QUATERPHENYL (B)	25.17	0.028	27.75869
4,4‴-DI(2-BUTYLOCTOXY-1)-P-QUATERPHENYL	25.93	0.026	33.18185
DIETHYL-P-QUINQUEPHENYL	24.48	0.030	23.56011
DI(3-ETHYLHEPTYL)-P-QUINQUEPHENYL	26.76	0.023	39.99070
TETRAMETHYL-P-SEXIPHENYL	24.27	0.031	22.41194
TETRAMETHYL-P-OCTAPHENYL (B)	28.42	0.019	57.34754
INDOLE (TRACE)	4.20	6.058	0.00061
2-PHENYL INDOLE	22.95	0.037	16.00024
1-METHYL-2-PHENYL INDOLE	20.31	0.053	7.68454
1-NEOPENTYL-2-PHENYL INDOLE	19.52	0.060	6.07461
1-N-BUTYL-2(BIPHENYLYL-4) INDOLE	23.62	0.034	18.90748
1,2-DIPHENYL INDOLE	21.07	0.048	9.59094
1-BIPHENYLYL-2-PHENYL INDOLE	22.35	0.040	13.58577
1,3-DIMETHYL-2-PHENYL INDOLE	19.95	0.056	6.90496
2,3-DIPHENYL INDOLE	22.10	0.041	12.68035
1-METHYL-2,3-DIPHENYL INDOLE	20.48	0.052	8.04320
2-PHENYL-3,2′-METHYLENE INDOLE (TRACE)	23.46	0.035	18.24293
2-PHENYL-3,2′-ETHYLENE INDOLE (TRACE)	23.59	0.034	18.75975
2-PHENYL-3,2′-TRIMETHYLENE INDOLE	23.21	0.036	17.08194
1-METHYL-3,2′-METHYLENE-2-PHENYL INDOLE (TRACE)	23.73	0.033	19.46855
1-METHYL-3,2′-ETHYLENE-2-PHENYL INDOLE	23.39	0.035	17.80690
1-METHYL-3,2′-TRIMETHYLENE-2-PHENYL INDOLE	22.08	0.042	12.66192
4,5-DIPHENYLIMIDAZOLE (A)	16.76	0.095	2.42870
PPF (TRACE)	25.00	0.029	26.63298
PPO (TRACE)	24.13	0.032	21.61278
2-(4-FLUOROPHENYL)-5-PHENYLOXAZOLE (TRACE)	23.12	0.036	16.71417
αNPO (TRACE)	24.00	0.032	20.72065
BBO (B)	26.72	0.023	39.37096
POPOP	24.97	0.029	26.19203
DIMETHYL POPOP (TRACE)	25.24	0.028	27.91518
PPD (TRACE)	18.12	0.075	3.91122
2,5-DI(2-METHYLPHENYL)OXADIAZOLE	18.14	0.075	3.94011
PBD	24.54	0.030	23.96323
BBD (B)	26.94	0.023	41.70774
αNPD (TRACE)	22.95	0.037	15.90914
βNPD	23.11	0.036	16.68532
αNND (TRACE)	23.17	0.036	16.80720
βNND (TRACE)	24.92	0.029	26.14526
TRANS-1-(4-BIPHENYLYL)-2-PHENYL ETHYLENE	26.21	0.025	35.23796
TRANS-1-(4-BIPHENYLYL)-2-(1-NAPHTHYL)-ETHYLENE	25.35	0.027	28.68071
DIPHENYLSTILBENE (B)	27.03	0.023	42.12203
1,1,4,4-TETRAPHENYLBUTADIENE	24.91	0.029	25.81753
1,6-DIPHENYLHEXATRIENE	26.80	0.023	40.01546
1,8-DIPHENYL-1,3,5,7-OCTATETRAENE	24.78	0.029	24.98345
BPSB (TRACE)	26.91	0.023	40.97033
BIS-MSB	26.82	0.023	40.17642
ESCULIN (A)	21.51	0.045	10.72830
AZULENE	16.64	0.097	2.29465
NAPHTHALENE	9.05	0.604	0.06055
DEUTERATED NAPHTHALENE-C10D8	10.16	0.427	0.12046
1-METHYLNAPHTHALENE 5ML/L	11.58	0.288	0.26547
2-METHYL NAPHTHALENE	10.80	0.355	0.17502
2,3-DIMETHYL NAPHTHALENE	11.14	0.323	0.21087
2,6-DIMETHYL NAPHTHALENE (TRACE)	12.50	0.229	0.41910
1-NAPHTHOL (TRACE)	15.43	0.122	1.48533
2-NAPHTHOL (TRACE)	14.61	0.143	1.06739
1-NAPHTHYLAMINE	18.27	0.073	4.05836
2-CHLORONAPHTHALENE	10.77	0.358	0.17187
ACENAPHTHENE	14.61	0.144	1.07052
1-PHENYL NAPHTHALENE	15.99	0.109	1.84197
2-PHENYL NAPHTHALENE	23.50	0.034	18.50027
1,4-DIPHENYL NAPHTHALENE	20.10	0.055	7.20925
1,5-DIPHENYL NAPHTHALENE	19.34	0.062	5.73323
1,7-DIPHENYL NAPHTHALENE	19.09	0.064	5.29949
1,4,5,8-TETRAPHENYL NAPHTHALENE	23.08	0.036	16.29935
1,1′-BINAPHTHYL	17.16	0.088	2.80855
4,4′-DI(N-BUTOXY)-1,1′-BINAPHTHYL	21.88	0.043	11.98009
2,2′-BINAPHTHYL	21.20	0.047	9.95156
ACRIDINE (A)	18.48	0.071	4.29366
ANTHRACENE (TRACE)	17.36	0.086	2.95687
DEUTERATED ANTHRACENE C14D10 (TRACE)	17.27	0.087	2.87569
9-METHYL ANTHRACENE (TRACE)	16.55	0.099	2.22013
1-AMINO ANTHRACENE	15.62	0.117	1.57074

ACCEPTOR MOLECULES	RO	CO	OVERLAP
2-AMINO ANTHRACENE	16.48	0.100	2.14577
9-VINYL ANTHRACENE	16.99	0.091	2.59936
9-PHENYL ANTHRACENE (TRACE)	17.47	0.084	3.07382
9,10-DIPHENYL ANTHRACENE	17.16	0.089	2.75534
9,10-DICHLOROANTHRACENE (TRACE)	16.24	0.104	1.97864
BIANTHRANYL (B)	18.51	0.071	4.33477
9,10-DINAPHTHYLANTHRACENE (B)	[illegible]	0.072	4.22132
9-(P-VINYLPHENYL)-10-PHENYL ANTHRACENE	17.59	0.082	3.19997
9,10-BIS(PHENYL-ETHYNYL)ANTHRACENE (TRACE)	21.45	0.045	10.50056
RUBICENE (B)	22.78	0.038	15.10051
DIBENZ(A,H)ANTHRACENE (B)	25.33	0.028	28.49595
TETRACENE (TRACE) (B)	17.69	0.081	3.30552
RUBRENE (B)	26.46	0.024	37.07758
TETRAHYDROCANNABINOL ACID METHYL ESTER	15.26	0.126	1.39745
PHENANTHRENE	12.50	0.229	0.41712
3,4-BENZOPHENANTHRENE	19.44	0.061	5.83197
2-PHENYL PHENANTHRENE	18.58	0.070	4.47809
CHRYSENE (TRACE)	20.01	0.056	6.97969
TRIPHENYLENE	13.59	0.178	0.68830
DEUTERATED TRIPHENYLENE-C18D12	13.49	0.182	0.65869
PYRENE	22.73	0.038	14.95986
1,3,6,8,-TETRAPHENYL PYRENE	22.77	0.038	15.05858
BIPYRENYL (B)	25.08	0.028	26.83961
O-PHENYLENE-PYRENE	23.17	0.036	16.68067
ANTHANTHRENE (B)	23.50	0.034	18.20219
FLUORANTHENE	19.18	0.063	5.37077
3-PHENYL FLUORANTHENE	21.04	0.048	9.35587
3-AMINO FLUORANTHENE	20.99	0.048	9.23506
3,3'-BIFLUORANTHENYL (B)	23.31	0.035	17.33388
BENZO(GHI)FLUORANTHENE	19.06	0.065	5.17768
9,12-DI(N-OCTADECOXY)-BENZ(E)ACEPHENANTHRYLENE	20.87	0.049	8.92718
DECACYCLENE (B)	24.92	0.029	25.87793
10-ETHOXY-BENZO(KL)XANTHENE	18.47	0.071	4.28453
PERYLENE	15.13	0.129	1.29291
BENZO(GHI)PERYLENE (B)	22.42	0.040	13.72044
BBOT (TRACE)	24.68	0.030	24.37616
QUININE SULFATE (A)	19.89	0.057	6.76251
ACRIDINE YELLOW (A)	15.57	0.119	1.53635
ACRIDINE ORANGE (A)	17.23	0.087	2.82876
ACRIDINE RED (A)	17.88	0.078	3.53282
URANINE (A)	18.11	0.075	3.81159
RHODAMINE B (A)	20.28	0.054	7.50761
RHODAMINE 6G (A)	19.69	0.059	6.29118
EOSIN B (A)	19.78	0.058	6.46287
EOSIN Y (A)	20.26	0.054	7.47065
ROSE BENGAL (A)	20.30	0.053	7.55940
INDENOINDENE	23.21	0.036	17.10317

8.15 Donor Molecule: Phenylcyclohexane (trace)

BENZENE (3 ML/L)	4.20	6.027	0.00030
DEUTERATED BENZENE-C6D6 5ML/L	3.80	8.121	0.00016
FLUOROBENZENE (TRACE)	7.83	0.932	0.01235
TOLUENE (TRACE)	7.35	1.125	0.00845
DEUTERATED TOLUENE-C7D8 5ML/L	6.86	1.388	0.00555
ETHYLBENZENE (TRACE)	7.20	1.201	0.00742
PROPYLBENZENE (TRACE)	6.95	1.334	0.00601
ISOPROPYLBENZENE (TRACE)	6.62	1.538	0.00452
SEC-BUTYLBENZENE (TRACE)	6.62	1.543	0.00450
DIPHENYLMETHANE (TRACE)	8.35	0.768	0.01813
BIBENZYL	7.89	0.910	0.01290
1,3,5-TRIPHENYLPENTANE 0.05ML/L	8.08	0.849	0.01480
POLYSTYRENE	7.18	1.207	0.00734
POLYVINYLTOLUENE	9.60	0.506	0.04125
PHENYLCYCLOHEXANE (TRACE)	6.59	1.561	0.00440
BENZYLALCOHOL 3ML/L	6.38	1.719	0.00362
BENZYLACETATE 5ML/L	6.38	1.724	0.00361
PHENOL (TRACE)	13.41	0.186	0.30878
ANISOLE (TRACE)	13.69	0.174	0.35033
PHENYL ETHER	14.28	0.154	0.44992
ANILINE 3ML/L	15.83	0.113	0.82834
N,N-DIMETHYLANILINE 3ML/L	16.91	0.093	1.21944
TRIPHENYLAMINE	24.79	0.029	12.07014
P-XYLENE (TRACE)	10.38	0.400	0.06658
DEUTERATED P-XYLENE C8D10 (TRACE)	10.36	0.403	0.06569
M-XYLENE (TRACE)	8.63	0.695	0.02206
O-XYLENE 3ML/L	8.31	0.779	0.01760
1-METHYL-4-ETHYLBENZENE (TRACE)	9.79	0.476	0.04709
1,4-BIS(TRIFLUOROMETHYL)BENZENE 5ML/L	8.31	0.779	0.01763
P-CRESOL	15.56	0.119	0.75155
P-METHYLANISOLE 3ML/L	15.03	0.132	0.61098
1,4-DIMETHOXY BENZENE	17.12	0.089	1.32161
1-METHYL-4-CHLOROBENZENE 5ML/L	10.61	0.375	0.07579
SODIUM SALICYLATE (A)	18.00	0.077	1.76943
ANTHRANILIC ACID (A)	15.02	0.132	0.59677
TMPD	20.90	0.049	4.32460
1,2,4-TRIMETHYLBENZENE (TRACE)	11.16	0.322	0.10261
MESITYLENE 5ML/L	8.62	0.697	0.02193
1,3,5-TRIETHYLBENZENE 5ML/L	8.33	0.775	0.01774
STYRENE	13.39	0.186	0.30417

ACCEPTOR MOLECULES	RO	CO	OVERLAP
1,1-DIPHENYLETHYLENE 3ML/L	15.65	0.117	0.77891
BIPHENYL	15.24	0.126	0.66669
DEUTERATED BIPHENYL-C12D10	15.30	0.125	0.68160
4-METHYL BIPHENYL	17.24	0.087	1.39081
MONOISOPROPYLBIPHENYL 5ML/L	16.73	0.095	1.16455
4-BENZYL BIPHENYL	18.91	0.066	2.41808
3,3'-DIMETHYL BIPHENYL	16.20	0.105	0.95613
2,2'-DIMETHYLBIPHENYL 2ML/L	11.19	0.319	0.10403
OCTAMETHYL BIPHENYL	12.64	0.221	0.21646
4-METHOXY BIPHENYL	20.92	0.049	4.41789
4-BIPHENYLYL PHENYL ETHER	21.51	0.045	5.21953
4,4'-DIHYDROXYBIPHENYL (A)	22.77	0.038	7.29259
4,4'-DIMETHOXY BIPHENYL	22.68	0.038	7.13835
4,4'-DIISOBUTOXYBIPHENYL	22.47	0.039	6.76345
BENZIDINE	25.97	0.026	15.95379
4-VINYL BIPHENYL	25.32	0.028	13.81343
M-TERPHENYL	20.02	0.056	3.37384
M-QUATERPHENYL	20.49	0.052	3.90332
1,3,5-TRIPHENYL BENZENE	21.44	0.045	5.06330
1,3-DIBIPHENYLYLBENZENE (B)	28.74	0.019	29.27650
1,3,5,-TRIBIPHENYLYLBENZENE (B)	31.51	0.014	50.85801
FLUORENE (TRACE)	21.01	0.048	4.51786
1-METHYL FLUORENE	18.23	0.074	1.91115
1,2-BENZOFLUORENE	23.03	0.037	7.73865
9,10-DIHYDROPHENANTHRENE	21.38	0.046	5.01733
CARBAZOLE (TRACE)	22.30	0.040	6.39031
N-METHYL CARBAZOLE (TRACE)	21.72	0.044	5.45261
N-VINYLCARBAZOLE	23.36	0.035	8.45222
N-PHENYL CARBAZOLE	23.46	0.035	8.66501
N-METHYL-2,3-BENZCARBAZOLE	27.75	0.021	23.70072
DIBENZOFURAN	22.48	0.039	6.79618
DIBENZOTHIOPHENE (TRACE)	20.06	0.055	3.38997
P-TERPHENYL	26.05	0.025	16.33912
DEUTERATED P-TERPHENYL-C18D14	25.75	0.026	15.25123
4-METHYL TERPHENYL	26.33	0.024	17.42467
2-METHYL-P-TERPHENYL	23.23	0.036	8.25828
3,3''-DIMETHYL-P-TERPHENYL	25.91	0.026	15.78011
4-(3,3-DIMETHYLBUTOXY)-P-TERPHENYL	26.74	0.023	18.99669
4,4''-DIHEXAHYDROFARNESOXY-P-TERPHENYL	27.62	0.021	23.08451
2,2'-METHYLENE-P-TERPHENYL	26.43	0.024	17.75609
2,2'-ETHYLENE-P-TERPHENYL	26.39	0.024	17.57134
2,2'-METHYLENE-5',6''-METHYLENE-P-TERPHENYL (TRACE)	25.39	0.027	13.91913
3-PHENYL DIBENZOFURAN	25.32	0.028	13.71732
BENZO(1,2-B-4,5-B')BISBENZOFURAN (TRACE) (B)	22.02	0.042	5.90045
SEXIPHENYL BENZENE	21.07	0.048	4.61549
P-QUATERPHENYL	27.44	0.022	22.19574
3,3'''-DIMETHYL-P-QUATERPHENYL	27.69	0.021	23.43560
TETRAMETHYL-P-QUATERPHENYL	25.47	0.027	14.30675
3,3''' -DIMETHYL-3',2''-METHYLENE-P-QUATERPHENYL	26.28	0.025	17.10478
3,3''' -DIMETHYL-3',2''-ETHYLENE-P-QUATERPHENYL	27.41	0.022	22.03996
4-(3-ETHYLBUTOXY)-P-QUATERPHENYL (B)	26.97	0.023	19.98916
4,4'''-DI(2-BUTYLOCTOXY-1)-P-QUATERPHENYL	28.03	0.020	25.19688
DIETHYL-P-QUINQUEPHENYL	29.10	0.018	31.54007
DI(3-ETHYLHEPTYL)-P-QUINQUEPHENYL	28.50	0.019	27.85061
TETRAMETHYL-P-SEXIPHENYL	29.62	0.017	35.08484
TETRAMETHYL-P-OCTAPHENYL (B)	28.94	0.018	30.50213
INDOLE (TRACE)	17.44	0.084	1.48958
2-PHENYL INDOLE	24.19	0.032	10.39867
1-METHYL-2-PHENYL INDOLE	23.64	0.034	9.07384
1-NEOPENTYL-2-PHENYL INDOLE	23.58	0.034	8.94285
1-N-BUTYL-2(BIPHENYLYL-4) INDOLE	24.65	0.030	11.65261
1,2-DIPHENYL INDOLE	24.49	0.030	11.21425
1-BIPHENYLYL-2-PHENYL INDOLE	26.22	0.025	16.88974
1,3-DIMETHYL-2-PHENYL INDOLE	22.67	0.038	7.05232
2,3-DIPHENYL INDOLE	23.40	0.035	8.52392
1-METHYL-2,3-DIPHENYL INDOLE	23.26	0.036	8.22368
2-PHENYL-3,2'-METHYLENE INDOLE (TRACE)	22.56	0.039	6.84585
2-PHENYL-3,2'-ETHYLENE INDOLE (TRACE)	22.01	0.042	5.90381
2-PHENYL-3,2'-TRIMETHYLENE INDOLE	23.14	0.036	7.96620
1-METHYL-3,2'-METHYLENE-2-PHENYL INDOLE (TRACE)	22.10	0.041	6.05622
1-METHYL-3,2'-ETHYLENE-2-PHENYL INDOLE	21.65	0.044	5.35391
1-METHYL-3,2'-TRIMETHYLENE-2-PHENYL INDOLE	22.44	0.040	6.63303
4,5-DIPHENYLIMIDAZOLE (A)	22.08	0.042	6.01562
PPF (TRACE)	23.65	0.034	9.08959
PPO (TRACE)	25.83	0.026	15.41665
2-(4-FLUOROPHENYL)-5-PHENYLOXAZOLE (TRACE)	25.01	0.029	12.70641
αNPO (TRACE)	23.14	0.036	7.96734
BBO (B)	24.42	0.031	11.01467
POPOP	22.55	0.039	6.83374
DIMETHYL POPOP (TRACE)	23.69	0.034	9.18499
PPD (TRACE)	25.73	0.026	15.18902
2,5-DI(2-METHYLPHENYL)OXADIAZOLE	25.06	0.028	12.96124
PBD	27.63	0.021	23.09390
BBD (B)	26.85	0.023	19.44522
αNPD (TRACE)	23.22	0.036	8.14887
βNPD	26.86	0.023	19.49071
αNND (TRACE)	21.63	0.044	5.31660
βNND (TRACE)	26.97	0.023	19.99034
TRANS-1-(4-BIPHENYLYL)-2-PHENYL-ETHYLENE	25.57	0.027	14.51187
TRANS-1-(4-BIPHENYLYL)-2-(1-NAPHTHYL)-ETHYLENE	24.30	0.031	10.69558
DIPHENYLSTILBENE (B)	23.12	0.036	7.93661
1.1.4.4-TETRAPHENYLBUTADIENE	21.40	0.046	4.98697
1,6-DIPHENYLHEXATRIENE	21.04	0.048	4.50205
1,8-DIPHENYL-1,3,5,7-OCTATETRAENE	19.82	0.057	3.15032
BPSB (TRACE)	22.02	0.042	5.92878
BIS-MSB	22.88	0.037	7.44474

ACCEPTOR MOLECULES	RO	CO	OVERLAP
ESCULIN (A)	20.30	0.053	3.63382
AZULENE	25.52	0.027	14.35666
NAPHTHALENE	18.13	0.075	1.85690
DEUTERATED NAPHTHALENE-C10D8	18.27	0.073	1.93500
1-METHYLNAPHTHALENE 5ML/L	19.34	0.062	2.72763
2-METHYL NAPHTHALENE	18.31	0.073	1.96496
2,3-DIMETHYL NAPHTHALENE	18.77	0.068	2.27818
2,6-DIMETHYL NAPHTHALENE (TRACE)	18.50	0.071	2.08942
1-NAPHTHOL (TRACE)	19.54	0.060	2.89347
2-NAPHTHOL (TRACE)	17.93	0.078	1.72791
1-NAPHTHYLAMINE	18.02	0.076	1.77678
2-CHLORONAPHTHALENE	18.31	0.073	1.96530
ACENAPHTHENE	20.14	0.055	3.47889
1-PHENYL NAPHTHALENE	21.91	0.043	5.76560
2-PHENYL NAPHTHALENE	32.54	0.013	61.67659
1,4-DIPHENYL NAPHTHALENE	22.77	0.038	7.23826
1,5-DIPHENYL NAPHTHALENE	22.42	0.040	6.60350
1,7-DIPHENYL NAPHTHALENE	22.55	0.039	6.82396
1,4,5,8-TETRAPHENYL NAPHTHALENE	21.98	0.042	5.85633
1,1'-BINAPHTHYL	23.10	0.036	7.91491
4,4'-DI(N-BUTOXY)-1,1'-BINAPHTHYL	22.63	0.039	6.97756
2,2'-BINAPHTHYL	25.52	0.027	14.35703
ACRIDINE (A)	14.90	0.135	0.56955
ANTHRACENE (TRACE)	13.21	0.194	0.27653
DEUTERATED ANTHRACENE C14D10 (TRACE)	13.93	0.165	0.37965
9-METHYL ANTHRACENE (TRACE)	13.61	0.177	0.33052
1-AMINO ANTHRACENE	22.31	0.040	6.40983
2-AMINO ANTHRACENE	11.63	0.285	0.12571
9-VINYL ANTHRACENE	15.50	0.120	0.72021
9-PHENYL ANTHRACENE (TRACE)	16.30	0.103	0.97571
9,10-DIPHENYL ANTHRACENE	19.16	0.064	2.57060
9,10-DICHLOROANTHRACENE (TRACE)	14.47	0.148	0.47733
BIANTHRANYL (B)	22.58	0.039	6.89267
9,10-DINAPHTHYLANTHRACENE (B)	23.72	0.034	9.25630
9-(P-VINYLPHENYL)-10-PHENYL ANTHRACENE	20.80	0.050	4.20538
9,10-BIS(PHENYL-ETHYNYL)ANTHRACENE (TRACE)	27.12	0.022	20.68834
RUBICENE (B)	27.06	0.023	20.39959
DIBENZ(A,H)ANTHRACENE (B)	29.56	0.017	34.67977
TETRACENE (TRACE) (B)	32.25	0.013	58.49670
RUBRENE (B)	29.01	0.018	30.94913
TETRAHYDROCANNABINOL ACID METHYL ESTER	23.85	0.033	9.62173
PHENANTHRENE	21.66	0.044	5.37088
3,4-BENZOPHENANTHRENE	27.65	0.021	23.23574
2-PHENYL PHENANTHRENE	27.82	0.021	24.08572
CHRYSENE (TRACE)	25.29	0.028	13.58380
TRIPHENYLENE	23.22	0.036	8.14759
DEUTERATED TRIPHENYLENE-C18D12	23.31	0.035	8.34231
PYRENE	22.44	0.040	6.62905
1,3,6,8,-TETRAPHENYL PYRENE	26.71	0.023	18.84778
BIPYRENYL (B)	27.27	0.022	21.36227
O-PHENYLENE-PYRENE	26.10	0.025	16.42803
ANTHANTHRENE (B)	25.60	0.027	14.62685
FLUORANTHENE	23.60	0.034	8.98271
3-PHENYL FLUORANTHENE	24.78	0.029	12.03981
3-AMINO FLUORANTHENE	23.01	0.037	7.71219
3,3'-BIFLUORANTHENYL (B)	26.50	0.024	17.97914
BENZO(GHI)FLUORANTHENE	22.84	0.038	7.37277
9,12-DI(N-OCTADECOXY)-BENZ(E)ACEPHENANTHRYLENE	26.58	0.024	18.32049
DECACYCLENE (B)	25.93	0.026	15.80378
10-ETHOXY-BENZO(KL)XANTHENE	19.43	0.061	2.79815
PERYLENE	16.96	0.092	1.23617
BENZO(GHI)PERYLENE (B)	26.86	0.023	19.52280
BBOT (TRACE)	20.34	0.053	3.67490
QUININE SULFATE (A)	20.75	0.050	4.14416
ACRIDINE YELLOW (A)	23.54	0.034	8.84075
ACRIDINE ORANGE (A)	23.79	0.033	9.42314
ACRIDINE RED (A)	20.61	0.051	3.98526
URANINE (A)	20.68	0.051	4.06749
RHODAMINE B (A)	23.73	0.033	9.27552
RHODAMINE 6G (A)	23.72	0.034	9.25686
EOSIN B (A)	21.24	0.047	4.77522
EOSIN Y (A)	21.87	0.043	5.68203
ROSE BENGAL (A)	22.59	0.039	6.90350
INDENOINDENE	23.25	0.036	8.19968

8.16 Donor Molecule: Benzyl Alcohol (3 ml/L)

BENZENE (3 ML/L)	3.20	12.857	0.00012
DEUTERATED BENZENE-C6D6 5ML/L	2.73	22.001	0.00004
FLUOROBENZENE (TRACE)	5.71	2.408	0.00346
TOLUENE (TRACE)	6.01	2.060	0.00471
DEUTERATED TOLUENE-C7D8 5ML/L	5.42	2.804	0.00255
ETHYLBENZENE (TRACE)	5.86	2.222	0.00405
PROPYLBENZENE (TRACE)	5.61	2.535	0.00312
ISOPROPYLBENZENE (TRACE)	5.25	3.099	0.00209
SEC-BUTYLBENZENE (TRACE)	5.26	3.070	0.00213
DIPHENYLMETHANE (TRACE)	6.94	1.340	0.01113
BIBENZYL	6.41	1.696	0.00696
1,3,5-TRIPHENYLPENTANE 0.05ML/L	6.58	1.571	0.00810
POLYSTYRENE	5.89	2.194	0.00416
POLYVINYLTOLUENE	8.34	0.771	0.03318

ACCEPTOR MOLECULES	RO	CO	OVERLAP
PHENYLCYCLOHEXANE (TRACE)	5.16	3.247	0.00190
BENZYLALCOHOL 3ML/L	5.14	3.287	0.00185
BENZYLACETATE 5ML/L	5.08	3.419	0.00171
PHENOL (TRACE)	11.78	0.274	0.26532
ANISOLE (TRACE)	12.07	0.254	0.30755
PHENYL ETHER	12.64	0.221	0.40481
ANILINE 3ML/L	14.27	0.154	0.82963
N,N-DIMETHYLANILINE 3ML/L	15.20	0.127	1.20569
TRIPHENYLAMINE	22.53	0.039	12.66237
P-XYLENE (TRACE)	9.02	0.609	0.05374
DEUTERATED P-XYLENE C8D10 (TRACE)	9.00	0.614	0.05280
M-XYLENE (TRACE)	7.43	1.091	0.01678
O-XYLENE 3ML/L	7.08	1.263	0.01251
1-METHYL-4-ETHYLBENZENE (TRACE)	8.49	0.731	0.03740
1,4-BIS(TRIFLUOROMETHYL)BENZENE 5ML/L	6.78	1.433	0.00975
P-CRESOL	13.91	0.166	0.71830
P-METHYLANISOLE 3ML/L	13.44	0.184	0.58466
1,4-DIMETHOXY BENZENE	15.47	0.121	1.34738
1-METHYL-4-CHLOROBENZENE 5ML/L	9.32	0.552	0.06523
SODIUM SALICYLATE (A)	16.35	0.102	1.85207
ANTHRANILIC ACID (A)	13.98	0.164	0.72395
TMPD	18.60	0.069	4.01253
1,2,4-TRIMETHYLBENZENE (TRACE)	9.76	0.481	0.08583
MESITYLENE 5ML/L	7.47	1.072	0.01734
1,3,5-TRIETHYLBENZENE 5ML/L	7.17	1.215	0.01350
STYRENE	11.87	0.267	0.27678
1,1-DIPHENYLETHYLENE 3ML/L	13.70	0.174	0.65543
BIPHENYL	13.22	0.194	0.53098
DEUTERATED BIPHENYL-C12D10	13.25	0.192	0.53898
4-METHYL BIPHENYL	15.09	0.130	1.17014
MONOISOPROPYLBIPHENYL 5ML/L	14.62	0.143	0.96750
4-BENZYL BIPHENYL	16.58	0.098	2.05988
3,3'-DIMETHYL BIPHENYL	14.26	0.154	0.83073
2,2'-DIMETHYLBIPHENYL 2ML/L	9.71	0.488	0.08330
OCTAMETHYL BIPHENYL	11.20	0.319	0.19530
4-METHOXY BIPHENYL	18.55	0.070	4.01263
4-BIPHENYLYL PHENYL ETHER	19.05	0.065	4.70854
4,4'-DIHYDROXYBIPHENYL (A)	20.30	0.053	6.85470
4,4'-DIMETHOXY BIPHENYL	20.21	0.054	6.67374
4,4'-DIISOBUTOXYBIPHENYL	20.04	0.056	6.35502
BENZIDINE	23.43	0.035	16.04009
4-VINYL BIPHENYL	22.69	0.038	13.36939
M-TERPHENYL	17.64	0.081	2.95670
M-QUATERPHENYL	17.95	0.077	3.30548
1,3,5-TRIPHENYL BENZENE	18.79	0.067	4.29847
1,3-DIBIPHENYLYLBENZENE (B)	25.88	0.026	29.22477
1,3,5,-TRIBIPHENYLYLBENZENE (B)	28.43	0.019	51.38231
FLUORENE (TRACE)	18.69	0.069	4.19256
1-METHYL FLUORENE	16.55	0.099	2.00572
1,2-BENZOFLUORENE	20.86	0.049	7.97962
9,10-DIHYDROPHENANTHRENE	19.02	0.065	4.65264
CARBAZOLE (TRACE)	20.17	0.054	6.53740
N-METHYL CARBAZOLE (TRACE)	19.61	0.059	5.50930
N-VINYLCARBAZOLE	20.98	0.048	8.25301
N-PHENYL CARBAZOLE	21.16	0.047	8.69785
N-METHYL-2,3-BENZCARBAZOLE	24.84	0.029	22.73470
DIBENZOFURAN	20.22	0.054	6.72735
DIBENZOTHIOPHENE (TRACE)	18.05	0.076	3.35779
P-TERPHENYL	23.35	0.035	15.86215
DEUTERATED P-TERPHENYL-C18D14	23.08	0.036	14.79132
4-METHYL TERPHENYL	23.65	0.034	17.10841
2-METHYL-P-TERPHENYL	20.68	0.051	7.68059
3,3''-DIMETHYL-P-TERPHENYL	23.25	0.036	15.44321
4-(3,3-DIMETHYLBUTOXY)-P-TERPHENYL	24.10	0.032	19.08373
4,4''-DIHEXAHYDROFARNESOXY-P-TERPHENYL	24.96	0.029	23.49209
2,2'-METHYLENE-P-TERPHENYL	23.88	0.033	18.07164
2,2'-ETHYLENE-P-TERPHENYL	23.86	0.033	17.92309
2,2'-METHYLENE-5',6''-METHYLENE-P-TERPHENYL (TRACE)	23.13	0.036	14.86057
3-PHENYL DIBENZOFURAN	22.96	0.037	14.27829
BENZO(1,2-B-4,5-B')BISBENZOFURAN (TRACE) (B)	20.42	0.053	6.99888
SEXIPHENYL BENZENE	18.74	0.068	4.26551
P-QUATERPHENYL	24.83	0.029	22.76462
3,3'''-DIMETHYL-P-QUATERPHENYL	25.08	0.028	24.12930
TETRAMETHYL-P-QUATERPHENYL	22.70	0.038	13.41275
3,3''' -DIMETHYL-3',2''-METHYLENE-P-QUATERPHENYL	24.01	0.032	18.53452
3,3''' -DIMETHYL-3',2''-ETHYLENE-P-QUATERPHENYL	24.91	0.029	23.14509
4-(3-ETHYLBUTOXY)-P-QUATERPHENYL (B)	24.55	0.030	21.18297
4,4'''-DI(2-BUTYLOCTOXY-1)-P-QUATERPHENYL	25.50	0.027	26.60076
DIETHYL-P-QUINQUEPHENYL	26.34	0.024	32.41599
DI(3-ETHYLHEPTYL)-P-QUINQUEPHENYL	25.95	0.026	29.56262
TETRAMETHYL-P-SEXIPHENYL	26.78	0.023	35.78358
TETRAMETHYL-P-OCTAPHENYL (B)	26.36	0.024	32.48293
INDOLE (TRACE)	15.47	0.121	1.35772
2-PHENYL INDOLE	22.04	0.042	11.12354
1-METHYL-2-PHENYL INDOLE	21.44	0.045	9.43439
1-NEOPENTYL-2-PHENYL INDOLE	21.35	0.046	9.22657
1-N-BUTYL-2(BIPHENYLYL-4) INDOLE	22.38	0.040	12.17416
1,2-DIPHENYL INDOLE	22.16	0.041	11.49589
1-BIPHENYLYL-2-PHENYL INDOLE	23.70	0.034	17.16222
1,3-DIMETHYL-2-PHENYL INDOLE	20.57	0.051	7.34197
2,3-DIPHENYL INDOLE	21.26	0.047	8.93338
1-METHYL-2,3-DIPHENYL INDOLE	21.01	0.048	8.32649
2-PHENYL-3,2'-METHYLENE INDOLE (TRACE)	20.68	0.051	7.57231
2-PHENYL-3,2'-ETHYLENE INDOLE (TRACE)	20.17	0.054	6.52757
2-PHENYL-3,2'-TRIMETHYLENE INDOLE	21.16	0.047	8.69049
1-METHYL-3,2'-METHYLENE-2-PHENYL INDOLE (TRACE)	20.27	0.054	6.71293

ACCEPTOR MOLECULES	RO	CO	OVERLAP
1-METHYL-3,2′-ETHYLENE-2-PHENYL INDOLE	19.83	0.057	5.89217
1-METHYL-3,2′-TRIMETHYLENE-2-PHENYL INDOLE	20.48	0.052	7.14686
4,5-DIPHENYLIMIDAZOLE (A)	19.88	0.057	6.00054
PPF (TRACE)	21.70	0.044	10.11916
PPO (TRACE)	23.50	0.034	16.33044
2-(4-FLUOROPHENYL)-5-PHENYLOXAZOLE (TRACE)	22.73	0.038	13.39240
αNPO (TRACE)	21.06	0.048	8.45457
BBO (B)	22.35	0.040	[illegible]
POPOP	20.64	0.051	7.48485
DIMETHYL POPOP (TRACE)	21.57	0.045	9.74935
PPD (TRACE)	23.13	0.036	15.01288
2,5-DI(2-METHYLPHENYL)OXADIAZOLE	22.54	0.039	12.81082
PBD	25.08	0.028	24.13885
BBD (B)	24.50	0.030	20.92574
αNPD (TRACE)	21.09	0.048	8.51082
βNPD	24.19	0.032	19.42577
αNND (TRACE)	19.82	0.057	5.87528
βNND (TRACE)	24.29	0.031	19.89392
TRANS-1-(4-BIPHENYLYL)-2-PHENYL-ETHYLENE	23.41	0.035	15.92510
TRANS-1-(4-BIPHENYLYL)-2-(1-NAPHTHYL)-ETHYLENE	22.17	0.041	11.48844
DIPHENYLSTILBENE (B)	21.41	0.046	9.31767
1,1,4,4-TETRAPHENYLBUTADIENE	19.71	0.058	5.67391
1,6-DIPHENYLHEXATRIENE	19.69	0.059	5.64443
1,8-DIPHENYL-1,3,5,7-OCTATETRAENE	18.30	0.073	3.63130
BPSB (TRACE)	20.45	0.052	7.08415
BIS-MSB	21.15	0.047	8.66849
ESCULIN (A)	18.59	0.070	3.99940
AZULENE	22.76	0.038	13.46093
NAPHTHALENE	16.15	0.106	1.73715
DEUTERATED NAPHTHALENE-C10D8	16.28	0.104	1.81100
1-METHYLNAPHTHALENE 5ML/L	17.34	0.086	2.65256
2-METHYL NAPHTHALENE	16.35	0.102	1.86067
2,3-DIMETHYL NAPHTHALENE	16.78	0.095	2.17557
2,6-DIMETHYL NAPHTHALENE (TRACE)	16.53	0.099	1.98818
1-NAPHTHOL (TRACE)	17.64	0.082	2.93284
2-NAPHTHOL (TRACE)	16.06	0.108	1.66496
1-NAPHTHYLAMINE	16.48	0.100	1.94207
2-CHLORONAPHTHALENE	16.35	0.102	1.86008
ACENAPHTHENE	18.15	0.075	3.48397
1-PHENYL NAPHTHALENE	19.75	0.058	5.78282
2-PHENYL NAPHTHALENE	29.26	0.018	61.05240
1,4-DIPHENYL NAPHTHALENE	20.67	0.051	7.56088
1,5-DIPHENYL NAPHTHALENE	20.34	0.053	6.87190
1,7-DIPHENYL NAPHTHALENE	20.30	0.053	6.79341
1,4,5,8-TETRAPHENYL NAPHTHALENE	19.99	0.056	6.17881
1,1′-BINAPHTHYL	20.83	0.050	7.94456
4,4′-DI(N-BUTOXY)-1,1′-BINAPHTHYL	20.63	0.051	7.45705
2,2′-BINAPHTHYL	22.93	0.037	14.12829
ACRIDINE (A)	13.78	0.171	0.66222
ANTHRACENE (TRACE)	12.34	0.238	0.34263
DEUTERATED ANTHRACENE C14D10 (TRACE)	12.89	0.209	0.44475
9-METHYL ANTHRACENE (TRACE)	12.31	0.239	0.33768
1-AMINO ANTHRACENE	19.56	0.060	5.41785
2-AMINO ANTHRACENE	11.17	0.321	0.18494
9-VINYL ANTHRACENE	13.71	0.174	0.64273
9-PHENYL ANTHRACENE (TRACE)	14.41	0.149	0.86804
9,10-DIPHENYL ANTHRACENE	16.44	0.101	1.90807
9,10-DICHLOROANTHRACENE (TRACE)	12.57	0.225	0.38237
BIANTHRANYL (B)	19.29	0.062	4.98298
9,10-DINAPHTHYLANTHRACENE (B)	21.08	0.048	8.49383
9-(P-VINYLPHENYL)-10-PHENYL ANTHRACENE	18.23	0.074	3.55285
9,10-BIS(PHENYL-ETHYNYL)ANTHRACENE (TRACE)	24.19	0.032	19.38996
RUBICENE (B)	24.43	0.031	20.57948
DIBENZ(A,H)ANTHRACENE (B)	26.85	0.023	36.29905
TETRACENE (TRACE) (B)	28.80	0.019	55.27696
RUBRENE (B)	26.43	0.024	33.00017
TETRAHYDROCANNABINOL ACID METHYL ESTER	21.31	0.046	9.18034
PHENANTHRENE	19.36	0.062	5.09267
3,4-BENZOPHENANTHRENE	24.81	0.029	22.59634
2-PHENYL PHENANTHRENE	24.81	0.029	22.56040
CHRYSENE (TRACE)	22.17	0.041	11.49206
TRIPHENYLENE	20.74	0.050	7.69404
DEUTERATED TRIPHENYLENE-C18D12	20.80	0.050	7.84134
PYRENE	20.22	0.054	6.61432
1,3,6,8,-TETRAPHENYL PYRENE	24.20	0.032	19.43304
BIPYRENYL (B)	24.60	0.030	21.43140
O-PHENYLENE-PYRENE	23.61	0.034	16.78090
ANTHANTHRENE (B)	23.27	0.035	15.38634
FLUORANTHENE	21.27	0.046	8.95767
3-PHENYL FLUORANTHENE	22.41	0.040	12.25298
3-AMINO FLUORANTHENE	20.82	0.050	7.88831
3,3′-BIFLUORANTHENYL (B)	23.96	0.033	18.31056
BENZO(GHI)FLUORANTHENE	20.60	0.051	7.40694
9,12-DI(N-OCTADECOXY)-BENZ(E)ACEPHENANTHRYLENE	23.87	0.033	17.91000
DECACYCLENE (B)	23.43	0.035	16.00193
10-ETHOXY-BENZO(KL)XANTHENE	17.60	0.082	2.87353
PERYLENE	15.02	0.132	1.10957
BENZO(GHI)PERYLENE (B)	24.30	0.031	19.93527
BBOT (TRACE)	18.75	0.068	4.20287
QUININE SULFATE (A)	18.80	0.067	4.27885
ACRIDINE YELLOW (A)	20.89	0.049	8.05415
ACRIDINE ORANGE (A)	21.27	0.047	8.95648
ACRIDINE RED (A)	18.59	0.070	4.00043
URANINE (A)	18.67	0.069	4.09749
RHODAMINE B (A)	21.39	0.046	9.27758
RHODAMINE 6G (A)	21.39	0.046	9.25944

ACCEPTOR MOLECULES	RO	CO	OVERLAP
EOSIN B (A)	19.17	0.064	4.80319
EOSIN Y (A)	19.74	0.058	5.72167
ROSE BENGAL (A)	20.27	0.054	6.70601
INDENOINDENE	21.27	0.047	8.96384

8.17 Donor Molecule: Benzyl Acetate (5 ml/L)

BENZENE (3 ML/L)	3.27	12.829	0.00014
DEUTERATED BENZENE-C6D6 5ML/L	2.79	20.617	0.00005
FLUOROBENZENE (TRACE)	5.84	2.246	0.00454
TOLUENE (TRACE)	5.98	2.092	0.00523
DEUTERATED TOLUENE-C7D8 5ML/L	5.44	2.780	0.00296
ETHYLBENZENE (TRACE)	5.83	2.256	0.00450
PROPYLBENZENE (TRACE)	5.59	2.566	0.00348
ISOPROPYLBENZENE (TRACE)	5.26	3.080	0.00241
SEC-BUTYLBENZENE (TRACE)	5.26	3.073	0.00243
DIPHENYLMETHANE (TRACE)	6.91	1.357	0.01241
BIBENZYL	6.38	1.718	0.00774
1,3,5-TRIPHENYLPENTANE 0.05ML/L	6.55	1.594	0.00899
POLYSTYRENE	5.86	2.222	0.00464
POLYVINYLTOLUENE	8.22	0.805	0.03480
PHENYLCYCLOHEXANE (TRACE)	5.18	3.218	0.00221
BENZYLALCOHOL 3ML/L	5.13	3.317	0.00208
BENZYLACETATE 5ML/L	5.07	3.428	0.00195
PHENOL (TRACE)	11.65	0.283	0.28413
ANISOLE (TRACE)	11.94	0.263	0.32910
PHENYL ETHER	12.47	0.231	0.42623
ANILINE 3ML/L	13.99	0.163	0.84450
N,N-DIMETHYLANILINE 3ML/L	14.88	0.136	1.21346
TRIPHENYLAMINE	21.99	0.042	12.55210
P-XYLENE (TRACE)	8.87	0.640	0.05561
DEUTERATED P-XYLENE C8D10 (TRACE)	8.85	0.645	0.05473
M-XYLENE (TRACE)	7.35	1.129	0.01791
O-XYLENE 3ML/L	7.02	1.291	0.01370
1-METHYL-4-ETHYLBENZENE (TRACE)	8.36	0.765	0.03896
1,4-BIS(TRIFLUOROMETHYL)BENZENE 5ML/L	6.77	1.442	0.01102
P-CRESOL	13.69	0.174	0.74459
P-METHYLANISOLE 3ML/L	13.23	0.193	0.60612
1,4-DIMETHOXY BENZENE	15.16	0.128	1.36259
1-METHYL-4-CHLOROBENZENE 5ML/L	9.22	0.570	0.06996
SODIUM SALICYLATE (A)	15.97	0.110	1.84188
ANTHRANILIC ACID (A)	13.38	0.187	0.63715
TMPD	18.26	0.073	4.11006
1,2,4-TRIMETHYLBENZENE (TRACE)	9.62	0.502	0.09030
MESITYLENE 5ML/L	7.38	1.112	0.01843
1,3,5-TRIETHYLBENZENE 5ML/L	7.10	1.252	0.01452
STYRENE	11.68	0.281	0.28602
1,1-DIPHENYLETHYLENE 3ML/L	13.51	0.181	0.68920
BIPHENYL	13.06	0.201	0.56461
DEUTERATED BIPHENYL-C12D10	13.10	0.199	0.57369
4-METHYL BIPHENYL	14.89	0.136	1.23329
MONOISOPROPYLBIPHENYL 5ML/L	14.42	0.149	1.02189
4-BENZYL BIPHENYL	16.35	0.102	2.16626
3,3'-DIMETHYL BIPHENYL	14.05	0.161	0.87030
2,2'-DIMETHYLBIPHENYL 2ML/L	9.60	0.506	0.08860
OCTAMETHYL BIPHENYL	11.04	0.333	0.20483
4-METHOXY BIPHENYL	18.26	0.073	4.17256
4-BIPHENYLYL PHENYL ETHER	18.75	0.068	4.90461
4,4'-DIHYDROXYBIPHENYL (A)	19.96	0.056	7.06961
4,4'-DIMETHOXY BIPHENYL	19.87	0.057	6.89259
4,4'-DIISOBUTOXYBIPHENYL	19.70	0.059	6.55720
BENZIDINE	22.95	0.037	16.19401
4-VINYL BIPHENYL	22.28	0.040	13.69624
M-TERPHENYL	17.37	0.085	3.08500
M-QUATERPHENYL	17.70	0.081	3.47482
1,3,5-TRIPHENYL BENZENE	18.53	0.070	4.51762
1,3-DIBIPHENYLYLBENZENE (B)	25.37	0.027	29.61698
1,3,5,-TRIBIPHENYLYLBENZENE (B)	27.85	0.021	51.83766
FLUORENE (TRACE)	18.36	0.072	4.31386
1-METHYL FLUORENE	16.19	0.105	2.00358
1,2-BENZOFLUORENE	20.38	0.053	7.93909
9,10-DIHYDROPHENANTHRENE	18.70	0.068	4.80637
CARBAZOLE (TRACE)	19.75	0.058	6.57999
N-METHYL CARBAZOLE (TRACE)	19.19	0.063	5.53067
N-VINYLCARBAZOLE	20.56	0.051	8.38735
N-PHENYL CARBAZOLE	20.73	0.050	8.79287
N-METHYL-2,3-BENZCARBAZOLE	24.36	0.031	23.20194
DIBENZOFURAN	19.85	0.057	6.88507
DIBENZOTHIOPHENE (TRACE)	17.69	0.081	3.40860
P-TERPHENYL	22.93	0.037	16.23432
DEUTERATED P-TERPHENYL-C18D14	22.66	0.038	15.14065
4-METHYL TERPHENYL	23.20	0.036	17.44378
2-METHYL-P-TERPHENYL	20.34	0.053	7.94626
3,3''-DIMETHYL-P-TERPHENYL	22.82	0.038	15.75875
4-(3,3-DIMETHYLBUTOXY)-P-TERPHENYL	23.61	0.034	19.28684
4,4''-DIHEXAHYDROFARNESOXY-P-TERPHENYL	24.42	0.031	23.59478
2,2'-METHYLENE-P-TERPHENYL	23.37	0.035	18.13174
2,2'-ETHYLENE-P-TERPHENYL	23.34	0.035	17.96648
2,2'-METHYLENE-5',6''-METHYLENE-P-TERPHENYL (TRACE)	22.52	0.039	14.47616

ACCEPTOR MOLECULES	RO	CO	OVERLAP
3-PHENYL DIBENZOFURAN	22.42	0.040	14.13576
BENZO(1,2-B-4,5-B')BISBENZOFURAN (TRACE) (B)	19.62	0.059	6.31226
SEXIPHENYL BENZENE	18.44	0.071	4.42643
P-QUATERPHENYL	24.29	0.031	22.77807
3,3'''-DIMETHYL-P-QUATERPHENYL	24.52	0.030	24.08127
TETRAMETHYL-P-QUATERPHENYL	22.31	0.040	13.84507
3,3''' -DIMETHYL-3',2''-METHYLENE-P-QUATERPHENYL	23.33	0.035	17.87074
3,3''' -DIMETHYL-4',2''-ETHYLENE-P-QUATERPHENYL	24.28	0.031	22.69488
4-(3-ETHYLBUTOXY)-P-QUATERPHENYL (B)	23.92	0.033	20.76310
4,4'''-DI(2-BUTYLOCTOXY-1)-P-QUATERPHENYL	24.85	0.029	26.12933
DIETHYL-P-QUINQUEPHENYL	25.76	0.026	32.38189
DI(3-ETHYLHEPTYL)-P-QUINQUEPHENYL	25.28	0.028	28.93876
TETRAMETHYL-P-SEXIPHENYL	26.20	0.025	35.87978
TETRAMETHYL-P-OCTAPHENYL (B)	25.64	0.027	31.50019
INDOLE (TRACE)	15.24	0.126	1.42251
2-PHENYL INDOLE	21.46	0.045	10.84808
1-METHYL-2-PHENYL INDOLE	20.95	0.049	9.38114
1-NEOPENTYL-2-PHENYL INDOLE	20.88	0.049	9.22063
1-N-BUTYL-2(BIPHENYLYL-4) INDOLE	21.80	0.043	11.89091
1,2-DIPHENYL INDOLE	21.66	0.044	11.45200
1-BIPHENYLYL-2-PHENYL INDOLE	23.17	0.036	17.15726
1,3-DIMETHYL-2-PHENYL INDOLE	20.08	0.055	7.27792
2,3-DIPHENYL INDOLE	20.71	0.050	8.75291
1-METHYL-2,3-DIPHENYL INDOLE	20.53	0.052	8.29498
2-PHENYL-3,2'-METHYLENE INDOLE (TRACE)	20.04	0.056	7.17789
2-PHENYL-3,2'-ETHYLENE INDOLE (TRACE)	19.52	0.060	6.12652
2-PHENYL-3,2'-TRIMETHYLENE INDOLE	20.55	0.052	8.34262
1-METHYL-3,2'-METHYLENE-2-PHENYL INDOLE (TRACE)	19.60	0.059	6.28372
1-METHYL-3,2'-ETHYLENE-2-PHENYL INDOLE	19.18	0.063	5.52683
1-METHYL-3,2'-TRIMETHYLENE-2-PHENYL INDOLE	19.91	0.057	6.90556
4,5-DIPHENYLIMIDAZOLE (A)	19.48	0.060	6.07401
PPF (TRACE)	21.01	0.048	9.54022
PPO (TRACE)	22.90	0.037	15.99988
2-(4-FLUOROPHENYL)-5-PHENYLOXAZOLE (TRACE)	22.17	0.041	13.15689
αNPO (TRACE)	20.45	0.052	8.11842
BBO (B)	21.62	0.044	11.33205
POPOP	19.98	0.056	7.04650
DIMETHYL POPOP (TRACE)	20.94	0.049	9.35738
PPD (TRACE)	22.69	0.038	15.27719
2,5-DI(2-METHYLPHENYL)OXADIAZOLE	22.10	0.041	13.01887
PBD	24.48	0.030	23.88897
BBD (B)	23.80	0.033	20.14964
αNPD (TRACE)	20.51	0.052	8.26641
βNPD	23.66	0.034	19.44770
αNND (TRACE)	19.19	0.063	5.53355
βNND (TRACE)	23.72	0.034	19.73118
TRANS-1-(4-BIPHENYLYL)-2-PHENYL-ETHYLENE	22.71	0.038	15.20451
TRANS-1-(4-BIPHENYLYL)-2-(1-NAPHTHYL)-ETHYLENE	21.51	0.045	10.99658
DIPHENYLSTILBENE (B)	20.57	0.051	8.40843
1,1,4,4-TETRAPHENYLBUTADIENE	18.96	0.066	5.16027
1,6-DIPHENYLHEXATRIENE	18.75	0.068	4.82338
1,8-DIPHENYL-1,3,5,7-OCTATETRAENE	17.56	0.083	3.25546
BPSB (TRACE)	19.59	0.059	6.27513
BIS-MSB	20.34	0.053	7.84870
ESCULIN (A)	18.02	0.076	3.79161
AZULENE	22.38	0.040	13.92436
NAPHTHALENE	15.88	0.112	1.79515
DEUTERATED NAPHTHALENE-C10D8	16.00	0.109	1.86836
1-METHYLNAPHTHALENE 5ML/L	17.03	0.091	2.72049
2-METHYL NAPHTHALENE	16.07	0.108	1.91859
2,3-DIMETHYL NAPHTHALENE	16.49	0.100	2.24189
2,6-DIMETHYL NAPHTHALENE (TRACE)	16.24	0.104	2.04217
1-NAPHTHOL (TRACE)	17.27	0.087	2.95378
2-NAPHTHOL (TRACE)	15.74	0.115	1.68701
1-NAPHTHYLAMINE	16.00	0.109	1.85964
2-CHLORONAPHTHALENE	16.07	0.108	1.92083
ACENAPHTHENE	17.79	0.079	3.53423
1-PHENYL NAPHTHALENE	19.35	0.062	5.85966
2-PHENYL NAPHTHALENE	28.69	0.019	62.00066
1,4-DIPHENYL NAPHTHALENE	20.18	0.054	7.48786
1,5-DIPHENYL NAPHTHALENE	19.87	0.057	6.82894
1,7-DIPHENYL NAPHTHALENE	19.86	0.057	6.80005
1,4,5,8-TETRAPHENYL NAPHTHALENE	19.40	0.061	5.90621
1,1'-BINAPHTHYL	20.41	0.053	8.03973
4,4'-DI(N-BUTOXY)-1,1'-BINAPHTHYL	20.07	0.055	7.24776
2,2'-BINAPHTHYL	22.45	0.040	14.22311
ACRIDINE (A)	13.19	0.195	0.58345
ANTHRACENE (TRACE)	11.72	0.278	0.28708
DEUTERATED ANTHRACENE C14D10 (TRACE)	12.34	0.238	0.39166
9-METHYL ANTHRACENE (TRACE)	11.84	0.269	0.30559
1-AMINO ANTHRACENE	19.26	0.063	5.66751
2-AMINO ANTHRACENE	10.41	0.396	0.13880
9-VINYL ANTHRACENE	13.33	0.189	0.62294
9-PHENYL ANTHRACENE (TRACE)	14.04	0.162	0.84830
9,10-DIPHENYL ANTHRACENE	16.19	0.105	1.99643
9,10-DICHLOROANTHRACENE (TRACE)	12.23	0.244	0.37164
BIANTHRANYL (B)	19.03	0.065	5.27386
9,10-DINAPHTHYLANTHRACENE (B)	20.69	0.051	8.69258
9-(P-VINYLPHENYL)-10-PHENYL ANTHRACENE	17.91	0.078	3.66561
9,10-BIS(PHENYL-ETHYNYL)ANTHRACENE (TRACE)	23.72	0.034	19.75410
RUBICENE (B)	23.89	0.033	20.62239
DIBENZ(A,H)ANTHRACENE (B)	26.22	0.025	36.04744
TETRACENE (TRACE) (B)	28.32	0.020	57.26651
RUBRENE (B)	25.76	0.026	32.37494
TETRAHYDROCANNABINOL ACID METHYL ESTER	20.93	0.049	9.41662
PHENANTHRENE	19.01	0.065	5.22889

ACCEPTOR MOLECULES	RO	CO	OVERLAP
3,4-BENZOPHENANTHRENE	24.35	0.031	23.13567
2-PHENYL PHENANTHRENE	24.37	0.031	23.21337
CHRYSENE (TRACE)	21.78	0.043	11.83303
TRIPHENYLENE	20.37	0.053	7.91272
DEUTERATED TRIPHENYLENE-C18D12	20.43	0.052	8.07032
PYRENE	19.66	0.059	6.40521
1,3,6,8,-TETRAPHENYL PYRENE	23.66	0.034	19.45122
BIPYRENYL (B)	24.05	0.032	21.47940
O-PHENYLENE-PYRENE	23.07	0.036	16.71130
ANTHANTHRENE (B)	22.69	0.038	15.15114
FLUORANTHENE	20.84	0.049	9.09512
3-PHENYL FLUORANTHENE	21.91	0.043	12.27581
3-AMINO FLUORANTHENE	20.32	0.053	7.80562
3,3'-BIFLUORANTHENYL (B)	23.42	0.035	18.28674
BENZO(GHI)FLUORANTHENE	20.16	0.055	7.44439
9,12-DI(N-OCTADECOXY)-BENZ(E)ACEPHENANTHRYLENE	23.41	0.035	18.23230
DECACYCLENE (B)	22.85	0.037	15.79736
10-ETHOXY-BENZO(KL)XANTHENE	17.17	0.088	2.84426
PERYLENE	14.73	0.140	1.13438
BENZO(GHI)PERYLENE (B)	23.76	0.033	19.96131
BBOT (TRACE)	18.02	0.076	3.79643
QUININE SULFATE (A)	18.32	0.073	4.19777
ACRIDINE YELLOW (A)	20.54	0.052	8.32871
ACRIDINE ORANGE (A)	20.87	0.049	9.15336
ACRIDINE RED (A)	18.18	0.074	4.00581
URANINE (A)	18.25	0.074	4.09400
RHODAMINE B (A)	20.92	0.049	9.30918
RHODAMINE 6G (A)	20.93	0.049	9.32784
EOSIN B (A)	18.70	0.068	4.74799
EOSIN Y (A)	19.26	0.063	5.65975
ROSE BENGAL (A)	19.82	0.057	6.71442
INDENOINDENE	20.66	0.051	8.61447

8.18 Donor Molecule: Phenol (trace)

ACCEPTOR MOLECULES	RO	CO	OVERLAP
BENZENE (3 ML/L)	0.00	0.000	0.00000
DEUTERATED BENZENE-C6D6 5ML/L	0.00	0.000	0.00000
FLUOROBENZENE (TRACE)	2.39	32.757	0.00002
TOLUENE (TRACE)	3.05	15.742	0.00008
DEUTERATED TOLUENE-C7D8 5ML/L	2.90	18.315	0.00006
ETHYLBENZENE (TRACE)	3.17	14.003	0.00010
PROPYLBENZENE (TRACE)	3.07	15.389	0.00008
ISOPROPYLBENZENE (TRACE)	2.79	20.576	0.00005
SEC-BUTYLBENZENE (TRACE)	2.63	24.565	0.00003
DIPHENYLMETHANE (TRACE)	3.94	7.320	0.00037
BIBENZYL	3.73	8.601	0.00027
1,3,5-TRIPHENYLPENTANE 0.05ML/L	4.00	7.000	0.00041
POLYSTYRENE	3.15	14.265	0.00010
POLYVINYLTOLUENE	6.53	1.603	0.00765
PHENYLCYCLOHEXANE (TRACE)	2.05	51.984	0.00001
BENZYLALCOHOL 3ML/L	2.90	18.337	0.00006
BENZYLACETATE 5ML/L	2.37	33.695	0.00002
PHENOL (TRACE)	9.63	0.501	0.07911
ANISOLE (TRACE)	10.04	0.442	0.10153
PHENYL ETHER	10.78	0.357	0.15502
ANILINE 3ML/L	14.12	0.159	0.77553
N,N-DIMETHYLANILINE 3ML/L	15.57	0.119	1.38557
TRIPHENYLAMINE	23.43	0.035	16.01690
P-XYLENE (TRACE)	6.55	1.590	0.00785
DEUTERATED P-XYLENE C8D10 (TRACE)	6.41	1.701	0.00687
M-XYLENE (TRACE)	4.91	3.779	0.00139
O-XYLENE 3ML/L	4.55	4.736	0.00089
1-METHYL-4-ETHYLBENZENE (TRACE)	5.86	2.227	0.00401
1,4-BIS(TRIFLUOROMETHYL)BENZENE 5ML/L	3.33	12.093	0.00014
P-CRESOL	12.79	0.214	0.43121
P-METHYLANISOLE 3ML/L	12.38	0.236	0.35584
1,4-DIMETHOXY BENZENE	15.63	0.117	1.43071
1-METHYL-4-CHLOROBENZENE 5ML/L	7.74	0.965	0.02130
SODIUM SALICYLATE (A)	17.00	0.091	2.33650
ANTHRANILIC ACID (A)	15.09	0.130	1.14097
TMPD	17.41	0.085	2.69345
1,2,4-TRIMETHYLBENZENE (TRACE)	7.64	1.002	0.01973
MESITYLENE 5ML/L	5.46	2.749	0.00263
1,3,5-TRIETHYLBENZENE 5ML/L	5.16	3.254	0.00188
STYRENE	10.98	0.338	0.17228
1,1-DIPHENYLETHYLENE 3ML/L	11.88	0.267	0.27710
BIPHENYL	10.62	0.373	0.14251
DEUTERATED BIPHENYL-C12D10	10.57	0.378	0.13850
4-METHYL BIPHENYL	12.75	0.216	0.42334
MONOISOPROPYLBIPHENYL 5ML/L	12.17	0.248	0.32151
4-BENZYL BIPHENYL	14.13	0.158	0.78703
3,3'-DIMETHYL BIPHENYL	12.55	0.227	0.38427
2,2'-DIMETHYLBIPHENYL 2ML/L	7.63	1.005	0.01956
OCTAMETHYL BIPHENYL	9.58	0.508	0.07667
4-METHOXY BIPHENYL	16.85	0.093	2.24727
4-BIPHENYLYL PHENYL ETHER	17.16	0.088	2.51291
4,4'-DIHYDROXYBIPHENYL (A)	19.13	0.064	4.78052
4,4'-DIMETHOXY BIPHENYL	18.91	0.066	4.46420
4,4'-DIISOBUTOXYBIPHENYL	18.83	0.067	4.36138
BENZIDINE	23.50	0.034	16.27082

ACCEPTOR MOLECULES	RO	CO	OVERLAP
4-VINYL BIPHENYL	21.90	0.043	10.77308
M-TERPHENYL	15.93	0.111	1.60162
M-QUATERPHENYL	15.59	0.118	1.41011
1,3,5-TRIPHENYL BENZENE	16.21	0.105	1.76190
1,3-DIBIPHENYLYLBENZENE (B)	25.80	0.026	28.55222
1,3,5,-TRIBIPHENYLYLBENZENE (B)	28.65	0.019	53.63833
FLUORENE (TRACE)	17.99	0.077	[illegible]
1-METHYL FLUORENE	17.12	0.089	2.45393
1,2-BENZOFLUORENE	21.41	0.046	9.32365
9,10-DIHYDROPHENANTHRENE	17.87	0.078	3.18519
CARBAZOLE (TRACE)	20.23	0.054	6.63836
N-METHYL CARBAZOLE (TRACE)	19.71	0.058	5.67911
N-VINYLCARBAZOLE	20.31	0.053	6.78166
N-PHENYL CARBAZOLE	20.97	0.049	8.21460
N-METHYL-2,3-BENZCARBAZOLE	23.79	0.033	17.51376
DIBENZOFURAN	19.66	0.059	5.66568
DIBENZOTHIOPHENE (TRACE)	17.61	0.082	2.89309
P-TERPHENYL	22.61	0.039	13.03739
DEUTERATED P-TERPHENYL-C18D14	22.34	0.040	12.13344
4-METHYL TERPHENYL	23.18	0.036	15.13029
2-METHYL-P-TERPHENYL	19.21	0.063	4.91675
3,3″-DIMETHYL-P-TERPHENYL	22.74	0.038	13.46794
4-(3,3-DIMETHYLBUTOXY)-P-TERPHENYL	24.17	0.032	19.34581
4,4″-DIHEXAHYDROFARNESOXY-P-TERPHENYL	25.33	0.028	25.60554
2,2′-METHYLENE-P-TERPHENYL	24.23	0.031	19.66306
2,2′-ETHYLENE-P-TERPHENYL	24.26	0.031	19.77458
2,2′-METHYLENE-5′,6″-METHYLENE-P-TERPHENYL (TRACE)	24.03	0.032	18.63504
3-PHENYL DIBENZOFURAN	24.06	0.032	18.83808
BENZO(1,2-B-4,5-B′)BISBENZOFURAN (TRACE) (B)	22.21	0.041	11.59757
SEXIPHENYL BENZENE	17.05	0.090	2.41599
P-QUATERPHENYL	25.37	0.027	25.81052
3,3‴-DIMETHYL-P-QUATERPHENYL	25.71	0.026	27.91673
TETRAMETHYL-P-QUATERPHENYL	21.30	0.046	9.14407
3,3‴-DIMETHYL-3′,2″-METHYLENE-P-QUATERPHENYL	25.28	0.028	25.22351
3,3‴-DIMETHYL-3′,2″-ETHYLENE-P-QUATERPHENYL	25.71	0.026	27.91498
4-(3-ETHYLBUTOXY)-P-QUATERPHENYL (B)	25.64	0.027	27.48021
4,4‴-DI(2-BUTYLOCTOXY-1)-P-QUATERPHENYL	26.57	0.024	34.04284
DIETHYL-P-QUINQUEPHENYL	26.96	0.023	37.12445
DI(3-ETHYLHEPTYL)-P-QUINQUEPHENYL	27.14	0.022	38.66634
TETRAMETHYL-P-SEXIPHENYL	27.25	0.022	39.64445
TETRAMETHYL-P-OCTAPHENYL (B)	27.54	0.021	42.23722
INDOLE (TRACE)	13.90	0.167	0.71266
2-PHENYL INDOLE	23.23	0.036	15.22235
1-METHYL-2-PHENYL INDOLE	22.19	0.041	11.54121
1-NEOPENTYL-2-PHENYL INDOLE	21.97	0.042	10.90999
1-N-BUTYL-2(BIPHENYLYL-4) INDOLE	23.31	0.035	15.51037
1,2-DIPHENYL INDOLE	22.83	0.038	13.68202
1-BIPHENYLYL-2-PHENYL INDOLE	24.09	0.032	18.92158
1,3-DIMETHYL-2-PHENYL INDOLE	21.31	0.046	9.05213
2,3-DIPHENYL INDOLE	22.08	0.042	11.19563
1-METHYL-2,3-DIPHENYL INDOLE	21.35	0.046	9.15081
2-PHENYL-3,2′-METHYLENE INDOLE (TRACE)	22.12	0.041	11.31996
2-PHENYL-3,2′-ETHYLENE INDOLE (TRACE)	21.38	0.046	9.24611
2-PHENYL-3,2′-TRIMETHYLENE INDOLE	22.52	0.039	12.60839
1-METHYL-3,2′-METHYLENE-2-PHENYL INDOLE (TRACE)	21.69	0.044	10.06020
1-METHYL-3,2′-ETHYLENE-2-PHENYL INDOLE	21.00	0.048	8.28924
1-METHYL-3,2′-TRIMETHYLENE-2-PHENYL INDOLE	21.70	0.044	10.08918
4,5-DIPHENYLIMIDAZOLE (A)	19.83	0.057	5.88484
PPF (TRACE)	23.10	0.036	14.69048
PPO (TRACE)	24.58	0.030	21.33884
2-(4-FLUOROPHENYL)-5-PHENYLOXAZOLE (TRACE)	23.70	0.034	17.14698
αNPO (TRACE)	21.50	0.045	9.56333
BBO (B)	23.12	0.036	14.75941
POPOP	21.23	0.047	8.84580
DIMETHYL POPOP (TRACE)	21.90	0.043	10.68043
PPD (TRACE)	22.84	0.038	13.88565
2,5-DI(2-METHYLPHENYL)OXADIAZOLE	22.31	0.040	12.03032
PBD	26.02	0.025	30.01483
BBD (B)	25.69	0.026	27.81283
αNPD (TRACE)	21.76	0.043	10.27815
βNPD	24.33	0.031	20.07092
αNND (TRACE)	20.84	0.049	7.92573
βNND (TRACE)	24.50	0.030	20.92800
TRANS-1-(4-BIPHENYLYL)-2-PHENYL-ETHYLENE	24.75	0.030	22.21563
TRANS-1-(4-BIPHENYLYL)-2-(1-NAPHTHYL)-ETHYLENE	22.82	0.038	13.67430
DIPHENYLSTILBENE (B)	22.85	0.037	13.76928
1,1,4,4-TETRAPHENYLBUTADIENE	20.72	0.050	7.65067
1,6-DIPHENYLHEXATRIENE	21.15	0.047	8.65078
1,8-DIPHENYL-1,3,5,7-OCTATETRAENE	18.78	0.068	4.24219
BPSB (TRACE)	21.69	0.044	10.06467
BIS-MSB	22.46	0.039	12.40651
ESCULIN (A)	19.43	0.061	5.20801
AZULENE	20.52	0.052	7.22044
NAPHTHALENE	14.84	0.137	1.03919
DEUTERATED NAPHTHALENE-C10D8	14.94	0.134	1.07905
1-METHYLNAPHTHALENE 5ML/L	16.65	0.097	2.07536
2-METHYL NAPHTHALENE	15.29	0.125	1.24221
2,3-DIMETHYL NAPHTHALENE	15.88	0.112	1.56084
2,6-DIMETHYL NAPHTHALENE (TRACE)	15.57	0.118	1.38747
1-NAPHTHOL (TRACE)	17.81	0.079	3.09565
2-NAPHTHOL (TRACE)	15.30	0.125	1.24312
1-NAPHTHYLAMINE	17.46	0.084	2.74007
2-CHLORONAPHTHALENE	15.32	0.124	1.25644
ACENAPHTHENE	18.08	0.076	3.39812
1-PHENYL NAPHTHALENE	19.72	0.058	5.72388
2-PHENYL NAPHTHALENE	29.07	0.018	58.56733

ACCEPTOR MOLECULES	RO	CO	OVERLAP
1,4-DIPHENYL NAPHTHALENE	21.46	0.045	9.45998
1,5-DIPHENYL NAPHTHALENE	21.06	0.048	8.43817
1,7-DIPHENYL NAPHTHALENE	20.79	0.050	7.79964
1,4,5,8-TETRAPHENYL NAPHTHALENE	20.57	0.051	7.31762
1,1'-BINAPHTHYL	20.83	0.049	7.93985
4,4' -DI(N-BUTOXY)-1,1' -BINAPHTHYL	21.66	0.044	9.98657
2,2'-BINAPHTHYL	23.06	0.036	14.55487
ACRIDINE (A)	14.29	0.153	0.82290
ANTHRACENE (TRACE)	13.17	0.196	0.50537
DEUTERATED ANTHRACENE C14D10 (TRACE)	13.57	0.179	0.60371
9-METHYL ANTHRACENE (TRACE)	12.29	0.241	0.33324
1-AMINO ANTHRACENE	17.21	0.088	2.51196
2-AMINO ANTHRACENE	12.31	0.240	0.33203
9-VINYL ANTHRACENE	13.18	0.195	0.50773
9-PHENYL ANTHRACENE (TRACE)	13.68	0.175	0.63420
9,10-DIPHENYL ANTHRACENE	14.30	0.153	0.82880
9,10-DICHLOROANTHRACENE (TRACE)	12.10	0.253	0.30304
BIANTHRANYL (B)	15.77	0.114	1.48669
9,10-DINAPHTHYLANTHRACENE (B)	20.43	0.052	7.04056
9-(P-VINYLPHENYL)-10-PHENYL ANTHRACENE	17.08	0.090	2.40322
9,10-BIS(PHENYL-ETHYNYL)ANTHRACENE (TRACE)	23.40	0.035	15.88272
RUBICENE (B)	24.87	0.029	22.86700
DIBENZ(A,H)ANTHRACENE (B)	27.90	0.021	45.66768
TETRACENE (TRACE) (B)	25.79	0.026	28.43720
RUBRENE (B)	28.13	0.020	47.89461
TETRAHYDROCANNABINOL ACID METHYL ESTER	20.58	0.051	7.40630
PHENANTHRENE	18.33	0.073	3.67217
3,4-BENZOPHENANTHRENE	23.47	0.035	16.18596
2-PHENYL PHENANTHRENE	23.68	0.034	17.05240
CHRYSENE (TRACE)	21.43	0.045	9.37383
TRIPHENYLENE	19.58	0.060	5.44604
DEUTERATED TRIPHENYLENE-C18D12	19.51	0.060	5.33126
PYRENE	19.58	0.060	5.45544
1,3,6,8,-TETRAPHENYL PYRENE	24.76	0.029	22.31297
BIPYRENYL (B)	23.75	0.033	17.37302
O-PHENYLENE-PYRENE	24.16	0.032	19.25617
ANTHANTHRENE (B)	24.43	0.031	20.55529
FLUORANTHENE	20.62	0.051	7.44009
3-PHENYL FLUORANTHENE	22.71	0.038	13.26564
3-AMINO FLUORANTHENE	21.53	0.045	9.62170
3,3'-BIFLUORANTHENYL (B)	24.26	0.031	19.72065
BENZO(GHI)FLUORANTHENE	20.46	0.052	7.09126
9,12-DI(N-OCTADECOXY)-BENZ(E)ACEPHENANTHRYLENE	23.08	0.036	14.62998
DECACYCLENE (B)	23.72	0.033	17.24608
10-ETHOXY-BENZO(KL)XANTHENE	18.00	0.077	3.28980
PERYLENE	14.29	0.153	0.82206
BENZO(GHI)PERYLENE (B)	24.86	0.029	22.84132
BBOT (TRACE)	19.49	0.060	5.30402
QUININE SULFATE (A)	18.94	0.066	4.46369
ACRIDINE YELLOW (A)	19.52	0.060	5.34636
ACRIDINE ORANGE (A)	20.75	0.050	7.72214
ACRIDINE RED (A)	18.72	0.068	4.16555
URANINE (A)	18.82	0.067	4.29769
RHODAMINE B (A)	21.49	0.045	9.52841
RHODAMINE 6G (A)	21.19	0.047	8.74480
EOSIN B (A)	19.60	0.059	5.47922
EOSIN Y (A)	20.20	0.054	6.57218
ROSE BENGAL (A)	20.05	0.056	6.27623
INDENOINDENE	22.66	0.038	13.09333

8.19 Donor Molecule: Anisole (trace)

BENZENE (3 ML/L)	0.00	0.000	0.00000
DEUTERATED BENZENE-C6D6 5ML/L	0.00	0.000	0.00000
FLUOROBENZENE (TRACE)	2.51	28.399	0.00001
TOLUENE (TRACE)	3.26	12.861	0.00003
DEUTERATED TOLUENE-C7D8 5ML/L	3.11	14.887	0.00002
ETHYLBENZENE (TRACE)	3.40	11.349	0.00004
PROPYLBENZENE (TRACE)	3.30	12.473	0.00004
ISOPROPYLBENZENE (TRACE)	2.99	16.693	0.00002
SEC-BUTYLBENZENE (TRACE)	2.82	19.984	0.00001
DIPHENYLMETHANE (TRACE)	4.26	5.771	0.00017
BIBENZYL	4.06	6.665	0.00012
1,3,5-TRIPHENYLPENTANE 0.05ML/L	4.46	5.036	0.00022
POLYSTYRENE	3.37	11.668	0.00004
POLYVINYLTOLUENE	7.71	0.975	0.00571
PHENYLCYCLOHEXANE (TRACE)	2.03	53.205	0.00000
BENZYLALCOHOL 3ML/L	3.11	14.892	0.00002
BENZYLACETATE 5ML/L	2.49	28.803	0.00001
PHENOL (TRACE)	10.93	0.343	0.04654
ANISOLE (TRACE)	11.53	0.292	0.06417
PHENYL ETHER	12.69	0.219	0.11394
ANILINE 3ML/L	17.37	0.085	0.74365
N,N-DIMETHYLANILINE 3ML/L	19.28	0.062	1.37798
TRIPHENYLAMINE	29.10	0.018	16.15496
P-XYLENE (TRACE)	7.32	1.140	0.00421
DEUTERATED P-XYLENE C8D10 (TRACE)	7.16	1.219	0.00369
M-XYLENE (TRACE)	5.45	2.762	0.00072
O-XYLENE 3ML/L	5.09	3.401	0.00047
1-METHYL-4-ETHYLBENZENE (TRACE)	6.48	1.646	0.00202
1,4-BIS(TRIFLUOROMETHYL)BENZENE 5ML/L	3.53	10.127	0.00005

ACCEPTOR MOLECULES	RO	CO	OVERLAP
P-CRESOL	15.47	0.121	0.37279
P-METHYLANISOLE 3ML/L	14.99	0.133	0.30902
1,4-DIMETHOXY BENZENE	19.30	0.062	1.39768
1-METHYL-4-CHLOROBENZENE 5ML/L	8.80	0.655	0.01274
SODIUM SALICYLATE (A)	21.10	0.048	2.34739
ANTHRANILIC ACID (A)	19.14	0.064	1.30798
TMPD	21.37	0.046	2.53618
1,2,4-TRIMETHYLBENZENE (TRACE)	8.67	0.688	0.01161
MESITYLENE 5ML/L	6.17	1.900	0.00152
1,3,5-TRIETHYLBENZENE 5ML/L	5.86	2.220	0.00111
STYRENE	13.38	0.187	0.15569
1,1-DIPHENYLETHYLENE 3ML/L	14.28	0.154	0.23038
BIPHENYL	12.44	0.232	0.10113
DEUTERATED BIPHENYL-C12D10	12.34	0.238	0.09661
4-METHYL BIPHENYL	15.17	0.128	0.33144
MONOISOPROPYLBIPHENYL 5ML/L	14.41	0.150	0.24388
4-BENZYL BIPHENYL	16.86	0.093	0.62604
3,3'-DIMETHYL BIPHENYL	15.11	0.130	0.32310
2,2'-DIMETHYLBIPHENYL 2ML/L	8.93	0.629	0.01379
OCTAMETHYL BIPHENYL	11.33	0.308	0.05753
4-METHOXY BIPHENYL	20.43	0.052	1.96729
4-BIPHENYLYL PHENYL ETHER	20.77	0.050	2.17565
4,4'-DIHYDROXYBIPHENYL (A)	23.37	0.035	4.38650
4,4'-DIMETHOXY BIPHENYL	23.07	0.036	4.06197
4,4'-DIISOBUTOXYBIPHENYL	22.99	0.037	3.98567
BENZIDINE	29.01	0.018	15.90162
4-VINYL BIPHENYL	26.86	0.023	10.11592
M-TERPHENYL	19.34	0.062	1.41215
M-QUATERPHENYL	18.73	0.068	1.17123
1,3,5-TRIPHENYL BENZENE	19.43	0.061	1.44471
1,3-DIBIPHENYLYLBENZENE (B)	31.81	0.014	27.69082
1,3,5,-TRIBIPHENYLYLBENZENE (B)	35.39	0.010	52.53716
FLUORENE (TRACE)	22.10	0.041	3.15046
1-METHYL FLUORENE	21.22	0.047	2.45878
1,2-BENZOFLUORENE	26.57	0.024	9.36444
9,10-DIHYDROPHENANTHRENE	21.82	0.043	2.91321
CARBAZOLE (TRACE)	25.02	0.029	6.53897
N-METHYL CARBAZOLE (TRACE)	24.43	0.031	5.65985
N-VINYLCARBAZOLE	25.02	0.029	6.53317
N-PHENYL CARBAZOLE	25.90	0.026	8.02591
N-METHYL-2,3-BENZCARBAZOLE	29.21	0.018	16.51527
DIBENZOFURAN	24.11	0.032	5.31532
DIBENZOTHIOPHENE (TRACE)	21.67	0.044	2.77004
P-TERPHENYL	27.75	0.021	12.28827
DEUTERATED P-TERPHENYL-C18D14	27.42	0.022	11.43395
4-METHYL TERPHENYL	28.51	0.019	14.44058
2-METHYL-P-TERPHENYL	23.40	0.035	4.42848
3,3''-DIMETHYL-P-TERPHENYL	27.96	0.020	12.82994
4-(3,3-DIMETHYLBUTOXY)-P-TERPHENYL	29.84	0.017	18.88035
4,4''-DIHEXAHYDROFARNESOXY-P-TERPHENYL	31.33	0.015	25.29745
2,2'-METHYLENE-P-TERPHENYL	29.98	0.017	19.44271
2,2'-ETHYLENE-P-TERPHENYL	30.03	0.017	19.60578
2,2'-METHYLENE-5',6''-METHYLENE-P-TERPHENYL (TRACE)	29.97	0.017	19.33009
3-PHENYL DIBENZOFURAN	29.89	0.017	19.08031
BENZO(1,2-B-4,5-B')BISBENZOFURAN (TRACE) (B)	28.02	0.020	12.86671
SEXIPHENYL BENZENE	20.64	0.051	2.09763
P-QUATERPHENYL	31.42	0.014	25.67287
3,3'''-DIMETHYL-P-QUATERPHENYL	31.86	0.014	27.88992
TETRAMETHYL-P-QUATERPHENYL	26.01	0.025	8.34991
3,3''' -DIMETHYL-3',2''-METHYLENE-P-QUATERPHENYL	31.58	0.014	26.40803
3,3''' -DIMETHYL-3',2''-ETHYLENE-P-QUATERPHENYL	32.01	0.014	28.63428
4-(3-ETHYLBUTOXY)-P-QUATERPHENYL (B)	31.92	0.014	28.16575
4,4'''-DI(2-BUTYLOCTOXY-1)-P-QUATERPHENYL	33.07	0.012	34.79741
DIETHYL-P-QUINQUEPHENYL	33.40	0.012	37.03019
DI(3-ETHYLHEPTYL)-P-QUINQUEPHENYL	33.80	0.012	39.69776
TETRAMETHYL-P-SEXIPHENYL	33.73	0.012	39.29968
TETRAMETHYL-P-OCTAPHENYL (B)	34.39	0.011	43.99151
INDOLE (TRACE)	16.75	0.095	0.60217
2-PHENYL INDOLE	28.94	0.018	15.67819
1-METHYL-2-PHENYL INDOLE	27.52	0.021	11.59363
1-NEOPENTYL-2-PHENYL INDOLE	27.22	0.022	10.87489
1-N-BUTYL-2(BIPHENYLYL-4) INDOLE	29.07	0.018	16.05325
1,2-DIPHENYL INDOLE	28.31	0.020	13.73673
1-BIPHENYLYL-2-PHENYL INDOLE	29.86	0.017	18.84625
1,3-DIMETHYL-2-PHENYL INDOLE	26.45	0.024	9.14122
2,3-DIPHENYL INDOLE	27.50	0.022	11.51181
1-METHYL-2,3-DIPHENYL INDOLE	26.50	0.024	9.20592
2-PHENYL-3,2'-METHYLENE INDOLE (TRACE)	27.70	0.021	12.04049
2-PHENYL-3,2'-ETHYLENE INDOLE (TRACE)	26.86	0.023	9.99732
2-PHENYL-3,2'-TRIMETHYLENE INDOLE	28.14	0.020	13.22245
1-METHYL-3,2'-METHYLENE-2-PHENYL INDOLE (TRACE)	27.25	0.022	10.88635
1-METHYL-3,2'-ETHYLENE-2-PHENYL INDOLE	26.40	0.024	9.00014
1-METHYL-3,2'-TRIMETHYLENE-2-PHENYL INDOLE	27.09	0.022	10.52725
4,5-DIPHENYLIMIDAZOLE (A)	24.47	0.031	5.72462
PPF (TRACE)	28.98	0.018	15.76014
PPO (TRACE)	30.60	0.016	21.88448
2-(4-FLUOROPHENYL)-5-PHENYLOXAZOLE (TRACE)	29.49	0.017	17.52654
αNPO (TRACE)	26.94	0.023	10.17694
BBO (B)	29.10	0.018	16.14355
POPOP	26.73	0.023	9.70190
DIMETHYL POPOP (TRACE)	27.47	0.022	11.44543
PPD (TRACE)	28.12	0.020	13.33170
2,5-DI(2-METHYLPHENYL)OXADIAZOLE	27.48	0.022	11.59290
PBD	32.31	0.013	30.35712
BBD (B)	32.12	0.014	29.20937
αNPD (TRACE)	27.18	0.022	10.73664

ACCEPTOR MOLECULES	RO	CO	OVERLAP
αNPD	30.14	0.016	19.96855
αNND (TRACE)	26.18	0.025	8.57577
βNND (TRACE)	30.47	0.016	21.30172
TRANS-1-(4-BIPHENYLYL)-2-PHENYL-ETHYLENE	30.98	0.015	23.52575
TRANS-1-(4-BIPHENYLYL)-2-(1-NAPHTHYL)-ETHYLENE	28.61	0.019	14.58113
DIPHENYLSTILBENE (B)	28.90	0.019	15.51718
1,1,4,4-TETRAPHENYLBUTADIENE	26.23	0.025	8.66096
1,6-DIPHENYLHEXATRIENE	26.98	0.023	10.26478
1,8-DIPHENYL-1,3,5,7-OCTATETRAENE	23.93	0.033	4.99049
BPSB (TRACE)	27.55	0.021	11.64084
BIS-MSB	28.41	0.020	13.99101
ESCULIN (A)	24.39	0.031	5.60080
AZULENE	24.83	0.029	6.22891
NAPHTHALENE	18.02	0.076	0.92116
DEUTERATED NAPHTHALENE-C10D8	18.17	0.075	0.96267
1-METHYLNAPHTHALENE 5ML/L	20.41	0.053	1.93666
2-METHYL NAPHTHALENE	18.64	0.069	1.12683
2,3-DIMETHYL NAPHTHALENE	19.40	0.061	1.43140
2,6-DIMETHYL NAPHTHALENE (TRACE)	19.05	0.065	1.27985
1-NAPHTHOL (TRACE)	22.01	0.042	3.04256
2-NAPHTHOL (TRACE)	18.84	0.067	1.19565
1-NAPHTHYLAMINE	21.84	0.043	2.88888
2-CHLORONAPHTHALENE	18.68	0.069	1.13769
ACENAPHTHENE	22.29	0.040	3.29052
1-PHENYL NAPHTHALENE	24.33	0.031	5.55439
2-PHENYL NAPHTHALENE	35.84	0.010	56.62102
1,4-DIPHENYL NAPHTHALENE	26.66	0.024	9.56296
1,5-DIPHENYL NAPHTHALENE	26.12	0.025	8.48678
1,7-DIPHENYL NAPHTHALENE	25.77	0.026	7.81295
1,4,5,8-TETRAPHENYL NAPHTHALENE	25.81	0.026	7.86430
1,1'-BINAPHTHYL	25.71	0.026	7.73050
4,4'-DI(N-BUTOXY)-1,1 -BINAPHTHYL	27.02	0.023	10.35150
2,2'-BINAPHTHYL	28.54	0.019	14.41545
ACRIDINE (A)	18.25	0.074	0.98184
ANTHRACENE (TRACE)	16.90	0.093	0.61967
DEUTERATED ANTHRACENE C14D10 (TRACE)	17.29	0.086	0.71190
9-METHYL ANTHRACENE (TRACE)	15.74	0.115	0.40499
1-AMINO ANTHRACENE	20.89	0.049	2.21109
2-AMINO ANTHRACENE	15.89	0.111	0.42249
9-VINYL ANTHRACENE	16.68	0.096	0.57263
9-PHENYL ANTHRACENE (TRACE)	17.28	0.087	0.70774
9,10-DIPHENYL ANTHRACENE	17.68	0.081	0.81349
9,10-DICHLOROANTHRACENE (TRACE)	15.35	0.124	0.34785
BIANTHRANYL (B)	19.30	0.062	1.37470
9,10-DINAPHTHYLANTHRACENE (B)	25.17	0.028	6.77054
9-(P-VINYLPHENYL)-10-PHENYL ANTHRACENE	21.07	0.048	2.32828
9,10-BIS(PHENYL-ETHYNYL)ANTHRACENE (TRACE)	28.88	0.019	15.44140
RUBICENE (B)	30.82	0.015	22.80110
DIBENZ(A,H)ANTHRACENE (B)	34.66	0.011	46.09356
TETRACENE (TRACE) (B)	30.81	0.015	22.76969
RUBRENE (B)	35.00	0.010	48.90265
TETRAHYDROCANNABINOL ACID METHYL ESTER	25.26	0.028	6.99226
PHENANTHRENE	22.46	0.039	3.41909
3,4-BENZOPHENANTHRENE	28.70	0.019	14.86416
2-PHENYL PHENANTHRENE	29.04	0.018	15.96860
CHRYSENE (TRACE)	26.53	0.024	9.27993
TRIPHENYLENE	0.00	0.000	0.00000
DEUTERATED TRIPHENYLENE-C18D12	23.86	0.033	4.91004
PYRENE	24.67	0.030	6.00448
1,3,6,8,-TETRAPHENYL PYRENE	30.70	0.015	22.28743
BIPYRENYL (B)	29.36	0.018	17.04700
O-PHENYLENE-PYRENE	29.99	0.017	19.34345
ANTHANTHRENE (B)	30.38	0.016	20.91801
FLUORANTHENE	25.42	0.027	7.18488
3-PHENYL FLUORANTHENE	28.15	0.020	13.24232
3-AMINO FLUORANTHENE	26.77	0.023	9.78791
3,3'-BIFLUORANTHENYL (B)	30.10	0.016	19.77107
BENZO(GHI)FLUORANTHENE	25.34	0.027	7.03952
9,12-DI(N-OCTADECOXY)-BENZ(E)ACEPHENANTHRYLENE	28.39	0.020	13.93328
DECACYCLENE (B)	29.56	0.017	17.74535
10-ETHOXY-BENZO(KL)XANTHENE	22.39	0.040	3.34930
PERYLENE	17.64	0.081	0.80295
BENZO(GHI)PERYLENE (B)	30.81	0.015	22.76592
BBOT (TRACE)	24.75	0.030	6.11043
QUININE SULFATE (A)	23.61	0.034	4.61210
ACRIDINE YELLOW (A)	23.88	0.033	4.93192
ACRIDINE ORANGE (A)	25.58	0.027	7.44605
ACRIDINE RED (A)	23.20	0.036	4.14654
URANINE (A)	23.34	0.035	4.30565
RHODAMINE B (A)	26.61	0.024	9.45003
RHODAMINE 6G (A)	26.19	0.025	8.58450
EOSIN B (A)	24.39	0.031	5.59658
EOSIN Y (A)	25.13	0.028	6.70311
ROSE BENGAL (A)	24.88	0.029	6.31048
INDENOINDENE	28.30	0.020	13.70224

8.20 Donor Molecule: Phenyl Ether

TOLUENE (TRACE)	1.25	******	0.00000
DEUTERATED TOLUENE-C7D8 5ML/L	1.35	******	0.00000
ETHYLBENZENE (TRACE)	1.46	******	0.00000

ACCEPTOR MOLECULES	RO	CO	OVERLAP
PROPYLBENZENE (TRACE)	1.43	******	0.00000
ISOPROPYLBENZENE (TRACE)	1.13	******	0.00000
SEC-BUTYLBENZENE (TRACE)	0.00	0.000	0.00000
DIPHENYLMETHANE (TRACE)	2.02	53.875	0.00002
BIBENZYL	1.97	58.588	0.00002
1,3,5-TRIPHENYLPENTANE 0.05ML/L	2.38	32.991	0.00005
POLYSTYRENE	1.41	******	0.00000
POLYVINYLTOLUENE	4.99	3.593	0.00606
PHENYLCYCLOHEXANE (TRACE)	0.00	0.000	0.00000
BENZYLALCOHOL 3ML/L	1.33	******	0.00000
BENZYLACETATE 5ML/L	0.00	0.000	0.00000
PHENOL (TRACE)	6.33	1.760	0.01705
ANISOLE (TRACE)	6.79	1.427	0.02591
PHENYL ETHER	7.93	0.896	0.06554
ANILINE 3ML/L	11.85	0.269	0.72369
N,N-DIMETHYLANILINE 3ML/L	13.26	0.192	1.41133
TRIPHENYLAMINE	20.05	0.056	16.65182
P-XYLENE (TRACE)	4.02	6.868	0.00112
DEUTERATED P-XYLENE C8D10 (TRACE)	3.91	7.509	0.00094
M-XYLENE (TRACE)	2.91	18.142	0.00016
O-XYLENE 3ML/L	2.78	20.868	0.00012
1-METHYL-4-ETHYLBENZENE (TRACE)	3.33	12.129	0.00036
1,4-BIS(TRIFLUOROMETHYL)BENZENE 5ML/L	1.11	******	0.00000
P-CRESOL	10.22	0.419	0.29986
P-METHYLANISOLE 3ML/L	9.91	0.459	0.24912
1,4-DIMETHOXY BENZENE	13.24	0.193	1.40094
1-METHYL-4-CHLOROBENZENE 5ML/L	5.10	3.371	0.00465
SODIUM SALICYLATE (A)	14.54	0.146	2.42104
ANTHRANILIC ACID (A)	13.27	0.192	1.39713
TMPD	14.46	0.148	2.33983
1,2,4-TRIMETHYLBENZENE (TRACE)	5.04	3.496	0.00431
MESITYLENE 5ML/L	3.51	10.312	0.00050
1,3,5-TRIETHYLBENZENE 5ML/L	3.41	11.266	0.00042
STYRENE	8.98	0.617	0.13755
1,1-DIPHENYLETHYLENE 3ML/L	9.38	0.541	0.17942
BIPHENYL	7.77	0.955	0.05782
DEUTERATED BIPHENYL-C12D10	7.66	0.995	0.05334
4-METHYL BIPHENYL	9.80	0.476	0.23208
MONOISOPROPYLBIPHENYL 5ML/L	9.20	0.575	0.15944
4-BENZYL BIPHENYL	10.94	0.342	0.45024
3,3′-DIMETHYL BIPHENYL	9.96	0.453	0.25616
2,2′-DIMETHYLBIPHENYL 2ML/L	5.60	2.541	0.00815
OCTAMETHYL BIPHENYL	7.14	1.227	0.03496
4-METHOXY BIPHENYL	13.62	0.177	1.66984
4-BIPHENYLYL PHENYL ETHER	13.81	0.170	1.81376
4,4′-DIHYDROXYBIPHENYL (A)	15.78	0.114	4.00388
4,4′-DIMETHOXY BIPHENYL	15.54	0.119	3.66050
4,4′-DIISOBUTOXYBIPHENYL	15.50	0.120	3.61720
BENZIDINE	19.86	0.057	15.81310
4-VINYL BIPHENYL	18.24	0.074	9.57107
M-TERPHENYL	12.95	0.206	1.22620
M-QUATERPHENYL	12.33	0.238	0.92172
1,3,5-TRIPHENYL BENZENE	12.74	0.216	1.10733
1,3-DIBIPHENYLYLBENZENE (B)	21.76	0.043	27.34112
1,3,5,-TRIBIPHENYLYLBENZENE (B)	24.25	0.031	52.49828
FLUORENE (TRACE)	15.08	0.130	3.07385
1-METHYL FLUORENE	14.60	0.144	2.51778
1,2-BENZOFLUORENE	18.28	0.073	9.58076
9,10-DIHYDROPHENANTHRENE	14.76	0.139	2.68841
CARBAZOLE (TRACE)	17.12	0.089	6.48507
N-METHYL CARBAZOLE (TRACE)	16.76	0.095	5.69219
N-VINYLCARBAZOLE	17.04	0.090	6.28451
N-PHENYL CARBAZOLE	17.70	0.081	7.90450
N-METHYL-2,3-BENZCARBAZOLE	19.84	0.057	15.62775
DIBENZOFURAN	16.35	0.102	4.99446
DIBENZOTHIOPHENE (TRACE)	14.68	0.141	2.57888
P-TERPHENYL	18.86	0.067	11.67955
DEUTERATED P-TERPHENYL-C18D14	18.63	0.069	10.86582
4-METHYL TERPHENYL	19.43	0.061	13.96903
2-METHYL-P-TERPHENYL	15.72	0.115	3.93258
3,3″-DIMETHYL-P-TERPHENYL	19.05	0.065	12.38114
4-(3,3-DIMETHYLBUTOXY)-P-TERPHENYL	20.43	0.052	18.77665
4,4″-DIHEXAHYDROFARNESOXY-P-TERPHENYL	21.50	0.045	25.48905
2,2′-METHYLENE-P-TERPHENYL	20.57	0.051	19.55809
2,2′-ETHYLENE-P-TERPHENYL	20.61	0.051	19.78597
2,2′-METHYLENE-5′,6′-METHYLENE-P-TERPHENYL (TRACE)	20.65	0.051	19.93157
3-PHENYL DIBENZOFURAN	20.61	0.051	19.82281
BENZO(1,2-B-4,5-B′)BISBENZOFURAN (TRACE) (B)	19.40	0.061	13.68304
SEXIPHENYL BENZENE	13.72	0.173	1.74096
P-QUATERPHENYL	21.58	0.044	26.03632
3,3‴-DIMETHYL-P-QUATERPHENYL	21.90	0.043	28.38327
TETRAMETHYL-P-QUATERPHENYL	17.54	0.083	7.57414
3,3‴-DIMETHYL-3′,2″-METHYLENE-P-QUATERPHENYL	21.80	0.043	27.50706
3,3‴-DIMETHYL-2′,3″-ETHYLENE-P-QUATERPHENYL	22.02	0.043	29.29106
4-(3-ETHYLBUTOXY)-P-QUATERPHENYL (B)	22.01	0.042	29.14483
4,4‴-DI(2-BUTYLOCTOXY-1)-P-QUATERPHENYL	22.79	0.038	35.93892
DIETHYL-P-QUINQUEPHENYL	22.95	0.037	37.62004
DI(3-ETHYLHEPTYL)-P-QUINQUEPHENYL	23.31	0.035	41.12579
TETRAMETHYL-P-SEXIPHENYL	23.16	0.036	39.71955
TETRAMETHYL-P-OCTAPHENYL (B)	23.72	0.034	45.62794
INDOLE (TRACE)	10.97	0.339	0.45791
2-PHENYL INDOLE	19.97	0.056	16.32512
1-METHYL-2-PHENYL INDOLE	18.94	0.066	11.89590
1-NEOPENTYL-2-PHENYL INDOLE	18.72	0.068	11.10292
1-N-BUTYL-2(BIPHENYLYL-4) INDOLE	20.05	0.055	16.64577
1,2-DIPHENYL INDOLE	19.48	0.060	14.07918

ACCEPTOR MOLECULES	RO	CO	OVERLAP
1-BIPHENYLYL-2-PHENYL INDOLE	20.50	0.052	19.03072
1,3-DIMETHYL-2-PHENYL INDOLE	18.21	0.074	9.39551
2,3-DIPHENYL INDOLE	18.96	0.066	11.88656
1-METHYL-2,3-DIPHENYL INDOLE	18.21	0.074	9.34134
2-PHENYL-3,2'-METHYLENE INDOLE (TRACE)	19.15	0.064	12.67607
2-PHENYL-3,2'-ETHYLENE INDOLE (TRACE)	18.57	0.070	10.50397
2-PHENYL-3,2'-TRIMETHYLENE INDOLE	19.44	0.061	13.86590
1-METHYL-3,2'-METHYLENE-2-PHENYL INDOLE (TRACE)	18.85	0.067	11.50536
1-METHYL-3,2'-ETHYLENE-2-PHENYL INDOLE	18.25	0.074	9.47897
1-METHYL-3,2'-TRIMETHYLENE-2-PHENYL INDOLE	18.71	0.068	11.02603
4,5-DIPHENYLIMIDAZOLE (A)	16.74	0.095	5.66193
PPF (TRACE)	20.03	0.056	16.55925
PPO (TRACE)	21.09	0.048	22.66865
2-(4-FLUOROPHENYL)-5-PHENYLOXAZOLE (TRACE)	20.32	0.053	18.10392
αNPO (TRACE)	18.54	0.070	10.39459
BBO (B)	20.08	0.055	16.78065
POPOP	18.44	0.071	10.08729
DIMETHYL POPOP (TRACE)	18.92	0.066	11.75719
PPD (TRACE)	19.20	0.063	13.02400
2,5-DI(2-METHYLPHENYL)OXADIAZOLE	18.78	0.068	11.37508
PBD	22.24	0.041	31.21072
BBD (B)	22.16	0.041	30.37847
αNPD (TRACE)	18.73	0.068	11.07609
βNPD	20.68	0.051	20.14758
αNND (TRACE)	18.09	0.076	8.97593
βNND (TRACE)	20.91	0.049	21.46069
TRANS-1-(4-BIPHENYLYL)-2-PHENYL-ETHYLENE	21.40	0.046	24.60617
TRANS-1-(4-BIPHENYLYL)-2-(1-NAPHTHYL)-ETHYLENE	19.71	0.058	15.00085
DIPHENYLSTILBENE (B)	20.02	0.056	16.49422
1,1,4,4-TETRAPHENYLBUTADIENE	18.15	0.075	9.15640
1,6-DIPHENYLHEXATRIENE	18.73	0.068	11.07316
1,8-DIPHENYL-1,3,5,7-OCTATETRAENE	16.57	0.098	5.29360
BPSB (TRACE)	19.10	0.064	12.42384
BIS-MSB	19.67	0.059	14.85678
ESCULIN (A)	16.84	0.094	5.83833
AZULENE	16.40	0.101	4.97673
NAPHTHALENE	12.04	0.256	0.78969
DEUTERATED NAPHTHALENE-C10D8	12.15	0.249	0.83065
1-METHYLNAPHTHALENE 5ML/L	13.83	0.169	1.80913
2-METHYL NAPHTHALENE	12.53	0.227	1.00081
2,3-DIMETHYL NAPHTHALENE	13.08	0.200	1.29692
2,6-DIMETHYL NAPHTHALENE (TRACE)	12.84	0.211	1.16083
1-NAPHTHOL (TRACE)	15.08	0.130	3.04402
2-NAPHTHOL (TRACE)	12.78	0.214	1.12235
1-NAPHTHYLAMINE	15.09	0.130	3.02305
2-CHLORONAPHTHALENE	12.55	0.226	1.00960
ACENAPHTHENE	15.23	0.127	3.23695
1-PHENYL NAPHTHALENE	16.64	0.097	5.49093
2-PHENYL NAPHTHALENE	24.50	0.030	55.74934
1,4-DIPHENYL NAPHTHALENE	18.36	0.072	9.84930
1,5-DIPHENYL NAPHTHALENE	17.98	0.077	8.71431
1,7-DIPHENYL NAPHTHALENE	17.73	0.080	7.98413
1,4,5,8-TETRAPHENYL NAPHTHALENE	17.80	0.079	8.13584
1,1'-BINAPHTHYL	17.60	0.082	7.67393
4,4'-DI(N-BUTOXY)-1,1'-BINAPHTHYL	18.64	0.069	10.76565
2,2'-BINAPHTHYL	19.58	0.060	14.53225
ACRIDINE (A)	12.65	0.221	1.04747
ANTHRACENE (TRACE)	11.75	0.276	0.67416
DEUTERATED ANTHRACENE C14D10 (TRACE)	12.00	0.259	0.76347
9-METHYL ANTHRACENE (TRACE)	10.89	0.346	0.42762
1-AMINO ANTHRACENE	13.96	0.165	1.89417
2-AMINO ANTHRACENE	11.05	0.332	0.45961
9-VINYL ANTHRACENE	11.48	0.296	0.58652
9-PHENYL ANTHRACENE (TRACE)	11.86	0.268	0.71455
9,10-DIPHENYL ANTHRACENE	11.97	0.261	0.75313
9,10-DICHLOROANTHRACENE (TRACE)	10.59	0.377	0.36159
BIANTHRANYL (B)	12.95	0.206	1.20859
9,10-DINAPHTHYLANTHRACENE (B)	17.16	0.089	6.53197
9-(P-VINYLPHENYL)-10-PHENYL ANTHRACENE	14.35	0.151	2.23388
9,10-BIS(PHENYL-ETHYNYL)ANTHRACENE (TRACE)	19.79	0.058	15.39197
RUBICENE (B)	21.20	0.047	23.22997
DIBENZ(A,H)ANTHRACENE (B)	23.89	0.033	47.66983
TETRACENE (TRACE) (B)	20.23	0.054	17.54143
RUBRENE (B)	24.20	0.032	51.48350
TETRAHYDROCANNABINOL ACID METHYL ESTER	17.17	0.088	6.66149
PHENANTHRENE	15.25	0.126	3.22863
3,4-BENZOPHENANTHRENE	19.30	0.062	13.25809
2-PHENYL PHENANTHRENE	19.75	0.058	15.19083
CHRYSENE (TRACE)	18.20	0.074	9.30522
TRIPHENYLENE	16.20	0.105	4.64301
DEUTERATED TRIPHENYLENE-C18D12	16.10	0.107	4.46411
PYRENE	17.03	0.091	6.25540
1,3,6,8,-TETRAPHENYL PYRENE	21.14	0.047	22.85651
BIPYRENYL (B)	19.87	0.057	15.78705
O-PHENYLENE-PYRENE	20.64	0.051	19.79541
ANTHANTHRENE (B)	20.97	0.049	21.76874
FLUORANTHENE	17.21	0.088	6.64661
3-PHENYL FLUORANTHENE	19.35	0.062	13.43791
3-AMINO FLUORANTHENE	18.47	0.071	10.16200
3,3'-BIFLUORANTHENYL (B)	20.68	0.051	20.04549
BENZO(GHI)FLUORANTHENE	17.34	0.086	6.95694
9,12-DI(N-OCTADECOXY)-BENZ(E)ACEPHENANTHRYLENE	19.27	0.063	13.10144
DECACYCLENE (B)	20.34	0.053	18.11684
10-ETHOXY-BENZO(KL)XANTHENE	15.42	0.122	3.44536
PERYLENE	12.02	0.257	0.77451
BENZO(GHI)PERYLENE (B)	21.21	0.047	23.30156
BBOT (TRACE)	17.13	0.089	6.47699

ACCEPTOR MOLECULES	RO	CO	OVERLAP
QUININE SULFATE (A)	16.20	0.105	4.63152
ACRIDINE YELLOW (A)	16.15	0.106	4.55343
ACRIDINE ORANGE (A)	17.49	0.084	7.32771
ACRIDINE RED (A)	15.92	0.111	4.16425
URANINE (A)	16.02	0.109	4.33598
RHODAMINE B (A)	18.25	0.074	9.45584
RHODAMINE 6G (A)	17.91	0.078	8.46866
EOSIN B (A)	16.78	0.095	5.12883
EOSIN Y (A)	17.30	0.086	6.86504
ROSE BENGAL (A)	17.05	0.090	6.30174
INDENOINDENE	19.55	0.060	14.37778

8.21 Donor Molecule: Aniline (3 ml/L)

POLYVINYLTOLUENE	2.72	22.259	0.00004
ANILINE 3ML/L	8.13	0.834	0.02786
N,N-DIMETHYLANILINE 3ML/L	13.14	0.197	0.49612
TRIPHENYLAMINE	21.74	0.044	10.07398
1,4-DIMETHOXY BENZENE	9.93	0.456	0.09287
1-METHYL-4-CHLOROBENZENE 5ML/L	0.00	0.000	0.00000
SODIUM SALICYLATE (A)	15.10	0.130	1.12850
ANTHRANILIC ACID (A)	18.78	0.068	4.14961
TMPD	16.94	0.092	2.24499
1,2,4-TRIMETHYLBENZENE (TRACE)	0.00	0.000	0.00000
MESITYLENE 5ML/L	0.00	0.000	0.00000
1,3,5-TRIETHYLBENZENE 5ML/L	0.00	0.000	0.00000
STYRENE	2.99	16.777	0.00007
3,3'-DIMETHYL BIPHENYL	3.76	8.418	0.00027
2,2'-DIMETHYLBIPHENYL 2ML/L	0.00	0.000	0.00000
OCTAMETHYL BIPHENYL	0.00	0.000	0.00000
4-METHOXY BIPHENYL	7.87	0.918	0.02301
4-BIPHENYLYL PHENYL ETHER	7.09	1.254	0.01235
4,4'-DIHYDROXYBIPHENYL (A)	12.21	0.246	0.31976
4,4'-DIMETHOXY BIPHENYL	11.12	0.325	0.18244
4,4'-DIISOBUTOXYBIPHENYL	11.01	0.335	0.17273
BENZIDINE	20.02	0.056	6.15509
4-VINYL BIPHENYL	14.91	0.135	1.06086
M-TERPHENYL	10.48	0.389	0.12768
M-QUATERPHENYL	6.64	1.526	0.00835
1,3,5-TRIPHENYL BENZENE	10.28	0.411	0.11370
1,3-DIBIPHENYLYLBENZENE (B)	20.70	0.050	7.53727
1,3,5,-TRIBIPHENYLYLBENZENE (B)	23.87	0.033	17.70754
FLUORENE (TRACE)	11.95	0.262	0.28208
1-METHYL FLUORENE	10.92	0.343	0.16412
1,2-BENZOFLUORENE	20.19	0.054	6.46277
9,10-DIHYDROPHENANTHRENE	12.00	0.259	0.28794
CARBAZOLE (TRACE)	17.51	0.083	2.75597
N-METHYL CARBAZOLE (TRACE)	17.51	0.083	2.74504
N-VINYLCARBAZOLE	17.88	0.078	3.12106
N-PHENYL CARBAZOLE	17.72	0.080	2.95266
N-METHYL-2,3-BENZCARBAZOLE	20.16	0.055	6.34387
DIBENZOFURAN	10.49	0.388	0.12899
DIBENZOTHIOPHENE (TRACE)	16.00	0.109	1.60698
P-TERPHENYL	15.85	0.112	1.52759
DEUTERATED P-TERPHENYL-C18D14	15.56	0.119	1.36853
4-METHYL TERPHENYL	17.19	0.088	2.48413
2-METHYL-P-TERPHENYL	10.74	0.361	0.14865
3,3''-DIMETHYL-P-TERPHENYL	16.73	0.095	2.11088
4-(3,3-DIMETHYLBUTOXY)-P-TERPHENYL	20.05	0.056	6.22372
4,4''-DIHEXAHYDROFARNESOXY-P-TERPHENYL	22.27	0.040	11.68755
2,2'-METHYLENE-P-TERPHENYL	21.89	0.043	10.56366
2,2'-ETHYLENE-P-TERPHENYL	21.94	0.042	10.69077
2,2'-METHYLENE-5',6''-METHYLENE-P-TERPHENYL (TRACE)	25.16	0.028	24.25688
3-PHENYL DIBENZOFURAN	22.41	0.040	12.15492
BENZO(1,2-B-4,5-B')BISBENZOFURAN (TRACE) (B)	26.43	0.024	32.53574
SEXIPHENYL BENZENE	7.67	0.991	0.01978
P-QUATERPHENYL	22.89	0.037	13.74519
3,3'''-DIMETHYL-P-QUATERPHENYL	23.67	0.034	16.79879
TETRAMETHYL-P-QUATERPHENYL	12.80	0.213	0.42589
3,3''' -DIMETHYL-3',2''-METHYLENE-P-QUATERPHENYL	27.07	0.023	37.47333
3,3''' -DIMETHYL-3',2''-ETHYLENE-P-QUATERPHENYL	26.86	0.023	35.73111
4-(3-ETHYLBUTOXY)-P-QUATERPHENYL (B)	25.88	0.026	28.58934
4,4'''-DI(2-BUTYLOCTOXY-1)-P-QUATERPHENYL	26.63	0.024	33.99613
DIETHYL-P-QUINQUEPHENYL	24.70	0.030	21.69469
DI(3-ETHYLHEPTYL)-P-QUINQUEPHENYL	27.53	0.021	41.31620
TETRAMETHYL-P-SEXIPHENYL	24.33	0.031	19.06002
TETRAMETHYL-P-OCTAPHENYL (B)	29.45	0.018	61.90034
INDOLE (TRACE)	0.00	0.000	0.00000
2-PHENYL INDOLE	23.65	0.034	16.72719
1-METHYL-2-PHENYL INDOLE	20.47	0.052	7.03497
1-NEOPENTYL-2-PHENYL INDOLE	19.42	0.061	5.14153
1-N-BUTYL-2(BIPHENYLYL-4) INDOLE	24.39	0.031	20.00884
1,2-DIPHENYL INDOLE	21.32	0.046	8.96563
1-BIPHENYLYL-2-PHENYL INDOLE	22.77	0.038	13.28367
1,3-DIMETHYL-2-PHENYL INDOLE	20.27	0.054	6.63454
2,3-DIPHENYL INDOLE	22.79	0.038	13.28651
1-METHYL-2,3-DIPHENYL INDOLE	20.99	0.048	8.12190
2-PHENYL-3,2'-METHYLENE INDOLE (TRACE)	24.44	0.031	20.34521
2-PHENYL-3,2'-ETHYLENE INDOLE (TRACE)	24.56	0.030	20.87181

ACCEPTOR MOLECULES	RO	CO	OVERLAP
2-PHENYL-3,2'-TRIMETHYLENE INDOLE	24.10	0.032	18.69123
1-METHYL-3,2'-METHYLENE-2-PHENYL INDOLE (TRACE)	24.71	0.030	21.69396
1-METHYL-3,2'-ETHYLENE-2-PHENYL INDOLE	24.31	0.031	19.62070
1-METHYL-3,2'-TRIMETHYLENE-2-PHENYL INDOLE	22.82	0.038	13.47700
4,5-DIPHENYLIMIDAZOLE (A)	16.54	0.099	1.95789
PPF (TRACE)	26.02	0.025	29.56997
PPO (TRACE)	24.83	0.029	22.41227
2-(4-FLUOROPHENYL)-5-PHENYLOXAZOLE (TRACE)	23.77	0.033	17.26400
αNPO (TRACE)	24.91	0.029	22.59657
BBO (B)	27.67	0.021	42.39035
POPOP	25.68	0.026	27.06077
DIMETHYL POPOP (TRACE)	25.86	0.026	28.21322
PPD (TRACE)	17.03	0.091	2.35465
2,5-DI(2-METHYLPHENYL)OXADIAZOLE	17.36	0.086	2.63537
PBD	25.03	0.029	23.50886
BBD (B)	27.97	0.020	45.55727
αNPD (TRACE)	23.82	0.033	17.34236
βNPD	23.66	0.034	16.77273
αNND (TRACE)	24.05	0.032	18.34817
βNND (TRACE)	25.87	0.026	28.60250
TRANS-1-(4-BIPHENYLYL)-2-PHENYL-ETHYLENE	27.19	0.022	38.35620
TRANS-1-(4-BIPHENYLYL)-2-(1-NAPHTHYL)-ETHYLENE	26.28	0.025	31.09628
DIPHENYLSTILBENE (B)	27.99	0.020	45.40584
1,1,4,4-TETRAPHENYLBUTADIENE	25.77	0.026	27.65770
1,6-DIPHENYLHEXATRIENE	27.65	0.021	42.15780
1,8-DIPHENYL-1,3,5,7-OCTATETRAENE	25.21	0.028	24.20405
BPSB (TRACE)	27.71	0.021	42.74212
BIS-MSB	27.70	0.021	42.58691
ESCULIN (A)	22.25	0.041	11.47486
AZULENE	17.14	0.089	2.39427
NAPHTHALENE	8.71	0.676	0.04216
DEUTERATED NAPHTHALENE-C10D8	10.02	0.444	0.09706
1-METHYLNAPHTHALENE 5ML/L	10.32	0.407	0.11596
2-METHYL NAPHTHALENE	10.83	0.352	0.15518
2,3-DIMETHYL NAPHTHALENE	10.86	0.350	0.15727
2,6-DIMETHYL NAPHTHALENE (TRACE)	12.83	0.212	0.42935
1-NAPHTHOL (TRACE)	15.40	0.122	1.28043
2-NAPHTHOL (TRACE)	15.31	0.125	1.23193
1-NAPHTHYLAMINE	18.96	0.066	4.43251
2-CHLORONAPHTHALENE	10.87	0.349	0.15814
ACENAPHTHENE	14.12	0.159	0.76292
1-PHENYL NAPHTHALENE	15.35	0.124	1.25414
2-PHENYL NAPHTHALENE	22.41	0.040	12.13004
1,4-DIPHENYL NAPHTHALENE	20.41	0.053	6.90875
1,5-DIPHENYL NAPHTHALENE	19.52	0.060	5.29988
1,7-DIPHENYL NAPHTHALENE	19.30	0.062	4.93658
1,4,5,8-TETRAPHENYL NAPHTHALENE	23.93	0.033	17.71055
1,1'-BINAPHTHYL	16.66	0.097	2.05120
4,4'-DI(N-BUTOXY)-1,1'-BINAPHTHYL	22.62	0.039	12.75829
2,2'-BINAPHTHYL	21.50	0.045	9.45629
ACRIDINE (A)	19.01	0.065	4.45435
ANTHRACENE (TRACE)	17.88	0.078	3.08461
DEUTERATED ANTHRACENE C14D10 (TRACE)	17.81	0.079	3.01552
9-METHYL ANTHRACENE (TRACE)	16.92	0.092	2.21735
1-AMINO ANTHRACENE	15.67	0.116	1.39787
2-AMINO ANTHRACENE	17.12	0.089	2.36267
9-VINYL ANTHRACENE	17.32	0.086	2.55037
9-PHENYL ANTHRACENE (TRACE)	17.86	0.079	3.06217
9,10-DIPHENYL ANTHRACENE	17.36	0.085	2.58245
9,10-DICHLOROANTHRACENE (TRACE)	16.40	0.101	1.83320
BIANTHRANYL (B)	18.85	0.067	4.22987
9,10-DINAPHTHYLANTHRACENE (B)	18.36	0.072	3.61337
9-(P-VINYLPHENYL)-10-PHENYL ANTHRACENE	17.65	0.081	2.85610
9,10-BIS(PHENYL-ETHYNYL)ANTHRACENE (TRACE)	21.57	0.045	9.51278
RUBICENE (B)	22.86	0.037	13.46019
DIBENZ(A,H)ANTHRACENE (B)	24.87	0.029	22.33841
TETRACENE (TRACE) (B)	15.77	0.114	1.45247
RUBRENE (B)	26.37	0.024	31.69656
TETRAHYDROCANNABINOL ACID METHYL ESTER	13.91	0.166	0.69963
PHENANTHRENE	11.21	0.318	0.18870
3,4-BENZOPHENANTHRENE	19.81	0.058	5.70156
2-PHENYL PHENANTHRENE	17.75	0.080	2.97416
CHRYSENE (TRACE)	20.54	0.052	7.12478
TRIPHENYLENE	13.01	0.203	0.46252
DEUTERATED TRIPHENYLENE-C18D12	12.97	0.205	0.45439
• PYRENE	23.85	0.033	17.40772
1,3,6,8,-TETRAPHENYL PYRENE	21.95	0.042	10.54576
BIPYRENYL (B)	25.87	0.026	28.29217
O-PHENYLENE-PYRENE	23.30	0.035	15.08149
ANTHANTHRENE (B)	23.74	0.033	16.90997
FLUORANTHENE	19.82	0.057	5.72346
3-PHENYL FLUORANTHENE	20.83	0.050	7.69928
3-AMINO FLUORANTHENE	21.14	0.047	8.42706
3,3'-BIFLUORANTHENYL (B)	23.46	0.035	15.74399
BENZO(GHI)FLUORANTHENE	19.52	0.060	5.22115
9,12-DI(N-OCTADECOXY)-BENZ(E)ACEPHENANTHRYLENE	21.03	0.048	8.16513
DECACYCLENE (B)	25.46	0.027	25.68975
10-ETHOXY-BENZO(KL)XANTHENE	18.42	0.072	3.68425
PERYLENE	15.06	0.131	1.09869
BENZO(GHI)PERYLENE (B)	21.85	0.043	10.26931
BBOT (TRACE)	25.22	0.028	24.26299
QUININE SULFATE (A)	20.73	0.050	7.56231
ACRIDINE YELLOW (A)	14.98	0.133	1.06444
ACRIDINE ORANGE (A)	16.66	0.097	2.01717
ACRIDINE RED (A)	18.12	0.075	3.33957
URANINE (A)	18.51	0.071	3.79328

ACCEPTOR MOLECULES	RO	CO	OVERLAP
RHODAMINE B (A)	20.50	0.052	6.99895
RHODAMINE 6G (A)	19.73	0.058	5.57171
EOSIN B (A)	20.37	0.053	6.73880
EOSIN Y (A)	20.84	0.049	7.72852
ROSE BENGAL (A)	20.97	0.049	8.01702
INDENOINDENE	24.07	0.032	18.55626

8.22 Donor Molecule: *N*,*N*-Dimethylaniline (3 ml/L)

ACCEPTOR MOLECULES	RO	CO	OVERLAP
ANILINE 3ML/L	3.39	11.492	0.00011
N,N-DIMETHYLANILINE 3ML/L	9.36	0.546	0.04683
TRIPHENYLAMINE	19.22	0.063	3.48335
1,4-DIMETHOXY BENZENE	4.02	6.871	0.00030
1-METHYL-4-CHLOROBENZENE 5ML/L	0.00	0.000	0.00000
SODIUM SALICYLATE (A)	12.91	0.208	0.31951
ANTHRANILIC ACID (A)	20.36	0.053	4.87770
TMPD	17.60	0.082	2.04031
4,4′-DIHYDROXYBIPHENYL (A)	8.75	0.667	0.03134
4,4′-DIMETHOXY BIPHENYL	6.88	1.372	0.00743
4,4′-DIISOBUTOXYBIPHENYL	5.77	2.328	0.00259
BENZIDINE	17.09	0.090	1.72427
4-VINYL BIPHENYL	9.32	0.553	0.04567
M-TERPHENYL	6.43	1.679	0.00497
M-QUATERPHENYL	0.00	0.000	0.00000
1,3,5-TRIPHENYL BENZENE	7.87	0.919	0.01648
1,3-DIBIPHENYLYLBENZENE (B)	16.50	0.100	1.39861
1,3,5,-TRIBIPHENYLYLBENZENE (B)	19.45	0.061	3.75832
FLUORENE (TRACE)	4.04	6.791	0.00030
1-METHYL FLUORENE	4.26	5.796	0.00042
1,2-BENZOFLUORENE	17.94	0.078	2.29942
9,10-DIHYDROPHENANTHRENE	5.75	2.358	0.00252
CARBAZOLE (TRACE)	16.91	0.092	1.62083
N-METHYL CARBAZOLE (TRACE)	18.55	0.070	2.81480
N-VINYLCARBAZOLE	18.71	0.068	2.96259
N-PHENYL CARBAZOLE	18.47	0.071	2.74126
N-METHYL-2,3-BENZCARBAZOLE	20.56	0.051	5.14372
DIBENZOFURAN	3.23	13.260	0.00008
DIBENZOTHIOPHENE (TRACE)	15.17	0.128	0.84351
P-TERPHENYL	10.99	0.337	0.12270
DEUTERATED P-TERPHENYL-C18D14	10.39	0.398	0.08812
4-METHYL TERPHENYL	12.43	0.233	0.25716
2-METHYL-P-TERPHENYL	5.48	2.719	0.00190
3,3″-DIMETHYL-P-TERPHENYL	12.23	0.245	0.23308
4-(3,3-DIMETHYLBUTOXY)-P-TERPHENYL	16.40	0.101	1.35282
4,4″-DIHEXAHYDROFARNESOXY-P-TERPHENYL	18.96	0.066	3.22539
2,2′-METHYLENE-P-TERPHENYL	18.19	0.074	2.51961
2,2′-ETHYLENE-P-TERPHENYL	18.80	0.067	3.06770
2,2′-METHYLENE-5′,6″-METHYLENE-P-TERPHENYL (TRACE)	25.73	0.026	20.05976
3-PHENYL DIBENZOFURAN	18.27	0.073	2.58391
BENZO(1,2-B-4,5-B′)BISBENZOFURAN (TRACE) (B)	26.86	0.023	25.92435
SEXIPHENYL BENZENE	0.00	0.000	0.00000
P-QUATERPHENYL	19.95	0.056	4.36858
3,3‴-DIMETHYL-P-QUATERPHENYL	21.02	0.048	5.97349
TETRAMETHYL-P-QUATERPHENYL	7.09	1.255	0.00889
3,3‴-DIMETHYL-3′,2″-METHYLENE-P-QUATERPHENYL	26.83	0.023	25.74475
3,3‴-DIMETHYL-3′,2″-ETHYLENE-P-QUATERPHENYL	26.47	0.024	23.69016
4-(3-ETHYLBUTOXY)-P-QUATERPHENYL (B)	24.56	0.030	15.12616
4,4‴-DI(2-BUTYLOCTOXY-1)-P-QUATERPHENYL	25.16	0.028	17.48650
DIETHYL-P-QUINQUEPHENYL	21.94	0.042	7.71211
DI(3-ETHYLHEPTYL)-P-QUINQUEPHENYL	26.19	0.025	22.19139
TETRAMETHYL-P-SEXIPHENYL	21.09	0.048	6.09126
TETRAMETHYL-P-OCTAPHENYL (B)	29.23	0.018	42.86228
INDOLE (TRACE)	0.00	0.000	0.00000
2-PHENYL INDOLE	21.82	0.043	7.48827
1-METHYL-2-PHENYL INDOLE	17.75	0.080	2.16461
1-NEOPENTYL-2-PHENYL INDOLE	15.91	0.111	1.12440
1-N-BUTYL-2(BIPHENYLYL-4) INDOLE	23.98	0.032	13.09819
1,2-DIPHENYL INDOLE	18.46	0.071	2.74258
1-BIPHENYLYL-2-PHENYL INDOLE	20.38	0.053	4.94276
1,3-DIMETHYL-2-PHENYL INDOLE	18.35	0.072	2.64459
2,3-DIPHENYL INDOLE	22.21	0.041	8.25783
1-METHYL-2,3-DIPHENYL INDOLE	20.21	0.054	4.69224
2-PHENYL-3,2′-METHYLENE INDOLE (TRACE)	23.95	0.033	13.05766
2-PHENYL-3,2′-ETHYLENE INDOLE (TRACE)	25.29	0.028	18.03013
2-PHENYL-3,2′-TRIMETHYLENE INDOLE	23.05	0.037	10.36506
1-METHYL-3,2′-METHYLENE-2-PHENYL INDOLE (TRACE)	25.09	0.028	17.21098
1-METHYL-3,2′-ETHYLENE-2-PHENYL INDOLE	25.39	0.027	18.42872
1-METHYL-3,2′-TRIMETHYLENE-2-PHENYL INDOLE	21.97	0.042	7.77554
4,5-DIPHENYLIMIDAZOLE (A)	14.11	0.159	0.54681
PPF (TRACE)	26.35	0.024	23.08550
PPO (TRACE)	23.31	0.035	11.10566
2-(4-FLUOROPHENYL)-5-PHENYLOXAZOLE (TRACE)	22.23	0.041	8.34894
αNPO (TRACE)	26.84	0.023	25.58956
BBO (B)	30.51	0.016	55.17824
POPOP	29.34	0.018	43.59596
DIMETHYL POPOP (TRACE)	29.85	0.017	48.21976
PPD (TRACE)	10.99	0.337	0.12325
2,5-DI(2-METHYLPHENYL)OXADIAZOLE	12.35	0.237	0.24802
PBD	22.57	0.039	9.16138
BBD (B)	28.00	0.020	33.19004
αNPD (TRACE)	24.13	0.032	13.59090

ACCEPTOR MOLECULES	RO	CO	OVERLAP
βNPD	21.59	0.044	7.00226
αNND (TRACE)	25.53	0.027	19.00938
βNND (TRACE)	25.64	0.027	19.63615
TRANS-1-(4-BIPHENYLYL)-2-PHENYL-ETHYLENE	27.58	0.021	30.26299
TRANS-1-(4-BIPHENYLYL)-2-(1-NAPHTHYL)-ETHYLENE	28.36	0.020	35.58731
DIPHENYLSTILBENE (B)	30.83	0.015	58.70169
1,1,4,4-TETRAPHENYLBUTADIENE	28.79	0.019	38.88065
1,6-DIPHENYLHEXATRIENE	31.60	0.014	67.92252
1,8-DIPHENYL-1,3,5,7-OCTATETRAENE	30.49	0.016	54.53108
BPSB (TRACE)	31.70	0.014	69.32358
BIS-MSB	31.01	0.015	60.68686
ESCULIN (A)	24.11	0.032	13.44985
AZULENE	18.30	0.073	2.54423
NAPHTHALENE	6.18	1.894	0.00389
DEUTERATED NAPHTHALENE-C10D8	8.14	0.830	0.02015
1-METHYLNAPHTHALENE 5ML/L	7.68	0.988	0.01426
2-METHYL NAPHTHALENE	9.30	0.555	0.04518
2,3-DIMETHYL NAPHTHALENE	9.14	0.586	0.04056
2,6-DIMETHYL NAPHTHALENE (TRACE)	11.92	0.264	0.19995
1-NAPHTHOL (TRACE)	13.18	0.195	0.36502
2-NAPHTHOL (TRACE)	14.93	0.134	0.76873
1-NAPHTHYLAMINE	18.81	0.067	3.06064
2-CHLORONAPHTHALENE	9.64	0.500	0.05576
ACENAPHTHENE	11.55	0.291	0.16506
1-PHENYL NAPHTHALENE	11.83	0.270	0.19094
2-PHENYL NAPHTHALENE	17.74	0.080	2.16332
1,4-DIPHENYL NAPHTHALENE	18.29	0.073	2.58890
1,5-DIPHENYL NAPHTHALENE	17.07	0.090	1.71436
1,7-DIPHENYL NAPHTHALENE	16.84	0.094	1.57840
1,4,5,8-TETRAPHENYL NAPHTHALENE	25.84	0.026	20.32884
1,1'-BINAPHTHYL	13.80	0.170	0.48069
4,4'-DI(N-BUTOXY)-1,1'-BINAPHTHYL	21.84	0.043	7.49447
2,2'-BINAPHTHYL	18.57	0.070	2.84337
ACRIDINE (A)	22.04	0.042	7.83006
ANTHRACENE (TRACE)	20.66	0.051	5.32329
DEUTERATED ANTHRACENE C14D10 (TRACE)	20.49	0.052	5.06176
9-METHYL ANTHRACENE (TRACE)	20.19	0.054	4.61806
1-AMINO ANTHRACENE	18.76	0.068	2.95696
2-AMINO ANTHRACENE	19.34	0.062	3.55401
9-VINYL ANTHRACENE	20.81	0.050	5.53037
9-PHENYL ANTHRACENE (TRACE)	21.25	0.047	6.28107
9,10-DIPHENYL ANTHRACENE	21.36	0.046	6.46539
9,10-DICHLOROANTHRACENE (TRACE)	20.30	0.053	4.76658
BIANTHRANYL (B)	22.65	0.038	9.18722
9,10-DINAPHTHYLANTHRACENE (B)	21.10	0.048	6.00448
9-(P-VINYLPHENYL)-10-PHENYL ANTHRACENE	21.42	0.046	6.56234
9,10-BIS(PHENYL-ETHYNYL)ANTHRACENE (TRACE)	19.37	0.062	3.58013
RUBICENE (B)	22.43	0.040	8.62871
DIBENZ(A,H)ANTHRACENE (B)	24.31	0.031	14.08766
TETRACENE (TRACE) (B)	15.81	0.113	1.05725
RUBRENE (B)	22.34	0.040	8.43085
TETRAHYDROCANNABINOL ACID METHYL ESTER	9.25	0.564	0.04386
PHENANTHRENE	11.38	0.303	0.15031
3,4-BENZOPHENANTHRENE	17.89	0.078	2.24249
2-PHENYL PHENANTHRENE	13.81	0.170	0.47788
CHRYSENE (TRACE)	18.56	0.070	2.80781
TRIPHENYLENE	12.17	0.248	0.22419
DEUTERATED TRIPHENYLENE-C18D12	12.05	0.256	0.21092
PYRENE	24.83	0.029	16.04811
1,3,6,8,-TETRAPHENYL PYRENE	26.41	0.024	23.02423
BIPYRENYL (B)	29.17	0.018	41.94418
O-PHENYLENE-PYRENE	24.30	0.031	13.95485
ANTHANTHRENE (B)	22.57	0.039	8.97433
FLUORANTHENE	21.99	0.042	7.69133
3-PHENYL FLUORANTHENE	23.63	0.034	11.82398
3-AMINO FLUORANTHENE	21.87	0.043	7.41778
3,3'-BIFLUORANTHENYL (B)	26.20	0.025	21.93519
BENZO(GHI)FLUORANTHENE	21.32	0.046	6.37946
9,12-DI(N-OCTADECOXY)-BENZ(E)ACEPHENANTHRYLENE	23.37	0.035	11.05246
DECACYCLENE (B)	28.54	0.019	36.61949
10-ETHOXY-BENZO(KL)XANTHENE	21.79	0.043	7.28700
PERYLENE	19.22	0.063	3.41742
BENZO(GHI)PERYLENE (B)	23.34	0.035	11.00109
BBOT (TRACE)	29.86	0.017	48.28112
QUININE SULFATE (A)	21.59	0.044	6.99247
ACRIDINE YELLOW (A)	16.83	0.094	1.53868
ACRIDINE ORANGE (A)	15.72	0.115	1.02244
ACRIDINE RED (A)	18.81	0.067	3.00381
URANINE (A)	19.07	0.065	3.26113
RHODAMINE B (A)	21.31	0.046	6.34900
RHODAMINE 6G (A)	21.71	0.044	7.09165
EOSIN B (A)	20.29	0.054	4.73859
EOSIN Y (A)	20.66	0.051	5.26713
ROSE BENGAL (A)	21.59	0.044	6.87774
INDENOINDENE	22.86	0.037	9.87119

8.23 Donor Molecule: Triphenylamine

TRIPHENYLAMINE	9.45	0.530	0.17870
SODIUM SALICYLATE (A)	7.40	1.102	0.04129
ANTHRANILIC ACID (A)	15.05	0.131	2.89518
TMPD	11.80	0.273	0.67302

ACCEPTOR MOLECULES	RO	CO	OVERLAP
BENZIDINE	6.52	1.613	0.01935
1,3-DIBIPHENYLYLBENZENE (B)	4.97	3.642	0.00380
1,3,5,-TRIBIPHENYLYLBENZENE (B)	5.17	3.241	0.00480
FLUORENE (TRACE)	0.00	0.000	0.00000
1-METHYL FLUORENE	0.00	0.000	0.00000
1,2-BENZOFLUORENE	9.54	0.515	0.18899
9,10-DIHYDROPHENANTHRENE	0.00	0.000	0.00000
CARBAZOLE (TRACE)	6.15	1.939	0.01338
N-METHYL CARBAZOLE (TRACE)	11.37	0.304	0.54215
N-VINYLCARBAZOLE	10.06	0.439	0.26074
N-PHENYL CARBAZOLE	10.25	0.415	0.29179
N-METHYL-2,3-BENZCARBAZOLE	15.99	0.109	4.14614
DIBENZOFURAN	0.00	0.000	0.00000
DIBENZOTHIOPHENE (TRACE)	3.87	7.723	0.00085
P-TERPHENYL	0.00	0.000	0.00000
DEUTERATED P-TERPHENYL-C18D14	0.00	0.000	0.00000
4-METHYL TERPHENYL	0.00	0.000	0.00000
2-METHYL-P-TERPHENYL	0.00	0.000	0.00000
3,3″-DIMETHYL-P-TERPHENYL	0.00	0.000	0.00000
4-(3,3-DIMETHYLBUTOXY)-P-TERPHENYL	3.64	9.276	0.00059
4,4″DIHEXAHYDROFARNESOXY-P-TERPHENYL	5.16	3.263	0.00474
2,2′-METHYLENE-P-TERPHENYL	2.82	20.029	0.00013
2,2′-ETHYLENE-P-TERPHENYL	4.40	5.268	0.00182
2,2′-METHYLENE-5′,6′-METHYLENE-P-TERPHENYL (TRACE)	10.99	0.337	0.44304
3-PHENYL DIBENZOFURAN	2.68	23.175	0.00009
BENZO(1,2-B-4,5-B′)BISBENZOFURAN (TRACE) (B)	13.89	0.167	1.80485
SEXIPHENYL BENZENE	0.00	0.000	0.00000
P-QUATERPHENYL	6.95	1.330	0.02848
3,3‴-DIMETHYL-P-QUATERPHENYL	8.66	0.690	0.10580
TETRAMETHYL-P-QUATERPHENYL	0.00	0.000	0.00000
3,3‴-DIMETHYL-3′,2″-METHYLENE-P-QUATERPHENYL	13.74	0.172	1.69072
3,3‴-DIMETHYL-3′,2″-ETHYLENE-P-QUATERPHENYL	14.65	0.142	2.47101
4-(3-ETHYLBUTOXY)-P-QUATERPHENYL (B)	13.00	0.203	1.21192
4,4‴-DI(2-BUTYLOCTOXY-1)-P-QUATERPHENYL	13.12	0.198	1.27955
DIETHYL-P-QUINQUEPHENYL	8.75	0.668	0.11290
DI(3-ETHYLHEPTYL)-P-QUINQUEPHENYL	14.24	0.155	2.08071
TETRAMETHYL-P-SEXIPHENYL	7.16	1.220	0.03383
TETRAMETHYL-P-OCTAPHENYL (B)	17.89	0.078	8.18698
INDOLE (TRACE)	0.00	0.000	0.00000
2-PHENYL INDOLE	7.22	1.190	0.03557
1-METHYL-2-PHENYL INDOLE	5.98	2.093	0.01149
1-NEOPENTYL-2-PHENYL INDOLE	3.39	11.479	0.00038
1-N-BUTYL-2(BIPHENYLYL-4) INDOLE	14.44	0.148	2.27161
1,2-DIPHENYL INDOLE	6.17	1.906	0.01385
1-BIPHENYLYL-2-PHENYL INDOLE	9.39	0.540	0.17222
1,3-DIMETHYL-2-PHENYL INDOLE	6.88	1.371	0.02678
2,3-DIPHENYL INDOLE	13.27	0.191	1.36615
1-METHYL-2,3-DIPHENYL INDOLE	11.83	0.270	0.68593
2-PHENYL-3,2′-METHYLENE INDOLE (TRACE)	10.41	0.396	0.32041
2-PHENYL-3,2′-ETHYLENE INDOLE (TRACE)	15.82	0.113	3.92839
2-PHENYL-3,2′-TRIMETHYLENE INDOLE	9.68	0.492	0.20728
1-METHYL-3,2′-METHYLENE-2-PHENYL INDOLE (TRACE)	14.72	0.140	2.54851
1-METHYL-3,2′-ETHYLENE-2-PHENYL INDOLE	17.56	0.083	7.33468
1-METHYL-3,2′-TRIMETHYLENE-2-PHENYL INDOLE	10.47	0.389	0.33134
4,5-DIPHENYLIMIDAZOLE (A)	5.48	2.716	0.00683
PPF (TRACE)	15.40	0.122	3.35032
PPO (TRACE)	9.11	0.592	0.14359
2-(4-FLUOROPHENYL)-5-PHENYLOXAZOLE (TRACE)	8.35	0.768	0.08546
αNPO (TRACE)	20.08	0.055	16.32840
BBO (B)	24.02	0.032	47.72522
POPOP	24.56	0.030	54.51909
DIMETHYL POPOP (TRACE)	25.58	0.027	69.47759
PPD (TRACE)	0.00	0.000	0.00000
2,5-DI(2-METHYLPHENYL)OXADIAZOLE	0.00	0.000	0.00000
PBD	7.48	1.067	0.04428
BBD (B)	16.78	0.095	5.58362
αNPD (TRACE)	14.83	0.137	2.65789
βNPD	9.10	0.594	0.14271
αNND (TRACE)	18.15	0.075	8.91933
βNND (TRACE)	13.88	0.167	1.79816
TRANS-1-(4-BIPHENYLYL)-2-PHENYL-ETHYLENE	17.34	0.086	6.80247
TRANS-1-(4-BIPHENYLYL)-2-(1-NAPHTHYL)-ETHYLENE	21.63	0.044	25.48124
DIPHENYLSTILBENE (B)	24.17	0.032	49.53277
1,1,4,4-TETRAPHENYLBUTADIENE	23.15	0.036	38.23023
1,6-DIPHENYLHEXATRIENE	26.36	0.024	83.29586
1,8-DIPHENYL-1,3,5,7-OCTATETRAENE	27.55	0.021	107.83160
BPSB (TRACE)	26.62	0.024	88.36275
BIS-MSB	25.31	0.028	65.26021
ESCULIN (A)	18.22	0.074	9.10419
AZULENE	11.56	0.290	0.58546
NAPHTHALENE	0.00	0.000	0.00000
DEUTERATED NAPHTHALENE-C10D8	2.18	42.928	0.00003
1-NAPHTHOL (TRACE)	1.78	79.733	0.00001
2-NAPHTHOL (TRACE)	4.31	5.573	0.00162
1 NAPHTHYLAMINE	9.94	0.455	0.24200
2-CHLORONAPHTHALENE	0.00	0.000	0.00000
ACENAPHTHENE	0.00	0.000	0.00000
1-PHENYL NAPHTHALENE	0.00	0.000	0.00000
2-PHENYL NAPHTHALENE	5.29	3.013	0.00556
1,4-DIPHENYL NAPHTHALENE	7.20	1.198	0.03508
1,5-DIPHENYL NAPHTHALENE	5.40	2.834	0.00627
1,7-DIPHENYL NAPHTHALENE	5.91	2.171	0.01069
1,4,5,8-TETRAPHENYL NAPHTHALENE	19.55	0.060	13.85030
1,1′-BINAPHTHYL	2.87	18.841	0.00014
4,4′-DI(N-BUTOXY)-1,1′-BINAPHTHYL	11.40	0.302	0.55119
2,2′-BINAPHTHYL	6.35	1.744	0.01657

ACCEPTOR MOLECULES	RO	CO	OVERLAP
ACRIDINE (A)	18.49	0.071	9.93445
ANTHRACENE (TRACE)	17.52	0.083	7.18887
DEUTERATED ANTHRACENE C14D10 (TRACE)	17.29	0.087	6.65406
9-METHYL ANTHRACENE (TRACE)	17.88	0.078	8.11148
1-AMINO ANTHRACENE	17.16	0.088	6.29499
2-AMINO ANTHRACENE	16.32	0.103	4.65622
9-VINYL ANTHRACENE	18.62	0.069	10.33761
9-PHENYL ANTHRACENE (TRACE)	18.78	0.068	10.87252
9,10-DIPHENYL ANTHRACENE	19.59	0.060	13.97206
9,10-DICHLOROANTHRACENE (TRACE)	18.77	0.068	10.80739
BIANTHRANYL (B)	20.50	0.052	18.35855
9,10-DINAPHTHYLANTHRACENE (B)	19.19	0.063	12.32530
9-(P-VINYLPHENYL)-10-PHENYL ANTHRACENE	19.69	0.059	14.39800
9,10-BIS(PHENYL-ETHYNYL)ANTHRACENE (TRACE)	17.93	0.078	8.16923
RUBICENE (B)	18.36	0.072	9.41168
DIBENZ(A,H)ANTHRACENE (B)	17.26	0.087	6.55805
TETRACENE (TRACE) (B)	13.99	0.163	1.84322
RUBRENE (B)	13.49	0.182	1.48306
TETRAHYDROCANNABINOL ACID METHYL ESTER	0.00	0.000	0.00000
PHENANTHRENE	6.89	1.366	0.02692
3,4-BENZOPHENANTHRENE	10.24	0.416	0.28698
2-PHENYL PHENANTHRENE	8.83	0.649	0.11878
CHRYSENE (TRACE)	10.51	0.385	0.33713
TRIPHENYLENE	6.16	1.914	0.01371
DEUTERATED TRIPHENYLENE-C18D12	6.08	1.990	0.01268
PYRENE	12.42	0.233	0.91618
1,3,6,8,-TETRAPHENYL PYRENE	24.36	0.031	51.63838
BIPYRENYL (B)	23.82	0.033	45.22433
O-PHENYLENE-PYRENE	21.11	0.048	21.76782
ANTHANTHRENE (B)	19.33	0.062	12.83165
FLUORANTHENE	17.65	0.081	7.48152
3-PHENYL FLUORANTHENE	20.38	0.053	17.72160
3-AMINO FLUORANTHENE	18.48	0.071	9.77566
3,3'-BIFLUORANTHENYL (B)	23.05	0.037	36.95622
BENZO(GHI)FLUORANTHENE	16.10	0.107	4.30107
9,12-DI(N-OCTADECOXY)-BENZ(E)ACEPHENANTHRYLENE	19.91	0.057	15.37417
DECACYCLENE (B)	24.67	0.030	55.35369
10-ETHOXY-BENZO(KL)XANTHENE	19.95	0.056	15.59102
PERYLENE	18.61	0.069	10.20766
BENZO(GHI)PERYLENE (B)	21.09	0.048	21.78690
BBOT (TRACE)	26.35	0.024	82.93269
QUININE SULFATE (A)	12.25	0.243	0.84972
ACRIDINE YELLOW (A)	16.27	0.104	4.56209
ACRIDINE ORANGE (A)	13.88	0.167	1.76114
ACRIDINE RED (A)	15.51	0.120	3.41752
URANINE (A)	15.04	0.131	2.84492
RHODAMINE B (A)	17.71	0.081	7.57790
RHODAMINE 6G (A)	17.48	0.084	7.01283
EOSIN B (A)	14.94	0.134	2.73061
EOSIN Y (A)	15.41	0.122	3.28682
ROSE BENGAL (A)	15.87	0.112	3.92572
INDENOINDENE	8.75	0.668	0.11269

8.24 Donor Molecule: *p*-Xylene (trace)

FLUOROBENZENE (TRACE)	4.15	6.256	0.00010
TOLUENE (TRACE)	5.18	3.217	0.00039
DEUTERATED TOLUENE-C7D8 5ML/L	4.78	4.094	0.00024
ETHYLBENZENE (TRACE)	5.26	3.077	0.00042
PROPYLBENZENE (TRACE)	5.08	3.415	0.00034
ISOPROPYLBENZENE (TRACE)	4.73	4.239	0.00022
SEC-BUTYLBENZENE (TRACE)	4.58	4.654	0.00018
DIPHENYLMETHANE (TRACE)	6.33	1.761	0.00129
BIBENZYL	5.93	2.144	0.00087
1,3,5-TRIPHENYLPENTANE 0.05ML/L	6.16	1.917	0.00108
POLYSTYRENE	5.27	3.061	0.00043
POLYVINYLTOLUENE	9.58	0.509	0.01517
PHENYLCYCLOHEXANE (TRACE)	3.48	10.582	0.00004
BENZYLALCOHOL 3ML/L	4.80	4.037	0.00024
BENZYLACETATE 5ML/L	4.13	6.329	0.00010
PHENOL (TRACE)	13.68	0.175	0.12990
ANISOLE (TRACE)	14.11	0.159	0.15621
PHENYL ETHER	15.18	0.128	0.24240
ANILINE 3ML/L	18.72	0.068	0.84432
N,N-DIMETHYLANILINE 3ML/L	20.19	0.054	1.31773
TRIPHENYLAMINE	30.23	0.016	14.76058
P-XYLENE (TRACE)	10.34	0.405	0.02420
DEUTERATED P-XYLENE C8D10 (TRACE)	10.16	0.427	0.02186
M-XYLENE (TRACE)	7.78	0.949	0.00442
O-XYLENE 3ML/L	7.12	1.238	0.00260
1-METHYL-4-ETHYLBENZENE (TRACE)	9.41	0.537	0.01382
1,4-BIS(TRIFLUOROMETHYL)BENZENE 5ML/L	5.76	2.342	0.00073
P-CRESOL	17.54	0.083	0.57463
P-METHYLANISOLE 3ML/L	16.97	0.092	0.47117
1,4-DIMETHOXY BENZENE	20.53	0.052	1.46743
1-METHYL-4-CHLOROBENZENE 5ML/L	10.87	0.348	0.03276
SODIUM SALICYLATE (A)	21.97	0.042	2.17091
ANTHRANILIC ACID (A)	19.01	0.065	0.91313
TMPD	23.38	0.035	3.15966
1,2,4-TRIMETHYLBENZENE (TRACE)	11.33	0.308	0.04186

ACCEPTOR MOLECULES	RO	CO	OVERLAP
MESITYLENE 5ML/L	8.26	0.793	0.00632
1,3,5-TRIETHYLBENZENE 5ML/L	7.77	0.952	0.00439
STYRENE	15.00	0.133	0.22407
1,1-DIPHENYLETHYLENE 3ML/L	16.55	0.099	0.40649
BIPHENYL	15.28	0.125	0.25248
DEUTERATED BIPHENYL-C12D10	15.26	0.126	0.24806
4-METHYL BIPHENYL	17.00	0.077	0.66763
MONOISOPROPYL BIPHENYL 5ML/L	17.28	0.087	0.52737
4-BENZYL BIPHENYL	19.88	0.057	1.21932
3,3'-DIMETHYL BIPHENYL	17.40	0.085	0.54637
2,2'-DIMETHYLBIPHENYL 2ML/L	11.07	0.330	0.03638
OCTAMETHYL BIPHENYL	13.53	0.181	0.12146
4-METHOXY BIPHENYL	23.08	0.036	2.96941
4-BIPHENYLYL PHENYL ETHER	23.59	0.034	3.38521
4,4'-DIHYDROXYBIPHENYL (A)	25.80	0.026	5.75220
4,4'-DIMETHOXY BIPHENYL	25.59	0.027	5.48786
4,4'-DIISOBUTOXYBIPHENYL	25.43	0.027	5.30479
BENZIDINE	30.81	0.015	16.54022
4-VINYL BIPHENYL	29.24	0.018	12.20801
M-TERPHENYL	21.75	0.043	2.07217
M-QUATERPHENYL	21.67	0.044	2.03810
1,3,5-TRIPHENYL BENZENE	22.60	0.039	2.59301
1,3-DIBIPHENYLYLBENZENE (B)	33.94	0.011	29.64087
1,3,5,-TRIBIPHENYLYLBENZENE (B)	37.52	0.008	54.08861
FLUORENE (TRACE)	23.84	0.033	3.60049
1-METHYL FLUORENE	22.35	0.040	2.42784
1,2-BENZOFLUORENE	27.72	0.021	8.77475
9,10-DIHYDROPHENANTHRENE	24.00	0.032	3.74220
CARBAZOLE (TRACE)	26.68	0.024	6.97694
N-METHYL CARBAZOLE (TRACE)	25.96	0.026	5.91548
N-VINYLCARBAZOLE	27.11	0.022	7.66483
N-PHENYL CARBAZOLE	27.78	0.021	8.87201
N-METHYL-2,3-BENZCARBAZOLE	31.86	0.014	20.21390
DIBENZOFURAN	26.25	0.025	6.42615
DIBENZOTHIOPHENE (TRACE)	23.48	0.035	3.24981
P-TERPHENYL	30.15	0.016	14.64653
DEUTERATED P-TERPHENYL-C18D14	29.79	0.017	13.64375
4-METHYL TERPHENYL	30.74	0.015	16.44486
2-METHYL-P-TERPHENYL	26.07	0.025	6.14740
3,3''-DIMETHYL-P-TERPHENYL	30.18	0.016	14.71534
4-(3,3-DIMETHYLBUTOXY)-P-TERPHENYL	31.71	0.014	19.73003
4,4''DIHEXAHYDROFARNESOXY-P-TERPHENYL	33.03	0.012	25.18734
2,2'-METHYLENE-P-TERPHENYL	31.58	0.014	19.27348
2,2'-ETHYLENE-P-TERPHENYL	31.59	0.014	19.29848
2,2'-METHYLENE-5',6'-METHYLENE-P-TERPHENYL (TRACE)	30.98	0.015	17.10970
3-PHENYL DIBENZOFURAN	30.90	0.015	16.91912
BENZO(1,2-B-4,5-B')BISBENZOFURAN (TRACE) (B)	27.89	0.021	9.08706
SEXIPHENYL BENZENE	23.41	0.035	3.23605
P-QUATERPHENYL	32.98	0.012	24.92279
3,3'''-DIMETHYL-P-QUATERPHENYL	33.36	0.012	26.65643
TETRAMETHYL-P-QUATERPHENYL	28.77	0.019	11.11045
3,3''' -DIMETHYL-3',2''-METHYLENE-P-QUATERPHENYL	32.34	0.013	22.11959
3,3''' -DIMETHYL-3',2''-ETHYLENE-P-QUATERPHENYL	33.16	0.012	25.69462
4-(3-ETHYLBUTOXY)-P-QUATERPHENYL (B)	32.95	0.013	24.73150
4,4'''-DI(2-BUTYLOCTOXY-1)-P-QUATERPHENYL	34.18	0.011	30.83580
DIETHYL-P-QUINQUEPHENYL	35.01	0.010	35.63406
DI(3-ETHYLHEPTYL)-P-QUINQUEPHENYL	34.86	0.011	34.64891
TETRAMETHYL-P-SEXIPHENYL	35.48	0.010	38.64781
TETRAMETHYL-P-OCTAPHENYL (B)	35.31	0.010	37.43967
INDOLE (TRACE)	19.14	0.064	0.97163
2-PHENYL INDOLE	29.72	0.017	13.34741
1-METHYL-2-PHENYL INDOLE	28.68	0.019	10.77174
1-NEOPENTYL-2-PHENYL INDOLE	28.48	0.019	10.36979
1-N-BUTYL-2(BIPHENYLYL-4) INDOLE	29.92	0.017	13.85614
1,2-DIPHENYL INDOLE	29.51	0.017	12.77787
1-BIPHENYLYL-2-PHENYL INDOLE	31.33	0.015	18.26141
1,3-DIMETHYL-2-PHENYL INDOLE	27.50	0.021	8.38212
2,3-DIPHENYL INDOLE	28.39	0.020	10.12121
1-METHYL-2,3-DIPHENYL INDOLE	27.72	0.021	8.76451
2-PHENYL-3,2'-METHYLENE INDOLE (TRACE)	28.05	0.020	9.42153
2-PHENYL-3,2'-ETHYLENE INDOLE (TRACE)	27.18	0.022	7.78224
2-PHENYL-3,2'-TRIMETHYLENE INDOLE	28.65	0.019	10.69616
1-METHYL-3,2'-METHYLENE-2-PHENYL INDOLE (TRACE)	27.46	0.022	8.28588
1-METHYL-3,2'-ETHYLENE-2-PHENYL INDOLE	26.67	0.024	6.95513
1-METHYL-3,2'-TRIMETHYLENE-2-PHENYL INDOLE	27.67	0.021	8.67492
4,5-DIPHENYLIMIDAZOLE (A)	26.07	0.025	6.07702
PPF (TRACE)	29.36	0.018	12.36622
PPO (TRACE)	31.56	0.014	19.10423
2-(4-FLUOROPHENYL)-5-PHENYLOXAZOLE (TRACE)	30.47	0.016	15.49889
αNPO (TRACE)	27.78	0.021	8.88301
BBO (B)	29.69	0.017	13.24113
POPOP	27.39	0.022	8.15783
DIMETHYL POPOP (TRACE)	28.44	0.019	10.21753
PPD (TRACE)	30.19	0.016	14.78970
2,5-DI(2-METHYLPHENYL)OXADIAZOLE	29.42	0.018	12.65505
PBD	33.56	0.012	27.65742
BBD (B)	32.87	0.013	24.36145
αNPD (TRACE)	27.97	0.020	9.24247
βNPD	31.66	0.014	19.49114
αNND (TRACE)	26.62	0.024	6.87833
βNND (TRACE)	31.80	0.014	19.98797
TRANS-1-(4-BIPHENYLYL)-2-PHENYL-ETHYLENE	31.59	0.014	19.19473
TRANS-1-(4-BIPHENYLYL)-2-(1-NAPHTHYL)-ETHYLENE	29.42	0.018	12.52999
DIPHENYLSTILBENE (B)	28.97	0.018	11.42228
1,1,4,4-TETRAPHENYLBUTADIENE	26.39	0.024	6.52983
1,6-DIPHENYLHEXATRIENE	26.70	0.024	6.99870

ACCEPTOR MOLECULES	RO	CO	OVERLAP
1,8-DIPHENYL-1,3,5,7-OCTATETRAENE	24.18	0.032	3.85849
BPSB (TRACE)	27.55	0.021	8.45396
BIS-MSB	28.54	0.019	10.43176
ESCULIN (A)	24.94	0.029	4.65016
AZULENE	28.28	0.020	9.89437
NAPHTHALENE	20.32	0.053	1.37263
DEUTERATED NAPHTHALENE-C10D8	20.46	0.052	1.42244
1-METHYLNAPHTHALENE 5ML/L	22.33	0.040	2.41614
2-METHYL NAPHTHALENE	20.73	0.050	1.54286
2,3-DIMETHYL NAPHTHALENE	21.40	0.046	1.87107
2,6-DIMETHYL NAPHTHALENE (TRACE)	21.01	0.048	1.67278
1-NAPHTHOL (TRACE)	23.29	0.035	3.10190
2-NAPHTHOL (TRACE)	20.45	0.052	1.41615
1-NAPHTHYLAMINE	22.26	0.041	2.35258
2-CHLORONAPHTHALENE	20.74	0.050	1.54810
ACENAPHTHENE	23.82	0.033	3.55401
1-PHENYL NAPHTHALENE	25.96	0.026	5.94769
2-PHENYL NAPHTHALENE	38.32	0.008	61.38140
1,4-DIPHENYL NAPHTHALENE	27.69	0.021	8.70738
1,5-DIPHENYL NAPHTHALENE	27.21	0.022	7.86174
1,7-DIPHENYL NAPHTHALENE	26.92	0.023	7.35927
1,4,5,8-TETRAPHENYL NAPHTHALENE	26.45	0.024	6.62168
1,1′-BINAPHTHYL	27.39	0.022	8.20557
4,4′-DI(N-BUTOXY)-1,1′-BINAPHTHYL	27.74	0.021	8.79637
2,2′-BINAPHTHYL	30.02	0.017	14.17231
ACRIDINE (A)	18.27	0.073	0.71840
ANTHRACENE (TRACE)	16.60	0.098	0.40387
DEUTERATED ANTHRACENE C14D10 (TRACE)	17.25	0.087	0.50825
9-METHYL ANTHRACENE (TRACE)	15.83	0.113	0.30442
1-AMINO ANTHRACENE	23.66	0.034	3.38929
2-AMINO ANTHRACENE	15.25	0.126	0.23872
9-VINYL ANTHRACENE	17.28	0.087	0.51460
9-PHENYL ANTHRACENE (TRACE)	18.01	0.077	0.65976
9,10-DIPHENYL ANTHRACENE	19.52	0.060	1.07007
9,10-DICHLOROANTHRACENE (TRACE)	15.83	0.113	0.30439
BIANTHRANYL (B)	22.00	0.042	2.18907
9,10-DINAPHTHYLANTHRACENE (B)	27.16	0.022	7.74806
9-(P-VINYLPHENYL)-10-PHENYL ANTHRACENE	22.83	0.038	2.73338
9,10-BIS(PHENYL-ETHYNYL)ANTHRACENE (TRACE)	30.78	0.015	16.42437
RUBICENE (B)	32.34	0.013	22.09640
DIBENZ(A,H)ANTHRACENE (B)	36.04	0.010	42.33781
TETRACENE (TRACE) (B)	35.74	0.010	40.27456
RUBRENE (B)	35.86	0.010	41.07213
TETRAHYDROCANNABINOL ACID METHYL ESTER	27.45	0.022	8.35033
PHENANTHRENE	24.69	0.030	4.37567
3,4-BENZOPHENANTHRENE	31.66	0.014	19.44852
2-PHENYL PHENANTHRENE	31.60	0.014	19.22115
CHRYSENE (TRACE)	28.03	0.020	9.36572
TRIPHENYLENE	26.44	0.024	6.60302
DEUTERATED TRIPHENYLENE-C18D12	26.43	0.024	6.58823
PYRENE	25.36	0.027	5.13815
1,3,6,8,-TETRAPHENYL PYRENE	32.29	0.013	21.92013
BIPYRENYL (B)	31.66	0.014	19.45996
O-PHENYLENE-PYRENE	31.30	0.015	18.18216
ANTHANTHRENE (B)	31.30	0.015	18.18378
FLUORANTHENE	27.58	0.021	8.49581
3-PHENYL FLUORANTHENE	29.70	0.017	13.26621
3-AMINO FLUORANTHENE	27.71	0.021	8.75108
3,3′-BIFLUORANTHENYL (B)	31.68	0.014	19.53874
BENZO(GHI)FLUORANTHENE	27.01	0.023	7.50133
9,12-DI(N-OCTADECOXY)-BENZ(E)ACEPHENANTHRYLENE	30.79	0.015	16.44797
DECACYCLENE (B)	30.82	0.015	16.57486
10-ETHOXY-BENZO(KL)XANTHENE	23.37	0.035	3.14419
PERYLENE	19.01	0.065	0.91201
BENZO(GHI)PERYLENE (B)	32.29	0.013	21.90188
BBOT (TRACE)	25.00	0.029	4.71376
QUININE SULFATE (A)	24.69	0.030	4.37363
ACRIDINE YELLOW (A)	26.29	0.025	6.37496
ACRIDINE ORANGE (A)	27.42	0.022	8.21957
ACRIDINE RED (A)	24.43	0.031	4.11246
URANINE (A)	24.56	0.030	4.23652
RHODAMINE B (A)	28.10	0.020	9.50632
RHODAMINE 6G (A)	27.92	0.021	9.15959
EOSIN B (A)	25.30	0.028	5.06278
EOSIN Y (A)	26.07	0.025	6.06353
ROSE BENGAL (A)	26.20	0.025	6.24989
INDENOINDENE	28.83	0.019	11.09575

8.25 Donor Molecule: Deuterated *p*-Xylene (C_8D_{10}) (trace)

BENZENE (3 ML/L)	2.13	46.385	0.00000
DEUTERATED BENZENE-C6D6 5ML/L	0.00	0.000	0.00000
FLUOROBENZENE (TRACE)	4.49	4.936	0.00025
TOLUENE (TRACE)	5.27	3.048	0.00066
DEUTERATED TOLUENE-C7D8 5ML/L	4.79	4.062	0.00037
ETHYLBENZENE (TRACE)	5.27	3.054	0.00066
PROPYLBENZENE (TRACE)	5.07	3.442	0.00052
ISOPROPYLBENZENE (TRACE)	4.72	4.245	0.00034
SEC-BUTYLBENZENE (TRACE)	4.65	4.462	0.00031
DIPHENYLMETHANE (TRACE)	6.42	1.689	0.00215
BIBENZYL	5.92	2.160	0.00132

ACCEPTOR MOLECULES	RO	CO	OVERLAP
1,3,5-TRIPHENYLPENTANE 0.05ML/L	6.11	1.958	0.00160
POLYSTYRENE	5.34	2.937	0.00071
POLYVINYLTOLUENE	9.15	0.584	0.01779
PHENYLCYCLOHEXANE (TRACE)	4.02	6.906	0.00013
BENZYLALCOHOL 3ML/L	4.80	4.057	0.00037
BENZYLACETATE 5ML/L	4.37	5.373	0.00021
PHENOL (TRACE)	13.11	0.198	[illegible]
ANISOLE (TRACE)	13.57	0.179	0.19094
PHENYL ETHER	14.48	0.147	0.28071
ANILINE 3ML/L	17.43	0.084	0.84695
N,N-DIMETHYLANILINE 3ML/L	18.73	0.068	1.29627
TRIPHENYLAMINE	28.02	0.020	14.39651
P-XYLENE (TRACE)	9.84	0.470	0.02773
DEUTERATED P-XYLENE C8D10 (TRACE)	9.75	0.483	0.02624
M-XYLENE (TRACE)	7.70	0.979	0.00639
O-XYLENE 3ML/L	7.11	1.245	0.00395
1-METHYL-4-ETHYLBENZENE (TRACE)	9.10	0.593	0.01741
1,4-BIS(TRIFLUOROMETHYL)BENZENE 5ML/L	6.01	2.061	0.00145
P-CRESOL	16.48	0.100	0.60964
P-METHYLANISOLE 3ML/L	15.94	0.110	0.49888
1,4-DIMETHOXY BENZENE	19.07	0.064	1.45244
1-METHYL-4-CHLOROBENZENE 5ML/L	10.43	0.394	0.03943
SODIUM SALICYLATE (A)	20.36	0.053	2.11619
ANTHRANILIC ACID (A)	17.53	0.083	0.86244
TMPD	21.94	0.042	3.31498
1,2,4-TRIMETHYLBENZENE (TRACE)	10.80	0.355	0.04845
MESITYLENE 5ML/L	8.01	0.871	0.00808
1,3,5-TRIETHYLBENZENE 5ML/L	7.56	1.036	0.00570
STYRENE	14.07	0.161	0.23524
1,1-DIPHENYLETHYLENE 3ML/L	15.66	0.116	0.44867
BIPHENYL	14.62	0.143	0.29784
DEUTERATED BIPHENYL-C12D10	14.61	0.144	0.29640
4-METHYL BIPHENYL	17.08	0.090	0.75541
MONOISOPROPYLBIPHENYL 5ML/L	16.45	0.101	0.60345
4-BENZYL BIPHENYL	18.86	0.067	1.36771
3,3′-DIMETHYL BIPHENYL	16.43	0.101	0.59810
2,2′-DIMETHYLBIPHENYL 2ML/L	10.63	0.373	0.04382
OCTAMETHYL BIPHENYL	12.86	0.210	0.13782
4-METHOXY BIPHENYL	21.72	0.044	3.17431
4-BIPHENYLYL PHENYL ETHER	22.22	0.041	3.64248
4,4′-DIHYDROXYBIPHENYL (A)	24.16	0.032	5.98168
4,4′-DIMETHOXY BIPHENYL	23.98	0.032	5.73107
4,4′-DIISOBUTOXYBIPHENYL	23.83	0.033	5.52434
BENZIDINE	28.66	0.019	16.51731
4-VINYL BIPHENYL	27.31	0.022	12.49548
M-TERPHENYL	20.49	0.052	2.22850
M-QUATERPHENYL	20.51	0.052	2.25589
1,3,5-TRIPHENYL BENZENE	21.41	0.046	2.88896
1,3-DIBIPHENYLYLBENZENE (B)	31.60	0.014	29.70058
1,3,5,-TRIBIPHENYLYLBENZENE (B)	34.88	0.011	53.81267
FLUORENE (TRACE)	22.29	0.040	3.70712
1-METHYL FLUORENE	20.69	0.051	2.35107
1,2-BENZOFLUORENE	25.73	0.026	8.62700
9,10-DIHYDROPHENANTHRENE	22.51	0.039	3.92576
CARBAZOLE (TRACE)	24.80	0.029	6.93273
N-METHYL CARBAZOLE (TRACE)	24.11	0.032	5.85079
N-VINYLCARBAZOLE	25.30	0.028	7.80357
N-PHENYL CARBAZOLE	25.85	0.026	8.87475
N-METHYL-2,3-BENZCARBAZOLE	29.78	0.017	20.74004
DIBENZOFURAN	24.51	0.030	6.54867
DIBENZOTHIOPHENE (TRACE)	21.88	0.043	3.27464
P-TERPHENYL	28.15	0.020	14.95577
DEUTERATED P-TERPHENYL-C18D14	27.82	0.021	13.93266
4-METHYL TERPHENYL	28.66	0.019	16.65360
2-METHYL-P-TERPHENYL	24.46	0.031	6.46273
3,3″-DIMETHYL-P-TERPHENYL	28.15	0.020	14.92509
4-(3,3-DIMETHYLBUTOXY)-P-TERPHENYL	29.50	0.017	19.69281
4,4″DIHEXAHYDROFARNESOXY-P-TERPHENYL	30.69	0.015	24.95730
2,2′-METHYLENE-P-TERPHENYL	29.35	0.018	19.11381
2,2′-ETHYLENE-P-TERPHENYL	29.35	0.018	19.10259
2,2′-METHYLENE-5′,6″-METHYLENE-P-TERPHENYL (TRACE)	28.71	0.019	16.65946
3-PHENYL DIBENZOFURAN	28.62	0.019	16.42865
BENZO(1,2-B-4,5-B′)BISBENZOFURAN (TRACE) (B)	25.72	0.026	8.60347
SEXIPHENYL BENZENE	22.02	0.042	3.45579
P-QUATERPHENYL	30.62	0.016	24.59327
3,3‴-DIMETHYL-P-QUATERPHENYL	30.97	0.015	26.25076
TETRAMETHYL-P-QUATERPHENYL	26.96	0.023	11.58892
3,3‴ -DIMETHYL-3′,2″-METHYLENE-P-QUATERPHENYL	29.93	0.017	21.39426
3,3‴ -DIMETHYL-3′,2″-ETHYLENE-P-QUATERPHENYL	30.77	0.015	25.24236
4-(3-ETHYLBUTOXY)-P-QUATERPHENYL (B)	30.53	0.016	24.06834
4,4‴-DI(2-BUTYLOCTOXY-1)-P-QUATERPHENYL	31.68	0.014	30.05333
DIETHYL-P-QUINQUEPHENYL	32.50	0.013	35.12686
DI(3-ETHYLHEPTYL)-P-QUINQUEPHENYL	32.29	0.013	[illegible]
TETRAMETHYL-P-SEXIPHENYL	32.96	0.012	38.23077
TETRAMETHYL-P-OCTAPHENYL (B)	32.71	0.013	36.41740
INDOLE (TRACE)	18.06	0.076	1.05794
2-PHENYL INDOLE	27.51	0.021	12.92562
1-METHYL-2-PHENYL INDOLE	26.60	0.024	10.54525
1-NEOPENTYL-2-PHENYL INDOLE	26.43	0.024	10.18626
1-N-BUTYL-2(BIPHENYLYL-4) INDOLE	27.72	0.021	13.50023
1,2-DIPHENYL INDOLE	27.38	0.022	12.55120
1-BIPHENYLYL-2-PHENYL INDOLE	29.11	0.018	18.09886
1,3-DIMETHYL-2-PHENYL INDOLE	25.50	0.027	8.19918
2,3-DIPHENYL INDOLE	26.32	0.025	9.88563
1-METHYL-2,3-DIPHENYL INDOLE	25.75	0.026	8.68016
2-PHENYL-3,2′-METHYLENE INDOLE (TRACE)	25.92	0.026	9.03129

ACCEPTOR MOLECULES	RO	CO	OVERLAP
2-PHENYL-3,2'-ETHYLENE INDOLE (TRACE)	25.12	0.028	7.48336
2-PHENYL-3,2'-TRIMETHYLENE INDOLE	26.50	0.024	10.29174
1-METHYL-3,2'-METHYLENE-2-PHENYL INDOLE (TRACE)	25.36	0.027	7.91755
1-METHYL-3,2'-ETHYLENE-2-PHENYL INDOLE	24.66	0.030	6.69436
1-METHYL-3,2'-TRIMETHYLENE-2-PHENYL INDOLE	25.60	0.027	8.36556
4,5-DIPHENYLIMIDAZOLE (A)	24.27	0.031	6.08753
PPF (TRACE)	27.13	0.022	11.87057
PPO (TRACE)	29.23	0.018	18.58233
2-(4-FLUOROPHENYL)-5-PHENYLOXAZOLE (TRACE)	28.24	0.020	15.10744
αNPO (TRACE)	25.79	0.026	8.75926
BBO (B)	27.51	0.021	12.89057
POPOP	25.38	0.027	7.94961
DIMETHYL POPOP (TRACE)	26.39	0.024	10.05348
PPD (TRACE)	28.13	0.020	14.90611
2,5-DI(2-METHYLPHENYL)OXADIAZOLE	27.41	0.022	12.74204
PBD	31.11	0.015	27.04190
BBD (B)	30.43	0.016	23.63623
αNPD (TRACE)	25.94	0.026	9.06529
βNPD	29.46	0.017	19.47016
αNND (TRACE)	24.63	0.030	6.63643
βNND (TRACE)	29.58	0.017	19.91447
TRANS-1-(4-BIPHENYLYL)-2-PHENYL-ETHYLENE	29.22	0.018	18.50406
TRANS-1-(4-BIPHENYLYL)-2-(1-NAPHTHYL)-ETHYLENE	27.28	0.022	12.26424
DIPHENYLSTILBENE (B)	26.74	0.023	10.86873
1,1,4,4-TETRAPHENYLBUTADIENE	24.39	0.031	6.26724
1,6-DIPHENYLHEXATRIENE	24.60	0.030	6.58602
1,8-DIPHENYL-1,3,5,7-OCTATETRAENE	22.38	0.040	3.73624
BPSB (TRACE)	25.43	0.027	8.04403
BIS-MSB	26.35	0.024	9.95436
ESCULIN (A)	23.08	0.036	4.49839
AZULENE	26.66	0.024	10.69304
NAPHTHALENE	19.06	0.065	1.44118
DEUTERATED NAPHTHALENE-C10D8	19.20	0.063	1.49498
1-METHYLNAPHTHALENE 5ML/L	20.88	0.049	2.48199
2-METHYL NAPHTHALENE	19.42	0.061	1.60996
2,3-DIMETHYL NAPHTHALENE	20.05	0.056	1.94487
2,6-DIMETHYL NAPHTHALENE (TRACE)	19.68	0.059	1.73878
1-NAPHTHOL (TRACE)	21.65	0.044	3.08435
2-NAPHTHOL (TRACE)	19.12	0.064	1.46020
1-NAPHTHYLAMINE	20.59	0.051	2.26766
2-CHLORONAPHTHALENE	19.44	0.061	1.61554
ACENAPHTHENE	22.17	0.041	3.56098
1-PHENYL NAPHTHALENE	24.15	0.032	5.94718
2-PHENYL NAPHTHALENE	35.68	0.010	61.58386
1,4-DIPHENYL NAPHTHALENE	25.66	0.026	8.50511
1,5-DIPHENYL NAPHTHALENE	25.23	0.028	7.69217
1,7-DIPHENYL NAPHTHALENE	24.98	0.029	7.24861
1,4,5,8-TETRAPHENYL NAPHTHALENE	24.53	0.030	6.48021
1,1'-BINAPHTHYL	25.49	0.027	8.19582
4,4'-DI(N-BUTOXY)-1,1'-BINAPHTHYL	25.68	0.026	8.53133
2,2'-BINAPHTHYL	27.93	0.021	14.16399
ACRIDINE (A)	16.89	0.093	0.68987
ANTHRACENE (TRACE)	15.29	0.125	0.38012
DEUTERATED ANTHRACENE C14D10 (TRACE)	15.92	0.111	0.48492
9-METHYL ANTHRACENE (TRACE)	14.69	0.141	0.29881
1-AMINO ANTHRACENE	22.36	0.040	3.71459
2-AMINO ANTHRACENE	14.00	0.163	0.21994
9-VINYL ANTHRACENE	16.12	0.107	0.52163
9-PHENYL ANTHRACENE (TRACE)	16.83	0.094	0.67649
9,10-DIPHENYL ANTHRACENE	18.46	0.071	1.17617
9,10 DICHLORDANTHRACENE (TRACE)	14.75	0.139	0.30689
BIANTHRANYL (B)	20.97	0.049	2.52902
9,10-DINAPHTHYLANTHRACENE (B)	25.34	0.027	7.88177
9-(P-VINYLPHENYL)-10-PHENYL ANTHRACENE	21.38	0.046	2.84263
9,10-BIS(PHENYL-ETHYNYL)ANTHRACENE (TRACE)	28.76	0.019	16.84087
RUBICENE (B)	30.03	0.017	21.83894
DIBENZ(A,H)ANTHRACENE (B)	33.40	0.012	41.26547
TETRACENE (TRACE) (B)	33.68	0.012	43.45249
RUBRENE (B)	33.16	0.012	39.56500
TETRAHYDROCANNABINOL ACID METHYL ESTER	25.63	0.027	8.52517
PHENANTHRENE	23.11	0.036	4.53703
3,4-BENZOPHENANTHRENE	29.68	0.017	20.32799
2-PHENYL PHENANTHRENE	29.57	0.017	19.87861
CHRYSENE (TRACE)	26.15	0.025	9.51513
TRIPHENYLENE	24.73	0.030	6.80196
DEUTERATED TRIPHENYLENE-C18D12	24.74	0.030	6.82542
PYRENE	23.74	0.033	5.33122
1,3,6,8,-TETRAPHENYL PYRENE	29.96	0.017	21.51655
BIPYRENYL (B)	29.61	0.017	20.06486
O-PHENYLENE-PYRENE	29.06	0.018	17.93009
ANTHANTHRENE (B)	29.00	0.018	17.69245
FLUORANTHENE	25.70	0.026	8.56844
3-PHENYL FLUORANTHENE	27.58	0.021	13.09603
3-AMINO FLUORANTHENE	25.70	0.026	8.56883
3,3'-BIFLUORANTHENYL (B)	29.43	0.018	19.33019
BENZO(GHI)FLUORANTHENE	25.14	0.028	7.50742
9,12-DI(N-OCTADECOXY)-BENZ(E)ACEPHENANTHRYLENE	28.76	0.019	16.84800
DECACYCLENE (B)	28.63	0.019	16.39616
10-ETHOXY-BENZO(KL)XANTHENE	21.68	0.044	3.08863
PERYLENE	17.78	0.080	0.94059
BENZO(GHI)PERYLENE (B)	29.97	0.017	21.57428
BBOT (TRACE)	23.11	0.036	4.53311
QUININE SULFATE (A)	22.95	0.037	4.34827
ACRIDINE YELLOW (A)	24.65	0.030	6.67445
ACRIDINE ORANGE (A)	25.58	0.027	8.34074
ACRIDINE RED (A)	22.72	0.038	4.09670

ACCEPTOR MOLECULES	RO	CO	OVERLAP
URANINE (A)	22.83	0.038	4.21205
RHODAMINE B (A)	26.13	0.025	9.47709
RHODAMINE 6G (A)	26.01	0.025	9.20645
EOSIN B (A)	23.49	0.035	4.99734
EOSIN Y (A)	24.20	0.032	5.98227
ROSE BENGAL (A)	24.42	0.031	6.30751
INDENOINDENE	28.65	0.024	10.67076

8.26 Donor Molecule: *m*-Xylene (trace)

ACCEPTOR MOLECULES	RO	CO	OVERLAP
BENZENE (3 ML/L)	2.56	26.727	0.00001
DEUTERATED BENZENE-C6D6 5ML/L	1.80	76.699	0.00000
FLUOROBENZENE (TRACE)	4.75	4.186	0.00054
TOLUENE (TRACE)	5.41	2.825	0.00118
DEUTERATED TOLUENE-C7D8 5ML/L	4.85	3.921	0.00061
ETHYLBENZENE (TRACE)	5.34	2.942	0.00109
PROPYLBENZENE (TRACE)	5.10	3.367	0.00083
ISOPROPYLBENZENE (TRACE)	4.75	4.166	0.00054
SEC-BUTYLBENZENE (TRACE)	4.72	4.256	0.00052
DIPHENYLMETHANE (TRACE)	6.50	1.632	0.00353
BIBENZYL	5.93	2.142	0.00205
1,3,5-TRIPHENYLPENTANE 0.05ML/L	6.10	1.971	0.00242
POLYSTYRENE	5.42	2.806	0.00119
POLYVINYLTOLUENE	8.74	0.670	0.02070
PHENYLCYCLOHEXANE (TRACE)	4.34	5.483	0.00031
BENZYLALCOHOL 3ML/L	4.80	4.034	0.00058
BENZYLACETATE 5ML/L	4.53	4.816	0.00041
PHENOL (TRACE)	12.48	0.230	0.17678
ANISOLE (TRACE)	12.89	0.209	0.21483
PHENYL ETHER	13.65	0.176	0.30137
ANILINE 3ML/L	16.23	0.105	0.84396
N,N-DIMETHYLANILINE 3ML/L	17.42	0.085	1.28113
TRIPHENYLAMINE	26.01	0.025	14.11467
P-XYLENE (TRACE)	9.37	0.544	0.03169
DEUTERATED P-XYLENE C8D10 (TRACE)	9.33	0.550	0.03097
M-XYLENE (TRACE)	7.56	1.037	0.00872
O-XYLENE 3ML/L	7.02	1.291	0.00563
1-METHYL-4-ETHYLBENZENE (TRACE)	8.78	0.662	0.02143
1,4-BIS(TRIFLUOROMETHYL)BENZENE 5ML/L	6.18	1.896	0.00262
P-CRESOL	15.44	0.122	0.63043
P-METHYLANISOLE 3ML/L	14.93	0.134	0.51569
1,4-DIMETHOXY BENZENE	17.73	0.080	1.43457
1-METHYL-4-CHLOROBENZENE 5ML/L	9.91	0.459	0.04445
SODIUM SALICYLATE (A)	18.89	0.066	2.07318
ANTHRANILIC ACID (A)	16.25	0.104	0.83787
TMPD	20.55	0.052	3.43383
1,2,4-TRIMETHYLBENZENE (TRACE)	10.27	0.413	0.05474
MESITYLENE 5ML/L	7.72	0.972	0.00992
1,3,5-TRIETHYLBENZENE 5ML/L	7.31	1.143	0.00717
STYRENE	13.15	0.197	0.23999
1,1-DIPHENYLETHYLENE 3ML/L	14.77	0.139	0.48262
BIPHENYL	13.91	0.166	0.33830
DEUTERATED BIPHENYL-C12D10	13.91	0.166	0.33842
4-METHYL BIPHENYL	16.15	0.106	0.82582
MONOISOPROPYLBIPHENYL 5ML/L	15.57	0.118	0.66545
4-BENZYL BIPHENYL	17.81	0.079	1.48544
3,3'-DIMETHYL BIPHENYL	15.47	0.121	0.63685
2,2'-DIMETHYLBIPHENYL 2ML/L	10.16	0.427	0.05120
OCTAMETHYL BIPHENYL	12.10	0.252	0.14648
4-METHOXY BIPHENYL	20.38	0.053	3.31357
4-BIPHENYLYL PHENYL ETHER	20.86	0.049	3.82145
4,4'-DIHYDROXYBIPHENYL (A)	22.60	0.039	6.12841
4,4'-DIMETHOXY BIPHENYL	22.44	0.040	5.88756
4,4'-DIISOBUTOXYBIPHENYL	22.29	0.040	5.66247
BENZIDINE	26.68	0.024	16.43941
4-VINYL BIPHENYL	25.49	0.027	12.63923
M-TERPHENYL	19.25	0.063	2.35090
M-QUATERPHENYL	19.35	0.062	2.43335
1,3,5-TRIPHENYL BENZENE	20.21	0.054	3.12871
1,3-DIBIPHENYLYLBENZENE (B)	29.42	0.018	29.62337
1,3,5,-TRIBIPHENYLYLBENZENE (B)	32.45	0.013	53.41254
FLUORENE (TRACE)	20.85	0.049	3.80077
1-METHYL FLUORENE	19.18	0.063	2.28481
1,2-BENZOFLUORENE	23.91	0.033	8.51976
9,10-DIHYDROPHENANTHRENE	21.08	0.048	4.05063
CARBAZOLE (TRACE)	23.05	0.037	6.84656
N-METHYL CARBAZOLE (TRACE)	22.41	0.040	5.77461
N-VINYLCARBAZOLE	23.59	0.034	7.85375
N-PHENYL CARBAZOLE	24.05	0.032	8.82781
N-METHYL-2,3-BENZCARBAZOLE	27.80	0.021	21.04538
DIBENZOFURAN	22.85	0.038	6.58260
DIBENZOTHIOPHENE (TRACE)	20.42	0.053	3.30936
P-TERPHENYL	26.27	0.025	15.10668
DEUTERATED P-TERPHENYL-C18D14	25.95	0.026	14.07504
4-METHYL TERPHENYL	26.72	0.023	16.73010
2-METHYL-P-TERPHENYL	22.90	0.037	6.66614
3,3''-DIMETHYL-P-TERPHENYL	26.25	0.025	15.01357
4-(3,3-DIMETHYLBUTOXY)-P-TERPHENYL	27.45	0.022	19.59431
4,4''-DIHEXAHYDROFARNESOXY-P-TERPHENYL	28.54	0.019	24.71880
2,2'-METHYLENE-P-TERPHENYL	27.30	0.022	18.94243

ACCEPTOR MOLECULES	RO	CO	OVERLAP
2,2'-ETHYLENE-P-TERPHENYL	27.30	0.022	18.90996
2,2'-METHYLENE-5',6'-METHYLENE-P-TERPHENYL (TRACE)	26.66	0.024	16.37762
3-PHENYL DIBENZOFURAN	26.56	0.024	16.07691
BENZO(1,2-B-4,5-B')BISBENZOFURAN (TRACE) (B)	23.83	0.033	8.33608
SEXIPHENYL BENZENE	20.64	0.051	3.58511
P-QUATERPHENYL	28.47	0.019	24.29597
3,3'''-DIMETHYL-P-QUATERPHENYL	28.78	0.019	25.90680
TETRAMETHYL-P-QUATERPHENYL	25.23	0.028	11.89314
3,3'''-DIMETHYL-3',2''-METHYLENE-P-QUATERPHENYL	27.78	0.021	20.93019
3,3'''-DIMETHYL-3',2''-ETHYLENE-P-QUATERPHENYL	28.59	0.019	24.87679
4-(3-ETHYLBUTOXY)-P-QUATERPHENYL (B)	28.34	0.020	23.59840
4,4'''-DI(2-BUTYLOCTOXY-1)-P-QUATERPHENYL	29.41	0.018	29.48946
DIETHYL-P-QUINQUEPHENYL	30.21	0.016	34.68648
DI(3-ETHYLHEPTYL)-P-QUINQUEPHENYL	29.97	0.017	33.01670
TETRAMETHYL-P-SEXIPHENYL	30.65	0.016	37.83495
TETRAMETHYL-P-OCTAPHENYL (B)	30.37	0.016	35.77584
INDOLE (TRACE)	16.97	0.092	1.11157
2-PHENYL INDOLE	25.53	0.027	12.63347
1-METHYL-2-PHENYL INDOLE	24.70	0.030	10.36336
1-NEOPENTYL-2-PHENYL INDOLE	24.55	0.030	10.02956
1-N-BUTYL-2(BIPHENYLYL-4) INDOLE	25.75	0.026	13.28516
1,2-DIPHENYL INDOLE	25.45	0.027	12.38238
1-BIPHENYLYL-2-PHENYL INDOLE	27.07	0.023	17.94550
1,3-DIMETHYL-2-PHENYL INDOLE	23.69	0.034	8.05970
2,3-DIPHENYL INDOLE	24.45	0.031	9.73181
1-METHYL-2,3-DIPHENYL INDOLE	23.96	0.033	8.62743
2-PHENYL-3,2'-METHYLENE INDOLE (TRACE)	24.04	0.032	8.79573
2-PHENYL-3,2'-ETHYLENE INDOLE (TRACE)	23.32	0.035	7.32311
2-PHENYL-3,2'-TRIMETHYLENE INDOLE	24.57	0.030	10.03093
1-METHYL-3,2'-METHYLENE-2-PHENYL INDOLE (TRACE)	23.52	0.034	7.71651
1-METHYL-3,2'-ETHYLENE-2-PHENYL INDOLE	22.89	0.037	6.56028
1-METHYL-3,2'-TRIMETHYLENE-2-PHENYL INDOLE	23.74	0.033	8.16723
4,5-DIPHENYLIMIDAZOLE (A)	22.59	0.039	6.07395
PPF (TRACE)	25.17	0.028	11.58409
PPO (TRACE)	27.14	0.022	18.21722
2-(4-FLUOROPHENYL)-5-PHENYLOXAZOLE (TRACE)	26.22	0.025	14.82919
αNPO (TRACE)	23.99	0.032	8.68884
BBO (B)	25.57	0.027	12.72961
POPOP	23.59	0.034	7.85281
DIMETHYL POPOP (TRACE)	24.55	0.030	9.98147
PPD (TRACE)	26.21	0.025	14.92814
2,5-DI(2-METHYLPHENYL)OXADIAZOLE	25.54	0.027	12.75336
PBD	28.89	0.019	26.56785
BBD (B)	28.26	0.020	23.19142
αNPD (TRACE)	24.12	0.032	8.97919
βNPD	27.45	0.022	19.49154
αNND (TRACE)	22.86	0.037	6.50639
βNND (TRACE)	27.54	0.021	19.89620
TRANS-1-(4-BIPHENYLYL)-2-PHENYL-ETHYLENE	27.11	0.022	18.07807
TRANS-1-(4-BIPHENYLYL)-2-(1-NAPHTHYL)-ETHYLENE	25.36	0.027	12.12368
DIPHENYLSTILBENE (B)	24.79	0.029	10.58997
1,1,4,4-TETRAPHENYLBUTADIENE	22.65	0.038	6.15296
1,6-DIPHENYLHEXATRIENE	22.80	0.038	6.39818
1,8-DIPHENYL-1,3,5,7-OCTATETRAENE	20.80	0.050	3.69440
BPSB (TRACE)	23.59	0.034	7.85304
BIS-MSB	24.44	0.031	9.71599
ESCULIN (A)	21.43	0.045	4.41132
AZULENE	24.99	0.029	11.09987
NAPHTHALENE	17.87	0.078	1.49778
DEUTERATED NAPHTHALENE-C10D8	17.99	0.077	1.55165
1-METHYLNAPHTHALENE 5ML/L	19.47	0.061	2.50253
2-METHYL NAPHTHALENE	18.17	0.075	1.65316
2,3-DIMETHYL NAPHTHALENE	18.73	0.068	1.97953
2,6-DIMETHYL NAPHTHALENE (TRACE)	18.41	0.072	1.78301
1-NAPHTHOL (TRACE)	20.14	0.055	3.05853
2-NAPHTHOL (TRACE)	17.89	0.078	1.50016
1-NAPHTHYLAMINE	19.10	0.064	2.21464
2-CHLORONAPHTHALENE	18.19	0.074	1.65933
ACENAPHTHENE	20.64	0.051	3.54835
1-PHENYL NAPHTHALENE	22.48	0.039	5.91906
2-PHENYL NAPHTHALENE	33.22	0.012	61.50444
1,4-DIPHENYL NAPHTHALENE	23.83	0.033	8.35166
1,5-DIPHENYL NAPHTHALENE	23.43	0.035	7.55878
1,7-DIPHENYL NAPHTHALENE	23.23	0.036	7.17187
1,4,5,8-TETRAPHENYL NAPHTHALENE	22.81	0.038	6.42520
1,1'-BINAPHTHYL	23.72	0.034	8.15015
4,4'-DI(N-BUTOXY)-1,1'-BINAPHTHYL	23.84	0.033	8.35763
2,2'-BINAPHTHYL	26.02	0.025	14.16760
ACRIDINE (A)	15.69	0.116	0.68121
ANTHRACENE (TRACE)	14.19	0.157	0.37200
DEUTERATED ANTHRACENE C14D10 (TRACE)	14.79	0.138	0.47619
9-METHYL ANTHRACENE (TRACE)	13.71	0.173	0.30298
1-AMINO ANTHRACENE	21.09	0.048	4.00928
2-AMINO ANTHRACENE	12.97	0.205	0.21323
9-VINYL ANTHRACENE	15.09	0.130	0.53811
9-PHENYL ANTHRACENE (TRACE)	15.78	0.114	0.70436
9,10-DIPHENYL ANTHRACENE	17.47	0.084	1.29476
9,10-DICHLOROANTHRACENE (TRACE)	13.81	0.170	0.31610
BIANTHRANYL (B)	20.03	0.056	2.94267
9,10-DINAPHTHYLANTHRACENE (B)	23.64	0.034	7.96165
9-(P-VINYLPHENYL)-10-PHENYL ANTHRACENE	20.04	0.056	2.95213
9,10-BIS(PHENYL-ETHYNYL)ANTHRACENE (TRACE)	26.93	0.023	17.36473
RUBICENE (B)	27.93	0.021	21.63433
DIBENZ(A,H)ANTHRACENE (B)	31.00	0.015	40.42150
TETRACENE (TRACE) (B)	31.61	0.014	45.45347
RUBRENE (B)	30.75	0.015	38.49249

ACCEPTOR MOLECULES	RO	CO	OVERLAP
TETRAHYDROCANNABINOL ACID METHYL ESTER	23.93	0.033	8.63913
PHENANTHRENE	21.58	0.044	4.60751
3,4-BENZOPHENANTHRENE	27.69	0.021	20.53632
2-PHENYL PHENANTHRENE	27.64	0.021	20.33502
CHRYSENE (TRACE)	24.46	0.031	9.76934
TRIPHENYLENE	23.12	0.036	6.96962
DEUTERATED TRIPHENYLENE-C18D12	23.14	0.036	[illegible]
PYRENE	22.36	0.040	5.69223
1,3,6,8,-TETRAPHENYL PYRENE	27.83	0.021	21.15694
BIPYRENYL (B)	27.60	0.021	20.12568
O-PHENYLENE-PYRENE	27.02	0.023	17.73522
ANTHANTHRENE (B)	26.90	0.023	17.27883
FLUORANTHENE	24.00	0.032	8.70872
3-PHENYL FLUORANTHENE	25.64	0.027	12.95357
3-AMINO FLUORANTHENE	23.89	0.033	8.46547
3,3'-BIFLUORANTHENYL (B)	27.37	0.022	19.14676
BENZO(GHI)FLUORANTHENE	23.39	0.035	7.47002
9,12-DI(N-OCTADECOXY)-BENZ(E)ACEPHENANTHRYLENE	26.82	0.023	16.97693
DECACYCLENE (B)	26.65	0.024	16.32659
10-ETHOXY-BENZO(KL)XANTHENE	20.15	0.055	3.05030
PERYLENE	16.64	0.097	0.96682
BENZO(GHI)PERYLENE (B)	27.85	0.021	21.28069
BBOT (TRACE)	21.46	0.045	4.45604
QUININE SULFATE (A)	21.37	0.046	4.33517
ACRIDINE YELLOW (A)	23.09	0.036	6.91203
ACRIDINE ORANGE (A)	23.88	0.033	8.45944
ACRIDINE RED (A)	21.15	0.047	4.07890
URANINE (A)	21.25	0.047	4.19419
RHODAMINE B (A)	24.32	0.031	9.43742
RHODAMINE 6G (A)	24.22	0.031	9.21031
EOSIN B (A)	21.85	0.043	4.96492
EOSIN Y (A)	22.52	0.039	5.93857
ROSE BENGAL (A)	22.79	0.038	6.37999
INDENOINDENE	24.72	0.030	10.39354

8.27 Donor Molecule: *o*-Xylene (trace)

ACCEPTOR MOLECULES	RO	CO	OVERLAP
BENZENE (3 ML/L)	2.81	20.067	0.00002
DEUTERATED BENZENE-C6D6 5ML/L	1.99	56.609	0.00000
FLUOROBENZENE (TRACE)	4.92	3.751	0.00060
TOLUENE (TRACE)	5.61	2.535	0.00131
DEUTERATED TOLUENE-C7D8 5ML/L	4.99	3.596	0.00065
ETHYLBENZENE (TRACE)	5.51	2.677	0.00117
PROPYLBENZENE (TRACE)	5.25	3.098	0.00088
ISOPROPYLBENZENE (TRACE)	4.87	3.866	0.00056
SEC-BUTYLBENZENE (TRACE)	4.84	3.942	0.00054
DIPHENYLMETHANE (TRACE)	6.69	1.491	0.00378
BIBENZYL	6.09	1.977	0.00215
1,3,5-TRIPHENYLPENTANE 0.05ML/L	6.27	1.811	0.00256
POLYSTYRENE	5.58	2.571	0.00127
POLYVINYLTOLUENE	8.85	0.646	0.01991
PHENYLCYCLOHEXANE (TRACE)	4.53	4.819	0.00036
BENZYLALCOHOL 3ML/L	4.91	3.783	0.00059
BENZYLACETATE 5ML/L	4.67	4.403	0.00043
PHENOL (TRACE)	12.74	0.216	0.17939
ANISOLE (TRACE)	13.15	0.197	0.21613
PHENYL ETHER	13.91	0.166	0.30248
ANILINE 3ML/L	16.53	0.099	0.84473
N,N-DIMETHYLANILINE 3ML/L	17.74	0.080	1.27870
TRIPHENYLAMINE	26.49	0.024	14.08984
P-XYLENE (TRACE)	9.48	0.525	0.03043
DEUTERATED P-XYLENE C8D10 (TRACE)	9.41	0.537	0.02908
M-XYLENE (TRACE)	7.59	1.022	0.00804
O-XYLENE 3ML/L	7.12	1.238	0.00548
1-METHYL-4-ETHYLBENZENE (TRACE)	8.82	0.652	0.01974
1,4-BIS(TRIFLUOROMETHYL)BENZENE 5ML/L	6.37	1.734	0.00280
P-CRESOL	15.74	0.115	0.63437
P-METHYLANISOLE 3ML/L	15.22	0.127	0.51874
1,4-DIMETHOXY BENZENE	18.06	0.076	1.43385
1-METHYL-4-CHLOROBENZENE 5ML/L	10.13	0.430	0.04526
SODIUM SALICYLATE (A)	19.24	0.063	2.06908
ANTHRANILIC ACID (A)	16.56	0.099	0.84012
TMPD	20.94	0.049	3.43352
1,2,4-TRIMETHYLBENZENE (TRACE)	10.44	0.393	0.05424
MESITYLENE 5ML/L	7.81	0.938	0.00953
1,3,5-TRIETHYLBENZENE 5ML/L	7.43	1.092	0.00702
STYRENE	13.40	0.186	0.24064
1,1-DIPHENYLETHYLENE 3ML/L	15.04	0.131	0.48288
BIPHENYL	14.15	0.158	0.33627
DEUTERATED BIPHENYL-C12D10	14.15	0.158	0.33633
4-METHYL BIPHENYL	16.45	0.101	0.82438
MONOISOPROPYLBIPHENYL 5ML/L	15.86	0.112	0.66370
4-BENZYL BIPHENYL	18.14	0.075	1.48318
3,3'-DIMETHYL BIPHENYL	15.76	0.114	0.63741
2,2'-DIMETHYLBIPHENYL 2ML/L	10.32	0.407	0.05035
OCTAMETHYL BIPHENYL	12.34	0.238	0.14749
4-METHOXY BIPHENYL	20.76	0.050	3.31820
4-BIPHENYLYL PHENYL ETHER	21.26	0.047	3.82587
4,4'-DIHYDROXYBIPHENYL (A)	23.03	0.037	6.13596
4,4'-DIMETHOXY BIPHENYL	22.87	0.037	5.89621

ACCEPTOR MOLECULES	RO	CO	OVERLAP
4,4'-DIISOBUTOXYBIPHENYL	22.71	0.038	5.67038
BENZIDINE	27.17	0.022	16.43422
4-VINYL BIPHENYL	25.97	0.026	12.64991
M-TERPHENYL	19.61	0.059	2.34934
M-QUATERPHENYL	19.71	0.058	2.43051
1,3,5-TRIPHENYL BENZENE	20.59	0.051	3.12364
1,3-DIBIPHENYLYLBENZENE (B)	29.97	0.017	29.61982
1,3,5,-TRIBIPHENYLYLBENZENE (B)	33.06	0.012	53.38959
FLUORENE (TRACE)	21.22	0.047	3.77777
1-METHYL FLUORENE	19.54	0.060	2.28285
1,2-BENZOFLUORENE	24.36	0.031	8.51177
9,10-DIHYDROPHENANTHRENE	21.47	0.045	4.04457
CARBAZOLE (TRACE)	23.49	0.035	6.85675
N-METHYL CARBAZOLE (TRACE)	22.84	0.038	5.78731
N-VINYLCARBAZOLE	24.04	0.032	7.86409
N-PHENYL CARBAZOLE	24.51	0.030	8.84415
N-METHYL-2,3-BENZCARBAZOLE	28.33	0.020	21.07927
DIBENZOFURAN	23.28	0.035	6.59774
DIBENZOTHIOPHENE (TRACE)	20.81	0.050	3.32379
P-TERPHENYL	26.76	0.023	15.11837
DEUTERATED P-TERPHENYL-C18D14	26.44	0.024	14.08599
4-METHYL TERPHENYL	27.22	0.022	16.73600
2-METHYL-P-TERPHENYL	23.34	0.035	6.67445
3,3''-DIMETHYL-P-TERPHENYL	26.74	0.023	15.01786
4-(3,3-DIMETHYLBUTOXY)-P-TERPHENYL	27.97	0.020	19.58770
4,4''DIHEXAHYDROFARNESOXY-P-TERPHENYL	29.07	0.018	24.69934
2,2'-METHYLENE-P-TERPHENYL	27.80	0.021	18.93204
2,2'-ETHYLENE-P-TERPHENYL	27.80	0.021	18.89690
2,2'-METHYLENE-5',6''-METHYLENE-P-TERPHENYL (TRACE)	27.16	0.022	16.35968
3-PHENYL DIBENZOFURAN	27.05	0.023	16.04077
BENZO(1,2-B-4,5-B')BISBENZOFURAN (TRACE) (B)	24.28	0.031	8.33751
SEXIPHENYL BENZENE	21.04	0.048	3.59740
P-QUATERPHENYL	29.00	0.018	24.27109
3,3'''-DIMETHYL-P-QUATERPHENYL	29.31	0.018	25.87679
TETRAMETHYL-P-QUATERPHENYL	25.70	0.026	11.90698
3,3'''-DIMETHYL-3',2''-METHYLENE-P-QUATERPHENYL	28.30	0.020	20.90876
3,3'''-DIMETHYL-3',2''-ETHYLENE-P-QUATERPHENYL	29.13	0.018	24.88589
4-(3-ETHYLBUTOXY)-P-QUATERPHENYL (B)	28.86	0.019	23.56577
4,4'''-DI(2-BUTYLOCTOXY-1)-P-QUATERPHENYL	29.96	0.017	29.45037
DIETHYL-P-QUINQUEPHENYL	30.77	0.015	34.65017
DI(3-ETHYLHEPTYL)-P-QUINQUEPHENYL	30.53	0.016	32.96974
TETRAMETHYL-P-SEXIPHENYL	31.22	0.015	37.79934
TETRAMETHYL-P-OCTAPHENYL (B)	30.94	0.015	35.74133
INDOLE (TRACE)	17.30	0.086	1.11592
2-PHENYL INDOLE	26.00	0.025	12.61083
1-METHYL-2-PHENYL INDOLE	25.16	0.028	10.34785
1-NEOPENTYL-2-PHENYL INDOLE	25.01	0.029	10.01510
1-N-BUTYL-2(BIPHENYLYL-4) INDOLE	26.23	0.025	13.26665
1,2-DIPHENYL INDOLE	25.91	0.026	12.35939
1-BIPHENYLYL-2-PHENYL INDOLE	27.58	0.021	17.93381
1,3-DIMETHYL-2-PHENYL INDOLE	24.12	0.032	8.04767
2,3-DIPHENYL INDOLE	24.90	0.029	9.71866
1-METHYL-2,3-DIPHENYL INDOLE	24.41	0.031	8.61654
2-PHENYL-3,2'-METHYLENE INDOLE (TRACE)	24.49	0.030	8.78173
2-PHENYL-3,2'-ETHYLENE INDOLE (TRACE)	23.76	0.033	7.32299
2-PHENYL-3,2'-TRIMETHYLENE INDOLE	25.03	0.029	10.01416
1-METHYL-3,2'-METHYLENE-2-PHENYL INDOLE (TRACE)	23.96	0.033	7.71224
1-METHYL-3,2'-ETHYLENE-2-PHENYL INDOLE	23.32	0.035	6.55783
1-METHYL-3,2'-TRIMETHYLENE-2-PHENYL INDOLE	24.18	0.032	8.15181
4,5-DIPHENYLIMIDAZOLE (A)	23.02	0.037	6.07257
PPF (TRACE)	25.64	0.027	11.57575
PPO (TRACE)	27.64	0.021	18.18706
2-(4-FLUOROPHENYL)-5-PHENYLOXAZOLE (TRACE)	26.71	0.023	14.80716
αNPO (TRACE)	24.45	0.031	8.71488
BBO (B)	26.06	0.025	12.76163
POPOP	24.05	0.032	7.87939
DIMETHYL POPOP (TRACE)	25.03	0.029	10.01031
PPD (TRACE)	26.70	0.024	14.92779
2,5-DI(2-METHYLPHENYL)OXADIAZOLE	26.01	0.025	12.75050
PBD	29.43	0.018	26.53150
BBD (B)	28.78	0.019	23.17374
αNPD (TRACE)	24.57	0.030	8.96921
βNPD	27.95	0.020	19.42692
αNND (TRACE)	23.29	0.035	6.51074
βNND (TRACE)	28.07	0.020	19.91400
TRANS-1-(4-BIPHENYLYL)-2-PHENYL-ETHYLENE	27.61	0.021	18.06612
TRANS-1-(4-BIPHENYLYL)-2-(1-NAPHTHYL)-ETHYLENE	25.85	0.026	12.14626
DIPHENYLSTILBENE (B)	25.27	0.028	10.61542
1,1,4,4-TETRAPHENYLBUTADIENE	23.09	0.036	6.17375
1,6-DIPHENYLHEXATRIENE	23.25	0.036	6.43657
1,8-DIPHENYL-1,3,5,7-OCTATETRAENE	21.23	0.047	3.72526
BPSB (TRACE)	24.05	0.032	7.89175
BIS-MSB	24.91	0.029	9.74447
ESCULIN (A)	21.84	0.043	4.41666
AZULENE	25.48	0.027	11.15251
NAPHTHALENE	18.20	0.074	1.49673
DEUTERATED NAPHTHALENE-C10D8	18.33	0.073	1.55115
1-METHYLNAPHTHALENE 5ML/L	19.85	0.057	2.51109
2-METHYL NAPHTHALENE	18.52	0.070	1.65628
2,3-DIMETHYL NAPHTHALENE	19.09	0.064	1.98614
2,6-DIMETHYL NAPHTHALENE (TRACE)	18.75	0.068	1.78457
1-NAPHTHOL (TRACE)	20.52	0.052	3.05835
2-NAPHTHOL (TRACE)	18.23	0.074	1.49835
1-NAPHTHYLAMINE	19.46	0.061	2.21194
2-CHLORONAPHTHALENE	18.54	0.070	1.66308
ACENAPHTHENE	21.03	0.048	3.54882

ACCEPTOR MOLECULES	RO	CO	OVERLAP
1-PHENYL NAPHTHALENE	22.90	0.037	5.91984
2-PHENYL NAPHTHALENE	33.85	0.012	61.49658
1,4-DIPHENYL NAPHTHALENE	24.27	0.031	8.33781
1,5-DIPHENYL NAPHTHALENE	23.87	0.033	7.54713
1,7-DIPHENYL NAPHTHALENE	23.66	0.034	7.16297
1,4,5,8-TETRAPHENYL NAPHTHALENE	23.25	0.036	6.43289
1,1'-BINAPHTHYL	24.16	0.032	8.15145
4,4'-DI(N-BUTOXY)-1,1'-BINAPHTHYL	[illegible]	0.031	8.34481
2,2'-BINAPHTHYL	26.50	0.024	14.13863
ACRIDINE (A)	16.01	0.109	0.68646
ANTHRACENE (TRACE)	14.47	0.147	0.37468
DEUTERATED ANTHRACENE C14D10 (TRACE)	15.08	0.130	0.47887
9-METHYL ANTHRACENE (TRACE)	13.99	0.163	0.30587
1-AMINO ANTHRACENE	21.47	0.045	3.99477
2-AMINO ANTHRACENE	13.23	0.193	0.21511
9-VINYL ANTHRACENE	15.39	0.123	0.54132
9-PHENYL ANTHRACENE (TRACE)	16.09	0.107	0.70806
9,10-DIPHENYL ANTHRACENE	17.80	0.079	1.29415
9,10-DICHLOROANTHRACENE (TRACE)	14.09	0.160	0.31857
BIANTHRANYL (B)	20.38	0.053	2.91845
9,10-DINAPHTHYLANTHRACENE (B)	24.10	0.032	7.98019
9-(P-VINYLPHENYL)-10-PHENYL ANTHRACENE	20.42	0.053	2.95638
9,10-BIS(PHENYL-ETHYNYL)ANTHRACENE (TRACE)	27.39	0.022	17.20083
RUBICENE (B)	28.45	0.019	21.60998
DIBENZ(A,H)ANTHRACENE (B)	31.56	0.014	40.27507
TETRACENE (TRACE) (B)	32.22	0.013	45.55509
RUBRENE (B)	31.30	0.015	38.34058
TETRAHYDROCANNABINOL ACID METHYL ESTER	24.38	0.031	8.64197
PHENANTHRENE	21.99	0.042	4.60280
3,4-BENZOPHENANTHRENE	28.25	0.020	20.68964
2-PHENYL PHENANTHRENE	28.16	0.020	20.30797
CHRYSENE (TRACE)	24.94	0.029	9.81232
TRIPHENYLENE	23.56	0.034	6.96829
DEUTERATED TRIPHENYLENE-C18D12	23.58	0.034	7.00999
PYRENE	22.66	0.038	5.51431
1,3,6,8,-TETRAPHENYL PYRENE	28.35	0.020	21.13668
BIPYRENYL (B)	28.17	0.020	20.34394
O-PHENYLENE-PYRENE	27.52	0.021	17.70331
ANTHANTHRENE (B)	27.40	0.022	17.25608
FLUORANTHENE	24.46	0.031	8.72517
3-PHENYL FLUORANTHENE	26.12	0.025	12.94727
3-AMINO FLUORANTHENE	24.32	0.031	8.43639
3,3'-BIFLUORANTHENYL (B)	27.88	0.021	19.14293
BENZO(GHI)FLUORANTHENE	23.84	0.033	7.48254
9,12-DI(N-OCTADECOXY)-BENZ(E)ACEPHENANTHRYLENE	27.35	0.022	17.04916
DECACYCLENE (B)	27.15	0.022	16.32785
10-ETHOXY-BENZO(KL)XANTHENE	20.53	0.052	3.04782
PERYLENE	16.96	0.092	0.96829
BENZO(GHI)PERYLENE (B)	28.37	0.020	21.23903
BBOT (TRACE)	21.89	0.043	4.48351
QUININE SULFATE (A)	21.77	0.043	4.34193
ACRIDINE YELLOW (A)	23.52	0.034	6.90574
ACRIDINE ORANGE (A)	24.32	0.031	8.43919
ACRIDINE RED (A)	21.55	0.045	4.07862
URANINE (A)	21.65	0.044	4.19256
RHODAMINE B (A)	24.78	0.029	9.44050
RHODAMINE 6G (A)	24.68	0.030	9.21585
EOSIN B (A)	22.26	0.041	4.96022
EOSIN Y (A)	22.94	0.037	5.93212
ROSE BENGAL (A)	23.21	0.036	6.37287
INDENOINDENE	25.17	0.028	10.37304

8.28 Donor Molecule: 1-Methyl-4-ethylbenzene (trace)

BENZENE (3 ML/L)	1.83	72.741	0.00000
DEUTERATED BENZENE-C6D6 5ML/L	0.00	0.000	0.00000
FLUOROBENZENE (TRACE)	4.34	5.490	0.00019
TOLUENE (TRACE)	5.17	3.245	0.00054
DEUTERATED TOLUENE-C7D8 5ML/L	4.73	4.240	0.00032
ETHYLBENZENE (TRACE)	5.19	3.194	0.00056
PROPYLBENZENE (TRACE)	5.00	3.572	0.00045
ISOPROPYLBENZENE (TRACE)	4.67	4.405	0.00029
SEC-BUTYLBENZENE (TRACE)	4.57	4.693	0.00026
DIPHENYLMETHANE (TRACE)	6.29	1.794	0.00177
BIBENZYL	5.84	2.250	0.00113
1,3,5-TRIPHENYLPENTANE 0.05ML/L	6.03	2.036	0.00137
POLYSTYRENE	5.24	3.102	0.00059
POLYVINYLTOLUENE	9.10	[illegible]	0.01680
PHENYLCYCLOHEXANE (TRACE)	3.81	8.073	0.00009
BENZYLALCOHOL 3ML/L	4.74	4.199	0.00032
BENZYLACETATE 5ML/L	4.24	5.863	0.00017
PHENOL (TRACE)	13.07	0.200	0.14169
ANISOLE (TRACE)	13.51	0.182	0.17205
PHENYL ETHER	14.48	0.147	0.26047
ANILINE 3ML/L	17.65	0.081	0.84616
N,N-DIMETHYLANILINE 3ML/L	18.99	0.065	1.30754
TRIPHENYLAMINE	28.42	0.019	14.57348
P-XYLENE (TRACE)	9.88	0.464	0.02641
DEUTERATED P-XYLENE C8D10 (TRACE)	9.77	0.479	0.02473
M-XYLENE (TRACE)	7.65	1.001	0.00568
O-XYLENE 3ML/L	7.02	1.292	0.00341

ACCEPTOR MOLECULES	RO	CO	OVERLAP
1-METHYL-4-ETHYLBENZENE (TRACE)	9.11	0.592	0.01622
1,4-BIS(TRIFLUOROMETHYL)BENZENE 5ML/L	5.85	2.235	0.00114
P-CRESOL	16.60	0.098	0.59163
P-METHYLANISOLE 3ML/L	16.06	0.108	0.48478
1,4-DIMETHOXY BENZENE	19.33	0.062	1.46102
1-METHYL-4-CHLOROBENZENE 5ML/L	10.39	0.399	0.03571
SODIUM SALICYLATE (A)	20.65	0.051	2.14346
ANTHRANILIC ACID (A)	17.83	0.079	0.88711
TMPD	22.12	0.041	3.24011
1,2,4-TRIMETHYLBENZENE (TRACE)	10.82	0.353	0.04539
MESITYLENE 5ML/L	7.99	0.878	0.00737
1,3,5-TRIETHYLBENZENE 5ML/L	7.52	1.051	0.00514
STYRENE	14.19	0.157	0.22966
1,1-DIPHENYLETHYLENE 3ML/L	15.74	0.115	0.42861
BIPHENYL	14.62	0.143	0.27672
DEUTERATED BIPHENYL-C12D10	14.60	0.144	0.27479
4-METHYL BIPHENYL	17.13	0.089	0.71353
MONOISOPROPYLBIPHENYL 5ML/L	16.48	0.100	0.56728
4-BENZYL BIPHENYL	18.92	0.066	1.29697
3,3'-DIMETHYL BIPHENYL	16.52	0.099	0.57317
2,2'-DIMETHYLBIPHENYL 2ML/L	10.62	0.373	0.04056
OCTAMETHYL BIPHENYL	12.88	0.209	0.12921
4-METHOXY BIPHENYL	21.87	0.043	3.07448
4-BIPHENYLYL PHENYL ETHER	22.37	0.040	3.51726
4,4'-DIHYDROXYBIPHENYL (A)	24.39	0.031	5.87126
4,4'-DIMETHOXY BIPHENYL	24.20	0.032	5.61380
4,4'-DIISOBUTOXYBIPHENYL	24.05	0.032	5.41842
BENZIDINE	29.02	0.018	16.52817
4-VINYL BIPHENYL	27.60	0.021	12.35650
M-TERPHENYL	20.63	0.051	2.15424
M-QUATERPHENYL	20.60	0.051	2.15278
1,3,5-TRIPHENYL BENZENE	21.50	0.045	2.74847
1,3-DIBIPHENYLYLBENZENE (B)	31.99	0.014	29.67600
1,3,5,-TRIBIPHENYLYLBENZENE (B)	35.33	0.010	53.95349
FLUORENE (TRACE)	22.52	0.039	3.66252
1-METHYL FLUORENE	21.01	0.048	2.39597
1,2-BENZOFLUORENE	26.08	0.025	8.69625
9,10-DIHYDROPHENANTHRENE	22.71	0.038	3.83692
CARBAZOLE (TRACE)	25.12	0.028	6.95123
N-METHYL CARBAZOLE (TRACE)	24.43	0.031	5.88132
N-VINYLCARBAZOLE	25.57	0.027	7.73158
N-PHENYL CARBAZOLE	26.17	0.025	8.87115
N-METHYL-2,3-BENZCARBAZOLE	30.08	0.016	20.47220
DIBENZOFURAN	24.78	0.029	6.48880
DIBENZOTHIOPHENE (TRACE)	22.14	0.041	3.26450
P-TERPHENYL	28.46	0.019	14.80733
DEUTERATED P-TERPHENYL-C18D14	28.12	0.020	13.79398
4-METHYL TERPHENYL	28.99	0.018	16.55504
2-METHYL-P-TERPHENYL	24.67	0.030	6.30961
3,3''-DIMETHYL-P-TERPHENYL	28.47	0.019	14.82671
4-(3,3-DIMETHYLBUTOXY)-P-TERPHENYL	29.87	0.017	19.71320
4,4''DIHEXAHYDROFARNESOXY-P-TERPHENYL	31.10	0.015	25.06869
2,2'-METHYLENE-P-TERPHENYL	29.73	0.017	19.18578
2,2'-ETHYLENE-P-TERPHENYL	29.74	0.017	19.19388
2,2'-METHYLENE-5',6''METHYLENE-P-TERPHENYL (TRACE)	29.13	0.018	16.88742
3-PHENYL DIBENZOFURAN	29.04	0.018	16.66653
BENZO(1,2-B-4,5-B')BISBENZOFURAN (TRACE) (B)	26.15	0.025	8.82836
SEXIPHENYL BENZENE	22.18	0.041	3.34521
P-QUATERPHENYL	31.04	0.015	24.75223
3,3'''-DIMETHYL-P-QUATERPHENYL	31.39	0.014	26.44551
TETRAMETHYL-P-QUATERPHENYL	27.21	0.022	11.35841
3,3'''-DIMETHYL-3',2''-METHYLENE-P-QUATERPHENYL	30.38	0.016	21.73684
3,3'''-DIMETHYL-3',2''-ETHYLENE-P-QUATERPHENYL	31.19	0.015	25.43465
4-(3-ETHYLBUTOXY)-P-QUATERPHENYL (B)	30.97	0.015	24.37967
4,4'''-DI(2-BUTYLOCTOXY-1)-P-QUATERPHENYL	32.13	0.013	30.42012
DIETHYL-P-QUINQUEPHENYL	32.95	0.013	35.37034
DI(3-ETHYLHEPTYL)-P-QUINQUEPHENYL	32.76	0.013	34.14026
TETRAMETHYL-P-SEXIPHENYL	33.40	0.012	38.43270
TETRAMETHYL-P-OCTAPHENYL (B)	33.18	0.012	36.89258
INDOLE (TRACE)	18.16	0.075	1.01507
2-PHENYL INDOLE	27.92	0.021	13.12028
1-METHYL-2-PHENYL INDOLE	26.97	0.023	10.65421
1-NEOPENTYL-2-PHENYL INDOLE	26.80	0.023	10.27602
1-N-BUTYL-2(BIPHENYLYL-4) INDOLE	28.12	0.020	13.66966
1,2-DIPHENYL INDOLE	27.76	0.021	12.66041
1-BIPHENYLYL-2-PHENYL INDOLE	29.49	0.017	18.16940
1,3-DIMETHYL-2-PHENYL INDOLE	25.87	0.026	8.28672
2,3-DIPHENYL INDOLE	26.69	0.024	9.99645
1-METHYL-2,3-DIPHENYL INDOLE	26.09	0.025	8.72121
2-PHENYL-3,2'-METHYLENE INDOLE (TRACE)	26.33	0.024	9.21177
2-PHENYL-3,2'-ETHYLENE INDOLE (TRACE)	25.51	0.027	7.62393
2-PHENYL-3,2'-TRIMETHYLENE INDOLE	26.90	0.023	10.47750
1-METHYL-3,2'-METHYLENE-2-PHENYL INDOLE (TRACE)	25.77	0.026	8.09398
1-METHYL-3,2'-ETHYLENE-2-PHENYL INDOLE	25.04	0.028	6.81845
1-METHYL-3,2'-TRIMETHYLENE-2-PHENYL INDOLE	25.99	0.025	8.51154
4,5-DIPHENYLIMIDAZOLE (A)	24.57	0.030	6.08319
PPF (TRACE)	27.56	0.021	12.10524
PPO (TRACE)	29.66	0.017	18.82713
2-(4-FLUOROPHENYL)-5-PHENYLOXAZOLE (TRACE)	28.65	0.019	15.29028
αNPO (TRACE)	26.14	0.025	8.81237
BBO (B)	27.91	0.021	13.06116
POPOP	25.75	0.026	8.05627
DIMETHYL POPOP (TRACE)	26.76	0.023	10.13815
PPD (TRACE)	28.46	0.019	14.85429
2,5-DI(2-METHYLPHENYL)OXADIAZOLE	27.74	0.021	12.70232
PBD	31.56	0.014	27.33111

ACCEPTOR MOLECULES	RO	CO	OVERLAP
BBD (B)	30.88	0.015	23.97325
αNPD (TRACE)	26.30	0.025	9.14855
βNPD	29.83	0.017	19.48168
αNND (TRACE)	25.00	0.029	6.75434
βNND (TRACE)	29.95	0.017	19.93597
TRANS-1-(4-BIPHENYLYL)-2-PHENYL-ETHYLENE	29.66	0.017	18.83131
TRANS-1-(4-BIPHENYLYL)-2-(1-NAPHTHYL)-ETHYLENE	27.66	0.021	12.38994
DIPHENYLSTILBENE (B)	27.18	0.022	11.13994
1,1,4,4-TETRAPHENYLBUTADIENE	24.78	0.029	6.39738
1,6-DIPHENYLHEXATRIENE	25.03	0.029	6.79066
1,8-DIPHENYL-1,3,5,7-OCTATETRAENE	22.72	0.038	3.79850
BPSB (TRACE)	25.85	0.026	8.24840
BIS-MSB	26.78	0.023	10.18915
ESCULIN (A)	23.43	0.035	4.57341
AZULENE	26.82	0.023	10.28450
NAPHTHALENE	19.23	0.063	1.41036
DEUTERATED NAPHTHALENE-C10D8	19.36	0.062	1.46218
1-METHYLNAPHTHALENE 5ML/L	21.09	0.048	2.44817
2-METHYL NAPHTHALENE	19.60	0.059	1.57745
2,3-DIMETHYL NAPHTHALENE	20.23	0.054	1.90673
2,6-DIMETHYL NAPHTHALENE (TRACE)	19.86	0.057	1.70676
1-NAPHTHOL (TRACE)	21.93	0.042	3.09208
2-NAPHTHOL (TRACE)	19.32	0.062	1.44002
1-NAPHTHYLAMINE	20.91	0.049	2.30765
2-CHLORONAPHTHALENE	19.61	0.059	1.58239
ACENAPHTHENE	22.44	0.040	3.55781
1-PHENYL NAPHTHALENE	24.46	0.031	5.94865
2-PHENYL NAPHTHALENE	36.11	0.009	61.50639
1,4-DIPHENYL NAPHTHALENE	26.03	0.025	8.60168
1,5-DIPHENYL NAPHTHALENE	25.59	0.027	7.77386
1,7-DIPHENYL NAPHTHALENE	25.33	0.028	7.30251
1,4,5,8-TETRAPHENYL NAPHTHALENE	24.88	0.029	6.54944
1,1′-BINAPHTHYL	25.81	0.026	8.20196
4,4′-DI(N-BUTOXY)-1,1′-BINAPHTHYL	26.06	0.025	8.65502
2,2′-BINAPHTHYL	28.28	0.020	14.16873
ACRIDINE (A)	17.16	0.089	0.70479
ANTHRACENE (TRACE)	15.56	0.119	0.39256
DEUTERATED ANTHRACENE C14D10 (TRACE)	16.19	0.105	0.49680
9-METHYL ANTHRACENE (TRACE)	14.89	0.135	0.30172
1-AMINO ANTHRACENE	22.48	0.039	3.56601
2-AMINO ANTHRACENE	14.27	0.154	0.22912
9-VINYL ANTHRACENE	16.30	0.103	0.51902
9-PHENYL ANTHRACENE (TRACE)	17.01	0.091	0.66974
9,10-DIPHENYL ANTHRACENE	18.56	0.070	1.13000
9,10-DICHLOROANTHRACENE (TRACE)	14.93	0.134	0.30666
BIANTHRANYL (B)	21.02	0.048	2.38684
9,10-DINAPHTHYLANTHRACENE (B)	25.62	0.027	7.81745
9-(P-VINYLPHENYL)-10-PHENYL ANTHRACENE	21.58	0.044	2.79396
9,10-BIS(PHENYL-ETHYNYL)ANTHRACENE (TRACE)	29.07	0.018	16.68918
RUBICENE (B)	30.44	0.016	21.96689
DIBENZ(A,H)ANTHRACENE (B)	33.88	0.011	41.82693
TETRACENE (TRACE) (B)	33.87	0.012	41.74626
RUBRENE (B)	33.67	0.012	40.29225
TETRAHYDROCANNABINOL ACID METHYL ESTER	25.91	0.026	8.44673
PHENANTHRENE	23.33	0.035	4.45620
3,4-BENZOPHENANTHRENE	29.92	0.017	19.84216
2-PHENYL PHENANTHRENE	29.85	0.017	19.56249
CHRYSENE (TRACE)	26.44	0.024	9.44634
TRIPHENYLENE	24.98	0.029	6.71645
DEUTERATED TRIPHENYLENE-C18D12	24.98	0.029	6.71897
PYRENE	24.01	0.032	5.29715
1,3,6,8,-TETRAPHENYL PYRENE	30.38	0.016	21.73387
BIPYRENYL (B)	29.89	0.017	19.69484
O-PHENYLENE-PYRENE	29.46	0.018	18.04973
ANTHANTHRENE (B)	29.42	0.018	17.91405
FLUORANTHENE	26.01	0.025	8.55942
3-PHENYL FLUORANTHENE	27.96	0.020	13.19192
3-AMINO FLUORANTHENE	26.06	0.025	8.65989
3,3′-BIFLUORANTHENYL (B)	29.82	0.017	19.43646
BENZO(GHI)FLUORANTHENE	25.45	0.027	7.50469
9,12-DI(N-OCTADECOXY)-BENZ(E)ACEPHENANTHRYLENE	29.06	0.018	16.63482
DECACYCLENE (B)	29.02	0.018	16.49199
10-ETHOXY-BENZO(KL)XANTHENE	21.98	0.042	3.11742
PERYLENE	17.96	0.077	0.92743
BENZO(GHI)PERYLENE (B)	30.38	0.016	21.73480
BBOT (TRACE)	23.47	0.035	4.62460
QUININE SULFATE (A)	23.25	0.036	4.36029
ACRIDINE YELLOW (A)	24.87	0.029	6.53625
ACRIDINE ORANGE (A)	25.88	0.026	8.29419
ACRIDINE RED (A)	23.01	0.037	4.10361
URANINE (A)	23.12	0.036	4.22427
RHODAMINE B (A)	26.46	0.024	9.48984
RHODAMINE 6G (A)	26.33	0.025	9.18523
EOSIN B (A)	23.80	0.033	5.02713
EOSIN Y (A)	24.53	0.030	6.01924
ROSE BENGAL (A)	24.70	0.030	6.28067
INDENOINDENE	27.07	0.023	10.86693

8.29 Donor Molecule: 1,4-Bis(trifluoromethyl)benzene (5 ml/L)

BENZENE (3 ML/L)	3.58	9.728	0.00011
DEUTERATED BENZENE-C6D6 5ML/L	2.92	17.956	0.00003

ACCEPTOR MOLECULES	RO	CO	OVERLAP
FLUOROBENZENE (TRACE)	6.22	1.855	0.00291
TOLUENE (TRACE)	6.64	1.525	0.00431
DEUTERATED TOLUENE-C7D8 5ML/L	5.96	2.112	0.00224
ETHYLBENZENE (TRACE)	6.47	1.648	0.00368
PROPYLBENZENE (TRACE)	6.18	1.895	0.00279
ISOPROPYLBENZENE (TRACE)	5.77	2.328	0.00185
SEC-BUTYLBENZENE (TRACE)	5.79	2.310	0.00188
DIPHENYLMETHANE (TRACE)	7.73	0.969	0.01064
BIBENZYL	7.09	1.256	0.00633
1,3,5-TRIPHENYLPENTANE 0.05ML/L	7.27	1.163	0.00738
POLYSTYRENE	6.52	1.611	0.00386
POLYVINYLTOLUENE	9.31	0.555	0.03202
PHENYLCYCLOHEXANE (TRACE)	5.66	2.465	0.00165
BENZYLALCOHOL 3ML/L	5.69	2.434	0.00169
BENZYLACETATE 5ML/L	5.60	2.547	0.00154
PHENOL (TRACE)	13.25	0.192	0.26970
ANISOLE (TRACE)	13.60	0.178	0.31450
PHENYL ETHER	14.22	0.156	0.41000
ANILINE 3ML/L	16.06	0.108	0.84502
N,N-DIMETHYLANILINE 3ML/L	17.09	0.090	1.21625
TRIPHENYLAMINE	25.31	0.028	12.76792
P-XYLENE (TRACE)	10.03	0.443	0.05071
DEUTERATED P-XYLENE C8D10 (TRACE)	10.00	0.448	0.04974
M-XYLENE (TRACE)	8.29	0.786	0.01615
O-XYLENE 3ML/L	7.91	0.904	0.01222
1-METHYL-4-ETHYLBENZENE (TRACE)	9.44	0.532	0.03524
1,4-BIS(TRIFLUOROMETHYL)BENZENE 5ML/L	7.52	1.052	0.00904
P-CRESOL	15.67	0.116	0.73311
P-METHYLANISOLE 3ML/L	15.14	0.129	0.59687
1,4-DIMETHOXY BENZENE	17.42	0.085	1.37149
1-METHYL-4-CHLOROBENZENE 5ML/L	10.50	0.387	0.06657
SODIUM SALICYLATE (A)	18.38	0.072	1.87231
ANTHRANILIC ACID (A)	15.49	0.120	0.67159
TMPD	20.87	0.049	4.00640
1,2,4-TRIMETHYLBENZENE (TRACE)	10.92	0.344	0.08417
MESITYLENE 5ML/L	8.35	0.769	0.01687
1,3,5-TRIETHYLBENZENE 5ML/L	8.02	0.868	0.01321
STYRENE	13.34	0.188	0.27883
1,1-DIPHENYLETHYLENE 3ML/L	15.37	0.123	0.65354
BIPHENYL	14.81	0.138	0.52423
DEUTERATED BIPHENYL-C12D10	14.84	0.137	0.53143
4-METHYL BIPHENYL	16.93	0.092	1.16538
MONOISOPROPYLBIPHENYL 5ML/L	16.39	0.102	0.96153
4-BENZYL BIPHENYL	18.61	0.069	2.05446
3,3'-DIMETHYL BIPHENYL	16.01	0.109	0.83232
2,2'-DIMETHYLBIPHENYL 2ML/L	10.88	0.348	0.08201
OCTAMETHYL BIPHENYL	12.59	0.224	0.19719
4-METHOXY BIPHENYL	20.85	0.049	4.04240
4-BIPHENYLYL PHENYL ETHER	21.40	0.046	4.73985
4,4'-DIHYDROXYBIPHENYL (A)	22.83	0.038	6.93027
4,4'-DIMETHOXY BIPHENYL	22.72	0.038	6.74482
4,4'-DIISOBUTOXYBIPHENYL	22.53	0.039	6.42650
BENZIDINE	26.35	0.024	16.23193
4-VINYL BIPHENYL	25.52	0.027	13.54915
M-TERPHENYL	19.80	0.058	2.95995
M-QUATERPHENYL	20.14	0.055	3.29603
1,3,5-TRIPHENYL BENZENE	21.08	0.048	4.28419
1,3-DIBIPHENYLYLBENZENE (B)	29.12	0.018	29.61663
1,3,5,-TRIBIPHENYLYLBENZENE (B)	31.98	0.014	52.05565
FLUORENE (TRACE)	20.99	0.048	4.20484
1-METHYL FLUORENE	18.63	0.069	2.03474
1,2-BENZOFLUORENE	23.43	0.035	8.02776
9,10-DIHYDROPHENANTHRENE	21.38	0.046	4.68337
CARBAZOLE (TRACE)	22.70	0.030	6.63289
N-METHYL CARBAZOLE (TRACE)	22.05	0.042	5.57504
N-VINYLCARBAZOLE	23.57	0.034	8.32151
N-PHENYL CARBAZOLE	23.80	0.033	8.81924
N-METHYL-2,3-BENZCARBAZOLE	27.91	0.021	22.94617
DIBENZOFURAN	22.77	0.038	6.86020
DIBENZOTHIOPHENE (TRACE)	20.31	0.053	3.41063
P-TERPHENYL	26.27	0.025	16.07732
DEUTERATED P-TERPHENYL-C18D14	25.96	0.026	14.99087
4-METHYL TERPHENYL	26.61	0.024	17.34342
2-METHYL-P-TERPHENYL	23.25	0.036	7.75786
3,3''-DIMETHYL-P-TERPHENYL	26.16	0.025	15.65098
4-(3,3-DIMETHYLBUTOXY)-P-TERPHENYL	27.11	0.022	19.32943
4,4''-DIHEXAHYDROFARNESOXY-P-TERPHENYL	28.06	0.020	23.75339
2,2'-METHYLENE-P-TERPHENYL	26.85	0.023	18.25565
2,2'-ETHYLENE-P-TERPHENYL	26.82	0.023	18.10493
2,2'-METHYLENE-5',6''-METHYLENE-P-TERPHENYL (TRACE)	25.94	0.026	14.77584
3-PHENYL DIBENZOFURAN	25.80	0.026	14.37621
BENZO(1,2-B-4,5-B')BISBENZOFURAN (TRACE) (B)	22.72	0.038	6.66513
SEXIPHENYL BENZENE	21.07	0.048	4.31204
P-QUATERPHENYL	27.92	0.021	22.99260
3,3'''-DIMETHYL-P-QUATERPHENYL	28.19	0.020	24.34049
TETRAMETHYL-P-QUATERPHENYL	25.52	0.027	13.55627
3,3'''-DIMETHYL-3',2''-METHYLENE-P-QUATERPHENYL	26.89	0.023	18.35749
3,3'''-DIMETHYL-3',2''-ETHYLENE-P-QUATERPHENYL	27.94	0.021	23.07383
4-(3-ETHYLBUTOXY)-P-QUATERPHENYL (B)	27.54	0.021	21.18853
4,4'''-DI(2-BUTYLOCTOXY-1)-P-QUATERPHENYL	28.61	0.019	26.63211
DIETHYL-P-QUINQUEPHENYL	29.61	0.017	32.71068
DI(3-ETHYLHEPTYL)-P-QUINQUEPHENYL	29.11	0.018	29.54827
TETRAMETHYL-P-SEXIPHENYL	30.11	0.016	36.15151
TETRAMETHYL-P-OCTAPHENYL (B)	29.53	0.017	32.17403
INDOLE (TRACE)	17.40	0.085	1.37715

ACCEPTOR MOLECULES	RO	CO	OVERLAP
2-PHENYL INDOLE	24.73	0.030	11.11500
1-METHYL-2-PHENYL INDOLE	24.10	0.032	9.51703
1-NEOPENTYL-2-PHENYL INDOLE	24.01	0.032	9.32864
1-N-BUTYL-2(BIPHENYLYL-4) INDOLE	25.08	0.028	12.08813
1,2-DIPHENYL INDOLE	24.90	0.029	11.56974
1-BIPHENYLYL-2-PHENYL INDOLE	26.62	0.024	17.27713
1,3-DIMETHYL-2-PHENYL INDOLE	23.11	0.036	7.38744
2,3-DIPHENYL INDOLE	23.83	0.033	8.89873
1-METHYL-2,3-DIPHENYL INDOLE	23.58	0.034	8.33832
2-PHENYL-3,2'-METHYLENE INDOLE (TRACE)	23.13	0.036	7.43452
2-PHENYL-3,2'-ETHYLENE INDOLE (TRACE)	22.51	0.039	6.32044
2-PHENYL-3,2'-TRIMETHYLENE INDOLE	23.70	0.034	8.60439
1-METHYL-3,2'-METHYLENE-2-PHENYL INDOLE (TRACE)	22.62	0.039	6.50479
1-METHYL-3,2'-ETHYLENE-2-PHENYL INDOLE	22.12	0.041	5.69150
1-METHYL-3,2'-TRIMETHYLENE-2-PHENYL INDOLE	22.95	0.037	7.09125
4,5-DIPHENYLIMIDAZOLE (A)	22.36	0.040	6.07329
PPF (TRACE)	24.25	0.031	9.87036
PPO (TRACE)	26.37	0.024	16.32988
2-(4-FLUOROPHENYL)-5-PHENYLOXAZOLE (TRACE)	25.52	0.027	13.40707
αNPO (TRACE)	23.53	0.034	8.24244
BBO (B)	24.91	0.029	11.58659
POPOP	23.00	0.037	7.18879
DIMETHYL POPOP (TRACE)	24.08	0.032	9.47031
PPD (TRACE)	26.03	0.025	15.22358
2,5-DI(2-METHYLPHENYL)OXADIAZOLE	25.35	0.027	12.97416
PBD	28.17	0.020	24.27784
BBD (B)	27.42	0.022	20.64013
αNPD (TRACE)	23.60	0.034	8.37907
βNPD	27.15	0.022	19.42409
αNND (TRACE)	22.13	0.041	5.69420
βNND (TRACE)	27.22	0.022	19.74878
TRANS-1-(4-BIPHENYLYL)-2-PHENYL-ETHYLENE	26.19	0.025	15.66223
TRANS-1-(4-BIPHENYLYL)-2-(1-NAPHTHYL)-ETHYLENE	24.77	0.029	11.20835
DIPHENYLSTILBENE (B)	23.78	0.033	8.77214
1,1,4,4-TETRAPHENYLBUTADIENE	21.88	0.043	5.32923
1,6-DIPHENYLHEXATRIENE	21.72	0.044	5.09371
1,8-DIPHENYL-1,3,5,7-OCTATETRAENE	20.24	0.054	3.34052
BPSB (TRACE)	22.64	0.039	6.54456
BIS-MSB	23.49	0.034	8.15993
ESCULIN (A)	20.77	0.050	3.89410
AZULENE	25.57	0.027	13.57849
NAPHTHALENE	18.16	0.075	1.75248
DEUTERATED NAPHTHALENE-C10D8	18.30	0.073	1.82488
1-METHYLNAPHTHALENE 5ML/L	19.52	0.060	2.69700
2-METHYL NAPHTHALENE	18.38	0.072	1.88237
2,3-DIMETHYL NAPHTHALENE	18.88	0.066	2.20707
2,6-DIMETHYL NAPHTHALENE (TRACE)	18.58	0.070	2.00719
1-NAPHTHOL (TRACE)	19.84	0.057	2.96995
2-NAPHTHOL (TRACE)	18.02	0.076	1.66552
1-NAPHTHYLAMINE	18.45	0.071	1.91577
2-CHLORONAPHTHALENE	18.39	0.072	1.88371
ACENAPHTHENE	20.42	0.053	3.53534
1-PHENYL NAPHTHALENE	22.22	0.041	5.86915
2-PHENYL NAPHTHALENE	32.92	0.013	61.87859
1,4-DIPHENYL NAPHTHALENE	23.22	0.036	7.60859
1,5-DIPHENYL NAPHTHALENE	22.86	0.037	6.92978
1,7-DIPHENYL NAPHTHALENE	22.81	0.038	6.82832
1,4,5,8-TETRAPHENYL NAPHTHALENE	22.31	0.040	5.99342
1,1'-BINAPHTHYL	23.43	0.035	8.05901
4,4'-DI(N-BUTOXY)-1,1'-BINAPHTHYL	23.12	0.036	7.41752
2,2'-BINAPHTHYL	25.76	0.026	14.18134
ACRIDINE (A)	15.21	0.127	0.60072
ANTHRACENE (TRACE)	13.56	0.179	0.30172
DEUTERATED ANTHRACENE C14D10 (TRACE)	14.25	0.155	0.40662
9-METHYL ANTHRACENE (TRACE)	13.58	0.179	0.30390
1-AMINO ANTHRACENE	21.91	0.043	5.36996
2-AMINO ANTHRACENE	12.16	0.249	0.15375
9-VINYL ANTHRACENE	15.22	0.127	0.60296
9-PHENYL ANTHRACENE (TRACE)	16.01	0.109	0.81724
9,10-DIPHENYL ANTHRACENE	18.33	0.073	1.84250
9,10-DICHLOROANTHRACENE (TRACE)	13.93	0.165	0.35522
BIANTHRANYL (B)	21.49	0.045	4.77968
9,10-DINAPHTHYLANTHRACENE (B)	23.67	0.034	8.53778
9-(P-VINYLPHENYL)-10-PHENYL ANTHRACENE	20.41	0.053	3.51002
9,10-BIS(PHENYL-ETHYNYL)ANTHRACENE (TRACE)	27.11	0.022	19.27805
RUBICENE (B)	27.44	0.022	20.74065
DIBENZ(A,H)ANTHRACENE (B)	30.15	0.016	36.48953
TETRACENE (TRACE) (B)	32.36	0.013	55.74435
RUBRENE (B)	29.67	0.017	33.12998
TETRAHYDROCANNABINOL ACID METHYL ESTER	23.97	0.032	9.28787
PHENANTHRENE	21.75	0.043	5.13898
3,4-BENZOPHENANTHRENE	27.91	0.021	22.92930
2-PHENYL PHENANTHRENE	27.86	0.021	22.72175
CHRYSENE (TRACE)	24.81	0.029	11.32274
TRIPHENYLENE	23.30	0.035	7.77060
DEUTERATED TRIPHENYLENE-C18D12	23.38	0.035	7.92011
PYRENE	22.50	0.039	6.29660
1,3,6,8,-TETRAPHENYL PYRENE	27.20	0.022	19.65299
BIPYRENYL (B)	27.60	0.021	21.46111
O-PHENYLENE-PYRENE	26.50	0.024	16.82358
ANTHANTHRENE (B)	26.13	0.025	15.45326
FLUORANTHENE	23.90	0.033	9.04308
3-PHENYL FLUORANTHENE	25.17	0.028	12.35060
3-AMINO FLUORANTHENE	23.35	0.035	7.87679
3,3'-BIFLUORANTHENYL (B)	26.90	0.023	18.40274
BENZO(GHI)FLUORANTHENE	23.14	0.036	7.45017
9,12-DI(N-OCTADECOXY)-BENZ(E)ACEPHENANTHRYLENE	26.83	0.023	18.10222

ACCEPTOR MOLECULES	RO	CO	OVERLAP
DECACYCLENE (B)	26.24	0.025	15.85635
10-ETHOXY-BENZO(KL)XANTHENE	19.73	0.058	2.86626
PERYLENE	16.82	0.094	1.10068
BENZO(GHI)PERYLENE (B)	27.30	0.022	20.09480
BBOT (TRACE)	20.78	0.050	3.90785
QUININE SULFATE (A)	21.06	0.048	4.23478
ACRIDINE YELLOW (A)	23.46	0.035	8.08747
ACRIDINE ORANGE (A)	23.89	0.033	9.01335
ACRIDINE RED (A)	20.87	0.049	4.01392
URANINE (A)	20.95	0.049	4.10772
RHODAMINE B (A)	24.02	0.032	9.32158
RHODAMINE 6G (A)	24.02	0.032	9.31388
EOSIN B (A)	21.48	0.045	4.77083
EOSIN Y (A)	22.12	0.041	5.68956
ROSE BENGAL (A)	22.71	0.038	6.65435
INDENOINDENE	23.83	0.033	8.88711

8.30 Donor Molecule: *p*-Cresol

POLYVINYLTOLUENE	5.27	3.050	0.00187
PHENYLCYCLOHEXANE (TRACE)	0.00	0.000	0.00000
BENZYLALCOHOL 3ML/L	0.00	0.000	0.00000
BENZYLACETATE 5ML/L	0.00	0.000	0.00000
PHENOL (TRACE)	4.10	6.487	0.00042
ANISOLE (TRACE)	4.56	4.704	0.00079
PHENYL ETHER	6.27	1.816	0.00530
ANILINE 3ML/L	13.12	0.198	0.44164
N,N-DIMETHYLANILINE 3ML/L	15.74	0.115	1.30882
TRIPHENYLAMINE	24.13	0.032	16.83109
P-XYLENE (TRACE)	2.55	27.019	0.00002
DEUTERATED P-XYLENE C8D10 (TRACE)	2.16	44.546	0.00001
M-XYLENE (TRACE)	0.00	0.000	0.00000
O-XYLENE 3ML/L	0.00	0.000	0.00000
1-METHYL-4-ETHYLBENZENE (TRACE)	0.00	0.000	0.00000
1,4-BIS(TRIFLUOROMETHYL)BENZENE 5ML/L	0.00	0.000	0.00000
P-CRESOL	9.56	0.513	0.06641
P-METHYLANISOLE 3ML/L	9.28	0.559	0.05576
1,4-DIMETHOXY BENZENE	15.03	0.132	0.99591
1-METHYL-4-CHLOROBENZENE 5ML/L	3.13	14.539	0.00008
SODIUM SALICYLATE (A)	17.36	0.085	2.33227
ANTHRANILIC ACID (A)	17.16	0.089	2.17199
TMPD	16.94	0.092	2.00937
1,2,4-TRIMETHYLBENZENE (TRACE)	3.77	8.349	0.00025
MESITYLENE 5ML/L	1.86	69.055	0.00000
1,3,5-TRIETHYLBENZENE 5ML/L	2.27	38.465	0.00001
STYRENE	9.14	0.586	0.05073
1,1-DIPHENYLETHYLENE 3ML/L	9.04	0.605	0.04765
BIPHENYL	6.29	1.797	0.00542
DEUTERATED BIPHENYL-C12D10	6.05	2.024	0.00428
4-METHYL BIPHENYL	9.03	0.608	0.04719
MONOISOPROPYLBIPHENYL 5ML/L	7.97	0.884	0.02236
4-BENZYL BIPHENYL	10.12	0.431	0.09383
3,3'-DIMETHYL BIPHENYL	9.85	0.468	0.07952
2,2'-DIMETHYLBIPHENYL 2ML/L	5.12	3.324	0.00158
OCTAMETHYL BIPHENYL	5.64	2.494	0.00281
4-METHOXY BIPHENYL	14.10	0.160	0.68132
4-BIPHENYLYL PHENYL ETHER	14.12	0.159	0.68895
4,4'-DIHYDROXYBIPHENYL (A)	17.13	0.089	2.17544
4,4'-DIMETHOXY BIPHENYL	16.65	0.097	1.83724
4,4'-DIISOBUTOXYBIPHENYL	16.68	0.096	1.86490
BENZIDINE	23.23	0.036	13.44253
4-VINYL BIPHENYL	20.34	0.053	6.11231
M-TERPHENYL	14.11	0.159	0.68302
M-QUATERPHENYL	12.53	0.227	0.33702
1,3,5-TRIPHENYL BENZENE	13.12	0.198	0.43907
1,3-DIBIPHENYLYLBENZENE (B)	25.14	0.028	21.64217
1,3,5,-TRIBIPHENYLYLBENZENE (B)	28.31	0.020	44.11562
FLUORENE (TRACE)	17.41	0.085	2.41100
1-METHYL FLUORENE	16.66	0.097	1.84824
1,2-BENZOFLUORENE	22.00	0.042	9.66060
9,10-DIHYDROPHENANTHRENE	16.70	0.096	1.87412
CARBAZOLE (TRACE)	19.42	0.061	4.58537
N-METHYL CARBAZOLE (TRACE)	19.08	0.064	4.11224
N-VINYLCARBAZOLE	19.15	0.064	4.20480
N-PHENYL CARBAZOLE	19.89	0.057	5.28412
N-METHYL-2,3-BENZCARBAZOLE	22.37	0.040	10.64903
DIBENZOFURAN	17.86	0.078	2.81801
DIBENZOTHIOPHENE (TRACE)	16.48	0.100	1.71630
P-TERPHENYL	21.10	0.048	7.61234
DEUTERATED P-TERPHENYL-C18D14	20.83	0.049	7.05007
4-METHYL TERPHENYL	22.04	0.042	9.87404
2-METHYL-P-TERPHENYL	16.75	0.095	1.91132
3,3''-DIMETHYL-P-TERPHENYL	21.57	0.045	8.66068
4-(3,3-DIMETHYLBUTOXY)-P-TERPHENYL	23.79	0.033	15.53309
4,4''DIHEXAHYDROFARNESOXY-P-TERPHENYL	25.41	0.027	23.05891
2,2'-METHYLENE-P-TERPHENYL	24.43	0.031	18.25585
2,2'-ETHYLENE-P-TERPHENYL	24.47	0.031	18.41184
2,2'-METHYLENE-5',6''METHYLENE P TERPHENYL (TRACE)	25.25	0.028	22.15480
3-PHENYL DIBENZOFURAN	24.96	0.029	20.77469
BENZO(1,2-B-4,5-B')BISBENZOFURAN (TRACE) (B)	24.97	0.029	20.66054
SEXIPHENYL BENZENE	13.87	0.168	0.61802

ACCEPTOR MOLECULES	RO	CO	OVERLAP
P-QUATERPHENYL	25.67	0.026	24.50081
3,3‴-DIMETHYL-P-QUATERPHENYL	26.18	0.025	27.51209
TETRAMETHYL-P-QUATERPHENYL	18.96	0.066	4.01912
3,3‴ -DIMETHYL-3′,2″-METHYLENE-P-QUATERPHENYL	27.02	0.023	33.20308
3,3‴ -DIMETHYL-3′,2″-ETHYLENE-P-QUATERPHENYL	27.08	0.023	33.56217
4-(3-ETHYLBUTOXY)-P-QUATERPHENYL (B)	26.94	0.023	32.52109
4,4‴-DI(2-BUTYLOCTOXY-1)-P-QUATERPHENYL	27.84	0.021	39.62308
DIETHYL-P-QUINQUEPHENYL	27.33	0.022	36.07820
DI(3-ETHYLHEPTYL)-P-QUINQUEPHENYL	28.57	0.019	46.29044
TETRAMETHYL-P-SEXIPHENYL	27.47	0.022	36.75186
TETRAMETHYL-P-OCTAPHENYL (B)	29.42	0.018	55.18733
INDOLE (TRACE)	10.34	0.404	0.10698
2-PHENYL INDOLE	24.59	0.030	18.90952
1-METHYL-2-PHENYL INDOLE	22.74	0.038	11.83633
1-NEOPENTYL-2-PHENYL INDOLE	22.29	0.040	10.50945
1-N-BUTYL-2(BIPHENYLYL-4) INDOLE	24.74	0.030	19.49929
1,2-DIPHENYL INDOLE	23.48	0.035	14.32053
1-BIPHENYLYL-2-PHENYL INDOLE	24.62	0.030	18.96155
1,3-DIMETHYL-2-PHENYL INDOLE	21.99	0.042	9.66130
2,3-DIPHENYL INDOLE	23.30	0.035	13.62271
1-METHYL-2,3-DIPHENYL INDOLE	22.02	0.042	9.69549
2-PHENYL-3,2′-METHYLENE INDOLE (TRACE)	24.09	0.032	16.68934
2-PHENYL-3,2′-ETHYLENE INDOLE (TRACE)	23.53	0.034	14.43705
2-PHENYL-3,2′-TRIMETHYLENE INDOLE	24.25	0.031	17.36456
1-METHYL-3,2′-METHYLENE-2-PHENYL INDOLE (TRACE)	23.87	0.033	15.77454
1-METHYL-3,2′-ETHYLENE-2-PHENYL INDOLE	23.17	0.036	13.17403
1-METHYL-3,2′-TRIMETHYLENE-2-PHENYL INDOLE	23.22	0.036	13.37395
4,5-DIPHENYLIMIDAZOLE (A)	19.44	0.061	4.62290
PPF (TRACE)	25.25	0.028	22.06573
PPO (TRACE)	25.85	0.026	25.51831
2-(4-FLUOROPHENYL)-5-PHENYLOXAZOLE (TRACE)	24.84	0.029	20.06603
αNPO (TRACE)	23.29	0.035	13.58723
BBO (B)	25.44	0.027	23.04175
POPOP	23.24	0.036	13.40084
DIMETHYL POPOP (TRACE)	23.57	0.034	14.57982
PPD (TRACE)	21.93	0.042	9.60438
2,5-DI(2-METHYLPHENYL)OXADIAZOLE	21.64	0.044	8.85404
PBD	26.93	0.023	32.63335
BBD (B)	27.64	0.021	37.91304
αNPD (TRACE)	23.38	0.035	13.88708
βNPD	25.11	0.028	21.40144
αNND (TRACE)	22.80	0.038	11.95001
βNND (TRACE)	25.72	0.026	24.69635
TRANS-1-(4-BIPHENYLYL)-2-PHENYL-ETHYLENE	26.72	0.023	30.95236
TRANS-1-(4-BIPHENYLYL)-2-(1-NAPHTHYL)-ETHYLENE	24.69	0.030	19.24507
DIPHENYLSTILBENE (B)	25.66	0.026	24.30398
1,1,4,4-TETRAPHENYLBUTADIENE	23.31	0.035	13.62751
1,6-DIPHENYLHEXATRIENE	24.41	0.031	17.97965
1,8-DIPHENYL-1,3,5,7-OCTATETRAENE	21.45	0.045	8.29020
BPSB (TRACE)	24.66	0.030	19.13760
BIS-MSB	25.21	0.028	21.80912
ESCULIN (A)	21.04	0.048	7.38457
AZULENE	17.45	0.084	2.40522
NAPHTHALENE	12.20	0.246	0.28404
DEUTERATED NAPHTHALENE-C10D8	12.62	0.223	0.34546
1-METHYLNAPHTHALENE 5ML/L	15.04	0.132	0.99446
2-METHYL NAPHTHALENE	13.36	0.188	0.48801
2,3-DIMETHYL NAPHTHALENE	14.11	0.159	0.67867
2,6-DIMETHYL NAPHTHALENE (TRACE)	14.18	0.157	0.69861
1-NAPHTHOL (TRACE)	17.68	0.081	2.62533
2-NAPHTHOL (TRACE)	15.07	0.131	1.00406
1-NAPHTHYLAMINE	18.83	0.067	3.79300
2-CHLORONAPHTHALENE	13.35	0.188	0.48623
ACENAPHTHENE	17.46	0.084	2.43459
1-PHENYL NAPHTHALENE	19.12	0.064	4.20152
2-PHENYL NAPHTHALENE	28.13	0.020	42.48613
1,4-DIPHENYL NAPHTHALENE	22.19	0.041	10.18902
1,5-DIPHENYL NAPHTHALENE	21.60	0.044	8.69899
1,7-DIPHENYL NAPHTHALENE	21.29	0.046	7.97261
1,4,5,8-TETRAPHENYL NAPHTHALENE	22.40	0.040	10.74629
1,1′-BINAPHTHYL	20.29	0.054	5.99037
4,4′ -DI(N-BUTOXY)-1,1′-BINAPHTHYL	23.03	0.037	12.72182
2,2′-BINAPHTHYL	23.55	0.034	14.60917
ACRIDINE (A)	16.48	0.100	1.70417
ANTHRACENE (TRACE)	15.50	0.120	1.18062
DEUTERATED ANTHRACENE C14D10 (TRACE)	15.61	0.118	1.23098
9-METHYL ANTHRACENE (TRACE)	14.32	0.152	0.73216
1-AMINO ANTHRACENE	15.67	0.116	1.25869
2-AMINO ANTHRACENE	14.83	0.137	0.89490
9-VINYL ANTHRACENE	14.77	0.139	0.88190
9-PHENYL ANTHRACENE (TRACE)	15.27	0.126	1.07840
9,10-DIPHENYL ANTHRACENE	14.73	0.140	0.86942
9,10-DICHLOROANTHRACENE (TRACE)	13.70	0.174	0.56329
DIANTHRANYL (B)	16.26	0.104	1.57356
9,10-DINAPHTHYLANTHRACENE (B)	19.62	0.059	4.85754
9-(P-VINYLPHENYL)-10-PHENYL ANTHRACENE	16.84	0.094	1.93961
9,10-BIS(PHENYL-ETHYNYL)ANTHRACENE (TRACE)	23.85	0.033	15.66819
RUBICENE (B)	25.24	0.028	21.99107
DIBENZ(A,H)ANTHRACENE (B)	28.57	0.019	46.26707
TETRACENE (TRACE) (B)	21.80	0.043	9.13239
RUBRENE (B)	29.67	0.017	57.97511
TETRAHYDROCANNABINOL ACID METHYL ESTER	19.11	0.064	4.20169
PHENANTHRENE	16.66	0.097	1.82319
3,4-BENZOPHENANTHRENE	21.67	0.044	8.79962
2-PHENYL PHENANTHRENE	22.49	0.039	10.99269
CHRYSENE (TRACE)	21.89	0.043	9.35163

ACCEPTOR MOLECULES	RO	CO	OVERLAP
TRIPHENYLENE	17.27	0.087	2.26491
DEUTERATED TRIPHENYLENE-C18D12	17.02	0.091	2.07461
PYRENE	22.03	0.042	9.70815
1,3,6,8,-TETRAPHENYL PYRENE	24.64	0.030	19.01936
BIPYRENYL (B)	23.75	0.033	15.25858
O-PHENYLENE-PYRENE	24.91	0.029	20.32891
ANTHANTHRENE (B)	25.69	0.026	24.44192
FLUORANTHENE	19.29	0.062	4.38045
3-PHENYL FLUORANTHENE	22.60	0.039	11.32285
3-AMINO FLUORANTHENE	22.56	0.039	11.19695
3,3'-BIFLUORANTHENYL (B)	24.49	0.030	18.33969
BENZO(GHI)FLUORANTHENE	20.00	0.056	5.44649
9,12-DI(N-OCTADECOXY)-BENZ(E)ACEPHENANTHRYLENE	22.06	0.042	9.79225
DECACYCLENE (B)	24.80	0.029	19.76230
10-ETHOXY-BENZO(KL)XANTHENE	18.58	0.070	3.49701
PERYLENE	14.18	0.157	0.69144
BENZO(GHI)PERYLENE (B)	25.17	0.028	21.64325
BBOT (TRACE)	22.00	0.042	9.64694
QUININE SULFATE (A)	19.86	0.057	5.23144
ACRIDINE YELLOW (A)	18.04	0.076	2.93022
ACRIDINE ORANGE (A)	20.18	0.054	5.74646
ACRIDINE RED (A)	19.07	0.064	4.09518
URANINE (A)	19.19	0.063	4.25041
RHODAMINE B (A)	21.75	0.043	8.99918
RHODAMINE 6G (A)	20.97	0.049	7.23080
EOSIN B (A)	20.67	0.051	6.62929
EOSIN Y (A)	21.28	0.046	7.88675
ROSE BENGAL (A)	20.84	0.049	6.96454
INDENOINDENE	24.37	0.031	17.87570

8.31 Donor Molecule: *p*-Methylanisole (3 ml/L)

POLYVINYLTOLUENE	6.18	1.898	0.00167
PHENYLCYCLOHEXANE (TRACE)	0.00	0.000	0.00000
BENZYLALCOHOL 3ML/L	0.00	0.000	0.00000
BENZYLACETATE 5ML/L	0.00	0.000	0.00000
PHENOL (TRACE)	4.35	5.428	0.00021
ANISOLE (TRACE)	4.93	3.724	0.00044
PHENYL ETHER	6.98	1.315	0.00350
ANILINE 3ML/L	15.43	0.122	0.40542
N,N-DIMETHYLANILINE 3ML/L	18.73	0.068	1.28547
TRIPHENYLAMINE	28.77	0.019	16.73110
P-XYLENE (TRACE)	2.67	23.555	0.00001
DEUTERATED P-XYLENE C8D10 (TRACE)	2.15	45.223	0.00000
M-XYLENE (TRACE)	0.00	0.000	0.00000
O-XYLENE 3ML/L	0.00	0.000	0.00000
1-METHYL-4-ETHYLBENZENE (TRACE)	0.00	0.000	0.00000
1,4-BIS(TRIFLUOROMETHYL)BENZENE 5ML/L	0.00	0.000	0.00000
P-CRESOL	10.79	0.356	0.04772
P-METHYLANISOLE 3ML/L	10.48	0.389	0.03990
1,4-DIMETHOXY BENZENE	17.75	0.080	0.93714
1-METHYL-4-CHLOROBENZENE 5ML/L	3.31	12.366	0.00004
SODIUM SALICYLATE (A)	20.67	0.051	2.30152
ANTHRANILIC ACID (A)	20.64	0.051	2.26368
TMPD	20.19	0.054	1.98844
1,2,4-TRIMETHYLBENZENE (TRACE)	4.16	6.225	0.00016
MESITYLENE 5ML/L	1.84	71.465	0.00000
1,3,5-TRIETHYLBENZENE 5ML/L	2.38	33.221	0.00001
STYRENE	10.58	0.378	0.04226
1,1-DIPHENYLETHYLENE 3ML/L	10.36	0.402	0.03742
BIPHENYL	6.94	1.341	0.00337
DEUTERATED BIPHENYL-C12D10	6.61	1.548	0.00253
4-METHYL BIPHENYL	10.29	0.410	0.03585
MONOISOPROPYLBIPHENYL 5ML/L	8.92	0.630	0.01526
4-BENZYL BIPHENYL	11.52	0.293	0.07052
3,3'-DIMETHYL BIPHENYL	11.38	0.304	0.06545
2,2'-DIMETHYLBIPHENYL 2ML/L	5.85	2.229	0.00122
OCTAMETHYL BIPHENYL	6.22	1.857	0.00175
4-METHOXY BIPHENYL	16.43	0.101	0.59106
4-BIPHENYLYL PHENYL ETHER	16.42	0.101	0.58929
4,4'-DIHYDROXYBIPHENYL (A)	20.13	0.055	1.98286
4,4'-DIMETHOXY BIPHENYL	19.51	0.060	1.65069
4,4'-DIISOBUTOXYBIPHENYL	19.57	0.060	1.68196
BENZIDINE	27.61	0.021	13.09848
4-VINYL BIPHENYL	24.00	0.032	5.71425
M-TERPHENYL	16.63	0.097	0.63321
M-QUATERPHENYL	14.60	0.144	0.29113
1,3,5-TRIPHENYL BENZENE	15.39	0.123	0.39529
1,3-DIBIPHENYLYLBENZENE (B)	29.82	0.017	20.86498
1,3,5,-TRIBIPHENYLYLBENZENE (B)	33.62	0.012	42.86465
FLUORENE (TRACE)	20.63	0.051	2.31515
1-METHYL FLUORENE	19.71	0.058	1.75719
1,2-BENZOFLUORENE	26.23	0.025	9.60934
9,10-DIHYDROPHENANTHRENE	19.77	0.058	1.78578
CARBAZOLE (TRACE)	22.95	0.037	4.31954
N-METHYL CARBAZOLE (TRACE)	22.53	0.039	3.86443
N-VINYLCARBAZOLE	22.60	0.039	3.93437
N-PHENYL CARBAZOLE	23.45	0.035	4.91374
N-METHYL-2,3-BENZCARBAZOLE	26.49	0.024	10.12840

ACCEPTOR MOLECULES	RO	CO	OVERLAP
DIBENZOFURAN	21.00	0.048	2.57081
DIBENZOTHIOPHENE (TRACE)	19.52	0.060	1.64132
P-TERPHENYL	24.92	0.029	7.13771
DEUTERATED P-TERPHENYL-C18D14	24.59	0.030	6.60523
4-METHYL TERPHENYL	26.07	0.025	9.36679
2-METHYL-P-TERPHENYL	19.62	0.059	1.71014
3,3''-DIMETHYL-P-TERPHENYL	25.51	0.027	8.28313
4-(3,3-DIMETHYLBUTOXY)-P-TERPHENYL	28.24	0.020	15.06604
4,4''-DIHEXAHYDROFARNESOXY-P-TERPHENYL	30.23	0.016	22.63473
2,2'-METHYLENE-P-TERPHENYL	29.09	0.018	18.01418
2,2'-ETHYLENE-P-TERPHENYL	29.13	0.018	18.14336
2,2'-METHYLENE-5',6''-METHYLENE-P-TERPHENYL (TRACE)	30.17	0.016	22.30392
3-PHENYL DIBENZOFURAN	29.78	0.017	20.72513
BENZO(1,2-B-4,5-B')BISBENZOFURAN (TRACE) (B)	29.98	0.017	21.42897
SEXIPHENYL BENZENE	16.11	0.107	0.52586
P-QUATERPHENYL	30.57	0.016	24.16562
3,3'''-DIMETHYL-P-QUATERPHENYL	31.19	0.015	27.23795
TETRAMETHYL-P-QUATERPHENYL	22.27	0.041	3.65017
3,3''' -DIMETHYL-3',2''-METHYLENE-P-QUATERPHENYL	32.33	0.013	33.70954
3,3''' -DIMETHYL-3',2''-ETHYLENE-P-QUATERPHENYL	32.38	0.013	33.94284
4-(3-ETHYLBUTOXY)-P-QUATERPHENYL (B)	32.18	0.013	32.73498
4,4'''-DI(2-BUTYLOCTOXY-1)-P-QUATERPHENYL	33.25	0.012	39.83340
DIETHYL-P-QUINQUEPHENYL	32.62	0.013	35.67443
DI(3-ETHYLHEPTYL)-P-QUINQUEPHENYL	34.15	0.011	46.63918
TETRAMETHYL-P-SEXIPHENYL	32.69	0.013	36.17893
TETRAMETHYL-P-OCTAPHENYL (B)	35.22	0.010	56.08341
INDOLE (TRACE)	11.72	0.278	0.07848
2-PHENYL INDOLE	29.39	0.018	19.10687
1-METHYL-2-PHENYL INDOLE	27.11	0.022	11.74739
1-NEOPENTYL-2-PHENYL INDOLE	26.53	0.024	10.36265
1-N-BUTYL-2(BIPHENYLYL-4) INDOLE	29.59	0.017	19.73829
1,2-DIPHENYL INDOLE	28.00	0.020	14.25723
1-BIPHENYLYL-2-PHENYL INDOLE	29.36	0.018	18.87524
1,3-DIMETHYL-2-PHENYL INDOLE	26.22	0.025	9.63014
2,3-DIPHENYL INDOLE	27.87	0.021	13.76278
1-METHYL-2,3-DIPHENYL INDOLE	26.28	0.025	9.69468
2-PHENYL-3,2'-METHYLENE INDOLE (TRACE)	28.86	0.019	17.08486
2-PHENYL-3,2'-ETHYLENE INDOLE (TRACE)	28.22	0.020	14.85300
2-PHENYL-3,2'-TRIMETHYLENE INDOLE	29.03	0.018	17.69099
1-METHYL-3,2'-METHYLENE-2-PHENYL INDOLE (TRACE)	28.62	0.019	16.21323
1-METHYL-3,2'-ETHYLENE-2-PHENYL INDOLE	27.80	0.021	13.57127
1-METHYL-3,2'-TRIMETHYLENE-2-PHENYL INDOLE	27.78	0.021	13.57430
4,5-DIPHENYLIMIDAZOLE (A)	23.08	0.036	4.47875
PPF (TRACE)	30.26	0.016	22.62484
PPO (TRACE)	30.89	0.015	25.70987
2-(4-FLUOROPHENYL)-5-PHENYLOXAZOLE (TRACE)	29.67	0.017	20.18660
αNPO (TRACE)	27.96	0.020	14.00262
BBO (B)	30.56	0.016	23.87484
POPOP	27.93	0.021	13.92541
DIMETHYL POPOP (TRACE)	28.31	0.020	15.10731
PPD (TRACE)	25.97	0.026	9.16310
2,5-DI(2-METHYLPHENYL)OXADIAZOLE	25.66	0.026	8.52585
PBD	32.13	0.013	32.60269
BBD (B)	33.09	0.012	38.65174
αNPD (TRACE)	28.00	0.020	14.17248
βNPD	29.98	0.017	21.46360
αNND (TRACE)	27.34	0.022	12.27180
βNND (TRACE)	30.77	0.015	25.04557
TRANS-1-(4-BIPHENYLYL)-2-PHENYL-ETHYLENE	32.00	0.014	31.56417
TRANS-1-(4-BIPHENYLYL)-2-(1-NAPHTHYL)-ETHYLENE	29.62	0.017	19.78107
DIPHENYLSTILBENE (B)	30.85	0.015	25.29234
1,1,4,4-TETRAPHENYLBUTADIENE	28.04	0.020	14.23891
1,6-DIPHENYLHEXATRIENE	29.42	0.018	19.01591
1,8-DIPHENYL-1,3,5,7-OCTATETRAENE	25.96	0.026	8.97982
BPSB (TRACE)	29.71	0.017	20.15945
BIS-MSB	30.32	0.016	22.76405
ESCULIN (A)	25.23	0.028	7.55979
AZULENE	20.63	0.051	2.25832
NAPHTHALENE	14.14	0.158	0.23817
DEUTERATED NAPHTHALENE-C10D8	14.74	0.140	0.30482
1-METHYLNAPHTHALENE 5ML/L	17.66	0.081	0.90219
2-METHYL NAPHTHALENE	15.65	0.117	0.43718
2,3-DIMETHYL NAPHTHALENE	16.55	0.099	0.61143
2,6-DIMETHYL NAPHTHALENE (TRACE)	16.74	0.095	0.65421
1-NAPHTHOL (TRACE)	21.02	0.048	2.56004
2-NAPHTHOL (TRACE)	17.99	0.077	1.00334
1-NAPHTHYLAMINE	22.54	0.039	3.86582
2-CHLORONAPHTHALENE	15.62	0.117	0.43223
ACENAPHTHENE	20.67	0.051	2.32422
1-PHENYL NAPHTHALENE	22.66	0.038	4.02472
2-PHENYL NAPHTHALENE	33.34	0.012	40.70918
1,4-DIPHENYL NAPHTHALENE	26.46	0.024	10.16069
1,5-DIPHENYL NAPHTHALENE	25.76	0.026	8.63620
1,7 DIPHENYL NAPHTHALENE	25.39	0.027	7.92064
1,4,5,8-TETRAPHENYL NAPHTHALENE	26.88	0.023	11.06404
1,1'-BINAPHTHYL	24.05	0.032	5.75426
4,4'-DI(N-BUTOXY)-1,1'-BINAPHTHYL	27.55	0.021	12.88673
2,2'-BINAPHTHYL	28.09	0.020	14.55492
ACRIDINE (A)	19.88	0.057	1.80706
ANTHRACENE (TRACE)	18.71	0.068	1.25744
DEUTERATED ANTHRACENE C14D10 (TRACE)	18.82	0.067	1.30406
9-METHYL ANTHRACENE (TRACE)	17.34	0.086	0.79811
1-AMINO ANTHRACENE	18.63	0.069	1.22640
2-AMINO ANTHRACENE	17.90	0.078	0.95649
9-VINYL ANTHRACENE	17.87	0.078	0.95336
9-PHENYL ANTHRACENE (TRACE)	18.46	0.071	1.16136

ACCEPTOR MOLECULES	RO	CO	OVERLAP
9,10-DIPHENYL ANTHRACENE	17.80	0.079	0.93392
9,10-DICHLOROANTHRACENE (TRACE)	16.61	0.098	0.61500
BIANTHRANYL (B)	19.65	0.059	1.68994
9,10-DINAPHTHYLANTHRACENE (B)	23.30	0.035	4.68995
9-(P-VINYLPHENYL)-10-PHENYL ANTHRACENE	20.13	0.055	1.95112
9,10-BIS(PHENYL-ETHYNYL)ANTHRACENE (TRACE)	28.48	0.019	15.63704
RUBICENE (B)	30.07	0.016	21.68853
DIBENZ(A,H)ANTHRACENE (B)	34.05	0.011	45.70834
TETRACENE (TRACE) (B)	25.70	0.026	8.45402
RUBRENE (B)	35.46	0.010	58.30481
TETRAHYDROCANNABINOL ACID METHYL ESTER	22.53	0.039	3.91002
PHENANTHRENE	19.53	0.060	1.63803
3,4-BENZOPHENANTHRENE	25.75	0.026	8.53955
2-PHENYL PHENANTHRENE	26.61	0.024	10.44780
CHRYSENE (TRACE)	26.12	0.025	9.32212
TRIPHENYLENE	20.21	0.054	2.01208
DEUTERATED TRIPHENYLENE-C18D12	19.92	0.057	1.84325
PYRENE	26.47	0.024	10.08198
1,3,6,8,-TETRAPHENYL PYRENE	29.28	0.018	18.49123
BIPYRENYL (B)	28.45	0.019	15.55230
O-PHENYLENE-PYRENE	29.75	0.017	20.33915
ANTHANTHRENE (B)	30.72	0.015	24.61844
FLUORANTHENE	22.84	0.038	4.16185
3-PHENYL FLUORANTHENE	26.87	0.023	11.02483
3-AMINO FLUORANTHENE	26.96	0.023	11.26646
3,3'-BIFLUORANTHENYL (B)	29.18	0.018	18.11226
BENZO(GHI)FLUORANTHENE	23.73	0.033	5.23660
9,12-DI(N-OCTADECOXY)-BENZ(E)ACEPHENANTHRYLENE	26.22	0.025	9.52438
DECACYCLENE (B)	29.67	0.017	20.01859
10-ETHOXY-BENZO(KL)XANTHENE	22.22	0.041	3.52561
PERYLENE	16.97	0.091	0.70099
BENZO(GHI)PERYLENE (B)	29.99	0.017	21.33520
BBOT (TRACE)	26.55	0.024	10.27535
QUININE SULFATE (A)	23.75	0.033	5.29424
ACRIDINE YELLOW (A)	21.33	0.046	2.76466
ACRIDINE ORANGE (A)	23.95	0.033	5.52907
ACRIDINE RED (A)	22.77	0.038	4.08128
URANINE (A)	22.90	0.037	4.22868
RHODAMINE B (A)	25.94	0.026	8.93498
RHODAMINE 6G (A)	24.95	0.029	7.07639
EOSIN B (A)	24.74	0.030	6.71719
EOSIN Y (A)	25.46	0.027	7.98452
ROSE BENGAL (A)	24.93	0.029	7.04210
INDENOINDENE	29.16	0.018	18.19863

8.32 Donor Molecule: 1,4-Dimethoxybenzene

POLYVINYLTOLUENE	3.17	14.010	0.00004
ANILINE 3ML/L	9.43	0.534	0.02590
N,N-DIMETHYLANILINE 3ML/L	15.26	0.126	0.46213
TRIPHENYLAMINE	25.35	0.027	9.63616
1,4-DIMETHOXY BENZENE	11.52	0.293	0.08620
1-METHYL-4-CHLOROBENZENE 5ML/L	0.00	0.000	0.00000
SODIUM SALICYLATE (A)	17.58	0.082	1.07138
ANTHRANILIC ACID (A)	22.12	0.041	4.22619
IMPD	19.91	0.057	2.24991
1,2,4-TRIMETHYLBENZENE (TRACE)	0.00	0.000	0.00000
MESITYLENE 5ML/L	0.00	0.000	0.00000
1,3,5-TRIETHYLBENZENE 5ML/L	0.00	0.000	0.00000
STYRENE	3.53	10.206	0.00007
3,3'-DIMETHYL BIPHENYL	4.42	5.189	0.00028
2,2'-DIMETHYLBIPHENYL 2ML/L	0.00	0.000	0.00000
OCTAMETHYL BIPHENYL	0.00	0.000	0.00000
4-METHOXY BIPHENYL	9.14	0.587	0.02148
4-BIPHENYLYL PHENYL ETHER	8.26	0.795	0.01171
4,4'-DIHYDROXYBIPHENYL (A)	14.18	0.157	0.29859
4,4'-DIMETHOXY BIPHENYL	12.90	0.209	0.16930
4,4'-DIISOBUTOXYBIPHENYL	12.76	0.215	0.15934
BENZIDINE	23.32	0.035	5.85367
4-VINYL BIPHENYL	17.29	0.087	0.98181
M-TERPHENYL	12.14	0.250	0.11804
M-QUATERPHENYL	7.73	0.967	0.00792
1,3,5-TRIPHENYL BENZENE	11.95	0.262	0.10659
1,3-DIBIPHENYLYLBENZENE (B)	24.07	0.032	7.09593
1,3,5,-TRIBIPHENYLYLBENZENE (B)	27.76	0.021	16.72334
FLUORENE (TRACE)	13.87	0.168	0.26232
1-METHYL FLUORENE	12.68	0.220	0.15293
1,2-BENZOFLUORENE	23.56	0.034	6.20636
9,10-DIHYDROPHENANTHRENE	13.90	0.167	0.26538
CARBAZOLE (TRACE)	20.50	0.052	2.70476
N-METHYL CARBAZOLE (TRACE)	20.61	0.051	2.78308
N-VINYLCARBAZOLE	21.03	0.048	3.14819
N-PHENYL CARBAZOLE	20.84	0.049	2.97376
N-METHYL-2,3-BENZCARBAZOLE	23.66	0.034	6.30224
DIBENZOFURAN	12.19	0.247	0.12122
DIBENZOTHIOPHENE (TRACE)	18.72	0.068	1.56993
P-TERPHENYL	18.39	0.072	1.42128
DEUTERATED P-TERPHENYL-C18D14	18.05	0.076	1.27113
4-METHYL TERPHENYL	19.96	0.056	2.31745
2-METHYL-P-TERPHENYL	12.45	0.232	0.13735

ACCEPTOR MOLECULES	RO	CO	OVERLAP
3,3''-DIMETHYL-P-TERPHENYL	19.43	0.061	1.97078
4-(3,3-DIMETHYLBUTOXY)-P-TERPHENYL	23.32	0.035	5.88234
4,4''-DIHEXAHYDROFARNESOXY-P-TERPHENYL	25.94	0.026	11.11671
2,2'-METHYLENE-P-TERPHENYL	25.49	0.027	10.02338
2,2'-ETHYLENE-P-TERPHENYL	25.56	0.027	10.19307
2,2'-METHYLENE-5',6''-METHYLENE-P-TERPHENYL (TRACE)	29.55	0.017	24.24087
3-PHENYL DIBENZOFURAN	26.01	0.025	11.48906
BENZO(1,2-B-4,5-B')BISBENZOFURAN (TRACE) (B)	31.03	0.015	32.46433
SEXIPHENYL BENZENE	8.91	0.632	0.01852
P-QUATERPHENYL	26.67	0.024	13.12779
3,3'''-DIMETHYL-P-QUATERPHENYL	27.60	0.021	16.10351
TETRAMETHYL-P-QUATERPHENYL	14.84	0.137	0.39350
3,3''' -DIMETHYL-3',2''-METHYLENE-P-QUATERPHENYL	31.72	0.014	37.01901
3,3''' -DIMETHYL-3',2''-ETHYLENE-P-QUATERPHENYL	31.47	0.014	35.20896
4-(3-ETHYLBUTOXY)-P-QUATERPHENYL (B)	30.26	0.016	27.84381
4,4'''-DI(2-BUTYLOCTOXY-1)-P-QUATERPHENYL	31.14	0.015	33.06952
DIETHYL-P-QUINQUEPHENYL	28.80	0.019	20.79610
DI(3-ETHYLHEPTYL)-P-QUINQUEPHENYL	32.19	0.013	40.26338
TETRAMETHYL-P-SEXIPHENYL	28.35	0.020	18.94617
TETRAMETHYL-P-OCTAPHENYL (B)	34.51	0.011	61.08014
INDOLE (TRACE)	0.00	0.000	0.00000
2-PHENYL INDOLE	27.62	0.021	16.19371
1-METHYL-2-PHENYL INDOLE	23.85	0.033	6.71304
1-NEOPENTYL-2-PHENYL INDOLE	22.59	0.039	4.86034
1-N-BUTYL-2(BIPHENYLYL-4) INDOLE	28.57	0.019	19.68027
1,2-DIPHENYL INDOLE	24.84	0.029	8.55548
1-BIPHENYLYL-2-PHENYL INDOLE	26.57	0.024	12.76021
1,3-DIMETHYL-2-PHENYL INDOLE	23.66	0.034	6.38718
2,3-DIPHENYL INDOLE	26.68	0.024	13.03126
1-METHYL-2,3-DIPHENYL INDOLE	24.55	0.030	7.93462
2-PHENYL-3,2'-METHYLENE INDOLE (TRACE)	28.63	0.019	20.03341
2-PHENYL-3,2'-ETHYLENE INDOLE (TRACE)	28.85	0.019	20.88172
2-PHENYL-3,2'-TRIMETHYLENE INDOLE	28.19	0.020	18.27269
1-METHYL-3,2'-METHYLENE-2-PHENYL INDOLE (TRACE)	29.00	0.018	21.60090
1-METHYL-3,2'-ETHYLENE-2-PHENYL INDOLE	28.57	0.019	19.68696
1-METHYL-3,2'-TRIMETHYLENE-2-PHENYL INDOLE	26.70	0.023	13.18480
4,5-DIPHENYLIMIDAZOLE (A)	19.26	0.063	1.86144
PPF (TRACE)	30.53	0.016	29.40914
PPO (TRACE)	29.03	0.018	21.78476
2-(4-FLUOROPHENYL)-5-PHENYLOXAZOLE (TRACE)	27.79	0.021	16.76501
αNPO (TRACE)	29.33	0.018	22.92497
BBO (B)	32.62	0.013	43.36271
POPOP	30.34	0.016	28.04274
DIMETHYL POPOP (TRACE)	30.57	0.016	29.31553
PPD (TRACE)	19.75	0.058	2.18079
2,5-DI(2-METHYLPHENYL)OXADIAZOLE	20.15	0.055	2.45979
PBD	29.21	0.018	22.62642
BBD (B)	32.79	0.013	45.10147
αNPD (TRACE)	27.95	0.020	17.23707
βNPD	27.63	0.021	16.18356
αNND (TRACE)	28.29	0.020	18.52763
βNND (TRACE)	30.32	0.016	28.24025
TRANS-1-(4-BIPHENYLYL)-2-PHENYL-ETHYLENE	31.90	0.014	38.12935
TRANS-1-(4-BIPHENYLYL)-2-(1-NAPHTHYL)-ETHYLENE	30.94	0.015	31.54066
DIPHENYLSTILBENE (B)	33.00	0.012	46.42921
1,1,4,4-TETRAPHENYLBUTADIENE	30.42	0.016	28.43345
1,6-DIPHENYLHEXATRIENE	32.68	0.013	43.71325
1,8-DIPHENYL-1,3,5,7-OCTATETRAENE	29.90	0.017	25.67426
BPSB (TRACE)	32.75	0.013	44.30852
BIS-MSB	32.68	0.013	43.76394
ESCULIN (A)	26.20	0.025	11.66727
AZULENE	20.19	0.054	2.43179
NAPHTHALENE	10.11	0.433	0.03924
DEUTERATED NAPHTHALENE-C10D8	11.66	0.283	0.09154
1-METHYLNAPHTHALENE 5ML/L	11.99	0.259	0.10910
2-METHYL NAPHTHALENE	12.62	0.223	0.14787
2,3-DIMETHYL NAPHTHALENE	12.64	0.221	0.14951
2,6-DIMETHYL NAPHTHALENE (TRACE)	15.00	0.133	0.41668
1-NAPHTHOL (TRACE)	17.94	0.077	1.21921
2-NAPHTHOL (TRACE)	17.93	0.078	1.21424
1-NAPHTHYLAMINE	22.23	0.041	4.37841
2-CHLORONAPHTHALENE	12.67	0.220	0.15163
ACENAPHTHENE	16.43	0.101	0.72100
1-PHENYL NAPHTHALENE	17.83	0.079	1.17684
2-PHENYL NAPHTHALENE	26.04	0.025	11.39679
1,4-DIPHENYL NAPHTHALENE	23.81	0.033	6.63775
1,5-DIPHENYL NAPHTHALENE	22.76	0.038	5.06693
1,7-DIPHENYL NAPHTHALENE	22.49	0.039	4.71677
1,4,5,8-TETRAPHENYL NAPHTHALENE	28.18	0.020	17.97799
1,1'-BINAPHTHYL	19.39	0.061	1.94363
4,4'-DI(N-BUTOXY)-1,1'-BINAPHTHYL	26.47	0.024	12.49234
2,2'-BINAPHTHYL	25.06	0.028	9.02037
ACRIDINE (A)	22.50	0.039	4.65459
ANTHRACENE (TRACE)	21.14	0.047	3.21118
DEUTERATED ANTHRACENE C14D10 (TRACE)	21.05	0.048	3.13291
9-METHYL ANTHRACENE (TRACE)	20.06	0.055	2.33942
1-AMINO ANTHRACENE	18.55	0.070	1.46446
2-AMINO ANTHRACENE	20.21	0.054	2.43399
9-VINYL ANTHRACENE	20.54	0.052	2.69872
9-PHENYL ANTHRACENE (TRACE)	21.16	0.047	3.22682
9,10-DIPHENYL ANTHRACENE	20.63	0.051	2.76500
9,10-DICHLOROANTHRACENE (TRACE)	19.48	0.060	1.96386
BIANTHRANYL (B)	22.35	0.040	4.47574
9,10-DINAPHTHYLANTHRACENE (B)	21.66	0.044	3.70523
9-(P-VINYLPHENYL)-10-PHENYL ANTHRACENE	20.93	0.049	3.02093
9,10-BIS(PHENYL-ETHYNYL)ANTHRACENE (TRACE)	25.10	0.028	8.97574

ACCEPTOR MOLECULES	RO	CO	OVERLAP
RUBICENE (B)	26.73	0.023	13.10246
DIBENZ(A,H)ANTHRACENE (B)	29.10	0.018	21.81530
TETRACENE (TRACE) (B)	18.45	0.071	1.41739
RUBRENE (B)	30.69	0.015	30.01579
TETRAHYDROCANNABINOL ACID METHYL ESTER	16.14	0.106	0.64960
PHENANTHRENE	13.16	0.196	0.18843
3,4-BENZOPHENANTHRENE	23.12	0.036	5.49421
2-PHENYL PHENANTHRENE	20.62	0.051	2.78536
CHRYSENE (TRACE)	23.97	0.032	6.85919
TRIPHENYLENE	15.20	0.127	0.44764
DEUTERATED TRIPHENYLENE-C18D12	15.15	0.129	0.43960
PYRENE	28.04	0.020	17.51764
1,3,6,8,-TETRAPHENYL PYRENE	25.99	0.025	11.05475
BIPYRENYL (B)	30.55	0.016	29.19388
O-PHENYLENE-PYRENE	27.31	0.022	14.91248
ANTHANTHRENE (B)	27.69	0.021	16.18366
FLUORANTHENE	23.38	0.035	5.85868
3-PHENYL FLUORANTHENE	24.57	0.030	7.90719
3-AMINO FLUORANTHENE	24.77	0.029	8.29766
3,3'-BIFLUORANTHENYL (B)	27.63	0.021	15.97088
BENZO(GHI)FLUORANTHENE	23.01	0.037	5.32853
9,12-DI(N-OCTADECOXY)-BENZ(E)ACEPHENANTHRYLENE	24.78	0.029	8.30781
DECACYCLENE (B)	30.01	0.017	26.22166
10-ETHOXY-BENZO(KL)XANTHENE	21.78	0.043	3.82792
PERYLENE	17.88	0.078	1.17478
BENZO(GHI)PERYLENE (B)	25.61	0.027	10.14046
BBOT (TRACE)	29.87	0.017	25.49943
QUININE SULFATE (A)	24.37	0.031	7.61265
ACRIDINE YELLOW (A)	17.58	0.082	1.06049
ACRIDINE ORANGE (A)	19.43	0.061	1.93389
ACRIDINE RED (A)	21.25	0.047	3.30250
URANINE (A)	21.72	0.044	3.76663
RHODAMINE B (A)	24.03	0.032	6.91744
RHODAMINE 6G (A)	23.24	0.036	5.66212
EOSIN B (A)	23.86	0.033	6.62389
EOSIN Y (A)	24.40	0.031	7.57969
ROSE BENGAL (A)	24.61	0.030	7.98165
INDENOINDENE	28.15	0.020	18.10984

8.36 Donor Molecule: TMPD

TRIPHENYLAMINE	4.86	3.889	0.00055
SODIUM SALICYLATE (A)	4.82	3.990	0.00052
ANTHRANILIC ACID (A)	13.88	0.167	0.29502
TMPD	9.29	0.557	0.02673
1,2-BENZOFLUORENE	4.40	5.236	0.00030
9,10-DIHYDROPHENANTHRENE	0.00	0.000	0.00000
CARBAZOLE (TRACE)	0.00	0.000	0.00000
N-METHYL CARBAZOLE (TRACE)	4.57	4.675	0.00038
N-VINYLCARBAZOLE	3.75	8.510	0.00012
N-PHENYL CARBAZOLE	3.93	7.378	0.00015
N-METHYL-2,3-BENZCARBAZOLE	22.28	0.040	5.03213
BENZO(1,2-B-4,5-B')BISBENZOFURAN (TRACE) (B)	4.82	3.998	0.00052
3,3''' -DIMETHYL-3',2''-METHYLENE-P-QUATERPHENYL	6.38	1.722	0.00281
3,3''' -DIMETHYL-3',2''-ETHYLENE-P-QUATERPHENYL	9.14	0.586	0.02422
4-(3-ETHYLBUTOXY)-P-QUATERPHENYL (B)	7.70	0.978	0.00869
4,4'''-DI(2-BUTYLOCTOXY-1)-P-QUATERPHENYL	7.43	1.092	0.00698
DIETHYL-P-QUINQUEPHENYL	0.00	0.000	0.00000
DI(3-ETHYLHEPTYL)-P-QUINQUEPHENYL	9.67	0.494	0.03395
TETRAMETHYL-P-SEXIPHENYL	0.00	0.000	0.00000
TETRAMETHYL-P-OCTAPHENYL (B)	13.66	0.175	0.26935
1-N-BUTYL-2(BIPHENYLYL-4) INDOLE	10.27	0.413	0.04869
1,2-DIPHENYL INDOLE	0.00	0.000	0.00000
1-BIPHENYLYL-2-PHENYL INDOLE	5.71	2.397	0.00145
1,3-DIMETHYL-2-PHENYL INDOLE	0.00	0.000	0.00000
2,3-DIPHENYL INDOLE	10.16	0.426	0.04569
1-METHYL-2,3-DIPHENYL INDOLE	8.26	0.793	0.01321
2-PHENYL-3,2'-METHYLENE INDOLE (TRACE)	0.00	0.000	0.00000
2-PHENYL-3,2'-ETHYLENE INDOLE (TRACE)	8.97	0.621	0.02156
2-PHENYL-3,2'-TRIMETHYLENE INDOLE	0.00	0.000	0.00000
1-METHYL-3,2'-METHYLENE-2-PHENYL INDOLE (TRACE)	7.13	1.235	0.00545
1-METHYL-3,2'-ETHYLENE-2-PHENYL INDOLE	12.93	0.207	0.19396
1-METHYL-3,2'-TRIMETHYLENE-2-PHENYL INDOLE	3.56	9.896	0.00009
4,5-DIPHENYLIMIDAZOLE (A)	0.00	0.000	0.00000
PPF (TRACE)	7.16	1.220	0.00559
αNPO (TRACE)	18.77	0.068	1.80615
BBO (B)	25.07	0.028	10.25242
POPOP	28.02	0.020	19.94678
DIMETHYL POPOP (TRACE)	30.95	0.015	36.14899
PPD (TRACE)	0.00	0.000	0.00000
2,5-DI(2-METHYLPHENYL)OXADIAZOLE	0.00	0.000	0.00000
PBD	0.00	0.000	0.00000
BBD (B)	10.11	0.432	0.04448
αNPD (TRACE)	10.78	0.358	0.06492
βNPD	0.00	0.000	0.00000
αNND (TRACE)	14.69	0.141	0.41634
βNND (TRACE)	5.09	3.300	0.00073
TRANS-1-(4-BIPHENYLYL)-2-PHENYL-ETHYLENE	11.69	0.280	0.10598
TRANS-1-(4-BIPHENYLYL)-2-(1-NAPHTHYL)-ETHYLENE	22.38	0.040	5.17750
DIPHENYLSTILBENE (B)	24.86	0.029	9.72518

ACCEPTOR MOLECULES	RO	CO	OVERLAP
1,1,4,4-TETRAPHENYLBUTADIENE	25.14	0.028	10.37928
1,6-DIPHENYLHEXATRIENE	29.55	0.017	27.43808
1,8-DIPHENYL-1,3,5,7-OCTATETRAENE	36.64	0.009	99.01136
BPSB (TRACE)	30.75	0.015	34.82495
BIS-MSB	28.47	0.019	21.94082
ESCULIN (A)	18.04	0.076	1.42205
AZULENE	9.23	0.570	0.00508
1-NAPHTHYLAMINE	4.97	3.645	0.00063
1,4,5,8-TETRAPHENYL NAPHTHALENE	19.82	0.057	2.49875
1,1'-BINAPHTHYL	0.00	0.000	0.00000
4,4'-DI(N-BUTOXY)-1,1'-BINAPHTHYL	6.58	1.573	0.00336
2,2'-BINAPHTHYL	0.00	0.000	0.00000
ACRIDINE (A)	20.71	0.050	3.25143
ANTHRACENE (TRACE)	20.21	0.054	2.80802
DEUTERATED ANTHRACENE C14D10 (TRACE)	19.72	0.058	2.42906
9-METHYL ANTHRACENE (TRACE)	22.46	0.039	5.29068
1-AMINO ANTHRACENE	24.22	0.031	8.24832
2-AMINO ANTHRACENE	23.23	0.036	6.42354
9-VINYL ANTHRACENE	24.05	0.032	7.96291
9-PHENYL ANTHRACENE (TRACE)	23.51	0.034	6.95348
9,10-DIPHENYL ANTHRACENE	26.45	0.024	14.05695
9,10-DICHLOROANTHRACENE (TRACE)	25.99	0.025	12.63825
BIANTHRANYL (B)	27.80	0.021	18.96166
9,10-DINAPHTHYLANTHRACENE (B)	26.33	0.025	13.66536
9-(P-VINYLPHENYL)-10-PHENYL ANTHRACENE	26.81	0.023	15.21853
9,10-BIS(PHENYL-ETHYNYL)ANTHRACENE (TRACE)	30.98	0.015	35.87507
RUBICENE (B)	26.45	0.024	13.90080
DIBENZ(A,H)ANTHRACENE (B)	16.20	0.105	0.74454
TETRACENE (TRACE) (B)	22.83	0.038	5.73927
RUBRENE (B)	20.44	0.052	2.96306
TETRAHYDROCANNABINOL ACID METHYL ESTER	0.00	0.000	0.00000
PHENANTHRENE	3.07	15.515	0.00003
3,4-BENZOPHENANTHRENE	10.70	0.366	0.06178
2-PHENYL PHENANTHRENE	6.73	1.465	0.00387
CHRYSENE (TRACE)	9.42	0.535	0.02897
TRIPHENYLENE	2.86	19.097	0.00002
DEUTERATED TRIPHENYLENE-C18D12	2.81	20.245	0.00002
PYRENE	10.14	0.429	0.04499
1,3,6,8,-TETRAPHENYL PYRENE	32.20	0.013	45.66790
BIPYRENYL (B)	27.28	0.022	16.94425
O-PHENYLENE-PYRENE	27.94	0.021	19.35279
ANTHANTHRENE (B)	29.76	0.017	28.35127
FLUORANTHENE	18.14	0.075	1.46185
3-PHENYL FLUORANTHENE	24.60	0.030	9.08778
3-AMINO FLUORANTHENE	26.25	0.025	13.29634
3,3'-BIFLUORANTHENYL (B)	30.37	0.016	32.11893
BENZO(GHI)FLUORANTHENE	17.94	0.077	1.36684
9,12-DI(N-OCTADECOXY)-BENZ(E)ACEPHENANTHRYLENE	27.77	0.021	18.75474
DECACYCLENE (B)	33.96	0.011	62.38646
10-ETHOXY-BENZO(KL)XANTHENE	26.90	0.023	15.56162
PERYLENE	30.60	0.016	33.50749
BENZO(GHI)PERYLENE (B)	27.54	0.021	17.93346
BBOT (TRACE)	33.63	0.012	59.48884
QUININE SULFATE (A)	6.00	2.067	0.00195
ACRIDINE YELLOW (A)	28.82	0.019	23.24963
ACRIDINE ORANGE (A)	24.22	0.031	8.19578
ACRIDINE RED (A)	20.54	0.052	3.04429
URANINE (A)	24.32	0.031	8.40222
RHODAMINE B (A)	23.80	0.033	7.38136
RHODAMINE 6G (A)	21.96	0.042	4.54587
EOSIN B (A)	21.50	0.045	4.01024
EOSIN Y (A)	22.67	0.038	5.51278
ROSE BENGAL (A)	21.50	0.045	4.01063

8.37 Donor Molecule: 1,2,4-Trimethylbenzene (trace)

ACCEPTOR MOLECULES	RO	CO	OVERLAP
FLUOROBENZENE (TRACE)	3.23	13.261	0.00002
TOLUENE (TRACE)	4.17	6.153	0.00010
DEUTERATED TOLUENE-C7D8 5ML/L	3.94	7.330	0.00007
ETHYLBENZENE (TRACE)	4.31	5.579	0.00013
PROPYLBENZENE (TRACE)	4.18	6.143	0.00010
ISOPROPYLBENZENE (TRACE)	3.83	7.991	0.00006
SEC-BUTYLBENZENE (TRACE)	3.62	9.424	0.00004
DIPHENYLMETHANE (TRACE)	5.31	2.992	0.00043
BIBENZYL	5.02	3.529	0.00031
1,3,5-TRIPHENYLPENTANE 0.05ML/L	5.36	2.910	0.00046
POLYSTYRENE	[illegible]	[illegible]	0.00012
POLYVINYLTOLUENE	8.73	0.672	0.00850
PHENYLCYCLOHEXANE (TRACE)	2.75	21.508	0.00001
BENZYLALCOHOL 3ML/L	3.94	7.314	0.00007
BENZYLACETATE 5ML/L	3.22	13.384	0.00002
PHENOL (TRACE)	12.78	0.214	0.08454
ANISOLE (TRACE)	13.33	0.189	0.10845
PHENYL ETHER	14.40	0.150	0.17203
ANILINE 3ML/L	18.66	0.069	0.80708
N,N-DIMETHYLANILINE 3ML/L	20.38	0.053	1.36284
TRIPHENYLAMINE	30.66	0.016	15.61372
P-XYLENE (TRACE)	8.86	0.642	0.00939
DEUTERATED P-XYLENE C8D10 (TRACE)	8.66	0.688	0.00820
M-XYLENE (TRACE)	6.61	1.546	0.00162

ACCEPTOR MOLECULES	RO	CO	OVERLAP
O-XYLENE 3ML/L	6.12	1.949	0.00102
1-METHYL-4-ETHYLBENZENE (TRACE)	7.92	0.899	0.00480
1,4-BIS(TRIFLUOROMETHYL)BENZENE 5ML/L	4.56	4.727	0.00017
P-CRESOL	17.07	0.090	0.47618
P-METHYLANISOLE 3ML/L	16.53	0.099	0.39232
1,4-DIMETHOXY BENZENE	20.60	0.051	1.45856
1-METHYL-4-CHLOROBENZENE 5ML/L	10.26	0.414	0.02255
SODIUM SALICYLATE (A)	22.26	0.041	2.28767
ANTHRANILIC ACID (A)	19.66	0.059	1.08777
TMPD	23.02	0.037	2.79721
1,2,4-TRIMETHYLBENZENE (TRACE)	10.23	0.418	0.02214
MESITYLENE 5ML/L	7.32	1.138	0.00299
1,3,5-TRIETHYLBENZENE 5ML/L	6.92	1.350	0.00213
STYRENE	14.64	0.143	0.18882
1,1-DIPHENYLETHYLENE 3ML/L	15.86	0.112	0.30694
BIPHENYL	14.22	0.156	0.15958
DEUTERATED BIPHENYL-C12D10	14.15	0.158	0.15518
4-METHYL BIPHENYL	17.04	0.090	0.47148
MONOISOPROPYLBIPHENYL 5ML/L	16.28	0.104	0.35983
4-BENZYL BIPHENYL	18.89	0.066	0.87628
3,3'-DIMETHYL BIPHENYL	16.74	0.095	0.42255
2,2'-DIMETHYLBIPHENYL 2ML/L	10.21	0.420	0.02184
OCTAMETHYL BIPHENYL	12.83	0.212	0.08587
4-METHOXY BIPHENYL	22.42	0.040	2.43595
4-BIPHENYLYL PHENYL ETHER	22.86	0.037	2.73464
4,4'-DIHYDROXYBIPHENYL (A)	25.37	0.027	5.06956
4,4'-DIMETHOXY BIPHENYL	25.10	0.028	4.76654
4,4'-DIISOBUTOXYBIPHENYL	24.98	0.029	4.64399
BENZIDINE	30.88	0.015	16.36440
4-VINYL BIPHENYL	28.95	0.018	11.22580
M-TERPHENYL	21.14	0.047	1.70372
M-QUATERPHENYL	20.77	0.050	1.54112
1,3,5-TRIPHENYL BENZENE	21.60	0.044	1.92544
1,3-DIBIPHENYLYLBENZENE (B)	33.95	0.011	28.95035
1,3,5,-TRIBIPHENYLYLBENZENE (B)	37.65	0.008	53.89481
FLUORENE (TRACE)	23.66	0.034	3.36079
1-METHYL FLUORENE	22.54	0.039	2.49543
1,2-BENZOFLUORENE	28.03	0.020	9.13318
9,10-DIHYDROPHENANTHRENE	23.60	0.034	3.29581
CARBAZOLE (TRACE)	26.72	0.023	6.87199
N-METHYL CARBAZOLE (TRACE)	26.05	0.025	5.87705
N-VINYLCARBAZOLE	26.90	0.023	7.12884
N-PHENYL CARBAZOLE	27.73	0.021	8.56465
N-METHYL-2,3-BENZCARBAZOLE	31.49	0.014	18.34874
DIBENZOFURAN	26.03	0.025	5.95070
DIBENZOTHIOPHENE (TRACE)	23.32	0.035	3.04426
P-TERPHENYL	29.89	0.017	13.55473
DEUTERATED P-TERPHENYL-C18D14	29.52	0.017	12.62028
4-METHYL TERPHENYL	30.59	0.016	15.57272
2-METHYL-P-TERPHENYL	25.51	0.027	5.26319
3,3''-DIMETHYL-P-TERPHENYL	30.01	0.017	13.88067
4-(3,3-DIMETHYLBUTOXY)-P-TERPHENYL	31.78	0.014	19.49868
4,4''-DIHEXAHYDROFARNESOXY-P-TERPHENYL	33.23	0.012	25.48204
2,2'-METHYLENE-P-TERPHENYL	31.78	0.014	19.51368
2,2'-ETHYLENE-P-TERPHENYL	31.82	0.014	19.62341
2,2'-METHYLENE-5',6''-METHYLENE-P-TERPHENYL (TRACE)	31.45	0.014	18.26524
3-PHENYL DIBENZOFURAN	31.40	0.014	18.17785
BENZO(1,2-B-4,5-B')BISBENZOFURAN (TRACE) (B)	28.83	0.019	10.81290
SEXIPHENYL BENZENE	22.72	0.038	2.64094
P-QUATERPHENYL	33.25	0.012	25.53349
3,3'''-DIMETHYL-P-QUATERPHENYL	33.68	0.012	27.50774
TETRAMETHYL-P-QUATERPHENYL	28.26	0.020	9.71637
3,3'''-DIMETHYL-3',2''-METHYLENE-P-QUATERPHENYL	32.99	0.012	24.26757
3,3'''-DIMETHYL-3',2''-ETHYLENE-P-QUATERPHENYL	33.62	0.012	27.17580
4-(3-ETHYLBUTOXY)-P-QUATERPHENYL (B)	33.50	0.012	26.57126
4,4'''-DI(2-BUTYLOCTOXY-1)-P-QUATERPHENYL	34.72	0.011	32.97828
DIETHYL-P-QUINQUEPHENYL	35.32	0.010	36.64978
DI(3-ETHYLHEPTYL)-P-QUINQUEPHENYL	35.45	0.010	37.33427
TETRAMETHYL-P-SEXIPHENYL	35.73	0.010	39.32971
TETRAMETHYL-P-OCTAPHENYL (B)	35.96	0.010	40.65758
INDOLE (TRACE)	18.54	0.070	0.78228
2-PHENYL INDOLE	30.28	0.016	14.57278
1-METHYL-2-PHENYL INDOLE	29.03	0.018	11.29041
1-NEOPENTYL-2-PHENYL INDOLE	28.77	0.019	10.73866
1-N-BUTYL-2(BIPHENYLYL-4) INDOLE	30.44	0.016	14.95851
1,2-DIPHENYL INDOLE	29.85	0.017	13.35099
1-BIPHENYLYL-2-PHENYL INDOLE	31.58	0.014	18.64870
1,3-DIMETHYL-2-PHENYL INDOLE	27.86	0.021	8.83144
2,3-DIPHENYL INDOLE	28.84	0.019	10.82942
1-METHYL-2,3-DIPHENYL INDOLE	27.97	0.020	9.00296
2-PHENYL-3,2'-METHYLENE INDOLE (TRACE)	28.77	0.019	10.68992
2-PHENYL-3,2'-ETHYLENE INDOLE (TRACE)	27.86	0.021	8.79845
2-PHENYL-3,2'-TRIMETHYLENE INDOLE	29.31	0.018	11.95850
1-METHYL-3,2'-METHYLENE-2-PHENYL INDOLE (TRACE)	28.22	0.020	9.50972
1-METHYL-3,2'-ETHYLENE-2-PHENYL INDOLE	27.36	0.022	7.88989
1-METHYL-3,2'-TRIMETHYLENE-2-PHENYL INDOLE	28.27	0.020	9.61474
4,5-DIPHENYLIMIDAZOLE (A)	26.09	0.025	5.95261
PPF (TRACE)	30.09	0.016	13.97246
PPO (TRACE)	32.08	0.014	20.58229
2-(4-FLUOROPHENYL)-5-PHENYLOXAZOLE (TRACE)	30.95	0.015	16.58872
αNPO (TRACE)	28.18	0.020	9.41276
BBO (B)	30.27	0.016	14.48244
POPOP	27.87	0.021	8.82446
DIMETHYL POPOP (TRACE)	28.80	0.019	10.73323
PPD (TRACE)	30.11	0.016	14.19410
2,5-DI(2-METHYLPHENYL)OXADIAZOLE	29.37	0.018	12.22434

ACCEPTOR MOLECULES	RO	CO	OVERLAP
PBD	34.01	0.011	29.22126
BBD (B)	33.52	0.012	26.69869
αNPD (TRACE)	28.42	0.019	9.91173
βNPD	31.84	0.014	19.63845
αNND (TRACE)	27.22	0.022	7.66031
βNND (TRACE)	32.08	0.014	20.33321
TRANS-1-(4-BIPHENYLYL)-2-PHENYL-ETHYLENE	32.29	0.013	21.31039
TRANS-1-(4-BIPHENYLYL)-2-(1-NAPHTHYL)-ETHYLENE	29.90	0.017	13.43142
DIPHENYLSTILBENE (B)	29.83	0.017	13.25914
1,1,4,4-TETRAPHENYLBUTADIENE	27.10	0.022	7.44979
1,6-DIPHENYLHEXATRIENE	27.66	0.021	8.42182
1,8-DIPHENYL-1,3,5,7-OCTATETRAENE	24.74	0.030	4.31425
BPSB (TRACE)	28.39	0.020	9.84029
BIS-MSB	29.35	0.018	12.02072
ESCULIN (A)	25.44	0.027	5.09860
AZULENE	27.35	0.022	7.88170
NAPHTHALENE	19.77	0.058	1.13549
DEUTERATED NAPHTHALENE-C10D8	19.90	0.057	1.17511
1-METHYLNAPHTHALENE 5ML/L	22.07	0.042	2.19593
2-METHYL NAPHTHALENE	20.31	0.053	1.33257
2,3-DIMETHYL NAPHTHALENE	21.07	0.048	1.65979
2,6-DIMETHYL NAPHTHALENE (TRACE)	20.65	0.051	1.47155
1-NAPHTHOL (TRACE)	23.39	0.035	3.10478
2-NAPHTHOL (TRACE)	20.21	0.054	1.28674
1-NAPHTHYLAMINE	22.76	0.038	2.61651
2-CHLORONAPHTHALENE	20.35	0.053	1.34589
ACENAPHTHENE	23.82	0.033	3.46466
1-PHENYL NAPHTHALENE	25.97	0.026	5.82073
2-PHENYL NAPHTHALENE	38.29	0.008	59.61779
1,4-DIPHENYL NAPHTHALENE	28.06	0.020	9.20982
1,5-DIPHENYL NAPHTHALENE	27.55	0.021	8.25258
1,7-DIPHENYL NAPHTHALENE	27.20	0.022	7.64038
1,4,5,8-TETRAPHENYL NAPHTHALENE	26.92	0.023	7.15429
1,1'-BINAPHTHYL	27.43	0.022	8.06068
4,4'-DI(N-BUTOXY)-1,1'-BINAPHTHYL	28.26	0.020	9.58842
2,2'-BINAPHTHYL	30.19	0.016	14.30523
ACRIDINE (A)	18.78	0.068	0.82459
ANTHRACENE (TRACE)	17.24	0.087	0.49415
DEUTERATED ANTHRACENE C14D10 (TRACE)	17.78	0.080	0.59459
9-METHYL ANTHRACENE (TRACE)	16.19	0.105	0.33848
1-AMINO ANTHRACENE	22.86	0.037	2.68348
2-AMINO ANTHRACENE	16.03	0.109	0.31477
9-VINYL ANTHRACENE	17.40	0.085	0.52297
9-PHENYL ANTHRACENE (TRACE)	18.06	0.076	0.65368
9,10-DIPHENYL ANTHRACENE	19.01	0.065	0.88807
9,10-DICHLOROANTHRACENE (TRACE)	15.99	0.109	0.31396
BIANTHRANYL (B)	20.92	0.049	1.57796
9,10-DINAPHTHYLANTHRACENE (B)	27.00	0.023	7.29531
9-(P-VINYLPHENYL)-10-PHENYL ANTHRACENE	22.59	0.039	2.50008
9,10-BIS(PHENYL-ETHYNYL)ANTHRACENE (TRACE)	30.66	0.016	15.61800
RUBICENE (B)	32.61	0.013	22.62235
DIBENZ(A,H)ANTHRACENE (B)	36.53	0.009	44.70245
TETRACENE (TRACE) (B)	34.25	0.011	30.35626
RUBRENE (B)	36.62	0.009	45.35758
TETRAHYDROCANNABINOL ACID METHYL ESTER	27.20	0.022	7.72040
PHENANTHRENE	24.31	0.031	3.87969
3,4-BENZOPHENANTHRENE	31.14	0.015	17.14514
2-PHENYL PHENANTHRENE	31.25	0.015	17.53712
CHRYSENE (TRACE)	28.11	0.020	9.27517
TRIPHENYLENE	26.02	0.025	5.83277
DEUTERATED TRIPHENYLENE-C18D12	25.95	0.026	5.74994
PYRENE	25.53	0.027	5.21321
1,3,6,8,-TETRAPHENYL PYRENE	32.57	0.013	22.44906
BIPYRENYL (B)	31.46	0.014	18.23736
O-PHENYLENE-PYRENE	31.64	0.014	18.85901
ANTHANTHRENE (B)	31.86	0.014	19.67066
FLUORANTHENE	27.37	0.022	7.90187
3-PHENYL FLUORANTHENE	29.89	0.017	13.41782
3-AMINO FLUORANTHENE	28.11	0.020	9.28683
3,3'-BIFLUORANTHENYL (B)	31.90	0.014	19.81636
BENZO(GHI)FLUORANTHENE	27.03	0.023	7.34394
9,12-DI(N-OCTADECOXY)-BENZ(E)ACEPHENANTHRYLENE	30.54	0.016	15.25668
DECACYCLENE (B)	31.13	0.015	17.13598
10-ETHOXY-BENZO(KL)XANTHENE	23.62	0.034	3.26634
PERYLENE	18.87	0.067	0.85024
BENZO(GHI)PERYLENE (B)	32.60	0.013	22.56370
BBOT (TRACE)	25.61	0.027	5.30964
QUININE SULFATE (A)	24.88	0.029	4.46603
ACRIDINE YELLOW (A)	25.84	0.026	5.60486
ACRIDINE ORANGE (A)	27.34	0.022	7.84929
ACRIDINE RED (A)	24.57	0.030	4.14388
URANINE (A)	24.71	0.030	4.28486
RHODAMINE B (A)	28.23	0.020	9.52316
RHODAMINE 6G (A)	27.92	0.021	8.91492
EOSIN B (A)	25.62	0.027	5.31425
EOSIN Y (A)	26.40	0.024	6.37034
ROSE BENGAL (A)	26.29	0.025	6.21812
INDENOINDENE	29.49	0.017	12.40903

8.38 Donor Molecule: Mesitylene (5 ml/L)

ACCEPTOR MOLECULES	RO	CO	OVERLAP
BENZENE (3 ML/L)	1.49	******	0.00000
DEUTERATED BENZENE-C6D6 5ML/L	0.00	0.000	0.00000
FLUOROBENZENE (TRACE)	3.20	13.700	0.00005
TOLUENE (TRACE)	3.86	7.788	0.00015
DEUTERATED TOLUENE-C7D8 5ML/L	3.56	9.910	0.00010
ETHYLBENZENE (TRACE)	3.91	7.497	0.00017
PROPYLBENZENE (TRACE)	3.76	8.383	0.00013
ISOPROPYLBENZENE (TRACE)	3.48	10.622	0.00008
SEC-BUTYLBENZENE (TRACE)	3.36	11.746	0.00007
DIPHENYLMETHANE (TRACE)	4.81	4.022	0.00058
BIBENZYL	4.48	4.971	0.00038
1,3,5-TRIPHENYLPENTANE 0.05ML/L	4.75	4.170	0.00054
POLYSTYRENE	3.93	7.361	0.00017
POLYVINYLTOLUENE	7.58	1.026	0.00880
PHENYLCYCLOHEXANE (TRACE)	2.86	19.198	0.00003
BENZYLALCOHOL 3ML/L	3.56	9.943	0.00010
BENZYLACETATE 5ML/L	3.12	14.733	0.00004
PHENOL (TRACE)	11.02	0.334	0.08362
ANISOLE (TRACE)	11.52	0.293	0.10915
PHENYL ETHER	12.48	0.230	0.17638
ANILINE 3ML/L	16.09	0.107	0.80172
N,N-DIMETHYLANILINE 3ML/L	17.58	0.082	1.35245
TRIPHENYLAMINE	26.44	0.024	15.50480
P-XYLENE (TRACE)	7.72	0.974	0.00986
DEUTERATED P-XYLENE C8D10 (TRACE)	7.58	1.028	0.00885
M-XYLENE (TRACE)	5.89	2.194	0.00195
O-XYLENE 3ML/L	5.47	2.733	0.00125
1-METHYL-4-ETHYLBENZENE (TRACE)	6.97	1.323	0.00536
1,4-BIS(TRIFLUOROMETHYL)BENZENE 5ML/L	4.34	5.478	0.00031
P-CRESOL	14.72	0.140	0.47294
P-METHYLANISOLE 3ML/L	14.25	0.155	0.38948
1,4-DIMETHOXY BENZENE	17.76	0.080	1.44760
1-METHYL-4-CHLOROBENZENE 5ML/L	8.84	0.647	0.02230
SODIUM SALICYLATE (A)	19.20	0.063	2.26968
ANTHRANILIC ACID (A)	17.04	0.090	1.11011
TMPD	19.89	0.057	2.81321
1,2,4-TRIMETHYLBENZENE (TRACE)	8.85	0.646	0.02237
MESITYLENE 5ML/L	6.40	1.703	0.00323
1,3,5-TRIETHYLBENZENE 5ML/L	6.06	2.015	0.00230
STYRENE	12.66	0.221	0.19053
1,1-DIPHENYLETHYLENE 3ML/L	13.72	0.173	0.31031
BIPHENYL	12.33	0.239	0.16405
DEUTERATED BIPHENYL-C12D10	12.28	0.242	0.15973
4-METHYL BIPHENYL	14.75	0.139	0.47918
MONOISOPROPYLBIPHENYL 5ML/L	14.10	0.159	0.36666
4-BENZYL BIPHENYL	16.35	0.102	0.88829
3,3'-DIMETHYL BIPHENYL	14.47	0.148	0.42598
2,2'-DIMETHYLBIPHENYL 2ML/L	8.87	0.641	0.02267
OCTAMETHYL BIPHENYL	11.11	0.326	0.08782
4-METHOXY BIPHENYL	19.37	0.062	2.44351
4-BIPHENYLYL PHENYL ETHER	19.75	0.058	2.74612
4,4'-DIHYDROXYBIPHENYL (A)	21.89	0.043	5.05973
4,4'-DIMETHOXY BIPHENYL	21.67	0.044	4.76049
4,4'-DIISOBUTOXYBIPHENYL	21.56	0.045	4.63596
BENZIDINE	26.64	0.024	16.27056
4-VINYL BIPHENYL	24.98	0.029	11.18234
M-TERPHENYL	18.27	0.073	1.71190
M-QUATERPHENYL	17.96	0.077	1.55720
1,3,5-TRIPHENYL BENZENE	18.68	0.069	1.94855
1,3-DIBIPHENYLYLBENZENE (B)	29.28	0.018	28.77723
1,3,5,-TRIBIPHENYLYLBENZENE (B)	32.47	0.013	53.54475
FLUORENE (TRACE)	20.43	0.052	3.35650
1-METHYL FLUORENE	19.44	0.061	2.47975
1,2-BENZOFLUORENE	24.18	0.032	9.08050
9,10-DIHYDROPHENANTHRENE	20.37	0.053	3.29682
CARBAZOLE (TRACE)	23.06	0.036	6.85805
N-METHYL CARBAZOLE (TRACE)	22.48	0.039	5.86376
N-VINYLCARBAZOLE	23.23	0.036	7.14557
N-PHENYL CARBAZOLE	23.93	0.033	8.53829
N-METHYL-2,3-BENZCARBAZOLE	27.18	0.022	18.30434
DIBENZOFURAN	22.45	0.040	5.91563
DIBENZOTHIOPHENE (TRACE)	20.10	0.055	3.00788
P-TERPHENYL	25.79	0.026	13.49743
DEUTERATED P-TERPHENYL-C18D14	25.47	0.027	12.56708
4-METHYL TERPHENYL	26.39	0.024	15.49322
2-METHYL-P-TERPHENYL	22.02	0.042	5.26284
3,3''-DIMETHYL-P-TERPHENYL	25.89	0.026	13.81213
4-(3,3-DIMETHYLBUTOXY)-P-TERPHENYL	27.41	0.022	19.37818
4,4''-DIHEXAHYDROFARNESOXY-P-TERPHENYL	28.66	0.019	25.32212
2,2'-METHYLENE-P-TERPHENYL	27.41	0.022	19.39532
2,2'-ETHYLENE-P-TERPHENYL	27.44	0.022	19.50589
2,2'-METHYLENE-5',6''-METHYLENE-P-TERPHENYL (TRACE)	27.16	0.022	18.26781
3-PHENYL DIBENZOFURAN	27.08	0.023	18.03440
BENZO(1,2-B-4,5-B')BISBENZOFURAN (TRACE) (B)	24.93	0.029	10.92217
SEXIPHENYL BENZENE	19.63	0.059	2.64836
P'-QUATERPHENYL	28.68	0.019	25.37408
3,3'''-DIMETHYL-P-QUATERPHENYL	29.05	0.018	27.34267
TETRAMETHYL-P-QUATERPHENYL	24.39	0.031	9.70433
3,3'''-DIMETHYL-3',2''-METHYLENE-P-QUATERPHENYL	28.49	0.019	24.26081
3,3'''-DIMETHYL-3',2''-ETHYLENE-P-QUATERPHENYL	29.03	0.018	27.17110
4-(3-ETHYLBUTOXY)-P-QUATERPHENYL (B)	28.91	0.019	26.46731
4,4'''-DI(2-BUTYLOCTOXY-1)-P-QUATERPHENYL	29.96	0.017	32.84395
DIETHYL-P-QUINQUEPHENYL	30.47	0.016	36.43039
DI(3-ETHYLHEPTYL)-P-QUINQUEPHENYL	30.59	0.016	37.19384

ACCEPTOR MOLECULES	RO	CO	OVERLAP
TETRAMETHYL-P-SEXIPHENYL	30.82	0.015	39.08644
TETRAMETHYL-P-OCTAPHENYL (B)	31.05	0.015	40.65658
INDOLE (TRACE)	16.02	0.109	0.78554
2-PHENYL INDOLE	26.13	0.025	14.50777
1-METHYL-2-PHENYL INDOLE	25.03	0.029	11.21489
1-NEOPENTYL-2-PHENYL INDOLE	24.81	0.029	10.66031
1-N-BUTYL-2(BIPHENYLYL-4) INDOLE	26.28	0.025	14.93957
1,2-DIPHENYL INDOLE	[illegible]	[illegible]	15.28943
1-BIPHENYLYL-2-PHENYL INDOLE	27.25	0.022	18.55969
1,3-DIMETHYL-2-PHENYL INDOLE	24.03	0.032	8.78017
2,3-DIPHENYL INDOLE	24.90	0.029	10.80724
1-METHYL-2,3-DIPHENYL INDOLE	24.14	0.032	8.97715
2-PHENYL-3,2'-METHYLENE INDOLE (TRACE)	24.85	0.029	10.69853
2-PHENYL-3,2'-ETHYLENE INDOLE (TRACE)	24.09	0.032	8.85738
2-PHENYL-3,2'-TRIMETHYLENE INDOLE	25.30	0.028	11.93805
1-METHYL-3,2'-METHYLENE-2-PHENYL INDOLE (TRACE)	24.39	0.031	9.55588
1-METHYL-3,2'-ETHYLENE-2-PHENYL INDOLE	23.66	0.034	7.95825
1-METHYL-3,2'-TRIMETHYLENE-2-PHENYL INDOLE	24.40	0.031	9.59186
4,5-DIPHENYLIMIDAZOLE (A)	22.50	0.039	5.91936
PPF (TRACE)	26.00	0.025	14.02506
PPO (TRACE)	27.68	0.021	20.50028
2-(4-FLUOROPHENYL)-5-PHENYLOXAZOLE (TRACE)	26.70	0.023	16.52018
αNPO (TRACE)	24.38	0.031	9.52436
BBO (B)	26.21	0.025	14.71002
POPOP	24.14	0.032	8.98646
DIMETHYL POPOP (TRACE)	24.93	0.029	10.89320
PPD (TRACE)	25.97	0.026	14.10949
2,5-DI(2-METHYLPHENYL)OXADIAZOLE	25.34	0.027	12.15727
PBD	29.34	0.018	29.05784
BBD (B)	28.95	0.018	26.73551
αNPD (TRACE)	24.55	0.030	9.94184
βNPD	27.47	0.022	19.57757
αNND (TRACE)	23.54	0.034	7.72691
βNND (TRACE)	27.71	0.021	20.56855
TRANS-1-(4-BIPHENYLYL)-2-PHENYL-ETHYLENE	27.89	0.021	21.35121
TRANS-1-(4-BIPHENYLYL)-2-(1-NAPHTHYL)-ETHYLENE	25.86	0.026	13.55998
DIPHENYLSTILBENE (B)	25.84	0.026	13.49651
1,1,4,4-TETRAPHENYLBUTADIENE	23.49	0.035	7.61458
1,6-DIPHENYLHEXATRIENE	24.02	0.032	8.70997
1,8-DIPHENYL-1,3,5,7-OCTATETRAENE	21.51	0.045	4.49621
BPSB (TRACE)	24.62	0.030	10.11468
BIS-MSB	25.42	0.027	12.25651
ESCULIN (A)	22.00	0.042	5.14191
AZULENE	23.67	0.034	7.97826
NAPHTHALENE	17.05	0.090	1.12821
DEUTERATED NAPHTHALENE-C10D8	17.17	0.088	1.17146
1-METHYLNAPHTHALENE 5ML/L	19.06	0.065	2.19342
2-METHYL NAPHTHALENE	17.53	0.083	1.32940
2,3-DIMETHYL NAPHTHALENE	18.19	0.074	1.65936
2,6-DIMETHYL NAPHTHALENE (TRACE)	17.83	0.079	1.46894
1-NAPHTHOL (TRACE)	20.18	0.054	3.08678
2-NAPHTHOL (TRACE)	17.45	0.084	1.28766
1-NAPHTHYLAMINE	19.66	0.059	2.61765
2-CHLORONAPHTHALENE	17.56	0.083	1.34118
ACENAPHTHENE	20.54	0.052	3.44441
1-PHENYL NAPHTHALENE	22.40	0.040	5.78396
2-PHENYL NAPHTHALENE	33.03	0.012	59.25229
1,4-DIPHENYL NAPHTHALENE	24.20	0.032	9.15424
1,5-DIPHENYL NAPHTHALENE	23.76	0.033	8.19795
1,7-DIPHENYL NAPHTHALENE	23.46	0.035	7.59604
1,4,5,8-TETRAPHENYL NAPHTHALENE	23.29	0.035	7.23548
1,1'-BINAPHTHYL	23.66	0.034	8.01303
4,4'-DI(N-BUTOXY)-1,1'-BINAPHTHYL	24.40	0.031	9.56729
2,2'-BINAPHTHYL	26.05	0.025	14.25140
ACRIDINE (A)	16.33	0.103	0.85952
ANTHRACENE (TRACE)	15.00	0.132	0.51722
DEUTERATED ANTHRACENE C14D10 (TRACE)	15.45	0.121	0.61652
9-METHYL ANTHRACENE (TRACE)	14.10	0.160	0.35607
1-AMINO ANTHRACENE	19.78	0.058	2.71802
2-AMINO ANTHRACENE	13.96	0.164	0.33069
9-VINYL ANTHRACENE	15.13	0.129	0.54427
9-PHENYL ANTHRACENE (TRACE)	15.70	0.116	0.67982
9,10-DIPHENYL ANTHRACENE	16.52	0.099	0.92136
9,10-DICHLOROANTHRACENE (TRACE)	13.91	0.166	0.32869
BIANTHRANYL (B)	18.20	0.074	1.65126
9,10-DINAPHTHYLANTHRACENE (B)	23.31	0.035	7.28760
9-(P-VINYLPHENYL)-10-PHENYL ANTHRACENE	19.54	0.060	2.52221
9,10-BIS(PHENYL-ETHYNYL)ANTHRACENE (TRACE)	26.47	0.024	15.59669
RUBICENE (B)	28.13	0.020	22.49716
DIBENZ(A,H)ANTHRACENE (B)	31.52	0.014	44.50719
TETRACENE (TRACE) (B)	29.53	0.017	30.11313
RUBRENE (B)	31.58	0.014	45.00645
TETRAHYDROCANNABINOL ACID METHYL ESTER	23.47	0.035	7.68535
PHENANTHRENE	21.00	0.048	3.89711
3,4-BENZOPHENANTHRENE	26.92	0.023	17.25398
2-PHENYL PHENANTHRENE	26.98	0.023	17.51290
CHRYSENE (TRACE)	24.27	0.031	9.27523
TRIPHENYLENE	22.44	0.040	5.80165
DEUTERATED TRIPHENYLENE-C18D12	22.39	0.040	5.73058
PYRENE	22.15	0.041	5.35492
1,3,6,8,-TETRAPHENYL PYRENE	28.10	0.020	22.33805
BIPYRENYL (B)	27.23	0.022	18.49353
O-PHENYLENE-PYRENE	27.30	0.022	18.79585
ANTHANTHRENE (B)	27.49	0.022	19.56599
FLUORANTHENE	23.59	0.034	7.82555
3-PHENYL FLUORANTHENE	25.80	0.026	13.36836

ACCEPTOR MOLECULES	RO	CO	OVERLAP
3-AMINO FLUORANTHENE	24.26	0.031	9.24974
3,3'-BIFLUORANTHENYL (B)	27.54	0.021	19.77731
BENZO(GHI)FLUORANTHENE	23.35	0.035	7.36076
9,12-DI(N-OCTADECOXY)-BENZ(E)ACEPHENANTHRYLENE	26.38	0.024	15.27481
DECACYCLENE (B)	26.90	0.023	17.19724
10-ETHOXY-BENZO(KL)XANTHENE	20.40	0.053	3.26733
PERYLENE	16.32	0.103	0.85647
BENZO(GHI)PERYLENE (B)	28.13	0.020	22.46995
BBOT (TRACE)	22.23	0.041	5.47582
QUININE SULFATE (A)	21.51	0.045	4.49196
ACRIDINE YELLOW (A)	22.32	0.040	5.60797
ACRIDINE ORANGE (A)	23.59	0.034	7.81876
ACRIDINE RED (A)	21.21	0.047	4.13402
URANINE (A)	21.33	0.046	4.27747
RHODAMINE B (A)	24.37	0.031	9.49650
RHODAMINE 6G (A)	24.11	0.032	8.90491
EOSIN B (A)	22.12	0.041	5.31029
EOSIN Y (A)	22.79	0.038	6.36283
ROSE BENGAL (A)	22.72	0.038	6.23860
INDENOINDENE	25.45	0.027	12.37996

8.39 Donor Molecule: 1,3,5-Triethylbenzene (5 ml/L)

FLUOROBENZENE (TRACE)	2.94	17.636	0.00004
TOLUENE (TRACE)	3.60	9.612	0.00014
DEUTERATED TOLUENE-C7D8 5ML/L	3.34	11.977	0.00009
ETHYLBENZENE (TRACE)	3.66	9.087	0.00016
PROPYLBENZENE (TRACE)	3.54	10.080	0.00013
ISOPROPYLBENZENE (TRACE)	3.27	12.791	0.00008
SEC-BUTYLBENZENE (TRACE)	3.15	14.282	0.00007
DIPHENYLMETHANE (TRACE)	4.50	4.917	0.00055
BIBENZYL	4.22	5.954	0.00038
1,3,5-TRIPHENYLPENTANE 0.05ML/L	4.47	5.000	0.00053
POLYSTYRENE	3.68	9.008	0.00016
POLYVINYLTOLUENE	7.15	1.226	0.00873
PHENYLCYCLOHEXANE (TRACE)	2.58	26.012	0.00002
BENZYLALCOHOL 3ML/L	3.35	11.919	0.00009
BENZYLACETATE 5ML/L	2.89	18.572	0.00004
PHENOL (TRACE)	10.41	0.396	0.08437
ANISOLE (TRACE)	10.88	0.348	0.10945
PHENYL ETHER	11.75	0.276	0.17339
ANILINE 3ML/L	15.06	0.131	0.76307
N,N-DIMETHYLANILINE 3ML/L	16.50	0.100	1.31135
TRIPHENYLAMINE	24.88	0.029	15.24755
P-XYLENE (TRACE)	7.33	1.138	0.01024
DEUTERATED P-XYLENE C8D10 (TRACE)	7.19	1.202	0.00917
M-XYLENE (TRACE)	5.56	2.596	0.00197
O-XYLENE 3ML/L	5.15	3.282	0.00123
1-METHYL-4-ETHYLBENZENE (TRACE)	6.62	1.544	0.00557
1,4-BIS(TRIFLUOROMETHYL)BENZENE 5ML/L	4.02	6.889	0.00028
P-CRESOL	13.81	0.170	0.45744
P-METHYLANISOLE 3ML/L	13.37	0.187	0.37675
1,4-DIMETHOXY BENZENE	16.62	0.097	1.37803
1-METHYL-4-CHLOROBENZENE 5ML/L	8.35	0.767	0.02248
SODIUM SALICYLATE (A)	18.04	0.076	2.21303
ANTHRANILIC ACID (A)	16.39	0.102	1.24577
TMPD	18.77	0.068	2.80871
1,2,4-TRIMETHYLBENZENE (TRACE)	8.37	0.762	0.02274
MESITYLENE 5ML/L	6.05	2.016	0.00326
1,3,5-TRIETHYLBENZENE 5ML/L	5.71	2.397	0.00231
STYRENE	11.85	0.269	0.18174
1,1-DIPHENYLETHYLENE 3ML/L	12.88	0.209	0.30050
BIPHENYL	11.61	0.285	0.16219
DEUTERATED BIPHENYL-C12D10	11.57	0.289	0.15823
4-METHYL BIPHENYL	13.87	0.168	0.46799
MONOISOPROPYLBIPHENYL 5ML/L	13.27	0.192	0.35974
4-BENZYL BIPHENYL	15.36	0.123	0.86612
3,3'-DIMETHYL BIPHENYL	13.58	0.179	0.41224
2,2'-DIMETHYLBIPHENYL 2ML/L	8.36	0.767	0.02242
OCTAMETHYL BIPHENYL	10.45	0.392	0.08608
4-METHOXY BIPHENYL	18.17	0.075	2.35358
4-BIPHENYLYL PHENYL ETHER	18.52	0.070	2.64817
4,4'-DIHYDROXYBIPHENYL (A)	20.52	0.052	4.85971
4,4'-DIMETHOXY BIPHENYL	20.31	0.053	4.56934
4,4'-DIISOBUTOXYBIPHENYL	20.21	0.054	4.44706
BENZIDINE	25.01	0.029	15.80650
4-VINYL BIPHENYL	23.41	0.035	10.73160
M-TERPHENYL	17.14	0.089	1.65630
M-QUATERPHENYL	16.87	0.093	1.51094
1,3,5-TRIPHENYL BENZENE	17.56	0.083	1.90183
1,3-DIBIPHENYLYLBENZENE (B)	27.47	0.022	27.78817
1,3,5,-TRIBIPHENYLYLBENZENE (B)	30.48	0.016	51.85408
FLUORENE (TRACE)	19.15	0.064	3.22582
1-METHYL FLUORENE	18.19	0.074	2.35566
1,2-BENZOFLUORENE	22.77	0.038	8.96066
9,10-DIHYDROPHENANTHRENE	19.11	0.064	3.18114
CARBAZOLE (TRACE)	21.64	0.044	6.62290
N-METHYL CARBAZOLE (TRACE)	21.09	0.048	5.66042
N-VINYLCARBAZOLE	21.81	0.043	6.93106
N-PHENYL CARBAZOLE	22.44	0.040	8.22050

ACCEPTOR MOLECULES	RO	CO	OVERLAP
N-METHYL-2,3-BENZCARBAZOLE	25.53	0.027	17.75926
DIBENZOFURAN	21.02	0.048	5.65142
DIBENZOTHIOPHENE (TRACE)	18.92	0.066	2.96206
P-TERPHENYL	24.17	0.032	12.96132
DEUTERATED P-TERPHENYL-C18D14	23.88	0.033	12.06486
4-METHYL TERPHENYL	24.74	0.030	14.89718
2-METHYL-P-TERPHENYL	20.64	0.051	5.05659
3,3''-DIMETHYL-P-TERPHENYL	24.27	0.031	13.37007
4-(3,3-DIMETHYLBUTOXY)-P-TERPHENYL	25.73	0.026	18.76352
4,4''-DIHEXAHYDROFARNESOXY-P-TERPHENYL	26.93	0.023	24.68828
2,2'-METHYLENE-P-TERPHENYL	25.78	0.026	19.00624
2,2'-ETHYLENE-P-TERPHENYL	25.81	0.026	19.10098
2,2'-METHYLENE-5',6''-METHYLENE-P-TERPHENYL (TRACE)	25.69	0.026	18.52406
3-PHENYL DIBENZOFURAN	25.49	0.027	17.76220
BENZO(1,2-B-4,5-B')BISBENZOFURAN (TRACE) (B)	23.89	0.033	11.95695
SEXIPHENYL BENZENE	18.41	0.072	2.55364
P-QUATERPHENYL	26.97	0.023	24.83198
3,3'''-DIMETHYL-P-QUATERPHENYL	27.33	0.022	26.85518
TETRAMETHYL-P-QUATERPHENYL	22.86	0.037	9.31660
3,3''' -DIMETHYL-3',2''-METHYLENE-P-QUATERPHENYL	26.99	0.023	24.88118
3,3''' -DIMETHYL-3',2''-ETHYLENE-P-QUATERPHENYL	27.48	0.022	27.60830
4-(3-ETHYLBUTOXY)-P-QUATERPHENYL (B)	27.30	0.022	26.58243
4,4'''-DI(2-BUTYLOCTOXY-1)-P-QUATERPHENYL	28.29	0.020	32.91810
DIETHYL-P-QUINQUEPHENYL	28.66	0.019	35.74729
DI(3-ETHYLHEPTYL)-P-QUINQUEPHENYL	28.90	0.019	37.41040
TETRAMETHYL-P-SEXIPHENYL	28.97	0.018	38.19526
TETRAMETHYL-P-OCTAPHENYL (B)	29.43	0.018	41.69226
INDOLE (TRACE)	15.04	0.131	0.76266
2-PHENYL INDOLE	24.69	0.030	14.62927
1-METHYL-2-PHENYL INDOLE	23.55	0.034	11.01874
1-NEOPENTYL-2-PHENYL INDOLE	23.32	0.035	10.39912
1-N-BUTYL-2(BIPHENYLYL-4) INDOLE	24.87	0.029	15.18713
1,2-DIPHENYL INDOLE	24.24	0.031	13.07939
1-BIPHENYLYL-2-PHENYL INDOLE	25.66	0.026	18.34917
1,3-DIMETHYL-2-PHENYL INDOLE	22.64	0.039	8.68136
2,3-DIPHENYL INDOLE	23.55	0.034	10.93510
1-METHYL-2,3-DIPHENYL INDOLE	22.77	0.038	8.95001
2-PHENYL-3,2'-METHYLENE INDOLE (TRACE)	23.61	0.034	11.16572
2-PHENYL-3,2'-ETHYLENE INDOLE (TRACE)	22.97	0.037	9.42230
2-PHENYL-3,2'-TRIMETHYLENE INDOLE	23.99	0.032	12.27204
1-METHYL-3,2'-METHYLENE-2-PHENYL INDOLE (TRACE)	23.24	0.036	10.12809
1-METHYL-3,2'-ETHYLENE-2-PHENYL INDOLE	22.58	0.039	8.49827
1-METHYL-3,2'-TRIMETHYLENE-2-PHENYL INDOLE	23.10	0.036	9.78240
4,5-DIPHENYLIMIDAZOLE (A)	21.12	0.047	5.73637
PPF (TRACE)	24.75	0.030	14.76125
PPO (TRACE)	26.15	0.025	20.60843
2-(4-FLUOROPHENYL)-5-PHENYLOXAZOLE (TRACE)	25.21	0.028	16.56983
αNPO (TRACE)	23.25	0.036	10.12458
BBO (B)	25.07	0.028	15.91507
POPOP	23.09	0.036	9.72064
DIMETHYL POPOP (TRACE)	23.77	0.033	11.56726
PPD (TRACE)	24.35	0.031	13.56662
2,5-DI(2-METHYLPHENYL)OXADIAZOLE	23.77	0.033	11.72508
PBD	27.64	0.021	28.81387
BBD (B)	27.48	0.022	27.64236
αNPD (TRACE)	23.32	0.035	10.31071
βNPD	25.92	0.026	19.53177
αNND (TRACE)	22.45	0.040	8.20725
βNND (TRACE)	26.24	0.025	21.02170
TRANS-1-(4-BIPHENYLYL)-2-PHENYL-ETHYLENE	26.48	0.024	22.13981
TRANS-1-(4-BIPHENYLYL)-2-(1-NAPHTHYL)-ETHYLENE	24.64	0.030	14.34631
DIPHENYLSTILBENE (B)	24.79	0.029	14.87865
1,1,4,4-TETRAPHENYLBUTADIENE	22.57	0.039	8.47194
1,6-DIPHENYLHEXATRIENE	23.23	0.036	10.08992
1,8-DIPHENYL-1,3,5,7-OCTATETRAENE	20.82	0.050	5.22307
BPSB (TRACE)	23.73	0.033	11.45997
BIS-MSB	24.40	0.031	13.53828
ESCULIN (A)	20.94	0.049	5.41277
AZULENE	22.27	0.040	7.82915
NAPHTHALENE	16.00	0.109	1.08984
DEUTERATED NAPHTHALENE-C10D8	16.12	0.107	1.13357
1-METHYLNAPHTHALENE 5ML/L	17.84	0.079	2.09466
2-METHYL NAPHTHALENE	16.44	0.101	1.28324
2,3-DIMETHYL NAPHTHALENE	17.05	0.090	1.59452
2,6-DIMETHYL NAPHTHALENE (TRACE)	16.75	0.095	1.42957
1-NAPHTHOL (TRACE)	18.95	0.066	3.00077
2-NAPHTHOL (TRACE)	16.48	0.100	1.29576
1-NAPHTHYLAMINE	18.65	0.069	2.70425
2-CHLORONAPHTHALENE	16.47	0.100	1.29492
ACENAPHTHENE	19.26	0.063	3.31685
1-PHENYL NAPHTHALENE	21.00	0.048	5.56681
2-PHENYL NAPHTHALENE	30.97	0.015	57.04373
1,4-DIPHENYL NAPHTHALENE	22.80	0.038	9.05054
1,5 DIPHENYL NAPHTHALENE	22.38	0.040	8.06106
1,7-DIPHENYL NAPHTHALENE	22.08	0.042	7.47612
1,4,5,8-TETRAPHENYL NAPHTHALENE	22.21	0.041	7.70512
1,1'-BINAPHTHYL	22.19	0.041	7.72148
4,4'-DI(N-BUTOXY)-1,1'-BINAPHTHYL	23.09	0.036	9.72509
2,2'-BINAPHTHYL	24.53	0.030	14.06883
ACRIDINE (A)	15.83	0.113	1.00812
ANTHRACENE (TRACE)	14.61	0.143	0.62473
DEUTERATED ANTHRACENE C14D10 (TRACE)	14.94	0.134	0.71477
9-METHYL ANTHRACENE (TRACE)	13.71	0.173	0.42677
1-AMINO ANTHRACENE	18.63	0.069	2.68343
2-AMINO ANTHRACENE	13.70	0.174	0.41805
9-VINYL ANTHRACENE	14.59	0.144	0.61967

ACCEPTOR MOLECULES	RO	CO	OVERLAP
9-PHENYL ANTHRACENE (TRACE)	15.14	0.129	0.77368
9,10-DIPHENYL ANTHRACENE	15.76	0.114	0.98427
9,10-DICHLOROANTHRACENE (TRACE)	13.48	0.183	0.38451
BIANTHRANYL (B)	17.39	0.085	1.77305
9,10-DINAPHTHYLANTHRACENE (B)	21.91	0.043	7.10035
9-(P-VINYLPHENYL)-10-PHENYL ANTHRACENE	18.44	0.071	2.52235
9,10-BIS(PHENYL-ETHYNYL)ANTHRACENE (TRACE)	24.93	0.029	15.41252
RUBICENE (B)	26.46	0.024	22.02853
DIBENZ(A,H)ANTHRACENE (B)	29.63	0.017	43.41785
TETRACENE (TRACE) (B)	27.79	0.021	29.57924
RUBRENE (B)	29.75	0.017	44.44916
TETRAHYDROCANNABINOL ACID METHYL ESTER	21.99	0.042	7.36695
PHENANTHRENE	19.66	0.059	3.70926
3,4-BENZOPHENANTHRENE	25.30	0.028	16.82389
2-PHENYL PHENANTHRENE	25.32	0.028	16.88769
CHRYSENE (TRACE)	22.87	0.037	9.18989
TRIPHENYLENE	21.03	0.048	5.57015
DEUTERATED TRIPHENYLENE-C18D12	20.99	0.048	5.50393
PYRENE	21.27	0.046	5.94457
1,3,6,8,-TETRAPHENYL PYRENE	26.37	0.024	21.56796
BIPYRENYL (B)	25.81	0.026	18.95679
O-PHENYLENE-PYRENE	25.73	0.026	18.61447
ANTHANTHRENE (B)	25.92	0.026	19.44078
FLUORANTHENE	22.24	0.041	7.77058
3-PHENYL FLUORANTHENE	24.24	0.031	13.01316
3-AMINO FLUORANTHENE	22.88	0.037	9.20336
3,3'-BIFLUORANTHENYL (B)	25.92	0.026	19.46929
BENZO(GHI)FLUORANTHENE	21.97	0.042	7.21393
9,12-DI(N-OCTADECOXY)-BENZ(E)ACEPHENANTHRYLENE	24.81	0.029	14.97251
DECACYCLENE (B)	25.47	0.027	17.49535
10-ETHOXY-BENZO(KL)XANTHENE	19.24	0.063	3.25697
PERYLENE	15.41	0.122	0.85857
BENZO(GHI)PERYLENE (B)	26.42	0.024	21.83038
BBOT (TRACE)	21.42	0.046	6.19135
QUININE SULFATE (A)	20.40	0.053	4.63635
ACRIDINE YELLOW (A)	20.95	0.049	5.42129
ACRIDINE ORANGE (A)	22.14	0.041	7.55012
ACRIDINE RED (A)	19.99	0.056	4.09913
URANINE (A)	20.12	0.055	4.25141
RHODAMINE B (A)	22.95	0.037	9.38309
RHODAMINE 6G (A)	22.68	0.038	8.73745
EOSIN B (A)	20.93	0.049	5.38978
EOSIN Y (A)	21.56	0.045	6.44159
ROSE BENGAL (A)	21.50	0.045	6.34150
INDENOINDENE	24.12	0.032	12.68879

8.41 Donor Molecule: 1,1-Diphenylethylene (3 ml/L)

POLYVINYLTOLUENE	2.48	29.189	0.00061
PHENYLCYCLOHEXANE (TRACE)	0.00	0.000	0.00000
BENZYLALCOHOL 3ML/L	0.00	0.000	0.00000
BENZYLACETATE 5ML/L	0.00	0.000	0.00000
PHENOL (TRACE)	1.45	******	0.00002
ANISOLE (TRACE)	1.77	81.342	0.00008
PHENYL ETHER	2.58	25.952	0.00078
ANILINE 3ML/L	6.33	1.762	0.16727
N,N DIMETHYLANILINE 3ML/L	8.27	0.791	0.82496
TRIPHENYLAMINE	13.08	0.200	12.76344
P-CRESOL	4.09	6.519	0.01230
P-METHYLANISOLE 3ML/L	3.97	7.130	0.01028
1,4-DIMETHOXY BENZENE	7.40	1.106	0.42438
1-METHYL-4-CHLOROBENZENE 5ML/L	1.07	******	0.00000
SODIUM SALICYLATE (A)	9.26	0.563	1.60911
ANTHRANILIC ACID (A)	10.46	0.391	3.31499
TMPD	9.66	0.496	2.06770
1,2,4-TRIMETHYLBENZENE (TRACE)	1.53	******	0.00003
MESITYLENE 5ML/L	0.00	0.000	0.00000
1,3,5-TRIETHYLBENZENE 5ML/L	0.00	0.000	0.00000
STYRENE	4.10	6.484	0.01240
1,1-DIPHENYLETHYLENE 3ML/L	3.98	7.121	0.01031
BIPHENYL	2.56	26.768	0.00073
DEUTERATED BIPHENYL-C12D10	2.41	31.773	0.00052
4-METHYL BIPHENYL	3.93	7.363	0.00964
MONOISOPROPYLBIPHENYL 5ML/L	3.32	12.207	0.00352
4-BENZYL BIPHENYL	4.38	5.313	0.01853
3,3'-DIMETHYL BIPHENYL	4.44	5.098	0.02008
2,2'-DIMETHYLBIPHENYL 2ML/L	2.24	39.986	0.00033
OCTAMETHYL BIPHENYL	2.28	37.764	0.00037
4-METHOXY BIPHENYL	6.59	1.560	0.21370
4-BIPHENYLYL PHENYL ETHER	6.52	1.610	0.20073
4,4'-DIHYDROXYBIPHENYL (A)	8.42	0.750	0.91872
4,4'-DIMETHOXY BIPHENYL	8.05	0.859	0.70195
4,4'-DIISOBUTOXYBIPHENYL	8.06	0.853	0.71320
BENZIDINE	12.30	0.240	8.87234
4-VINYL BIPHENYL	10.16	0.427	2.83984
M-TERPHENYL	7.05	1.278	0.31755
M-QUATERPHENYL	5.85	2.238	0.10406
1,3,5-TRIPHENYL BENZENE	6.60	1.552	0.21387
1,3-DIBIPHENYLYLBENZENE (B)	13.07	0.200	12.77948
1,3,5,-TRIBIPHENYLYLBENZENE (B)	14.86	0.136	27.67313

ACCEPTOR MOLECULES	RO	CO	OVERLAP
FLUORENE (TRACE)	8.71	0.677	1.13484
1-METHYL FLUORENE	8.24	0.798	0.81379
1,2-BENZOFLUORENE	12.01	0.258	7.68179
9,10-DIHYDROPHENANTHRENE	8.38	0.760	0.89844
CARBAZOLE (TRACE)	10.35	0.403	3.14841
N-METHYL CARBAZOLE (TRACE)	10.24	0.416	2.94856
N-VINYLCARBAZOLE	10.36	0.402	3.16902
N-PHENYL CARBAZOLE	10.60	[illegible]	[illegible]
N-METHYL-2,3-BENZCARBAZOLE	12.00	0.259	7.55464
DIBENZOFURAN	8.57	0.711	1.02830
DIBENZOTHIOPHENE (TRACE)	9.18	0.577	1.54057
P-TERPHENYL	10.60	0.375	3.66588
DEUTERATED P-TERPHENYL-C18D14	10.45	0.392	3.36624
4-METHYL TERPHENYL	11.22	0.317	5.14899
2-METHYL-P-TERPHENYL	8.04	0.860	0.70143
3,3''-DIMETHYL-P-TERPHENYL	10.96	0.340	4.46368
4-(3,3-DIMETHYLBUTOXY)-P-TERPHENYL	12.48	0.230	9.70048
4,4''-DIHEXAHYDROFARNESOXY-P-TERPHENYL	13.57	0.179	16.00256
2,2'-METHYLENE-P-TERPHENYL	13.18	0.196	13.45506
2,2'-ETHYLENE-P-TERPHENYL	13.19	0.195	13.54169
2,2'-METHYLENE-5',6''-METHYLENE-P-TERPHENYL (TRACE)	14.40	0.150	22.87221
3-PHENYL DIBENZOFURAN	13.52	0.181	15.68018
BENZO(1,2-B-4,5-B')BISBENZOFURAN (TRACE) (B)	14.83	0.137	27.18570
SEXIPHENYL BENZENE	6.45	1.666	0.18759
P-QUATERPHENYL	13.82	0.170	17.82911
3,3'''-DIMETHYL-P-QUATERPHENYL	14.18	0.157	20.85310
TETRAMETHYL-P-QUATERPHENYL	9.23	0.569	1.59870
3,3''' -DIMETHYL-3',2''-METHYLENE-P-QUATERPHENYL	15.48	0.121	35.10040
3,3''' -DIMETHYL-3',2''-ETHYLENE-P-QUATERPHENYL	15.42	0.122	34.21531
4-(3-ETHYLBUTOXY)-P-QUATERPHENYL (B)	15.07	0.131	29.84150
4,4'''-DI(2-BUTYLOCTOXY-1)-P-QUATERPHENYL	15.53	0.119	35.85718
DIETHYL-P-QUINQUEPHENYL	14.82	0.137	27.11706
DI(3-ETHYLHEPTYL)-P-QUINQUEPHENYL	16.01	0.109	42.86068
TETRAMETHYL-P-SEXIPHENYL	14.74	0.140	26.27081
TETRAMETHYL-P-OCTAPHENYL (B)	16.86	0.093	58.35476
INDOLE (TRACE)	4.45	5.092	0.02018
2-PHENYL INDOLE	13.77	0.171	17.47134
1-METHYL-2-PHENYL INDOLE	12.31	0.240	8.91089
1-NEOPENTYL-2-PHENYL INDOLE	11.89	0.266	7.26297
1-N-BUTYL-2(BIPHENYLYL-4) INDOLE	14.05	0.161	19.54773
1,2-DIPHENYL INDOLE	12.76	0.215	11.07380
1-BIPHENYLYL-2-PHENYL INDOLE	13.49	0.182	15.38983
1,3-DIMETHYL-2-PHENYL INDOLE	12.03	0.257	7.78342
2,3-DIPHENYL INDOLE	13.17	0.196	13.26724
1-METHYL-2,3-DIPHENYL INDOLE	12.25	0.243	8.62611
2-PHENYL-3,2'-METHYLENE INDOLE (TRACE)	13.92	0.166	18.57905
2-PHENYL-3,2'-ETHYLENE INDOLE (TRACE)	13.85	0.168	17.96015
2-PHENYL-3,2'-TRIMETHYLENE INDOLE	13.84	0.169	17.94379
1-METHYL-3,2'-METHYLENE-2-PHENYL INDOLE (TRACE)	13.97	0.164	18.98810
1-METHYL-3,2'-ETHYLENE-2-PHENYL INDOLE	13.70	0.174	16.79176
1-METHYL-3,2'-TRIMETHYLENE-2-PHENYL INDOLE	13.16	0.196	13.30016
4,5-DIPHENYLIMIDAZOLE (A)	10.23	0.418	2.93502
PPF (TRACE)	14.73	0.140	26.07750
PPO (TRACE)	14.46	0.148	23.41991
2-(4-FLUOROPHENYL)-5-PHENYLOXAZOLE (TRACE)	13.86	0.168	18.18602
αNPO (TRACE)	13.96	0.164	18.76362
BBO (B)	15.46	0.121	34.57740
POPOP	14.36	0.151	22.14568
DIMETHYL POPOP (TRACE)	14.53	0.146	23.78098
PPD (TRACE)	11.17	0.321	5.02853
2,5-DI(2-METHYLPHENYL)OXADIAZOLE	11.16	0.322	4.99509
PBD	14.80	0.138	26.93465
BBD (B)	15.94	0.110	41.81261
αNPD (TRACE)	13.54	0.180	15.69741
βNPD	13.90	0.167	18.46768
αNND (TRACE)	13.52	0.181	15.53031
βNND (TRACE)	14.77	0.139	26.50067
TRANS-1-(4-BIPHENYLYL)-2-PHENYL-ETHYLENE	15.46	0.121	34.80052
TRANS-1-(4-BIPHENYLYL)-2-(1-NAPHTHYL)-ETHYLENE	14.76	0.139	26.11512
DIPHENYLSTILBENE (B)	15.64	0.117	36.94066
1,1,4,4-TETRAPHENYLBUTADIENE	14.37	0.151	22.19815
1,6-DIPHENYLHEXATRIENE	15.38	0.123	33.35728
1,8-DIPHENYL-1,3,5,7-OCTATETRAENE	14.18	0.157	20.49304
BPSB (TRACE)	15.45	0.121	34.38532
BIS-MSB	15.48	0.120	34.78320
ESCULIN (A)	12.51	0.228	9.71980
AZULENE	9.77	0.480	2.19597
NAPHTHALENE	5.85	2.235	0.10339
DEUTERATED NAPHTHALENE-C10D8	6.34	1.757	0.16644
1-METHYLNAPHTHALENE 5ML/L	7.30	1.151	0.38882
2-METHYL NAPHTHALENE	6.75	1.454	0.24368
2,3-DIMETHYL NAPHTHALENE	7.01	1.299	0.30544
2,6-DIMETHYL NAPHTHALENE (TRACE)	7.57	1.030	0.48572
1-NAPHTHOL (TRACE)	9.40	0.539	1.76923
2-NAPHTHOL (TRACE)	8.68	0.684	1.09599
1-NAPHTHYLAMINE	10.83	0.353	4.11052
2-CHLORONAPHTHALENE	6.73	1.467	0.23917
ACENAPHTHENE	9.00	0.615	1.36553
1-PHENYL NAPHTHALENE	9.84	0.470	2.33588
2-PHENYL NAPHTHALENE	14.45	0.148	23.41743
1,4-DIPHENYL NAPHTHALENE	12.13	0.250	8.16652
1,5-DIPHENYL NAPHTHALENE	11.71	0.278	6.61537
1,7-DIPHENYL NAPHTHALENE	11.56	0.290	6.11457
1,4,5,8-TETRAPHENYL NAPHTHALENE	13.42	0.185	14.76772
1,1'-BINAPHTHYL	10.51	0.385	3.47056
4,4'-DI(N-BUTOXY)-1,1'-BINAPHTHYL	13.05	0.201	12.59247

ACCEPTOR MOLECULES	RO	CO	OVERLAP
2,2'-BINAPHTHYL	12.84	0.211	11.47359
ACRIDINE (A)	10.54	0.382	3.46046
ANTHRACENE (TRACE)	9.92	0.458	2.41744
DEUTERATED ANTHRACENE C14D10 (TRACE)	9.90	0.461	2.38048
9-METHYL ANTHRACENE (TRACE)	9.45	0.530	1.79471
1-AMINO ANTHRACENE	9.13	0.589	1.45775
2-AMINO ANTHRACENE	9.51	0.521	1.84595
9-VINYL ANTHRACENE	9.71	0.488	2.11943
9-PHENYL ANTHRACENE (TRACE)	9.98	0.450	2.49151
9,10-DIPHENYL ANTHRACENE	9.82	0.473	2.26087
9,10-DICHLOROANTHRACENE (TRACE)	9.29	0.559	1.61861
BIANTHRANYL (B)	10.64	0.372	3.65373
9,10-DINAPHTHYLANTHRACENE (B)	10.91	0.344	4.25878
9-(P-VINYLPHENYL)-10-PHENYL ANTHRACENE	10.20	0.422	2.83814
9,10-BIS(PHENYL-ETHYNYL)ANTHRACENE (TRACE)	13.01	0.203	12.25436
RUBICENE (B)	13.71	0.174	16.76061
DIBENZ(A,H)ANTHRACENE (B)	15.30	0.125	32.38198
TETRACENE (TRACE) (B)	10.84	0.351	4.10237
RUBRENE (B)	16.08	0.108	43.63364
TETRAHYDROCANNABINOL ACID METHYL ESTER	9.51	0.520	1.91376
PHENANTHRENE	7.97	0.882	0.65629
3,4-BENZOPHENANTHRENE	11.76	0.275	6.68317
2-PHENYL PHENANTHRENE	11.48	0.296	5.82786
CHRYSENE (TRACE)	12.08	0.254	7.89958
TRIPHENYLENE	8.52	0.723	0.97823
DEUTERATED TRIPHENYLENE-C18D12	8.43	0.746	0.91865
PYRENE	13.29	0.191	13.93321
1,3,6,8,-TETRAPHENYL PYRENE	13.44	0.184	14.88963
BIPYRENYL (B)	14.47	0.148	23.14800
O-PHENYLENE-PYRENE	13.84	0.169	17.72727
ANTHANTHRENE (B)	14.18	0.157	20.52829
FLUORANTHENE	11.18	0.320	4.93072
3-PHENYL FLUORANTHENE	12.39	0.235	9.15136
3-AMINO FLUORANTHENE	12.56	0.226	9.89595
3,3'-BIFLUORANTHENYL (B)	13.77	0.171	17.23041
BENZO(GHI)FLUORANTHENE	11.20	0.318	4.98818
9,12-DI(N-OCTADECOXY)-BENZ(E)ACEPHENANTHRYLENE	12.33	0.239	8.85215
DECACYCLENE (B)	14.58	0.144	24.25253
10-ETHOXY-BENZO(KL)XANTHENE	10.82	0.353	4.04858
PERYLENE	8.85	0.645	1.21383
BENZO(GHI)PERYLENE (B)	13.55	0.180	15.64804
BBOT (TRACE)	14.14	0.158	20.14618
QUININE SULFATE (A)	11.65	0.283	6.38351
ACRIDINE YELLOW (A)	9.45	0.529	1.80248
ACRIDINE ORANGE (A)	10.52	0.384	3.42664
ACRIDINE RED (A)	10.64	0.371	3.67069
URANINE (A)	10.77	0.358	3.93764
RHODAMINE B (A)	12.08	0.254	7.85537
RHODAMINE 6G (A)	11.60	0.286	6.15681
EOSIN B (A)	11.76	0.275	6.68980
EOSIN Y (A)	12.07	0.254	7.80677
ROSE BENGAL (A)	11.99	0.259	7.50951
INDENOINDENE	13.85	0.168	18.09687

8.42 Donor Molecule: Biphenyl

POLYVINYLTOLUENE	4.51	4.884	0.00036
PHENYL ETHER	3.84	7.912	0.00014
ANILINE 3ML/L	11.79	0.273	0.11612
N,N-DIMETHYLANILINE 3ML/L	16.22	0.105	0.78233
TRIPHENYLAMINE	25.88	0.026	12.76989
P-CRESOL	6.20	1.874	0.00248
P-METHYLANISOLE 3ML/L	5.93	2.146	0.00189
1,4-DIMETHOXY BENZENE	13.99	0.163	0.32422
1-METHYL-4-CHLOROBENZENE 5ML/L	0.00	0.000	0.00000
SODIUM SALICYLATE (A)	18.24	0.074	1.56631
ANTHRANILIC ACID (A)	20.92	0.049	3.54094
TMPD	19.23	0.063	2.14147
1,2,4-TRIMETHYLBENZENE (TRACE)	1.91	64.144	0.00000
MESITYLENE 5ML/L	0.00	0.000	0.00000
1,3,5-TRIETHYLBENZENE 5ML/L	0.00	0.000	0.00000
STYRENE	6.98	1.318	0.00500
1,1-DIPHENYLETHYLENE 3ML/L	6.54	1.597	0.00341
BIPHENYL	3.11	14.819	0.00004
DEUTERATED BIPHENYL-C12D10	0.00	0.000	0.00000
4-METHYL BIPHENYL	6.41	1.695	0.00303
MONOISOPROPYLBIPHENYL 5ML/L	4.78	4.091	0.00052
4'-BENZYL BIPHENYL	7.02	1.292	0.00521
3,3'-DIMETHYL BIPHENYL	7.67	0.993	0.00681
2,2'-DIMETHYLBIPHENYL 2ML/L	3.64	9.281	0.00010
OCTAMETHYL BIPHENYL	3.15	14.293	0.00004
4-METHOXY BIPHENYL	11.97	0.261	0.12730
4-BIPHENYLYL PHENYL ETHER	11.63	0.284	0.10737
4,4'-DIHYDROXYBIPHENYL (A)	15.95	0.110	0.70861
4,4'-DIMETHOXY BIPHENYL	15.06	0.131	0.50248
4,4'-DIISOBUTOXYBIPHENYL	15.09	0.130	0.50937
BENZIDINE	24.17	0.032	8.50643
4-VINYL BIPHENYL	19.45	0.061	2.33389
M-TERPHENYL	13.49	0.182	0.26047

ACCEPTOR MOLECULES	RO	CO	OVERLAP
M-QUATERPHENYL	10.54	0.382	0.05944
1,3,5-TRIPHENYL BENZENE	12.75	0.216	0.18413
1,3-DIBIPHENYLYLBENZENE (B)	25.51	0.027	11.78090
1,3,5,-TRIBIPHENYLYLBENZENE (B)	29.13	0.018	26.11064
FLUORENE (TRACE)	16.64	0.097	0.91804
1-METHYL FLUORENE	15.59	0.118	0.62049
1,2-BENZOFLUORENE	23.88	0.033	7.88283
9,10-DIHYDROPHENANTHRENE	16.12	0.107	0.76710
CARBAZOLE (TRACE)	20.34	0.053	3.02415
N-METHYL CARBAZOLE (TRACE)	20.05	0.055	2.76630
N-VINYLCARBAZOLE	20.35	0.053	3.02563
N-PHENYL CARBAZOLE	20.45	0.052	3.11543
N-METHYL-2,3-BENZCARBAZOLE	23.49	0.035	7.08547
DIBENZOFURAN	15.85	0.112	0.68425
DIBENZOTHIOPHENE (TRACE)	18.27	0.073	1.58947
P-TERPHENYL	20.37	0.053	3.06878
DEUTERATED P-TERPHENYL-C18D14	20.07	0.055	2.80890
4-METHYL TERPHENYL	21.71	0.044	4.49604
2-METHYL-P-TERPHENYL	14.96	0.134	0.48416
3,3''-DIMETHYL-P-TERPHENYL	21.18	0.047	3.87272
4-(3,3-DIMETHYLBUTOXY)-P-TERPHENYL	24.45	0.031	9.13232
4,4''-DIHEXAHYDROFARNESOXY-P-TERPHENYL	26.74	0.023	15.61300
2,2'-METHYLENE-P-TERPHENYL	26.05	0.025	13.39077
2,2'-ETHYLENE-P-TERPHENYL	26.08	0.025	13.44226
2,2'-METHYLENE-5',6''-METHYLENE-P-TERPHENYL (TRACE)	28.75	0.019	24.05001
3-PHENYL DIBENZOFURAN	26.76	0.023	15.74050
BENZO(1,2-B-4,5-B')BISBENZOFURAN (TRACE) (B)	29.79	0.017	29.73334
SEXIPHENYL BENZENE	11.67	0.282	0.10922
P-QUATERPHENYL	27.29	0.022	17.62333
3,3'''-DIMETHYL-P-QUATERPHENYL	28.07	0.020	20.83439
TETRAMETHYL-P-QUATERPHENYL	17.36	0.085	1.18153
3,3''' -DIMETHYL-3',2''-METHYLENE-P-QUATERPHENYL	30.93	0.015	37.20916
3,3''' -DIMETHYL-3',2''-ETHYLENE-P-QUATERPHENYL	30.78	0.015	36.05303
4-(3-ETHYLBUTOXY)-P-QUATERPHENYL (B)	30.01	0.017	31.02233
4,4'''-DI(2-BUTYLOCTOXY-1)-P-QUATERPHENYL	30.93	0.015	37.19102
DIETHYL-P-QUINQUEPHENYL	29.31	0.018	27.02422
DI(3-ETHYLHEPTYL)-P-QUINQUEPHENYL	31.90	0.014	44.62044
TETRAMETHYL-P-SEXIPHENYL	29.09	0.018	25.84107
TETRAMETHYL-P-OCTAPHENYL (B)	33.68	0.012	61.76591
INDOLE (TRACE)	6.49	1.638	0.00325
2-PHENYL INDOLE	27.46	0.022	18.29163
1-METHYL-2-PHENYL INDOLE	24.36	0.031	8.90693
1-NEOPENTYL-2-PHENYL INDOLE	23.44	0.035	7.09520
1-N-BUTYL-2(BIPHENYLYL-4) INDOLE	28.02	0.020	20.53515
1,2-DIPHENYL INDOLE	25.29	0.028	11.15704
1-BIPHENYLYL-2-PHENYL INDOLE	26.78	0.023	15.66600
1,3-DIMETHYL-2-PHENYL INDOLE	23.87	0.033	7.89670
2,3-DIPHENYL INDOLE	26.26	0.025	13.88870
1-METHYL-2,3-DIPHENYL INDOLE	24.37	0.031	8.87856
2-PHENYL-3,2'-METHYLENE INDOLE (TRACE)	27.86	0.021	19.92685
2-PHENYL-3,2'-ETHYLENE INDOLE (TRACE)	27.75	0.021	19.35874
2-PHENYL-3,2'-TRIMETHYLENE INDOLE	27.67	0.021	19.12950
1-METHYL-3,2'-METHYLENE-2-PHENYL INDOLE (TRACE)	28.00	0.020	20.46577
1-METHYL-3,2'-ETHYLENE-2-PHENYL INDOLE	27.40	0.022	17.93715
1-METHYL-3,2'-TRIMETHYLENE-2-PHENYL INDOLE	26.28	0.025	14.02712
4,5-DIPHENYLIMIDAZOLE (A)	20.05	0.055	2.77261
PPF (TRACE)	29.49	0.017	27.97375
PPO (TRACE)	28.82	0.019	24.44425
2-(4-FLUOROPHENYL)-5-PHENYLOXAZOLE (TRACE)	27.62	0.021	18.93540
αNPO (TRACE)	27.89	0.021	19.85450
BBO (B)	30.80	0.015	36.03227
POPOP	28.38	0.020	22.04896
DIMETHYL POPOP (TRACE)	28.58	0.019	23.00928
PPD (TRACE)	21.58	0.045	4.34845
2,5-DI(2-METHYLPHENYL)OXADIAZOLE	21.74	0.044	4.54179
PBD	29.39	0.018	27.50698
BBD (B)	31.87	0.014	44.52232
αNPD (TRACE)	27.08	0.023	16.71376
βNPD	27.64	0.021	19.01245
αNND (TRACE)	27.03	0.023	16.51230
βNND (TRACE)	29.52	0.017	28.14148
TRANS-1-(4-BIPHENYLYL)-2-PHENYL-ETHYLENE	30.92	0.015	36.99510
TRANS-1-(4-BIPHENYLYL)-2-(1-NAPHTHYL)-ETHYLENE	29.43	0.018	27.41708
DIPHENYLSTILBENE (B)	31.16	0.015	38.64074
1,1,4,4-TETRAPHENYLBUTADIENE	28.57	0.019	22.95147
1,6-DIPHENYLHEXATRIENE	30.45	0.016	33.65696
1,8-DIPHENYL-1,3,5,7-OCTATETRAENE	27.39	0.022	17.82762
BPSB (TRACE)	30.56	0.016	34.39443
BIS-MSB	30.74	0.015	35.60976
ESCULIN (A)	24.94	0.029	10.15883
AZULENE	19.33	0.062	2.19938
NAPHTHALENE	10.93	0.342	0.07335
DEUTERATED NAPHTHALENE-C10D8	12.20	0.246	0.14083
1-METHYLNAPHTHALENE 5ML/L	13.68	0.176	0.28104
2-METHYL NAPHTHALENE	13.03	0.202	0.21009
2,3-DIMETHYL NAPHTHALENE	13.38	0.187	0.24624
2,6-DIMETHYL NAPHTHALENE (TRACE)	14.87	0.136	0.46322
1-NAPHTHOL (TRACE)	18.46	0.071	1.69524
2-NAPHTHOL (TRACE)	17.39	0.085	1.17979
1-NAPHTHYLAMINE	21.65	0.044	4.37883
2-CHLORONAPHTHALENE	12.94	0.206	0.20156
ACENAPHTHENE	17.50	0.084	1.23041
1-PHENYL NAPHTHALENE	19.13	0.064	2.09899
2-PHENYL NAPHTHALENE	28.08	0.020	20.94665
1,4-DIPHENYL NAPHTHALENE	24.07	0.032	8.28150
1,5-DIPHENYL NAPHTHALENE	23.18	0.036	6.62481

ACCEPTOR MOLECULES	RO	CO	OVERLAP
1,7-DIPHENYL NAPHTHALENE	22.89	0.037	6.14176
1,4,5,8-TETRAPHENYL NAPHTHALENE	26.77	0.023	15.54786
1,1'-BINAPHTHYL	20.48	0.052	3.16202
4,4'-DI(N-BUTOXY)-1,1'-BINAPHTHYL	26.05	0.025	13.27438
2,2'-BINAPHTHYL	25.46	0.027	11.61662
ACRIDINE (A)	20.83	0.049	3.44977
ANTHRACENE (TRACE)	19.59	0.060	2.38406
DEUTERATED ANTHRACENE C14D10 (TRACE)	19.56	0.060	2.36393
9-METHYL ANTHRACENE (TRACE)	18.42	0.072	1.65116
1-AMINO ANTHRACENE	17.52	0.083	1.21959
2-AMINO ANTHRACENE	18.79	0.067	1.84694
9-VINYL ANTHRACENE	18.84	0.067	1.89065
9-PHENYL ANTHRACENE (TRACE)	19.47	0.061	2.29694
9,10-DIPHENYL ANTHRACENE	18.77	0.068	1.84379
9,10-DICHLOROANTHRACENE (TRACE)	17.70	0.081	1.29698
BIANTHRANYL (B)	20.54	0.052	3.17046
9,10-DINAPHTHYLANTHRACENE (B)	21.02	0.048	3.64159
9-(P-VINYLPHENYL)-10-PHENYL ANTHRACENE	19.51	0.060	2.33094
9,10-BIS(PHENYL-ETHYNYL)ANTHRACENE (TRACE)	25.60	0.027	11.87149
RUBICENE (B)	26.95	0.023	16.17092
DIBENZ(A,H)ANTHRACENE (B)	30.03	0.017	30.96968
TETRACENE (TRACE) (B)	20.50	0.052	3.13271
RUBRENE (B)	31.85	0.014	44.09280
TETRAHYDROCANNABINOL ACID METHYL ESTER	18.18	0.074	1.55336
PHENANTHRENE	14.67	0.142	0.42467
3,4-BENZOPHENANTHRENE	23.28	0.035	6.72294
2-PHENYL PHENANTHRENE	22.24	0.041	5.12301
CHRYSENE (TRACE)	24.04	0.032	8.16030
TRIPHENYLENE	16.20	0.105	0.76767
DEUTERATED TRIPHENYLENE-C18D12	16.09	0.107	0.73700
PYRENE	26.69	0.024	15.27034
1,3,6,8,-TETRAPHENYL PYRENE	25.55	0.027	11.75147
BIPYRENYL (B)	28.67	0.019	23.46904
O-PHENYLENE-PYRENE	27.29	0.022	17.42646
ANTHANTHRENE (B)	28.00	0.020	20.34000
FLUORANTHENE	22.07	0.042	4.87869
3-PHENYL FLUORANTHENE	24.04	0.032	8.15293
3-AMINO FLUORANTHENE	24.81	0.029	9.84714
3,3'-BIFLUORANTHENYL (B)	26.85	0.023	15.83113
BENZO(GHI)FLUORANTHENE	22.07	0.042	4.87454
9,12-DI(N-OCTADECOXY)-BENZ(E)ACEPHENANTHRYLENE	24.10	0.032	8.26621
DECACYCLENE (B)	28.63	0.019	23.26450
10-ETHOXY-BENZO(KL)XANTHENE	20.89	0.049	3.50625
PERYLENE	16.52	0.099	0.85785
BENZO(GHI)PERYLENE (B)	26.45	0.024	14.45895
BBOT (TRACE)	27.59	0.021	18.63245
QUININE SULFATE (A)	23.30	0.035	6.81188
ACRIDINE YELLOW (A)	18.07	0.076	1.46823
ACRIDINE ORANGE (A)	20.42	0.052	3.06511
ACRIDINE RED (A)	21.01	0.048	3.63013
URANINE (A)	21.29	0.046	3.93512
RHODAMINE B (A)	23.82	0.033	7.71289
RHODAMINE 6G (A)	22.77	0.038	5.88689
EOSIN B (A)	23.42	0.035	6.95985
EOSIN Y (A)	24.01	0.032	8.09508
ROSE BENGAL (A)	23.87	0.033	7.81666
INDENOINDENE	27.70	0.021	19.24938

8.43 Donor Molecule: Deuterated Biphenyl ($C_{12}D_{10}$)

POLYVINYLTOLUENE	4.68	4.376	0.00048
ANISOLE (TRACE)	2.41	31.826	0.00001
PHENYL ETHER	4.35	5.424	0.00031
ANILINE 3ML/L	12.12	0.252	0.14472
N,N-DIMETHYLANILINE 3ML/L	16.26	0.104	0.84020
TRIPHENYLAMINE	25.75	0.026	13.12947
P-CRESOL	6.92	1.351	0.00505
P-METHYLANISOLE 3ML/L	6.67	1.509	0.00405
1,4-DIMETHOXY BENZENE	14.33	0.152	0.39596
1-METHYL-4-CHLOROBENZENE 5ML/L	0.00	0.000	0.00000
SODIUM SALICYLATE (A)	18.21	0.074	1.64107
ANTHRANILIC ACID (A)	20.54	0.052	3.35708
TMPD	18.96	0.066	2.08452
1,2,4-TRIMETHYLBENZENE (TRACE)	2.39	32.665	0.00001
MESITYLENE 5ML/L	0.00	0.000	0.00000
1,3,5-TRIETHYLBENZENE 5ML/L	0.00	0.000	0.00000
STYRENE	7.48	1.071	0.00802
1,1-DIPHENYLETHYLENE 3ML/L	7.09	1.254	0.00586
BIPHENYL	3.94	7.293	0.00017
DEUTERATED BIPHENYL-C12D10	3.48	10.584	0.00008
4-METHYL BIPHENYL	6.95	1.332	0.00519
MONOISOPROPYLBIPHENYL 5ML/L	5.53	2.646	0.00132
4-BENZYL BIPHENYL	7.69	0.983	0.00955
3,3'-DIMETHYL BIPHENYL	8.10	0.841	0.01301
2,2'-DIMETHYLBIPHENYL 2ML/L	3.97	7.155	0.00018
OCTAMETHYL BIPHENYL	3.68	8.994	0.00011
4-METHOXY BIPHENYL	12.41	0.234	0.16746
4-BIPHENYLYL PHENYL ETHER	12.16	0.249	0.14837
4,4'-DIHYDROXYBIPHENYL (A)	16.21	0.105	0.82728
4,4'-DIMETHOXY BIPHENYL	15.40	0.123	0.60775
4,4'-DIISOBUTOXYBIPHENYL	15.44	0.122	0.61870

ACCEPTOR MOLECULES	RO	CO	OVERLAP
BENZIDINE	24.14	0.032	8.94897
4-VINYL BIPHENYL	19.71	0.058	2.67275
M-TERPHENYL	13.67	0.175	0.29797
M-QUATERPHENYL	10.99	0.337	0.08073
1,3,5-TRIPHENYL BENZENE	12.83	0.212	0.20252
1,3-DIBIPHENYLYLBENZENE (B)	25.59	0.027	12.69878
1,3,5,-TRIBIPHENYLYLBENZENE (B)	29.15	0.018	27.79903
FLUORENE (TRACE)	17.05	0.090	1.12612
1-METHYL FLUORENE	15.99	0.110	0.76292
1,2-BENZOFLUORENE	23.70	0.034	7.98083
9,10-DIHYDROPHENANTHRENE	16.36	0.102	0.87634
CARBAZOLE (TRACE)	20.20	0.054	3.06831
N-METHYL CARBAZOLE (TRACE)	19.88	0.057	2.78308
N-VINYLCARBAZOLE	20.15	0.055	3.02148
N-PHENYL CARBAZOLE	20.34	0.053	3.19122
N-METHYL-2,3-BENZCARBAZOLE	23.37	0.035	7.25443
DIBENZOFURAN	16.35	0.102	0.87477
DIBENZOTHIOPHENE (TRACE)	18.05	0.076	1.56457
P-TERPHENYL	20.60	0.051	3.47543
DEUTERATED P-TERPHENYL-C18D14	20.30	0.053	3.18885
4-METHYL TERPHENYL	21.88	0.043	4.99141
2-METHYL-P-TERPHENYL	15.35	0.124	0.59754
3,3''-DIMETHYL-P-TERPHENYL	21.35	0.046	4.31070
4-(3,3-DIMETHYLBUTOXY)-P-TERPHENYL	24.47	0.031	9.72336
4,4''-DIHEXAHYDROFARNESOXY-P-TERPHENYL	26.67	0.024	16.27950
2,2'-METHYLENE-P-TERPHENYL	25.94	0.026	13.81691
2,2'-ETHYLENE-P-TERPHENYL	25.96	0.026	13.85033
2,2'-METHYLENE-5',6''-METHYLENE-P-TERPHENYL (TRACE)	28.34	0.020	23.38774
3-PHENYL DIBENZOFURAN	26.64	0.024	16.22453
BENZO(1,2-B-4,5-B')BISBENZOFURAN (TRACE) (B)	29.25	0.018	28.21402
SEXIPHENYL BENZENE	12.14	0.250	0.14676
P-QUATERPHENYL	27.18	0.022	18.21394
3,3'''-DIMETHYL-P-QUATERPHENYL	27.92	0.021	21.37138
TETRAMETHYL-P-QUATERPHENYL	17.72	0.080	1.41676
3,3'''-DIMETHYL-3',2''-METHYLENE-P-QUATERPHENYL	30.49	0.016	36.19862
3,3'''-DIMETHYL-3',2''-ETHYLENE-P-QUATERPHENYL	30.37	0.016	35.25184
4-(3-ETHYLBUTOXY)-P-QUATERPHENYL (B)	29.70	0.017	30.84817
4,4'''-DI(2-BUTYLOCTOXY-1)-P-QUATERPHENYL	30.62	0.016	37.05748
DIETHYL-P-QUINQUEPHENYL	29.16	0.018	27.75758
DI(3-ETHYLHEPTYL)-P-QUINQUEPHENYL	31.56	0.014	44.32414
TETRAMETHYL-P-SEXIPHENYL	28.98	0.018	26.79366
TETRAMETHYL-P-OCTAPHENYL (B)	33.22	0.012	60.25555
INDOLE (TRACE)	7.51	1.056	0.00828
2-PHENYL INDOLE	27.17	0.022	18.17170
1-METHYL-2-PHENYL INDOLE	24.24	0.031	9.16352
1-NEOPENTYL-2-PHENYL INDOLE	23.39	0.035	7.42725
1-N-BUTYL-2(BIPHENYLYL-4) INDOLE	27.67	0.021	20.17251
1,2-DIPHENYL INDOLE	25.15	0.028	11.42898
1-BIPHENYLYL-2-PHENYL INDOLE	26.58	0.024	15.87891
1,3-DIMETHYL-2-PHENYL INDOLE	23.70	0.034	8.01793
2,3-DIPHENYL INDOLE	25.95	0.026	13.70004
1-METHYL-2,3-DIPHENYL INDOLE	24.13	0.032	8.86214
2-PHENYL-3,2'-METHYLENE INDOLE (TRACE)	27.45	0.022	19.28228
2-PHENYL-3,2'-ETHYLENE INDOLE (TRACE)	27.28	0.022	18.51035
2-PHENYL-3,2'-TRIMETHYLENE INDOLE	27.30	0.022	18.68432
1-METHYL-3,2'-METHYLENE-2-PHENYL INDOLE (TRACE)	27.54	0.021	19.62207
1-METHYL-3,2'-ETHYLENE-2-PHENYL INDOLE	26.95	0.023	17.20518
1-METHYL-3,2'-TRIMETHYLENE-2-PHENYL INDOLE	25.96	0.026	13.79210
4,5-DIPHENYLIMIDAZOLE (A)	20.05	0.056	2.93637
PPF (TRACE)	29.03	0.018	26.91287
PPO (TRACE)	28.52	0.019	24.27975
2-(4-FLUOROPHENYL)-5-PHENYLOXAZOLE (TRACE)	27.33	0.022	18.83778
αNPO (TRACE)	27.42	0.022	19.00470
BBO (B)	30.31	0.016	34.59061
POPOP	28.01	0.020	21.54745
DIMETHYL POPOP (TRACE)	28.29	0.020	22.84000
PPD (TRACE)	21.78	0.043	4.87269
2,5-DI(2-METHYLPHENYL)OXADIAZOLE	21.87	0.043	4.97704
PBD	29.16	0.018	27.82217
BBD (B)	31.41	0.014	43.17730
αNPD (TRACE)	26.68	0.024	16.17067
βNPD	27.41	0.022	19.12641
αNND (TRACE)	26.58	0.024	15.81188
βNND (TRACE)	29.10	0.018	27.35176
TRANS-1-(4-BIPHENYLYL)-2-PHENYL-ETHYLENE	30.46	0.016	35.82154
TRANS-1-(4-BIPHENYLYL)-2-(1-NAPHTHYL)-ETHYLENE	28.97	0.018	26.36036
DIPHENYLSTILBENE (B)	30.66	0.016	37.04832
1,1,4,4-TETRAPHENYLBUTADIENE	28.12	0.020	22.05316
1,6-DIPHENYLHEXATRIENE	30.01	0.017	32.58510
1,8-DIPHENYL-1,3,5,7-OCTATETRAENE	27.28	0.022	18.36852
BPSB (TRACE)	30.15	0.016	33.50058
BIS-MSB	30.29	0.016	34.45568
ESCULIN (A)	24.55	0.030	9.78012
AZULENE	19.06	0.065	2.13613
NAPHTHALENE	11.15	0.323	0.08745
DEUTERATED NAPHTHALENE-C10D8	12.27	0.242	0.15483
1-METHYLNAPHTHALENE 5ML/L	13.95	0.165	0.33544
2-METHYL NAPHTHALENE	13.07	0.200	0.22625
2,3-DIMETHYL NAPHTHALENE	13.50	0.182	0.27446
2,6-DIMETHYL NAPHTHALENE (TRACE)	14.79	0.138	0.47552
1-NAPHTHOL (TRACE)	18.43	0.071	1.77943
2-NAPHTHOL (TRACE)	17.11	0.089	1.13419
1-NAPHTHYLAMINE	21.34	0.046	4.24897
2-CHLORONAPHTHALENE	12.99	0.204	0.21797
ACENAPHTHENE	17.57	0.082	1.33739
1-PHENYL NAPHTHALENE	19.23	0.063	2.29268
2-PHENYL NAPHTHALENE	28.25	0.020	23.01028

ACCEPTOR MOLECULES	RO	CO	OVERLAP
1,4-DIPHENYL NAPHTHALENE	23.90	0.033	8.42298
1,5-DIPHENYL NAPHTHALENE	23.06	0.036	6.80312
1,7-DIPHENYL NAPHTHALENE	22.77	0.038	6.29542
1,4,5,8-TETRAPHENYL NAPHTHALENE	26.35	0.024	14.91433
1,1'-BINAPHTHYL	20.55	0.052	3.40989
4,4'-DI(N-BUTOXY)-1,1'-BINAPHTHYL	25.72	0.026	13.04206
2,2'-BINAPHTHYL	25.30	0.028	11.86098
ACRIDINE (A)	20.53	0.052	3.34246
ANTHRACENE (TRACE)	19.32	0.062	2.32240
DEUTERATED ANTHRACENE C14D10 (TRACE)	19.29	0.062	2.29873
9-METHYL ANTHRACENE (TRACE)	18.29	0.073	1.67037
1-AMINO ANTHRACENE	17.56	0.083	1.30494
2-AMINO ANTHRACENE	18.51	0.071	1.77621
9-VINYL ANTHRACENE	18.75	0.068	1.93484
9-PHENYL ANTHRACENE (TRACE)	19.32	0.062	2.32092
9,10-DIPHENYL ANTHRACENE	18.78	0.068	1.95335
9,10-DICHLOROANTHRACENE (TRACE)	17.73	0.080	1.38353
BIANTHRANYL (B)	20.46	0.052	3.27342
9,10-DINAPHTHYLANTHRACENE (B)	21.07	0.048	3.89626
9-(P-VINYLPHENYL)-10-PHENYL ANTHRACENE	19.57	0.060	2.50099
9,10-BIS(PHENYL-ETHYNYL)ANTHRACENE (TRACE)	25.56	0.027	12.43247
RUBICENE (B)	26.84	0.023	16.68204
DIBENZ(A,H)ANTHRACENE (B)	30.12	0.016	33.25179
TETRACENE (TRACE) (B)	20.85	0.049	3.66587
RUBRENE (B)	31.76	0.014	45.72817
TETRAHYDROCANNABINOL ACID METHYL ESTER	18.43	0.071	1.78831
PHENANTHRENE	15.04	0.132	0.52100
3,4-BENZOPHENANTHRENE	23.13	0.036	6.83880
2-PHENYL PHENANTHRENE	22.38	0.040	5.63508
CHRYSENE (TRACE)	23.80	0.033	8.14227
TRIPHENYLENE	16.38	0.102	0.87037
DEUTERATED TRIPHENYLENE-C18D12	16.25	0.104	0.83007
PYRENE	26.15	0.025	14.27192
1,3,6,8,-TETRAPHENYL PYRENE	25.85	0.026	13.30827
BIPYRENYL (B)	28.26	0.020	22.71623
O-PHENYLENE-PYRENE	27.19	0.022	18.02404
ANTHANTHRENE (B)	27.85	0.021	20.82079
FLUORANTHENE	21.79	0.043	4.76801
3-PHENYL FLUORANTHENE	24.05	0.032	8.63548
3-AMINO FLUORANTHENE	24.65	0.030	9.99738
3,3'-BIFLUORANTHENYL (B)	26.79	0.023	16.47089
BENZO(GHI)FLUORANTHENE	21.83	0.043	4.82504
9,12-DI(N-OCTADECOXY)-BENZ(E)ACEPHENANTHRYLENE	24.00	0.032	8.52546
DECACYCLENE (B)	28.41	0.020	23.45086
10-ETHOXY-BENZO(KL)XANTHENE	20.93	0.049	3.74976
PERYLENE	16.76	0.095	0.98706
BENZO(GHI)PERYLENE (B)	26.59	0.024	15.74646
BBOT (TRACE)	27.37	0.022	18.75110
QUININE SULFATE (A)	22.89	0.037	6.48212
ACRIDINE YELLOW (A)	18.25	0.074	1.64806
ACRIDINE ORANGE (A)	20.54	0.052	3.34292
ACRIDINE RED (A)	20.88	0.049	3.69326
URANINE (A)	21.11	0.048	3.94812
RHODAMINE B (A)	23.69	0.034	7.87823
RHODAMINE 6G (A)	22.65	0.038	6.02123
EOSIN B (A)	23.16	0.036	6.88345
EOSIN Y (A)	23.77	0.033	8.03618
ROSE BENGAL (A)	23.58	0.034	7.65398
INDENOINDENE	27.34	0.022	18.85966

8.44 Donor Molecule: 4-Methylbiphenyl

POLYVINYLTOLUENE	3.92	7.416	0.00017
PHENYL ETHER	2.58	25.909	0.00001
ANILINE 3ML/L	10.47	0.389	0.06032
N,N-DIMETHYLANILINE 3ML/L	15.24	0.126	0.56837
TRIPHENYLAMINE	24.87	0.029	10.63399
P-CRESOL	4.65	4.443	0.00047
P-METHYLANISOLE 3ML/L	4.40	5.253	0.00033
1,4-DIMETHOXY BENZENE	12.52	0.228	0.17600
1-METHYL-4-CHLOROBENZENE 5ML/L	0.00	0.000	0.00000
SODIUM SALICYLATE (A)	17.38	0.085	1.23609
ANTHRANILIC ACID (A)	21.16	0.047	4.00036
TMPD	19.15	0.064	2.20976
1,2,4-TRIMETHYLBENZENE (TRACE)	0.00	0.000	0.00000
MESITYLENE 5ML/L	0.00	0.000	0.00000
1,3,5-TRIETHYLBENZENE 5ML/L	0.00	0.000	0.00000
STYRENE	5.62	2.522	0.00144
1,1-DIPHENYLETHYLENE 3ML/L	5.14	3.294	0.00085
BIPHENYL	0.00	0.000	0.00000
DEUTERATED BIPHENYL-C12D10	0.00	0.000	0.00000
4-METHYL BIPHENYL	5.06	3.458	0.00077
MONOISOPROPYLBIPHENYL 5ML/L	0.00	0.000	0.00000
4-BENZYL BIPHENYL	5.40	2.846	0.00114
3,3'-DIMETHYL BIPHENYL	6.41	1.696	0.00319
2,2'-DIMETHYLBIPHENYL 2ML/L	2.83	19.776	0.00002
OCTAMETHYL BIPHENYL	1.93	62.192	0.00000
4-METHOXY BIPHENYL	10.43	0.394	0.05894
4-BIPHENYLYL PHENYL ETHER	10.02	0.445	0.04626

ACCEPTOR MOLECULES	RO	CO	OVERLAP
4,4'-DIHYDROXYBIPHENYL (A)	14.62	0.143	0.44404
4,4'-DIMETHOXY BIPHENYL	13.60	0.178	0.28810
4,4'-DIISOBUTOXYBIPHENYL	13.59	0.178	0.28699
BENZIDINE	23.03	0.037	6.72901
4-VINYL BIPHENYL	17.83	0.079	1.46282
M-TERPHENYL	12.43	0.233	0.16799
M-QUATERPHENYL	9.18	0.577	0.02750
1,3,5-TRIPHENYL BENZENE	11.95	0.262	0.13313
1,3-DIBIPHENYLYLBENZENE (B)	24.03	0.032	8.68988
1,3,5,-TRIBIPHENYLYLBENZENE (B)	27.57	0.021	19.86432
FLUORENE (TRACE)	14.97	0.133	0.51374
1-METHYL FLUORENE	13.98	0.164	0.34005
1,2-BENZOFLUORENE	23.01	0.037	6.67374
9,10-DIHYDROPHENANTHRENE	14.61	0.143	0.44362
CARBAZOLE (TRACE)	19.89	0.057	2.78839
N-METHYL CARBAZOLE (TRACE)	19.87	0.057	2.77057
N-VINYLCARBAZOLE	20.24	0.054	3.09400
N-PHENYL CARBAZOLE	20.15	0.055	3.01524
N-METHYL-2,3-BENZCARBAZOLE	22.97	0.037	6.53286
DIBENZOFURAN	13.99	0.163	0.34260
DIBENZOTHIOPHENE (TRACE)	18.02	0.076	1.54554
P-TERPHENYL	18.80	0.067	2.00384
DEUTERATED P-TERPHENYL-C18D14	18.48	0.071	1.81122
4-METHYL TERPHENYL	20.19	0.054	3.07446
2-METHYL-P-TERPHENYL	13.40	0.186	0.26403
3,3''-DIMETHYL-P-TERPHENYL	19.68	0.059	2.63481
4-(3,3-DIMETHYLBUTOXY)-P-TERPHENYL	23.15	0.036	6.96973
4,4''-DIHEXAHYDROFARNESOXY-P-TERPHENYL	25.55	0.027	12.57119
2,2'-METHYLENE-P-TERPHENYL	24.99	0.029	11.03336
2,2'-ETHYLENE-P-TERPHENYL	25.06	0.028	11.19216
2,2'-METHYLENE-5',6''-METHYLENE-P-TERPHENYL (TRACE)	28.53	0.019	24.29237
3-PHENYL DIBENZOFURAN	25.62	0.027	12.80097
BENZO(1,2-B-4,5-B')BISBENZOFURAN (TRACE) (B)	29.74	0.017	31.13358
SEXIPHENYL BENZENE	10.17	0.426	0.05057
P-QUATERPHENYL	26.19	0.025	14.56047
3,3'''-DIMETHYL-P-QUATERPHENYL	27.03	0.023	17.58788
TETRAMETHYL-P-QUATERPHENYL	15.68	0.116	0.67789
3,3''' -DIMETHYL-3',2''-METHYLENE-P-QUATERPHENYL	30.60	0.016	36.88751
3,3''' -DIMETHYL-3',2''-ETHYLENE-P-QUATERPHENYL	30.38	0.016	35.28444
4-(3-ETHYLBUTOXY)-P-QUATERPHENYL (B)	29.35	0.018	28.70265
4,4'''-DI(2-BUTYLOCTOXY-1)-P-QUATERPHENYL	30.22	0.016	34.19929
DIETHYL-P-QUINQUEPHENYL	28.22	0.020	22.75775
DI(3-ETHYLHEPTYL)-P-QUINQUEPHENYL	31.21	0.015	41.42356
TETRAMETHYL-P-SEXIPHENYL	27.87	0.021	21.15979
TETRAMETHYL-P-OCTAPHENYL (B)	33.29	0.012	60.93817
INDOLE (TRACE)	4.27	5.753	0.00028
2-PHENYL INDOLE	26.81	0.023	16.74854
1-METHYL-2-PHENYL INDOLE	23.40	0.035	7.40130
1-NEOPENTYL-2-PHENYL INDOLE	22.32	0.040	5.58681
1-N-BUTYL-2(BIPHENYLYL-4) INDOLE	27.61	0.021	19.84826
1,2-DIPHENYL INDOLE	24.33	0.031	9.35656
1-BIPHENYLYL-2-PHENYL INDOLE	25.92	0.026	13.61506
1,3-DIMETHYL-2-PHENYL INDOLE	23.10	0.036	6.84810
2,3-DIPHENYL INDOLE	25.81	0.026	13.23594
1-METHYL-2,3-DIPHENYL INDOLE	23.83	0.033	8.20644
2-PHENYL-3,2'-METHYLENE INDOLE (TRACE)	27.60	0.021	19.88564
2-PHENYL-3,2'-ETHYLENE INDOLE (TRACE)	27.70	0.021	20.26362
2-PHENYL-3,2'-TRIMETHYLENE INDOLE	27.24	0.022	18.40074
1-METHYL-3,2'-METHYLENE-2-PHENYL INDOLE (TRACE)	27.88	0.021	21.09671
1-METHYL-3,2'-ETHYLENE-2-PHENYL INDOLE	27.43	0.022	19.06988
1-METHYL-3,2'-TRIMETHYLENE-2-PHENYL INDOLE	25.83	0.026	13.37706
4,5-DIPHENYLIMIDAZOLE (A)	19.06	0.065	2.16666
PPF (TRACE)	29.36	0.018	28.75492
PPO (TRACE)	28.17	0.020	22.51471
2-(4-FLUOROPHENYL)-5-PHENYLOXAZOLE (TRACE)	26.97	0.023	17.36003
αNPO (TRACE)	28.09	0.020	21.90076
BBO (B)	31.19	0.015	41.02554
POPOP	28.96	0.018	26.23934
DIMETHYL POPOP (TRACE)	29.18	0.018	27.45129
PPD (TRACE)	20.05	0.056	2.95323
2,5-DI(2-METHYLPHENYL)OXADIAZOLE	20.27	0.054	3.15523
PBD	28.48	0.019	24.06371
BBD (B)	31.59	0.014	44.64379
αNPD (TRACE)	26.90	0.023	16.96787
βNPD	26.87	0.023	16.94127
αNND (TRACE)	27.13	0.022	17.83689
βNND (TRACE)	29.22	0.018	28.00614
TRANS-1-(4-BIPHENYLYL)-2-PHENYL-ETHYLENE	30.71	0.015	37.57515
TRANS-1-(4-BIPHENYLYL)-2-(1-NAPHTHYL)-ETHYLENE	29.64	0.017	30.17435
DIPHENYLSTILBENE (B)	31.56	0.014	43.94379
1,1,4,4-TETRAPHENYLBUTADIENE	29.04	0.018	26.71311
1,6-DIPHENYLHEXATRIENE	31.15	0.015	40.67273
1,8-DIPHENYL-1,3,5,7-OCTATETRAENE	28.40	0.020	23.36699
BPSB (TRACE)	31.23	0.015	41.30461
BIS-MSB	31.22	0.015	41.22203
ESCULIN (A)	25.11	0.028	11.18530
AZULENE	19.38	0.061	2.35565
NAPHTHALENE	10.13	0.431	0.04897
DEUTERATED NAPHTHALENE-C10D8	11.55	0.290	0.10739
1-METHYLNAPHTHALENE 5ML/L	12.45	0.232	0.16856
2-METHYL NAPHTHALENE	12.44	0.232	0.16785
2,3-DIMETHYL NAPHTHALENE	12.60	0.224	0.18134
2,6-DIMETHYL NAPHTHALENE (TRACE)	14.54	0.146	0.42728
1-NAPHTHOL (TRACE)	17.68	0.081	1.38218
2-NAPHTHOL (TRACE)	17.23	0.087	1.18347
1-NAPHTHYLAMINE	21.43	0.045	4.35523

ACCEPTOR MOLECULES	RO	CO	OVERLAP
2-CHLORONAPHTHALENE	12.42	0.233	0.16653
ACENAPHTHENE	16.45	0.100	0.90021
1-PHENYL NAPHTHALENE	17.91	0.078	1.49490
2-PHENYL NAPHTHALENE	26.23	0.025	14.73553
1,4-DIPHENYL NAPHTHALENE	23.26	0.036	7.13840
1,5-DIPHENYL NAPHTHALENE	22.30	0.040	5.55335
1,7-DIPHENYL NAPHTHALENE	22.03	0.042	5.15792
1,4,5,8-TETRAPHENYL NAPHTHALENE	26.98	0.023	17.17452
1,1'-BINAPHTHYL	19.33	0.062	2.35833
4,4'-DI(N-BUTOXY)-1,1'-BINAPHTHYL	25.61	0.027	12.66414
2,2'-BINAPHTHYL	24.52	0.030	9.80210
ACRIDINE (A)	21.41	0.046	4.29162
ANTHRACENE (TRACE)	20.13	0.055	2.96223
DEUTERATED ANTHRACENE C14D10 (TRACE)	20.06	0.055	2.90792
9-METHYL ANTHRACENE (TRACE)	19.07	0.065	2.13840
1-AMINO ANTHRACENE	17.77	0.080	1.40147
2-AMINO ANTHRACENE	19.27	0.063	2.26164
9-VINYL ANTHRACENE	19.52	0.060	2.46337
9-PHENYL ANTHRACENE (TRACE)	20.12	0.055	2.95063
9,10-DIPHENYL ANTHRACENE	19.57	0.060	2.49755
9,10-DICHLOROANTHRACENE (TRACE)	18.47	0.071	1.76488
BIANTHRANYL (B)	21.24	0.047	4.08687
9,10-DINAPHTHYLANTHRACENE (B)	20.91	0.049	3.71573
9-(P-VINYLPHENYL)-10-PHENYL ANTHRACENE	19.98	0.056	2.83326
9,10-BIS(PHENYL-ETHYNYL)ANTHRACENE (TRACE)	24.60	0.030	9.86139
RUBICENE (B)	26.13	0.025	14.16900
DIBENZ(A,H)ANTHRACENE (B)	28.76	0.019	25.16274
TETRACENE (TRACE) (B)	18.99	0.065	2.08846
RUBRENE (B)	30.32	0.016	34.57216
TETRAHYDROCANNABINOL ACID METHYL ESTER	16.66	0.097	0.97130
PHENANTHRENE	13.49	0.182	0.27143
3,4-BENZOPHENANTHRENE	22.53	0.039	5.81836
2-PHENYL PHENANTHRENE	20.75	0.050	3.57837
CHRYSENE (TRACE)	23.37	0.035	7.28411
TRIPHENYLENE	15.20	0.127	0.55427
DEUTERATED TRIPHENYLENE-C18D12	15.13	0.129	0.53848
PYRENE	26.92	0.023	16.96494
1,3,6,8,-TETRAPHENYL PYRENE	25.29	0.028	11.63817
BIPYRENYL (B)	29.17	0.018	27.39581
O-PHENYLENE-PYRENE	26.53	0.024	15.52917
ANTHANTHRENE (B)	27.11	0.022	17.65527
FLUORANTHENE	22.34	0.040	5.53866
3-PHENYL FLUORANTHENE	23.82	0.033	8.12702
3-AMINO FLUORANTHENE	24.13	0.032	8.78717
3,3'-BIFLUORANTHENYL (B)	26.68	0.024	16.05606
BENZO(GHI)FLUORANTHENE	22.11	0.041	5.19948
9,12-DI(N-OCTADECOXY)-BENZ(E)ACEPHENANTHRYLENE	23.93	0.033	8.35801
DECACYCLENE (B)	28.80	0.019	25.39256
10-ETHOXY-BENZO(KL)XANTHENE	20.96	0.049	3.76955
PERYLENE	17.03	0.091	1.08668
BENZO(GHI)PERYLENE (B)	25.28	0.028	11.61629
BBOT (TRACE)	28.43	0.019	23.47249
QUININE SULFATE (A)	23.38	0.035	7.33763
ACRIDINE YELLOW (A)	17.34	0.086	1.21105
ACRIDINE ORANGE (A)	19.34	0.062	2.32753
ACRIDINE RED (A)	20.61	0.051	3.41062
URANINE (A)	21.00	0.048	3.82118
RHODAMINE B (A)	23.34	0.035	7.18635
RHODAMINE 6G (A)	22.52	0.039	5.79708
EOSIN B (A)	23.06	0.036	6.69175
EOSIN Y (A)	23.61	0.034	7.69867
ROSE BENGAL (A)	23.71	0.034	7.89938
INDENOINDENE	27.22	0.022	18.31501

8.45 Donor Molecule: Monoisopropylbiphenyl (5 ml/L)

POLYVINYLTOLUENE	4.53	4.813	0.00027
PHENYL ETHER	3.51	10.336	0.00006
ANILINE 3ML/L	12.02	0.258	0.09355
N,N-DIMETHYLANILINE 3ML/L	16.89	0.093	0.71646
TRIPHENYLAMINE	27.11	0.022	12.13588
P-CRESOL	5.70	2.411	0.00108
P-METHYLANISOLE 3ML/L	5.40	2.839	0.00078
1,4-DIMETHOXY BENZENE	14.37	0.151	0.27386
1-METHYL-4-CHLOROBENZENE 5ML/L	0.00	0.000	0.00000
SODIUM SALICYLATE (A)	19.06	0.065	1.46455
ANTHRANILIC ACID (A)	22.25	0.041	3.68892
TMPD	20.34	0.053	2.15957
1,2,4-TRIMETHYLBENZENE (TRACE)	0.00	0.000	0.00000
MESITYLENE 5ML/L	0.00	0.000	0.00000
1,3,5-TRIETHYLBENZENE 5ML/L	0.00	0.000	0.00000
STYRENE	6.70	1.484	0.00284
1,1-DIPHENYLETHYLENE 3ML/L	6.15	1.922	0.00169
BIPHENYL	0.00	0.000	0.00000
DEUTERATED BIPHENYL-C12D10	0.00	0.000	0.00000
4-METHYL BIPHENYL	6.03	2.043	0.00150
MONOISOPROPYLBIPHENYL 5ML/L	4.20	6.023	0.00017
4-BENZYL BIPHENYL	6.54	1.597	0.00246
3,3'-DIMETHYL BIPHENYL	7.47	1.073	0.00543
2,2'-DIMETHYLBIPHENYL 2ML/L	3.40	11.345	0.00005

ACCEPTOR MOLECULES	RO	CO	OVERLAP
OCTAMETHYL BIPHENYL	2.76	21.338	0.00001
4-METHOXY BIPHENYL	12.04	0.256	0.09471
4-BIPHENYLYL PHENYL ETHER	11.63	0.285	0.07693
4,4'-DIHYDROXYBIPHENYL (A)	16.40	0.101	0.60279
4,4'-DIMETHOXY BIPHENYL	15.39	0.123	0.41203
4,4'-DIISOBUTOXYBIPHENYL	15.42	0.122	0.41723
BENZIDINE	25.24	0.028	7.93459
4-VINYL BIPHENYL	20.05	0.055	[illegible]
M-TERPHENYL	13.93	0.165	0.22666
M-QUATERPHENYL	10.67	0.369	0.04594
1,3,5-TRIPHENYL BENZENE	13.24	0.193	0.16586
1,3-DIBIPHENYLYLBENZENE (B)	26.54	0.024	10.75090
1,3,5,-TRIBIPHENYLYLBENZENE (B)	30.36	0.016	24.08508
FLUORENE (TRACE)	17.27	0.087	0.82405
1-METHYL FLUORENE	16.13	0.107	0.54714
1,2-BENZOFLUORENE	25.01	0.029	7.48829
9,10-DIHYDROPHENANTHRENE	16.61	0.098	0.65064
CARBAZOLE (TRACE)	21.35	0.046	2.90980
N-METHYL CARBAZOLE (TRACE)	21.09	0.048	2.69287
N-VINYLCARBAZOLE	21.46	0.045	2.99176
N-PHENYL CARBAZOLE	21.46	0.045	2.99363
N-METHYL-2,3-BENZCARBAZOLE	24.63	0.030	6.77806
DIBENZOFURAN	16.14	0.106	0.55063
DIBENZOTHIOPHENE (TRACE)	19.28	0.062	1.57955
P-TERPHENYL	21.04	0.048	2.68122
DEUTERATED P-TERPHENYL-C18D14	20.71	0.050	2.44504
4-METHYL TERPHENYL	22.49	0.039	3.99818
2-METHYL-P-TERPHENYL	15.26	0.126	0.39140
3,3''-DIMETHYL-P-TERPHENYL	21.93	0.042	3.43573
4-(3,3-DIMETHYLBUTOXY)-P-TERPHENYL	25.48	0.027	8.42579
4,4''DIHEXAHYDROFARNESOXY-P-TERPHENYL	27.95	0.020	14.66275
2,2'-METHYLENE-P-TERPHENYL	27.28	0.022	12.69154
2,2'-ETHYLENE-P-TERPHENYL	27.31	0.022	12.75452
2,2'-METHYLENE-5',6''-METHYLENE-P-TERPHENYL (TRACE)	30.38	0.016	24.12145
3-PHENYL DIBENZOFURAN	28.02	0.020	14.89534
BENZO(1,2-B-4,5-B')BISBENZOFURAN (TRACE) (B)	31.55	0.014	30.18665
SEXIPHENYL BENZENE	11.77	0.274	0.08295
P-QUATERPHENYL	28.57	0.019	16.68395
3,3'''-DIMETHYL-P-QUATERPHENYL	29.41	0.018	19.85337
TETRAMETHYL-P-QUATERPHENYL	17.78	0.080	0.97967
3,3''' -DIMETHYL-3',2''-METHYLENE-P-QUATERPHENYL	32.67	0.013	37.18310
3,3''' -DIMETHYL-3',2''-ETHYLENE-P-QUATERPHENYL	32.49	0.013	35.88257
4-(3-ETHYLBUTOXY)-P-QUATERPHENYL (B)	31.60	0.014	30.39464
4,4'''-DI(2-BUTYLOCTOXY-1)-P-QUATERPHENYL	32.55	0.013	36.37073
DIETHYL-P-QUINQUEPHENYL	30.71	0.015	25.72541
DI(3-ETHYLHEPTYL)-P-QUINQUEPHENYL	33.59	0.012	43.76661
TETRAMETHYL-P-SEXIPHENYL	30.43	0.016	24.39363
TETRAMETHYL-P-OCTAPHENYL (B)	35.57	0.010	61.68376
INDOLE (TRACE)	5.64	2.488	0.00101
2-PHENYL INDOLE	28.90	0.019	17.88654
1-METHYL-2-PHENYL INDOLE	25.51	0.027	8.45862
1-NEOPENTYL-2-PHENYL INDOLE	24.49	0.030	6.63486
1-N-BUTYL-2(BIPHENYLYL-4) INDOLE	29.57	0.017	20.37715
1,2-DIPHENYL INDOLE	26.50	0.024	10.63279
1-BIPHENYLYL-2-PHENYL INDOLE	28.11	0.020	15.07349
1,3-DIMETHYL-2-PHENYL INDOLE	25.05	0.028	7.58869
2,3-DIPHENYL INDOLE	27.69	0.021	13.72613
1-METHYL-2,3-DIPHENYL INDOLE	25.65	0.027	8.68583
2-PHENYL-3,2'-METHYLENE INDOLE (TRACE)	29.45	0.018	19.99084
2-PHENYL-3,2'-ETHYLENE INDOLE (TRACE)	29.39	0.018	19.66736
2-PHENYL-3,2'-TRIMETHYLENE INDOLE	29.20	0.018	18.97599
1-METHYL-3,2'-METHYLENE-2-PHENYL INDOLE (TRACE)	29.63	0.017	20.70102
1-METHYL-3,2'-ETHYLENE-2-PHENYL INDOLE	29.06	0.018	18.34453
1-METHYL-3,2'-TRIMETHYLENE-2-PHENYL INDOLE	27.71	0.021	13.87758
4,5-DIPHENYLIMIDAZOLE (A)	20.92	0.049	2.57295
PPF (TRACE)	31.22	0.015	28.29009
PPO (TRACE)	30.34	0.016	23.92834
2-(4-FLUOROPHENYL)-5-PHENYLOXAZOLE (TRACE)	29.07	0.018	18.50764
αNPO (TRACE)	29.62	0.017	20.52180
BBO (B)	32.77	0.013	37.63558
POPOP	30.27	0.016	23.36563
DIMETHYL POPOP (TRACE)	30.49	0.016	24.38350
PPD (TRACE)	22.38	0.040	3.88827
2,5-DI(2-METHYLPHENYL)OXADIAZOLE	22.57	0.039	4.08378
PBD	30.86	0.015	26.51784
BBD (B)	33.69	0.012	44.67893
αNPD (TRACE)	28.64	0.019	16.83026
βNPD	29.05	0.018	18.44531
αNND (TRACE)	28.68	0.019	16.94519
βNND (TRACE)	31.19	0.015	28.17392
TRANS-1-(4-BIPHENYLYL)-2-PHENYL-ETHYLENE	32.70	0.013	37.25771
TRANS-1-(4-BIPHENYLYL)-2-(1-NAPHTHYL)-ETHYLENE	31.25	0.015	28.31724
DIPHENYLSTILBENE (B)	33.16	0.012	40.35909
1,1,4,4-TETRAPHENYLBUTADIENE	30.44	0.016	24.15574
1,6-DIPHENYLHEXATRIENE	32.52	0.013	35.92468
1,8-DIPHENYL-1,3,5,7-OCTATETRAENE	29.38	0.018	19.55677
BPSB (TRACE)	32.62	0.013	36.58444
BIS-MSB	32.74	0.013	37.41182
ESCULIN (A)	26.48	0.024	10.48354
AZULENE	20.46	0.052	2.22839
NAPHTHALENE	11.28	0.312	0.06362
DEUTERATED NAPHTHALENE-C10D8	12.70	0.218	0.12934
1-METHYLNAPHTHALENE 5ML/L	13.96	0.164	0.22816
2-METHYL NAPHTHALENE	13.55	0.180	0.19127
2,3-DIMETHYL NAPHTHALENE	13.84	0.169	0.21691
2,6-DIMETHYL NAPHTHALENE (TRACE)	15.64	0.117	0.45131

ACCEPTOR MOLECULES	RO	CO	OVERLAP
1-NAPHTHOL (TRACE)	19.31	0.062	1.59778
2-NAPHTHOL (TRACE)	18.37	0.072	1.18191
1-NAPHTHYLAMINE	22.88	0.037	4.38347
2-CHLORONAPHTHALENE	13.50	0.182	0.18630
ACENAPHTHENE	18.20	0.074	1.12009
1-PHENYL NAPHTHALENE	19.86	0.057	1.89071
2-PHENYL NAPHTHALENE	29.14	0.018	18.83680
1,4-DIPHENYL NAPHTHALENE	25.25	0.028	7.94777
1,5-DIPHENYL NAPHTHALENE	24.29	0.031	6.30595
1,7-DIPHENYL NAPHTHALENE	23.99	0.032	5.85098
1,4,5,8-TETRAPHENYL NAPHTHALENE	28.44	0.019	16.07420
1,1'-BINAPHTHYL	21.30	0.046	2.87705
4,4'-DI(N-BUTOXY)-1,1'-BINAPHTHYL	27.47	0.022	13.12258
2,2'-BINAPHTHYL	26.69	0.024	11.09432
ACRIDINE (A)	22.26	0.040	3.71369
ANTHRACENE (TRACE)	20.96	0.049	2.57259
DEUTERATED ANTHRACENE C14D10 (TRACE)	20.92	0.049	2.54319
9-METHYL ANTHRACENE (TRACE)	19.74	0.058	1.79626
1-AMINO ANTHRACENE	18.62	0.069	1.26573
2-AMINO ANTHRACENE	20.09	0.055	1.98352
9-VINYL ANTHRACENE	20.20	0.054	2.06476
9-PHENYL ANTHRACENE (TRACE)	20.84	0.049	2.49117
9,10-DIPHENYL ANTHRACENE	20.17	0.054	2.04697
9,10-DICHLOROANTHRACENE (TRACE)	19.02	0.065	1.43683
BIANTHRANYL (B)	22.01	0.042	3.45433
9,10-DINAPHTHYLANTHRACENE (B)	22.19	0.041	3.62951
9-(P-VINYLPHENYL)-10-PHENYL ANTHRACENE	20.83	0.050	2.47870
9,10-BIS(PHENYL-ETHYNYL)ANTHRACENE (TRACE)	26.84	0.023	11.34637
RUBICENE (B)	28.26	0.020	15.47817
DIBENZ(A,H)ANTHRACENE (B)	31.50	0.014	29.65635
TETRACENE (TRACE) (B)	21.15	0.047	2.71953
RUBRENE (B)	33.29	0.012	41.32234
TETRAHYDROCANNABINOL ACID METHYL ESTER	18.73	0.068	1.33612
PHENANTHRENE	14.84	0.137	0.32738
3,4-BENZOPHENANTHRENE	24.41	0.031	6.43361
2-PHENYL PHENANTHRENE	23.03	0.037	4.55041
CHRYSENE (TRACE)	25.25	0.028	7.88041
TRIPHENYLENE	16.74	0.095	0.67334
DEUTERATED TRIPHENYLENE-C18D12	16.66	0.097	0.65417
PYRENE	28.36	0.020	15.81625
1,3,6,8,-TETRAPHENYL PYRENE	26.91	0.023	11.53513
BIPYRENYL (B)	30.55	0.016	24.68008
O-PHENYLENE-PYRENE	28.65	0.019	16.79269
ANTHANTHRENE (B)	29.41	0.018	19.65026
FLUORANTHENE	23.47	0.035	5.07572
3-PHENYL FLUORANTHENE	25.36	0.027	8.08348
3-AMINO FLUORANTHENE	26.04	0.025	9.46196
3,3'-BIFLUORANTHENYL (B)	28.37	0.020	15.83113
BENZO(GHI)FLUORANTHENE	23.35	0.035	4.92644
9,12-DI(N-OCTADECOXY)-BENZ(E)ACEPHENANTHRYLENE	25.46	0.027	8.27262
DECACYCLENE (B)	30.38	0.016	23.88815
10-ETHOXY-BENZO(KL)XANTHENE	22.14	0.041	3.57650
PERYLENE	17.63	0.082	0.91181
BENZO(GHI)PERYLENE (B)	27.62	0.021	13.48629
BBOT (TRACE)	29.53	0.017	20.14611
QUININE SULFATE (A)	24.71	0.030	6.96724
ACRIDINE YELLOW (A)	18.86	0.067	1.36517
ACRIDINE ORANGE (A)	21.27	0.046	2.81100
ACRIDINE RED (A)	22.13	0.041	3.56996
URANINE (A)	22.46	0.040	3.89476
RHODAMINE B (A)	25.08	0.028	7.56393
RHODAMINE 6G (A)	24.02	0.032	5.83392
EOSIN B (A)	24.70	0.030	6.90301
EOSIN Y (A)	25.32	0.028	8.00437
ROSE BENGAL (A)	25.24	0.028	7.85659
INDENOINDENE	29.21	0.018	19.04687

8.46 Donor Molecule: 4-Benzylbiphenyl

	RO	CO	OVERLAP
POLYVINYLTOLUENE	4.17	6.175	0.00020
PHENYL ETHER	3.34	12.027	0.00005
ANILINE 3ML/L	11.04	0.333	0.07018
N,N-DIMETHYLANILINE 3ML/L	15.80	0.113	0.60076
TRIPHENYLAMINE	25.68	0.026	10.96362
P-CRESOL	5.39	2.857	0.00096
P-METHYLANISOLE 3ML/L	5.11	3.354	0.00070
1,4-DIMETHOXY BENZENE	13.14	0.197	0.20021
1-METHYL-4-CHLOROBENZENE 5ML/L	0.00	0.000	0.00000
SODIUM SALICYLATE (A)	17.97	0.077	1.28670
ANTHRANILIC ACID (A)	21.68	0.044	3.93833
TMPD	19.67	0.059	2.20565
1,2,4-TRIMETHYLBENZENE (TRACE)	0.00	0.000	0.00000
MESITYLENE 5ML/L	0.00	0.000	0.00000
1,3,5-TRIETHYLBENZENE 5ML/L	0.00	0.000	0.00000
STYRENE	6.29	1.796	0.00242
1,1-DIPHENYLETHYLENE 3ML/L	5.80	2.289	0.00149
BIPHENYL	0.00	0.000	0.00000
DEUTERATED BIPHENYL-C12D10	0.00	0.000	0.00000
4-METHYL BIPHENYL	5.68	2.446	0.00131
MONOISOPROPYLBIPHENYL 5ML/L	4.02	6.898	0.00016
4-BENZYL BIPHENYL	6.18	1.892	0.00219

ACCEPTOR MOLECULES	RO	CO	OVERLAP
3,3'-DIMETHYL BIPHENYL	6.96	1.327	0.00444
2,2'-DIMETHYLBIPHENYL 2ML/L	3.22	13.447	0.00004
OCTAMETHYL BIPHENYL	2.63	24.586	0.00001
4-METHOXY BIPHENYL	11.07	0.329	0.07180
4-BIPHENYLYL PHENYL ETHER	10.71	0.364	0.05870
4,4'-DIHYDROXYBIPHENYL (A)	15.27	0.126	0.48979
4,4'-DIMETHOXY BIPHENYL	14.26	0.154	0.32606
4,4'-DIISOBUTOXYBIPHENYL	14.26	0.154	0.32635
BENZIDINE	23.82	0.033	7.00071
[illegible] VINYL BIPHENYL	18.60	0.070	1.60375
M-TERPHENYL	12.96	0.206	0.18342
M-QUATERPHENYL	9.77	0.480	0.03382
1,3,5-TRIPHENYL BENZENE	12.41	0.234	0.14057
1,3-DIBIPHENYLYLBENZENE (B)	24.91	0.029	9.17684
1,3,5,-TRIBIPHENYLYLBENZENE (B)	28.55	0.019	20.84197
FLUORENE (TRACE)	15.68	0.116	0.57837
1-METHYL FLUORENE	14.67	0.142	0.38668
1,2-BENZOFLUORENE	23.74	0.033	6.83908
9,10-DIHYDROPHENANTHRENE	15.26	0.126	0.49067
CARBAZOLE (TRACE)	20.50	0.052	2.84283
N-METHYL CARBAZOLE (TRACE)	20.43	0.052	2.78513
N-VINYLCARBAZOLE	20.81	0.050	3.10958
N-PHENYL CARBAZOLE	20.75	0.050	3.05416
N-METHYL-2,3-BENZCARBAZOLE	23.64	0.034	6.61329
DIBENZOFURAN	14.78	0.139	0.40468
DIBENZOTHIOPHENE (TRACE)	18.53	0.070	1.55710
P-TERPHENYL	19.58	0.060	2.17714
DEUTERATED P-TERPHENYL-C18D14	19.25	0.063	1.97242
4-METHYL TERPHENYL	20.99	0.048	3.30192
2-METHYL-P-TERPHENYL	14.09	0.160	0.30356
3,3''-DIMETHYL-P-TERPHENYL	20.46	0.052	2.83353
4-(3,3-DIMETHYLBUTOXY)-P-TERPHENYL	23.98	0.032	7.31056
4,4''DIHEXAHYDROFARNESOXY-P-TERPHENYL	26.41	0.024	13.04830
2,2'-METHYLENE-P-TERPHENYL	25.82	0.026	11.40530
2,2'-ETHYLENE-P-TERPHENYL	25.88	0.026	11.55125
2,2'-METHYLENE-5',6''-METHYLENE-P-TERPHENYL (TRACE)	29.31	0.018	24.29793
3-PHENYL DIBENZOFURAN	26.47	0.024	13.25140
BENZO(1,2-B-4,5-B')BISBENZOFURAN (TRACE) (B)	30.52	0.016	30.95226
SEXIPHENYL BENZENE	10.80	0.355	0.06167
P-QUATERPHENYL	27.05	0.023	15.04344
3,3'''-DIMETHYL-P-QUATERPHENYL	27.91	0.021	18.10174
TETRAMETHYL-P-QUATERPHENYL	16.44	0.101	0.76452
3,3'''-DIMETHYL-3',2''-METHYLENE-P-QUATERPHENYL	31.45	0.014	37.01294
3,3'''-DIMETHYL-3',2''-ETHYLENE-P-QUATERPHENYL	31.24	0.015	35.47347
4-(3-ETHYLBUTOXY)-P-QUATERPHENYL (B)	30.22	0.016	29.09755
4,4'''-DI(2-BUTYLOCTOXY-1)-P-QUATERPHENYL	31.12	0.015	34.70357
DIETHYL-P-QUINQUEPHENYL	29.13	0.018	23.43417
DI(3-ETHYLHEPTYL)-P-QUINQUEPHENYL	32.14	0.013	41.97020
TETRAMETHYL-P-SEXIPHENYL	28.80	0.019	21.89621
TETRAMETHYL-P-OCTAPHENYL (B)	34.22	0.011	61.17522
INDOLE (TRACE)	5.38	2.880	0.00094
2-PHENYL INDOLE	27.61	0.021	17.00352
1-METHYL-2-PHENYL INDOLE	24.16	0.032	7.63578
1-NEOPENTYL-2-PHENYL INDOLE	23.08	0.036	5.81870
1-N-BUTYL-2(BIPHENYLYL-4) INDOLE	28.40	0.020	19.98639
1,2-DIPHENYL INDOLE	25.12	0.028	9.63680
1-BIPHENYLYL-2-PHENYL INDOLE	26.73	0.023	13.94216
1,3-DIMETHYL-2-PHENYL INDOLE	23.82	0.033	7.01447
2,3-DIPHENYL INDOLE	26.56	0.024	13.35555
1-METHYL-2,3-DIPHENYL INDOLE	24.54	0.030	8.32144
2-PHENYL-3,2'-METHYLENE INDOLE (TRACE)	28.36	0.020	19.93724
2-PHENYL-3,2'-ETHYLENE INDOLE (TRACE)	28.43	0.019	20.15998
2-PHENYL-3,2'-TRIMETHYLENE INDOLE	28.02	0.020	18.53645
1-METHYL-3,2'-METHYLENE-2-PHENYL INDOLE (TRACE)	28.63	0.019	21.03779
1-METHYL-3,2'-ETHYLENE-2-PHENYL INDOLE	28.15	0.020	18.94373
1-METHYL-3,2'-TRIMETHYLENE-2-PHENYL INDOLE	26.58	0.024	13.49842
4,5-DIPHENYLIMIDAZOLE (A)	19.73	0.058	2.26349
PPF (TRACE)	30.15	0.016	28.70293
PPO (TRACE)	29.01	0.018	22.83149
2-(4-FLUOROPHENYL)-5-PHENYLOXAZOLE (TRACE)	27.78	0.021	17.61805
αNPO (TRACE)	28.80	0.019	21.62897
BBO (B)	31.95	0.014	40.31921
POPOP	29.62	0.017	25.58853
DIMETHYL POPOP (TRACE)	29.83	0.017	26.69949
PPD (TRACE)	20.85	0.049	3.18480
2,5-DI(2-METHYLPHENYL)OXADIAZOLE	21.05	0.048	3.36589
PBD	29.37	0.018	24.61305
BBD (B)	32.47	0.013	44.71126
αNPD (TRACE)	27.63	0.021	16.96268
βNPD	27.69	0.021	17.27803
αNND (TRACE)	27.83	0.021	17.67140
βNND (TRACE)	30.03	0.017	28.09136
TRANS-1-(4-BIPHENYLYL)-2-PHENYL-ETHYLENE	31.55	0.014	37.56395
TRANS-1-(4-BIPHENYLYL)-2-(1-NAPHTHYL)-ETHYLENE	30.38	0.016	29.80243
DIPHENYLSTILBENE (B)	32.32	0.013	43.19428
1,1,4,4-TETRAPHENYLBUTADIENE	29.73	0.017	26.16086
1,6-DIPHENYLHEXATRIENE	31.85	0.014	39.59990
1,8-DIPHENYL-1,3,5,7-OCTATETRAENE	28.95	0.018	22.30918
BPSB (TRACE)	31.93	0.014	40.20929
BIS-MSB	31.96	0.014	40.37223
ESCULIN (A)	25.75	0.026	11.04974
AZULENE	19.88	0.057	2.34269
NAPHTHALENE	10.56	0.380	0.05348
DEUTERATED NAPHTHALENE-C10D8	11.97	0.261	0.11327
1-METHYLNAPHTHALENE 5ML/L	13.03	0.202	0.18863
2-METHYL NAPHTHALENE	12.88	0.209	0.17582

ACCEPTOR MOLECULES	RO	CO	OVERLAP
2,3-DIMETHYL NAPHTHALENE	13.09	0.200	0.19386
2,6-DIMETHYL NAPHTHALENE (TRACE)	14.99	0.133	0.43688
1-NAPHTHOL (TRACE)	18.28	0.073	1.43604
2-NAPHTHOL (TRACE)	17.71	0.081	1.18394
1-NAPHTHYLAMINE	22.03	0.042	4.36687
2-CHLORONAPHTHALENE	12.86	0.210	0.17443
ACENAPHTHENE	17.07	0.090	0.95391
1-PHENYL NAPHTHALENE	18.59	0.070	1.59060
2-PHENYL NAPHTHALENE	27.24	0.022	15.71429
1,4-DIPHENYL NAPHTHALENE	24.00	0.032	7.31952
1,5-DIPHENYL NAPHTHALENE	23.03	0.037	5.72168
1,7-DIPHENYL NAPHTHALENE	22.75	0.038	5.31189
1,4,5,8-TETRAPHENYL NAPHTHALENE	27.65	0.021	16.95589
1,1'-BINAPHTHYL	20.03	0.056	2.48727
4,4'-DI(N-BUTOXY)-1,1'-BINAPHTHYL	26.34	0.024	12.77526
2,2'-BINAPHTHYL	25.31	0.028	10.08442
ACRIDINE (A)	21.88	0.043	4.15883
ANTHRACENE (TRACE)	20.57	0.051	2.87338
DEUTERATED ANTHRACENE C14D10 (TRACE)	20.51	0.052	2.82561
9-METHYL ANTHRACENE (TRACE)	19.44	0.061	2.04457
1-AMINO ANTHRACENE	18.16	0.075	1.35951
2-AMINO ANTHRACENE	19.69	0.059	2.19524
9-VINYL ANTHRACENE	19.90	0.057	2.35265
9-PHENYL ANTHRACENE (TRACE)	20.52	0.052	2.82843
9,10-DIPHENYL ANTHRACENE	19.92	0.057	2.36632
9,10-DICHLOROANTHRACENE (TRACE)	18.79	0.067	1.67053
BIANTHRANYL (B)	21.65	0.044	3.89949
9,10-DINAPHTHYLANTHRACENE (B)	21.44	0.045	3.68640
9-(P-VINYLPHENYL)-10-PHENYL ANTHRACENE	20.39	0.053	2.72485
9,10-BIS(PHENYL-ETHYNYL)ANTHRACENE (TRACE)	25.40	0.027	10.18241
RUBICENE (B)	26.94	0.023	14.48078
DIBENZ(A,H)ANTHRACENE (B)	29.71	0.017	26.07150
TETRACENE (TRACE) (B)	19.77	0.058	2.26633
RUBRENE (B)	31.34	0.015	35.89447
TETRAHYDROCANNABINOL ACID METHYL ESTER	17.38	0.085	1.06637
PHENANTHRENE	14.11	0.159	0.30218
3,4-BENZOPHENANTHRENE	23.23	0.036	5.95649
2-PHENYL PHENANTHRENE	21.56	0.045	3.82621
CHRYSENE (TRACE)	24.09	0.032	7.42536
TRIPHENYLENE	15.79	0.114	0.59284
DEUTERATED TRIPHENYLENE-C18D12	15.71	0.115	0.57408
PYRENE	27.60	0.021	16.76890
1,3,6,8,-TETRAPHENYL PYRENE	25.94	0.026	11.54839
BIPYRENYL (B)	29.84	0.017	26.78915
O-PHENYLENE-PYRENE	27.30	0.022	15.70523
ANTHANTHRENE (B)	27.95	0.020	18.07735
FLUORANTHENE	22.89	0.037	5.44854
3-PHENYL FLUORANTHENE	24.47	0.031	8.13318
3-AMINO FLUORANTHENE	24.83	0.029	8.89651
3,3'-BIFLUORANTHENYL (B)	27.38	0.022	15.97699
BENZO(GHI)FLUORANTHENE	22.69	0.038	5.17283
9,12-DI(N-OCTADECOXY)-BENZ(E)ACEPHENANTHRYLENE	24.56	0.030	8.32684
DECACYCLENE (B)	29.50	0.017	24.98716
10-ETHOXY-BENZO(KL)XANTHENE	21.45	0.045	3.68866
PERYLENE	17.30	0.086	1.01722
BENZO(GHI)PERYLENE (B)	26.07	0.025	11.91116
BBOT (TRACE)	29.01	0.018	22.60012
QUININE SULFATE (A)	23.99	0.032	7.28160
ACRIDINE YELLOW (A)	17.89	0.078	1.24323
ACRIDINE ORANGE (A)	20.02	0.056	2.44144
ACRIDINE RED (A)	21.20	0.047	3.44302
URANINE (A)	21.59	0.044	3.84145
RHODAMINE B (A)	24.01	0.032	7.26384
RHODAMINE 6G (A)	23.14	0.036	5.81688
EOSIN B (A)	23.71	0.034	6.73703
EOSIN Y (A)	24.28	0.031	7.76235
ROSE BENGAL (A)	24.35	0.031	7.89686
INDENOINDENE	28.00	0.020	18.47439

8.47 Donor Molecule: 3,3′-Dimethylbiphenyl

POLYVINYLTOLUENE	4.12	6.405	0.00018
PHENYL ETHER	2.83	19.666	0.00002
ANILINE 3ML/L	11.29	0.311	0.07670
N,N-DIMETHYLANILINE 3ML/L	16.46	0.100	0.73066
TRIPHENYLAMINE	26.42	0.024	12.39073
P-CRESOL	4.76	4.140	0.00043
P-METHYLANISOLE 3ML/L	4.50	4.910	0.00031
1,4-DIMETHOXY BENZENE	13.67	0.175	0.24079
1-METHYL-4-CHLOROBENZENE 5ML/L	0.00	0.000	0.00000
SODIUM SALICYLATE (A)	18.57	0.070	1.49006
ANTHRANILIC ACID (A)	21.55	0.045	3.62285
TMPD	19.72	0.058	2.13277
1,2,4-TRIMETHYLBENZENE (TRACE)	0.00	0.000	0.00000
MESITYLENE 5ML/L	0.00	0.000	0.00000
1,3,5-TRIETHYLBENZENE 5ML/L	0.00	0.000	0.00000
STYRENE	5.72	2.390	0.00130
1,1-DIPHENYLETHYLENE 3ML/L	5.17	3.245	0.00071
BIPHENYL	0.00	0.000	0.00000
DEUTERATED BIPHENYL-C12D10	0.00	0.000	0.00000

ACCEPTOR MOLECULES	RO	CO	OVERLAP
4-METHYL BIPHENYL	5.10	3.380	0.00065
MONOISOPROPYLBIPHENYL 5ML/L	3.30	12.471	0.00005
4-BENZYL BIPHENYL	5.47	2.738	0.00099
3,3'-DIMETHYL BIPHENYL	6.55	1.593	0.00293
2,2'-DIMETHYLBIPHENYL 2ML/L	2.84	19.571	0.00002
OCTAMETHYL BIPHENYL	2.20	41.913	0.00000
4-METHOXY BIPHENYL	11.22	0.317	0.07384
4-BIPHENYLYL PHENYL ETHER	10.59	0.377	0.05228
4,4' DIHYDROXYBIPHENYL (A)	15.74	0.113	0.33912
4,4'-DIMETHOXY BIPHENYL	14.67	0.142	0.36731
4,4'-DIISOBUTOXYBIPHENYL	14.69	0.141	0.37086
BENZIDINE	24.56	0.030	8.01253
4-VINYL BIPHENYL	19.34	0.062	1.93278
M-TERPHENYL	13.45	0.184	0.21871
M-QUATERPHENYL	9.75	0.483	0.03188
1,3,5-TRIPHENYL BENZENE	12.86	0.210	0.16595
1,3-DIBIPHENYLYLBENZENE (B)	25.79	0.026	10.76842
1,3,5,-TRIBIPHENYLYLBENZENE (B)	29.53	0.017	24.29123
FLUORENE (TRACE)	16.44	0.101	0.73150
1-METHYL FLUORENE	15.17	0.128	0.45050
1,2-BENZOFLUORENE	24.42	0.031	7.72077
9,10-DIHYDROPHENANTHRENE	16.04	0.108	0.62863
CARBAZOLE (TRACE)	20.70	0.050	2.87406
N-METHYL CARBAZOLE (TRACE)	20.29	0.054	2.54229
N-VINYLCARBAZOLE	20.70	0.050	2.86470
N-PHENYL CARBAZOLE	20.64	0.051	2.82183
N-METHYL-2,3-BENZCARBAZOLE	23.84	0.033	6.60940
DIBENZOFURAN	15.00	0.133	0.42139
DIBENZOTHIOPHENE (TRACE)	18.77	0.068	1.59771
P-TERPHENYL	20.32	0.053	2.58888
DEUTERATED P-TERPHENYL-C18D14	20.00	0.056	2.35976
4-METHYL TERPHENYL	21.79	0.043	3.93768
2-METHYL-P-TERPHENYL	14.47	0.148	0.33926
3,3''-DIMETHYL-P-TERPHENYL	21.23	0.047	3.36929
4-(3,3-DIMETHYLBUTOXY)-P-TERPHENYL	24.78	0.029	8.48458
4,4''DIHEXAHYDROFARNESOXY-P-TERPHENYL	27.22	0.022	14.89397
2,2'-METHYLENE-P-TERPHENYL	26.60	0.024	12.99749
2,2'-ETHYLENE-P-TERPHENYL	26.61	0.024	12.99861
2,2'-METHYLENE-5',6''METHYLENE-P-TERPHENYL (TRACE)	29.47	0.017	23.92574
3-PHENYL DIBENZOFURAN	27.35	0.022	15.35143
BENZO(1,2-B-4,5-B')BISBENZOFURAN (TRACE) (B)	30.63	0.016	30.09437
SEXIPHENYL BENZENE	10.99	0.337	0.06538
P-QUATERPHENYL	27.82	0.021	16.96394
3,3'''-DIMETHYL-P-QUATERPHENYL	28.65	0.019	20.19611
TETRAMETHYL-P-QUATERPHENYL	16.99	0.091	0.88823
3,3'''-DIMETHYL-3',2''-METHYLENE-P-QUATERPHENYL	31.75	0.014	37.33417
3,3'''-DIMETHYL-3',2''-ETHYLENE-P-QUATERPHENYL	31.59	0.014	36.09423
4-(3-ETHYLBUTOXY)-P-QUATERPHENYL (B)	30.77	0.015	30.84891
4,4'''-DI(2-BUTYLOCTOXY-1)-P-QUATERPHENYL	31.70	0.014	36.93035
DIETHYL-P-QUINQUEPHENYL	29.91	0.017	26.15136
DI(3-ETHYLHEPTYL)-P-QUINQUEPHENYL	32.71	0.013	44.41575
TETRAMETHYL-P-SEXIPHENYL	29.64	0.017	24.78987
TETRAMETHYL-P-OCTAPHENYL (B)	34.59	0.011	62.13614
INDOLE (TRACE)	4.57	4.689	0.00034
2-PHENYL INDOLE	28.18	0.020	18.28894
1-METHYL-2-PHENYL INDOLE	24.87	0.029	8.64016
1-NEOPENTYL-2-PHENYL INDOLE	23.87	0.033	6.77643
1-N-BUTYL-2(BIPHENYLYL-4) INDOLE	28.76	0.019	20.55442
1,2-DIPHENYL INDOLE	25.85	0.026	10.89126
1-BIPHENYLYL-2-PHENYL INDOLE	27.40	0.022	15.40831
1,3-DIMETHYL-2-PHENYL INDOLE	24.41	0.031	7.72725
2,3-DIPHENYL INDOLE	26.95	0.023	13.87771
1-METHYL-2,3-DIPHENYL INDOLE	24.96	0.029	8.77756
2-PHENYL-3,2'-METHYLENE INDOLE (TRACE)	28.66	0.019	20.20774
2-PHENYL-3,2'-ETHYLENE INDOLE (TRACE)	28.53	0.019	19.59063
2-PHENYL-3,2'-TRIMETHYLENE INDOLE	28.44	0.019	19.29764
1-METHYL-3,2'-METHYLENE-2-PHENYL INDOLE (TRACE)	28.79	0.019	20.71570
1-METHYL-3,2'-ETHYLENE-2-PHENYL INDOLE	28.20	0.020	18.24843
1-METHYL-3,2'-TRIMETHYLENE-2-PHENYL INDOLE	26.98	0.023	14.07672
4,5-DIPHENYLIMIDAZOLE (A)	20.33	0.053	2.58375
PPF (TRACE)	30.33	0.016	28.30333
PPO (TRACE)	29.56	0.017	24.34237
2-(4-FLUOROPHENYL)-5-PHENYLOXAZOLE (TRACE)	28.31	0.020	18.82179
αNPO (TRACE)	28.71	0.019	20.23221
BBO (B)	31.75	0.014	36.92453
POPOP	29.31	0.018	22.83143
DIMETHYL POPOP (TRACE)	29.55	0.017	23.96015
PPD (TRACE)	21.64	0.044	3.78544
2,5-DI(2-METHYLPHENYL)OXADIAZOLE	21.92	0.042	4.08003
PBD	30.07	0.016	27.04590
BBD (B)	32.75	0.013	44.89238
αNPD (TRACE)	27.83	0.021	16.84191
βNPD	28.31	0.020	18.78727
αNND (TRACE)	27.80	0.021	16.74531
βNND (TRACE)	30.33	0.016	28.35988
TRANS-1-(4-BIPHENYLYL)-2-PHENYL-ETHYLENE	31.77	0.014	37.27930
TRANS-1-(4-BIPHENYLYL)-2-(1-NAPHTHYL)-ETHYLENE	30.31	0.016	27.95794
DIPHENYLSTILBENE (B)	32.12	0.013	39.61539
1,1,4,4-TETRAPHENYLBUTADIENE	29.49	0.017	23.65392
1,6-DIPHENYLHEXATRIENE	31.49	0.014	35.11976
1,8-DIPHENYL-1,3,5,7-OCTATETRAENE	28.57	0.019	19.55610
BPSB (TRACE)	31.60	0.014	35.86753
BIS-MSB	31.73	0.014	36.76292
ESCULIN (A)	25.65	0.027	10.29033
AZULENE	19.77	0.058	2.14836
NAPHTHALENE	10.91	0.344	0.06223

ACCEPTOR MOLECULES	RO	CO	OVERLAP
DEUTERATED NAPHTHALENE-C10D8	12.35	0.238	0.12973
1-METHYLNAPHTHALENE 5ML/L	13.28	0.191	0.20177
2-METHYL NAPHTHALENE	13.19	0.195	0.19339
2,3-DIMETHYL NAPHTHALENE	13.37	0.187	0.20970
2,6-DIMETHYL NAPHTHALENE (TRACE)	15.18	0.128	0.44835
1-NAPHTHOL (TRACE)	18.79	0.067	1.61193
2-NAPHTHOL (TRACE)	17.86	0.078	1.18937
1-NAPHTHYLAMINE	22.24	0.041	4.40946
2-CHLORONAPHTHALENE	13.09	0.199	0.18456
ACENAPHTHENE	17.64	0.081	1.10883
1-PHENYL NAPHTHALENE	19.26	0.063	1.87516
2-PHENYL NAPHTHALENE	28.26	0.020	18.63690
1,4-DIPHENYL NAPHTHALENE	24.61	0.030	8.10708
1,5-DIPHENYL NAPHTHALENE	23.67	0.034	6.43358
1,7-DIPHENYL NAPHTHALENE	23.39	0.035	5.98153
1,4,5,8-TETRAPHENYL NAPHTHALENE	27.58	0.021	15.84455
1,1'-BINAPHTHYL	20.65	0.051	2.84430
4,4'-DI(N-BUTOXY)-1,1'-BINAPHTHYL	26.74	0.023	13.30094
2,2'-BINAPHTHYL	26.04	0.025	11.38845
ACRIDINE (A)	21.55	0.045	3.60599
ANTHRACENE (TRACE)	20.29	0.054	2.51545
DEUTERATED ANTHRACENE C14D10 (TRACE)	20.25	0.054	2.49012
9-METHYL ANTHRACENE (TRACE)	19.16	0.064	1.77813
1-AMINO ANTHRACENE	18.11	0.075	1.27025
2-AMINO ANTHRACENE	19.50	0.060	1.96571
9-VINYL ANTHRACENE	19.62	0.059	2.05380
9-PHENYL ANTHRACENE (TRACE)	20.22	0.054	2.46334
9,10-DIPHENYL ANTHRACENE	19.64	0.059	2.06365
9,10-DICHLOROANTHRACENE (TRACE)	18.54	0.070	1.46394
BIANTHRANYL (B)	21.42	0.045	3.48128
9,10-DINAPHTHYLANTHRACENE (B)	21.56	0.045	3.61123
9-(P-VINYLPHENYL)-10-PHENYL ANTHRACENE	20.27	0.054	2.49492
9,10-BIS(PHENYL-ETHYNYL)ANTHRACENE (TRACE)	26.26	0.025	11.80073
RUBICENE (B)	27.50	0.022	15.56403
DIBENZ(A,H)ANTHRACENE (B)	30.53	0.016	29.16454
TETRACENE (TRACE) (B)	20.13	0.055	2.39356
RUBRENE (B)	32.58	0.013	43.02446
TETRAHYDROCANNABINOL ACID METHYL ESTER	18.05	0.076	1.27355
PHENANTHRENE	13.85	0.169	0.25662
3,4-BENZOPHENANTHRENE	23.80	0.033	6.55074
2-PHENYL PHENANTHRENE	22.34	0.040	4.51368
CHRYSENE (TRACE)	24.66	0.030	8.14472
TRIPHENYLENE	16.12	0.107	0.63954
DEUTERATED TRIPHENYLENE-C18D12	16.05	0.108	0.62349
PYRENE	27.55	0.021	15.79304
1,3,6,8,-TETRAPHENYL PYRENE	25.99	0.025	11.09224
BIPYRENYL (B)	29.57	0.017	24.07816
O-PHENYLENE-PYRENE	27.95	0.020	17.16498
ANTHANTHRENE (B)	28.77	0.019	20.39735
FLUORANTHENE	22.74	0.038	4.97587
3-PHENYL FLUORANTHENE	24.45	0.031	7.69032
3-AMINO FLUORANTHENE	25.45	0.027	9.78321
3,3'-BIFLUORANTHENYL (B)	27.55	0.021	15.74002
BENZO(GHI)FLUORANTHENE	22.56	0.039	4.74186
9,12-DI(N-OCTADECOXY)-BENZ(E)ACEPHENANTHRYLENE	24.69	0.030	8.15417
DECACYCLENE (B)	29.54	0.017	23.91217
10-ETHOXY-BENZO(KL)XANTHENE	21.59	0.044	3.64438
PERYLENE	17.36	0.085	0.98636
BENZO(GHI)PERYLENE (B)	27.02	0.023	14.00767
BBOT (TRACE)	28.65	0.019	19.92638
QUININE SULFATE (A)	23.94	0.033	6.85800
ACRIDINE YELLOW (A)	18.30	0.073	1.35388
ACRIDINE ORANGE (A)	20.64	0.051	2.78174
ACRIDINE RED (A)	21.55	0.045	3.60269
URANINE (A)	21.83	0.043	3.89166
RHODAMINE B (A)	24.42	0.031	7.63535
RHODAMINE 6G (A)	23.24	0.036	5.66690
EOSIN B (A)	24.08	0.032	7.01340
EOSIN Y (A)	24.69	0.030	8.15411
ROSE BENGAL (A)	24.54	0.030	7.86730
INDENOINDENE	28.45	0.019	19.35151

8.48 Donor Molecule: 2,2'-Dimethylbiphenyl (2 ml/L)

POLYVINYLTOLUENE	2.61	25.273	0.00041
ANISOLE (TRACE)	1.59	******	0.00002
PHENYL ETHER	2.51	28.324	0.00033
ANILINE 3ML/L	6.77	1.442	0.12482
N,N-DIMETHYLANILINE 3ML/L	9.21	0.572	0.78811
TRIPHENYLAMINE	14.67	0.142	12.69982
P-CRESOL	4.00	6.987	0.00535
P-METHYLANISOLE 3ML/L	3.87	7.700	0.00441
1,4-DIMETHOXY BENZENE	8.00	0.874	0.33926
1-METHYL-4-CHLOROBENZENE 5ML/L	0.00	0.000	0.00000
SODIUM SALICYLATE (A)	10.35	0.404	1.56739
ANTHRANILIC ACID (A)	11.81	0.272	3.43348
TMPD	10.87	0.348	2.09356
1,2,4-TRIMETHYLBENZENE (TRACE)	1.45	******	0.00001
MESITYLENE 5ML/L	0.00	0.000	0.00000
1,3,5-TRIETHYLBENZENE 5ML/L	0.00	0.000	0.00000

ACCEPTOR MOLECULES	RO	CO	OVERLAP
STYRENE	4.17	6.186	0.00681
1,1-DIPHENYLETHYLENE 3ML/L	3.99	7.064	0.00524
BIPHENYL	2.41	32.116	0.00025
DEUTERATED BIPHENYL-C12D10	2.23	40.377	0.00016
4-METHYL BIPHENYL	3.93	7.373	0.00481
MONOISOPROPYLBIPHENYL 5ML/L	3.22	13.395	0.00146
4-BENZYL BIPHENYL	4.36	5.407	0.00894
3,3' DIMETHYL BIPHENYL	4.54	4.774	0.01144
2,2'-DIMETHYLBIPHENYL 2ML/L	2.23	40.232	0.00016
OCTAMETHYL BIPHENYL	2.17	43.655	0.00014
4-METHOXY BIPHENYL	6.94	1.336	0.14571
4-BIPHENYLYL PHENYL ETHER	6.79	1.429	0.12734
4,4'-DIHYDROXYBIPHENYL (A)	9.14	0.587	0.75015
4,4'-DIMETHOXY BIPHENYL	8.65	0.691	0.54255
4,4'-DIISOBUTOXYBIPHENYL	8.67	0.687	0.54967
BENZIDINE	13.72	0.173	8.53991
4-VINYL BIPHENYL	11.11	0.326	2.42671
M-TERPHENYL	7.71	0.978	0.27106
M-QUATERPHENYL	6.12	1.952	0.06832
1,3,5-TRIPHENYL BENZENE	7.28	1.161	0.19099
1,3-DIBIPHENYLYLBENZENE (B)	14.50	0.147	11.93603
1,3,5,-TRIBIPHENYLYLBENZENE (B)	16.54	0.099	26.31979
FLUORENE (TRACE)	9.48	0.524	0.94417
1-METHYL FLUORENE	8.90	0.635	0.64243
1,2-BENZOFLUORENE	13.53	0.181	7.82832
9,10-DIHYDROPHENANTHRENE	9.19	0.577	0.77888
CARBAZOLE (TRACE)	11.54	0.291	3.03135
N-METHYL CARBAZOLE (TRACE)	11.36	0.305	2.74714
N-VINYLCARBAZOLE	11.53	0.292	2.99948
N-PHENYL CARBAZOLE	11.61	0.286	3.12171
N-METHYL-2,3-BENZCARBAZOLE	13.36	0.188	7.17805
DIBENZOFURAN	9.11	0.592	0.74184
DIBENZOTHIOPHENE (TRACE)	10.34	0.405	1.56643
P-TERPHENYL	11.62	0.285	3.17725
DEUTERATED P-TERPHENYL-C18D14	11.45	0.298	2.91113
4-METHYL TERPHENYL	12.37	0.237	4.61195
2-METHYL-P-TERPHENYL	8.61	0.701	0.52755
3,3''-DIMETHYL-P-TERPHENYL	12.07	0.255	3.97696
4-(3,3-DIMETHYLBUTOXY)-P-TERPHENYL	13.89	0.167	9.20986
4,4''-DIHEXAHYDROFARNESOXY-P-TERPHENYL	15.17	0.128	15.62315
2,2'-METHYLENE-P-TERPHENYL	14.77	0.139	13.37172
2,2'-ETHYLENE-P-TERPHENYL	14.79	0.138	13.41322
2,2'-METHYLENE-5',6''-METHYLENE-P-TERPHENYL (TRACE)	16.23	0.105	23.37718
3-PHENYL DIBENZOFURAN	15.17	0.128	15.69091
BENZO(1,2-B-4,5-B')BISBENZOFURAN (TRACE) (B)	16.80	0.094	28.68266
SEXIPHENYL BENZENE	6.78	1.435	0.12637
P-QUATERPHENYL	15.47	0.121	17.57614
3,3'''-DIMETHYL-P-QUATERPHENYL	15.91	0.111	20.72208
TETRAMETHYL-P-QUATERPHENYL	9.96	0.453	1.26258
3,3'''-DIMETHYL-3',2''-METHYLENE-P-QUATERPHENYL	17.47	0.084	36.27249
3,3'''-DIMETHYL-3',2''-ETHYLENE-P-QUATERPHENYL	17.39	0.085	35.23624
4-(3-ETHYLBUTOXY)-P-QUATERPHENYL (B)	16.98	0.091	30.50990
4,4'''-DI(2-BUTYLOCTOXY-1)-P-QUATERPHENYL	17.50	0.083	36.60472
DIETHYL-P-QUINQUEPHENYL	16.61	0.098	26.88997
DI(3-ETHYLHEPTYL)-P-QUINQUEPHENYL	18.04	0.076	43.86656
TETRAMETHYL-P-SEXIPHENYL	16.49	0.100	25.80198
TETRAMETHYL-P-OCTAPHENYL (B)	19.03	0.065	60.33726
INDOLE (TRACE)	4.36	5.381	0.00903
2-PHENYL INDOLE	15.53	0.119	18.00042
1-METHYL-2-PHENYL INDOLE	13.81	0.170	8.86616
1-NEOPENTYL-2-PHENYL INDOLE	13.30	0.190	7.10430
1-N-BUTYL-2(BIPHENYLYL-4) INDOLE	15.84	0.113	20.10495
1,2-DIPHENYL INDOLE	14.33	0.152	11.09226
1-BIPHENYLYL-2-PHENYL INDOLE	15.17	0.128	15.53192
1,3-DIMETHYL-2-PHENYL INDOLE	13.52	0.181	7.82226
2,3-DIPHENYL INDOLE	14.85	0.137	13.61939
1-METHYL-2,3-DIPHENYL INDOLE	13.79	0.171	8.74323
2-PHENYL-3,2'-METHYLENE INDOLE (TRACE)	15.74	0.115	19.40959
2-PHENYL-3,2'-ETHYLENE INDOLE (TRACE)	15.66	0.117	18.75260
2-PHENYL-3,2'-TRIMETHYLENE INDOLE	15.64	0.117	18.71013
1-METHYL-3,2'-METHYLENE-2-PHENYL INDOLE (TRACE)	15.80	0.113	19.85433
1-METHYL-3,2'-ETHYLENE-2-PHENYL INDOLE	15.47	0.121	17.42043
1-METHYL-3,2'-TRIMETHYLENE-2-PHENYL INDOLE	14.86	0.136	13.73669
4,5-DIPHENYLIMIDAZOLE (A)	11.39	0.303	2.79186
PPF (TRACE)	16.65	0.097	27.15839
PPO (TRACE)	16.30	0.103	24.04264
2-(4-FLUOROPHENYL)-5-PHENYLOXAZOLE (TRACE)	15.63	0.117	18.63598
αNPO (TRACE)	15.75	0.114	19.33410
BBO (B)	17.42	0.085	35.34356
POPOP	16.12	0.107	22.14625
DIMETHYL POPOP (TRACE)	16.29	0.104	23.56169
PPD (TRACE)	12.29	0.241	4.45596
2,5-DI(2-METHYLPHENYL)OXADIAZOLE	12.37	0.237	4.61410
PBD	16.64	0.097	27.22115
BBD (B)	18.00	0.077	43.38270
αNPD (TRACE)	15.29	0.125	16.26345
βNPD	15.64	0.117	18.76304
αNND (TRACE)	15.26	0.126	16.03399
βNND (TRACE)	16.67	0.096	27.44980
TRANS-1-(4-BIPHENYLYL)-2-PHENYL-ETHYLENE	17.46	0.084	36.01635
TRANS-1-(4-BIPHENYLYL)-2-(1-NAPHTHYL)-ETHYLENE	16.64	0.097	26.81690
DIPHENYLSTILBENE (B)	17.62	0.082	37.84058
1,1,4,4-TETRAPHENYLBUTADIENE	16.18	0.106	22.62448
1,6-DIPHENYLHEXATRIENE	17.29	0.087	33.65769
1,8-DIPHENYL-1,3,5,7-OCTATETRAENE	15.82	0.113	19.76949
BPSB (TRACE)	17.37	0.085	34.59564

ACCEPTOR MOLECULES	RO	CO	OVERLAP
BIS-MSB	17.43	0.085	35.33299
ESCULIN (A)	14.09	0.160	9.91998
AZULENE	10.93	0.342	2.15438
NAPHTHALENE	6.31	1.779	0.08159
DEUTERATED NAPHTHALENE-C10D8	6.97	1.321	0.14723
1-METHYLNAPHTHALENE 5ML/L	7.85	0.924	0.30187
2-METHYL NAPHTHALENE	7.44	1.084	0.21915
2,3-DIMETHYL NAPHTHALENE	7.66	0.995	0.26034
2,6-DIMETHYL NAPHTHALENE (TRACE)	8.44	0.744	0.46512
1-NAPHTHOL (TRACE)	10.48	0.389	1.70156
2-NAPHTHOL (TRACE)	9.82	0.473	1.14796
1-NAPHTHYLAMINE	12.23	0.245	4.26637
2-CHLORONAPHTHALENE	7.40	1.104	0.21120
ACENAPHTHENE	9.96	0.453	1.25409
1-PHENYL NAPHTHALENE	10.88	0.347	2.13813
2-PHENYL NAPHTHALENE	15.97	0.110	21.32704
1,4-DIPHENYL NAPHTHALENE	13.63	0.177	8.20923
1,5-DIPHENYL NAPHTHALENE	13.14	0.197	6.58921
1,7-DIPHENYL NAPHTHALENE	12.97	0.205	6.10778
1,4,5,8-TETRAPHENYL NAPHTHALENE	15.14	0.129	15.17844
1,1'-BINAPHTHYL	11.65	0.283	3.20613
4,4'-DI(N-BUTOXY)-1,1'-BINAPHTHYL	14.72	0.140	13.00017
2,2'-BINAPHTHYL	14.42	0.149	11.54496
ACRIDINE (A)	11.83	0.270	3.46165
ANTHRACENE (TRACE)	11.14	0.324	2.41203
DEUTERATED ANTHRACENE C14D10 (TRACE)	11.11	0.326	2.38480
9-METHYL ANTHRACENE (TRACE)	10.58	0.378	1.76659
1-AMINO ANTHRACENE	10.13	0.431	1.36030
2-AMINO ANTHRACENE	10.70	0.365	1.88304
9-VINYL ANTHRACENE	10.85	0.350	2.06218
9-PHENYL ANTHRACENE (TRACE)	11.17	0.321	2.44778
9,10-DIPHENYL ANTHRACENE	10.92	0.344	2.13664
9,10-DICHLOROANTHRACENE (TRACE)	10.32	0.407	1.52701
BIANTHRANYL (B)	11.88	0.267	3.54419
9,10-DINAPHTHYLANTHRACENE (B)	12.09	0.253	3.94348
9-(P-VINYLPHENYL)-10-PHENYL ANTHRACENE	11.32	0.309	2.65161
9,10-BIS(PHENYL-ETHYNYL)ANTHRACENE (TRACE)	14.57	0.145	12.06096
RUBICENE (B)	15.32	0.124	16.32153
DIBENZ(A,H)ANTHRACENE (B)	17.04	0.090	30.86164
TETRACENE (TRACE) (B)	11.74	0.276	3.30753
RUBRENE (B)	18.07	0.076	43.87593
TETRAHYDROCANNABINOL ACID METHYL ESTER	10.39	0.399	1.62054
PHENANTHRENE	8.45	0.741	0.46555
3,4-BENZOPHENANTHRENE	13.20	0.194	6.68801
2-PHENYL PHENANTHRENE	12.67	0.220	5.25326
CHRYSENE (TRACE)	13.62	0.177	8.09792
TRIPHENYLENE	9.27	0.561	0.81255
DEUTERATED TRIPHENYLENE-C18D12	9.20	0.574	0.77484
PYRENE	15.05	0.131	14.72491
1,3,6,8,-TETRAPHENYL PYRENE	14.76	0.139	13.06772
BIPYRENYL (B)	16.25	0.104	23.32675
O-PHENYLENE-PYRENE	15.53	0.119	17.67989
ANTHANTHRENE (B)	15.92	0.111	20.51576
FLUORANTHENE	12.53	0.228	4.87291
3-PHENYL FLUORANTHENE	13.72	0.173	8.40954
3-AMINO FLUORANTHENE	14.11	0.159	9.94391
3,3'-BIFLUORANTHENYL (B)	15.35	0.124	16.48288
BENZO(GHI)FLUORANTHENE	12.51	0.228	4.83453
9,12-DI(N-OCTADECOXY)-BENZ(E)ACEPHENANTHRYLENE	13.74	0.172	8.48674
DECACYCLENE (B)	16.34	0.103	24.00157
10-ETHOXY-BENZO(KL)XANTHENE	12.04	0.256	3.84394
PERYLENE	9.80	0.475	1.11921
BENZO(GHI)PERYLENE (B)	15.11	0.130	15.00691
BBOT (TRACE)	15.82	0.113	19.75657
QUININE SULFATE (A)	13.15	0.197	6.58840
ACRIDINE YELLOW (A)	10.40	0.397	1.59818
ACRIDINE ORANGE (A)	11.64	0.283	3.14530
ACRIDINE RED (A)	11.93	0.263	3.64113
URANINE (A)	12.08	0.254	3.92498
RHODAMINE B (A)	13.53	0.180	7.75442
RHODAMINE 6G (A)	12.93	0.207	5.88975
EOSIN B (A)	13.26	0.192	6.87050
EOSIN Y (A)	13.61	0.178	8.00711
ROSE BENGAL (A)	13.51	0.181	7.68486
INDENOINDENE	15.66	0.117	18.84120

8.49 Donor Molecule: Octamethylbiphenyl

DIPHENYLMETHANE (TRACE)	1.73	86.769	0.00000
BIBENZYL	1.78	79.944	0.00000
1,3,5-TRIPHENYLPENTANE 0.05ML/L	2.56	26.782	0.00001
POLYSTYRENE	0.00	0.000	0.00000
POLYVINYLTOLUENE	6.47	1.648	0.00321
PHENYLCYCLOHEXANE (TRACE)	0.00	0.000	0.00000
BENZYLALCOHOL 3ML/L	0.00	0.000	0.00000
BENZYLACETATE 5ML/L	0.00	0.000	0.00000
PHENOL (TRACE)	7.24	1.177	0.00634
ANISOLE (TRACE)	7.86	0.921	0.01036
PHENYL ETHER	9.62	0.502	0.03473
ANILINE 3ML/L	15.68	0.116	0.64488
N,N-DIMETHYLANILINE 3ML/L	17.91	0.078	1.42129

ACCEPTOR MOLECULES	RO	CO	OVERLAP
TRIPHENYLAMINE	27.15	0.022	17.10780
P-XYLENE (TRACE)	4.62	4.535	0.00043
DEUTERATED P-XYLENE C8D10 (TRACE)	4.44	5.100	0.00034
M-XYLENE (TRACE)	3.21	13.472	0.00005
O-XYLENE 3ML/L	3.09	15.167	0.00004
1-METHYL-4-ETHYLBENZENE (TRACE)	3.61	9.545	0.00010
1,4-BIS(TRIFLUOROMETHYL)BENZENE 5ML/L	0.00	0.000	0.00000
P-CRESOL	13.02	0.203	0.21314
P-METHYLANISOLE 3ML/L	12.64	0.222	0.17800
1,4-DIMETHOXY BENZENE	17.64	0.081	1.30778
1-METHYL-4-CHLOROBENZENE 5ML/L	5.79	2.308	0.00165
SODIUM SALICYLATE (A)	19.65	0.059	2.46052
ANTHRANILIC ACID (A)	18.26	0.074	1.57787
TMPD	19.26	0.063	2.17417
1,2,4-TRIMETHYLBENZENE (TRACE)	5.92	2.151	0.00190
MESITYLENE 5ML/L	4.00	6.965	0.00018
1,3,5-TRIETHYLBENZENE 5ML/L	3.99	7.052	0.00018
STYRENE	11.60	0.287	0.10619
1,1-DIPHENYLETHYLENE 3ML/L	11.95	0.262	0.12736
BIPHENYL	9.51	0.521	0.03236
DEUTERATED BIPHENYL-C12D10	9.33	0.550	0.02902
4-METHYL BIPHENYL	12.31	0.240	0.15248
MONOISOPROPYLBIPHENYL 5ML/L	11.43	0.299	0.09791
4-BENZYL BIPHENYL	13.79	0.171	0.30059
3,3'-DIMETHYL BIPHENYL	12.75	0.216	0.18733
2,2'-DIMETHYLBIPHENYL 2ML/L	6.98	1.315	0.00506
OCTAMETHYL BIPHENYL	8.71	0.677	0.01913
4-METHOXY BIPHENYL	17.64	0.081	1.31012
4-BIPHENYLYL PHENYL ETHER	17.83	0.079	1.39664
4,4'-DIHYDROXYBIPHENYL (A)	20.71	0.050	3.40396
4,4'-DIMETHOXY BIPHENYL	20.32	0.053	3.04823
4,4'-DIISOBUTOXYBIPHENYL	20.30	0.053	3.03695
BENZIDINE	26.64	0.024	15.33121
4-VINYL BIPHENYL	24.13	0.032	8.53865
M-TERPHENYL	16.96	0.092	1.03315
M-QUATERPHENYL	15.86	0.112	0.69220
1,3,5-TRIPHENYL BENZENE	16.38	0.102	0.83190
1,3-DIBIPHENYLYLBENZENE (B)	29.09	0.018	26.01348
1,3,5,-TRIBIPHENYLYLBENZENE (B)	32.53	0.013	50.89366
FLUORENE (TRACE)	20.18	0.054	2.93716
1-METHYL FLUORENE	19.49	0.060	2.37616
1,2-BENZOFLUORENE	24.74	0.030	9.78861
9,10-DIHYDROPHENANTHRENE	19.58	0.060	2.44289
CARBAZOLE (TRACE)	22.73	0.038	5.91610
N-METHYL CARBAZOLE (TRACE)	22.26	0.040	5.22463
N-VINYLCARBAZOLE	22.52	0.039	5.57426
N-PHENYL CARBAZOLE	23.45	0.035	7.10368
N-METHYL-2,3-BENZCARBAZOLE	26.25	0.025	13.93762
DIBENZOFURAN	21.53	0.045	4.33141
DIBENZOTHIOPHENE (TRACE)	19.38	0.061	2.27587
P-TERPHENYL	24.98	0.029	10.48035
DEUTERATED P-TERPHENYL-C18D14	24.67	0.030	9.74044
4-METHYL TERPHENYL	25.84	0.026	12.84854
2-METHYL-P-TERPHENYL	20.52	0.052	3.23976
3,3''-DIMETHYL-P-TERPHENYL	25.32	0.028	11.35185
4-(3,3-DIMETHYLBUTOXY)-P-TERPHENYL	27.38	0.022	18.09697
4,4''-DIHEXAHYDROFARNESOXY-P-TERPHENYL	28.94	0.018	25.22494
2,2'-METHYLENE-P-TERPHENYL	27.71	0.021	19.49293
2,2'-ETHYLENE-P-TERPHENYL	27.77	0.021	19.70679
2,2'-METHYLENE-5',6''-METHYLENE-P-TERPHENYL (TRACE)	27.99	0.020	20.59641
3-PHENYL DIBENZOFURAN	27.99	0.020	20.67885
BENZO(1,2-B-4,5-B')BISBENZOFURAN (TRACE) (B)	26.71	0.023	15.50895
SEXIPHENYL BENZENE	17.65	0.081	1.31799
P-QUATERPHENYL	29.11	0.018	26.07171
3,3'''-DIMETHYL-P-QUATERPHENYL	29.58	0.017	28.65704
TETRAMETHYL-P-QUATERPHENYL	22.99	0.037	6.39782
3,3'''-DIMETHYL-3',2''-METHYLENE-P-QUATERPHENYL	29.70	0.017	29.31268
3,3'''-DIMETHYL-3',2''-ETHYLENE-P-QUATERPHENYL	29.93	0.017	30.63337
4-(3-ETHYLBUTOXY)-P-QUATERPHENYL (B)	29.92	0.017	30.58244
4,4'''-DI(2-BUTYLOCTOXY-1)-P-QUATERPHENYL	30.96	0.015	37.57798
DIETHYL-P-QUINQUEPHENYL	30.98	0.015	37.84993
DI(3-ETHYLHEPTYL)-P-QUINQUEPHENYL	31.70	0.014	43.26914
TETRAMETHYL-P-SEXIPHENYL	31.20	0.015	39.56340
TETRAMETHYL-P-OCTAPHENYL (B)	32.33	0.013	48.68208
INDOLE (TRACE)	13.94	0.165	0.32149
2-PHENYL INDOLE	27.20	0.022	17.37871
1-METHYL-2-PHENYL INDOLE	25.63	0.027	12.14701
1-NEOPENTYL-2-PHENYL INDOLE	25.27	0.028	11.19276
1-N-BUTYL-2(BIPHENYLYL-4) INDOLE	27.30	0.022	17.64592
1,2-DIPHENYL INDOLE	26.39	0.024	14.46253
1-BIPHENYLYL-2-PHENYL INDOLE	27.70	0.021	19.28178
1,3-DIMETHYL-2-PHENYL INDOLE	24.67	0.030	9.66138
2,3-DIPHENYL INDOLE	25.78	0.026	12.52551
1-METHYL-2,3-DIPHENYL INDOLE	24.64	0.030	9.54257
2-PHENYL-3,2'-METHYLENE INDOLE (TRACE)	26.22	0.025	13.90979
2-PHENYL-3,2'-ETHYLENE INDOLE (TRACE)	25.44	0.027	11.56564
2-PHENYL-3,2'-TRIMETHYLENE INDOLE	26.57	0.024	15.04854
1-METHYL-3,2'-METHYLENE-2-PHENYL INDOLE (TRACE)	25.84	0.026	12.71838
1-METHYL-3,2'-ETHYLENE-2-PHENYL INDOLE	25.03	0.029	10.47220
1-METHYL-3,2'-TRIMETHYLENE-2-PHENYL INDOLE	25.54	0.027	11.86866
4,5-DIPHENYLIMIDAZOLE (A)	22.40	0.040	5.42327
PPF (TRACE)	27.42	0.022	18.12979
PPO (TRACE)	28.69	0.019	23.86468
2-(4-FLUOROPHENYL)-5-PHENYLOXAZOLE (TRACE)	27.60	0.021	18.96135
αNPO (TRACE)	25.26	0.028	11.06011
BBO (B)	27.42	0.022	18.12168

ACCEPTOR MOLECULES	RO	CO	OVERLAP
POPOP	25.13	0.028	10.72440
DIMETHYL POPOP (TRACE)	25.69	0.026	12.25310
PPD (TRACE)	25.59	0.027	12.15546
2,5-DI(2-METHYLPHENYL)OXADIAZOLE	25.09	0.028	10.77331
PBD	30.15	0.016	32.22985
BBD (B)	30.25	0.016	32.65179
αNPD (TRACE)	25.56	0.027	11.87880
βNPD	28.04	0.020	20.79001
αNND (TRACE)	24.72	0.030	9.72452
βNND (TRACE)	28.37	0.020	22.30887
TRANS-1-(4-BIPHENYLYL)-2-PHENYL-ETHYLENE	29.22	0.018	26.50222
TRANS-1-(4-BIPHENYLYL)-2-(1-NAPHTHYL)-ETHYLENE	26.84	0.023	15.93701
DIPHENYLSTILBENE (B)	27.48	0.022	18.33952
1,1,4,4-TETRAPHENYLBUTADIENE	24.89	0.029	10.13374
1,6-DIPHENYLHEXATRIENE	25.79	0.026	12.54792
1,8-DIPHENYL-1,3,5,7-OCTATETRAENE	22.70	0.038	5.82459
BPSB (TRACE)	26.23	0.025	13.87021
BIS-MSB	26.99	0.023	16.46260
ESCULIN (A)	22.94	0.037	6.20552
AZULENE	21.16	0.047	3.82843
NAPHTHALENE	15.55	0.119	0.61143
DEUTERATED NAPHTHALENE-C10D8	15.75	0.114	0.65679
1-METHYLNAPHTHALENE 5ML/L	18.16	0.075	1.54655
2-METHYL NAPHTHALENE	16.35	0.102	0.82209
2,3-DIMETHYL NAPHTHALENE	17.13	0.089	1.08791
2,6-DIMETHYL NAPHTHALENE (TRACE)	16.88	0.093	0.99664
1-NAPHTHOL (TRACE)	20.25	0.054	2.96452
2-NAPHTHOL (TRACE)	17.05	0.090	1.05472
1-NAPHTHYLAMINE	20.60	0.051	3.26375
2-CHLORONAPHTHALENE	16.37	0.102	0.82815
ACENAPHTHENE	20.32	0.053	3.03700
1-PHENYL NAPHTHALENE	22.22	0.041	5.18295
2-PHENYL NAPHTHALENE	32.70	0.013	52.54502
1,4-DIPHENYL NAPHTHALENE	24.88	0.029	10.15840
1,5-DIPHENYL NAPHTHALENE	24.33	0.031	8.90317
1,7-DIPHENYL NAPHTHALENE	23.98	0.032	8.14746
1,4,5,8-TETRAPHENYL NAPHTHALENE	24.28	0.031	8.72841
1,1'-BINAPHTHYL	23.52	0.034	7.27497
4,4'-DI(N-BUTOXY)-1,1'-BINAPHTHYL	25.40	0.027	11.46583
2,2'-BINAPHTHYL	26.49	0.024	14.80306
ACRIDINE (A)	17.37	0.085	1.17078
ANTHRACENE (TRACE)	16.23	0.105	0.77821
DEUTERATED ANTHRACENE C14D10 (TRACE)	16.49	0.100	0.85768
9-METHYL ANTHRACENE (TRACE)	14.98	0.133	0.48204
1-AMINO ANTHRACENE	18.35	0.072	1.62776
2-AMINO ANTHRACENE	15.33	0.124	0.54720
9-VINYL ANTHRACENE	15.67	0.116	0.63147
9-PHENYL ANTHRACENE (TRACE)	16.19	0.105	0.76826
9,10-DIPHENYL ANTHRACENE	16.05	0.108	0.72980
9,10-DICHLOROANTHRACENE (TRACE)	14.49	0.147	0.39394
BIANTHRANYL (B)	17.48	0.084	1.21745
9,10-DINAPHTHYLANTHRACENE (B)	22.81	0.038	6.00177
9-(P-VINYLPHENYL)-10-PHENYL ANTHRACENE	19.15	0.064	2.10354
9,10-BIS(PHENYL-ETHYNYL)ANTHRACENE (TRACE)	26.80	0.023	15.80065
RUBICENE (B)	28.59	0.019	23.28344
DIBENZ(A,H)ANTHRACENE (B)	32.32	0.013	48.58677
TETRACENE (TRACE) (B)	26.32	0.025	14.17736
RUBRENE (B)	33.03	0.012	55.31801
TETRAHYDROCANNABINOL ACID METHYL ESTER	22.72	0.038	5.94101
PHENANTHRENE	20.06	0.055	2.78247
3,4-BENZOPHENANTHRENE	25.40	0.027	11.44631
2-PHENYL PHENANTHRENE	26.27	0.025	14.00905
CHRYSENE (TRACE)	24.68	0.030	9.40957
TRIPHENYLENE	21.19	0.047	3.86834
DEUTERATED TRIPHENYLENE-C18D12	20.99	0.048	3.65360
PYRENE	23.42	0.035	7.03719
1,3,6,8,-TETRAPHENYL PYRENE	28.36	0.020	22.16989
BIPYRENYL (B)	26.49	0.024	14.72587
O-PHENYLENE-PYRENE	27.96	0.020	20.34761
ANTHANTHRENE (B)	28.57	0.019	23.16559
FLUORANTHENE	22.69	0.038	5.82055
3-PHENYL FLUORANTHENE	25.93	0.026	12.96709
3-AMINO FLUORANTHENE	25.12	0.028	10.71569
3,3'-BIFLUORANTHENYL (B)	27.81	0.021	19.70927
BENZO(GHI)FLUORANTHENE	23.08	0.036	6.44299
9,12-DI(N-OCTADECOXY)-BENZ(E)ACEPHENANTHRYLENE	25.56	0.027	11.89405
DECACYCLENE (B)	27.52	0.021	18.53298
10-ETHOXY-BENZO(KL)XANTHENE	20.85	0.049	3.50390
PERYLENE	16.08	0.108	0.73564
BENZO(GHI)PERYLENE (B)	28.62	0.019	23.41217
BBOT (TRACE)	23.45	0.035	7.09117
QUININE SULFATE (A)	21.89	0.043	4.70027
ACRIDINE YELLOW (A)	21.34	0.046	4.02956
ACRIDINE ORANGE (A)	23.38	0.035	6.95416
ACRIDINE RED (A)	21.48	0.045	4.18340
URANINE (A)	21.60	0.044	4.32968
RHODAMINE B (A)	24.58	0.030	9.40696
RHODAMINE 6G (A)	23.99	0.032	8.11541
EOSIN B (A)	22.84	0.038	6.04406
EOSIN Y (A)	23.53	0.034	7.23854
ROSE BENGAL (A)	23.09	0.036	6.46112
INDENOINDENE	26.72	0.023	15.59949

8.50 Donor Molecule: 4-Methoxybiphenyl

ACCEPTOR MOLECULES	RO	CO	OVERLAP
POLYVINYLTOLUENE	2.44	30.680	0.00001
ANILINE 3ML/L	7.75	0.960	0.00646
N,N-DIMETHYLANILINE 3ML/L	13.85	0.168	0.20901
TRIPHENYLAMINE	24.37	0.031	6.13727
1,4-DIMETHOXY BENZENE	9.47	0.526	0.02149
1-METHYL-4-CHLOROBENZENE 5ML/L	0.00	0.000	0.00000
SODIUM SALICYLATE (A)	16.67	0.097	0.62719
ANTHRANILIC ACID (A)	23.39	0.035	4.75684
TMPD	20.58	0.051	2.21882
1,2,4-TRIMETHYLBENZENE (TRACE)	0.00	0.000	0.00000
MESITYLENE 5ML/L	0.00	0.000	0.00000
1,3,5-TRIETHYLBENZENE 5ML/L	0.00	0.000	0.00000
STYRENE	2.54	27.367	0.00001
4-METHOXY BIPHENYL	7.36	1.121	0.00475
4-BIPHENYLYL PHENYL ETHER	6.48	1.646	0.00220
4,4'-DIHYDROXYBIPHENYL (A)	12.74	0.217	0.12636
4,4'-DIMETHOXY BIPHENYL	11.28	0.312	0.06106
4,4'-DIISOBUTOXYBIPHENYL	10.94	0.342	0.05099
BENZIDINE	22.13	0.041	3.44783
4-VINYL BIPHENYL	15.19	0.128	0.36391
M-TERPHENYL	10.76	0.359	0.04607
M-QUATERPHENYL	6.09	1.978	0.00153
1,3,5-TRIPHENYL BENZENE	10.97	0.338	0.05162
1,3-DIBIPHENYLYLBENZENE (B)	22.33	0.040	3.64387
1,3,5,-TRIBIPHENYLYLBENZENE (B)	25.95	0.026	8.99465
FLUORENE (TRACE)	11.30	0.310	0.06198
1-METHYL FLUORENE	10.32	0.407	0.03585
1,2-BENZOFLUORENE	22.70	0.038	4.00507
9,10-DIHYDROPHENANTHRENE	11.69	0.280	0.07598
CARBAZOLE (TRACE)	20.50	0.052	2.18350
N-METHYL CARBAZOLE (TRACE)	21.58	0.045	2.95609
N-VINYLCARBAZOLE	21.87	0.043	3.20965
N-PHENYL CARBAZOLE	21.62	0.044	2.99371
N-METHYL-2,3-BENZCARBAZOLE	24.16	0.032	5.74643
DIBENZOFURAN	9.76	0.481	0.02578
DIBENZOTHIOPHENE (TRACE)	18.62	0.069	1.22791
P-TERPHENYL	16.45	0.100	0.58724
DEUTERATED P-TERPHENYL-C18D14	16.06	0.108	0.50858
4-METHYL TERPHENYL	18.05	0.076	1.02349
2-METHYL-P-TERPHENYL	10.59	0.376	0.04203
3,3''-DIMETHYL-P-TERPHENYL	17.59	0.082	0.87605
4-(3,3-DIMETHYLBUTOXY)-P-TERPHENYL	21.83	0.043	3.19076
4,4''-DIHEXAHYDROFARNESOXY-P-TERPHENYL	24.62	0.030	6.55989
2,2'-METHYLENE-P-TERPHENYL	24.09	0.032	5.76418
2,2'-ETHYLENE-P-TERPHENYL	24.37	0.031	6.16855
2,2'-METHYLENE-5',6'-METHYLENE-P-TERPHENYL (TRACE)	30.33	0.016	22.83477
3-PHENYL DIBENZOFURAN	24.49	0.030	6.35628
BENZO(1,2-B-4,5-B')BISBENZOFURAN (TRACE) (B)	31.85	0.014	30.58288
SEXIPHENYL BENZENE	7.13	1.232	0.00393
P-QUATERPHENYL	25.53	0.027	8.15074
3,3'''-DIMETHYL-P-QUATERPHENYL	26.61	0.024	10.42956
TETRAMETHYL-P-QUATERPHENYL	12.78	0.214	0.12947
3,3'''-DIMETHYL-3',2''-METHYLENE-P-QUATERPHENYL	32.10	0.014	32.05501
3,3'''-DIMETHYL-3',2''-ETHYLENE-P-QUATERPHENYL	31.75	0.014	29.91635
4-(3-ETHYLBUTOXY)-P-QUATERPHENYL (B)	29.99	0.017	21.29713
4,4'''-DI(2-BUTYLOCTOXY-1)-P-QUATERPHENYL	30.80	0.015	24.98822
DIETHYL-P-QUINQUEPHENYL	27.77	0.021	13.46139
DI(3-ETHYLHEPTYL)-P-QUINQUEPHENYL	31.94	0.014	30.99449
TETRAMETHYL-P-SEXIPHENYL	27.08	0.023	11.59889
TETRAMETHYL-P-OCTAPHENYL (B)	34.91	0.011	52.82332
INDOLE (TRACE)	0.00	0.000	0.00000
2-PHENYL INDOLE	27.08	0.023	11.60410
1-METHYL-2-PHENYL INDOLE	22.80	0.038	4.13183
1-NEOPENTYL-2-PHENYL INDOLE	21.19	0.047	2.66575
1-N-BUTYL-2(BIPHENYLYL-4) INDOLE	28.77	0.019	16.55185
1,2-DIPHENYL INDOLE	23.75	0.033	5.27139
1-BIPHENYLYL-2-PHENYL INDOLE	25.71	0.026	8.45995
1,3-DIMETHYL-2-PHENYL INDOLE	22.99	0.037	4.33452
2,3-DIPHENYL INDOLE	26.75	0.023	10.68876
1-METHYL-2,3-DIPHENYL INDOLE	24.47	0.031	6.27719
2-PHENYL-3,2'-METHYLENE INDOLE (TRACE)	28.82	0.019	16.81416
2-PHENYL-3,2'-ETHYLENE INDOLE (TRACE)	29.74	0.017	20.20788
2-PHENYL-3,2'-TRIMETHYLENE INDOLE	28.08	0.020	14.37175
1-METHYL-3,2'-METHYLENE-2-PHENYL INDOLE (TRACE)	29.69	0.017	20.06001
1-METHYL-3,2'-ETHYLENE-2-PHENYL INDOLE	29.62	0.017	19.72119
1-METHYL-3,2'-TRIMETHYLENE-2-PHENYL INDOLE	26.65	0.024	10.50492
4,5-DIPHENYLIMIDAZOLE (A)	18.26	0.073	1.09085
PPF (TRACE)	31.22	0.015	27.09609
PPO (TRACE)	28.65	0.019	16.22671
2-(4-FLUOROPHENYL)-5-PHENYLOXAZOLE (TRACE)	27.30	0.022	12.37304
αNPO (TRACE)	30.88	0.015	25.20669
BBO (B)	34.75	0.011	51.09028
POPOP	32.90	0.013	36.78830
DIMETHYL POPOP (TRACE)	33.29	0.012	39.40011
PPD (TRACE)	17.67	0.081	0.90188
2,5-DI(2-METHYLPHENYL)OXADIAZOLE	18.24	0.074	1.09184
PBD	28.36	0.020	15.29391
BBD (B)	33.32	0.012	40.04270
αNPD (TRACE)	28.57	0.019	15.85831
βNPD	26.98	0.023	11.31582
αNND (TRACE)	29.58	0.017	19.51576
βNND (TRACE)	30.68	0.015	24.44372
TRANS-1-(4-BIPHENYLYL)-2-PHENYL-ETHYLENE	32.62	0.013	35.13625

ACCEPTOR MOLECULES	RO	CO	OVERLAP
TRANS-1-(4-BIPHENYLYL)-2-(1-NAPHTHYL)-ETHYLENE	32.59	0.013	34.75239
DIPHENYLSTILBENE (B)	35.13	0.010	54.53818
1,1,4,4-TETRAPHENYLBUTADIENE	32.62	0.013	34.85003
1,6-DIPHENYLHEXATRIENE	35.44	0.010	57.41148
1,8-DIPHENYL-1,3,5,7-OCTATETRAENE	33.31	0.012	39.52334
BPSB (TRACE)	35.53	0.010	58.27929
BIS-MSB	35.06	0.010	53.80148
ESCULIN (A)	27.67	0.021	13.04122
AZULENE	21.19	0.047	2.61913
NAPHTHALENE	9.24	0.568	0.01836
DEUTERATED NAPHTHALENE-C10D8	10.92	0.344	0.04990
1-METHYLNAPHTHALENE 5ML/L	10.93	0.343	0.05028
2-METHYL NAPHTHALENE	11.99	0.260	0.08760
2,3-DIMETHYL NAPHTHALENE	11.92	0.264	0.08484
2,6-DIMETHYL NAPHTHALENE (TRACE)	14.79	0.138	0.30892
1-NAPHTHOL (TRACE)	17.05	0.090	0.72331
2-NAPHTHOL (TRACE)	18.07	0.076	1.02511
1-NAPHTHYLAMINE	22.50	0.039	3.79815
2-CHLORONAPHTHALENE	12.21	0.246	0.09758
ACENAPHTHENE	15.26	0.126	0.37383
1-PHENYL NAPHTHALENE	16.38	0.102	0.56982
2-PHENYL NAPHTHALENE	24.00	0.032	5.63045
1,4-DIPHENYL NAPHTHALENE	23.05	0.037	4.39993
1,5-DIPHENYL NAPHTHALENE	21.82	0.043	3.17568
1,7-DIPHENYL NAPHTHALENE	21.56	0.045	2.94740
1,4,5,8-TETRAPHENYL NAPHTHALENE	29.70	0.017	19.87327
1,1'-BINAPHTHYL	18.21	0.074	1.07331
4,4'-DI(N-BUTOXY)-1,1'-BINAPHTHYL	26.46	0.024	10.03794
2,2'-BINAPHTHYL	23.95	0.033	5.54320
ACRIDINE (A)	24.61	0.030	6.43862
ANTHRACENE (TRACE)	23.07	0.036	4.37100
DEUTERATED ANTHRACENE C14D10 (TRACE)	22.90	0.037	4.19145
9-METHYL ANTHRACENE (TRACE)	22.23	0.041	3.49798
1-AMINO ANTHRACENE	20.45	0.052	2.11731
2-AMINO ANTHRACENE	21.70	0.044	3.01161
9-VINYL ANTHRACENE	22.83	0.038	4.09764
9-PHENYL ANTHRACENE (TRACE)	23.43	0.035	4.79131
9,10-DIPHENYL ANTHRACENE	23.18	0.036	4.48765
9,10-DICHLOROANTHRACENE (TRACE)	21.96	0.042	3.24343
BIANTHRANYL (B)	24.78	0.029	6.69695
9,10-DINAPHTHYLANTHRACENE (B)	23.33	0.035	4.65623
9-(P-VINYLPHENYL)-10-PHENYL ANTHRACENE	23.29	0.035	4.61756
9,10-BIS(PHENYL-ETHYNYL)ANTHRACENE (TRACE)	23.96	0.033	5.46833
RUBICENE (B)	26.64	0.024	10.32786
DIBENZ(A,H)ANTHRACENE (B)	28.95	0.018	17.03358
TETRACENE (TRACE) (B)	18.24	0.074	1.06582
RUBRENE (B)	28.92	0.018	16.93051
TETRAHYDROCANNABINOL ACID METHYL ESTER	14.26	0.154	0.24959
PHENANTHRENE	13.43	0.184	0.17224
3,4-BENZOPHENANTHRENE	22.46	0.039	3.72961
2-PHENYL PHENANTHRENE	18.96	0.066	1.35554
CHRYSENE (TRACE)	23.22	0.036	4.56112
TRIPHENYLENE	14.85	0.137	0.31399
DEUTERATED TRIPHENYLENE-C18D12	14.77	0.139	0.30406
PYRENE	29.02	0.018	17.35523
1,3,6,8,-TETRAPHENYL PYRENE	28.72	0.019	16.23647
BIPYRENYL (B)	32.92	0.013	36.78347
O-PHENYLENE-PYRENE	27.96	0.020	13.81899
ANTHANTHRENE (B)	27.10	0.022	11.44441
FLUORANTHENE	24.97	0.029	7.00447
3-PHENYL FLUORANTHENE	26.43	0.024	9.85304
3-AMINO FLUORANTHENE	25.23	0.028	7.45735
3,3'-BIFLUORANTHENYL (B)	29.37	0.018	18.56259
BENZO(GHI)FLUORANTHENE	24.39	0.031	6.08045
9,12-DI(N-OCTADECOXY)-BENZ(E)ACEPHENANTHRYLENE	26.29	0.025	9.54503
DECACYCLENE (B)	32.01	0.014	31.10976
10-ETHOXY-BENZO(KL)XANTHENE	23.85	0.033	5.31874
PERYLENE	20.12	0.055	1.91579
BENZO(GHI)PERYLENE (B)	26.31	0.025	9.58640
BBOT (TRACE)	32.95	0.013	36.99525
QUININE SULFATE (A)	25.28	0.028	7.64791
ACRIDINE YELLOW (A)	18.29	0.073	1.08189
ACRIDINE ORANGE (A)	18.85	0.067	1.29503
ACRIDINE RED (A)	21.78	0.043	3.08944
URANINE (A)	22.24	0.041	3.49408
RHODAMINE B (A)	24.64	0.030	6.46699
RHODAMINE 6G (A)	24.63	0.030	6.45335
EOSIN B (A)	24.09	0.032	5.65313
EOSIN Y (A)	24.56	0.030	6.34955
ROSE BENGAL (A)	25.28	0.028	7.54208
INDENOINDENE	27.96	0.020	14.01792

8.51 Donor Molecule: 4-Biphenylyl Phenyl Ether

POLYVINYLTOLUENE	2.42	31.668	0.00001
ANILINE 3ML/L	7.25	1.174	0.00937
N,N-DIMETHYLANILINE 3ML/L	12.47	0.231	0.24090
TRIPHENYLAMINE	21.70	0.044	6.63117
1,4-DIMETHOXY BENZENE	8.81	0.655	0.03008
1-METHYL-4-CHLOROBENZENE 5ML/L	0.00	0.000	0.00000
SODIUM SALICYLATE (A)	14.87	0.136	0.68698

ACCEPTOR MOLECULES	RO	CO	OVERLAP
ANTHRANILIC ACID (A)	20.52	0.052	4.70578
TMPD	18.12	0.075	2.23997
1,2,4-TRIMETHYLBENZENE (TRACE)	0.00	0.000	0.00000
MESITYLENE 5ML/L	0.00	0.000	0.00000
1,3,5-TRIETHYLBENZENE 5ML/L	0.00	0.000	0.00000
STYRENE	2.71	22.409	0.00003
3,3'-DIMETHYL BIPHENYL	3.46	10.799	0.00011
2,2'-DIMETHYLBIPHENYL 2ML/L	0.00	0.000	0.00000
OCTAMETHYL BIPHENYL	0.00	0.000	0.00000
4-METHOXY BIPHENYL	6.95	1.332	0.00729
4-BIPHENYLYL PHENYL ETHER	6.28	1.810	0.00395
4,4'-DIHYDROXYBIPHENYL (A)	11.51	0.293	0.14942
4,4'-DIMETHOXY BIPHENYL	10.27	0.413	0.07552
4,4'-DIISOBUTOXYBIPHENYL	10.02	0.445	0.06529
BENZIDINE	19.75	0.058	3.78178
4-VINYL BIPHENYL	13.79	0.171	0.44256
M-TERPHENYL	9.74	0.484	0.05509
M-QUATERPHENYL	5.87	2.213	0.00265
1,3,5-TRIPHENYL BENZENE	9.86	0.467	0.05879
1,3-DIBIPHENYLYLBENZENE (B)	20.01	0.056	4.10323
1,3,5,-TRIBIPHENYLYLBENZENE (B)	23.23	0.036	10.03567
FLUORENE (TRACE)	10.52	0.384	0.08741
1-METHYL FLUORENE	9.70	0.490	0.05368
1,2-BENZOFLUORENE	20.20	0.054	4.32311
9,10-DIHYDROPHENANTHRENE	10.75	0.360	0.09970
CARBAZOLE (TRACE)	18.16	0.075	2.28239
N-METHYL CARBAZOLE (TRACE)	18.97	0.066	2.95810
N-VINYLCARBAZOLE	19.25	0.063	3.23574
N-PHENYL CARBAZOLE	19.03	0.065	3.02251
N-METHYL-2,3-BENZCARBAZOLE	21.30	0.046	5.86265
DIBENZOFURAN	9.27	0.561	0.04104
DIBENZOTHIOPHENE (TRACE)	16.50	0.100	1.28819
P-TERPHENYL	14.87	0.136	0.69523
DEUTERATED P-TERPHENYL-C18D14	14.54	0.146	0.60652
4-METHYL TERPHENYL	16.27	0.104	1.19085
2-METHYL-P-TERPHENYL	9.74	0.485	0.05492
3,3''-DIMETHYL-P-TERPHENYL	15.85	0.112	1.01842
4-(3,3-DIMETHYLBUTOXY)-P-TERPHENYL	19.54	0.060	3.55548
4,4''-DIHEXAHYDROFARNESOXY-P-TERPHENYL	21.98	0.042	7.19534
2,2' -METHYLENE-P-TERPHENYL	21.52	0.045	6.35191
2,2' -ETHYLENE-P-TERPHENYL	21.74	0.044	6.74021
2,2'-METHYLENE-5',6''-METHYLENE-P-TERPHENYL (TRACE)	26.75	0.023	23.30958
3-PHENYL DIBENZOFURAN	21.89	0.043	7.04811
BENZO(1,2-B-4,5-B')BISBENZOFURAN (TRACE) (B)	28.08	0.020	31.19112
SEXIPHENYL BENZENE	6.74	1.463	0.00605
P-QUATERPHENYL	22.76	0.038	8.85884
3,3'''-DIMETHYL-P-QUATERPHENYL	23.69	0.034	11.25254
TETRAMETHYL-P-QUATERPHENYL	11.68	0.280	0.16401
3,3''' -DIMETHYL-3',2''-METHYLENE-P-QUATERPHENYL	28.36	0.020	33.07272
3,3''' -DIMETHYL-3',2''-ETHYLENE-P-QUATERPHENYL	28.06	0.020	30.92583
4-(3-ETHYLBUTOXY)-P-QUATERPHENYL (B)	26.58	0.024	22.35830
4,4'''-DI(2-BUTYLOCTOXY-1)-P-QUATERPHENYL	27.30	0.022	26.28496
DIETHYL-P-QUINQUEPHENYL	24.72	0.030	14.52529
DI(3-ETHYLHEPTYL)-P-QUINQUEPHENYL	28.30	0.020	32.50251
TETRAMETHYL-P-SEXIPHENYL	24.15	0.032	12.63787
TETRAMETHYL-P-OCTAPHENYL (B)	30.84	0.015	54.39547
INDOLE (TRACE)	0.00	0.000	0.00000
2-PHENYL INDOLE	24.05	0.032	12.32805
1-METHYL-2-PHENYL INDOLE	20.33	0.053	4.49718
1-NEOPENTYL-2-PHENYL INDOLE	18.95	0.066	2.96186
1-N-BUTYL-2(BIPHENYLYL-4) INDOLE	25.42	0.027	17.11265
1,2-DIPHENYL INDOLE	21.17	0.047	5.73633
1-BIPHENYLYL-2-PHENYL INDOLE	22.87	0.037	9.09205
1,3-DIMETHYL-2-PHENYL INDOLE	20.43	0.052	4.64151
2,3-DIPHENYL INDOLE	23.66	0.034	11.09142
1-METHYL-2,3-DIPHENYL INDOLE	21.66	0.044	6.54869
2-PHENYL-3,2'-METHYLENE INDOLE (TRACE)	25.49	0.027	17.43829
2-PHENYL-3,2'-ETHYLENE INDOLE (TRACE)	26.19	0.025	20.47905
2-PHENYL-3,2' -TRIMETHYLENE INDOLE	24.87	0.029	15.05095
1-METHYL-3,2'-METHYLENE-2-PHENYL INDOLE (TRACE)	26.18	0.025	20.45803
1-METHYL-3,2'-ETHYLENE-2-PHENYL INDOLE	26.06	0.025	19.84219
1-METHYL-3,2'-TRIMETHYLENE-2-PHENYL INDOLE	23.59	0.034	10.96496
4,5-DIPHENYLIMIDAZOLE (A)	16.30	0.103	1.19771
PPF (TRACE)	27.53	0.021	27.66029
PPO (TRACE)	25.40	0.027	17.12745
2-(4-FLUOROPHENYL)-5-PHENYLOXAZOLE (TRACE)	24.29	0.031	13.08258
αNPO (TRACE)	27.10	0.022	24.99529
BBO (B)	30.44	0.016	50.10050
POPOP	28.74	0.019	35.44030
DIMETHYL POPOP (TRACE)	29.06	0.018	37.76619
PPD (TRACE)	15.96	0.110	1.06322
2,5-DI(2-METHYLPHENYL)OXADIAZOLE	16.44	0.101	1.26744
PBD	25.22	0.028	16.39450
BBD (B)	29.41	0.018	41.09067
αNPD (TRACE)	25.19	0.028	16.17945
βNPD	23.96	0.033	12.05504
αNND (TRACE)	25.99	0.025	19.48995
βNND (TRACE)	27.10	0.022	25.18609
TRANS-1-(4-BIPHENYLYL)-2-PHENYL-ETHYLENE	28.76	0.019	35.84002
TRANS-1-(4-BIPHENYLYL)-2-(1-NAPHTHYL)-ETHYLENE	28.60	0.019	34.42514
DIPHENYLSTILBENE (B)	30.78	0.015	53.50808
1,1,4,4-TETRAPHENYLBUTADIENE	28.55	0.019	33.96474
1,6-DIPHENYLHEXATRIENE	30.96	0.015	55.32956
1,8-DIPHENYL-1,3,5,7-OCTATETRAENE	28.96	0.018	37.03020
BPSB (TRACE)	31.03	0.015	56.11687
BIS-MSB	30.68	0.015	52.35057
ESCULIN (A)	24.28	0.031	12.89972

ACCEPTOR MOLECULES	RO	CO	OVERLAP
AZULENE	18.62	0.069	2.61622
NAPHTHALENE	8.30	0.783	0.02098
DEUTERATED NAPHTHALENE-C10D8	9.77	0.480	0.05541
1-METHYLNAPHTHALENE 5ML/L	9.85	0.469	0.05840
2-METHYL NAPHTHALENE	10.70	0.365	0.09636
2,3-DIMETHYL NAPHTHALENE	10.66	0.369	0.09408
2,6-DIMETHYL NAPHTHALENE (TRACE)	13.12	0.198	0.32655
1-NAPHTHOL (TRACE)	15.21	0.127	0.79314
2-NAPHTHOL (TRACE)	15.98	0.110	1.06481
1-NAPHTHYLAMINE	19.88	0.057	3.91590
2-CHLORONAPHTHALENE	10.87	0.348	0.10560
ACENAPHTHENE	13.69	0.174	0.42142
1-PHENYL NAPHTHALENE	14.72	0.140	0.65010
2-PHENYL NAPHTHALENE	21.55	0.045	6.39151
1,4-DIPHENYL NAPHTHALENE	20.50	0.052	4.73020
1,5-DIPHENYL NAPHTHALENE	19.44	0.061	3.44696
1,7-DIPHENYL NAPHTHALENE	19.21	0.063	3.19958
1,4,5,8-TETRAPHENYL NAPHTHALENE	26.06	0.025	19.68527
1,1'-BINAPHTHYL	16.29	0.104	1.19381
4,4' -DI(N-BUTOXY)-1,1' -BINAPHTHYL	23.42	0.035	10.46621
2,2'-BINAPHTHYL	21.35	0.046	6.03377
ACRIDINE (A)	21.47	0.045	6.16094
ANTHRACENE (TRACE)	20.13	0.055	4.18374
DEUTERATED ANTHRACENE C14D10 (TRACE)	19.99	0.056	4.02496
9-METHYL ANTHRACENE (TRACE)	19.36	0.062	3.30437
1-AMINO ANTHRACENE	17.79	0.079	1.99151
2-AMINO ANTHRACENE	18.98	0.065	2.92355
9-VINYL ANTHRACENE	19.87	0.057	3.85493
9-PHENYL ANTHRACENE (TRACE)	20.40	0.053	4.52752
9,10-DIPHENYL ANTHRACENE	20.13	0.055	4.16813
9,10-DICHLOROANTHRACENE (TRACE)	19.05	0.065	2.99795
BIANTHRANYL (B)	21.56	0.045	6.29652
9,10-DINAPHTHYLANTHRACENE (B)	20.35	0.053	4.45221
9-(P-VINYLPHENYL)-10-PHENYL ANTHRACENE	20.24	0.054	4.31035
9,10-BIS(PHENYL-ETHYNYL)ANTHRACENE (TRACE)	21.30	0.046	5.85071
RUBICENE (B)	23.55	0.034	10.69361
DIBENZ(A,H)ANTHRACENE (B)	25.61	0.027	17.69590
TETRACENE (TRACE) (B)	16.15	0.106	1.11457
RUBRENE (B)	25.83	0.026	18.64630
TETRAHYDROCANNABINOL ACID METHYL ESTER	12.94	0.207	0.30136
PHENANTHRENE	11.86	0.268	0.17666
3,4-BENZOPHENANTHRENE	19.97	0.056	3.99303
2-PHENYL PHENANTHRENE	17.02	0.091	1.53696
CHRYSENE (TRACE)	20.65	0.051	4.90650
TRIPHENYLENE	13.19	0.195	0.33401
DEUTERATED TRIPHENYLENE-C18D12	13.12	0.198	0.32450
PYRENE	25.56	0.027	17.59216
1,3,6,8,-TETRAPHENYL PYRENE	24.97	0.029	15.21870
BIPYRENYL (B)	28.79	0.019	35.70264
O-PHENYLENE-PYRENE	24.58	0.030	13.84838
ANTHANTHRENE (B)	23.99	0.032	11.96572
FLUORANTHENE	21.86	0.043	6.84295
3-PHENYL FLUORANTHENE	23.10	0.036	9.52459
3-AMINO FLUORANTHENE	22.21	0.041	7.52828
3,3' -BIFLUORANTHENYL (B)	25.69	0.026	18.03639
BENZO(GHI)FLUORANTHENE	21.39	0.046	6.00232
9,12-DI(N-OCTADECOXY)-BENZ(E)ACEPHENANTHRYLENE	23.01	0.037	9.32072
DECACYCLENE (B)	28.00	0.020	30.24628
10-ETHOXY-BENZO(KL)XANTHENE	20.75	0.050	5.01390
PERYLENE	17.42	0.085	1.75150
BENZO(GHI)PERYLENE (B)	23.08	0.036	9.49084
BBOT (TRACE)	28.70	0.019	35.02714
QUININE SULFATE (A)	22.26	0.041	7.72634
ACRIDINE YELLOW (A)	16.00	0.109	1.05330
ACRIDINE ORANGE (A)	16.72	0.096	1.37169
ACRIDINE RED (A)	19.16	0.064	3.10332
URANINE (A)	19.59	0.060	3.54423
RHODAMINE B (A)	21.67	0.044	6.49046
RHODAMINE 6G (A)	21.57	0.045	6.32546
EOSIN B (A)	21.27	0.047	5.80159
EOSIN Y (A)	21.69	0.044	6.53088
ROSE BENGAL (A)	22.26	0.041	7.64068
INDENOINDENE	24.77	0.029	14.71825

8.52 Donor Molecule: 4,4'-Dihydroxybiphenyl (A)

N,N-DIMETHYLANILINE 3ML/L	0.00	0.000	0.00045
TRIPHENYLAMINE	0.00	0.000	0.65042
SODIUM SALICYLATE (A)	0.00	0.000	0.07886
ANTHRANILIC ACID (A)	0.00	0.000	3.52855
TMPD	0.00	0.000	1.03978
4,4'-DIHYDROXYBIPHENYL (A)	0.00	0.000	0.00039
4,4'-DIMETHOXY BIPHENYL	0.00	0.000	0.00000
4,4'-DIISOBUTOXYBIPHENYL	0.00	0.000	0.00000
BENZIDINE	0.00	0.000	0.20200
M-QUATERPHENYL	0.00	0.000	0.00000
1,3,5-TRIPHENYL BENZENE	0.00	0.000	0.00046
1,3-DIBIPHENYLYLBENZENE (B)	0.00	0.000	0.09901
1,3,5,-TRIBIPHENYLYLBENZENE (B)	0.00	0.000	0.26967
FLUORENE (TRACE)	0.00	0.000	0.00000
1-METHYL FLUORENE	0.00	0.000	0.00000

ACCEPTOR MOLECULES	RO	CO	OVERLAP
1,2-BENZOFLUORENE	0.00	0.000	0.45028
9,10-DIHYDROPHENANTHRENE	0.00	0.000	0.00000
CARBAZOLE (TRACE)	0.00	0.000	0.27925
N-METHYL CARBAZOLE (TRACE)	0.00	0.000	1.26437
N-VINYLCARBAZOLE	0.00	0.000	1.05189
N-PHENYL CARBAZOLE	0.00	0.000	1.00544
N-METHYL-2,3-BENZCARBAZOLE	0.00	0.000	4.11656
DIBENZOFURAN	0.00	0.000	0.00000
DIBENZOTHIOPHENE (TRACE)	0.00	0.000	0.08950
P-TERPHENYL	0.00	0.000	0.00114
DEUTERATED P-TERPHENYL-C18D14	0.00	0.000	0.00000
4-METHYL TERPHENYL	0.00	0.000	0.00376
2-METHYL-P-TERPHENYL	0.00	0.000	0.00000
3,3''-DIMETHYL-P-TERPHENYL	0.00	0.000	0.00519
4-(3,3-DIMETHYLBUTOXY)-P-TERPHENYL	0.00	0.000	0.08732
4,4''-DIHEXAHYDROFARNESOXY-P-TERPHENYL	0.00	0.000	0.27833
2,2'-METHYLENE-P-TERPHENYL	0.00	0.000	0.11341
2,2'-ETHYLENE-P-TERPHENYL	0.00	0.000	0.20350
2,2'-METHYLENE-5',6''-METHYLENE-P-TERPHENYL (TRACE)	0.00	0.000	5.23108
3-PHENYL DIBENZOFURAN	0.00	0.000	0.11196
BENZO(1,2-B-4,5-B')BISBENZOFURAN (TRACE) (B)	0.00	0.000	7.66542
SEXIPHENYL BENZENE	0.00	0.000	0.00000
P-QUATERPHENYL	0.00	0.000	0.49668
3,3'''-DIMETHYL-P-QUATERPHENYL	0.00	0.000	0.83511
TETRAMETHYL-P-QUATERPHENYL	0.00	0.000	0.00000
3,3''' -DIMETHYL-3',2''-METHYLENE-P-QUATERPHENYL	0.00	0.000	6.79866
3,3''' -DIMETHYL-3',2''-ETHYLENE-P-QUATERPHENYL	0.00	0.000	6.92563
4-(3-ETHYLBUTOXY)-P-QUATERPHENYL (B)	0.00	0.000	3.70018
4,4'''-DI(2-BUTYLOCTOXY-1)-P-QUATERPHENYL	0.00	0.000	4.09664
DIETHYL-P-QUINQUEPHENYL	0.00	0.000	1.06030
DI(3-ETHYLHEPTYL)-P-QUINQUEPHENYL	0.00	0.000	5.69265
TETRAMETHYL-P-SEXIPHENYL	0.00	0.000	0.66329
TETRAMETHYL-P-OCTAPHENYL (B)	0.00	0.000	15.67602
INDOLE (TRACE)	0.00	0.000	0.00000
2-PHENYL INDOLE	0.00	0.000	1.00558
1-METHYL-2-PHENYL INDOLE	0.00	0.000	0.22359
1-NEOPENTYL-2-PHENYL INDOLE	0.00	0.000	0.06584
1-N-BUTYL-2(BIPHENYLYL-4) INDOLE	0.00	0.000	4.57249
1,2-DIPHENYL INDOLE	0.00	0.000	0.26910
1-BIPHENYLYL-2-PHENYL INDOLE	0.00	0.000	0.72592
1,3-DIMETHYL-2-PHENYL INDOLE	0.00	0.000	0.38047
2,3-DIPHENYL INDOLE	0.00	0.000	2.78119
1-METHYL-2,3-DIPHENYL INDOLE	0.00	0.000	1.48733
2-PHENYL-3,2'-METHYLENE INDOLE (TRACE)	0.00	0.000	2.83396
2-PHENYL-3,2'-ETHYLENE INDOLE (TRACE)	0.00	0.000	7.67329
2-PHENYL-3,2'-TRIMETHYLENE INDOLE	0.00	0.000	1.92648
1-METHYL-3,2'-METHYLENE-2-PHENYL INDOLE (TRACE)	0.00	0.000	6.24967
1-METHYL-3,2'-ETHYLENE-2-PHENYL INDOLE	0.00	0.000	10.34504
1-METHYL-3,2'-TRIMETHYLENE-2-PHENYL INDOLE	0.00	0.000	1.80712
4,5-DIPHENYLIMIDAZOLE (A)	0.00	0.000	0.07042
PPF (TRACE)	0.00	0.000	8.25748
PPO (TRACE)	0.00	0.000	1.94300
2-(4-FLUOROPHENYL)-5-PHENYLOXAZOLE (TRACE)	0.00	0.000	1.37438
αNPO (TRACE)	0.00	0.000	19.18376
BBO (B)	0.00	0.000	50.65139
POPOP	0.00	0.000	52.38412
DIMETHYL POPOP (TRACE)	0.00	0.000	64.65835
PPD (TRACE)	0.00	0.000	0.00000
2,5-DI(2-METHYLPHENYL)OXADIAZOLE	0.00	0.000	0.00097
PBD	0.00	0.000	1.13379
BBD (B)	0.00	0.000	11.93600
αNPD (TRACE)	0.00	0.000	5.41539
βNPD	0.00	0.000	1.06211
αNND (TRACE)	0.00	0.000	11.91611
βNND (TRACE)	0.00	0.000	5.89041
TRANS-1-(4-BIPHENYLYL)-2-PHENYL-ETHYLENE	0.00	0.000	12.53131
TRANS-1-(4-BIPHENYLYL)-2-(1-NAPHTHYL)-ETHYLENE	0.00	0.000	28.55618
DIPHENYLSTILBENE (B)	0.00	0.000	53.09844
1,1,4,4-TETRAPHENYLBUTADIENE	0.00	0.000	39.14100
1,6-DIPHENYLHEXATRIENE	0.00	0.000	80.18364
1,8-DIPHENYL-1,3,5,7-OCTATETRAENE	0.00	0.000	94.89543
BPSB (TRACE)	0.00	0.000	84.51379
BIS-MSB	0.00	0.000	65.07148
ESCULIN (A)	0.00	0.000	10.49588
AZULENE	0.00	0.000	1.20868
NAPHTHALENE	0.00	0.000	0.00004
DEUTERATED NAPHTHALENE-C10D8	0.00	0.000	0.00153
1-METHYLNAPHTHALENE 5ML/L	0.00	0.000	0.00028
2-METHYL NAPHTHALENE	0.00	0.000	0.00121
2,3-DIMETHYL NAPHTHALENE	0.00	0.000	0.00123
2,6-DIMETHYL NAPHTHALENE (TRACE)	0.00	0.000	0.01626
1-NAPHTHOL (TRACE)	0.00	0.000	0.01903
2-NAPHTHOL (TRACE)	0.00	0.000	0.11482
1-NAPHTHYLAMINE	0.00	0.000	0.88349
2-CHLORONAPHTHALENE	0.00	0.000	0.00300
ACENAPHTHENE	0.00	0.000	0.00608
1-PHENYL NAPHTHALENE	0.00	0.000	0.00740
2-PHENYL NAPHTHALENE	0.00	0.000	0.21707
1,4-DIPHENYL NAPHTHALENE	0.00	0.000	0.33820
1,5-DIPHENYL NAPHTHALENE	0.00	0.000	0.16863
1,7-DIPHENYL NAPHTHALENE	0.00	0.000	0.17109
1,4,5,8-TETRAPHENYL NAPHTHALENE	0.00	0.000	15.85205
1,1'-BINAPHTHYL	0.00	0.000	0.03364
4,4'-DI(N-BUTOXY)-1,1'-BINAPHTHYL	0.00	0.000	1.84138
2,2'-BINAPHTHYL	0.00	0.000	0.28625
ACRIDINE (A)	0.00	0.000	9.57788
ANTHRACENE (TRACE)	0.00	0.000	6.76449

ACCEPTOR MOLECULES	RO	CO	OVERLAP
DEUTERATED ANTHRACENE C14D10 (TRACE)	0.00	0.000	6.28199
9-METHYL ANTHRACENE (TRACE)	0.00	0.000	7.27182
1-AMINO ANTHRACENE	0.00	0.000	5.53025
2-AMINO ANTHRACENE	0.00	0.000	4.49586
9-VINYL ANTHRACENE	0.00	0.000	9.16829
9-PHENYL ANTHRACENE (TRACE)	0.00	0.000	9.77927
9,10-DIPHENYL ANTHRACENE	0.00	0.000	12.13221
9,10-DICHLOROANTHRACENE (TRACE)	0.00	0.000	9.37014
BIANTHRANYL (B)	0.00	0.000	16.15125
9,10-DINAPHTHYLANTHRACENE (B)	0.00	0.000	10.74140
9-(P-VINYLPHENYL)-10-PHENYL ANTHRACENE	0.00	0.000	12.47712
9,10-BIS(PHENYL-ETHYNYL)ANTHRACENE (TRACE)	0.00	0.000	6.99355
RUBICENE (B)	0.00	0.000	8.76343
DIBENZ(A,H)ANTHRACENE (B)	0.00	0.000	8.39627
TETRACENE (TRACE) (B)	0.00	0.000	1.63872
RUBRENE (B)	0.00	0.000	2.28920
TETRAHYDROCANNABINOL ACID METHYL ESTER	0.00	0.000	0.00018
PHENANTHRENE	0.00	0.000	0.06006
3,4-BENZOPHENANTHRENE	0.00	0.000	0.46210
2-PHENYL PHENANTHRENE	0.00	0.000	0.16200
CHRYSENE (TRACE)	0.00	0.000	0.49583
TRIPHENYLENE	0.00	0.000	0.06290
DEUTERATED TRIPHENYLENE-C18D12	0.00	0.000	0.05425
PYRENE	0.00	0.000	5.05107
1,3,6,8,-TETRAPHENYL PYRENE	0.00	0.000	44.38346
BIPYRENYL (B)	0.00	0.000	45.12543
O-PHENYLENE-PYRENE	0.00	0.000	19.63795
ANTHANTHRENE (B)	0.00	0.000	11.97942
FLUORANTHENE	0.00	0.000	7.64134
3-PHENYL FLUORANTHENE	0.00	0.000	16.22250
3-AMINO FLUORANTHENE	0.00	0.000	9.17547
3,3'-BIFLUORANTHENYL (B)	0.00	0.000	33.11330
BENZO(GHI)FLUORANTHENE	0.00	0.000	5.03364
9,12-DI(N-OCTADECOXY)-BENZ(E)ACEPHENANTHRYLENE	0.00	0.000	14.62989
DECACYCLENE (B)	0.00	0.000	51.07751
10-ETHOXY-BENZO(KL)XANTHENE	0.00	0.000	13.55788
PERYLENE	0.00	0.000	8.94179
BENZO(GHI)PERYLENE (B)	0.00	0.000	18.96830
BBOT (TRACE)	0.00	0.000	74.65239
QUININE SULFATE (A)	0.00	0.000	2.53497
ACRIDINE YELLOW (A)	0.00	0.000	3.75132
ACRIDINE ORANGE (A)	0.00	0.000	1.49591
ACRIDINE RED (A)	0.00	0.000	3.34227
URANINE (A)	0.00	0.000	2.80699
RHODAMINE B (A)	0.00	0.000	7.34218
RHODAMINE 6G (A)	0.00	0.000	7.34818
EOSIN B (A)	0.00	0.000	3.18181
EOSIN Y (A)	0.00	0.000	3.71295
ROSE BENGAL (A)	0.00	0.000	4.60411
INDENOINDENE	0.00	0.000	1.69451

8.53 Donor Molecule: 4,4′-Dimethoyxbiphenyl

N,N-DIMETHYLANILINE 3ML/L	8.09	0.846	0.01018
TRIPHENYLAMINE	18.89	0.066	1.64242
1,4-DIMETHOXY BENZENE	2.95	17.496	0.00002
1-METHYL 4 CHLOROBENZENE 5ML/L	0.00	0.000	0.00000
SODIUM SALICYLATE (A)	12.86	0.210	0.16347
ANTHRANILIC ACID (A)	22.41	0.040	4.54649
TMPD	18.82	0.067	1.59803
4,4'-DIHYDROXYBIPHENYL (A)	7.61	1.017	0.00706
4,4'-DIMETHOXY BIPHENYL	5.78	2.317	0.00136
4,4'-DIISOBUTOXYBIPHENYL	4.62	4.544	0.00035
BENZIDINE	16.28	0.104	0.67447
4-VINYL BIPHENYL	7.82	0.937	0.00834
M-TERPHENYL	5.36	2.913	0.00086
1,3,5-TRIPHENYL BENZENE	7.00	1.304	0.00428
1,3-DIBIPHENYLYLBENZENE (B)	15.23	0.127	0.45193
1,3,5,-TRIBIPHENYLYLBENZENE (B)	17.99	0.077	1.22941
FLUORENE (TRACE)	0.00	0.000	0.00000
1-METHYL FLUORENE	3.28	12.700	0.00005
1,2-BENZOFLUORENE	17.60	0.082	1.07337
9,10-DIHYDROPHENANTHRENE	4.49	4.951	0.00030
CARBAZOLE (TRACE)	16.60	0.098	0.75851
N-METHYL CARBAZOLE (TRACE)	19.88	0.057	2.22797
N-VINYLCARBAZOLE	19.67	0.059	2.09462
N-PHENYL CARBAZOLE	19.46	0.061	1.96061
N-METHYL-2,3-BENZCARBAZOLE	22.05	0.042	4.09456
DIBENZOFURAN	0.00	0.000	0.00000
DIBENZOTHIOPHENE (TRACE)	14.48	0.147	0.33466
P-TERPHENYL	9.48	0.526	0.02642
DEUTERATED P-TERPHENYL-C18D14	8.82	0.652	0.01719
4-METHYL TERPHENYL	10.84	0.351	0.05915
2-METHYL-P-TERPHENYL	4.34	5.465	0.00025
3,3''-DIMETHYL-P-TERPHENYL	10.79	0.356	0.05741
4-(3,3-DIMETHYLBUTOXY)-P-TERPHENYL	15.10	0.130	0.43124
4,4''-DIHEXAHYDROFARNESOXY-P-TERPHENYL	17.75	0.080	1.13342
2,2'-METHYLENE-P-TERPHENYL	16.50	0.100	0.73350
2,2'-ETHYLENE-P-TERPHENYL	17.37	0.085	0.99674
2,2'-METHYLENE-5',6''-METHYLENE-P-TERPHENYL (TRACE)	26.20	0.025	11.71473

ACCEPTOR MOLECULES	RO	CO	OVERLAP
3-PHENYL DIBENZOFURAN	16.53	0.099	0.74233
BENZO(1,2-B-4,5-B')BISBENZOFURAN (TRACE) (B)	27.68	0.021	16.24555
SEXIPHENYL BENZENE	0.00	0.000	0.00000
P-QUATERPHENYL	18.99	0.065	1.70157
3,3'''-DIMETHYL-P-QUATERPHENYL	20.29	0.054	2.52242
TETRAMETHYL-P-QUATERPHENYL	5.80	2.296	0.00139
3,3''' -DIMETHYL-3',2''-METHYLENE-P-QUATERPHENYL	27.28	0.022	14.89195
3,3''' -DIMETHYL-3',2''-ETHYLENE-P-QUATERPHENYL	27.07	[illegible]	[illegible]
4-(3-ETHYLBUTOXY)-P-QUATERPHENYL (B)	24.70	0.030	8.19282
4,4'''-DI(2-BUTYLOCTOXY-1)-P-QUATERPHENYL	25.22	0.028	9.27633
DIETHYL-P-QUINQUEPHENYL	21.15	0.047	3.24381
DI(3-ETHYLHEPTYL)-P-QUINQUEPHENYL	26.42	0.024	12.22579
TETRAMETHYL-P-SEXIPHENYL	20.02	0.056	2.33114
TETRAMETHYL-P-OCTAPHENYL (B)	30.34	0.016	28.03506
INDOLE (TRACE)	0.00	0.000	0.00000
2-PHENYL INDOLE	21.04	0.048	3.14423
1-METHYL-2-PHENYL INDOLE	16.79	0.094	0.81204
1-NEOPENTYL-2-PHENYL INDOLE	14.56	0.144	0.34844
1-N-BUTYL-2(BIPHENYLYL-4) INDOLE	24.81	0.029	8.39011
1,2-DIPHENYL INDOLE	17.41	0.085	1.00912
1-BIPHENYLYL-2-PHENYL INDOLE	19.67	0.059	2.09303
1,3-DIMETHYL-2-PHENYL INDOLE	17.77	0.080	1.14175
2,3-DIPHENYL INDOLE	22.89	0.037	5.17030
1-METHYL-2,3-DIPHENYL INDOLE	20.73	0.050	2.85745
2-PHENYL-3,2'-METHYLENE INDOLE (TRACE)	23.99	0.032	6.89712
2-PHENYL-3,2'-ETHYLENE INDOLE (TRACE)	26.74	0.023	13.18681
2-PHENYL-3,2'-TRIMETHYLENE INDOLE	22.77	0.038	5.03993
1-METHYL-3,2'-METHYLENE-2-PHENYL INDOLE (TRACE)	26.22	0.025	11.73372
1-METHYL-3,2'-ETHYLENE-2-PHENYL INDOLE	27.36	0.022	15.09850
1-METHYL-3,2'-TRIMETHYLENE-2-PHENYL INDOLE	22.11	0.041	4.22789
4,5-DIPHENYLIMIDAZOLE (A)	13.53	0.181	0.22244
PPF (TRACE)	27.50	0.021	15.61876
PPO (TRACE)	22.92	0.037	5.24264
2-(4-FLUOROPHENYL)-5-PHENYLOXAZOLE (TRACE)	21.75	0.043	3.82989
αNPO (TRACE)	29.59	0.017	24.07294
BBO (B)	34.18	0.011	57.02015
POPOP	33.60	0.012	51.43747
DIMETHYL POPOP (TRACE)	34.43	0.011	59.38904
PPD (TRACE)	9.23	0.569	0.02255
2,5-DI(2-METHYLPHENYL)OXADIAZOLE	10.59	0.377	0.05139
PBD	21.63	0.044	3.71580
BBD (B)	29.11	0.018	21.92930
αNPD (TRACE)	25.36	0.027	9.56828
βNPD	21.00	0.048	3.10073
αNND (TRACE)	27.80	0.021	16.58041
βNND (TRACE)	26.40	0.024	12.23425
TRANS-1-(4-BIPHENYLYL)-2-PHENYL-ETHYLENE	29.00	0.018	21.42124
TRANS-1-(4-BIPHENYLYL)-2-(1-NAPHTHYL)-ETHYLENE	31.38	0.014	34.17139
DIPHENYLSTILBENE (B)	34.51	0.011	60.42645
1,1,4,4-TETRAPHENYLBUTADIENE	32.50	0.013	42.03311
1,6-DIPHENYLHEXATRIENE	36.12	0.009	79.35438
1,8-DIPHENYL-1,3,5,7-OCTATETRAENE	35.82	0.010	74.96516
BPSB (TRACE)	36.33	0.009	82.14238
BIS-MSB	35.15	0.010	67.30178
ESCULIN (A)	26.67	0.024	12.88256
AZULENE	19.64	0.059	2.03470
NAPHTHALENE	5.36	2.906	0.00086
DEUTERATED NAPHTHALENE-C10D8	7.54	1.042	0.00669
1-METHYLNAPHTHALENE 5ML/L	6.75	1.456	0.00343
2-METHYL NAPHTHALENE	8.30	0.784	0.01186
2,3-DIMETHYL NAPHTHALENE	8.18	0.817	0.01090
2,6-DIMETHYL NAPHTHALENE (TRACE)	11.26	0.314	0.07401
1-NAPHTHOL (TRACE)	12.09	0.253	0.11353
2-NAPHTHOL (TRACE)	14.54	0.145	0.34291
1-NAPHTHYLAMINE	19.22	0.063	1.82233
2-CHLORONAPHTHALENE	8.86	0.643	0.01760
ACENAPHTHENE	10.40	0.397	0.04620
1-PHENYL NAPHTHALENE	10.64	0.371	0.05278
2-PHENYL NAPHTHALENE	16.72	0.096	0.79327
1,4-DIPHENYL NAPHTHALENE	17.58	0.082	1.06794
1,5-DIPHENYL NAPHTHALENE	16.12	0.107	0.63569
1,7-DIPHENYL NAPHTHALENE	15.98	0.110	0.60314
1,4,5,8-TETRAPHENYL NAPHTHALENE	28.55	0.019	19.35033
1,1'-BINAPHTHYL	12.78	0.214	0.15827
4,4'-DI(N-BUTOXY)-1,1'-BINAPHTHYL	22.00	0.042	4.09122
2,2'-BINAPHTHYL	17.53	0.083	1.05116
ACRIDINE (A)	25.34	0.027	9.46564
ANTHRACENE (TRACE)	23.78	0.033	6.46362
DEUTERATED ANTHRACENE C14D10 (TRACE)	23.51	0.034	6.03898
9-METHYL ANTHRACENE (TRACE)	23.57	0.034	6.12761
1-AMINO ANTHRACENE	22.09	0.042	4.11673
2-AMINO ANTHRACENE	21.94	0.042	3.95379
9-VINYL ANTHRACENE	24.37	0.031	7.47470
9-PHENYL ANTHRACENE (TRACE)	[illegible]	0.029	8.32811
9,10-DIPHENYL ANTHRACENE	25.23	0.028	9.18991
9,10-DICHLOROANTHRACENE (TRACE)	24.07	0.032	6.91699
BIANTHRANYL (B)	26.55	0.024	12.45361
9,10-DINAPHTHYLANTHRACENE (B)	24.74	0.030	8.14932
9-(P-VINYLPHENYL)-10-PHENYL ANTHRACENE	25.31	0.028	9.34990
9,10-BIS(PHENYL-ETHYNYL)ANTHRACENE (TRACE)	22.05	0.042	4.07360
RUBICENE (B)	24.69	0.030	8.03285
DIBENZ(A,H)ANTHRACENE (B)	26.22	0.025	11.58613
TETRACENE (TRACE) (B)	17.98	0.077	1.19747
RUBRENE (B)	22.19	0.041	4.23735
TETRAHYDROCANNABINOL ACID METHYL ESTER	7.90	0.908	0.00886
PHENANTHRENE	12.00	0.259	0.10766
3,4-BENZOPHENANTHRENE	17.46	0.084	1.01275

ACCEPTOR MOLECULES	RO	CO	OVERLAP
2-PHENYL PHENANTHRENE	14.00	0.163	0.27088
CHRYSENE (TRACE)	17.87	0.078	1.17094
TRIPHENYLENE	12.45	0.232	0.13464
DEUTERATED TRIPHENYLENE-C18D12	12.21	0.246	0.11967
PYRENE	25.68	0.026	10.27083
1,3,6,8,-TETRAPHENYL PYRENE	31.33	0.015	33.56650
BIPYRENYL (B)	33.06	0.012	46.53814
O-PHENYLENE-PYRENE	27.92	0.021	16.80106
ANTHANTHRENE (B)	25.39	0.027	9.50023
FLUORANTHENE	24.82	0.029	8.32035
3-PHENYL FLUORANTHENE	27.28	0.022	14.64565
3-AMINO FLUORANTHENE	24.66	0.030	7.98744
3,3'-BIFLUORANTHENYL (B)	30.31	0.016	27.49403
BENZO(GHI)FLUORANTHENE	23.61	0.034	6.14627
9,12-DI(N-OCTADECOXY)-BENZ(E)ACEPHENANTHRYLENE	26.70	0.023	12.86135
DECACYCLENE (B)	32.71	0.013	43.41613
10-ETHOXY-BENZO(KL)XANTHENE	25.72	0.026	10.32137
PERYLENE	22.86	0.037	5.06417
BENZO(GHI)PERYLENE (B)	27.34	0.022	14.88129
BBOT (TRACE)	34.80	0.011	63.23901
QUININE SULFATE (A)	22.62	0.039	4.83871
ACRIDINE YELLOW (A)	19.80	0.058	2.13858
ACRIDINE ORANGE (A)	17.61	0.082	1.05943
ACRIDINE RED (A)	21.28	0.046	3.29186
URANINE (A)	20.73	0.050	2.81644
RHODAMINE B (A)	24.17	0.032	7.08161
RHODAMINE 6G (A)	24.66	0.030	7.98622
EOSIN B (A)	21.90	0.043	3.91333
EOSIN Y (A)	22.32	0.040	4.38326
ROSE BENGAL (A)	23.32	0.035	5.70068
INDENOINDENE	22.44	0.040	4.62128

8.54 Donor Molecule: 4,4′-Diisobutoxybiphenyl

N,N-DIMETHYLANILINE 3ML/L	8.07	0.851	0.00845
TRIPHENYLAMINE	19.16	0.064	1.49802
1,4-DIMETHOXY BENZENE	2.60	25.541	0.00001
1-METHYL-4-CHLOROBENZENE 5ML/L	0.00	0.000	0.00000
SODIUM SALICYLATE (A)	13.07	0.200	0.15099
ANTHRANILIC ACID (A)	22.95	0.037	4.39479
TMPD	19.22	0.063	1.52135
4,4′-DIHYDROXYBIPHENYL (A)	7.60	1.017	0.00593
4,4′-DIMETHOXY BIPHENYL	5.71	2.398	0.00107
4,4′-DIISOBUTOXYBIPHENYL	4.25	5.839	0.00018
BENZIDINE	16.45	0.100	0.60303
4-VINYL BIPHENYL	7.74	0.964	0.00661
M-TERPHENYL	5.27	3.055	0.00066
1,3,5-TRIPHENYL BENZENE	7.02	1.292	0.00366
1,3-DIBIPHENYLYLBENZENE (B)	15.34	0.124	0.39631
1,3,5,-TRIBIPHENYLYLBENZENE (B)	18.13	0.075	1.08090
FLUORENE (TRACE)	0.00	0.000	0.00000
1-METHYL FLUORENE	2.99	16.743	0.00002
1,2-BENZOFLUORENE	17.88	0.078	0.98935
9,10-DIHYDROPHENANTHRENE	4.28	5.712	0.00019
CARBAZOLE (TRACE)	16.82	0.094	0.68830
N-METHYL CARBAZOLE (TRACE)	20.29	0.054	2.11812
N-VINYLCARBAZOLE	19.99	0.056	1.93901
N-PHENYL CARBAZOLE	19.79	0.058	1.82513
N-METHYL-2,3-BENZCARBAZOLE	22.66	0.038	4.07700
DIBENZOFURAN	0.00	0.000	0.00000
DIBENZOTHIOPHENE (TRACE)	14.64	0.143	0.29946
P-TERPHENYL	9.46	0.529	0.02193
DEUTERATED P-TERPHENYL-C18D14	8.79	0.660	0.01411
4-METHYL TERPHENYL	10.84	0.351	0.04961
2-METHYL-P-TERPHENYL	3.99	7.053	0.00012
3,3″-DIMETHYL-P-TERPHENYL	10.80	0.355	0.04863
4-(3,3-DIMETHYLBUTOXY)-P-TERPHENYL	15.21	0.127	0.37826
4,4″-DIHEXAHYDROFARNESOXY-P-TERPHENYL	17.90	0.078	1.00370
2,2′-METHYLENE-P-TERPHENYL	16.60	0.098	0.63785
2,2′-ETHYLENE-P-TERPHENYL	17.50	0.083	0.87648
2,2′-METHYLENE-5′,6′-METHYLENE-P-TERPHENYL (TRACE)	26.59	0.024	10.73902
3-PHENYL DIBENZOFURAN	16.62	0.097	0.64414
BENZO(1,2-B-4,5-B′)BISBENZOFURAN (TRACE) (B)	28.13	0.020	15.03400
SEXIPHENYL BENZENE	0.00	0.000	0.00000
P-QUATERPHENYL	19.19	0.063	1.51908
3,3‴-DIMETHYL-P-QUATERPHENYL	20.53	0.052	2.27227
TETRAMETHYL-P-QUATERPHENYL	5.64	2.493	0.00099
3,3‴-DIMETHYL-3′,2″-METHYLENE-P-QUATERPHENYL	27.71	0.021	13.74143
3,3‴-DIMETHYL-3′,2″-ETHYLENE-P-QUATERPHENYL	27.52	0.021	13.15010
4-(3-ETHYLBUTOXY)-P-QUATERPHENYL (B)	25.09	0.028	7.55173
4,4‴-DI(2-BUTYLOCTOXY-1)-P-QUATERPHENYL	25.60	0.027	8.53635
DIETHYL-P-QUINQUEPHENYL	21.40	0.046	2.91853
DI(3-ETHYLHEPTYL)-P-QUINQUEPHENYL	26.84	0.023	11.29383
TETRAMETHYL-P-SEXIPHENYL	20.22	0.054	2.07813
TETRAMETHYL-P-OCTAPHENYL (B)	30.90	0.015	26.30010
INDOLE (TRACE)	0.00	0.000	0.00000
2-PHENYL INDOLE	21.28	0.046	2.82461
1-METHYL-2-PHENYL INDOLE	16.96	0.092	0.72320
1-NEOPENTYL-2-PHENYL INDOLE	14.68	0.142	0.30474
1-N-BUTYL-2(BIPHENYLYL-4) INDOLE	25.26	0.028	7.85682

ACCEPTOR MOLECULES	RO	CO	OVERLAP
1,2-DIPHENYL INDOLE	17.58	0.082	0.89728
1-BIPHENYLYL-2-PHENYL INDOLE	19.92	0.057	1.89081
1,3-DIMETHYL-2-PHENYL INDOLE	17.98	0.077	1.02782
2,3-DIPHENYL INDOLE	23.30	0.035	4.83300
1-METHYL-2,3-DIPHENYL INDOLE	21.09	0.048	2.66436
2-PHENYL-3,2'-METHYLENE INDOLE (TRACE)	24.32	0.031	6.28806
2-PHENYL-3,2'-ETHYLENE INDOLE (TRACE)	27.30	0.022	12.51552
2-PHENYL-3,2'-TRIMETHYLENE INDOLE	23.07	0.034	4.31919
1-METHYL-3,2'-METHYLENE-2-PHENYL INDOLE (TRACE)	26.73	0.023	11.04886
1-METHYL-3,2'-ETHYLENE-2-PHENYL INDOLE	27.96	0.020	14.42100
1-METHYL-3,2'-TRIMETHYLENE-2-PHENYL INDOLE	22.44	0.040	3.87184
4,5-DIPHENYLIMIDAZOLE (A)	13.68	0.175	0.19907
PPF (TRACE)	28.02	0.020	14.67675
PPO (TRACE)	23.21	0.036	4.74772
2-(4-FLUOROPHENYL)-5-PHENYLOXAZOLE (TRACE)	22.02	0.042	3.46319
αNPO (TRACE)	30.31	0.016	23.32018
BBO (B)	35.05	0.010	55.68943
POPOP	34.53	0.011	50.91825
DIMETHYL POPOP (TRACE)	35.45	0.010	59.40137
PPD (TRACE)	9.10	0.594	0.01740
2,5-DI(2-METHYLPHENYL)OXADIAZOLE	10.58	0.378	0.04285
PBD	21.87	0.043	3.33030
BBD (B)	29.65	0.017	20.57648
αNPD (TRACE)	25.85	0.026	9.02021
βNPD	21.26	0.047	2.80441
αNND (TRACE)	28.43	0.019	15.92963
βNND (TRACE)	26.85	0.023	11.36872
TRANS-1-(4-BIPHENYLYL)-2-PHENYL-ETHYLENE	29.58	0.017	20.23017
TRANS-1-(4-BIPHENYLYL)-2-(1-NAPHTHYL)-ETHYLENE	32.15	0.013	33.18986
DIPHENYLSTILBENE (B)	35.39	0.010	58.98399
1,1,4,4-TETRAPHENYLBUTADIENE	33.35	0.012	41.25363
1,6-DIPHENYLHEXATRIENE	37.11	0.009	78.27885
1,8-DIPHENYL-1,3,5,7-OCTATETRAENE	37.04	0.009	76.95226
BPSB (TRACE)	37.35	0.009	81.46306
BIS-MSB	36.10	0.010	66.31476
ESCULIN (A)	27.31	0.022	12.49481
AZULENE	20.02	0.056	1.91831
NAPHTHALENE	5.31	2.991	0.00068
DEUTERATED NAPHTHALENE-C10D8	7.60	1.020	0.00586
1-METHYLNAPHTHALENE 5ML/L	6.77	1.443	0.00294
2-METHYL NAPHTHALENE	8.33	0.774	0.01021
2,3-DIMETHYL NAPHTHALENE	8.21	0.808	0.00936
2,6-DIMETHYL NAPHTHALENE (TRACE)	11.35	0.306	0.06545
1-NAPHTHOL (TRACE)	12.15	0.249	0.09824
2-NAPHTHOL (TRACE)	14.73	0.140	0.31059
1-NAPHTHYLAMINE	19.54	0.060	1.68662
2-CHLORONAPHTHALENE	8.90	0.634	0.01520
ACENAPHTHENE	10.43	0.394	0.03946
1-PHENYL NAPHTHALENE	10.69	0.366	0.04560
2-PHENYL NAPHTHALENE	16.87	0.093	0.70354
1,4-DIPHENYL NAPHTHALENE	17.78	0.080	0.95982
1,5-DIPHENYL NAPHTHALENE	16.27	0.104	0.56525
1,7-DIPHENYL NAPHTHALENE	16.14	0.106	0.53797
1,4,5,8-TETRAPHENYL NAPHTHALENE	29.25	0.018	18.77387
1,1'-BINAPHTHYL	12.88	0.209	0.13937
4,4'-DI(N-BUTOXY)-1,1'-BINAPHTHYL	22.34	0.040	3.77032
2,2'-BINAPHTHYL	17.70	0.081	0.93441
ACRIDINE (A)	26.05	0.025	9.37586
ANTHRACENE (TRACE)	24.42	0.031	6.37303
DEUTERATED ANTHRACENE C14D10 (TRACE)	24.14	0.032	5.94424
9-METHYL ANTHRACENE (TRACE)	24.34	0.031	6.23643
1-AMINO ANTHRACENE	22.93	0.037	4.33266
2-AMINO ANTHRACENE	22.69	0.038	4.06081
9-VINYL ANTHRACENE	25.18	0.028	7.62960
9-PHENYL ANTHRACENE (TRACE)	25.61	0.027	8.45495
9,10-DIPHENYL ANTHRACENE	26.12	0.025	9.49515
9,10-DICHLOROANTHRACENE (TRACE)	24.97	0.029	7.23925
BIANTHRANYL (B)	27.48	0.022	12.85842
9,10-DINAPHTHYLANTHRACENE (B)	25.62	0.027	8.44981
9-(P-VINYLPHENYL)-10-PHENYL ANTHRACENE	26.22	0.025	9.69800
9,10-BIS(PHENYL-ETHYNYL)ANTHRACENE (TRACE)	23.39	0.035	4.88426
RUBICENE (B)	25.47	0.027	8.12379
DIBENZ(A,H)ANTHRACENE (B)	26.79	0.023	11.08222
TETRACENE (TRACE) (B)	18.78	0.068	1.30731
RUBRENE (B)	22.59	0.039	3.95930
TETRAHYDROCANNABINOL ACID METHYL ESTER	7.87	0.916	0.00731
PHENANTHRENE	12.24	0.244	0.10172
3,4-BENZOPHENANTHRENE	17.72	0.080	0.92910
2-PHENYL PHENANTHRENE	14.26	0.154	0.25420
CHRYSENE (TRACE)	18.12	0.075	1.06798
TRIPHENYLENE	12.65	0.221	0.12404
DEUTERATED TRIPHENYLENE-C18D12	12.40	0.235	0.11031
PYRENE	26.07	0.025	9.45632
1,3,6,8,-TETRAPHENYL PYRENE	32.42	0.013	34.63530
BIPYRENYL (B)	33.98	0.011	46.01615
O-PHENYLENE-PYRENE	28.83	0.019	17.09381
ANTHANTHRENE (B)	26.44	0.024	10.17162
FLUORANTHENE	25.47	0.027	8.16147
3-PHENYL FLUORANTHENE	28.09	0.020	14.64840
3-AMINO FLUORANTHENE	25.50	0.027	8.19834
3,3'-BIFLUORANTHENYL (B)	31.30	0.015	28.01272
BENZO(GHI)FLUORANTHENE	24.21	0.032	6.00894
9,12-DI(N-OCTADECOXY)-BENZ(E)ACEPHENANTHRYLENE	27.61	0.021	13.20410
DECACYCLENE (B)	33.79	0.012	44.38236
10-ETHOXY-BENZO(KL)XANTHENE	26.64	0.024	10.68068
PERYLENE	24.24	0.031	6.04602
BENZO(GHI)PERYLENE (B)	28.28	0.020	15.30179

ACCEPTOR MOLECULES	RO	CO	OVERLAP
BBOT (TRACE)	35.90	0.010	64.06105
QUININE SULFATE (A)	23.03	0.037	4.52034
ACRIDINE YELLOW (A)	21.02	0.048	2.57337
ACRIDINE ORANGE (A)	18.44	0.071	1.16985
ACRIDINE RED (A)	21.92	0.042	3.30521
URANINE (A)	21.37	0.046	2.83839
RHODAMINE B (A)	24.93	0.029	7.14481
RHODAMINE 6G (A)	25.36	0.027	7.92208
EOSIN B (A)	22.48	0.039	3.84223
EOSIN Y (A)	22.93	0.037	4.33212
ROSE BENGAL (A)	23.91	0.033	5.56400
INDENOINDENE	22.73	0.038	4.18879

8.56 Donor Molecule: 4-Vinylbiphenyl

ANILINE 3ML/L	6.64	1.527	0.00109
N,N-DIMETHYLANILINE 3ML/L	14.86	0.136	0.13541
TRIPHENYLAMINE	27.43	0.022	5.30758
1,4-DIMETHOXY BENZENE	7.89	0.909	0.00306
1-METHYL-4-CHLOROBENZENE 5ML/L	0.00	0.000	0.00000
SODIUM SALICYLATE (A)	18.58	0.070	0.51267
ANTHRANILIC ACID (A)	27.00	0.023	4.79719
TMPD	23.68	0.034	2.19039
4-METHOXY BIPHENYL	5.86	2.227	0.00051
4-BIPHENYLYL PHENYL ETHER	4.09	6.551	0.00006
4,4'-DIHYDROXYBIPHENYL (A)	13.64	0.176	0.08114
4,4'-DIMETHOXY BIPHENYL	11.46	0.297	0.02856
4,4'-DIISOBUTOXYBIPHENYL	10.33	0.406	0.01534
BENZIDINE	24.76	0.029	2.88074
4-VINYL BIPHENYL	15.48	0.121	0.17347
M-TERPHENYL	10.86	0.349	0.02075
1,3,5-TRIPHENYL BENZENE	12.00	0.259	0.03749
1,3-DIBIPHENYLYLBENZENE (B)	24.59	0.030	2.76990
1,3,5,-TRIBIPHENYLYLBENZENE (B)	28.77	0.019	7.10874
FLUORENE (TRACE)	8.58	0.708	0.00505
1-METHYL FLUORENE	8.19	0.813	0.00383
1,2-BENZOFLUORENE	25.91	0.026	3.77186
9,10-DIHYDROPHENANTHRENE	10.61	0.374	0.01806
CARBAZOLE (TRACE)	23.57	0.034	2.14379
N-METHYL CARBAZOLE (TRACE)	24.81	0.029	2.90645
N-VINYLCARBAZOLE	25.13	0.028	3.13780
N-PHENYL CARBAZOLE	24.82	0.029	2.91885
N-METHYL-2,3-BENZCARBAZOLE	27.93	0.021	5.83371
DIBENZOFURAN	7.24	1.181	0.00182
DIBENZOTHIOPHENE (TRACE)	21.20	0.047	1.13634
P-TERPHENYL	17.41	0.085	0.35107
DEUTERATED P-TERPHENYL-C18D14	16.85	0.093	0.28906
4-METHYL TERPHENYL	19.40	0.061	0.67070
2-METHYL-P-TERPHENYL	9.87	0.465	0.01172
3,3''-DIMETHYL-P-TERPHENYL	18.91	0.066	0.57568
4-(3,3-DIMETHYLBUTOXY)-P-TERPHENYL	24.24	0.031	2.54404
4,4''-DIHEXAHYDROFARNESOXY-P-TERPHENYL	27.60	0.021	5.53818
2,2'-METHYLENE-P-TERPHENYL	27.05	0.023	4.91935
2,2'-ETHYLENE-P-TERPHENYL	27.48	0.022	5.39758
2,2'-METHYLENE-5',6''-METHYLENE-P-TERPHENYL (TRACE)	34.96	0.010	22.79838
3-PHENYL DIBENZOFURAN	27.31	0.022	5.20764
BENZO(1,2-B-4,5-B')BISBENZOFURAN (TRACE) (B)	36.56	0.009	29.78311
SEXIPHENYL BENZENE	4.92	3.755	0.00018
P-QUATERPHENYL	28.74	0.019	7.05234
3,3'''-DIMETHYL-P-QUATERPHENYL	30.06	0.016	9.21131
TETRAMETHYL-P-QUATERPHENYL	12.32	0.239	0.04427
3,3''' -DIMETHYL-3',2''-METHYLENE-P-QUATERPHENYL	36.80	0.009	30.96845
3,3''' -DIMETHYL-3',2''-ETHYLENE-P-QUATERPHENYL	36.38	0.009	28.80949
4-(3-ETHYLBUTOXY)-P-QUATERPHENYL (B)	34.21	0.011	19.96625
4,4'''-DI(2-BUTYLOCTOXY-1)-P-QUATERPHENYL	35.12	0.010	23.35505
DIETHYL-P-QUINQUEPHENYL	31.35	0.015	11.86700
DI(3-ETHYLHEPTYL)-P-QUINQUEPHENYL	36.45	0.009	29.10942
TETRAMETHYL-P-SEXIPHENYL	30.45	0.016	9.96415
TETRAMETHYL-P-OCTAPHENYL (B)	40.03	0.007	51.03654
INDOLE (TRACE)	0.00	0.000	0.00000
2-PHENYL INDOLE	30.87	0.015	10.81630
1-METHYL-2-PHENYL INDOLE	25.66	0.026	3.56922
1-NEOPENTYL-2-PHENYL INDOLE	23.57	0.034	2.15110
1-N-BUTYL-2(BIPHENYLYL-4) INDOLE	32.94	0.013	15.86553
1,2-DIPHENYL INDOLE	26.74	0.023	4.57270
1-BIPHENYLYL-2-PHENYL INDOLE	29.13	0.018	7.61355
1,3-DIMETHYL-2-PHENYL INDOLE	26.04	0.025	3.90170
2,3-DIPHENYL INDOLE	30.60	0.016	10.18109
1-METHYL-2,3-DIPHENYL INDOLE	27.93	0.021	5.89867
2-PHENYL-3,2'-METHYLENE INDOLE (TRACE)	33.08	0.012	16.34408
2-PHENYL-3,2'-ETHYLENE INDOLE (TRACE)	34.20	0.011	19.90372
2-PHENYL-3,2' -TRIMETHYLENE INDOLE	32.16	0.013	13.80212
1-METHYL-3,2'-METHYLENE-2-PHENYL INDOLE (TRACE)	34.11	0.011	19.62308
1-METHYL-3,2'-ETHYLENE-2-PHENYL INDOLE	34.10	0.011	19.53469
1-METHYL-3,2'-TRIMETHYLENE-2-PHENYL INDOLE	30.47	0.016	9.98105
4,5-DIPHENYLIMIDAZOLE (A)	20.42	0.053	0.90719
PPF (TRACE)	35.82	0.010	26.31653
PPO (TRACE)	32.73	0.013	15.35011
2-(4-FLUOROPHENYL)-5-PHENYLOXAZOLE (TRACE)	31.26	0.015	11.66292
αNPO (TRACE)	35.64	0.010	25.33021
BBO (B)	40.19	0.007	52.02244

ACCEPTOR MOLECULES	RO	CO	OVERLAP
POPOP	38.17	0.008	38.13485
DIMETHYL POPOP (TRACE)	38.69	0.008	41.26822
PPD (TRACE)	18.25	0.074	0.46671
2,5-DI(2-METHYLPHENYL)OXADIAZOLE	19.69	0.059	0.73456
PBD	32.16	0.013	13.82792
BBD (B)	38.24	0.008	38.88925
αNPD (TRACE)	32.78	0.013	15.39728
βNPD	30.57	0.016	10.19537
αNND (TRACE)	34.08	0.011	19.44097
βNND (TRACE)	35.14	0.010	23.46469
TRANS-1-(4-BIPHENYLYL)-2-PHENYL-ETHYLENE	37.45	0.009	34.27056
TRANS-1-(4-BIPHENYLYL)-2-(1-NAPHTHYL)-ETHYLENE	37.62	0.008	35.01307
DIPHENYLSTILBENE (B)	40.62	0.007	55.43974
1,1,4,4-TETRAPHENYLBUTADIENE	37.77	0.008	35.73186
1,6-DIPHENYLHEXATRIENE	41.15	0.006	59.83434
1,8-DIPHENYL-1,3,5,7-OCTATETRAENE	39.00	0.008	43.21608
BPSB (TRACE)	41.24	0.006	60.59027
BIS-MSB	40.61	0.007	55.28038
ESCULIN (A)	31.94	0.014	13.14030
AZULENE	24.40	0.031	2.59093
NAPHTHALENE	9.84	0.469	0.01146
DEUTERATED NAPHTHALENE-C10D8	12.05	0.256	0.03833
1-METHYLNAPHTHALENE 5ML/L	11.92	0.264	0.03602
2-METHYL NAPHTHALENE	13.64	0.176	0.08084
2,3-DIMETHYL NAPHTHALENE	13.49	0.182	0.07576
2,6-DIMETHYL NAPHTHALENE (TRACE)	16.72	0.096	0.27484
1-NAPHTHOL (TRACE)	19.10	0.064	0.60909
2-NAPHTHOL (TRACE)	20.72	0.050	0.99059
1-NAPHTHYLAMINE	25.81	0.026	3.68181
2-CHLORONAPHTHALENE	13.77	0.171	0.08578
ACENAPHTHENE	16.95	0.092	0.29834
1-PHENYL NAPHTHALENE	17.92	0.078	0.41553
2-PHENYL NAPHTHALENE	26.12	0.025	3.97497
1,4-DIPHENYL NAPHTHALENE	26.09	0.025	3.93587
1,5-DIPHENYL NAPHTHALENE	24.60	0.030	2.77217
1,7-DIPHENYL NAPHTHALENE	24.29	0.031	2.56937
1,4,5,8-TETRAPHENYL NAPHTHALENE	34.29	0.011	20.02318
1,1'-BINAPHTHYL	20.26	0.054	0.86867
4,4'-DI(N-BUTOXY)-1,1'-BINAPHTHYL	30.25	0.016	9.54120
2,2'-BINAPHTHYL	26.99	0.023	4.83178
ACRIDINE (A)	28.57	0.019	6.70457
ANTHRACENE (TRACE)	26.77	0.023	4.54176
DEUTERATED ANTHRACENE C14D10 (TRACE)	26.61	0.024	4.39015
9-METHYL ANTHRACENE (TRACE)	25.95	0.026	3.76149
1-AMINO ANTHRACENE	23.96	0.033	2.32315
2-AMINO ANTHRACENE	25.31	0.028	3.22464
9-VINYL ANTHRACENE	26.69	0.024	4.44448
9-PHENYL ANTHRACENE (TRACE)	27.31	0.022	5.10982
9,10-DIPHENYL ANTHRACENE	27.22	0.022	4.99281
9,10-DICHLOROANTHRACENE (TRACE)	25.78	0.026	3.60231
BIANTHRANYL (B)	29.05	0.018	7.38109
9,10-DINAPHTHYLANTHRACENE (B)	27.20	0.022	4.97046
9-(P-VINYLPHENYL)-10-PHENYL ANTHRACENE	27.31	0.022	5.09906
9,10-BIS(PHENYL-ETHYNYL)ANTHRACENE (TRACE)	26.66	0.024	4.40630
RUBICENE (B)	30.49	0.016	9.86878
DIBENZ(A,H)ANTHRACENE (B)	32.84	0.013	15.43761
TETRACENE (TRACE) (B)	20.85	0.049	1.00965
RUBRENE (B)	32.14	0.013	13.54729
TETRAHYDROCANNABINOL ACID METHYL ESTER	14.86	0.136	0.13551
PHENANTHRENE	15.40	0.122	0.16640
3,4-BENZOPHENANTHRENE	25.56	0.027	3.44375
2-PHENYL PHENANTHRENE	20.54	0.052	0.93241
CHRYSENE (TRACE)	26.55	0.024	4.34131
TRIPHENYLENE	16.77	0.095	0.27705
DEUTERATED TRIPHENYLENE-C18D12	16.72	0.096	0.27283
PYRENE	33.66	0.012	17.97527
1,3,6,8,-TETRAPHENYL PYRENE	33.59	0.012	17.62679
BIPYRENYL (B)	38.17	0.008	37.98210
O-PHENYLENE-PYRENE	32.29	0.013	13.93181
ANTHANTHRENE (B)	30.54	0.016	9.95684
FLUORANTHENE	28.86	0.019	7.08931
3-PHENYL FLUORANTHENE	30.66	0.016	10.20390
3-AMINO FLUORANTHENE	28.92	0.018	7.18185
3,3'-BIFLUORANTHENYL (B)	34.07	0.011	19.19967
BENZO(GHI)FLUORANTHENE	28.22	0.020	6.19959
9,12-DI(N-OCTADECOXY)-BENZ(E)ACEPHENANTHRYLENE	30.49	0.016	9.87592
DECACYCLENE (B)	37.20	0.009	32.55919
10-ETHOXY-BENZO(KL)XANTHENE	27.84	0.021	5.71931
PERYLENE	24.08	0.032	2.39680
BENZO(GHI)PERYLENE (B)	30.22	0.016	9.36521
BBOT (TRACE)	38.43	0.008	39.62098
QUININE SULFATE (A)	29.12	0.018	7.60016
ACRIDINE YELLOW (A)	21.40	0.046	1.18025
ACRIDINE ORANGE (A)	21.31	0.046	1.16060
ACRIDINE RED (A)	25.01	0.029	3.00849
URANINE (A)	25.59	0.027	3.45296
RHODAMINE B (A)	28.29	0.020	6.29344
RHODAMINE 6G (A)	28.41	0.020	6.45283
EOSIN B (A)	27.61	0.021	5.44269
EOSIN Y (A)	28.14	0.020	6.09205
ROSE BENGAL (A)	29.09	0.018	7.44846
INDENOINDENE	32.01	0.014	13.43764

8.57 Donor Molecule: *m*-Terphenyl

ACCEPTOR MOLECULES	RO	CO	OVERLAP
ANILINE 3ML/L	4.81	4.022	0.00033
N,N-DIMETHYLANILINE 3ML/L	11.17	0.321	0.05133
TRIPHENYLAMINE	22.52	0.039	3.41901
1,4-DIMETHOXY BENZENE	5.65	2.481	0.00086
1-METHYL-4-CHLOROBENZENE 5ML/L	0.00	0.000	0.00000
SODIUM SALICYLATE (A)	15.16	0.128	0.31816
ANTHRANILIC ACID (A)	23.90	0.033	4.84936
TMPD	20.62	0.051	2.00707
4-METHOXY BIPHENYL	3.89	7.594	0.00009
4-BIPHENYLYL PHENYL ETHER	0.00	0.000	0.00000
4,4'-DIHYDROXYBIPHENYL (A)	10.37	0.401	0.03292
4,4'-DIMETHOXY BIPHENYL	8.44	0.744	0.00960
4,4'-DIISOBUTOXYBIPHENYL	7.63	1.006	0.00525
BENZIDINE	20.03	0.056	1.69556
4-VINYL BIPHENYL	11.43	0.300	0.05906
M-TERPHENYL	8.03	0.865	0.00710
1,3,5-TRIPHENYL BENZENE	9.28	0.560	0.01687
1,3-DIBIPHENYLYLBENZENE (B)	19.41	0.061	1.40521
1,3,5,-TRIBIPHENYLYLBENZENE (B)	22.84	0.038	3.73668
FLUORENE (TRACE)	5.96	2.094	0.00121
1-METHYL FLUORENE	5.88	2.204	0.00109
1,2-BENZOFLUORENE	20.94	0.049	2.21018
9,10-DIHYDROPHENANTHRENE	7.73	0.969	0.00566
CARBAZOLE (TRACE)	19.72	0.058	1.54394
N-METHYL CARBAZOLE (TRACE)	21.76	0.043	2.78489
N-VINYLCARBAZOLE	21.93	0.042	2.91566
N-PHENYL CARBAZOLE	21.65	0.044	2.69994
N-METHYL-2,3-BENZCARBAZOLE	24.08	0.032	5.03372
DIBENZOFURAN	5.01	3.561	0.00042
DIBENZOTHIOPHENE (TRACE)	17.74	0.080	0.82000
P-TERPHENYL	13.12	0.198	0.13513
DEUTERATED P-TERPHENYL-C18D14	12.51	0.229	0.10140
4-METHYL TERPHENYL	14.75	0.139	0.27291
2-METHYL-P-TERPHENYL	7.28	1.159	0.00396
3,3''-DIMETHYL-P-TERPHENYL	14.50	0.147	0.24613
4-(3,3-DIMETHYLBUTOXY)-P-TERPHENYL	19.25	0.063	1.34178
4,4''-DIHEXAHYDROFARNESOXY-P-TERPHENYL	22.21	0.041	3.15894
2,2'-METHYLENE-P-TERPHENYL	21.31	0.046	2.46800
2,2'-ETHYLENE-P-TERPHENYL	22.00	0.042	2.98298
2,2'-METHYLENE-5',6''-METHYLENE-P-TERPHENYL (TRACE)	30.02	0.017	19.20511
3-PHENYL DIBENZOFURAN	21.44	0.045	2.56135
BENZO(1,2-B-4,5-B')BISBENZOFURAN (TRACE) (B)	31.43	0.014	25.26978
SEXIPHENYL BENZENE	0.00	0.000	0.00000
P-QUATERPHENYL	23.35	0.035	4.26493
3,3'''-DIMETHYL-P-QUATERPHENYL	24.60	0.030	5.81898
TETRAMETHYL-P-QUATERPHENYL	9.08	0.597	0.01490
3,3'''-DIMETHYL-3',2''-METHYLENE-P-QUATERPHENYL	31.37	0.014	24.96328
3,3'''-DIMETHYL-3',2''-ETHYLENE-P-QUATERPHENYL	30.96	0.015	23.03112
4-(3-ETHYLBUTOXY)-P-QUATERPHENYL (B)	28.73	0.019	14.70369
4,4'''-DI(2-BUTYLOCTOXY-1)-P-QUATERPHENYL	29.43	0.018	16.99214
DIETHYL-P-QUINQUEPHENYL	25.67	0.026	7.51406
DI(3-ETHYLHEPTYL)-P-QUINQUEPHENYL	30.64	0.016	21.58194
TETRAMETHYL-P-SEXIPHENYL	24.69	0.030	5.95230
TETRAMETHYL-P-OCTAPHENYL (B)	34.22	0.011	41.84846
INDOLE (TRACE)	0.00	0.000	0.00000
2-PHENYL INDOLE	25.49	0.027	7.20907
1-METHYL-2-PHENYL INDOLE	20.78	0.050	2.11370
1-NEOPENTYL-2-PHENYL INDOLE	18.66	0.069	1.11411
1-N-BUTYL-2(BIPHENYLYL-4) INDOLE	28.07	0.020	12.78805
1,2-DIPHENYL INDOLE	21.61	0.044	2.67644
1-BIPHENYLYL-2-PHENYL INDOLE	23.84	0.033	4.80294
1,3-DIMETHYL-2-PHENYL INDOLE	21.46	0.045	2.56601
2,3-DIPHENYL INDOLE	26.00	0.025	8.05930
1-METHYL-2,3-DIPHENYL INDOLE	23.66	0.034	4.58222
2-PHENYL-3,2'-METHYLENE INDOLE (TRACE)	27.98	0.020	12.58160
2-PHENYL-3,2'-ETHYLENE INDOLE (TRACE)	29.62	0.017	17.65430
2-PHENYL-3,2'-TRIMETHYLENE INDOLE	26.91	0.023	9.97130
1-METHYL-3,2'-METHYLENE-2-PHENYL INDOLE (TRACE)	29.38	0.018	16.82332
1-METHYL-3,2'-ETHYLENE-2-PHENYL INDOLE	29.77	0.017	18.18385
1-METHYL-3,2'-TRIMETHYLENE-2-PHENYL INDOLE	25.69	0.026	7.53987
4,5-DIPHENYLIMIDAZOLE (A)	16.54	0.099	0.53853
PPF (TRACE)	30.86	0.015	22.60583
PPO (TRACE)	27.22	0.022	10.68718
2-(4-FLUOROPHENYL)-5-PHENYLOXAZOLE (TRACE)	25.96	0.026	8.03265
αNPO (TRACE)	31.51	0.014	25.46572
BBO (B)	35.87	0.010	55.29526
POPOP	34.57	0.011	44.25298
DIMETHYL POPOP (TRACE)	35.18	0.010	49.07121
PPD (TRACE)	13.50	0.182	0.16025
2,5-DI(2-METHYLPHENYL)OXADIAZOLE	14.68	0.141	0.26459
PBD	26.38	0.024	8.86986
BBD (B)	32.77	0.013	32.38886
αNPD (TRACE)	28.27	0.020	13.33419
βNPD	25.26	0.028	6.82664
αNND (TRACE)	29.96	0.017	18.84035
βNND (TRACE)	30.02	0.017	19.20046
TRANS-1-(4-BIPHENYLYL)-2-PHENYL-ETHYLENE	32.30	0.013	29.63520
TRANS-1-(4-BIPHENYLYL)-2-(1-NAPHTHYL)-ETHYLENE	33.32	0.012	35.48111
DIPHENYLSTILBENE (B)	36.24	0.009	58.80760
1,1,4,4-TETRAPHENYLBUTADIENE	33.87	0.012	39.10429
1,6-DIPHENYLHEXATRIENE	37.21	0.009	68.76801
1,8-DIPHENYL-1,3,5,7-OCTATETRAENE	35.99	0.010	56.04253
BPSB (TRACE)	37.35	0.009	70.33740

ACCEPTOR MOLECULES	RO	CO	OVERLAP
BIS-MSB	36.50	0.009	61.19346
ESCULIN (A)	28.32	0.020	13.40373
AZULENE	21.46	0.045	2.51660
NAPHTHALENE	7.41	1.101	0.00436
DEUTERATED NAPHTHALENE-C10D8	9.57	0.510	0.02028
1-METHYLNAPHTHALENE 5ML/L	9.07	0.600	0.01467
2-METHYL NAPHTHALENE	10.82	0.354	0.04227
2,3-DIMETHYL NAPHTHALENE	10.66	0.369	0.03874
2,6-DIMETHYL NAPHTHALENE (TRACE)	13.99	0.163	0.19770
1-NAPHTHOL (TRACE)	15.45	0.121	0.35900
2-NAPHTHOL (TRACE)	17.43	0.085	0.73658
1-NAPHTHYLAMINE	22.00	0.042	2.96768
2-CHLORONAPHTHALENE	11.28	0.312	0.05438
ACENAPHTHENE	13.52	0.181	0.16145
1-PHENYL NAPHTHALENE	13.94	0.165	0.19403
2-PHENYL NAPHTHALENE	20.91	0.049	2.20211
1,4-DIPHENYL NAPHTHALENE	21.39	0.046	2.51518
1,5-DIPHENYL NAPHTHALENE	19.96	0.056	1.67033
1,7-DIPHENYL NAPHTHALENE	19.71	0.058	1.53925
1,4,5,8-TETRAPHENYL NAPHTHALENE	30.35	0.016	20.25149
1,1'-BINAPHTHYL	16.19	0.105	0.47472
4,4'-DI(N-BUTOXY)-1,1'-BINAPHTHYL	25.54	0.027	7.26333
2,2'-BINAPHTHYL	21.74	0.044	2.77396
ACRIDINE (A)	25.98	0.026	7.96577
ANTHRACENE (TRACE)	24.38	0.031	5.44798
DEUTERATED ANTHRACENE C14D10 (TRACE)	24.15	0.032	5.14760
9-METHYL ANTHRACENE (TRACE)	23.80	0.033	4.70789
1-AMINO ANTHRACENE	22.13	0.041	3.03057
2-AMINO ANTHRACENE	22.73	0.038	3.55290
9-VINYL ANTHRACENE	24.55	0.030	5.66062
9-PHENYL ANTHRACENE (TRACE)	25.07	0.028	6.42855
9,10-DIPHENYL ANTHRACENE	25.23	0.028	6.66507
9,10-DICHLOROANTHRACENE (TRACE)	24.02	0.032	4.95150
BIANTHRANYL (B)	26.72	0.023	9.39413
9,10-DINAPHTHYLANTHRACENE (B)	24.93	0.029	6.19239
9-(P-VINYLPHENYL)-10-PHENYL ANTHRACENE	25.31	0.028	6.77407
9,10-BIS(PHENYL-ETHYNYL)ANTHRACENE (TRACE)	22.83	0.038	3.64845
RUBICENE (B)	26.35	0.024	8.63295
DIBENZ(A,H)ANTHRACENE (B)	28.54	0.019	13.98802
TETRACENE (TRACE) (B)	18.57	0.070	1.05602
RUBRENE (B)	26.23	0.025	8.39626
TETRAHYDROCANNABINOL ACID METHYL ESTER	11.11	0.326	0.04985
PHENANTHRENE	13.34	0.189	0.14741
3,4-BENZOPHENANTHRENE	20.92	0.049	2.17368
2-PHENYL PHENANTHRENE	16.37	0.102	0.50278
CHRYSENE (TRACE)	21.66	0.044	2.68856
TRIPHENYLENE	14.25	0.155	0.21920
DEUTERATED TRIPHENYLENE-C18D12	14.09	0.160	0.20485
PYRENE	28.97	0.018	15.36212
1,3,6,8,-TETRAPHENYL PYRENE	31.21	0.015	23.84202
BIPYRENYL (B)	34.32	0.011	42.19785
O-PHENYLENE-PYRENE	28.63	0.019	14.19588
ANTHANTHRENE (B)	26.59	0.024	9.11491
FLUORANTHENE	25.89	0.026	7.76706
3-PHENYL FLUORANTHENE	27.86	0.021	12.04386
3-AMINO FLUORANTHENE	25.73	0.026	7.47407
3,3'-BIFLUORANTHENYL (B)	30.88	0.015	22.37705
BENZO(GHI)FLUORANTHENE	25.02	0.029	6.32135
9,12-DI(N-OCTADECOXY)-BENZ(E)ACEPHENANTHRYLENE	27.50	0.022	11.15746
DECACYCLENE (B)	33.60	0.012	37.08266
10-ETHOXY-BENZO(KL)XANTHENE	25.75	0.026	7.52913
PERYLENE	22.57	0.039	3.41037
BENZO(GHI)PERYLENE (B)	27.56	0.021	11.31882
BBOT (TRACE)	35.23	0.010	49.37886
QUININE SULFATE (A)	25.27	0.028	6.82061
ACRIDINE YELLOW (A)	19.71	0.058	1.51312
ACRIDINE ORANGE (A)	18.47	0.071	1.02415
ACRIDINE RED (A)	22.14	0.041	3.03596
URANINE (A)	22.36	0.040	3.21922
RHODAMINE B (A)	25.08	0.028	6.42094
RHODAMINE 6G (A)	25.54	0.027	7.15530
EOSIN B (A)	23.81	0.033	4.69550
EOSIN Y (A)	24.23	0.031	5.22212
ROSE BENGAL (A)	25.32	0.028	6.78733
INDENOINDENE	26.68	0.024	9.47788

8.58 Donor Molecule: *m*-Quaterphenyl

ANILINE 3ML/L	4.27	5.750	0.00025
N,N-DIMETHYLANILINE 3ML/L	9.95	0.454	0.03911
TRIPHENYLAMINE	20.79	0.050	3.23363
1,4-DIMETHOXY BENZENE	5.05	3.478	0.00067
1-METHYL-4-CHLOROBENZENE 5ML/L	0.00	0.000	0.00000
SODIUM SALICYLATE (A)	13.96	0.164	0.29621
ANTHRANILIC ACID (A)	22.35	0.040	4.94700
TMPD	19.27	0.063	2.03707
4-METHOXY BIPHENYL	3.72	8.659	0.00011
4-BIPHENYLYL PHENYL ETHER	0.00	0.000	0.00000
4,4'-DIHYDROXYBIPHENYL (A)	9.29	0.558	0.02593
4,4'-DIMETHOXY BIPHENYL	7.45	1.084	0.00690
4,4'-DIISOBUTOXYBIPHENYL	6.67	1.507	0.00357

ACCEPTOR MOLECULES	RO	CO	OVERLAP
BENZIDINE	18.41	0.072	1.56121
4-VINYL BIPHENYL	10.09	0.436	0.04268
M-TERPHENYL	7.01	1.298	0.00481
1,3,5-TRIPHENYL BENZENE	8.38	0.761	0.01392
1,3-DIBIPHENYLYLBENZENE (B)	17.68	0.081	1.22598
1,3,5,-TRIBIPHENYLYLBENZENE (B)	20.86	0.049	3.30831
FLUORENE (TRACE)	5.43	2.801	0.00104
1-METHYL FLUORENE	5.24	3.111	0.00084
1,2-BENZOFLUORENE	19.26	0.063	2.04093
9,10-DIHYDROPHENANTHRENE	6.84	1.398	0.00415
CARBAZOLE (TRACE)	18.36	0.072	1.53781
N-METHYL CARBAZOLE (TRACE)	20.37	0.053	2.85644
N-VINYLCARBAZOLE	20.49	0.052	2.96065
N-PHENYL CARBAZOLE	20.23	0.054	2.74442
N-METHYL-2,3-BENZCARBAZOLE	22.44	0.040	5.04414
DIBENZOFURAN	4.59	4.625	0.00038
DIBENZOTHIOPHENE (TRACE)	16.46	0.100	0.79839
P-TERPHENYL	11.70	0.279	0.10383
DEUTERATED P-TERPHENYL-C18D14	11.09	0.328	0.07528
4-METHYL TERPHENYL	13.21	0.194	0.21488
2-METHYL-P-TERPHENYL	6.37	1.728	0.00272
3,3''-DIMETHYL-P-TERPHENYL	13.04	0.202	0.19840
4-(3,3-DIMETHYLBUTOXY)-P-TERPHENYL	17.58	0.082	1.18770
4,4''-DIHEXAHYDROFARNESOXY-P-TERPHENYL	20.38	0.053	2.88019
2,2' -METHYLENE-P-TERPHENYL	19.38	0.061	2.13520
2,2' -ETHYLENE-P-TERPHENYL	20.13	0.055	2.67683
2,2'-METHYLENE-5',6''-METHYLENE-P-TERPHENYL (TRACE)	28.11	0.020	19.78337
3-PHENYL DIBENZOFURAN	19.48	0.061	2.19851
BENZO(1,2-B-4,5-B')BISBENZOFURAN (TRACE) (B)	29.37	0.018	25.66963
SEXIPHENYL BENZENE	0.00	0.000	0.00000
P-QUATERPHENYL	21.50	0.045	3.96309
3,3'''-DIMETHYL-P-QUATERPHENYL	22.70	0.038	5.48071
TETRAMETHYL-P-QUATERPHENYL	7.96	0.888	0.01028
3,3''' -DIMETHYL-3',2''-METHYLENE-P-QUATERPHENYL	29.25	0.018	25.05542
3,3''' -DIMETHYL-3',2''-ETHYLENE-P-QUATERPHENYL	28.85	0.019	22.99116
4-(3-ETHYLBUTOXY)-P-QUATERPHENYL (B)	26.68	0.024	14.40585
4,4'''-DI(2-BUTYLOCTOXY-1)-P-QUATERPHENYL	27.32	0.022	16.60401
DIETHYL-P-QUINQUEPHENYL	23.69	0.034	7.08321
DI(3-ETHYLHEPTYL)-P-QUINQUEPHENYL	28.46	0.019	21.16542
TETRAMETHYL-P-SEXIPHENYL	22.71	0.038	5.51050
TETRAMETHYL-P-OCTAPHENYL (B)	31.88	0.014	41.81124
INDOLE (TRACE)	0.00	0.000	0.00000
2-PHENYL INDOLE	23.60	0.034	6.92953
1-METHYL-2-PHENYL INDOLE	19.11	0.064	1.95118
1-NEOPENTYL-2-PHENYL INDOLE	17.02	0.091	0.97787
1-N-BUTYL-2(BIPHENYLYL-4) INDOLE	26.14	0.025	12.73790
1,2-DIPHENYL INDOLE	19.86	0.057	2.45838
1-BIPHENYLYL-2-PHENYL INDOLE	21.98	0.042	4.51273
1,3-DIMETHYL-2-PHENYL INDOLE	19.85	0.057	2.45336
2,3-DIPHENYL INDOLE	24.20	0.032	8.00056
1-METHYL-2,3-DIPHENYL INDOLE	22.01	0.042	4.53277
2-PHENYL-3,2'-METHYLENE INDOLE (TRACE)	26.05	0.025	12.51320
2-PHENYL-3,2'-ETHYLENE INDOLE (TRACE)	27.68	0.021	17.95416
2-PHENYL-3,2' -TRIMETHYLENE INDOLE	25.01	0.029	9.80486
1-METHYL-3,2'-METHYLENE-2-PHENYL INDOLE (TRACE)	27.42	0.022	17.00414
1-METHYL-3,2'-ETHYLENE-2-PHENYL INDOLE	27.80	0.021	18.41374
1-METHYL-3,2'-TRIMETHYLENE-2-PHENYL INDOLE	23.89	0.033	7.44390
4,5-DIPHENYLIMIDAZOLE (A)	15.21	0.127	0.49665
PPF (TRACE)	28.80	0.019	22.79128
PPO (TRACE)	25.26	0.028	10.42118
2-(4-FLUOROPHENYL)-5-PHENYLOXAZOLE (TRACE)	24.08	0.032	7.82112
αNPO (TRACE)	29.45	0.018	25.91296
BBO (B)	33.52	0.012	56.22399
POPOP	32.28	0.013	44.80131
DIMETHYL POPOP (TRACE)	32.84	0.013	49.54098
PPD (TRACE)	11.84	0.269	0.11175
2,5-DI(2-METHYLPHENYL)OXADIAZOLE	13.06	0.201	0.20068
PBD	24.38	0.031	8.42550
BBD (B)	30.56	0.016	32.51655
αNPD (TRACE)	26.38	0.024	13.44294
βNPD	23.38	0.035	6.54603
αNND (TRACE)	27.99	0.020	19.15979
βNND (TRACE)	27.97	0.020	19.14548
TRANS-1-(4-BIPHENYLYL)-2-PHENYL-ETHYLENE	30.14	0.016	29.88767
TRANS-1-(4-BIPHENYLYL)-2-(1-NAPHTHYL)-ETHYLENE	31.13	0.015	36.02342
DIPHENYLSTILBENE (B)	33.87	0.012	59.82968
1,1,4,4-TETRAPHENYLBUTADIENE	31.65	0.014	39.74210
1,6-DIPHENYLHEXATRIENE	34.75	0.011	69.64044
1,8-DIPHENYL-1,3,5,7-OCTATETRAENE	33.52	0.012	55.92175
BPSB (TRACE)	34.88	0.011	71.18896
BIS-MSB	34.09	0.011	62.04427
ESCULIN (A)	26.47	0.024	13.64542
AZULENE	20.06	0.055	2.56849
NAPHTHALENE	6.57	1.578	0.00325
DEUTERATED NAPHTHALENE-C10D8	8.74	0.671	0.01789
1-METHYLNAPHTHALENE 5ML/L	8.14	0.830	0.01171
2-METHYL NAPHTHALENE	9.82	0.472	0.03619
2,3-DIMETHYL NAPHTHALENE	9.68	0.493	0.03321
2,6-DIMETHYL NAPHTHALENE (TRACE)	12.93	0.207	0.18848
1-NAPHTHOL (TRACE)	14.12	0.159	0.31897
2-NAPHTHOL (TRACE)	16.25	0.104	0.73927
1-NAPHTHYLAMINE	20.52	0.052	2.98131
2-CHLORONAPHTHALENE	10.32	0.407	0.04877
ACENAPHTHENE	12.29	0.241	0.13934
1-PHENYL NAPHTHALENE	12.63	0.222	0.16370
2-PHENYL NAPHTHALENE	19.17	0.064	1.99438
1,4-DIPHENYL NAPHTHALENE	19.74	0.058	2.37529

ACCEPTOR MOLECULES	RO	CO	OVERLAP
1,5-DIPHENYL NAPHTHALENE	18.38	0.072	1.54648
1,7-DIPHENYL NAPHTHALENE	18.13	0.075	1.42490
1,4,5,8-TETRAPHENYL NAPHTHALENE	28.36	0.020	20.58227
1,1'-BINAPHTHYL	14.82	0.137	0.42668
4,4'-DI(N-BUTOXY)-1,1'-BINAPHTHYL	23.77	0.033	7.20281
2,2'-BINAPHTHYL	19.95	0.056	2.53201
ACRIDINE (A)	24.27	0.031	8.07406
ANTHRACENE (TRACE)	22.73	0.038	5.47001
DEUTERATED ANTHRACENE C14D10 (TRACE)	22.52	0.039	5.17918
9-METHYL ANTHRACENE (TRACE)	22.22	0.041	4.76155
1-AMINO ANTHRACENE	20.59	0.051	3.00552
2-AMINO ANTHRACENE	21.18	0.047	3.55962
9-VINYL ANTHRACENE	22.90	0.037	5.68668
9-PHENYL ANTHRACENE (TRACE)	23.40	0.035	6.49021
9,10-DIPHENYL ANTHRACENE	23.49	0.035	6.62012
9,10-DICHLOROANTHRACENE (TRACE)	22.34	0.040	4.89622
BIANTHRANYL (B)	24.89	0.029	9.37212
9,10-DINAPHTHYLANTHRACENE (B)	23.17	0.036	6.10019
9-(P-VINYLPHENYL)-10-PHENYL ANTHRACENE	23.54	0.034	6.71106
9,10-BIS(PHENYL-ETHYNYL)ANTHRACENE (TRACE)	20.89	0.049	3.27812
RUBICENE (B)	24.42	0.031	8.35364
DIBENZ(A,H)ANTHRACENE (B)	26.61	0.024	14.01972
TETRACENE (TRACE) (B)	17.21	0.088	1.02325
RUBRENE (B)	24.15	0.032	7.81395
TETRAHYDROCANNABINOL ACID METHYL ESTER	9.89	0.463	0.03777
PHENANTHRENE	12.46	0.231	0.14958
3,4-BENZOPHENANTHRENE	19.30	0.062	2.04632
2-PHENYL PHENANTHRENE	14.96	0.133	0.44767
CHRYSENE (TRACE)	19.88	0.057	2.45712
TRIPHENYLENE	13.27	0.191	0.21862
DEUTERATED TRIPHENYLENE-C18D12	13.11	0.199	0.20314
PYRENE	27.07	0.023	15.61769
1,3,6,8,-TETRAPHENYL PYRENE	29.05	0.018	23.70011
BIPYRENYL (B)	32.08	0.014	42.99268
O-PHENYLENE-PYRENE	26.59	0.024	13.94057
ANTHANTHRENE (B)	24.62	0.030	8.77755
FLUORANTHENE	24.17	0.032	7.86346
3-PHENYL FLUORANTHENE	25.98	0.026	12.12723
3-AMINO FLUORANTHENE	23.95	0.033	7.44074
3,3'-BIFLUORANTHENYL (B)	28.76	0.019	22.30852
BENZO(GHI)FLUORANTHENE	23.40	0.035	6.47811
9,12-DI(N-OCTADECOXY)-BENZ(E)ACEPHENANTHRYLENE	25.62	0.027	11.15902
DECACYCLENE (B)	31.28	0.015	36.94702
10-ETHOXY-BENZO(KL)XANTHENE	23.97	0.032	7.47799
PERYLENE	20.89	0.049	3.27627
BENZO(GHI)PERYLENE (B)	25.67	0.026	11.27739
BBOT (TRACE)	32.85	0.013	49.60659
QUININE SULFATE (A)	23.63	0.034	6.95654
ACRIDINE YELLOW (A)	18.21	0.074	1.43513
ACRIDINE ORANGE (A)	17.05	0.090	0.96987
ACRIDINE RED (A)	20.61	0.051	3.01709
URANINE (A)	20.81	0.050	3.20296
RHODAMINE B (A)	23.34	0.035	6.37832
RHODAMINE 6G (A)	23.87	0.033	7.28621
EOSIN B (A)	22.14	0.041	4.63957
EOSIN Y (A)	22.52	0.039	5.14008
ROSE BENGAL (A)	23.60	0.034	6.80502
INDENOINDENE	24.80	0.029	9.32744

8.59 Donor Molecule: 1,3,5-Triphenylbenzene

N,N-DIMETHYLANILINE 3ML/L	5.02	3.542	0.00045
TRIPHENYLAMINE	17.56	0.083	0.82188
SODIUM SALICYLATE (A)	12.26	0.242	0.09522
ANTHRANILIC ACID (A)	22.88	0.037	3.98946
TMPD	18.83	0.067	1.24275
4,4'-DIHYDROXYBIPHENYL (A)	4.86	3.907	0.00037
4,4'-DIMETHOXY BIPHENYL	0.00	0.000	0.00000
4,4'-DIISOBUTOXYBIPHENYL	0.00	0.000	0.00000
BENZIDINE	14.43	0.149	0.25304
4-VINYL BIPHENYL	0.00	0.000	0.00000
M-TERPHENYL	0.00	0.000	0.00000
1,3,5-TRIPHENYL BENZENE	5.07	3.442	0.00048
1,3-DIBIPHENYLYLBENZENE (B)	12.73	0.217	0.11974
1,3,5,-TRIBIPHENYLYLBENZENE (B)	14.99	0.133	0.31895
FLUORENE (TRACE)	0.00	0.000	0.00000
1-METHYL FLUORENE	0.00	0.000	0.00000
1,2-BENZOFLUORENE	16.50	0.099	0.56593
9,10-DIHYDROPHENANTHRENE	0.00	0.000	0.00000
CARBAZOLE (TRACE)	15.26	0.126	0.35493
N-METHYL CARBAZOLE (TRACE)	19.73	0.058	1.64982
N-VINYLCARBAZOLE	19.25	0.063	1.42667
N-PHENYL CARBAZOLE	19.07	0.064	1.34904
N-METHYL-2,3-BENZCARBAZOLE	22.79	0.038	3.87565
DIBENZOFURAN	0.00	0.000	0.00000
DIBENZOTHIOPHENE (TRACE)	12.41	0.234	0.10248
P-TERPHENYL	5.80	2.288	0.00108
DEUTERATED P-TERPHENYL-C18D14	0.00	0.000	0.00000
4-METHYL TERPHENYL	7.11	1.247	0.00364
2-METHYL-P-TERPHENYL	0.00	0.000	0.00000
3,3''-DIMETHYL-P-TERPHENYL	7.55	1.040	0.00523

ACCEPTOR MOLECULES	RO	CO	OVERLAP
4-(3,3-DIMETHYLBUTOXY)-P-TERPHENYL	12.35	0.238	0.09986
4,4''-DIHEXAHYDROFARNESOXY-P-TERPHENYL	15.08	0.130	0.33141
2,2'-METHYLENE-P-TERPHENYL	12.78	0.214	0.12257
2,2'-ETHYLENE-P-TERPHENYL	14.21	0.156	0.23150
2,2'-METHYLENE-5',6''-METHYLENE-P-TERPHENYL (TRACE)	25.09	0.028	7.00604
3-PHENYL DIBENZOFURAN	12.74	0.216	0.12075
BENZO(1,2-B-4,5-B')BISBENZOFURAN (TRACE) (B)	26.68	0.024	10.11276
SEXIPHENYL BENZENE	0.00	0.000	0.00000
P-QUATERPHENYL	16.74	0.095	0.61731
3,3'''-DIMETHYL-P-QUATERPHENYL	18.30	0.073	1.05273
TETRAMETHYL-P-QUATERPHENYL	0.00	0.000	0.00000
3,3'''-DIMETHYL-3',2''-METHYLENE-P-QUATERPHENYL	26.08	0.025	8.81935
3,3'''-DIMETHYL-3',2''-ETHYLENE-P-QUATERPHENYL	26.08	0.025	8.79585
4-(3-ETHYLBUTOXY)-P-QUATERPHENYL (B)	23.47	0.035	4.67527
4,4'''-DI(2-BUTYLOCTOXY-1)-P-QUATERPHENYL	23.87	0.033	5.17897
DIETHYL-P-QUINQUEPHENYL	19.04	0.065	1.33862
DI(3-ETHYLHEPTYL)-P-QUINQUEPHENYL	25.19	0.028	7.13135
TETRAMETHYL-P-SEXIPHENYL	17.55	0.083	0.82012
TETRAMETHYL-P-OCTAPHENYL (B)	29.71	0.017	19.16307
INDOLE (TRACE)	0.00	0.000	0.00000
2-PHENYL INDOLE	18.84	0.067	1.25695
1-METHYL-2-PHENYL INDOLE	14.61	0.143	0.27355
1-NEOPENTYL-2-PHENYL INDOLE	11.76	0.275	0.07442
1-N-BUTYL-2(BIPHENYLYL-4) INDOLE	24.22	0.031	5.63119
1,2-DIPHENYL INDOLE	15.06	0.131	0.32735
1-BIPHENYLYL-2-PHENYL INDOLE	17.80	0.079	0.88985
1,3-DIMETHYL-2-PHENYL INDOLE	16.06	0.108	0.48251
2,3-DIPHENYL INDOLE	22.29	0.040	3.41802
1-METHYL-2,3-DIPHENYL INDOLE	20.10	0.055	1.84077
2-PHENYL-3,2'-METHYLENE INDOLE (TRACE)	22.61	0.039	3.74481
2-PHENYL-3,2'-ETHYLENE INDOLE (TRACE)	26.46	0.024	9.58845
2-PHENYL-3,2'-TRIMETHYLENE INDOLE	21.12	0.047	2.49107
1-METHYL-3,2'-METHYLENE-2-PHENYL INDOLE (TRACE)	25.68	0.026	8.02674
1-METHYL-3,2'-ETHYLENE-2-PHENYL INDOLE	27.53	0.021	12.15564
1-METHYL-3,2'-TRIMETHYLENE-2-PHENYL INDOLE	20.96	0.049	2.37908
4,5-DIPHENYLIMIDAZOLE (A)	12.15	0.250	0.09019
PPF (TRACE)	26.90	0.023	10.60518
PPO (TRACE)	21.17	0.047	2.52929
2-(4-FLUOROPHENYL)-5-PHENYLOXAZOLE (TRACE)	19.95	0.056	1.77064
αNPO (TRACE)	30.28	0.016	21.40537
BBO (B)	35.36	0.010	54.23114
POPOP	35.27	0.010	53.33986
DIMETHYL POPOP (TRACE)	36.37	0.009	64.01138
PPD (TRACE)	0.00	0.000	0.00000
2,5-DI(2-METHYLPHENYL)OXADIAZOLE	5.65	2.474	0.00092
PBD	19.19	0.063	1.40504
BBD (B)	28.48	0.019	14.90974
αNPD (TRACE)	24.98	0.029	6.78404
βNPD	19.03	0.065	1.33272
αNND (TRACE)	28.15	0.020	13.86474
βNND (TRACE)	25.51	0.027	7.72576
TRANS-1-(4-BIPHENYLYL)-2-PHENYL-ETHYLENE	28.64	0.019	15.40154
TRANS-1-(4-BIPHENYLYL)-2-(1-NAPHTHYL)-ETHYLENE	32.24	0.013	31.15218
DIPHENYLSTILBENE (B)	35.68	0.010	57.16420
1,1,4,4-TETRAPHENYLBUTADIENE	33.79	0.012	41.19144
1,6-DIPHENYLHEXATRIENE	37.88	0.008	81.79420
1,8-DIPHENYL-1,3,5,7-OCTATETRAENE	38.42	0.008	88.47376
BPSB (TRACE)	38.17	0.008	85.65394
BIS-MSB	36.68	0.009	67.42051
ESCULIN (A)	27.34	0.022	11.61247
AZULENE	19.56	0.060	1.54079
NAPHTHALENE	3.26	12.949	0.00003
DEUTERATED NAPHTHALENE-C10D8	6.34	1.756	0.00183
1-METHYLNAPHTHALENE 5ML/L	4.64	4.473	0.00028
2-METHYL NAPHTHALENE	5.91	2.169	0.00120
2,3-DIMETHYL NAPHTHALENE	5.93	2.150	0.00122
2,6-DIMETHYL NAPHTHALENE (TRACE)	9.23	0.569	0.01742
1-NAPHTHOL (TRACE)	9.41	0.537	0.01958
2-NAPHTHOL (TRACE)	13.06	0.201	0.13964
1-NAPHTHYLAMINE	18.49	0.071	1.11804
2-CHLORONAPHTHALENE	6.89	1.368	0.00302
ACENAPHTHENE	7.70	0.978	0.00590
1-PHENYL NAPHTHALENE	8.10	0.843	0.00795
2-PHENYL NAPHTHALENE	14.55	0.145	0.26656
1,4-DIPHENYL NAPHTHALENE	15.70	0.115	0.42091
1,5-DIPHENYL NAPHTHALENE	13.91	0.166	0.20349
1,7-DIPHENYL NAPHTHALENE	13.99	0.163	0.21094
1,4,5,8-TETRAPHENYL NAPHTHALENE	29.28	0.018	17.45668
1,1'-BINAPHTHYL	10.53	0.384	0.03830
4,4'-DI(N-BUTOXY)-1,1'-BINAPHTHYL	20.89	0.049	2.32618
2,2'-BINAPHTHYL	15.26	0.126	0.35489
ACRIDINE (A)	26.60	0.024	9.81786
ANTHRACENE (TRACE)	25.02	0.029	6.80865
DEUTERATED ANTHRACENE C14D10 (TRACE)	24.71	0.030	6.32797
9-METHYL ANTHRACENE (TRACE)	25.13	0.028	6.96810
1-AMINO ANTHRACENE	23.82	0.033	5.02593
2-AMINO ANTHRACENE	23.21	0.036	4.30338
9-VINYL ANTHRACENE	26.06	0.025	8.65868
9-PHENYL ANTHRACENE (TRACE)	26.42	0.024	9.42150
9,10-DIPHENYL ANTHRACENE	27.18	0.022	11.12472
9,10-DICHLOROANTHRACENE (TRACE)	26.00	0.025	8.53372
BIANTHRANYL (B)	28.52	0.019	14.84898
9,10-DINAPHTHYLANTHRACENE (B)	26.63	0.024	9.82090
9-(P-VINYLPHENYL)-10-PHENYL ANTHRACENE	27.20	0.022	11.39277
9,10-BIS(PHENYL-ETHYNYL)ANTHRACENE (TRACE)	24.29	0.031	5.65027
RUBICENE (B)	25.90	0.026	8.29702
DIBENZ(A,H)ANTHRACENE (B)	26.49	0.024	9.55351

ACCEPTOR MOLECULES	RO	CO	OVERLAP
TETRACENE (TRACE) (B)	19.34	0.062	1.43907
RUBRENE (B)	21.28	0.046	2.55122
TETRAHYDROCANNABINOL ACID METHYL ESTER	4.29	5.658	0.00018
PHENANTHRENE	11.83	0.271	0.07655
3,4-BENZOPHENANTHRENE	16.23	0.105	0.50660
2-PHENYL PHENANTHRENE	13.75	0.172	0.18874
CHRYSENE (TRACE)	16.42	0.101	0.54528
TRIPHENYLENE	12.00	0.259	0.08352
DEUTERATED TRIPHENYLENE C18D12	11.60	0.281	0.07107
PYRENE	25.03	0.029	6.82857
1,3,6,8,-TETRAPHENYL PYRENE	33.76	0.012	40.79983
BIPYRENYL (B)	34.51	0.011	46.64273
O-PHENYLENE-PYRENE	29.67	0.017	18.76720
ANTHANTHRENE (B)	27.10	0.022	10.89085
FLUORANTHENE	25.78	0.026	8.09746
3-PHENYL FLUORANTHENE	28.86	0.019	15.91575
3-AMINO FLUORANTHENE	26.08	0.025	8.65993
3,3'-BIFLUORANTHENYL (B)	32.31	0.013	31.25659
BENZO(GHI)FLUORANTHENE	24.22	0.031	5.56750
9,12-DI(N-OCTADECOXY)-BENZ(E)ACEPHENANTHRYLENE	28.28	0.020	14.07053
DECACYCLENE (B)	34.76	0.011	48.47565
10-ETHOXY-BENZO(KL)XANTHENE	27.70	0.021	12.47279
PERYLENE	25.36	0.027	7.31714
BENZO(GHI)PERYLENE (B)	29.33	0.018	17.60109
BBOT (TRACE)	37.06	0.009	71.55210
QUININE SULFATE (A)	22.16	0.041	3.31968
ACRIDINE YELLOW (A)	21.92	0.042	3.05014
ACRIDINE ORANGE (A)	18.97	0.065	1.28336
ACRIDINE RED (A)	22.30	0.040	3.37755
URANINE (A)	21.50	0.045	2.71671
RHODAMINE B (A)	25.39	0.027	7.36046
RHODAMINE 6G (A)	25.66	0.026	7.85528
EOSIN B (A)	22.34	0.040	3.41440
EOSIN Y (A)	22.84	0.038	3.90769
ROSE BENGAL (A)	23.75	0.033	4.93955
INDENOINDENE	20.65	0.051	2.17307

8.62 Donor Molecule: Fluorene (trace)

POLYVINYLTOLUENE	5.20	3.185	0.00019
ANILINE 3ML/L	14.86	0.136	0.10440
N,N-DIMETHYLANILINE 3ML/L	21.62	0.044	0.98177
TRIPHENYLAMINE	34.13	0.011	15.06032
1,4-DIMETHOXY BENZENE	18.39	0.072	0.37486
1-METHYL-4-CHLOROBENZENE 5ML/L	0.00	0.000	0.00000
SODIUM SALICYLATE (A)	24.12	0.032	1.87701
ANTHRANILIC ACID (A)	26.28	0.025	3.12860
TMPD	24.49	0.030	2.04963
1,2,4-TRIMETHYLBENZENE (TRACE)	0.00	0.000	0.00000
MESITYLENE 5ML/L	0.00	0.000	0.00000
1,3,5-TRIETHYLBENZENE 5ML/L	0.00	0.000	0.00000
STYRENE	5.66	2.462	0.00032
3,3'-DIMETHYL BIPHENYL	6.92	1.351	0.00107
2,2'-DIMETHYLBIPHENYL 2ML/L	0.00	0.000	0.00000
OCTAMETHYL BIPHENYL	0.00	0.000	0.00000
4-METHOXY BIPHENYL	14.55	0.145	0.09199
4-BIPHENYLYL PHENYL ETHER	13.55	0.180	0.05996
4,4'-DIHYDROXYBIPHENYL (A)	20.58	0.051	0.73257
4,4'-DIMETHOXY BIPHENYL	19.20	0.063	0.48373
4,4'-DIISOBUTOXYBIPHENYL	19.28	0.062	0.49753
BENZIDINE	31.84	0.014	9.96114
4-VINYL BIPHENYL	25.50	0.027	2.65618
M-TERPHENYL	17.69	0.081	0.29669
M-QUATERPHENYL	12.90	0.209	0.04469
1,3,5-TRIPHENYL BENZENE	16.81	0.094	0.21658
1,3-DIBIPHENYLYLBENZENE (B)	33.66	0.012	13.93122
1,3,5,-TRIBIPHENYLYLBENZENE (B)	38.46	0.008	31.01383
FLUORENE (TRACE)	22.64	0.039	1.30462
1-METHYL FLUORENE	20.54	0.052	0.72659
1,2-BENZOFLUORENE	31.41	0.014	9.16468
9,10-DIHYDROPHENANTHRENE	21.44	0.045	0.93805
CARBAZOLE (TRACE)	26.02	0.025	2.97145
N-METHYL CARBAZOLE (TRACE)	24.82	0.029	2.22730
N-VINYLCARBAZOLE	25.39	0.027	2.55750
N-PHENYL CARBAZOLE	25.37	0.027	2.54534
N-METHYL-2,3-BENZCARBAZOLE	29.80	0.017	6.64625
DIBENZOFURAN	19.85	0.057	0.59291
DIBENZOTHIOPHENE (TRACE)	23.67	0.034	1.68611
P-TERPHENYL	26.70	0.026	3.48936
DEUTERATED P-TERPHENYL-C18D14	26.31	0.025	3.20296
4-METHYL TERPHENYL	28.59	0.019	5.25937
2-METHYL-P-TERPHENYL	19.00	0.065	0.45490
3,3''-DIMETHYL-P-TERPHENYL	27.85	0.021	4.49371
4-(3,3-DIMETHYLBUTOXY)-P-TERPHENYL	32.24	0.013	10.77932
4,4''-DIHEXAHYDROFARNESOXY-P-TERPHENYL	35.25	0.010	18.39168
2,2'-METHYLENE-P-TERPHENYL	34.40	0.011	15.90774
2,2'-ETHYLENE-P-TERPHENYL	34.32	0.011	15.66477
2,2'-METHYLENE-5',6''-METHYLENE-P-TERPHENYL (TRACE)	36.88	0.009	24.03883
3-PHENYL DIBENZOFURAN	35.47	0.010	19.12133
BENZO(1,2-B-4,5-B')BISBENZOFURAN (TRACE) (B)	38.04	0.008	28.92278
SEXIPHENYL BENZENE	14.58	0.144	0.09307

ACCEPTOR MOLECULES	RO	CO	OVERLAP
P-QUATERPHENYL	35.93	0.010	20.58783
3,3'''-DIMETHYL-P-QUATERPHENYL	36.90	0.009	24.13679
TETRAMETHYL-P-QUATERPHENYL	22.36	0.040	1.20928
3,3''' -DIMETHYL-3',2''-METHYLENE-P-QUATERPHENYL	39.88	0.007	38.38571
3,3''' -DIMETHYL-3',2''-ETHYLENE-P-QUATERPHENYL	39.76	0.007	37.57408
4-(3-ETHYLBUTOXY)-P-QUATERPHENYL (B)	39.13	0.007	34.18308
4,4'''-DI(2-BUTYLOCTOXY-1)-P-QUATERPHENYL	40.35	0.007	41.15108
DIETHYL-P-QUINQUEPHENYL	38.53	0.008	31.28184
DI(3-ETHYLHEPTYL)-P-QUINQUEPHENYL	41.58	0.006	49.07867
TETRAMETHYL-P-SEXIPHENYL	38.31	0.008	30.26689
TETRAMETHYL-P-OCTAPHENYL (B)	43.51	0.005	64.41743
INDOLE (TRACE)	0.00	0.000	0.00000
2-PHENYL INDOLE	35.92	0.010	20.56188
1-METHYL-2-PHENYL INDOLE	32.12	0.013	10.50263
1-NEOPENTYL-2-PHENYL INDOLE	31.03	0.015	8.56952
1-N-BUTYL-2(BIPHENYLYL-4) INDOLE	36.30	0.009	21.76225
1,2-DIPHENYL INDOLE	33.36	0.012	13.18790
1-BIPHENYLYL-2-PHENYL INDOLE	35.18	0.010	18.07441
1,3-DIMETHYL-2-PHENYL INDOLE	31.34	0.015	9.06161
2,3-DIPHENYL INDOLE	34.10	0.011	14.92858
1-METHYL-2,3-DIPHENYL INDOLE	31.72	0.014	9.68783
2-PHENYL-3,2'-METHYLENE INDOLE (TRACE)	36.01	0.010	20.82321
2-PHENYL-3,2'-ETHYLENE INDOLE (TRACE)	35.45	0.010	18.87068
2-PHENYL-3,2' -TRIMETHYLENE INDOLE	35.97	0.010	20.67178
1-METHYL-3,2'-METHYLENE-2-PHENYL INDOLE (TRACE)	35.90	0.010	20.38649
1-METHYL-3,2'-ETHYLENE-2-PHENYL INDOLE	34.95	0.010	17.31082
1-METHYL-3,2'-TRIMETHYLENE-2-PHENYL INDOLE	34.17	0.011	15.20352
4,5-DIPHENYLIMIDAZOLE (A)	26.37	0.024	3.21769
PPF (TRACE)	37.85	0.008	28.01855
PPO (TRACE)	37.64	0.008	27.18083
2-(4-FLUOROPHENYL)-5-PHENYLOXAZOLE (TRACE)	36.08	0.010	21.08029
αNPO (TRACE)	35.20	0.010	18.06230
BBO (B)	38.57	0.008	31.27098
POPOP	35.11	0.010	17.77976
DIMETHYL POPOP (TRACE)	35.34	0.010	18.50715
PPD (TRACE)	28.48	0.019	5.14993
2,5-DI(2-METHYLPHENYL)OXADIAZOLE	28.83	0.019	5.52873
PBD	38.58	0.008	31.54481
BBD (B)	41.07	0.006	45.69825
αNPD (TRACE)	34.80	0.011	16.87198
βNPD	36.24	0.009	21.64447
αNND (TRACE)	34.28	0.011	15.40824
βNND (TRACE)	38.09	0.008	29.12861
TRANS-1-(4-BIPHENYLYL)-2-PHENYL-ETHYLENE	39.72	0.007	37.29685
TRANS-1-(4-BIPHENYLYL)-2-(1-NAPHTHYL)-ETHYLENE	37.16	0.009	25.00639
DIPHENYLSTILBENE (B)	39.06	0.008	33.71061
1,1,4,4-TETRAPHENYLBUTADIENE	35.60	0.010	19.33311
1,6-DIPHENYLHEXATRIENE	37.61	0.008	26.86597
1,8-DIPHENYL-1,3,5,7-OCTATETRAENE	33.15	0.012	12.59192
BPSB (TRACE)	37.80	0.008	27.68813
BIS-MSB	38.38	0.008	30.32510
ESCULIN (A)	31.42	0.014	9.12937
AZULENE	24.14	0.032	1.87886
NAPHTHALENE	14.29	0.153	0.08207
DEUTERATED NAPHTHALENE-C10D8	16.08	0.108	0.16541
1-METHYLNAPHTHALENE 5ML/L	17.11	0.089	0.24141
2-METHYL NAPHTHALENE	16.84	0.094	0.21891
2,3-DIMETHYL NAPHTHALENE	17.11	0.089	0.24117
2,6-DIMETHYL NAPHTHALENE (TRACE)	19.24	0.063	0.48795
1-NAPHTHOL (TRACE)	24.28	0.031	1.96614
2-NAPHTHOL (TRACE)	22.38	0.040	1.20459
1-NAPHTHYLAMINE	27.94	0.021	4.53511
2-CHLORONAPHTHALENE	16.64	0.097	0.20409
ACENAPHTHENE	23.01	0.037	1.42647
1-PHENYL NAPHTHALENE	25.18	0.028	2.45136
2-PHENYL NAPHTHALENE	37.04	0.009	24.74230
1,4-DIPHENYL NAPHTHALENE	31.63	0.014	9.56726
1,5-DIPHENYL NAPHTHALENE	30.54	0.016	7.75843
1,7-DIPHENYL NAPHTHALENE	30.17	0.016	7.21447
1,4,5,8-TETRAPHENYL NAPHTHALENE	33.77	0.012	14.08619
1,1'-BINAPHTHYL	26.81	0.023	3.56654
4,4' -DI(N-BUTOXY)-1,1' -BINAPHTHYL	33.84	0.012	14.30957
2,2'-BINAPHTHYL	33.59	0.012	13.74065
ACRIDINE (A)	25.48	0.027	2.59618
ANTHRACENE (TRACE)	24.03	0.032	1.82943
DEUTERATED ANTHRACENE C14D10 (TRACE)	24.08	0.032	1.84871
9-METHYL ANTHRACENE (TRACE)	22.35	0.040	1.18277
1-AMINO ANTHRACENE	21.71	0.044	0.99298
2-AMINO ANTHRACENE	23.30	0.035	1.51317
9-VINYL ANTHRACENE	22.82	0.038	1.34159
9-PHENYL ANTHRACENE (TRACE)	23.65	0.034	1.66022
9,10-DIPHENYL ANTHRACENE	22.47	0.039	1.22093
9,10-DICHLOROANTHRACENE (TRACE)	21.19	0.047	0.86012
BIANTHRANYL (B)	25.02	0.029	2.32867
9,10-DINAPHTHYLANTHRACENE (B)	26.46	0.024	3.26071
9-(P-VINYLPHENYL)-10-PHENYL ANTHRACENE	23.94	0.033	1.78519
9,10-BIS(PHENYL-ETHYNYL)ANTHRACENE (TRACE)	33.96	0.011	14.56586
RUBICENE (B)	34.98	0.010	17.38885
DIBENZ(A,H)ANTHRACENE (B)	39.88	0.007	38.18746
TETRACENE (TRACE) (B)	25.91	0.026	2.87088
RUBRENE (B)	42.48	0.006	55.80500
TETRAHYDROCANNABINOL ACID METHYL ESTER	23.77	0.033	1.74059
PHENANTHRENE	16.71	0.096	0.20794
3,4-BENZOPHENANTHRENE	30.49	0.016	7.62328
2-PHENYL PHENANTHRENE	29.18	0.018	5.86577
CHRYSENE (TRACE)	31.45	0.014	9.18448
TRIPHENYLENE	20.75	0.050	0.76167

ACCEPTOR MOLECULES	RO	CO	OVERLAP
DEUTERATED TRIPHENYLENE-C18D12	20.72	0.050	0.75478
PYRENE	33.81	0.012	14.17251
1,3,6,8,-TETRAPHENYL PYRENE	31.76	0.014	9.74998
BIPYRENYL (B)	35.61	0.010	19.34191
O-PHENYLENE-PYRENE	35.63	0.010	19.41837
ANTHANTHRENE (B)	36.98	0.009	24.28136
FLUORANTHENE	27.58	0.021	4.17730
8 PHENYL FLUORANTHENE	30.11	0.016	7.07567
3-AMINO FLUORANTHENE	32.33	0.013	10.83718
3,3'-BIFLUORANTHENYL (B)	34.10	0.011	14.92987
BENZO(GHI)FLUORANTHENE	27.51	0.021	4.11321
9,12-DI(N-OCTADECOXY)-BENZ(E)ACEPHENANTHRYLENE	30.66	0.016	7.88879
DECACYCLENE (B)	36.15	0.009	21.19548
10-ETHOXY-BENZO(KL)XANTHENE	26.45	0.024	3.25293
PERYLENE	20.36	0.053	0.67570
BENZO(GHI)PERYLENE (B)	35.05	0.010	17.60239
BBOT (TRACE)	33.72	0.012	13.94654
QUININE SULFATE (A)	29.49	0.017	6.27454
ACRIDINE YELLOW (A)	23.30	0.035	1.51862
ACRIDINE ORANGE (A)	26.75	0.023	3.47853
ACRIDINE RED (A)	27.25	0.022	3.89082
URANINE (A)	27.38	0.022	4.00126
RHODAMINE B (A)	30.93	0.015	6.30386
RHODAMINE 6G (A)	28.94	0.018	5.57803
EOSIN B (A)	30.40	0.016	7.49766
EOSIN Y (A)	31.24	0.015	8.82908
ROSE BENGAL (A)	30.65	0.016	7.87589
INDENOINDENE	36.07	0.010	21.02944

8.63 Donor Molecule: 1-Methylfluorene

ACCEPTOR MOLECULES	RO	CO	OVERLAP
POLYVINYLTOLUENE	5.29	3.015	0.00030
ANILINE 3ML/L	14.67	0.142	0.13311
N,N-DIMETHYLANILINE 3ML/L	20.54	0.052	0.99827
TRIPHENYLAMINE	32.33	0.013	15.04370
P-CRESOL	4.30	5.623	0.00009
P-METHYLANISOLE 3ML/L	4.02	6.890	0.00006
1,4-DIMETHOXY BENZENE	17.93	0.078	0.44346
1-METHYL-4-CHLOROBENZENE 5ML/L	0.00	0.000	0.00000
SODIUM SALICYLATE (A)	22.89	0.037	1.89556
ANTHRANILIC ACID (A)	24.85	0.029	3.08007
TMPD	23.18	0.036	2.03245
1,2,4-TRIMETHYLBENZENE (TRACE)	0.00	0.000	0.00000
MESITYLENE 5ML/L	0.00	0.000	0.00000
1,3,5-TRIETHYLBENZENE 5ML/L	0.00	0.000	0.00000
STYRENE	6.28	1.803	0.00083
1,1-DIPHENYLETHYLENE 3ML/L	4.67	4.389	0.00014
BIPHENYL	0.00	0.000	0.00000
DEUTERATED BIPHENYL-C12D10	0.00	0.000	0.00000
4-METHYL BIPHENYL	4.75	4.182	0.00015
MONOISOPROPYLBIPHENYL 5ML/L	0.00	0.000	0.00000
4-BENZYL BIPHENYL	4.88	3.852	0.00018
3,3'-DIMETHYL BIPHENYL	7.73	0.969	0.00287
2,2'-DIMETHYLBIPHENYL 2ML/L	2.49	29.143	0.00000
OCTAMETHYL BIPHENYL	0.00	0.000	0.00000
4-METHOXY BIPHENYL	14.37	0.151	0.11811
4-BIPHENYLYL PHENYL ETHER	13.64	0.176	0.08624
4,4'-DIHYDROXYBIPHENYL (A)	19.81	0.058	0.80524
4,4'-DIMETHOXY BIPHENYL	18.58	0.070	0.54886
4,4'-DIISOBUTOXYBIPHENYL	18.68	0.069	0.56767
BENZIDINE	30.24	0.016	10.10295
4-VINYL BIPHENYL	24.47	0.031	2.86386
M-TERPHENYL	16.95	0.092	0.31662
M-QUATERPHENYL	12.96	0.206	0.06349
1,3,5-TRIPHENYL BENZENE	15.99	0.109	0.22232
1,3-DIBIPHENYLYLBENZENE (B)	32.05	0.014	14.34501
1,3,5,-TRIBIPHENYLYLBENZENE (B)	36.57	0.009	31.65291
FLUORENE (TRACE)	22.07	0.042	1.54495
1-METHYL FLUORENE	20.50	0.052	0.99316
1,2-BENZOFLUORENE	29.70	0.017	9.03311
9,10-DIHYDROPHENANTHRENE	20.63	0.051	1.03072
CARBAZOLE (TRACE)	24.62	0.030	2.94142
N-METHYL CARBAZOLE (TRACE)	23.55	0.034	2.24930
N-VINYLCARBAZOLE	24.05	0.032	2.55350
N-PHENYL CARBAZOLE	24.11	0.032	2.58689
N-METHYL-2,3-BENZCARBAZOLE	28.32	0.020	6.73997
DIBENZOFURAN	19.73	0.058	0.79028
DIBENZOTHIOPHENE (TRACE)	22.34	0.040	1.64273
P-TERPHENYL	25.59	0.027	3.73499
DEUTERATED P-TERPHENYL-C18D14	25.22	0.028	3.43290
4-METHYL TERPHENYL	27.31	0.022	5.52056
2-METHYL-P-TERPHENYL	18.47	0.071	0.53161
3,3''-DIMETHYL-P-TERPHENYL	26.63	0.024	4.74023
4-(3,3-DIMETHYLBUTOXY)-P-TERPHENYL	30.66	0.016	11.00711
4,4''-DIHEXAHYDROFARNESOXY-P-TERPHENYL	33.44	0.012	18.52179
2,2'-METHYLENE-P-TERPHENYL	32.56	0.013	15.81428
2,2'-ETHYLENE-P-TERPHENYL	32.51	0.013	15.63804
2,2'-METHYLENE-5',6'-METHYLENE-P-TERPHENYL (TRACE)	34.90	0.011	23.85380
3-PHENYL DIBENZOFURAN	33.57	0.012	18.97566
BENZO(1,2-B-4,5-B')BISBENZOFURAN (TRACE) (B)	35.92	0.010	28.29913

ACCEPTOR MOLECULES	RO	CO	OVERLAP
SEXIPHENYL BENZENE	14.29	0.153	0.11431
P-QUATERPHENYL	34.06	0.011	20.64768
3,3'''-DIMETHYL-P-QUATERPHENYL	34.96	0.010	24.12068
TETRAMETHYL-P-QUATERPHENYL	21.64	0.044	1.37123
3,3''' -DIMETHYL-3',2''-METHYLENE-P-QUATERPHENYL	37.69	0.008	37.76077
3,3''' -DIMETHYL-3',2''-ETHYLENE-P-QUATERPHENYL	37.57	0.008	36.97151
4-(3-ETHYLBUTOXY)-P-QUATERPHENYL (B)	37.00	0.009	33.71718
4,4'''-DI(2-BUTYLOCTOXY-1)-P-QUATERPHENYL	38.16	0.008	40.61713
DIETHYL-P-QUINQUEPHENYL	36.51	0.009	31.28697
DI(3-ETHYLHEPTYL)-P-QUINQUEPHENYL	39.31	0.007	48.39255
TETRAMETHYL-P-SEXIPHENYL	36.33	0.009	30.40457
TETRAMETHYL-P-OCTAPHENYL (B)	41.10	0.006	63.23755
INDOLE (TRACE)	3.57	9.852	0.00003
2-PHENYL INDOLE	33.93	0.011	20.18543
1-METHYL-2-PHENYL INDOLE	30.43	0.016	10.48437
1-NEOPENTYL-2-PHENYL INDOLE	29.44	0.018	8.62188
1-N-BUTYL-2(BIPHENYLYL-4) INDOLE	34.31	0.011	21.42501
1,2-DIPHENYL INDOLE	31.58	0.014	13.11210
1-BIPHENYLYL-2-PHENYL INDOLE	33.28	0.012	17.88971
1,3-DIMETHYL-2-PHENYL INDOLE	29.67	0.017	9.01302
2,3-DIPHENYL INDOLE	32.23	0.013	14.70320
1-METHYL-2,3-DIPHENYL INDOLE	30.01	0.017	9.59693
2-PHENYL-3,2'-METHYLENE INDOLE (TRACE)	33.99	0.011	20.33369
2-PHENYL-3,2'-ETHYLENE INDOLE (TRACE)	33.47	0.012	18.48460
2-PHENYL-3,2'-TRIMETHYLENE INDOLE	33.95	0.011	20.19782
1-METHYL-3,2'-METHYLENE-2-PHENYL INDOLE (TRACE)	33.89	0.011	19.94010
1-METHYL-3,2'-ETHYLENE-2-PHENYL INDOLE	33.01	0.012	16.98054
1-METHYL-3,2'-TRIMETHYLENE-2-PHENYL INDOLE	32.28	0.013	14.92025
4,5-DIPHENYLIMIDAZOLE (A)	25.07	0.028	3.28260
PPF (TRACE)	35.74	0.010	27.41684
PPO (TRACE)	35.57	0.010	26.76711
2-(4-FLUOROPHENYL)-5-PHENYLOXAZOLE (TRACE)	34.10	0.011	20.78109
αNPO (TRACE)	33.30	0.012	17.81856
BBO (B)	36.53	0.009	31.04129
POPOP	33.35	0.012	17.96557
DIMETHYL POPOP (TRACE)	33.60	0.012	18.81606
PPD (TRACE)	27.25	0.022	5.46535
2,5-DI(2-METHYLPHENYL)OXADIAZOLE	27.50	0.022	5.75773
PBD	36.50	0.009	31.27475
BBD (B)	38.79	0.008	44.81300
αNPD (TRACE)	32.88	0.013	16.57124
βNPD	34.27	0.011	21.38871
αNND (TRACE)	32.41	0.013	15.18169
βNND (TRACE)	35.97	0.010	28.53255
TRANS-1-(4-BIPHENYLYL)-2-PHENYL-ETHYLENE	37.53	0.008	36.64420
TRANS-1-(4-BIPHENYLYL)-2-(1-NAPHTHYL)-ETHYLENE	35.16	0.010	24.70415
DIPHENYLSTILBENE (B)	36.98	0.009	33.42356
1,1,4,4-TETRAPHENYLBUTADIENE	33.73	0.012	19.24984
1,6-DIPHENYLHEXATRIENE	35.69	0.010	26.98778
1,8-DIPHENYL-1,3,5,7-OCTATETRAENE	31.67	0.014	13.17121
BPSB (TRACE)	35.88	0.010	27.85367
BIS-MSB	36.37	0.009	30.24509
ESCULIN (A)	29.76	0.017	9.07721
AZULENE	22.91	0.037	1.88718
NAPHTHALENE	13.61	0.177	0.08467
DEUTERATED NAPHTHALENE-C10D8	15.25	0.126	0.16634
1-METHYLNAPHTHALENE 5ML/L	16.50	0.100	0.26771
2-METHYL NAPHTHALENE	15.93	0.111	0.21747
2,3-DIMETHYL NAPHTHALENE	16.27	0.104	0.24597
2,6-DIMETHYL NAPHTHALENE (TRACE)	18.17	0.075	0.47786
1-NAPHTHOL (TRACE)	23.04	0.037	1.98554
2-NAPHTHOL (TRACE)	21.12	0.047	1.17546
1-NAPHTHYLAMINE	26.39	0.024	4.44005
2-CHLORONAPHTHALENE	15.78	0.114	0.20495
ACENAPHTHENE	22.00	0.042	1.50804
1-PHENYL NAPHTHALENE	24.03	0.032	2.55322
2-PHENYL NAPHTHALENE	35.37	0.010	25.93924
1,4-DIPHENYL NAPHTHALENE	29.94	0.017	9.51378
1,5-DIPHENYL NAPHTHALENE	28.93	0.018	7.74861
1,7-DIPHENYL NAPHTHALENE	28.57	0.019	7.18277
1,4,5,8-TETRAPHENYL NAPHTHALENE	31.96	0.014	13.91796
1,1'-BINAPHTHYL	25.57	0.027	3.70117
4,4'-DI(N-BUTOXY)-1,1'-BINAPHTHYL	31.97	0.014	14.04364
2,2'-BINAPHTHYL	31.78	0.014	13.61035
ACRIDINE (A)	24.20	0.032	2.62577
ANTHRACENE (TRACE)	22.83	0.038	1.85052
DEUTERATED ANTHRACENE C14D10 (TRACE)	22.86	0.037	1.86393
9-METHYL ANTHRACENE (TRACE)	21.29	0.046	1.21793
1-AMINO ANTHRACENE	20.75	0.050	1.04303
2-AMINO ANTHRACENE	22.07	0.042	1.49969
9-VINYL ANTHRACENE	21.78	0.043	1.39380
9-PHENYL ANTHRACENE (TRACE)	22.54	0.039	1.71211
9,10-DIPHENYL ANTHRACENE	21.54	0.045	1.30563
9,10-DICHLOROANTHRACENE (TRACE)	20.34	0.053	0.92616
BIANTHRANYL (B)	23.87	0.033	2.41387
9,10-DINAPHTHYLANTHRACENE (B)	25.29	0.028	3.41761
9-(P-VINYLPHENYL)-10-PHENYL ANTHRACENE	22.94	0.037	1.90230
9,10-BIS(PHENYL-ETHYNYL)ANTHRACENE (TRACE)	32.10	0.014	14.29460
RUBICENE (B)	33.22	0.012	17.55945
DIBENZ(A,H)ANTHRACENE (B)	38.18	0.008	40.48523
TETRACENE (TRACE) (B)	25.37	0.027	3.47982
RUBRENE (B)	40.25	0.007	55.54021
TETRAHYDROCANNABINOL ACID METHYL ESTER	22.03	0.030	1.00007
PHENANTHRENE	16.22	0.105	0.24006
3,4-BENZOPHENANTHRENE	28.87	0.019	7.55804
2-PHENYL PHENANTHRENE	27.76	0.021	6.01363

ACCEPTOR MOLECULES	RO	CO	OVERLAP
CHRYSENE (TRACE)	29.69	0.017	8.96991
TRIPHENYLENE	19.84	0.057	0.80273
DEUTERATED TRIPHENYLENE-C18D12	19.83	0.057	0.80030
PYRENE	31.89	0.014	13.75443
1,3,6,8,-TETRAPHENYL PYRENE	30.81	0.015	11.18480
BIPYRENYL (B)	33.77	0.012	19.39468
O-PHENYLENE-PYRENE	33.82	0.012	19.53481
ANTHANTHRENE (B)	[illegible]	0.011	23.82124
FLUORANTHENE	26.12	0.025	4.14518
3-PHENYL FLUORANTHENE	28.97	0.018	7.71654
3-AMINO FLUORANTHENE	30.64	0.016	10.81600
3,3'-BIFLUORANTHENYL (B)	32.47	0.013	15.31620
BENZO(GHI)FLUORANTHENE	26.12	0.025	4.14890
9,12-DI(N-OCTADECOXY)-BENZ(E)ACEPHENANTHRYLENE	29.24	0.018	8.16049
DECACYCLENE (B)	34.33	0.011	21.39756
10-ETHOXY-BENZO(KL)XANTHENE	25.23	0.028	3.37062
PERYLENE	19.51	0.060	0.71959
BENZO(GHI)PERYLENE (B)	33.28	0.012	17.76219
BBOT (TRACE)	32.12	0.014	14.33957
QUININE SULFATE (A)	27.88	0.021	6.17991
ACRIDINE YELLOW (A)	22.33	0.040	1.61844
ACRIDINE ORANGE (A)	25.55	0.027	3.63808
ACRIDINE RED (A)	25.83	0.026	3.88415
URANINE (A)	25.98	0.026	4.01717
RHODAMINE B (A)	29.32	0.018	8.30009
RHODAMINE 6G (A)	27.60	0.021	5.77064
EOSIN B (A)	28.74	0.019	7.35617
EOSIN Y (A)	29.53	0.017	8.66097
ROSE BENGAL (A)	29.01	0.018	7.78400
INDENOINDENE	34.05	0.011	20.58312

8.64 Donor Molecule: 1,2-Benzofluorene

TRIPHENYLAMINE	15.72	0.115	0.22273
SODIUM SALICYLATE (A)	11.94	0.263	0.04267
ANTHRANILIC ACID (A)	24.04	0.032	2.82658
TMPD	19.03	0.065	0.69819
BENZIDINE	10.84	0.351	0.02397
1,3-DIBIPHENYLYLBENZENE (B)	7.51	1.056	0.00266
1,3,5,-TRIBIPHENYLYLBENZENE (B)	6.59	1.566	0.00121
FLUORENE (TRACE)	0.00	0.000	0.00000
1-METHYL FLUORENE	0.00	0.000	0.00000
1,2-BENZOFLUORENE	16.17	0.106	0.26325
9,10-DIHYDROPHENANTHRENE	0.00	0.000	0.00000
CARBAZOLE (TRACE)	9.72	0.487	0.01245
N-METHYL CARBAZOLE (TRACE)	19.35	0.062	0.77328
N-VINYLCARBAZOLE	17.09	0.090	0.36815
N-PHENYL CARBAZOLE	17.46	0.084	0.41862
N-METHYL-2,3-BENZCARBAZOLE	25.59	0.027	4.09140
DIBENZOFURAN	0.00	0.000	0.00000
DIBENZOTHIOPHENE (TRACE)	5.66	2.471	0.00049
4,4''-DIHEXAHYDROFARNESOXY-P-TERPHENYL	6.44	1.672	0.00106
2,2'-METHYLENE-P-TERPHENYL	0.00	0.000	0.00000
2,2'-ETHYLENE-P-TERPHENYL	5.47	2.736	0.00040
2,2'-METHYLENE-5',6''-METHYLENE-P-TERPHENYL (TRACE)	17.27	0.087	0.39146
3-PHENYL DIBENZOFURAN	0.00	0.000	0.00000
BENZO(1,2-B-4,5-B')BISBENZOFURAN (TRACE) (B)	23.73	0.033	2.63190
SEXIPHENYL BENZENE	0.00	0.000	0.00000
P-QUATERPHENYL	11.38	0.304	0.03206
3,3'''-DIMETHYL-P-QUATERPHENYL	14.59	0.144	0.14233
TETRAMETHYL-P-QUATERPHENYL	0.00	0.000	0.00000
3,3''' -DIMETHYL-3',2''-METHYLENE-P-QUATERPHENYL	22.99	0.037	2.17632
3,3''' -DIMETHYL-3',2''-ETHYLENE-P-QUATERPHENYL	24.15	0.032	2.92131
4-(3-ETHYLBUTOXY)-P-QUATERPHENYL (B)	21.44	0.045	1.43035
4,4'''-DI(2-BUTYLOCTOXY-1)-P-QUATERPHENYL	21.67	0.044	1.52398
DIETHYL-P-QUINQUEPHENYL	14.62	0.143	0.14453
DI(3-ETHYLHEPTYL)-P-QUINQUEPHENYL	23.35	0.035	2.38431
TETRAMETHYL-P-SEXIPHENYL	11.61	0.286	0.03627
TETRAMETHYL-P-OCTAPHENYL (B)	29.02	0.018	8.76878
INDOLE (TRACE)	0.00	0.000	0.00000
2-PHENYL INDOLE	10.66	0.369	0.02177
1-METHYL-2-PHENYL INDOLE	9.68	0.493	0.01215
1-NEOPENTYL-2-PHENYL INDOLE	0.00	0.000	0.00000
1-N-BUTYL-2(BIPHENYLYL-4) INDOLE	23.49	0.035	2.47069
1,2-DIPHENYL INDOLE	10.01	0.446	0.01486
1-BIPHENYLYL-2-PHENYL INDOLE	15.30	0.125	0.18896
1,3-DIMETHYL-2-PHENYL INDOLE	11.22	0.317	0.02953
2,3 DIPHENYL INDOLE	[illegible]	0.045	[illegible]
1-METHYL-2,3-DIPHENYL INDOLE	19.27	0.063	0.75257
2-PHENYL-3,2'-METHYLENE INDOLE (TRACE)	16.85	0.093	0.33822
2-PHENYL-3,2'-ETHYLENE INDOLE (TRACE)	26.33	0.024	4.90556
2-PHENYL-3,2'-TRIMETHYLENE INDOLE	15.96	0.110	0.24378
1-METHYL-3,2'-METHYLENE-2-PHENYL INDOLE (TRACE)	24.81	0.029	3.43510
1-METHYL-3,2'-ETHYLENE-2-PHENYL INDOLE	28.02	0.020	7.11858
1-METHYL-3,2'-TRIMETHYLENE-2-PHENYL INDOLE	17.56	0.083	0.43320
4,5-DIPHENYLIMIDAZOLE (A)	9.22	0.572	0.00906
PPF (TRACE)	26.00	0.025	4.55547
PPO (TRACE)	14.40	0.150	0.13187
2-(4-FLUOROPHENYL)-5-PHENYLOXAZOLE (TRACE)	13.03	0.202	0.07225
αNPO (TRACE)	32.12	0.013	16.06853
BBO (B)	38.20	0.008	45.42171

ACCEPTOR MOLECULES	RO	CO	OVERLAP
POPOP	38.96	0.008	51.08848
DIMETHYL POPOP (TRACE)	40.75	0.007	66.75498
PPD (TRACE)	0.00	0.000	0.00000
2,5-DI(2-METHYLPHENYL)OXADIAZOLE	0.00	0.000	0.00000
PBD	11.62	0.285	0.03634
BBD (B)	27.56	0.021	6.44637
αNPD (TRACE)	24.49	0.030	3.16935
βNPD	15.50	0.120	0.20485
αNND (TRACE)	29.07	0.018	8.86782
βNND (TRACE)	23.61	0.034	2.55610
TRANS-1-(4-BIPHENYLYL)-2-PHENYL-ETHYLENE	28.29	0.020	7.53372
TRANS-1-(4-BIPHENYLYL)-2-(1-NAPHTHYL)-ETHYLENE	34.47	0.011	24.50784
DIPHENYLSTILBENE (B)	38.48	0.008	47.42025
1,1,4,4-TETRAPHENYLBUTADIENE	36.82	0.009	36.34589
1,6-DIPHENYLHEXATRIENE	41.66	0.006	76.22281
1,8-DIPHENYL-1,3,5,7-OCTATETRAENE	43.92	0.005	104.09116
BPSB (TRACE)	42.29	0.006	83.49996
BIS-MSB	40.25	0.007	62.03795
ESCULIN (A)	29.05	0.018	8.80493
AZULENE	18.91	0.066	0.66191
NAPHTHALENE	0.00	0.000	0.00000
DEUTERATED NAPHTHALENE-C10D8	2.74	21.766	0.00001
2-NAPHTHOL (TRACE)	5.60	2.553	0.00045
1-NAPHTHYLAMINE	16.55	0.099	0.30254
2-CHLORONAPHTHALENE	0.00	0.000	0.00000
ACENAPHTHENE	0.00	0.000	0.00000
1-PHENYL NAPHTHALENE	0.00	0.000	0.00000
2-PHENYL NAPHTHALENE	6.78	1.437	0.00144
1,4-DIPHENYL NAPHTHALENE	12.10	0.253	0.04636
1,5-DIPHENYL NAPHTHALENE	8.64	0.694	0.00615
1,7-DIPHENYL NAPHTHALENE	9.66	0.496	0.01205
1,4,5,8-TETRAPHENYL NAPHTHALENE	31.18	0.015	13.39560
1,1'-BINAPHTHYL	0.00	0.000	0.00000
4,4'-DI(N-BUTOXY)-1,1'-BINAPHTHYL	18.76	0.068	0.64144
2,2'-BINAPHTHYL	10.35	0.403	0.01820
ACRIDINE (A)	29.30	0.018	9.23283
ANTHRACENE (TRACE)	27.48	0.022	6.29221
DEUTERATED ANTHRACENE C14D10 (TRACE)	27.13	0.022	5.83377
9-METHYL ANTHRACENE (TRACE)	28.79	0.019	8.30109
1-AMINO ANTHRACENE	27.61	0.021	6.40485
2-AMINO ANTHRACENE	26.21	0.025	4.68931
9-VINYL ANTHRACENE	29.81	0.017	10.22782
9-PHENYL ANTHRACENE (TRACE)	30.16	0.016	10.97814
9,10-DIPHENYL ANTHRACENE	31.21	0.015	13.43097
9,10-DICHLOROANTHRACENE (TRACE)	29.98	0.017	10.56002
BIANTHRANYL (B)	32.79	0.013	18.07534
9,10-DINAPHTHYLANTHRACENE (B)	30.58	0.016	11.87480
9-(P-VINYLPHENYL)-10-PHENYL ANTHRACENE	31.40	0.014	13.92053
9,10-BIS(PHENYL-ETHYNYL)ANTHRACENE (TRACE)	29.44	0.018	9.40237
RUBICENE (B)	29.40	0.018	9.33469
DIBENZ(A,H)ANTHRACENE (B)	27.24	0.022	5.95089
TETRACENE (TRACE) (B)	22.70	0.038	1.98043
RUBRENE (B)	21.83	0.043	1.56395
TETRAHYDROCANNABINOL ACID METHYL ESTER	0.00	0.000	0.00000
PHENANTHRENE	11.72	0.278	0.03824
3,4-BENZOPHENANTHRENE	16.23	0.105	0.26704
2-PHENYL PHENANTHRENE	14.16	0.157	0.11865
CHRYSENE (TRACE)	16.72	0.096	0.32012
TRIPHENYLENE	10.25	0.416	0.01707
DEUTERATED TRIPHENYLENE-C18D12	10.18	0.424	0.01644
PYRENE	19.63	0.059	0.83765
1,3,6,8,-TETRAPHENYL PYRENE	38.93	0.008	50.51746
BIPYRENYL (B)	37.91	0.008	43.19851
O-PHENYLENE-PYRENE	33.71	0.012	21.21940
ANTHANTHRENE (B)	31.40	0.014	13.86015
FLUORANTHENE	28.04	0.020	7.07305
3-PHENYL FLUORANTHENE	32.38	0.013	16.74448
3-AMINO FLUORANTHENE	29.72	0.017	9.96250
3,3'-BIFLUORANTHENYL (B)	36.81	0.009	36.05237
BENZO(GHI)FLUORANTHENE	25.80	0.026	4.28489
9,12-DI(N-OCTADECOXY)-BENZ(E)ACEPHENANTHRYLENE	31.99	0.014	15.53020
DECACYCLENE (B)	39.55	0.007	55.30773
10-ETHOXY-BENZO(KL)XANTHENE	31.85	0.014	15.18043
PERYLENE	30.61	0.016	11.88511
BENZO(GHI)PERYLENE (B)	33.90	0.011	22.06925
BBOT (TRACE)	41.99	0.006	79.81426
QUININE SULFATE (A)	20.51	0.052	1.09895
ACRIDINE YELLOW (A)	26.57	0.024	5.08739
ACRIDINE ORANGE (A)	22.51	0.039	1.88094
ACRIDINE RED (A)	24.77	0.029	3.33553
URANINE (A)	24.19	0.032	2.89923
RHODAMINE B (A)	28.26	0.020	7.36209
RHODAMINE 6G (A)	27.93	0.021	6.85890
EOSIN B (A)	24.20	0.032	2.90039
EOSIN Y (A)	24.97	0.029	3.50177
ROSE BENGAL (A)	25.48	0.027	3.95984
INDENOINDENE	14.04	0.162	0.11323

8.65 Donor Molecule: 9,10-Dihydrophenanthrene

ACCEPTOR MOLECULES	RO	CO	OVERLAP
POLYVINYLTOLUENE	3.35	11.923	0.00002
ANILINE 3ML/L	10.50	0.386	0.01891
N,N-DIMETHYLANILINE 3ML/L	17.84	0.079	0.45063
TRIPHENYLAMINE	29.70	0.017	9.51477
1,4 DIMETHOXY BENZENE	12.74	0.216	0.06022
1-METHYL-4-CHLOROBENZENE 5ML/L	0.00	0.000	0.00000
SODIUM SALICYLATE (A)	20.58	0.051	1.05178
ANTHRANILIC ACID (A)	25.94	0.026	4.19581
TMPD	23.33	0.035	2.22388
1,2,4-TRIMETHYLBENZENE (TRACE)	0.00	0.000	0.00000
MESITYLENE 5ML/L	0.00	0.000	0.00000
1,3,5-TRIETHYLBENZENE 5ML/L	0.00	0.000	0.00000
STYRENE	3.66	9.088	0.00003
3,3'-DIMETHYL BIPHENYL	4.62	4.535	0.00014
2,2'-DIMETHYLBIPHENYL 2ML/L	0.00	0.000	0.00000
OCTAMETHYL BIPHENYL	0.00	0.000	0.00000
4-METHOXY BIPHENYL	10.02	0.445	0.01425
4-BIPHENYLYL PHENYL ETHER	8.76	0.665	0.00640
4,4'-DIHYDROXYBIPHENYL (A)	16.44	0.101	0.27633
4,4'-DIMETHOXY BIPHENYL	14.85	0.136	0.15086
4,4'-DIISOBUTOXYBIPHENYL	14.68	0.141	0.14074
BENZIDINE	27.29	0.022	5.74049
4-VINYL BIPHENYL	20.01	0.056	0.90145
M-TERPHENYL	14.14	0.158	0.11246
M-QUATERPHENYL	8.12	0.837	0.00404
1,3,5-TRIPHENYL BENZENE	13.96	0.164	0.10347
1,3-DIBIPHENYLYLBENZENE (B)	28.12	0.020	6.88394
1,3,5,-TRIBIPHENYLYLBENZENE (B)	32.47	0.013	16.32517
FLUORENE (TRACE)	15.07	0.131	0.16486
1-METHYL FLUORENE	13.76	0.172	0.09567
1,2-BENZOFLUORENE	27.53	0.021	6.04155
9,10-DIHYDROPHENANTHRENE	15.76	0.114	0.21589
CARBAZOLE (TRACE)	24.00	0.032	2.65988
N-METHYL CARBAZOLE (TRACE)	24.11	0.032	2.72762
N-VINYLCARBAZOLE	24.66	0.030	3.12039
N-PHENYL CARBAZOLE	24.41	0.031	2.93317
N-METHYL-2,3-BENZCARBAZOLE	27.68	0.021	6.16191
DIBENZOFURAN	13.15	0.197	0.07290
DIBENZOTHIOPHENE (TRACE)	21.99	0.042	1.57676
P-TERPHENYL	21.36	0.046	1.32979
DEUTERATED P-TERPHENYL-C18D14	20.93	0.049	1.18114
4-METHYL TERPHENYL	23.24	0.036	2.20723
2-METHYL-P-TERPHENYL	14.23	0.155	0.11708
3,3''-DIMETHYL-P-TERPHENYL	22.61	0.039	1.87134
4-(3,3-DIMETHYLBUTOXY)-P-TERPHENYL	27.28	0.022	5.74776
4,4''DIHEXAHYDROFARNESOXY-P-TERPHENYL	30.38	0.016	10.95266
2,2'-METHYLENE-P-TERPHENYL	29.90	0.017	9.98152
2,2'-ETHYLENE-P-TERPHENYL	29.98	0.017	10.11737
2,2'-METHYLENE-5',6''-METHYLENE-P-TERPHENYL (TRACE)	34.52	0.011	23.50872
3-PHENYL DIBENZOFURAN	30.60	0.016	11.45771
BENZO(1,2-B-4,5-B')BISBENZOFURAN (TRACE) (B)	36.34	0.009	31.95722
SEXIPHENYL BENZENE	9.57	0.510	0.01085
P-QUATERPHENYL	31.25	0.015	12.96659
3,3'''-DIMETHYL-P-QUATERPHENYL	32.35	0.013	15.92221
TETRAMETHYL-P-QUATERPHENYL	17.05	0.090	0.34512
3,3''' -DIMETHYL-3',2''-METHYLENE-P-QUATERPHENYL	37.18	0.009	36.63162
3,3''' -DIMETHYL-3',2''-ETHYLENE-P-QUATERPHENYL	36.89	0.009	34.90368
4-(3-ETHYLBUTOXY)-P-QUATERPHENYL (B)	35.48	0.010	27.63512
4,4'''-DI(2-BUTYLOCTOXY-1)-P-QUATERPHENYL	36.51	0.009	32.81929
DIETHYL-P-QUINQUEPHENYL	33.75	0.012	20.55266
DI(3-ETHYLHEPTYL)-P-QUINQUEPHENYL	37.75	0.008	39.96727
TETRAMETHYL-P-SEXIPHENYL	33.21	0.012	18.68944
TETRAMETHYL-P-OCTAPHENYL (B)	40.47	0.007	60.62143
INDOLE (TRACE)	0.00	0.000	0.00000
2-PHENYL INDOLE	32.40	0.013	16.11618
1-METHYL-2-PHENYL INDOLE	27.96	0.020	6.64362
1-NEOPENTYL-2-PHENYL INDOLE	26.46	0.024	4.78946
1-N-BUTYL-2(BIPHENYLYL-4) INDOLE	33.49	0.012	19.51906
1,2-DIPHENYL INDOLE	29.12	0.018	8.48962
1-BIPHENYLYL-2-PHENYL INDOLE	31.16	0.015	12.68333
1,3-DIMETHYL-2-PHENYL INDOLE	27.73	0.021	6.32052
2,3-DIPHENYL INDOLE	31.28	0.015	12.92465
1-METHYL-2,3-DIPHENYL INDOLE	28.78	0.019	7.85486
2-PHENYL-3,2'-METHYLENE INDOLE (TRACE)	33.58	0.012	19.91766
2-PHENYL-3,2'-ETHYLENE INDOLE (TRACE)	33.79	0.012	20.60031
2-PHENYL-3,2'-TRIMETHYLENE INDOLE	33.06	0.012	18.13791
1-METHYL-3,2'-METHYLENE-2-PHENYL INDOLE (TRACE)	33.99	0.011	21.35960
1-METHYL-3,2'-ETHYLENE-2-PHENYL INDOLE	33.51	0.012	19.57599
1-METHYL-3,2'-TRIMETHYLENE-2-PHENYL INDOLE	31.32	0.015	13.09744
4,5-DIPHENYLIMIDAZOLE (A)	22.53	0.039	1.82172
PPP (TRACE)	35.00	0.010	30.18078
PPO (TRACE)	34.02	0.011	21.55502
2-(4-FLUOROPHENYL)-5-PHENYLOXAZOLE (TRACE)	32.57	0.013	16.59096
αNPO (TRACE)	34.40	0.011	22.78197
BBO (B)	38.29	0.008	43.27496
POPOP	35.68	0.010	28.31899
DIMETHYL POPOP (TRACE)	35.99	0.010	29.79446
PPD (TRACE)	23.07	0.036	2.11644
2,5-DI(2-METHYLPHENYL)OXADIAZOLE	23.44	0.035	2.32467
PBD	34.24	0.011	22.44356
BBD (B)	38.44	0.008	44.71253
αNPD (TRACE)	32.76	0.013	17.08048
βNPD	32.41	0.013	16.10779

ACCEPTOR MOLECULES	RO	CO	OVERLAP
αNND (TRACE)	33.16	0.012	18.36509
βNND (TRACE)	35.57	0.010	28.09199
TRANS-1-(4-BIPHENYLYL)-2-PHENYL-ETHYLENE	37.39	0.009	37.77564
TRANS-1-(4-BIPHENYLYL)-2-(1-NAPHTHYL)-ETHYLENE	36.30	0.009	31.41227
DIPHENYLSTILBENE (B)	38.73	0.008	46.31378
1,1,4,4-TETRAPHENYLBUTADIENE	35.72	0.010	28.46492
1,6-DIPHENYLHEXATRIENE	38.42	0.008	44.11109
1,8-DIPHENYL-1,3,5,7-OCTATETRAENE	35.40	0.010	26.96945
BPSB (TRACE)	38.52	0.008	44.80956
BIS-MSB	38.40	0.008	43.94542
ESCULIN (A)	30.73	0.015	11.58336
AZULENE	23.65	0.034	2.39950
NAPHTHALENE	11.83	0.270	0.03841
DEUTERATED NAPHTHALENE-C10D8	13.67	0.175	0.09117
1-METHYLNAPHTHALENE 5ML/L	13.92	0.166	0.10158
2-METHYL NAPHTHALENE	14.81	0.138	0.14780
2,3-DIMETHYL NAPHTHALENE	14.82	0.137	0.14813
2,6-DIMETHYL NAPHTHALENE (TRACE)	17.68	0.081	0.42637
1-NAPHTHOL (TRACE)	21.06	0.048	1.21906
2-NAPHTHOL (TRACE)	21.01	0.048	1.19719
1-NAPHTHYLAMINE	26.05	0.025	4.33292
2-CHLORONAPHTHALENE	14.93	0.134	0.15499
ACENAPHTHENE	19.16	0.064	0.69291
1-PHENYL NAPHTHALENE	20.80	0.050	1.13108
2-PHENYL NAPHTHALENE	30.32	0.016	10.83941
1,4-DIPHENYL NAPHTHALENE	27.92	0.021	6.57796
1,5-DIPHENYL NAPHTHALENE	26.68	0.024	5.01596
1,7-DIPHENYL NAPHTHALENE	26.37	0.024	4.67585
1,4,5,8-TETRAPHENYL NAPHTHALENE	33.06	0.012	17.88566
1,1'-BINAPHTHYL	22.65	0.039	1.88372
4,4'-DI(N-BUTOXY)-1,1'-BINAPHTHYL	31.04	0.015	12.38803
2,2'-BINAPHTHYL	29.39	0.018	8.97236
ACRIDINE (A)	26.47	0.024	4.71186
ANTHRACENE (TRACE)	24.92	0.029	3.29089
DEUTERATED ANTHRACENE C14D10 (TRACE)	24.79	0.029	3.19282
9-METHYL ANTHRACENE (TRACE)	23.66	0.034	2.40624
1-AMINO ANTHRACENE	21.96	0.042	1.53595
2-AMINO ANTHRACENE	23.81	0.033	2.48646
9-VINYL ANTHRACENE	24.27	0.031	2.80531
9-PHENYL ANTHRACENE (TRACE)	24.97	0.029	3.32383
9,10-DIPHENYL ANTHRACENE	24.48	0.030	2.94955
9,10-DICHLOROANTHRACENE (TRACE)	23.16	0.036	2.11569
BIANTHRANYL (B)	26.45	0.024	4.69294
9,10-DINAPHTHYLANTHRACENE (B)	25.59	0.027	3.84654
9-(P-VINYLPHENYL)-10-PHENYL ANTHRACENE	24.82	0.029	3.20424
9,10-BIS(PHENYL-ETHYNYL)ANTHRACENE (TRACE)	29.69	0.017	9.38656
RUBICENE (B)	31.39	0.014	13.12083
DIBENZ(A,H)ANTHRACENE (B)	33.81	0.012	20.46914
TETRACENE (TRACE) (B)	21.36	0.046	1.30225
RUBRENE (B)	35.86	0.010	29.16701
TETRAHYDROCANNABINOL ACID METHYL ESTER	18.65	0.069	0.59125
PHENANTHRENE	15.37	0.123	0.18259
3,4-BENZOPHENANTHRENE	27.07	0.023	5.40515
2-PHENYL PHENANTHRENE	24.09	0.032	2.69884
CHRYSENE (TRACE)	28.15	0.020	6.85850
TRIPHENYLENE	17.68	0.081	0.42387
DEUTERATED TRIPHENYLENE-C18D12	17.60	0.082	0.41287
PYRENE	32.85	0.013	17.28691
1,3,6,8,-TETRAPHENYL PYRENE	30.67	0.016	11.41092
BIPYRENYL (B)	35.86	0.010	29.16791
O-PHENYLENE-PYRENE	31.95	0.014	14.58277
ANTHANTHRENE (B)	32.70	0.013	16.75620
FLUORANTHENE	27.48	0.022	5.89843
3-PHENYL FLUORANTHENE	28.84	0.019	7.88470
3-AMINO FLUORANTHENE	29.05	0.018	8.24246
3,3'-BIFLUORANTHENYL (B)	32.54	0.013	16.28612
BENZO(GHI)FLUORANTHENE	26.95	0.023	5.25421
9,12-DI(N-OCTADECOXY)-BENZ(E)ACEPHENANTHRYLENE	29.09	0.018	8.31451
DECACYCLENE (B)	35.34	0.010	26.72561
10-ETHOXY-BENZO(KL)XANTHENE	25.74	0.026	3.98324
PERYLENE	21.39	0.046	1.31134
BENZO(GHI)PERYLENE (B)	29.86	0.017	9.72448
BBOT (TRACE)	35.25	0.010	26.31177
QUININE SULFATE (A)	28.55	0.019	7.51022
ACRIDINE YELLOW (A)	20.71	0.050	1.08288
ACRIDINE ORANGE (A)	22.72	0.038	1.88686
ACRIDINE RED (A)	24.96	0.029	3.31343
URANINE (A)	25.48	0.027	3.74955
RHODAMINE B (A)	28.23	0.020	6.94516
RHODAMINE 6G (A)	27.22	0.022	5.58129
EOSIN B (A)	28.01	0.020	6.62368
EOSIN Y (A)	28.65	0.019	7.58527
ROSE BENGAL (A)	28.87	0.019	7.93737
INDENOINDENE	32.99	0.012	17.92714

8.66 Donor Molecule: Carbazole (trace)

N,N-DIMETHYLANILINE 3ML/L	3.77	0.320	0.00006
TRIPHENYLAMINE	20.04	0.056	1.28824
SODIUM SALICYLATE (A)	13.77	0.171	0.13523
ANTHRANILIC ACID (A)	25.09	0.028	4.91751

ACCEPTOR MOLECULES	RO	CO	OVERLAP
TMPD	21.02	0.048	1.70408
BENZIDINE	16.57	0.098	0.41201
1,3,5-TRIPHENYL BENZENE	4.35	5.432	0.00014
1,3-DIBIPHENYLYLBENZENE (B)	14.43	0.149	0.18035
1,3,5,-TRIBIPHENYLYLBENZENE (B)	17.10	0.089	0.49970
FLUORENE (TRACE)	0.00	0.000	0.00000
1-METHYL FLUORENE	0.00	0.000	0.00000
1,2 BENZOFLUORENE	18.68	0.071	0.79158
9,10-DIHYDROPHENANTHRENE	0.00	0.000	0.00000
CARBAZOLE (TRACE)	18.57	0.070	0.81640
N-METHYL CARBAZOLE (TRACE)	21.95	0.042	2.21951
N-VINYLCARBAZOLE	21.70	0.044	2.07642
N-PHENYL CARBAZOLE	21.43	0.045	1.92315
N-METHYL-2,3-BENZCARBAZOLE	24.30	0.031	4.03411
DIBENZOFURAN	0.00	0.000	0.00000
DIBENZOTHIOPHENE (TRACE)	13.48	0.182	0.11988
P-TERPHENYL	0.00	0.000	0.00000
DEUTERATED P-TERPHENYL-C18D14	0.00	0.000	0.00000
4-METHYL TERPHENYL	4.66	4.431	0.00020
2-METHYL-P-TERPHENYL	0.00	0.000	0.00000
3,3''-DIMETHYL-P-TERPHENYL	6.05	2.023	0.00098
4-(3,3-DIMETHYLBUTOXY)-P-TERPHENYL	13.90	0.166	0.14425
4,4''-DIHEXAHYDROFARNESOXY-P-TERPHENYL	17.39	0.085	0.55153
2,2'-METHYLENE-P-TERPHENYL	13.19	0.195	0.10504
2,2'-ETHYLENE-P-TERPHENYL	15.67	0.116	0.29574
2,2'-METHYLENE-5',6''-METHYLENE-P-TERPHENYL (TRACE)	30.38	0.016	15.63853
3-PHENYL DIBENZOFURAN	13.04	0.202	0.09841
BENZO(1,2-B-4,5-B')BISBENZOFURAN (TRACE) (B)	30.46	0.016	15.86246
SEXIPHENYL BENZENE	0.00	0.000	0.00000
P-QUATERPHENYL	19.33	0.062	1.03877
3,3'''-DIMETHYL-P-QUATERPHENYL	21.08	0.048	1.74624
TETRAMETHYL-P-QUATERPHENYL	0.00	0.000	0.00000
3,3'''-DIMETHYL-3',2''-METHYLENE-P-QUATERPHENYL	30.11	0.016	14.80231
3,3'''-DIMETHYL-3',2''-ETHYLENE-P-QUATERPHENYL	29.80	0.017	13.87002
4-(3-ETHYLBUTOXY)-P-QUATERPHENYL (B)	26.83	0.023	7.39873
4,4'''-DI(2-BUTYLOCTOXY-1)-P-QUATERPHENYL	27.31	0.022	8.23804
DIETHYL-P-QUINQUEPHENYL	21.97	0.042	2.24231
DI(3-ETHYLHEPTYL)-P-QUINQUEPHENYL	28.72	0.019	11.10925
TETRAMETHYL-P-SEXIPHENYL	20.28	0.054	1.38703
TETRAMETHYL-P-OCTAPHENYL (B)	33.50	0.012	27.98351
INDOLE (TRACE)	0.00	0.000	0.00000
2-PHENYL INDOLE	22.36	0.040	2.49027
1-METHYL-2-PHENYL INDOLE	16.82	0.094	0.45133
1-NEOPENTYL-2-PHENYL INDOLE	13.04	0.202	0.09814
1-N-BUTYL-2(BIPHENYLYL-4) INDOLE	27.36	0.022	8.30042
1,2-DIPHENYL INDOLE	17.25	0.087	0.52404
1-BIPHENYLYL-2-PHENYL INDOLE	20.31	0.053	1.39355
1,3-DIMETHYL-2-PHENYL INDOLE	18.77	0.068	0.87021
2,3-DIPHENYL INDOLE	25.21	0.026	5.07267
1-METHYL-2,3-DIPHENYL INDOLE	22.77	0.038	2.75786
2-PHENYL-3,2'-METHYLENE INDOLE (TRACE)	26.56	0.024	6.98748
2-PHENYL-3,2'-ETHYLENE INDOLE (TRACE)	29.81	0.017	13.90693
2-PHENYL-3,2'-TRIMETHYLENE INDOLE	24.89	0.029	4.72449
1-METHYL-3,2'-METHYLENE-2-PHENYL INDOLE (TRACE)	28.97	0.018	11.73870
1-METHYL-3,2'-ETHYLENE-2-PHENYL INDOLE	30.59	0.016	16.22198
1-METHYL-3,2'-TRIMETHYLENE-2-PHENYL INDOLE	24.28	0.031	4.07563
4,5-DIPHENYLIMIDAZOLE (A)	13.96	0.164	0.14786
PPF (TRACE)	30.30	0.016	15.35077
PPO (TRACE)	25.10	0.028	4.97438
2-(4-FLUOROPHENYL)-5-PHENYLOXAZOLE (TRACE)	23.73	0.033	3.56115
αNPO (TRACE)	33.06	0.012	25.73299
BBO (B)	38.16	0.008	60.78760
POPOP	37.34	0.009	53.31022
DIMETHYL POPOP (TRACE)	38.19	0.008	60.87442
PPD (TRACE)	0.00	0.000	0.00000
2,5-DI(2-METHYLPHENYL)OXADIAZOLE	0.00	0.000	0.00000
PBD	22.46	0.039	2.56209
BBD (B)	32.25	0.013	22.30184
αNPD (TRACE)	28.02	0.020	9.58083
βNPD	21.58	0.044	2.01199
αNND (TRACE)	31.04	0.015	17.69183
βNND (TRACE)	28.76	0.019	11.24575
TRANS-1-(4-BIPHENYLYL)-2-PHENYL-ETHYLENE	32.21	0.013	22.11748
TRANS-1-(4-BIPHENYLYL)-2-(1-NAPHTHYL)-ETHYLENE	35.02	0.010	36.28381
DIPHENYLSTILBENE (B)	38.55	0.008	64.56677
1,1,4,4-TETRAPHENYLBUTADIENE	36.25	0.009	44.57863
1,6-DIPHENYLHEXATRIENE	40.29	0.007	84.10532
1,8-DIPHENYL-1,3,5,7-OCTATETRAENE	39.45	0.007	73.88082
BPSB (TRACE)	40.38	0.007	85.24482
BIS-MSB	39.14	0.007	70.65779
ESCULIN (A)	29.80	0.017	13.80833
AZULENE	21.79	0.043	2.09902
NAPHTHALENE	0.00	0.000	0.00000
DEUTERATED NAPHTHALENE-C10D8	7.31	1.147	0.00304
1-METHYLNAPHTHALENE 5ML/L	3.88	7.675	0.00007
2-METHYL NAPHTHALENE	4.49	4.931	0.00016
2,3-DIMETHYL NAPHTHALENE	4.61	4.578	0.00019
2,6-DIMETHYL NAPHTHALENE (TRACE)	8.63	0.695	0.00829
1-NAPHTHOL (TRACE)	8.79	0.658	0.00925
2-NAPHTHOL (TRACE)	15.47	0.121	0.27337
1-NAPHTHYLAMINE	21.38	0.046	1.89799
2-CHLORONAPHTHALENE	5.95	2.121	0.00089
ACENAPHTHENE	5.64	2.491	0.00065
1-PHENYL NAPHTHALENE	8.15	0.826	0.00587
2-PHENYL NAPHTHALENE	17.02	0.091	0.48445
1,4-DIPHENYL NAPHTHALENE	18.07	0.076	0.69415

ACCEPTOR MOLECULES	RO	CO	OVERLAP
1,5-DIPHENYL NAPHTHALENE	15.89	0.112	0.32046
1,7-DIPHENYL NAPHTHALENE	16.09	0.107	0.34538
1,4,5,8-TETRAPHENYL NAPHTHALENE	31.88	0.014	20.62520
1,1'-BINAPHTHYL	11.82	0.271	0.05439
4,4'-DI(N-BUTOXY)-1,1'-BINAPHTHYL	24.09	0.032	3.88049
2,2'-BINAPHTHYL	17.56	0.083	0.58545
ACRIDINE (A)	28.20	0.020	9.88125
ANTHRACENE (TRACE)	26.30	0.025	6.50843
DEUTERATED ANTHRACENE C14D10 (TRACE)	26.17	0.025	6.32377
9-METHYL ANTHRACENE (TRACE)	26.13	0.025	6.25932
1-AMINO ANTHRACENE	24.19	0.032	3.92728
2-AMINO ANTHRACENE	24.36	0.031	4.09710
9-VINYL ANTHRACENE	26.96	0.023	7.52323
9-PHENYL ANTHRACENE (TRACE)	27.40	0.022	8.31073
9,10-DIPHENYL ANTHRACENE	27.77	0.021	8.99060
9,10-DICHLOROANTHRACENE (TRACE)	26.21	0.025	6.36031
BIANTHRANYL (B)	29.30	0.018	12.39810
9,10-DINAPHTHYLANTHRACENE (B)	27.01	0.023	7.61803
9-(P-VINYLPHENYL)-10-PHENYL ANTHRACENE	27.79	0.021	9.02280
9,10-BIS(PHENYL-ETHYNYL)ANTHRACENE (TRACE)	23.34	0.035	3.17012
RUBICENE (B)	26.77	0.023	7.21377
DIBENZ(A,H)ANTHRACENE (B)	29.29	0.018	12.38463
TETRACENE (TRACE) (B)	19.48	0.060	1.07140
RUBRENE (B)	23.54	0.034	3.33163
TETRAHYDROCANNABINOL ACID METHYL ESTER	0.00	0.000	0.00000
PHENANTHRENE	13.29	0.190	0.10956
3,4-BENZOPHENANTHRENE	17.76	0.080	0.61725
2-PHENYL PHENANTHRENE	15.21	0.127	0.24479
CHRYSENE (TRACE)	17.53	0.083	0.57420
TRIPHENYLENE	13.85	0.168	0.14006
DEUTERATED TRIPHENYLENE-C18D12	13.57	0.179	0.12388
PYRENE	29.68	0.017	13.47692
1,3,6,8,-TETRAPHENYL PYRENE	34.38	0.011	32.36942
BIPYRENYL (B)	36.95	0.009	49.87910
O-PHENYLENE-PYRENE	30.67	0.016	16.30455
ANTHANTHRENE (B)	27.82	0.021	9.09276
FLUORANTHENE	27.51	0.021	8.49805
3-PHENYL FLUORANTHENE	30.23	0.016	14.96052
3-AMINO FLUORANTHENE	27.06	0.023	7.69332
3,3'-BIFLUORANTHENYL (B)	33.38	0.012	27.13227
BENZO(GHI)FLUORANTHENE	26.65	0.024	7.02902
9,12-DI(N-OCTADECOXY)-BENZ(E)ACEPHENANTHRYLENE	29.45	0.018	12.79643
DECACYCLENE (B)	35.95	0.010	42.27526
10-ETHOXY-BENZO(KL)XANTHENE	28.26	0.020	9.97518
PERYLENE	24.45	0.031	4.18376
BENZO(GHI)PERYLENE (B)	30.06	0.016	14.47143
BBOT (TRACE)	38.50	0.008	63.87174
QUININE SULFATE (A)	25.33	0.028	5.24559
ACRIDINE YELLOW (A)	21.20	0.047	1.78089
ACRIDINE ORANGE (A)	18.88	0.066	0.88751
ACRIDINE RED (A)	23.58	0.034	3.36953
URANINE (A)	22.67	0.038	2.65753
RHODAMINE B (A)	26.83	0.023	7.30582
RHODAMINE 6G (A)	27.63	0.021	8.72034
EOSIN B (A)	24.06	0.032	3.80525
EOSIN Y (A)	24.48	0.030	4.22028
ROSE BENGAL (A)	25.80	0.026	5.77554
INDENOINDENE	24.71	0.030	4.53512

8.67 Donor Molecule: *N*-Methylcarbazole (trace)

TRIPHENYLAMINE	16.86	0.093	0.33910
SODIUM SALICYLATE (A)	12.57	0.225	0.05802
ANTHRANILIC ACID (A)	24.96	0.029	3.54052
TMPD	19.98	0.056	0.93474
BENZIDINE	11.88	0.267	0.04156
1,3-DIBIPHENYLYLBENZENE (B)	7.75	0.960	0.00321
1,3,5,-TRIBIPHENYLYLBENZENE (B)	7.60	1.019	0.00286
FLUORENE (TRACE)	0.00	0.000	0.00000
1-METHYL FLUORENE	0.00	0.000	0.00000
1,2-BENZOFLUORENE	17.27	0.087	0.39081
9,10-DIHYDROPHENANTHRENE	0.00	0.000	0.00000
CARBAZOLE (TRACE)	10.74	0.361	0.02275
N-METHYL CARBAZOLE (TRACE)	21.34	0.046	1.39334
N-VINYLCARBAZOLE	18.11	0.075	0.52003
N-PHENYL CARBAZOLE	18.76	0.068	0.64323
N-METHYL-2,3-BENZCARBAZOLE	25.00	0.029	3.55465
DIBENZOFURAN	0.00	0.000	0.00000
DIBENZOTHIOPHENE (TRACE)	5.90	2.176	0.00063
4-(3,3-DIMETHYLBUTOXY)-P-TERPHENYL	4.81	4.030	0.00018
4,4''-DIHEXAHYDROFARNESOXY-P-TERPHENYL	7.52	1.052	0.00268
2,2'-METHYLENE-P-TERPHENYL	0.00	0.000	0.00000
2,2'-ETHYLENE-P-TERPHENYL	6.36	1.736	0.00098
2,2'-METHYLENE-5',6''-METHYLENE-P-TERPHENYL (TRACE)	18.63	0.069	0.61713
3-PHENYL DIBENZOFURAN	0.00	0.000	0.00000
BENZO(1,2-B-4,5-B')BISBENZOFURAN (TRACE) (B)	25.48	0.027	4.04093
SEXIPHENYL BENZENE	0.00	0.000	0.00000
P-QUATERPHENYL	11.78	0.273	0.03964
3,3'''-DIMETHYL-P-QUATERPHENYL	15.77	0.114	0.22754
TETRAMETHYL-P-QUATERPHENYL	0.00	0.000	0.00000
3,3'''-DIMETHYL-3',2''-METHYLENE-P-QUATERPHENYL	24.70	0.030	3.34552

ACCEPTOR MOLECULES	RO	CO	OVERLAP
3,3'''-DIMETHYL-3',2''-ETHYLENE-P-QUATERPHENYL	25.75	0.026	4.29354
4-(3-ETHYLBUTOXY)-P-QUATERPHENYL (B)	22.86	0.037	2.09982
4,4'''-DI(2-BUTYLOCTOXY-1)-P-QUATERPHENYL	23.12	0.036	2.25055
DIETHYL-P-QUINQUEPHENYL	15.94	0.110	0.24293
DI(3-ETHYLHEPTYL)-P-QUINQUEPHENYL	24.81	0.029	3.42961
TETRAMETHYL-P-SEXIPHENYL	11.96	0.261	0.04333
TETRAMETHYL-P-OCTAPHENYL (B)	30.59	0.016	12.04110
INDOLE (TRACE)	0.00	0.000	0.00000
2-PHENYL INDOLE	11.27	0.312	0.03040
1-METHYL-2-PHENYL INDOLE	10.43	0.394	0.01906
1-NEOPENTYL-2-PHENYL INDOLE	4.50	4.896	0.00012
1-N-BUTYL-2(BIPHENYLYL-4) INDOLE	24.84	0.029	3.45158
1,2-DIPHENYL INDOLE	10.84	0.351	0.02406
1-BIPHENYLYL-2-PHENYL INDOLE	16.21	0.105	0.26730
1,3-DIMETHYL-2-PHENYL INDOLE	11.94	0.263	0.04280
2,3-DIPHENYL INDOLE	22.77	0.038	2.04534
1-METHYL-2,3-DIPHENYL INDOLE	20.40	0.053	1.05945
2-PHENYL-3,2'-METHYLENE INDOLE (TRACE)	18.06	0.076	0.51262
2-PHENYL-3,2'-ETHYLENE INDOLE (TRACE)	28.15	0.020	7.32432
2-PHENYL-3,2'-TRIMETHYLENE INDOLE	17.14	0.089	0.37425
1-METHYL-3,2'-METHYLENE-2-PHENYL INDOLE (TRACE)	27.05	0.023	5.77319
1-METHYL-3,2'-ETHYLENE-2-PHENYL INDOLE	29.27	0.018	9.23602
1-METHYL-3,2'-TRIMETHYLENE-2-PHENYL INDOLE	18.88	0.066	0.66952
4,5-DIPHENYLIMIDAZOLE (A)	9.53	0.517	0.01109
PPF (TRACE)	28.08	0.020	7.22352
PPO (TRACE)	15.41	0.122	0.19807
2-(4-FLUOROPHENYL)-5-PHENYLOXAZOLE (TRACE)	14.09	0.160	0.11562
αNPO (TRACE)	33.26	0.012	19.83693
BBO (B)	39.16	0.007	52.74535
POPOP	39.53	0.007	55.73945
DIMETHYL POPOP (TRACE)	41.00	0.006	69.24684
PPD (TRACE)	0.00	0.000	0.00000
2,5-DI(2-METHYLPHENYL)OXADIAZOLE	0.00	0.000	0.00000
PBD	12.06	0.255	0.04553
BBD (B)	29.26	0.018	9.23439
αNPD (TRACE)	26.21	0.025	4.76379
βNPD	16.69	0.096	0.31926
αNND (TRACE)	30.44	0.016	11.68657
βNND (TRACE)	25.74	0.026	4.29168
TRANS-1-(4-BIPHENYLYL)-2-PHENYL-ETHYLENE	29.93	0.017	10.56473
TRANS-1-(4-BIPHENYLYL)-2-(1-NAPHTHYL)-ETHYLENE	35.45	0.010	29.00156
DIPHENYLSTILBENE (B)	39.51	0.007	55.56793
1,1,4,4-TETRAPHENYLBUTADIENE	37.60	0.008	41.23146
1,6-DIPHENYLHEXATRIENE	42.19	0.006	82.33902
1,8-DIPHENYL-1,3,5,7-OCTATETRAENE	43.42	0.005	97.17509
BPSB (TRACE)	42.83	0.006	90.13935
BIS-MSB	40.94	0.007	68.73835
ESCULIN (A)	30.00	0.017	10.67624
AZULENE	20.09	0.055	0.95243
NAPHTHALENE	0.00	0.000	0.00000
DEUTERATED NAPHTHALENE-C10D8	3.19	13.759	0.00002
2-NAPHTHOL (TRACE)	6.23	1.851	0.00086
1-NAPHTHYLAMINE	17.73	0.080	0.45859
2-CHLORONAPHTHALENE	0.00	0.000	0.00000
ACENAPHTHENE	0.00	0.000	0.00000
1-PHENYL NAPHTHALENE	0.00	0.000	0.00000
2-PHENYL NAPHTHALENE	7.81	0.938	0.00337
1,4-DIPHENYL NAPHTHALENE	13.06	0.201	0.07329
1,5-DIPHENYL NAPHTHALENE	8.97	0.620	0.00771
1,7-DIPHENYL NAPHTHALENE	10.42	0.396	0.01891
1,4,5,8-TETRAPHENYL NAPHTHALENE	32.14	0.013	16.08177
1,1'-BINAPHTHYL	0.00	0.000	0.00000
4,4'-DI(N-BUTOXY)-1,1'-BINAPHTHYL	19.97	0.056	0.93402
2,2'-BINAPHTHYL	10.74	0.361	0.02266
ACRIDINE (A)	29.77	0.017	10.16265
ANTHRACENE (TRACE)	27.71	0.021	6.62430
DEUTERATED ANTHRACENE C14D10 (TRACE)	27.27	0.022	6.02007
9-METHYL ANTHRACENE (TRACE)	28.75	0.019	8.24120
1-AMINO ANTHRACENE	27.06	0.023	5.69183
2-AMINO ANTHRACENE	25.72	0.026	4.19586
9-VINYL ANTHRACENE	29.60	0.017	9.80603
9-PHENYL ANTHRACENE (TRACE)	30.22	0.016	11.11179
9,10-DIPHENYL ANTHRACENE	30.66	0.016	12.08255
9,10-DICHLOROANTHRACENE (TRACE)	29.55	0.017	9.68959
BIANTHRANYL (B)	32.17	0.013	16.13922
9,10-DINAPHTHYLANTHRACENE (B)	30.07	0.016	10.74531
9-(P-VINYLPHENYL)-10-PHENYL ANTHRACENE	30.84	0.015	12.49458
9,10-BIS(PHENYL-ETHYNYL)ANTHRACENE (TRACE)	27.54	0.021	6.33024
RUBICENE (B)	28.99	0.018	8.60108
DIBENZ(A,H)ANTHRACENE (B)	28.01	0.020	7.04259
TETRACENE (TRACE) (B)	21.75	0.043	1.53383
RUBRENE (B)	22.01	0.042	1.64984
TETRAHYDROCANNABINOL ACID METHYL ESTER	0.00	0.000	0.00000
PHENANTHRENE	12.42	0.234	0.05405
3,4-BENZOPHENANTHRENE	16.57	0.098	0.30231
2-PHENYL PHENANTHRENE	14.82	0.137	0.15601
CHRYSENE (TRACE)	17.50	0.084	0.42084
TRIPHENYLENE	11.01	0.335	0.02623
DEUTERATED TRIPHENYLENE-C18D12	10.98	0.338	0.02579
PYRENE	20.75	0.050	1.16930
1,3,6,8,-TETRAPHENYL PYRENE	38.45	0.008	46.92837
BIPYRENYL (B)	38.52	0.008	47.57727
O-PHENYLENE-PYRENE	33.57	0.012	20.74941
ANTHANTHRENE (B)	30.41	0.016	11.48048
FLUORANTHENE	28.84	0.019	8.37749
3-PHENYL FLUORANTHENE	32.53	0.013	17.20540

ACCEPTOR MOLECULES	RO	CO	OVERLAP
3-AMINO FLUORANTHENE	29.25	0.018	9.09124
3,3'-BIFLUORANTHENYL (B)	36.49	0.009	34.26062
BENZO(GHI)FLUORANTHENE	26.48	0.024	5.00762
9,12-DI(N-OCTADECOXY)-BENZ(E)ACEPHENANTHRYLENE	31.67	0.014	14.62498
DECACYCLENE (B)	39.06	0.008	51.48058
10-ETHOXY-BENZO(KL)XANTHENE	31.40	0.014	13.96525
PERYLENE	28.88	0.019	8.42075
BENZO(GHI)PERYLENE (B)	33.58	0.012	20.87655
BBOT (TRACE)	41.85	0.006	78.21616
QUININE SULFATE (A)	22.03	0.042	1.68444
ACRIDINE YELLOW (A)	24.85	0.029	3.41345
ACRIDINE ORANGE (A)	21.42	0.045	1.40199
ACRIDINE RED (A)	24.95	0.029	3.49823
URANINE (A)	23.72	0.034	2.58316
RHODAMINE B (A)	28.38	0.020	7.57156
RHODAMINE 6G (A)	28.60	0.019	7.93081
EOSIN B (A)	24.54	0.030	3.16964
EOSIN Y (A)	25.16	0.028	3.67901
ROSE BENGAL (A)	25.83	0.026	4.30289
INDENOINDENE	15.17	0.128	0.17980

8.68 Donor Molecule: *N*-Vinylcarbazole

ACCEPTOR MOLECULES	RO	CO	OVERLAP
TRIPHENYLAMINE	17.00	0.091	0.53569
SODIUM SALICYLATE (A)	12.34	0.238	0.07812
ANTHRANILIC ACID (A)	23.90	0.033	4.10561
TMPD	19.41	0.061	1.18043
BENZIDINE	12.91	0.208	0.10291
1,3-DIBIPHENYLYLBENZENE (B)	10.26	0.414	0.02601
1,3,5,-TRIBIPHENYLYLBENZENE (B)	10.46	0.391	0.02915
FLUORENE (TRACE)	0.00	0.000	0.00000
1-METHYL FLUORENE	0.00	0.000	0.00000
1,2-BENZOFLUORENE	16.85	0.094	0.50674
9,10-DIHYDROPHENANTHRENE	0.00	0.000	0.00000
CARBAZOLE (TRACE)	12.20	0.246	0.07338
N-METHYL CARBAZOLE (TRACE)	20.80	0.050	1.79515
N-VINYLCARBAZOLE	19.81	0.058	1.34138
N-PHENYL CARBAZOLE	19.76	0.058	1.32163
N-METHYL-2,3-BENZCARBAZOLE	22.96	0.037	3.20918
DIBENZOFURAN	0.00	0.000	0.00000
DIBENZOTHIOPHENE (TRACE)	8.05	0.859	0.00603
4-(3,3-DIMETHYLBUTOXY)-P-TERPHENYL	7.39	1.106	0.00364
4,4''-DIHEXAHYDROFARNESOXY-P-TERPHENYL	10.43	0.394	0.02865
2,2'-METHYLENE-P-TERPHENYL	6.26	1.820	0.00135
2,2'-ETHYLENE-P-TERPHENYL	8.98	0.618	0.01166
2,2'-METHYLENE-5',6''-METHYLENE-P-TERPHENYL (TRACE)	21.85	0.043	2.41913
3-PHENYL DIBENZOFURAN	6.11	1.964	0.00116
BENZO(1,2-B-4,5-B)BISBENZOFURAN (TRACE) (B)	26.89	0.023	8.38189
SEXIPHENYL BENZENE	0.00	0.000	0.00000
P-QUATERPHENYL	14.21	0.156	0.18272
3,3'''-DIMETHYL-P-QUATERPHENYL	16.60	0.098	0.46422
TETRAMETHYL-P-QUATERPHENYL	0.00	0.000	0.00000
3,3'''-DIMETHYL-3',2''-METHYLENE-P-QUATERPHENYL	25.36	0.027	5.88733
3,3'''-DIMETHYL-3',2''-ETHYLENE-P-QUATERPHENYL	25.86	0.026	6.60502
4-(3-ETHYLBUTOXY)-P-QUATERPHENYL (B)	22.99	0.037	3.26548
4,4'''-DI(2-BUTYLOCTOXY-1)-P-QUATERPHENYL	23.27	0.035	3.51651
DIETHYL-P-QUINQUEPHENYL	17.15	0.089	0.56444
DI(3-ETHYLHEPTYL)-P-QUINQUEPHENYL	24.79	0.029	5.12588
TETRAMETHYL-P-SEXIPHENYL	14.66	0.142	0.22054
TETRAMETHYL-P-OCTAPHENYL (B)	30.06	0.016	16.26883
INDOLE (TRACE)	0.00	0.000	0.00000
2-PHENYL INDOLE	14.69	0.141	0.22381
1-METHYL-2-PHENYL INDOLE	12.03	0.257	0.06747
1-NEOPENTYL-2-PHENYL INDOLE	6.90	1.361	0.00241
1-N-BUTYL-2(BIPHENYLYL-4) INDOLE	24.45	0.031	4.72504
1,2-DIPHENYL INDOLE	12.39	0.235	0.08036
1-BIPHENYLYL-2-PHENYL INDOLE	16.45	0.100	0.43886
1,3-DIMETHYL-2-PHENYL INDOLE	13.94	0.165	0.16292
2,3-DIPHENYL INDOLE	22.43	0.040	2.80660
1-METHYL-2,3-DIPHENYL INDOLE	20.16	0.055	1.48474
2-PHENYL-3,2'-METHYLENE INDOLE (TRACE)	20.49	0.052	1.64044
2-PHENYL-3,2'-ETHYLENE INDOLE (TRACE)	27.13	0.022	8.81847
2-PHENYL-3,2'-TRIMETHYLENE INDOLE	18.92	0.066	1.01836
1-METHYL-3,2'-METHYLENE-2-PHENYL INDOLE (TRACE)	26.28	0.025	7.30161
1-METHYL-3,2'-ETHYLENE-2-PHENYL INDOLE	28.48	0.019	11.79179
1-METHYL-3,2'-TRIMETHYLENE-2-PHENYL INDOLE	20.18	0.054	1.49809
4,5-DIPHENYLIMIDAZOLE (A)	11.19	0.319	0.04365
PPF (TRACE)	27.54	0.021	9.65874
PPO (TRACE)	18.38	0.072	0.85514
2-(4-FLUOROPHENYL)-5-PHENYLOXAZOLE (TRACE)	16.84	0.094	0.50591
αNPO (TRACE)	31.67	0.014	22.20516
BBO (B)	37.07	0.009	57.01748
POPOP	37.13	0.009	57.52053
DIMETHYL POPOP (TRACE)	38.19	0.008	67.91102
PPD (TRACE)	0.00	0.000	0.00000
2,5-DI(2-METHYLPHENYL)OXADIAZOLE	0.00	0.000	0.00000
PBD	15.32	0.124	0.28799
BBD (B)	28.79	0.019	12.59247
αNPD (TRACE)	25.70	0.026	6.35730
βNPD	17.70	0.081	0.68320
αNND (TRACE)	29.39	0.018	14.20754

ACCEPTOR MOLECULES	RO	CO	OVERLAP
(3NND (TRACE)	26.05	0.025	6.93587
TRANS-1-(4-BIPHENYLYL)-2-PHENYL-ETHYLENE	29.25	0.018	13.81862
TRANS-1-(4-BIPHENYLYL)-2-(1-NAPHTHYL)-ETHYLENE	33.70	0.012	32.16477
DIPHENYLSTILBENE (B)	37.42	0.009	60.22034
1,1,4,4-TETRAPHENYLBUTADIENE	35.47	0.010	43.62033
1,6-DIPHENYLHEXATRIENE	39.75	0.007	86.47885
1,8-DIPHENYL-1,3,5,7-OCTATETRAENE	40.12	0.007	90.78972
BPSB (TRACE)	40.15	0.007	91.77884
BIS-MSB	38.51	0.008	71.51728
ESCULIN (A)	28.60	0.019	[illegible]
AZULENE	20.41	0.053	1.57454
NAPHTHALENE	0.00	0.000	0.00000
DEUTERATED NAPHTHALENE-C10D8	4.41	5.201	0.00016
1-METHYLNAPHTHALENE 5ML/L	0.00	0.000	0.00000
2-METHYL NAPHTHALENE	0.00	0.000	0.00000
2,3-DIMETHYL NAPHTHALENE	0.00	0.000	0.00000
2,6-DIMETHYL NAPHTHALENE (TRACE)	3.18	13.936	0.00002
1-NAPHTHOL (TRACE)	4.05	6.751	0.00010
2-NAPHTHOL (TRACE)	8.89	0.636	0.01100
1-NAPHTHYLAMINE	18.05	0.076	0.76486
2-CHLORONAPHTHALENE	2.40	32.228	0.00000
ACENAPHTHENE	0.00	0.000	0.00000
1-PHENYL NAPHTHALENE	3.64	9.276	0.00005
2-PHENYL NAPHTHALENE	10.68	0.367	0.03302
1,4-DIPHENYL NAPHTHALENE	13.85	0.168	0.15657
1,5-DIPHENYL NAPHTHALENE	11.06	0.331	0.04068
1,7-DIPHENYL NAPHTHALENE	11.86	0.268	0.06172
1,4,5,8-TETRAPHENYL NAPHTHALENE	30.61	0.016	18.02928
1,1'-BINAPHTHYL	6.01	2.061	0.00105
4,4'-DI(N-BUTOXY)-1,1'-BINAPHTHYL	20.13	0.055	1.47428
2,2'-BINAPHTHYL	12.98	0.204	0.10642
ACRIDINE (A)	28.07	0.020	10.72874
ANTHRACENE (TRACE)	26.42	0.024	7.46644
DEUTERATED ANTHRACENE C14D10 (TRACE)	25.92	0.026	6.66519
9-METHYL ANTHRACENE (TRACE)	26.29	0.025	7.22952
1-AMINO ANTHRACENE	24.79	0.029	5.05345
2-AMINO ANTHRACENE	23.92	0.033	4.07882
9-VINYL ANTHRACENE	27.22	0.022	8.90552
9-PHENYL ANTHRACENE (TRACE)	27.75	0.021	10.01001
9,10-DIPHENYL ANTHRACENE	28.30	0.020	11.22980
9,10-DICHLOROANTHRACENE (TRACE)	27.25	0.022	8.93662
BIANTHRANYL (B)	29.57	0.017	14.61389
9,10-DINAPHTHYLANTHRACENE (B)	27.80	0.021	10.07741
9-(P-VINYLPHENYL)-10-PHENYL ANTHRACENE	28.44	0.019	11.54313
9,10-BIS(PHENYL-ETHYNYL)ANTHRACENE (TRACE)	24.65	0.030	4.88785
RUBICENE (B)	26.88	0.023	8.20854
DIBENZ(A,H)ANTHRACENE (B)	27.29	0.022	9.03322
TETRACENE (TRACE) (B)	19.86	0.057	1.33618
RUBRENE (B)	21.17	0.047	1.95913
TETRAHYDROCANNABINOL ACID METHYL ESTER	0.00	0.000	0.00000
PHENANTHRENE	12.02	0.257	0.06692
3,4-BENZOPHENANTHRENE	15.79	0.114	0.34079
2-PHENYL PHENANTHRENE	14.28	0.154	0.18722
CHRYSENE (TRACE)	16.62	0.097	0.46505
TRIPHENYLENE	11.65	0.283	0.05543
DEUTERATED TRIPHENYLENE-C18D12	11.34	0.307	0.04712
PYRENE	23.21	0.036	3.44191
1,3,6,8,-TETRAPHENYL PYRENE	35.31	0.010	42.29290
BIPYRENYL (B)	36.12	0.009	48.54919
O-PHENYLENE-PYRENE	31.13	0.015	19.83375
ANTHANTHRENE (B)	27.82	0.021	10.10326
FLUORANTHENE	27.28	0.022	8.99724
3-PHENYL FLUORANTHENE	30.28	0.016	16.80347
3-AMINO FLUORANTHENE	27.00	0.023	8.43882
3,3'-BIFLUORANTHENYL (B)	33.71	0.012	31.95018
BENZO(GHI)FLUORANTHENE	25.06	0.028	5.40276
9,12-DI(N-OCTADECOXY)-BENZ(E)ACEPHENANTHRYLENE	29.45	0.018	14.20887
DECACYCLENE (B)	36.20	0.009	48.99422
10-ETHOXY-BENZO(KL)XANTHENE	28.92	0.018	12.79376
PERYLENE	25.86	0.026	6.51667
BENZO(GHI)PERYLENE (B)	30.65	0.016	18.12619
BBOT (TRACE)	38.79	0.008	74.50712
QUININE SULFATE (A)	22.36	0.040	2.76562
ACRIDINE YELLOW (A)	22.32	0.040	2.69390
ACRIDINE ORANGE (A)	19.43	0.061	1.17291
ACRIDINE RED (A)	23.38	0.035	3.55965
URANINE (A)	22.03	0.042	2.48675
RHODAMINE B (A)	26.59	0.024	7.69814
RHODAMINE 6G (A)	26.97	0.023	8.37965
EOSIN B (A)	23.20	0.036	3.39348
EOSIN Y (A)	23.70	0.034	3.86007
ROSE BENGAL (A)	24.52	0.030	4.72865
INDENOINDENE	17.33	0.086	0.60126

8.69 Donor Molecule: *N*-Phenylcarbazole

ACCEPTOR MOLECULES	RO	CO	OVERLAP
TRIPHENYLAMINE	16.95	0.092	0.48260
SODIUM SALICYLATE (A)	12.36	0.237	0.07256
ANTHRANILIC ACID (A)	24.09	0.032	3.95020
TMPD	19.49	0.060	1.11121
BENZIDINE	12.77	0.215	0.08851

ACCEPTOR MOLECULES	RO	CO	OVERLAP
1,3-DIBIPHENYLYLBENZENE (B)	9.97	0.451	0.02011
1,3,5,-TRIBIPHENYLYLBENZENE (B)	10.10	0.434	0.02170
FLUORENE (TRACE)	0.00	0.000	0.00000
1-METHYL FLUORENE	0.00	0.000	0.00000
1,2-BENZOFLUORENE	16.88	0.093	0.47088
9,10-DIHYDROPHENANTHRENE	0.00	0.000	0.00000
CARBAZOLE (TRACE)	11.98	0.260	0.06046
N-METHYL CARBAZOLE (TRACE)	20.88	0.049	1.68592
N-VINYLCARBAZOLE	19.61	0.059	1.15987
N-PHENYL CARBAZOLE	19.65	0.059	1.17176
N-METHYL-2,3-BENZCARBAZOLE	23.42	0.035	3.32108
DIBENZOFURAN	0.00	0.000	0.00000
DIBENZOTHIOPHENE (TRACE)	7.82	0.936	0.00467
4-(3,3-DIMETHYLBUTOXY)-P-TERPHENYL	7.15	1.221	0.00275
4,4''-DIHEXAHYDROFARNESOXY-P-TERPHENYL	10.07	0.438	0.02133
2,2'-METHYLENE-P-TERPHENYL	6.08	1.987	0.00104
2,2'-ETHYLENE-P-TERPHENYL	8.68	0.685	0.00873
2,2'-METHYLENE-5',6''-METHYLENE-P-TERPHENYL (TRACE)	21.37	0.046	1.94574
3-PHENYL DIBENZOFURAN	5.94	2.135	0.00090
BENZO(1,2-B-4,5-B')BISBENZOFURAN (TRACE) (B)	26.73	0.023	7.43515
SEXIPHENYL BENZENE	0.00	0.000	0.00000
P-QUATERPHENYL	13.93	0.165	0.14933
3,3'''-DIMETHYL-P-QUATERPHENYL	16.45	0.101	0.40404
TETRAMETHYL-P-QUATERPHENYL	0.00	0.000	0.00000
3,3''' -DIMETHYL-3',2''-METHYLENE-P-QUATERPHENYL	25.21	0.028	5.22994
3,3''' -DIMETHYL-3',2''-ETHYLENE-P-QUATERPHENYL	25.80	0.026	5.98810
4-(3-ETHYLBUTOXY)-P-QUATERPHENYL (B)	22.93	0.037	2.95413
4,4'''-DI(2-BUTYLOCTOXY-1)-P-QUATERPHENYL	23.20	0.036	3.17399
DIETHYL-P-QUINQUEPHENYL	16.96	0.092	0.48660
DI(3-ETHYLHEPTYL)-P-QUINQUEPHENYL	24.75	0.029	4.66834
TETRAMETHYL-P-SEXIPHENYL	14.35	0.151	0.17866
TETRAMETHYL-P-OCTAPHENYL (B)	30.11	0.016	15.11311
INDOLE (TRACE)	0.00	0.000	0.00000
2-PHENYL INDOLE	14.28	0.154	0.17338
1-METHYL-2-PHENYL INDOLE	11.82	0.271	0.05571
1-NEOPENTYL-2-PHENYL INDOLE	6.68	1.499	0.00182
1-N-BUTYL-2(BIPHENYLYL-4) INDOLE	24.48	0.030	4.37438
1,2-DIPHENYL INDOLE	12.18	0.248	0.06670
1-BIPHENYLYL-2-PHENYL INDOLE	16.38	0.102	0.39334
1,3-DIMETHYL-2-PHENYL INDOLE	13.69	0.174	0.13441
2,3-DIPHENYL INDOLE	22.45	0.040	2.59696
1-METHYL-2,3-DIPHENYL INDOLE	20.17	0.054	1.36908
2-PHENYL-3,2'-METHYLENE INDOLE (TRACE)	20.11	0.055	1.34930
2-PHENYL-3,2'-ETHYLENE INDOLE (TRACE)	27.21	0.022	8.25713
2-PHENYL-3,2'-TRIMETHYLENE INDOLE	18.65	0.069	0.85738
1-METHYL-3,2'-METHYLENE-2-PHENYL INDOLE (TRACE)	26.34	0.024	6.80198
1-METHYL-3,2'-ETHYLENE-2-PHENYL INDOLE	28.60	0.019	11.11847
1-METHYL-3,2'-TRIMETHYLENE-2-PHENYL INDOLE	19.98	0.056	1.29732
4,5-DIPHENYLIMIDAZOLE (A)	11.01	0.335	0.03646
PPF (TRACE)	27.56	0.021	8.92491
PPO (TRACE)	17.97	0.077	0.68844
2-(4-FLUOROPHENYL)-5-PHENYLOXAZOLE (TRACE)	16.46	0.100	0.40578
αNPO (TRACE)	31.95	0.014	21.49467
BBO (B)	37.47	0.009	55.88446
POPOP	37.64	0.008	57.35080
DIMETHYL POPOP (TRACE)	38.77	0.008	68.30685
PPD (TRACE)	0.00	0.000	0.00000
2,5-DI(2-METHYLPHENYL)OXADIAZOLE	0.00	0.000	0.00000
PBD	14.94	0.134	0.22731
BBD (B)	28.80	0.019	11.60614
αNPD (TRACE)	25.75	0.026	5.91826
βNPD	17.57	0.083	0.59941
αNND (TRACE)	29.57	0.017	13.55094
βNND (TRACE)	26.00	0.025	6.29618
TRANS-1-(4-BIPHENYLYL)-2-PHENYL-ETHYLENE	29.31	0.018	12.85619
TRANS-1-(4-BIPHENYLYL)-2-(1-NAPHTHYL)-ETHYLENE	34.03	0.011	31.32390
DIPHENYLSTILBENE (B)	37.81	0.008	58.92667
1,1,4,4-TETRAPHENYLBUTADIENE	35.89	0.010	43.00085
1,6-DIPHENYLHEXATRIENE	40.27	0.007	85.92858
1,8-DIPHENYL-1,3,5,7-OCTATETRAENE	40.84	0.007	92.94312
BPSB (TRACE)	40.70	0.007	91.62364
BIS-MSB	39.00	0.008	70.86411
ESCULIN (A)	28.85	0.019	11.66590
AZULENE	20.38	0.053	1.43672
NAPHTHALENE	0.00	0.000	0.00000
DEUTERATED NAPHTHALENE-C10D8	4.26	5.783	0.00012
1-METHYLNAPHTHALENE 5ML/L	0.00	0.000	0.00000
2-METHYL NAPHTHALENE	0.00	0.000	0.00000
2,3-DIMETHYL NAPHTHALENE	0.00	0.000	0.00000
2,6-DIMETHYL NAPHTHALENE (TRACE)	3.22	13.345	0.00002
1-NAPHTHOL (TRACE)	3.93	7.366	0.00008
2-NAPHTHOL (TRACE)	8.60	0.702	0.00829
1-NAPHTHYLAMINE	17.96	0.077	0.68320
2-CHLORONAPHTHALENE	2.41	31.785	0.00000
ACENAPHTHENE	0.00	0.000	0.00000
1-PHENYL NAPHTHALENE	3.59	9.650	0.00004
2-PHENYL NAPHTHALENE	10.31	0.408	0.02455
1,4-DIPHENYL NAPHTHALENE	13.71	0.174	0.13540
1,5-DIPHENYL NAPHTHALENE	10.82	0.353	0.03273
1,7-DIPHENYL NAPHTHALENE	11.67	0.282	0.05149
1,4,5,8-TETRAPHENYL NAPHTHALENE	30.89	0.015	17.50157
1,1'-BINAPHTHYL	5.81	2.279	0.00079
4,4' DI(N BUTOXY) 1,1' BINAPHTHYL	20.06	0.055	1.32760
2,2'-BINAPHTHYL	12.72	0.217	0.08670
ACRIDINE (A)	28.44	0.019	10.65500
ANTHRACENE (TRACE)	26.77	0.023	7.43318

ACCEPTOR MOLECULES	RO	CO	OVERLAP
DEUTERATED ANTHRACENE C14D10 (TRACE)	26.25	0.025	6.61387
9-METHYL ANTHRACENE (TRACE)	26.75	0.023	7.37873
1-AMINO ANTHRACENE	25.27	0.028	5.21944
2-AMINO ANTHRACENE	24.29	0.031	4.11609
9-VINYL ANTHRACENE	27.71	0.021	9.09649
9-PHENYL ANTHRACENE (TRACE)	28.24	0.020	10.20591
9,10-DIPHENYL ANTHRACENE	28.83	0.019	11.53737
9,10 DICHLOROANTHRACENE (TRACE)	27.70	0.021	[illegible]
BIANTHRANYL (B)	30.12	0.016	15.00236
9,10-DINAPHTHYLANTHRACENE (B)	28.34	0.020	10.38854
9-(P-VINYLPHENYL)-10-PHENYL ANTHRACENE	28.98	0.018	11.88639
9,10-BIS(PHENYL-ETHYNYL)ANTHRACENE (TRACE)	25.30	0.028	5.25198
RUBICENE (B)	27.35	0.022	8.37774
DIBENZ(A,H)ANTHRACENE (B)	27.41	0.022	8.52961
TETRACENE (TRACE) (B)	20.27	0.054	1.38982
RUBRENE (B)	21.31	0.046	1.87630
TETRAHYDROCANNABINOL ACID METHYL ESTER	0.00	0.000	0.00000
PHENANTHRENE	12.02	0.258	0.06138
3,4-BENZOPHENANTHRENE	15.94	0.110	0.33147
2-PHENYL PHENANTHRENE	14.36	0.151	0.17821
CHRYSENE (TRACE)	16.78	0.095	0.45308
TRIPHENYLENE	11.54	0.291	0.04808
DEUTERATED TRIPHENYLENE-C18D12	11.28	0.312	0.04193
PYRENE	22.81	0.038	2.84406
1,3,6,8,-TETRAPHENYL PYRENE	36.00	0.010	43.60198
BIPYRENYL (B)	36.58	0.009	48.15613
O-PHENYLENE-PYRENE	31.66	0.014	20.16334
ANTHANTHRENE (B)	28.35	0.020	10.39078
FLUORANTHENE	27.60	0.021	8.87874
3-PHENYL FLUORANTHENE	30.75	0.015	16.94847
3-AMINO FLUORANTHENE	27.47	0.022	8.60386
3,3'-BIFLUORANTHENYL (B)	34.30	0.011	32.62552
BENZO(GHI)FLUORANTHENE	25.26	0.028	5.21244
9,12-DI(N-OCTADECOXY)-BENZ(E)ACEPHENANTHRYLENE	29.92	0.017	14.35102
DECACYCLENE (B)	36.81	0.009	49.81090
10-ETHOXY-BENZO(KL)XANTHENE	29.47	0.017	13.17302
PERYLENE	26.52	0.024	6.97136
BENZO(GHI)PERYLENE (B)	31.24	0.015	18.68475
BBOT (TRACE)	39.44	0.007	75.62506
QUININE SULFATE (A)	22.27	0.040	2.48437
ACRIDINE YELLOW (A)	22.89	0.037	2.87686
ACRIDINE ORANGE (A)	19.87	0.057	1.23201
ACRIDINE RED (A)	23.69	0.034	3.54303
URANINE (A)	22.38	0.040	2.51377
RHODAMINE B (A)	26.95	0.023	7.67078
RHODAMINE 6G (A)	27.25	0.022	8.20216
EOSIN B (A)	23.43	0.035	3.31092
EOSIN Y (A)	23.96	0.033	3.78636
ROSE BENGAL (A)	24.76	0.029	4.61440
INDENOINDENE	17.01	0.091	0.49352

8.70 Donor Molecule: *N*-Methyl-2,3-benzcarbazole

N-METHYL-2,3-BENZCARBAZOLE	24.64	0.030	2.73901
BBO (B)	9.14	0.586	0.00715
POPOP	13.95	0.165	0.09041
DIMETHYL POPOP (TRACE)	25.51	0.027	3.37388
TRANS-1-(4-BIPHENYLYL)-2-(1-NAPHTHYL)-ETHYLENE	9.90	0.461	0.01154
DIPHENYLSTILBENE (B)	10.27	0.413	0.01440
1,1,4,4-TETRAPHENYLBUTADIENE	17.35	0.086	0.33420
1,6-DIPHENYLHEXATRIENE	16.97	0.092	0.29287
1,8-DIPHENYL-1,3,5,7-OCTATETRAENE	40.06	0.007	50.34086
BPSB (TRACE)	19.59	0.059	0.69276
BIS-MSB	19.38	0.061	0.64916
ESCULIN (A)	0.00	0.000	0.00000
AZULENE	12.46	0.231	0.04503
1,4,5,8-TETRAPHENYL NAPHTHALENE	12.32	0.239	0.04281
1,1'-BINAPHTHYL	0.00	0.000	0.00000
4,4'-DI(N-BUTOXY)-1,1'-BINAPHTHYL	0.00	0.000	0.00000
2,2'-BINAPHTHYL	0.00	0.000	0.00000
ACRIDINE (A)	14.25	0.155	0.10253
ANTHRACENE (TRACE)	6.67	1.509	0.00108
DEUTERATED ANTHRACENE C14D10 (TRACE)	5.88	2.202	0.00051
9-METHYL ANTHRACENE (TRACE)	15.29	0.125	0.15671
1-AMINO ANTHRACENE	28.66	0.019	6.73432
2-AMINO ANTHRACENE	28.30	0.020	6.25017
9-VINYL ANTHRACENE	22.85	0.038	1.74091
9-PHENYL ANTHRACENE (TRACE)	17.21	0.088	0.31841
9,10-DIPHENYL ANTHRACENE	28.77	0.019	6.93836
9,10-DICHLOROANTHRACENE (TRACE)	30.36	0.016	9.56591
BIANTHRANYL (B)	29.13	0.018	7.46565
9,10-DINAPHTHYLANTHRACENE (B)	30.09	0.016	9.06932
9-(P-VINYLPHENYL)-10-PHENYL ANTHRACENE	29.69	0.017	8.36236
9,10-BIS(PHENYL-ETHYNYL)ANTHRACENE (TRACE)	42.13	0.006	67.64638
RUBICENE (B)	33.63	0.012	17.44010
DIBENZ(A,H)ANTHRACENE (B)	16.17	0.106	0.21917
TETRACENE (TRACE) (B)	30.50	0.016	9.73986
RUBRENE (B)	28.25	0.020	6.12830
TETRAHYDROCANNABINOL ACID METHYL ESTER	0.00	0.000	0.00000
PHENANTHRENE	0.00	0.000	0.00000
3,4-BENZOPHENANTHRENE	3.77	8.332	0.00004

ACCEPTOR MOLECULES	RO	CO	OVERLAP
1,3,6,8,-TETRAPHENYL PYRENE	31.90	0.014	12.84977
BIPYRENYL (B)	25.02	0.029	3.00006
O-PHENYLENE-PYRENE	32.56	0.013	14.42720
ANTHANTHRENE (B)	38.81	0.008	41.53944
FLUORANTHENE	13.98	0.164	0.09136
3-PHENYL FLUORANTHENE	22.86	0.037	1.73979
3-AMINO FLUORANTHENE	32.50	0.013	14.27456
3,3'-BIFLUORANTHENYL (B)	32.81	0.013	15.20548
BENZO(GHI)FLUORANTHENE	20.13	0.055	0.81165
9,12-DI(N-OCTADECOXY)-BENZ(E)ACEPHENANTHRYLENE	33.41	0.012	16.93895
DECACYCLENE (B)	39.84	0.007	48.44837
10-ETHOXY-BENZO(KL)XANTHENE	29.57	0.017	8.17476
PERYLENE	40.08	0.007	50.40552
BENZO(GHI)PERYLENE (B)	23.17	0.036	1.89027
BBOT (TRACE)	34.39	0.011	20.25371
QUININE SULFATE (A)	0.00	0.000	0.00000
ACRIDINE YELLOW (A)	39.91	0.007	48.76113
ACRIDINE ORANGE (A)	33.67	0.012	17.58672
ACRIDINE RED (A)	26.37	0.024	4.05143
URANINE (A)	33.74	0.012	17.78343
RHODAMINE B (A)	30.78	0.015	10.24834
RHODAMINE 6G (A)	29.60	0.017	8.10390
EOSIN B (A)	29.23	0.018	7.52252
EOSIN Y (A)	30.75	0.015	10.20379
ROSE BENGAL (A)	28.43	0.019	6.36997

8.71 Donor Molecule: Dibenzofuran

ACCEPTOR MOLECULES	RO	CO	OVERLAP
POLYVINYLTOLUENE	4.13	6.365	0.00007
ANILINE 3ML/L	11.99	0.259	0.04349
N,N-DIMETHYLANILINE 3ML/L	18.71	0.068	0.62272
TRIPHENYLAMINE	30.54	0.016	11.68780
1,4-DIMETHOXY BENZENE	14.74	0.140	0.14952
1-METHYL-4-CHLOROBENZENE 5ML/L	0.00	0.000	0.00000
SODIUM SALICYLATE (A)	21.30	0.046	1.34136
ANTHRANILIC ACID (A)	25.42	0.027	3.85360
TMPD	23.19	0.036	2.22772
1,2,4-TRIMETHYLBENZENE (TRACE)	0.00	0.000	0.00000
MESITYLENE 5ML/L	0.00	0.000	0.00000
1,3,5-TRIETHYLBENZENE 5ML/L	0.00	0.000	0.00000
STYRENE	4.55	4.738	0.00013
3,3'-DIMETHYL BIPHENYL	5.77	2.334	0.00054
2,2'-DIMETHYLBIPHENYL 2ML/L	0.00	0.000	0.00000
OCTAMETHYL BIPHENYL	0.00	0.000	0.00000
4-METHOXY BIPHENYL	11.72	0.278	0.03794
4-BIPHENYLYL PHENYL ETHER	10.73	0.362	0.02235
4,4'-DIHYDROXYBIPHENYL (A)	17.52	0.083	0.42002
4,4'-DIMETHOXY BIPHENYL	16.05	0.108	0.24959
4,4'-DIISOBUTOXYBIPHENYL	15.91	0.111	0.23671
BENZIDINE	28.24	0.020	7.31875
4-VINYL BIPHENYL	21.44	0.045	1.41579
M-TERPHENYL	14.96	0.134	0.16377
M-QUATERPHENYL	10.04	0.442	0.01505
1,3,5-TRIPHENYL BENZENE	14.62	0.143	0.14168
1,3-DIBIPHENYLYLBENZENE (B)	29.38	0.018	9.29043
1,3,5,-TRIBIPHENYLYLBENZENE (B)	33.79	0.012	21.55164
FLUORENE (TRACE)	17.87	0.078	0.47610
1-METHYL FLUORENE	16.24	0.104	0.26810
1,2 BENZOFLUORENE	28.50	0.019	7.71787
9,10-DIHYDROPHENANTHRENE	17.53	0.083	0.42321
CARBAZOLE (TRACE)	24.32	0.031	2.98465
N-METHYL CARBAZOLE (TRACE)	23.84	0.033	2.64206
N-VINYLCARBAZOLE	24.34	0.031	2.99221
N-PHENYL CARBAZOLE	24.18	0.032	2.87693
N-METHYL-2,3-BENZCARBAZOLE	27.95	0.020	6.80442
DIBENZOFURAN	15.72	0.115	0.22095
DIBENZOTHIOPHENE (TRACE)	22.16	0.041	1.71339
P-TERPHENYL	22.71	0.038	1.99422
DEUTERATED P-TERPHENYL-C18D14	22.33	0.040	1.80504
4-METHYL TERPHENYL	24.55	0.030	3.17933
2-METHYL-P-TERPHENYL	15.59	0.118	0.20967
3,3''-DIMETHYL-P-TERPHENYL	23.89	0.033	2.70146
4-(3,3-DIMETHYLBUTOXY)-P-TERPHENYL	28.37	0.020	7.55647
4,4''-DIHEXAHYDROFARNESOXY-P-TERPHENYL	31.39	0.014	13.85252
2,2'-METHYLENE-P-TERPHENYL	30.89	0.015	12.57862
2,2'-ETHYLENE-P-TERPHENYL	30.89	0.015	12.57371
2,2'-METHYLENE-5',6''-METHYLENE-P-TERPHENYL (TRACE)	34.56	0.011	24.58577
3-PHENYL DIBENZOFURAN	31.63	0.014	14.52217
BENZO(1,2-B-4,5-B')BISBENZOFURAN (TRACE) (B)	36.32	0.009	33.07734
SEXIPHENYL BENZENE	11.59	0.288	0.03547
P-QUATERPHENYL	32.18	0.013	16.05398
3,3'''-DIMETHYL-P-QUATERPHENYL	33.22	0.012	19.39781
TETRAMETHYL-P-QUATERPHENYL	18.51	0.071	0.58727
3,3''' -DIMETHYL-3',2''-METHYLENE-P-QUATERPHENYL	37.34	0.009	39.04144
3,3''' -DIMETHYL-3',2''-ETHYLENE-P-QUATERPHENYL	37.12	0.009	37.57348
4-(3-ETHYLBUTOXY)-P-QUATERPHENYL (B)	36.00	0.010	31.28314
4,4'''-DI(2-BUTYLOCTOXY-1)-P-QUATERPHENYL	37.08	0.009	37.35981
DIETHYL-P-QUINQUEPHENYL	34.67	0.011	25.05520
DI(3-ETHYLHEPTYL)-P-QUINQUEPHENYL	38.28	0.008	45.11686
TETRAMETHYL-P-SEXIPHENYL	34.25	0.011	23.30240
TETRAMETHYL-P-OCTAPHENYL (B)	40.66	0.007	64.76121

ACCEPTOR MOLECULES	RO	CO	OVERLAP
2-PHENYL INDOLE	32.99	0.012	18.63845
1-METHYL-2-PHENYL INDOLE	28.79	0.019	8.22818
1-NEOPENTYL-2-PHENYL INDOLE	27.45	0.022	6.19438
1-N-BUTYL-2(BIPHENYLYL-4) INDOLE	33.74	0.012	21.18121
1,2-DIPHENYL INDOLE	29.97	0.017	10.46822
1-BIPHENYLYL-2-PHENYL INDOLE	31.92	0.014	15.20426
1,3-DIMETHYL-2-PHENYL INDOLE	28.38	0.020	7.54984
2,3-DIPHENYL INDOLE	31.58	0.014	14.20141
1-METHYL-2,3-DIPHENYL INDOLE	29.15	0.018	8.80389
2-PHENYL-3,2'-METHYLENE INDOLE (TRACE)	33.76	0.012	21.35012
2-PHENYL-3,2'-ETHYLENE INDOLE (TRACE)	33.67	0.012	20.93621
2-PHENYL-3,2'-TRIMETHYLENE INDOLE	33.46	0.012	20.24449
1-METHYL-3,2'-METHYLENE-2-PHENYL INDOLE (TRACE)	33.97	0.011	22.09001
1-METHYL-3,2'-ETHYLENE-2-PHENYL INDOLE	33.21	0.012	19.23089
1-METHYL-3,2'-TRIMETHYLENE-2-PHENYL INDOLE	31.65	0.014	14.49217
4,5-DIPHENYLIMIDAZOLE (A)	23.34	0.035	2.33321
PPF (TRACE)	35.75	0.010	30.02389
PPO (TRACE)	34.62	0.011	24.83898
2-(4-FLUOROPHENYL)-5-PHENYLOXAZOLE (TRACE)	33.16	0.012	19.18711
αNPO (TRACE)	33.81	0.012	21.31661
BBO (B)	37.32	0.009	38.60451
POPOP	34.29	0.011	23.19700
DIMETHYL POPOP (TRACE)	34.47	0.011	23.96653
PPD (TRACE)	24.26	0.031	2.97049
2,5-DI(2-METHYLPHENYL)OXADIAZOLE	24.85	0.029	3.42721
PBD	35.05	0.010	26.76472
BBD (B)	38.55	0.008	47.15859
αNPD (TRACE)	32.75	0.013	17.69778
βNPD	33.07	0.012	18.85913
αNND (TRACE)	32.74	0.013	17.65187
βNND (TRACE)	35.69	0.010	29.76126
TRANS-1-(4-BIPHENYLYL)-2-PHENYL-ETHYLENE	37.38	0.009	39.11767
TRANS-1-(4-BIPHENYLYL)-2-(1-NAPHTHYL)-ETHYLENE	35.65	0.010	29.32187
DIPHENYLSTILBENE (B)	37.76	0.008	41.41059
1,1,4,4-TETRAPHENYLBUTADIENE	34.63	0.011	24.63983
1,6-DIPHENYLHEXATRIENE	36.90	0.009	36.01693
1,8-DIPHENYL-1,3,5,7-OCTATETRAENE	33.08	0.012	18.72504
BPSB (TRACE)	36.99	0.009	36.58197
BIS-MSB	37.22	0.009	37.97775
ESCULIN (A)	30.16	0.016	10.74290
AZULENE	23.25	0.036	2.25772
NAPHTHALENE	12.35	0.237	0.05171
DEUTERATED NAPHTHALENE-C10D8	14.14	0.158	0.11556
1-METHYLNAPHTHALENE 5ML/L	14.80	0.138	0.15278
2-METHYL NAPHTHALENE	15.21	0.127	0.17937
2,3-DIMETHYL NAPHTHALENE	15.31	0.125	0.18703
2,6-DIMETHYL NAPHTHALENE (TRACE)	17.75	0.080	0.45426
1-NAPHTHOL (TRACE)	21.63	0.044	1.48492
2-NAPHTHOL (TRACE)	21.19	0.047	1.30902
1-NAPHTHYLAMINE	26.17	0.025	4.62323
2-CHLORONAPHTHALENE	15.12	0.129	0.17359
ACENAPHTHENE	20.02	0.056	0.93388
1-PHENYL NAPHTHALENE	21.84	0.043	1.57425
2-PHENYL NAPHTHALENE	31.88	0.014	15.18548
1,4-DIPHENYL NAPHTHALENE	28.61	0.019	7.91012
1,5-DIPHENYL NAPHTHALENE	27.44	0.022	6.16488
1,7-DIPHENYL NAPHTHALENE	27.12	0.022	5.74503
1,4,5,8-TETRAPHENYL NAPHTHALENE	32.45	0.013	16.66396
1,1'-BINAPHTHYL	23.60	0.034	2.50113
4,4'-DI(N-BUTOXY)-1,1'-BINAPHTHYL	31.37	0.014	13.69816
2,2'-BINAPHTHYL	30.24	0.016	11.04800
ACRIDINE (A)	25.21	0.028	3.66551
ANTHRACENE (TRACE)	23.69	0.034	2.52602
DEUTERATED ANTHRACENE C14D10 (TRACE)	23.67	0.034	2.51197
9-METHYL ANTHRACENE (TRACE)	22.33	0.040	1.76867
1-AMINO ANTHRACENE	20.81	0.050	1.15925
2-AMINO ANTHRACENE	22.96	0.037	2.08106
9-VINYL ANTHRACENE	22.79	0.038	2.00012
9-PHENYL ANTHRACENE (TRACE)	23.57	0.034	2.44702
9,10-DIPHENYL ANTHRACENE	22.61	0.039	1.90908
9,10-DICHLOROANTHRACENE (TRACE)	21.34	0.046	1.34947
BIANTHRANYL (B)	24.66	0.029	3.36873
9,10-DINAPHTHYLANTHRACENE (B)	24.78	0.029	3.30659
9-(P-VINYLPHENYL)-10-PHENYL ANTHRACENE	23.26	0.036	2.25922
9,10-BIS(PHENYL-ETHYNYL)ANTHRACENE (TRACE)	30.14	0.016	10.70710
RUBICENE (B)	31.80	0.014	14.75841
DIBENZ(A,H)ANTHRACENE (B)	34.84	0.011	25.51921
TETRACENE (TRACE) (B)	22.29	0.040	1.75054
RUBRENE (B)	37.27	0.009	38.28785
TETRAHYDROCANNABINOL ACID METHYL ESTER	20.02	0.056	0.93784
PHENANTHRENE	15.48	0.121	0.19777
3,4-BENZOPHENANTHRENE	27.81	0.021	6.61077
2-PHENYL PHENANTHRENE	25.23	0.028	3.70117
CHRYSENE (TRACE)	28.75	0.019	8.08719
TRIPHENYLENE	18.28	0.073	0.53731
DEUTERATED TRIPHENYLENE-C18D12	18.28	0.073	0.53645
PYRENE	32.52	0.013	16.90233
1,3,6,8,-TETRAPHENYL PYRENE	29.27	0.018	8.97181
BIPYRENYL (B)	34.72	0.011	25.00185
O-PHENYLENE-PYRENE	32.48	0.013	16.75920
ANTHANTHRENE (B)	32.97	0.012	18.33814
FLUORANTHENE	26.67	0.024	5.13393
3-PHENYL FLUORANTHENE	28.12	0.020	7.06532
3-AMINO FLUORANTHENE	29.20	0.018	8.85663
3,3'-BIFLUORANTHENYL (B)	31.82	0.014	14.81358
BENZO(GHI)FLUORANTHENE	26.43	0.024	4.86977
9,12-DI(N-OCTADECOXY)-BENZ(E)ACEPHENANTHRYLENE	28.57	0.019	7.76563

ACCEPTOR MOLECULES	RO	CO	OVERLAP
DECACYCLENE (B)	34.40	0.011	23.65528
10-ETHOXY-BENZO(KL)XANTHENE	24.62	0.030	3.18255
PERYLENE	19.80	0.058	0.86053
BENZO(GHI)PERYLENE (B)	30.49	0.016	11.47352
BBOT (TRACE)	33.33	0.012	19.57884
QUININE SULFATE (A)	28.28	0.020	7.36767
ACRIDINE YELLOW (A)	20.68	0.051	1.11817
ACRIDINE ORANGE (A)	23.46	0.035	2.37980
ACRIDINE RED (A)	24.96	0.029	3.45422
URANINE (A)	25.47	0.027	3.89334
RHODAMINE B (A)	28.24	0.020	7.24351
RHODAMINE 6G (A)	26.80	0.023	5.29350
EOSIN B (A)	28.17	0.020	7.13420
EOSIN Y (A)	28.86	0.019	8.24221
ROSE BENGAL (A)	28.80	0.019	8.14412
INDENOINDENE	33.48	0.012	20.32350

8.72 Donor Molecule: Dibenzothiophene (trace)

N,N-DIMETHYLANILINE 3ML/L	5.30	3.012	0.00187
TRIPHENYLAMINE	17.39	0.085	2.32452
SODIUM SALICYLATE (A)	11.58	0.288	0.20269
ANTHRANILIC ACID (A)	19.94	0.056	5.24366
TMPD	17.07	0.090	2.06838
4,4'-DIHYDROXYBIPHENYL (A)	4.50	4.911	0.00070
4,4'-DIMETHOXY BIPHENYL	0.00	0.000	0.00000
4,4'-DIISOBUTOXYBIPHENYL	0.00	0.000	0.00000
BENZIDINE	14.95	0.134	0.94189
1,3,5-TRIPHENYL BENZENE	5.65	2.484	0.00275
1,3-DIBIPHENYLYLBENZENE (B)	13.46	0.183	0.50241
1,3,5,-TRIBIPHENYLYLBENZENE (B)	16.17	0.106	1.51084
FLUORENE (TRACE)	0.00	0.000	0.00000
1-METHYL FLUORENE	0.00	0.000	0.00000
1,2-BENZOFLUORENE	15.68	0.116	1.24963
9,10-DIHYDROPHENANTHRENE	0.00	0.000	0.00000
CARBAZOLE (TRACE)	15.86	0.112	1.34125
N-METHYL CARBAZOLE (TRACE)	18.30	0.073	3.15414
N-VINYLCARBAZOLE	18.12	0.075	2.97940
N-PHENYL CARBAZOLE	17.95	0.077	2.81604
N-METHYL-2,3-BENZCARBAZOLE	19.78	0.058	4.96032
DIBENZOFURAN	0.00	0.000	0.00000
DIBENZOTHIOPHENE (TRACE)	14.09	0.160	0.66118
P-TERPHENYL	5.44	2.783	0.00219
DEUTERATED P-TERPHENYL-C18D14	0.00	0.000	0.00000
4-METHYL TERPHENYL	7.08	1.261	0.01067
2-METHYL-P-TERPHENYL	0.00	0.000	0.00000
3,3''-DIMETHYL-P-TERPHENYL	8.27	0.791	0.02711
4-(3,3-DIMETHYLBUTOXY)-P-TERPHENYL	13.60	0.178	0.53437
4,4''-DIHEXAHYDROFARNESOXY-P-TERPHENYL	16.31	0.103	1.58656
2,2'-METHYLENE-P-TERPHENYL	14.28	0.153	0.71795
2,2'-ETHYLENE-P-TERPHENYL	15.67	0.116	1.24804
2,2'-METHYLENE-5',6''-METHYLENE-P-TERPHENYL (TRACE)	24.76	0.029	19.37128
3-PHENYL DIBENZOFURAN	14.29	0.153	0.72064
BENZO(1,2-B-4,5-B')BISBENZOFURAN (TRACE) (B)	26.08	0.025	26.43610
SEXIPHENYL BENZENE	0.00	0.000	0.00000
P-QUATERPHENYL	17.57	0.082	2.48217
3,3'''-DIMETHYL-P-QUATERPHENYL	18.81	0.067	3.73168
TETRAMETHYL-P-QUATERPHENYL	0.00	0.000	0.00000
3,3''' -DIMETHYL-3',2''-METHYLENE-P-QUATERPHENYL	25.55	0.027	23.38889
3,3''' -DIMETHYL-3',2''-ETHYLENE-P-QUATERPHENYL	25.13	0.028	21.13326
4-(3-ETHYLBUTOXY)-P-QUATERPHENYL (B)	22.91	0.037	12.13927
4,4'''-DI(2-BUTYLOCTOXY-1)-P-QUATERPHENYL	23.41	0.035	13.80012
DIETHYL-P-QUINQUEPHENYL	19.65	0.059	4.84711
DI(3-ETHYLHEPTYL)-P-QUINQUEPHENYL	24.47	0.031	17.94688
TETRAMETHYL-P-SEXIPHENYL	18.49	0.071	3.37202
TETRAMETHYL-P-OCTAPHENYL (B)	27.85	0.021	39.05729
INDOLE (TRACE)	0.00	0.000	0.00000
2-PHENYL INDOLE	19.79	0.058	5.06840
1-METHYL-2-PHENYL INDOLE	15.51	0.120	1.17200
1-NEOPENTYL-2-PHENYL INDOLE	12.97	0.205	0.40292
1-N-BUTYL-2(BIPHENYLYL-4) INDOLE	22.79	0.038	11.74323
1,2-DIPHENYL INDOLE	16.01	0.109	1.42052
1-BIPHENYLYL-2-PHENYL INDOLE	18.16	0.075	3.00572
1,3-DIMETHYL-2-PHENYL INDOLE	16.66	0.097	1.80388
2,3-DIPHENYL INDOLE	21.04	0.048	7.25228
1-METHYL-2,3-DIPHENYL INDOLE	19.09	0.064	4.04934
2-PHENYL-3,2'-METHYLENE INDOLE (TRACE)	22.44	0.040	10.74104
2-PHENYL-3,2'-ETHYLENE INDOLE (TRACE)	24.57	0.030	18.45366
2-PHENYL-3,2'-TRIMETHYLENE INDOLE	21.43	0.045	8.13667
1-METHYL-3,2'-METHYLENE-2-PHENYL INDOLE (TRACE)	24.23	0.031	16.98937
1-METHYL-3,2'-ETHYLENE-2-PHENYL INDOLE	24.60	0.030	18.52959
1-METHYL-3,2'-TRIMETHYLENE-2-PHENYL INDOLE	20.58	0.051	6.38977
4,5-DIPHENYLIMIDAZOLE (A)	12.37	0.236	0.30246
PPF (TRACE)	25.38	0.027	22.44709
PPO (TRACE)	21.48	0.045	8.26204
2-(4-FLUOROPHENYL)-5-PHENYLOXAZOLE (TRACE)	20.45	0.052	6.16920
αNPO (TRACE)	26.26	0.025	27.30569
BBO (B)	29.92	0.017	59.76759
POPOP	28.85	0.019	47.95037
DIMETHYL POPOP (TRACE)	29.33	0.018	52.83507

ACCEPTOR MOLECULES	RO	CO	OVERLAP
2,5-DI(2-METHYLPHENYL)OXADIAZOLE	5.16	3.261	0.00160
PBD	20.33	0.053	5.95451
BBD (B)	26.79	0.023	31.03553
αNPD (TRACE)	23.28	0.035	13.30969
βNPD	19.84	0.057	5.13437
αNND (TRACE)	24.91	0.029	19.95256
βNND (TRACE)	24.43	0.031	17.87609
TRANS-1-(4-BIPHENYLYL)-2-PHENYL-ETHYLENE	[illegible]	[illegible]	[illegible]
TRANS-1-(4-BIPHENYLYL)-2-(1-NAPHTHYL)-ETHYLENE	27.71	0.021	37.68187
DIPHENYLSTILBENE (B)	30.25	0.016	63.74597
1,1,4,4-TETRAPHENYLBUTADIENE	28.28	0.020	42.47547
1,6-DIPHENYLHEXATRIENE	31.01	0.015	73.90230
1,8-DIPHENYL-1,3,5,7-OCTATETRAENE	29.82	0.017	58.12505
BPSB (TRACE)	31.16	0.015	76.08031
BIS-MSB	30.44	0.016	66.02802
ESCULIN (A)	23.60	0.034	14.40914
AZULENE	17.88	0.078	2.69732
NAPHTHALENE	3.05	15.706	0.00007
DEUTERATED NAPHTHALENE-C10D8	6.82	1.409	0.00851
1-METHYLNAPHTHALENE 5ML/L	5.11	3.344	0.00151
2-METHYL NAPHTHALENE	6.12	1.947	0.00447
2,3-DIMETHYL NAPHTHALENE	6.24	1.840	0.00500
2,6-DIMETHYL NAPHTHALENE (TRACE)	10.74	0.361	0.12981
1-NAPHTHOL (TRACE)	10.28	0.412	0.09974
2-NAPHTHOL (TRACE)	14.27	0.154	0.71360
1-NAPHTHYLAMINE	17.93	0.078	2.79298
2-CHLORONAPHTHALENE	7.42	1.093	0.01418
ACENAPHTHENE	7.95	0.890	0.02140
1-PHENYL NAPHTHALENE	9.10	0.594	0.04798
2-PHENYL NAPHTHALENE	15.49	0.120	1.16684
1,4-DIPHENYL NAPHTHALENE	16.38	0.102	1.62290
1,5-DIPHENYL NAPHTHALENE	14.92	0.135	0.92996
1,7-DIPHENYL NAPHTHALENE	14.75	0.139	0.86700
1,4,5,8-TETRAPHENYL NAPHTHALENE	25.26	0.028	21.58674
1,1'-BINAPHTHYL	11.71	0.278	0.21798
4,4'-DI(N-BUTOXY)-1,1'-BINAPHTHYL	20.59	0.051	6.39230
2,2'-BINAPHTHYL	15.96	0.110	1.39235
ACRIDINE (A)	21.72	0.044	8.72293
ANTHRACENE (TRACE)	20.22	0.054	5.69543
DEUTERATED ANTHRACENE C14D10 (TRACE)	20.00	0.056	5.33874
9-METHYL ANTHRACENE (TRACE)	19.91	0.057	5.17962
1-AMINO ANTHRACENE	18.30	0.073	3.10049
2-AMINO ANTHRACENE	18.72	0.068	3.55687
9-VINYL ANTHRACENE	20.43	0.052	6.03706
9-PHENYL ANTHRACENE (TRACE)	20.97	0.049	7.05736
9,10-DIPHENYL ANTHRACENE	20.84	0.049	6.77525
9,10-DICHLOROANTHRACENE (TRACE)	19.83	0.057	5.03537
BIANTHRANYL (B)	22.10	0.041	9.65656
9,10-DINAPHTHYLANTHRACENE (B)	20.46	0.052	6.06595
9-(P-VINYLPHENYL)-10-PHENYL ANTHRACENE	20.87	0.049	6.82980
9,10-BIS(PHENYL-ETHYNYL)ANTHRACENE (TRACE)	17.78	0.080	2.61410
RUBICENE (B)	21.08	0.048	7.24933
DIBENZ(A,H)ANTHRACENE (B)	23.34	0.035	13.41967
TETRACENE (TRACE) (B)	15.09	0.130	0.97707
RUBRENE (B)	19.99	0.056	5.27217
TETRAHYDROCANNABINOL ACID METHYL ESTER	3.68	8.949	0.00021
PHENANTHRENE	11.02	0.334	0.15071
3,4-BENZOPHENANTHRENE	16.09	0.107	1.44453
2-PHENYL PHENANTHRENE	12.25	0.244	0.28233
CHRYSENE (TRACE)	15.46	0.121	1.13984
TRIPHENYLENE	11.54	0.291	0.19821
DEUTERATED TRIPHENYLENE-C18D12	11.37	0.304	0.18189
PYRENE	23.45	0.035	13.85971
1,3,6,8,-TETRAPHENYL PYRENE	25.84	0.026	24.61715
BIPYRENYL (B)	28.74	0.019	46.65123
O-PHENYLENE-PYRENE	23.42	0.035	13.63953
ANTHANTHRENE (B)	21.39	0.046	7.92450
FLUORANTHENE	21.58	0.045	8.35413
3-PHENYL FLUORANTHENE	23.19	0.036	12.83792
3-AMINO FLUORANTHENE	21.21	0.047	7.52060
3,3'-BIFLUORANTHENYL (B)	25.52	0.027	22.85103
BENZO(GHI)FLUORANTHENE	20.92	0.049	6.93556
9,12-DI(N-OCTADECOXY)-BENZ(E)ACEPHENANTHRYLENE	22.76	0.038	11.48023
DECACYCLENE (B)	27.69	0.021	37.23979
10-ETHOXY-BENZO(KL)XANTHENE	21.29	0.046	7.70688
PERYLENE	18.37	0.072	3.17294
BENZO(GHI)PERYLENE (B)	22.83	0.038	11.71804
BBOT (TRACE)	29.30	0.018	52.40993
QUININE SULFATE (A)	20.90	0.049	7.00582
ACRIDINE YELLOW (A)	16.03	0.109	1.40014
ACRIDINE ORANGE (A)	14.71	0.141	0.83618
ACRIDINE RED (A)	18.22	0.074	3.02252
URANINE (A)	18.28	0.073	3.07956
RHODAMINE B (A)	20.84	0.051	6.39044
RHODAMINE 6G (A)	21.41	0.046	7.95792
EOSIN B (A)	19.36	0.062	4.34691
EOSIN Y (A)	19.64	0.059	4.74826
ROSE BENGAL (A)	20.78	0.050	6.66362
INDENOINDENE	21.27	0.046	7.80206

8.73 Donor Molecule: *p*-Terphenyl

ACCEPTOR MOLECULES	RO	CO	OVERLAP
ANILINE 3ML/L	4.89	3.823	0.00011
N,N-DIMETHYLANILINE 3ML/L	12.70	0.218	0.03460
TRIPHENYLAMINE	26.71	0.023	2.96793
1,4-DIMETHOXY BENZENE	5.79	2.306	0.00031
1-METHYL-4-CHLOROBENZENE 5ML/L	0.00	0.000	0.00000
SODIUM SALICYLATE (A)	17.96	0.077	0.27358
ANTHRANILIC ACID (A)	28.85	0.019	4.67219
TMPD	24.78	0.029	1.88164
4-METHOXY BIPHENYL	3.89	7.621	0.00003
4-BIPHENYLYL PHENYL ETHER	0.00	0.000	0.00000
4,4'-DIHYDROXYBIPHENYL (A)	11.95	0.262	0.02398
4,4'-DIMETHOXY BIPHENYL	9.29	0.558	0.00531
4,4'-DIISOBUTOXYBIPHENYL	7.98	0.880	0.00214
BENZIDINE	23.67	0.034	1.44026
4-VINYL BIPHENYL	12.59	0.224	0.03295
M-TERPHENYL	8.67	0.687	0.00351
1,3,5-TRIPHENYL BENZENE	10.80	0.355	0.01309
1,3-DIBIPHENYLYLBENZENE (B)	22.75	0.038	1.13694
1,3,5,-TRIBIPHENYLYLBENZENE (B)	26.83	0.023	3.06614
FLUORENE (TRACE)	6.04	2.033	0.00040
1-METHYL FLUORENE	6.06	2.009	0.00041
1,2-BENZOFLUORENE	24.84	0.029	1.91741
9,10-DIHYDROPHENANTHRENE	8.05	0.858	0.00225
CARBAZOLE (TRACE)	23.41	0.035	1.34991
N-METHYL CARBAZOLE (TRACE)	26.11	0.025	2.59084
N-VINYLCARBAZOLE	26.36	0.024	2.73946
N-PHENYL CARBAZOLE	26.00	0.025	2.52464
N-METHYL-2,3-BENZCARBAZOLE	29.01	0.018	4.80314
DIBENZOFURAN	5.03	3.505	0.00014
DIBENZOTHIOPHENE (TRACE)	21.13	0.047	0.73100
P-TERPHENYL	14.95	0.134	0.09225
DEUTERATED P-TERPHENYL-C18D14	14.03	0.162	0.06307
4-METHYL TERPHENYL	16.98	0.091	0.19774
2-METHYL-P-TERPHENYL	7.60	1.021	0.00159
3,3''-DIMETHYL-P-TERPHENYL	16.77	0.095	0.18319
4-(3,3-DIMETHYLBUTOXY)-P-TERPHENYL	22.63	0.039	1.10497
4,4''DIHEXAHYDROFARNESOXY-P-TERPHENYL	26.22	0.025	2.66840
2,2'-METHYLENE-P-TERPHENYL	25.06	0.028	2.03805
2,2'-ETHYLENE-P-TERPHENYL	25.99	0.025	2.53312
2,2'-METHYLENE-5',6'-METHYLENE-P-TERPHENYL (TRACE)	35.85	0.010	17.37357
3-PHENYL DIBENZOFURAN	25.15	0.028	2.08215
BENZO(1,2-B-4,5-B')BISBENZOFURAN (TRACE) (B)	37.54	0.008	22.86854
SEXIPHENYL BENZENE	0.00	0.000	0.00000
P-QUATERPHENYL	27.64	0.021	3.65822
3,3'''-DIMETHYL-P-QUATERPHENYL	29.17	0.018	5.04433
TETRAMETHYL-P-QUATERPHENYL	9.67	0.495	0.00676
3,3''' -DIMETHYL-3',2''-METHYLENE-P-QUATERPHENYL	37.47	0.009	22.62008
3,3''' -DIMETHYL-3',2''-ETHYLENE-P-QUATERPHENYL	37.00	0.009	20.92686
4-(3-ETHYLBUTOXY)-P-QUATERPHENYL (B)	34.25	0.011	13.17509
4,4'''-DI(2-BUTYLOCTOXY-1)-P-QUATERPHENYL	35.07	0.010	15.18926
DIETHYL-P-QUINQUEPHENYL	30.44	0.016	6.51304
DI(3-ETHYLHEPTYL)-P-QUINQUEPHENYL	36.54	0.009	19.37579
TETRAMETHYL-P-SEXIPHENYL	29.21	0.018	5.08780
TETRAMETHYL-P-OCTAPHENYL (B)	40.95	0.007	38.36949
INDOLE (TRACE)	0.00	0.000	0.00000
2-PHENYL INDOLE	30.28	0.016	6.31680
1-METHYL-2-PHENYL INDOLE	24.57	0.030	1.80435
1-NEOPENTYL-2-PHENYL INDOLE	21.94	0.042	0.91571
1-N-BUTYL-2(BIPHENYLYL-4) INDOLE	33.58	0.012	11.68063
1,2-DIPHENYL INDOLE	25.56	0.027	2.28329
1-BIPHENYLYL-2-PHENYL INDOLE	28.29	0.020	4.18233
1,3-DIMETHYL-2-PHENYL INDOLE	25.47	0.027	2.23698
2,3-DIPHENYL INDOLE	31.08	0.015	7.33706
1-METHYL-2,3-DIPHENYL INDOLE	28.26	0.020	4.15137
2-PHENYL-3,2'-METHYLENE INDOLE (TRACE)	33.41	0.012	11.38710
2-PHENYL-3,2'-ETHYLENE INDOLE (TRACE)	35.49	0.010	16.27592
2-PHENYL-3,2'-TRIMETHYLENE INDOLE	32.05	0.014	8.86424
1-METHYL-3,2'-METHYLENE-2-PHENYL INDOLE (TRACE)	35.16	0.010	15.41946
1-METHYL-3,2'-ETHYLENE-2-PHENYL INDOLE	35.82	0.010	17.19882
1-METHYL-3,2'-TRIMETHYLENE-2-PHENYL INDOLE	30.66	0.016	6.78834
4,5-DIPHENYLIMIDAZOLE (A)	19.56	0.060	0.45907
PPF (TRACE)	36.96	0.009	20.80213
PPO (TRACE)	32.39	0.013	9.44897
2-(4-FLUOROPHENYL)-5-PHENYLOXAZOLE (TRACE)	30.85	0.015	7.06864
αNPO (TRACE)	38.04	0.008	24.57054
BBO (B)	43.45	0.005	54.48100
POPOP	42.12	0.006	45.16446
DIMETHYL POPOP (TRACE)	42.98	0.006	50.86968
PPD (TRACE)	14.79	0.138	0.08660
2,5-DI(2-METHYLPHENYL)OXADIAZOLE	16.68	0.096	0.17753
PBD	31.29	0.015	7.69419
BBD (B)	39.22	0.007	29.69015
αNPD (TRACE)	33.88	0.012	12.31081
βNPD	29.99	0.017	5.95767
αNND (TRACE)	36.08	0.010	17.94696
βNND (TRACE)	35.92	0.010	17.56613
TRANS-1-(4-BIPHENYLYL)-2-PHENYL-ETHYLENE	38.71	0.008	27.39117
TRANS-1-(4-BIPHENYLYL)-2-(1-NAPHTHYL)-ETHYLENE	40.27	0.007	34.48790
DIPHENYLSTILBENE (B)	43.89	0.005	57.82873
1,1,4,4-TETRAPHENYLBUTADIENE	41.11	0.006	38.94489
1,6-DIPHENYLHEXATRIENE	45.32	0.005	69.98728
1,8-DIPHENYL-1,3,5,7-OCTATETRAENE	44.36	0.005	61.17793

ACCEPTOR MOLECULES	RO	CO	OVERLAP
BPSB (TRACE)	45.52	0.005	71.90152
BIS-MSB	44.35	0.005	61.44052
ESCULIN (A)	34.21	0.011	12.99901
AZULENE	25.83	0.026	2.38329
NAPHTHALENE	8.38	0.759	0.00286
DEUTERATED NAPHTHALENE-C10D8	11.23	0.316	0.01644
1-METHYLNAPHTHALENE 5ML/L	10.47	0.390	0.01084
2-METHYL NAPHTHALENE	12.84	0.211	0.03694
2,3-DIMETHYL NAPHTHALENE	12.62	0.222	0.03335
2,6-DIMETHYL NAPHTHALENE (TRACE)	16.72	0.096	0.18013
1-NAPHTHOL (TRACE)	18.37	0.072	0.31650
2-NAPHTHOL (TRACE)	20.68	0.051	0.64148
1-NAPHTHYLAMINE	26.28	0.025	2.68877
2-CHLORONAPHTHALENE	13.48	0.183	0.04931
ACENAPHTHENE	16.03	0.109	0.13982
1-PHENYL NAPHTHALENE	16.26	0.104	0.15211
2-PHENYL NAPHTHALENE	24.53	0.030	1.78925
1,4-DIPHENYL NAPHTHALENE	25.38	0.027	2.18979
1,5-DIPHENYL NAPHTHALENE	23.66	0.034	1.43768
1,7-DIPHENYL NAPHTHALENE	23.33	0.035	1.32008
1,4,5,8-TETRAPHENYL NAPHTHALENE	36.67	0.009	19.63650
1,1'-BINAPHTHYL	19.07	0.065	0.39525
4,4'-DI(N-BUTOXY)-1,1'-BINAPHTHYL	30.45	0.016	6.51063
2,2'-BINAPHTHYL	25.70	0.026	2.36114
ACRIDINE (A)	31.67	0.014	8.15499
ANTHRACENE (TRACE)	29.79	0.017	5.65901
DEUTERATED ANTHRACENE C14D10 (TRACE)	29.48	0.017	5.31297
9-METHYL ANTHRACENE (TRACE)	29.17	0.018	4.98106
1-AMINO ANTHRACENE	27.38	0.022	3.37924
2-AMINO ANTHRACENE	27.82	0.021	3.71876
9-VINYL ANTHRACENE	30.17	0.016	6.08277
9-PHENYL ANTHRACENE (TRACE)	30.74	0.015	6.80878
9,10-DIPHENYL ANTHRACENE	31.18	0.015	7.40274
9,10-DICHLOROANTHRACENE (TRACE)	29.76	0.017	5.58983
BIANTHRANYL (B)	32.93	0.013	10.26541
9,10-DINAPHTHYLANTHRACENE (B)	30.78	0.015	6.83204
9-(P-VINYLPHENYL)-10-PHENYL ANTHRACENE	31.29	0.015	7.55051
9,10-BIS(PHENYL-ETHYNYL)ANTHRACENE (TRACE)	28.32	0.020	4.13202
RUBICENE (B)	32.02	0.014	8.64066
DIBENZ(A,H)ANTHRACENE (B)	34.36	0.011	13.27491
TETRACENE (TRACE) (B)	22.90	0.037	1.15695
RUBRENE (B)	31.11	0.015	7.26729
TETRAHYDROCANNABINOL ACID METHYL ESTER	12.56	0.226	0.03243
PHENANTHRENE	15.98	0.110	0.13604
3,4-BENZOPHENANTHRENE	24.84	0.029	1.90316
2-PHENYL PHENANTHRENE	19.17	0.064	0.40389
CHRYSENE (TRACE)	25.85	0.026	2.42611
TRIPHENYLENE	17.02	0.091	0.19828
DEUTERATED TRIPHENYLENE-C18D12	16.79	0.094	0.18319
PYRENE	34.77	0.011	14.31853
1,3,6,8,-TETRAPHENYL PYRENE	38.58	0.008	26.48755
BIPYRENYL (B)	41.67	0.006	42.20357
O-PHENYLENE-PYRENE	35.01	0.010	14.76180
ANTHANTHRENE (B)	32.50	0.013	9.44049
FLUORANTHENE	31.42	0.014	7.74119
3-PHENYL FLUORANTHENE	34.04	0.011	12.50159
3-AMINO FLUORANTHENE	31.40	0.014	7.68389
3,3'-BIFLUORANTHENYL (B)	37.88	0.008	23.68988
BENZO(GHI)FLUORANTHENE	30.21	0.016	6.10455
9,12-DI(N-OCTADECOXY)-BENZ(E)ACEPHENANTHRYLENE	33.67	0.012	11.68206
DECACYCLENE (B)	41.20	0.006	39.20467
10-ETHOXY-BENZO(KL)XANTHENE	31.80	0.014	8.32770
PERYLENE	28.63	0.019	4.41215
BENZO(GHI)PERYLENE (B)	33.87	0.012	12.17084
BBOT (TRACE)	43.22	0.006	52.51785
QUININE SULFATE (A)	30.30	0.016	6.32025
ACRIDINE YELLOW (A)	24.97	0.029	1.94318
ACRIDINE ORANGE (A)	22.72	0.038	1.10200
ACRIDINE RED (A)	26.92	0.023	3.05114
URANINE (A)	27.13	0.022	3.19393
RHODAMINE B (A)	30.53	0.016	6.49346
RHODAMINE 6G (A)	31.01	0.015	7.12595
EOSIN B (A)	28.70	0.019	4.48265
EOSIN Y (A)	29.24	0.018	5.01111
ROSE BENGAL (A)	30.54	0.016	6.50672
INDENOINDENE	31.72	0.014	8.33276

8.74 Donor Molecule: Deuterated *p*-Terphenyl ($C_{18}D_{14}$)

ANILINE 3ML/L	4.29	5.649	0.00005
N,N-DIMETHYLANILINE 3ML/L	12.67	0.220	0.03442
TRIPHENYLAMINE	26.73	0.023	3.01197
1,4-DIMETHOXY BENZENE	5.15	3.278	0.00016
1-METHYL-4-CHLOROBENZENE 5ML/L	0.00	0.000	0.00000
SODIUM SALICYLATE (A)	17.96	0.077	0.27681
ANTHRANILIC ACID (A)	28.82	0.019	4.69155
TMPD	24.76	0.029	1.89471
4,4'-DIHYDROXYBIPHENYL (A)	11.95	0.262	0.02430
4,4'-DIMETHOXY BIPHENYL	9.15	0.584	0.00490
4,4'-DIISOBUTOXYBIPHENYL	7.58	1.026	0.00159
BENZIDINE	23.70	0.034	1.46522

ACCEPTOR MOLECULES	RO	CO	OVERLAP
4-VINYL BIPHENYL	12.40	0.235	0.03029
M-TERPHENYL	8.51	0.726	0.00317
1,3,5-TRIPHENYL BENZENE	10.81	0.354	0.01328
1,3-DIBIPHENYLYLBENZENE (B)	22.77	0.038	1.15439
1,3,5,-TRIBIPHENYLYLBENZENE (B)	26.87	0.023	3.12102
FLUORENE (TRACE)	4.85	3.925	0.00011
1-METHYL FLUORENE	5.49	2.699	0.00023
1,2-BENZOFLUORENE	24.89	0.029	1.96275
9,10-DIHYDROPHENANTHRENE	7.46	1.077	0.00144
CARBAZOLE (TRACE)	23.46	0.035	1.38014
N-METHYL CARBAZOLE (TRACE)	26.10	0.025	2.60707
N-VINYLCARBAZOLE	26.33	0.025	2.74948
N-PHENYL CARBAZOLE	25.98	0.026	2.53597
N-METHYL-2,3-BENZCARBAZOLE	29.00	0.018	4.84113
DIBENZOFURAN	0.00	0.000	0.00000
DIBENZOTHIOPHENE (TRACE)	21.14	0.047	0.73948
P-TERPHENYL	14.91	0.135	0.09169
DEUTERATED P-TERPHENYL-C18D14	13.93	0.165	0.06105
4-METHYL TERPHENYL	16.97	0.091	0.19926
2-METHYL-P-TERPHENYL	7.19	1.202	0.00116
3,3″-DIMETHYL-P-TERPHENYL	16.76	0.095	0.18428
4-(3,3-DIMETHYLBUTOXY)-P-TERPHENYL	22.67	0.038	1.12628
4,4″-DIHEXAHYDROFARNESOXY-P-TERPHENYL	26.27	0.025	2.72212
2,2′-METHYLENE-P-TERPHENYL	25.13	0.028	2.09017
2,2′-ETHYLENE-P-TERPHENYL	26.06	0.025	2.59535
2,2′-METHYLENE-5′,6″-METHYLENE-P-TERPHENYL (TRACE)	35.86	0.010	17.57020
3-PHENYL DIBENZOFURAN	25.20	0.028	2.12815
BENZO(1,2-B-4,5-B′)BISBENZOFURAN (TRACE) (B)	37.53	0.008	23.08116
SEXIPHENYL BENZENE	0.00	0.000	0.00000
P-QUATERPHENYL	27.68	0.021	3.72655
3,3‴-DIMETHYL-P-QUATERPHENYL	29.21	0.018	5.13460
TETRAMETHYL-P-QUATERPHENYL	9.33	0.550	0.00553
3,3‴-DIMETHYL-3′,2″-METHYLENE-P-QUATERPHENYL	37.49	0.008	22.88620
3,3‴-DIMETHYL-3′,2″-ETHYLENE-P-QUATERPHENYL	37.02	0.009	21.17343
4-(3-ETHYLBUTOXY)-P-QUATERPHENYL (B)	34.28	0.011	13.36039
4,4‴-DI(2-BUTYLOCTOXY-1)-P-QUATERPHENYL	35.10	0.010	15.40881
DIETHYL-P-QUINQUEPHENYL	30.48	0.016	6.62870
DI(3-ETHYLHEPTYL)-P-QUINQUEPHENYL	36.57	0.009	19.64424
TETRAMETHYL-P-SEXIPHENYL	29.25	0.018	5.18387
TETRAMETHYL-P-OCTAPHENYL (B)	40.96	0.007	38.77952
INDOLE (TRACE)	0.00	0.000	0.00000
2-PHENYL INDOLE	30.33	0.016	6.44552
1-METHYL-2-PHENYL INDOLE	24.61	0.030	1.83956
1-NEOPENTYL-2-PHENYL INDOLE	21.98	0.042	0.93457
1-N-BUTYL-2(BIPHENYLYL-4) INDOLE	33.59	0.012	11.80897
1,2-DIPHENYL INDOLE	25.60	0.027	2.32988
1-BIPHENYLYL-2-PHENYL INDOLE	28.33	0.020	4.26252
1,3-DIMETHYL-2-PHENYL INDOLE	25.50	0.027	2.27561
2,3-DIPHENYL INDOLE	31.09	0.015	7.42090
1-METHYL-2,3-DIPHENYL INDOLE	28.27	0.020	4.19940
2-PHENYL-3,2′-METHYLENE INDOLE (TRACE)	33.43	0.012	11.53509
2-PHENYL-3,2′-ETHYLENE INDOLE (TRACE)	35.48	0.010	16.42107
2-PHENYL-3,2′-TRIMETHYLENE INDOLE	32.09	0.014	9.01461
1-METHYL-3,2′-METHYLENE-2-PHENYL INDOLE (TRACE)	35.16	0.010	15.57650
1-METHYL-3,2′-ETHYLENE-2-PHENYL INDOLE	35.80	0.010	17.29952
1-METHYL-3,2′-TRIMETHYLENE-2-PHENYL INDOLE	30.67	0.015	6.87782
4,5-DIPHENYLIMIDAZOLE (A)	19.58	0.060	0.46647
PPF (TRACE)	36.95	0.009	20.99151
PPO (TRACE)	32.42	0.013	9.60688
2-(4-FLUOROPHENYL)-5-PHENYLOXAZOLE (TRACE)	30.89	0.015	7.19254
αNPO (TRACE)	38.01	0.008	24.68599
BBO (B)	43.40	0.005	54.63725
POPOP	42.05	0.006	45.16708
DIMETHYL POPOP (TRACE)	42.91	0.006	50.81586
PPD (TRACE)	14.61	0.143	0.08111
2,5-DI(2-METHYLPHENYL)OXADIAZOLE	16.64	0.097	0.17716
PBD	31.34	0.015	7.84093
BBD (B)	39.22	0.007	29.99569
αNPD (TRACE)	33.87	0.012	12.41815
βNPD	30.01	0.017	6.04392
αNND (TRACE)	36.06	0.010	18.04048
βNND (TRACE)	35.93	0.010	17.75326
TRANS-1-(4-BIPHENYLYL)-2-PHENYL-ETHYLENE	38.71	0.008	27.63377
TRANS-1-(4-BIPHENYLYL)-2-(1-NAPHTHYL)-ETHYLENE	40.23	0.007	34.62634
DIPHENYLSTILBENE (B)	43.84	0.005	58.00178
1,1,4,4-TETRAPHENYLBUTADIENE	41.05	0.006	39.02333
1,6-DIPHENYLHEXATRIENE	45.25	0.005	69.98949
1,8-DIPHENYL-1,3,5,7-OCTATETRAENE	44.24	0.005	60.76418
BPSB (TRACE)	45.45	0.005	71.90709
BIS-MSB	44.29	0.005	61.52784
ESCULIN (A)	34.18	0.011	13.04949
AZULENE	25.80	0.026	2.39104
NAPHTHALENE	8.37	0.762	0.00287
DEUTERATED NAPHTHALENE-C10D8	11.24	0.315	0.01673
1-METHYLNAPHTHALENE 5ML/L	10.47	0.390	0.01095
2-METHYL NAPHTHALENE	12.94	0.207	0.03897
2,3-DIMETHYL NAPHTHALENE	12.69	0.219	0.03477
2,6-DIMETHYL NAPHTHALENE (TRACE)	16.72	0.096	0.18143
1-NAPHTHOL (TRACE)	18.43	0.071	0.32568
2-NAPHTHOL (TRACE)	20.70	0.050	0.65180
1-NAPHTHYLAMINE	26.29	0.025	2.72021
2-CHLORONAPHTHALENE	13.53	0.180	0.05107
ACENAPHTHENE	16.11	0.107	0.14535
1-PHENYL NAPHTHALENE	16.27	0.104	0.15432
2-PHENYL NAPHTHALENE	24.52	0.030	1.80100
1,4-DIPHENYL NAPHTHALENE	25.42	0.027	2.23083

ACCEPTOR MOLECULES	RO	CO	OVERLAP
1,5-DIPHENYL NAPHTHALENE	23.70	0.034	1.46651
1,7-DIPHENYL NAPHTHALENE	23.36	0.035	1.34600
1,4,5,8-TETRAPHENYL NAPHTHALENE	36.63	0.009	19.71682
1,1'-BINAPHTHYL	19.10	0.064	0.40239
4,4'-DI(N-BUTOXY)-1,1'-BINAPHTHYL	30.47	0.016	6.60196
2,2'-BINAPHTHYL	25.74	0.026	2.41007
ACRIDINE (A)	31.62	0.014	8.16681
ANTHRACENE (TRACE)	29.73	0.017	5.64234
DEUTERATED ANTHRACENE C14D10 (TRACE)	29.42	0.018	5.30193
9-METHYL ANTHRACENE (TRACE)	29.12	0.018	4.97569
1-AMINO ANTHRACENE	27.28	0.022	3.34300
2-AMINO ANTHRACENE	27.74	0.021	3.69701
9-VINYL ANTHRACENE	30.10	0.016	6.05813
9-PHENYL ANTHRACENE (TRACE)	30.68	0.015	6.79674
9,10-DIPHENYL ANTHRACENE	31.09	0.015	7.33622
9,10-DICHLOROANTHRACENE (TRACE)	29.66	0.017	5.53311
BIANTHRANYL (B)	32.83	0.013	10.18817
9,10-DINAPHTHYLANTHRACENE (B)	30.68	0.015	6.77426
9-(P-VINYLPHENYL)-10-PHENYL ANTHRACENE	31.19	0.015	7.47981
9,10-BIS(PHENYL-ETHYNYL)ANTHRACENE (TRACE)	28.00	0.020	3.91120
RUBICENE (B)	31.96	0.014	8.63954
DIBENZ(A,H)ANTHRACENE (B)	34.33	0.011	13.32520
TETRACENE (TRACE) (B)	22.77	0.038	1.13077
RUBRENE (B)	31.10	0.015	7.33424
TETRAHYDROCANNABINOL ACID METHYL ESTER	12.49	0.229	0.03173
PHENANTHRENE	15.97	0.110	0.13705
3,4-BENZOPHENANTHRENE	24.88	0.029	1.93653
2-PHENYL PHENANTHRENE	19.12	0.064	0.40202
CHRYSENE (TRACE)	25.95	0.026	2.50292
TRIPHENYLENE	17.01	0.091	0.19990
DEUTERATED TRIPHENYLENE-C18D12	16.80	0.094	0.18529
PYRENE	34.78	0.011	14.47730
1,3,6,8,-TETRAPHENYL PYRENE	38.47	0.008	26.29829
BIPYRENYL (B)	41.61	0.006	42.25343
O-PHENYLENE-PYRENE	34.92	0.011	14.70601
ANTHANTHRENE (B)	32.33	0.013	9.25852
FLUORANTHENE	31.38	0.014	7.76216
3-PHENYL FLUORANTHENE	33.98	0.011	12.48740
3-AMINO FLUORANTHENE	31.31	0.015	7.63579
3,3'-BIFLUORANTHENYL (B)	37.78	0.008	23.59580
BENZO(GHI)FLUORANTHENE	30.18	0.016	6.12640
9,12-DI(N-OCTADECOXY)-BENZ(E)ACEPHENANTHRYLENE	33.58	0.012	11.61582
DECACYCLENE (B)	41.09	0.006	39.00311
10-ETHOXY-BENZO(KL)XANTHENE	31.70	0.014	8.25912
PERYLENE	28.32	0.020	4.18747
BENZO(GHI)PERYLENE (B)	33.80	0.012	12.11977
BBOT (TRACE)	43.12	0.006	52.32119
QUININE SULFATE (A)	30.29	0.016	6.36912
ACRIDINE YELLOW (A)	24.64	0.030	1.81294
ACRIDINE ORANGE (A)	22.57	0.039	1.07063
ACRIDINE RED (A)	26.86	0.023	3.04636
URANINE (A)	27.06	0.023	3.18425
RHODAMINE B (A)	30.46	0.016	6.47291
RHODAMINE 6G (A)	30.94	0.015	7.11814
EOSIN B (A)	28.67	0.019	4.50340
EOSIN Y (A)	29.20	0.018	5.02906
ROSE BENGAL (A)	30.51	0.016	6.53635
INDENOINDENE	31.76	0.014	8.47793

8.75 Donor Molecule: 4-Methylterphenyl

N,N-DIMETHYLANILINE 3ML/L	10.33	0.406	0.00986
TRIPHENYLAMINE	25.27	0.028	2.09954
SODIUM SALICYLATE (A)	16.96	0.092	0.19165
ANTHRANILIC ACID (A)	28.73	0.019	4.50159
TMPD	24.42	0.031	1.70348
4,4'-DIHYDROXYBIPHENYL (A)	9.95	0.454	0.00792
4,4'-DIMETHOXY BIPHENYL	6.78	1.436	0.00079
4,4'-DIISOBUTOXYBIPHENYL	0.00	0.000	0.00000
BENZIDINE	22.03	0.042	0.92348
4-VINYL BIPHENYL	9.30	0.557	0.00526
M-TERPHENYL	5.34	2.945	0.00019
1,3,5-TRIPHENYL BENZENE	9.42	0.536	0.00566
1,3-DIBIPHENYLYLBENZENE (B)	20.56	0.051	0.61195
1,3,5,-TRIBIPHENYLYLBENZENE (B)	24.43	0.031	1.72400
FLUORENE (TRACE)	0.00	0.000	0.00000
1-METHYL FLUORENE	0.00	0.000	0.00000
1,2-BENZOFLUORENE	23.25	0.036	1.27228
9,10-DIHYDROPHENANTHRENE	4.03	6.830	0.00004
CARBAZOLE (TRACE)	22.46	0.039	1.03809
N-METHYL CARBAZOLE (TRACE)	25.88	0.026	2.42349
N-VINYLCARBAZOLE	25.82	0.026	2.38794
N-PHENYL CARBAZOLE	25.52	0.027	2.22845
N-METHYL-2,3-BENZCARBAZOLE	28.88	0.019	4.61613
DIBENZOFURAN	0.00	0.000	0.00000
DIBENZOTHIOPHENE (TRACE)	20.20	0.054	0.54963
P-TERPHENYL	12.20	0.246	0.02681
DEUTERATED P-TERPHENYL-C18D14	10.89	0.346	0.01361
4-METHYL TERPHENYL	14.21	0.156	0.06695
2-METHYL-P-TERPHENYL	0.00	0.000	0.00000

ACCEPTOR MOLECULES	RO	CO	OVERLAP
3,3''-DIMETHYL-P-TERPHENYL	14.36	0.151	0.07143
4-(3,3-DIMETHYLBUTOXY)-P-TERPHENYL	20.60	0.051	0.62008
4,4''-DIHEXAHYDROFARNESOXY-P-TERPHENYL	24.23	0.031	1.63716
2,2'-METHYLENE-P-TERPHENYL	22.48	0.039	1.04673
2,2'-ETHYLENE-P-TERPHENYL	23.79	0.033	1.46981
2,2'-METHYLENE-5',6'-METHYLENE-P-TERPHENYL (TRACE)	34.85	0.011	14.47230
3-PHENYL DIBENZOFURAN	22.49	0.039	1.04852
BENZO(1,2-B-4,5-B')BISBENZOFURAN (TRACE) (B)	36.73	0.009	19.81704
SEXIPHENYL BENZENE	0.00	0.000	0.00000
P-QUATERPHENYL	25.81	0.026	2.38878
3,3'''-DIMETHYL-P-QUATERPHENYL	27.44	0.022	3.44696
TETRAMETHYL-P-QUATERPHENYL	5.61	2.534	0.00025
3,3''' -DIMETHYL-3',2''-METHYLENE-P-QUATERPHENYL	36.35	0.009	18.58234
3,3''' -DIMETHYL-3',2''-ETHYLENE-P-QUATERPHENYL	35.89	0.010	17.18985
4-(3-ETHYLBUTOXY)-P-QUATERPHENYL (B)	32.92	0.013	10.24738
4,4'''-DI(2-BUTYLOCTOXY-1)-P-QUATERPHENYL	33.66	0.012	11.70504
DIETHYL-P-QUINQUEPHENYL	28.64	0.019	4.45564
DI(3-ETHYLHEPTYL)-P-QUINQUEPHENYL	35.16	0.010	15.16679
TETRAMETHYL-P-SEXIPHENYL	27.21	0.022	3.28329
TETRAMETHYL-P-OCTAPHENYL (B)	39.89	0.007	32.33470
INDOLE (TRACE)	0.00	0.000	0.00000
2-PHENYL INDOLE	28.63	0.019	4.45827
1-METHYL-2-PHENYL INDOLE	22.86	0.037	1.15426
1-NEOPENTYL-2-PHENYL INDOLE	19.87	0.057	0.49860
1-N-BUTYL-2(BIPHENYLYL-4) INDOLE	32.65	0.013	9.73885
1,2-DIPHENYL INDOLE	23.72	0.034	1.43771
1-BIPHENYLYL-2-PHENYL INDOLE	26.59	0.024	2.84657
1,3-DIMETHYL-2-PHENYL INDOLE	24.09	0.032	1.58112
2,3-DIPHENYL INDOLE	30.17	0.016	6.04908
1-METHYL-2,3-DIPHENYL INDOLE	27.37	0.022	3.38049
2-PHENYL-3,2'-METHYLENE INDOLE (TRACE)	32.12	0.013	8.86602
2-PHENYL-3,2'-ETHYLENE INDOLE (TRACE)	34.93	0.010	14.59648
2-PHENYL-3,2'-TRIMETHYLENE INDOLE	30.67	0.016	6.71754
1-METHYL-3,2'-METHYLENE-2-PHENYL INDOLE (TRACE)	34.44	0.011	13.44970
1-METHYL-3,2'-ETHYLENE-2-PHENYL INDOLE	35.34	0.010	15.63364
1-METHYL-3,2'-TRIMETHYLENE-2-PHENYL INDOLE	29.49	0.017	5.30265
4,5-DIPHENYLIMIDAZOLE (A)	18.23	0.074	0.29646
PPF (TRACE)	36.16	0.009	18.01545
PPO (TRACE)	30.85	0.015	6.96363
2-(4-FLUOROPHENYL)-5-PHENYLOXAZOLE (TRACE)	29.36	0.018	5.18081
αNPO (TRACE)	37.92	0.008	23.77245
BBO (B)	43.55	0.005	54.51155
POPOP	42.57	0.006	47.50529
DIMETHYL POPOP (TRACE)	43.62	0.005	54.82942
PPD (TRACE)	10.38	0.400	0.01021
2,5-DI(2-METHYLPHENYL)OXADIAZOLE	13.31	0.190	0.04534
PBD	29.48	0.017	5.31336
BBD (B)	38.26	0.008	25.24844
αNPD (TRACE)	33.21	0.012	10.76614
βNPD	28.52	0.019	4.34833
αNND (TRACE)	35.78	0.010	16.83060
βNND (TRACE)	34.91	0.011	14.60580
TRANS-1-(4-BIPHENYLYL)-2-PHENYL-ETHYLENE	37.93	0.008	23.91439
TRANS-1-(4-BIPHENYLYL)-2-(1-NAPHTHYL)-ETHYLENE	40.18	0.007	33.57378
DIPHENYLSTILBENE (B)	43.98	0.005	57.79077
1,1,4,4-TETRAPHENYLBUTADIENE	41.33	0.006	39.68000
1,6-DIPHENYLHEXATRIENE	45.73	0.005	72.86424
1,8-DIPHENYL-1,3,5,7-OCTATETRAENE	45.39	0.005	69.21792
BPSB (TRACE)	46.04	0.005	75.88806
BIS-MSB	44.66	0.005	63.22500
ESCULIN (A)	34.13	0.011	12.64775
AZULENE	25.50	0.027	2.17197
NAPHTHALENE	6.65	1.524	0.00070
DEUTERATED NAPHTHALENE-C10D8	10.23	0.418	0.00929
1-METHYLNAPHTHALENE 5ML/L	8.89	0.637	0.00400
2-METHYL NAPHTHALENE	11.30	0.310	0.01690
2,3-DIMETHYL NAPHTHALENE	11.15	0.322	0.01563
2,6-DIMETHYL NAPHTHALENE (TRACE)	15.84	0.113	0.12809
1-NAPHTHOL (TRACE)	16.70	0.096	0.17567
2-NAPHTHOL (TRACE)	19.99	0.056	0.51679
1-NAPHTHYLAMINE	25.51	0.027	2.21841
2-CHLORONAPHTHALENE	12.27	0.242	0.02775
ACENAPHTHENE	14.24	0.155	0.06779
1-PHENYL NAPHTHALENE	14.40	0.150	0.07224
2-PHENYL NAPHTHALENE	22.68	0.038	1.10378
1,4-DIPHENYL NAPHTHALENE	23.89	0.033	1.50047
1,5-DIPHENYL NAPHTHALENE	22.03	0.042	0.92529
1,7-DIPHENYL NAPHTHALENE	21.73	0.044	0.85185
1,4,5,8-TETRAPHENYL NAPHTHALENE	36.57	0.009	19.05363
1,1'-BINAPHTHYL	17.51	0.083	0.23363
4,4'-DI(N-BUTOXY)-1,1'-BINAPHTHYL	29.39	0.018	5.18810
2,2'-BINAPHTHYL	23.76	0.033	1.45618
ACRIDINE (A)	32.04	0.014	8.61422
ANTHRACENE (TRACE)	30.06	0.016	5.89146
DEUTERATED ANTHRACENE C14D10 (TRACE)	29.70	0.017	5.48552
9-METHYL ANTHRACENE (TRACE)	29.88	0.017	5.66853
1-AMINO ANTHRACENE	28.13	0.020	3.91981
2-AMINO ANTHRACENE	28.10	0.020	3.89340
9-VINYL ANTHRACENE	30.88	0.015	6.89204
9-PHENYL ANTHRACENE (TRACE)	31.45	0.014	7.70062
9,10-DIPHENYL ANTHRACENE	31.96	0.014	8.47086
9,10-DICHLOROANTHRACENE (TRACE)	30.58	0.016	6.49848
BIANTHRANYL (B)	33.71	0.012	11.65745
9,10-DINAPHTHYLANTHRACENE (B)	31.45	0.014	7.67119
9-(P-VINYLPHENYL)-10-PHENYL ANTHRACENE	32.09	0.014	8.66295
9,10-BIS(PHENYL-ETHYNYL)ANTHRACENE (TRACE)	28.97	0.018	4.67118

ACCEPTOR MOLECULES	RO	CO	OVERLAP
RUBICENE (B)	31.86	0.014	8.27233
DIBENZ(A,H)ANTHRACENE (B)	33.80	0.012	11.85964
TETRACENE (TRACE) (B)	23.32	0.035	1.27254
RUBRENE (B)	29.44	0.018	5.15208
TETRAHYDROCANNABINOL ACID METHYL ESTER	10.01	0.445	0.00822
PHENANTHRENE	15.70	0.116	0.12066
3,4-BENZOPHENANTHRENE	23.51	0.034	1.34858
2-PHENYL PHENANTHRENE	17.98	0.077	0.27119
CHRYSENE (TRACE)	23.98	0.032	1.52605
TRIPHENYLENE	16.46	0.100	0.16034
DEUTERATED TRIPHENYLENE-C18D12	16.20	0.105	0.14573
PYRENE	33.63	0.012	11.56205
1,3,6,8,-TETRAPHENYL PYRENE	39.65	0.007	30.79862
BIPYRENYL (B)	42.03	0.006	43.82892
O-PHENYLENE-PYRENE	35.42	0.010	15.62593
ANTHANTHRENE (B)	32.76	0.013	9.77541
FLUORANTHENE	31.56	0.014	7.85454
3-PHENYL FLUORANTHENE	34.55	0.011	13.47928
3-AMINO FLUORANTHENE	31.74	0.014	8.08493
3,3'-BIFLUORANTHENYL (B)	38.54	0.008	25.92471
BENZO(GHI)FLUORANTHENE	30.18	0.016	5.99167
9,12-DI(N-OCTADECOXY)-BENZ(E)ACEPHENANTHRYLENE	34.12	0.011	12.48014
DECACYCLENE (B)	41.78	0.006	42.07641
10-ETHOXY-BENZO(KL)XANTHENE	32.61	0.013	9.55968
PERYLENE	29.90	0.017	5.64527
BENZO(GHI)PERYLENE (B)	34.71	0.011	13.91173
BBOT (TRACE)	44.06	0.005	58.16126
QUININE SULFATE (A)	29.69	0.017	5.51940
ACRIDINE YELLOW (A)	26.01	0.025	2.44687
ACRIDINE ORANGE (A)	22.99	0.037	1.16658
ACRIDINE RED (A)	27.04	0.023	3.09274
URANINE (A)	27.04	0.023	3.09268
RHODAMINE B (A)	30.71	0.015	6.62762
RHODAMINE 6G (A)	31.24	0.015	7.35465
EOSIN B (A)	28.33	0.020	4.08886
EOSIN Y (A)	28.88	0.019	4.58580
ROSE BENGAL (A)	30.23	0.016	6.03141
INDENOINDENE	30.33	0.016	6.28194

8.76 Donor Molecule: 2-Methyl-*p*-terphenyl

ACCEPTOR MOLECULES	RO	CO	OVERLAP
POLYVINYLTOLUENE	2.18	43.132	0.00000
ANILINE 3ML/L	7.60	1.018	0.00223
N,N-DIMETHYLANILINE 3ML/L	14.78	0.139	0.11946
TRIPHENYLAMINE	26.95	0.023	4.34912
1,4-DIMETHOXY BENZENE	9.23	0.568	0.00715
1-METHYL-4-CHLOROBENZENE 5ML/L	0.00	0.000	0.00000
SODIUM SALICYLATE (A)	18.35	0.072	0.43278
ANTHRANILIC ACID (A)	27.16	0.022	4.52840
TMPD	23.57	0.034	1.94141
4-METHOXY BIPHENYL	7.10	1.248	0.00148
4-BIPHENYLYL PHENYL ETHER	5.89	2.185	0.00048
4,4'-DIHYDROXYBIPHENYL (A)	13.57	0.179	0.07170
4,4'-DIMETHOXY BIPHENYL	11.73	0.277	0.03004
4,4'-DIISOBUTOXYBIPHENYL	11.14	0.324	0.02202
BENZIDINE	24.29	0.031	2.34052
4-VINYL BIPHENYL	15.85	0.112	0.18208
M-TERPHENYL	11.22	0.317	0.02294
M-QUATERPHENYL	5.36	2.899	0.00028
1,3,5-TRIPHENYL BENZENE	11.83	0.270	0.03143
1,3-DIBIPHENYLYLBENZENE (B)	24.22	0.031	2.30325
1,3,5,-TRIBIPHENYLYLBENZENE (B)	28.26	0.020	5.81566
FLUORENE (TRACE)	10.72	0.363	0.01752
1-METHYL FLUORENE	9.71	0.489	0.00964
1,2-BENZOFLUORENE	25.22	0.028	2.92394
9,10-DIHYDROPHENANTHRENE	11.76	0.275	0.03052
CARBAZOLE (TRACE)	22.99	0.037	1.68024
N-METHYL CARBAZOLE (TRACE)	24.70	0.030	2.58238
N-VINYLCARBAZOLE	24.94	0.029	2.73773
N-PHENYL CARBAZOLE	24.64	0.030	2.54594
N-METHYL-2,3-BENZCARBAZOLE	27.91	0.021	5.29722
DIBENZOFURAN	9.14	0.586	0.00672
DIBENZOTHIOPHENE (TRACE)	20.71	0.050	0.90108
P-TERPHENYL	17.44	0.084	0.32308
DEUTERATED P-TERPHENYL-C18D14	16.94	0.092	0.27171
4-METHYL TERPHENYL	19.28	0.062	0.58913
2-METHYL-P-TERPHENYL	10.73	0.362	0.01756
3,3''-DIMETHYL-P-TERPHENYL	18.80	0.067	0.50683
4-(3,3-DIMETHYLBUTOXY)-P-TERPHENYL	23.79	0.033	2.06989
4,4''DIHEXAHYDROFARNESOXY-P-TERPHENYL	27.01	0.023	4.43547
2,2'-METHYLENE-P-TERPHENYL	26.36	0.024	3.84066
2,2'-ETHYLENE-P-TERPHENYL	26.79	0.023	4.21973
2,2'-METHYLENE-5',6''-METHYLENE-P-TERPHENYL (TRACE)	34.38	0.011	18.80549
3-PHENYL DIBENZOFURAN	26.70	0.023	4.14355
BENZO(1,2-B-4,5-B')BISBENZOFURAN (TRACE) (B)	36.04	0.010	24.90743
SEXIPHENYL BENZENE	6.70	1.489	0.00104
P-QUATERPHENYL	28.13	0.020	5.65233
3,3'''-DIMETHYL-P-QUATERPHENYL	29.42	0.018	7.38954
TETRAMETHYL-P-QUATERPHENYL	13.09	0.199	0.05790

ACCEPTOR MOLECULES	RO	CO	OVERLAP
3,3''' -DIMETHYL-3',2''-METHYLENE-P-QUATERPHENYL	36.20	0.009	25.56997
3,3''' -DIMETHYL-3',2''-ETHYLENE-P-QUATERPHENYL	35.81	0.010	23.90350
4-(3-ETHYLBUTOXY)-P-QUATERPHENYL (B)	33.60	0.012	16.31712
4,4'''-DI(2-BUTYLOCTOXY-1)-P-QUATERPHENYL	34.47	0.011	19.02884
DIETHYL-P-QUINQUEPHENYL	30.70	0.015	9.52794
DI(3-ETHYLHEPTYL)-P-QUINQUEPHENYL	35.81	0.010	23.85039
TETRAMETHYL-P-SEXIPHENYL	29.80	0.017	7.99281
TETRAMETHYL-P-OCTAPHENYL (B)	39.52	0.007	43.10706
INDOLE (TRACE)	0.00	0.000	0.00000
2-PHENYL INDOLE	30.13	0.016	8.52454
1-METHYL-2-PHENYL INDOLE	25.10	0.028	2.84914
1-NEOPENTYL-2-PHENYL INDOLE	23.10	0.036	1.73647
1-N-BUTYL-2(BIPHENYLYL-4) INDOLE	32.50	0.013	13.35056
1,2-DIPHENYL INDOLE	26.14	0.025	3.63523
1-BIPHENYLYL-2-PHENYL INDOLE	28.48	0.019	6.05891
1,3-DIMETHYL-2-PHENYL INDOLE	25.49	0.027	3.12325
2,3-DIPHENYL INDOLE	30.17	0.016	8.53449
1-METHYL-2,3-DIPHENYL INDOLE	27.53	0.021	4.93373
2-PHENYL-3,2'-METHYLENE INDOLE (TRACE)	32.44	0.013	13.24695
2-PHENYL-3,2'-ETHYLENE INDOLE (TRACE)	33.90	0.011	17.20665
2-PHENYL-3,2'-TRIMETHYLENE INDOLE	31.45	0.014	10.99908
1-METHYL-3,2'-METHYLENE-2-PHENYL INDOLE (TRACE)	33.71	0.012	16.67122
1-METHYL-3,2'-ETHYLENE-2-PHENYL INDOLE	34.02	0.011	17.55289
1-METHYL-3,2'-TRIMETHYLENE-2-PHENYL INDOLE	29.92	0.017	8.15338
4,5-DIPHENYLIMIDAZOLE (A)	20.05	0.055	0.74147
PPF (TRACE)	35.42	0.010	22.42388
PPO (TRACE)	31.99	0.014	12.21386
2-(4-FLUOROPHENYL)-5-PHENYLOXAZOLE (TRACE)	30.54	0.016	9.25047
αNPO (TRACE)	35.87	0.010	23.99529
BBO (B)	40.77	0.007	51.68548
POPOP	39.27	0.007	41.20732
DIMETHYL POPOP (TRACE)	40.04	0.007	46.23140
PPD (TRACE)	18.55	0.070	0.46895
2,5-DI(2-METHYLPHENYL)OXADIAZOLE	19.48	0.061	0.62715
PBD	31.41	0.014	10.95615
BBD (B)	37.76	0.008	32.88212
αNPD (TRACE)	32.46	0.013	13.24824
βNPD	29.93	0.017	8.17677
αNND (TRACE)	34.12	0.011	17.85287
βNND (TRACE)	34.64	0.011	19.65726
TRANS-1-(4-BIPHENYLYL)-2-PHENYL-ETHYLENE	37.10	0.009	29.49798
TRANS-1-(4-BIPHENYLYL)-2-(1-NAPHTHYL)-ETHYLENE	37.95	0.008	33.58746
DIPHENYLSTILBENE (B)	41.19	0.006	54.92317
1,1,4,4-TETRAPHENYLBUTADIENE	38.48	0.008	36.43545
1,6-DIPHENYLHEXATRIENE	42.27	0.006	64.05977
1,8-DIPHENYL-1,3,5,7-OCTATETRAENE	41.16	0.006	54.26548
BPSB (TRACE)	42.45	0.006	65.67525
BIS-MSB	41.50	0.006	57.32940
ESCULIN (A)	32.21	0.013	12.57943
AZULENE	24.36	0.031	2.32888
NAPHTHALENE	9.83	0.470	0.01038
DEUTERATED NAPHTHALENE-C10D8	11.87	0.268	0.03190
1-METHYLNAPHTHALENE 5ML/L	11.74	0.276	0.02998
2-METHYL NAPHTHALENE	13.23	0.193	0.06150
2,3-DIMETHYL NAPHTHALENE	13.11	0.199	0.05808
2,6-DIMETHYL NAPHTHALENE (TRACE)	16.38	0.102	0.22108
1-NAPHTHOL (TRACE)	18.71	0.068	0.49140
2-NAPHTHOL (TRACE)	20.21	0.054	0.77847
1-NAPHTHYLAMINE	25.40	0.027	3.04835
2-CHLORONAPHTHALENE	13.49	0.182	0.06891
ACENAPHTHENE	16.65	0.097	0.24416
1-PHENYL NAPHTHALENE	17.67	0.081	0.34833
2-PHENYL NAPHTHALENE	25.94	0.026	3.47785
1,4-DIPHENYL NAPHTHALENE	25.51	0.027	3.13825
1,5-DIPHENYL NAPHTHALENE	24.05	0.032	2.20520
1,7-DIPHENYL NAPHTHALENE	23.76	0.033	2.04779
1,4,5,8-TETRAPHENYL NAPHTHALENE	34.55	0.011	19.11289
1,1'-BINAPHTHYL	19.87	0.057	0.70343
4,4'-DI(N-BUTOXY)-1,1'-BINAPHTHYL	29.70	0.017	7.79742
2,2'-BINAPHTHYL	26.36	0.024	3.82703
ACRIDINE (A)	29.45	0.018	7.33140
ANTHRACENE (TRACE)	27.67	0.021	5.05011
DEUTERATED ANTHRACENE C14D10 (TRACE)	27.43	0.022	4.80030
9-METHYL ANTHRACENE (TRACE)	27.13	0.022	4.48145
1-AMINO ANTHRACENE	25.44	0.027	3.02331
2-AMINO ANTHRACENE	26.06	0.025	3.49033
9-VINYL ANTHRACENE	28.03	0.020	5.43335
9-PHENYL ANTHRACENE (TRACE)	28.56	0.019	6.09184
9,10-DIPHENYL ANTHRACENE	28.90	0.019	6.52059
9,10-DICHLOROANTHRACENE (TRACE)	27.53	0.021	4.86974
BIANTHRANYL (B)	30.62	0.016	9.22848
9,10-DINAPHTHYLANTHRACENE (B)	28.69	0.019	6.22967
9-(P-VINYLPHENYL)-10-PHENYL ANTHRACENE	29.03	0.018	6.68903
9,10-BIS(PHENYL-ETHYNYL)ANTHRACENE (TRACE)	27.98	0.020	5.34821
RUBICENE (B)	30.85	0.015	9.62281
DIBENZ(A,H)ANTHRACENE (B)	32.93	0.013	14.29165
TETRACENE (TRACE) (B)	21.74	0.044	1.17651
RUBRENE (B)	31.83	0.014	11.60213
TETRAHYDROCANNABINOL ACID METHYL ESTER	15.01	0.132	0.13117
PHENANTHRENE	15.28	0.125	0.14432
3,4 BENZOPHENANTHRENE	24.96	0.029	2.72126
2-PHENYL PHENANTHRENE	20.54	0.052	0.84963
CHRYSENE (TRACE)	25.87	0.026	3.38464
TRIPHENYLENE	16.61	0.098	0.23833
DEUTERATED TRIPHENYLENE-C18D12	16.49	0.100	0.22846
PYRENE	33.16	0.012	14.99144

ACCEPTOR MOLECULES	RO	CO	OVERLAP
1,3,6,8,-TETRAPHENYL PYRENE	35.77	0.010	23.39247
BIPYRENYL (B)	38.99	0.008	39.38011
O-PHENYLENE-PYRENE	33.22	0.012	15.00107
ANTHANTHRENE (B)	31.49	0.014	10.88830
FLUORANTHENE	29.40	0.018	7.22502
3-PHENYL FLUORANTHENE	31.77	0.014	11.4[illegible]
3-AMINO FLUORANTHENE	29.75	0.017	7.73976
3,3'-BIFLUORANTHENYL (B)	35.47	0.010	22.20637
BENZO(GHI)FLUORANTHENE	28.46	0.019	5.93160
9,12-DI(N-OCTADECOXY)-BENZ(E)ACEPHENANTHRYLENE	31.53	0.014	10.96203
DECACYCLENE (B)	38.55	0.008	36.61754
10-ETHOXY-BENZO(KL)XANTHENE	29.54	0.017	7.43768
PERYLENE	26.53	0.024	3.88971
BENZO(GHI)PERYLENE (B)	31.80	0.014	11.57970
BBOT (TRACE)	40.17	0.007	47.05241
QUININE SULFATE (A)	28.83	0.019	6.51669
ACRIDINE YELLOW (A)	23.32	0.035	1.79569
ACRIDINE ORANGE (A)	22.00	0.042	1.26571
ACRIDINE RED (A)	25.55	0.027	3.10005
URANINE (A)	25.82	0.026	3.30899
RHODAMINE B (A)	28.95	0.018	6.57208
RHODAMINE 6G (A)	29.06	0.018	6.71493
EOSIN B (A)	27.62	0.021	4.95213
EOSIN Y (A)	28.17	0.020	5.58060
ROSE BENGAL (A)	29.12	0.018	6.80714
INDENOINDENE	31.26	0.015	10.61900

8.77 Donor Molecule: 3,3″-Dimethyl-*p*-terphenyl

N,N-DIMETHYLANILINE 3ML/L	10.91	0.345	0.01430
TRIPHENYLAMINE	25.23	0.028	2.17424
1,4-DIMETHOXY BENZENE	3.61	9.495	0.00002
1-METHYL-4-CHLOROBENZENE 5ML/L	0.00	0.000	0.00000
SODIUM SALICYLATE (A)	16.95	0.092	0.19980
ANTHRANILIC ACID (A)	28.46	0.019	4.44409
TMPD	24.20	0.032	1.68730
4,4'-DIHYDROXYBIPHENYL (A)	10.36	0.402	0.01053
4,4'-DIMETHOXY BIPHENYL	7.58	1.026	0.00162
4,4'-DIISOBUTOXYBIPHENYL	5.72	2.384	0.00030
BENZIDINE	22.08	0.042	0.97804
4-VINYL BIPHENYL	10.30	0.409	0.01017
M-TERPHENYL	6.89	1.369	0.00091
1,3,5-TRIPHENYL BENZENE	9.63	0.501	0.00677
1,3-DIBIPHENYLYLBENZENE (B)	20.78	0.050	0.68080
1,3,5,-TRIBIPHENYLYLBENZENE (B)	24.64	0.030	1.89424
FLUORENE (TRACE)	0.00	0.000	0.00000
1-METHYL FLUORENE	4.05	6.712	0.00004
1,2-BENZOFLUORENE	23.29	0.035	1.34526
9,10-DIHYDROPHENANTHRENE	5.67	2.450	0.00028
CARBAZOLE (TRACE)	22.35	0.040	1.05358
N-METHYL CARBAZOLE (TRACE)	25.62	0.027	2.38476
N-VINYLCARBAZOLE	25.56	0.027	2.35205
N-PHENYL CARBAZOLE	25.27	0.028	2.19625
N-METHYL-2,3-BENZCARBAZOLE	28.71	0.019	4.65861
DIBENZOFURAN	0.00	0.000	0.00000
DIBENZOTHIOPHENE (TRACE)	20.10	0.055	0.55831
P-TERPHENYL	12.82	0.212	0.03780
DEUTERATED P-TERPHENYL-C18D14	11.76	0.275	0.02251
4-METHYL TERPHENYL	14.77	0.139	0.08819
2-METHYL-P-TERPHENYL	5.38	2.871	0.00021
3,3″-DIMETHYL-P-TERPHENYL	14.78	0.138	0.08868
4-(3,3-DIMETHYLBUTOXY)-P-TERPHENYL	20.78	0.050	0.68158
4,4″-DIHEXAHYDROFARNESOXY-P-TERPHENYL	24.32	0.031	1.75353
2,2'-METHYLENE-P-TERPHENYL	22.77	0.038	1.18282
2,2'-ETHYLENE-P-TERPHENYL	23.95	0.033	1.59791
2,2'-METHYLENE-5',6″-METHYLENE-P-TERPHENYL (TRACE)	34.54	0.011	14.34660
3-PHENYL DIBENZOFURAN	22.80	0.038	1.19259
BENZO(1,2-B-4,5-B')BISBENZOFURAN (TRACE) (B)	36.42	0.009	19.66279
SEXIPHENYL BENZENE	0.00	0.000	0.00000
P-QUATERPHENYL	25.84	0.026	2.51573
3,3‴-DIMETHYL-P-QUATERPHENYL	27.42	0.022	3.59127
TETRAMETHYL-P-QUATERPHENYL	7.41	1.099	0.00141
3,3‴ -DIMETHYL-3',2″-METHYLENE-P-QUATERPHENYL	36.08	0.010	18.57731
3,3‴ -DIMETHYL-3',2″-ETHYLENE-P-QUATERPHENYL	35.64	0.010	17.21910
4-(3-ETHYLBUTOXY)-P-QUATERPHENYL (B)	32.75	0.013	10.38125
4,4‴-DI(2-BUTYLOCTOXY-1)-P-QUATERPHENYL	33.69	0.012	11.8[illegible]
DIETHYL-P-QUINQUEPHENYL	28.62	0.019	4.63963
DI(3-ETHYLHEPTYL)-P-QUINQUEPHENYL	34.97	0.010	15.35311
TETRAMETHYL-P-SEXIPHENYL	27.26	0.022	3.46860
TETRAMETHYL-P-OCTAPHENYL (B)	39.60	0.007	32.33823
INDOLE (TRACE)	0.00	0.000	0.00000
2-PHENYL INDOLE	28.57	0.019	4.60127
1-METHYL-2-PHENYL INDOLE	22.91	0.037	1.22264
1-NEOPENTYL-2-PHENYL INDOLE	20.06	0.055	0.55299
1-N-BUTYL-2(BIPHENYLYL-4) INDOLE	32.42	0.013	9.75609
1,2-DIPHENYL INDOLE	23.78	0.033	1.52929
1-BIPHENYLYL-2-PHENYL INDOLE	26.58	0.024	2.97112
1,3-DIMETHYL-2-PHENYL INDOLE	24.04	0.032	1.63236
2,3-DIPHENYL INDOLE	29.96	0.017	6.07248
1-METHYL-2,3-DIPHENYL INDOLE	27.20	0.022	3.40013

ACCEPTOR MOLECULES	RO	CO	OVERLAP
2-PHENYL-3,2'-METHYLENE INDOLE (TRACE)	31.92	0.014	8.92125
2-PHENYL-3,2'-ETHYLENE INDOLE (TRACE)	34.62	0.011	14.46688
2-PHENYL-3,2'-TRIMETHYLENE INDOLE	30.52	0.016	6.82130
1-METHYL-3,2'-METHYLENE-2-PHENYL INDOLE (TRACE)	34.16	0.011	13.36952
1-METHYL-3,2'-ETHYLENE-2-PHENYL INDOLE	35.02	0.010	15.48548
1-METHYL-3,2'-TRIMETHYLENE-2-PHENYL INDOLE	29.31	0.018	5.34856
4,5-DIPHENYLIMIDAZOLE (A)	18.26	0.074	0.31306
PPF (TRACE)	35.86	0.010	17.90683
PPO (TRACE)	30.73	0.015	7.10754
2-(4-FLUOROPHENYL)-5-PHENYLOXAZOLE (TRACE)	29.26	0.018	5.29825
αNPO (TRACE)	37.57	0.008	23.49827
BBO (B)	43.15	0.006	53.91211
POPOP	42.20	0.006	47.09811
DIMETHYL POPOP (TRACE)	43.25	0.006	54.48888
PPD (TRACE)	12.02	0.258	0.02572
2,5-DI(2-METHYLPHENYL)OXADIAZOLE	14.20	0.156	0.06980
PBD	29.45	0.018	5.51857
BBD (B)	37.96	0.008	25.18799
αNPD (TRACE)	32.93	0.013	10.70052
βNPD	28.44	0.019	4.46637
αNND (TRACE)	35.45	0.010	16.63461
βNND (TRACE)	34.65	0.011	14.59602
TRANS-1-(4-BIPHENYLYL)-2-PHENYL-ETHYLENE	37.61	0.008	23.78031
TRANS-1-(4-BIPHENYLYL)-2-(1-NAPHTHYL)-ETHYLENE	39.81	0.007	33.23299
DIPHENYLSTILBENE (B)	43.58	0.005	57.13112
1,1,4,4-TETRAPHENYLBUTADIENE	40.95	0.007	39.26038
1,6-DIPHENYLHEXATRIENE	45.32	0.005	72.21617
1,8-DIPHENYL-1,3,5,7-OCTATETRAENE	45.06	0.005	69.31725
BPSB (TRACE)	45.64	0.005	75.26531
BIS-MSB	44.26	0.005	62.64149
ESCULIN (A)	33.82	0.012	12.50040
AZULENE	25.25	0.028	2.13651
NAPHTHALENE	7.16	1.216	0.00115
DEUTERATED NAPHTHALENE-C10D8	10.31	0.408	0.01020
1-METHYLNAPHTHALENE 5ML/L	9.21	0.573	0.00517
2-METHYL NAPHTHALENE	11.50	0.294	0.01962
2,3-DIMETHYL NAPHTHALENE	11.33	0.308	0.01797
2,6-DIMETHYL NAPHTHALENE (TRACE)	15.77	0.114	0.13065
1-NAPHTHOL (TRACE)	16.81	0.094	0.19170
2-NAPHTHOL (TRACE)	19.87	0.057	0.52037
1-NAPHTHYLAMINE	25.32	0.028	2.21742
2-CHLORONAPHTHALENE	12.35	0.237	0.03013
ACENAPHTHENE	14.44	0.149	0.07702
1-PHENYL NAPHTHALENE	14.64	0.142	0.08371
2-PHENYL NAPHTHALENE	22.77	0.038	1.18061
1,4-DIPHENYL NAPHTHALENE	23.87	0.033	1.56128
1,5-DIPHENYL NAPHTHALENE	22.07	0.042	0.97748
1,7-DIPHENYL NAPHTHALENE	21.77	0.043	0.90000
1,4,5,8-TETRAPHENYL NAPHTHALENE	36.23	0.009	18.84818
1,1'-BINAPHTHYL	17.61	0.082	0.25301
4,4'-DI(N-BUTOXY)-1,1'-BINAPHTHYL	29.21	0.018	5.22792
2,2'-BINAPHTHYL	23.85	0.033	1.55723
ACRIDINE (A)	31.74	0.014	8.52579
ANTHRACENE (TRACE)	29.80	0.017	5.84880
DEUTERATED ANTHRACENE C14D10 (TRACE)	29.44	0.018	5.44459
9-METHYL ANTHRACENE (TRACE)	29.64	0.017	5.64480
1-AMINO ANTHRACENE	27.96	0.020	3.94185
2-AMINO ANTHRACENE	27.91	0.021	3.89919
9-VINYL ANTHRACENE	30.64	0.016	6.88023
9-PHENYL ANTHRACENE (TRACE)	31.19	0.015	7.66880
9,10-DIPHENYL ANTHRACENE	31.75	0.014	8.50080
9,10-DICHLOROANTHRACENE (TRACE)	30.39	0.016	6.54596
BIANTHRANYL (B)	33.48	0.012	11.68674
9,10-DINAPHTHYLANTHRACENE (B)	31.25	0.015	7.72879
9-(P-VINYLPHENYL)-10-PHENYL ANTHRACENE	31.88	0.014	8.70562
9,10-BIS(PHENYL-ETHYNYL)ANTHRACENE (TRACE)	29.02	0.018	4.93200
RUBICENE (B)	31.72	0.014	8.41302
DIBENZ(A,H)ANTHRACENE (B)	33.51	0.012	11.79103
TETRACENE (TRACE) (B)	23.25	0.036	1.30485
RUBRENE (B)	29.45	0.018	5.38144
TETRAHYDROCANNABINOL ACID METHYL ESTER	10.63	0.373	0.01227
PHENANTHRENE	15.56	0.119	0.11956
3,4-BENZOPHENANTHRENE	23.49	0.034	1.40375
2-PHENYL PHENANTHRENE	18.04	0.076	0.28915
CHRYSENE (TRACE)	24.05	0.032	1.62378
TRIPHENYLENE	16.34	0.102	0.16053
DEUTERATED TRIPHENYLENE-C18D12	16.10	0.107	0.14670
PYRENE	33.35	0.012	11.49860
1,3,6,8,-TETRAPHENYL PYRENE	39.37	0.007	30.88420
BIPYRENYL (B)	41.65	0.006	43.39530
O-PHENYLENE-PYRENE	35.21	0.010	15.72614
ANTHANTHRENE (B)	32.62	0.013	9.95340
FLUORANTHENE	31.28	0.015	7.77250
3-PHENYL FLUORANTHENE	34.26	0.011	13.41193
3-AMINO FLUORANTHENE	31.54	0.014	8.11555
3,3'-BIFLUORANTHENYL (B)	38.26	0.008	25.96012
BENZO(GHI)FLUORANTHENE	29.90	0.017	5.91991
9,12-DI(N-OCTADECOXY)-BENZ(E)ACEPHENANTHRYLENE	33.87	0.012	12.48017
DECACYCLENE (B)	41.50	0.006	42.17126
10-ETHOXY-BENZO(KL)XANTHENE	32.39	0.013	9.59889
PERYLENE	29.82	0.017	5.79765
BENZO(GHI)PERYLENE (B)	34.47	0.011	13.94005
BBOT (TRACE)	43.71	0.005	58.00247
QUININE SULFATE (A)	29.41	0.018	5.45688
ACRIDINE YELLOW (A)	26.03	0.025	2.56543
ACRIDINE ORANGE (A)	22.97	0.037	1.21269

ACCEPTOR MOLECULES	RO	CO	OVERLAP
ACRIDINE RED (A)	26.85	0.023	3.09254
URANINE (A)	26.89	0.023	3.12228
RHODAMINE B (A)	30.49	0.016	6.63072
RHODAMINE 6G (A)	30.96	0.015	7.27232
EOSIN B (A)	28.15	0.020	4.10932
EOSIN Y (A)	28.71	0.019	4.62001
ROSE BENGAL (A)	30.01	0.017	6.02362
INDENOINDENE	30.19	0.016	6.39066

8.78 Donor Molecule: 4-(3,3-Dimethylbutoxy)-*p*-terphenyl

N,N-DIMETHYLANILINE 3ML/L	6.29	1.799	0.00053
TRIPHENYLAMINE	21.74	0.044	0.89767
SODIUM SALICYLATE (A)	15.01	0.132	0.09716
ANTHRANILIC ACID (A)	27.62	0.021	3.74746
TMPD	22.87	0.037	1.20947
4,4'-DIHYDROXYBIPHENYL (A)	6.09	1.978	0.00044
4,4'-DIMETHOXY BIPHENYL	0.00	0.000	0.00000
4,4'-DIISOBUTOXYBIPHENYL	0.00	0.000	0.00000
BENZIDINE	18.10	0.075	0.29994
1,3,5-TRIPHENYL BENZENE	6.35	1.749	0.00056
1,3-DIBIPHENYLYLBENZENE (B)	16.08	0.108	0.14724
1,3,5,-TRIBIPHENYLYLBENZENE (B)	19.03	0.065	0.40578
FLUORENE (TRACE)	0.00	0.000	0.00000
1-METHYL FLUORENE	0.00	0.000	0.00000
1,2-BENZOFLUORENE	20.10	0.055	0.56040
9,10-DIHYDROPHENANTHRENE	0.00	0.000	0.00000
CARBAZOLE (TRACE)	19.44	0.061	0.46039
N-METHYL CARBAZOLE (TRACE)	23.79	0.033	1.53979
N-VINYLCARBAZOLE	23.55	0.034	1.45085
N-PHENYL CARBAZOLE	23.26	0.036	1.34548
N-METHYL-2,3-BENZCARBAZOLE	28.22	0.020	4.24012
DIBENZOFURAN	0.00	0.000	0.00000
DIBENZOTHIOPHENE (TRACE)	15.72	0.115	0.12865
P-TERPHENYL	7.30	1.149	0.00130
DEUTERATED P-TERPHENYL-C18D14	5.18	3.219	0.00017
4-METHYL TERPHENYL	8.88	0.639	0.00420
2-METHYL-P-TERPHENYL	0.00	0.000	0.00000
3,3''-DIMETHYL-P-TERPHENYL	9.43	0.534	0.00601
4-(3,3-DIMETHYLBUTOXY)-P-TERPHENYL	15.69	0.116	0.12767
4,4''-DIHEXAHYDROFARNESOXY-P-TERPHENYL	19.18	0.063	0.42496
2,2'-METHYLENE-P-TERPHENYL	16.14	0.106	0.15115
2,2'-ETHYLENE-P-TERPHENYL	18.01	0.077	0.29171
2,2'-METHYLENE-5',6''-METHYLENE-P-TERPHENYL (TRACE)	31.76	0.014	8.73884
3-PHENYL DIBENZOFURAN	16.10	0.107	0.14873
BENZO(1,2-B-4,5-B')BISBENZOFURAN (TRACE) (B)	32.86	0.013	10.70144
SEXIPHENYL BENZENE	0.00	0.000	0.00000
P-QUATERPHENYL	21.12	0.047	0.75667
3,3'''-DIMETHYL-P-QUATERPHENYL	22.92	0.037	1.23463
TETRAMETHYL-P-QUATERPHENYL	0.00	0.000	0.00000
3,3''' -DIMETHYL-3',2''-METHYLENE-P-QUATERPHENYL	32.33	0.013	9.70423
3,3''' -DIMETHYL-3',2''-ETHYLENE-P-QUATERPHENYL	32.13	0.013	9.33198
4-(3-ETHYLBUTOXY)-P-QUATERPHENYL (B)	28.99	0.018	5.04108
4,4'''-DI(2-BUTYLOCTOXY-1)-P-QUATERPHENYL	29.52	0.017	5.61864
DIETHYL-P-QUINQUEPHENYL	23.89	0.033	1.58412
DI(3-ETHYLHEPTYL)-P-QUINQUEPHENYL	31.07	0.015	7.62195
TETRAMETHYL-P-SEXIPHENYL	22.17	0.041	1.01171
TETRAMETHYL-P-OCTAPHENYL (B)	36.34	0.009	19.47263
INDOLE (TRACE)	0.00	0.000	0.00000
2-PHENYL INDOLE	23.95	0.033	1.61015
1-METHYL-2-PHENYL INDOLE	18.47	0.071	0.33802
1-NEOPENTYL-2-PHENYL INDOLE	14.92	0.135	0.09441
1-N-BUTYL-2(BIPHENYLYL-4) INDOLE	29.64	0.017	5.74378
1,2-DIPHENYL INDOLE	19.02	0.065	0.40316
1-BIPHENYLYL-2-PHENYL INDOLE	22.20	0.041	1.01750
1,3-DIMETHYL-2-PHENYL INDOLE	20.26	0.054	0.59012
2,3-DIPHENYL INDOLE	27.31	0.022	3.51123
1-METHYL-2,3-DIPHENYL INDOLE	24.65	0.030	1.90257
2-PHENYL-3,2'-METHYLENE INDOLE (TRACE)	28.40	0.020	4.46820
2-PHENYL-3,2'-ETHYLENE INDOLE (TRACE)	32.16	0.013	9.37892
2-PHENYL-3,2'-TRIMETHYLENE INDOLE	26.58	0.024	3.00098
1-METHYL-3,2'-METHYLENE-2-PHENYL INDOLE (TRACE)	31.28	0.015	7.95836
1-METHYL-3,2'-ETHYLENE-2-PHENYL INDOLE	33.39	0.012	11.72872
1-METHYL-3,2'-TRIMETHYLENE-2-PHENYL INDOLE	26.07	0.025	2.66827
4,5-DIPHENYLIMIDAZOLE (A)	15.21	0.127	0.10539
PPF (TRACE)	32.77	0.013	10.51219
PPO (TRACE)	26.75	0.023	3.11831
2-(4-FLUOROPHENYL)-5-PHENYLOXAZOLE (TRACE)	25.27	0.028	2.21633
αNPO (TRACE)	36.51	0.009	19.97025
BBO (B)	42.64	0.006	50.59423
POPOP	42.52	0.006	49.71075
DIMETHYL POPOP (TRACE)	43.95	0.005	60.52794
PPD (TRACE)	0.00	0.000	0.00000
2,5-DI(2-METHYLPHENYL)OXADIAZOLE	7.20	1.198	0.00120
PBD	24.35	0.031	1.77555
BBD (B)	34.83	0.011	15.15445
αNPD (TRACE)	30.40	0.016	6.67796
βNPD	23.75	0.033	1.52891
αNND (TRACE)	34.00	0.011	13.07400
βNND (TRACE)	31.19	0.015	7.82578

ACCEPTOR MOLECULES	RO	CO	OVERLAP
TRANS-1-(4-BIPHENYLYL)-2-PHENYL-ETHYLENE	34.90	0.011	15.30141
TRANS-1-(4-BIPHENYLYL)-2-(1-NAPHTHYL)-ETHYLENE	38.93	0.008	29.31838
DIPHENYLSTILBENE (B)	43.00	0.006	53.19985
1,1,4,4-TETRAPHENYLBUTADIENE	40.74	0.007	36.40371
1,6-DIPHENYLHEXATRIENE	45.73	0.005	76.89965
1,8-DIPHENYL-1,3,5,7-OCTATETRAENE	46.83	0.004	88.02707
BPSB (TRACE)	46.04	0.005	80.07813
BIS-MSB	44.25	0.005	63.05516
ESCULIN (A)	33.00	0.012	10.89028
AZULENE	23.77	0.033	1.49791
NAPHTHALENE	4.07	6.655	0.00004
DEUTERATED NAPHTHALENE-C10D8	8.06	0.855	0.00234
1-METHYLNAPHTHALENE 5ML/L	5.81	2.285	0.00033
2-METHYL NAPHTHALENE	7.42	1.094	0.00143
2,3-DIMETHYL NAPHTHALENE	7.40	1.102	0.00141
2,6-DIMETHYL NAPHTHALENE (TRACE)	11.56	0.290	0.02042
1-NAPHTHOL (TRACE)	11.77	0.274	0.02277
2-NAPHTHOL (TRACE)	16.70	0.096	0.18492
1-NAPHTHYLAMINE	22.89	0.037	1.22267
2-CHLORONAPHTHALENE	8.59	0.705	0.00344
ACENAPHTHENE	9.56	0.512	0.00654
1-PHENYL NAPHTHALENE	10.24	0.417	0.00985
2-PHENYL NAPHTHALENE	18.50	0.071	0.34203
1,4-DIPHENYL NAPHTHALENE	19.73	0.058	0.50233
1,5-DIPHENYL NAPHTHALENE	17.60	0.082	0.25301
1,7-DIPHENYL NAPHTHALENE	17.65	0.081	0.25755
1,4,5,8-TETRAPHENYL NAPHTHALENE	35.35	0.010	16.41030
1,1'-BINAPHTHYL	13.38	0.187	0.04901
4,4'-DI(N-BUTOXY)-1,1'-BINAPHTHYL	25.92	0.026	2.57350
2,2'-BINAPHTHYL	19.24	0.063	0.43276
ACRIDINE (A)	32.04	0.014	9.08576
ANTHRACENE (TRACE)	30.22	0.016	6.41994
DEUTERATED ANTHRACENE C14D10 (TRACE)	29.91	0.017	6.03220
9-METHYL ANTHRACENE (TRACE)	30.38	0.016	6.61265
1-AMINO ANTHRACENE	29.22	0.018	5.18047
2-AMINO ANTHRACENE	28.58	0.019	4.53598
9-VINYL ANTHRACENE	31.65	0.014	8.42634
9-PHENYL ANTHRACENE (TRACE)	31.92	0.014	8.88361
9,10-DIPHENYL ANTHRACENE	33.22	0.012	11.26527
9,10-DICHLOROANTHRACENE (TRACE)	31.78	0.014	8.62611
BIANTHRANYL (B)	34.90	0.011	15.14150
9,10-DINAPHTHYLANTHRACENE (B)	32.57	0.013	9.98168
9-(P-VINYLPHENYL)-10-PHENYL ANTHRACENE	33.38	0.012	11.56575
9,10-BIS(PHENYL-ETHYNYL)ANTHRACENE (TRACE)	31.25	0.015	7.74472
RUBICENE (B)	31.83	0.014	8.63692
DIBENZ(A,H)ANTHRACENE (B)	32.18	0.013	9.32070
TETRACENE (TRACE) (B)	24.37	0.031	1.74076
RUBRENE (B)	26.37	0.024	2.79652
TETRAHYDROCANNABINOL ACID METHYL ESTER	5.56	2.608	0.00025
PHENANTHRENE	14.39	0.150	0.07546
3,4-BENZOPHENANTHRENE	20.01	0.056	0.54111
2-PHENYL PHENANTHRENE	16.63	0.097	0.17939
CHRYSENE (TRACE)	19.95	0.056	0.53336
TRIPHENYLENE	14.89	0.135	0.09268
DEUTERATED TRIPHENYLENE-C18D12	14.50	0.147	0.07904
PYRENE	31.38	0.014	8.05256
1,3,6,8,-TETRAPHENYL PYRENE	41.08	0.006	40.19608
BIPYRENYL (B)	41.68	0.006	43.94057
O-PHENYLENE-PYRENE	36.08	0.010	18.33375
ANTHANTHRENE (B)	33.69	0.012	12.18181
FLUORANTHENE	30.95	0.015	7.36592
3-PHENYL FLUORANTHENE	34.90	0.011	15.09515
3-AMINO FLUORANTHENE	32.05	0.014	9.01633
3,3'-BIFLUORANTHENYL (B)	39.33	0.007	30.90145
BENZO(GHI)FLUORANTHENE	29.35	0.018	5.34431
9,12-DI(N-OCTADECOXY)-BENZ(E)ACEPHENANTHRYLENE	34.60	0.011	14.32324
DECACYCLENE (B)	42.51	0.006	49.07775
10-ETHOXY-BENZO(KL)XANTHENE	33.80	0.012	12.49125
PERYLENE	32.14	0.013	9.17610
BENZO(GHI)PERYLENE (B)	35.62	0.010	17.11724
BBOT (TRACE)	44.98	0.005	69.39365
QUININE SULFATE (A)	27.20	0.022	3.44423
ACRIDINE YELLOW (A)	28.36	0.020	4.32474
ACRIDINE ORANGE (A)	24.11	0.032	1.63373
ACRIDINE RED (A)	27.06	0.023	3.26269
URANINE (A)	26.72	0.023	3.02384
RHODAMINE B (A)	30.87	0.015	7.18814
RHODAMINE 6G (A)	30.98	0.015	7.34191
EOSIN B (A)	27.26	0.022	3.40762
EOSIN Y (A)	27.95	0.020	3.96106
ROSE BENGAL (A)	29.03	0.018	4.97360
INDENOINDENE	26.17	0.025	2.73517

8.79 Donor Molecule: 4,4″-Dihexahydrofarnesoxy-*p*-terphenyl

N,N-DIMETHYLANILINE 3ML/L	3.79	8.224	0.00002
TRIPHENYLAMINE	20.28	0.054	0.54183
SODIUM SALICYLATE (A)	14.37	0.151	0.06847
ANTHRANILIC ACID (A)	27.47	0.022	3.32085
TMPD	22.38	0.040	0.97362
BENZIDINE	16.24	0.105	0.14291

ACCEPTOR MOLECULES	RO	CO	OVERLAP
1,3,5-TRIPHENYL BENZENE	4.40	5.244	0.00006
1,3-DIBIPHENYLYLBENZENE (B)	13.91	0.166	0.05666
1,3,5,-TRIBIPHENYLYLBENZENE (B)	16.02	0.109	0.13231
FLUORENE (TRACE)	0.00	0.000	0.00000
1-METHYL FLUORENE	0.00	0.000	0.00000
1,2-BENZOFLUORENE	19.29	0.062	0.40128
9,10-DIHYDROPHENANTHRENE	0.00	0.000	0.00000
CARBAZOLE (TRACE)	17.12	0.089	0.19635
N-METHYL CARBAZOLE (TRACE)	23.41	0.035	1.28082
N-VINYLCARBAZOLE	22.90	0.037	1.12164
N-PHENYL CARBAZOLE	22.66	0.038	1.05451
N-METHYL-2,3-BENZCARBAZOLE	28.41	0.020	4.03796
DIBENZOFURAN	0.00	0.000	0.00000
DIBENZOTHIOPHENE (TRACE)	12.86	0.210	0.03541
P-TERPHENYL	0.00	0.000	0.00000
DEUTERATED P-TERPHENYL-C18D14	0.00	0.000	0.00000
4-METHYL TERPHENYL	0.00	0.000	0.00000
2-METHYL-P-TERPHENYL	0.00	0.000	0.00000
3,3″-DIMETHYL-P-TERPHENYL	6.13	1.945	0.00041
4-(3,3-DIMETHYLBUTOXY)-P-TERPHENYL	12.78	0.214	0.03403
4,4″-DIHEXAHYDROFARNESOXY-P-TERPHENYL	16.19	0.105	0.14086
2,2′-METHYLENE-P-TERPHENYL	12.37	0.237	0.02800
2,2′-ETHYLENE-P-TERPHENYL	14.66	0.142	0.07751
2,2′-METHYLENE-5′,6″-METHYLENE-P-TERPHENYL (TRACE)	29.04	0.018	4.67285
3-PHENYL DIBENZOFURAN	12.30	0.240	0.02710
BENZO(1,2-B-4,5-B′)BISBENZOFURAN (TRACE) (B)	31.32	0.015	7.34355
SEXIPHENYL BENZENE	0.00	0.000	0.00000
P-QUATERPHENYL	18.65	0.069	0.32891
3,3‴-DIMETHYL-P-QUATERPHENYL	20.72	0.050	0.61755
TETRAMETHYL-P-QUATERPHENYL	0.00	0.000	0.00000
3,3‴-DIMETHYL-3′,2″-METHYLENE-P-QUATERPHENYL	30.37	0.016	6.10331
3,3‴-DIMETHYL-3′,2″-ETHYLENE-P-QUATERPHENYL	30.51	0.016	6.26180
4-(3-ETHYLBUTOXY)-P-QUATERPHENYL (B)	27.30	0.022	3.21579
4,4‴-DI(2-BUTYLOCTOXY-1)-P-QUATERPHENYL	27.71	0.021	3.52003
DIETHYL-P-QUINQUEPHENYL	21.54	0.045	0.77937
DI(3-ETHYLHEPTYL)-P-QUINQUEPHENYL	29.36	0.018	4.96417
TETRAMETHYL-P-SEXIPHENYL	19.47	0.061	0.42582
TETRAMETHYL-P-OCTAPHENYL (B)	35.03	0.010	14.30252
INDOLE (TRACE)	0.00	0.000	0.00000
2-PHENYL INDOLE	20.94	0.049	0.65813
1-METHYL-2-PHENYL INDOLE	16.11	0.107	0.13621
1-NEOPENTYL-2-PHENYL INDOLE	12.03	0.257	0.02373
1-N-BUTYL-2(BIPHENYLYL-4) INDOLE	28.51	0.019	4.16497
1,2-DIPHENYL INDOLE	16.54	0.099	0.15969
1-BIPHENYLYL-2-PHENYL INDOLE	20.20	0.054	0.52806
1,3-DIMETHYL-2-PHENYL INDOLE	18.13	0.075	0.27690
2,3-DIPHENYL INDOLE	26.22	0.025	2.51839
1-METHYL-2,3-DIPHENYL INDOLE	23.60	0.034	1.34038
2-PHENYL-3,2-METHYLENE INDOLE (TRACE)	26.15	0.025	2.49189
2-PHENYL-3,2-ETHYLENE INDOLE (TRACE)	31.17	0.015	7.11907
2-PHENYL-3,2 -TRIMETHYLENE INDOLE	24.16	0.032	1.55029
1-METHYL-3,2-METHYLENE-2-PHENYL INDOLE (TRACE)	30.23	0.016	5.92927
1-METHYL-3,2-ETHYLENE-2-PHENYL INDOLE	32.79	0.013	9.62984
1-METHYL-3,2 TRIMETHYLENE-2-PHENYL INDOLE	24.41	0.031	1.64713
4,5-DIPHENYLIMIDAZOLE (A)	13.86	0.168	0.05526
PPF (TRACE)	31.67	0.014	7.84026
PPO (TRACE)	24.20	0.032	1.56611
2-(4-FLUOROPHENYL)-5-PHENYLOXAZOLE (TRACE)	22.64	0.039	1.04970
αNPO (TRACE)	36.41	0.009	17.98568
BBO (B)	42.86	0.006	47.80408
POPOP	43.26	0.006	50.49773
DIMETHYL POPOP (TRACE)	44.87	0.005	62.74108
PPD (TRACE)	0.00	0.000	0.00000
2,5-DI(2-METHYLPHENYL)OXADIAZOLE	0.00	0.000	0.00000
PBD	21.26	0.047	0.72130
BBD (B)	33.48	0.012	10.94136
αNPD (TRACE)	29.56	0.017	5.15691
βNPD	21.56	0.045	0.78291
αNND (TRACE)	33.67	0.012	11.28818
βNND (TRACE)	30.03	0.017	5.70957
TRANS-1-(4-BIPHENYLYL)-2-PHENYL-ETHYLENE	33.81	0.012	11.58092
TRANS-1-(4-BIPHENYLYL)-2-(1-NAPHTHYL)-ETHYLENE	38.96	0.008	26.94618
DIPHENYLSTILBENE (B)	43.20	0.006	50.06757
1,1,4,4-TETRAPHENYLBUTADIENE	41.11	0.006	37.13176
1,6-DIPHENYLHEXATRIENE	46.35	0.004	76.30714
1,8-DIPHENYL-1,3,5,7-OCTATETRAENE	48.17	0.004	95.52750
BPSB (TRACE)	46.85	0.004	81.36459
BIS-MSB	44.79	0.005	62.14241
ESCULIN (A)	32.96	0.012	9.90061
AZULENE	22.95	0.037	1.11121
NAPHTHALENE	0.00	0.000	0.00000
DEUTERATED NAPHTHALENE-C10D8	6.81	1.417	0.00078
1-METHYL NAPHTHALENE 5ML/L	3.89	7.603	0.00003
2-METHYL NAPHTHALENE	4.54	4.787	0.00007
2,3-DIMETHYL NAPHTHALENE	4.61	4.558	0.00008
2,6-DIMETHYL NAPHTHALENE (TRACE)	8.62	0.697	0.00322
1-NAPHTHOL (TRACE)	8.31	0.779	0.00258
2-NAPHTHOL (TRACE)	14.22	0.156	0.06452
1-NAPHTHYLAMINE	21.58	0.045	0.78493
2-CHLORONAPHTHALENE	5.78	2.318	0.00029
ACENAPHTHENE	5.64	2.494	0.00025
1-PHENYL NAPHTHALENE	7.79	0.945	0.00175
2-PHENYL NAPHTHALENE	15.97	0.110	0.12967
1,4-DIPHENYL NAPHTHALENE	17.61	0.082	0.23245
1,5-DIPHENYL NAPHTHALENE	15.16	0.128	0.09475
1,7-DIPHENYL NAPHTHALENE	15.53	0.119	0.10945

ACCEPTOR MOLECULES	RO	CO	OVERLAP
1,4,5,8-TETRAPHENYL NAPHTHALENE	35.32	0.010	14.94778
1,1'-BINAPHTHYL	10.83	0.352	0.01260
4,4'-DI(N-BUTOXY)-1,1'-BINAPHTHYL	24.20	0.032	1.56314
2,2'-BINAPHTHYL	16.99	0.091	0.18768
ACRIDINE (A)	32.57	0.013	9.19461
ANTHRACENE (TRACE)	30.84	0.015	6.63845
DEUTERATED ANTHRACENE C14D10 (TRACE)	30.36	0.016	6.03849
9-METHYL ANTHRACENE (TRACE)	31.16	0.015	7.05090
1-AMINO ANTHRACENE	30.17	0.016	5.75512
2-AMINO ANTHRACENE	29.13	0.018	4.65875
9-VINYL ANTHRACENE	32.48	0.013	9.02531
9-PHENYL ANTHRACENE (TRACE)	32.81	0.013	9.59845
9,10-DIPHENYL ANTHRACENE	34.20	0.011	12.27672
9,10-DICHLOROANTHRACENE (TRACE)	32.99	0.012	9.88518
BIANTHRANYL (B)	35.81	0.010	16.17622
9,10-DINAPHTHYLANTHRACENE (B)	33.66	0.012	11.13569
9-(P-VINYLPHENYL)-10-PHENYL ANTHRACENE	34.42	0.011	12.73800
9,10-BIS(PHENYL-ETHYNYL)ANTHRACENE (TRACE)	32.82	0.013	9.50097
RUBICENE (B)	32.61	0.013	9.14005
DIBENZ(A,H)ANTHRACENE (B)	31.64	0.014	7.71286
TETRACENE (TRACE) (B)	25.24	0.028	1.96702
RUBRENE (B)	25.36	0.027	2.02071
TETRAHYDROCANNABINOL ACID METHYL ESTER	0.00	0.000	0.00000
PHENANTHRENE	13.91	0.166	0.05645
3,4-BENZOPHENANTHRENE	18.91	0.066	0.35206
2-PHENYL PHENANTHRENE	16.38	0.102	0.14963
CHRYSENE (TRACE)	19.20	0.063	0.38773
TRIPHENYLENE	14.08	0.160	0.06061
DEUTERATED TRIPHENYLENE-C18D12	13.64	0.176	0.05001
PYRENE	29.49	0.017	5.07403
1,3,6,8,-TETRAPHENYL PYRENE	42.43	0.006	44.67540
BIPYRENYL (B)	42.08	0.006	42.64538
O-PHENYLENE-PYRENE	37.08	0.009	19.76951
ANTHANTHRENE (B)	34.60	0.011	13.07773
FLUORANTHENE	31.35	0.015	7.28914
3-PHENYL FLUORANTHENE	35.63	0.010	15.65533
3-AMINO FLUORANTHENE	32.80	0.013	9.46062
3,3'-BIFLUORANTHENYL (B)	40.37	0.007	33.07875
BENZO(GHI)FLUORANTHENE	29.09	0.018	4.64102
9,12-DI(N-OCTADECOXY)-BENZ(E)ACEPHENANTHRYLENE	35.39	0.010	15.02540
DECACYCLENE (B)	43.64	0.005	52.50864
10-ETHOXY-BENZO(KL)XANTHENE	34.88	0.011	13.81347
PERYLENE	33.53	0.012	10.82775
BENZO(GHI)PERYLENE (B)	36.72	0.009	18.82800
BBOT (TRACE)	46.08	0.005	73.54897
QUININE SULFATE (A)	26.14	0.025	2.48390
ACRIDINE YELLOW (A)	29.78	0.017	5.30029
ACRIDINE ORANGE (A)	25.13	0.028	1.91545
ACRIDINE RED (A)	27.43	0.022	3.23966
URANINE (A)	27.11	0.022	3.01640
RHODAMINE B (A)	31.31	0.015	7.15934
RHODAMINE 6G (A)	31.03	0.015	6.78944
EOSIN B (A)	27.22	0.022	3.09135
EOSIN Y (A)	28.04	0.020	3.69174
ROSE BENGAL (A)	28.79	0.019	4.33159
INDENOINDENE	23.36	0.035	1.26663

8.80 Donor Molecule: 2,2′-Methylene-*p*-terphenyl

N,N-DIMETHYLANILINE 3ML/L	9.63	0.501	0.00671
TRIPHENYLAMINE	25.32	0.028	2.19288
SODIUM SALICYLATE (A)	16.94	0.092	0.19698
ANTHRANILIC ACID (A)	28.92	0.018	4.84202
TMPD	24.66	0.030	1.86570
4,4'-DIHYDROXYBIPHENYL (A)	9.20	0.574	0.00511
4,4'-DIMETHOXY BIPHENYL	6.21	1.866	0.00048
4,4'-DIISOBUTOXYBIPHENYL	0.00	0.000	0.00000
BENZIDINE	21.93	0.042	0.92897
4-VINYL BIPHENYL	8.51	0.727	0.00319
M-TERPHENYL	5.43	2.792	0.00022
1,3,5-TRIPHENYL BENZENE	9.00	0.614	0.00446
1,3-DIBIPHENYLYLBENZENE (B)	20.19	0.054	0.56680
1,3,5,-TRIBIPHENYLYLBENZENE (B)	24.08	0.032	1.63161
FLUORENE (TRACE)	0.00	0.000	0.00000
1-METHYL FLUORENE	0.00	0.000	0.00000
1,2-BENZOFLUORENE	23.09	0.036	1.26232
9,10-DIHYDROPHENANTHRENE	4.26	5.782	0.00005
CARBAZOLE (TRACE)	22.77	0.038	1.16356
N-METHYL CARBAZOLE (TRACE)	26.24	0.025	2.71962
N-VINYLCARBAZOLE	26.12	0.025	2.64516
N-PHENYL CARBAZOLE	25.83	0.026	2.47594
N-METHYL-2,3-BENZCARBAZOLE	28.84	0.019	4.72591
DIBENZOFURAN	0.00	0.000	0.00000
DIBENZOTHIOPHENE (TRACE)	20.27	0.054	0.58066
P-TERPHENYL	11.27	0.313	0.01720
DEUTERATED P-TERPHENYL-C18D14	9.92	0.459	0.00801
4-METHYL TERPHENYL	13.24	0.193	0.04521
2-METHYL-P-TERPHENYL	0.00	0.000	0.00000
3,3''-DIMETHYL-P-TERPHENYL	13.61	0.177	0.05349
4-(3,3-DIMETHYLBUTOXY)-P-TERPHENYL	20.27	0.054	0.58140

ACCEPTOR MOLECULES	RO	CO	OVERLAP
4,4'-DIHEXAHYDROFARNESOXY-P-TERPHENYL	24.02	0.032	1.60887
2,2' -METHYLENE-P-TERPHENYL	21.83	0.043	0.90639
2,2' -ETHYLENE-P-TERPHENYL	23.39	0.035	1.36993
2,2'-METHYLENE-5',6'-METHYLENE-P-TERPHENYL (TRACE)	35.48	0.010	16.65866
3-PHENYL DIBENZOFURAN	21.83	0.043	0.90776
BENZO(1,2-B-4,5-B')BISBENZOFURAN (TRACE) (B)	37.32	0.009	22.51738
SEXIPHENYL BENZENE	0.00	0.000	0.00000
P-QUATERPHENYL	25.72	0.026	2.42037
3,3'''-DIMETHYL-P-QUATERPHENYL	27.43	0.022	3.55630
TETRAMETHYL-P-QUATERPHENYL	5.75	2.347	0.00031
3,3''' -DIMETHYL-3',2''-METHYLENE-P-QUATERPHENYL	36.78	0.009	20.60829
3,3''' -DIMETHYL-3',2''-ETHYLENE-P-QUATERPHENYL	36.25	0.009	18.86449
4-(3-ETHYLBUTOXY)-P-QUATERPHENYL (B)	33.15	0.012	11.04191
4,4'''-DI(2-BUTYLOCTOXY-1)-P-QUATERPHENYL	33.88	0.012	12.58067
DIETHYL-P-QUINQUEPHENYL	28.64	0.019	4.60573
DI(3-ETHYLHEPTYL)-P-QUINQUEPHENYL	35.40	0.010	16.33779
TETRAMETHYL-P-SEXIPHENYL	27.10	0.022	3.30960
TETRAMETHYL-P-OCTAPHENYL (B)	40.25	0.007	35.23605
INDOLE (TRACE)	0.00	0.000	0.00000
2-PHENYL INDOLE	28.73	0.019	4.70407
1-METHYL-2-PHENYL INDOLE	22.75	0.038	1.15662
1-NEOPENTYL-2-PHENYL INDOLE	19.47	0.061	0.45685
1-N-BUTYL-2(BIPHENYLYL-4) INDOLE	32.94	0.013	10.60402
1,2-DIPHENYL INDOLE	23.55	0.034	1.42449
1-BIPHENYLYL-2-PHENYL INDOLE	26.53	0.024	2.90463
1,3-DIMETHYL-2-PHENYL INDOLE	24.18	0.032	1.66928
2,3-DIPHENYL INDOLE	30.42	0.016	6.56867
1-METHYL-2,3-DIPHENYL INDOLE	27.60	0.021	3.66752
2-PHENYL-3,2'-METHYLENE INDOLE (TRACE)	32.44	0.013	9.71180
2-PHENYL-3,2'-ETHYLENE INDOLE (TRACE)	35.36	0.010	16.25600
2-PHENYL-3,2' -TRIMETHYLENE INDOLE	30.94	0.015	7.31416
1-METHYL-3,2'-METHYLENE-2-PHENYL INDOLE (TRACE)	34.86	0.011	14.94181
1-METHYL-3,2'-ETHYLENE-2-PHENYL INDOLE	35.63	0.010	16.96629
1-METHYL-3,2'-TRIMETHYLENE-2-PHENYL INDOLE	29.75	0.017	5.77514
4,5-DIPHENYLIMIDAZOLE (A)	18.15	0.075	0.29905
PPF (TRACE)	36.56	0.009	19.87704
PPO (TRACE)	31.08	0.015	7.52133
2-(4-FLUOROPHENYL)-5-PHENYLOXAZOLE (TRACE)	29.58	0.017	5.59535
αNPO (TRACE)	38.13	0.008	25.39431
BBO (B)	43.64	0.005	57.02693
POPOP	42.39	0.006	47.84711
DIMETHYL POPOP (TRACE)	43.29	0.006	54.10414
PPD (TRACE)	9.73	0.486	0.00714
2,5-DI(2-METHYLPHENYL)OXADIAZOLE	12.18	0.247	0.02750
PBD	29.54	0.017	5.55402
BBD (B)	38.66	0.008	27.75487
αNPD (TRACE)	33.56	0.012	11.84702
βNPD	28.66	0.019	4.62651
αNND (TRACE)	36.06	0.010	18.23443
βNND (TRACE)	35.24	0.010	15.95806
TRANS-1-(4-BIPHENYLYL)-2-PHENYL-ETHYLENE	38.32	0.008	26.28104
TRANS-1-(4-BIPHENYLYL)-2-(1-NAPHTHYL)-ETHYLENE	40.33	0.007	35.47634
DIPHENYLSTILBENE (B)	44.10	0.005	60.63099
1,1,4,4-TETRAPHENYLBUTADIENE	41.34	0.006	41.05930
1,6-DIPHENYLHEXATRIENE	45.57	0.005	73.71391
1,8-DIPHENYL-1,3,5,7-OCTATETRAENE	44.60	0.005	64.40022
BPSB (TRACE)	45.82	0.005	76.22699
BIS-MSB	44.59	0.005	64.67964
ESCULIN (A)	34.30	0.011	13.46113
AZULENE	25.78	0.026	2.40050
NAPHTHALENE	6.21	1.870	0.00048
DEUTERATED NAPHTHALENE-C10D8	10.11	0.432	0.00896
1-METHYLNAPHTHALENE 5ML/L	8.43	0.746	0.00302
2-METHYL NAPHTHALENE	10.61	0.374	0.01199
2,3-DIMETHYL NAPHTHALENE	10.53	0.383	0.01145
2,6-DIMETHYL NAPHTHALENE (TRACE)	15.68	0.116	0.12486
1-NAPHTHOL (TRACE)	16.13	0.107	0.14752
2-NAPHTHOL (TRACE)	20.34	0.053	0.59171
1-NAPHTHYLAMINE	25.82	0.026	2.46571
2-CHLORONAPHTHALENE	11.83	0.270	0.02297
ACENAPHTHENE	13.49	0.182	0.05057
1-PHENYL NAPHTHALENE	13.95	0.165	0.06191
2-PHENYL NAPHTHALENE	22.64	0.039	1.12656
1,4-DIPHENYL NAPHTHALENE	23.87	0.033	1.54495
1,5-DIPHENYL NAPHTHALENE	21.90	0.043	0.92199
1,7-DIPHENYL NAPHTHALENE	21.63	0.044	0.85496
1,4,5,8-TETRAPHENYL NAPHTHALENE	36.73	0.009	20.22320
1,1'-BINAPHTHYL	17.30	0.086	0.22490
4,4' -DI(N-BUTOXY)-1,1' -BINAPHTHYL	29.68	0.017	5.69133
2,2'-BINAPHTHYL	23.56	0.034	1.42778
ACRIDINE (A)	31.91	0.014	8.69901
ANTHRACENE (TRACE)	29.84	0.017	5.82205
DEUTERATED ANTHRACENE C14D10 (TRACE)	29.52	0.017	5.46129
9-METHYL ANTHRACENE (TRACE)	29.54	0.017	5.46930
1-AMINO ANTHRACENE	27.52	0.021	3.55066
2-AMINO ANTHRACENE	27.78	0.021	3.75994
9-VINYL ANTHRACENE	30.43	0.016	6.53128
9-PHENYL ANTHRACENE (TRACE)	31.08	0.015	7.42160
9,10-DIPHENYL ANTHRACENE	31.31	0.015	7.73395
9,10-DICHLOROANTHRACENE (TRACE)	29.87	0.017	5.82374
BIANTHRANYL (B)	33.10	0.012	10.79419
9,10-DINAPHTHYLANTHRACENE (B)	30.76	0.015	6.94570
9-(P-VINYLPHENYL)-10-PHENYL ANTHRACENE	31.40	0.014	7.85607
9,10-BIS(PHENYL-ETHYNYL)ANTHRACENE (TRACE)	27.69	0.021	3.68623
RUBICENE (B)	31.37	0.014	7.80261
DIBENZ(A,H)ANTHRACENE (B)	33.97	0.011	12.63456

ACCEPTOR MOLECULES	RO	CO	OVERLAP
TETRACENE (TRACE) (B)	22.73	0.038	1.12927
RUBRENE (B)	29.32	0.018	5.20131
TETRAHYDROCANNABINOL ACID METHYL ESTER	9.15	0.584	0.00493
PHENANTHRENE	15.90	0.111	0.13436
3,4-BENZOPHENANTHRENE	23.45	0.035	1.37363
2-PHENYL PHENANTHRENE	17.95	0.077	0.27779
CHRYSENE (TRACE)	23.44	0.035	1.37262
TRIPHENYLENE	16.66	0.097	0.17804
DEUTERATED TRIPHENYLENE-C18D12	16.38	0.102	0.16110
PYRENE	34.04	0.011	12.85584
1,3,6,8,-TETRAPHENYL PYRENE	38.83	0.008	28.08125
BIPYRENYL (B)	42.02	0.006	45.21032
O-PHENYLENE-PYRENE	34.89	0.011	14.75989
ANTHANTHRENE (B)	32.10	0.014	8.95816
FLUORANTHENE	31.54	0.014	8.08408
3-PHENYL FLUORANTHENE	34.25	0.011	13.23465
3-AMINO FLUORANTHENE	31.38	0.014	7.80939
3,3'-BIFLUORANTHENYL (B)	37.99	0.008	24.58761
BENZO(GHI)FLUORANTHENE	30.39	0.016	6.44999
9,12-DI(N-OCTADECOXY)-BENZ(E)ACEPHENANTHRYLENE	33.73	0.012	12.04133
DECACYCLENE (B)	41.19	0.006	39.95042
10-ETHOXY-BENZO(KL)XANTHENE	31.95	0.014	8.73927
PERYLENE	28.59	0.019	4.47017
BENZO(GHI)PERYLENE (B)	34.11	0.011	12.94534
BBOT (TRACE)	43.51	0.005	55.76288
QUININE SULFATE (A)	30.09	0.016	6.17500
ACRIDINE YELLOW (A)	24.86	0.029	1.93291
ACRIDINE ORANGE (A)	22.30	0.040	1.00568
ACRIDINE RED (A)	26.86	0.023	3.07377
URANINE (A)	26.85	0.023	3.06653
RHODAMINE B (A)	30.47	0.016	6.55283
RHODAMINE 6G (A)	31.28	0.015	7.66424
EOSIN B (A)	28.30	0.020	4.20320
EOSIN Y (A)	28.79	0.019	4.65728
ROSE BENGAL (A)	30.28	0.016	6.30620
INDENOINDENE	30.64	0.016	6.90095

8.81 Donor Molecule: 2,2′-Ethylene-*p*-terphenyl

ACCEPTOR MOLECULES	RO	CO	OVERLAP
N,N-DIMETHYLANILINE 3ML/L	7.07	1.266	0.00113
TRIPHENYLAMINE	22.71	0.038	1.23607
SODIUM SALICYLATE (A)	15.49	0.120	0.12435
ANTHRANILIC ACID (A)	27.91	0.021	4.22584
TMPD	23.34	0.035	1.44950
4,4'-DIHYDROXYBIPHENYL (A)	6.85	1.394	0.00094
4,4'-DIMETHOXY BIPHENYL	0.00	0.000	0.00000
4,4'-DIISOBUTOXYBIPHENYL	0.00	0.000	0.00000
BENZIDINE	19.15	0.064	0.44499
4-VINYL BIPHENYL	5.05	3.475	0.00015
1,3,5-TRIPHENYL BENZENE	7.02	1.293	0.00109
1,3-DIBIPHENYLYLBENZENE (B)	17.14	0.089	0.22924
1,3,5,-TRIBIPHENYLYLBENZENE (B)	20.41	0.053	0.65428
FLUORENE (TRACE)	0.00	0.000	0.00000
1-METHYL FLUORENE	0.00	0.000	0.00000
1,2-BENZOFLUORENE	20.89	0.049	0.74914
9,10-DIHYDROPHENANTHRENE	0.00	0.000	0.00000
CARBAZOLE (TRACE)	20.57	0.051	0.68450
N-METHYL CARBAZOLE (TRACE)	24.52	0.030	1.95717
N-VINYLCARBAZOLE	24.19	0.032	1.80771
N-PHENYL CARBAZOLE	23.93	0.033	1.69296
N-METHYL-2,3-BENZCARBAZOLE	28.07	0.020	4.34843
DIBENZOFURAN	0.00	0.000	0.00000
DIBENZOTHIOPHENE (TRACE)	17.03	0.090	0.22100
P-TERPHENYL	8.25	0.796	0.00287
DEUTERATED P-TERPHENYL-C18D14	6.60	1.555	0.00075
4-METHYL TERPHENYL	9.91	0.459	0.00863
2-METHYL-P-TERPHENYL	0.00	0.000	0.00000
3,3''-DIMETHYL-P-TERPHENYL	10.46	0.391	0.01189
4-(3,3-DIMETHYLBUTOXY)-P-TERPHENYL	16.98	0.091	0.21711
4,4''-DIHEXAHYDROFARNESOXY-P-TERPHENYL	20.55	0.052	0.68090
2,2'-METHYLENE-P-TERPHENYL	17.68	0.081	0.27651
2,2'-ETHYLENE-P-TERPHENYL	19.50	0.060	0.49716
2,2'-METHYLENE-5',6''-METHYLENE-P-TERPHENYL (TRACE)	32.97	0.012	11.58698
3-PHENYL DIBENZOFURAN	17.64	0.081	0.27349
BENZO(1,2-B-4,5-B')BISBENZOFURAN (TRACE) (B)	34.18	0.011	14.36776
SEXIPHENYL BENZENE	0.00	0.000	0.00000
P-QUATERPHENYL	22.39	0.040	1.13938
3,3'''-DIMETHYL-P-QUATERPHENYL	24.17	0.032	1.79999
TETRAMETHYL-P-QUATERPHENYL	0.00	0.000	0.00000
3,3'''-DIMETHYL-3',2''-METHYLENE-P-QUATERPHENYL	33.61	0.012	12.99115
3,3'''-DIMETHYL-3',2''-ETHYLENE-P-QUATERPHENYL	33.29	0.012	12.22319
4-(3-ETHYLBUTOXY)-P-QUATERPHENYL (B)	30.14	0.016	6.74603
4,4'''-DI(2-BUTYLOCTOXY-1)-P-QUATERPHENYL	30.73	0.015	7.57166
DIETHYL-P-QUINQUEPHENYL	25.21	0.028	2.31803
DI(3-ETHYLHEPTYL)-P-QUINQUEPHENYL	32.26	0.013	10.11482
TETRAMETHYL-P-SEXIPHENYL	23.54	0.034	1.53676
TETRAMETHYL-P-OCTAPHENYL (B)	37.38	0.009	24.48146
INDOLE (TRACE)	0.00	0.000	0.00000
2-PHENYL INDOLE	25.37	0.027	2.40898
1-METHYL-2-PHENYL INDOLE	19.66	0.059	0.52179

ACCEPTOR MOLECULES	RO	CO	OVERLAP
1-NEOPENTYL-2-PHENYL INDOLE	16.18	0.106	0.16271
1-N-BUTYL-2(BIPHENYLYL-4) INDOLE	30.53	0.016	7.27235
1,2-DIPHENYL INDOLE	20.27	0.054	0.62702
1-BIPHENYLYL-2-PHENYL INDOLE	23.38	0.035	1.46812
1,3-DIMETHYL-2-PHENYL INDOLE	21.38	0.046	0.86321
2,3-DIPHENYL INDOLE	28.15	0.020	4.45995
1-METHYL-2,3-DIPHENYL INDOLE	25.45	0.027	2.44034
2-PHENYL-3,2'-METHYLENE INDOLE (TRACE)	29.52	0.017	5.97097
2-PHENYL-3,2'-ETHYLENE INDOLE (TRACE)	33.12	0.012	11.85259
2-PHENYL-3,2'-TRIMETHYLENE INDOLE	27.86	0.021	4.21968
1-METHYL-3,2'-METHYLENE-2-PHENYL INDOLE (TRACE)	32.34	0.013	10.28647
1-METHYL-3,2'-ETHYLENE-2-PHENYL INDOLE	33.94	0.011	13.70855
1-METHYL-3,2'-TRIMETHYLENE-2-PHENYL INDOLE	27.09	0.022	3.55436
4,5-DIPHENYLIMIDAZOLE (A)	15.98	0.110	0.15062
PPF (TRACE)	33.84	0.012	13.52345
PPO (TRACE)	28.00	0.020	4.34818
2-(4-FLUOROPHENYL)-5-PHENYLOXAZOLE (TRACE)	26.54	0.024	3.15614
αNPO (TRACE)	36.85	0.009	22.37425
BBO (B)	42.71	0.006	54.19669
POPOP	42.18	0.006	50.19119
DIMETHYL POPOP (TRACE)	43.41	0.005	59.53208
PPD (TRACE)	5.08	3.417	0.00016
2,5-DI(2-METHYLPHENYL)OXADIAZOLE	8.41	0.752	0.00322
PBD	25.86	0.026	2.69976
BBD (B)	35.90	0.010	19.26344
αNPD (TRACE)	31.27	0.015	8.38559
βNPD	25.13	0.028	2.27554
αNND (TRACE)	34.50	0.011	15.12151
βNND (TRACE)	32.30	0.013	10.22832
TRANS-1-(4-BIPHENYLYL)-2-PHENYL-ETHYLENE	35.84	0.010	19.02590
TRANS-1-(4-BIPHENYLYL)-2-(1-NAPHTHYL)-ETHYLENE	39.15	0.007	32.11405
DIPHENYLSTILBENE (B)	43.11	0.006	57.28365
1,1,4,4-TETRAPHENYLBUTADIENE	40.68	0.007	40.36496
1,6-DIPHENYLHEXATRIENE	45.37	0.005	77.73041
1,8-DIPHENYL-1,3,5,7-OCTATETRAENE	45.66	0.005	80.23212
BPSB (TRACE)	45.65	0.005	80.60424
BIS-MSB	44.06	0.005	65.17742
ESCULIN (A)	33.24	0.012	12.05590
AZULENE	24.27	0.031	1.80326
NAPHTHALENE	4.53	4.797	0.00008
DEUTERATED NAPHTHALENE-C10D8	8.63	0.696	0.00374
1-METHYLNAPHTHALENE 5ML/L	6.44	1.678	0.00064
2-METHYL NAPHTHALENE	8.26	0.792	0.00289
2,3-DIMETHYL NAPHTHALENE	8.19	0.815	0.00273
2,6-DIMETHYL NAPHTHALENE (TRACE)	12.69	0.219	0.03780
1-NAPHTHOL (TRACE)	12.87	0.210	0.04129
2-NAPHTHOL (TRACE)	17.89	0.078	0.29632
1-NAPHTHYLAMINE	23.75	0.033	1.61467
2-CHLORONAPHTHALENE	9.39	0.540	0.00623
ACENAPHTHENE	10.49	0.388	0.01209
1-PHENYL NAPHTHALENE	11.25	0.314	0.01836
2-PHENYL NAPHTHALENE	19.66	0.059	0.52305
1,4-DIPHENYL NAPHTHALENE	20.89	0.049	0.75004
1,5-DIPHENYL NAPHTHALENE	18.79	0.067	0.39795
1,7-DIPHENYL NAPHTHALENE	18.75	0.068	0.39217
1,4,5,8-TETRAPHENYL NAPHTHALENE	35.60	0.010	18.11689
1,1'-BINAPHTHYL	14.51	0.147	0.08444
4,4'-DI(N-BUTOXY)-1,1'-BINAPHTHYL	27.01	0.023	3.49176
2,2'-BINAPHTHYL	20.40	0.053	0.65130
ACRIDINE (A)	31.78	0.014	9.18176
ANTHRACENE (TRACE)	29.82	0.017	6.26931
DEUTERATED ANTHRACENE C14D10 (TRACE)	29.54	0.017	5.93397
9-METHYL ANTHRACENE (TRACE)	29.92	0.017	6.38572
1-AMINO ANTHRACENE	28.35	0.020	4.58941
2-AMINO ANTHRACENE	28.00	0.020	4.25603
9-VINYL ANTHRACENE	31.01	0.015	7.89853
9-PHENYL ANTHRACENE (TRACE)	31.42	0.014	8.55864
9,10-DIPHENYL ANTHRACENE	32.28	0.013	10.04661
9,10-DICHLOROANTHRACENE (TRACE)	30.79	0.015	7.55600
BIANTHRANYL (B)	33.99	0.011	13.69830
9,10-DINAPHTHYLANTHRACENE (B)	31.60	0.014	8.82900
9-(P-VINYLPHENYL)-10-PHENYL ANTHRACENE	32.40	0.013	10.25264
9,10-BIS(PHENYL-ETHYNYL)ANTHRACENE (TRACE)	29.45	0.018	5.75780
RUBICENE (B)	31.18	0.015	8.11272
DIBENZ(A,H)ANTHRACENE (B)	32.53	0.013	10.54488
TETRACENE (TRACE) (B)	23.39	0.035	1.44778
RUBRENE (B)	26.91	0.023	3.35501
TETRAHYDROCANNABINOL ACID METHYL ESTER	6.53	1.606	0.00071
PHENANTHRENE	14.83	0.137	0.09563
3,4-BENZOPHENANTHRENE	20.83	0.050	0.72852
2-PHENYL PHENANTHRENE	16.93	0.092	0.21118
CHRYSENE (TRACE)	20.65	0.051	0.69667
TRIPHENYLENE	15.37	0.123	0.11881
DEUTERATED TRIPHENYLENE-C18D12	15.04	0.131	0.10430
PYRENE	32.14	0.013	9.84515
1,3,6,8,-TETRAPHENYL PYRENE	39.98	0.007	36.20835
BIPYRENYL (B)	41.54	0.006	45.66579
O-PHENYLENE-PYRENE	35.33	0.010	17.17739
ANTHANTHRENE (B)	32.70	0.013	10.80084
FLUORANTHENE	30.93	0.015	7.77860
3-PHENYL FLUORANTHENE	34.42	0.011	14.73670
3-AMINO FLUORANTHENE	31.42	0.014	8.50172
3,3'-BIFLUORANTHENYL (B)	38.52	0.008	28.88415
BENZO(GHI)FLUORANTHENE	29.59	0.017	5.94415
9,12-DI(N-OCTADECOXY)-BENZ(E)ACEPHENANTHRYLENE	33.94	0.011	13.51771
DECACYCLENE (B)	41.59	0.006	45.73235

ACCEPTOR MOLECULES	RO	CO	OVERLAP
10-ETHOXY-BENZO(KL)XANTHENE	32.87	0.013	11.19698
PERYLENE	30.49	0.016	7.09836
BENZO(GHI)PERYLENE (B)	34.84	0.011	15.87457
BBOT (TRACE)	44.13	0.005	65.61223
QUININE SULFATE (A)	28.03	0.020	4.36593
ACRIDINE YELLOW (A)	26.61	0.024	3.13900
ACRIDINE ORANGE (A)	22.97	0.037	1.29887
ACRIDINE RED (A)	26.76	0.023	3.24234
URANINE (A)	26.29	0.025	2.91674
RHODAMINE B (A)	30.46	0.016	7.06130
RHODAMINE 6G (A)	30.92	0.015	7.71239
EOSIN B (A)	27.29	0.022	3.64894
EOSIN Y (A)	27.88	0.021	4.14510
ROSE BENGAL (A)	29.14	0.018	5.40653
INDENOINDENE	27.55	0.021	3.94565

8.82 Donor Molecule: 2,2′-Methylene-5′,6″-methylene-*p*-terphenyl (trace)

ACCEPTOR MOLECULES	RO	CO	OVERLAP
N,N-DIMETHYLANILINE 3ML/L	2.66	23.878	0.00000
TRIPHENYLAMINE	22.31	0.040	1.03484
SODIUM SALICYLATE (A)	15.54	0.119	0.11776
ANTHRANILIC ACID (A)	28.78	0.019	4.72709
TMPD	23.96	0.033	1.57735
BENZIDINE	18.05	0.076	0.29102
1,3,5-TRIPHENYL BENZENE	3.47	10.739	0.00001
1,3-DIBIPHENYLYLBENZENE (B)	15.60	0.118	0.12153
1,3,5,-TRIBIPHENYLYLBENZENE (B)	17.96	0.077	0.28309
FLUORENE (TRACE)	0.00	0.000	0.00000
1-METHYL FLUORENE	0.00	0.000	0.00000
1,2-BENZOFLUORENE	20.78	0.050	0.67426
9,10-DIHYDROPHENANTHRENE	0.00	0.000	0.00000
CARBAZOLE (TRACE)	19.39	0.061	0.44685
N-METHYL CARBAZOLE (TRACE)	25.15	0.028	2.12400
N-VINYLCARBAZOLE	25.12	0.028	2.10737
N-PHENYL CARBAZOLE	24.72	0.030	1.91303
N-METHYL-2,3-BENZCARBAZOLE	27.10	0.022	3.28010
DIBENZOFURAN	0.00	0.000	0.00000
DIBENZOTHIOPHENE (TRACE)	13.32	0.189	0.04695
P-TERPHENYL	0.00	0.000	0.00000
DEUTERATED P-TERPHENYL-C18D14	0.00	0.000	0.00000
4-METHYL TERPHENYL	0.00	0.000	0.00000
2-METHYL-P-TERPHENYL	0.00	0.000	0.00000
3,3″-DIMETHYL-P-TERPHENYL	4.38	5.311	0.00006
4-(3,3-DIMETHYLBUTOXY)-P-TERPHENYL	13.88	0.167	0.06034
4,4″-DIHEXAHYDROFARNESOXY-P-TERPHENYL	18.15	0.075	0.30118
2,2′-METHYLENE-P-TERPHENYL	12.16	0.249	0.02727
2,2′-ETHYLENE-P-TERPHENYL	15.83	0.113	0.13260
2,2′-METHYLENE-5′,6″-METHYLENE-P-TERPHENYL (TRACE)	33.71	0.012	12.32433
3-PHENYL DIBENZOFURAN	11.91	0.265	0.02406
BENZO(1,2-B-4,5-B′)BISBENZOFURAN (TRACE) (B)	34.09	0.011	13.17123
SEXIPHENYL BENZENE	0.00	0.000	0.00000
P-QUATERPHENYL	20.87	0.049	0.69590
3,3‴-DIMETHYL-P-QUATERPHENYL	23.05	0.037	1.26057
TETRAMETHYL-P-QUATERPHENYL	0.00	0.000	0.00000
3,3‴-DIMETHYL-3′,2″-METHYLENE-P-QUATERPHENYL	33.69	0.012	12.25819
3,3‴-DIMETHYL-3′,2″-ETHYLENE-P-QUATERPHENYL	33.50	0.012	11.82416
4-(3-ETHYLBUTOXY)-P-QUATERPHENYL (B)	30.01	0.017	6.11131
4,4‴-DI(2-BUTYLOCTOXY-1)-P-QUATERPHENYL	30.50	0.016	6.73743
DIETHYL-P-QUINQUEPHENYL	24.00	0.032	1.60507
DI(3-ETHYLHEPTYL)-P-QUINQUEPHENYL	32.16	0.013	9.24262
TETRAMETHYL-P-SEXIPHENYL	21.78	0.043	0.89897
TETRAMETHYL-P-OCTAPHENYL (B)	37.88	0.008	24.67497
INDOLE (TRACE)	0.00	0.000	0.00000
2-PHENYL INDOLE	23.86	0.033	1.55112
1-METHYL-2-PHENYL INDOLE	17.95	0.077	0.28071
1-NEOPENTYL-2-PHENYL INDOLE	12.96	0.205	0.04003
1-N-BUTYL-2(BIPHENYLYL-4) INDOLE	30.92	0.015	7.29756
1,2-DIPHENYL INDOLE	18.37	0.072	0.32316
1-BIPHENYLYL-2-PHENYL INDOLE	22.22	0.041	1.00683
1,3-DIMETHYL-2-PHENYL INDOLE	20.39	0.053	0.60355
2,3-DIPHENYL INDOLE	28.45	0.019	4.42604
1-METHYL-2,3-DIPHENYL INDOLE	25.67	0.026	2.39064
2-PHENYL-3,2′-METHYLENE INDOLE (TRACE)	29.91	0.017	6.01317
2-PHENYL-3,2′-ETHYLENE INDOLE (TRACE)	33.70	0.012	12.25912
2-PHENYL-3,2′-TRIMETHYLENE INDOLE	27.31	0.022	3.48196
1-METHYL-3,2′-METHYLENE-2-PHENYL INDOLE (TRACE)	32.84	0.013	10.49735
1-METHYL-3,2′-ETHYLENE-2-PHENYL INDOLE	35.01	0.010	15.37204
1-METHYL-3,2′-TRIMETHYLENE-2-PHENYL INDOLE	27.30	0.022	3.47448
4,5-DIPHENYLIMIDAZOLE (A)	15.47	0.121	0.11534
PPF (TRACE)	34.43	0.011	13.95179
PPO (TRACE)	27.69	0.021	3.78504
2-(4-FLUOROPHENYL)-5-PHENYLOXAZOLE (TRACE)	25.90	0.026	2.53544
αNPO (TRACE)	37.91	0.008	24.69052
BBO (B)	43.92	0.005	59.62505
POPOP	43.25	0.006	64.33629
DIMETHYL POPOP (TRACE)	44.25	0.005	62.20213
PPD (TRACE)	0.00	0.000	0.00000

ACCEPTOR MOLECULES	RO	CO	OVERLAP
PBD	23.92	0.033	1.57570
BBD (B)	36.44	0.009	19.60802
αNPD (TRACE)	31.90	0.014	8.79573
βNPD	23.44	0.035	1.39454
αNND (TRACE)	35.59	0.010	16.95252
βNND (TRACE)	32.75	0.013	10.34377
TRANS-1-(4-BIPHENYLYL)-2-PHENYL-ETHYLENE	36.54	0.009	19.88968
TRANS-1-(4-BIPHENYLYL)-2-(1-NAPHTHYL)-ETHYLENE	40.22	0.007	35.16742
DIPHENYLSTILBENE (B)	44.35	0.005	63.21767
1,1,4,4-TETRAPHENYLBUTADIENE	41.79	0.006	44.14304
1,6-DIPHENYLHEXATRIENE	46.62	0.004	85.15759
1,8-DIPHENYL-1,3,5,7-OCTATETRAENE	46.07	0.005	78.76167
BPSB (TRACE)	46.75	0.004	86.61355
BIS-MSB	45.19	0.005	70.59264
ESCULIN (A)	34.22	0.011	13.36292
AZULENE	25.25	0.028	2.12732
NAPHTHALENE	0.00	0.000	0.00000
DEUTERATED NAPHTHALENE-C10D8	7.66	0.996	0.00170
1-METHYLNAPHTHALENE 5ML/L	3.01	16.412	0.00001
2-METHYL NAPHTHALENE	3.51	10.365	0.00002
2,3-DIMETHYL NAPHTHALENE	3.60	9.617	0.00002
2,6-DIMETHYL NAPHTHALENE (TRACE)	7.05	1.274	0.00104
1-NAPHTHOL (TRACE)	7.89	0.911	0.00203
2-NAPHTHOL (TRACE)	15.59	0.118	0.12094
1-NAPHTHYLAMINE	23.96	0.032	1.58707
2-CHLORONAPHTHALENE	5.00	3.582	0.00013
ACENAPHTHENE	4.29	5.659	0.00005
1-PHENYL NAPHTHALENE	7.13	1.235	0.00111
2-PHENYL NAPHTHALENE	18.20	0.074	0.30597
1,4-DIPHENYL NAPHTHALENE	19.56	0.060	0.47006
1,5-DIPHENYL NAPHTHALENE	16.83	0.094	0.19124
1,7-DIPHENYL NAPHTHALENE	17.31	0.086	0.22649
1,4,5,8-TETRAPHENYL NAPHTHALENE	36.61	0.009	19.95136
1,1'-BINAPHTHYL	11.47	0.297	0.01918
4,4'-DI(N-BUTOXY)-1,1'-BINAPHTHYL	26.70	0.024	3.03269
2,2'-BINAPHTHYL	19.06	0.065	0.40381
ACRIDINE (A)	32.70	0.013	10.13938
ANTHRACENE (TRACE)	30.74	0.015	7.01403
DEUTERATED ANTHRACENE C14D10 (TRACE)	30.46	0.016	6.64347
9-METHYL ANTHRACENE (TRACE)	30.15	0.016	6.22438
1-AMINO ANTHRACENE	28.32	0.020	4.24318
2-AMINO ANTHRACENE	28.36	0.020	4.27375
9-VINYL ANTHRACENE	31.28	0.015	7.74686
9-PHENYL ANTHRACENE (TRACE)	31.74	0.014	8.47106
9,10-DIPHENYL ANTHRACENE	32.49	0.013	9.72574
9,10-DICHLOROANTHRACENE (TRACE)	30.87	0.015	7.15716
BIANTHRANYL (B)	34.10	0.011	12.99744
9,10-DINAPHTHYLANTHRACENE (B)	31.74	0.014	8.43678
9-(P-VINYLPHENYL)-10-PHENYL ANTHRACENE	32.56	0.013	9.83420
9,10-BIS(PHENYL-ETHYNYL)ANTHRACENE (TRACE)	27.94	0.021	3.89767
RUBICENE (B)	31.17	0.015	7.52608
DIBENZ(A,H)ANTHRACENE (B)	33.63	0.012	11.97848
TETRACENE (TRACE) (B)	22.91	0.037	1.18473
RUBRENE (B)	26.54	0.024	2.86836
TETRAHYDROCANNABINOL ACID METHYL ESTER	0.00	0.000	0.00000
PHENANTHRENE	15.12	0.130	0.09995
3,4-BENZOPHENANTHRENE	19.51	0.060	0.45886
2-PHENYL PHENANTHRENE	17.37	0.085	0.22970
CHRYSENE (TRACE)	19.87	0.057	0.51243
TRIPHENYLENE	15.85	0.112	0.13308
DEUTERATED TRIPHENYLENE-C18D12	15.29	0.125	0.10721
PYRENE	34.10	0.011	13.07009
1,3,6,8,-TETRAPHENYL PYRENE	40.19	0.007	34.75975
BIPYRENYL (B)	42.46	0.006	48.42828
O-PHENYLENE-PYRENE	35.79	0.010	17.24612
ANTHANTHRENE (B)	32.41	0.013	9.51428
FLUORANTHENE	31.81	0.014	8.56287
3-PHENYL FLUORANTHENE	35.03	0.010	15.25463
3-AMINO FLUORANTHENE	31.38	0.014	7.83609
3,3'-BIFLUORANTHENYL (B)	38.83	0.008	28.22783
BENZO(GHI)FLUORANTHENE	30.36	0.016	6.46536
9,12-DI(N-OCTADECOXY)-BENZ(E)ACEPHENANTHRYLENE	34.29	0.011	13.38490
DECACYCLENE (B)	41.97	0.006	44.81775
10-ETHOXY-BENZO(KL)XANTHENE	33.04	0.012	10.76543
PERYLENE	29.18	0.018	5.07067
BENZO(GHI)PERYLENE (B)	34.88	0.011	14.88393
BBOT (TRACE)	44.75	0.005	66.41213
QUININE SULFATE (A)	28.84	0.019	4.82754
ACRIDINE YELLOW (A)	25.32	0.028	2.16115
ACRIDINE ORANGE (A)	22.22	0.041	0.98663
ACRIDINE RED (A)	27.33	0.022	3.42124
URANINE (A)	26.09	0.025	2.58579
RHODAMINE B (A)	31.10	0.015	7.42488
RHODAMINE 6G (A)	31.86	0.014	8.56973
EOSIN B (A)	27.71	0.021	3.71289
EOSIN Y (A)	28.24	0.020	4.16078
ROSE BENGAL (A)	29.58	0.017	5.48902
INDENOINDENE	26.62	0.024	2.99082

8.83 Donor Molecule: 3-Phenyldibenzofuran

ACCEPTOR MOLECULES	RO	CO	OVERLAP
N,N-DIMETHYLANILINE 3ML/L	9.46	0.528	0.00547
TRIPHENYLAMINE	25.69	0.026	2.17823
SODIUM SALICYLATE (A)	17.19	0.088	0.19539
ANTHRANILIC ACID (A)	29.44	0.018	4.89383
TMPD	25.10	0.028	1.88571
4,4-DIHYDROXYBIPHENYL (A)	9.12	0.589	0.00441
4,4'-DIMETHOXY BIPHENYL	5.78	2.315	0.00029
4,4'-DIISOBUTOXYBIPHENYL	0.00	0.000	0.00000
BENZIDINE	22.21	0.041	0.91249
4-VINYL BIPHENYL	8.02	0.867	0.00204
1,3,5-TRIPHENYL BENZENE	9.00	0.614	0.00405
1,3-DIBIPHENYLYLBENZENE (B)	20.37	0.053	0.54395
1,3,5,-TRIBIPHENYLYLBENZENE (B)	24.32	0.031	1.57533
FLUORENE (TRACE)	0.00	0.000	0.00000
1-METHYL FLUORENE	0.00	0.000	0.00000
1,2-BENZOFLUORENE	23.43	0.035	1.25119
9,10-DIHYDROPHENANTHRENE	0.00	0.000	0.00000
CARBAZOLE (TRACE)	23.18	0.036	1.17733
N-METHYL CARBAZOLE (TRACE)	26.72	0.023	2.75622
N-VINYLCARBAZOLE	26.57	0.024	2.66785
N-PHENYL CARBAZOLE	26.29	0.025	2.49962
N-METHYL-2,3-BENZCARBAZOLE	29.33	0.018	4.75628
DIBENZOFURAN	0.00	0.000	0.00000
DIBENZOTHIOPHENE (TRACE)	20.54	0.052	0.57031
P-TERPHENYL	11.10	0.327	0.01432
DEUTERATED P-TERPHENYL-C18D14	9.58	0.508	0.00593
4-METHYL TERPHENYL	13.11	0.198	0.03886
2-METHYL-P-TERPHENYL	0.00	0.000	0.00000
3,3''-DIMETHYL-P-TERPHENYL	13.56	0.180	0.04740
4-(3,3-DIMETHYLBUTOXY)-P-TERPHENYL	20.47	0.052	0.55967
4,4''-DIHEXAHYDROFARNESOXY-P-TERPHENYL	24.31	0.031	1.56964
2,2'-METHYLENE-P-TERPHENYL	21.96	0.042	0.85398
2,2'-ETHYLENE-P-TERPHENYL	23.61	0.034	1.31894
2,2'-METHYLENE-5',6''-METHYLENE-P-TERPHENYL (TRACE)	36.13	0.009	16.88678
3-PHENYL DIBENZOFURAN	21.96	0.042	0.85367
BENZO(1,2-B-4,5-B')BISBENZOFURAN (TRACE) (B)	38.01	0.008	22.85929
SEXIPHENYL BENZENE	0.00	0.000	0.00000
P-QUATERPHENYL	26.06	0.025	2.38164
3,3'''-DIMETHYL-P-QUATERPHENYL	27.82	0.021	3.51769
TETRAMETHYL-P-QUATERPHENYL	0.00	0.000	0.00000
3,3''' -DIMETHYL-3',2''-METHYLENE-P-QUATERPHENYL	37.41	0.009	20.76745
3,3''' -DIMETHYL-3',2''-ETHYLENE-P-QUATERPHENYL	36.87	0.009	18.98539
4-(3-ETHYLBUTOXY)-P-QUATERPHENYL (B)	33.69	0.012	11.05959
4,4'''-DI(2-BUTYLOCTOXY-1)-P-QUATERPHENYL	34.42	0.011	12.59038
DIETHYL-P-QUINQUEPHENYL	29.04	0.018	4.55643
DI(3-ETHYLHEPTYL)-P-QUINQUEPHENYL	35.98	0.010	16.36872
TETRAMETHYL-P-SEXIPHENYL	27.45	0.022	3.25223
TETRAMETHYL-P-OCTAPHENYL (B)	40.93	0.007	35.47596
INDOLE (TRACE)	0.00	0.000	0.00000
2-PHENYL INDOLE	29.16	0.018	4.67704
1-METHYL-2-PHENYL INDOLE	23.04	0.037	1.13469
1-NEOPENTYL-2-PHENYL INDOLE	19.64	0.059	0.43710
1-N-BUTYL-2(BIPHENYLYL-4) INDOLE	33.50	0.012	10.67137
1,2-DIPHENYL INDOLE	23.84	0.033	1.39370
1-BIPHENYLYL-2-PHENYL INDOLE	26.90	0.023	2.86695
1,3-DIMETHYL-2-PHENYL INDOLE	24.54	0.030	1.65973
2,3-DIPHENYL INDOLE	30.93	0.015	6.60490
1-METHYL-2,3-DIPHENYL INDOLE	28.06	0.020	3.68527
2-PHENYL-3,2'-METHYLENE INDOLE (TRACE)	32.97	0.012	9.74738
2-PHENYL-3,2'-ETHYLENE INDOLE (TRACE)	36.01	0.010	16.46611
2-PHENYL-3,2'-TRIMETHYLENE INDOLE	31.45	0.014	7.33572
1-METHYL-3,2'-METHYLENE-2-PHENYL INDOLE (TRACE)	35.48	0.010	15.10342
1-METHYL-3,2'-ETHYLENE-2-PHENYL INDOLE	36.25	0.009	17.13066
1-METHYL-3,2'-TRIMETHYLENE-2-PHENYL INDOLE	30.24	0.016	5.79437
4,5-DIPHENYLIMIDAZOLE (A)	18.39	0.072	0.29418
PPF (TRACE)	37.20	0.009	20.04153
PPO (TRACE)	31.58	0.014	7.52623
2-(4-FLUOROPHENYL)-5-PHENYLOXAZOLE (TRACE)	30.05	0.016	5.59650
αNPO (TRACE)	38.80	0.008	25.64193
BBO (B)	44.40	0.005	57.47873
POPOP	43.09	0.006	48.01975
DIMETHYL POPOP (TRACE)	43.98	0.005	54.14552
PPD (TRACE)	8.74	0.671	0.00340
2,5-DI(2-METHYLPHENYL)OXADIAZOLE	11.86	0.268	0.02125
PBD	29.97	0.017	5.50668
BBD (B)	39.33	0.007	27.99220
αNPD (TRACE)	34.14	0.011	11.95585
βNPD	29.09	0.018	4.60052
αNND (TRACE)	36.70	0.009	18.42827
βNND (TRACE)	35.83	0.010	16.04075
TRANS-1-(4-BIPHENYLYL)-2-PHENYL-ETHYLENE	39.00	0.008	26.53244
TRANS-1-(4-BIPHENYLYL)-2-(1-NAPHTHYL)-ETHYLENE	41.03	0.006	35.77092
DIPHENYLSTILBENE (B)	44.86	0.005	61.13521
1,1,4,4-TETRAPHENYLBUTADIENE	42.04	0.006	41.33972
1,6-DIPHENYLHEXATRIENE	46.33	0.004	74.04756
1,8-DIPHENYL-1,3,5,7-OCTATETRAENE	45.26	0.005	63.92589
BPSB (TRACE)	46.58	0.004	76.47364
BIS-MSB	45.34	0.005	65.02274
ESCULIN (A)	34.90	0.011	13.58799
AZULENE	26.24	0.025	2.42684
NAPHTHALENE	6.06	2.011	0.00038
DEUTERATED NAPHTHALENE-C10D8	10.22	0.419	0.00868
1-METHYLNAPHTHALENE 5ML/L	8.35	0.768	0.00259

ACCEPTOR MOLECULES	RO	CO	OVERLAP
2-METHYL NAPHTHALENE	10.64	0.372	0.01107
2,3-DIMETHYL NAPHTHALENE	10.55	0.381	0.01054
2,6-DIMETHYL NAPHTHALENE (TRACE)	15.80	0.113	0.11884
1-NAPHTHOL (TRACE)	16.19	0.105	0.13750
2-NAPHTHOL (TRACE)	20.71	0.050	0.60028
1-NAPHTHYLAMINE	26.27	0.025	2.48793
2-CHLORONAPHTHALENE	11.87	0.267	0.03137
ACENAPHTHENE	13.50	0.182	0.04623
1-PHENYL NAPHTHALENE	14.03	0.162	0.05813
2-PHENYL NAPHTHALENE	22.92	0.037	1.10394
1,4-DIPHENYL NAPHTHALENE	24.21	0.032	1.52635
1,5-DIPHENYL NAPHTHALENE	22.17	0.041	0.90233
1,7-DIPHENYL NAPHTHALENE	21.91	0.043	0.83925
1,4,5,8-TETRAPHENYL NAPHTHALENE	37.37	0.009	20.40246
1,1'-BINAPHTHYL	17.48	0.084	0.21762
4,4'-DI(N-BUTOXY)-1,1'-BINAPHTHYL	30.19	0.016	5.72482
2,2'-BINAPHTHYL	23.84	0.033	1.39384
ACRIDINE (A)	32.45	0.013	8.73876
ANTHRACENE (TRACE)	30.31	0.016	5.82378
DEUTERATED ANTHRACENE C14D10 (TRACE)	30.00	0.017	5.47265
9-METHYL ANTHRACENE (TRACE)	30.00	0.017	5.45753
1-AMINO ANTHRACENE	27.90	0.021	3.50882
2-AMINO ANTHRACENE	28.20	0.020	3.74551
9-VINYL ANTHRACENE	30.90	0.015	6.50102
9-PHENYL ANTHRACENE (TRACE)	31.57	0.014	7.40381
9,10-DIPHENYL ANTHRACENE	31.76	0.014	7.65556
9,10-DICHLOROANTHRACENE (TRACE)	30.27	0.016	5.74224
BIANTHRANYL (B)	33.58	0.012	10.70759
9,10-DINAPHTHYLANTHRACENE (B)	31.19	0.015	6.85829
9-(P-VINYLPHENYL)-10-PHENYL ANTHRACENE	31.84	0.014	7.76609
9,10-BIS(PHENYL-ETHYNYL)ANTHRACENE (TRACE)	27.94	0.020	3.54288
RUBICENE (B)	31.82	0.014	7.72051
DIBENZ(A,H)ANTHRACENE (B)	34.54	0.011	12.70356
TETRACENE (TRACE) (B)	23.03	0.037	1.11088
RUBRENE (B)	29.74	0.017	5.14509
TETRAHYDROCANNABINOL ACID METHYL ESTER	8.98	0.618	0.00401
PHENANTHRENE	16.19	0.105	0.13616
3,4-BENZOPHENANTHRENE	23.78	0.033	1.35802
2-PHENYL PHENANTHRENE	18.21	0.074	0.27533
CHRYSENE (TRACE)	23.68	0.034	1.32760
TRIPHENYLENE	16.95	0.092	0.17988
DEUTERATED TRIPHENYLENE-C18D12	16.67	0.097	0.16231
PYRENE	34.65	0.011	13.00209
1,3,6,8,-TETRAPHENYL PYRENE	39.38	0.007	27.77919
BIPYRENYL (B)	42.74	0.006	45.54405
O-PHENYLENE-PYRENE	35.41	0.010	14.65950
ANTHANTHRENE (B)	32.54	0.013	8.83564
FLUORANTHENE	32.07	0.014	8.12506
3-PHENYL FLUORANTHENE	34.80	0.011	13.23832
3-AMINO FLUORANTHENE	31.86	0.014	7.77595
3,3'-BIFLUORANTHENYL (B)	38.56	0.008	24.44857
BENZO(GHI)FLUORANTHENE	30.94	0.015	6.53437
9,12-DI(N-OCTADECOXY)-BENZ(E)ACEPHENANTHRYLENE	34.24	0.011	11.99473
DECACYCLENE (B)	41.80	0.006	39.68703
10-ETHOXY-BENZO(KL)XANTHENE	32.41	0.013	8.64979
PERYLENE	28.87	0.019	4.30946
BENZO(GHI)PERYLENE (B)	34.61	0.011	12.83814
BBOT (TRACE)	44.18	0.005	55.59636
QUININE SULFATE (A)	30.62	0.016	6.24721
ACRIDINE YELLOW (A)	25.12	0.028	1.87128
ACRIDINE ORANGE (A)	22.56	0.039	0.98199
ACRIDINE RED (A)	27.29	0.022	3.07586
URANINE (A)	27.27	0.022	3.05687
RHODAMINE B (A)	30.96	0.015	6.55478
RHODAMINE 6G (A)	31.83	0.014	7.73516
EOSIN B (A)	28.76	0.019	4.20927
EOSIN Y (A)	29.24	0.018	4.65498
ROSE BENGAL (A)	30.79	0.015	6.33381
INDENOINDENE	31.15	0.015	6.92696

8.85 Donor Molecule: Sexiphenylbenzene

ANILINE 3ML/L	2.24	39.532	0.00010
N,N-DIMETHYLANILINE 3ML/L	5.71	2.404	0.02654
TRIPHENYLAMINE	12.26	0.243	2.57943
1,4-DIMETHOXY BENZENE	2.66	23.854	0.00027
1-METHYL-4-CHLOROBENZENE 5ML/L	0.00	0.000	0.00000
SODIUM SALICYLATE (A)	8.24	0.799	0.23764
ANTHRANILIC ACID (A)	13.48	0.183	4.52388
TMPD	11.55	0.290	1.79666
4-METHOXY BIPHENYL	1.75	83.497	0.00002
4-BIPHENYLYL PHENYL ETHER	0.00	0.000	0.00000
4,4'-DIHYDROXYBIPHENYL (A)	5.34	2.944	0.01772
4,4'-DIMETHOXY BIPHENYL	4.22	5.965	0.00433
4,4'-DIISOBUTOXYBIPHENYL	3.69	8.936	0.00193
BENZIDINE	10.79	0.356	1.20368
4-VINYL BIPHENYL	5.71	2.406	0.02658
M-TERPHENYL	3.97	7.158	0.00301
1,3,5-TRIPHENYL BENZENE	4.84	3.950	0.00982
1,3-DIBIPHENYLYLBENZENE (B)	10.29	0.610	0.00641

ACCEPTOR MOLECULES	RO	CO	OVERLAP
1,3,5,-TRIBIPHENYLYLBENZENE (B)	12.15	0.249	2.46045
FLUORENE (TRACE)	2.75	21.561	0.00033
1-METHYL FLUORENE	2.78	20.730	0.00036
1,2-BENZOFLUORENE	11.34	0.306	1.61943
9,10-DIHYDROPHENANTHRENE	3.69	8.893	0.00195
CARBAZOLE (TRACE)	10.86	0.349	1.25199
N-METHYL CARBAZOLE (TRACE)	12.21	0.246	2.51268
N-VINYLCARBAZOLE	12.26	0.243	2.58209
N-PHENYL CARBAZOLE	12.10	0.252	2.38941
N-METHYL-2,3-BENZCARBAZOLE	13.66	0.175	4.87293
DIBENZOFURAN	2.28	37.961	0.00011
DIBENZOTHIOPHENE (TRACE)	9.65	0.498	0.61501
P-TERPHENYL	6.70	1.485	0.06961
DEUTERATED P-TERPHENYL-C18D14	6.32	1.770	0.04910
4-METHYL TERPHENYL	7.60	1.021	0.14729
2-METHYL-P-TERPHENYL	3.51	10.369	0.00143
3,3″-DIMETHYL-P-TERPHENYL	7.51	1.057	0.13736
4-(3,3-DIMETHYLBUTOXY)-P-TERPHENYL	10.23	0.417	0.87837
4,4″-DIHEXAHYDROFARNESOXY-P-TERPHENYL	11.91	0.265	2.18072
2,2′-METHYLENE-P-TERPHENYL	11.26	0.313	1.55744
2,2′-ETHYLENE-P-TERPHENYL	11.74	0.277	1.99543
2,2′-METHYLENE-5′,6″-METHYLENE-P-TERPHENYL (TRACE)	16.78	0.095	16.99816
3-PHENYL DIBENZOFURAN	11.30	0.310	1.59498
BENZO(1,2-B-4,5-B′)BISBENZOFURAN (TRACE) (B)	17.50	0.084	21.81562
SEXIPHENYL BENZENE	0.00	0.000	0.00000
P-QUATERPHENYL	12.61	0.223	3.06357
3,3‴-DIMETHYL-P-QUATERPHENYL	13.34	0.188	4.30032
TETRAMETHYL-P-QUATERPHENYL	4.43	5.134	0.00585
3,3‴ -DIMETHYL-3′,2″-METHYLENE-P-QUATERPHENYL	17.39	0.085	21.00985
3,3‴ -DIMETHYL-3′,2″-ETHYLENE-P-QUATERPHENYL	17.16	0.089	19.32424
4-(3-ETHYLBUTOXY)-P-QUATERPHENYL (B)	15.81	0.113	11.85475
4,4‴-DI(2-BUTYLOCTOXY-1)-P-QUATERPHENYL	16.18	0.106	13.60888
DIETHYL-P-QUINQUEPHENYL	13.93	0.166	5.55635
DI(3-ETHYLHEPTYL)-P-QUINQUEPHENYL	16.87	0.093	17.46277
TETRAMETHYL-P-SEXIPHENYL	13.31	0.190	4.24417
TETRAMETHYL-P-OCTAPHENYL (B)	19.00	0.065	35.61645
INDOLE (TRACE)	0.00	0.000	0.00000
2-PHENYL INDOLE	13.89	0.167	5.47851
1-METHYL-2-PHENYL INDOLE	11.19	0.319	1.49635
1-NEOPENTYL-2-PHENYL INDOLE	9.90	0.461	0.71953
1-N-BUTYL-2(BIPHENYLYL-4) INDOLE	15.57	0.118	10.80421
1,2-DIPHENYL INDOLE	11.62	0.285	1.87774
1-BIPHENYLYL-2-PHENYL INDOLE	12.92	0.207	3.53505
1,3-DIMETHYL-2-PHENYL INDOLE	11.69	0.280	1.94220
2,3-DIPHENYL INDOLE	14.41	0.150	6.75895
1-METHYL-2,3-DIPHENYL INDOLE	13.09	0.200	3.80647
2-PHENYL-3,2′-METHYLENE INDOLE (TRACE)	15.46	0.121	10.40469
2-PHENYL-3,2′-ETHYLENE INDOLE (TRACE)	16.56	0.099	15.61004
2-PHENYL-3,2′-TRIMETHYLENE INDOLE	14.80	0.138	7.98921
1-METHYL-3,2′-METHYLENE-2-PHENYL INDOLE (TRACE)	16.37	0.102	14.59632
1-METHYL-3,2′-ETHYLENE-2-PHENYL INDOLE	16.69	0.096	16.35409
1-METHYL-3,2′-TRIMETHYLENE-2-PHENYL INDOLE	14.18	0.157	6.17831
4,5-DIPHENYLIMIDAZOLE (A)	8.93	0.629	0.38600
PPF (TRACE)	17.18	0.088	19.51584
PPO (TRACE)	14.94	0.134	8.46133
2-(4-FLUOROPHENYL)-5-PHENYLOXAZOLE (TRACE)	14.22	0.155	6.31121
αNPO (TRACE)	17.77	0.080	23.73290
BBO (B)	20.32	0.053	53.05746
POPOP	19.74	0.058	44.49429
DIMETHYL POPOP (TRACE)	20.20	0.054	50.92216
PPD (TRACE)	6.74	1.463	0.07192
2,5-DI(2-METHYLPHENYL)OXADIAZOLE	7.49	1.065	0.13550
PBD	14.33	0.152	6.59191
BBD (B)	18.22	0.074	27.77647
αNPD (TRACE)	15.76	0.114	11.60233
βNPD	13.76	0.172	5.16965
αNND (TRACE)	16.84	0.094	17.22359
βNND (TRACE)	16.64	0.097	16.15361
TRANS-1-(4-BIPHENYLYL)-2-PHENYL-ETHYLENE	18.01	0.077	25.82896
TRANS-1-(4-BIPHENYLYL)-2-(1-NAPHTHYL)-ETHYLENE	18.81	0.067	33.37276
DIPHENYLSTILBENE (B)	20.53	0.052	56.31029
1,1,4,4-TETRAPHENYLBUTADIENE	19.24	0.063	38.12760
1,6-DIPHENYLHEXATRIENE	21.23	0.047	68.82903
1,8-DIPHENYL-1,3,5,7-OCTATETRAENE	20.99	0.048	63.87369
BPSB (TRACE)	21.35	0.046	71.06151
BIS-MSB	20.77	0.050	60.33850
ESCULIN (A)	15.99	0.109	12.59218
AZULENE	12.05	0.255	2.28242
NAPHTHALENE	3.78	8.289	0.00223
DEUTERATED NAPHTHALENE-C10D8	5.10	3.380	0.01337
1-METHYLNAPHTHALENE 5ML/L	4.69	4.330	0.00817
2-METHYL NAPHTHALENE	5.69	2.423	0.02611
2,3-DIMETHYL NAPHTHALENE	5.61	2.532	0.02389
2,6-DIMETHYL NAPHTHALENE (TRACE)	7.55	1.039	0.14177
1-NAPHTHOL (TRACE)	8.21	0.808	0.23419
2-NAPHTHOL (TRACE)	9.59	0.508	0.59235
1-NAPHTHYLAMINE	12.21	0.246	2.51736
2-CHLORONAPHTHALENE	6.01	2.064	0.03593
ACENAPHTHENE	7.12	1.238	0.10005
1-PHENYL NAPHTHALENE	7.31	1.144	0.11693
2-PHENYL NAPHTHALENE	11.20	0.318	1.50812
1,4-DIPHENYL NAPHTHALENE	11.60	0.286	1.85599
1,5-DIPHENYL NAPHTHALENE	10.76	0.359	1.18280
1,7-DIPHENYL NAPHTHALENE	10.63	0.373	1.09683
1,4,5,8-TETRAPHENYL NAPHTHALENE	17.13	0.089	18.99537
1,1′-BINAPHTHYL	8.64	0.694	0.31745

ACCEPTOR MOLECULES	RO	CO	OVERLAP
4,4'-DI(N-BUTOXY)-1,1'-BINAPHTHYL	14.10	0.160	5.95510
2,2'-BINAPHTHYL	11.68	0.281	1.93798
ACRIDINE (A)	14.84	0.137	8.01981
ANTHRACENE (TRACE)	13.93	0.166	5.49770
DEUTERATED ANTHRACENE C14D10 (TRACE)	13.79	0.170	5.18528
9-METHYL ANTHRACENE (TRACE)	13.78	0.171	5.14527
1-AMINO ANTHRACENE	13.05	0.201	3.67487
2-AMINO ANTHRACENE	13.18	0.195	3.90699
9-VINYL ANTHRACENE	14.26	0.154	6.30737
9-PHENYL ANTHRACENE (TRACE)	14.50	0.147	6.97100
9,10-DIPHENYL ANTHRACENE	14.79	0.138	7.82402
9,10-DICHLOROANTHRACENE (TRACE)	14.13	0.159	5.95823
BIANTHRANYL (B)	15.63	0.117	10.90676
9,10-DINAPHTHYLANTHRACENE (B)	14.57	0.145	7.16231
9-(P-VINYLPHENYL)-10-PHENYL ANTHRACENE	14.85	0.137	8.00617
9,10-BIS(PHENYL-ETHYNYL)ANTHRACENE (TRACE)	14.02	0.162	5.66083
RUBICENE (B)	15.03	0.132	8.56413
DIBENZ(A,H)ANTHRACENE (B)	15.98	0.110	12.47426
TETRACENE (TRACE) (B)	11.09	0.328	1.38181
RUBRENE (B)	14.32	0.152	6.40730
TETRAHYDROCANNABINOL ACID METHYL ESTER	5.64	2.495	0.02470
PHENANTHRENE	7.45	1.082	0.12982
3,4-BENZOPHENANTHRENE	11.36	0.305	1.61523
2-PHENYL PHENANTHRENE	8.83	0.650	0.35880
CHRYSENE (TRACE)	11.66	0.282	1.89743
TRIPHENYLENE	7.91	0.905	0.18540
DEUTERATED TRIPHENYLENE-C18D12	7.79	0.946	0.16991
PYRENE	16.22	0.105	13.74292
1,3,6,8,-TETRAPHENYL PYRENE	18.27	0.073	27.78787
BIPYRENYL (B)	19.56	0.060	41.92985
O-PHENYLENE-PYRENE	16.49	0.100	14.94944
ANTHANTHRENE (B)	15.51	0.120	10.37749
FLUORANTHENE	14.67	0.142	7.45246
3-PHENYL FLUORANTHENE	15.99	0.109	12.48993
3-AMINO FLUORANTHENE	14.86	0.136	8.02011
3,3'-BIFLUORANTHENYL (B)	17.88	0.078	24.34079
BENZO(GHI)FLUORANTHENE	14.17	0.157	6.05455
9,12-DI(N-OCTADECOXY)-BENZ(E)ACEPHENANTHRYLENE	15.92	0.111	12.15978
DECACYCLENE (B)	19.47	0.061	40.57588
10-ETHOXY-BENZO(KL)XANTHENE	15.07	0.131	8.78505
PERYLENE	14.16	0.158	5.98601
BENZO(GHI)PERYLENE (B)	16.02	0.109	12.67313
BBOT (TRACE)	20.38	0.053	53.69346
QUININE SULFATE (A)	14.13	0.159	6.04999
ACRIDINE YELLOW (A)	12.57	0.225	2.93660
ACRIDINE ORANGE (A)	11.02	0.334	1.33122
ACRIDINE RED (A)	12.64	0.221	3.03756
URANINE (A)	12.81	0.213	3.29102
RHODAMINE B (A)	14.36	0.151	6.52446
RHODAMINE 6G (A)	14.57	0.145	7.11585
EOSIN B (A)	13.41	0.186	4.32440
EOSIN Y (A)	13.68	0.175	4.87062
ROSE BENGAL (A)	14.26	0.154	6.31041
INDENOINDENE	14.66	0.142	7.56380

8.86 Donor Molecule: *p*-Quaterphenyl

TRIPHENYLAMINE	18.01	0.077	0.28897
P-XYLENE (TRACE)	0.00	0.000	0.00000
SODIUM SALICYLATE (A)	13.24	0.193	0.04559
ANTHRANILIC ACID (A)	26.16	0.025	2.69714
TMPD	20.91	0.049	0.70423
BENZIDINE	13.69	0.174	0.05595
1,3,5-TRIPHENYL BENZENE	2.67	23.452	0.00000
1,3-DIBIPHENYLYLBENZENE (B)	11.34	0.306	0.01813
1,3,5,-TRIBIPHENYLYLBENZENE (B)	12.75	0.216	0.03663
FLUORENE (TRACE)	0.00	0.000	0.00000
1-METHYL FLUORENE	0.00	0.000	0.00000
1,2-BENZOFLUORENE	17.72	0.080	0.26181
9,10-DIHYDROPHENANTHRENE	0.00	0.000	0.00000
CARBAZOLE (TRACE)	14.04	0.162	0.06495
N-METHYL CARBAZOLE (TRACE)	21.44	0.045	0.82332
N-VINYLCARBAZOLE	20.07	0.055	0.55418
N-PHENYL CARBAZOLE	20.09	0.055	0.55689
N-METHYL-2,3-BENZCARBAZOLE	28.20	0.020	4.20259
DIBENZOFURAN	0.00	0.000	0.00000
DIBENZOTHIOPHENE (TRACE)	9.82	0.472	0.00764
P-TERPHENYL	0.00	0.000	0.00000
DEUTERATED P-TERPHENYL-C18D14	0.00	0.000	0.00000
4-METHYL TERPHENYL	0.00	0.000	0.00000
2-METHYL-P-TERPHENYL	0.00	0.000	0.00000
3,3''-DIMETHYL-P-TERPHENYL	0.00	0.000	0.00000
4-(3,3-DIMETHYLBUTOXY)-P-TERPHENYL	9.91	0.460	0.00807
4,4''-DIHEXAHYDROFARNESOXY-P-TERPHENYL	12.87	0.210	0.03872
2,2'-METHYLENE-P-TERPHENYL	9.13	0.588	0.00493
2,2'-ETHYLENE-P-TERPHENYL	11.40	0.302	0.01870
2,2'-METHYLENE-5',6''-METHYLENE-P-TERPHENYL (TRACE)	24.20	0.032	1.70604
3-PHENYL DIBENZOFURAN	9.01	0.611	0.00457
BENZO(1,2-B-4,5-B')BISBENZOFURAN (TRACE) (B)	27.50	0.022	3.66664
SEXIPHENYL BENZENE	0.00	0.000	0.00000

ACCEPTOR MOLECULES	RO	CO	OVERLAP
P-QUATERPHENYL	15.39	0.123	0.11295
3,3‴-DIMETHYL-P-QUATERPHENYL	17.69	0.081	0.26013
TETRAMETHYL-P-QUATERPHENYL	0.00	0.000	0.00000
3,3‴ -DIMETHYL-3′,2″-METHYLENE-P-QUATERPHENYL	26.73	0.023	3.08740
3,3‴ -DIMETHYL-3′,2″-ETHYLENE-P-QUATERPHENYL	27.40	0.022	3.57297
4-(3-ETHYLBUTOXY)-P-QUATERPHENYL (B)	24.41	0.031	1.78874
4,4‴-DI(2-BUTYLOCTOXY-1)-P-QUATERPHENYL	24.72	0.030	1.93249
DIETHYL-P-QUINQUEPHENYL	18.23	0.074	0.31212
DI(3-ETHYLHEPTYL)-P-QUINQUEPHENYL	26.42	0.024	2.86740
TETRAMETHYL-P-SEXIPHENYL	16.00	0.109	0.14273
TETRAMETHYL-P-OCTAPHENYL (B)	32.22	0.013	9.43754
INDOLE (TRACE)	0.00	0.000	0.00000
2-PHENYL INDOLE	17.00	0.091	0.20548
1-METHYL-2-PHENYL INDOLE	13.24	0.193	0.04569
1-NEOPENTYL-2-PHENYL INDOLE	9.29	0.559	0.00546
1-N-BUTYL-2(BIPHENYLYL-4) INDOLE	26.14	0.025	2.69392
1,2-DIPHENYL INDOLE	13.60	0.178	0.05368
1-BIPHENYLYL-2-PHENYL INDOLE	17.74	0.080	0.26383
1,3-DIMETHYL-2-PHENYL INDOLE	15.06	0.131	0.09902
2,3-DIPHENYL INDOLE	24.03	0.032	1.62490
1-METHYL-2,3-DIPHENYL INDOLE	21.53	0.045	0.84125
2-PHENYL-3,2′-METHYLENE INDOLE (TRACE)	22.07	0.042	0.98111
2-PHENYL-3,2′-ETHYLENE INDOLE (TRACE)	28.89	0.019	4.91048
2-PHENYL-3,2′ -TRIMETHYLENE INDOLE	20.39	0.053	0.60956
1-METHYL-3,2′-METHYLENE-2-PHENYL INDOLE (TRACE)	27.58	0.021	3.72785
1-METHYL-3,2′-ETHYLENE-2-PHENYL INDOLE	30.75	0.015	7.13657
1-METHYL-3,2′-TRIMETHYLENE-2-PHENYL INDOLE	21.12	0.047	0.75221
4,5-DIPHENYLIMIDAZOLE (A)	11.67	0.281	0.02147
PPF (TRACE)	28.87	0.019	4.90281
PPO (TRACE)	20.11	0.055	0.56185
2-(4-FLUOROPHENYL)-5-PHENYLOXAZOLE (TRACE)	18.70	0.068	0.36339
αNPO (TRACE)	34.85	0.011	15.05305
BBO (B)	41.43	0.006	42.44585
POPOP	42.24	0.006	47.67310
DIMETHYL POPOP (TRACE)	44.22	0.005	62.53955
PPD (TRACE)	0.00	0.000	0.00000
2,5-DI(2-METHYLPHENYL)OXADIAZOLE	0.00	0.000	0.00000
PBD	17.31	0.086	0.22833
BBD (B)	30.66	0.016	7.01973
αNPD (TRACE)	27.14	0.022	3.37148
βNPD	18.47	0.071	0.33653
αNND (TRACE)	31.75	0.014	8.63075
βNND (TRACE)	26.83	0.023	3.16486
TRANS-1-(4-BIPHENYLYL)-2-PHENYL-ETHYLENE	31.25	0.015	7.86659
TRANS-1-(4-BIPHENYLYL)-2-(1-NAPHTHYL)-ETHYLENE	37.46	0.009	23.17465
DIPHENYLSTILBENE (B)	41.72	0.006	44.21938
1,1,4,4-TETRAPHENYLBUTADIENE	39.92	0.007	33.87909
1,6-DIPHENYLHEXATRIENE	45.22	0.005	71.63623
1,8-DIPHENYL-1,3,5,7-OCTATETRAENE	48.00	0.004	101.81463
BPSB (TRACE)	45.85	0.005	77.87808
BIS-MSB	43.65	0.005	57.93263
ESCULIN (A)	31.57	0.014	8.32959
AZULENE	21.23	0.047	0.75916
NAPHTHALENE	0.00	0.000	0.00000
DEUTERATED NAPHTHALENE-C10D8	5.42	2.803	0.00022
1-METHYLNAPHTHALENE 5ML/L	2.31	36.180	0.00000
2-METHYL NAPHTHALENE	2.63	24.675	0.00000
2,3-DIMETHYL NAPHTHALENE	2.76	21.223	0.00000
2,6-DIMETHYL NAPHTHALENE (TRACE)	5.49	2.700	0.00023
1-NAPHTHOL (TRACE)	5.93	2.140	0.00037
2-NAPHTHOL (TRACE)	11.27	0.313	0.01742
1-NAPHTHYLAMINE	19.10	0.064	0.41164
2-CHLORONAPHTHALENE	4.01	6.962	0.00004
1-PHENYL NAPHTHALENE	5.71	2.405	0.00030
2-PHENYL NAPHTHALENE	12.04	0.211	0.03809
1,4-DIPHENYL NAPHTHALENE	14.89	0.135	0.09265
1,5-DIPHENYL NAPHTHALENE	12.31	0.240	0.02962
1,7-DIPHENYL NAPHTHALENE	12.86	0.210	0.03840
1,4,5,8-TETRAPHENYL NAPHTHALENE	33.89	0.011	12.68619
1,1′-BINAPHTHYL	8.31	0.780	0.00280
4,4′ -DI(N-BUTOXY)-1,1′ -BINAPHTHYL	21.51	0.045	0.83853
2,2′-BINAPHTHYL	14.03	0.162	0.06492
ACRIDINE (A)	31.76	0.014	8.59222
ANTHRACENE (TRACE)	29.96	0.017	6.06952
DEUTERATED ANTHRACENE C14D10 (TRACE)	29.56	0.017	5.60112
9-METHYL ANTHRACENE (TRACE)	31.11	0.015	7.60278
1-AMINO ANTHRACENE	30.30	0.016	6.43008
2-AMINO ANTHRACENE	28.96	0.018	4.89763
9-VINYL ANTHRACENE	32.41	0.013	9.68939
9-PHENYL ANTHRACENE (TRACE)	32.62	0.013	10.09128
9,10-DIPHENYL ANTHRACENE	34.18	0.011	13.31861
9,10-DICHLOROANTHRACENE (TRACE)	32.88	0.013	10.55950
BIANTHRANYL (B)	35.90	0.010	17.88641
9,10-DINAPHTHYLANTHRACENE (B)	33.55	0.012	11.89664
9-(P-VINYLPHENYL)-10-PHENYL ANTHRACENE	34.41	0.011	13.84034
9,10-BIS(PHENYL-ETHYNYL)ANTHRACENE (TRACE)	33.60	0.012	11.93247
RUBICENE (B)	32.45	0.013	9.68114
DIBENZ(A,H)ANTHRACENE (B)	29.92	0.017	6.00512
TETRACENE (TRACE) (B)	25.59	0.027	2.32877
RUBRENE (B)	24.57	0.030	1.82344
TETRAHYDROCANNABINOL ACID METHYL ESTER	0.00	0.000	0.00000
PHENANTHRENE	12.80	0.213	0.03726
3,4-BENZOPHENANTHRENE	17.80	0.079	0.26741
2-PHENYL PHENANTHRENE	15.45	0.121	0.11483
CHRYSENE (TRACE)	10.21	0.074	0.30735
TRIPHENYLENE	12.25	0.243	0.02868

ACCEPTOR MOLECULES	RO	CO	OVERLAP
DEUTERATED TRIPHENYLENE-C18D12	11.94	0.263	0.02450
PYRENE	25.22	0.028	2.16262
1,3,6,8,-TETRAPHENYL PYRENE	42.42	0.006	48.55809
BIPYRENYL (B)	41.11	0.006	40.35789
O-PHENYLENE-PYRENE	36.81	0.009	20.62358
ANTHANTHRENE (B)	34.95	0.010	15.13708
FLUORANTHENE	30.34	0.016	6.51410
3-PHENYL FLUORANTHENE	35.14	0.010	15.69625
3-AMINO FLUORANTHENE	32.73	0.013	[illegible]
3,3'-BIFLUORANTHENYL (B)	40.17	0.007	34.95002
BENZO(GHI)FLUORANTHENE	28.09	0.020	4.09248
9,12-DI(N-OCTADECOXY)-BENZ(E)ACEPHENANTHRYLENE	35.18	0.010	15.78327
DECACYCLENE (B)	43.40	0.005	55.41373
10-ETHOXY-BENZO(KL)XANTHENE	34.82	0.011	14.89826
PERYLENE	34.42	0.011	13.80195
BENZO(GHI)PERYLENE (B)	36.79	0.009	20.70916
BBOT (TRACE)	45.72	0.005	76.34689
QUININE SULFATE (A)	23.39	0.035	1.38596
ACRIDINE YELLOW (A)	30.58	0.016	6.77814
ACRIDINE ORANGE (A)	25.62	0.027	2.34617
ACRIDINE RED (A)	27.07	0.023	3.26497
URANINE (A)	27.10	0.022	3.28767
RHODAMINE B (A)	30.95	0.015	7.29332
RHODAMINE 6G (A)	30.39	0.016	6.52817
EOSIN B (A)	26.68	0.024	2.99289
EOSIN Y (A)	27.61	0.021	3.66902
ROSE BENGAL (A)	28.09	0.020	4.07530
INDENOINDENE	19.33	0.062	0.44344

8.87 Donor Molecule: 3,3‴-Dimethyl-*p*-quaterphenyl

TRIPHENYLAMINE	16.80	0.094	0.16952
P-XYLENE (TRACE)	0.00	0.000	0.00000
SODIUM SALICYLATE (A)	12.76	0.215	0.03248
ANTHRANILIC ACID (A)	25.72	0.026	2.16884
TMPD	20.25	0.054	0.51792
BENZIDINE	12.30	0.240	0.02620
1,3-DIBIPHENYLYLBENZENE (B)	9.97	0.451	0.00744
1,3,5,-TRIBIPHENYLYLBENZENE (B)	11.01	0.336	0.01345
FLUORENE (TRACE)	0.00	0.000	0.00000
1-METHYL FLUORENE	0.00	0.000	0.00000
1,2-BENZOFLUORENE	16.71	0.096	0.16430
9,10-DIHYDROPHENANTHRENE	0.00	0.000	0.00000
CARBAZOLE (TRACE)	12.34	0.238	0.02664
N-METHYL CARBAZOLE (TRACE)	20.07	0.055	0.49297
N-VINYLCARBAZOLE	18.40	0.072	0.29273
N-PHENYL CARBAZOLE	18.53	0.070	0.30541
N-METHYL-2,3-BENZCARBAZOLE	29.06	0.018	4.47922
DIBENZOFURAN	0.00	0.000	0.00000
DIBENZOTHIOPHENE (TRACE)	8.29	0.785	0.00246
P-TERPHENYL	0.00	0.000	0.00000
DEUTERATED P-TERPHENYL-C18D14	0.00	0.000	0.00000
4-METHYL TERPHENYL	0.00	0.000	0.00000
2-METHYL-P-TERPHENYL	0.00	0.000	0.00000
3,3″-DIMETHYL-P-TERPHENYL	0.00	0.000	0.00000
4-(3,3-DIMETHYLBUTOXY)-P-TERPHENYL	8.37	0.762	0.00261
4,4″-DIHEXAHYDROFARNESOXY-P-TERPHENYL	11.08	0.328	0.01405
2,2'-METHYLENE-P-TERPHENYL	7.41	1.100	0.00125
2,2'-ETHYLENE-P-TERPHENYL	9.69	0.492	0.00625
2,2'-METHYLENE-5',6″-METHYLENE-P-TERPHENYL (TRACE)	21.54	0.045	0.75421
3-PHENYL DIBENZOFURAN	7.26	1.167	0.00111
BENZO(1,2-B-4,5-B')BISBENZOFURAN (TRACE) (B)	25.25	0.028	1.95357
SEXIPHENYL BENZENE	0.00	0.000	0.00000
P-QUATERPHENYL	13.63	0.177	0.04850
3,3‴-DIMETHYL-P-QUATERPHENYL	16.04	0.108	0.12842
TETRAMETHYL-P-QUATERPHENYL	0.00	0.000	0.00000
3,3‴-DIMETHYL-3',2″-METHYLENE-P-QUATERPHENYL	24.72	0.030	1.72008
3,3‴-DIMETHYL-3',2″-ETHYLENE-P-QUATERPHENYL	25.78	0.026	2.20842
4-(3-ETHYLBUTOXY)-P-QUATERPHENYL (B)	22.94	0.037	1.09536
4,4‴-DI(2-BUTYLOCTOXY-1)-P-QUATERPHENYL	23.19	0.036	1.17121
DIETHYL-P-QUINQUEPHENYL	16.43	0.101	0.14836
DI(3-ETHYLHEPTYL)-P-QUINQUEPHENYL	24.96	0.029	1.81609
TETRAMETHYL-P-SEXIPHENYL	14.13	0.158	0.06027
TETRAMETHYL-P-OCTAPHENYL (B)	30.93	0.015	6.57295
INDOLE (TRACE)	0.00	0.000	0.00000
2-PHENYL INDOLE	14.86	0.136	0.08136
1-METHYL-2-PHENYL INDOLE	11.70	0.279	0.01942
1-NEOPENTYL-2-PHENYL INDOLE	7.83	0.932	0.00175
1-N-BUTYL-2(BIPHENYLYL-4) INDOLE	25.01	0.029	1.84092
1,2-DIPHENYL INDOLE	12.03	0.257	0.02292
1-BIPHENYLYL-2-PHENYL INDOLE	16.62	0.097	0.15861
1,3-DIMETHYL-2-PHENYL INDOLE	13.38	0.187	0.04345
2,3-DIPHENYL INDOLE	23.01	0.037	1.11462
1-METHYL-2,3-DIPHENYL INDOLE	20.54	0.052	0.56464
2-PHENYL-3,2'-METHYLENE INDOLE (TRACE)	19.83	0.057	0.45904
2-PHENYL-3,2'-ETHYLENE INDOLE (TRACE)	27.50	0.022	3.25161
2-PHENYL-3,2'-TRIMETHYLENE INDOLE	18.33	0.073	0.28640
1-METHYL-3,2'-METHYLENE-2-PHENYL INDOLE (TRACE)	25.91	0.026	2.27652
1-METHYL-3,2'-ETHYLENE-2-PHENYL INDOLE	29.98	0.017	5.45667
1-METHYL-3,2'-TRIMETHYLENE-2-PHENYL INDOLE	19.27	0.063	0.38606

ACCEPTOR MOLECULES	RO	CO	OVERLAP
4,5-DIPHENYLIMIDAZOLE (A)	10.46	0.390	0.00993
PPF (TRACE)	27.12	0.022	2.99826
PPO (TRACE)	17.86	0.078	0.24546
2-(4-FLUOROPHENYL)-5-PHENYLOXAZOLE (TRACE)	16.55	0.099	0.15507
αNPO (TRACE)	34.32	0.011	12.21280
BBO (B)	41.24	0.006	36.75106
POPOP	42.47	0.006	43.77542
DIMETHYL POPOP (TRACE)	44.70	0.005	59.41376
PPD (TRACE)	0.00	0.000	0.00000
2,5-DI(2-METHYLPHENYL)OXADIAZOLE	0.00	0.000	0.00000
PBD	15.16	0.128	0.09183
BBD (B)	29.19	0.018	4.65646
αNPD (TRACE)	25.81	0.026	2.21915
βNPD	16.77	0.095	0.16771
αNND (TRACE)	30.96	0.015	6.60674
βNND (TRACE)	24.87	0.029	1.78318
TRANS-1-(4-BIPHENYLYL)-2-PHENYL-ETHYLENE	29.98	0.017	5.45295
TRANS-1-(4-BIPHENYLYL)-2-(1-NAPHTHYL)-ETHYLENE	37.16	0.009	19.67006
DIPHENYLSTILBENE (B)	41.46	0.006	37.92137
1,1,4,4-TETRAPHENYLBUTADIENE	39.88	0.007	29.97730
1,6-DIPHENYLHEXATRIENE	45.52	0.005	66.35665
1,8-DIPHENYL-1,3,5,7-OCTATETRAENE	49.23	0.004	105.49042
BPSB (TRACE)	46.14	0.005	71.94086
BIS-MSB	43.76	0.005	52.31326
ESCULIN (A)	31.21	0.015	6.90958
AZULENE	20.22	0.054	0.50249
NAPHTHALENE	0.00	0.000	0.00000
DEUTERATED NAPHTHALENE-C10D8	4.68	4.368	0.00008
1-METHYLNAPHTHALENE 5ML/L	0.00	0.000	0.00000
2-METHYL NAPHTHALENE	0.00	0.000	0.00000
2,3-DIMETHYL NAPHTHALENE	0.00	0.000	0.00000
2,6-DIMETHYL NAPHTHALENE (TRACE)	4.10	6.508	0.00004
1-NAPHTHOL (TRACE)	4.80	4.034	0.00009
2-NAPHTHOL (TRACE)	9.58	0.508	0.00586
1-NAPHTHYLAMINE	17.76	0.080	0.23670
2-CHLORONAPHTHALENE	3.06	15.637	0.00001
ACENAPHTHENE	0.00	0.000	0.00000
1-PHENYL NAPHTHALENE	4.52	4.830	0.00007
2-PHENYL NAPHTHALENE	11.15	0.323	0.01452
1,4-DIPHENYL NAPHTHALENE	13.44	0.184	0.04448
1,5-DIPHENYL NAPHTHALENE	10.80	0.355	0.01202
1,7-DIPHENYL NAPHTHALENE	11.43	0.300	0.01683
1,4,5,8-TETRAPHENYL NAPHTHALENE	33.54	0.012	10.62031
1,1'-BINAPHTHYL	6.94	1.337	0.00085
4,4'-DI(N-BUTOXY)-1,1'-BINAPHTHYL	20.17	0.055	0.50672
2,2'-BINAPHTHYL	12.44	0.232	0.02807
ACRIDINE (A)	31.87	0.014	7.81720
ANTHRACENE (TRACE)	30.31	0.016	5.78817
DEUTERATED ANTHRACENE C14D10 (TRACE)	29.92	0.017	5.36569
9-METHYL ANTHRACENE (TRACE)	31.50	0.014	7.28504
1-AMINO ANTHRACENE	31.25	0.015	6.87627
2-AMINO ANTHRACENE	29.82	0.017	5.19870
9-VINYL ANTHRACENE	33.04	0.012	9.68229
9-PHENYL ANTHRACENE (TRACE)	33.02	0.012	9.65548
9,10-DIPHENYL ANTHRACENE	35.19	0.010	14.10476
9,10-DICHLOROANTHRACENE (TRACE)	33.87	0.012	11.21094
BIANTHRANYL (B)	36.95	0.009	18.89909
9,10-DINAPHTHYLANTHRACENE (B)	34.58	0.011	12.68833
9-(P-VINYLPHENYL)-10-PHENYL ANTHRACENE	35.44	0.010	14.70354
9,10-BIS(PHENYL-ETHYNYL)ANTHRACENE (TRACE)	35.91	0.010	15.76028
RUBICENE (B)	33.57	0.012	10.53007
DIBENZ(A,H)ANTHRACENE (B)	29.54	0.017	4.95195
TETRACENE (TRACE) (B)	27.11	0.022	2.92129
RUBRENE (B)	25.21	0.028	1.88861
TETRAHYDROCANNABINOL ACID METHYL ESTER	0.00	0.000	0.00000
PHENANTHRENE	12.08	0.254	0.02338
3,4-BENZOPHENANTHRENE	17.64	0.081	0.22521
2-PHENYL PHENANTHRENE	15.07	0.131	0.08782
CHRYSENE (TRACE)	17.88	0.078	0.24512
TRIPHENYLENE	11.24	0.315	0.01521
DEUTERATED TRIPHENYLENE-C18D12	11.00	0.336	0.01336
PYRENE	23.04	0.037	1.11969
1,3,6,8,-TETRAPHENYL PYRENE	43.44	0.005	49.83385
BIPYRENYL (B)	41.27	0.006	36.76349
O-PHENYLENE-PYRENE	37.61	0.008	20.82031
ANTHANTHRENE (B)	36.40	0.009	17.19175
FLUORANTHENE	30.17	0.016	5.60480
3-PHENYL FLUORANTHENE	35.59	0.010	15.08352
3-AMINO FLUORANTHENE	33.72	0.012	10.81045
3,3'-BIFLUORANTHENYL (B)	41.07	0.006	35.51579
BENZO(GHI)FLUORANTHENE	27.92	0.021	3.51468
9,12-DI(N-OCTADECOXY)-BENZ(E)ACEPHENANTHRYLENE	36.11	0.010	16.40178
DECACYCLENE (B)	44.54	0.005	57.43462
10-ETHOXY-BENZO(KL)XANTHENE	35.77	0.010	15.58399
PERYLENE	36.26	0.009	16.78760
BENZO(GHI)PERYLENE (B)	37.49	0.008	20.66005
BBOT (TRACE)	46.59	0.004	76.11904
QUININE SULFATE (A)	21.81	0.043	0.81201
ACRIDINE YELLOW (A)	33.00	0.012	9.50164
ACRIDINE ORANGE (A)	27.74	0.021	3.34904
ACRIDINE RED (A)	27.58	0.021	3.23442
URANINE (A)	28.72	0.019	4.12516
RHODAMINE B (A)	31.62	0.014	7.34474
RHODAMINE 6G (A)	30.66	0.016	6.10167
EOSIN B (A)	27.30	0.022	3.03999
EOSIN Y (A)	28.39	0.020	3.84860

ACCEPTOR MOLECULES	RO	CO	OVERLAP
ROSE BENGAL (A)	28.48	0.019	3.92447
INDENOINDENE	17.15	0.089	0.19180

8.88 Donor Molecule: Tetramethyl-*p*-quaterphenyl

ACCEPTOR MOLECULES	RO	CO	OVERLAP
N,N-DIMETHYLANILINE 3ML/L	9.68	0.493	0.00864
TRIPHENYLAMINE	22.45	0.040	1.33058
1,4-DIMETHOXY BENZENE	3.62	9.411	0.00002
1-METHYL-4-CHLOROBENZENE 5ML/L	0.00	0.000	0.00000
SODIUM SALICYLATE (A)	15.29	0.125	0.13278
ANTHRANILIC ACID (A)	26.77	0.023	3.79937
TMPD	22.39	0.040	1.30181
4,4'-DIHYDROXYBIPHENYL (A)	9.06	0.601	0.00582
4,4'-DIMETHOXY BIPHENYL	6.99	1.312	0.00122
4,4'-DIISOBUTOXYBIPHENYL	5.67	2.457	0.00035
BENZIDINE	19.37	0.062	0.54979
4-VINYL BIPHENYL	9.44	0.531	0.00745
M-TERPHENYL	6.55	1.589	0.00083
1,3,5-TRIPHENYL BENZENE	8.32	0.777	0.00347
1,3-DIBIPHENYLYLBENZENE (B)	18.12	0.075	0.36948
1,3,5,-TRIBIPHENYLYLBENZENE (B)	21.42	0.046	1.00786
FLUORENE (TRACE)	0.00	0.000	0.00000
1-METHYL FLUORENE	4.03	6.859	0.00004
1,2-BENZOFLUORENE	20.86	0.049	0.85670
9,10-DIHYDROPHENANTHRENE	5.50	2.682	0.00029
CARBAZOLE (TRACE)	19.86	0.057	0.63930
N-METHYL CARBAZOLE (TRACE)	23.46	0.035	1.73458
N-VINYLCARBAZOLE	23.27	0.036	1.64909
N-PHENYL CARBAZOLE	23.00	0.037	1.54022
N-METHYL-2,3-BENZCARBAZOLE	27.49	0.022	4.42846
DIBENZOFURAN	0.00	0.000	0.00000
DIBENZOTHIOPHENE (TRACE)	17.21	0.088	0.27111
P-TERPHENYL	11.33	0.308	0.02216
DEUTERATED P-TERPHENYL-C18D14	10.61	0.375	0.01497
4-METHYL TERPHENYL	12.92	0.208	0.04870
2-METHYL-P-TERPHENYL	5.33	2.955	0.00024
3,3''-DIMETHYL-P-TERPHENYL	12.83	0.212	0.04682
4-(3,3-DIMETHYLBUTOXY)-P-TERPHENYL	17.97	0.077	0.35226
4,4''-DIHEXAHYDROFARNESOXY-P-TERPHENYL	21.13	0.047	0.92958
2,2'-METHYLENE-P-TERPHENYL	19.60	0.059	0.59263
2,2'-ETHYLENE-P-TERPHENYL	20.64	0.051	0.80784
2,2'-METHYLENE-5',6''-METHYLENE-P-TERPHENYL (TRACE)	31.33	0.015	9.84192
3-PHENYL DIBENZOFURAN	19.65	0.059	0.60113
BENZO(1,2-B-4,5-B')BISBENZOFURAN (TRACE) (B)	32.84	0.013	13.04365
SEXIPHENYL BENZENE	0.00	0.000	0.00000
P-QUATERPHENYL	22.61	0.039	1.39189
3,3'''-DIMETHYL-P-QUATERPHENYL	24.13	0.032	2.05545
TETRAMETHYL-P-QUATERPHENYL	7.10	1.249	0.00135
3,3''' -DIMETHYL-3',2''-METHYLENE-P-QUATERPHENYL	32.43	0.013	12.08542
3,3''' -DIMETHYL-3',2''-ETHYLENE-P-QUATERPHENYL	32.16	0.013	11.46783
4-(3-ETHYLBUTOXY)-P-QUATERPHENYL (B)	29.36	0.018	6.63995
4,4'''-DI(2-BUTYLOCTOXY-1)-P-QUATERPHENYL	29.97	0.017	7.52005
DIETHYL-P-QUINQUEPHENYL	25.17	0.028	2.64580
DI(3-ETHYLHEPTYL)-P-QUINQUEPHENYL	31.40	0.014	9.92119
TETRAMETHYL-P-SEXIPHENYL	23.83	0.033	1.90813
TETRAMETHYL-P-OCTAPHENYL (B)	36.07	0.010	22.79203
INDOLE (TRACE)	0.00	0.000	0.00000
2-PHENYL INDOLE	25.09	0.028	2.60278
1-METHYL-2-PHENYL INDOLE	19.99	0.056	0.66414
1-NEOPENTYL-2-PHENYL INDOLE	17.34	0.086	0.28382
1-N-BUTYL-2(BIPHENYLYL-4) INDOLE	29.48	0.017	6.80277
1,2-DIPHENYL INDOLE	20.72	0.050	0.82368
1-BIPHENYLYL-2-PHENYL INDOLE	23.40	0.035	1.70411
1,3-DIMETHYL-2-PHENYL INDOLE	21.17	0.047	0.93818
2,3-DIPHENYL INDOLE	27.22	0.022	4.20345
1-METHYL-2,3-DIPHENYL INDOLE	24.64	0.030	2.31779
2-PHENYL-3,2'-METHYLENE INDOLE (TRACE)	28.58	0.019	5.67032
2-PHENYL-3,2'-ETHYLENE INDOLE (TRACE)	31.71	0.014	10.54190
2-PHENYL-3,2'-TRIMETHYLENE INDOLE	27.14	0.022	4.15303
1-METHYL-3,2'-METHYLENE-2-PHENYL INDOLE (TRACE)	31.07	0.015	9.32961
1-METHYL-3,2'-ETHYLENE-2-PHENYL INDOLE	32.56	0.013	12.32556
1-METHYL-3,2'-TRIMETHYLENE-2-PHENYL INDOLE	26.29	0.025	3.43087
4,5-DIPHENYLIMIDAZOLE (A)	16.09	0.107	0.18091
PPF (TRACE)	32.57	0.013	12.39474
PPO (TRACE)	27.33	0.022	4.33788
2-(4-FLUOROPHENYL)-5-PHENYLOXAZOLE (TRACE)	25.94	0.026	3.17755
αNPO (TRACE)	35.40	0.010	30.27140
BBO (B)	41.14	0.006	49.88266
POPOP	40.85	0.007	47.79635
DIMETHYL POPOP (TRACE)	42.19	0.006	57.84775
PPD (TRACE)	11.20	0.318	0.02075
2,5-DI(2-METHYLPHENYL)OXADIAZOLE	12.70	0.219	0.04396
PBD	25.78	0.026	3.06290
BBD (B)	34.56	0.011	17.67281
αNPD (TRACE)	30.08	0.016	7.66442
βNPD	24.96	0.029	2.51690
αNND (TRACE)	33.10	0.012	13.59189
βNND (TRACE)	31.26	0.015	9.69286
TRANS-1-(4-BIPHENYLYL)-2-PHENYL-ETHYLENE	34.44	0.011	17.27904
TRANS-1-(4-BIPHENYLYL)-2-(1-NAPHTHYL)-ETHYLENE	37.69	0.008	29.48341
DIPHENYLSTILBENE (B)	41.50	0.006	52.54001

ACCEPTOR MOLECULES	RO	CO	OVERLAP
1,1,4,4-TETRAPHENYLBUTADIENE	39.24	0.007	37.49293
1,6-DIPHENYLHEXATRIENE	43.88	0.005	73.30201
1,8-DIPHENYL-1,3,5,7-OCTATETRAENE	44.81	0.005	82.63267
BPSB (TRACE)	44.23	0.005	76.91378
BIS-MSB	42.59	0.006	61.28341
ESCULIN (A)	31.95	0.014	10.96178
AZULENE	23.28	0.035	1.61458
NAPHTHALENE	6.44	1.674	0.00075
DEUTERATED NAPHTHALENE-C10D8	8.99	0.616	0.00551
1-METHYLNAPHTHALENE 5ML/L	8.04	0.859	0.00284
2-METHYL NAPHTHALENE	9.81	0.473	0.00935
2,3-DIMETHYL NAPHTHALENE	9.67	0.494	0.00858
2,6-DIMETHYL NAPHTHALENE (TRACE)	13.30	0.190	0.05798
1-NAPHTHOL (TRACE)	14.28	0.153	0.08880
2-NAPHTHOL (TRACE)	17.39	0.085	0.28832
1-NAPHTHYLAMINE	22.86	0.037	1.48115
2-CHLORONAPHTHALENE	10.45	0.392	0.01363
ACENAPHTHENE	12.29	0.241	0.03610
1-PHENYL NAPHTHALENE	12.66	0.221	0.04301
2-PHENYL NAPHTHALENE	19.95	0.056	0.65741
1,4-DIPHENYL NAPHTHALENE	20.91	0.049	0.87071
1,5-DIPHENYL NAPHTHALENE	19.18	0.063	0.51828
1,7-DIPHENYL NAPHTHALENE	19.01	0.065	0.49241
1,4,5,8-TETRAPHENYL NAPHTHALENE	34.24	0.011	16.54870
1,1'-BINAPHTHYL	15.21	0.127	0.12923
4,4'-DI(N-BUTOXY)-1,1'-BINAPHTHYL	26.18	0.025	3.33880
2,2'-BINAPHTHYL	20.85	0.049	0.85771
ACRIDINE (A)	30.75	0.015	8.67733
ANTHRACENE (TRACE)	28.97	0.018	6.09045
DEUTERATED ANTHRACENE C14D10 (TRACE)	28.63	0.019	5.67626
9-METHYL ANTHRACENE (TRACE)	29.16	0.018	6.31487
1-AMINO ANTHRACENE	28.01	0.020	4.91197
2-AMINO ANTHRACENE	27.46	0.022	4.36521
9-VINYL ANTHRACENE	30.31	0.016	7.95492
9-PHENYL ANTHRACENE (TRACE)	30.65	0.016	8.51170
9,10-DIPHENYL ANTHRACENE	31.75	0.014	10.48999
9,10-DICHLOROANTHRACENE (TRACE)	30.45	0.016	8.16541
BIANTHRANYL (B)	33.38	0.012	14.16543
9,10-DINAPHTHYLANTHRACENE (B)	31.20	0.015	9.43760
9-(P-VINYLPHENYL)-10-PHENYL ANTHRACENE	31.92	0.014	10.80758
9,10-BIS(PHENYL-ETHYNYL)ANTHRACENE (TRACE)	30.37	0.016	7.95255
RUBICENE (B)	30.93	0.015	8.88079
DIBENZ(A,H)ANTHRACENE (B)	31.30	0.015	9.65378
TETRACENE (TRACE) (B)	23.59	0.034	1.74742
RUBRENE (B)	26.86	0.023	3.80748
TETRAHYDROCANNABINOL ACID METHYL ESTER	9.46	0.528	0.00755
PHENANTHRENE	14.21	0.156	0.08566
3,4-BENZOPHENANTHRENE	20.87	0.049	0.85039
2-PHENYL PHENANTHRENE	16.68	0.096	0.22313
CHRYSENE (TRACE)	21.24	0.047	0.94829
TRIPHENYLENE	14.80	0.138	0.10930
DEUTERATED TRIPHENYLENE-C18D12	14.52	0.146	0.09743
PYRENE	30.63	0.016	8.51650
1,3,6,8,-TETRAPHENYL PYRENE	39.31	0.007	37.72948
BIPYRENYL (B)	40.10	0.007	42.58500
O-PHENYLENE-PYRENE	34.70	0.011	17.71229
ANTHANTHRENE (B)	32.53	0.013	12.05243
FLUORANTHENE	29.88	0.017	7.28200
3-PHENYL FLUORANTHENE	33.47	0.012	14.37691
3-AMINO FLUORANTHENE	30.94	0.015	8.89013
3,3'-BIFLUORANTHENYL (B)	37.75	0.008	29.50381
BENZO(GHI)FLUORANTHENE	28.34	0.020	5.29620
9,12-DI(N-OCTADECOXY)-BENZ(E)ACEPHENANTHRYLENE	33.28	0.012	13.85001
DECACYCLENE (B)	40.90	0.007	47.46479
10-ETHOXY-BENZO(KL)XANTHENE	32.34	0.013	11.73168
PERYLENE	30.95	0.015	8.94725
BENZO(GHI)PERYLENE (B)	34.18	0.011	16.33752
BBOT (TRACE)	43.08	0.006	65.53204
QUININE SULFATE (A)	26.83	0.023	3.87780
ACRIDINE YELLOW (A)	27.64	0.021	4.51794
ACRIDINE ORANGE (A)	23.63	0.034	1.76370
ACRIDINE RED (A)	26.09	0.025	3.19834
URANINE (A)	26.17	0.025	3.26054
RHODAMINE B (A)	29.73	0.017	7.00236
RHODAMINE 6G (A)	29.83	0.017	7.13921
EOSIN B (A)	26.68	0.024	3.65715
EOSIN Y (A)	27.34	0.022	4.23698
ROSE BENGAL (A)	28.34	0.020	5.24968
INDENOINDENE	26.79	0.023	3.85044

8.89 Donor Molecule: 3,3'''-Dimethyl-3',2''-methylene-*p*-quaterphenyl

ACCEPTOR MOLECULES	RO	CO	OVERLAP
TRIPHENYLAMINE	16.12	0.107	0.13190
P-XYLENE (TRACE)	0.00	0.000	0.00000
SODIUM SALICYLATE (A)	12.69	0.219	0.03139
ANTHRANILIC ACID (A)	25.90	0.026	2.25432
TMPD	20.21	0.054	0.51019
BENZIDINE	11.05	0.332	0.01373

ACCEPTOR MOLECULES	RO	CO	OVERLAP
1,3-DIBIPHENYLYLBENZENE (B)	8.51	0.726	0.00287
1,3,5,-TRIBIPHENYLYLBENZENE (B)	8.87	0.640	0.00369
FLUORENE (TRACE)	0.00	0.000	0.00000
1-METHYL FLUORENE	0.00	0.000	0.00000
1,2-BENZOFLUORENE	16.24	0.104	0.13817
9,10-DIHYDROPHENANTHRENE	0.00	0.000	0.00000
CARBAZOLE (TRACE)	10.30	0.400	0.00943
N-METHYL CARBAZOLE (TRACE)	19.27	0.063	0.38517
N-VINYLCARBAZOLE	17.07	0.090	0.18646
N-PHENYL CARBAZOLE	17.38	0.085	0.20769
N-METHYL-2,3-BENZCARBAZOLE	29.11	0.018	4.52144
DIBENZOFURAN	0.00	0.000	0.00000
DIBENZOTHIOPHENE (TRACE)	6.59	1.565	0.00062
P-TERPHENYL	0.00	0.000	0.00000
DEUTERATED P-TERPHENYL-C18D14	0.00	0.000	0.00000
4-METHYL TERPHENYL	0.00	0.000	0.00000
2-METHYL-P-TERPHENYL	0.00	0.000	0.00000
3,3''-DIMETHYL-P-TERPHENYL	0.00	0.000	0.00000
4-(3,3-DIMETHYLBUTOXY)-P-TERPHENYL	6.15	1.922	0.00041
4,4''-DIHEXAHYDROFARNESOXY-P-TERPHENYL	8.84	0.647	0.00361
2,2'-METHYLENE-P-TERPHENYL	0.00	0.000	0.00000
2,2'-ETHYLENE-P-TERPHENYL	7.51	1.057	0.00135
2,2'-METHYLENE-5',6''-METHYLENE-P-TERPHENYL (TRACE)	18.77	0.068	0.33025
3-PHENYL DIBENZOFURAN	0.00	0.000	0.00000
BENZO(1,2-B-4,5-B')BISBENZOFURAN (TRACE) (B)	23.53	0.034	1.27826
SEXIPHENYL BENZENE	0.00	0.000	0.00000
P-QUATERPHENYL	11.83	0.270	0.02069
3,3'''-DIMETHYL-P-QUATERPHENYL	14.69	0.141	0.07585
TETRAMETHYL-P-QUATERPHENYL	0.00	0.000	0.00000
3,3''' -DIMETHYL-3',2''-METHYLENE-P-QUATERPHENYL	23.40	0.035	1.23576
3,3''' -DIMETHYL-3',2''-ETHYLENE-P-QUATERPHENYL	25.00	0.029	1.83127
4-(3-ETHYLBUTOXY)-P-QUATERPHENYL (B)	22.20	0.041	0.90050
4,4'''-DI(2-BUTYLOCTOXY-1)-P-QUATERPHENYL	22.41	0.040	0.95135
DIETHYL-P-QUINQUEPHENYL	14.84	0.137	0.08056
DI(3-ETHYLHEPTYL)-P-QUINQUEPHENYL	24.32	0.031	1.55194
TETRAMETHYL-P-SEXIPHENYL	12.18	0.247	0.02467
TETRAMETHYL-P-OCTAPHENYL (B)	30.62	0.016	6.17422
INDOLE (TRACE)	0.00	0.000	0.00000
2-PHENYL INDOLE	12.34	0.238	0.02664
1-METHYL-2-PHENYL INDOLE	10.16	0.427	0.00829
1-NEOPENTYL-2-PHENYL INDOLE	5.75	2.357	0.00027
1-N-BUTYL-2(BIPHENYLYL-4) INDOLE	24.70	0.030	1.70335
1,2-DIPHENYL INDOLE	10.47	0.389	0.00996
1-BIPHENYLYL-2-PHENYL INDOLE	16.07	0.108	0.12937
1,3-DIMETHYL-2-PHENYL INDOLE	11.70	0.279	0.01934
2,3-DIPHENYL INDOLE	22.72	0.038	1.03128
1-METHYL-2,3-DIPHENYL INDOLE	20.22	0.054	0.51347
2-PHENYL-3,2'-METHYLENE INDOLE (TRACE)	17.77	0.080	0.23759
2-PHENYL-3,2'-ETHYLENE INDOLE (TRACE)	27.01	0.023	2.91700
2-PHENYL-3,2'-TRIMETHYLENE INDOLE	16.47	0.100	0.15067
1-METHYL-3,2'-METHYLENE-2-PHENYL INDOLE (TRACE)	25.03	0.029	1.85026
1-METHYL-3,2'-ETHYLENE-2-PHENYL INDOLE	30.10	0.016	5.58184
1-METHYL-3,2'-TRIMETHYLENE-2-PHENYL INDOLE	17.79	0.079	0.23907
4,5-DIPHENYLIMIDAZOLE (A)	9.30	0.556	0.00490
PPF (TRACE)	26.21	0.025	2.44023
PPO (TRACE)	15.54	0.119	0.10641
2-(4-FLUOROPHENYL)-5-PHENYLOXAZOLE (TRACE)	14.25	0.155	0.06324
αNPO (TRACE)	34.57	0.011	12.74388
BBO (B)	41.66	0.006	38.98206
POPOP	42.92	0.006	46.59976
DIMETHYL POPOP (TRACE)	45.09	0.005	62.51431
PPD (TRACE)	0.00	0.000	0.00000
2,5-DI(2-METHYLPHENYL)OXADIAZOLE	0.00	0.000	0.00000
PBD	12.77	0.215	0.03277
BBD (B)	28.68	0.019	4.18591
αNPD (TRACE)	25.29	0.028	1.96227
βNPD	15.41	0.122	0.10095
αNND (TRACE)	31.06	0.015	6.72590
βNND (TRACE)	23.54	0.034	1.28068
TRANS-1-(4-BIPHENYLYL)-2-PHENYL-ETHYLENE	29.67	0.017	5.11804
TRANS-1-(4-BIPHENYLYL)-2-(1-NAPHTHYL)-ETHYLENE	37.48	0.008	20.66340
DIPHENYLSTILBENE (B)	41.87	0.006	40.15625
1,1,4,4-TETRAPHENYLBUTADIENE	40.29	0.007	31.81851
1,6-DIPHENYLHEXATRIENE	46.09	0.005	71.38714
1,8-DIPHENYL-1,3,5,7-OCTATETRAENE	49.51	0.004	109.00286
BPSB (TRACE)	46.61	0.004	76.37071
BIS-MSB	44.19	0.005	55.43767
ESCULIN (A)	31.46	0.014	7.23646
AZULENE	19.79	0.058	0.44012
NAPHTHALENE	0.00	0.000	0.00000
DEUTERATED NAPHTHALENE-C10D8	3.75	8.511	0.00002
1-METHYLNAPHTHALENE 5ML/L	0.00	0.000	0.00000
2-METHYL NAPHTHALENE	0.00	0.000	0.00000
2,3-DIMETHYL NAPHTHALENE	0.00	0.000	0.00000
2,6-DIMETHYL NAPHTHALENE (TRACE)	0.00	0.000	0.00000
1-NAPHTHOL (TRACE)	0.00	0.000	0.00000
2-NAPHTHOL (TRACE)	7.30	1.150	0.00114
1-NAPHTHYLAMINE	16.95	0.092	0.17833
2-PHENYL NAPHTHALENE	9.11	0.592	0.00431
1,4-DIPHENYL NAPHTHALENE	12.23	0.245	0.02518
1,5-DIPHENYL NAPHTHALENE	9.20	0.574	0.00458
1,7-DIPHENYL NAPHTHALENE	10.03	0.443	0.00770
1,4,5,8-TETRAPHENYL NAPHTHALENE	33.81	0.012	11.12032
1,1'-BINAPHTHYL	4.66	4.407	0.00008
4,4'-DI(N-BUTOXY)-1,1'-BINAPHTHYL	19.48	0.060	0.41126
2,2'-BINAPHTHYL	10.81	0.354	0.01208

ACCEPTOR MOLECULES	RO	CO	OVERLAP
ACRIDINE (A)	32.24	0.013	8.35595
ANTHRACENE (TRACE)	30.70	0.015	6.24900
DEUTERATED ANTHRACENE C14D10 (TRACE)	30.35	0.016	5.82979
9-METHYL ANTHRACENE (TRACE)	31.69	0.014	7.54331
1-AMINO ANTHRACENE	31.25	0.015	6.87417
2-AMINO ANTHRACENE	29.78	0.017	5.14317
9-VINYL ANTHRACENE	33.25	0.012	10.03735
9-PHENYL ANTHRACENE (TRACE)	33.23	0.012	10.01346
9,10-DIPHENYL ANTHRACENE	35.38	0.010	14.54682
9,10-DICHLOROANTHRACENE (TRACE)	33.97	0.011	11.39567
BIANTHRANYL (B)	37.11	0.009	19.37609
9,10-DINAPHTHYLANTHRACENE (B)	34.72	0.011	12.98541
9-(P-VINYLPHENYL)-10-PHENYL ANTHRACENE	35.60	0.010	15.10089
9,10-BIS(PHENYL-ETHYNYL)ANTHRACENE (TRACE)	35.06	0.010	13.62167
RUBICENE (B)	33.45	0.012	10.25737
DIBENZ(A,H)ANTHRACENE (B)	29.82	0.017	5.23373
TETRACENE (TRACE) (B)	26.62	0.024	2.61191
RUBRENE (B)	24.68	0.030	1.65417
TETRAHYDROCANNABINOL ACID METHYL ESTER	0.00	0.000	0.00000
PHENANTHRENE	11.75	0.276	0.01983
3,4-BENZOPHENANTHRENE	17.78	0.080	0.23592
2-PHENYL PHENANTHRENE	15.12	0.129	0.08978
CHRYSENE (TRACE)	18.01	0.077	0.25538
TRIPHENYLENE	10.50	0.387	0.01006
DEUTERATED TRIPHENYLENE-C18D12	10.35	0.404	0.00924
PYRENE	21.33	0.046	0.70310
1,3,6,8,-TETRAPHENYL PYRENE	43.69	0.005	51.51476
BIPYRENYL (B)	41.67	0.006	38.89912
O-PHENYLENE-PYRENE	37.76	0.008	21.31041
ANTHANTHRENE (B)	36.06	0.010	16.23330
FLUORANTHENE	30.49	0.016	5.96132
3-PHENYL FLUORANTHENE	35.95	0.010	15.98581
3-AMINO FLUORANTHENE	33.65	0.012	10.67340
3,3'-BIFLUORANTHENYL (B)	41.28	0.006	36.61390
BENZO(GHI)FLUORANTHENE	28.09	0.020	3.63956
9,12-DI(N-OCTADECOXY)-BENZ(E)ACEPHENANTHRYLENE	36.09	0.010	16.33042
DECACYCLENE (B)	44.58	0.005	57.68674
10-ETHOXY-BENZO(KL)XANTHENE	35.95	0.010	16.02306
PERYLENE	35.68	0.010	15.23173
BENZO(GHI)PERYLENE (B)	37.68	0.008	21.25526
BBOT (TRACE)	46.94	0.004	79.44176
QUININE SULFATE (A)	20.88	0.049	0.62280
ACRIDINE YELLOW (A)	32.03	0.014	7.90816
ACRIDINE ORANGE (A)	26.87	0.023	2.75659
ACRIDINE RED (A)	27.63	0.021	3.26105
URANINE (A)	28.08	0.020	3.59185
RHODAMINE B (A)	31.67	0.014	7.38495
RHODAMINE 6G (A)	30.72	0.015	6.15701
EOSIN B (A)	26.97	0.023	2.81703
EOSIN Y (A)	28.02	0.020	3.54460
ROSE BENGAL (A)	28.35	0.020	3.80446
INDENOINDENE	14.88	0.136	0.08184

8.90 Donor Molecule: 3,3‴-Dimethyl-3′,2″-ethylene-*p*-quaterphenyl

TRIPHENYLAMINE	12.56	0.226	0.03937
P-XYLENE (TRACE)	0.00	0.000	0.00000
SODIUM SALICYLATE (A)	10.55	0.381	0.01378
ANTHRANILIC ACID (A)	22.71	0.038	1.36529
TMPD	17.17	0.088	0.25606
BENZIDINE	7.83	0.932	0.00231
1,3-DIBIPHENYLYLBENZENE (B)	5.38	2.874	0.00024
1,3,5,-TRIBIPHENYLYLBENZENE (B)	0.00	0.000	0.00000
FLUORENE (TRACE)	0.00	0.000	0.00000
1-METHYL FLUORENE	0.00	0.000	0.00000
1,2-BENZOFLUORENE	12.65	0.221	0.04099
9,10-DIHYDROPHENANTHRENE	0.00	0.000	0.00000
CARBAZOLE (TRACE)	7.02	1.292	0.00121
N-METHYL CARBAZOLE (TRACE)	14.69	0.141	0.10072
N-VINYLCARBAZOLE	12.52	0.228	0.03871
N-PHENYL CARBAZOLE	12.94	0.207	0.04701
N-METHYL-2,3-BENZCARBAZOLE	28.15	0.020	4.92589
DIBENZOFURAN	0.00	0.000	0.00000
DIBENZOTHIOPHENE (TRACE)	4.04	6.790	0.00004
P-TERPHENYL	0.00	0.000	0.00000
DEUTERATED P-TERPHENYL-C18D14	0.00	0.000	0.00000
4-METHYL TERPHENYL	0.00	0.000	0.00000
2-METHYL-P-TERPHENYL	0.00	0.000	0.00000
3,3"-DIMETHYL-P-TERPHENYL	0.00	0.000	0.00000
4-(3,3-DIMETHYLBUTOXY)-P-TERPHENYL	0.00	0.000	0.00000
4,4"-DIHEXAHYDROFARNESOXY-P-TERPHENYL	0.00	0.000	0.00000
2,2'-METHYLENE-P-TERPHENYL	0.00	0.000	0.00000
2,2'-ETHYLENE-P-TERPHENYL	0.00	0.000	0.00000
2,2'-METHYLENE-5',6"-METHYLENE-P-TERPHENYL (TRACE)	12.42	0.234	0.03682
3-PHENYL DIBENZOFURAN	0.00	0.000	0.00000
BENZO(1,2-B-4,5-B')BISBENZOFURAN (TRACE) (B)	17.44	0.084	0.28173
SEXIPHENYL BENZENE	0.00	0.000	0.00000
P-QUATERPHENYL	7.90	0.907	0.00244

ACCEPTOR MOLECULES	RO	CO	OVERLAP
3,3'''-DIMETHYL-P-QUATERPHENYL	10.70	0.365	0.01510
TETRAMETHYL-P-QUATERPHENYL	0.00	0.000	0.00000
3,3''' -DIMETHYL-3',2''-METHYLENE-P-QUATERPHENYL	17.86	0.078	0.32543
3,3''' -DIMETHYL-3',2''-ETHYLENE-P-QUATERPHENYL	20.08	0.055	0.65636
4-(3-ETHYLBUTOXY)-P-QUATERPHENYL (B)	17.75	0.080	0.31783
4,4'''-DI(2-BUTYLOCTOXY-1)-P-QUATERPHENYL	17.70	0.080	0.31672
DIETHYL-P-QUINQUEPHENYL	10.59	0.376	0.01420
DI(3-ETHYLHEPTYL)-P-QUINQUEPHENYL	19.81	0.058	0.60415
TETRAMETHYL-P-SEXIPHENYL	8.06	0.854	0.00276
TETRAMETHYL-P-OCTAPHENYL (B)	25.72	0.026	2.88770
INDOLE (TRACE)	0.00	0.000	0.00000
2-PHENYL INDOLE	7.39	1.107	0.00164
1-METHYL-2-PHENYL INDOLE	6.90	1.362	0.00108
1-NEOPENTYL-2-PHENYL INDOLE	0.00	0.000	0.00000
1-N-BUTYL-2(BIPHENYLYL-4) INDOLE	20.55	0.052	0.75340
1,2-DIPHENYL INDOLE	7.16	1.217	0.00136
1-BIPHENYLYL-2-PHENYL INDOLE	12.97	0.205	0.04773
1,3-DIMETHYL-2-PHENYL INDOLE	7.92	0.899	0.00249
2,3-DIPHENYL INDOLE	19.00	0.065	0.46983
1-METHYL-2,3-DIPHENYL INDOLE	16.76	0.095	0.22131
2-PHENYL-3,2'-METHYLENE INDOLE (TRACE)	12.38	0.236	0.03616
2-PHENYL-3,2'-ETHYLENE INDOLE (TRACE)	21.56	0.045	1.00384
2-PHENYL-3,2'-TRIMETHYLENE INDOLE	11.73	0.277	0.02620
1-METHYL-3,2'-METHYLENE-2-PHENYL INDOLE (TRACE)	19.42	0.061	0.53653
1-METHYL-3,2'-ETHYLENE-2-PHENYL INDOLE	25.65	0.027	2.84295
1-METHYL-3,2' TRIMETHYLENE-2-PHENYL INDOLE	13.05	0.201	0.04944
4,5-DIPHENYLIMIDAZOLE (A)	6.41	1.699	0.00070
PPF (TRACE)	20.30	0.053	0.70060
PPO (TRACE)	10.24	0.417	0.01157
2-(4-FLUOROPHENYL)-5-PHENYLOXAZOLE (TRACE)	9.28	0.559	0.00643
αNPO (TRACE)	30.46	0.016	7.93768
BBO (B)	37.50	0.008	27.61255
POPOP	39.58	0.007	38.15378
DIMETHYL POPOP (TRACE)	42.03	0.006	54.58525
PPD (TRACE)	0.00	0.000	0.00000
2,5-DI(2-METHYLPHENYL)OXADIAZOLE	0.00	0.000	0.00000
PBD	8.07	0.851	0.00278
BBD (B)	23.21	0.036	1.56446
αNPD (TRACE)	20.83	0.049	0.81579
βNPD	11.26	0.313	0.02048
αNND (TRACE)	26.86	0.023	3.74277
βNND (TRACE)	17.69	0.081	0.30741
TRANS-1-(4-BIPHENYLYL)-2-PHENYL-ETHYLENE	24.47	0.031	2.14145
TRANS-1-(4-BIPHENYLYL)-2-(1-NAPHTHYL)-ETHYLENE	33.60	0.012	14.29372
DIPHENYLSTILBENE (B)	37.56	0.008	27.88425
1,1,4,4-TETRAPHENYLBUTADIENE	36.57	0.009	23.70421
1,6-DIPHENYLHEXATRIENE	42.24	0.006	56.27871
1,8-DIPHENYL-1,3,5,7-OCTATETRAENE	47.25	0.004	109.61922
BPSB (TRACE)	43.03	0.006	62.96034
BIS-MSB	40.46	0.007	43.45653
ESCULIN (A)	27.97	0.020	4.76198
AZULENE	16.32	0.103	0.18422
NAPHTHALENE	0.00	0.000	0.00000
DEUTERATED NAPHTHALENE-C10D8	0.00	0.000	0.00000
1-METHYLNAPHTHALENE 5ML/L	0.00	0.000	0.00000
2-METHYL NAPHTHALENE	0.00	0.000	0.00000
2,3-DIMETHYL NAPHTHALENE	0.00	0.000	0.00000
2,6-DIMETHYL NAPHTHALENE (TRACE)	0.00	0.000	0.00000
1-NAPHTHOL (TRACE)	0.00	0.000	0.00000
2-NAPHTHOL (TRACE)	3.89	7.584	0.00003
1-NAPHTHYLAMINE	13.11	0.198	0.05087
1,4-DIPHENYL NAPHTHALENE	8.87	0.641	0.00489
1,5-DIPHENYL NAPHTHALENE	6.02	2.054	0.00048
1,7-DIPHENYL NAPHTHALENE	6.90	1.363	0.00108
1,4,5,8-TETRAPHENYL NAPHTHALENE	30.16	0.016	7.46453
1,1'-BINAPHTHYL	0.00	0.000	0.00000
4,4'-DI(N-BUTOXY)-1,1'-BINAPHTHYL	15.53	0.119	0.14052
2,2'-BINAPHTHYL	7.21	1.195	0.00141
ACRIDINE (A)	29.57	0.017	6.63372
ANTHRACENE (TRACE)	28.51	0.019	5.32981
DEUTERATED ANTHRACENE C14D10 (TRACE)	27.97	0.020	4.75539
9-METHYL ANTHRACENE (TRACE)	29.79	0.017	6.92357
1-AMINO ANTHRACENE	30.28	0.016	7.58249
2-AMINO ANTHRACENE	28.82	0.019	5.63638
9-VINYL ANTHRACENE	31.45	0.014	9.58002
9-PHENYL ANTHRACENE (TRACE)	31.30	0.015	9.32177
9,10-DIPHENYL ANTHRACENE	33.86	0.012	14.89061
9,10-DICHLOROANTHRACENE (TRACE)	32.97	0.012	12.68326
BIANTHRANYL (B)	35.43	0.010	19.56073
9,10-DINAPHTHYLANTHRACENE (B)	33.51	0.012	13.97890
9-(P-VINYLPHENYL)-10-PHENYL ANTHRACENE	34.19	0.011	15.77533
9,10-BIS(PHENYL-ETHYNYL)ANTHRACENE (TRACE)	35.95	0.010	21.08972
RUBICENE (B)	32.65	0.013	11.83772
DIBENZ(A,H)ANTHRACENE (B)	25.94	0.026	3.01344
TETRACENE (TRACE) (B)	26.82	0.023	3.63704
RUBRENE (B)	24.20	0.032	1.96208
TETRAHYDROCANNABINOL ACID METHYL ESTER	0.00	0.000	0.00000
PHENANTHRENE	9.12	0.589	0.00578
3,4-BENZOPHENANTHRENE	16.02	0.109	0.16805
2-PHENYL PHENANTHRENE	12.94	0.206	0.04689
CHRYSENE (TRACE)	15.96	0.110	0.16446
TRIPHENYLENE	7.86	0.922	0.00236
DEUTERATED TRIPHENYLENE-C18D12	7.83	0.933	0.00230
PYRENE	17.01	0.091	0.24091
1,3,6,8,-TETRAPHENYL PYRENE	41.73	0.006	52.06451
BIPYRENYL (B)	38.16	0.008	30.51669

ACCEPTOR MOLECULES	RO	CO	OVERLAP
O-PHENYLENE-PYRENE	36.01	0.010	21.34122
ANTHANTHRENE (B)	35.72	0.010	20.41179
FLUORANTHENE	27.57	0.021	4.34099
3-PHENYL FLUORANTHENE	33.50	0.012	13.95307
3-AMINO FLUORANTHENE	32.60	0.013	11.74754
3,3'-BIFLUORANTHENYL (B)	39.32	0.007	36.39446
BENZO(GHI)FLUORANTHENE	25.03	0.029	2.42315
9,12-DI(N-OCTADECOXY)-BENZ(E)ACEPHENANTHRYLENE	34.74	0.011	17.30119
DECACYCLENE (B)	42.93	0.006	61.21070
10-ETHOXY-BENZO(KL)XANTHENE	34.49	0.011	16.63777
PERYLENE	36.10	0.010	21.75495
BENZO(GHI)PERYLENE (B)	35.86	0.010	21.01712
BBOT (TRACE)	44.28	0.005	74.61825
QUININE SULFATE (A)	16.11	0.107	0.17540
ACRIDINE YELLOW (A)	33.11	0.012	12.87895
ACRIDINE ORANGE (A)	27.76	0.021	4.46583
ACRIDINE RED (A)	26.11	0.025	3.09396
URANINE (A)	28.32	0.020	5.03740
RHODAMINE B (A)	30.04	0.017	7.17733
RHODAMINE 6G (A)	28.31	0.020	5.02931
EOSIN B (A)	25.95	0.026	2.98142
EOSIN Y (A)	27.20	0.022	3.95648
ROSE BENGAL (A)	26.87	0.023	3.67712
INDENOINDENE	10.12	0.431	0.01080

8.92 Donor Molecule: 4,4'''-Di(2-butyloctoxy-1)-*p*-quaterphenyl

ACCEPTOR MOLECULES	RO	CO	OVERLAP
TRIPHENYLAMINE	11.93	0.263	0.02331
P-XYLENE (TRACE)	0.00	0.000	0.00000
SODIUM SALICYLATE (A)	10.26	0.414	0.00941
ANTHRANILIC ACID (A)	22.92	0.037	1.16143
TMPD	17.08	0.090	0.19992
BENZIDINE	7.15	1.224	0.00108
1,2-BENZOFLUORENE	11.99	0.260	0.02396
9,10-DIHYDROPHENANTHRENE	0.00	0.000	0.00000
CARBAZOLE (TRACE)	6.41	1.699	0.00056
N-METHYL CARBAZOLE (TRACE)	13.86	0.168	0.05738
N-VINYLCARBAZOLE	11.47	0.297	0.01836
N-PHENYL CARBAZOLE	12.01	0.258	0.02430
N-METHYL-2,3-BENZCARBAZOLE	29.17	0.018	4.91736
2,2'-METHYLENE-5',6''-METHYLENE-P-TERPHENYL (TRACE)	11.08	0.329	0.01493
3-PHENYL DIBENZOFURAN	0.00	0.000	0.00000
BENZO(1,2-B-4,5-B')BISBENZOFURAN (TRACE) (B)	16.16	0.106	0.14397
SEXIPHENYL BENZENE	0.00	0.000	0.00000
P-QUATERPHENYL	6.86	1.386	0.00084
3,3'''-DIMETHYL-P-QUATERPHENYL	9.91	0.459	0.00767
TETRAMETHYL-P-QUATERPHENYL	0.00	0.000	0.00000
3,3''' -DIMETHYL-3',2''-METHYLENE-P-QUATERPHENYL	16.83	0.094	0.18377
3,3''' -DIMETHYL-3',2''-ETHYLENE-P-QUATERPHENYL	19.44	0.061	0.43551
4-(3-ETHYLBUTOXY)-P-QUATERPHENYL (B)	17.08	0.090	0.20037
4,4'''-DI(2-BUTYLOCTOXY-1)-P-QUATERPHENYL	17.03	0.090	0.19716
DIETHYL-P-QUINQUEPHENYL	9.72	0.487	0.00684
DI(3-ETHYLHEPTYL)-P-QUINQUEPHENYL	19.34	0.062	0.42140
TETRAMETHYL-P-SEXIPHENYL	6.90	1.363	0.00087
TETRAMETHYL-P-OCTAPHENYL (B)	25.47	0.027	2.19444
INDOLE (TRACE)	0.00	0.000	0.00000
2-PHENYL INDOLE	5.95	2.122	0.00036
1-METHYL-2-PHENYL INDOLE	6.16	1.917	0.00044
1-NEOPENTYL-2-PHENYL INDOLE	0.00	0.000	0.00000
1-N-BUTYL-2(BIPHENYLYL-4) INDOLE	20.23	0.054	0.55222
1,2-DIPHENYL INDOLE	6.42	1.693	0.00057
1-BIPHENYLYL-2-PHENYL INDOLE	12.52	0.228	0.03105
1,3-DIMETHYL-2-PHENYL INDOLE	7.01	1.301	0.00096
2,3-DIPHENYL INDOLE	18.79	0.067	0.35403
1-METHYL-2,3-DIPHENYL INDOLE	16.46	0.100	0.16001
2-PHENYL-3,2'-METHYLENE INDOLE (TRACE)	11.25	0.314	0.01638
2-PHENYL-3,2'-ETHYLENE INDOLE (TRACE)	20.69	0.051	0.63183
2-PHENYL-3,2' -TRIMETHYLENE INDOLE	10.76	0.359	0.01254
1-METHYL-3,2'-METHYLENE-2-PHENYL INDOLE (TRACE)	18.44	0.071	0.31678
1-METHYL-3,2'-ETHYLENE-2-PHENYL INDOLE	25.34	0.027	2.13194
1-METHYL-3,2' TRIMETHYLENE-2-PHENYL INDOLE	12.07	0.254	0.02500
4,5-DIPHENYLIMIDAZOLE (A)	5.64	2.495	0.00026
PPF (TRACE)	19.23	0.063	0.40808
PPO (TRACE)	8.92	0.630	0.00408
2-(4-FLUOROPHENYL)-5-PHENYLOXAZOLE (TRACE)	8.10	0.841	0.00229
αNPO (TRACE)	30.90	0.015	6.98015
BBO (B)	38.22	0.008	24.96638
POPOP	40.66	0.007	36.13798
DIMETHYL POPOP (TRACE)	43.37	0.005	53.12215
PPD (TRACE)	0.00	0.000	0.00000
2,5-DI(2-METHYLPHENYL)OXADIAZOLE	0.00	0.000	0.00000
PBD	6.68	1.502	0.00072
BBD (B)	22.38	0.040	1.01440
αNPD (TRACE)	20.55	0.052	0.60575
βNPD	10.37	0.401	0.01008
αNND (TRACE)	26.93	0.023	3.06474
βNND (TRACE)	16.57	0.098	0.16711
TRANS-1-(4-BIPHENYLYL) 2 PHENYL-ETHYLENE	23.88	0.033	1.49066
TRANS-1-(4-BIPHENYLYL)-2-(1-NAPHTHYL)-ETHYLENE	34.21	0.011	12.83410
DIPHENYLSTILBENE (B)	38.27	0.008	25.14942
1,1,4,4-TETRAPHENYLBUTADIENE	37.40	0.009	21.85386

ACCEPTOR MOLECULES	RO	CO	OVERLAP
1,6-DIPHENYLHEXATRIENE	43.14	0.006	51.54131
1,8-DIPHENYL-1,3,5,7-OCTATETRAENE	48.85	0.004	107.97658
BPSB (TRACE)	44.25	0.005	59.97286
BIS-MSB	41.49	0.006	40.76664
ESCULIN (A)	28.40	0.020	4.20866
AZULENE	15.92	0.111	0.12829
NAPHTHALENE	0.00	0.000	0.00000
DEUTERATED NAPHTHALENE-C10D8	0.00	0.000	0.00000
1-METHYLNAPHTHALENE 5ML/L	0.00	0.000	0.00000
2-METHYL NAPHTHALENE	0.00	0.000	0.00000
2,3-DIMETHYL NAPHTHALENE	0.00	0.000	0.00000
2,6-DIMETHYL NAPHTHALENE (TRACE)	0.00	0.000	0.00000
1-NAPHTHOL (TRACE)	0.00	0.000	0.00000
2-NAPHTHOL (TRACE)	0.00	0.000	0.00000
1-NAPHTHYLAMINE	12.42	0.233	0.02968
2-CHLORONAPHTHALENE	0.00	0.000	0.00000
ACENAPHTHENE	0.00	0.000	0.00000
1-PHENYL NAPHTHALENE	0.00	0.000	0.00000
2-PHENYL NAPHTHALENE	0.00	0.000	0.00000
1,4-DIPHENYL NAPHTHALENE	8.20	0.810	0.00246
1,5-DIPHENYL NAPHTHALENE	5.13	3.315	0.00015
1,7-DIPHENYL NAPHTHALENE	6.19	1.886	0.00046
1,4,5,8-TETRAPHENYL NAPHTHALENE	30.66	0.016	6.64335
1,1'-BINAPHTHYL	0.00	0.000	0.00000
4,4'-DI(N-BUTOXY)-1,1'-BINAPHTHYL	14.90	0.135	0.08841
2,2'-BINAPHTHYL	6.22	1.857	0.00047
ACRIDINE (A)	30.33	0.016	6.22497
ANTHRACENE (TRACE)	29.10	0.018	4.85990
DEUTERATED ANTHRACENE C14D10 (TRACE)	28.46	0.019	4.25788
9-METHYL ANTHRACENE (TRACE)	31.02	0.015	7.11494
1-AMINO ANTHRACENE	31.51	0.014	7.75599
2-AMINO ANTHRACENE	29.95	0.017	5.71846
9-VINYL ANTHRACENE	32.60	0.013	9.57113
9-PHENYL ANTHRACENE (TRACE)	32.56	0.013	9.51641
9,10-DIPHENYL ANTHRACENE	34.99	0.010	14.61571
9,10-DICHLOROANTHRACENE (TRACE)	34.23	0.011	12.81789
BIANTHRANYL (B)	36.63	0.009	19.23682
9,10-DINAPHTHYLANTHRACENE (B)	34.69	0.011	13.86759
9-(P-VINYLPHENYL)-10-PHENYL ANTHRACENE	35.38	0.010	15.61173
9,10-BIS(PHENYL-ETHYNYL)ANTHRACENE (TRACE)	37.73	0.008	22.73432
RUBICENE (B)	33.95	0.011	12.08184
DIBENZ(A,H)ANTHRACENE (B)	25.81	0.026	2.35751
TETRACENE (TRACE) (B)	28.03	0.020	3.82036
RUBRENE (B)	25.15	0.028	1.99472
TETRAHYDROCANNABINOL ACID METHYL ESTER	0.00	0.000	0.00000
PHENANTHRENE	8.62	0.698	0.00332
3,4-BENZOPHENANTHRENE	16.33	0.103	0.15172
2-PHENYL PHENANTHRENE	12.87	0.210	0.03654
CHRYSENE (TRACE)	16.25	0.104	0.14791
TRIPHENYLENE	7.37	1.119	0.00129
DEUTERATED TRIPHENYLENE-C18D12	7.37	1.117	0.00130
PYRENE	16.96	0.092	0.19074
1,3,6,8,-TETRAPHENYL PYRENE	43.29	0.006	52.33628
BIPYRENYL (B)	39.12	0.007	28.56598
O-PHENYLENE-PYRENE	37.35	0.009	21.41131
ANTHANTHRENE (B)	37.35	0.009	21.51410
FLUORANTHENE	28.23	0.020	4.02817
3-PHENYL FLUORANTHENE	34.51	0.011	13.43568
3-AMINO FLUORANTHENE	33.92	0.011	12.00894
3,3'-BIFLUORANTHENYL (B)	40.74	0.007	36.30945
BENZO(GHI)FLUORANTHENE	25.35	0.027	2.11141
9,12-DI(N-OCTADECOXY)-BENZ(E)ACEPHENANTHRYLENE	36.07	0.010	17.47630
DECACYCLENE (B)	44.60	0.005	62.04681
10-ETHOXY-BENZO(KL)XANTHENE	35.72	0.010	16.55928
PERYLENE	37.92	0.008	23.55674
BENZO(GHI)PERYLENE (B)	37.35	0.009	21.64345
BBOT (TRACE)	45.73	0.005	72.96618
QUININE SULFATE (A)	15.24	0.126	0.10135
ACRIDINE YELLOW (A)	34.78	0.011	13.94760
ACRIDINE ORANGE (A)	29.03	0.018	4.72220
ACRIDINE RED (A)	26.94	0.023	3.01692
URANINE (A)	29.48	0.017	5.17316
RHODAMINE B (A)	31.03	0.015	7.03083
RHODAMINE 6G (A)	28.91	0.019	4.60755
EOSIN B (A)	26.81	0.023	2.93007
EOSIN Y (A)	28.18	0.020	3.94824
ROSE BENGAL (A)	27.76	0.021	3.60968
INDENOINDENE	9.17	0.580	0.00481

8.93 Donor Molecule: Diethyl-*p*-quinquephenyl

TRIPHENYLAMINE	16.30	0.103	0.15347
P-XYLENE (TRACE)	0.00	0.000	0.00000
SODIUM SALICYLATE (A)	12.39	0.235	0.02968
ANTHRANILIC ACID (A)	25.15	0.028	2.06093
TMPD	19.73	0.058	0.48088
BENZIDINE	12.02	0.257	0.02478
1,3-DIBIPHENYLYLBENZENE (B)	9.83	0.471	0.00743
1,3,5,-TRIBIPHENYLYLBENZENE (B)	10.93	0.342	0.01407
FLUORENE (TRACE)	0.00	0.000	0.00000
1-METHYL FLUORENE	0.00	0.000	0.00000
1,2-BENZOFLUORENE	16.15	0.106	0.14518

ACCEPTOR MOLECULES	RO	CO	OVERLAP
CARBAZOLE (TRACE)	12.19	0.247	0.02695
N-METHYL CARBAZOLE (TRACE)	19.39	0.061	0.43547
N-VINYLCARBAZOLE	17.85	0.079	0.26495
N-PHENYL CARBAZOLE	17.95	0.077	0.27388
N-METHYL-2,3-BENZCARBAZOLE	28.73	0.019	4.55006
DIBENZOFURAN	0.00	0.000	0.00000
DIBENZOTHIOPHENE (TRACE)	8.25	0.797	0.00259
P-TERPHENYL	0.00	0.000	0.00000
DEUTERATED P-TERPHENYL-C18D14	0.00	0.000	0.00000
4-METHYL TERPHENYL	0.00	0.000	0.00000
2-METHYL-P-TERPHENYL	0.00	0.000	0.00000
3,3''-DIMETHYL-P-TERPHENYL	0.00	0.000	0.00000
4-(3,3-DIMETHYLBUTOXY)-P-TERPHENYL	8.38	0.760	0.00285
4,4''-DIHEXAHYDROFARNESOXY-P-TERPHENYL	11.03	0.334	0.01479
2,2'-METHYLENE-P-TERPHENYL	7.48	1.070	0.00144
2,2'-ETHYLENE-P-TERPHENYL	9.66	0.496	0.00669
2,2'-METHYLENE-5',6''-METHYLENE-P-TERPHENYL (TRACE)	21.18	0.047	0.74087
3-PHENYL DIBENZOFURAN	7.33	1.135	0.00128
BENZO(1,2-B-4,5-B')BISBENZOFURAN (TRACE) (B)	24.48	0.030	1.76470
SEXIPHENYL BENZENE	0.00	0.000	0.00000
P-QUATERPHENYL	13.39	0.186	0.04728
3,3'''-DIMETHYL-P-QUATERPHENYL	15.63	0.117	0.11983
TETRAMETHYL-P-QUATERPHENYL	0.00	0.000	0.00000
3,3''' -DIMETHYL-3',2''-METHYLENE-P-QUATERPHENYL	24.00	0.032	1.56527
3,3''' -DIMETHYL-3',2''-ETHYLENE-P-QUATERPHENYL	25.04	0.029	2.01302
4-(3-ETHYLBUTOXY)-P-QUATERPHENYL (B)	22.27	0.041	0.99746
4,4'''-DI(2-BUTYLOCTOXY-1)-P-QUATERPHENYL	22.51	0.039	1.06422
DIETHYL-P-QUINQUEPHENYL	16.04	0.108	0.13984
DI(3-ETHYLHEPTYL)-P-QUINQUEPHENYL	24.26	0.031	1.66386
TETRAMETHYL-P-SEXIPHENYL	13.90	0.167	0.05922
TETRAMETHYL-P-OCTAPHENYL (B)	30.11	0.016	6.07730
INDOLE (TRACE)	0.00	0.000	0.00000
2-PHENYL INDOLE	14.70	0.141	0.08310
1-METHYL-2-PHENYL INDOLE	11.51	0.294	0.01905
1-NEOPENTYL-2-PHENYL INDOLE	7.84	0.929	0.00191
1-N-BUTYL-2(BIPHENYLYL-4) INDOLE	24.33	0.031	1.69342
1,2-DIPHENYL INDOLE	11.82	0.271	0.02243
1-BIPHENYLYL-2-PHENYL INDOLE	16.17	0.106	0.14642
1,3-DIMETHYL-2-PHENYL INDOLE	13.13	0.197	0.04214
2,3-DIPHENYL INDOLE	22.39	0.040	1.02838
1-METHYL-2,3-DIPHENYL INDOLE	19.98	0.056	0.51907
2-PHENYL-3,2'-METHYLENE INDOLE (TRACE)	19.41	0.061	0.43871
2-PHENYL-3,2'-ETHYLENE INDOLE (TRACE)	26.60	0.024	2.89680
2-PHENYL-3,2'-TRIMETHYLENE INDOLE	17.93	0.078	0.27292
1-METHYL-3,2'-METHYLENE-2-PHENYL INDOLE (TRACE)	25.06	0.028	2.02856
1-METHYL-3,2'-ETHYLENE-2-PHENYL INDOLE	29.21	0.018	5.07032
1-METHYL-3,2'-TRIMETHYLENE-2-PHENYL INDOLE	18.75	0.068	0.35601
4,5-DIPHENYLIMIDAZOLE (A)	10.22	0.419	0.00937
PPF (TRACE)	26.23	0.025	2.66488
PPO (TRACE)	17.56	0.083	0.24042
2-(4-FLUOROPHENYL)-5-PHENYLOXAZOLE (TRACE)	16.29	0.103	0.15354
αNPO (TRACE)	33.58	0.012	11.65988
BBO (B)	40.46	0.007	35.63109
POPOP	41.82	0.006	43.41674
DIMETHYL POPOP (TRACE)	44.04	0.005	59.08357
PPD (TRACE)	0.00	0.000	0.00000
2,5-DI(2-METHYLPHENYL)OXADIAZOLE	0.00	0.000	0.00000
PBD	14.97	0.133	0.09248
BBD (B)	28.30	0.020	4.19953
αNPD (TRACE)	25.08	0.028	2.02929
βNPD	16.31	0.103	0.15452
αNND (TRACE)	30.24	0.016	6.23578
βNND (TRACE)	24.07	0.032	1.59556
TRANS-1-(4-BIPHENYLYL)-2-PHENYL-ETHYLENE	29.10	0.018	4.95550
TRANS-1-(4-BIPHENYLYL)-2-(1-NAPHTHYL)-ETHYLENE	36.44	0.009	19.00364
DIPHENYLSTILBENE (B)	40.66	0.007	36.67057
1,1,4,4-TETRAPHENYLBUTADIENE	39.17	0.007	29.24945
1,6-DIPHENYLHEXATRIENE	44.77	0.005	65.22622
1,8-DIPHENYL-1,3,5,7-OCTATETRAENE	48.61	0.004	106.25255
BPSB (TRACE)	45.43	0.005	71.24829
BIS-MSB	43.02	0.006	51.38348
ESCULIN (A)	30.57	0.016	6.63955
AZULENE	19.63	0.059	0.45795
NAPHTHALENE	0.00	0.000	0.00000
DEUTERATED NAPHTHALENE-C10D8	4.65	4.443	0.00008
1-METHYLNAPHTHALENE 5ML/L	0.00	0.000	0.00000
2-METHYL NAPHTHALENE	0.00	0.000	0.00000
2,3-DIMETHYL NAPHTHALENE	0.00	0.000	0.00000
2,6-DIMETHYL NAPHTHALENE (TRACE)	4.17	6.177	0.00004
1-NAPHTHOL (TRACE)	4.85	3.924	0.00011
2-NAPHTHOL (TRACE)	9.57	0.511	0.00630
1-NAPHTHYLAMINE	17.24	0.087	0.21500
2-CHLORONAPHTHALENE	3.11	14.938	0.00001
ACENAPHTHENE	0.00	0.000	0.00000
1-PHENYL NAPHTHALENE	4.58	4.662	0.00008
2-PHENYL NAPHTHALENE	11.06	0.331	0.01506
1,4-DIPHENYL NAPHTHALENE	13.12	0.198	0.04189
1,5-DIPHENYL NAPHTHALENE	10.65	0.371	0.01197
1,7-DIPHENYL NAPHTHALENE	11.21	0.317	0.01631
1,4,5,8-TETRAPHENYL NAPHTHALENE	32.87	0.013	10.21886
1,1'-BINAPHTHYL	6.97	1.319	0.00095
4,4'-DI(N-BUTOXY)-1,1'-BINAPHTHYL	19.59	0.059	0.46262
2,2'-BINAPHTHYL	12.22	0.245	0.02734
ACRIDINE (A)	31.37	0.014	7.72352
ANTHRACENE (TRACE)	29.89	0.017	5.78784
DEUTERATED ANTHRACENE C14D10 (TRACE)	29.45	0.018	5.29831
9-METHYL ANTHRACENE (TRACE)	31.06	0.015	7.27329
1-AMINO ANTHRACENE	30.90	0.015	6.98711

ACCEPTOR MOLECULES	RO	CO	OVERLAP
2-AMINO ANTHRACENE	29.47	0.017	5.25709
9-VINYL ANTHRACENE	32.59	0.013	9.68302
9-PHENYL ANTHRACENE (TRACE)	32.59	0.013	9.70217
9,10-DIPHENYL ANTHRACENE	34.74	0.011	14.19349
9,10-DICHLOROANTHRACENE (TRACE)	33.57	0.012	11.55001
BIANTHRANYL (B)	36.42	0.009	18.86152
9,10-DINAPHTHYLANTHRACENE (B)	34.21	0.011	12.93601
9-(P-VINYLPHENYL)-10-PHENYL ANTHRACENE	35.01	0.010	14.86414
9,10-BIS(PHENYL-ETHYNYL)ANTHRACENE (TRACE)	35.59	0.010	16.26509
RUBICENE (B)	33.17	0.012	10.66544
DIBENZ(A,H)ANTHRACENE (B)	28.82	0.019	4.64039
TETRACENE (TRACE) (B)	26.78	0.023	2.95121
RUBRENE (B)	24.73	0.030	1.83130
TETRAHYDROCANNABINOL ACID METHYL ESTER	0.00	0.000	0.00000
PHENANTHRENE	11.67	0.281	0.02070
3,4-BENZOPHENANTHRENE	17.32	0.086	0.21930
2-PHENYL PHENANTHRENE	14.69	0.141	0.08192
CHRYSENE (TRACE)	17.54	0.083	0.23750
TRIPHENYLENE	10.93	0.342	0.01399
DEUTERATED TRIPHENYLENE-C18D12	10.68	0.367	0.01219
PYRENE	22.59	0.039	1.08017
1,3,6,8,-TETRAPHENYL PYRENE	42.92	0.006	50.38849
BIPYRENYL (B)	40.56	0.007	35.97458
O-PHENYLENE-PYRENE	37.15	0.009	21.02539
ANTHANTHRENE (B)	36.05	0.010	17.61830
FLUORANTHENE	29.66	0.017	5.49879
3-PHENYL FLUORANTHENE	35.07	0.010	15.00502
3-AMINO FLUORANTHENE	33.32	0.012	10.94492
3,3'-BIFLUORANTHENYL (B)	40.55	0.007	35.78718
BENZO(GHI)FLUORANTHENE	27.28	0.022	3.32045
9,12-DI(N-OCTADECOXY)-BENZ(E)ACEPHENANTHRYLENE	35.65	0.010	16.52841
DECACYCLENE (B)	44.01	0.005	58.17287
10-ETHOXY-BENZO(KL)XANTHENE	35.36	0.010	15.79077
PERYLENE	36.00	0.010	17.47373
BENZO(GHI)PERYLENE (B)	37.03	0.009	20.83013
BBOT (TRACE)	45.95	0.005	76.08890
QUININE SULFATE (A)	21.14	0.047	0.73091
ACRIDINE YELLOW (A)	32.63	0.013	9.66819
ACRIDINE ORANGE (A)	27.22	0.022	3.25534
ACRIDINE RED (A)	27.12	0.022	3.18238
URANINE (A)	28.16	0.020	3.99111
RHODAMINE B (A)	31.11	0.015	7.25443
RHODAMINE 6G (A)	29.95	0.017	5.78144
EOSIN B (A)	26.71	0.023	2.90494
EOSIN Y (A)	27.82	0.021	3.70880
ROSE BENGAL (A)	27.98	0.020	3.84249
INDENOINDENE	16.87	0.093	0.18919

8.94 Donor Molecule: Di(3-ethylheptyl)-*p*-quinquephenyl

TRIPHENYLAMINE	10.62	0.374	0.01077
P-XYLENE (TRACE)	0.00	0.000	0.00000
SODIUM SALICYLATE (A)	9.40	0.539	0.00517
ANTHRANILIC ACID (A)	22.08	0.042	0.86378
TMPD	16.12	0.107	0.13132
BENZIDINE	5.94	2.139	0.00033
1,2-BENZOFLUORENE	10.60	0.376	0.01064
9,10-DIHYDROPHENANTHRENE	0.00	0.000	0.00000
CARBAZOLE (TRACE)	5.34	2.934	0.00018
N METHYL CARBAZOLE (TRACE)	12.05	0.256	0.02297
N-VINYLCARBAZOLE	9.73	0.485	0.00639
N-PHENYL CARBAZOLE	10.29	0.411	0.00892
N-METHYL-2,3-BENZCARBAZOLE	29.58	0.017	4.97344
2,2'-METHYLENE-5',6'-METHYLENE-P-TERPHENYL (TRACE)	9.34	0.549	0.00500
3-PHENYL DIBENZOFURAN	0.00	0.000	0.00000
BENZO(1,2-B-4,5-B')BISBENZOFURAN (TRACE) (B)	13.76	0.172	0.05094
SEXIPHENYL BENZENE	0.00	0.000	0.00000
P-QUATERPHENYL	5.37	2.888	0.00018
3,3'''-DIMETHYL-P-QUATERPHENYL	8.50	0.729	0.00283
TETRAMETHYL-P-QUATERPHENYL	0.00	0.000	0.00000
3,3''' -DIMETHYL-3',2''-METHYLENE-P-QUATERPHENYL	14.83	0.137	0.07985
3,3''' -DIMETHYL-3',2''-ETHYLENE-P-QUATERPHENYL	17.74	0.080	0.23392
4-(3-ETHYLBUTOXY)-P-QUATERPHENYL (B)	15.47	0.121	0.10278
4,4'''-DI(2-BUTYLOCTOXY-1)-P-QUATERPHENYL	15.34	0.124	0.09761
DIETHYL-P-QUINQUEPHENYL	8.19	0.815	0.00226
DI(3-ETHYLHEPTYL)-P-QUINQUEPHENYL	17.86	0.079	0.24273
TETRAMETHYL-P-SEXIPHENYL	5.35	2.927	0.00018
TETRAMETHYL-P-OCTAPHENYL (B)	23.92	0.033	1.40235
INDOLE (TRACE)	0.00	0.000	0.00000
2-PHENYL INDOLE	4.54	4.779	0.00007
1-METHYL-2-PHENYL INDOLE	5.04	3.492	0.00012
1-NEOPENTYL-2-PHENYL INDOLE	0.00	0.000	0.00000
1-N-BUTYL-2(BIPHENYLYL-4) INDOLE	18.81	0.067	0.33162
1,2-DIPHENYL INDOLE	5.27	3.054	0.00016
1-BIPHENYLYL-2-PHENYL INDOLE	11.38	0.304	0.01627
1,3-DIMETHYL-2-PHENYL INDOLE	5.64	2.488	0.00024
2,3-DIPHENYL INDOLE	17.66	0.081	0.22670
1-METHYL-2,3-DIPHENYL INDOLE	15.26	0.126	0.09458
2-PHENYL-3,2'-METHYLENE INDOLE (TRACE)	9.63	0.501	0.00599
2-PHENYL 3,2' ETHYLENE INDOLE (TRACE)	18.64	0.069	0.31465
2-PHENYL-3,2'-TRIMETHYLENE INDOLE	9.19	0.576	0.00454

ACCEPTOR MOLECULES	RO	CO	OVERLAP
1-METHYL-3,2'-METHYLENE-2-PHENYL INDOLE (TRACE)	16.32	0.103	0.14159
1-METHYL-3,2'-ETHYLENE-2-PHENYL INDOLE	23.63	0.034	1.30321
1-METHYL-3,2'-TRIMETHYLENE-2-PHENYL INDOLE	10.37	0.401	0.00937
4,5-DIPHENYLIMIDAZOLE (A)	4.44	5.120	0.00006
PPF (TRACE)	16.99	0.091	0.18021
PPO (TRACE)	7.31	1.144	0.00115
2-(4-FLUOROPHENYL)-5-PHENYLOXAZOLE (TRACE)	6.63	1.532	0.00064
αNPO (TRACE)	29.97	0.017	5.40073
BBO (B)	37.47	0.008	20.61930
POPOP	40.19	0.007	31.36569
DIMETHYL POPOP (TRACE)	43.34	0.005	49.16393
PPD (TRACE)	0.00	0.000	0.00000
2,5-DI(2-METHYLPHENYL)OXADIAZOLE	0.00	0.000	0.00000
PBD	5.12	3.332	0.00014
BBD (B)	20.31	0.053	0.52671
αNPD (TRACE)	19.19	0.063	0.37351
βNPD	8.78	0.661	0.00345
αNND (TRACE)	25.52	0.027	2.06600
βNND (TRACE)	14.30	0.153	0.06435
TRANS-1-(4-BIPHENYLYL)-2-PHENYL-ETHYLENE	21.99	0.042	0.84711
TRANS-1-(4-BIPHENYLYL)-2-(1-NAPHTHYL)-ETHYLENE	33.47	0.012	10.47092
DIPHENYLSTILBENE (B)	37.48	0.008	20.63697
1,1,4,4-TETRAPHENYLBUTADIENE	36.85	0.009	18.58746
1,6-DIPHENYLHEXATRIENE	42.58	0.006	44.30441
1,8-DIPHENYL-1,3,5,7-OCTATETRAENE	49.29	0.004	105.90168
BPSB (TRACE)	43.88	0.005	53.04096
BIS-MSB	41.05	0.006	35.53020
ESCULIN (A)	27.62	0.021	3.31301
AZULENE	14.77	0.139	0.07592
NAPHTHALENE	0.00	0.000	0.00000
DEUTERATED NAPHTHALENE-C10D8	0.00	0.000	0.00000
1-METHYLNAPHTHALENE 5ML/L	0.00	0.000	0.00000
2-METHYL NAPHTHALENE	0.00	0.000	0.00000
2,3-DIMETHYL NAPHTHALENE	0.00	0.000	0.00000
2,6-DIMETHYL NAPHTHALENE (TRACE)	0.00	0.000	0.00000
1-NAPHTHOL (TRACE)	0.00	0.000	0.00000
2-NAPHTHOL (TRACE)	0.00	0.000	0.00000
1-NAPHTHYLAMINE	11.03	0.333	0.01352
2-CHLORONAPHTHALENE	0.00	0.000	0.00000
ACENAPHTHENE	0.00	0.000	0.00000
1-PHENYL NAPHTHALENE	0.00	0.000	0.00000
2-PHENYL NAPHTHALENE	0.00	0.000	0.00000
1,4-DIPHENYL NAPHTHALENE	7.03	1.286	0.00091
1,5-DIPHENYL NAPHTHALENE	4.00	6.999	0.00003
1,7-DIPHENYL NAPHTHALENE	5.08	3.416	0.00013
1,4,5,8-TETRAPHENYL NAPHTHALENE	29.93	0.017	5.34109
1,1'-BINAPHTHYL	0.00	0.000	0.00000
4,4'-DI(N-BUTOXY)-1,1'-BINAPHTHYL	13.45	0.184	0.04444
2,2'-BINAPHTHYL	4.87	3.864	0.00010
ACRIDINE (A)	29.93	0.017	5.34660
ANTHRACENE (TRACE)	28.64	0.019	4.10888
DEUTERATED ANTHRACENE C14D10 (TRACE)	28.07	0.020	3.64118
9-METHYL ANTHRACENE (TRACE)	31.25	0.015	6.92154
1-AMINO ANTHRACENE	32.03	0.014	7.95924
2-AMINO ANTHRACENE	30.47	0.016	5.89693
9-VINYL ANTHRACENE	32.89	0.013	9.39654
9-PHENYL ANTHRACENE (TRACE)	32.69	0.013	9.06311
9,10-DIPHENYL ANTHRACENE	35.40	0.010	14.57032
9,10-DICHLOROANTHRACENE (TRACE)	34.49	0.011	12.46374
BIANTHRANYL (B)	37.21	0.009	19.67176
9,10-DINAPHTHYLANTHRACENE (B)	35.01	0.010	13.61723
9-(P-VINYLPHENYL)-10-PHENYL ANTHRACENE	35.79	0.010	15.54287
9,10-BIS(PHENYL-ETHYNYL)ANTHRACENE (TRACE)	39.09	0.007	26.13860
RUBICENE (B)	34.56	0.011	12.46786
DIBENZ(A,H)ANTHRACENE (B)	24.63	0.030	1.65501
TETRACENE (TRACE) (B)	28.99	0.018	4.34585
RUBRENE (B)	25.92	0.026	2.21884
TETRAHYDROCANNABINOL ACID METHYL ESTER	0.00	0.000	0.00000
PHENANTHRENE	7.65	0.999	0.00151
3,4-BENZOPHENANTHRENE	16.04	0.108	0.12675
2-PHENYL PHENANTHRENE	12.08	0.254	0.02324
CHRYSENE (TRACE)	15.65	0.117	0.10980
TRIPHENYLENE	6.47	1.654	0.00055
DEUTERATED TRIPHENYLENE-C18D12	6.46	1.659	0.00055
PYRENE	16.26	0.104	0.13800
1,3,6,8,-TETRAPHENYL PYRENE	43.68	0.005	51.37867
BIPYRENYL (B)	38.78	0.008	25.21047
O-PHENYLENE-PYRENE	37.66	0.008	20.91760
ANTHANTHRENE (B)	38.28	0.008	23.18550
FLUORANTHENE	27.61	0.021	3.28404
3-PHENYL FLUORANTHENE	34.45	0.011	12.35098
3-AMINO FLUORANTHENE	34.53	0.011	12.44850
3,3'-BIFLUORANTHENYL (B)	41.09	0.006	35.53056
BENZO(GHI)FLUORANTHENE	24.95	0.029	1.78361
9,12-DI(N-OCTADECOXY)-BENZ(E)ACEPHENANTHRYLENE	36.64	0.009	17.86795
DECACYCLENE (B)	45.19	0.005	62.43023
10-ETHOXY-BENZO(KL)XANTHENE	36.05	0.010	16.28006
PERYLENE	39.08	0.007	26.21726
BENZO(GHI)PERYLENE (B)	37.74	0.008	21.42424
BBOT (TRACE)	45.98	0.005	70.07781
QUININE SULFATE (A)	13.49	0.182	0.04528
ACRIDINE YELLOW (A)	36.16	0.009	16.35372
ACRIDINE ORANGE (A)	30.24	0.016	5.59664
ACRIDINE RED (A)	27.22	0.022	2.97563
URANINE (A)	30.56	0.016	5.96106
RHODAMINE B (A)	31.38	0.014	6.99519
RHODAMINE 6G (A)	29.02	0.018	4.37420

ACCEPTOR MOLECULES	RO	CO	OVERLAP
EOSIN B (A)	27.47	0.022	3.14557
EOSIN Y (A)	28.93	0.018	4.29195
ROSE BENGAL (A)	28.15	0.020	3.64578
INDENOINDENE	7.73	0.968	0.00161

8.95 Donor Molecule: Tetramethyl-*p*-sexiphenyl

TRIPHENYLAMINE	14.61	0.143	0.07801
P-XYLENE (TRACE)	0.00	0.000	0.00000
SODIUM SALICYLATE (A)	11.48	0.295	0.01837
ANTHRANILIC ACID (A)	24.09	0.032	1.55686
TMPD	18.52	0.070	0.32241
BENZIDINE	10.46	0.391	0.01049
1,3-DIBIPHENYLYLBENZENE (B)	8.43	0.746	0.00290
1,3,5,-TRIBIPHENYLYLBENZENE (B)	9.22	0.570	0.00495
FLUORENE (TRACE)	0.00	0.000	0.00000
1-METHYL FLUORENE	0.00	0.000	0.00000
1,2-BENZOFLUORENE	14.53	0.146	0.07546
9,10-DIHYDROPHENANTHRENE	0.00	0.000	0.00000
CARBAZOLE (TRACE)	10.43	0.395	0.01032
N-METHYL CARBAZOLE (TRACE)	17.32	0.086	0.21632
N-VINYLCARBAZOLE	15.75	0.114	0.12253
N-PHENYL CARBAZOLE	15.90	0.111	0.12945
N-METHYL-2,3-BENZCARBAZOLE	29.03	0.018	4.73725
DIBENZOFURAN	0.00	0.000	0.00000
DIBENZOTHIOPHENE (TRACE)	6.84	1.396	0.00083
P-TERPHENYL	0.00	0.000	0.00000
DEUTERATED P-TERPHENYL-C18D14	0.00	0.000	0.00000
4-METHYL TERPHENYL	0.00	0.000	0.00000
2-METHYL-P-TERPHENYL	0.00	0.000	0.00000
3,3''-DIMETHYL-P-TERPHENYL	0.00	0.000	0.00000
4-(3,3-DIMETHYLBUTOXY)-P-TERPHENYL	6.92	1.353	0.00088
4,4''-DIHEXAHYDROFARNESOXY-P-TERPHENYL	9.28	0.560	0.00514
2,2'-METHYLENE-P-TERPHENYL	5.91	2.164	0.00034
2,2'-ETHYLENE-P-TERPHENYL	8.01	0.870	0.00213
2,2'-METHYLENE-5',6''-METHYLENE-P-TERPHENYL (TRACE)	18.33	0.073	0.30489
3-PHENYL DIBENZOFURAN	5.74	2.360	0.00029
BENZO(1,2-B-4,5-B')BISBENZOFURAN (TRACE) (B)	21.60	0.044	0.81502
SEXIPHENYL BENZENE	0.00	0.000	0.00000
P-QUATERPHENYL	11.53	0.292	0.01886
3,3'''-DIMETHYL-P-QUATERPHENYL	13.68	0.175	0.05271
TETRAMETHYL-P-QUATERPHENYL	0.00	0.000	0.00000
3,3''' -DIMETHYL-3',2''-METHYLENE-P-QUATERPHENYL	21.35	0.046	0.75963
3,3''' -DIMETHYL-3',2''-ETHYLENE-P-QUATERPHENYL	22.74	0.038	1.10538
4-(3-ETHYLBUTOXY)-P-QUATERPHENYL (B)	20.17	0.055	0.53873
4,4'''-DI(2-BUTYLOCTOXY-1)-P-QUATERPHENYL	20.33	0.053	0.56460
DIETHYL-P-QUINQUEPHENYL	13.98	0.164	0.05985
DI(3-ETHYLHEPTYL)-P-QUINQUEPHENYL	22.18	0.041	0.94932
TETRAMETHYL-P-SEXIPHENYL	11.94	0.263	0.02334
TETRAMETHYL-P-OCTAPHENYL (B)	28.03	0.020	3.87053
INDOLE (TRACE)	0.00	0.000	0.00000
2-PHENYL INDOLE	12.54	0.227	0.03122
1-METHYL-2-PHENYL INDOLE	9.89	0.462	0.00753
1-NEOPENTYL-2-PHENYL INDOLE	6.44	1.677	0.00057
1-N-BUTYL-2(BIPHENYLYL-4) INDOLE	22.53	0.039	1.04419
1,2-DIPHENYL INDOLE	10.17	0.425	0.00889
1-BIPHENYLYL-2-PHENYL INDOLE	14.66	0.142	0.07952
1,3-DIMETHYL-2-PHENYL INDOLE	11.34	0.307	0.01709
2,3-DIPHENYL INDOLE	20.79	0.050	0.64444
1 METHYL-2,3-DIPHENYL INDOLE	18.44	0.071	0.31438
2-PHENYL-3,2'-METHYLENE INDOLE (TRACE)	16.91	0.093	0.18726
2-PHENYL-3,2'-ETHYLENE INDOLE (TRACE)	24.20	0.032	1.60586
2-PHENYL-3,2'-TRIMETHYLENE INDOLE	15.62	0.117	0.11655
1-METHYL-3,2'-METHYLENE-2-PHENYL INDOLE (TRACE)	22.50	0.039	1.03990
1-METHYL-3,2'-ETHYLENE-2-PHENYL INDOLE	27.49	0.022	3.44235
1-METHYL-3,2'-TRIMETHYLENE-2-PHENYL INDOLE	16.50	0.100	0.16157
4,5-DIPHENYLIMIDAZOLE (A)	8.88	0.638	0.00395
PPF (TRACE)	23.55	0.034	1.36450
PPO (TRACE)	15.16	0.128	0.09767
2-(4-FLUOROPHENYL)-5-PHENYLOXAZOLE (TRACE)	14.03	0.162	0.06125
αNPO (TRACE)	32.29	0.013	9.01532
BBO (B)	39.40	0.007	29.69152
POPOP	41.27	0.006	39.20379
DIMETHYL POPOP (TRACE)	43.75	0.005	55.49663
PPD (TRACE)	0.00	0.000	0.00000
2,5-DI(2-METHYLPHENYL)OXADIAZOLE	0.00	0.000	0.00000
PBD	12.78	0.214	0.03505
BBD (B)	25.86	0.026	2.39335
αNPD (TRACE)	23.09	0.036	1.21105
βNPD	14.29	0.153	0.06827
αNND (TRACE)	28.69	0.019	4.44319
βNND (TRACE)	21.37	0.046	0.76455
TRANS-1-(4-BIPHENYLYL)-2-PHENYL-ETHYLENE	26.86	0.023	2.99948
TRANS-1-(4-BIPHENYLYL)-2-(1-NAPHTHYL)-ETHYLENE	35.37	0.010	15.54410
DIPHENYLSTILBENE (B)	39.52	0.007	30.23706
1,1,4,4-TETRAPHENYLBUTADIENE	38.33	0.008	25.10406
1,6-DIPHENYLHEXATRIENE	44.04	0.005	57.86628
1,8-DIPHENYL-1,3,5,7-OCTATETRAENE	48.85	0.004	106.99019
BPSB (TRACE)	44.87	0.005	64.72047
BIS-MSB	42.30	0.006	45.36061
ESCULIN (A)	29.53	0.017	5.27174

ACCEPTOR MOLECULES	RO	CO	OVERLAP
AZULENE	18.09	0.076	0.27344
NAPHTHALENE	0.00	0.000	0.00000
DEUTERATED NAPHTHALENE-C10D8	3.92	7.424	0.00003
1-METHYLNAPHTHALENE 5ML/L	0.00	0.000	0.00000
2-METHYL NAPHTHALENE	0.00	0.000	0.00000
2,3-DIMETHYL NAPHTHALENE	0.00	0.000	0.00000
2,6-DIMETHYL NAPHTHALENE (TRACE)	0.00	0.000	0.00000
1-NAPHTHOL (TRACE)	3.80	8.132	0.00002
2-NAPHTHOL (TRACE)	7.95	0.892	0.00202
1-NAPHTHYLAMINE	15.42	0.122	0.10747
2-PHENYL NAPHTHALENE	9.38	0.541	0.00550
1,4-DIPHENYL NAPHTHALENE	11.45	0.298	0.01810
1,5-DIPHENYL NAPHTHALENE	9.10	0.593	0.00457
1,7-DIPHENYL NAPHTHALENE	9.68	0.494	0.00659
1,4,5,8-TETRAPHENYL NAPHTHALENE	31.81	0.014	8.21044
1,1'-BINAPHTHYL	5.69	2.429	0.00027
4,4'-DI(N-BUTOXY)-1,1'-BINAPHTHYL	17.70	0.081	0.24649
2,2'-BINAPHTHYL	10.53	0.383	0.01098
ACRIDINE (A)	30.87	0.015	6.86449
ANTHRACENE (TRACE)	29.57	0.017	5.30475
DEUTERATED ANTHRACENE C14D10 (TRACE)	29.05	0.018	4.77552
9-METHYL ANTHRACENE (TRACE)	31.03	0.015	7.07882
1-AMINO ANTHRACENE	31.28	0.015	7.35874
2-AMINO ANTHRACENE	29.78	0.017	5.47876
9-VINYL ANTHRACENE	32.64	0.013	9.56629
9-PHENYL ANTHRACENE (TRACE)	32.57	0.013	9.45170
9,10-DIPHENYL ANTHRACENE	34.97	0.010	14.44189
9,10-DICHLOROANTHRACENE (TRACE)	33.95	0.011	12.10400
BIANTHRANYL (B)	36.65	0.009	19.13214
9,10-DINAPHTHYLANTHRACENE (B)	34.54	0.011	13.39028
9-(P-VINYLPHENYL)-10-PHENYL ANTHRACENE	35.29	0.010	15.25184
9,10-BIS(PHENYL-ETHYNYL)ANTHRACENE (TRACE)	37.06	0.009	20.25229
RUBICENE (B)	33.80	0.012	11.63304
DIBENZ(A,H)ANTHRACENE (B)	27.47	0.022	3.40215
TETRACENE (TRACE) (B)	27.72	0.021	3.53894
RUBRENE (B)	25.36	0.027	2.07854
TETRAHYDROCANNABINOL ACID METHYL ESTER	0.00	0.000	0.00000
PHENANTHRENE	10.50	0.386	0.01074
3,4-BENZOPHENANTHRENE	16.84	0.094	0.18143
2-PHENYL PHENANTHRENE	13.86	0.168	0.05650
CHRYSENE (TRACE)	16.89	0.093	0.18526
TRIPHENYLENE	9.67	0.494	0.00657
DEUTERATED TRIPHENYLENE-C18D12	9.48	0.525	0.00582
PYRENE	20.30	0.053	0.55650
1,3,6,8,-TETRAPHENYL PYRENE	43.18	0.006	51.11785
BIPYRENYL (B)	39.87	0.007	31.77160
O-PHENYLENE-PYRENE	37.32	0.009	21.12091
ANTHANTHRENE (B)	36.84	0.009	19.64766
FLUORANTHENE	28.95	0.018	4.65022
3-PHENYL FLUORANTHENE	34.84	0.011	14.10795
3-AMINO FLUORANTHENE	33.72	0.012	11.49528
3,3'-BIFLUORANTHENYL (B)	40.72	0.007	35.87578
BENZO(GHI)FLUORANTHENE	26.36	0.024	2.64495
9,12-DI(N-OCTADECOXY)-BENZ(E)ACEPHENANTHRYLENE	35.93	0.010	16.94932
DECACYCLENE (B)	44.41	0.005	59.98526
10-ETHOXY-BENZO(KL)XANTHENE	35.61	0.010	16.13386
PERYLENE	37.17	0.009	20.70430
BENZO(GHI)PERYLENE (B)	37.22	0.009	21.00797
BBOT (TRACE)	45.92	0.005	74.18638
QUININE SULFATE (A)	18.92	0.066	0.36795
ACRIDINE YELLOW (A)	34.25	0.011	12.60884
ACRIDINE ORANGE (A)	28.93	0.018	4.57573
ACRIDINE RED (A)	27.20	0.022	3.16334
URANINE (A)	29.63	0.017	5.28464
RHODAMINE B (A)	31.25	0.015	7.26675
RHODAMINE 6G (A)	29.79	0.017	5.46115
EOSIN B (A)	27.21	0.022	3.17091
EOSIN Y (A)	28.46	0.019	4.14576
ROSE BENGAL (A)	28.08	0.020	3.82546
INDENOINDENE	14.55	0.145	0.07628

8.97 Donor Molecule: Indole (trace)

POLYVINYLTOLUENE	6.92	1.349	0.00221
PHENYLCYCLOHEXANE (TRACE)	0.00	0.000	0.00000
BENZYLALCOHOL 3ML/L	0.00	0.000	0.00000
BENZYLACETATE 5ML/L	0.00	0.000	0.00000
PHENOL (TRACE)	5.59	2.557	0.00062
ANISOLE (TRACE)	6.45	1.669	0.00145
PHENYL ETHER	8.71	0.677	0.00882
ANILINE 3ML/L	17.09	0.090	0.49921
N,N-DIMETHYLANILINE 3ML/L	20.21	0.054	1.35253
TRIPHENYLAMINE	30.88	0.015	17.06876
P-XYLENE (TRACE)	3.43	11.110	0.00003
DEUTERATED P-XYLENE C8D10 (TRACE)	3.11	14.897	0.00002
M-XYLENE (TRACE)	2.00	55.999	0.00000
O-XYLENE 3ML/L	1.91	64.028	0.00000
1-METHYL-4-ETHYLBENZENE (TRACE)	2.16	44.140	0.00000
1,4-BIS(TRIFLUOROMETHYL)BENZENE 5ML/L	0.00	0.000	0.00000
P-CRESOL	13.39	0.186	0.11600
P-METHYLANISOLE 3ML/L	13.07	0.200	0.10059
1,4-DIMETHOXY BENZENE	19.46	0.061	1.08717

ACCEPTOR MOLECULES	RO	CO	OVERLAP
1-METHYL-4-CHLOROBENZENE 5ML/L	4.29	5.675	0.00013
SODIUM SALICYLATE (A)	22.25	0.041	2.38652
ANTHRANILIC ACID (A)	21.57	0.045	1.97713
TMPD	21.69	0.044	2.04112
1,2,4-TRIMETHYLBENZENE (TRACE)	5.39	2.860	0.00049
MESITYLENE 5ML/L	2.75	21.509	0.00001
1,3,5-TRIETHYLBENZENE 5ML/L	3.01	16.408	0.00002
STYRENE	12.03	0.257	0.06099
1,1-DIPHENYLETHYLENE 3ML/L	12.21	0.245	0.06690
BIPHENYL	8.97	0.620	0.01052
DEUTERATED BIPHENYL-C12D10	8.58	0.709	0.00805
4-METHYL BIPHENYL	12.31	0.240	0.07013
MONOISOPROPYLBIPHENYL 5ML/L	11.20	0.319	0.03977
4-BENZYL BIPHENYL	13.88	0.167	0.14382
3,3'-DIMETHYL BIPHENYL	13.21	0.194	0.10704
2,2'-DIMETHYLBIPHENYL 2ML/L	6.96	1.328	0.00229
OCTAMETHYL BIPHENYL	7.91	0.905	0.00492
4-METHOXY BIPHENYL	18.66	0.069	0.84689
4-BIPHENYLYL PHENYL ETHER	18.76	0.068	0.87418
4,4'-DIHYDROXYBIPHENYL (A)	22.41	0.040	2.51892
4,4'-DIMETHOXY BIPHENYL	21.84	0.043	2.16596
4,4'-DIISOBUTOXYBIPHENYL	21.86	0.043	2.18341
BENZIDINE	29.88	0.017	14.05775
4-VINYL BIPHENYL	26.42	0.024	6.78742
M-TERPHENYL	18.39	0.072	0.77195
M-QUATERPHENYL	16.62	0.097	0.42363
1,3,5-TRIPHENYL BENZENE	17.31	0.086	0.53451
1,3-DIBIPHENYLYLBENZENE (B)	32.42	0.013	22.98246
1,3,5,-TRIBIPHENYLYLBENZENE (B)	36.44	0.009	46.30864
FLUORENE (TRACE)	22.53	0.039	2.61535
1-METHYL FLUORENE	21.68	0.044	2.07567
1,2-BENZOFLUORENE	28.13	0.020	9.75442
9,10-DIHYDROPHENANTHRENE	21.60	0.044	2.02602
CARBAZOLE (TRACE)	25.06	0.028	4.88952
N-METHYL CARBAZOLE (TRACE)	24.48	0.031	4.23451
N-VINYLCARBAZOLE	24.63	0.030	4.39729
N-PHENYL CARBAZOLE	25.64	0.027	5.59340
N-METHYL-2,3-BENZCARBAZOLE	28.75	0.019	11.08397
DIBENZOFURAN	23.36	0.035	3.25245
DIBENZOTHIOPHENE (TRACE)	21.55	0.045	1.98276
P-TERPHENYL	27.40	0.022	8.42839
DEUTERATED P-TERPHENYL-C18D14	27.05	0.023	7.81444
4-METHYL TERPHENYL	28.54	0.019	10.73868
2-METHYL-P-TERPHENYL	22.01	0.042	2.27134
3,3''-DIMETHYL-P-TERPHENYL	27.93	0.021	9.44132
4-(3,3-DIMETHYLBUTOXY)-P-TERPHENYL	30.63	0.016	16.34783
4,4''-DIHEXAHYDROFARNESOXY-P-TERPHENYL	32.62	0.013	23.84352
2,2'-METHYLENE-P-TERPHENYL	31.35	0.015	18.80388
2,2'-ETHYLENE-P-TERPHENYL	31.39	0.014	18.93518
2,2'-METHYLENE-5',6''-METHYLENE-P-TERPHENYL (TRACE)	32.13	0.013	21.72588
3-PHENYL DIBENZOFURAN	31.93	0.014	20.99039
BENZO(1,2-B-4,5-B')BISBENZOFURAN (TRACE) (B)	31.50	0.014	19.25561
SEXIPHENYL BENZENE	18.49	0.071	0.80208
P-QUATERPHENYL	32.92	0.013	25.14150
3,3'''-DIMETHYL-P-QUATERPHENYL	33.53	0.012	28.06094
TETRAMETHYL-P-QUATERPHENYL	24.82	0.029	4.67379
3,3''' -DIMETHYL-3',2''-METHYLENE-P-QUATERPHENYL	34.33	0.011	32.21825
3,3''' -DIMETHYL-3',2''-ETHYLENE-P-QUATERPHENYL	34.45	0.011	32.80403
4-(3-ETHYLBUTOXY)-P-QUATERPHENYL (B)	34.34	0.011	32.22028
4,4'''-DI(2-BUTYLOCTOXY-1)-P-QUATERPHENYL	35.50	0.010	39.35207
DIETHYL-P-QUINQUEPHENYL	35.09	0.010	36.86489
DI(3-ETHYLHEPTYL)-P-QUINQUEPHENYL	36.42	0.009	45.79575
TETRAMETHYL-P-SEXIPHENYL	35.24	0.010	37.83496
TETRAMETHYL-P-OCTAPHENYL (B)	37.39	0.009	53.60295
INDOLE (TRACE)	14.78	0.139	0.21019
2-PHENYL INDOLE	31.32	0.015	18.66159
1-METHYL-2-PHENYL INDOLE	29.12	0.018	12.03200
1-NEOPENTYL-2-PHENYL INDOLE	28.58	0.019	10.79645
1-N-BUTYL-2(BIPHENYLYL-4) INDOLE	31.47	0.014	19.07738
1,2-DIPHENYL INDOLE	30.04	0.017	14.50261
1-BIPHENYLYL-2-PHENYL INDOLE	31.50	0.014	19.23236
1,3-DIMETHYL-2-PHENYL INDOLE	28.11	0.020	9.74494
2,3-DIPHENYL INDOLE	29.67	0.017	13.39662
1-METHYL-2,3-DIPHENYL INDOLE	28.12	0.020	9.71343
2-PHENYL-3,2'-METHYLENE INDOLE (TRACE)	30.54	0.016	15.99937
2-PHENYL-3,2'-ETHYLENE INDOLE (TRACE)	29.75	0.017	13.60872
2-PHENYL-3,2'-TRIMETHYLENE INDOLE	30.81	0.015	16.85390
1-METHYL-3,2'-METHYLENE-2-PHENYL INDOLE (TRACE)	30.20	0.016	14.92733
1-METHYL-3,2'-ETHYLENE-2-PHENYL INDOLE	29.28	0.018	12.37819
1-METHYL-3,2'-TRIMETHYLENE-2-PHENYL INDOLE	29.52	0.017	13.04547
4,5-DIPHENYLIMIDAZOLE (A)	25.04	0.028	4.87280
PPF (TRACE)	31.97	0.014	20.99484
PPO (TRACE)	32.95	0.013	25.27418
2-(4-FLUOROPHENYL)-5-PHENYLOXAZOLE (TRACE)	31.67	0.014	19.92734
αNPO (TRACE)	29.43	0.018	12.74130
BBO (B)	32.07	0.014	21.34732
POPOP	29.28	0.018	12.37241
DIMETHYL POPOP (TRACE)	29.77	0.017	13.64818
PPD (TRACE)	28.36	0.020	10.36611
2,5-DI(2-METHYLPHENYL)OXADIAZOLE	27.93	0.021	9.45462
PBD	34.41	0.011	32.82625
BBD (B)	35.07	0.010	36.55473
αNPD (TRACE)	29.65	0.017	13.34140
βNPD	32.06	0.014	21.43096
αNND (TRACE)	28.82	0.019	11.25453
βNND (TRACE)	32.71	0.013	24.13057
TRANS-1-(4-BIPHENYLYL)-2-PHENYL-ETHYLENE	33.89	0.011	29.75364

ACCEPTOR MOLECULES	RO	CO	OVERLAP
TRANS-1-(4-BIPHENYLYL)-2-(1-NAPHTHYL)-ETHYLENE	31.22	0.015	18.17921
DIPHENYLSTILBENE (B)	32.31	0.013	22.31068
1,1,4,4-TETRAPHENYLBUTADIENE	29.30	0.018	12.41245
1,6-DIPHENYLHEXATRIENE	30.57	0.016	16.00561
1,8-DIPHENYL-1,3,5,7-OCTATETRAENE	26.81	0.023	7.28999
BPSB (TRACE)	30.94	0.015	17.23046
BIS-MSB	31.72	0.014	19.97575
ESCULIN (A)	26.62	0.024	6.98385
AZULENE	22.72	0.038	2.69591
NAPHTHALENE	16.53	0.099	0.40615
DEUTERATED NAPHTHALENE-C10D8	16.83	0.094	0.45006
1-METHYLNAPHTHALENE 5ML/L	19.57	0.060	1.11649
2-METHYL NAPHTHALENE	17.63	0.082	0.59637
2,3-DIMETHYL NAPHTHALENE	18.48	0.071	0.79114
2,6-DIMETHYL NAPHTHALENE (TRACE)	18.49	0.071	0.79230
1-NAPHTHOL (TRACE)	22.73	0.038	2.73195
2-NAPHTHOL (TRACE)	19.32	0.062	1.02930
1-NAPHTHYLAMINE	23.89	0.033	3.65948
2-CHLORONAPHTHALENE	17.68	0.081	0.60599
ACENAPHTHENE	22.58	0.039	2.63328
1-PHENYL NAPHTHALENE	24.69	0.030	4.49254
2-PHENYL NAPHTHALENE	36.33	0.009	45.54921
1,4-DIPHENYL NAPHTHALENE	28.36	0.020	10.27248
1,5-DIPHENYL NAPHTHALENE	27.65	0.021	8.83735
1,7-DIPHENYL NAPHTHALENE	27.26	0.022	8.09578
1,4,5,8-TETRAPHENYL NAPHTHALENE	28.30	0.020	10.08528
1,1'-BINAPHTHYL	26.15	0.025	6.33426
4,4'-DI(N-BUTOXY)-1,1'-BINAPHTHYL	29.30	0.018	12.45044
2,2'-BINAPHTHYL	30.13	0.016	14.79412
ACRIDINE (A)	20.59	0.051	1.49701
ANTHRACENE (TRACE)	19.35	0.062	1.02892
DEUTERATED ANTHRACENE C14D10 (TRACE)	19.53	0.060	1.08894
9-METHYL ANTHRACENE (TRACE)	17.86	0.078	0.63718
1-AMINO ANTHRACENE	20.20	0.054	1.33407
2-AMINO ANTHRACENE	18.50	0.071	0.77807
9-VINYL ANTHRACENE	18.47	0.071	0.77930
9-PHENYL ANTHRACENE (TRACE)	19.11	0.064	0.95628
9,10-DIPHENYL ANTHRACENE	18.51	0.070	0.79040
9,10-DICHLOROANTHRACENE (TRACE)	17.11	0.089	0.49175
BIANTHRANYL (B)	20.41	0.053	1.41750
9,10-DINAPHTHYLANTHRACENE (B)	25.28	0.028	5.12694
9-(P-VINYLPHENYL)-10-PHENYL ANTHRACENE	21.52	0.045	1.94846
9,10-BIS(PHENYL-ETHYNYL)ANTHRACENE (TRACE)	30.53	0.016	15.89947
RUBICENE (B)	32.28	0.013	22.22289
DIBENZ(A,H)ANTHRACENE (B)	36.70	0.009	47.99049
TETRACENE (TRACE) (B)	28.40	0.020	10.30553
RUBRENE (B)	37.87	0.008	57.91764
TETRAHYDROCANNABINOL ACID METHYL ESTER	24.87	0.029	4.71268
PHENANTHRENE	21.43	0.045	1.90532
3,4-BENZOPHENANTHRENE	27.92	0.021	9.29623
2-PHENYL PHENANTHRENE	29.01	0.018	11.70296
CHRYSENE (TRACE)	27.95	0.020	9.34850
TRIPHENYLENE	22.81	0.038	2.76938
DEUTERATED TRIPHENYLENE-C18D12	22.45	0.040	2.52425
PYRENE	27.68	0.021	8.82533
1,3,6,8,-TETRAPHENYL PYRENE	31.61	0.014	19.59384
BIPYRENYL (B)	30.15	0.016	14.72757
O-PHENYLENE-PYRENE	31.86	0.014	20.51047
ANTHANTHRENE (B)	32.72	0.013	24.07997
FLUORANTHENE	25.49	0.027	5.38154
3-PHENYL FLUORANTHENE	29.00	0.018	11.67407
3-AMINO FLUORANTHENE	28.74	0.019	11.06949
3,3'-BIFLUORANTHENYL (B)	31.32	0.016	18.53551
BENZO(GHI)FLUORANTHENE	25.68	0.026	5.62833
9,12-DI(N-OCTADECOXY)-BENZ(E)ACEPHENANTHRYLENE	28.43	0.019	10.35345
DECACYCLENE (B)	31.53	0.014	19.26467
10-ETHOXY-BENZO(KL)XANTHENE	23.70	0.034	3.47641
PERYLENE	18.12	0.075	0.69389
BENZO(GHI)PERYLENE (B)	32.28	0.013	22.18517
BBOT (TRACE)	27.59	0.021	8.65352
QUININE SULFATE (A)	25.20	0.028	5.03675
ACRIDINE YELLOW (A)	23.41	0.035	3.23200
ACRIDINE ORANGE (A)	26.03	0.025	6.09970
ACRIDINE RED (A)	24.40	0.031	4.13750
URANINE (A)	24.53	0.030	4.27404
RHODAMINE B (A)	27.85	0.021	9.15923
RHODAMINE 6G (A)	26.90	0.023	7.42906
EOSIN B (A)	26.31	0.025	6.50898
EOSIN Y (A)	27.09	0.022	7.76219
ROSE BENGAL (A)	26.52	0.024	6.82931
INDENOINDENE	30.97	0.015	17.40229

8.98 Donor Molecule: 2-Phenylindole

N,N-DIMETHYLANILINE 3ML/L	2.80	20.471	0.00000
TRIPHENYLAMINE	19.37	0.062	0.46323
SODIUM SALICYLATE (A)	13.80	0.170	0.06050
ANTHRANILIC ACID (A)	26.38	0.024	2.93464
TMPD	21.43	0.045	0.84543
BENZIDINE	15.56	0.119	0.12500
1,3,5-TRIPHENYL BENZENE	3.60	9.588	0.00002

ACCEPTOR MOLECULES	RO	CO	OVERLAP
1,3-DIBIPHENYLYLBENZENE (B)	13.46	0.184	0.05232
1,3,5,-TRIBIPHENYLYLBENZENE (B)	15.54	0.119	0.12434
FLUORENE (TRACE)	0.00	0.000	0.00000
1-METHYL FLUORENE	0.00	0.000	0.00000
1,2-BENZOFLUORENE	18.16	0.075	0.31433
9,10-DIHYDROPHENANTHRENE	0.00	0.000	0.00000
CARBAZOLE (TRACE)	16.71	0.096	0.19133
N-METHYL CARBAZOLE (TRACE)	21.81	0.043	0.94541
N-VINYLCARBAZOLE	21.64	0.044	0.90188
N-PHENYL CARBAZOLE	21.33	0.046	0.82624
N-METHYL-2,3-BENZCARBAZOLE	28.11	0.020	4.26962
DIBENZOFURAN	0.00	0.000	0.00000
DIBENZOTHIOPHENE (TRACE)	11.93	0.263	0.02545
P-TERPHENYL	0.00	0.000	0.00000
DEUTERATED P-TERPHENYL-C18D14	0.00	0.000	0.00000
4-METHYL TERPHENYL	0.00	0.000	0.00000
2-METHYL-P-TERPHENYL	0.00	0.000	0.00000
3,3''-DIMETHYL-P-TERPHENYL	4.61	4.553	0.00009
4-(3,3-DIMETHYLBUTOXY)-P-TERPHENYL	12.19	0.247	0.02888
4,4''-DIHEXAHYDROFARNESOXY-P-TERPHENYL	15.71	0.115	0.13269
2,2'-METHYLENE-P-TERPHENYL	11.23	0.316	0.01767
2,2'-ETHYLENE-P-TERPHENYL	13.93	0.166	0.06436
2,2'-METHYLENE-5',6''-METHYLENE-P-TERPHENYL (TRACE)	28.67	0.019	4.87985
3-PHENYL DIBENZOFURAN	11.09	0.328	0.01643
BENZO(1,2-B-4,5-B')BISBENZOFURAN (TRACE) (B)	29.50	0.017	5.78414
SEXIPHENYL BENZENE	0.00	0.000	0.00000
P-QUATERPHENYL	18.00	0.077	0.29941
3,3'''-DIMETHYL-P-QUATERPHENYL	19.88	0.057	0.54228
TETRAMETHYL-P-QUATERPHENYL	0.00	0.000	0.00000
3,3''' -DIMETHYL-3',2''-METHYLENE-P-QUATERPHENYL	29.08	0.018	5.30792
3,3''' -DIMETHYL-3',2''-ETHYLENE-P-QUATERPHENYL	29.18	0.018	5.39893
4-(3-ETHYLBUTOXY)-P-QUATERPHENYL (B)	26.13	0.025	2.78679
4,4'''-DI(2-BUTYLOCTOXY-1)-P-QUATERPHENYL	26.53	0.024	3.05523
DIETHYL-P-QUINQUEPHENYL	20.68	0.051	0.68752
DI(3-ETHYLHEPTYL)-P-QUINQUEPHENYL	28.10	0.020	4.30623
TETRAMETHYL-P-SEXIPHENYL	18.80	0.067	0.38854
TETRAMETHYL-P-OCTAPHENYL (B)	33.53	0.012	12.40875
INDOLE (TRACE)	0.00	0.000	0.00000
2-PHENYL INDOLE	20.49	0.052	0.65146
1-METHYL-2-PHENYL INDOLE	15.51	0.120	0.12266
1-NEOPENTYL-2-PHENYL INDOLE	11.42	0.300	0.01954
1-N-BUTYL-2(BIPHENYLYL-4) INDOLE	27.28	0.022	3.60279
1,2-DIPHENYL INDOLE	15.90	0.111	0.14209
1-BIPHENYLYL-2-PHENYL INDOLE	19.40	0.061	0.46697
1,3-DIMETHYL-2-PHENYL INDOLE	17.54	0.083	0.25613
2,3-DIPHENYL INDOLE	25.11	0.028	2.18928
1-METHYL-2,3-DIPHENYL INDOLE	22.59	0.039	1.16150
2-PHENYL-3,2'-METHYLENE INDOLE (TRACE)	25.55	0.027	2.44442
2-PHENYL-3,2'-ETHYLENE INDOLE (TRACE)	29.58	0.017	5.86469
2-PHENYL-3,2'-TRIMETHYLENE INDOLE	23.44	0.035	1.45631
1-METHYL-3,2'-METHYLENE-2-PHENYL INDOLE (TRACE)	28.53	0.019	4.72794
1-METHYL-3,2'-ETHYLENE-2-PHENYL INDOLE	31.55	0.014	8.61702
1-METHYL-3,2'-TRIMETHYLENE-2-PHENYL INDOLE	23.45	0.035	1.45913
4,5-DIPHENYLIMIDAZOLE (A)	13.30	0.190	0.04882
PPF (TRACE)	29.91	0.017	6.27997
PPO (TRACE)	23.67	0.034	1.54808
2-(4-FLUOROPHENYL)-5-PHENYLOXAZOLE (TRACE)	22.16	0.041	1.04240
αNPO (TRACE)	34.92	0.011	15.77817
BBO (B)	41.34	0.006	43.41614
POPOP	41.89	0.006	46.93740
DIMETHYL POPOP (TRACE)	43.62	0.005	59.67802
PPD (TRACE)	0.00	0.000	0.00000
2,5-DI(2-METHYLPHENYL)OXADIAZOLE	0.00	0.000	0.00000
PBD	20.64	0.051	0.67997
BBD (B)	31.99	0.014	9.38270
αNPD (TRACE)	28.02	0.020	4.22837
βNPD	20.37	0.053	0.62751
αNND (TRACE)	32.21	0.013	9.75138
βNND (TRACE)	28.28	0.020	4.49373
TRANS-1-(4-BIPHENYLYL)-2-PHENYL-ETHYLENE	32.31	0.013	9.93833
TRANS-1-(4-BIPHENYLYL)-2-(1-NAPHTHYL)-ETHYLENE	37.53	0.008	24.29422
DIPHENYLSTILBENE (B)	41.62	0.006	45.17135
1,1,4,4-TETRAPHENYLBUTADIENE	39.71	0.007	34.02011
1,6-DIPHENYLHEXATRIENE	45.07	0.005	72.72604
1,8-DIPHENYL-1,3,5,7-OCTATETRAENE	47.44	0.004	98.27849
BPSB (TRACE)	45.40	0.005	76.01997
BIS-MSB	43.34	0.005	57.48331
ESCULIN (A)	31.70	0.014	8.82746
AZULENE	22.20	0.041	1.02455
NAPHTHALENE	0.00	0.000	0.00000
DEUTERATED NAPHTHALENE-C10D8	6.62	1.541	0.00074
1-METHYLNAPHTHALENE 5ML/L	3.12	14.699	0.00001
2-METHYL NAPHTHALENE	3.65	9.181	0.00002
2,3-DIMETHYL NAPHTHALENE	3.73	8.603	0.00002
2,6-DIMETHYL NAPHTHALENE (TRACE)	7.14	1.228	0.00117
1-NAPHTHOL (TRACE)	7.36	1.122	0.00140
2-NAPHTHOL (TRACE)	13.68	0.175	0.05778
1-NAPHTHYLAMINE	20.71	0.050	0.69110
2-CHLORONAPHTHALENE	4.97	3.652	0.00013
ACENAPHTHENE	4.49	4.929	0.00007
1-PHENYL NAPHTHALENE	6.93	1.343	0.00098
2-PHENYL NAPHTHALENE	15.64	0.117	0.12899
1,4-DIPHENYL NAPHTHALENE	16.89	0.093	0.20433
1,5-DIPHENYL NAPHTHALENE	14.58	0.144	0.08464
1,7-DIPHENYL NAPHTHALENE	14.95	0.134	0.09809
1,4,5,8-TETRAPHENYL NAPHTHALENE	33.99	0.011	13.39207
1,1'-BINAPHTHYL	10.20	0.421	0.00996

ACCEPTOR MOLECULES	RO	CO	OVERLAP
4,4'-DI(N-BUTOXY)-1,1'-BINAPHTHYL	23.24	0.036	1.37921
2,2'-BINAPHTHYL	16.41	0.101	0.17206
ACRIDINE (A)	31.53	0.014	8.53056
ANTHRACENE (TRACE)	30.06	0.016	6.41265
DEUTERATED ANTHRACENE C14D10 (TRACE)	29.70	0.017	5.97333
9-METHYL ANTHRACENE (TRACE)	30.26	0.016	6.65624
1-AMINO ANTHRACENE	29.84	0.017	6.07363
2-AMINO ANTHRACENE	28.89	0.019	4.99766
9-VINYL ANTHRACENE	31.80	0.014	8.95315
9-PHENYL ANTHRACENE (TRACE)	31.81	0.014	8.99066
9,10-DIPHENYL ANTHRACENE	33.83	0.012	12.97074
9,10-DICHLOROANTHRACENE (TRACE)	32.50	0.013	10.18759
BIANTHRANYL (B)	35.45	0.010	17.15796
9,10-DINAPHTHYLANTHRACENE (B)	33.25	0.012	11.68219
9-(P-VINYLPHENYL)-10-PHENYL ANTHRACENE	34.04	0.011	13.43473
9,10-BIS(PHENYL-ETHYNYL)ANTHRACENE (TRACE)	33.55	0.012	12.22397
RUBICENE (B)	32.28	0.013	9.69368
DIBENZ(A,H)ANTHRACENE (B)	30.74	0.015	7.31507
TETRACENE (TRACE) (B)	25.57	0.027	2.39408
RUBRENE (B)	25.23	0.028	2.21255
TETRAHYDROCANNABINOL ACID METHYL ESTER	0.00	0.000	0.00000
PHENANTHRENE	13.19	0.195	0.04624
3,4-BENZOPHENANTHRENE	18.20	0.074	0.31585
2-PHENYL PHENANTHRENE	15.70	0.116	0.13094
CHRYSENE (TRACE)	18.29	0.073	0.32717
TRIPHENYLENE	13.60	0.178	0.05541
DEUTERATED TRIPHENYLENE-C18D12	13.12	0.198	0.04478
PYRENE	29.08	0.018	5.26239
1,3,6,8,-TETRAPHENYL PYRENE	41.65	0.006	45.08858
BIPYRENYL (B)	40.81	0.007	39.98213
O-PHENYLENE-PYRENE	36.33	0.009	19.71186
ANTHANTHRENE (B)	34.67	0.011	14.94403
FLUORANTHENE	30.09	0.016	6.41707
3-PHENYL FLUORANTHENE	34.71	0.011	15.08588
3-AMINO FLUORANTHENE	32.41	0.013	9.92787
3,3'-BIFLUORANTHENYL (B)	39.62	0.007	33.31590
BENZO(GHI)FLUORANTHENE	28.23	0.020	4.36493
9,12-DI(N-OCTADECOXY)-BENZ(E)ACEPHENANTHRYLENE	34.89	0.011	15.55675
DECACYCLENE (B)	42.95	0.006	53.81521
10-ETHOXY-BENZO(KL)XANTHENE	34.37	0.011	14.25943
PERYLENE	34.00	0.011	13.27780
BENZO(GHI)PERYLENE (B)	35.85	0.010	18.36761
BBOT (TRACE)	45.13	0.005	73.09959
QUININE SULFATE (A)	24.94	0.029	2.10870
ACRIDINE YELLOW (A)	30.72	0.015	7.20556
ACRIDINE ORANGE (A)	25.76	0.026	2.50303
ACRIDINE RED (A)	26.92	0.023	3.26036
URANINE (A)	27.28	0.022	3.53528
RHODAMINE B (A)	30.82	0.015	7.34200
RHODAMINE 6G (A)	30.33	0.016	6.67462
EOSIN B (A)	26.77	0.023	3.15160
EOSIN Y (A)	27.66	0.021	3.83530
ROSE BENGAL (A)	28.30	0.020	4.40622
INDENOINDENE	22.82	0.038	1.24217

8.99 Donor Molecule: 1-Methyl-2-phenylindole

N,N-DIMETHYLANILINE 3ML/L	2.67	23.547	0.00000
TRIPHENYLAMINE	16.89	0.093	0.20566
SODIUM SALICYLATE (A)	12.35	0.238	0.03144
ANTHRANILIC ACID (A)	24.58	0.030	1.94623
TMPD	19.53	0.060	0.48969
BENZIDINE	13.08	0.200	0.04460
1,3,5-TRIPHENYL BENZENE	3.28	12.653	0.00001
1,3-DIBIPHENYLYLBENZENE (B)	11.03	0.334	0.01600
1,3,5,-TRIBIPHENYLYLBENZENE (B)	12.57	0.225	0.03518
FLUORENE (TRACE)	0.00	0.000	0.00000
1-METHYL FLUORENE	0.00	0.000	0.00000
1,2-BENZOFLUORENE	16.42	0.101	0.17370
9,10-DIHYDROPHENANTHRENE	0.00	0.000	0.00000
CARBAZOLE (TRACE)	13.58	0.179	0.05582
N-METHYL CARBAZOLE (TRACE)	19.85	0.057	0.54360
N-VINYLCARBAZOLE	18.84	0.067	0.39728
N-PHENYL CARBAZOLE	18.79	0.067	0.39037
N-METHYL-2,3-BENZCARBAZOLE	28.18	0.020	4.38333
DIBENZOFURAN	0.00	0.000	0.00000
DIBENZOTHIOPHENE (TRACE)	9.99	0.449	0.00886
P-TERPHENYL	0.00	0.000	0.00000
DEUTERATED P-TERPHENYL-C18D14	0.00	0.000	0.00000
4-METHYL TERPHENYL	0.00	0.000	0.00000
2-METHYL-P-TERPHENYL	0.00	0.000	0.00000
3,3''-DIMETHYL-P-TERPHENYL	4.38	5.321	0.00006
4-(3,3-DIMETHYLBUTOXY)-P-TERPHENYL	9.93	0.457	0.00855
4,4''-DIHEXAHYDROFARNESOXY-P-TERPHENYL	12.69	0.219	0.03731
2,2'-METHYLENE-P-TERPHENYL	9.48	0.524	0.00650
2,2'-ETHYLENE-P-TERPHENYL	11.41	0.301	0.01967
2,2'-METHYLENE-5',6''-METHYLENE-P-TERPHENYL (TRACE)	23.24	0.036	1.39994
3-PHENYL DIBENZOFURAN	9.42	0.535	0.00624
BENZO(1,2-B-4,5-B')BISBENZOFURAN (TRACE) (B)	25.81	0.026	2.62479
SEXIPHENYL BENZENE	0.00	0.000	0.00000
P-QUATERPHENYL	14.86	0.136	0.09590
3,3'''-DIMETHYL-P-QUATERPHENYL	16.82	0.094	0.20122

ACCEPTOR MOLECULES	RO	CO	OVERLAP
3,3'' -DIMETHYL-3',2''-METHYLENE-P-QUATERPHENYL	25.11	0.028	2.22401
3,3''' -DIMETHYL-3',2''-ETHYLENE-P-QUATERPHENYL	25.64	0.027	2.51531
4-(3-ETHYLBUTOXY)-P-QUATERPHENYL (B)	22.87	0.037	1.26689
4,4'''-DI(2-BUTYLOCTOXY-1)-P-QUATERPHENYL	23.17	0.036	1.37066
DIETHYL-P-QUINQUEPHENYL	17.39	0.085	0.24642
DI(3-ETHYLHEPTYL)-P-QUINQUEPHENYL	24.75	0.030	2.03116
TETRAMETHYL-P-SEXIPHENYL	15.48	0.121	0.12268
TETRAMETHYL-P-OCTAPHENYL (B)	30.13	0.016	6.61149
INDOLE (TRACE)	0.00	0.000	0.00000
2-PHENYL INDOLE	16.55	0.099	0.18310
1-METHYL-2-PHENYL INDOLE	12.81	0.213	0.03936
1-NEOPENTYL-2-PHENYL INDOLE	9.34	0.550	0.00591
1-N-BUTYL-2(BIPHENYLYL-4) INDOLE	24.42	0.031	1.87560
1,2-DIPHENYL INDOLE	13.16	0.196	0.04623
1-BIPHENYLYL-2-PHENYL INDOLE	16.75	0.095	0.19571
1,3-DIMETHYL-2-PHENYL INDOLE	14.49	0.147	0.08225
2,3-DIPHENYL INDOLE	22.48	0.039	1.14072
1-METHYL-2,3-DIPHENYL INDOLE	20.13	0.055	0.58817
2-PHENYL-3,2'-METHYLENE INDOLE (TRACE)	21.07	0.048	0.77847
2-PHENYL-3,2'-ETHYLENE INDOLE (TRACE)	26.76	0.023	3.24850
2-PHENYL-3,2'-TRIMETHYLENE INDOLE	19.47	0.061	0.48435
1-METHYL-3,2'-METHYLENE-2-PHENYL INDOLE (TRACE)	25.61	0.027	2.49969
1-METHYL-3,2'-ETHYLENE-2-PHENYL INDOLE	28.70	0.019	4.93933
1-METHYL-3,2'-TRIMETHYLENE-2-PHENYL INDOLE	19.95	0.056	0.55978
4,5-DIPHENYLIMIDAZOLE (A)	11.15	0.323	0.01708
PPF (TRACE)	26.81	0.023	3.29082
PPO (TRACE)	19.34	0.062	0.46530
2-(4-FLUOROPHENYL)-5-PHENYLOXAZOLE (TRACE)	18.04	0.076	0.30652
αNPO (TRACE)	32.79	0.013	10.94692
BBO (B)	39.33	0.007	32.56028
POPOP	40.61	0.007	39.38972
DIMETHYL POPOP (TRACE)	42.86	0.006	54.32178
PPD (TRACE)	0.00	0.000	0.00000
2,5-DI(2-METHYLPHENYL)OXADIAZOLE	0.00	0.000	0.00000
PBD	16.83	0.094	0.20231
BBD (B)	28.53	0.019	4.78249
αNPD (TRACE)	25.30	0.028	2.31940
βNPD	17.53	0.083	0.25816
αNND (TRACE)	29.71	0.017	6.07056
βNND (TRACE)	25.04	0.028	2.19020
TRANS-1-(4-BIPHENYLYL)-2-PHENYL-ETHYLENE	29.09	0.018	5.35671
TRANS-1-(4-BIPHENYLYL)-2-(1-NAPHTHYL)-ETHYLENE	35.51	0.010	17.62861
DIPHENYLSTILBENE (B)	39.54	0.007	33.60494
1,1,4,4-TETRAPHENYLBUTADIENE	38.06	0.008	26.67270
1,6-DIPHENYLHEXATRIENE	43.34	0.005	58.20240
1,8-DIPHENYL-1,3,5,7-OCTATETRAENE	47.44	0.004	99.48806
BPSB (TRACE)	44.13	0.005	64.84517
BIS-MSB	41.81	0.006	46.91430
ESCULIN (A)	29.82	0.017	6.19575
AZULENE	19.88	0.057	0.53372
NAPHTHALENE	0.00	0.000	0.00000
DEUTERATED NAPHTHALENE-C10D8	5.34	2.933	0.00021
1-METHYLNAPHTHALENE 5ML/L	2.85	19.229	0.00000
2-METHYL NAPHTHALENE	3.35	11.878	0.00001
2,3-DIMETHYL NAPHTHALENE	3.41	11.315	0.00001
2,6-DIMETHYL NAPHTHALENE (TRACE)	6.42	1.694	0.00062
1-NAPHTHOL (TRACE)	6.29	1.800	0.00055
2-NAPHTHOL (TRACE)	11.13	0.325	0.01692
1-NAPHTHYLAMINE	17.92	0.078	0.29359
2-CHLORONAPHTHALENE	4.36	5.414	0.00006
ACENAPHTHENE	4.16	6.226	0.00005
1-PHENYL NAPHTHALENE	5.96	2.113	0.00040
2-PHENYL NAPHTHALENE	12.58	0.225	0.03528
1,4-DIPHENYL NAPHTHALENE	14.22	0.156	0.07357
1,5-DIPHENYL NAPHTHALENE	12.00	0.259	0.02653
1,7-DIPHENYL NAPHTHALENE	12.40	0.235	0.03231
1,4,5,8-TETRAPHENYL NAPHTHALENE	32.06	0.014	9.53183
1,1'-BINAPHTHYL	8.39	0.758	0.00311
4,4'-DI(N-BUTOXY)-1,1'-BINAPHTHYL	20.19	0.054	0.60065
2,2'-BINAPHTHYL	13.54	0.180	0.05487
ACRIDINE (A)	30.45	0.016	6.99639
ANTHRACENE (TRACE)	28.91	0.019	5.13411
DEUTERATED ANTHRACENE C14D10 (TRACE)	28.44	0.019	4.65494
9-METHYL ANTHRACENE (TRACE)	30.34	0.016	6.84794
1-AMINO ANTHRACENE	30.41	0.016	6.88626
2-AMINO ANTHRACENE	29.10	0.018	5.28240
9-VINYL ANTHRACENE	31.79	0.014	9.04950
9-PHENYL ANTHRACENE (TRACE)	31.82	0.014	9.10632
9,10-DIPHENYL ANTHRACENE	33.90	0.011	13.28878
9,10-DICHLOROANTHRACENE (TRACE)	32.89	0.013	11.07585
BIANTHRANYL (B)	35.58	0.010	17.74932
9,10-DINAPHTHYLANTHRACENE (B)	33.45	0.012	12.25226
9-(P-VINYLPHENYL)-10-PHENYL ANTHRACENE	34.21	0.011	14.01146
9,10-BIS(PHENYL-ETHYNYL)ANTHRACENE (TRACE)	36.37	0.009	20.01361
RUBICENE (B)	33.07	0.012	11.30220
DIBENZ(A,H)ANTHRACENE (B)	28.12	0.020	4.33445
TETRACENE (TRACE) (B)	27.20	0.022	3.49881
RUBRENE (B)	25.53	0.027	2.39599
TETRAHYDROCANNABINOL ACID METHYL ESTER	0.00	0.000	0.00000
PHENANTHRENE	11.86	0.268	0.02465
3,4-BENZOPHENANTHRENE	16.99	0.091	0.21158
2-PHENYL PHENANTHRENE	14.42	0.149	0.07938
CHRYSENE (TRACE)	17.17	0.088	0.22623
TRIPHENYLENE	11.56	0.290	0.02113
DEUTERATED TRIPHENYLENE-C18D12	11.23	0.315	0.01785
PYRENE	24.03	0.032	1.69694
1,3,6,8,-TETRAPHENYL PYRENE	41.89	0.006	47.22892

ACCEPTOR MOLECULES	RO	CO	OVERLAP
BIPYRENYL (B)	39.43	0.007	32.92684
O-PHENYLENE-PYRENE	36.42	0.009	20.23048
ANTHANTHRENE (B)	36.04	0.010	19.06922
FLUORANTHENE	28.83	0.019	5.01989
3-PHENYL FLUORANTHENE	34.07	0.011	13.65624
3-AMINO FLUORANTHENE	32.94	0.013	11.06865
3,3'-BIFLUORANTHENYL (B)	39.63	0.007	33.79230
BENZO(GHI)FLUORANTHENE	26.66	0.024	3.13717
9,12-DI(N-OCTADECOXY)-BENZ(E)ACEPHENANTHRYLENE	35.14	0.010	16.42754
DECACYCLENE (B)	43.32	0.006	57.19964
10-ETHOXY-BENZO(KL)XANTHENE	34.54	0.011	14.87392
PERYLENE	36.27	0.009	19.79853
BENZO(GHI)PERYLENE (B)	36.20	0.009	19.71395
BBOT (TRACE)	44.76	0.005	70.44100
QUININE SULFATE (A)	21.85	0.043	0.96451
ACRIDINE YELLOW (A)	33.79	0.012	12.86362
ACRIDINE ORANGE (A)	28.67	0.019	4.80383
ACRIDINE RED (A)	26.87	0.023	3.25552
URANINE (A)	29.44	0.018	5.62437
RHODAMINE B (A)	30.83	0.015	7.42210
RHODAMINE 6G (A)	29.86	0.017	6.15601
EOSIN B (A)	27.25	0.022	3.53604
EOSIN Y (A)	28.39	0.020	4.52304
ROSE BENGAL (A)	28.02	0.020	4.18332
INDENOINDENE	18.63	0.069	0.37226

8.100 Donor Molecule: 1-Neopentyl-2-phenylindole

N,N-DIMETHYLANILINE 3ML/L	3.58	9.725	0.00002
TRIPHENYLAMINE	17.50	0.084	0.27048
SODIUM SALICYLATE (A)	12.58	0.225	0.03729
ANTHRANILIC ACID (A)	24.61	0.030	2.08335
TMPD	19.70	0.058	0.54947
BENZIDINE	13.95	0.165	0.06974
1,3,5-TRIPHENYL BENZENE	4.12	6.415	0.00005
1,3-DIBIPHENYLYLBENZENE (B)	12.02	0.257	0.02858
1,3,5,-TRIBIPHENYLYLBENZENE (B)	14.03	0.162	0.07217
FLUORENE (TRACE)	0.00	0.000	0.00000
1-METHYL FLUORENE	0.00	0.000	0.00000
1,2-BENZOFLUORENE	16.70	0.096	0.20477
9,10-DIHYDROPHENANTHRENE	0.00	0.000	0.00000
CARBAZOLE (TRACE)	14.82	0.137	0.10017
N-METHYL CARBAZOLE (TRACE)	20.06	0.055	0.61526
N-VINYLCARBAZOLE	19.31	0.062	0.48887
N-PHENYL CARBAZOLE	19.19	0.063	0.47093
N-METHYL-2,3-BENZCARBAZOLE	27.93	0.021	4.42099
DIBENZOFURAN	0.00	0.000	0.00000
DIBENZOTHIOPHENE (TRACE)	11.53	0.292	0.02224
P-TERPHENYL	0.00	0.000	0.00000
DEUTERATED P-TERPHENYL-C18D14	0.00	0.000	0.00000
4-METHYL TERPHENYL	0.00	0.000	0.00000
2-METHYL-P-TERPHENYL	0.00	0.000	0.00000
3,3''-DIMETHYL-P-TERPHENYL	5.79	2.307	0.00036
4-(3,3-DIMETHYLBUTOXY)-P-TERPHENYL	11.36	0.305	0.02038
4,4''-DIHEXAHYDROFARNESOXY-P-TERPHENYL	14.18	0.157	0.07722
2,2'-METHYLENE-P-TERPHENYL	11.23	0.316	0.01900
2,2'-ETHYLENE-P-TERPHENYL	13.01	0.203	0.04603
2,2'-METHYLENE-5',6''-METHYLENE-P-TERPHENYL (TRACE)	24.75	0.029	2.17498
3-PHENYL DIBENZOFURAN	11.19	0.319	0.01864
BENZO(1,2-B-4,5-B')BISBENZOFURAN (TRACE) (B)	26.65	0.024	3.37933
SEXIPHENYL BENZENE	0.00	0.000	0.00000
P-QUATERPHENYL	16.07	0.108	0.16317
3,3'''-DIMETHYL-P-QUATERPHENYL	17.83	0.079	0.30326
TETRAMETHYL-P-QUATERPHENYL	0.00	0.000	0.00000
3,3'''-DIMETHYL-3',2''-METHYLENE-P-QUATERPHENYL	26.04	0.025	2.94053
3,3'''-DIMETHYL-3',2''-ETHYLENE-P-QUATERPHENYL	26.34	0.024	3.14601
4-(3-ETHYLBUTOXY)-P-QUATERPHENYL (B)	23.59	0.034	1.62131
4,4'''-DI(2-BUTYLOCTOXY-1)-P-QUATERPHENYL	23.93	0.033	1.77186
DIETHYL-P-QUINQUEPHENYL	18.50	0.071	0.37967
DI(3-ETHYLHEPTYL)-P-QUINQUEPHENYL	25.45	0.027	2.55167
TETRAMETHYL-P-SEXIPHENYL	16.81	0.094	0.21341
TETRAMETHYL-P-OCTAPHENYL (B)	30.62	0.016	7.74884
INDOLE (TRACE)	0.00	0.000	0.00000
2-PHENYL INDOLE	18.12	0.075	0.33477
1-METHYL-2-PHENYL INDOLE	13.95	0.165	0.06977
1-NEOPENTYL-2-PHENYL INDOLE	10.72	0.363	0.01443
1-N-BUTYL-2(BIPHENYLYL-4) INDOLE	24.86	0.029	2.21893
1,2-DIPHENYL INDOLE	14.34	0.152	0.08220
1-BIPHENYLYL-2-PHENYL INDOLE	17.56	0.083	0.27609
1,3-DIMETHYL-2-PHENYL INDOLE	15.57	0.118	0.13486
2,3-DIPHENYL INDOLE	22.90	0.037	1.35293
1-METHYL-2,3-DIPHENYL INDOLE	20.54	0.052	0.70662
2-PHENYL-3,2'-METHYLENE INDOLE (TRACE)	22.27	0.041	1.15128
2-PHENYL-3,2'-ETHYLENE INDOLE (TRACE)	27.13	0.022	3.75687
2-PHENYL-3,2'-TRIMETHYLENE INDOLE	20.70	0.050	0.74321
1-METHYL-3,2'-METHYLENE-2-PHENYL INDOLE (TRACE)	26.05	0.025	2.94766
1-METHYL-3,2'-ETHYLENE-2-PHENYL INDOLE	28.93	0.018	5.51105
1-METHYL-3,2'-TRIMETHYLENE-2-PHENYL INDOLE	20.79	0.050	0.76254
4,5-DIPHENYLIMIDAZOLE (A)	11.81	0.271	0.02571
PPF (TRACE)	27.28	0.022	3.88861
PPO (TRACE)	20.69	0.051	0.74140

ACCEPTOR MOLECULES	RO	CO	OVERLAP
2-(4-FLUOROPHENYL)-5-PHENYLOXAZOLE (TRACE)	19.42	0.061	0.50787
αNPO (TRACE)	32.77	0.013	11.59813
BBO (B)	39.19	0.007	33.85256
POPOP	40.29	0.007	39.92857
DIMETHYL POPOP (TRACE)	42.45	0.006	54.50564
PPD (TRACE)	0.00	0.000	0.00000
2,5-DI(2-METHYLPHENYL)OXADIAZOLE	0.00	0.000	0.00000
PBD	18.41	0.072	0.36894
BBD (B)	29.09	0.018	5.70381
αNPD (TRACE)	25.65	0.027	2.67675
βNPD	18.56	0.070	0.38608
αNND (TRACE)	29.82	0.017	6.60460
βNND (TRACE)	25.62	0.027	2.66836
TRANS-1-(4-BIPHENYLYL)-2-PHENYL-ETHYLENE	29.52	0.017	6.21563
TRANS-1-(4-BIPHENYLYL)-2-(1-NAPHTHYL)-ETHYLENE	35.43	0.010	18.49716
DIPHENYLSTILBENE (B)	39.41	0.007	34.99456
1,1,4,4-TETRAPHENYLBUTADIENE	37.86	0.008	27.46988
1,6-DIPHENYLHEXATRIENE	43.08	0.006	59.63040
1,8-DIPHENYL-1,3,5,7-OCTATETRAENE	46.91	0.004	98.75898
BPSB (TRACE)	43.77	0.005	65.63051
BIS-MSB	41.54	0.006	47.93973
ESCULIN (A)	29.79	0.017	6.54224
AZULENE	20.14	0.055	0.61298
NAPHTHALENE	0.00	0.000	0.00000
DEUTERATED NAPHTHALENE-C10D8	5.96	2.116	0.00042
1-METHYLNAPHTHALENE 5ML/L	3.65	9.236	0.00002
2-METHYL NAPHTHALENE	4.26	5.806	0.00006
2,3-DIMETHYL NAPHTHALENE	4.32	5.548	0.00006
2,6-DIMETHYL NAPHTHALENE (TRACE)	8.04	0.862	0.00256
1-NAPHTHOL (TRACE)	7.61	1.014	0.00185
2-NAPHTHOL (TRACE)	12.51	0.228	0.03635
1-NAPHTHYLAMINE	18.53	0.070	0.38207
2-CHLORONAPHTHALENE	5.34	2.943	0.00022
ACENAPHTHENE	5.30	3.003	0.00021
1-PHENYL NAPHTHALENE	7.11	1.243	0.00123
2-PHENYL NAPHTHALENE	13.88	0.167	0.06767
1,4-DIPHENYL NAPHTHALENE	15.20	0.127	0.11657
1,5-DIPHENYL NAPHTHALENE	13.18	0.195	0.04961
1,7-DIPHENYL NAPHTHALENE	13.41	0.185	0.05504
1,4,5,8-TETRAPHENYL NAPHTHALENE	32.01	0.014	10.04346
1,1'-BINAPHTHYL	9.68	0.493	0.00781
4,4'-DI(N-BUTOXY)-1,1'-BINAPHTHYL	20.93	0.049	0.79278
2,2'-BINAPHTHYL	14.62	0.143	0.09235
ACRIDINE (A)	30.22	0.016	7.10731
ANTHRACENE (TRACE)	28.71	0.019	5.23250
DEUTERATED ANTHRACENE C14D10 (TRACE)	28.28	0.020	4.78574
9-METHYL ANTHRACENE (TRACE)	29.96	0.017	6.74963
1-AMINO ANTHRACENE	30.01	0.017	6.75845
2-AMINO ANTHRACENE	28.78	0.019	5.25223
9-VINYL ANTHRACENE	31.43	0.014	8.97091
9-PHENYL ANTHRACENE (TRACE)	31.42	0.014	8.97408
9,10-DIPHENYL ANTHRACENE	33.52	0.012	13.19504
9,10-DICHLOROANTHRACENE (TRACE)	32.45	0.013	10.85542
BIANTHRANYL (B)	35.19	0.010	17.66407
9,10-DINAPHTHYLANTHRACENE (B)	33.05	0.012	12.09863
9-(P-VINYLPHENYL)-10-PHENYL ANTHRACENE	33.81	0.012	13.86821
9,10-BIS(PHENYL-ETHYNYL)ANTHRACENE (TRACE)	35.75	0.010	19.19179
RUBICENE (B)	32.60	0.013	11.03513
DIBENZ(A,H)ANTHRACENE (B)	28.30	0.020	4.78387
TETRACENE (TRACE) (B)	26.76	0.023	3.37677
RUBRENE (B)	25.27	0.028	2.39225
TETRAHYDROCANNABINOL ACID METHYL ESTER	0.00	0.000	0.00000
PHENANTHRENE	12.05	0.256	0.02881
3,4-BENZOPHENANTHRENE	17.22	0.088	0.24402
2-PHENYL PHENANTHRENE	14.50	0.147	0.08728
CHRYSENE (TRACE)	17.24	0.087	0.24655
TRIPHENYLENE	11.98	0.260	0.02791
DEUTERATED TRIPHENYLENE-C18D12	11.66	0.282	0.02375
PYRENE	25.06	0.028	2.32101
1,3,6,8,-TETRAPHENYL PYRENE	41.37	0.006	46.51893
BIPYRENYL (B)	39.19	0.007	33.69415
O-PHENYLENE-PYRENE	35.99	0.010	20.01646
ANTHANTHRENE (B)	35.54	0.010	18.63331
FLUORANTHENE	28.65	0.019	5.14473
3-PHENYL FLUORANTHENE	33.75	0.012	13.72528
3-AMINO FLUORANTHENE	32.55	0.013	10.94985
3,3'-BIFLUORANTHENYL (B)	39.18	0.007	33.52572
BENZO(GHI)FLUORANTHENE	26.65	0.024	3.32807
9,12-DI(N-OCTADECOXY)-BENZ(E)ACEPHENANTHRYLENE	34.75	0.011	16.30748
DECACYCLENE (B)	42.80	0.006	56.56100
10-ETHOXY-BENZO(KL)XANTHENE	34.12	0.011	14.69085
PERYLENE	35.68	0.010	19.07762
BENZO(GHI)PERYLENE (B)	35.72	0.010	19.33912
BBOT (TRACE)	44.29	0.005	70.29653
QUININE SULFATE (A)	22.38	0.040	1.18523
ACRIDINE YELLOW (A)	33.15	0.012	12.19015
ACRIDINE ORANGE (A)	27.95	0.020	4.38053
ACRIDINE RED (A)	26.54	0.024	3.21438
URANINE (A)	28.73	0.019	5.17197
RHODAMINE B (A)	30.46	0.016	7.33991
RHODAMINE 6G (A)	29.52	0.017	6.08290
EOSIN B (A)	26.81	0.023	3.41394
EOSIN Y (A)	27.92	0.021	4.34917
ROSE BENGAL (A)	27.76	0.021	4.20689
INDENOINDENE	20.11	0.055	0.62502

8.101 Donor Molecule: 1-*n*-Butyl-2-(biphenylyl-4)indole

ACCEPTOR MOLECULES	RO	CO	OVERLAP
TRIPHENYLAMINE	5.45	2.763	0.00020
P-XYLENE (TRACE)	0.00	0.000	0.00000
SODIUM SALICYLATE (A)	5.94	2.135	0.00034
ANTHRANILIC ACID (A)	17.57	0.082	0.22558
TMPD	11.74	0.276	0.02015
1,2-BENZOFLUORENE	4.74	4.188	0.00009
9,10-DIHYDROPHENANTHRENE	0.00	0.000	0.00000
CARBAZOLE (TRACE)	0.00	0.000	0.00000
N-METHYL CARBAZOLE (TRACE)	4.71	4.280	0.00008
N-VINYLCARBAZOLE	0.00	0.000	0.00000
N-PHENYL CARBAZOLE	0.00	0.000	0.00000
N-METHYL-2,3-BENZCARBAZOLE	28.61	0.019	4.18394
3,3''' -DIMETHYL-3',2''-METHYLENE-P-QUATERPHENYL	6.86	1.386	0.00080
3,3''' -DIMETHYL-3',2''-ETHYLENE-P-QUATERPHENYL	11.38	0.303	0.01674
4-(3-ETHYLBUTOXY)-P-QUATERPHENYL (B)	9.41	0.536	0.00536
4,4'''-DI(2-BUTYLOCTOXY-1)-P-QUATERPHENYL	8.90	0.635	0.00382
DIETHYL-P-QUINQUEPHENYL	0.00	0.000	0.00000
DI(3-ETHYLHEPTYL)-P-QUINQUEPHENYL	12.12	0.251	0.02443
TETRAMETHYL-P-SEXIPHENYL	0.00	0.000	0.00000
TETRAMETHYL-P-OCTAPHENYL (B)	17.24	0.087	0.20185
INDOLE (TRACE)	0.00	0.000	0.00000
2-PHENYL INDOLE	0.00	0.000	0.00000
1-METHYL-2-PHENYL INDOLE	0.00	0.000	0.00000
1-NEOPENTYL-2-PHENYL INDOLE	0.00	0.000	0.00000
1-N-BUTYL-2(BIPHENYLYL-4) INDOLE	12.93	0.207	0.03602
1,2-DIPHENYL INDOLE	0.00	0.000	0.00000
1-BIPHENYLYL-2-PHENYL INDOLE	6.96	1.326	0.00088
1,3-DIMETHYL-2-PHENYL INDOLE	0.00	0.000	0.00000
2,3-DIPHENYL INDOLE	12.82	0.213	0.03405
1-METHYL-2,3-DIPHENYL INDOLE	10.40	0.398	0.00973
2-PHENYL-3,2'-METHYLENE INDOLE (TRACE)	0.00	0.000	0.00000
2-PHENYL-3,2'-ETHYLENE INDOLE (TRACE)	10.80	0.355	0.01221
2-PHENYL-3,2'-TRIMETHYLENE INDOLE	0.00	0.000	0.00000
1-METHYL-3,2'-METHYLENE-2-PHENYL INDOLE (TRACE)	8.33	0.773	0.00258
1-METHYL-3,2'-ETHYLENE-2-PHENYL INDOLE	16.27	0.104	0.14271
1-METHYL-3,2'-TRIMETHYLENE-2-PHENYL INDOLE	0.00	0.000	0.00000
4,5-DIPHENYLIMIDAZOLE (A)	0.00	0.000	0.00000
PPF (TRACE)	8.06	0.856	0.00211
αNPO (TRACE)	23.77	0.033	1.38230
BBO (B)	31.74	0.014	7.81671
POPOP	35.45	0.010	15.16356
DIMETHYL POPOP (TRACE)	39.26	0.007	27.95060
PPD (TRACE)	0.00	0.000	0.00000
2,5-DI(2-METHYLPHENYL)OXADIAZOLE	0.00	0.000	0.00000
PBD	0.00	0.000	0.00000
BBD (B)	12.25	0.243	0.02604
αNPD (TRACE)	13.65	0.176	0.04971
βNPD	0.00	0.000	0.00000
αNND (TRACE)	18.65	0.069	0.32336
βNND (TRACE)	0.00	0.000	0.00000
TRANS-1-(4-BIPHENYLYL)-2-PHENYL-ETHYLENE	14.57	0.145	0.07353
TRANS-1-(4-BIPHENYLYL)-2-(1-NAPHTHYL)-ETHYLENE	28.32	0.020	3.94528
DIPHENYLSTILBENE (B)	31.47	0.014	7.42179
1,1,4,4-TETRAPHENYLBUTADIENE	31.84	0.014	7.94781
1,6-DIPHENYLHEXATRIENE	37.41	0.009	20.94733
1,8-DIPHENYL-1,3,5,7-OCTATETRAENE	47.03	0.004	82.13523
BPSB (TRACE)	38.92	0.008	26.55533
BIS-MSB	36.06	0.010	16.78144
ESCULIN (A)	22.83	0.038	1.08437
AZULENE	13.12	0.198	0.03835
NAPHTHALENE	0.00	0.000	0.00000
DEUTERATED NAPHTHALENE-C10D8	0.00	0.000	0.00000
1-METHYLNAPHTHALENE 5ML/L	0.00	0.000	0.00000
2-METHYL NAPHTHALENE	0.00	0.000	0.00000
2,3-DIMETHYL NAPHTHALENE	0.00	0.000	0.00000
2,6-DIMETHYL NAPHTHALENE (TRACE)	0.00	0.000	0.00000
1-NAPHTHOL (TRACE)	0.00	0.000	0.00000
2-NAPHTHOL (TRACE)	0.00	0.000	0.00000
1-NAPHTHYLAMINE	5.55	2.618	0.00023
2-CHLORONAPHTHALENE	0.00	0.000	0.00000
ACENAPHTHENE	0.00	0.000	0.00000
1-PHENYL NAPHTHALENE	0.00	0.000	0.00000
2-PHENYL NAPHTHALENE	0.00	0.000	0.00000
1,4-DIPHENYL NAPHTHALENE	0.00	0.000	0.00000
1,5-DIPHENYL NAPHTHALENE	0.00	0.000	0.00000
1,7-DIPHENYL NAPHTHALENE	0.00	0.000	0.00000
1,4,5,8-TETRAPHENYL NAPHTHALENE	25.10	0.028	1.91089
1,1'-BINAPHTHYL	0.00	0.000	0.00000
4,4' -DI(N-BUTOXY)-1,1'-BINAPHTHYL	7.90	0.907	0.00187
2,2'-BINAPHTHYL	0.00	0.000	0.00000
ACRIDINE (A)	26.23	0.025	2.48757
ANTHRACENE (TRACE)	25.59	0.027	2.14729
DEUTERATED ANTHRACENE C14D10 (TRACE)	24.97	0.029	1.85470
9-METHYL ANTHRACENE (TRACE)	28.40	0.020	4.00802
1-AMINO ANTHRACENE	31.52	0.014	7.42776
2-AMINO ANTHRACENE	30.40	0.016	5.98116
9-VINYL ANTHRACENE	30.62	0.016	6.28614
9-PHENYL ANTHRACENE (TRACE)	29.75	0.017	5.29601
9,10-DIPHENYL ANTHRACENE	33.95	0.011	11.65074
9,10-DICHLOROANTHRACENE (TRACE)	33.51	0.012	10.77205
BIANTHRANYL (B)	35.62	0.010	15.55590
9,10-DINAPHTHYLANTHRACENE (B)	33.90	0.011	11.54793
9-(P-VINYLPHENYL)-10-PHENYL ANTHRACENE	34.45	0.011	12.71137

ACCEPTOR MOLECULES	RO	CO	OVERLAP
9,10-BIS(PHENYL-ETHYNYL)ANTHRACENE (TRACE)	42.77	0.006	46.07119
RUBICENE (B)	35.93	0.010	16.17969
DIBENZ(A,H)ANTHRACENE (B)	20.63	0.051	0.58784
TETRACENE (TRACE) (B)	31.50	0.014	7.35551
RUBRENE (B)	29.34	0.018	4.79563
TETRAHYDROCANNABINOL ACID METHYL ESTER	0.00	0.000	0.00000
PHENANTHRENE	3.30	12.460	0.00001
3,4-BENZOPHENANTHRENE	13.54	0.180	0.04713
2-PHENYL PHENANTHRENE	8.52	0.722	0.00395
CHRYSENE (TRACE)	11.00	0.259	0.02283
TRIPHENYLENE	3.18	13.925	0.00001
DEUTERATED TRIPHENYLENE-C18D12	2.82	19.872	0.00000
PYRENE	12.85	0.211	0.03454
1,3,6,8,-TETRAPHENYL PYRENE	41.07	0.006	36.45553
BIPYRENYL (B)	34.69	0.011	13.28570
O-PHENYLENE-PYRENE	36.44	0.009	17.67233
ANTHANTHRENE (B)	39.79	0.007	30.05443
FLUORANTHENE	23.03	0.037	1.13556
3-PHENYL FLUORANTHENE	31.32	0.015	7.17545
3-AMINO FLUORANTHENE	34.64	0.011	13.02927
3,3'-BIFLUORANTHENYL (B)	39.02	0.008	26.79765
BENZO(GHI)FLUORANTHENE	23.09	0.036	1.15325
9,12-DI(N-OCTADECOXY)-BENZ(E)ACEPHENANTHRYLENE	36.26	0.009	17.25562
DECACYCLENE (B)	44.32	0.005	57.10424
10-ETHOXY-BENZO(KL)XANTHENE	34.50	0.011	12.84942
PERYLENE	41.05	0.006	36.22050
BENZO(GHI)PERYLENE (B)	34.93	0.010	13.84474
BBOT (TRACE)	42.96	0.006	47.90468
QUININE SULFATE (A)	6.68	1.499	0.00069
ACRIDINE YELLOW (A)	40.62	0.007	33.78389
ACRIDINE ORANGE (A)	34.98	0.010	13.77556
ACRIDINE RED (A)	28.21	0.020	3.78973
URANINE (A)	35.52	0.010	15.10763
RHODAMINE B (A)	32.56	0.013	8.96086
RHODAMINE 6G (A)	31.67	0.014	7.58351
EOSIN B (A)	30.90	0.015	6.54667
EOSIN Y (A)	32.39	0.013	8.67239
ROSE BENGAL (A)	29.96	0.017	5.43678

8.102 Donor Molecule: 1,2-Diphenylindole

ACCEPTOR MOLECULES	RO	CO	OVERLAP
N,N-DIMETHYLANILINE 3ML/L	3.37	11.652	0.00001
TRIPHENYLAMINE	18.11	0.075	0.29533
SODIUM SALICYLATE (A)	13.03	0.202	0.04094
ANTHRANILIC ACID (A)	25.52	0.027	2.30065
TMPD	20.47	0.052	0.61453
BENZIDINE	14.30	0.153	0.07184
1,3,5-TRIPHENYL BENZENE	3.87	7.715	0.00003
1,3-DIBIPHENYLYLBENZENE (B)	12.17	0.248	0.02732
1,3,5,-TRIBIPHENYLYLBENZENE (B)	13.95	0.165	0.06204
FLUORENE (TRACE)	0.00	0.000	0.00000
1-METHYL FLUORENE	0.00	0.000	0.00000
1,2-BENZOFLUORENE	17.40	0.085	0.23292
9,10-DIHYDROPHENANTHRENE	0.00	0.000	0.00000
CARBAZOLE (TRACE)	14.94	0.134	0.09349
N-METHYL CARBAZOLE (TRACE)	21.11	0.048	0.74286
N-VINYLCARBAZOLE	20.37	0.053	0.59958
N-PHENYL CARBAZOLE	20.23	0.054	0.57491
N-METHYL-2,3-BENZCARBAZOLE	28.36	0.020	4.31063
DIBENZOFURAN	0.00	0.000	0.00000
DIBENZOTHIOPHENE (TRACE)	11.21	0.317	0.01673
P-TERPHENYL	0.00	0.000	0.00000
DEUTERATED P-TERPHENYL-C18D14	0.00	0.000	0.00000
4-METHYL TERPHENYL	0.00	0.000	0.00000
2-METHYL-P-TERPHENYL	0.00	0.000	0.00000
3,3''-DIMETHYL-P-TERPHENYL	5.45	2.767	0.00022
4-(3,3-DIMETHYLBUTOXY)-P-TERPHENYL	11.10	0.327	0.01580
4,4''-DIHEXAHYDROFARNESOXY-P-TERPHENYL	14.09	0.160	0.06584
2,2'-METHYLENE-P-TERPHENYL	10.77	0.359	0.01313
2,2'-ETHYLENE-P-TERPHENYL	12.75	0.216	0.03625
2,2'-METHYLENE-5',6'-METHYLENE-P-TERPHENYL (TRACE)	25.43	0.027	2.27191
3-PHENYL DIBENZOFURAN	10.71	0.364	0.01273
BENZO(1,2-B-4,5-B')BISBENZOFURAN (TRACE) (B)	27.87	0.021	3.93220
SEXIPHENYL BENZENE	0.00	0.000	0.00000
P-QUATERPHENYL	16.34	0.102	0.16028
3,3'''-DIMETHYL-P-QUATERPHENYL	18.30	0.073	0.31515
TETRAMETHYL-P-QUATERPHENYL	0.00	0.000	0.00000
3,3''' -DIMETHYL-3',2''-METHYLENE-P-QUATERPHENYL	27.02	0.023	3.26440
3,3''' -DIMETHYL-3',2''-ETHYLENE-P-QUATERPHENYL	27.36	0.022	3.50919
4-(3-ETHYLBUTOXY)-P-QUATERPHENYL (B)	24.44	0.031	1.78354
4,4'''-DI(2-BUTYLOCTOXY-1)-P-QUATERPHENYL	24.78	0.029	1.94006
DIETHYL-P-QUINQUEPHENYL	18.98	0.065	0.39310
DI(3-ETHYLHEPTYL)-P-QUINQUEPHENYL	26.37	0.024	2.80764
TETRAMETHYL-P-SEXIPHENYL	17.05	0.090	0.20647
TETRAMETHYL-P-OCTAPHENYL (B)	31.79	0.014	8.62232
INDOLE (TRACE)	0.00	0.000	0.00000
2-PHENYL INDOLE	18.26	0.073	0.31166
1-METHYL-2-PHENYL INDOLE	14.10	0.159	0.06617
1-NEOPENTYL-2-PHENYL INDOLE	10.46	0.391	0.01104
1-N-BUTYL-2(BIPHENYLYL-4) INDOLE	25.82	0.026	2.47511
1,2-DIPHENYL INDOLE	14.49	0.147	0.07775

ACCEPTOR MOLECULES	RO	CO	OVERLAP
1-BIPHENYLYL-2-PHENYL INDOLE	18.00	0.077	0.28467
1,3-DIMETHYL-2-PHENYL INDOLE	15.89	0.111	0.13562
2,3-DIPHENYL INDOLE	23.76	0.033	1.50082
1-METHYL-2,3-DIPHENYL INDOLE	21.32	0.046	0.78584
2-PHENYL-3,2'-METHYLENE INDOLE (TRACE)	22.99	0.037	1.23921
2-PHENYL-3,2'-ETHYLENE INDOLE (TRACE)	28.22	0.020	4.22759
2-PHENYL-3,2'-TRIMETHYLENE INDOLE	21.25	0.047	0.77291
1-METHYL-3,2'-METHYLENE-2-PHENYL INDOLE (TRACE)	27.21	0.022	3.40357
1-METHYL-3,2'-ETHYLENE-2-PHENYL INDOLE	30.04	0.017	6.13674
1-METHYL-3,2'-TRIMETHYLENE-2-PHENYL INDOLE	21.61	0.044	0.85591
4,5-DIPHENYLIMIDAZOLE (A)	12.20	0.247	0.02770
PPF (TRACE)	28.50	0.019	4.48837
PPO (TRACE)	21.20	0.047	0.75261
2-(4-FLUOROPHENYL)-5-PHENYLOXAZOLE (TRACE)	19.80	0.058	0.50701
αNPO (TRACE)	33.95	0.011	12.74680
BBO (B)	40.47	0.007	36.49374
POPOP	41.52	0.006	42.52687
DIMETHYL POPOP (TRACE)	43.55	0.005	56.52894
PPD (TRACE)	0.00	0.000	0.00000
2,5-DI(2-METHYLPHENYL)OXADIAZOLE	0.00	0.000	0.00000
PBD	18.56	0.070	0.34428
BBD (B)	30.21	0.016	6.36874
αNPD (TRACE)	26.77	0.023	3.07665
βNPD	19.07	0.064	0.40451
αNND (TRACE)	31.03	0.015	7.44654
βNND (TRACE)	26.86	0.023	3.15343
TRANS-1-(4-BIPHENYLYL)-2-PHENYL-ETHYLENE	30.67	0.016	6.95271
TRANS-1-(4-BIPHENYLYL)-2-(1-NAPHTHYL)-ETHYLENE	36.62	0.009	20.04782
DIPHENYLSTILBENE (B)	40.71	0.007	37.82701
1,1,4,4-TETRAPHENYLBUTADIENE	39.05	0.008	29.38262
1,6-DIPHENYLHEXATRIENE	44.34	0.005	62.99010
1,8-DIPHENYL-1,3,5,7-OCTATETRAENE	47.83	0.004	98.69274
BPSB (TRACE)	45.05	0.005	69.35372
BIS-MSB	42.79	0.006	50.91580
ESCULIN (A)	30.84	0.015	7.15888
AZULENE	21.11	0.048	0.72351
NAPHTHALENE	0.00	0.000	0.00000
DEUTERATED NAPHTHALENE-C10D8	5.92	2.151	0.00036
1-METHYLNAPHTHALENE 5ML/L	3.43	11.085	0.00001
2-METHYL NAPHTHALENE	4.00	6.984	0.00003
2,3-DIMETHYL NAPHTHALENE	4.06	6.665	0.00004
2,6-DIMETHYL NAPHTHALENE (TRACE)	7.58	1.025	0.00161
1-NAPHTHOL (TRACE)	7.27	1.166	0.00124
2-NAPHTHOL (TRACE)	12.34	0.238	0.02975
1-NAPHTHYLAMINE	19.23	0.063	0.42408
2-CHLORONAPHTHALENE	5.05	3.466	0.00014
ACENAPHTHENE	4.98	3.621	0.00013
1-PHENYL NAPHTHALENE	6.78	1.437	0.00082
2-PHENYL NAPHTHALENE	13.91	0.166	0.06092
1,4-DIPHENYL NAPHTHALENE	15.51	0.120	0.11719
1,5-DIPHENYL NAPHTHALENE	13.25	0.192	0.04556
1,7-DIPHENYL NAPHTHALENE	13.62	0.177	0.05368
1,4,5,8-TETRAPHENYL NAPHTHALENE	33.11	0.012	10.93175
1,1'-BINAPHTHYL	9.41	0.537	0.00585
4,4'-DI(N-BUTOXY)-1,1'-BINAPHTHYL	21.62	0.044	0.85474
2,2'-BINAPHTHYL	14.89	0.136	0.09166
ACRIDINE (A)	31.16	0.015	7.59365
ANTHRACENE (TRACE)	29.61	0.017	5.60141
DEUTERATED ANTHRACENE C14D10 (TRACE)	29.09	0.018	5.03716
9-METHYL ANTHRACENE (TRACE)	30.63	0.016	6.84879
1-AMINO ANTHRACENE	30.48	0.016	6.59329
2-AMINO ANTHRACENE	29.22	0.018	5.11476
9-VINYL ANTHRACENE	32.07	0.014	9.00128
9-PHENYL ANTHRACENE (TRACE)	32.20	0.013	9.24029
9,10-DIPHENYL ANTHRACENE	34.11	0.011	13.02330
9,10-DICHLOROANTHRACENE (TRACE)	33.14	0.012	10.95557
BIANTHRANYL (B)	35.73	0.010	17.19559
9,10-DINAPHTHYLANTHRACENE (B)	33.70	0.012	12.08913
9-(P-VINYLPHENYL)-10-PHENYL ANTHRACENE	34.41	0.011	13.71659
9,10-BIS(PHENYL-ETHYNYL)ANTHRACENE (TRACE)	35.75	0.010	17.07117
RUBICENE (B)	33.14	0.012	10.81130
DIBENZ(A,H)ANTHRACENE (B)	29.23	0.018	5.17154
TETRACENE (TRACE) (B)	26.86	0.023	3.06862
RUBRENE (B)	25.69	0.026	2.34837
TETRAHYDROCANNABINOL ACID METHYL ESTER	0.00	0.000	0.00000
PHENANTHRENE	12.55	0.226	0.03271
3,4-BENZOPHENANTHRENE	17.62	0.082	0.24881
2-PHENYL PHENANTHRENE	15.06	0.131	0.09740
CHRYSENE (TRACE)	17.87	0.078	0.27171
TRIPHENYLENE	12.48	0.230	0.03171
DEUTERATED TRIPHENYLENE-C18D12	12.11	0.252	0.02649
PYRENE	26.04	0.025	2.59209
1,3,6,8,-TETRAPHENYL PYRENE	42.21	0.006	46.64512
BIPYRENYL (B)	40.29	0.007	35.36420
O-PHENYLENE-PYRENE	36.76	0.009	20.18118
ANTHANTHRENE (B)	35.85	0.010	17.45017
FLUORANTHENE	29.67	0.017	5.64514
3-PHENYL FLUORANTHENE	34.61	0.011	14.18830
3-AMINO FLUORANTHENE	33.03	0.012	10.62600
3,3'-BIFLUORANTHENYL (B)	39.98	0.007	33.65509
BENZO(GHI)FLUORANTHENE	27.39	0.022	3.48442
9,12-DI(N-OCTADECOXY)-BENZ(E)ACEPHENANTHRYLENE	35.34	0.010	16.04863
DECACYCLENE (B)	43.59	0.005	56.11527
10-ETHOXY BENZO(KL)XANTHENE	34.79	0.011	14.67703
PERYLENE	35.84	0.010	17.40069
BENZO(GHI)PERYLENE (B)	36.45	0.009	19.41728
BBOT (TRACE)	45.27	0.005	71.23872

ACCEPTOR MOLECULES	RO	CO	OVERLAP
QUININE SULFATE (A)	23.38	0.035	1.36974
ACRIDINE YELLOW (A)	33.11	0.012	10.76255
ACRIDINE ORANGE (A)	28.15	0.020	4.06662
ACRIDINE RED (A)	27.15	0.022	3.27071
URANINE (A)	29.15	0.018	5.01297
RHODAMINE B (A)	31.10	0.015	7.38490
RHODAMINE 6G (A)	30.35	0.016	6.38297
EOSIN B (A)	27.42	0.022	3.46892
EOSIN Y (A)	28.48	0.019	4.35500
ROSE BENGAL (A)	28.40	0.020	4.28177
INDENOINDENE	20.44	0.052	0.61346

8.103 Donor Molecule: 1-Biphenylyl-2-phenylindole

N,N-DIMETHYLANILINE 3ML/L	3.02	16.161	0.00001
TRIPHENYLAMINE	17.60	0.082	0.27031
SODIUM SALICYLATE (A)	12.73	0.217	0.03875
ANTHRANILIC ACID (A)	25.04	0.028	2.22765
TMPD	20.05	0.055	0.58872
BENZIDINE	13.77	0.171	0.05214
1,3,5-TRIPHENYL BENZENE	3.54	10.083	0.00002
1,3-DIBIPHENYLYLBENZENE (B)	11.64	0.283	0.02274
1,3,5,-TRIBIPHENYLYLBENZENE (B)	13.29	0.191	0.05032
FLUORENE (TRACE)	0.00	0.000	0.00000
1-METHYL FLUORENE	0.00	0.000	0.00000
1,2-BENZOFLUORENE	17.03	0.091	0.22177
9,10-DIHYDROPHENANTHRENE	0.00	0.000	0.00000
CARBAZOLE (TRACE)	14.33	0.152	0.07885
N-METHYL CARBAZOLE (TRACE)	20.66	0.051	0.70724
N-VINYLCARBAZOLE	19.77	0.058	0.54267
N-PHENYL CARBAZOLE	19.67	0.059	0.52692
N-METHYL-2,3-BENZCARBAZOLE	27.87	0.021	4.20580
DIBENZOFURAN	0.00	0.000	0.00000
DIBENZOTHIOPHENE (TRACE)	10.58	0.378	0.01279
P-TERPHENYL	0.00	0.000	0.00000
DEUTERATED P-TERPHENYL-C18D14	0.00	0.000	0.00000
4-METHYL TERPHENYL	0.00	0.000	0.00000
2-METHYL-P-TERPHENYL	0.00	0.000	0.00000
3,3''-DIMETHYL-P-TERPHENYL	4.90	3.794	0.00013
4-(3,3-DIMETHYLBUTOXY)-P-TERPHENYL	10.52	0.385	0.01237
4,4''-DIHEXAHYDROFARNESOXY-P-TERPHENYL	13.42	0.185	0.05337
2,2' -METHYLENE-P-TERPHENYL	10.08	0.436	0.00961
2,2' -ETHYLENE-P-TERPHENYL	12.08	0.254	0.02844
2,2'-METHYLENE-5',6''-METHYLENE-P-TERPHENYL (TRACE)	24.48	0.031	1.95860
3-PHENYL DIBENZOFURAN	10.02	0.445	0.00925
BENZO(1,2-B-4,5-B')BISBENZOFURAN (TRACE) (B)	27.05	0.023	3.56400
SEXIPHENYL BENZENE	0.00	0.000	0.00000
P-QUATERPHENYL	15.68	0.116	0.13543
3,3'''-DIMETHYL-P-QUATERPHENYL	17.66	0.081	0.27648
TETRAMETHYL-P-QUATERPHENYL	0.00	0.000	0.00000
3,3''' -DIMETHYL-3',2''-METHYLENE-P-QUATERPHENYL	26.24	0.025	2.96657
3,3''' -DIMETHYL-3',2''-ETHYLENE-P-QUATERPHENYL	26.65	0.024	3.24924
4-(3-ETHYLBUTOXY)-P-QUATERPHENYL (B)	23.78	0.033	1.64361
4,4'''-DI(2-BUTYLOCTOXY-1)-P-QUATERPHENYL	24.11	0.032	1.78407
DIETHYL-P-QUINQUEPHENYL	18.30	0.073	0.34183
DI(3-ETHYLHEPTYL)-P-QUINQUEPHENYL	25.68	0.026	2.60077
TETRAMETHYL-P-SEXIPHENYL	16.34	0.103	0.17359
TETRAMETHYL-P-OCTAPHENYL (B)	31.07	0.015	8.14823
INDOLE (TRACE)	0.00	0.000	0.00000
2-PHENYL INDOLE	17.47	0.084	0.25974
1-METHYL-2-PHENYL INDOLE	13.52	0.181	0.05557
1-NEOPENTYL-2-PHENYL INDOLE	9.89	0.462	0.00856
1-N-BUTYL-2(BIPHENYLYL-4) INDOLE	25.22	0.028	2.33380
1,2-DIPHENYL INDOLE	13.88	0.167	0.06525
1-BIPHENYLYL-2-PHENYL INDOLE	17.45	0.084	0.25615
1,3-DIMETHYL-2-PHENYL INDOLE	15.27	0.126	0.11568
2,3-DIPHENYL INDOLE	23.21	0.036	1.41391
1-METHYL-2,3-DIPHENYL INDOLE	20.82	0.050	0.73747
2-PHENYL-3,2'-METHYLENE INDOLE (TRACE)	22.17	0.041	1.08101
2-PHENYL-3,2'-ETHYLENE INDOLE (TRACE)	27.66	0.021	4.06099
2-PHENYL-3,2' -TRIMETHYLENE INDOLE	20.48	0.052	0.67192
1-METHYL-3,2'-METHYLENE-2-PHENYL INDOLE (TRACE)	26.61	0.024	3.22308
1-METHYL-3,2'-ETHYLENE-2-PHENYL INDOLE	29.43	0.018	5.89161
1-METHYL-3,2'-TRIMETHYLENE-2-PHENYL INDOLE	20.93	0.049	0.76452
4,5-DIPHENYLIMIDAZOLE (A)	11.75	0.276	0.02397
PPF (TRACE)	27.86	0.021	4.24381
PPO (TRACE)	20.38	0.053	0.65352
2-(4-FLUOROPHENYL)-5-PHENYLOXAZOLE (TRACE)	19.02	0.065	0.43145
αNPO (TRACE)	33.34	0.012	12.38664
BBO (B)	39.74	0.007	35.51498
POPOP	40.77	0.007	41.36457
DIMETHYL POPOP (TRACE)	42.81	0.006	55.32395
PPD (TRACE)	0.00	0.000	0.00000
2,5-DI(2-METHYLPHENYL)OXADIAZOLE	0.00	0.000	0.00000
PBD	17.77	0.080	0.28697
BBD (B)	29.53	0.017	6.02108
αNPD (TRACE)	26.17	0.025	2.91370
βNPD	18.41	0.072	0.35500
αNND (TRACE)	30.42	0.016	7.16535
βNND (TRACE)	26.16	0.025	2.91701
TRANS-1-(4-BIPHENYLYL)-2-PHENYL-ETHYLENE	30.01	0.017	6.62410

ACCEPTOR MOLECULES	RO	CO	OVERLAP
TRANS-1-(4-BIPHENYLYL)-2-(1-NAPHTHYL)-ETHYLENE	35.95	0.010	19.45791
DIPHENYLSTILBENE (B)	39.99	0.007	36.83094
1,1,4,4-TETRAPHENYLBUTADIENE	38.36	0.008	28.63599
1,6-DIPHENYLHEXATRIENE	43.54	0.005	61.29054
1,8-DIPHENYL-1,3,5,7-OCTATETRAENE	47.04	0.004	96.84375
BPSB (TRACE)	44.26	0.005	67.61919
BIS-MSB	42.04	0.006	49.64041
ESCULIN (A)	30.27	0.016	6.94215
AZULENE	20.63	0.051	0.68110
NAPHTHALENE	0.00	0.000	0.00000
DEUTERATED NAPHTHALENE-C10D8	5.65	2.484	0.00030
1-METHYLNAPHTHALENE 5ML/L	3.12	14.721	0.00001
2-METHYL NAPHTHALENE	3.65	9.230	0.00002
2,3-DIMETHYL NAPHTHALENE	3.71	8.793	0.00002
2,6-DIMETHYL NAPHTHALENE (TRACE)	6.95	1.332	0.00103
1-NAPHTHOL (TRACE)	6.75	1.454	0.00087
2-NAPHTHOL (TRACE)	11.75	0.276	0.02406
1-NAPHTHYLAMINE	18.69	0.068	0.38792
2-CHLORONAPHTHALENE	4.68	4.374	0.00010
ACENAPHTHENE	4.53	4.805	0.00008
1-PHENYL NAPHTHALENE	6.34	1.753	0.00060
2-PHENYL NAPHTHALENE	13.28	0.191	0.05019
1,4-DIPHENYL NAPHTHALENE	14.95	0.134	0.10162
1,5-DIPHENYL NAPHTHALENE	12.67	0.220	0.03769
1,7-DIPHENYL NAPHTHALENE	13.07	0.200	0.04547
1,4,5,8-TETRAPHENYL NAPHTHALENE	32.51	0.013	10.60860
1,1'-BINAPHTHYL	8.89	0.638	0.00450
4,4'-DI(N-BUTOXY)-1,1'-BINAPHTHYL	21.02	0.048	0.78257
2,2'-BINAPHTHYL	14.29	0.153	0.07754
ACRIDINE (A)	30.61	0.016	7.39551
ANTHRACENE (TRACE)	29.03	0.018	5.39179
DEUTERATED ANTHRACENE C14D10 (TRACE)	28.54	0.019	4.87170
9-METHYL ANTHRACENE (TRACE)	30.16	0.016	6.77290
1-AMINO ANTHRACENE	30.04	0.017	6.55064
2-AMINO ANTHRACENE	28.79	0.019	5.07878
9-VINYL ANTHRACENE	31.56	0.014	8.86732
9-PHENYL ANTHRACENE (TRACE)	31.68	0.014	9.08572
9,10-DIPHENYL ANTHRACENE	33.55	0.012	12.79181
9,10-DICHLOROANTHRACENE (TRACE)	32.57	0.013	10.69197
BIANTHRANYL (B)	35.17	0.010	16.97315
9,10-DINAPHTHYLANTHRACENE (B)	33.12	0.012	11.81686
9-(P-VINYLPHENYL)-10-PHENYL ANTHRACENE	33.85	0.012	13.46626
9,10-BIS(PHENYL-ETHYNYL)ANTHRACENE (TRACE)	35.80	0.010	18.65656
RUBICENE (B)	32.90	0.013	11.20415
DIBENZ(A,H)ANTHRACENE (B)	28.66	0.019	4.98250
TETRACENE (TRACE) (B)	26.87	0.023	3.33247
RUBRENE (B)	25.82	0.026	2.62196
TETRAHYDROCANNABINOL ACID METHYL ESTER	0.00	0.000	0.00000
PHENANTHRENE	12.29	0.241	0.03129
3,4-BENZOPHENANTHRENE	17.23	0.087	0.23575
2-PHENYL PHENANTHRENE	14.77	0.139	0.09400
CHRYSENE (TRACE)	17.50	0.083	0.26012
TRIPHENYLENE	12.10	0.252	0.02857
DEUTERATED TRIPHENYLENE-C18D12	11.75	0.276	0.02397
PYRENE	25.18	0.028	2.29811
1,3,6,8,-TETRAPHENYL PYRENE	41.51	0.006	45.80175
BIPYRENYL (B)	39.60	0.007	34.60210
O-PHENYLENE-PYRENE	36.20	0.009	19.97241
ANTHANTHRENE (B)	35.51	0.010	17.86863
FLUORANTHENE	29.14	0.018	5.48721
3-PHENYL FLUORANTHENE	34.03	0.011	13.88456
3-AMINO FLUORANTHENE	32.60	0.013	10.64867
3,3'-BIFLUORANTHENYL (B)	39.34	0.007	33.08596
BENZO(GHI)FLUORANTHENE	26.94	0.023	3.42188
9,12-DI(N-OCTADECOXY) BENZ(E)ACEPHENANTHRYLENE	34.82	0.011	15.92456
DECACYCLENE (B)	42.94	0.006	55.61433
10-ETHOXY-BENZO(KL)XANTHENE	34.21	0.011	14.38602
PERYLENE	35.58	0.010	18.06091
BENZO(GHI)PERYLENE (B)	35.88	0.010	19.14702
BBOT (TRACE)	44.52	0.005	69.86959
QUININE SULFATE (A)	22.77	0.038	1.26829
ACRIDINE YELLOW (A)	33.44	0.012	12.36490
ACRIDINE ORANGE (A)	28.78	0.019	5.02896
ACRIDINE RED (A)	27.03	0.023	3.44635
URANINE (A)	29.88	0.017	6.29273
RHODAMINE B (A)	30.93	0.015	7.74557
RHODAMINE 6G (A)	30.44	0.016	7.02749
EOSIN B (A)	27.72	0.021	4.00715
EOSIN Y (A)	28.78	0.019	5.02587
ROSE BENGAL (A)	28.31	0.020	4.54887
INDENOINDENE	19.64	0.059	0.52213

8.104 Donor Molecule: 1,3-Dimethyl-2-phenylindole

TRIPHENYLAMINE	16.24	0.105	0.18924
P-XYLENE (TRACE)	0.00	0.000	0.00000
SODIUM SALICYLATE (A)	11.94	0.262	0.02997
ANTHRANILIC ACID (A)	23.90	0.033	1.91298
TMPD	18.94	0.066	0.47415
BENZIDINE	12.46	0.231	0.03874
1,3,5-TRIPHENYL BENZENE	2.65	24.168	0.00000

ACCEPTOR MOLECULES	RO	CO	OVERLAP
1,3-DIBIPHENYLYLBENZENE (B)	10.39	0.398	0.01307
1,3,5,-TRIBIPHENYLYLBENZENE (B)	11.66	0.282	0.02603
FLUORENE (TRACE)	0.00	0.000	0.00000
1-METHYL FLUORENE	0.00	0.000	0.00000
1,2-BENZOFLUORENE	15.86	0.112	0.16425
9,10-DIHYDROPHENANTHRENE	0.00	0.000	0.00000
CARBAZOLE (TRACE)	12.75	0.216	0.04455
N-METHYL CARBAZOLE (TRACE)	19.23	0.063	0.52183
N-VINYLCARBAZOLE	18.25	0.074	0.38248
N-PHENYL CARBAZOLE	18.20	0.074	0.37560
N-METHYL-3,3 BENZCARBAZOLE	27.49	0.022	4.40300
DIBENZOFURAN	0.00	0.000	0.00000
DIBENZOTHIOPHENE (TRACE)	9.00	0.614	0.00551
P-TERPHENYL	0.00	0.000	0.00000
DEUTERATED P-TERPHENYL-C18D14	0.00	0.000	0.00000
4-METHYL TERPHENYL	0.00	0.000	0.00000
2-METHYL-P-TERPHENYL	0.00	0.000	0.00000
3,3''-DIMETHYL-P-TERPHENYL	0.00	0.000	0.00000
4-(3,3-DIMETHYLBUTOXY)-P-TERPHENYL	9.02	0.610	0.00558
4,4''-DIHEXAHYDROFARNESOXY-P-TERPHENYL	11.75	0.276	0.02735
2,2'-METHYLENE-P-TERPHENYL	8.32	0.776	0.00345
2,2'-ETHYLENE-P-TERPHENYL	10.40	0.398	0.01314
2,2'-METHYLENE-5',6''-METHYLENE-P-TERPHENYL (TRACE)	22.13	0.041	1.21526
3-PHENYL DIBENZOFURAN	8.23	0.803	0.00323
BENZO(1,2-B-4,5-B')BISBENZOFURAN (TRACE) (B)	24.91	0.029	2.46719
SEXIPHENYL BENZENE	0.00	0.000	0.00000
P-QUATERPHENYL	14.07	0.161	0.08028
3,3'''-DIMETHYL-P-QUATERPHENYL	16.04	0.108	0.17655
TETRAMETHYL-P-QUATERPHENYL	0.00	0.000	0.00000
3,3''' -DIMETHYL-3',2''-METHYLENE-P-QUATERPHENYL	24.15	0.032	2.04889
3,3''' -DIMETHYL-3',2''-ETHYLENE-P-QUATERPHENYL	24.73	0.030	2.35786
4-(3-ETHYLBUTOXY)-P-QUATERPHENYL (B)	22.03	0.042	1.17815
4,4'''-DI(2-BUTYLOCTOXY-1)-P-QUATERPHENYL	22.30	0.040	1.26948
DIETHYL-P-QUINQUEPHENYL	16.58	0.098	0.21497
DI(3-ETHYLHEPTYL)-P-QUINQUEPHENYL	23.86	0.033	1.90029
TETRAMETHYL-P-SEXIPHENYL	14.63	0.143	0.10144
TETRAMETHYL-P-OCTAPHENYL (B)	29.16	0.018	6.32297
INDOLE (TRACE)	0.00	0.000	0.00000
2-PHENYL INDOLE	15.54	0.119	0.14580
1-METHYL-2-PHENYL INDOLE	12.07	0.254	0.03207
1-NEOPENTYL-2-PHENYL INDOLE	8.45	0.742	0.00378
1-N-BUTYL-2(BIPHENYLYL-4) INDOLE	23.62	0.034	1.78851
1,2-DIPHENYL INDOLE	12.40	0.235	0.03761
1-BIPHENYLYL-2-PHENYL INDOLE	16.04	0.108	0.17584
1,3-DIMETHYL-2-PHENYL INDOLE	13.75	0.172	0.05997
2,3-DIPHENYL INDOLE	21.74	0.044	1.08650
1-METHYL-2,3-DIPHENYL INDOLE	19.45	0.061	0.55837
2-PHENYL-3,2'-METHYLENE INDOLE (TRACE)	20.15	0.055	0.69315
2-PHENYL-3,2'-ETHYLENE INDOLE (TRACE)	25.86	0.026	3.08221
2-PHENYL-3,2' -TRIMETHYLENE INDOLE	18.56	0.070	0.42295
1-METHYL-3,2'-METHYLENE-2-PHENYL INDOLE (TRACE)	24.74	0.030	2.36417
1-METHYL-3,2'-ETHYLENE-2-PHENYL INDOLE	27.85	0.021	4.80797
1-METHYL-3,2' TRIMETHYLENE-2-PHENYL INDOLE	19.18	0.063	0.51418
4,5-DIPHENYLIMIDAZOLE (A)	10.67	0.369	0.01527
PPF (TRACE)	25.90	0.026	3.11675
PPO (TRACE)	18.39	0.072	0.40068
2-(4-FLUOROPHENYL)-5-PHENYLOXAZOLE (TRACE)	17.09	0.090	0.25822
αNPO (TRACE)	31.88	0.014	10.76367
BBO (B)	38.29	0.008	32.28766
POPOP	39.59	0.007	39.40847
DIMETHYL POPOP (TRACE)	41.78	0.006	54.27991
PPD (TRACE)	0.00	0.000	0.00000
2,5-DI(2-METHYLPHENYL)OXADIAZOLE	0.00	0.000	0.00000
PBD	15.81	0.113	0.16190
BBD (B)	27.57	0.021	4.53149
αNPD (TRACE)	24.50	0.030	2.22434
βNPD	16.75	0.095	0.22836
αNND (TRACE)	28.86	0.019	5.94749
βNND (TRACE)	24.20	0.032	2.07694
TRANS-1-(4-BIPHENYLYL)-2-PHENYL-ETHYLENE	28.15	0.020	5.12190
TRANS-1-(4-BIPHENYLYL)-2-(1-NAPHTHYL)-ETHYLENE	34.56	0.011	17.43469
DIPHENYLSTILBENE (B)	38.49	0.008	33.28046
1,1,4,4-TETRAPHENYLBUTADIENE	37.07	0.009	26.51117
1,6-DIPHENYLHEXATRIENE	42.26	0.006	58.19955
1,8-DIPHENYL-1,3,5,7-OCTATETRAENE	46.31	0.005	100.20568
BPSB (TRACE)	43.02	0.006	64.78996
BIS-MSB	40.74	0.007	46.72836
ESCULIN (A)	29.01	0.018	6.11502
AZULENE	19.29	0.062	0.51904
NAPHTHALENE	0.00	0.000	0.00000
DEUTERATED NAPHTHALENE-C10D8	4.95	3.680	0.00015
1-METHYLNAPHTHALENE 5ML/L	2.27	38.333	0.00000
2-METHYL NAPHTHALENE	2.64	24.190	0.00000
2,3-DIMETHYL NAPHTHALENE	2.72	22.155	0.00000
2,6-DIMETHYL NAPHTHALENE (TRACE)	5.19	3.191	0.00020
1-NAPHTHOL (TRACE)	5.43	2.795	0.00027
2-NAPHTHOL (TRACE)	10.24	0.417	0.01194
1-NAPHTHYLAMINE	17.24	0.087	0.27149
2-CHLORONAPHTHALENE	3.69	8.911	0.00003
ACENAPHTHENE	3.22	13.437	0.00001
1-PHENYL NAPHTHALENE	5.17	3.246	0.00020
2-PHENYL NAPHTHALENE	11.74	0.276	0.02723
1,4-DIPHENYL NAPHTHALENE	13.52	0.181	0.06311
1,5-DIPHENYL NAPHTHALENE	11.25	0.314	0.02101
1,7-DIPHENYL NAPHTHALENE	11.73	0.277	0.02695
1,4,5,8-TETRAPHENYL NAPHTHALENE	31.19	0.015	9.41086
1,1'-BINAPHTHYL	7.54	1.043	0.00191

ACCEPTOR MOLECULES	RO	CO	OVERLAP
4,4'-DI(N-BUTOXY)-1,1'-BINAPHTHYL	19.41	0.061	0.55203
2,2'-BINAPHTHYL	12.83	0.212	0.04625
ACRIDINE (A)	29.68	0.017	6.99065
ANTHRACENE (TRACE)	28.24	0.020	5.19172
DEUTERATED ANTHRACENE C14D10 (TRACE)	27.76	0.021	4.68434
9-METHYL ANTHRACENE (TRACE)	29.54	0.017	6.79492
1-AMINO ANTHRACENE	29.68	0.017	6.92189
2-AMINO ANTHRACENE	28.39	0.020	5.29864
9-VINYL ANTHRACENE	31.00	0.015	9.04486
9-PHENYL ANTHRACENE (TRACE)	31.01	0.015	9.07908
9,10-DIPHENYL ANTHRACENE	33.10	0.012	13.39536
9,10-DICHLOROANTHRACENE (TRACE)	32.15	0.013	11.23706
BIANTHRANYL (B)	34.71	0.011	17.81376
9,10-DINAPHTHYLANTHRACENE (B)	32.69	0.013	12.41279
9-(P-VINYLPHENYL)-10-PHENYL ANTHRACENE	33.41	0.012	14.13761
9,10-BIS(PHENYL-ETHYNYL)ANTHRACENE (TRACE)	35.50	0.010	20.16202
RUBICENE (B)	32.27	0.013	11.37252
DIBENZ(A,H)ANTHRACENE (B)	27.32	0.022	4.24462
TETRACENE (TRACE) (B)	26.55	0.024	3.52262
RUBRENE (B)	24.86	0.029	2.37319
TETRAHYDROCANNABINOL ACID METHYL ESTER	0.00	0.000	0.00000
PHENANTHRENE	11.45	0.298	0.02327
3,4-BENZOPHENANTHRENE	16.46	0.100	0.20392
2-PHENYL PHENANTHRENE	14.00	0.163	0.07738
CHRYSENE (TRACE)	16.69	0.096	0.22191
TRIPHENYLENE	11.13	0.325	0.01961
DEUTERATED TRIPHENYLENE-C18D12	10.81	0.354	0.01646
PYRENE	23.11	0.036	1.56020
1,3,6,8,-TETRAPHENYL PYRENE	40.88	0.007	47.44443
BIPYRENYL (B)	38.41	0.008	32.73484
O-PHENYLENE-PYRENE	35.54	0.010	20.31368
ANTHANTHRENE (B)	35.16	0.010	19.14009
FLUORANTHENE	28.09	0.020	5.00242
3-PHENYL FLUORANTHENE	33.22	0.012	13.67113
3-AMINO FLUORANTHENE	32.13	0.013	11.09976
3,3'-BIFLUORANTHENYL (B)	38.67	0.008	33.91334
BENZO(GHI)FLUORANTHENE	25.91	0.026	3.07583
9,12-DI(N-OCTADECOXY)-BENZ(E)ACEPHENANTHRYLENE	34.28	0.011	16.47655
DECACYCLENE (B)	42.26	0.006	57.43943
10-ETHOXY-BENZO(KL)XANTHENE	33.73	0.012	15.00364
PERYLENE	35.40	0.010	19.91066
BENZO(GHI)PERYLENE (B)	35.28	0.010	19.66763
BBOT (TRACE)	43.66	0.005	70.66811
QUININE SULFATE (A)	21.10	0.048	0.91059
ACRIDINE YELLOW (A)	32.98	0.012	12.94841
ACRIDINE ORANGE (A)	27.98	0.020	4.83161
ACRIDINE RED (A)	26.20	0.025	3.25585
URANINE (A)	28.72	0.019	5.64586
RHODAMINE B (A)	30.06	0.016	7.42883
RHODAMINE 6G (A)	29.11	0.018	6.11772
EOSIN B (A)	26.55	0.024	3.52407
EOSIN Y (A)	27.67	0.021	4.51542
ROSE BENGAL (A)	27.29	0.022	4.16110
INDENOINDENE	17.62	0.082	0.31014

8.105 Donor Molecule: 2,3-Diphenylindole

ACCEPTOR MOLECULES	RO	CO	OVERLAP
TRIPHENYLAMINE	4.43	5.139	0.00009
P-XYLENE (TRACE)	0.00	0.000	0.00000
SODIUM SALICYLATE (A)	4.75	4.163	0.00013
ANTHRANILIC ACID (A)	14.01	0.163	0.00793
TMPD	9.30	0.557	0.00752
1,2-BENZOFLUORENE	3.86	7.755	0.00004
9,10-DIHYDROPHENANTHRENE	0.00	0.000	0.00000
CARBAZOLE (TRACE)	0.00	0.000	0.00000
N-METHYL CARBAZOLE (TRACE)	3.85	7.813	0.00004
N-VINYLCARBAZOLE	0.00	0.000	0.00000
N-PHENYL CARBAZOLE	0.00	0.000	0.00000
N-METHYL-2,3-BENZCARBAZOLE	24.36	0.031	2.41516
3,3'''-DIMETHYL-3 ,2 -METHYLENE-P-QUATERPHENYL	5.59	2.559	0.00036
3,3'''-DIMETHYL-3 ,2 -ETHYLENE-P-QUATERPHENYL	9.06	0.601	0.00647
4-(3-ETHYLBUTOXY)-P-QUATERPHENYL (B)	7.53	1.048	0.00213
4,4'''-DI(2-BUTYLOCTOXY-1)-P-QUATERPHENYL	7.14	1.228	0.00155
DIETHYL-P-QUINQUEPHENYL	0.00	0.000	0.00000
DI(3-ETHYLHEPTYL)-P-QUINQUEPHENYL	9.61	0.504	0.00920
TETRAMETHYL-P-SEXIPHENYL	0.00	0.000	0.00000
TETRAMETHYL-P-OCTAPHENYL (B)	13.65	0.176	0.07551
INDOLE (TRACE)	0.00	0.000	0.00000
2-PHENYL INDOLE	0.00	0.000	0.00000
1-METHYL-2-PHENYL INDOLE	0.00	0.000	0.00000
1-NEOPENTYL-2-PHENYL INDOLE	0.00	0.000	0.00000
1-N-BUTYL-2(BIPHENYLYL-4) INDOLE	10.26	0.415	0.01357
1,2-DIPHENYL INDOLE	0.00	0.000	0.00000
1-BIPHENYLYL-2-PHENYL INDOLE	5.58	2.580	0.00035
1,3-DIMETHYL-2-PHENYL INDOLE	0.00	0.000	0.00000
2,3-DIPHENYL INDOLE	10.15	0.427	0.01276
1-METHYL-2,3-DIPHENYL INDOLE	8.25	0.796	0.00368
2-PHENYL-3,2'-METHYLENE INDOLE (TRACE)	0.00	0.000	0.00000
2-PHENYL-3,2'-ETHYLENE INDOLE (TRACE)	8.63	0.697	0.00481
2-PHENYL-3,2'-TRIMETHYLENE INDOLE	0.00	0.000	0.00000
1-METHYL-3,2'-METHYLENE-2-PHENYL INDOLE (TRACE)	6.70	1.488	0.00106
1-METHYL-3,2'-ETHYLENE-2-PHENYL INDOLE	12.93	0.207	0.05444

ACCEPTOR MOLECULES	RO	CO	OVERLAP
PPF (TRACE)	6.48	1.642	0.00087
αNPO (TRACE)	18.86	0.067	0.52290
BBO (B)	25.52	0.027	3.20403
POPOP	28.90	0.019	6.75338
DIMETHYL POPOP (TRACE)	32.58	0.013	13.82546
PPD (TRACE)	0.00	0.000	0.00000
2,5-DI(2-METHYLPHENYL)OXADIAZOLE	0.00	0.000	0.00000
PBD	0.00	0.000	0.00000
BBD (B)	9.81	0.474	0.01041
αNPD (TRACE)	10.70	[illegible]	0.01842
βNND	0.00	0.000	0.00000
αNND (TRACE)	14.76	0.139	0.12063
βNND (TRACE)	0.00	0.000	0.00000
TRANS-1-(4-BIPHENYLYL)-2-PHENYL-ETHYLENE	11.58	0.288	0.02812
TRANS-1-(4-BIPHENYLYL)-2-(1-NAPHTHYL)-ETHYLENE	22.86	0.037	1.65265
DIPHENYLSTILBENE (B)	25.27	0.028	3.01576
1,1,4,4-TETRAPHENYLBUTADIENE	25.87	0.026	3.47042
1,6-DIPHENYLHEXATRIENE	30.29	0.016	8.94052
1,8-DIPHENYL-1,3,5,7-OCTATETRAENE	40.02	0.007	47.27611
BPSB (TRACE)	31.90	0.014	12.18673
BIS-MSB	29.49	0.017	7.61263
ESCULIN (A)	18.34	0.073	0.44135
AZULENE	15.40	0.122	0.15115
NAPHTHALENE	0.00	0.000	0.00000
DEUTERATED NAPHTHALENE-C10D8	0.00	0.000	0.00000
1-METHYLNAPHTHALENE 5ML/L	0.00	0.000	0.00000
2-METHYL NAPHTHALENE	0.00	0.000	0.00000
2,3-DIMETHYL NAPHTHALENE	0.00	0.000	0.00000
2,6-DIMETHYL NAPHTHALENE (TRACE)	0.00	0.000	0.00000
1-NAPHTHOL (TRACE)	0.00	0.000	0.00000
2-NAPHTHOL (TRACE)	0.00	0.000	0.00000
1-NAPHTHYLAMINE	4.49	4.933	0.00010
2-CHLORONAPHTHALENE	0.00	0.000	0.00000
ACENAPHTHENE	0.00	0.000	0.00000
1-PHENYL NAPHTHALENE	0.00	0.000	0.00000
2-PHENYL NAPHTHALENE	0.00	0.000	0.00000
1,4-DIPHENYL NAPHTHALENE	0.00	0.000	0.00000
1,5-DIPHENYL NAPHTHALENE	0.00	0.000	0.00000
1,7-DIPHENYL NAPHTHALENE	0.00	0.000	0.00000
1,4,5,8-TETRAPHENYL NAPHTHALENE	20.31	0.053	0.81215
1,1'-BINAPHTHYL	0.00	0.000	0.00000
4,4'-DI(N-BUTOXY)-1,1'-BINAPHTHYL	6.34	1.755	0.00076
2,2'-BINAPHTHYL	0.00	0.000	0.00000
ACRIDINE (A)	21.57	0.045	1.16429
ANTHRACENE (TRACE)	20.71	0.050	0.91485
DEUTERATED ANTHRACENE C14D10 (TRACE)	20.15	0.055	0.77566
9-METHYL ANTHRACENE (TRACE)	23.68	0.034	2.04026
1-AMINO ANTHRACENE	27.77	0.021	5.26013
2-AMINO ANTHRACENE	27.05	0.023	4.49142
9-VINYL ANTHRACENE	25.75	0.026	3.36516
9-PHENYL ANTHRACENE (TRACE)	24.78	0.029	2.67677
9,10-DIPHENYL ANTHRACENE	28.88	0.019	6.69570
9,10-DICHLOROANTHRACENE (TRACE)	28.73	0.019	6.48385
BIANTHRANYL (B)	30.27	0.016	8.87800
9,10-DINAPHTHYLANTHRACENE (B)	28.98	0.018	6.82109
9-(P-VINYLPHENYL)-10-PHENYL ANTHRACENE	29.40	0.018	7.44490
9,10-BIS(PHENYL-ETHYNYL)ANTHRACENE (TRACE)	41.69	0.006	59.95493
RUBICENE (B)	35.72	0.010	23.54501
DIBENZ(A,H)ANTHRACENE (B)	17.26	0.087	0.30616
TETRACENE (TRACE) (B)	30.97	0.015	10.07499
RUBRENE (B)	32.10	0.014	12.40233
TETRAHYDROCANNABINOL ACID METHYL ESTER	0.00	0.000	0.00000
PHENANTHRENE	2.68	23.112	0.00000
3,4-BENZOPHENANTHRENE	10.85	0.350	0.01894
2-PHENYL PHENANTHRENE	6.75	1.452	0.00111
CHRYSENE (TRACE)	9.50	0.522	0.00855
TRIPHENYLENE	2.59	25.844	0.00000
DEUTERATED TRIPHENYLENE-C18D12	2.30	36.649	0.00000
PYRENE	10.23	0.418	0.01331
1,3,6,8,-TETRAPHENYL PYRENE	34.65	0.011	19.92542
BIPYRENYL (B)	28.84	0.019	6.64509
O-PHENYLENE-PYRENE	32.44	0.013	13.31429
ANTHANTHRENE (B)	36.76	0.009	28.32584
FLUORANTHENE	18.69	0.069	0.49168
3-PHENYL FLUORANTHENE	26.09	0.025	3.63579
3-AMINO FLUORANTHENE	31.41	0.014	10.98067
3,3'-BIFLUORANTHENYL (B)	33.45	0.012	16.09423
BENZO(GHI)FLUORANTHENE	19.87	0.057	0.70976
9,12-DI(N-OCTADECOXY)-BENZ(E)ACEPHENANTHRYLENE	32.11	0.014	12.61404
DECACYCLENE (B)	39.48	0.007	43.24663
10-ETHOXY-BENZO(KL)XANTHENE	29.31	0.018	7.32275
PERYLENE	38.17	0.008	35.46992
BENZO(GHI)PERYLENE (B)	29.31	0.018	7.32457
BBOT (TRACE)	36.16	0.009	25.83814
QUININE SULFATE (A)	5.43	2.795	0.00030
ACRIDINE YELLOW (A)	40.89	0.007	53.22347
ACRIDINE ORANGE (A)	37.36	0.009	30.89873
ACRIDINE RED (A)	32.91	0.013	14.37021
URANINE (A)	39.20	0.007	41.20821
RHODAMINE B (A)	37.63	0.008	32.14138
RHODAMINE 6G (A)	38.83	0.008	38.81935
EOSIN B (A)	36.95	0.009	28.83385
EOSIN Y (A)	38.12	0.008	34.76715
ROSE BENGAL (A)	35.73	0.010	23.55080

8.106 Donor Molecule: 1-Methyl-2,3-diphenylindole

ACCEPTOR MOLECULES	RO	CO	OVERLAP
TRIPHENYLAMINE	3.16	14.121	0.00002
P-XYLENE (TRACE)	0.00	0.000	0.00000
SODIUM SALICYLATE (A)	3.81	8.104	0.00006
ANTHRANILIC ACID (A)	12.12	0.251	0.06207
TMPD	7.88	0.913	0.00470
1,2-BENZOFLUORENE	2.68	23.289	0.00001
N-METHYL-2,3-BENZCARBAZOLE	21.98	0.042	2.19471
3,3''' -DIMETHYL-3',2''-METHYLENE-P-QUATERPHENYL	3.86	7.758	0.00007
3,3''' -DIMETHYL-3',2''-ETHYLENE-P-QUATERPHENYL	7.46	1.076	0.00340
4-(3-ETHYLBUTOXY)-P-QUATERPHENYL (B)	6.07	1.998	0.00099
4,4'''-DI(2-BUTYLOCTOXY-1)-P-QUATERPHENYL	5.62	2.518	0.00062
DIETHYL-P-QUINQUEPHENYL	0.00	0.000	0.00000
DI(3-ETHYLHEPTYL)-P-QUINQUEPHENYL	8.05	0.859	0.00533
TETRAMETHYL-P-SEXIPHENYL	0.00	0.000	0.00000
TETRAMETHYL-P-OCTAPHENYL (B)	11.56	0.289	0.04682
INDOLE (TRACE)	0.00	0.000	0.00000
2-PHENYL INDOLE	0.00	0.000	0.00000
1-METHYL-2-PHENYL INDOLE	0.00	0.000	0.00000
1-NEOPENTYL-2-PHENYL INDOLE	0.00	0.000	0.00000
1-N-BUTYL-2(BIPHENYLYL-4) INDOLE	8.58	0.709	0.00783
1,2-DIPHENYL INDOLE	0.00	0.000	0.00000
1-BIPHENYLYL-2-PHENYL INDOLE	4.44	5.101	0.00015
1,3-DIMETHYL-2-PHENYL INDOLE	0.00	0.000	0.00000
2,3-DIPHENYL INDOLE	8.61	0.700	0.00801
1-METHYL-2,3-DIPHENYL INDOLE	6.88	1.371	0.00209
2-PHENYL-3,2'-METHYLENE INDOLE (TRACE)	0.00	0.000	0.00000
2-PHENYL-3,2'-ETHYLENE INDOLE (TRACE)	6.97	1.320	0.00226
2-PHENYL-3,2'-TRIMETHYLENE INDOLE	0.00	0.000	0.00000
1-METHYL-3,2'-METHYLENE-2-PHENYL INDOLE (TRACE)	5.21	3.155	0.00040
1-METHYL-3,2'-ETHYLENE-2-PHENYL INDOLE	10.74	0.361	0.03012
1-METHYL-3,2'-TRIMETHYLENE-2-PHENYL INDOLE	0.00	0.000	0.00000
4,5-DIPHENYLIMIDAZOLE (A)	0.00	0.000	0.00000
PPF (TRACE)	5.00	3.582	0.00031
αNPO (TRACE)	16.26	0.104	0.36073
BBO (B)	22.34	0.040	2.42808
POPOP	25.60	0.027	5.48942
DIMETHYL POPOP (TRACE)	29.10	0.018	11.80991
PPD (TRACE)	0.00	0.000	0.00000
2,5-DI(2-METHYLPHENYL)OXADIAZOLE	0.00	0.000	0.00000
PBD	0.00	0.000	0.00000
BBD (B)	7.85	0.923	0.00461
αNPD (TRACE)	9.14	0.586	0.01143
βNPD	0.00	0.000	0.00000
αNND (TRACE)	12.50	0.229	0.07477
βNND (TRACE)	0.00	0.000	0.00000
TRANS-1-(4-BIPHENYLYL)-2-PHENYL-ETHYLENE	9.59	0.507	0.01530
TRANS-1-(4-BIPHENYLYL)-2-(1-NAPHTHYL)-ETHYLENE	20.06	0.055	1.27066
DIPHENYLSTILBENE (B)	22.08	0.042	2.26152
1,1,4,4-TETRAPHENYLBUTADIENE	22.82	0.038	2.75238
1,6-DIPHENYLHEXATRIENE	26.69	0.024	7.03749
1,8-DIPHENYL-1,3,5,7-OCTATETRAENE	36.09	0.010	42.81852
BPSB (TRACE)	28.31	0.020	10.04314
BIS-MSB	26.13	0.025	6.20479
ESCULIN (A)	16.02	0.109	0.32996
AZULENE	14.21	0.156	0.15691
NAPHTHALENE	0.00	0.000	0.00000
DEUTERATED NAPHTHALENE-C10D8	0.00	0.000	0.00000
1-METHYLNAPHTHALENE 5ML/L	0.00	0.000	0.00000
2-METHYL NAPHTHALENE	0.00	0.000	0.00000
2,3-DIMETHYL NAPHTHALENE	0.00	0.000	0.00000
2,6-DIMETHYL NAPHTHALENE (TRACE)	0.00	0.000	0.00000
1-NAPHTHOL (TRACE)	0.00	0.000	0.00000
2-NAPHTHOL (TRACE)	0.00	0.000	0.00000
1-NAPHTHYLAMINE	3.30	12.400	0.00003
2-CHLORONAPHTHALENE	0.00	0.000	0.00000
ACENAPHTHENE	0.00	0.000	0.00000
1-PHENYL NAPHTHALENE	0.00	0.000	0.00000
2-PHENYL NAPHTHALENE	0.00	0.000	0.00000
1,4-DIPHENYL NAPHTHALENE	0.00	0.000	0.00000
1,5-DIPHENYL NAPHTHALENE	0.00	0.000	0.00000
1,7-DIPHENYL NAPHTHALENE	0.00	0.000	0.00000
1,4,5,8-TETRAPHENYL NAPHTHALENE	17.83	0.079	0.62642
1,1'-BINAPHTHYL	0.00	0.000	0.00000
4,4'-DI(N-BUTOXY)-1,1'-BINAPHTHYL	5.00	3.582	0.00031
2,2'-BINAPHTHYL	0.00	0.000	0.00000
ACRIDINE (A)	19.16	0.064	0.96381
ANTHRACENE (TRACE)	18.28	0.073	0.72826
DEUTERATED ANTHRACENE C14D10 (TRACE)	17.72	0.080	0.60448
9-METHYL ANTHRACENE (TRACE)	21.21	0.047	1.77422
1-AMINO ANTHRACENE	25.27	0.028	5.02953
2-AMINO ANTHRACENE	24.66	0.030	4.35052
9-VINYL ANTHRACENE	23.12	0.036	2.97242
9-PHENYL ANTHRACENE (TRACE)	22.19	0.041	2.32437
9,10-DIPHENYL ANTHRACENE	26.04	0.025	6.05741
9,10-DICHLOROANTHRACENE (TRACE)	26.00	0.025	5.99840
BIANTHRANYL (B)	27.27	0.022	7.99559
9,10-DINAPHTHYLANTHRACENE (B)	26.18	0.025	6.25124
9-(P-VINYLPHENYL)-10-PHENYL ANTHRACENE	26.54	0.024	6.78469
9,10-BIS(PHENYL-ETHYNYL)ANTHRACENE (TRACE)	38.46	0.008	62.23583
RUBICENE (B)	32.94	0.013	24.37007
DIBENZ(A,H)ANTHRACENE (B)	15.38	0.123	0.25765
TETRACENE (TRACE) (B)	28.58	0.019	10.46227
RUBRENE (B)	29.72	0.017	13.15756

ACCEPTOR MOLECULES	RO	CO	OVERLAP
TETRAHYDROCANNABINOL ACID METHYL ESTER	0.00	0.000	0.00000
PHENANTHRENE	1.88	67.413	0.00000
3,4-BENZOPHENANTHRENE	9.46	0.528	0.01402
2-PHENYL PHENANTHRENE	5.67	2.455	0.00065
CHRYSENE (TRACE)	8.11	0.838	0.00559
TRIPHENYLENE	1.78	79.645	0.00000
DEUTERATED TRIPHENYLENE-C18D12	0.00	0.000	0.00000
PYRENE	8.84	0.648	0.00932
1,3,6,8,-TETRAPHENYL PYRENE	31.16	0.015	17.77012
BIPYRENYL (B)	25.75	0.026	5.67140
O-PHENYLENE-PYRENE	29.56	0.017	12.82796
ANTHANTHRENE (B)	33.74	0.012	28.51121
FLUORANTHENE	16.43	0.101	0.38229
3-PHENYL FLUORANTHENE	23.33	0.035	3.12621
3-AMINO FLUORANTHENE	28.73	0.019	10.82284
3,3'-BIFLUORANTHENYL (B)	30.22	0.016	14.74220
BENZO(GHI)FLUORANTHENE	17.97	0.077	0.65317
9,12-DI(N-OCTADECOXY)-BENZ(E)ACEPHENANTHRYLENE	29.26	0.018	12.14294
DECACYCLENE (B)	35.99	0.010	41.77915
10-ETHOXY-BENZO(KL)XANTHENE	26.43	0.024	6.63455
PERYLENE	35.06	0.010	35.87816
BENZO(GHI)PERYLENE (B)	26.32	0.025	6.46583
BBOT (TRACE)	32.49	0.013	22.85514
QUININE SULFATE (A)	3.91	7.499	0.00007
ACRIDINE YELLOW (A)	37.80	0.008	55.90269
ACRIDINE ORANGE (A)	34.60	0.011	32.81551
ACRIDINE RED (A)	30.38	0.016	14.98282
URANINE (A)	36.33	0.009	43.97053
RHODAMINE B (A)	34.70	0.011	33.28616
RHODAMINE 6G (A)	35.96	0.010	41.26249
EOSIN B (A)	34.24	0.011	30.70911
EOSIN Y (A)	35.31	0.010	37.00542
ROSE BENGAL (A)	32.96	0.012	24.45250

8.107 Donor Molecule: 2-Phenyl-3,2′-methyleneindole (trace)

N,N-DIMETHYLANILINE 3ML/L	2.46	29.893	0.00000
TRIPHENYLAMINE	18.90	0.066	0.39944
SODIUM SALICYLATE (A)	13.59	0.178	0.05519
ANTHRANILIC ACID (A)	26.29	0.025	2.87922
TMPD	21.25	0.047	0.80472
BENZIDINE	14.91	0.135	0.09671
1,3,5-TRIPHENYL BENZENE	3.15	14.325	0.00001
1,3-DIBIPHENYLYLBENZENE (B)	12.74	0.216	0.03767
1,3,5,-TRIBIPHENYLYLBENZENE (B)	14.41	0.149	0.07895
FLUORENE (TRACE)	0.00	0.000	0.00000
1-METHYL FLUORENE	0.00	0.000	0.00000
1,2-BENZOFLUORENE	18.01	0.077	0.29902
9,10-DIHYDROPHENANTHRENE	0.00	0.000	0.00000
CARBAZOLE (TRACE)	15.46	0.121	0.12015
N-METHYL CARBAZOLE (TRACE)	21.84	0.043	0.95299
N-VINYLCARBAZOLE	21.43	0.045	0.85113
N-PHENYL CARBAZOLE	21.18	0.047	0.79198
N-METHYL-2,3-BENZCARBAZOLE	27.95	0.020	4.13014
DIBENZOFURAN	0.00	0.000	0.00000
DIBENZOTHIOPHENE (TRACE)	10.83	0.352	0.01421
P-TERPHENYL	0.00	0.000	0.00000
DEUTERATED P-TERPHENYL-C18D14	0.00	0.000	0.00000
4-METHYL TERPHENYL	0.00	0.000	0.00000
2-METHYL-P-TERPHENYL	0.00	0.000	0.00000
3,3''-DIMETHYL-P-TERPHENYL	4.07	6.649	0.00004
4-(3,3-DIMETHYLBUTOXY)-P-TERPHENYL	10.93	0.343	0.01501
4,4''-DIHEXAHYDROFARNESOXY-P-TERPHENYL	14.50	0.147	0.08197
2,2'-METHYLENE-P-TERPHENYL	9.76	0.481	0.00764
2,2'-ETHYLENE-P-TERPHENYL	12.68	0.219	0.03669
2,2'-METHYLENE-5',6'-METHYLENE-P-TERPHENYL (TRACE)	27.30	0.022	3.63557
3-PHENYL DIBENZOFURAN	9.63	0.502	0.00702
BENZO(1,2-B-4,5-B')BISBENZOFURAN (TRACE) (B)	28.98	0.018	5.19602
SEXIPHENYL BENZENE	0.00	0.000	0.00000
P-QUATERPHENYL	17.07	0.090	0.21806
3,3'''-DIMETHYL-P-QUATERPHENYL	19.11	0.064	0.42767
TETRAMETHYL-P-QUATERPHENYL	0.00	0.000	0.00000
3,3''' -DIMETHYL-3',2''-METHYLENE-P-QUATERPHENYL	28.35	0.020	4.55463
3,3''' -DIMETHYL-3',2''-ETHYLENE-P-QUATERPHENYL	28.61	0.019	4.79663
4-(3-ETHYLBUTOXY)-P-QUATERPHENYL (B)	25.54	0.027	2.43337
4,4'''-DI(2-BUTYLOCTOXY-1)-P-QUATERPHENYL	25.91	0.026	2.64953
DIETHYL-P-QUINQUEPHENYL	19.83	0.057	0.53514
DI(3-ETHYLHEPTYL)-P-QUINQUEPHENYL	27.52	0.021	3.79767
TETRAMETHYL-P-SEXIPHENYL	17.77	0.080	0.27724
TETRAMETHYL-P-OCTAPHENYL (B)	33.08	0.012	11.44616
INDOLE (TRACE)	0.00	0.000	0.00000
2-PHENYL INDOLE	19.14	0.064	0.43259
1-METHYL-2-PHENYL INDOLE	14.63	0.143	0.08613
1-NEOPENTYL-2-PHENYL INDOLE	10.23	0.418	0.01011
1-N-BUTYL-2(BIPHENYLYL-4) INDOLE	26.89	0.023	3.30824
1,2-DIPHENYL INDOLE	14.99	0.133	0.09999
1-BIPHENYLYL-2-PHENYL INDOLE	18.73	0.068	0.37823
1,3-DIMETHYL-2-PHENYL INDOLE	16.68	0.096	0.18939
2,3-DIPHENYL INDOLE	24.74	0.030	2.00157
1-METHYL-2,3-DIPHENYL INDOLE	22.23	0.041	1.05516
2-PHENYL-3,2'-METHYLENE INDOLE (TRACE)	24.62	0.030	1.95816

ACCEPTOR MOLECULES	RO	CO	OVERLAP
2-PHENYL-3,2'-ETHYLENE INDOLE (TRACE)	29.31	0.018	5.54966
2-PHENYL-3,2'-TRIMETHYLENE INDOLE	22.41	0.040	1.11338
1-METHYL-3,2'-METHYLENE-2-PHENYL INDOLE (TRACE)	28.29	0.020	4.49596
1-METHYL-3,2'-ETHYLENE-2-PHENYL INDOLE	31.27	0.015	8.17359
1-METHYL-3,2'-TRIMETHYLENE-2-PHENYL INDOLE	22.81	0.038	1.23763
4,5-DIPHENYLIMIDAZOLE (A)	12.83	0.212	0.03921
PPF (TRACE)	29.66	0.017	5.97006
PPO (TRACE)	22.57	0.039	1.16102
2-(4-FLUOROPHENYL)-5-PHENYLOXAZOLE (TRACE)	20.99	0.048	0.75241
αNPO (TRACE)	34.87	0.011	15.65621
BBO (B)	41.32	0.006	43.28393
POPOP	41.99	0.006	47.60830
DIMETHYL POPOP (TRACE)	43.75	0.005	60.75378
PPD (TRACE)	0.00	0.000	0.00000
2,5-DI(2-METHYLPHENYL)OXADIAZOLE	0.00	0.000	0.00000
PBD	19.34	0.062	0.46067
BBD (B)	31.52	0.014	8.59583
αNPD (TRACE)	27.81	0.021	4.04308
βNPD	19.67	0.059	0.50966
αNND (TRACE)	32.09	0.014	9.53118
βNND (TRACE)	28.01	0.020	4.23816
TRANS-1-(4-BIPHENYLYL)-2-PHENYL-ETHYLENE	31.94	0.014	9.27504
TRANS-1-(4-BIPHENYLYL)-2-(1-NAPHTHYL)-ETHYLENE	37.47	0.009	24.06049
DIPHENYLSTILBENE (B)	41.61	0.006	45.07472
1,1,4,4-TETRAPHENYLBUTADIENE	39.73	0.007	34.11798
1,6-DIPHENYLHEXATRIENE	45.04	0.005	72.41130
1,8-DIPHENYL-1,3,5,7-OCTATETRAENE	47.48	0.004	98.84491
BPSB (TRACE)	45.51	0.005	77.09978
BIS-MSB	43.39	0.005	57.89717
ESCULIN (A)	31.64	0.014	8.73072
AZULENE	22.06	0.042	0.98747
NAPHTHALENE	0.00	0.000	0.00000
DEUTERATED NAPHTHALENE-C10D8	6.12	1.949	0.00046
1-METHYLNAPHTHALENE 5ML/L	2.73	22.064	0.00000
2-METHYL NAPHTHALENE	3.20	13.643	0.00001
2,3-DIMETHYL NAPHTHALENE	3.26	12.896	0.00001
2,6-DIMETHYL NAPHTHALENE (TRACE)	6.20	1.875	0.00050
1-NAPHTHOL (TRACE)	6.39	1.712	0.00060
2-NAPHTHOL (TRACE)	12.38	0.236	0.03172
1-NAPHTHYLAMINE	20.20	0.054	0.59488
2-CHLORONAPHTHALENE	4.29	5.685	0.00005
ACENAPHTHENE	3.96	7.221	0.00003
1-PHENYL NAPHTHALENE	5.96	2.117	0.00039
2-PHENYL NAPHTHALENE	14.60	0.144	0.08554
1,4-DIPHENYL NAPHTHALENE	16.14	0.106	0.15521
1,5-DIPHENYL NAPHTHALENE	13.68	0.175	0.05770
1,7-DIPHENYL NAPHTHALENE	14.17	0.157	0.07132
1,4,5,8-TETRAPHENYL NAPHTHALENE	33.93	0.011	13.24379
1,1'-BINAPHTHYL	9.00	0.614	0.00469
4,4'-DI(N-BUTOXY)-1,1'-BINAPHTHYL	22.59	0.039	1.16499
2,2'-BINAPHTHYL	15.60	0.118	0.12679
ACRIDINE (A)	31.61	0.014	8.65590
ANTHRACENE (TRACE)	30.05	0.016	6.40385
DEUTERATED ANTHRACENE C14D10 (TRACE)	29.62	0.017	5.87457
9-METHYL ANTHRACENE (TRACE)	30.47	0.016	6.95055
1-AMINO ANTHRACENE	29.91	0.017	6.15885
2-AMINO ANTHRACENE	28.83	0.019	4.93449
9-VINYL ANTHRACENE	31.90	0.014	9.13149
9-PHENYL ANTHRACENE (TRACE)	32.05	0.014	9.40577
9,10-DIPHENYL ANTHRACENE	33.82	0.012	12.94142
9,10-DICHLOROANTHRACENE (TRACE)	32.61	0.013	10.39606
BIANTHRANYL (B)	35.41	0.010	17.05007
9,10-DINAPHTHYLANTHRACENE (B)	33.29	0.012	11.75276
9-(P-VINYLPHENYL)-10-PHENYL ANTHRACENE	34.05	0.011	13.46319
9,10-BIS(PHENYL-ETHYNYL)ANTHRACENE (TRACE)	33.53	0.012	12.17567
RUBICENE (B)	32.32	0.013	9.76906
DIBENZ(A,H)ANTHRACENE (B)	30.41	0.016	6.84797
TETRACENE (TRACE) (B)	25.53	0.027	2.37165
RUBRENE (B)	25.01	0.029	2.09579
TETRAHYDROCANNABINOL ACID METHYL ESTER	0.00	0.000	0.00000
PHENANTHRENE	13.09	0.199	0.04417
3,4-BENZOPHENANTHRENE	17.93	0.078	0.28858
2-PHENYL PHENANTHRENE	15.61	0.118	0.12628
CHRYSENE (TRACE)	18.25	0.074	0.32285
TRIPHENYLENE	13.29	0.191	0.04824
DEUTERATED TRIPHENYLENE-C18D12	12.77	0.215	0.03805
PYRENE	28.18	0.020	4.36128
1,3,6,8,-TETRAPHENYL PYRENE	41.81	0.006	46.13781
BIPYRENYL (B)	40.80	0.007	39.91976
O-PHENYLENE-PYRENE	36.45	0.009	20.10414
ANTHANTHRENE (B)	34.65	0.011	14.86887
FLUORANTHENE	30.20	0.016	6.56478
3-PHENYL FLUORANTHENE	34.78	0.011	15.26899
3-AMINO FLUORANTHENE	32.42	0.013	9.94833
3,3'-BIFLUORANTHENYL (B)	39.70	0.007	33.74084
BENZO(GHI)FLUORANTHENE	28.02	0.020	4.17575
9,12-DI(N-OCTADECOXY)-BENZ(E)ACEPHENANTHRYLENE	34.90	0.011	15.58112
DECACYCLENE (B)	43.03	0.006	54.40345
10-ETHOXY-BENZO(KL)XANTHENE	34.43	0.011	14.41700
PERYLENE	34.05	0.011	13.38816
BENZO(GHI)PERYLENE (B)	36.09	0.010	19.11031
BBOT (TRACE)	45.20	0.005	73.83963
QUININE SULFATE (A)	24.55	0.030	1.92207
ACRIDINE YELLOW (A)	30.73	0.015	7.21883
ACRIDINE ORANGE (A)	25.82	0.026	2.53860
ACRIDINE RED (A)	26.93	0.023	3.27030
URANINE (A)	27.29	0.022	3.54367

ACCEPTOR MOLECULES	RO	CO	OVERLAP
RHODAMINE B (A)	30.81	0.015	7.32611
RHODAMINE 6G (A)	30.30	0.016	6.63543
EOSIN B (A)	26.74	0.023	3.13340
EOSIN Y (A)	27.64	0.021	3.81990
ROSE BENGAL (A)	28.21	0.020	4.31593
INDENOINDENE	21.50	0.045	0.86758

8.108 Donor Molecule: 2-Phenyl-3,2′-ethyleneindole (trace)

ACCEPTOR MOLECULES	RO	CO	OVERLAP
TRIPHENYLAMINE	13.86	0.168	0.05065
P-XYLENE (TRACE)	0.00	0.000	0.00000
SODIUM SALICYLATE (A)	11.20	0.318	0.01585
ANTHRANILIC ACID (A)	23.45	0.035	1.40998
TMPD	18.00	0.077	0.28946
BENZIDINE	8.62	0.697	0.00352
1,3-DIBIPHENYLYLBENZENE (B)	5.86	2.226	0.00035
1,3,5,-TRIBIPHENYLYLBENZENE (B)	5.14	3.296	0.00016
FLUORENE (TRACE)	0.00	0.000	0.00000
1-METHYL FLUORENE	0.00	0.000	0.00000
1,2-BENZOFLUORENE	14.14	0.158	0.06835
9,10-DIHYDROPHENANTHRENE	0.00	0.000	0.00000
CARBAZOLE (TRACE)	7.77	0.954	0.00188
N-METHYL CARBAZOLE (TRACE)	16.43	0.101	0.16785
N-VINYLCARBAZOLE	13.77	0.171	0.05837
N-PHENYL CARBAZOLE	14.31	0.153	0.07351
N-METHYL-2,3-BENZCARBAZOLE	28.65	0.019	4.66919
DIBENZOFURAN	0.00	0.000	0.00000
DIBENZOTHIOPHENE (TRACE)	4.41	5.214	0.00006
P-TERPHENYL	0.00	0.000	0.00000
DEUTERATED P-TERPHENYL-C18D14	0.00	0.000	0.00000
4-METHYL TERPHENYL	0.00	0.000	0.00000
2-METHYL-P-TERPHENYL	0.00	0.000	0.00000
3,3''-DIMETHYL-P-TERPHENYL	0.00	0.000	0.00000
4-(3,3-DIMETHYLBUTOXY)-P-TERPHENYL	0.00	0.000	0.00000
4,4''-DIHEXAHYDROFARNESOXY-P-TERPHENYL	4.99	3.598	0.00013
2,2' -METHYLENE-P-TERPHENYL	0.00	0.000	0.00000
2,2' -ETHYLENE-P-TERPHENYL	4.25	5.833	0.00005
2,2'-METHYLENE-5',6'-METHYLENE-P-TERPHENYL (TRACE)	13.83	0.169	0.06001
3-PHENYL DIBENZOFURAN	0.00	0.000	0.00000
BENZO(1,2-B-4,5-B')BISBENZOFURAN (TRACE) (B)	19.30	0.062	0.44165
SEXIPHENYL BENZENE	0.00	0.000	0.00000
P-QUATERPHENYL	8.57	0.711	0.00339
3,3'''-DIMETHYL-P-QUATERPHENYL	12.00	0.259	0.02551
TETRAMETHYL-P-QUATERPHENYL	0.00	0.000	0.00000
3,3''' -DIMETHYL-3',2''-METHYLENE-P-QUATERPHENYL	19.84	0.057	0.52102
3,3''' -DIMETHYL-3',2''-ETHYLENE-P-QUATERPHENYL	21.70	0.044	0.88919
4-(3-ETHYLBUTOXY)-P-QUATERPHENYL (B)	19.25	0.063	0.43389
4,4'''-DI(2-BUTYLOCTOXY-1)-P-QUATERPHENYL	19.39	0.061	0.45427
DIETHYL-P-QUINQUEPHENYL	11.76	0.275	0.02270
DI(3-ETHYLHEPTYL)-P-QUINQUEPHENYL	21.24	0.047	0.78110
TETRAMETHYL-P-SEXIPHENYL	8.73	0.672	0.00380
TETRAMETHYL-P-OCTAPHENYL (B)	27.11	0.022	3.37846
INDOLE (TRACE)	0.00	0.000	0.00000
2-PHENYL INDOLE	8.13	0.834	0.00247
1-METHYL-2-PHENYL INDOLE	7.57	1.030	0.00161
1-NEOPENTYL-2-PHENYL INDOLE	0.00	0.000	0.00000
1-N-BUTYL-2(BIPHENYLYL-4) INDOLE	21.77	0.043	0.90673
1,2-DIPHENYL INDOLE	7.86	0.919	0.00203
1-BIPHENYLYL-2-PHENYL INDOLE	13.95	0.165	0.06303
1,3-DIMETHYL-2-PHENYL INDOLE	8.64	0.693	0.00357
2,3-DIPHENYL INDOLE	20.10	0.055	0.56113
1-METHYL-2,3-DIPHENYL INDOLE	17.78	0.080	0.26949
2-PHENYL-3,2'-METHYLENE INDOLE (TRACE)	13.81	0.170	0.05934
2-PHENYL-3,2'-ETHYLENE INDOLE (TRACE)	23.73	0.033	1.52346
2-PHENYL-3,2' -TRIMETHYLENE INDOLE	13.07	0.200	0.04266
1-METHYL-3,2'-METHYLENE-2-PHENYL INDOLE (TRACE)	21.61	0.044	0.86824
1-METHYL-3,2'-ETHYLENE-2-PHENYL INDOLE	26.78	0.023	3.14055
1-METHYL-3,2'-TRIMETHYLENE-2-PHENYL INDOLE	14.52	0.146	0.08024
4,5-DIPHENYLIMIDAZOLE (A)	6.93	1.344	0.00095
PPF (TRACE)	22.66	0.038	1.15528
PPO (TRACE)	11.29	0.311	0.01770
2-(4-FLUOROPHENYL)-5-PHENYLOXAZOLE (TRACE)	10.23	0.417	0.00985
αNPO (TRACE)	31.42	0.014	8.16451
BBO (B)	38.42	0.008	27.25313
POPOP	40.17	0.007	35.56861
DIMETHYL POPOP (TRACE)	42.80	0.006	51.90690
PPD (TRACE)	0.00	0.000	0.00000
2,5-DI(2-METHYLPHENYL)OXADIAZOLE	0.00	0.000	0.00000
PBD	8.80	0.656	0.00399
BBD (B)	25.16	0.028	2.16506
αNPD (TRACE)	22.17	0.041	1.01132
βNPD	12.52	0.228	0.03290
αNND (TRACE)	27.76	0.021	3.89141
βNND (TRACE)	19.75	0.058	0.50761
TRANS-1-(4-BIPHENYLYL)-2-PHENYL-ETHYLENE	26.18	0.025	2.74125
TRANS-1-(4-BIPHENYLYL)-2-(1-NAPHTHYL)-ETHYLENE	34.47	0.011	14.19141
DIPHENYLSTILBENE (B)	38.54	0.008	27.73379
1,1,4,4-TETRAPHENYLBUTADIENE	37.40	0.009	23.11399
1,6-DIPHENYLHEXATRIENE	43.04	0.006	53.76846
1,8-DIPHENYL-1,3,5,7-OCTATETRAENE	48.12	0.004	104.22948
BPSB (TRACE)	43.77	0.005	59.40235

ACCEPTOR MOLECULES	RO	CO	OVERLAP
BIS-MSB	41.28	0.006	41.80038
ESCULIN (A)	28.74	0.019	4.77462
AZULENE	17.27	0.087	0.22045
NAPHTHALENE	0.00	0.000	0.00000
DEUTERATED NAPHTHALENE-C10D8	2.12	47.041	0.00000
1-METHYLNAPHTHALENE 5ML/L	0.00	0.000	0.00000
2-METHYL NAPHTHALENE	0.00	0.000	0.00000
2,3-DIMETHYL NAPHTHALENE	0.00	0.000	0.00000
2,6-DIMETHYL NAPHTHALENE (TRACE)	0.00	0.000	0.00000
1-NAPHTHOL (TRACE)	0.00	0.000	0.00000
2-NAPHTHOL (TRACE)	4.40	5.261	0.00006
1-NAPHTHYLAMINE	14.50	0.147	0.07946
2-PHENYL NAPHTHALENE	5.29	3.018	0.00019
1,4-DIPHENYL NAPHTHALENE	9.94	0.456	0.00825
1,5-DIPHENYL NAPHTHALENE	6.53	1.603	0.00067
1,7-DIPHENYL NAPHTHALENE	7.57	1.031	0.00161
1,4,5,8-TETRAPHENYL NAPHTHALENE	30.99	0.015	7.48722
1,1'-BINAPHTHYL	0.00	0.000	0.00000
4,4'-DI(N-BUTOXY)-1,1'-BINAPHTHYL	16.89	0.093	0.19809
2,2'-BINAPHTHYL	7.82	0.935	0.00196
ACRIDINE (A)	30.05	0.016	6.22469
ANTHRACENE (TRACE)	28.74	0.019	4.77330
DEUTERATED ANTHRACENE C14D10 (TRACE)	28.40	0.020	4.44493
9-METHYL ANTHRACENE (TRACE)	30.41	0.016	6.67901
1-AMINO ANTHRACENE	30.93	0.015	7.33900
2-AMINO ANTHRACENE	29.51	0.017	5.53798
9-VINYL ANTHRACENE	32.11	0.014	9.24053
9-PHENYL ANTHRACENE (TRACE)	31.80	0.014	8.74175
9,10-DIPHENYL ANTHRACENE	34.53	0.011	14.29144
9,10-DICHLOROANTHRACENE (TRACE)	33.30	0.012	11.48132
BIANTHRANYL (B)	36.32	0.009	19.34945
9,10-DINAPHTHYLANTHRACENE (B)	33.98	0.011	12.95907
9-(P-VINYLPHENYL)-10-PHENYL ANTHRACENE	34.82	0.011	14.99727
9,10-BIS(PHENYL-ETHYNYL)ANTHRACENE (TRACE)	37.62	0.008	23.61944
RUBICENE (B)	33.66	0.012	12.08594
DIBENZ(A,H)ANTHRACENE (B)	26.96	0.023	3.23916
TETRACENE (TRACE) (B)	28.12	0.020	4.11608
RUBRENE (B)	25.81	0.026	2.45790
TETRAHYDROCANNABINOL ACID METHYL ESTER	0.00	0.000	0.00000
PHENANTHRENE	10.28	0.412	0.01006
3,4-BENZOPHENANTHRENE	16.42	0.101	0.16573
2-PHENYL PHENANTHRENE	13.49	0.182	0.05128
CHRYSENE (TRACE)	16.27	0.104	0.15748
TRIPHENYLENE	8.68	0.684	0.00366
DEUTERATED TRIPHENYLENE-C18D12	8.67	0.687	0.00363
PYRENE	17.77	0.080	0.26722
1,3,6,8,-TETRAPHENYL PYRENE	42.41	0.006	48.93590
BIPYRENYL (B)	39.05	0.008	29.88704
O-PHENYLENE-PYRENE	36.70	0.009	20.39215
ANTHANTHRENE (B)	36.80	0.009	20.80942
FLUORANTHENE	28.04	0.020	4.09525
3-PHENYL FLUORANTHENE	34.12	0.011	13.27018
3-AMINO FLUORANTHENE	33.45	0.012	11.69514
3,3'-BIFLUORANTHENYL (B)	40.04	0.007	34.60825
BENZO(GHI)FLUORANTHENE	26.08	0.025	2.64399
9,12-DI(N-OCTADECOXY)-BENZ(E)ACEPHENANTHRYLENE	35.56	0.010	16.98853
DECACYCLENE (B)	43.82	0.005	59.02339
10-ETHOXY-BENZO(KL)XANTHENE	35.05	0.010	15.61967
PERYLENE	37.24	0.009	22.32768
BENZO(GHI)PERYLENE (B)	36.50	0.009	19.92576
BBOT (TRACE)	45.14	0.005	71.36302
QUININE SULFATE (A)	17.84	0.079	0.27541
ACRIDINE YELLOW (A)	35.10	0.010	15.55540
ACRIDINE ORANGE (A)	29.89	0.017	5.93318
ACRIDINE RED (A)	27.06	0.023	3.26524
URANINE (A)	30.48	0.016	6.67080
RHODAMINE B (A)	31.11	0.015	7.53741
RHODAMINE 6G (A)	29.90	0.017	5.94252
EOSIN B (A)	27.60	0.021	3.67181
EOSIN Y (A)	28.86	0.019	4.80564
ROSE BENGAL (A)	28.03	0.020	4.03031
INDENOINDENE	11.26	0.313	0.01746

8.109 Donor Molecule: 2-Phenyl-3,2′-trimethyleneindole

TRIPHENYLAMINE	16.46	0.100	0.19723
P-XYLENE (TRACE)	0.00	0.000	0.00000
SODIUM SALICYLATE (A)	12.16	0.248	0.03210
ANTHRANILIC ACID (A)	24.36	0.031	2.05646
TMPD	19.30	0.062	0.51105
BENZIDINE	12.49	0.229	0.03778
1,3-DIBIPHENYLYLBENZENE (B)	10.30	0.409	0.01192
1,3,5,-TRIBIPHENYLYLBENZENE (B)	11.32	0.308	0.02101
FLUORENE (TRACE)	0.00	0.000	0.00000
1-METHYL FLUORENE	0.00	0.000	0.00000
1,2-BENZOFLUORENE	16.17	0.106	0.17723
9,10-DIHYDROPHENANTHRENE	0.00	0.000	0.00000
CARBAZOLE (TRACE)	12.54	0.227	0.03873
N-METHYL CARBAZOLE (TRACE)	19.66	0.059	0.57192
N-VINYLCARBAZOLE	18.58	0.070	0.40805
N-PHENYL CARBAZOLE	18.54	0.070	0.40330

ACCEPTOR MOLECULES	RO	CO	OVERLAP
N-METHYL-2,3-BENZCARBAZOLE	27.63	0.021	4.35810
DIBENZOFURAN	0.00	0.000	0.00000
DIBENZOTHIOPHENE (TRACE)	8.49	0.732	0.00372
P-TERPHENYL	0.00	0.000	0.00000
DEUTERATED P-TERPHENYL-C18D14	0.00	0.000	0.00000
4-METHYL TERPHENYL	0.00	0.000	0.00000
2-METHYL-P-TERPHENYL	0.00	0.000	0.00000
3,3″-DIMETHYL-P-TERPHENYL	0.00	0.000	0.00000
4-(3,3-DIMETHYLBUTOXY)-P-TERPHENYL	8.50	0.728	0.00376
4,4″-DIHEXAHYDROFARNESOXY-P-TERPHENYL	11.38	0.303	0.02169
2,2′-METHYLENE-P-TERPHENYL	7.49	1.069	0.0017[illegible]
2,2′-ETHYLENE-P-TERPHENYL	9.91	0.459	0.00945
2,2′-METHYLENE-5′,6″-METHYLENE-P-TERPHENYL (TRACE)	22.09	0.041	1.15475
3-PHENYL DIBENZOFURAN	7.32	1.140	0.00154
BENZO(1,2-B-4,5-B)BISBENZOFURAN (TRACE) (B)	25.28	0.028	2.59116
SEXIPHENYL BENZENE	0.00	0.000	0.00000
P-QUATERPHENYL	13.99	0.163	0.07452
3,3‴-DIMETHYL-P-QUATERPHENYL	16.13	0.107	0.17469
TETRAMETHYL-P-QUATERPHENYL	0.00	0.000	0.00000
3,3‴-DIMETHYL-3′,2″-METHYLENE-P-QUATERPHENYL	24.47	0.031	2.12670
3,3‴-DIMETHYL-3′,2″-ETHYLENE-P-QUATERPHENYL	25.12	0.028	2.48876
4-(3-ETHYLBUTOXY)-P-QUATERPHENYL (B)	22.36	0.040	1.23596
4,4‴-DI(2-BUTYLOCTOXY-1)-P-QUATERPHENYL	22.62	0.039	1.32806
DIETHYL-P-QUINQUEPHENYL	16.63	0.097	0.21057
DI(3-ETHYLHEPTYL)-P-QUINQUEPHENYL	24.23	0.031	2.00029
TETRAMETHYL-P-SEXIPHENYL	14.51	0.146	0.09295
TETRAMETHYL-P-OCTAPHENYL (B)	29.68	0.017	6.75655
INDOLE (TRACE)	0.00	0.000	0.00000
2-PHENYL INDOLE	15.28	0.125	0.12646
1-METHYL-2-PHENYL INDOLE	11.96	0.261	0.02913
1-NEOPENTYL-2-PHENYL INDOLE	7.95	0.892	0.00251
1-N-BUTYL-2(BIPHENYLYL-4) INDOLE	24.05	0.032	1.91154
1,2-DIPHENYL INDOLE	12.29	0.241	0.03421
1-BIPHENYLYL-2-PHENYL INDOLE	16.19	0.105	0.17853
1,3-DIMETHYL-2-PHENYL INDOLE	13.71	0.174	0.06600
2,3-DIPHENYL INDOLE	22.12	0.041	1.15757
1-METHYL-2,3-DIPHENYL INDOLE	19.79	0.058	0.59467
2-PHENYL-3,2′-METHYLENE INDOLE (TRACE)	20.25	0.054	0.68413
2-PHENYL-3,2′-ETHYLENE INDOLE (TRACE)	26.38	0.024	3.33437
2-PHENYL-3,2′-TRIMETHYLENE INDOLE	18.60	0.070	0.41073
1-METHYL-3,2′-METHYLENE-2-PHENYL INDOLE (TRACE)	25.23	0.028	2.55504
1-METHYL-3,2′-ETHYLENE-2-PHENYL INDOLE	28.41	0.019	5.20007
1-METHYL-3,2′ TRIMETHYLENE-2-PHENYL INDOLE	19.39	0.061	0.52799
4,5-DIPHENYLIMIDAZOLE (A)	10.72	0.363	0.01510
PPF (TRACE)	26.41	0.024	3.36269
PPO (TRACE)	18.36	0.072	0.38042
2-(4-FLUOROPHENYL)-5-PHENYLOXAZOLE (TRACE)	16.99	0.091	0.23939
αNPO (TRACE)	32.48	0.013	11.55616
BBO (B)	38.94	0.008	34.27489
POPOP	40.18	0.007	41.29637
DIMETHYL POPOP (TRACE)	42.31	0.006	56.16647
PPD (TRACE)	0.00	0.000	0.00000
2,5-DI(2-METHYLPHENYL)OXADIAZOLE	0.00	0.000	0.00000
PBD	15.58	0.118	0.14248
BBD (B)	28.08	0.020	4.85309
αNPD (TRACE)	24.98	0.029	2.39840
βNPD	16.86	0.093	0.22832
αNND (TRACE)	29.45	0.018	6.43658
βNND (TRACE)	24.65	0.030	2.22615
TRANS-1-(4-BIPHENYLYL)-2-PHENYL-ETHYLENE	28.69	0.019	5.51386
TRANS-1-(4-BIPHENYLYL)-2-(1-NAPHTHYL)-ETHYLENE	35.15	0.010	18.53005
DIPHENYLSTILBENE (B)	39.16	0.007	35.40069
1,1,4,4-TETRAPHENYLBUTADIENE	37.67	0.008	27.99587
1,6-DIPHENYLHEXATRIENE	42.89	0.006	61.04732
1,8-DIPHENYL-1,3,5,7-OCTATETRAENE	46.69	0.004	100.94152
BPSB (TRACE)	43.64	0.005	67.71729
BIS-MSB	41.36	0.006	49.06931
ESCULIN (A)	29.54	0.017	6.53408
AZULENE	19.67	0.059	0.55941
NAPHTHALENE	0.00	0.000	0.00000
DEUTERATED NAPHTHALENE-C10D8	4.81	4.023	0.00012
1-METHYLNAPHTHALENE 5ML/L	0.00	0.000	0.00000
2-METHYL NAPHTHALENE	0.00	0.000	0.00000
2,3-DIMETHYL NAPHTHALENE	0.00	0.000	0.00000
2,6-DIMETHYL NAPHTHALENE (TRACE)	4.13	6.359	0.00005
1-NAPHTHOL (TRACE)	4.84	3.947	0.00013
2-NAPHTHOL (TRACE)	9.78	0.478	0.00871
1-NAPHTHYLAMINE	17.48	0.084	0.28288
2-CHLORONAPHTHALENE	3.07	15.449	0.00001
ACENAPHTHENE	0.00	0.000	0.00000
1-PHENYL NAPHTHALENE	4.51	4.874	0.00008
2-PHENYL NAPHTHALENE	11.49	0.295	0.02297
1,4-DIPHENYL NAPHTHALENE	13.53	0.180	0.06104
1,5-DIPHENYL NAPHTHALENE	11.09	0.328	0.01849
1,7-DIPHENYL NAPHTHALENE	11.67	0.282	0.02509
1,4,5,8-TETRAPHENYL NAPHTHALENE	31.74	0.014	10.02880
1,1′-BINAPHTHYL	7.02	1.293	0.00119
4,4′-DI(N-BUTOXY)-1,1′-BINAPHTHYL	19.66	0.059	0.57237
2,2′-BINAPHTHYL	12.78	0.214	0.04332
ACRIDINE (A)	30.14	0.016	7.35701
ANTHRACENE (TRACE)	28.64	0.019	5.42476
DEUTERATED ANTHRACENE C14D10 (TRACE)	28.15	0.020	4.89320
9-METHYL ANTHRACENE (TRACE)	29.88	0.017	6.97496
1-AMINO ANTHRACENE	29.83	0.017	6.84703
2-AMINO ANTHRACENE	28.51	0.019	5.22205
9-VINYL ANTHRACENE	31.30	0.015	9.20033
9-PHENYL ANTHRACENE (TRACE)	31.37	0.014	9.33872

ACCEPTOR MOLECULES	RO	CO	OVERLAP
9,10-DIPHENYL ANTHRACENE	33.34	0.012	13.42284
9,10-DICHLOROANTHRACENE (TRACE)	32.35	0.013	11.20855
BIANTHRANYL (B)	34.95	0.010	17.82379
9,10-DINAPHTHYLANTHRACENE (B)	32.90	0.013	12.39393
9-(P-VINYLPHENYL)-10-PHENYL ANTHRACENE	33.63	0.012	14.13937
9,10-BIS(PHENYL-ETHYNYL)ANTHRACENE (TRACE)	35.22	0.010	18.44572
RUBICENE (B)	32.34	0.013	11.05449
DIBENZ(A,H)ANTHRACENE (B)	27.82	0.021	4.53973
TETRACENE (TRACE) (B)	26.40	0.024	3.27312
RUBRENE (B)	24.75	0.029	2.22162
TETRAHYDROCANNABINOL ACID METHYL ESTER	0.00	0.000	0.00000
PHENANTHRENE	11.68	0.281	0.02517
3,4-BENZOPHENANTHRENE	16.68	0.096	0.21166
2-PHENYL PHENANTHRENE	14.28	0.154	0.08383
CHRYSENE (TRACE)	17.00	0.091	0.23792
TRIPHENYLENE	11.26	0.313	0.02026
DEUTERATED TRIPHENYLENE-C18D12	10.93	0.342	0.01693
PYRENE	23.29	0.035	1.57064
1,3,6,8,-TETRAPHENYL PYRENE	41.23	0.006	47.95901
BIPYRENYL (B)	38.97	0.008	34.24896
O-PHENYLENE-PYRENE	35.83	0.010	20.48549
ANTHANTHRENE (B)	35.15	0.010	18.33458
FLUORANTHENE	28.58	0.019	5.33311
3-PHENYL FLUORANTHENE	33.63	0.012	14.13041
3-AMINO FLUORANTHENE	32.26	0.013	10.92277
3,3'-BIFLUORANTHENYL (B)	39.00	0.008	34.26520
BENZO(GHI)FLUORANTHENE	26.30	0.025	3.23325
9,12-DI(N-OCTADECOXY)-BENZ(E)ACEPHENANTHRYLENE	34.48	0.011	16.36866
DECACYCLENE (B)	42.53	0.006	57.24936
10-ETHOXY-BENZO(KL)XANTHENE	33.98	0.011	15.06554
PERYLENE	35.27	0.010	18.70403
BENZO(GHI)PERYLENE (B)	35.62	0.010	19.97912
BBOT (TRACE)	44.11	0.005	72.05482
QUININE SULFATE (A)	21.46	0.045	0.96851
ACRIDINE YELLOW (A)	32.62	0.013	11.64718
ACRIDINE ORANGE (A)	27.61	0.021	4.28255
ACRIDINE RED (A)	26.37	0.024	3.25010
URANINE (A)	28.42	0.019	5.08981
RHODAMINE B (A)	30.24	0.016	7.38753
RHODAMINE 6G (A)	29.32	0.018	6.13799
EOSIN B (A)	26.54	0.024	3.37257
EOSIN Y (A)	27.63	0.021	4.29407
ROSE BENGAL (A)	27.43	0.022	4.11164
INDENOINDENE	17.49	0.084	0.28469

8.110 Donor Molecule: 1-Methyl-3,2'-methylene-2-phenylindole (trace)

TRIPHENYLAMINE	15.41	0.122	0.11060
P-XYLENE (TRACE)	0.00	0.000	0.00000
SODIUM SALICYLATE (A)	11.94	0.263	0.02395
ANTHRANILIC ACID (A)	24.65	0.030	1.84006
TMPD	19.18	0.063	0.40974
BENZIDINE	10.47	0.389	0.01093
1,3-DIBIPHENYLYLBENZENE (B)	7.40	1.102	0.00137
1,3,5,-TRIBIPHENYLYLBENZENE (B)	7.39	1.109	0.00135
FLUORENE (TRACE)	0.00	0.000	0.00000
1-METHYL FLUORENE	0.00	0.000	0.00000
1,2-BENZOFLUORENE	15.72	0.113	0.12455
9,10-DIHYDROPHENANTHRENE	0.00	0.000	0.00000
CARBAZOLE (TRACE)	9.55	0.514	0.00628
N-METHYL CARBAZOLE (TRACE)	18.94	0.066	0.38220
N-VINYLCARBAZOLE	16.15	0.106	0.14669
N-PHENYL CARBAZOLE	16.67	0.097	0.17737
N-METHYL-2,3-BENZCARBAZOLE	28.71	0.019	4.56982
DIBENZOFURAN	0.00	0.000	0.00000
DIBENZOTHIOPHENE (TRACE)	5.64	2.488	0.00027
P-TERPHENYL	0.00	0.000	0.00000
DEUTERATED P-TERPHENYL-C18D14	0.00	0.000	0.00000
4-METHYL TERPHENYL	0.00	0.000	0.00000
2-METHYL-P-TERPHENYL	0.00	0.000	0.00000
3,3''-DIMETHYL-P-TERPHENYL	0.00	0.000	0.00000
4-(3,3-DIMETHYLBUTOXY)-P-TERPHENYL	4.59	4.633	0.00008
4,4''-DIHEXAHYDROFARNESOXY-P-TERPHENYL	7.30	1.152	0.00125
2,2'-METHYLENE-P-TERPHENYL	0.00	0.000	0.00000
2,2'-ETHYLENE-P-TERPHENYL	6.18	1.898	0.00046
2,2'-METHYLENE-5',6''-METHYLENE-P-TERPHENYL (TRACE)	16.97	0.092	0.19776
3-PHENYL DIBENZOFURAN	0.00	0.000	0.00000
BENZO(1,2-B-4,5-B')BISBENZOFURAN (TRACE) (B)	22.58	0.039	1.09665
SEXIPHENYL BENZENE	0.00	0.000	0.00000
P-QUATERPHENYL	10.61	0.374	0.01185
3,3'''-DIMETHYL-P-QUATERPHENYL	14.09	0.160	0.06483
TETRAMETHYL-P-QUATERPHENYL	0.00	0.000	0.00000
3,3''' -DIMETHYL-3',2''-METHYLENE-P-QUATERPHENYL	22.39	0.040	1.04014
3,3'''-DIMETHYL-3',2''-ETHYLENE-P-QUATERPHENYL	23.81	0.033	1.50358
4-(3-ETHYLBUTOXY)-P-QUATERPHENYL (B)	21.12	0.047	0.73262
4,4'''-DI(2-BUTYLOCTOXY-1)-P-QUATERPHENYL	21.32	0.046	0.77570
DIETHYL-P-QUINQUEPHENYL	14.13	0.158	0.06607
DI(3-ETHYLHEPTYL)-P-QUINQUEPHENYL	23.14	0.036	1.26398
TETRAMETHYL-P-SEXIPHENYL	10.85	0.350	0.01353

ACCEPTOR MOLECULES	RO	CO	OVERLAP
TETRAMETHYL-P-OCTAPHENYL (B)	29.08	0.018	4.97976
INDOLE (TRACE)	0.00	0.000	0.00000
2-PHENYL INDOLE	10.56	0.380	0.01152
1-METHYL-2-PHENYL INDOLE	9.32	0.552	0.00544
1-NEOPENTYL-2-PHENYL INDOLE	4.30	5.630	0.00005
1-N-BUTYL-2(BIPHENYLYL-4) INDOLE	23.45	0.035	1.37046
1,2-DIPHENYL INDOLE	9.66	0.496	0.00675
1-BIPHENYLYL-2-PHENYL INDOLE	15.15	0.129	0.09991
1,3-DIMETHYL-2-PHENYL INDOLE	10.67	0.369	0.01221
2,3-DIPHENYL INDOLE	21.59	0.044	0.83461
1-METHYL-2,3-DIPHENYL INDOLE	19.20	0.063	0.41291
2-PHENYL-3,2'-METHYLENE INDOLE (TRACE)	16.41	0.101	0.16189
2-PHENYL-3,2'-ETHYLENE INDOLE (TRACE)	25.97	0.026	2.52897
2-PHENYL-3,2'-TRIMETHYLENE INDOLE	15.42	0.122	0.11137
1-METHYL-3,2'-METHYLENE-2-PHENYL INDOLE (TRACE)	24.40	0.031	1.74316
1-METHYL-3,2'-ETHYLENE-2-PHENYL INDOLE	28.30	0.020	4.23362
1-METHYL-3,2'-TRIMETHYLENE-2-PHENYL INDOLE	16.92	0.092	0.19404
4,5-DIPHENYLIMIDAZOLE (A)	8.48	0.733	0.00309
PPF (TRACE)	25.42	0.027	2.23002
PPO (TRACE)	14.01	0.163	0.06277
2-(4-FLUOROPHENYL)-5-PHENYLOXAZOLE (TRACE)	12.80	0.213	0.03639
αNPO (TRACE)	33.03	0.012	10.65408
BBO (B)	39.90	0.007	33.06122
POPOP	41.42	0.006	41.31597
DIMETHYL POPOP (TRACE)	43.83	0.005	57.94663
PPD (TRACE)	0.00	0.000	0.00000
2,5-DI(2-METHYLPHENYL)OXADIAZOLE	0.00	0.000	0.00000
PBD	11.15	0.323	0.01593
BBD (B)	27.27	0.022	3.39442
αNPD (TRACE)	24.31	0.031	1.70169
βNPD	14.80	0.138	0.08694
αNND (TRACE)	29.48	0.017	5.40665
βNND (TRACE)	22.88	0.037	1.18779
TRANS-1-(4-BIPHENYLYL)-2-PHENYL-ETHYLENE	28.17	0.020	4.11556
TRANS-1-(4-BIPHENYLYL)-2-(1-NAPHTHYL)-ETHYLENE	35.88	0.010	17.47897
DIPHENYLSTILBENE (B)	40.09	0.007	34.01215
1,1,4,4-TETRAPHENYLBUTADIENE	38.71	0.008	27.49799
1,6-DIPHENYLHEXATRIENE	44.20	0.005	60.96286
1,8-DIPHENYL-1,3,5,7-OCTATETRAENE	48.50	0.004	105.82572
BPSB (TRACE)	45.05	0.005	68.37688
BIS-MSB	42.60	0.006	48.84042
ESCULIN (A)	30.05	0.016	6.04320
AZULENE	18.72	0.068	0.34673
NAPHTHALENE	0.00	0.000	0.00000
DEUTERATED NAPHTHALENE-C10D8	3.10	15.051	0.00001
1-METHYLNAPHTHALENE 5ML/L	0.00	0.000	0.00000
2-METHYL NAPHTHALENE	0.00	0.000	0.00000
2,3-DIMETHYL NAPHTHALENE	0.00	0.000	0.00000
2,6-DIMETHYL NAPHTHALENE (TRACE)	0.00	0.000	0.00000
1-NAPHTHOL (TRACE)	0.00	0.000	0.00000
2-NAPHTHOL (TRACE)	6.01	2.057	0.00039
1-NAPHTHYLAMINE	16.18	0.106	0.14838
2-PHENYL NAPHTHALENE	7.60	1.020	0.00160
1,4-DIPHENYL NAPHTHALENE	11.68	0.281	0.02105
1,5-DIPHENYL NAPHTHALENE	8.16	0.823	0.00245
1,7-DIPHENYL NAPHTHALENE	9.28	0.560	0.00529
1,4,5,8-TETRAPHENYL NAPHTHALENE	32.34	0.013	9.34962
1,1'-BINAPHTHYL	0.00	0.000	0.00000
4,4'-DI(N-BUTOXY)-1,1'-BINAPHTHYL	18.49	0.071	0.32993
2,2'-BINAPHTHYL	9.69	0.491	0.00688
ACRIDINE (A)	31.02	0.015	7.28982
ANTHRACENE (TRACE)	29.39	0.018	5.27914
DEUTERATED ANTHRACENE C14D10 (TRACE)	28.97	0.018	4.84419
9-METHYL ANTHRACENE (TRACE)	31.19	0.015	7.53502
1-AMINO ANTHRACENE	30.99	0.015	7.18570
2-AMINO ANTHRACENE	29.48	0.017	5.31702
9-VINYL ANTHRACENE	32.60	0.013	9.80417
9-PHENYL ANTHRACENE (TRACE)	32.67	0.013	9.93809
9,10-DIPHENYL ANTHRACENE	34.66	0.011	14.14076
9,10-DICHLOROANTHRACENE (TRACE)	33.53	0.012	11.58783
BIANTHRANYL (B)	36.41	0.009	18.99669
9,10-DINAPHTHYLANTHRACENE (B)	34.13	0.011	12.87895
9-(P-VINYLPHENYL)-10-PHENYL ANTHRACENE	34.96	0.010	14.86600
9,10-BIS(PHENYL-ETHYNYL)ANTHRACENE (TRACE)	36.08	0.010	17.77048
RUBICENE (B)	33.28	0.012	10.94104
DIBENZ(A,H)ANTHRACENE (B)	27.96	0.020	3.90370
TETRACENE (TRACE) (B)	27.07	0.023	3.17254
RUBRENE (B)	24.85	0.029	1.89542
TETRAHYDROCANNABINOL ACID METHYL ESTER	0.00	0.000	0.00000
PHENANTHRENE	11.41	0.301	0.01820
3,4-BENZOPHENANTHRENE	17.00	0.091	0.19769
2-PHENYL PHENANTHRENE	14.31	0.153	0.07072
CHRYSENE (TRACE)	17.21	0.088	0.21394
TRIPHENYLENE	9.95	0.454	0.00804
DEUTERATED TRIPHENYLENE-C18D12	9.88	0.464	0.00769
PYRENE	19.77	0.058	0.49027
1,3,6,8,-TETRAPHENYL PYRENE	42.94	0.006	51.00278
BIPYRENYL (B)	40.19	0.007	34.35610
O-PHENYLENE-PYRENE	37.10	0.009	21.06793
ANTHANTHRENE (B)	36.29	0.009	18.52674
FLUORANTHENE	29.27	0.018	5.13055
3-PHENYL FLUORANTHENE	34.86	0.011	14.61484
3-AMINO FLUORANTHENE	33.40	0.012	11.20360
3,3'-BIFLUORANTHENYL (B)	40.51	0.007	35.88431
BENZO(GHI)FLUORANTHENE	26.88	0.023	3.07209
9,12-DI(N-OCTADECOXY)-BENZ(E)ACEPHENANTHRYLENE	35.66	0.010	16.69815
DECACYCLENE (B)	44.04	0.005	58.88786

ACCEPTOR MOLECULES	RO	CO	OVERLAP
10-ETHOXY-BENZO(KL)XANTHENE	35.32	0.010	15.84060
PERYLENE	36.43	0.009	18.94642
BENZO(GHI)PERYLENE (B)	37.19	0.009	21.57544
BBOT (TRACE)	45.80	0.005	75.31203
QUININE SULFATE (A)	20.01	0.056	0.53012
ACRIDINE YELLOW (A)	33.18	0.012	10.74295
ACRIDINE ORANGE (A)	27.87	0.021	3.77829
ACRIDINE RED (A)	27.07	0.023	3.17016
URANINE (A)	28.68	0.019	4.48097
RHODAMINE B (A)	31.05	0.015	7.22015
RHODAMINE 6G (A)	29.84	0.017	5.68283
EOSIN B (A)	26.88	0.023	3.03870
EOSIN Y (A)	28.04	0.020	3.91637
ROSE BENGAL (A)	27.91	0.021	3.80553
INDENOINDENE	13.66	0.176	0.05375

8.111 Donor Molecule: 1-Methyl-3,2′-ethylene-2-phenylindole

ACCEPTOR MOLECULES	RO	CO	OVERLAP
TRIPHENYLAMINE	8.53	0.721	0.00335
P-XYLENE (TRACE)	0.00	0.000	0.00000
SODIUM SALICYLATE (A)	8.45	0.741	0.00317
ANTHRANILIC ACID (A)	20.47	0.052	0.63738
TMPD	14.72	0.140	0.08837
1,2-BENZOFLUORENE	7.92	0.901	0.00215
9,10-DIHYDROPHENANTHRENE	0.00	0.000	0.00000
CARBAZOLE (TRACE)	0.00	0.000	0.00000
N-METHYL CARBAZOLE (TRACE)	8.16	0.822	0.00258
N-VINYLCARBAZOLE	6.65	1.522	0.00075
N-PHENYL CARBAZOLE	6.99	1.307	0.00102
N-METHYL-2,3-BENZCARBAZOLE	28.70	0.019	4.80950
BENZO(1,2-B-4,5-B')BISBENZOFURAN (TRACE) (B)	8.82	0.651	0.00411
SEXIPHENYL BENZENE	0.00	0.000	0.00000
P-QUATERPHENYL	0.00	0.000	0.00000
3,3'''-DIMETHYL-P-QUATERPHENYL	0.00	0.000	0.00000
TETRAMETHYL-P-QUATERPHENYL	0.00	0.000	0.00000
3,3''' -DIMETHYL-3' ,2''-METHYLENE-P-QUATERPHENYL	11.29	0.310	0.01809
3,3''' -DIMETHYL-3' ,2''-ETHYLENE-P-QUATERPHENYL	15.60	0.118	0.12514
4-(3-ETHYLBUTOXY)-P-QUATERPHENYL (B)	13.43	0.185	0.05096
4,4'''-DI(2-BUTYLOCTOXY-1)-P-QUATERPHENYL	13.05	0.201	0.04301
DIETHYL-P-QUINQUEPHENYL	0.00	0.000	0.00000
DI(3-ETHYLHEPTYL)-P-QUINQUEPHENYL	15.91	0.111	0.14073
TETRAMETHYL-P-SEXIPHENYL	0.00	0.000	0.00000
TETRAMETHYL-P-OCTAPHENYL (B)	21.75	0.043	0.91830
INDOLE (TRACE)	0.00	0.000	0.00000
2-PHENYL INDOLE	0.00	0.000	0.00000
1-METHYL-2-PHENYL INDOLE	0.00	0.000	0.00000
1-NEOPENTYL-2-PHENYL INDOLE	0.00	0.000	0.00000
1-N-BUTYL-2(BIPHENYLYL-4) INDOLE	17.03	0.091	0.21186
1,2-DIPHENYL INDOLE	0.00	0.000	0.00000
1-BIPHENYLYL-2-PHENYL INDOLE	10.02	0.445	0.00880
1,3-DIMETHYL-2-PHENYL INDOLE	0.00	0.000	0.00000
2,3-DIPHENYL INDOLE	15.99	0.109	0.14535
1-METHYL-2,3-DIPHENYL INDOLE	13.79	0.171	0.05976
2-PHENYL-3,2'-METHYLENE INDOLE (TRACE)	6.06	2.012	0.00043
2-PHENYL-3,2'-ETHYLENE INDOLE (TRACE)	15.56	0.119	0.12333
2-PHENYL-3,2' -TRIMETHYLENE INDOLE	5.64	2.497	0.00028
1-METHYL-3,2'-METHYLENE-2-PHENYL INDOLE (TRACE)	12.62	0.223	0.03512
1-METHYL-3,2'-ETHYLENE-2-PHENYL INDOLE	21.67	0.044	0.90063
1-METHYL-3,2'-TRIMETHYLENE-2-PHENYL INDOLE	6.74	1.469	0.00082
4,5-DIPHENYLIMIDAZOLE (A)	0.00	0.000	0.00000
PPF (TRACE)	12.82	0.212	0.03859
αNPO (TRACE)	27.63	0.021	3.85165
BBO (B)	35.12	0.010	16.20875
POPOP	38.19	0.008	26.76924
DIMETHYL POPOP (TRACE)	41.17	0.006	41.97459
PPD (TRACE)	0.00	0.000	0.00000
2,5-DI(2-METHYLPHENYL)OXADIAZOLE	0.00	0.000	0.00000
PBD	0.00	0.000	0.00000
BBD (B)	17.64	0.082	0.26192
αNPD (TRACE)	17.30	0.086	0.23249
βNPD	0.00	0.000	0.00000
αNND (TRACE)	23.57	0.034	1.48612
βNND (TRACE)	9.29	0.559	0.00559
TRANS-1-(4-BIPHENYLYL)-2-PHENYL-ETHYLENE	19.63	0.059	0.49656
TRANS-1-(4-BIPHENYLYL)-2-(1-NAPHTHYL)-ETHYLENE	31.38	0.014	8.24090
DIPHENYLSTILBENE (B)	35.02	0.010	15.92024
1,1,4,4-TETRAPHENYLBUTADIENE	34.67	0.011	14.95985
1,6-DIPHENYLHEXATRIENE	40.40	0.007	37.47335
1,8-DIPHENYL-1,3,5,7-OCTATETRAENE	47.61	0.004	99.81402
BPSB (TRACE)	41.62	0.006	44.83459
BIS-MSB	38.78	0.008	29.34087
ESCULIN (A)	25.79	0.026	2.54373
AZULENE	13.75	0.172	0.05754
1-NAPHTHYLAMINE	8.75	0.667	0.00392
1,4,5,8-TETRAPHENYL NAPHTHALENE	27.98	0.020	4.13549
1,1'-BINAPHTHYL	0.00	0.000	0.00000
4,4' -DI(N-BUTOXY)-1,1' -BINAPHTHYL	11.54	0.291	0.02062
2,2'-BINAPHTHYL	0.00	0.000	0.00000
ACRIDINE (A)	28.33	0.020	4.45953
ANTHRACENE (TRACE)	27.59	0.021	3.81155

ACCEPTOR MOLECULES	RO	CO	OVERLAP
DEUTERATED ANTHRACENE C14D10 (TRACE)	26.85	0.023	3.23822
9-METHYL ANTHRACENE (TRACE)	29.51	0.017	5.69191
1-AMINO ANTHRACENE	31.15	0.015	7.81797
2-AMINO ANTHRACENE	29.78	0.017	5.95880
9-VINYL ANTHRACENE	31.37	0.014	8.20097
9-PHENYL ANTHRACENE (TRACE)	31.03	0.015	7.69782
9,10-DIPHENYL ANTHRACENE	34.20	0.011	13.76302
9,10-DICHLOROANTHRACENE (TRACE)	33.74	0.012	12.68419
BIANTHRANYL (B)	35.75	0.010	17.94585
9,10-DINAPHTHYLANTHRACENE (B)	34.12	0.011	13.55258
9-(P-VINYLPHENYL)-10-PHENYL ANTHRACENE	34.66	0.011	14.00774
9,10-BIS(PHENYL-ETHYNYL)ANTHRACENE (TRACE)	39.59	0.007	32.76373
RUBICENE (B)	34.32	0.011	13.88394
DIBENZ(A,H)ANTHRACENE (B)	23.04	0.037	1.28978
TETRACENE (TRACE) (B)	29.23	0.018	5.29589
RUBRENE (B)	26.54	0.024	2.96835
PHENANTHRENE	5.53	2.640	0.00025
3,4-BENZOPHENANTHRENE	14.99	0.133	0.09787
2-PHENYL PHENANTHRENE	11.10	0.327	0.01624
CHRYSENE (TRACE)	14.52	0.146	0.08136
TRIPHENYLENE	5.00	3.578	0.00014
DEUTERATED TRIPHENYLENE-C18D12	4.88	3.843	0.00012
PYRENE	14.99	0.133	0.09829
1,3,6,8,-TETRAPHENYL PYRENE	41.94	0.006	46.70731
BIPYRENYL (B)	36.73	0.009	21.13924
O-PHENYLENE-PYRENE	36.48	0.009	20.07724
ANTHANTHRENE (B)	37.97	0.008	25.64164
FLUORANTHENE	25.86	0.026	2.57272
3-PHENYL FLUORANTHENE	32.81	0.013	10.70901
3-AMINO FLUORANTHENE	33.77	0.012	12.62594
3,3'-BIFLUORANTHENYL (B)	39.57	0.007	32.90612
BENZO(GHI)FLUORANTHENE	23.77	0.033	1.54688
9,12-DI(N-OCTADECOXY)-BENZ(E)ACEPHENANTHRYLENE	35.67	0.010	17.65539
DECACYCLENE (B)	43.95	0.005	61.31756
10-ETHOXY-BENZO(KL)XANTHENE	34.91	0.011	15.56176
PERYLENE	38.91	0.008	29.64931
BENZO(GHI)PERYLENE (B)	35.90	0.010	18.42605
BBOT (TRACE)	44.06	0.005	62.96012
QUININE SULFATE (A)	10.58	0.378	0.01219
ACRIDINE YELLOW (A)	37.06	0.009	22.01672
ACRIDINE ORANGE (A)	31.34	0.015	8.05613
ACRIDINE RED (A)	26.78	0.023	3.13468
URANINE (A)	31.59	0.014	8.44460
RHODAMINE B (A)	30.94	0.015	7.44952
RHODAMINE 6G (A)	28.99	0.018	5.04610
EOSIN B (A)	27.88	0.021	3.98831
EOSIN Y (A)	29.32	0.018	5.40255
ROSE BENGAL (A)	27.97	0.020	4.07084

8.112 Donor Molecule: 1-Methyl-3,2′-trimethylene-2-phenylindole

TRIPHENYLAMINE	14.92	0.135	0.14301
P-XYLENE (TRACE)	0.00	0.000	0.00000
SODIUM SALICYLATE (A)	11.24	0.315	0.02613
ANTHRANILIC ACID (A)	22.89	0.037	1.85402
TMPD	17.96	0.077	0.43464
BENZIDINE	10.93	0.342	0.02223
1,3-DIBIPHENYLYLBENZENE (B)	8.67	0.687	0.00553
1,3,5,-TRIBIPHENYLYLBENZENE (B)	9.14	0.586	0.00760
FLUORENE (TRACE)	0.00	0.000	0.00000
1-METHYL FLUORENE	0.00	0.000	0.00000
1,2-BENZOFLUORENE	14.90	0.135	0.14207
9,10-DIHYDROPHENANTHRENE	0.00	0.000	0.00000
CARBAZOLE (TRACE)	10.48	0.388	0.01729
N-METHYL CARBAZOLE (TRACE)	18.10	0.075	0.45619
N-VINYLCARBAZOLE	16.65	0.097	0.27682
N-PHENYL CARBAZOLE	16.75	0.095	0.28734
N-METHYL-2,3-BENZCARBAZOLE	26.51	0.024	4.44700
DIBENZOFURAN	0.00	0.000	0.00000
DIBENZOTHIOPHENE (TRACE)	6.82	1.413	0.00131
P-TERPHENYL	0.00	0.000	0.00000
DEUTERATED P-TERPHENYL-C18D14	0.00	0.000	0.00000
4-METHYL TERPHENYL	0.00	0.000	0.00000
2-METHYL-P-TERPHENYL	0.00	0.000	0.00000
3,3''-DIMETHYL-P-TERPHENYL	0.00	0.000	0.00000
4-(3,3-DIMETHYLBUTOXY)-P-TERPHENYL	6.54	1.600	0.00102
4,4''-DIHEXAHYDROFARNESOXY-P-TERPHENYL	9.14	0.586	0.00759
2,2'-METHYLENE-P-TERPHENYL	5.27	3.056	0.00028
2,2'-ETHYLENE-P-TERPHENYL	7.82	0.937	0.00298
2,2'-METHYLENE-5',6''-METHYLENE-P-TERPHENYL (TRACE)	18.79	0.067	0.57202
3-PHENYL DIBENZOFURAN	5.08	3.414	0.00022
BENZO(1,2-B-4,5-B')BISBENZOFURAN (TRACE) (B)	22.76	0.038	1.80667
SEXIPHENYL BENZENE	0.00	0.000	0.00000
P-QUATERPHENYL	11.95	0.262	0.03787
3,3'''-DIMETHYL-P-QUATERPHENYL	14.24	0.155	0.10839
TETRAMETHYL-P-QUATERPHENYL	0.00	0.000	0.00000
3,3''' -DIMETHYL-3',2''-METHYLENE-P-QUATERPHENYL	22.01	0.042	1.47614
3,3''' -DIMETHYL-3',2''-ETHYLENE-P-QUATERPHENYL	22.90	0.037	1.87021
4-(3-ETHYLBUTOXY)-P-QUATERPHENYL (B)	20.34	0.053	0.91710

ACCEPTOR MOLECULES	RO	CO	OVERLAP
4,4'''-DI(2-BUTYLOCTOXY-1)-P-QUATERPHENYL	20.55	0.052	0.97727
DIETHYL-P-QUINQUEPHENYL	14.59	0.144	0.12538
DI(3-ETHYLHEPTYL)-P-QUINQUEPHENYL	22.14	0.041	1.52424
TETRAMETHYL-P-SEXIPHENYL	12.33	0.238	0.04586
TETRAMETHYL-P-OCTAPHENYL (B)	27.44	0.022	5.52362
INDOLE (TRACE)	0.00	0.000	0.00000
2-PHENYL INDOLE	12.62	0.222	0.05268
1-METHYL-2-PHENYL INDOLE	10.20	0.422	0.01465
1-NEOPENTYL-2-PHENYL INDOLE	6.09	1.978	0.00067
1-N-BUTYL-2(BIPHENYLYL-4) INDOLE	22.18	0.041	1.54180
1,2-DIPHENYL INDOLE	10.50	0.387	0.01743
1-BIPHENYLYL-2-PHENYL INDOLE	14.63	0.143	0.12706
1,3-DIMETHYL-2-PHENYL INDOLE	11.76	0.275	0.03446
2,3-DIPHENYL INDOLE	20.41	0.053	0.93597
1-METHYL-2,3-DIPHENYL INDOLE	18.21	0.074	0.47281
2-PHENYL-3,2'-METHYLENE INDOLE (TRACE)	17.52	0.083	0.37554
2-PHENYL-3,2'-ETHYLENE INDOLE (TRACE)	24.39	0.031	2.73003
2-PHENYL-3,2'-TRIMETHYLENE INDOLE	16.17	0.106	0.23202
1-METHYL-3,2'-METHYLENE-2-PHENYL INDOLE (TRACE)	23.17	0.036	2.00678
1-METHYL-3,2'-ETHYLENE-2-PHENYL INDOLE	26.46	0.024	4.44331
1-METHYL-3,2'-TRIMETHYLENE-2-PHENYL INDOLE	17.22	0.088	0.33902
4,5-DIPHENYLIMIDAZOLE (A)	9.34	0.549	0.00864
PPF (TRACE)	24.24	0.031	2.63043
PPO (TRACE)	15.64	0.117	0.19015
2-(4-FLUOROPHENYL)-5-PHENYLOXAZOLE (TRACE)	14.38	0.151	0.11492
αNPO (TRACE)	30.60	0.016	10.57373
BBO (B)	36.86	0.009	32.30297
POPOP	38.22	0.008	40.10706
DIMETHYL POPOP (TRACE)	40.38	0.007	55.64951
PPD (TRACE)	0.00	0.000	0.00000
2,5-DI(2-METHYLPHENYL)OXADIAZOLE	0.00	0.000	0.00000
PBD	13.01	0.203	0.06330
BBD (B)	25.83	0.026	3.85232
αNPD (TRACE)	23.05	0.037	1.94220
βNPD	15.01	0.132	0.14851
αNND (TRACE)	27.53	0.021	5.63322
βNND (TRACE)	22.40	0.040	1.64141
TRANS-1-(4-BIPHENYLYL)-2-PHENYL-ETHYLENE	26.53	0.024	4.51392
TRANS-1-(4-BIPHENYLYL)-2-(1-NAPHTHYL)-ETHYLENE	33.21	0.012	17.25926
DIPHENYLSTILBENE (B)	37.05	0.009	33.25524
1,1,4,4-TETRAPHENYLBUTADIENE	35.73	0.010	26.72131
1,6-DIPHENYLHEXATRIENE	40.77	0.007	58.95418
1,8-DIPHENYL-1,3,5,7-OCTATETRAENE	44.75	0.005	102.44803
BPSB (TRACE)	41.55	0.006	66.06699
BIS-MSB	39.30	0.007	47.33507
ESCULIN (A)	27.85	0.021	6.01212
AZULENE	18.06	0.076	0.43804
NAPHTHALENE	0.00	0.000	0.00000
DEUTERATED NAPHTHALENE-C10D8	3.87	7.732	0.00004
1-METHYLNAPHTHALENE 5ML/L	0.00	0.000	0.00000
2-METHYL NAPHTHALENE	0.00	0.000	0.00000
2,3-DIMETHYL NAPHTHALENE	0.00	0.000	0.00000
2,6-DIMETHYL NAPHTHALENE (TRACE)	0.00	0.000	0.00000
1-NAPHTHOL (TRACE)	3.36	11.760	0.00002
2-NAPHTHOL (TRACE)	7.68	0.986	0.00268
1-NAPHTHYLAMINE	15.78	0.114	0.20032
2-PHENYL NAPHTHALENE	9.35	0.548	0.00870
1,4-DIPHENYL NAPHTHALENE	11.88	0.267	0.03655
1,5-DIPHENYL NAPHTHALENE	9.34	0.549	0.00864
1,7-DIPHENYL NAPHTHALENE	10.03	0.443	0.01327
1,4,5,8-TETRAPHENYL NAPHTHALENE	29.95	0.017	9.27354
1,1'-BINAPHTHYL	5.24	3.111	0.00027
4,4'-DI(N-BUTOXY)-1,1'-BINAPHTHYL	17.82	0.079	0.41604
2,2'-BINAPHTHYL	10.92	0.343	0.02210
ACRIDINE (A)	28.64	0.019	7.09599
ANTHRACENE (TRACE)	27.21	0.022	5.22808
DEUTERATED ANTHRACENE C14D10 (TRACE)	26.75	0.023	4.71390
9-METHYL ANTHRACENE (TRACE)	28.63	0.019	7.07577
1-AMINO ANTHRACENE	28.65	0.019	7.04961
2-AMINO ANTHRACENE	27.33	0.022	5.30494
9-VINYL ANTHRACENE	29.99	0.017	9.33278
9-PHENYL ANTHRACENE (TRACE)	30.03	0.017	9.42320
9,10-DIPHENYL ANTHRACENE	31.97	0.014	13.68436
9,10-DICHLOROANTHRACENE (TRACE)	31.04	0.015	11.45060
BIANTHRANYL (B)	33.54	0.012	18.23436
9,10-DINAPHTHYLANTHRACENE (B)	31.56	0.014	12.63407
9-(P-VINYLPHENYL)-10-PHENYL ANTHRACENE	32.27	0.013	14.43766
9,10-BIS(PHENYL-ETHYNYL)ANTHRACENE (TRACE)	33.98	0.011	19.49745
RUBICENE (B)	31.03	0.015	11.26781
DIBENZ(A,H)ANTHRACENE (B)	26.02	0.025	3.98527
TETRACENE (TRACE) (B)	25.43	0.027	3.42164
RUBRENE (B)	23.57	0.034	2.16691
TETRAHYDROCANNABINOL ACID METHYL ESTER	0.00	0.000	0.00000
PHENANTHRENE	10.74	0.361	0.01990
3,4-BENZOPHENANTHRENE	15.73	0.115	0.19470
2-PHENYL PHENANTHRENE	13.34	0.188	0.07295
CHRYSENE (TRACE)	16.00	0.109	0.21700
TRIPHENYLENE	10.03	0.444	0.01320
DEUTERATED TRIPHENYLENE-C18D12	9.81	0.474	0.01155
PYRENE	20.38	0.053	0.92402
1,3,6,8,-TETRAPHENYL PYRENE	39.55	0.007	48.94426
BIPYRENYL (B)	37.05	0.009	33.15730
O-PHENYLENE-PYRENE	34.31	0.011	20.68186
ANTHANTHRENE (B)	33.81	0.012	19.02884
FLUORANTHENE	27.08	0.023	5.05264
3-PHENYL FLUORANTHENE	32.11	0.014	14.01872

ACCEPTOR MOLECULES	RO	CO	OVERLAP
3-AMINO FLUORANTHENE	30.96	0.015	11.17246
3,3'-BIFLUORANTHENYL (B)	37.37	0.009	34.71746
BENZO(GHI)FLUORANTHENE	24.86	0.029	3.01344
9,12-DI(N-OCTADECOXY)-BENZ(E)ACEPHENANTHRYLENE	33.04	0.012	16.59583
DECACYCLENE (B)	40.77	0.007	58.16980
10-ETHOXY-BENZO(KL)XANTHENE	32.59	0.013	15.35927
PERYLENE	34.03	0.011	19.76589
BENZO(GHI)PERYLENE (B)	34.18	0.011	20.44328
BBOT (TRACE)	42.20	0.006	72.41925
QUININE SULFATE (A)	19.47	0.061	0.70800
ACRIDINE YELLOW (A)	31.47	0.014	12.27834
ACRIDINE ORANGE (A)	26.58	0.024	4.45721
ACRIDINE RED (A)	25.18	0.028	3.31063
URANINE (A)	27.28	0.022	5.20219
RHODAMINE B (A)	28.89	0.019	7.34754
RHODAMINE 6G (A)	27.85	0.021	5.89898
EOSIN B (A)	25.30	0.028	3.31689
EOSIN Y (A)	26.39	0.024	4.26852
ROSE BENGAL (A)	26.11	0.025	4.00003
INDENOINDENE	14.86	0.136	0.14034

8.114 Donor Molecule: PPF (trace)

ACCEPTOR MOLECULES	RO	CO	OVERLAP
TRIPHENYLAMINE	16.07	0.108	0.12961
P-XYLENE (TRACE)	0.00	0.000	0.00000
SODIUM SALICYLATE (A)	12.39	0.235	0.02719
ANTHRANILIC ACID (A)	25.50	0.027	2.05520
TMPD	19.92	0.057	0.46856
BENZIDINE	10.72	0.363	0.01146
1,3-DIBIPHENYLYLBENZENE (B)	7.52	1.052	0.00137
1,3,5,-TRIBIPHENYLYLBENZENE (B)	7.11	1.245	0.00097
FLUORENE (TRACE)	0.00	0.000	0.00000
1-METHYL FLUORENE	0.00	0.000	0.00000
1,2-BENZOFLUORENE	16.51	0.099	0.15221
9,10-DIHYDROPHENANTHRENE	0.00	0.000	0.00000
CARBAZOLE (TRACE)	9.69	0.491	0.00625
N-METHYL CARBAZOLE (TRACE)	19.64	0.059	0.43142
N-VINYLCARBAZOLE	16.76	0.095	0.16673
N-PHENYL CARBAZOLE	17.29	0.086	0.20129
N-METHYL-2,3-BENZCARBAZOLE	29.05	0.018	4.46566
DIBENZOFURAN	0.00	0.000	0.00000
DIBENZOTHIOPHENE (TRACE)	5.69	2.432	0.00026
P-TERPHENYL	0.00	0.000	0.00000
DEUTERATED P-TERPHENYL-C18D14	0.00	0.000	0.00000
4-METHYL TERPHENYL	0.00	0.000	0.00000
2-METHYL-P-TERPHENYL	0.00	0.000	0.00000
3,3''-DIMETHYL-P-TERPHENYL	0.00	0.000	0.00000
4-(3,3-DIMETHYLBUTOXY)-P-TERPHENYL	0.00	0.000	0.00000
4,4''-DIHEXAHYDROFARNESOXY-P-TERPHENYL	6.96	1.327	0.00086
2,2'-METHYLENE-P-TERPHENYL	0.00	0.000	0.00000
2,2'-ETHYLENE-P-TERPHENYL	5.91	2.172	0.00032
2,2'-METHYLENE-5',6''-METHYLENE-P-TERPHENYL (TRACE)	17.27	0.087	0.19987
3-PHENYL DIBENZOFURAN	0.00	0.000	0.00000
BENZO(1,2-B-4,5-B')BISBENZOFURAN (TRACE) (B)	23.42	0.035	1.24212
SEXIPHENYL BENZENE	0.00	0.000	0.00000
P-QUATERPHENYL	10.88	0.348	0.01249
3,3'''-DIMETHYL-P-QUATERPHENYL	14.59	0.144	0.07268
TETRAMETHYL-P-QUATERPHENYL	0.00	0.000	0.00000
3,3''' -DIMETHYL-3',2''-METHYLENE-P-QUATERPHENYL	23.34	0.035	1.21516
3,3''' -DIMETHYL-3',2''-ETHYLENE-P-QUATERPHENYL	24.82	0.029	1.75352
4-(3-ETHYLBUTOXY)-P-QUATERPHENYL (B)	22.01	0.042	0.85309
4,4'''-DI(2-BUTYLOCTOXY-1)-P-QUATERPHENYL	22.23	0.041	0.90578
DIETHYL-P-QUINQUEPHENYL	14.52	0.146	0.07051
DI(3-ETHYLHEPTYL)-P-QUINQUEPHENYL	24.09	0.032	1.46518
TETRAMETHYL-P-SEXIPHENYL	11.11	0.326	0.01420
TETRAMETHYL-P-OCTAPHENYL (B)	30.23	0.016	5.71492
INDOLE (TRACE)	0.00	0.000	0.00000
2-PHENYL INDOLE	10.55	0.381	0.01041
1-METHYL-2-PHENYL INDOLE	9.51	0.520	0.00558
1-NEOPENTYL-2-PHENYL INDOLE	0.00	0.000	0.00000
1-N-BUTYL-2(BIPHENYLYL-4) INDOLE	24.39	0.031	1.57770
1,2-DIPHENYL INDOLE	9.86	0.466	0.00694
1-BIPHENYLYL-2-PHENYL INDOLE	15.75	0.114	0.11468
1,3-DIMETHYL-2-PHENYL INDOLE	10.90	0.345	0.01268
2,3-DIPHENYL INDOLE	22.46	0.039	0.96161
1-METHYL-2,3-DIPHENYL INDOLE	19.97	0.056	0.47578
2-PHENYL-3,2'-METHYLENE INDOLE (TRACE)	16.91	0.093	0.17591
2-PHENYL-3,2'-ETHYLENE INDOLE (TRACE)	27.22	0.022	3.05533
2-PHENYL-3,2'-TRIMETHYLENE INDOLE	15.93	0.111	0.12317
1-METHYL-3,2'-METHYLENE-2-PHENYL INDOLE (TRACE)	25.40	0.027	2.01755
1-METHYL-3,2'-ETHYLENE-2-PHENYL INDOLE	29.32	0.018	4.75768
1-METHYL-3,2'-TRIMETHYLENE-2-PHENYL INDOLE	17.55	0.083	0.21992
4,5-DIPHENYLIMIDAZOLE (A)	8.75	0.669	0.00338
PPF (TRACE)	26.58	0.024	2.65148
PPO (TRACE)	14.25	0.154	0.06318
2-(4-FLUOROPHENYL)-5-PHENYLOXAZOLE (TRACE)	12.96	0.206	0.03571
αNPO (TRACE)	34.21	0.011	11.97559
BBO (B)	41.13	0.006	36.08346
POPOP	42.45	0.006	43.55802
DIMETHYL POPOP (TRACE)	44.86	0.005	60.59866
PPD (TRACE)	0.00	0.000	0.00000

ACCEPTOR MOLECULES	RO	CO	OVERLAP
PBD	11.30	0.310	0.01572
BBD (B)	28.48	0.019	4.00644
αNPD (TRACE)	25.28	0.028	1.95507
βNPD	15.31	0.125	0.09721
αNND (TRACE)	30.55	0.016	6.08654
βNND (TRACE)	23.74	0.033	1.34761
TRANS-1-(4-BIPHENYLYL)-2-PHENYL-ETHYLENE	29.37	0.018	4.80859
TRANS-1-(4-BIPHENYLYL)-2-(1-NAPHTHYL)-ETHYLENE	37.00	0.009	19.11534
DIPHENYLSTILBENE (B)	41.37	0.006	37.34492
1,1,4,4-TETRAPHENYLBUTADIENE	39.84	0.007	29.72128
1,6-DIPHENYLHEXATRIENE	45.31	0.005	64.37448
1,8-DIPHENYL-1,3,5,7-OCTATETRAENE	49.27	0.004	105.77302
BPSB (TRACE)	46.18	0.005	72.16675
BIS-MSB	43.75	0.005	52.17967
ESCULIN (A)	31.04	0.015	6.67035
AZULENE	19.23	0.063	0.37289
NAPHTHALENE	0.00	0.000	0.00000
DEUTERATED NAPHTHALENE-C10D8	2.96	17.257	0.00001
1-METHYLNAPHTHALENE 5ML/L	0.00	0.000	0.00000
2-METHYL NAPHTHALENE	0.00	0.000	0.00000
2,3-DIMETHYL NAPHTHALENE	0.00	0.000	0.00000
2,6-DIMETHYL NAPHTHALENE (TRACE)	0.00	0.000	0.00000
1-NAPHTHOL (TRACE)	0.00	0.000	0.00000
2-NAPHTHOL (TRACE)	5.84	2.241	0.00030
1-NAPHTHYLAMINE	16.89	0.093	0.17427
2-PHENYL NAPHTHALENE	7.32	1.143	0.00116
1,4-DIPHENYL NAPHTHALENE	12.09	0.253	0.02358
1,5-DIPHENYL NAPHTHALENE	8.33	0.775	0.00251
1,7-DIPHENYL NAPHTHALENE	9.49	0.524	0.00549
1,4,5,8-TETRAPHENYL NAPHTHALENE	33.37	0.012	10.28274
1,1'-BINAPHTHYL	0.00	0.000	0.00000
4,4'-DI(N-BUTOXY)-1,1'-BINAPHTHYL	19.27	0.063	0.38464
2,2'-BINAPHTHYL	9.93	0.457	0.00722
ACRIDINE (A)	31.81	0.014	7.71069
ANTHRACENE (TRACE)	29.91	0.017	5.33703
DEUTERATED ANTHRACENE C14D10 (TRACE)	29.53	0.017	4.94695
9-METHYL ANTHRACENE (TRACE)	31.98	0.014	7.95422
1-AMINO ANTHRACENE	31.39	0.014	7.06276
2-AMINO ANTHRACENE	29.80	0.017	5.16817
9-VINYL ANTHRACENE	33.28	0.012	10.08649
9-PHENYL ANTHRACENE (TRACE)	33.41	0.012	10.34829
9,10-DIPHENYL ANTHRACENE	35.17	0.010	14.02994
9,10-DICHLOROANTHRACENE (TRACE)	33.88	0.012	11.20609
BIANTHRANYL (B)	37.02	0.009	19.10581
9,10-DINAPHTHYLANTHRACENE (B)	34.51	0.011	12.51937
9-(P-VINYLPHENYL)-10-PHENYL ANTHRACENE	35.43	0.010	14.65935
9,10-BIS(PHENYL-ETHYNYL)ANTHRACENE (TRACE)	35.76	0.010	15.39800
RUBICENE (B)	33.45	0.012	10.32022
DIBENZ(A,H)ANTHRACENE (B)	28.81	0.019	4.25277
TETRACENE (TRACE) (B)	26.95	0.023	2.81991
RUBRENE (B)	24.72	0.030	1.67975
TETRAHYDROCANNABINOL ACID METHYL ESTER	0.00	0.000	0.00000
PHENANTHRENE	12.01	0.258	0.02257
3,4-BENZOPHENANTHRENE	17.50	0.083	0.21446
2-PHENYL PHENANTHRENE	14.85	0.137	0.08050
CHRYSENE (TRACE)	17.80	0.079	0.23800
TRIPHENYLENE	10.31	0.408	0.00905
DEUTERATED TRIPHENYLENE-C18D12	10.25	0.415	0.00874
PYRENE	20.28	0.054	0.51893
1,3,6,8,-TETRAPHENYL PYRENE	43.67	0.005	51.37965
BIPYRENYL (B)	41.28	0.006	36.71634
O-PHENYLENE-PYRENE	37.70	0.008	21.15736
ANTHANTHRENE (B)	36.51	0.009	17.46377
FLUORANTHENE	30.16	0.016	5.58595
3-PHENYL FLUORANTHENE	35.67	0.010	15.24654
3-AMINO FLUORANTHENE	33.81	0.012	11.00829
3,3'-BIFLUORANTHENYL (B)	41.21	0.006	36.21915
BENZO(GHI)FLUORANTHENE	27.78	0.021	3.40566
9,12-DI(N-OCTADECOXY)-BENZ(E)ACEPHENANTHRYLENE	36.15	0.009	16.49364
DECACYCLENE (B)	44.63	0.005	58.22057
10-ETHOXY-BENZO(KL)XANTHENE	35.83	0.010	15.70142
PERYLENE	36.47	0.009	17.33948
BENZO(GHI)PERYLENE (B)	37.95	0.008	22.16214
BBOT (TRACE)	46.70	0.004	77.01084
QUININE SULFATE (A)	20.86	0.049	0.62045
ACRIDINE YELLOW (A)	32.59	0.013	8.81509
ACRIDINE ORANGE (A)	27.10	0.022	2.91428
ACRIDINE RED (A)	27.46	0.022	3.15520
URANINE (A)	28.13	0.020	3.64482
RHODAMINE B (A)	31.46	0.014	7.13937
RHODAMINE 6G (A)	30.33	0.016	5.73273
EOSIN B (A)	26.95	0.023	2.82045
EOSIN Y (A)	28.06	0.020	3.59727
ROSE BENGAL (A)	28.24	0.020	3.73817
INDENOINDENE	13.96	0.164	0.05587

8.115 Donor Molecule: PPO (trace)

N,N-DIMETHYLANILINE 3ML/L	2.57	26.385	0.00000
TRIPHENYLAMINE	18.61	0.069	0.31321
SODIUM SALICYLATE (A)	13.41	0.185	0.04386

ACCEPTOR MOLECULES	RO	CO	OVERLAP
ANTHRANILIC ACID (A)	26.14	0.025	2.39204
TMPD	21.01	0.048	0.64592
BENZIDINE	14.75	0.139	0.07805
1,3,5-TRIPHENYL BENZENE	3.27	12.764	0.00001
1,3-DIBIPHENYLYLBENZENE (B)	12.67	0.220	0.03129
1,3,5,-TRIBIPHENYLYLBENZENE (B)	14.30	0.153	0.06472
FLUORENE (TRACE)	0.00	0.000	0.00000
1-METHYL FLUORENE	0.00	0.000	0.00000
1,2-BENZOFLUORENE	17.63	0.082	0.22667
9,10-DIHYDROPHENANTHRENE	0.00	0.000	0.00000
CARBAZOLE (TRACE)	15.36	0.123	0.09928
N-METHYL CARBAZOLE (TRACE)	21.29	0.046	0.70288
N-VINYLCARBAZOLE	21.23	0.047	0.69123
N-PHENYL CARBAZOLE	20.90	0.049	0.62872
N-METHYL-2,3-BENZCARBAZOLE	28.87	0.019	4.31634
DIBENZOFURAN	0.00	0.000	0.00000
DIBENZOTHIOPHENE (TRACE)	10.92	0.343	0.01288
P-TERPHENYL	0.00	0.000	0.00000
DEUTERATED P-TERPHENYL-C18D14	0.00	0.000	0.00000
4-METHYL TERPHENYL	0.00	0.000	0.00000
2-METHYL-P-TERPHENYL	0.00	0.000	0.00000
3,3''-DIMETHYL-P-TERPHENYL	4.24	5.868	0.00004
4-(3,3-DIMETHYLBUTOXY)-P-TERPHENYL	10.96	0.340	0.01316
4,4''-DIHEXAHYDROFARNESOXY-P-TERPHENYL	14.40	0.150	0.05753
2,2'-METHYLENE-P-TERPHENYL	10.02	0.444	0.00770
2,2'-ETHYLENE-P-TERPHENYL	12.69	0.219	0.03163
2,2'-METHYLENE-5',6''-METHYLENE-P-TERPHENYL (TRACE)	26.89	0.023	2.86066
3-PHENYL DIBENZOFURAN	9.91	0.460	0.00718
BENZO(1,2-B-4,5-B')BISBENZOFURAN (TRACE) (B)	28.64	0.019	4.16535
SEXIPHENYL BENZENE	0.00	0.000	0.00000
P-QUATERPHENYL	16.95	0.092	0.17951
3,3'''-DIMETHYL-P-QUATERPHENYL	18.86	0.067	0.34070
TETRAMETHYL-P-QUATERPHENYL	0.00	0.000	0.00000
3,3'''-DIMETHYL-3',2''-METHYLENE-P-QUATERPHENYL	27.93	0.021	3.57873
3,3'''-DIMETHYL-3',2''-ETHYLENE-P-QUATERPHENYL	28.19	0.020	3.77633
4-(3-ETHYLBUTOXY)-P-QUATERPHENYL (B)	25.17	0.028	1.91706
4,4'''-DI(2-BUTYLOCTOXY-1)-P-QUATERPHENYL	25.52	0.027	2.08361
DIETHYL-P-QUINQUEPHENYL	19.61	0.059	0.42961
DI(3-ETHYLHEPTYL)-P-QUINQUEPHENYL	27.14	0.022	3.00269
TETRAMETHYL-P-SEXIPHENYL	17.66	0.081	0.22941
TETRAMETHYL-P-OCTAPHENYL (B)	32.67	0.013	9.13622
INDOLE (TRACE)	0.00	0.000	0.00000
2-PHENYL INDOLE	18.97	0.066	0.35249
1-METHYL-2-PHENYL INDOLE	14.52	0.146	0.07090
1-NEOPENTYL-2-PHENYL INDOLE	10.27	0.413	0.00891
1-N-BUTYL-2(BIPHENYLYL-4) INDOLE	26.54	0.024	2.62619
1,2-DIPHENYL INDOLE	14.89	0.136	0.08239
1-BIPHENYLYL-2-PHENYL INDOLE	18.51	0.071	0.30319
1,3-DIMETHYL-2-PHENYL INDOLE	16.53	0.099	0.15451
2,3-DIPHENYL INDOLE	24.42	0.031	1.59424
1-METHYL-2,3-DIPHENYL INDOLE	21.93	0.042	0.83682
2-PHENYL-3,2'-METHYLENE INDOLE (TRACE)	24.28	0.031	1.54727
2-PHENYL-3,2'-ETHYLENE INDOLE (TRACE)	28.71	0.019	4.21535
2-PHENYL-3,2'-TRIMETHYLENE INDOLE	22.15	0.041	0.89273
1-METHYL-3,2'-METHYLENE-2-PHENYL INDOLE (TRACE)	27.66	0.021	3.37499
1-METHYL-3,2'-ETHYLENE-2-PHENYL INDOLE	30.96	0.015	6.62203
1-METHYL-3,2'-TRIMETHYLENE-2-PHENYL INDOLE	22.51	0.039	0.98267
4,5-DIPHENYLIMIDAZOLE (A)	12.73	0.217	0.03215
PPF (TRACE)	29.04	0.018	4.52086
PPO (TRACE)	22.30	0.040	0.93097
2-(4-FLUOROPHENYL)-5-PHENYLOXAZOLE (TRACE)	20.75	0.050	0.60466
αNPO (TRACE)	34.69	0.011	13.05331
BBO (B)	41.38	0.006	37.54486
POPOP	42.38	0.006	43.27016
DIMETHYL POPOP (TRACE)	44.35	0.005	56.67544
PPD (TRACE)	0.00	0.000	0.00000
2,5-DI(2-METHYLPHENYL)OXADIAZOLE	0.00	0.000	0.00000
PBD	19.21	0.063	0.38069
BBD (B)	31.02	0.015	6.71334
αNPD (TRACE)	27.36	0.022	3.15049
βNPD	19.49	0.060	0.41476
αNND (TRACE)	31.80	0.014	7.77236
βNND (TRACE)	27.51	0.021	3.26854
TRANS-1-(4-BIPHENYLYL)-2-PHENYL-ETHYLENE	31.46	0.014	7.29155
TRANS-1-(4-BIPHENYLYL)-2-(1-NAPHTHYL)-ETHYLENE	37.47	0.009	20.67122
DIPHENYLSTILBENE (B)	41.62	0.006	38.84037
1,1,4,4-TETRAPHENYLBUTADIENE	39.89	0.007	30.05643
1,6-DIPHENYLHEXATRIENE	45.43	0.005	65.59607
1,8-DIPHENYL-1,3,5,7-OCTATETRAENE	48.76	0.004	99.64662
BPSB (TRACE)	45.96	0.005	70.28780
BIS-MSB	43.68	0.005	51.83444
ESCULIN (A)	31.56	0.014	7.39379
AZULENE	21.78	0.043	0.78630
NAPHTHALENE	0.00	0.000	0.00000
DEUTERATED NAPHTHALENE-C10D8	6.07	1.997	0.00038
1-METHYLNAPHTHALENE 5ML/L	2.84	19.583	0.00000
2-METHYL NAPHTHALENE	3.33	12.148	0.00001
2,3-DIMETHYL NAPHTHALENE	3.39	11.456	0.00001
2,6-DIMETHYL NAPHTHALENE (TRACE)	6.45	1.666	0.00055
1-NAPHTHOL (TRACE)	6.58	1.569	0.00062
2-NAPHTHOL (TRACE)	12.41	0.234	0.02773
1-NAPHTHYLAMINE	19.88	0.057	0.46542
2-CHLORONAPHTHALENE	4.46	5.055	0.00006
ACENAPHTHENE	4.11	6.455	0.00004
1-PHENYL NAPHTHALENE	6.19	1.888	0.00043
2-PHENYL NAPHTHALENE	14.44	0.149	0.06873
1,4-DIPHENYL NAPHTHALENE	15.96	0.110	0.12473

ACCEPTOR MOLECULES	RO	CO	OVERLAP
1,5-DIPHENYL NAPHTHALENE	13.61	0.177	0.04810
1,7-DIPHENYL NAPHTHALENE	14.06	0.161	0.05837
1,4,5,8-TETRAPHENYL NAPHTHALENE	33.86	0.012	11.27990
1,1'-BINAPHTHYL	9.12	0.590	0.00436
4,4'-DI(N-BUTOXY)-1,1'-BINAPHTHYL	22.28	0.040	0.92095
2,2'-BINAPHTHYL	15.48	0.121	0.10409
ACRIDINE (A)	31.85	0.014	7.78891
ANTHRACENE (TRACE)	30.46	0.016	5.97126
DEUTERATED ANTHRACENE C14D10 (TRACE)	29.97	0.017	5.42157
9-METHYL ANTHRACENE (TRACE)	30.93	0.015	6.53965
1-AMINO ANTHRACENE	30.96	0.015	6.51789
2-AMINO ANTHRACENE	29.79	0.017	5.17342
9-VINYL ANTHRACENE	32.57	0.013	8.89360
9-PHENYL ANTHRACENE (TRACE)	32.56	0.013	8.88256
9,10-DIPHENYL ANTHRACENE	34.82	0.011	13.24928
9,10-DICHLOROANTHRACENE (TRACE)	33.71	0.012	10.91234
BIANTHRANYL (B)	36.43	0.009	17.38514
9,10-DINAPHTHYLANTHRACENE (B)	34.36	0.011	12.23125
9-(P-VINYLPHENYL)-10-PHENYL ANTHRACENE	35.10	0.010	13.88233
9,10-BIS(PHENYL-ETHYNYL)ANTHRACENE (TRACE)	36.22	0.009	16.63061
RUBICENE (B)	33.61	0.012	10.61791
DIBENZ(A,H)ANTHRACENE (B)	30.27	0.016	5.73360
TETRACENE (TRACE) (B)	27.25	0.022	3.01250
RUBRENE (B)	25.92	0.026	2.23346
TETRAHYDROCANNABINOL ACID METHYL ESTER	0.00	0.000	0.00000
PHENANTHRENE	12.83	0.212	0.03359
3,4-BENZOPHENANTHRENE	17.98	0.077	0.25258
2-PHENYL PHENANTHRENE	15.44	0.121	0.10205
CHRYSENE (TRACE)	18.18	0.074	0.27053
TRIPHENYLENE	13.07	0.200	0.03765
DEUTERATED TRIPHENYLENE-C18D12	12.55	0.226	0.02949
PYRENE	27.74	0.021	3.41462
1,3,6,8,-TETRAPHENYL PYRENE	42.89	0.006	46.19808
BIPYRENYL (B)	41.11	0.006	35.90405
O-PHENYLENE-PYRENE	37.39	0.009	20.12654
ANTHANTHRENE (B)	36.41	0.009	17.23377
FLUORANTHENE	30.27	0.016	5.71663
3-PHENYL FLUORANTHENE	35.29	0.010	14.34926
3-AMINO FLUORANTHENE	33.58	0.012	10.56199
3,3'-BIFLUORANTHENYL (B)	40.69	0.007	33.61689
BENZO(GHI)FLUORANTHENE	28.02	0.020	3.59273
9,12-DI(N-OCTADECOXY)-BENZ(E)ACEPHENANTHRYLENE	35.98	0.010	16.08476
DECACYCLENE (B)	44.34	0.005	55.95879
10-ETHOXY-BENZO(KL)XANTHENE	35.42	0.010	14.70810
PERYLENE	36.30	0.009	16.91341
BENZO(GHI)PERYLENE (B)	36.87	0.009	18.69491
BBOT (TRACE)	46.13	0.005	71.75033
QUININE SULFATE (A)	24.15	0.032	1.49560
ACRIDINE YELLOW (A)	33.46	0.012	10.32289
ACRIDINE ORANGE (A)	28.02	0.020	3.56268
ACRIDINE RED (A)	27.55	0.021	3.21741
URANINE (A)	28.99	0.018	4.37321
RHODAMINE B (A)	31.60	0.014	7.33194
RHODAMINE 6G (A)	30.69	0.015	6.14674
EOSIN B (A)	27.52	0.021	3.19970
EOSIN Y (A)	28.60	0.019	4.02484
ROSE BENGAL (A)	28.83	0.019	4.22374
INDENOINDENE	21.27	0.046	0.69952

8.116 Donor Molecule: 2-(4-Fluorophenyl)-5-phenyloxazole (trace)

N,N-DIMETHYLANILINE 3ML/L	2.58	26.014	0.00000
TRIPHENYLAMINE	18.69	0.069	0.33841
SODIUM SALICYLATE (A)	13.42	0.185	0.04645
ANTHRANILIC ACID (A)	26.01	0.025	2.44187
TMPD	20.96	0.049	0.67108
BENZIDINE	14.89	0.135	0.08680
1,3,5-TRIPHENYL BENZENE	3.32	12.240	0.00001
1,3-DIBIPHENYLYLBENZENE (B)	12.85	0.211	0.03588
1,3,5,-TRIBIPHENYLYLBENZENE (B)	14.70	0.141	0.08038
FLUORENE (TRACE)	0.00	0.000	0.00000
1-METHYL FLUORENE	0.00	0.000	0.00000
1,2-BENZOFLUORENE	17.63	0.082	0.23838
9,10-DIHYDROPHENANTHRENE	0.00	0.000	0.00000
CARBAZOLE (TRACE)	15.78	0.114	0.12282
N-METHYL CARBAZOLE (TRACE)	21.16	0.047	0.71365
N-VINYLCARBAZOLE	21.04	0.048	0.68892
N-PHENYL CARBAZOLE	20.72	0.050	0.62852
N-METHYL-2,3-BENZCARBAZOLE	28.61	0.019	4.29570
DIBENZOFURAN	0.00	0.000	0.00000
DIBENZOTHIOPHENE (TRACE)	11.21	0.317	0.01585
P-TERPHENYL	0.00	0.000	0.00000
DEUTERATED P-TERPHENYL-C18D14	0.00	0.000	0.00000
4-METHYL TERPHENYL	0.00	0.000	0.00000
2-METHYL-P-TERPHENYL	0.00	0.000	0.00000
3,3''-DIMETHYL-P-TERPHENYL	4.25	5.832	0.00005
4-(3,3-DIMETHYLBUTOXY)-P-TERPHENYL	11.39	0.303	0.01741
4,4''-DIHEXAHYDROFARNESOXY-P-TERPHENYL	14.83	0.137	0.08488
2,2'-METHYLENE-P-TERPHENYL	10.41	0.397	0.01016
2,2'-ETHYLENE-P-TERPHENYL	13.09	0.199	0.04017
2,2'-METHYLENE-5',6''-METHYLENE-P-TERPHENYL (TRACE)	27.36	0.022	3.34094

ACCEPTOR MOLECULES	RO	CO	OVERLAP
3-PHENYL DIBENZOFURAN	10.28	0.411	0.00945
BENZO(1,2-B-4,5-B')BISBENZOFURAN (TRACE) (B)	28.51	0.019	4.26934
SEXIPHENYL BENZENE	0.00	0.000	0.00000
P-QUATERPHENYL	17.18	0.088	0.20467
3,3'''-DIMETHYL-P-QUATERPHENYL	19.04	0.065	0.37957
TETRAMETHYL-P-QUATERPHENYL	0.00	0.000	0.00000
3,3''' -DIMETHYL-3',2''-METHYLENE-P-QUATERPHENYL	28.04	0.020	3.86230
3,3''' -DIMETHYL-3',2''-ETHYLENE-P-QUATERPHENYL	28.25	0.020	4.02606
4-(3-ETHYLBUTOXY)-P-QUATERPHENYL (B)	25.26	0.028	2.06008
4,4''' DI(3-BUTYLOCTOXY-1)-P-QUATERPHENYL	25.63	0.027	2.24840
DIETHYL-P-QUINQUEPHENYL	19.79	0.058	0.47859
DI(3-ETHYLHEPTYL)-P-QUINQUEPHENYL	27.21	0.022	3.21395
TETRAMETHYL-P-SEXIPHENYL	17.92	0.078	0.26376
TETRAMETHYL-P-OCTAPHENYL (B)	32.66	0.013	9.59988
INDOLE (TRACE)	0.00	0.000	0.00000
2-PHENYL INDOLE	19.43	0.061	0.42810
1-METHYL-2-PHENYL INDOLE	14.76	0.139	0.08244
1-NEOPENTYL-2-PHENYL INDOLE	10.67	0.368	0.01178
1-N-BUTYL-2(BIPHENYLYL-4) INDOLE	26.54	0.024	2.76672
1,2-DIPHENYL INDOLE	15.13	0.129	0.09556
1-BIPHENYLYL-2-PHENYL INDOLE	18.66	0.069	0.33542
1,3-DIMETHYL-2-PHENYL INDOLE	16.75	0.095	0.17578
2,3-DIPHENYL INDOLE	24.43	0.031	1.68280
1-METHYL-2,3-DIPHENYL INDOLE	21.95	0.042	0.88512
2-PHENYL-3,2'-METHYLENE INDOLE (TRACE)	24.53	0.030	1.73416
2-PHENYL-3,2'-ETHYLENE INDOLE (TRACE)	28.76	0.019	4.48457
2-PHENYL-3,2' -TRIMETHYLENE INDOLE	22.42	0.040	1.01101
1-METHYL-3,2'-METHYLENE-2-PHENYL INDOLE (TRACE)	27.66	0.021	3.54988
1-METHYL-3,2'-ETHYLENE-2-PHENYL INDOLE	30.89	0.015	6.86896
1-METHYL-3,2'-TRIMETHYLENE-2-PHENYL INDOLE	22.59	0.039	1.05642
4,5-DIPHENYLIMIDAZOLE (A)	12.78	0.214	0.03474
PPF (TRACE)	29.02	0.018	4.74252
PPO (TRACE)	22.63	0.039	1.06869
2-(4-FLUOROPHENYL)-5-PHENYLOXAZOLE (TRACE)	21.12	0.047	0.70739
αNPO (TRACE)	34.50	0.011	13.29786
BBO (B)	41.09	0.006	37.86673
POPOP	41.96	0.006	42.91694
DIMETHYL POPOP (TRACE)	43.92	0.005	56.31426
PPD (TRACE)	0.00	0.000	0.00000
2,5-DI(2-METHYLPHENYL)OXADIAZOLE	0.00	0.000	0.00000
PBD	19.60	0.059	0.45227
BBD (B)	31.07	0.015	7.13165
αNPD (TRACE)	27.28	0.022	3.26394
βNPD	19.56	0.060	0.44502
αNND (TRACE)	31.64	0.014	7.93580
βNND (TRACE)	27.39	0.022	3.35795
TRANS-1-(4-BIPHENYLYL)-2-PHENYL-ETHYLENE	31.47	0.014	7.68008
TRANS-1-(4-BIPHENYLYL)-2-(1-NAPHTHYL)-ETHYLENE	37.22	0.009	20.91700
DIPHENYLSTILBENE (B)	41.33	0.006	39.22966
1,1,4,4-TETRAPHENYLBUTADIENE	39.58	0.007	30.19053
1,6-DIPHENYLHEXATRIENE	45.04	0.005	65.55218
1,8-DIPHENYL-1,3,5,7-OCTATETRAENE	48.24	0.004	98.42413
BPSB (TRACE)	45.52	0.005	69.87981
BIS-MSB	43.31	0.006	51.85263
ESCULIN (A)	31.36	0.014	7.50096
AZULENE	21.65	0.044	0.79637
NAPHTHALENE	0.00	0.000	0.00000
DEUTERATED NAPHTHALENE-C10D8	6.25	1.831	0.00048
1-METHYLNAPHTHALENE 5ML/L	2.88	18.701	0.00000
2-METHYL NAPHTHALENE	3.36	11.745	0.00001
2,3-DIMETHYL NAPHTHALENE	3.44	10.973	0.00001
2,6-DIMETHYL NAPHTHALENE (TRACE)	6.59	1.560	0.00066
1-NAPHTHOL (TRACE)	6.82	1.409	0.00081
2-NAPHTHOL (TRACE)	12.84	0.211	0.03576
1-NAPHTHYLAMINE	19.98	0.056	0.50430
2-CHLORONAPHTHALENE	4.59	4.612	0.00008
ACENAPHTHENE	4.13	6.341	0.00004
1-PHENYL NAPHTHALENE	6.42	1.687	0.00056
2-PHENYL NAPHTHALENE	14.82	0.137	0.08460
1,4-DIPHENYL NAPHTHALENE	16.15	0.106	0.14106
1,5-DIPHENYL NAPHTHALENE	13.86	0.168	0.05641
1,7-DIPHENYL NAPHTHALENE	14.25	0.154	0.06679
1,4,5,8-TETRAPHENYL NAPHTHALENE	33.67	0.012	11.43968
1,1'-BINAPHTHYL	9.49	0.523	0.00584
4,4' -DI(N-BUTOXY)-1,1' -BINAPHTHYL	22.41	0.040	1.00523
2,2'-BINAPHTHYL	15.68	0.116	0.11822
ACRIDINE (A)	31.54	0.014	7.73831
ANTHRACENE (TRACE)	30.10	0.016	5.85077
DEUTERATED ANTHRACENE C14D10 (TRACE)	29.68	0.017	5.38781
9-METHYL ANTHRACENE (TRACE)	30.67	0.016	6.53493
1-AMINO ANTHRACENE	30.64	0.016	6.44083
2-AMINO ANTHRACENE	29.53	0.017	5.16386
9-VINYL ANTHRACENE	32.26	0.013	8.84469
9-PHENYL ANTHRACENE (TRACE)	32.23	0.013	8.79363
9,10-DIPHENYL ANTHRACENE	34.45	0.011	13.09930
9,10-DICHLOROANTHRACENE (TRACE)	33.27	0.012	10.61096
BIANTHRANYL (B)	36.10	0.010	17.32344
9,10-DINAPHTHYLANTHRACENE (B)	33.95	0.011	11.97412
9-(P-VINYLPHENYL)-10-PHENYL ANTHRACENE	34.71	0.011	13.68637
9,10-BIS(PHENYL-ETHYNYL)ANTHRACENE (TRACE)	35.93	0.010	16.66525
RUBICENE (B)	33.32	0.012	10.58550
DIBENZ(A,H)ANTHRACENE (B)	30.20	0.016	5.95539
TETRACENE (TRACE) (B)	27.06	0.023	3.03699
RUBRENE (B)	25.94	0.026	2.35925
TETRAHYDROCANNABINOL ACID METHYL ESTER	0.00	0.000	0.00000
PHENANTHRENE	12.84	0.211	0.03564
3,4-BENZOPHENANTHRENE	17.93	0.078	0.26175

ACCEPTOR MOLECULES	RO	CO	OVERLAP
2-PHENYL PHENANTHRENE	15.40	0.123	0.10547
CHRYSENE (TRACE)	18.05	0.076	0.27320
TRIPHENYLENE	13.12	0.198	0.04056
DEUTERATED TRIPHENYLENE-C18D12	12.62	0.222	0.03216
PYRENE	28.01	0.020	3.80915
1,3,6,8,-TETRAPHENYL PYRENE	42.42	0.006	45.50606
BIPYRENYL (B)	40.80	0.007	36.15598
O-PHENYLENE-PYRENE	36.99	0.009	19.86887
ANTHANTHRENE (B)	36.08	0.010	17.16425
FLUORANTHENE	29.97	0.017	5.67622
3-PHENYL FLUORANTHENE	34.96	0.010	14.25602
3-AMINO FLUORANTHENE	33.26	0.012	10.50285
3,3'-BIFLUORANTHENYL (B)	40.27	0.007	33.26144
BENZO(GHI)FLUORANTHENE	27.97	0.020	3.74100
9,12-DI(N-OCTADECOXY)-BENZ(E)ACEPHENANTHRYLENE	35.64	0.010	15.98503
DECACYCLENE (B)	43.88	0.005	55.32986
10-ETHOXY-BENZO(KL)XANTHENE	35.03	0.010	14.47345
PERYLENE	35.92	0.010	16.72076
BENZO(GHI)PERYLENE (B)	36.49	0.009	18.51119
BBOT (TRACE)	45.68	0.005	71.16225
QUININE SULFATE (A)	24.15	0.032	1.57545
ACRIDINE YELLOW (A)	33.28	0.012	10.50080
ACRIDINE ORANGE (A)	28.09	0.020	3.79891
ACRIDINE RED (A)	27.39	0.022	3.26367
URANINE (A)	29.10	0.018	4.69390
RHODAMINE B (A)	31.41	0.014	7.42144
RHODAMINE 6G (A)	30.66	0.016	6.42233
EOSIN B (A)	27.53	0.021	3.36573
EOSIN Y (A)	28.58	0.019	4.21555
ROSE BENGAL (A)	28.71	0.019	4.33034
INDENOINDENE	21.69	0.044	0.82938

8.117 Donor Molecule: α-NPO (trace)

TRIPHENYLAMINE	4.62	4.540	0.00008
P-XYLENE (TRACE)	0.00	0.000	0.00000
SODIUM SALICYLATE (A)	5.19	3.192	0.00016
ANTHRANILIC ACID (A)	17.57	0.082	0.23274
TMPD	11.78	0.273	0.02124
1,2-BENZOFLUORENE	4.01	6.957	0.00003
9,10-DIHYDROPHENANTHRENE	0.00	0.000	0.00000
CARBAZOLE (TRACE)	0.00	0.000	0.00000
N-METHYL CARBAZOLE (TRACE)	3.95	7.277	0.00003
N-VINYLCARBAZOLE	0.00	0.000	0.00000
N-PHENYL CARBAZOLE	0.00	0.000	0.00000
N-METHYL-2,3-BENZCARBAZOLE	28.09	0.020	3.86173
3,3''' -DIMETHYL-3',2''-METHYLENE-P-QUATERPHENYL	5.79	2.300	0.00030
3,3''' -DIMETHYL-3',2''-ETHYLENE-P-QUATERPHENYL	10.63	0.372	0.01146
4-(3-ETHYLBUTOXY)-P-QUATERPHENYL (B)	8.41	0.752	0.00281
4,4'''-DI(2-BUTYLOCTOXY-1)-P-QUATERPHENYL	7.73	0.968	0.00170
DIETHYL-P-QUINQUEPHENYL	0.00	0.000	0.00000
DI(3-ETHYLHEPTYL)-P-QUINQUEPHENYL	11.90	0.266	0.02252
TETRAMETHYL-P-SEXIPHENYL	0.00	0.000	0.00000
TETRAMETHYL-P-OCTAPHENYL (B)	17.24	0.087	0.20848
INDOLE (TRACE)	0.00	0.000	0.00000
2-PHENYL INDOLE	0.00	0.000	0.00000
1-METHYL-2-PHENYL INDOLE	0.00	0.000	0.00000
1-NEOPENTYL-2-PHENYL INDOLE	0.00	0.000	0.00000
1-N-BUTYL-2(BIPHENYLYL-4) INDOLE	12.55	0.226	0.03107
1,2-DIPHENYL INDOLE	0.00	0.000	0.00000
1-BIPHENYLYL-2-PHENYL INDOLE	6.08	1.994	0.00040
1,3-DIMETHYL-2-PHENYL INDOLE	0.00	0.000	0.00000
2,3-DIPHENYL INDOLE	12.89	0.209	0.03641
1-METHYL-2,3-DIPHENYL INDOLE	10.02	0.445	0.00802
2-PHENYL-3,2'-METHYLENE INDOLE (TRACE)	0.00	0.000	0.00000
2-PHENYL-3,2'-ETHYLENE INDOLE (TRACE)	9.89	0.463	0.00743
2-PHENYL-3,2'-TRIMETHYLENE INDOLE	0.00	0.000	0.00000
1-METHYL-3,2'-METHYLENE-2-PHENYL INDOLE (TRACE)	7.22	1.189	0.00113
1-METHYL-3,2'-ETHYLENE-2-PHENYL INDOLE	15.55	0.119	0.11236
1-METHYL-3,2'-TRIMETHYLENE-2-PHENYL INDOLE	0.00	0.000	0.00000
4,5-DIPHENYLIMIDAZOLE (A)	0.00	0.000	0.00000
PPF (TRACE)	6.96	1.325	0.00091
αNPO (TRACE)	24.22	0.031	1.59588
BBO (B)	31.56	0.014	7.80034
POPOP	34.80	0.011	14.00907
DIMETHYL POPOP (TRACE)	38.88	0.008	27.18572
PPD (TRACE)	0.00	0.000	0.00000
2,5-DI(2-METHYLPHENYL)OXADIAZOLE	0.00	0.000	0.00000
PBD	0.00	0.000	0.00000
BBD (B)	10.88	0.347	0.01322
αNPD (TRACE)	13.63	0.177	0.05089
βNPD	0.00	0.000	0.00000
αNND (TRACE)	18.52	0.070	0.31941
βNND (TRACE)	0.00	0.000	0.00000
TRANS-1-(4-BIPHENYLYL)-2-PHENYL-ETHYLENE	13.96	0.164	0.05878
TRANS-1-(4-BIPHENYLYL)-2-(1-NAPHTHYL)-ETHYLENE	28.12	0.020	3.90232
DIPHENYLSTILBENE (B)	31.43	0.014	7.60848
1,1,4,4-TETRAPHENYLBUTADIENE	31.60	0.014	7.84510
1,6-DIPHENYLHEXATRIENE	36.73	0.009	19.34400
1,8-DIPHENYL-1,3,5,7-OCTATETRAENE	46.22	0.005	76.36232
BPSB (TRACE)	38.39	0.008	25.23985

ACCEPTOR MOLECULES	RO	CO	OVERLAP
BIS-MSB	35.69	0.010	16.28718
ESCULIN (A)	22.73	0.038	1.08907
AZULENE	13.42	0.185	0.04522
NAPHTHALENE	0.00	0.000	0.00000
DEUTERATED NAPHTHALENE-C10D8	0.00	0.000	0.00000
1-METHYLNAPHTHALENE 5ML/L	0.00	0.000	0.00000
2-METHYL NAPHTHALENE	0.00	0.000	0.00000
2,3-DIMETHYL NAPHTHALENE	0.00	0.000	0.00000
2,6-DIMETHYL NAPHTHALENE (TRACE)	0.00	0.000	0.00000
1-NAPHTHOL (TRACE)	0.00	0.000	0.00000
2-NAPHTHOL (TRACE)	0.00	0.000	0.00000
1-NAPHTHYLAMINE	4.72	4.259	0.00009
2-CHLORONAPHTHALENE	0.00	0.000	0.00000
ACENAPHTHENE	0.00	0.000	0.00000
1-PHENYL NAPHTHALENE	0.00	0.000	0.00000
2-PHENYL NAPHTHALENE	0.00	0.000	0.00000
1,4-DIPHENYL NAPHTHALENE	0.00	0.000	0.00000
1,5-DIPHENYL NAPHTHALENE	0.00	0.000	0.00000
1,7-DIPHENYL NAPHTHALENE	0.00	0.000	0.00000
1,4,5,8-TETRAPHENYL NAPHTHALENE	25.01	0.029	1.92769
1,1'-BINAPHTHYL	0.00	0.000	0.00000
4,4'-DI(N-BUTOXY)-1,1'-BINAPHTHYL	6.87	1.380	0.00084
2,2'-BINAPHTHYL	0.00	0.000	0.00000
ACRIDINE (A)	25.93	0.026	2.39680
ANTHRACENE (TRACE)	24.64	0.030	1.76679
DEUTERATED ANTHRACENE C14D10 (TRACE)	24.23	0.031	1.59880
9-METHYL ANTHRACENE (TRACE)	28.40	0.020	4.13210
1-AMINO ANTHRACENE	31.17	0.015	7.16266
2-AMINO ANTHRACENE	30.06	0.016	5.76963
9-VINYL ANTHRACENE	30.38	0.016	6.18801
9-PHENYL ANTHRACENE (TRACE)	29.54	0.017	5.23012
9,10-DIPHENYL ANTHRACENE	33.40	0.012	10.90683
9,10-DICHLOROANTHRACENE (TRACE)	32.58	0.013	9.40017
BIANTHRANYL (B)	35.28	0.010	15.15642
9,10-DINAPHTHYLANTHRACENE (B)	33.10	0.012	10.32169
9-(P-VINYLPHENYL)-10-PHENYL ANTHRACENE	33.84	0.012	11.78241
9,10-BIS(PHENYL-ETHYNYL)ANTHRACENE (TRACE)	42.78	0.006	47.60810
RUBICENE (B)	35.94	0.010	16.69567
DIBENZ(A,H)ANTHRACENE (B)	20.12	0.055	0.52168
TETRACENE (TRACE) (B)	31.61	0.014	7.74653
RUBRENE (B)	29.91	0.017	5.54869
TETRAHYDROCANNABINOL ACID METHYL ESTER	0.00	0.000	0.00000
PHENANTHRENE	2.79	20.635	0.00000
3,4-BENZOPHENANTHRENE	13.54	0.180	0.04860
2-PHENYL PHENANTHRENE	8.31	0.779	0.00262
CHRYSENE (TRACE)	12.03	0.257	0.02400
TRIPHENYLENE	2.68	23.272	0.00000
DEUTERATED TRIPHENYLENE-C18D12	2.35	34.373	0.00000
PYRENE	12.90	0.208	0.03642
1,3,6,8,-TETRAPHENYL PYRENE	40.51	0.007	34.65273
BIPYRENYL (B)	34.34	0.011	12.88652
O-PHENYLENE-PYRENE	36.08	0.010	17.15668
ANTHANTHRENE (B)	39.48	0.007	29.59717
FLUORANTHENE	23.01	0.037	1.16692
3-PHENYL FLUORANTHENE	30.92	0.015	6.85679
3-AMINO FLUORANTHENE	34.38	0.011	12.85921
3,3'-BIFLUORANTHENYL (B)	38.50	0.008	25.51593
BENZO(GHI)FLUORANTHENE	22.72	0.038	1.07902
9,12-DI(N-OCTADECOXY)-BENZ(E)ACEPHENANTHRYLENE	35.92	0.010	16.82628
DECACYCLENE (B)	43.83	0.005	55.17354
10-ETHOXY-BENZO(KL)XANTHENE	33.82	0.012	11.77197
PERYLENE	40.83	0.007	36.17273
BENZO(GHI)PERYLENE (B)	34.74	0.011	13.83031
BBOT (TRACE)	42.37	0.006	45.49498
QUININE SULFATE (A)	5.67	2.458	0.00026
ACRIDINE YELLOW (A)	40.84	0.007	35.95581
ACRIDINE ORANGE (A)	35.46	0.010	15.40528
ACRIDINE RED (A)	28.99	0.018	4.59402
URANINE (A)	36.32	0.009	17.77977
RHODAMINE B (A)	33.26	0.012	10.47633
RHODAMINE 6G (A)	33.39	0.012	10.72950
EOSIN B (A)	32.42	0.013	8.99462
EOSIN Y (A)	33.82	0.012	11.59450
ROSE BENGAL (A)	30.87	0.015	6.69860

8.119 Donor Molecule: POPOP

ANTHRANILIC ACID (A)	10.02	0.445	0.00805
TMPD	0.00	0.000	0.00000
N-METHYL-2,3-BENZCARBAZOLE	25.76	0.026	2.31626
αNPO (TRACE)	11.41	0.301	0.01752
BBO (B)	21.67	0.044	0.82383
POPOP	28.09	0.020	3.90418
DIMETHYL POPOP (TRACE)	33.76	0.012	11.72860
αNND (TRACE)	4.39	5.301	0.00006
βNND (TRACE)	0.00	0.000	0.00000
TRANS-1-(4-BIPHENYLYL)-2-PHENYL-ETHYLENE	0.00	0.000	0.00000
TRANS-1-(4-BIPHENYLYL)-2-(1-NAPHTHYL)-ETHYLENE	20.46	0.052	0.58324
DIPHENYLSTILBENE (B)	21.10	0.048	0.70139
1,1,4,4-TETRAPHENYLBUTADIENE	24.30	0.031	1.63281
1,6-DIPHENYLHEXATRIENE	27.07	0.023	3.12173
1,8-DIPHENYL-1,3,5,7-OCTATETRAENE	42.35	0.006	45.52314

ACCEPTOR MOLECULES	RO	CO	OVERLAP
BPSB (TRACE)	31.80	0.014	8.19607
BIS-MSB	28.95	0.018	4.66387
ESCULIN (A)	15.54	0.119	0.11191
AZULENE	14.20	0.156	0.06383
1,4,5,8-TETRAPHENYL NAPHTHALENE	18.55	0.070	0.32319
1,1'-BINAPHTHYL	0.00	0.000	0.00000
4,4'-DI(N-BUTOXY)-1,1'-BINAPHTHYL	0.00	0.000	0.00000
2,2'-BINAPHTHYL	0.00	0.000	0.00000
ACRIDINE (A)	21.88	0.043	0.87042
ANTHRACENE (TRACE)	18.10	0.075	0.27908
DEUTERATED ANTHRACENE C14D10 (TRACE)	16.82	0.094	0.18004
9-METHYL ANTHRACENE (TRACE)	25.74	0.026	2.30516
1-AMINO ANTHRACENE	30.41	0.016	6.22333
2-AMINO ANTHRACENE	29.75	0.017	5.45762
9-VINYL ANTHRACENE	27.40	0.022	3.34757
9-PHENYL ANTHRACENE (TRACE)	26.56	0.024	2.78246
9,10-DIPHENYL ANTHRACENE	30.58	0.016	6.46649
9,10-DICHLOROANTHRACENE (TRACE)	30.86	0.015	6.82280
BIANTHRANYL (B)	32.05	0.014	8.57937
9,10-DINAPHTHYLANTHRACENE (B)	30.81	0.015	6.75603
9-(P-VINYLPHENYL)-10-PHENYL ANTHRACENE	31.33	0.015	7.46961
9,10-BIS(PHENYL-ETHYNYL)ANTHRACENE (TRACE)	44.90	0.005	64.10523
RUBICENE (B)	36.96	0.009	19.88815
DIBENZ(A,H)ANTHRACENE (B)	17.49	0.084	0.22653
TETRACENE (TRACE) (B)	32.87	0.013	9.86826
RUBRENE (B)	31.77	0.014	8.01313
TETRAHYDROCANNABINOL ACID METHYL ESTER	0.00	0.000	0.00000
PHENANTHRENE	0.00	0.000	0.00000
3,4-BENZOPHENANTHRENE	8.63	0.696	0.00328
2-PHENYL PHENANTHRENE	0.00	0.000	0.00000
CHRYSENE (TRACE)	4.53	4.804	0.00007
TRIPHENYLENE	0.00	0.000	0.00000
DEUTERATED TRIPHENYLENE-C18D12	0.00	0.000	0.00000
PYRENE	7.05	1.274	0.00098
1,3,6,8,-TETRAPHENYL PYRENE	36.92	0.009	19.99108
BIPYRENYL (B)	29.68	0.017	5.40581
O-PHENYLENE-PYRENE	35.13	0.010	14.73717
ANTHANTHRENE (B)	40.76	0.007	36.08947
FLUORANTHENE	17.45	0.084	0.22353
3-PHENYL FLUORANTHENE	26.71	0.023	2.86838
3-AMINO FLUORANTHENE	34.31	0.011	12.79149
3,3'-BIFLUORANTHENYL (B)	35.95	0.010	17.01673
BENZO(GHI)FLUORANTHENE	21.40	0.046	0.75877
9,12-DI(N-OCTADECOXY)-BENZ(E)ACEPHENANTHRYLENE	35.34	0.010	15.34705
DECACYCLENE (B)	42.85	0.006	48.44700
10-ETHOXY-BENZO(KL)XANTHENE	31.10	0.015	7.15979
PERYLENE	42.42	0.006	45.84993
BENZO(GHI)PERYLENE (B)	31.76	0.014	8.11742
BBOT (TRACE)	37.90	0.008	23.46442
QUININE SULFATE (A)	0.00	0.000	0.00000
ACRIDINE YELLOW (A)	43.32	0.006	51.56237
ACRIDINE ORANGE (A)	37.85	0.008	22.92540
ACRIDINE RED (A)	29.90	0.017	5.57454
URANINE (A)	38.77	0.008	26.48849
RHODAMINE B (A)	34.29	0.011	12.68150
RHODAMINE 6G (A)	35.12	0.010	14.62070
EOSIN B (A)	34.15	0.011	12.36221
EOSIN Y (A)	35.59	0.010	15.84326
ROSE BENGAL (A)	32.16	0.013	8.63310

8.120 Donor Molecule: Dimethyl POPOP (trace)

ACCEPTOR MOLECULES	RO	CO	OVERLAP
ANTHRANILIC ACID (A)	5.61	2.530	0.00025
TMPD	0.00	0.000	0.00000
N-METHYL-2,3-BENZCARBAZOLE	24.64	0.030	1.76760
αNPO (TRACE)	6.31	1.776	0.00050
BBO (B)	13.70	0.174	0.05239
POPOP	19.13	0.064	0.38856
DIMETHYL POPOP (TRACE)	28.01	0.020	3.81780
TRANS-1-(4-BIPHENYLYL)-2-(1-NAPHTHYL)-ETHYLENE	13.82	0.170	0.05521
DIPHENYLSTILBENE (B)	14.11	0.159	0.06257
1,1,4,4-TETRAPHENYLBUTADIENE	18.99	0.065	0.37113
1,6-DIPHENYLHEXATRIENE	19.62	0.059	0.45213
1,8-DIPHENYL-1,3,5,7-OCTATETRAENE	39.94	0.007	31.94154
BPSB (TRACE)	23.79	0.033	1.43463
BIS-MSB	22.48	0.039	1.02073
ESCULIN (A)	9.56	0.511	0.00607
AZULENE	15.74	0.115	0.11814
1,4,5,8-TETRAPHENYL NAPHTHALENE	14.27	0.154	0.06681
1,1'-BINAPHTHYL	0.00	0.000	0.00000
4,4'-DI(N-BUTOXY)-1,1'-BINAPHTHYL	0.00	0.000	0.00000
2,2'-BINAPHTHYL	0.00	0.000	0.00000
ACRIDINE (A)	16.85	0.094	0.18115
ANTHRACENE (TRACE)	10.68	0.367	0.01180
DEUTERATED ANTHRACENE C14D10 (TRACE)	9.69	0.492	0.00657
9-METHYL ANTHRACENE (TRACE)	19.25	0.063	0.40303
1-AMINO ANTHRACENE	29.12	0.018	4.79057
2-AMINO ANTHRACENE	28.81	0.019	4.49276
9-VINYL ANTHRACENE	24.01	0.032	1.51586
9-PHENYL ANTHRACENE (TRACE)	20.43	0.052	0.57576
9,10-DIPHENYL ANTHRACENE	28.86	0.019	4.56429

ACCEPTOR MOLECULES	RO	CO	OVERLAP
9,10-DICHLOROANTHRACENE (TRACE)	28.93	0.018	4.62790
BIANTHRANYL (B)	30.24	0.016	6.03404
9,10-DINAPHTHYLANTHRACENE (B)	29.33	0.018	5.01860
9-(P-VINYLPHENYL)-10-PHENYL ANTHRACENE	29.51	0.017	5.20514
9,10-BIS(PHENYL-ETHYNYL)ANTHRACENE (TRACE)	45.88	0.005	72.89580
RUBICENE (B)	38.24	0.008	24.30605
DIBENZ(A,H)ANTHRACENE (B)	16.76	0.095	0.17533
TETRACENE (TRACE) (B)	34.16	0.011	12.41489
RUBRENE (B)	34.19	0.011	12.41071
TETRAHYDROCANNABINOL ACID METHYL ESTER	0.00	0.000	0.00000
PHENANTHRENE	0.00	0.000	0.00000
3,4-BENZOPHENANTHRENE	5.48	2.713	0.00022
2-PHENYL PHENANTHRENE	0.00	0.000	0.00000
CHRYSENE (TRACE)	0.00	0.000	0.00000
TRIPHENYLENE	0.00	0.000	0.00000
DEUTERATED TRIPHENYLENE-C18D12	0.00	0.000	0.00000
PYRENE	4.06	6.685	0.00004
1,3,6,8,-TETRAPHENYL PYRENE	32.72	0.013	9.67465
BIPYRENYL (B)	25.71	0.026	2.28229
O-PHENYLENE-PYRENE	33.92	0.011	11.92319
ANTHANTHRENE (B)	39.75	0.007	30.95563
FLUORANTHENE	14.44	0.149	0.07159
3-PHENYL FLUORANTHENE	23.56	0.034	1.34682
3-AMINO FLUORANTHENE	33.77	0.012	11.60171
3,3'-BIFLUORANTHENYL (B)	33.27	0.012	10.66305
BENZO(GHI)FLUORANTHENE	20.18	0.054	0.53169
9,12-DI(N-OCTADECOXY)-BENZ(E)ACEPHENANTHRYLENE	34.23	0.011	12.64898
DECACYCLENE (B)	41.45	0.006	39.66059
10-ETHOXY-BENZO(KL)XANTHENE	29.07	0.018	4.76722
PERYLENE	41.42	0.006	39.67182
BENZO(GHI)PERYLENE (B)	25.86	0.026	2.36172
BBOT (TRACE)	34.70	0.011	13.78394
QUININE SULFATE (A)	0.00	0.000	0.00000
ACRIDINE YELLOW (A)	44.89	0.005	63.80115
ACRIDINE ORANGE (A)	40.29	0.007	33.29799
ACRIDINE RED (A)	33.36	0.012	10.70874
URANINE (A)	41.76	0.006	41.26389
RHODAMINE B (A)	37.96	0.008	23.24799
RHODAMINE 6G (A)	39.95	0.007	31.59414
EOSIN B (A)	38.32	0.008	24.61984
EOSIN Y (A)	39.67	0.007	30.26802
ROSE BENGAL (A)	35.93	0.010	16.71214

8.121 Donor Molecule: PPD (trace)

ACCEPTOR MOLECULES	RO	CO	OVERLAP
ANILINE 3ML/L	5.90	2.180	0.00036
N,N-DIMETHYLANILINE 3ML/L	13.81	0.170	0.05976
TRIPHENYLAMINE	26.12	0.025	2.71190
1,4-DIMETHOXY BENZENE	7.00	1.303	0.00102
1-METHYL-4-CHLOROBENZENE 5ML/L	0.00	0.000	0.00000
SODIUM SALICYLATE (A)	17.75	0.080	0.26693
ANTHRANILIC ACID (A)	27.99	0.020	4.07344
TMPD	23.89	0.033	1.57879
4-METHOXY BIPHENYL	4.82	3.995	0.00011
4-BIPHENYLYL PHENYL ETHER	0.00	0.000	0.00000
4,4'-DIHYDROXYBIPHENYL (A)	12.52	0.228	0.03330
4,4'-DIMETHOXY BIPHENYL	10.68	0.368	0.01281
4,4'-DIISOBUTOXYBIPHENYL	9.64	0.499	0.00696
BENZIDINE	23.29	0.035	1.36705
4-VINYL BIPHENYL	14.39	0.150	0.07662
M-TERPHENYL	10.33	0.406	0.01053
1,3,5-TRIPHENYL BENZENE	11.08	0.329	0.01593
1,3-DIBIPHENYLYLBENZENE (B)	22.89	0.037	1.23306
1,3,5,-TRIBIPHENYLYLBENZENE (B)	26.81	0.023	3.18851
FLUORENE (TRACE)	7.31	1.143	0.00133
1-METHYL FLUORENE	7.33	1.134	0.00135
1,2-BENZOFLUORENE	24.40	0.031	1.80331
9,10-DIHYDROPHENANTHRENE	9.68	0.494	0.00711
CARBAZOLE (TRACE)	22.59	0.039	1.13664
N-METHYL CARBAZOLE (TRACE)	25.09	0.028	2.12966
N-VINYLCARBAZOLE	25.03	0.029	2.09999
N-PHENYL CARBAZOLE	24.77	0.029	1.97327
N-METHYL-2,3-BENZCARBAZOLE	28.96	0.018	4.96829
DIBENZOFURAN	6.08	1.995	0.00044
DIBENZOTHIOPHENE (TRACE)	20.21	0.054	0.58449
P-TERPHENYL	16.06	0.108	0.14825
DEUTERATED P-TERPHENYL-C18D14	15.59	0.118	0.12423
4-METHYL TERPHENYL	17.89	0.078	0.28235
2-METHYL-P-TERPHENYL	9.18	0.578	0.00519
3,3''-DIMETHYL-P-TERPHENYL	17.47	0.084	0.24485
4-(3,3-DIMETHYLBUTOXY)-P-TERPHENYL	22.56	0.039	1.13380
4,4''-DIHEXAHYDROFARNESOXY-P-TERPHENYL	25.82	0.026	2.54429
2,2'-METHYLENE-P-TERPHENYL	25.03	0.029	2.11578
2,2'-ETHYLENE-P-TERPHENYL	25.56	0.027	2.39127
2,2'-METHYLENE-5,6'-METHYLENE-P-TERPHENYL (TRACE)	34.21	0.011	13.70670
3-PHENYL DIBENZOFURAN	25.29	0.028	2.24921
BENZO(1,2-B-4,5-B')BISBENZOFURAN (TRACE) (B)	36.07	0.010	18.82876
SEXIPHENYL BENZENE	0.00	0.000	0.00000
P-QUATERPHENYL	27.05	0.023	3.35593
3,3'''-DIMETHYL-P-QUATERPHENYL	28.43	0.019	4.51815
TETRAMETHYL-P-QUATERPHENYL	11.56	0.289	0.02070

ACCEPTOR MOLECULES	RO	CO	OVERLAP
3,3''' -DIMETHYL-3',2''-METHYLENE-P-QUATERPHENYL	35.90	0.010	18.27857
3,3''' -DIMETHYL-3',2''-ETHYLENE-P-QUATERPHENYL	35.54	0.010	17.15628
4-(3-ETHYLBUTOXY)-P-QUATERPHENYL (B)	33.04	0.012	11.09755
4,4'''-DI(2-BUTYLOCTOXY-1)-P-QUATERPHENYL	33.85	0.012	12.83123
DIETHYL-P-QUINQUEPHENYL	29.66	0.017	5.82276
DI(3-ETHYLHEPTYL)-P-QUINQUEPHENYL	35.25	0.010	16.33631
TETRAMETHYL-P-SEXIPHENYL	28.62	0.019	4.70931
TETRAMETHYL-P-OCTAPHENYL (B)	39.44	0.007	32.00517
INDOLE (TRACE)	0.00	0.000	0.00000
2-PHENYL INDOLE	29.25	0.018	5.36949
1-METHYL-2-PHENYL INDOLE	24.10	0.032	1.67596
1-NEOPENTYL-2-PHENYL INDOLE	21.90	0.043	0.94851
1-N-BUTYL-2(BIPHENYLYL-4) INDOLE	32.36	0.013	9.77549
1,2-DIPHENYL INDOLE	25.07	0.028	2.12768
1-BIPHENYLYL-2-PHENYL INDOLE	27.54	0.021	3.72657
1,3-DIMETHYL-2-PHENYL INDOLE	24.72	0.030	1.95308
2,3-DIPHENYL INDOLE	29.98	0.017	6.17382
1-METHYL-2,3-DIPHENYL INDOLE	27.28	0.022	3.50780
2-PHENYL-3,2'-METHYLENE INDOLE (TRACE)	31.92	0.014	9.04804
2-PHENYL-3,2'-ETHYLENE INDOLE (TRACE)	34.20	0.011	13.62912
2-PHENYL-3,2'-TRIMETHYLENE INDOLE	30.82	0.015	7.33339
1-METHYL-3,2'-METHYLENE-2-PHENYL INDOLE (TRACE)	33.80	0.012	12.71836
1-METHYL-3,2'-ETHYLENE-2-PHENYL INDOLE	34.55	0.011	14.47532
1-METHYL-3,2'-TRIMETHYLENE-2-PHENYL INDOLE	29.44	0.018	5.56964
4,5-DIPHENYLIMIDAZOLE (A)	19.23	0.063	0.43384
PPF (TRACE)	35.47	0.010	16.98293
PPO (TRACE)	31.23	0.015	7.94957
2-(4-FLUOROPHENYL)-5-PHENYLOXAZOLE (TRACE)	29.80	0.017	5.99551
αNPO (TRACE)	36.99	0.009	21.71623
BBO (B)	42.52	0.006	49.97469
POPOP	41.64	0.006	44.08394
DIMETHYL POPOP (TRACE)	42.83	0.006	52.02315
PPD (TRACE)	17.22	0.088	0.22547
2,5-DI(2-METHYLPHENYL)OXADIAZOLE	18.22	0.074	0.31566
PBD	30.40	0.016	6.76945
BBD (B)	37.73	0.008	24.58452
αNPD (TRACE)	32.59	0.013	10.19560
βNPD	29.14	0.018	5.23770
αNND (TRACE)	34.89	0.011	15.32505
βNND (TRACE)	34.43	0.011	14.22554
TRANS-1-(4-BIPHENYLYL)-2-PHENYL-ETHYLENE	37.26	0.009	22.77202
TRANS-1-(4-BIPHENYLYL)-2-(1-NAPHTHYL)-ETHYLENE	39.26	0.007	30.95348
DIPHENYLSTILBENE (B)	42.92	0.006	52.86811
1,1,4,4-TETRAPHENYLBUTADIENE	40.36	0.007	36.48307
1,6-DIPHENYLHEXATRIENE	44.74	0.005	67.70074
1,8-DIPHENYL-1,3,5,7-OCTATETRAENE	44.97	0.005	69.36738
BPSB (TRACE)	45.07	0.005	70.75727
BIS-MSB	43.69	0.005	58.66523
ESCULIN (A)	33.29	0.012	11.53519
AZULENE	24.71	0.030	1.90056
NAPHTHALENE	9.27	0.562	0.00546
DEUTERATED NAPHTHALENE-C10D8	11.27	0.312	0.01763
1-METHYLNAPHTHALENE 5ML/L	11.06	0.331	0.01576
2-METHYL NAPHTHALENE	12.32	0.239	0.03010
2,3-DIMETHYL NAPHTHALENE	12.26	0.243	0.02919
2,6-DIMETHYL NAPHTHALENE (TRACE)	15.85	0.112	0.13620
1-NAPHTHOL (TRACE)	17.71	0.081	0.26481
2-NAPHTHOL (TRACE)	19.95	0.056	0.54030
1-NAPHTHYLAMINE	25.22	0.028	2.19680
2-CHLORONAPHTHALENE	12.77	0.215	0.03723
ACENAPHTHENE	15.51	0.120	0.11970
1-PHENYL NAPHTHALENE	16.57	0.098	0.17822
2-PHENYL NAPHTHALENE	24.57	0.030	1.88785
1,4-DIPHENYL NAPHTHALENE	24.66	0.030	1.92304
1,5-DIPHENYL NAPHTHALENE	23.09	0.036	1.29886
1,7-DIPHENYL NAPHTHALENE	22.83	0.038	1.21233
1,4,5,8-TETRAPHENYL NAPHTHALENE	35.70	0.010	17.48707
1,1'-BINAPHTHYL	18.91	0.066	0.39314
4,4'-DI(N-BUTOXY)-1,1'-BINAPHTHYL	29.31	0.018	5.41358
2,2'-BINAPHTHYL	25.27	0.028	2.23379
ACRIDINE (A)	31.28	0.015	7.90763
ANTHRACENE (TRACE)	29.38	0.018	5.44274
DEUTERATED ANTHRACENE C14D10 (TRACE)	29.07	0.018	5.11013
9-METHYL ANTHRACENE (TRACE)	29.45	0.018	5.51231
1-AMINO ANTHRACENE	28.07	0.020	4.09330
2-AMINO ANTHRACENE	27.92	0.021	3.95662
9-VINYL ANTHRACENE	30.51	0.016	6.80018
9-PHENYL ANTHRACENE (TRACE)	30.96	0.015	7.42731
9,10-DIPHENYL ANTHRACENE	31.76	0.014	8.63512
9,10-DICHLOROANTHRACENE (TRACE)	30.42	0.016	6.66631
BIANTHRANYL (B)	33.51	0.012	11.92161
9,10-DINAPHTHYLANTHRACENE (B)	31.33	0.015	7.94719
9-(P-VINYLPHENYL)-10-PHENYL ANTHRACENE	31.93	0.014	8.89930
9,10-BIS(PHENYL-ETHYNYL)ANTHRACENE (TRACE)	30.92	0.015	7.29012
RUBICENE (B)	32.15	0.013	9.22890
DIBENZ(A,H)ANTHRACENE (B)	33.25	0.012	11.38779
TETRACENE (TRACE) (B)	23.85	0.033	1.53720
RUBRENE (B)	30.88	0.015	7.23338
TETRAHYDROCANNABINOL ACID METHYL ESTER	13.66	0.175	0.05616
PHENANTHRENE	15.37	0.123	0.11256
3,4-BENZOPHENANTHRENE	24.31	0.031	1.74782
2-PHENYL PHENANTHRENE	19.71	0.058	0.49894
CHRYSENE (TRACE)	24.79	0.029	1.97129
TRIPHENYLENE	16.35	0.102	0.16290
DEUTERATED TRIPHENYLENE-C18D12	16.20	0.105	0.15427
PYRENE	32.93	0.013	10.81039
1,3,6,8,-TETRAPHENYL PYRENE	39.34	0.007	31.12359

ACCEPTOR MOLECULES	RO	CO	OVERLAP
BIPYRENYL (B)	41.13	0.006	40.78150
O-PHENYLENE-PYRENE	35.44	0.010	16.55599
ANTHANTHRENE (B)	33.47	0.012	11.76114
FLUORANTHENE	30.78	0.015	7.14949
3-PHENYL FLUORANTHENE	33.98	0.011	12.91986
3-AMINO FLUORANTHENE	31.68	0.014	8.43439
3,3'-BIFLUORANTHENYL (B)	38.22	0.008	26.13282
BENZO(GHI)FLUORANTHENE	29.50	0.017	5.53262
9,12-DI(N-OCTADECOXY)-BENZ(E)ACEPHENANTHRYLENE	33.82	0.012	12.54926
DECACYCLENE (B)	41.48	0.006	42.50185
10-ETHOXY BENZO(KL)XANTHENE	32.40	0.013	9.74320
PERYLENE	30.77	0.015	7.10228
BENZO(GHI)PERYLENE (B)	34.49	0.011	14.17381
BBOT (TRACE)	43.45	0.005	56.68442
QUININE SULFATE (A)	28.94	0.018	5.02065
ACRIDINE YELLOW (A)	27.37	0.022	3.50668
ACRIDINE ORANGE (A)	23.96	0.033	1.57795
ACRIDINE RED (A)	26.86	0.023	3.13226
URANINE (A)	27.10	0.022	3.30754
RHODAMINE B (A)	30.53	0.016	6.76413
RHODAMINE 6G (A)	30.59	0.016	6.83748
EOSIN B (A)	28.25	0.020	4.24697
EOSIN Y (A)	28.89	0.019	4.85790
ROSE BENGAL (A)	29.89	0.017	5.95166
INDENOINDENE	30.63	0.016	7.06287

8.122 Donor Molecule: 2,5-Di(2-methylphenyl) oxadiazole

ACCEPTOR MOLECULES	RO	CO	OVERLAP
ANILINE 3ML/L	4.44	5.120	0.00007
N,N-DIMETHYLANILINE 3ML/L	12.29	0.241	0.02962
TRIPHENYLAMINE	24.91	0.029	2.03765
1,4-DIMETHOXY BENZENE	5.28	3.036	0.00019
1-METHYL-4-CHLOROBENZENE 5ML/L	0.00	0.000	0.00000
SODIUM SALICYLATE (A)	16.89	0.093	0.19777
ANTHRANILIC ACID (A)	27.70	0.021	3.82856
TMPD	23.47	0.035	1.41946
4,4'-DIHYDROXYBIPHENYL (A)	11.36	0.304	0.01871
4,4'-DIMETHOXY BIPHENYL	9.16	0.581	0.00512
4,4'-DIISOBUTOXYBIPHENYL	7.58	1.028	0.00164
BENZIDINE	22.01	0.042	0.97168
4-VINYL BIPHENYL	12.38	0.236	0.03106
M-TERPHENYL	8.50	0.729	0.00325
1,3,5-TRIPHENYL BENZENE	10.16	0.427	0.00945
1,3-DIBIPHENYLYLBENZENE (B)	21.28	0.046	0.79567
1,3,5,-TRIBIPHENYLYLBENZENE (B)	25.03	0.029	2.10940
FLUORENE (TRACE)	5.33	2.960	0.00020
1-METHYL FLUORENE	5.60	2.552	0.00027
1,2-BENZOFLUORENE	23.42	0.035	1.40914
9,10-DIHYDROPHENANTHRENE	7.56	1.036	0.00161
CARBAZOLE (TRACE)	21.87	0.043	0.93754
N-METHYL CARBAZOLE (TRACE)	24.56	0.030	1.87129
N-VINYLCARBAZOLE	24.47	0.031	1.83436
N-PHENYL CARBAZOLE	24.20	0.032	1.71574
N-METHYL-2,3-BENZCARBAZOLE	28.78	0.019	4.78689
DIBENZOFURAN	4.29	5.648	0.00005
DIBENZOTHIOPHENE (TRACE)	19.21	0.063	0.43051
P-TERPHENYL	14.37	0.151	0.07595
DEUTERATED P-TERPHENYL-C18D14	13.77	0.171	0.05891
4-METHYL TERPHENYL	16.17	0.106	0.15406
2-METHYL-P-TERPHENYL	7.19	1.202	0.00120
3,3''-DIMETHYL-P-TERPHENYL	15.85	0.112	0.13625
4-(3,3-DIMETHYLBUTOXY)-P-TERPHENYL	21.07	0.048	0.75013
4,4''-DIHEXAHYDROFARNESOXY-P-TERPHENYL	24.33	0.031	1.77988
2,2'-METHYLENE-P-TERPHENYL	23.37	0.035	1.39929
2,2'-ETHYLENE-P-TERPHENYL	24.07	0.032	1.66802
2,2'-METHYLENE-5',6''-METHYLENE-P-TERPHENYL (TRACE)	33.59	0.012	12.29134
3-PHENYL DIBENZOFURAN	23.50	0.034	1.44515
BENZO(1,2-B-4,5-B')BISBENZOFURAN (TRACE) (B)	35.02	0.010	15.75374
SEXIPHENYL BENZENE	0.00	0.000	0.00000
P-QUATERPHENYL	25.64	0.027	2.43165
3,3'''-DIMETHYL-P-QUATERPHENYL	27.07	0.023	3.36707
TETRAMETHYL-P-QUATERPHENYL	9.35	0.548	0.00577
3,3''' -DIMETHYL-3',2''-METHYLENE-P-QUATERPHENYL	34.88	0.011	15.37359
3,3''' -DIMETHYL-3',2''-ETHYLENE-P-QUATERPHENYL	34.53	0.011	14.43875
4-(3-ETHYLBUTOXY)-P-QUATERPHENYL (B)	31.91	0.014	9.00039
4,4'''-DI(2-BUTYLOCTOXY-1)-P-QUATERPHENYL	32.66	0.013	10.34164
DIETHYL-P-QUINQUEPHENYL	28.24	0.020	4.33484
DI(3-ETHYLHEPTYL)-P-QUINQUEPHENYL	34.07	0.011	13.30467
TETRAMETHYL-P-SEXIPHENYL	27.09	0.022	3.38842
TETRAMETHYL-P-OCTAPHENYL (B)	38.44	0.008	27.42330
INDOLE (TRACE)	0.00	0.000	0.00000
2-PHENYL INDOLE	28.04	0.020	4.16173
1-METHYL-2-PHENYL INDOLE	22.79	0.038	1.20019
1-NEOPENTYL-2-PHENYL INDOLE	20.43	0.052	0.62402
1-N-BUTYL-2(BIPHENYLYL-4) INDOLE	31.49	0.014	8.30369
1,2-DIPHENYL INDOLE	23.70	0.034	1.51693
1-BIPHENYLYL-2-PHENYL INDOLE	26.26	0.025	2.79459
1,3-DIMETHYL-2-PHENYL INDOLE	23.62	0.034	1.48574
2,3-DIPHENYL INDOLE	29.14	0.018	5.20718
1-METHYL-2,3-DIPHENYL INDOLE	26.47	0.024	2.92337
2-PHENYL-3,2'-METHYLENE INDOLE (TRACE)	31.01	0.015	7.60581

ACCEPTOR MOLECULES	RO	CO	OVERLAP
2-PHENYL-3,2'-ETHYLENE INDOLE (TRACE)	33.46	0.012	11.95170
2-PHENYL-3,2'-TRIMETHYLENE INDOLE	29.78	0.017	5.95779
1-METHYL-3,2'-METHYLENE-2-PHENYL INDOLE (TRACE)	32.97	0.012	10.95633
1-METHYL-3,2'-ETHYLENE-2-PHENYL INDOLE	34.02	0.011	13.17975
1-METHYL-3,2'-TRIMETHYLENE-2-PHENYL INDOLE	28.51	0.019	4.58660
4,5-DIPHENYLIMIDAZOLE (A)	18.20	0.074	0.31101
PPF (TRACE)	34.57	0.011	14.55648
PPO (TRACE)	30.13	0.016	6.40683
2-(4-FLUOROPHENYL)-5-PHENYLOXAZOLE (TRACE)	28.70	0.019	4.78370
αNPO (TRACE)	36.61	0.009	20.39498
BBO (B)	42.29	0.006	48.37073
POPOP	41.68	0.006	44.30463
DIMETHYL POPOP (TRACE)	43.00	0.006	53.26126
PPD (TRACE)	14.59	0.144	0.08316
2,5-DI(2-METHYLPHENYL)OXADIAZOLE	16.33	0.103	0.16357
PBD	28.99	0.018	5.07853
BBD (B)	36.79	0.009	21.12688
αNPD (TRACE)	31.82	0.014	8.82622
βNPD	27.76	0.021	3.91019
αNND (TRACE)	34.41	0.011	14.09765
βNND (TRACE)	33.43	0.012	11.91412
TRANS-1-(4-BIPHENYLYL)-2-PHENYL-ETHYLENE	36.43	0.009	19.88140
TRANS-1-(4-BIPHENYLYL)-2-(1-NAPHTHYL)-ETHYLENE	38.93	0.008	29.40063
DIPHENYLSTILBENE (B)	42.67	0.006	51.02590
1,1,4,4-TETRAPHENYLBUTADIENE	40.24	0.007	35.78787
1,6-DIPHENYLHEXATRIENE	44.81	0.005	68.27714
1,8-DIPHENYL-1,3,5,7-OCTATETRAENE	45.57	0.005	75.07262
BPSB (TRACE)	45.13	0.005	71.33478
BIS-MSB	43.61	0.005	58.04833
ESCULIN (A)	33.00	0.012	10.92973
AZULENE	24.29	0.031	1.71263
NAPHTHALENE	8.09	0.846	0.00241
DEUTERATED NAPHTHALENE-C10D8	10.47	0.390	0.01129
1-METHYLNAPHTHALENE 5ML/L	10.01	0.446	0.00867
2-METHYL NAPHTHALENE	11.81	0.272	0.02333
2,3-DIMETHYL NAPHTHALENE	11.65	0.283	0.02150
2,6-DIMETHYL NAPHTHALENE (TRACE)	14.97	0.133	0.09656
1-NAPHTHOL (TRACE)	16.63	0.097	0.18157
2-NAPHTHOL (TRACE)	19.16	0.064	0.42419
1-NAPHTHYLAMINE	24.54	0.030	1.86376
2-CHLORONAPHTHALENE	12.05	0.255	0.02637
ACENAPHTHENE	14.60	0.144	0.08326
1-PHENYL NAPHTHALENE	15.27	0.126	0.10891
2-PHENYL NAPHTHALENE	22.86	0.037	1.22249
1,4-DIPHENYL NAPHTHALENE	23.49	0.035	1.43672
1,5-DIPHENYL NAPHTHALENE	21.86	0.043	0.93517
1,7-DIPHENYL NAPHTHALENE	21.63	0.044	0.87562
1,4,5,8-TETRAPHENYL NAPHTHALENE	35.38	0.010	16.55676
1,1'-BINAPHTHYL	17.70	0.081	0.26423
4,4'-DI(N-BUTOXY)-1,1'-BINAPHTHYL	28.37	0.020	4.44941
2,2'-BINAPHTHYL	23.89	0.033	1.59350
ACRIDINE (A)	31.33	0.015	7.97612
ANTHRACENE (TRACE)	29.49	0.017	5.55639
DEUTERATED ANTHRACENE C14D10 (TRACE)	29.19	0.018	5.23403
9-METHYL ANTHRACENE (TRACE)	29.66	0.017	5.74712
1-AMINO ANTHRACENE	28.56	0.019	4.53470
2-AMINO ANTHRACENE	28.24	0.020	4.23787
9-VINYL ANTHRACENE	30.83	0.015	7.22755
9-PHENYL ANTHRACENE (TRACE)	31.16	0.015	7.71543
9,10-DIPHENYL ANTHRACENE	32.27	0.013	9.50099
9,10-DICHLOROANTHRACENE (TRACE)	30.93	0.015	7.36165
BIANTHRANYL (B)	34.00	0.011	13.00123
9,10-DINAPHTHYLANTHRACENE (B)	31.78	0.014	8.65034
9-(P-VINYLPHENYL)-10-PHENYL ANTHRACENE	32.45	0.013	9.80120
9,10-BIS(PHENYL-ETHYNYL)ANTHRACENE (TRACE)	31.94	0.014	8.84152
RUBICENE (B)	32.22	0.013	9.32536
DIBENZ(A,H)ANTHRACENE (B)	32.71	0.013	10.32350
TETRACENE (TRACE) (B)	24.58	0.030	1.83818
RUBRENE (B)	29.61	0.017	5.61267
TETRAHYDROCANNABINOL ACID METHYL ESTER	12.16	0.249	0.02793
PHENANTHRENE	15.00	0.132	0.09731
3,4-BENZOPHENANTHRENE	23.25	0.036	1.33521
2-PHENYL PHENANTHRENE	18.42	0.072	0.33274
CHRYSENE (TRACE)	23.92	0.033	1.58932
TRIPHENYLENE	15.87	0.112	0.13646
DEUTERATED TRIPHENYLENE-C18D12	15.71	0.115	0.12818
PYRENE	32.60	0.013	10.16820
1,3,6,8,-TETRAPHENYL PYRENE	39.89	0.007	33.80644
BIPYRENYL (B)	41.08	0.006	40.46186
O-PHENYLENE-PYRENE	35.60	0.010	16.96554
ANTHANTHRENE (B)	33.77	0.012	12.39777
FLUORANTHENE	30.62	0.016	6.92516
3-PHENYL FLUORANTHENE	34.12	0.011	13.24923
3-AMINO FLUORANTHENE	31.89	0.014	8.75508
3,3'-BIFLUORANTHENYL (B)	38.53	0.008	27.42655
BENZO(GHI)FLUORANTHENE	29.31	0.018	5.31819
9,12-DI(N-OCTADECOXY)-BENZ(E)ACEPHENANTHRYLENE	34.13	0.011	13.24662
DECACYCLENE (B)	41.89	0.006	44.99869
10-ETHOXY-BENZO(KL)XANTHENE	32.87	0.013	10.61746
PERYLENE	31.97	0.014	8.92553
BENZO(GHI)PERYLENE (B)	34.79	0.011	14.90771
BBOT (TRACE)	43.84	0.005	59.79005
QUININE SULFATE (A)	28.36	0.020	4.44079
ACRIDINE YELLOW (A)	29.11	0.018	5.07275
ACRIDINE ORANGE (A)	24.96	0.029	2.01513
ACRIDINE RED (A)	26.88	0.023	3.14001
URANINE (A)	27.50	0.022	3.60381

ACCEPTOR MOLECULES	RO	CO	OVERLAP
RHODAMINE B (A)	30.62	0.016	6.86170
RHODAMINE 6G (A)	30.62	0.016	6.86051
EOSIN B (A)	28.01	0.020	4.01821
EOSIN Y (A)	28.70	0.019	4.65855
ROSE BENGAL (A)	29.64	0.017	5.64490
INDENOINDENE	29.55	0.017	5.68864

8.123 Donor Molecule: PBD

N,N-DIMETHYLANILINE 3ML/L	3.29	12.605	0.00001
TRIPHENYLAMINE	18.57	0.070	0.37338
SODIUM SALICYLATE (A)	13.29	0.191	0.04996
ANTHRANILIC ACID (A)	25.70	0.026	2.60423
TMPD	20.76	0.050	0.72407
BENZIDINE	14.81	0.138	0.09633
1,3,5-TRIPHENYL BENZENE	3.81	8.057	0.00003
1,3-DIBIPHENYLYLBENZENE (B)	12.72	0.217	0.03865
1,3,5,-TRIBIPHENYLYLBENZENE (B)	14.58	0.144	0.08763
FLUORENE (TRACE)	0.00	0.000	0.00000
1-METHYL FLUORENE	0.00	0.000	0.00000
1,2-BENZOFLUORENE	17.63	0.082	0.27254
9,10-DIHYDROPHENANTHRENE	0.00	0.000	0.00000
CARBAZOLE (TRACE)	15.62	0.117	0.13219
N-METHYL CARBAZOLE (TRACE)	21.29	0.046	0.84767
N-VINYLCARBAZOLE	20.98	0.048	0.77593
N-PHENYL CARBAZOLE	20.72	0.050	0.72005
N-METHYL-2,3-BENZCARBAZOLE	27.92	0.021	4.25062
DIBENZOFURAN	0.00	0.000	0.00000
DIBENZOTHIOPHENE (TRACE)	11.49	0.295	0.02102
P-TERPHENYL	0.00	0.000	0.00000
DEUTERATED P-TERPHENYL-C18D14	0.00	0.000	0.00000
4-METHYL TERPHENYL	0.00	0.000	0.00000
2-METHYL-P-TERPHENYL	0.00	0.000	0.00000
3,3″-DIMETHYL-P-TERPHENYL	5.31	2.986	0.00021
4-(3,3-DIMETHYLBUTOXY)-P-TERPHENYL	11.49	0.295	0.02103
4,4″-DIHEXAHYDROFARNESOXY-P-TERPHENYL	14.72	0.140	0.09287
2,2′-METHYLENE-P-TERPHENYL	10.93	0.342	0.01563
2,2′-ETHYLENE-P-TERPHENYL	13.20	0.195	0.04830
2,2′-METHYLENE-5′,6″-METHYLENE-P-TERPHENYL (TRACE)	26.75	0.023	3.34165
3-PHENYL DIBENZOFURAN	10.85	0.350	0.01495
BENZO(1,2-B-4,5-B′)BISBENZOFURAN (TRACE) (B)	28.57	0.019	4.94268
SEXIPHENYL BENZENE	0.00	0.000	0.00000
P-QUATERPHENYL	17.04	0.090	0.22307
3,3‴-DIMETHYL-P-QUATERPHENYL	18.92	0.066	0.41812
TETRAMETHYL-P-QUATERPHENYL	0.00	0.000	0.00000
3,3‴-DIMETHYL-3′,2″-METHYLENE-P-QUATERPHENYL	27.82	0.021	4.21785
3,3‴-DIMETHYL-3′,2″-ETHYLENE-P-QUATERPHENYL	28.03	0.020	4.39790
4-(3-ETHYLBUTOXY)-P-QUATERPHENYL (B)	25.06	0.028	2.24989
4,4‴-DI(2-BUTYLOCTOXY-1)-P-QUATERPHENYL	25.43	0.027	2.45521
DIETHYL-P-QUINQUEPHENYL	19.67	0.059	0.52745
DI(3-ETHYLHEPTYL)-P-QUINQUEPHENYL	26.99	0.023	3.50402
TETRAMETHYL-P-SEXIPHENYL	17.78	0.080	0.28800
TETRAMETHYL-P-OCTAPHENYL (B)	32.36	0.013	10.40517
INDOLE (TRACE)	0.00	0.000	0.00000
2-PHENYL INDOLE	19.15	0.064	0.44986
1-METHYL-2-PHENYL INDOLE	14.68	0.142	0.09106
1-NEOPENTYL-2-PHENYL INDOLE	10.80	0.355	0.01452
1-N-BUTYL-2(BIPHENYLYL-4) INDOLE	26.31	0.025	3.00571
1,2-DIPHENYL INDOLE	15.06	0.131	0.10631
1-BIPHENYLYL-2-PHENYL INDOLE	18.52	0.070	0.36658
1,3-DIMETHYL-2-PHENYL INDOLE	16.58	0.098	0.18934
2,3-DIPHENYL INDOLE	24.21	0.032	1.82306
1-METHYL-2,3-DIPHENYL INDOLE	21.76	0.043	0.96197
2-PHENYL-3,2′-METHYLENE INDOLE (TRACE)	24.07	0.032	1.77189
2-PHENYL-3,2′-ETHYLENE INDOLE (TRACE)	28.60	0.019	4.96933
2-PHENYL-3,2′-TRIMETHYLENE INDOLE	22.14	0.041	1.07282
1-METHYL-3,2′-METHYLENE-2-PHENYL INDOLE (TRACE)	27.61	0.021	4.02635
1-METHYL-3,2′-ETHYLENE-2-PHENYL INDOLE	30.51	0.016	7.30853
1-METHYL-3,2′-TRIMETHYLENE-2-PHENYL INDOLE	22.38	0.040	1.14264
4,5-DIPHENYLIMIDAZOLE (A)	12.68	0.219	0.03800
PPF (TRACE)	28.95	0.018	5.35283
PPO (TRACE)	22.23	0.041	1.10096
2-(4-FLUOROPHENYL)-5-PHENYLOXAZOLE (TRACE)	20.77	0.050	0.73194
αNPO (TRACE)	34.10	0.011	14.19938
BBO (B)	40.50	0.007	39.76964
POPOP	41.29	0.006	44.62802
DIMETHYL POPOP (TRACE)	43.13	0.006	57.79386
PPD (TRACE)	0.00	0.000	0.00000
2,5-DI(2-METHYLPHENYL)OXADIAZOLE	0.00	0.000	0.00000
PBD	19.41	0.061	0.48740
BBD (B)	30.81	0.015	7.77084
αNPD (TRACE)	27.17	0.022	3.64255
βNPD	19.61	0.059	0.51734
αNND (TRACE)	31.34	0.015	8.57987
βNND (TRACE)	27.41	0.022	3.85418
TRANS-1-(4-BIPHENYLYL)-2-PHENYL-ETHYLENE	31.20	0.015	8.34981
TRANS-1-(4-BIPHENYLYL)-2-(1-NAPHTHYL)-ETHYLENE	36.71	0.009	22.06228
DIPHENYLSTILBENE (B)	40.76	0.007	41.31355
1,1,4,4-TETRAPHENYLBUTADIENE	38.99	0.008	31.55933
1,6-DIPHENYLHEXATRIENE	44.24	0.005	67.47064
1,8-DIPHENYL-1,3,5,7-OCTATETRAENE	47.13	0.004	97.98645

ACCEPTOR MOLECULES	RO	CO	OVERLAP
BPSB (TRACE)	44.77	0.005	72.41425
BIS-MSB	42.63	0.006	53.95826
ESCULIN (A)	30.97	0.015	7.95730
AZULENE	21.53	0.045	0.88316
NAPHTHALENE	0.00	0.000	0.00000
DEUTERATED NAPHTHALENE-C10D8	6.20	1.881	0.00052
1-METHYLNAPHTHALENE 5ML/L	3.38	11.630	0.00001
2-METHYL NAPHTHALENE	3.93	7.346	0.00003
2,3-DIMETHYL NAPHTHALENE	4.00	6.974	0.00004
2,6-DIMETHYL NAPHTHALENE (TRACE)	7.50	1.059	0.00163
1-NAPHTHOL (TRACE)	7.32	1.139	0.00141
2-NAPHTHOL (TRACE)	12.85	0.211	0.04104
1-NAPHTHYLAMINE	19.79	0.058	0.54637
2-CHLORONAPHTHALENE	5.05	3.473	0.00015
ACENAPHTHENE	4.89	3.837	0.00012
1-PHENYL NAPHTHALENE	6.84	1.397	0.00094
2-PHENYL NAPHTHALENE	14.60	0.144	0.08842
1,4-DIPHENYL NAPHTHALENE	16.06	0.108	0.15624
1,5-DIPHENYL NAPHTHALENE	13.79	0.170	0.06286
1,7-DIPHENYL NAPHTHALENE	14.16	0.157	0.07359
1,4,5,8-TETRAPHENYL NAPHTHALENE	33.23	0.012	12.10303
1,1'-BINAPHTHYL	9.68	0.492	0.00755
4,4'-DI(N-BUTOXY)-1,1'-BINAPHTHYL	22.21	0.041	1.09190
2,2'-BINAPHTHYL	15.53	0.119	0.12804
ACRIDINE (A)	31.04	0.015	8.04948
ANTHRACENE (TRACE)	29.56	0.017	6.01614
DEUTERATED ANTHRACENE C14D10 (TRACE)	29.10	0.018	5.47584
9-METHYL ANTHRACENE (TRACE)	30.10	0.016	6.68417
1-AMINO ANTHRACENE	29.87	0.017	6.33105
2-AMINO ANTHRACENE	28.76	0.019	5.04665
9-VINYL ANTHRACENE	31.58	0.014	8.89815
9-PHENYL ANTHRACENE (TRACE)	31.66	0.014	9.05452
9,10-DIPHENYL ANTHRACENE	33.61	0.012	12.91386
9,10-DICHLOROANTHRACENE (TRACE)	32.52	0.013	10.59239
BIANTHRANYL (B)	35.18	0.010	17.00223
9,10-DINAPHTHYLANTHRACENE (B)	33.14	0.012	11.86895
9-(P-VINYLPHENYL)-10-PHENYL ANTHRACENE	33.87	0.012	13.50899
9,10-BIS(PHENYL-ETHYNYL)ANTHRACENE (TRACE)	34.56	0.011	15.12969
RUBICENE (B)	32.40	0.013	10.27702
DIBENZ(A,H)ANTHRACENE (B)	29.70	0.017	6.16079
TETRACENE (TRACE) (B)	26.08	0.025	2.79412
RUBRENE (B)	25.15	0.028	2.24495
TETRAHYDROCANNABINOL ACID METHYL ESTER	0.00	0.000	0.00000
PHENANTHRENE	12.76	0.215	0.03918
3,4-BENZOPHENANTHRENE	17.73	0.080	0.27975
2-PHENYL PHENANTHRENE	15.23	0.127	0.11328
CHRYSENE (TRACE)	17.93	0.078	0.30025
TRIPHENYLENE	12.94	0.206	0.04274
DEUTERATED TRIPHENYLENE-C18D12	12.50	0.229	0.03466
PYRENE	27.32	0.022	3.75005
1,3,6,8,-TETRAPHENYL PYRENE	41.48	0.006	45.59096
BIPYRENYL (B)	40.12	0.007	37.38423
O-PHENYLENE-PYRENE	36.20	0.009	19.97590
ANTHANTHRENE (B)	35.00	0.010	16.38983
FLUORANTHENE	29.62	0.017	6.05314
3-PHENYL FLUORANTHENE	34.30	0.011	14.56162
3-AMINO FLUORANTHENE	32.42	0.013	10.31308
3,3'-BIFLUORANTHENYL (B)	39.39	0.007	33.33567
BENZO(GHI)FLUORANTHENE	27.48	0.022	3.85723
9,12-DI(N-OCTADECOXY)-BENZ(E)ACEPHENANTHRYLENE	34.78	0.011	15.81811
DECACYCLENE (B)	42.86	0.006	54.99235
10-ETHOXY-BENZO(KL)XANTHENE	34.22	0.011	14.40027
PERYLENE	34.74	0.011	15.65985
BENZO(GHI)PERYLENE (B)	35.75	0.010	18.72891
BBOT (TRACE)	44.72	0.005	71.73134
QUININE SULFATE (A)	23.98	0.032	1.72814
ACRIDINE YELLOW (A)	31.88	0.014	9.31442
ACRIDINE ORANGE (A)	26.70	0.023	3.21563
ACRIDINE RED (A)	26.72	0.023	3.23220
URANINE (A)	27.81	0.021	4.10440
RHODAMINE B (A)	30.62	0.016	7.31495
RHODAMINE 6G (A)	29.91	0.017	6.35285
EOSIN B (A)	26.69	0.024	3.20905
EOSIN Y (A)	27.67	0.021	3.98544
ROSE BENGAL (A)	28.03	0.020	4.30262
INDENOINDENE	21.39	0.046	0.87218

8.125 Donor Molecule: α-NPD (trace)

ACCEPTOR MOLECULES	RO	CO	OVERLAP
TRIPHENYLAMINE	14.20	0.156	0.08816
P-XYLENE (TRACE)	0.00	0.000	0.00000
SODIUM SALICYLATE (A)	10.99	0.337	0.01887
ANTHRANILIC ACID (A)	22.96	0.037	1.56260
TMPD	17.78	0.080	0.33836
BENZIDINE	9.51	0.520	0.00797
1,3-DIBIPHENYLYLBENZENE (B)	6.58	1.571	0.00087
1,3,5,-TRIBIPHENYLYLBENZENE (B)	5.93	2.146	0.00047
FLUORENE (TRACE)	0.00	0.000	0.00000
1-METHYL FLUORENE	0.00	0.000	0.00000
1,2-BENZOFLUORENE	14.59	0.144	0.10347
9,10-DIHYDROPHENANTHRENE	0.00	0.000	0.00000

ACCEPTOR MOLECULES	RO	CO	OVERLAP
CARBAZOLE (TRACE)	8.57	0.712	0.00425
N-METHYL CARBAZOLE (TRACE)	17.42	0.085	0.30041
N-VINYLCARBAZOLE	14.86	0.136	0.11578
N-PHENYL CARBAZOLE	15.34	0.124	0.13985
N-METHYL-2,3-BENZCARBAZOLE	27.37	0.022	4.45613
DIBENZOFURAN	0.00	0.000	0.00000
DIBENZOTHIOPHENE (TRACE)	4.96	3.669	0.00016
P-TERPHENYL	0.00	0.000	0.00000
DEUTERATED P-TERPHENYL-C18D14	0.00	0.000	0.00000
4-METHYL TERPHENYL	0.00	0.000	0.00000
2-METHYL-P-TERPHENYL	0.00	0.000	0.00000
3,3'' DIMETHYL-P-TERPHENYL	0.00	0.000	0.00000
4-(3,3-DIMETHYLBUTOXY)-P-TERPHENYL	0.00	0.000	0.00000
4,4''-DIHEXAHYDROFARNESOXY-P-TERPHENYL	5.77	2.333	0.00040
2,2'-METHYLENE-P-TERPHENYL	0.00	0.000	0.00000
2,2'-ETHYLENE-P-TERPHENYL	4.91	3.784	0.00015
2,2'-METHYLENE-5',6''-METHYLENE-P-TERPHENYL (TRACE)	15.18	0.128	0.13153
3-PHENYL DIBENZOFURAN	0.00	0.000	0.00000
BENZO(1,2-B-4,5-B')BISBENZOFURAN (TRACE) (B)	20.78	0.050	0.86572
SEXIPHENYL BENZENE	0.00	0.000	0.00000
P-QUATERPHENYL	9.60	0.506	0.00843
3,3'''-DIMETHYL-P-QUATERPHENYL	12.94	0.207	0.05040
TETRAMETHYL-P-QUATERPHENYL	0.00	0.000	0.00000
3,3''' -DIMETHYL-3',2''-METHYLENE-P-QUATERPHENYL	20.63	0.051	0.82761
3,3''' -DIMETHYL-3',2''-ETHYLENE-P-QUATERPHENYL	21.99	0.042	1.21044
4-(3-ETHYLBUTOXY)-P-QUATERPHENYL (B)	19.47	0.061	0.58477
4,4'''-DI(2-BUTYLOCTOXY-1)-P-QUATERPHENYL	19.65	0.059	0.61819
DIETHYL-P-QUINQUEPHENYL	12.88	0.209	0.04913
DI(3-ETHYLHEPTYL)-P-QUINQUEPHENYL	21.38	0.046	1.02155
TETRAMETHYL-P-SEXIPHENYL	9.80	0.476	0.00953
TETRAMETHYL-P-OCTAPHENYL (B)	26.94	0.023	4.08456
INDOLE (TRACE)	0.00	0.000	0.00000
2-PHENYL INDOLE	9.16	0.582	0.00637
1-METHYL-2-PHENYL INDOLE	8.40	0.754	0.00379
1-NEOPENTYL-2-PHENYL INDOLE	0.00	0.000	0.00000
1-N-BUTYL-2(BIPHENYLYL-4) INDOLE	21.69	0.044	1.11382
1,2-DIPHENYL INDOLE	8.72	0.675	0.00473
1-BIPHENYLYL-2-PHENYL INDOLE	13.94	0.165	0.07865
1,3-DIMETHYL-2-PHENYL INDOLE	9.63	0.500	0.00860
2,3-DIPHENYL INDOLE	20.00	0.056	0.68471
1-METHYL-2,3-DIPHENYL INDOLE	17.75	0.080	0.33457
2-PHENYL-3,2'-METHYLENE INDOLE (TRACE)	14.89	0.135	0.11722
2-PHENYL-3,2'-ETHYLENE INDOLE (TRACE)	24.05	0.032	2.07527
2-PHENYL-3,2'-TRIMETHYLENE INDOLE	14.08	0.160	0.08370
1-METHYL-3,2'-METHYLENE-2-PHENYL INDOLE (TRACE)	22.50	0.039	1.39266
1-METHYL-3,2'-ETHYLENE-2-PHENYL INDOLE	26.12	0.025	3.40065
1-METHYL-3,2'-TRIMETHYLENE-2-PHENYL INDOLE	15.53	0.119	0.15085
4,5-DIPHENYLIMIDAZOLE (A)	7.75	0.962	0.00233
PPF (TRACE)	23.51	0.034	1.81227
PPO (TRACE)	12.54	0.227	0.04188
2-(4-FLUOROPHENYL)-5-PHENYLOXAZOLE (TRACE)	11.39	0.303	0.02348
αNPO (TRACE)	30.86	0.015	9.21600
BBO (B)	37.37	0.009	28.97950
POPOP	38.96	0.008	37.22556
DIMETHYL POPOP (TRACE)	41.43	0.006	53.66621
PPD (TRACE)	0.00	0.000	0.00000
2,5-DI(2-METHYLPHENYL)OXADIAZOLE	0.00	0.000	0.00000
PBD	9.90	0.461	0.01015
BBD (B)	25.18	0.028	2.73274
αNPD (TRACE)	22.51	0.039	1.39412
βNPD	13.60	0.178	0.06811
αNND (TRACE)	27.39	0.022	4.50975
βNND (TRACE)	21.05	0.048	0.93576
TRANS-1-(4-BIPHENYLYL)-2-PHENYL-ETHYLENE	26.04	0.025	3.33327
TRANS-1-(4-BIPHENYLYL)-2-(1-NAPHTHYL)-ETHYLENE	33.58	0.012	15.25579
DIPHENYLSTILBENE (B)	37.54	0.008	29.78687
1,1,4,4-TETRAPHENYLBUTADIENE	36.33	0.009	24.41138
1,6-DIPHENYLHEXATRIENE	41.44	0.006	53.81481
1,8-DIPHENYL-1,3,5,7-OCTATETRAENE	46.09	0.005	101.26877
BPSB (TRACE)	42.43	0.006	61.99923
BIS-MSB	40.05	0.007	43.85908
ESCULIN (A)	28.08	0.020	5.22636
AZULENE	17.19	0.088	0.27072
NAPHTHALENE	0.00	0.000	0.00000
DEUTERATED NAPHTHALENE-C10D8	2.45	30.417	0.00000
1-METHYLNAPHTHALENE 5ML/L	0.00	0.000	0.00000
2-METHYL NAPHTHALENE	0.00	0.000	0.00000
2,3-DIMETHYL NAPHTHALENE	0.00	0.000	0.00000
2,6-DIMETHYL NAPHTHALENE (TRACE)	0.00	0.000	0.00000
1-NAPHTHOL (TRACE)	0.00	0.000	0.00000
2-NAPHTHOL (TRACE)	4.98	3.619	0.00016
1-NAPHTHYLAMINE	14.92	0.135	0.11829
2-PHENYL NAPHTHALENE	6.11	1.965	0.00056
1,4-DIPHENYL NAPHTHALENE	10.72	0.363	0.01632
1,5-DIPHENYL NAPHTHALENE	7.33	1.138	0.00166
1,7-DIPHENYL NAPHTHALENE	8.39	0.757	0.00376
1,4,5,8-TETRAPHENYL NAPHTHALENE	30.24	0.016	8.12534
1,1'-BINAPHTHYL	0.00	0.000	0.00000
4,4'-DI(N-BUTOXY)-1,1'-BINAPHTHYL	17.04	0.090	0.26237
2,2'-BINAPHTHYL	8.76	0.666	0.00486
ACRIDINE (A)	29.16	0.018	6.53288
ANTHRACENE (TRACE)	27.51	0.021	4.61355
DEUTERATED ANTHRACENE C14D10 (TRACE)	27.07	0.023	4.19239
9-METHYL ANTHRACENE (TRACE)	29.67	0.017	7.25012
1-AMINO ANTHRACENE	29.73	0.017	7.27337
2-AMINO ANTHRACENE	28.30	0.020	5.40853
9-VINYL ANTHRACENE	30.97	0.015	9.36164

ACCEPTOR MOLECULES	RO	CO	OVERLAP
9-PHENYL ANTHRACENE (TRACE)	31.03	0.015	9.48386
9,10-DIPHENYL ANTHRACENE	32.95	0.013	13.56085
9,10-DICHLOROANTHRACENE (TRACE)	32.00	0.014	11.38102
BIANTHRANYL (B)	34.64	0.011	18.29522
9,10-DINAPHTHYLANTHRACENE (B)	32.50	0.013	12.47443
9-(P-VINYLPHENYL)-10-PHENYL ANTHRACENE	33.27	0.012	14.35878
9,10-BIS(PHENYL-ETHYNYL)ANTHRACENE (TRACE)	35.88	0.010	22.37289
RUBICENE (B)	32.14	0.013	11.56900
DIBENZ(A,H)ANTHRACENE (B)	25.84	0.026	3.16186
TETRACENE (TRACE) (B)	26.70	0.024	3.79998
RUBRENE (B)	24.30	0.031	2.15915
TETRAHYDROCANNABINOL ACID METHYL ESTER	0.00	0.000	0.00000
PHENANTHRENE	10.62	0.374	0.01540
3,4-BENZOPHENANTHRENE	15.91	0.111	0.17274
2-PHENYL PHENANTHRENE	13.26	0.192	0.05816
CHRYSENE (TRACE)	16.10	0.107	0.18580
TRIPHENYLENE	9.12	0.590	0.00617
DEUTERATED TRIPHENYLENE-C18D12	9.07	0.599	0.00599
PYRENE	18.08	0.076	0.37263
1,3,6,8,-TETRAPHENYL PYRENE	40.84	0.007	49.07109
BIPYRENYL (B)	37.82	0.008	31.00674
O-PHENYLENE-PYRENE	35.41	0.010	20.69744
ANTHANTHRENE (B)	35.36	0.010	20.59812
FLUORANTHENE	27.47	0.022	4.55243
3-PHENYL FLUORANTHENE	32.93	0.013	13.47209
3-AMINO FLUORANTHENE	32.15	0.013	11.57975
3,3'-BIFLUORANTHENYL (B)	38.54	0.008	34.60167
BENZO(GHI)FLUORANTHENE	25.21	0.028	2.71627
9,12-DI(N-OCTADECOXY)-BENZ(E)ACEPHENANTHRYLENE	34.18	0.011	16.82029
DECACYCLENE (B)	42.16	0.006	58.90672
10-ETHOXY-BENZO(KL)XANTHENE	33.61	0.012	15.27338
PERYLENE	35.79	0.010	22.14341
BENZO(GHI)PERYLENE (B)	35.42	0.010	20.93760
BBOT (TRACE)	43.40	0.005	70.89561
QUININE SULFATE (A)	18.45	0.071	0.42335
ACRIDINE YELLOW (A)	33.29	0.012	14.28748
ACRIDINE ORANGE (A)	27.76	0.021	4.80014
ACRIDINE RED (A)	25.83	0.026	3.11872
URANINE (A)	28.22	0.020	5.29563
RHODAMINE B (A)	29.70	0.017	7.20779
RHODAMINE 6G (A)	28.28	0.020	5.36474
EOSIN B (A)	25.88	0.026	3.15524
EOSIN Y (A)	27.10	0.022	4.15663
ROSE BENGAL (A)	26.75	0.023	3.84666
INDENOINDENE	12.32	0.239	0.03770

8.126 Donor Molecule: β-NPD

TRIPHENYLAMINE	15.54	0.119	0.48214
P-XYLENE (TRACE)	0.00	0.000	0.00000
SODIUM SALICYLATE (A)	11.14	0.324	0.06544
ANTHRANILIC ACID (A)	21.58	0.045	3.43571
TMPD	17.50	0.084	0.97891
BENZIDINE	12.15	0.249	0.11056
1,3,5-TRIPHENYL BENZENE	2.75	21.409	0.00002
1,3-DIBIPHENYLYLBENZENE (B)	10.12	0.432	0.03697
1,3,5,-TRIBIPHENYLYLBENZENE (B)	11.04	0.332	0.06242
FLUORENE (TRACE)	0.00	0.000	0.00000
1-METHYL FLUORENE	0.00	0.000	0.00000
1,2-BENZOFLUORENE	16.10	0.130	0.40682
9,10 DIHYDROPHENANTHRENE	0.00	0.000	0.00000
CARBAZOLE (TRACE)	12.20	0.246	0.11323
N-METHYL CARBAZOLE (TRACE)	18.44	0.071	1.34872
N-VINYLCARBAZOLE	18.17	0.075	1.23408
N-PHENYL CARBAZOLE	17.96	0.077	1.15145
N-METHYL-2,3-BENZCARBAZOLE	22.01	0.042	3.84387
DIBENZOFURAN	0.00	0.000	0.00000
DIBENZOTHIOPHENE (TRACE)	8.90	0.634	0.01713
P-TERPHENYL	0.00	0.000	0.00000
DEUTERATED P-TERPHENYL-C18D14	0.00	0.000	0.00000
4-METHYL TERPHENYL	0.00	0.000	0.00000
2-METHYL-P-TERPHENYL	0.00	0.000	0.00000
3,3''-DIMETHYL-P-TERPHENYL	0.00	0.000	0.00000
4-(3,3-DIMETHYLBUTOXY)-P-TERPHENYL	8.68	0.683	0.01477
4,4''-DIHEXAHYDROFARNESOXY-P-TERPHENYL	11.14	0.324	0.06567
2,2'-METHYLENE-P-TERPHENYL	8.34	0.771	0.01160
2,2'-ETHYLENE-P-TERPHENYL	10.03	0.443	0.03506
2,2'-METHYLENE-5',6''-METHYLENE-P-TERPHENYL (TRACE)	21.01	0.048	2.95039
3-PHENYL DIBENZOFURAN	8.28	0.787	0.01115
BENZO(1,2-B-4,5-B')BISBENZOFURAN (TRACE) (B)	24.61	0.030	7.61175
SEXIPHENYL BENZENE	0.00	0.000	0.00000
P-QUATERPHENYL	13.74	0.172	0.23187
3,3'''-DIMETHYL-P-QUATERPHENYL	15.53	0.119	0.48260
TETRAMETHYL-P-QUATERPHENYL	0.00	0.000	0.00000
3,3''' -DIMETHYL-3',2''-METHYLENE-P-QUATERPHENYL	23.26	0.036	5.42311
3,3''' -DIMETHYL-3',2''-ETHYLENE-P-QUATERPHENYL	23.54	0.034	5.81716
4-(3-ETHYLBUTOXY)-P-QUATERPHENYL (B)	20.97	0.048	2.91329
4,4'''-DI(2-BUTYLOCTOXY-1)-P-QUATERPHENYL	21.25	0.047	3.15001
DIETHYL-P-QUINQUEPHENYL	16.12	0.107	0.60232
DI(3-ETHYLHEPTYL)-P-QUINQUEPHENYL	22.59	0.039	4.54017
TETRAMETHYL-P-SEXIPHENYL	14.28	0.153	0.29213

ACCEPTOR MOLECULES	RO	CO	OVERLAP
TETRAMETHYL-P-OCTAPHENYL (B)	27.21	0.022	13.85799
INDOLE (TRACE)	0.00	0.000	0.00000
2-PHENYL INDOLE	14.82	0.137	0.36501
1-METHYL-2-PHENYL INDOLE	11.73	0.277	0.08947
1-NEOPENTYL-2-PHENYL INDOLE	8.16	0.822	0.01020
1-N-BUTYL-2(BIPHENYLYL-4) INDOLE	22.13	0.041	4.01674
1,2-DIPHENYL INDOLE	12.05	0.255	0.10544
1-BIPHENYLYL-2-PHENYL INDOLE	15.24	0.126	0.42838
1,3-DIMETHYL-2-PHENYL INDOLE	13.38	0.187	0.19758
2,3-DIPHENYL INDOLE	20.33	0.053	2.40816
1-METHYL-2,3 DIPHENYL INDOLE	18.28	0.073	1.27506
2-PHENYL-3,2'-METHYLENE INDOLE (TRACE)	19.38	0.062	1.80713
2-PHENYL-3,2'-ETHYLENE INDOLE (TRACE)	24.26	0.031	6.96811
2-PHENYL-3,2'-TRIMETHYLENE INDOLE	17.93	0.078	1.14123
1-METHYL-3,2'-METHYLENE-2-PHENYL INDOLE (TRACE)	23.50	0.034	5.77556
1-METHYL-3,2'-ETHYLENE-2-PHENYL INDOLE	25.66	0.026	9.74816
1-METHYL-3,2'-TRIMETHYLENE-2-PHENYL INDOLE	18.66	0.069	1.44955
4,5-DIPHENYLIMIDAZOLE (A)	10.55	0.381	0.04730
PPF (TRACE)	24.70	0.030	7.77536
PPO (TRACE)	17.69	0.081	1.05237
2-(4-FLUOROPHENYL)-5-PHENYLOXAZOLE (TRACE)	16.37	0.102	0.66233
αNPO (TRACE)	28.60	0.019	18.63358
BBO (B)	33.69	0.012	49.64133
POPOP	34.06	0.011	52.99761
DIMETHYL POPOP (TRACE)	35.22	0.010	64.61876
PPD (TRACE)	0.00	0.000	0.00000
2,5-DI(2-METHYLPHENYL)OXADIAZOLE	0.00	0.000	0.00000
PBD	15.28	0.125	0.43767
BBD (B)	25.97	0.026	10.50955
αNPD (TRACE)	23.11	0.036	5.20767
βNPD	16.49	0.100	0.69036
αNND (TRACE)	26.48	0.024	11.75847
βNND (TRACE)	23.52	0.034	5.80238
TRANS-1-(4-BIPHENYLYL)-2-PHENYL-ETHYLENE	26.33	0.025	11.38035
TRANS-1-(4-BIPHENYLYL)-2-(1-NAPHTHYL)-ETHYLENE	30.59	0.016	27.85295
DIPHENYLSTILBENE (B)	33.95	0.011	52.02780
1,1,4,4-TETRAPHENYLBUTADIENE	32.31	0.013	38.58403
1,6-DIPHENYLHEXATRIENE	36.42	0.009	79.18510
1,8-DIPHENYL-1,3,5,7-OCTATETRAENE	37.68	0.008	96.43651
BPSB (TRACE)	36.84	0.009	84.78662
BIS-MSB	35.20	0.010	64.53310
ESCULIN (A)	25.90	0.026	10.26101
AZULENE	18.41	0.072	1.31105
NAPHTHALENE	0.00	0.000	0.00000
DEUTERATED NAPHTHALENE-C10D8	4.68	4.351	0.00036
1-METHYLNAPHTHALENE 5ML/L	2.37	33.642	0.00001
2-METHYL NAPHTHALENE	2.77	21.120	0.00002
2,3-DIMETHYL NAPHTHALENE	2.84	19.432	0.00002
2,6-DIMETHYL NAPHTHALENE (TRACE)	5.41	2.822	0.00087
1-NAPHTHOL (TRACE)	5.47	2.736	0.00092
2-NAPHTHOL (TRACE)	9.89	0.462	0.03226
1-NAPHTHYLAMINE	16.51	0.099	0.69417
2-CHLORONAPHTHALENE	3.81	8.072	0.00011
ACENAPHTHENE	3.37	11.642	0.00005
1-PHENYL NAPHTHALENE	5.30	3.010	0.00076
2-PHENYL NAPHTHALENE	11.03	0.333	0.06218
1,4-DIPHENYL NAPHTHALENE	13.08	0.200	0.17239
1,5-DIPHENYL NAPHTHALENE	10.96	0.340	0.05967
1,7-DIPHENYL NAPHTHALENE	11.42	0.301	0.07614
1,4,5,8-TETRAPHENYL NAPHTHALENE	27.74	0.021	15.45887
1,1'-BINAPHTHYL	7.37	1.117	0.00553
4,4'-DI(N-BUTOXY)-1,1'-BINAPHTHYL	18.46	0.071	1.35500
2,2'-BINAPHTHYL	12.53	0.227	0.13303
ACRIDINE (A)	25.68	0.026	9.71657
ANTHRACENE (TRACE)	24.39	0.031	7.15448
DEUTERATED ANTHRACENE C14D10 (TRACE)	23.88	0.033	6.30457
9-METHYL ANTHRACENE (TRACE)	24.34	0.031	7.04795
1-AMINO ANTHRACENE	23.52	0.034	5.69577
2-AMINO ANTHRACENE	22.62	0.039	4.50615
9-VINYL ANTHRACENE	25.38	0.027	9.04214
9-PHENYL ANTHRACENE (TRACE)	25.70	0.026	9.76542
9,10-DIPHENYL ANTHRACENE	26.71	0.023	12.27969
9,10-DICHLOROANTHRACENE (TRACE)	25.89	0.026	10.16841
BIANTHRANYL (B)	27.88	0.021	15.85947
9,10-DINAPHTHYLANTHRACENE (B)	26.36	0.024	11.33760
9-(P-VINYLPHENYL)-10-PHENYL ANTHRACENE	26.90	0.023	12.77752
9,10-BIS(PHENYL-ETHYNYL)ANTHRACENE (TRACE)	24.84	0.029	7.90475
RUBICENE (B)	25.41	0.027	9.06553
DIBENZ(A,H)ANTHRACENE (B)	24.69	0.030	7.67683
TETRACENE (TRACE) (B)	19.32	0.062	1.75061
RUBRENE (B)	19.70	0.058	1.96987
TETRAHYDROCANNABINOL ACID METHYL ESTER	0.00	0.000	0.00000
PHENANTHRENE	10.84	0.351	0.05561
3,4-BENZOPHENANTHRENE	14.57	0.145	0.32528
2-PHENYL PHENANTHRENE	12.84	0.211	0.15316
CHRYSENE (TRACE)	15.04	0.131	0.39510
TRIPHENYLENE	10.73	0.362	0.05245
DEUTERATED TRIPHENYLENE-C18D12	10.40	0.398	0.04327
PYRENE	21.80	0.043	3.65167
1,3,6,8,-TETRAPHENYL PYRENE	33.20	0.012	45.13760
BIPYRENYL (B)	33.00	0.012	43.70641
O-PHENYLENE-PYRENE	29.05	0.018	20.23981
ANTHANTHRENE (B)	26.75	0.023	12.32868
FLUORANTHENE	24.78	0.029	7.82598
3-PHENYL FLUORANTHENE	27.96	0.020	16.11000
3-AMINO FLUORANTHENE	25.50	0.027	9.25280

ACCEPTOR MOLECULES	RO	CO	OVERLAP
3,3'-BIFLUORANTHENYL (B)	31.57	0.014	33.33996
BENZO(GHI)FLUORANTHENE	22.67	0.038	4.58103
9,12-DI(N-OCTADECOXY)-BENZ(E)ACEPHENANTHRYLENE	27.64	0.021	15.00657
DECACYCLENE (B)	34.07	0.011	52.62534
10-ETHOXY-BENZO(KL)XANTHENE	27.28	0.022	13.95185
PERYLENE	25.81	0.026	9.94708
BENZO(GHI)PERYLENE (B)	28.68	0.019	18.80769
BBOT (TRACE)	36.09	0.010	74.74373
QUININE SULFATE (A)	20.29	0.054	2.39184
ACRIDINE YELLOW (A)	22.39	0.040	4.23567
ACRIDINE ORANGE (A)	19.05	0.065	1.61074
ACRIDINE RED (A)	21.54	0.045	3.35778
URANINE (A)	20.89	0.049	2.79424
RHODAMINE B (A)	24.55	0.030	7.36891
RHODAMINE 6G (A)	24.44	0.031	7.18026
EOSIN B (A)	21.32	0.046	3.15898
EOSIN Y (A)	21.90	0.043	3.71261
ROSE BENGAL (A)	22.58	0.039	4.46553
INDENOINDENE	16.87	0.093	0.79192

8.127 Donor Molecule: α-NND (trace)

TRIPHENYLAMINE	6.62	1.543	0.00091
P-XYLENE (TRACE)	0.00	0.000	0.00000
SODIUM SALICYLATE (A)	6.78	1.437	0.00105
ANTHRANILIC ACID (A)	18.70	0.068	0.46156
TMPD	13.16	0.196	0.05618
1,2-BENZOFLUORENE	5.97	2.101	0.00049
9,10-DIHYDROPHENANTHRENE	0.00	0.000	0.00000
CARBAZOLE (TRACE)	0.00	0.000	0.00000
N-METHYL CARBAZOLE (TRACE)	6.17	1.903	0.00060
N-VINYLCARBAZOLE	5.05	3.481	0.00018
N-PHENYL CARBAZOLE	5.29	3.019	0.00024
N-METHYL-2,3-BENZCARBAZOLE	27.18	0.022	4.32781
BENZO(1,2-B-4,5-B')BISBENZOFURAN (TRACE) (B)	6.49	1.635	0.00081
SEXIPHENYL BENZENE	0.00	0.000	0.00000
P-QUATERPHENYL	0.00	0.000	0.00000
3,3'''-DIMETHYL-P-QUATERPHENYL	0.00	0.000	0.00000
TETRAMETHYL-P-QUATERPHENYL	0.00	0.000	0.00000
3,3''' -DIMETHYL-3',2''-METHYLENE-P-QUATERPHENYL	8.65	0.691	0.00454
3,3''' -DIMETHYL-3',2''-ETHYLENE-P-QUATERPHENYL	13.23	0.193	0.05816
4-(3-ETHYLBUTOXY)-P-QUATERPHENYL (B)	10.91	0.345	0.01827
4,4'''-DI(2-BUTYLOCTOXY-1)-P-QUATERPHENYL	10.34	0.405	0.01323
DIETHYL-P-QUINQUEPHENYL	0.00	0.000	0.00000
DI(3-ETHYLHEPTYL)-P-QUINQUEPHENYL	13.86	0.168	0.07661
TETRAMETHYL-P-SEXIPHENYL	0.00	0.000	0.00000
TETRAMETHYL-P-OCTAPHENYL (B)	19.35	0.062	0.56826
INDOLE (TRACE)	0.00	0.000	0.00000
2-PHENYL INDOLE	0.00	0.000	0.00000
1-METHYL-2-PHENYL INDOLE	0.00	0.000	0.00000
1-NEOPENTYL-2-PHENYL INDOLE	0.00	0.000	0.00000
1-N-BUTYL-2(BIPHENYLYL-4) INDOLE	14.90	0.135	0.11839
1,2-DIPHENYL INDOLE	0.00	0.000	0.00000
1-BIPHENYLYL-2-PHENYL INDOLE	8.00	0.873	0.00285
1,3-DIMETHYL-2-PHENYL INDOLE	0.00	0.000	0.00000
2,3-DIPHENYL INDOLE	14.27	0.154	0.09124
1-METHYL-2,3-DIPHENYL INDOLE	11.98	0.260	0.03206
2-PHENYL-3,2'-METHYLENE INDOLE (TRACE)	0.00	0.000	0.00000
2-PHENYL-3,2'-ETHYLENE INDOLE (TRACE)	12.72	0.217	0.04598
2-PHENYL-3,2'-TRIMETHYLENE INDOLE	0.00	0.000	0.00000
1-METHYL-3,2'-METHYLENE-2-PHENYL INDOLE (TRACE)	9.86	0.466	0.00997
1-METHYL-3,2'-ETHYLENE-2-PHENYL INDOLE	18.61	0.069	0.44977
1-METHYL-3,2'-TRIMETHYLENE-2-PHENYL INDOLE	4.80	4.053	0.00013
4,5-DIPHENYLIMIDAZOLE (A)	0.00	0.000	0.00000
PPF (TRACE)	9.84	0.469	0.00985
αNPO (TRACE)	25.57	0.027	3.01412
BBO (B)	32.53	0.013	12.76110
POPOP	35.62	0.010	21.95854
DIMETHYL POPOP (TRACE)	38.84	0.008	36.83461
PPD (TRACE)	0.00	0.000	0.00000
2,5-DI(2-METHYLPHENYL)OXADIAZOLE	0.00	0.000	0.00000
PBD	0.00	0.000	0.00000
BBD (B)	14.25	0.155	0.09073
αNPD (TRACE)	15.50	0.120	0.15019
βNPD	0.00	0.000	0.00000
αNND (TRACE)	21.15	0.047	0.96855
βNND (TRACE)	6.86	1.385	0.00113
TRANS-1-(4-BIPHENYLYL)-2-PHENYL-ETHYLENE	16.72	0.096	0.23661
TRANS-1-(4-BIPHENYLYL)-2-(1-NAPHTHYL)-ETHYLENE	29.05	0.018	6.46896
DIPHENYLSTILBENE (B)	32.46	0.013	12.58855
1,1,4,4-TETRAPHENYLBUTADIENE	32.28	0.013	12.15264
1,6-DIPHENYLHEXATRIENE	37.35	0.009	29.15027
1,8-DIPHENYL-1,3,5,7-OCTATETRAENE	45.03	0.005	89.00194
BPSB (TRACE)	38.97	0.008	37.62285
BIS-MSB	36.25	0.009	24.36954
ESCULIN (A)	23.79	0.033	1.95554
AZULENE	12.42	0.233	0.03886
1-NAPHTHYLAMINE	6.78	1.435	0.00105
2-CHLORONAPHTHALENE	0.00	0.000	0.00000
ACENAPHTHENE	0.00	0.000	0.00000
1-PHENYL NAPHTHALENE	0.00	0.000	0.00000

ACCEPTOR MOLECULES	RO	CO	OVERLAP
1,4,5,8-TETRAPHENYL NAPHTHALENE	25.89	0.026	3.23704
1,1'-BINAPHTHYL	0.00	0.000	0.00000
4,4'-DI(N-BUTOXY)-1,1'-BINAPHTHYL	9.16	0.582	0.00641
2,2'-BINAPHTHYL	0.00	0.000	0.00000
ACRIDINE (A)	26.44	0.024	3.66976
ANTHRACENE (TRACE)	25.20	0.028	2.75885
DEUTERATED ANTHRACENE C14D10 (TRACE)	24.54	0.030	2.35113
9-METHYL ANTHRACENE (TRACE)	28.29	0.020	5.50341
1-AMINO ANTHRACENE	29.91	0.017	7.62820
2-AMINO ANTHRACENE	28.65	0.019	5.89120
9-VINYL ANTHRACENE	29.83	0.017	7.55349
9-PHENYL ANTHRACENE (TRACE)	29.61	0.017	7.24106
9,10-DIPHENYL ANTHRACENE	32.34	0.013	12.26431
9,10-DICHLOROANTHRACENE (TRACE)	31.94	0.014	11.36232
BIANTHRANYL (B)	33.93	0.011	16.34772
9,10-DINAPHTHYLANTHRACENE (B)	32.23	0.013	11.98851
9-(P-VINYLPHENYL)-10-PHENYL ANTHRACENE	32.82	0.013	13.37266
9,10-BIS(PHENYL-ETHYNYL)ANTHRACENE (TRACE)	39.22	0.007	38.52821
RUBICENE (B)	33.41	0.012	14.72987
DIBENZ(A,H)ANTHRACENE (B)	20.60	0.051	0.81995
TETRACENE (TRACE) (B)	28.87	0.019	6.12459
RUBRENE (B)	26.56	0.024	3.71955
PHENANTHRENE	4.16	6.215	0.00006
3,4-BENZOPHENANTHRENE	13.88	0.167	0.07692
2-PHENYL PHENANTHRENE	9.77	0.479	0.00944
CHRYSENE (TRACE)	13.42	0.185	0.06317
TRIPHENYLENE	3.88	7.652	0.00004
DEUTERATED TRIPHENYLENE-C18D12	3.78	8.250	0.00003
PYRENE	13.82	0.170	0.07493
1,3,6,8,-TETRAPHENYL PYRENE	39.81	0.007	42.54438
BIPYRENYL (B)	34.46	0.011	17.93096
O-PHENYLENE-PYRENE	34.91	0.011	19.19875
ANTHANTHRENE (B)	37.11	0.009	27.83703
FLUORANTHENE	24.06	0.032	2.07763
3-PHENYL FLUORANTHENE	30.85	0.015	9.21227
3-AMINO FLUORANTHENE	32.63	0.013	12.80648
3,3'-BIFLUORANTHENYL (B)	37.62	0.008	30.23336
BENZO(GHI)FLUORANTHENE	22.31	0.040	1.31887
9,12-DI(N-OCTADECOXY)-BENZ(E)ACEPHENANTHRYLENE	34.29	0.011	17.34620
DECACYCLENE (B)	42.15	0.006	59.46608
10-ETHOXY-BENZO(KL)XANTHENE	33.03	0.012	13.91200
PERYLENE	38.21	0.008	33.11472
BENZO(GHI)PERYLENE (B)	34.34	0.011	17.57158
BBOT (TRACE)	41.62	0.006	55.73163
QUININE SULFATE (A)	8.17	0.819	0.00324
ACRIDINE YELLOW (A)	37.00	0.009	27.16259
ACRIDINE ORANGE (A)	31.55	0.014	10.44465
ACRIDINE RED (A)	26.08	0.025	3.33158
URANINE (A)	31.86	0.014	11.06762
RHODAMINE B (A)	30.12	0.016	7.89565
RHODAMINE 6G (A)	28.65	0.019	5.85754
EOSIN B (A)	27.80	0.021	4.88477
EOSIN Y (A)	29.21	0.018	6.57092
ROSE BENGAL (A)	27.46	0.022	4.53884

8.128 Donor Molecule: β-NND (trace)

TRIPHENYLAMINE	16.71	0.096	0.17455
P-XYLENE (TRACE)	0.00	0.000	0.00000
SODIUM SALICYLATE (A)	12.49	0.230	0.03033
ANTHRANILIC ACID (A)	25.34	0.027	2.10739
TMPD	20.00	0.056	0.51004
BENZIDINE	12.14	0.250	0.02576
1,3-DIBIPHENYLYLBENZENE (B)	8.99	0.616	0.00424
1,3,5,-TRIBIPHENYLYLBENZENE (B)	9.03	0.607	0.00437
FLUORENE (TRACE)	0.00	0.000	0.00000
1-METHYL FLUORENE	0.00	0.000	0.00000
1,2-BENZOFLUORENE	16.90	0.093	0.18637
9,10-DIHYDROPHENANTHRENE	0.00	0.000	0.00000
CARBAZOLE (TRACE)	11.19	0.320	0.01573
N-METHYL CARBAZOLE (TRACE)	20.83	0.050	0.65324
N-VINYLCARBAZOLE	18.63	0.069	0.33455
N-PHENYL CARBAZOLE	18.93	0.066	0.36889
N-METHYL-2,3-BENZCARBAZOLE	28.49	0.019	4.22830
DIBENZOFURAN	0.00	0.000	0.00000
DIBENZOTHIOPHENE (TRACE)	6.91	1.357	0.00087
P-TERPHENYL	0.00	0.000	0.00000
DEUTERATED P-TERPHENYL-C18D14	0.00	0.000	0.00000
4-METHYL TERPHENYL	0.00	0.000	0.00000
2-METHYL-P-TERPHENYL	0.00	0.000	0.00000
3,3''-DIMETHYL-P-TERPHENYL	0.00	0.000	0.00000
4-(3,3-DIMETHYLBUTOXY)-P-TERPHENYL	6.14	1.929	0.00043
4,4''-DIHEXAHYDROFARNESOXY-P-TERPHENYL	8.98	0.618	0.00421
2,2'-METHYLENE-P-TERPHENYL	0.00	0.000	0.00000
2,2'-ETHYLENE-P-TERPHENYL	7.63	1.009	0.00158
2,2'-METHYLENE-5',6'-METHYLENE-P-TERPHENYL (TRACE)	19.93	0.057	0.50209
3-PHENYL DIBENZOFURAN	0.00	0.000	0.00000
BENZO(1,2-B-4,5-B')BISBENZOFURAN (TRACE) (B)	25.73	0.026	2.32350
SEXIPHENYL BENZENE	0.00	0.000	0.00000
P-QUATERPHENYL	12.83	0.212	0.03577
3,3'''-DIMETHYL-P-QUATERPHENYL	15.88	0.112	0.12869

ACCEPTOR MOLECULES	RO	CO	OVERLAP
3,3''' -DIMETHYL-3',2''-METHYLENE-P-QUATERPHENYL	24.64	0.030	1.79019
3,3''' -DIMETHYL-3',2''-ETHYLENE-P-QUATERPHENYL	25.62	0.027	2.25923
4-(3-ETHYLBUTOXY)-P-QUATERPHENYL (B)	22.72	0.038	1.10024
4,4'''-DI(2-BUTYLOCTOXY-1)-P-QUATERPHENYL	22.96	0.037	1.17191
DIETHYL-P-QUINQUEPHENYL	16.22	0.105	0.14592
DI(3-ETHYLHEPTYL)-P-QUINQUEPHENYL	24.72	0.030	1.82015
TETRAMETHYL-P-SEXIPHENYL	13.15	0.197	0.04161
TETRAMETHYL-P-OCTAPHENYL (B)	30.58	0.016	6.52181
INDOLE (TRACE)	0.00	0.000	0.00000
2-PHENYL INDOLE	12.92	0.207	0.03739
1-METHYL-2-PHENYL INDOLE	11.01	0.335	0.01433
1-NEOPENTYL-2-PHENYL INDOLE	5.74	2.368	0.00029
1-N-BUTYL-2(BIPHENYLYL-4) INDOLE	24.75	0.029	1.83588
1,2-DIPHENYL INDOLE	11.38	0.303	0.01745
1-BIPHENYLYL-2-PHENYL INDOLE	16.20	0.105	0.14479
1,3-DIMETHYL-2-PHENYL INDOLE	12.72	0.217	0.03398
2,3-DIPHENYL INDOLE	22.74	0.038	1.10386
1-METHYL-2,3-DIPHENYL INDOLE	20.33	0.053	0.56335
2-PHENYL-3,2'-METHYLENE INDOLE (TRACE)	18.93	0.066	0.36881
2-PHENYL-3,2'-ETHYLENE INDOLE (TRACE)	27.50	0.021	3.45795
2-PHENYL-3,2' -TRIMETHYLENE INDOLE	17.68	0.081	0.24496
1-METHYL-3,2'-METHYLENE-2-PHENYL INDOLE (TRACE)	26.40	0.024	2.70789
1-METHYL-3,2'-ETHYLENE-2-PHENYL INDOLE	29.33	0.018	5.08044
1-METHYL-3,2'-TRIMETHYLENE-2-PHENYL INDOLE	19.17	0.063	0.39820
4,5-DIPHENYLIMIDAZOLE (A)	10.22	0.419	0.00914
PPF (TRACE)	27.51	0.021	3.46681
PPO (TRACE)	16.64	0.097	0.17025
2-(4-FLUOROPHENYL)-5-PHENYLOXAZOLE (TRACE)	15.22	0.127	0.09971
αNPO (TRACE)	33.88	0.012	12.00813
BBO (B)	40.53	0.007	35.19207
POPOP	41.84	0.006	42.52340
DIMETHYL POPOP (TRACE)	43.99	0.005	57.34610
PPD (TRACE)	0.00	0.000	0.00000
2,5-DI(2-METHYLPHENYL)OXADIAZOLE	0.00	0.000	0.00000
PBD	13.57	0.179	0.05019
BBD (B)	28.88	0.019	4.64125
αNPD (TRACE)	25.99	0.025	2.46012
βNPD	16.87	0.093	0.18503
αNND (TRACE)	30.67	0.016	6.63492
βNND (TRACE)	25.50	0.027	2.20133
TRANS-1-(4-BIPHENYLYL)-2-PHENYL-ETHYLENE	29.63	0.017	5.40194
TRANS-1-(4-BIPHENYLYL)-2-(1-NAPHTHYL)-ETHYLENE	36.57	0.009	18.97135
DIPHENYLSTILBENE (B)	40.78	0.007	36.46427
1,1,4,4-TETRAPHENYLBUTADIENE	39.21	0.007	28.76002
1,6-DIPHENYLHEXATRIENE	44.50	0.005	61.46687
1,8-DIPHENYL-1,3,5,7-OCTATETRAENE	48.30	0.004	99.98232
BPSB (TRACE)	45.42	0.005	69.53854
BIS-MSB	43.04	0.006	50.33691
ESCULIN (A)	30.75	0.015	6.71692
AZULENE	20.23	0.054	0.53531
NAPHTHALENE	0.00	0.000	0.00000
DEUTERATED NAPHTHALENE-C10D8	3.80	8.123	0.00002
1-METHYLNAPHTHALENE 5ML/L	0.00	0.000	0.00000
2-METHYL NAPHTHALENE	0.00	0.000	0.00000
2,3-DIMETHYL NAPHTHALENE	0.00	0.000	0.00000
2,6-DIMETHYL NAPHTHALENE (TRACE)	0.00	0.000	0.00000
1-NAPHTHOL (TRACE)	0.00	0.000	0.00000
2-NAPHTHOL (TRACE)	7.48	1.070	0.00141
1-NAPHTHYLAMINE	17.64	0.081	0.24113
2-PHENYL NAPHTHALENE	9.27	0.561	0.00511
1,4-DIPHENYL NAPHTHALENE	13.19	0.195	0.04229
1,5-DIPHENYL NAPHTHALENE	9.89	0.462	0.00752
1,7-DIPHENYL NAPHTHALENE	10.92	0.343	0.01362
1,4,5,8-TETRAPHENYL NAPHTHALENE	33.03	0.012	10.28062
1,1'-BINAPHTHYL	4.67	4.384	0.00008
4,4'-DI(N-BUTOXY)-1,1'-BINAPHTHYL	19.86	0.057	0.49026
2,2'-BINAPHTHYL	11.71	0.279	0.02072
ACRIDINE (A)	31.39	0.014	7.57714
ANTHRACENE (TRACE)	29.68	0.017	5.42331
DEUTERATED ANTHRACENE C14D10 (TRACE)	29.07	0.018	4.79202
9-METHYL ANTHRACENE (TRACE)	31.14	0.015	7.22547
1-AMINO ANTHRACENE	30.89	0.015	6.81664
2-AMINO ANTHRACENE	29.43	0.018	5.09659
9-VINYL ANTHRACENE	32.47	0.013	9.26050
9-PHENYL ANTHRACENE (TRACE)	32.74	0.013	9.74237
9,10-DIPHENYL ANTHRACENE	34.41	0.011	13.10578
9,10-DICHLOROANTHRACENE (TRACE)	33.55	0.012	11.25620
BIANTHRANYL (B)	36.03	0.010	17.28616
9,10-DINAPHTHYLANTHRACENE (B)	34.03	0.011	12.25174
9-(P-VINYLPHENYL)-10-PHENYL ANTHRACENE	34.75	0.011	13.87759
9,10-BIS(PHENYL-ETHYNYL)ANTHRACENE (TRACE)	36.35	0.009	18.03430
RUBICENE (B)	33.46	0.012	10.95906
DIBENZ(A,H)ANTHRACENE (B)	28.49	0.019	4.22976
TETRACENE (TRACE) (B)	27.25	0.022	3.20103
RUBRENE (B)	25.40	0.027	2.09661
TETRAHYDROCANNABINOL ACID METHYL ESTER	0.00	0.000	0.00000
PHENANTHRENE	12.11	0.252	0.02527
3,4-BENZOPHENANTHRENE	17.23	0.087	0.20770
2-PHENYL PHENANTHRENE	14.84	0.137	0.08523
CHRYSENE (TRACE)	17.83	0.079	0.25594
TRIPHENYLENE	11.14	0.324	0.01530
DEUTERATED TRIPHENYLENE-C18D12	10.99	0.337	0.01409
PYRENE	21.93	0.042	0.88376
1,3,6,8,-TETRAPHENYL PYRENE	42.76	0.006	48.17695
BIPYRENYL (B)	40.51	0.007	34.90159
O-PHENYLENE-PYRENE	37.22	0.009	20.77896
ANTHANTHRENE (B)	36.33	0.009	18.04379

ACCEPTOR MOLECULES	RO	CO	OVERLAP
FLUORANTHENE	29.92	0.017	5.66706
3-PHENYL FLUORANTHENE	34.94	0.010	14.34245
3-AMINO FLUORANTHENE	33.40	0.012	10.84923
3,3'-BIFLUORANTHENYL (B)	40.42	0.007	34.33010
BENZO(GHI)FLUORANTHENE	27.19	0.022	3.18152
9,12-DI(N-OCTADECOXY)-BENZ(E)ACEPHENANTHRYLENE	35.69	0.010	16.26304
DECACYCLENE (B)	44.05	0.005	57.09095
10-ETHOXY-BENZO(KL)XANTHENE	35.18	0.010	14.97183
PERYLENE	36.47	0.009	18.45059
BENZO(GHI)PERYLENE (B)	37.05	0.009	20.43924
BBOT (TRACE)	45.72	0.005	72.17931
QUININE SULFATE (A)	21.88	0.043	0.87906
ACRIDINE YELLOW (A)	33.67	0.012	11.38579
ACRIDINE ORANGE (A)	28.34	0.020	4.04570
ACRIDINE RED (A)	27.30	0.022	3.23560
URANINE (A)	29.10	0.018	4.74562
RHODAMINE B (A)	31.29	0.015	7.33026
RHODAMINE 6G (A)	30.30	0.016	6.05007
EOSIN B (A)	27.31	0.022	3.24523
EOSIN Y (A)	28.43	0.019	4.13038
ROSE BENGAL (A)	28.33	0.020	4.04248
INDENOINDENE	15.93	0.111	0.13105

8.129 Donor Molecule: *trans*-1-(4-Biphenylyl)-2-phenylethylene

ACCEPTOR MOLECULES	RO	CO	OVERLAP
TRIPHENYLAMINE	10.71	0.364	0.01618
P-XYLENE (TRACE)	0.00	0.000	0.00000
SODIUM SALICYLATE (A)	9.65	0.498	0.00863
ANTHRANILIC ACID (A)	21.55	0.045	1.06778
TMPD	16.00	0.109	0.17891
BENZIDINE	5.38	2.878	0.00026
1,2-BENZOFLUORENE	10.59	0.376	0.01514
9,10-DIHYDROPHENANTHRENE	0.00	0.000	0.00000
CARBAZOLE (TRACE)	4.89	3.829	0.00015
N-METHYL CARBAZOLE (TRACE)	11.77	0.274	0.02850
N-VINYLCARBAZOLE	9.31	0.554	0.00699
N-PHENYL CARBAZOLE	9.92	0.459	0.01021
N-METHYL-2,3-BENZCARBAZOLE	27.86	0.021	4.95323
2,2'-METHYLENE-5',6''-METHYLENE-P-TERPHENYL (TRACE)	8.78	0.660	0.00494
3-PHENYL DIBENZOFURAN	0.00	0.000	0.00000
BENZO(1,2-B-4,5-B')BISBENZOFURAN (TRACE) (B)	13.13	0.197	0.05509
SEXIPHENYL BENZENE	0.00	0.000	0.00000
P-QUATERPHENYL	0.00	0.000	0.00000
3,3'''-DIMETHYL-P-QUATERPHENYL	8.14	0.829	0.00313
TETRAMETHYL-P-QUATERPHENYL	0.00	0.000	0.00000
3,3''' -DIMETHYL-3',2''-METHYLENE-P-QUATERPHENYL	14.79	0.138	0.11241
3,3''' -DIMETHYL-3',2''-ETHYLENE-P-QUATERPHENYL	17.89	0.078	0.35054
4-(3-ETHYLBUTOXY)-P-QUATERPHENYL (B)	15.70	0.116	0.16046
4,4'''-DI(2-BUTYLOCTOXY-1)-P-QUATERPHENYL	15.58	0.118	0.15322
DIETHYL-P-QUINQUEPHENYL	7.62	1.011	0.00211
DI(3-ETHYLHEPTYL)-P-QUINQUEPHENYL	17.89	0.078	0.35055
TETRAMETHYL-P-SEXIPHENYL	0.00	0.000	0.00000
TETRAMETHYL-P-OCTAPHENYL (B)	23.78	0.033	1.93262
INDOLE (TRACE)	0.00	0.000	0.00000
2-PHENYL INDOLE	0.00	0.000	0.00000
1-METHYL-2-PHENYL INDOLE	4.45	5.073	0.00008
1-NEOPENTYL-2-PHENYL INDOLE	0.00	0.000	0.00000
1-N-BUTYL-2(BIPHENYLYL-4) INDOLE	18.88	0.067	0.48354
1,2-DIPHENYL INDOLE	4.68	4.374	0.00011
1-BIPHENYLYL-2-PHENYL INDOLE	11.63	0.284	0.02651
1,3-DIMETHYL-2-PHENYL INDOLE	4.70	4.304	0.00012
2,3-DIPHENYL INDOLE	17.52	0.083	0.30842
1-METHYL-2,3-DIPHENYL INDOLE	15.34	0.124	0.13952
2-PHENYL-3,2'-METHYLENE INDOLE (TRACE)	9.28	0.560	0.00685
2-PHENYL-3,2'-ETHYLENE INDOLE (TRACE)	18.82	0.067	0.47611
2-PHENYL-3,2'-TRIMETHYLENE INDOLE	8.83	0.650	0.00508
1-METHYL-3,2'-METHYLENE-2-PHENYL INDOLE (TRACE)	16.29	0.103	0.20042
1-METHYL-3,2'-ETHYLENE-2-PHENYL INDOLE	23.91	0.033	1.99738
1-METHYL-3,2'-TRIMETHYLENE-2-PHENYL INDOLE	10.04	0.443	0.01096
4,5-DIPHENYLIMIDAZOLE (A)	0.00	0.000	0.00000
PPF (TRACE)	16.94	0.092	0.25332
PPO (TRACE)	6.54	1.596	0.00084
2-(4-FLUOROPHENYL)-5-PHENYLOXAZOLE (TRACE)	5.91	2.172	0.00046
αNPO (TRACE)	28.95	0.018	6.26935
BBO (B)	36.05	0.010	23.31544
POPOP	38.49	0.008	34.54157
DIMETHYL POPOP (TRACE)	41.02	0.006	50.48736
PPD (TRACE)	0.00	0.000	0.00000
2,5-DI(2-METHYLPHENYL)OXADIAZOLE	0.00	0.000	0.00000
PBD	0.00	0.000	0.00000
BBD (B)	20.68	0.051	0.83821
αNPD (TRACE)	18.99	0.065	0.50082
βNPD	8.27	0.792	0.00343
αNND (TRACE)	25.31	0.028	2.80694
βNND (TRACE)	13.87	0.168	0.07643
TRANS-1-(4-BIPHENYLYL)-2-PHENYL-ETHYLENE	22.23	0.041	1.28774
TRANS-1-(4-BIPHENYLYL)-2-(1-NAPHTHYL)-ETHYLENE	32.26	0.013	11.97342
DIPHENYLSTILBENE (B)	36.04	0.010	23.29564
1,1,4,4-TETRAPHENYLBUTADIENE	35.30	0.010	20.50354
1,6-DIPHENYLHEXATRIENE	40.93	0.007	49.89167
1,8-DIPHENYL-1,3,5,7-OCTATETRAENE	46.57	0.004	107.49931

ACCEPTOR MOLECULES	RO	CO	OVERLAP
BPSB (TRACE)	41.85	0.006	57.02690
BIS-MSB	39.20	0.007	38.49604
ESCULIN (A)	26.74	0.023	3.89441
AZULENE	14.92	0.135	0.11520
NAPHTHALENE	0.00	0.000	0.00000
DEUTERATED NAPHTHALENE-C10D8	0.00	0.000	0.00000
1-METHYLNAPHTHALENE 5ML/L	0.00	0.000	0.00000
2-METHYL NAPHTHALENE	0.00	0.000	0.00000
2,3-DIMETHYL NAPHTHALENE	0.00	0.000	0.00000
2,6-DIMETHYL NAPHTHALENE (TRACE)	0.00	0.000	0.00000
1-NAPHTHOL (TRACE)	0.00	0.000	0.00000
2-NAPHTHOL (TRACE)	0.00	0.000	0.00000
1-NAPHTHYLAMINE	11.09	0.328	0.01997
2-CHLORONAPHTHALENE	0.00	0.000	0.00000
ACENAPHTHENE	0.00	0.000	0.00000
1-PHENYL NAPHTHALENE	0.00	0.000	0.00000
2-PHENYL NAPHTHALENE	0.00	0.000	0.00000
1,4-DIPHENYL NAPHTHALENE	6.74	1.461	0.00101
1,5-DIPHENYL NAPHTHALENE	0.00	0.000	0.00000
1,7-DIPHENYL NAPHTHALENE	4.52	4.853	0.00009
1,4,5,8-TETRAPHENYL NAPHTHALENE	28.88	0.019	6.16192
1,1'-BINAPHTHYL	0.00	0.000	0.00000
4,4'-DI(N-BUTOXY)-1,1'-BINAPHTHYL	13.69	0.174	0.07058
2,2'-BINAPHTHYL	0.00	0.000	0.00000
ACRIDINE (A)	28.69	0.019	5.92291
ANTHRACENE (TRACE)	27.82	0.021	4.92430
DEUTERATED ANTHRACENE C14D10 (TRACE)	27.17	0.022	4.27995
9-METHYL ANTHRACENE (TRACE)	29.14	0.018	6.50017
1-AMINO ANTHRACENE	30.06	0.016	7.76446
2-AMINO ANTHRACENE	28.65	0.019	5.81020
9-VINYL ANTHRACENE	30.86	0.015	9.13905
9-PHENYL ANTHRACENE (TRACE)	30.68	0.015	8.83963
9,10-DIPHENYL ANTHRACENE	33.39	0.012	14.65091
9,10-DICHLOROANTHRACENE (TRACE)	32.75	0.013	13.05230
BIANTHRANYL (B)	34.87	0.011	19.00919
9,10-DINAPHTHYLANTHRACENE (B)	33.19	0.012	14.12794
9-(P-VINYLPHENYL)-10-PHENYL ANTHRACENE	33.78	0.012	15.69127
9,10-BIS(PHENYL-ETHYNYL)ANTHRACENE (TRACE)	36.49	0.009	24.68117
RUBICENE (B)	32.60	0.013	12.53218
DIBENZ(A,H)ANTHRACENE (B)	24.50	0.030	2.28973
TETRACENE (TRACE) (B)	27.06	0.023	4.10800
RUBRENE (B)	24.38	0.031	2.19375
TETRAHYDROCANNABINOL ACID METHYL ESTER	0.00	0.000	0.00000
PHENANTHRENE	7.66	0.995	0.00217
3,4-BENZOPHENANTHRENE	15.40	0.122	0.14176
2-PHENYL PHENANTHRENE	12.10	0.252	0.03361
CHRYSENE (TRACE)	15.24	0.126	0.13352
TRIPHENYLENE	6.44	1.678	0.00076
DEUTERATED TRIPHENYLENE-C18D12	6.42	1.694	0.00075
PYRENE	15.78	0.114	0.16455
1,3,6,8,-TETRAPHENYL PYRENE	41.10	0.006	50.90928
BIPYRENYL (B)	36.99	0.009	27.10903
O-PHENYLENE-PYRENE	35.55	0.010	21.15212
ANTHANTHRENE (B)	35.87	0.010	22.41319
FLUORANTHENE	26.57	0.024	3.72429
3-PHENYL FLUORANTHENE	32.70	0.013	12.91863
3-AMINO FLUORANTHENE	32.40	0.013	12.12281
3,3'-BIFLUORANTHENYL (B)	38.74	0.008	35.62475
BENZO(GHI)FLUORANTHENE	24.02	0.032	2.02556
9,12-DI(N-OCTADECOXY)-BENZ(E)ACEPHENANTHRYLENE	34.44	0.011	17.60051
DECACYCLENE (B)	42.53	0.006	61.93807
10-ETHOXY-BENZO(KL)XANTHENE	34.06	0.011	16.52110
PERYLENE	36.47	0.009	24.73503
BENZO(GHI)PERYLENE (B)	35.25	0.010	20.29006
BBOT (TRACE)	43.44	0.005	71.09345
QUININE SULFATE (A)	13.52	0.181	0.06560
ACRIDINE YELLOW (A)	33.74	0.012	15.39983
ACRIDINE ORANGE (A)	28.32	0.020	5.39351
ACRIDINE RED (A)	25.80	0.026	3.08160
URANINE (A)	28.76	0.019	5.91300
RHODAMINE B (A)	29.73	0.017	7.22156
RHODAMINE 6G (A)	27.80	0.021	4.82455
EOSIN B (A)	25.96	0.026	3.19470
EOSIN Y (A)	27.27	0.022	4.29451
ROSE BENGAL (A)	26.65	0.024	3.74409
INDENOINDENE	7.25	1.175	0.00156

8.130 Donor Molecule: *trans*-1-(4-Biphenylyl)-2-(1-naphthyl) ethylene

ANTHRANILIC ACID (A)	10.90	0.346	0.01240
TMPD	0.00	0.000	0.00000
N-METHYL-2,3-BENZCARBAZOLE	26.86	0.023	2.76572
αNPO (TRACE)	12.63	0.222	0.03007
BBO (B)	22.72	0.038	1.01775
POPOP	28.85	0.019	4.25415
DIMETHYL POPOP (TRACE)	34.70	0.011	12.85244
PPD (TRACE)	0.00	0.000	0.00000
2,5-DI(2-METHYLPHENYL)OXADIAZOLE	0.00	0.000	0.00000
PBD	0.00	0.000	0.00000
BBD (B)	0.00	0.000	0.00000

ACCEPTOR MOLECULES	RO	CO	OVERLAP
αNND (TRACE)	5.74	2.369	0.00026
βNND (TRACE)	0.00	0.000	0.00000
TRANS-1-(4-BIPHENYLYL)-2-PHENYL-ETHYLENE	0.00	0.000	0.00000
TRANS-1-(4-BIPHENYLYL)-2-(1-NAPHTHYL)-ETHYLENE	21.20	0.047	0.67148
DIPHENYLSTILBENE (B)	22.16	0.041	0.87495
1,1,4,4-TETRAPHENYLBUTADIENE	25.14	0.028	1.86113
1,6-DIPHENYLHEXATRIENE	28.45	0.019	3.91458
1,8-DIPHENYL-1,3,5,7-OCTATETRAENE	43.84	0.005	52.04303
BPSB (TRACE)	32.61	0.013	8.86657
BIS-MSB	29.79	0.017	5.15900
ESCULIN (A)	16.23	0.105	0.13514
AZULENE	14.21	0.156	0.05949
1,4,5,8-TETRAPHENYL NAPHTHALENE	19.21	0.063	0.37015
1,1'-BINAPHTHYL	0.00	0.000	0.00000
4,4'-DI(N-BUTOXY)-1,1'-BINAPHTHYL	0.00	0.000	0.00000
2,2'-BINAPHTHYL	0.00	0.000	0.00000
ACRIDINE (A)	22.38	0.040	0.92671
ANTHRACENE (TRACE)	19.33	0.062	0.38496
DEUTERATED ANTHRACENE C14D10 (TRACE)	18.12	0.075	0.26162
9-METHYL ANTHRACENE (TRACE)	25.97	0.026	2.26388
1-AMINO ANTHRACENE	31.02	0.015	6.52023
2-AMINO ANTHRACENE	30.35	0.016	5.71737
9-VINYL ANTHRACENE	28.20	0.020	3.70535
9-PHENYL ANTHRACENE (TRACE)	26.97	0.023	2.83899
9,10-DIPHENYL ANTHRACENE	31.74	0.014	7.52396
9,10-DICHLOROANTHRACENE (TRACE)	31.65	0.014	7.39306
BIANTHRANYL (B)	33.45	0.012	10.31104
9,10-DINAPHTHYLANTHRACENE (B)	31.85	0.014	7.67471
9-(P-VINYLPHENYL)-10-PHENYL ANTHRACENE	32.39	0.013	8.49062
9,10-BIS(PHENYL-ETHYNYL)ANTHRACENE (TRACE)	45.13	0.005	61.50743
RUBICENE (B)	37.14	0.009	18.99803
DIBENZ(A,H)ANTHRACENE (B)	18.36	0.072	0.28202
TETRACENE (TRACE) (B)	33.22	0.012	9.78267
RUBRENE (B)	31.72	0.014	7.36887
TETRAHYDROCANNABINOL ACID METHYL ESTER	0.00	0.000	0.00000
PHENANTHRENE	0.00	0.000	0.00000
3,4-BENZOPHENANTHRENE	9.16	0.583	0.00436
2-PHENYL PHENANTHRENE	0.00	0.000	0.00000
CHRYSENE (TRACE)	5.43	2.790	0.00019
TRIPHENYLENE	0.00	0.000	0.00000
DEUTERATED TRIPHENYLENE-C18D12	0.00	0.000	0.00000
PYRENE	7.70	0.980	0.00154
1,3,6,8,-TETRAPHENYL PYRENE	37.97	0.008	21.99273
BIPYRENYL (B)	30.51	0.016	5.93051
O-PHENYLENE-PYRENE	35.65	0.010	14.96775
ANTHANTHRENE (B)	41.00	0.006	34.73044
FLUORANTHENE	17.96	0.077	0.24709
3-PHENYL FLUORANTHENE	27.47	0.022	3.15671
3-AMINO FLUORANTHENE	34.84	0.011	13.04038
3,3'-BIFLUORANTHENYL (B)	36.80	0.009	18.20522
BENZO(GHI)FLUORANTHENE	21.96	0.042	0.82441
9,12-DI(N-OCTADECOXY)-BENZ(E)ACEPHENANTHRYLENE	36.14	0.009	16.31847
DECACYCLENE (B)	43.58	0.005	49.88343
10-ETHOXY-BENZO(KL)XANTHENE	32.17	0.013	8.15361
PERYLENE	42.62	0.006	43.81791
BENZO(GHI)PERYLENE (B)	32.36	0.013	8.45284
BBOT (TRACE)	39.23	0.007	26.83512
QUININE SULFATE (A)	0.00	0.000	0.00000
ACRIDINE YELLOW (A)	43.35	0.005	48.19263
ACRIDINE ORANGE (A)	37.71	0.008	20.84407
ACRIDINE RED (A)	30.21	0.016	5.50267
URANINE (A)	38.53	0.008	23.68204
RHODAMINE B (A)	34.69	0.011	12.60960
RHODAMINE 6G (A)	35.27	0.010	13.93857
EOSIN B (A)	34.25	0.011	11.67586
EOSIN Y (A)	35.69	0.010	14.96266
ROSE BENGAL (A)	32.42	0.013	8.40727

8.132 Donor Molecule: 1,1,4,4-Tetraphenylbutadiene

ANTHRANILIC ACID (A)	4.28	5.723	0.00008
TMPD	0.00	0.000	0.00000
N-METHYL-2,3-BENZCARBAZOLE	16.93	0.092	0.28804
αNPO (TRACE)	4.80	4.056	0.00015
BBO (B)	10.24	0.417	0.01417
POPOP	14.00	0.163	0.09258
DIMETHYL POPOP (TRACE)	19.43	0.061	0.66059
TRANS-1-(4-BIPHENYLYL)-2-(1-NAPHTHYL)-ETHYLENE	10.13	0.430	0.01330
DIPHENYLSTILBENE (B)	10.28	0.412	0.01450
1,1,4,4-TETRAPHENYLBUTADIENE	13.32	0.189	0.06837
1,6-DIPHENYLHEXATRIENE	13.82	0.170	0.08540
1,8-DIPHENYL-1,3,5,7-OCTATETRAENE	28.51	0.019	6.55932
BPSB (TRACE)	16.83	0.094	0.27929
BIS-MSB	15.71	0.115	0.18469
ESCULIN (A)	7.23	1.184	0.00176
AZULENE	18.50	0.071	0.47993
1,4,5,8-TETRAPHENYL NAPHTHALENE	9.97	0.451	0.01206
1,1'-BINAPHTHYL	0.00	0.000	0.00000
4,4'-DI(N-BUTOXY)-1,1'-BINAPHTHYL	0.00	0.000	0.00000
2,2'-BINAPHTHYL	0.00	0.000	0.00000
ACRIDINE (A)	11.84	0.270	0.03375

ACCEPTOR MOLECULES	RO	CO	OVERLAP
ANTHRACENE (TRACE)	8.03	0.864	0.00330
DEUTERATED ANTHRACENE C14D10 (TRACE)	7.32	1.140	0.00189
9-METHYL ANTHRACENE (TRACE)	13.75	0.172	0.08302
1-AMINO ANTHRACENE	22.77	0.038	1.69620
2-AMINO ANTHRACENE	22.61	0.039	1.62703
9-VINYL ANTHRACENE	16.62	0.097	0.25813
9-PHENYL ANTHRACENE (TRACE)	14.41	0.149	0.10989
9,10-DIPHENYL ANTHRACENE	20.24	0.054	0.84066
9,10-DICHLOROANTHRACENE (TRACE)	20.92	0.049	1.02427
BIANTHRANYL (B)	20.78	0.050	0.98655
9,10-DINAPHTHYLANTHRACENE (B)	20.83	0.050	0.99805
9-(P-VINYLPHENYL)-10-PHENYL ANTHRACENE	20.96	0.049	1.03646
9,10-BIS(PHENYL-ETHYNYL)ANTHRACENE (TRACE)	41.90	0.006	65.52494
RUBICENE (B)	37.93	0.008	35.83548
DIBENZ(A,H)ANTHRACENE (B)	11.29	0.310	0.02544
TETRACENE (TRACE) (B)	31.70	0.014	12.27805
RUBRENE (B)	36.16	0.009	26.89714
TETRAHYDROCANNABINOL ACID METHYL ESTER	0.00	0.000	0.00000
PHENANTHRENE	0.00	0.000	0.00000
3,4-BENZOPHENANTHRENE	4.08	6.606	0.00006
2-PHENYL PHENANTHRENE	0.00	0.000	0.00000
CHRYSENE (TRACE)	0.00	0.000	0.00000
TRIPHENYLENE	0.00	0.000	0.00000
DEUTERATED TRIPHENYLENE-C18D12	0.00	0.000	0.00000
PYRENE	3.08	15.379	0.00001
1,3,6,8,-TETRAPHENYL PYRENE	23.14	0.036	1.87745
BIPYRENYL (B)	18.25	0.074	0.45199
O-PHENYLENE-PYRENE	28.47	0.019	6.45842
ANTHANTHRENE (B)	33.49	0.012	17.18073
FLUORANTHENE	10.34	0.404	0.01500
3-PHENYL FLUORANTHENE	17.36	0.086	0.33413
3-AMINO FLUORANTHENE	28.18	0.020	6.07713
3,3'-BIFLUORANTHENYL (B)	25.10	0.028	3.05215
BENZO(GHI)FLUORANTHENE	14.99	0.133	0.13882
9,12-DI(N-OCTADECOXY)-BENZ(E)ACEPHENANTHRYLENE	26.34	0.024	4.06890
DECACYCLENE (B)	34.52	0.011	20.51637
10-ETHOXY-BENZO(KL)XANTHENE	20.36	0.053	0.87342
PERYLENE	35.12	0.010	22.84232
BENZO(GHI)PERYLENE (B)	18.13	0.075	0.43505
BBOT (TRACE)	24.14	0.032	2.42353
QUININE SULFATE (A)	0.00	0.000	0.00000
ACRIDINE YELLOW (A)	42.64	0.006	72.57720
ACRIDINE ORANGE (A)	41.21	0.006	59.09731
ACRIDINE RED (A)	38.84	0.008	41.18161
URANINE (A)	44.17	0.005	89.51520
RHODAMINE B (A)	44.44	0.005	92.47834
RHODAMINE 6G (A)	45.85	0.005	111.67430
EOSIN B (A)	43.29	0.006	79.11752
EOSIN Y (A)	44.46	0.005	92.87334
ROSE BENGAL (A)	42.46	0.006	70.28966

8.133 Donor Molecule: 1,6-Diphenylhexatriene

ACCEPTOR MOLECULES	RO	CO	OVERLAP
ANTHRANILIC ACID (A)	6.71	1.483	0.00084
TMPD	0.00	0.000	0.00000
N-METHYL-2,3-BENZCARBAZOLE	20.19	0.054	0.62260
αNPO (TRACE)	7.46	1.076	0.00160
BBO (B)	14.03	0.162	0.07048
POPOP	18.03	0.076	0.31647
DIMETHYL POPOP (TRACE)	23.41	0.035	1.51410
TRANS-1-(4-BIPHENYLYL)-2-(1-NAPHTHYL)-ETHYLENE	13.19	0.195	0.04851
DIPHENYLSTILBENE (B)	13.75	0.172	0.06241
1,1,4,4-TETRAPHENYLBUTADIENE	16.50	0.100	0.18556
1,6-DIPHENYLHEXATRIENE	18.15	0.075	0.32983
1,8-DIPHENYL-1,3,5,7-OCTATETRAENE	33.27	0.012	12.42857
BPSB (TRACE)	20.86	0.049	0.75992
BIS-MSB	19.36	0.062	0.48532
ESCULIN (A)	10.01	0.445	0.00932
AZULENE	19.05	0.065	0.42974
1,4,5,8-TETRAPHENYL NAPHTHALENE	12.42	0.234	0.03376
1,1-BINAPHTHYL	0.00	0.000	0.00000
4,4'-DI(N-BUTOXY)-1,1'-BINAPHTHYL	0.00	0.000	0.00000
2,2'-BINAPHTHYL	0.00	0.000	0.00000
ACRIDINE (A)	14.47	0.148	0.08465
ANTHRACENE (TRACE)	12.15	0.249	0.02966
DEUTERATED ANTHRACENE C14D10 (TRACE)	11.18	0.320	0.01802
9-METHYL ANTHRACENE (TRACE)	16.49	0.100	0.18551
1-AMINO ANTHRACENE	25.20	0.028	2.33854
2-AMINO ANTHRACENE	24.97	0.029	2.21389
9-VINYL ANTHRACENE	19.81	0.058	0.55560
9-PHENYL ANTHRACENE (TRACE)	17.43	0.084	0.25805
9,10-DIPHENYL ANTHRACENE	23.82	0.033	1.67879
9,10-DICHLOROANTHRACENE (TRACE)	24.48	0.031	1.97482
BIANTHRANYL (B)	24.61	0.030	2.04198
9,10-DINAPHTHYLANTHRACENE (B)	24.46	0.031	1.96602
9-(P-VINYLPHENYL)-10-PHENYL ANTHRACENE	24.52	0.030	1.99294
9,10-BIS(PHENYL-ETHYNYL)ANTHRACENE (TRACE)	43.96	0.005	65.59053
RUBICENE (B)	39.31	0.007	33.33121
DIBENZ(A,H)ANTHRACENE (B)	13.56	0.179	0.05720
TETRACENE (TRACE) (B)	33.07	0.012	11.87714
RUBRENE (B)	37.27	0.009	24.19201

ACCEPTOR MOLECULES	RO	CO	OVERLAP
3,4-BENZOPHENANTHRENE	5.55	2.614	0.00027
2-PHENYL PHENANTHRENE	0.00	0.000	0.00000
CHRYSENE (TRACE)	0.00	0.000	0.00000
TRIPHENYLENE	0.00	0.000	0.00000
DEUTERATED TRIPHENYLENE-C18D12	0.00	0.000	0.00000
PYRENE	4.60	4.599	0.00009
1,3,6,8,-TETRAPHENYL PYRENE	27.27	0.022	3.76774
BIPYRENYL (B)	21.57	0.045	0.92691
O-PHENYLENE-PYRENE	30.66	0.016	7.55546
ANTHANTHRENE (B)	36.04	0.010	20.00822
FLUORANTHENE	12.35	0.237	0.03260
3-PHENYL FLUORANTHENE	20.02	0.056	0.59094
3-AMINO FLUORANTHENE	30.38	0.016	7.12554
3,3'-BIFLUORANTHENYL (B)	28.36	0.020	4.75995
BENZO(GHI)FLUORANTHENE	17.06	0.090	0.22590
9,12-DI(N-OCTADECOXY)-BENZ(E)ACEPHENANTHRYLENE	29.23	0.018	5.71482
DECACYCLENE (B)	37.26	0.009	24.35271
10-ETHOXY-BENZO(KL)XANTHENE	24.15	0.032	1.82248
PERYLENE	37.61	0.008	25.86762
BENZO(GHI)PERYLENE (B)	21.61	0.044	0.93644
BBOT (TRACE)	28.69	0.019	5.13040
QUININE SULFATE (A)	0.00	0.000	0.00000
ACRIDINE YELLOW (A)	44.36	0.005	69.06818
ACRIDINE ORANGE (A)	42.48	0.006	53.15394
ACRIDINE RED (A)	40.02	0.007	36.98819
URANINE (A)	45.39	0.005	79.14194
RHODAMINE B (A)	45.84	0.005	83.53640
RHODAMINE 6G (A)	47.15	0.004	99.21533
EOSIN B (A)	44.53	0.005	70.35174
EOSIN Y (A)	45.74	0.005	82.68205
ROSE BENGAL (A)	43.77	0.005	63.33542

8.134 Donor Molecule: 1,8-Diphenyl-1,3,5,7-octatetraene

ACCEPTOR MOLECULES	RO	CO	OVERLAP
1,8-DIPHENYL-1,3,5,7-OCTATETRAENE	5.51	2.672	0.00226
BPSB (TRACE)	0.00	0.000	0.00000
BIS-MSB	0.00	0.000	0.00000
ESCULIN (A)	0.00	0.000	0.00000
AZULENE	16.70	0.096	1.71135
1-AMINO ANTHRACENE	7.94	0.892	0.02024
2-AMINO ANTHRACENE	7.53	1.048	0.01466
9,10-BIS(PHENYL-ETHYNYL)ANTHRACENE (TRACE)	23.82	0.033	14.63942
RUBICENE (B)	27.22	0.022	32.35805
DIBENZ(A,H)ANTHRACENE (B)	0.00	0.000	0.00000
TETRACENE (TRACE) (B)	19.59	0.060	4.52342
RUBRENE (B)	27.63	0.021	35.38836
O-PHENYLENE-PYRENE	14.04	0.162	0.61500
ANTHANTHRENE (B)	12.47	0.231	0.30249
FLUORANTHENE	0.00	0.000	0.00000
3-PHENYL FLUORANTHENE	0.00	0.000	0.00000
3-AMINO FLUORANTHENE	13.54	0.180	0.49351
3,3'-BIFLUORANTHENYL (B)	6.00	2.066	0.00378
BENZO(GHI)FLUORANTHENE	0.00	0.000	0.00000
9,12-DI(N-OCTADECOXY)-BENZ(E)ACEPHENANTHRYLENE	5.41	2.821	0.00203
DECACYCLENE (B)	17.25	0.087	2.11360
10-ETHOXY-BENZO(KL)XANTHENE	0.00	0.000	0.00000
PERYLENE	14.09	0.160	0.62910
BENZO(GHI)PERYLENE (B)	0.00	0.000	0.00000
BBOT (TRACE)	0.00	0.000	0.00000
QUININE SULFATE (A)	0.00	0.000	0.00000
ACRIDINE YELLOW (A)	26.08	0.025	25.13919
ACRIDINE ORANGE (A)	29.00	0.018	47.42978
ACRIDINE RED (A)	33.17	0.012	105.87397
URANINE (A)	32.45	0.013	93.10694
RHODAMINE B (A)	38.56	0.008	261.20213
RHODAMINE 6G (A)	37.66	0.008	227.09642
EOSIN B (A)	35.15	0.010	150.03277
EOSIN Y (A)	35.93	0.010	171.16178
ROSE BENGAL (A)	36.95	0.009	202.27311

8.135 Donor Molecule: BPSB (trace)

ACCEPTOR MOLECULES	RO	CO	OVERLAP
ANTHRANILIC ACID (A)	9.08	0.597	0.00441
TMPD	0.00	0.000	0.00000
N-METHYL-2,3-BENZCARBAZOLE	26.25	0.025	2.56278
αNPO (TRACE)	10.28	0.412	0.00928
BBO (B)	20.06	0.055	0.51206
POPOP	26.40	0.024	2.65842
DIMETHYL POPOP (TRACE)	33.31	0.012	10.70821
TRANS-1-(4-BIPHENYLYL)-2-(1-NAPHTHYL)-ETHYLENE	19.23	0.063	0.39695
DIPHENYLSTILBENE (B)	19.73	0.058	0.46372
1,1,4,4-TETRAPHENYLBUTADIENE	23.43	0.035	1.29647
1,6-DIPHENYLHEXATRIENE	25.59	0.027	2.20014
1,8-DIPHENYL-1,3,5,7-OCTATETRAENE	42.68	0.006	47.15353
BPSB (TRACE)	30.56	0.016	6.38433

ACCEPTOR MOLECULES	RO	CO	OVERLAP
BIS-MSB	28.04	0.020	3.80856
ESCULIN (A)	14.26	0.154	0.06613
AZULENE	14.52	0.146	0.07231
1,4,5,8-TETRAPHENYL NAPHTHALENE	17.89	0.078	0.25718
1,1'-BINAPHTHYL	0.00	0.000	0.00000
4,4'-DI(N-BUTOXY)-1,1'-BINAPHTHYL	0.00	0.000	0.00000
2,2'-BINAPHTHYL	0.00	0.000	0.00000
ACRIDINE (A)	21.20	0.047	0.71177
ANTHRACENE (TRACE)	16.66	0.097	0.16794
DEUTERATED ANTHRACENE C14D10 (TRACE)	15.42	0.122	0.10586
9-METHYL ANTHRACENE (TRACE)	24.90	0.029	1.86798
1-AMINO ANTHRACENE	30.40	0.016	6.13952
2-AMINO ANTHRACENE	29.82	0.017	5.46505
9-VINYL ANTHRACENE	27.40	0.022	3.31576
9-PHENYL ANTHRACENE (TRACE)	25.76	0.026	2.28799
9,10-DIPHENYL ANTHRACENE	31.02	0.015	6.96242
9,10-DICHLOROANTHRACENE (TRACE)	30.51	0.016	6.30498
BIANTHRANYL (B)	32.93	0.013	9.96552
9,10-DINAPHTHYLANTHRACENE (B)	30.92	0.015	6.83063
9-(P-VINYLPHENYL)-10-PHENYL ANTHRACENE	31.58	0.014	7.74695
9,10-BIS(PHENYL-ETHYNYL)ANTHRACENE (TRACE)	44.98	0.005	64.09562
RUBICENE (B)	37.11	0.009	20.10981
DIBENZ(A,H)ANTHRACENE (B)	17.95	0.077	0.26192
TETRACENE (TRACE) (B)	33.31	0.012	10.55863
RUBRENE (B)	32.17	0.013	8.53679
TETRAHYDROCANNABINOL ACID METHYL ESTER	0.00	0.000	0.00000
PHENANTHRENE	0.00	0.000	0.00000
3,4-BENZOPHENANTHRENE	7.99	0.877	0.00204
2-PHENYL PHENANTHRENE	0.00	0.000	0.00000
CHRYSENE (TRACE)	3.83	7.933	0.00003
TRIPHENYLENE	0.00	0.000	0.00000
DEUTERATED TRIPHENYLENE-C18D12	0.00	0.000	0.00000
PYRENE	6.39	1.718	0.00053
1,3,6,8,-TETRAPHENYL PYRENE	36.77	0.009	19.28680
BIPYRENYL (B)	29.32	0.018	4.97324
O-PHENYLENE-PYRENE	34.95	0.010	14.13216
ANTHANTHRENE (B)	40.32	0.007	33.43080
FLUORANTHENE	16.98	0.091	0.18740
3-PHENYL FLUORANTHENE	26.34	0.024	2.60690
3-AMINO FLUORANTHENE	34.35	0.011	12.73741
3,3'-BIFLUORANTHENYL (B)	35.78	0.010	16.35474
BENZO(GHI)FLUORANTHENE	21.40	0.046	0.75054
9,12-DI(N-OCTADECOXY)-BENZ(E)ACEPHENANTHRYLENE	35.54	0.010	15.71136
DECACYCLENE (B)	42.77	0.006	47.37783
10-ETHOXY-BENZO(KL)XANTHENE	31.22	0.015	7.25122
PERYLENE	41.98	0.006	42.53289
BENZO(GHI)PERYLENE (B)	31.26	0.015	7.29569
BBOT (TRACE)	38.09	0.008	23.89187
QUININE SULFATE (A)	0.00	0.000	0.00000
ACRIDINE YELLOW (A)	43.43	0.005	51.80470
ACRIDINE ORANGE (A)	38.16	0.008	23.77083
ACRIDINE RED (A)	30.82	0.015	6.59928
URANINE (A)	39.16	0.007	27.76056
RHODAMINE B (A)	35.21	0.010	14.68463
RHODAMINE 6G (A)	36.47	0.009	18.13691
EOSIN B (A)	35.24	0.010	14.74620
EOSIN Y (A)	36.62	0.009	18.58985
ROSE BENGAL (A)	33.05	0.012	10.04516

8.136 Donor Molecule: Bis-MSB

ACCEPTOR MOLECULES	RO	CO	OVERLAP
ANTHRANILIC ACID (A)	8.57	0.710	0.00313
TMPD	0.00	0.000	0.00000
N-METHYL-2,3-BENZCARBAZOLE	25.66	0.026	2.23350
αNPO (TRACE)	9.80	0.475	0.00697
BBO (B)	18.47	0.071	0.31171
POPOP	24.09	0.032	1.53658
DIMETHYL POPOP (TRACE)	31.14	0.015	7.14665
TRANS-1-(4-BIPHENYLYL)-2-(1-NAPHTHYL)-ETHYLENE	17.61	0.082	0.23447
DIPHENYLSTILBENE (B)	18.21	0.074	0.28674
1,1,4,4-TETRAPHENYLBUTADIENE	21.82	0.043	0.84683
1,6-DIPHENYLHEXATRIENE	23.81	0.033	1.43084
1,8-DIPHENYL-1,3,5,7-OCTATETRAENE	41.67	0.006	40.82365
BPSB (TRACE)	28.09	0.020	3.85381
BIS-MSB	25.95	0.026	2.39253
ESCULIN (A)	13.08	0.200	0.03948
AZULENE	15.02	0.132	0.08834
1,4,5,8-TETRAPHENYL NAPHTHALENE	16.59	0.098	0.16329
1,1'-BINAPHTHYL	0.00	0.000	0.00000
4,4'-DI(N-BUTOXY)-1,1'-BINAPHTHYL	0.00	0.000	0.00000
2,2'-BINAPHTHYL	0.00	0.000	0.00000
ACRIDINE (A)	19.53	0.060	0.43515
ANTHRACENE (TRACE)	15.49	0.120	0.10865
DEUTERATED ANTHRACENE C14D10 (TRACE)	14.45	0.148	0.07149
9-METHYL ANTHRACENE (TRACE)	22.70	0.038	1.07389
1-AMINO ANTHRACENE	29.91	0.017	5.56713
2-AMINO ANTHRACENE	29.45	0.018	5.07842
9-VINYL ANTHRACENE	25.96	0.026	2.39909
9-PHENYL ANTHRACENE (TRACE)	23.65	0.034	1.37211
9,10-DIPHENYL ANTHRACENE	30.19	0.016	5.92526

ACCEPTOR MOLECULES	RO	CO	OVERLAP
9,10-DICHLOROANTHRACENE (TRACE)	30.15	0.016	5.87008
BIANTHRANYL (B)	31.78	0.014	8.05992
9,10-DINAPHTHYLANTHRACENE (B)	30.45	0.016	6.23394
9-(P-VINYLPHENYL)-10-PHENYL ANTHRACENE	30.82	0.015	6.69115
9,10-BIS(PHENYL-ETHYNYL)ANTHRACENE (TRACE)	45.53	0.005	68.91749
RUBICENE (B)	37.60	0.008	21.76181
DIBENZ(A,H)ANTHRACENE (B)	17.46	0.084	0.22188
TETRACENE (TRACE) (B)	33.64	0.012	11.22127
RUBRENE (B)	33.03	0.012	9.99044
TETRAHYDROCANNABINOL ACID METHYL ESTER	0.00	0.000	0.00000
PHENANTHRENE	0.00	0.000	0.00000
3,4-BENZOPHENANTHRENE	7.41	1.101	0.00130
2-PHENYL PHENANTHRENE	0.00	0.000	0.00000
CHRYSENE (TRACE)	4.03	6.822	0.00003
TRIPHENYLENE	0.00	0.000	0.00000
DEUTERATED TRIPHENYLENE-C18D12	0.00	0.000	0.00000
PYRENE	6.03	2.036	0.00038
1,3,6,8,-TETRAPHENYL PYRENE	35.09	0.010	14.56836
BIPYRENYL (B)	27.86	0.021	3.65608
O-PHENYLENE-PYRENE	34.57	0.011	13.22268
ANTHANTHRENE (B)	40.32	0.007	33.44386
FLUORANTHENE	16.01	0.109	0.13159
3-PHENYL FLUORANTHENE	25.20	0.028	1.99858
3-AMINO FLUORANTHENE	34.18	0.011	12.36596
3,3'-BIFLUORANTHENYL (B)	34.75	0.011	13.72379
BENZO(GHI)FLUORANTHENE	20.92	0.049	0.65536
9,12-DI(N-OCTADECOXY)-BENZ(E)ACEPHENANTHRYLENE	34.98	0.010	14.26922
DECACYCLENE (B)	42.26	0.006	44.12874
10-ETHOXY-BENZO(KL)XANTHENE	30.47	0.016	6.25748
PERYLENE	41.90	0.006	42.09343
BENZO(GHI)PERYLENE (B)	29.01	0.018	4.66345
BBOT (TRACE)	36.74	0.009	19.25469
QUININE SULFATE (A)	0.00	0.000	0.00000
ACRIDINE YELLOW (A)	44.17	0.005	57.31275
ACRIDINE ORANGE (A)	39.06	0.008	27.35189
ACRIDINE RED (A)	31.77	0.014	7.91754
URANINE (A)	40.23	0.007	32.64926
RHODAMINE B (A)	36.26	0.009	17.48055
RHODAMINE 6G (A)	37.83	0.008	22.56997
EOSIN B (A)	36.47	0.009	18.11009
EOSIN Y (A)	37.84	0.008	22.60819
ROSE BENGAL (A)	34.17	0.011	12.25588

8.138 Donor Molecule: Azulene

TRIPHENYLAMINE	5.32	2.964	0.00569
P-XYLENE (TRACE)	0.00	0.000	0.00000
SODIUM SALICYLATE (A)	5.09	3.389	0.00435
ANTHRANILIC ACID (A)	12.07	0.254	0.76848
TMPD	8.49	0.730	0.09335
1,2-BENZOFLUORENE	4.92	3.753	0.00355
9,10-DIHYDROPHENANTHRENE	0.00	0.000	0.00000
CARBAZOLE (TRACE)	0.00	0.000	0.00000
N-METHYL CARBAZOLE (TRACE)	5.08	3.403	0.00432
N-VINYLCARBAZOLE	4.13	6.338	0.00124
N-PHENYL CARBAZOLE	4.34	5.454	0.00168
N-METHYL-2,3-BENZCARBAZOLE	17.14	0.089	6.26766
BENZO(1,2-B-4,5-B')BISBENZOFURAN (TRACE) (B)	5.44	2.776	0.00649
SEXIPHENYL BENZENE	0.00	0.000	0.00000
P-QUATERPHENYL	0.00	0.000	0.00000
3,3'''-DIMETHYL-P-QUATERPHENYL	0.00	0.000	0.00000
TETRAMETHYL-P-QUATERPHENYL	0.00	0.000	0.00000
3,3''' -DIMETHYL-3',2''-METHYLENE-P-QUATERPHENYL	7.04	1.284	0.03033
3,3''' -DIMETHYL-3',2''-ETHYLENE-P-QUATERPHENYL	9.14	0.585	0.14553
4-(3-ETHYLBUTOXY)-P-QUATERPHENYL (B)	8.03	0.863	0.06709
4,4'''-DI(2-BUTYLOCTOXY-1)-P-QUATERPHENYL	7.96	0.888	0.06336
DIETHYL-P-QUINQUEPHENYL	0.00	0.000	0.00000
DI(3-ETHYLHEPTYL)-P-QUINQUEPHENYL	9.28	0.559	0.15925
TETRAMETHYL-P-SEXIPHENYL	0.00	0.000	0.00000
TETRAMETHYL-P-OCTAPHENYL (B)	12.57	0.225	0.98381
INDOLE (TRACE)	0.00	0.000	0.00000
2-PHENYL INDOLE	0.00	0.000	0.00000
1-METHYL-2-PHENYL INDOLE	0.00	0.000	0.00000
1-NEOPENTYL-2-PHENYL INDOLE	0.00	0.000	0.00000
1-N-BUTYL-2(BIPHENYLYL-4) INDOLE	9.85	0.468	0.22723
1,2-DIPHENYL INDOLE	0.00	0.000	0.00000
1-BIPHENYLYL-2-PHENYL INDOLE	6.06	2.007	0.01239
1,3-DIMETHYL-2-PHENYL INDOLE	0.00	0.000	0.00000
2,3-DIPHENYL INDOLE	9.29	0.558	0.15971
1-METHYL-2,3-DIPHENYL INDOLE	7.99	0.877	0.06488
2-PHENYL-3,2'-METHYLENE INDOLE (TRACE)	3.68	8.960	0.00062
2-PHENYL-3,2'-ETHYLENE INDOLE (TRACE)	9.35	0.547	0.16665
2-PHENYL-3,2'-TRIMETHYLENE INDOLE	3.43	11.128	0.00040
1-METHYL-3,2'-METHYLENE-2-PHENYL INDOLE (TRACE)	7.73	0.969	0.05315
1-METHYL-3,2'-ETHYLENE-2-PHENYL INDOLE	12.64	0.221	1.01560
1-METHYL-3,2'-TRIMETHYLENE-2-PHENYL INDOLE	4.13	6.368	0.00123
4,5-DIPHENYLIMIDAZOLE (A)	0.00	0.000	0.00000
PPF (TRACE)	7.87	0.919	0.05912
αNPO (TRACE)	16.11	0.107	4.33231
BBO (B)	21.21	0.047	22.52403

ACCEPTOR MOLECULES	RO	CO	OVERLAP
POPOP	23.11	0.036	37.70549
DIMETHYL POPOP (TRACE)	24.79	0.029	57.34113
PPD (TRACE)	0.00	0.000	0.00000
2,5-DI(2-METHYLPHENYL)OXADIAZOLE	0.00	0.000	0.00000
PBD	0.00	0.000	0.00000
BBD (B)	10.62	0.373	0.35864
αNPD (TRACE)	9.86	0.466	0.22894
βNPD	0.00	0.000	0.00000
αNND (TRACE)	13.40	0.186	1.44013
βNND (TRACE)	5.73	2.377	0.00885
TRANS-1-(4-BIPHENYLYL)-2-PHENYL-ETHYLENE	11.56	0.290	0.59347
TRANS-1-(4-BIPHENYLYL)-2-(1-NAPHTHYL)-ETHYLENE	18.87	0.067	11.18254
DIPHENYLSTILBENE (B)	21.05	0.048	21.51070
1,1,4,4-TETRAPHENYLBUTADIENE	20.92	0.049	20.72601
1,6-DIPHENYLHEXATRIENE	24.87	0.029	58.57118
1,8-DIPHENYL-1,3,5,7-OCTATETRAENE	28.50	0.019	131.79259
BPSB (TRACE)	25.16	0.028	62.79228
BIS-MSB	23.42	0.035	40.83270
ESCULIN (A)	15.34	0.124	3.23342
AZULENE	7.92	0.900	0.06051
1-NAPHTHYLAMINE	5.44	2.775	0.00649
2-CHLORONAPHTHALENE	0.00	0.000	0.00000
ACENAPHTHENE	0.00	0.000	0.00000
1-PHENYL NAPHTHALENE	0.00	0.000	0.00000
2-PHENYL NAPHTHALENE	0.00	0.000	0.00000
1,4-DIPHENYL NAPHTHALENE	0.00	0.000	0.00000
1,5-DIPHENYL NAPHTHALENE	0.00	0.000	0.00000
1,7-DIPHENYL NAPHTHALENE	0.00	0.000	0.00000
1,4,5,8-TETRAPHENYL NAPHTHALENE	16.71	0.096	5.37516
1,1'-BINAPHTHYL	0.00	0.000	0.00000
4,4'-DI(N-BUTOXY)-1,1'-BINAPHTHYL	7.03	1.285	0.03023
2,2'-BINAPHTHYL	0.00	0.000	0.00000
ACRIDINE (A)	16.94	0.092	5.84079
ANTHRACENE (TRACE)	17.15	0.089	6.30294
DEUTERATED ANTHRACENE C14D10 (TRACE)	16.90	0.093	5.76614
9-METHYL ANTHRACENE (TRACE)	17.48	0.084	7.06254
1-AMINO ANTHRACENE	18.05	0.076	8.48621
2-AMINO ANTHRACENE	17.04	0.090	5.99701
9-VINYL ANTHRACENE	18.79	0.067	10.85658
9-PHENYL ANTHRACENE (TRACE)	18.42	0.072	9.64573
9,10-DIPHENYL ANTHRACENE	20.55	0.052	18.55879
9,10-DICHLOROANTHRACENE (TRACE)	19.84	0.057	15.03100
BIANTHRANYL (B)	21.50	0.045	24.32738
9,10-DINAPHTHYLANTHRACENE (B)	20.31	0.053	17.27948
9-(P-VINYLPHENYL)-10-PHENYL ANTHRACENE	20.72	0.050	19.49155
9,10-BIS(PHENYL-ETHYNYL)ANTHRACENE (TRACE)	20.37	0.053	17.50033
RUBICENE (B)	19.25	0.063	12.48270
DIBENZ(A,H)ANTHRACENE (B)	13.87	0.168	1.76038
TETRACENE (TRACE) (B)	15.39	0.123	3.25472
RUBRENE (B)	13.38	0.187	1.40661
PHENANTHRENE	3.43	11.039	0.00041
3,4-BENZOPHENANTHRENE	9.10	0.594	0.14058
2-PHENYL PHENANTHRENE	6.37	1.734	0.01657
CHRYSENE (TRACE)	8.23	0.802	0.07741
TRIPHENYLENE	3.13	14.551	0.00024
DEUTERATED TRIPHENYLENE-C18D12	3.05	15.696	0.00020
PYRENE	8.79	0.658	0.11448
1,3,6,8,-TETRAPHENYL PYRENE	25.12	0.028	61.88943
BIPYRENYL (B)	22.04	0.042	28.24945
O-PHENYLENE-PYRENE	21.23	0.047	22.46212
ANTHANTHRENE (B)	20.72	0.050	19.40249
FLUORANTHENE	15.16	0.128	2.99760
3-PHENYL FLUORANTHENE	19.86	0.057	15.08037
3-AMINO FLUORANTHENE	19.18	0.063	12.20336
3,3'-BIFLUORANTHENYL (B)	23.54	0.034	41.75573
BENZO(GHI)FLUORANTHENE	14.04	0.162	1.88097
9,12-DI(N-OCTADECOXY)-BENZ(E)ACEPHENANTHRYLENE	20.44	0.052	17.93731
DECACYCLENE (B)	25.56	0.027	68.26513
10-ETHOXY-BENZO(KL)XANTHENE	20.83	0.050	20.13084
PERYLENE	20.95	0.049	20.71883
BENZO(GHI)PERYLENE (B)	21.35	0.046	23.35918
BBOT (TRACE)	26.50	0.024	85.40727
QUININE SULFATE (A)	6.59	1.564	0.02044
ACRIDINE YELLOW (A)	18.32	0.073	9.24969
ACRIDINE ORANGE (A)	15.28	0.125	3.12177
ACRIDINE RED (A)	14.86	0.136	2.64291
URANINE (A)	15.77	0.114	3.76779
RHODAMINE B (A)	17.09	0.090	6.09670
RHODAMINE 6G (A)	15.51	0.120	3.41091
EOSIN B (A)	14.63	0.143	2.39906
EOSIN Y (A)	15.39	0.123	3.25070
ROSE BENGAL (A)	15.17	0.128	2.98412

8.139 Donor Molecule: Naphthalene

ANILINE 3ML/L	3.77	8.353	0.00010
N,N-DIMETHYLANILINE 3ML/L	10.84	0.351	0.05403
TRIPHENYLAMINE	22.41	0.040	4.17553
1,4-DIMETHOXY BENZENE	4.70	4.305	0.00036
1-METHYL-4-CHLOROBENZENE 5ML/L	0.00	0.000	0.00000
SODIUM SALICYLATE (A)	15.00	0.133	0.37536

ACCEPTOR MOLECULES	RO	CO	OVERLAP
ANTHRANILIC ACID (A)	23.25	0.036	5.17038
TMPD	20.26	0.054	2.27443
4,4'-DIHYDROXYBIPHENYL (A)	10.19	0.423	0.03735
4,4'-DIMETHOXY BIPHENYL	7.92	0.899	0.00827
4,4'-DIISOBUTOXYBIPHENYL	7.05	1.278	0.00410
BENZIDINE	19.98	0.056	2.10391
4-VINYL BIPHENYL	10.69	0.366	0.04982
M-TERPHENYL	7.66	0.996	0.00675
1,3,5-TRIPHENYL BENZENE	9.19	0.577	0.01998
1,3-DIBIPHENYLYLBENZENE (B)	19.28	0.062	1.70100
1,3,5,-TRIBIPHENYLYLBENZENE (B)	22.75	0.038	4.59832
FLUORENE (TRACE)	4.06	6.672	0.00015
1-METHYL FLUORENE	5.07	3.438	0.00057
1,2-BENZOFLUORENE	20.69	0.050	2.59070
9,10-DIHYDROPHENANTHRENE	6.77	1.444	0.00321
CARBAZOLE (TRACE)	19.67	0.059	1.91706
N-METHYL CARBAZOLE (TRACE)	21.31	0.046	3.09115
N-VINYLCARBAZOLE	21.71	0.044	3.45261
N-PHENYL CARBAZOLE	21.38	0.046	3.15792
N-METHYL-2,3-BENZCARBAZOLE	23.27	0.035	5.17993
DIBENZOFURAN	0.00	0.000	0.00000
DIBENZOTHIOPHENE (TRACE)	17.85	0.079	1.07219
P-TERPHENYL	12.74	0.216	0.14227
DEUTERATED P-TERPHENYL-C18D14	11.84	0.269	0.09199
4-METHYL TERPHENYL	14.48	0.147	0.30679
2-METHYL-P-TERPHENYL	6.66	1.512	0.00293
3,3''-DIMETHYL-P-TERPHENYL	14.28	0.154	0.28229
4-(3,3-DIMETHYLBUTOXY)-P-TERPHENYL	19.20	0.063	1.66306
4,4''-DIHEXAHYDROFARNESOXY-P-TERPHENYL	22.21	0.041	3.97767
2,2'-METHYLENE-P-TERPHENYL	21.28	0.046	3.08725
2,2'-ETHYLENE-P-TERPHENYL	22.05	0.042	3.81129
2,2'-METHYLENE-5',6''-METHYLENE-P-TERPHENYL (TRACE)	29.92	0.017	23.70483
3-PHENYL DIBENZOFURAN	21.36	0.046	3.15575
BENZO(1,2-B-4,5-B')BISBENZOFURAN (TRACE) (B)	31.07	0.015	29.70642
SEXIPHENYL BENZENE	0.00	0.000	0.00000
P-QUATERPHENYL	23.35	0.035	5.36222
3,3'''-DIMETHYL-P-QUATERPHENYL	24.57	0.030	7.27928
TETRAMETHYL-P-QUATERPHENYL	8.44	0.744	0.01210
3,3''' -DIMETHYL-3',2''-METHYLENE-P-QUATERPHENYL	31.21	0.015	30.51652
3,3''' -DIMETHYL-3',2''-ETHYLENE-P-QUATERPHENYL	30.76	0.015	27.89456
4-(3-ETHYLBUTOXY)-P-QUATERPHENYL (B)	28.60	0.019	18.01748
4,4'''-DI(2-BUTYLOCTOXY-1)-P-QUATERPHENYL	29.30	0.018	20.88078
DIETHYL-P-QUINQUEPHENYL	25.65	0.027	9.41292
DI(3-ETHYLHEPTYL)-P-QUINQUEPHENYL	30.48	0.016	26.36995
TETRAMETHYL-P-SEXIPHENYL	24.68	0.030	7.47693
TETRAMETHYL-P-OCTAPHENYL (B)	33.87	0.012	49.58361
INDOLE (TRACE)	0.00	0.000	0.00000
2-PHENYL INDOLE	25.55	0.027	9.20929
1-METHYL-2-PHENYL INDOLE	20.77	0.050	2.65832
1-NEOPENTYL-2-PHENYL INDOLE	18.62	0.069	1.38332
1-N-BUTYL-2(BIPHENYLYL-4) INDOLE	27.80	0.021	15.20046
1,2-DIPHENYL INDOLE	21.62	0.044	3.37556
1-BIPHENYLYL-2-PHENYL INDOLE	23.82	0.033	6.02921
1,3-DIMETHYL-2-PHENYL INDOLE	21.45	0.045	3.22246
2,3-DIPHENYL INDOLE	25.77	0.026	9.61297
1-METHYL-2,3-DIPHENYL INDOLE	23.46	0.035	5.48586
2-PHENYL-3,2'-METHYLENE INDOLE (TRACE)	27.99	0.020	15.88975
2-PHENYL-3,2'-ETHYLENE INDOLE (TRACE)	29.16	0.018	20.23507
2-PHENYL-3,2'-TRIMETHYLENE INDOLE	26.89	0.023	12.49260
1-METHYL-3,2'-METHYLENE-2-PHENYL INDOLE (TRACE)	29.00	0.018	19.60603
1-METHYL-3,2'-ETHYLENE-2-PHENYL INDOLE	29.22	0.018	20.44479
1-METHYL-3,2'-TRIMETHYLENE-2-PHENYL INDOLE	25.61	0.027	9.31488
4,5-DIPHENYLIMIDAZOLE (A)	16.49	0.100	0.66547
PPF (TRACE)	30.52	0.016	26.62268
PPO (TRACE)	27.19	0.022	13.37362
2-(4-FLUOROPHENYL)-5-PHENYLOXAZOLE (TRACE)	25.93	0.026	10.06393
αNPO (TRACE)	30.62	0.016	26.97161
BBO (B)	34.59	0.011	56.03135
POPOP	32.90	0.013	41.39976
DIMETHYL POPOP (TRACE)	33.27	0.012	44.24134
PPD (TRACE)	12.87	0.210	0.15157
2,5-DI(2-METHYLPHENYL)OXADIAZOLE	14.11	0.159	0.26365
PBD	26.40	0.024	11.21506
BBD (B)	32.43	0.013	38.34253
αNPD (TRACE)	27.89	0.021	15.48069
βNPD	25.20	0.028	8.45879
αNND (TRACE)	29.26	0.018	20.59199
βNND (TRACE)	29.79	0.017	23.06023
TRANS-1-(4-BIPHENYLYL)-2-PHENYL-ETHYLENE	31.88	0.014	34.51020
TRANS-1-(4-BIPHENYLYL)-2-(1-NAPHTHYL)-ETHYLENE	32.31	0.013	37.15155
DIPHENYLSTILBENE (B)	34.97	0.010	59.74247
1,1,4,4-TETRAPHENYLBUTADIENE	32.53	0.013	38.65878
1,6-DIPHENYLHEXATRIENE	35.48	0.010	65.10628
1,8-DIPHENYL-1,3,5,7-OCTATETRAENE	33.39	0.012	45.17734
BPSB (TRACE)	35.52	0.010	65.56051
BIS-MSB	34.95	0.010	59.44479
ESCULIN (A)	27.48	0.022	14.09075
AZULENE	21.07	0.048	2.85494
NAPHTHALENE	7.35	1.126	0.00526
DEUTERATED NAPHTHALENE-C10D8	9.55	0.514	0.02515
1-METHYLNAPHTHALENE 5ML/L	8.89	0.637	0.01641
2-METHYL NAPHTHALENE	11.02	0.335	0.05944
2,3-DIMETHYL NAPHTHALENE	10.79	0.356	0.05261
2,6-DIMETHYL NAPHTHALENE (TRACE)	14.16	0.158	0.26778
1-NAPHTHOL (TRACE)	15.77	0.114	0.51000
2-NAPHTHOL (TRACE)	17.36	0.085	0.90860

ACCEPTOR MOLECULES	RO	CO	OVERLAP
1-NAPHTHYLAMINE	21.86	0.043	3.59615
2-CHLORONAPHTHALENE	11.64	0.284	0.08251
ACENAPHTHENE	13.86	0.168	0.23587
1-PHENYL NAPHTHALENE	13.80	0.170	0.22929
2-PHENYL NAPHTHALENE	20.77	0.050	2.66202
1,4-DIPHENYL NAPHTHALENE	21.41	0.046	3.18270
1,5-DIPHENYL NAPHTHALENE	20.02	0.056	2.13004
1,7-DIPHENYL NAPHTHALENE	19.71	0.058	1.94110
1,4,5,8-TETRAPHENYL NAPHTHALENE	29.46	0.017	21.31924
1,1'-BINAPHTHYL	16.15	0.106	0.59005
4,4'-DI(N-BUTOXY)-1,1'-BINAPHTHYL	25.42	0.027	8.90544
2,2'-BINAPHTHYL	21.72	0.044	3.47829
ACRIDINE (A)	24.69	0.030	7.38614
ANTHRACENE (TRACE)	23.16	0.036	5.04342
DEUTERATED ANTHRACENE C14D10 (TRACE)	23.00	0.037	4.84054
9-METHYL ANTHRACENE (TRACE)	22.25	0.041	3.95440
1-AMINO ANTHRACENE	20.39	0.053	2.34184
2-AMINO ANTHRACENE	21.65	0.044	3.35448
9-VINYL ANTHRACENE	22.87	0.037	4.66730
9-PHENYL ANTHRACENE (TRACE)	23.43	0.035	5.39418
9,10-DIPHENYL ANTHRACENE	23.27	0.035	5.18119
9,10-DICHLOROANTHRACENE (TRACE)	21.97	0.042	3.66991
BIANTHRANYL (B)	24.80	0.029	7.58474
9,10-DINAPHTHYLANTHRACENE (B)	23.07	0.036	4.91252
9-(P-VINYLPHENYL)-10-PHENYL ANTHRACENE	23.29	0.035	5.20779
9,10-BIS(PHENYL-ETHYNYL)ANTHRACENE (TRACE)	20.94	0.049	2.74775
RUBICENE (B)	25.34	0.027	8.63965
DIBENZ(A,H)ANTHRACENE (B)	28.03	0.020	15.82004
TETRACENE (TRACE) (B)	17.42	0.085	0.91239
RUBRENE (B)	25.91	0.026	9.86250
TETRAHYDROCANNABINOL ACID METHYL ESTER	10.63	0.373	0.04806
PHENANTHRENE	13.16	0.196	0.17111
3,4-BENZOPHENANTHRENE	20.78	0.050	2.63459
2-PHENYL PHENANTHRENE	15.86	0.112	0.52404
CHRYSENE (TRACE)	21.81	0.043	3.53444
TRIPHENYLENE	14.18	0.157	0.26829
DEUTERATED TRIPHENYLENE-C18D12	13.99	0.164	0.24694
PYRENE	28.97	0.018	19.33569
1,3,6,8,-TETRAPHENYL PYRENE	28.66	0.019	18.06990
BIPYRENYL (B)	32.82	0.013	40.75176
O-PHENYLENE-PYRENE	26.86	0.023	12.25081
ANTHANTHRENE (B)	25.19	0.028	8.32554
FLUORANTHENE	24.87	0.029	7.70829
3-PHENYL FLUORANTHENE	26.29	0.025	10.77237
3-AMINO FLUORANTHENE	24.45	0.031	6.97226
3,3'-BIFLUORANTHENYL (B)	28.98	0.018	19.30848
BENZO(GHI)FLUORANTHENE	24.26	0.031	6.64973
9,12-DI(N-OCTADECOXY)-BENZ(E)ACEPHENANTHRYLENE	26.04	0.025	10.17026
DECACYCLENE (B)	31.75	0.014	33.37210
10-ETHOXY-BENZO(KL)XANTHENE	23.73	0.033	5.82819
PERYLENE	19.96	0.056	2.06008
BENZO(GHI)PERYLENE (B)	25.54	0.027	9.05478
BBOT (TRACE)	33.00	0.012	42.08533
QUININE SULFATE (A)	25.01	0.029	8.07372
ACRIDINE YELLOW (A)	17.68	0.081	0.99692
ACRIDINE ORANGE (A)	17.41	0.085	0.90938
ACRIDINE RED (A)	21.13	0.047	2.90273
URANINE (A)	21.70	0.044	3.40690
RHODAMINE B (A)	23.90	0.033	6.08222
RHODAMINE 6G (A)	24.44	0.031	6.94563
EOSIN B (A)	23.21	0.036	5.09536
EOSIN Y (A)	23.60	0.034	5.62647
ROSE BENGAL (A)	24.72	0.030	7.43215
INDENOINDENE	26.63	0.024	11.78486

8.140 Donor Molecule: Deuterated Naphthalene $C_{10}D_8$)

ANILINE 3ML/L	4.40	5.243	0.00021
N,N-DIMETHYLANILINE 3ML/L	11.25	0.314	0.05754
TRIPHENYLAMINE	22.82	0.038	3.97197
1,4-DIMETHOXY BENZENE	5.24	3.117	0.00059
1-METHYL-4-CHLOROBENZENE 5ML/L	0.00	0.000	0.00000
SODIUM SALICYLATE (A)	15.31	0.125	0.36231
ANTHRANILIC ACID (A)	23.68	0.034	4.91711
TMPD	20.59	0.051	2.13409
4,4'-DIHYDROXYBIPHENYL (A)	10.54	0.382	0.03906
4,4'-DIMETHOXY BIPHENYL	8.35	0.770	0.00962
4,4'-DIISOBUTOXYBIPHENYL	7.59	1.023	0.00545
BENZIDINE	20.37	0.053	2.01459
4-VINYL BIPHENYL	11.27	0.313	0.05830
M-TERPHENYL	8.06	0.854	0.00781
1,3,5-TRIPHENYL BENZENE	9.44	0.532	0.02007
1,3-DIBIPHENYLYLBENZENE (B)	19.74	0.058	1.67068
1,3,5,-TRIBIPHENYLYLBENZENE (B)	23.26	0.036	4.47616
FLUORENE (TRACE)	5.04	3.496	0.00047
1-METHYL FLUORENE	5.54	2.628	0.00083
1,2-BENZOFLUORENE	21.14	0.047	2.51115
9,10-DIHYDROPHENANTHRENE	7.37	1.119	0.00455
CARBAZOLE (TRACE)	19.98	0.056	1.79481
N-METHYL CARBAZOLE (TRACE)	21.64	0.044	2.88689
N-VINYLCARBAZOLE	22.00	0.042	3.19236

ACCEPTOR MOLECULES	RO	CO	OVERLAP
N-PHENYL CARBAZOLE	21.68	0.044	2.92423
N-METHYL-2,3-BENZCARBAZOLE	23.94	0.033	5.22731
DIBENZOFURAN	3.82	8.054	0.00009
DIBENZOTHIOPHENE (TRACE)	18.06	0.076	0.98065
P-TERPHENYL	13.24	0.193	0.15283
DEUTERATED P-TERPHENYL-C18D14	12.40	0.235	0.10326
4-METHYL TERPHENYL	14.97	0.133	0.32001
2-METHYL-P-TERPHENYL	7.21	1.196	0.00399
3,3''-DIMETHYL-P-TERPHENYL	14.72	0.140	0.28874
4-(3,3 DIMETHYLBUTOXY)-P-TERPHENYL	19.63	0.059	1.61659
4,4''-DIHEXAHYDROFARNESOXY-P-TERPHENYL	22.66	0.039	3.81087
2,2'-METHYLENE-P-TERPHENYL	21.77	0.043	3.01107
2,2'-ETHYLENE-P-TERPHENYL	22.49	0.039	3.65702
2,2'-METHYLENE-5',6''-METHYLENE-P-TERPHENYL (TRACE)	30.27	0.016	21.67495
3-PHENYL DIBENZOFURAN	21.87	0.043	3.09852
BENZO(1,2-B-4,5-B')BISBENZOFURAN (TRACE) (B)	31.55	0.014	27.73848
SEXIPHENYL BENZENE	0.00	0.000	0.00000
P-QUATERPHENYL	23.78	0.033	5.10302
3,3'''-DIMETHYL-P-QUATERPHENYL	25.01	0.029	6.89980
TETRAMETHYL-P-QUATERPHENYL	8.99	0.616	0.01504
3,3''' -DIMETHYL-3',2''-METHYLENE-P-QUATERPHENYL	31.66	0.014	28.32494
3,3''' -DIMETHYL-3',2''-ETHYLENE-P-QUATERPHENYL	31.22	0.015	25.99277
4-(3-ETHYLBUTOXY)-P-QUATERPHENYL (B)	29.05	0.018	16.88088
4,4'''-DI(2-BUTYLOCTOXY-1)-P-QUATERPHENYL	29.77	0.017	19.57267
DIETHYL-P-QUINQUEPHENYL	26.10	0.025	8.91552
DI(3-ETHYLHEPTYL)-P-QUINQUEPHENYL	30.97	0.015	24.71388
TETRAMETHYL-P-SEXIPHENYL	25.14	0.028	7.12775
TETRAMETHYL-P-OCTAPHENYL (B)	34.40	0.011	46.40638
INDOLE (TRACE)	0.00	0.000	0.00000
2-PHENYL INDOLE	25.96	0.026	8.63377
1-METHYL-2-PHENYL INDOLE	21.17	0.047	2.53542
1-NEOPENTYL-2-PHENYL INDOLE	19.04	0.065	1.34796
1-N-BUTYL-2(BIPHENYLYL-4) INDOLE	28.24	0.020	14.23401
1,2-DIPHENYL INDOLE	22.03	0.042	3.22301
1-BIPHENYLYL-2-PHENYL INDOLE	24.25	0.031	5.71283
1,3-DIMETHYL-2-PHENYL INDOLE	21.81	0.043	3.03477
2,3-DIPHENYL INDOLE	26.18	0.025	9.00957
1-METHYL-2,3-DIPHENYL INDOLE	23.84	0.033	5.14438
2-PHENYL-3,2'-METHYLENE INDOLE (TRACE)	28.37	0.020	14.68048
2-PHENYL-3,2'-ETHYLENE INDOLE (TRACE)	29.62	0.017	18.95026
2-PHENYL-3,2'-TRIMETHYLENE INDOLE	27.30	0.022	11.65359
1-METHYL-3,2'-METHYLENE-2-PHENYL INDOLE (TRACE)	29.45	0.018	18.34506
1-METHYL-3,2'-ETHYLENE-2-PHENYL INDOLE	29.70	0.017	19.21037
1-METHYL-3,2'-TRIMETHYLENE-2-PHENYL INDOLE	25.99	0.025	8.67897
4,5-DIPHENYLIMIDAZOLE (A)	16.81	0.094	0.63729
PPF (TRACE)	30.97	0.015	24.80358
PPO (TRACE)	27.61	0.021	12.50106
2-(4-FLUOROPHENYL)-5-PHENYLOXAZOLE (TRACE)	26.34	0.024	9.41350
αNPO (TRACE)	31.21	0.015	25.76220
BBO (B)	35.35	0.010	54.37051
POPOP	33.81	0.012	41.55902
DIMETHYL POPOP (TRACE)	34.33	0.011	45.48144
PPD (TRACE)	13.52	0.181	0.17401
2,5-DI(2-METHYLPHENYL)OXADIAZOLE	14.67	0.142	0.28270
PBD	26.85	0.023	10.57870
BBD (B)	32.93	0.013	35.81099
αNPD (TRACE)	28.33	0.020	14.48349
βNPD	25.62	0.027	7.96214
αNND (TRACE)	29.76	0.017	19.45728
βNND (TRACE)	30.23	0.016	21.48751
TRANS-1-(4-BIPHENYLYL)-2-PHENYL-ETHYLENE	32.38	0.013	32.27662
TRANS-1-(4-BIPHENYLYL)-2-(1-NAPHTHYL)-ETHYLENE	32.96	0.012	35.71349
DIPHENYLSTILBENE (B)	35.72	0.010	57.87527
1,1,4,4-TETRAPHENYLBUTADIENE	33.30	0.012	37.88393
1,6-DIPHENYLHEXATRIENE	36.43	0.009	64.98327
1,8-DIPHENYL-1,3,5,7-OCTATETRAENE	34.91	0.011	50.12629
BPSB (TRACE)	36.53	0.009	66.02707
BIS-MSB	35.83	0.010	58.80882
ESCULIN (A)	28.02	0.020	13.48916
AZULENE	21.39	0.046	2.65344
NAPHTHALENE	7.60	1.020	0.00546
DEUTERATED NAPHTHALENE-C10D8	9.78	0.478	0.02470
1-METHYLNAPHTHALENE 5ML/L	9.16	0.582	0.01675
2-METHYL NAPHTHALENE	11.30	0.310	0.05901
2,3-DIMETHYL NAPHTHALENE	11.05	0.332	0.05155
2,6-DIMETHYL NAPHTHALENE (TRACE)	14.29	0.153	0.24105
1-NAPHTHOL (TRACE)	16.03	0.109	0.47945
2-NAPHTHOL (TRACE)	17.64	0.082	0.84988
1-NAPHTHYLAMINE	22.18	0.041	3.34324
2-CHLORONAPHTHALENE	11.81	0.272	0.07677
ACENAPHTHENE	14.13	0.158	0.22586
1-PHENYL NAPHTHALENE	14.17	0.157	0.22944
2-PHENYL NAPHTHALENE	21.22	0.047	2.58036
1,4-DIPHENYL NAPHTHALENE	21.78	0.043	3.00548
1,5-DIPHENYL NAPHTHALENE	20.38	0.053	2.02159
1,7-DIPHENYL NAPHTHALENE	20.08	0.055	1.84821
1,4,5,8-TETRAPHENYL NAPHTHALENE	30.04	0.016	20.43278
1,1'-BINAPHTHYL	16.49	0.100	0.56855
4,4'-DI(N-BUTOXY)-1,1'-BINAPHTHYL	25.82	0.026	8.31821
2,2'-BINAPHTHYL	22.15	0.041	3.33011
ACRIDINE (A)	25.37	0.027	7.41381
ANTHRACENE (TRACE)	23.82	0.033	5.09164
DEUTERATED ANTHRACENE C14D10 (TRACE)	23.64	0.034	4.87028
9-METHYL ANTHRACENE (TRACE)	23.11	0.036	4.24012
1-AMINO ANTHRACENE	21.46	0.045	2.70544
2-AMINO ANTHRACENE	22.38	0.040	3.47725
9-VINYL ANTHRACENE	23.83	0.033	5.08522

ACCEPTOR MOLECULES	RO	CO	OVERLAP
9-PHENYL ANTHRACENE (TRACE)	24.34	0.031	5.77435
9,10-DIPHENYL ANTHRACENE	24.44	0.031	5.90922
9,10-DICHLOROANTHRACENE (TRACE)	23.20	0.036	4.31747
BIANTHRANYL (B)	25.97	0.026	6.50744
9,10-DINAPHTHYLANTHRACENE (B)	24.21	0.032	5.58530
9-(P-VINYLPHENYL)-10-PHENYL ANTHRACENE	24.50	0.030	5.99548
9,10-BIS(PHENYL-ETHYNYL)ANTHRACENE (TRACE)	22.42	0.040	3.52294
RUBICENE (B)	26.15	0.025	8.85855
DIBENZ(A,H)ANTHRACENE (B)	28.51	0.019	14.90173
TETRACENE (TRACE) (B)	18.23	0.074	1.01744
RUBRENE (B)	26.48	0.024	9.56213
TETRAHYDROCANNABINOL ACID METHYL ESTER	11.12	0.326	0.05364
PHENANTHRENE	13.37	0.187	0.16074
3,4-BENZOPHENANTHRENE	21.17	0.047	2.51066
2-PHENYL PHENANTHRENE	16.32	0.103	0.53009
CHRYSENE (TRACE)	22.24	0.041	3.38092
TRIPHENYLENE	14.40	0.150	0.25025
DEUTERATED TRIPHENYLENE-C18D12	14.21	0.156	0.23128
PYRENE	29.32	0.018	17.73043
1,3,6,8,-TETRAPHENYL PYRENE	30.14	0.016	20.76393
BIPYRENYL (B)	33.67	0.012	40.37206
O-PHENYLENE-PYRENE	27.90	0.021	13.06082
ANTHANTHRENE (B)	26.24	0.025	9.04948
FLUORANTHENE	25.45	0.027	7.52621
3-PHENYL FLUORANTHENE	27.16	0.022	11.12702
3-AMINO FLUORANTHENE	25.31	0.028	7.27723
3,3'-BIFLUORANTHENYL (B)	30.12	0.016	20.70932
BENZO(GHI)FLUORANTHENE	24.75	0.030	6.36745
9,12-DI(N-OCTADECOXY)-BENZ(E)ACEPHENANTHRYLENE	26.96	0.023	10.64235
DECACYCLENE (B)	32.92	0.013	35.27281
10-ETHOXY-BENZO(KL)XANTHENE	24.93	0.029	6.65164
PERYLENE	21.87	0.043	3.03251
BENZO(GHI)PERYLENE (B)	26.73	0.023	10.11378
BBOT (TRACE)	34.26	0.011	44.85848
QUININE SULFATE (A)	25.36	0.027	7.47676
ACRIDINE YELLOW (A)	19.11	0.064	1.34828
ACRIDINE ORANGE (A)	18.22	0.074	1.01424
ACRIDINE RED (A)	21.75	0.043	2.93474
URANINE (A)	22.26	0.041	3.37095
RHODAMINE B (A)	24.62	0.030	6.17794
RHODAMINE 6G (A)	25.04	0.029	6.82465
EOSIN B (A)	23.73	0.033	4.94914
EOSIN Y (A)	24.15	0.032	5.50018
ROSE BENGAL (A)	25.23	0.028	7.14519
INDENOINDENE	27.03	0.023	10.99373

8.141 Donor Molecule: 1-Methylnaphthalene (5 ml/L)

N,N-DIMETHYLANILINE 3ML/L	9.77	0.479	0.02668
TRIPHENYLAMINE	21.77	0.043	3.22675
1,4-DIMETHOXY BENZENE	2.73	21.935	0.00001
1-METHYL-4-CHLOROBENZENE 5ML/L	0.00	0.000	0.00000
SODIUM SALICYLATE (A)	14.55	0.145	0.28758
ANTHRANILIC ACID (A)	23.54	0.034	5.12956
TMPD	20.33	0.053	2.13129
4,4-DIHYDROXYBIPHENYL (A)	9.22	0.570	0.01887
4,4'-DIMETHOXY BIPHENYL	6.86	1.389	0.00319
4,4'-DIISOBUTOXYBIPHENYL	4.55	4.746	0.00027
BENZIDINE	19.20	0.063	1.52478
4-VINYL BIPHENYL	9.32	0.552	0.02011
M-TERPHENYL	6.21	1.866	0.00177
1,3,5-TRIPHENYL BENZENE	8.53	0.720	0.01180
1,3-DIBIPHENYLYLBENZENE (B)	18.23	0.074	1.11617
1,3,5,-TRIBIPHENYLYLBENZENE (B)	21.58	0.045	3.07861
FLUORENE (TRACE)	0.00	0.000	0.00000
1-METHYL FLUORENE	3.19	13.741	0.00003
1,2-BENZOFLUORENE	20.07	0.055	1.98462
9,10-DIHYDROPHENANTHRENE	4.92	3.749	0.00044
CARBAZOLE (TRACE)	19.30	0.062	1.57299
N-METHYL CARBAZOLE (TRACE)	21.49	0.045	2.99099
N-VINYLCARBAZOLE	21.76	0.043	3.22085
N-PHENYL CARBAZOLE	21.45	0.045	2.95318
N-METHYL-2,3-BENZCARBAZOLE	23.42	0.035	4.93768
DIBENZOFURAN	0.00	0.000	0.00000
DIBENZOTHIOPHENE (TRACE)	17.42	0.085	0.85227
P-TERPHENYL	11.47	0.297	0.06961
DEUTERATED P-TERPHENYL-C18D14	10.65	0.371	0.04466
4-METHYL TERPHENYL	13.17	0.196	0.15981
2-METHYL-P-TERPHENYL	4.27	5.742	0.00019
3,3''-DIMETHYL-P-TERPHENYL	13.14	0.197	0.15749
4-(3,3-DIMETHYLBUTOXY)-P-TERPHENYL	18.21	0.074	1.11149
4,4''-DIHEXAHYDROFARNESOXY-P-TERPHENYL	21.24	0.047	2.79692
2,2'-METHYLENE-P-TERPHENYL	20.06	0.055	1.98641
2,2'-ETHYLENE-P-TERPHENYL	20.98	0.048	2.59863
2,2'-METHYLENE-5',6''-METHYLENE-P-TERPHENYL (TRACE)	29.71	0.017	20.91012
3-PHENYL DIBENZOFURAN	20.09	0.055	2.00634
BENZO(1,2-B-4,5-B')BISBENZOFURAN (TRACE) (B)	31.02	0.015	27.03071
SEXIPHENYL BENZENE	0.00	0.000	0.00000
P-QUATERPHENYL	22.48	0.039	3.92848
3,3'''-DIMETHYL-P-QUATERPHENYL	23.78	0.033	5.49619
TETRAMETHYL-P-QUATERPHENYL	6.60	1.554	0.00255

ACCEPTOR MOLECULES	RO	CO	OVERLAP
3,3''' -DIMETHYL-3',2''-METHYLENE-P-QUATERPHENYL	30.89	0.015	26.36014
3,3''' -DIMETHYL-3',2''-ETHYLENE-P-QUATERPHENYL	30.44	0.016	24.07441
4-(3-ETHYLBUTOXY)-P-QUATERPHENYL (B)	28.10	0.020	14.89347
4,4'''-DI(2-BUTYLOCTOXY-1)-P-QUATERPHENYL	28.76	0.019	17.14029
DIETHYL-P-QUINQUEPHENYL	24.82	0.029	7.10795
DI(3-ETHYLHEPTYL)-P-QUINQUEPHENYL	29.97	0.017	21.88220
TETRAMETHYL-P-SEXIPHENYL	23.73	0.033	5.43889
TETRAMETHYL-P-OCTAPHENYL (B)	33.62	0.012	43.60248
INDOLE (TRACE)	0.00	0.000	0.00000
2-PHENYL INDOLE	24.80	0.029	7.07719
1-METHYL-2-PHENYL INDOLE	19.96	0.056	1.92303
1-NEOPENTYL-2-PHENYL INDOLE	17.61	0.082	0.90920
1-N-BUTYL-2(BIPHENYLYL-4) INDOLE	27.56	0.021	13.23889
1,2-DIPHENYL INDOLE	20.73	0.050	2.41677
1-BIPHENYLYL-2-PHENYL INDOLE	23.04	0.037	4.53331
1,3-DIMETHYL-2-PHENYL INDOLE	20.84	0.049	2.49053
2,3-DIPHENYL INDOLE	25.50	0.027	8.30736
1-METHYL-2,3-DIPHENYL INDOLE	23.16	0.036	4.69513
2-PHENYL-3,2'-METHYLENE INDOLE (TRACE)	27.52	0.021	13.19453
2-PHENYL-3,2'-ETHYLENE INDOLE (TRACE)	29.19	0.018	18.72393
2-PHENYL-3,2'-TRIMETHYLENE INDOLE	26.35	0.024	10.16135
1-METHYL-3,2'-METHYLENE-2-PHENYL INDOLE (TRACE)	28.93	0.018	17.75849
1-METHYL-3,2'-ETHYLENE-2-PHENYL INDOLE	29.34	0.018	19.29324
1-METHYL-3,2'-TRIMETHYLENE-2-PHENYL INDOLE	25.20	0.028	7.77453
4,5-DIPHENYLIMIDAZOLE (A)	15.87	0.112	0.48620
PPF (TRACE)	30.41	0.016	23.95748
PPO (TRACE)	26.58	0.024	10.72870
2-(4-FLUOROPHENYL)-5-PHENYLOXAZOLE (TRACE)	25.33	0.028	8.02926
αNPO (TRACE)	31.00	0.015	26.71609
BBO (B)	35.22	0.010	57.35896
POPOP	33.79	0.012	44.67177
DIMETHYL POPOP (TRACE)	34.29	0.011	48.65385
PPD (TRACE)	10.84	0.351	0.04990
2,5-DI(2-METHYLPHENYL)OXADIAZOLE	12.82	0.212	0.13615
PBD	25.58	0.027	8.51681
BBD (B)	32.24	0.013	34.01729
αNPD (TRACE)	27.84	0.021	14.07134
βNPD	24.57	0.030	6.69080
αNND (TRACE)	29.52	0.017	19.95485
βNND (TRACE)	29.53	0.017	20.12941
TRANS-1-(4-BIPHENYLYL)-2-PHENYL-ETHYLENE	31.81	0.014	31.28554
TRANS-1-(4-BIPHENYLYL)-2-(1-NAPHTHYL)-ETHYLENE	32.74	0.013	36.98686
DIPHENYLSTILBENE (B)	35.60	0.010	61.09939
1,1,4,4-TETRAPHENYLBUTADIENE	33.22	0.012	40.27285
1,6-DIPHENYLHEXATRIENE	36.41	0.009	69.87221
1,8-DIPHENYL-1,3,5,7-OCTATETRAENE	34.79	0.011	53.03889
BPSB (TRACE)	36.49	0.009	70.82516
BIS-MSB	35.74	0.010	62.44642
ESCULIN (A)	27.86	0.021	14.06464
AZULENE	21.24	0.047	2.74247
NAPHTHALENE	6.35	1.751	0.00200
DEUTERATED NAPHTHALENE-C10D8	9.02	0.610	0.01638
1-METHYLNAPHTHALENE 5ML/L	8.25	0.798	0.00961
2-METHYL NAPHTHALENE	10.14	0.429	0.03322
2,3-DIMETHYL NAPHTHALENE	9.99	0.448	0.03046
2,6-DIMETHYL NAPHTHALENE (TRACE)	13.81	0.170	0.21207
1-NAPHTHOL (TRACE)	14.77	0.139	0.31683
2-NAPHTHOL (TRACE)	17.11	0.089	0.76527
1-NAPHTHYLAMINE	21.65	0.044	3.12491
2-CHLORONAPHTHALENE	10.82	0.353	0.04898
ACENAPHTHENE	12.66	0.221	0.12563
1-PHENYL NAPHTHALENE	12.91	0.208	0.14168
2-PHENYL NAPHTHALENE	19.84	0.057	1.86109
1,4-DIPHENYL NAPHTHALENE	20.71	0.050	2.39958
1,5-DIPHENYL NAPHTHALENE	19.23	0.063	1.54074
1,7-DIPHENYL NAPHTHALENE	18.95	0.066	1.41025
1,4,5,8-TETRAPHENYL NAPHTHALENE	29.84	0.017	21.19037
1,1'-BINAPHTHYL	15.41	0.122	0.40876
4,4'-DI(N-BUTOXY)-1,1'-BINAPHTHYL	25.05	0.028	7.48280
2,2'-BINAPHTHYL	20.82	0.050	2.47552
ACRIDINE (A)	25.42	0.027	8.09135
ANTHRACENE (TRACE)	23.83	0.033	5.49985
DEUTERATED ANTHRACENE C14D10 (TRACE)	23.63	0.034	5.23134
9-METHYL ANTHRACENE (TRACE)	23.07	0.036	4.51881
1-AMINO ANTHRACENE	21.32	0.046	2.80586
2-AMINO ANTHRACENE	22.15	0.041	3.53415
9-VINYL ANTHRACENE	23.78	0.033	5.40583
9-PHENYL ANTHRACENE (TRACE)	24.30	0.031	6.17766
9,10-DIPHENYL ANTHRACENE	24.35	0.031	6.22777
9,10-DICHLOROANTHRACENE (TRACE)	23.07	0.036	4.50722
BIANTHRANYL (B)	25.83	0.026	8.87376
9,10-DINAPHTHYLANTHRACENE (B)	24.00	0.032	5.70477
9-(P-VINYLPHENYL)-10-PHENYL ANTHRACENE	24.38	0.031	6.27285
9,10-BIS(PHENYL-ETHYNYL)ANTHRACENE (TRACE)	21.26	0.047	2.76350
RUBICENE (B)	25.43	0.027	8.08150
DIBENZ(A,H)ANTHRACENE (B)	28.07	0.020	14.65557
TETRACENE (TRACE) (B)	17.87	0.078	0.97440
RUBRENE (B)	25.14	0.028	7.54989
TETRAHYDROCANNABINOL ACID METHYL ESTER	9.54	0.514	0.02318
PHENANTHRENE	13.17	0.196	0.15808
3,4-BENZOPHENANTHRENE	20.21	0.054	2.04773
2-PHENYL PHENANTHRENE	15.33	0.124	0.39215
CHRYSENE (TRACE)	20.79	0.050	2.43644
TRIPHENYLENE	13.98	0.164	0.22683
DEUTERATED TRIPHENYLENE-C18D12	13.79	0.171	0.20851
PYRENE	28.64	0.019	16.59790

ACCEPTOR MOLECULES	RO	CO	OVERLAP
1,3,6,8,-TETRAPHENYL PYRENE	30.03	0.017	21.91650
BIPYRENYL (B)	33.62	0.012	43.19880
O-PHENYLENE-PYRENE	27.62	0.021	13.27427
ANTHANTHRENE (B)	25.51	0.027	8.24424
FLUORANTHENE	25.37	0.027	7.97605
3-PHENYL FLUORANTHENE	27.11	0.022	11.86163
3-AMINO FLUORANTHENE	24.97	0.029	7.25238
3,3'-BIFLUORANTHENYL (B)	29.90	0.017	21.36096
BENZO(GHI)FLUORANTHENE	24.63	0.030	6.66615
9,12-DI(N-OCTADECOXY)-BENZ(E)ACEPHENANTHRYLENE	26.77	0.023	10.99716
DECACYCLENE (B)	32.62	0.013	35.98204
10-ETHOXY-BENZO(KL)XANTHENE	24.80	0.029	6.96551
PERYLENE	21.43	0.045	2.89369
BENZO(GHI)PERYLENE (B)	26.53	0.024	10.42302
BBOT (TRACE)	34.20	0.011	47.89548
QUININE SULFATE (A)	25.00	0.029	7.39446
ACRIDINE YELLOW (A)	18.74	0.068	1.29419
ACRIDINE ORANGE (A)	17.64	0.081	0.90077
ACRIDINE RED (A)	21.54	0.045	2.98462
URANINE (A)	21.82	0.043	3.22903
RHODAMINE B (A)	24.40	0.031	6.30121
RHODAMINE 6G (A)	25.05	0.028	7.38516
EOSIN B (A)	23.23	0.036	4.70141
EOSIN Y (A)	23.62	0.034	5.18406
ROSE BENGAL (A)	24.82	0.029	6.98399
INDENOINDENE	26.10	0.025	9.60992

8.142 Donor Molecule: 2-Methylnaphthalene

ACCEPTOR MOLECULES	RO	CO	OVERLAP
N,N-DIMETHYLANILINE 3ML/L	10.20	0.422	0.02687
TRIPHENYLAMINE	23.24	0.036	3.73376
1,4-DIMETHOXY BENZENE	2.64	24.225	0.00001
1-METHYL-4-CHLOROBENZENE 5ML/L	0.00	0.000	0.00000
SODIUM SALICYLATE (A)	15.50	0.120	0.32874
ANTHRANILIC ACID (A)	24.65	0.030	5.27845
TMPD	21.42	0.046	2.27615
4,4'-DIHYDROXYBIPHENYL (A)	10.14	0.430	0.02596
4,4'-DIMETHOXY BIPHENYL	6.16	1.917	0.00131
4,4'-DIISOBUTOXYBIPHENYL	4.32	5.556	0.00016
BENZIDINE	20.59	0.051	1.80974
4-VINYL BIPHENYL	8.76	0.666	0.01081
M-TERPHENYL	5.39	2.858	0.00059
1,3,5-TRIPHENYL BENZENE	9.18	0.577	0.01433
1,3-DIBIPHENYLYLBENZENE (B)	19.58	0.060	1.34251
1,3,5,-TRIBIPHENYLYLBENZENE (B)	23.21	0.036	3.72586
FLUORENE (TRACE)	0.00	0.000	0.00000
1-METHYL FLUORENE	3.04	15.923	0.00002
1,2-BENZOFLUORENE	21.58	0.045	2.39131
9,10-DIHYDROPHENANTHRENE	4.36	5.389	0.00017
CARBAZOLE (TRACE)	20.64	0.051	1.83454
N-METHYL CARBAZOLE (TRACE)	22.62	0.039	3.17100
N-VINYLCARBAZOLE	22.90	0.037	3.42147
N-PHENYL CARBAZOLE	22.59	0.039	3.14850
N-METHYL-2,3-BENZCARBAZOLE	24.43	0.031	4.98546
DIBENZOFURAN	0.00	0.000	0.00000
DIBENZOTHIOPHENE (TRACE)	18.52	0.070	0.96151
P-TERPHENYL	12.33	0.238	0.08420
DEUTERATED P-TERPHENYL-C18D14	10.79	0.356	0.03774
4-METHYL TERPHENYL	14.21	0.156	0.19666
2-METHYL-P-TERPHENYL	4.05	6.711	0.00011
3,3''-DIMETHYL-P-TERPHENYL	14.11	0.159	0.18835
4-(3,3-DIMETHYLBUTOXY)-P-TERPHENYL	19.60	0.059	1.35269
4,4''-DIHEXAHYDROFARNESOXY-P-TERPHENYL	22.84	0.038	3.38130
2,2'-METHYLENE-P-TERPHENYL	21.67	0.044	2.47023
2,2'-ETHYLENE-P-TERPHENYL	22.65	0.038	3.21889
2,2'-METHYLENE-5',6'-METHYLENE-P-TERPHENYL (TRACE)	31.69	0.014	24.03300
3-PHENYL DIBENZOFURAN	21.67	0.044	2.46837
BENZO(1,2-B-4,5-B')BISBENZOFURAN (TRACE) (B)	32.73	0.013	29.15441
SEXIPHENYL BENZENE	0.00	0.000	0.00000
P-QUATERPHENYL	24.12	0.032	4.68339
3,3'''-DIMETHYL-P-QUATERPHENYL	25.47	0.027	6.48669
TETRAMETHYL-P-QUATERPHENYL	5.76	2.338	0.00088
3,3''' -DIMETHYL-3',2''-METHYLENE-P-QUATERPHENYL	32.82	0.013	29.59571
3,3''' -DIMETHYL-3',2''-ETHYLENE-P-QUATERPHENYL	32.32	0.013	26.95590
4-(3-ETHYLBUTOXY)-P-QUATERPHENYL (B)	29.92	0.017	16.96341
4,4'''-DI(2-BUTYLOCTOXY-1)-P-QUATERPHENYL	30.64	0.016	19.58092
DIETHYL-P-QUINQUEPHENYL	26.59	0.024	8.38517
DI(3-ETHYLHEPTYL)-P-QUINQUEPHENYL	31.90	0.014	24.86072
TETRAMETHYL-P-SEXIPHENYL	25.47	0.027	6.49157
TETRAMETHYL-P-OCTAPHENYL (B)	35.62	0.010	48.17829
INDOLE (TRACE)	0.00	0.000	0.00000
2-PHENYL INDOLE	26.60	0.024	8.41364
1-METHYL-2-PHENYL INDOLE	21.43	0.045	2.29859
1-NEOPENTYL-2-PHENYL INDOLE	18.99	0.065	1.11599
1-N-BUTYL-2(BIPHENYLYL-4) INDOLE	29.22	0.018	14.69970
1,2-DIPHENYL INDOLE	22.28	0.040	2.90527
1-BIPHENYLYL-2-PHENYL INDOLE	24.70	0.030	5.36928
1,3-DIMETHYL-2-PHENYL INDOLE	22.29	0.040	2.91890
2,3-DIPHENYL INDOLE	27.05	0.023	9.24436
1-METHYL-2,3-DIPHENYL INDOLE	24.61	0.030	5.24362
2-PHENYL-3,2'-METHYLENE INDOLE (TRACE)	29.39	0.018	15.27758

ACCEPTOR MOLECULES	RO	CO	OVERLAP
2-PHENYL-3,2'-ETHYLENE INDOLE (TRACE)	30.80	0.015	20.19393
2-PHENYL-3,2'-TRIMETHYLENE INDOLE	28.16	0.020	11.81944
1-METHYL-3,2'-METHYLENE-2-PHENYL INDOLE (TRACE)	30.61	0.016	19.45643
1-METHYL-3,2'-ETHYLENE-2-PHENYL INDOLE	30.88	0.015	20.47872
1-METHYL-3,2'-TRIMETHYLENE-2-PHENYL INDOLE	26.84	0.023	8.86568
4,5-DIPHENYLIMIDAZOLE (A)	17.01	0.091	0.57535
PPF (TRACE)	32.16	0.013	26.18486
PPO (TRACE)	28.45	0.019	12.58333
2-(4-FLUOROPHENYL)-5-PHENYLOXAZOLE (TRACE)	27.10	0.022	9.40084
αNPO (TRACE)	32.45	0.013	27.42650
BBO (B)	36.71	0.009	57.47212
POPOP	34.97	0.010	42.91307
DIMETHYL POPOP (TRACE)	35.36	0.010	[illegible]
PPD (TRACE)	9.79	0.476	0.02117
2,5-DI(2-METHYLPHENYL)OXADIAZOLE	13.16	0.196	0.12436
PBD	27.40	0.022	10.06251
BBD (B)	34.14	0.011	37.46226
αNPD (TRACE)	29.41	0.018	15.28232
βNPD	26.17	0.025	7.62232
αNND (TRACE)	30.98	0.015	20.85567
βNND (TRACE)	31.32	0.015	22.37144
TRANS-1-(4-BIPHENYLYL)-2-PHENYL-ETHYLENE	33.61	0.012	34.02642
TRANS-1-(4-BIPHENYLYL)-2-(1-NAPHTHYL)-ETHYLENE	34.23	0.011	37.77477
DIPHENYLSTILBENE (B)	37.11	0.009	61.27853
1,1,4,4-TETRAPHENYLBUTADIENE	34.54	0.011	39.80993
1,6-DIPHENYLHEXATRIENE	37.73	0.008	67.54927
1,8-DIPHENYL-1,3,5,7-OCTATETRAENE	35.50	0.010	46.88712
BPSB (TRACE)	37.76	0.008	67.92709
BIS-MSB	37.11	0.009	61.20688
ESCULIN (A)	29.13	0.018	14.37726
AZULENE	22.33	0.040	2.90106
NAPHTHALENE	6.76	1.445	0.00229
DEUTERATED NAPHTHALENE-C10D8	9.72	0.488	0.02002
1-METHYLNAPHTHALENE 5ML/L	8.63	0.695	0.00990
2-METHYL NAPHTHALENE	11.75	0.276	0.06291
2,3-DIMETHYL NAPHTHALENE	11.23	0.316	0.04789
2,6-DIMETHYL NAPHTHALENE (TRACE)	14.49	0.147	0.22125
1-NAPHTHOL (TRACE)	15.98	0.110	0.39677
2-NAPHTHOL (TRACE)	18.21	0.074	0.86766
1-NAPHTHYLAMINE	23.00	0.037	3.50408
2-CHLORONAPHTHALENE	11.85	0.269	0.06591
ACENAPHTHENE	14.20	0.156	0.19584
1-PHENYL NAPHTHALENE	13.91	0.166	0.17274
2-PHENYL NAPHTHALENE	21.18	0.047	2.14862
1,4-DIPHENYL NAPHTHALENE	22.20	0.041	2.83933
1,5-DIPHENYL NAPHTHALENE	20.66	0.051	1.84914
1,7-DIPHENYL NAPHTHALENE	20.35	0.053	1.68912
1,4,5,8-TETRAPHENYL NAPHTHALENE	31.22	0.015	21.69143
1,1'-BINAPHTHYL	16.52	0.099	0.48451
4,4'-DI(N-BUTOXY)-1,1'-BINAPHTHYL	26.65	0.024	8.48431
2,2'-BINAPHTHYL	22.37	0.040	2.97652
ACRIDINE (A)	26.26	0.025	7.68779
ANTHRACENE (TRACE)	24.63	0.030	5.23797
DEUTERATED ANTHRACENE C14D10 (TRACE)	24.45	0.031	5.01612
9-METHYL ANTHRACENE (TRACE)	23.63	0.034	4.07966
1-AMINO ANTHRACENE	21.64	0.044	2.40698
2-AMINO ANTHRACENE	22.97	0.037	3.44079
9-VINYL ANTHRACENE	24.31	0.031	4.83460
9-PHENYL ANTHRACENE (TRACE)	24.91	0.029	5.59593
9,10-DIPHENYL ANTHRACENE	24.75	0.030	5.38007
9,10-DICHLOROANTHRACENE (TRACE)	23.36	0.035	3.80739
BIANTHRANYL (B)	26.34	0.024	7.82578
9,10-DINAPHTHYLANTHRACENE (B)	24.45	0.031	5.01097
9-(P-VINYLPHENYL)-10-PHENYL ANTHRACENE	24.75	0.029	5.38813
9,10-BIS(PHENYL-ETHYNYL)ANTHRACENE (TRACE)	21.19	0.047	2.12179
RUBICENE (B)	26.61	0.024	8.32576
DIBENZ(A,H)ANTHRACENE (B)	29.53	0.017	15.53314
TETRACENE (TRACE) (B)	18.35	0.072	0.89528
RUBRENE (B)	26.78	0.023	8.64273
TETRAHYDROCANNABINOL ACID METHYL ESTER	10.33	0.406	0.02907
PHENANTHRENE	13.82	0.169	0.16528
3,4-BENZOPHENANTHRENE	21.57	0.045	2.36217
2-PHENYL PHENANTHRENE	16.01	0.109	0.39830
CHRYSENE (TRACE)	22.80	0.038	3.31113
TRIPHENYLENE	14.92	0.135	0.26176
DEUTERATED TRIPHENYLENE-C18D12	14.77	0.139	0.24557
PYRENE	30.74	0.015	19.84084
1,3,6,8,-TETRAPHENYL PYRENE	30.47	0.016	18.73213
BIPYRENYL (B)	34.87	0.011	42.08802
O-PHENYLENE-PYRENE	28.31	0.020	12.05409
ANTHANTHRENE (B)	26.31	0.025	7.77796
FLUORANTHENE	26.37	0.024	7.88593
3-PHENYL FLUORANTHENE	27.93	0.021	11.11224
3-AMINO FLUORANTHENE	25.83	0.026	6.95497
3,3'-BIFLUORANTHENYL (B)	30.71	0.015	19.63921
BENZO(GHI)FLUORANTHENE	25.73	0.026	6.80089
9,12-DI(N-OCTADECOXY)-BENZ(E)ACEPHENANTHRYLENE	27.61	0.021	10.37731
DECACYCLENE (B)	33.63	0.012	33.90976
10-ETHOXY-BENZO(KL)XANTHENE	25.22	0.028	6.02219
PERYLENE	21.08	0.048	2.05592
BENZO(GHI)PERYLENE (B)	27.05	0.023	9.17529
BBOT (TRACE)	35.08	0.010	43.67585
QUININE SULFATE (A)	26.44	0.024	8.08640
ACRIDINE YELLOW (A)	18.63	0.069	0.97927
ACRIDINE ORANGE (A)	18.13	0.075	0.83161
ACRIDINE RED (A)	22.32	0.040	2.90056

ACCEPTOR MOLECULES	RO	CO	OVERLAP
URANINE (A)	22.86	0.037	3.34052
RHODAMINE B (A)	25.25	0.028	6.07519
RHODAMINE 6G (A)	25.97	0.026	7.18397
EOSIN B (A)	24.41	0.031	4.95420
EOSIN Y (A)	24.80	0.029	5.44652
ROSE BENGAL (A)	26.07	0.025	7.35819
INDENOINDENE	27.85	0.021	11.07545

8.143 Donor Molecule: 2,3-Dimethylnaphthalene

ACCEPTOR MOLECULES	RO	CO	OVERLAP
N,N-DIMETHYLANILINE 3ML/L	10.01	0.447	0.02017
TRIPHENYLAMINE	23.63	0.034	3.47158
SODIUM SALICYLATE (A)	15.74	0.115	0.30287
ANTHRANILIC ACID (A)	25.39	0.027	5.30421
TMPD	22.01	0.042	2.25695
4,4'-DIHYDROXYBIPHENYL (A)	9.95	0.454	0.01958
4,4'-DIMETHOXY BIPHENYL	6.08	1.988	0.00102
4,4'-DIISOBUTOXYBIPHENYL	0.00	0.000	0.00000
BENZIDINE	20.85	0.049	1.64179
4-VINYL BIPHENYL	8.47	0.736	0.00745
M-TERPHENYL	4.57	4.697	0.00018
1,3,5-TRIPHENYL BENZENE	9.14	0.587	0.01169
1,3-DIBIPHENYLYLBENZENE (B)	19.68	0.059	1.16327
1,3,5,-TRIBIPHENYLYLBENZENE (B)	23.37	0.035	3.26625
FLUORENE (TRACE)	0.00	0.000	0.00000
1-METHYL FLUORENE	0.00	0.000	0.00000
1,2-BENZOFLUORENE	21.74	0.044	2.10273
9,10-DIHYDROPHENANTHRENE	3.43	11.065	0.00003
CARBAZOLE (TRACE)	21.11	0.048	1.76730
N-METHYL CARBAZOLE (TRACE)	23.29	0.035	3.18312
N-VINYLCARBAZOLE	23.54	0.034	3.39640
N-PHENYL CARBAZOLE	23.22	0.036	3.12436
N-METHYL-2,3-BENZCARBAZOLE	25.22	0.028	5.07226
DIBENZOFURAN	0.00	0.000	0.00000
DIBENZOTHIOPHENE (TRACE)	18.89	0.066	0.90962
P-TERPHENYL	12.06	0.255	0.06202
DEUTERATED P-TERPHENYL-C18D14	10.31	0.408	0.02423
4-METHYL TERPHENYL	14.02	0.162	0.15290
2-METHYL-P-TERPHENYL	0.00	0.000	0.00000
3,3''-DIMETHYL-P-TERPHENYL	13.98	0.164	0.15043
4-(3,3-DIMETHYLBUTOXY)-P-TERPHENYL	19.73	0.058	1.18253
4,4'-DIHEXAHYDROFARNESOXY-P-TERPHENYL	23.08	0.036	3.03144
2,2'-METHYLENE-P-TERPHENYL	21.69	0.044	2.09002
2,2'-ETHYLENE-P-TERPHENYL	22.81	0.038	2.82197
2,2'-METHYLENE-5',6''-METHYLENE-P-TERPHENYL (TRACE)	32.44	0.013	23.26325
3-PHENYL DIBENZOFURAN	21.68	0.044	2.08260
BENZO(1,2-B-4,5-B')BISBENZOFURAN (TRACE) (B)	33.69	0.012	29.14743
SEXIPHENYL BENZENE	0.00	0.000	0.00000
P-QUATERPHENYL	24.45	0.031	4.26900
3,3'''-DIMETHYL-P-QUATERPHENYL	25.86	0.026	5.97766
TETRAMETHYL-P-QUATERPHENYL	4.80	4.056	0.00025
3,3''' -DIMETHYL-3',2''-METHYLENE-P-QUATERPHENYL	33.62	0.012	28.79910
3,3''' -DIMETHYL-3',2''-ETHYLENE-P-QUATERPHENYL	33.10	0.012	26.14280
4-(3-ETHYLBUTOXY)-P-QUATERPHENYL (B)	30.55	0.016	16.18547
4,4'''-DI(2-BUTYLOCTOXY-1)-P-QUATERPHENYL	31.27	0.015	18.63694
DIETHYL-P-QUINQUEPHENYL	27.00	0.023	7.74027
DI(3-ETHYLHEPTYL)-P-QUINQUEPHENYL	32.58	0.013	23.75754
TETRAMETHYL-P-SEXIPHENYL	25.80	0.026	5.90116
TETRAMETHYL-P-OCTAPHENYL (B)	36.49	0.009	46.88736
INDOLE (TRACE)	0.00	0.000	0.00000
2-PHENYL INDOLE	27.06	0.023	7.85817
1-METHYL-2-PHENYL INDOLE	21.69	0.044	2.08367
1-NEOPENTYL-2-PHENYL INDOLE	19.07	0.064	0.96578
1-N-BUTYL-2(BIPHENYLYL-4) INDOLE	29.92	0.017	14.25892
1,2-DIPHENYL INDOLE	22.54	0.039	2.61972
1-BIPHENYLYL-2-PHENYL INDOLE	25.06	0.028	4.92983
1,3-DIMETHYL-2-PHENYL INDOLE	22.68	0.038	2.72441
2,3-DIPHENYL INDOLE	27.69	0.021	8.94586
1-METHYL-2,3-DIPHENYL INDOLE	25.18	0.028	5.06068
2-PHENYL-3,2'-METHYLENE INDOLE (TRACE)	30.02	0.017	14.60307
2-PHENYL-3,2'-ETHYLENE INDOLE (TRACE)	31.66	0.014	20.05169
2-PHENYL-3,2'-TRIMETHYLENE INDOLE	28.73	0.019	11.23441
1-METHYL-3,2'-METHYLENE-2-PHENYL INDOLE (TRACE)	31.42	0.014	19.17946
1-METHYL-3,2'-ETHYLENE-2-PHENYL INDOLE	31.73	0.014	20.27762
1-METHYL-3,2'-TRIMETHYLENE-2-PHENYL INDOLE	27.42	0.022	8.49053
4,5-DIPHENYLIMIDAZOLE (A)	17.22	0.088	0.52218
PPF (TRACE)	33.00	0.012	25.72390
PPO (TRACE)	28.97	0.018	11.81341
2-(4-FLUOROPHENYL)-5-PHENYLOXAZOLE (TRACE)	27.60	0.021	8.83766
αNPO (TRACE)	33.42	0.012	27.55476
BBO (B)	37.85	0.008	58.09092
POPOP	36.12	0.009	43.82807
DIMETHYL POPOP (TRACE)	36.55	0.009	46.96544
PPD (TRACE)	9.32	0.552	0.01326
2,5-DI(2-METHYLPHENYL)OXADIAZOLE	12.82	0.212	0.08948
PBD	27.85	0.021	9.33632
BBD (B)	35.00	0.010	36.61847
αNPD (TRACE)	30.19	0.016	15.04871
βNPD	26.66	0.024	7.17564
αNND (TRACE)	31.88	0.014	20.82894
βNND (TRACE)	32.08	0.014	21.73765

ACCEPTOR MOLECULES	RO	CO	OVERLAP
TRANS-1-(4-BIPHENYLYL)-2-PHENYL-ETHYLENE	34.49	0.011	33.46298
TRANS-1-(4-BIPHENYLYL)-2-(1-NAPHTHYL)-ETHYLENE	35.26	0.010	37.93697
DIPHENYLSTILBENE (B)	38.26	0.008	61.95050
1,1,4,4-TETRAPHENYLBUTADIENE	35.64	0.010	40.39064
1,6-DIPHENYLHEXATRIENE	38.94	0.008	68.72773
1,8-DIPHENYL-1,3,5,7-OCTATETRAENE	36.74	0.009	48.46573
BPSB (TRACE)	39.00	0.008	69.37269
BIS-MSB	38.29	0.008	62.16423
ESCULIN (A)	30.01	0.017	14.45813
AZULENE	22.96	0.037	2.89027
NAPHTHALENE	6.61	1.548	0.00168
DEUTERATED NAPHTHALENE-C10D8	9.79	0.477	0.01759
1-METHYLNAPHTHALENE 5ML/L	8.57	0.710	0.00799
2-METHYL NAPHTHALENE	11.47	0.296	0.04574
2,3-DIMETHYL NAPHTHALENE	11.14	0.324	0.03836
2,6-DIMETHYL NAPHTHALENE (TRACE)	14.76	0.139	0.20780
1-NAPHTHOL (TRACE)	16.26	0.104	0.37133
2-NAPHTHOL (TRACE)	18.69	0.069	0.85330
1-NAPHTHYLAMINE	23.56	0.034	3.40810
2-CHLORONAPHTHALENE	12.08	0.254	0.06247
ACENAPHTHENE	14.29	0.153	0.17133
1-PHENYL NAPHTHALENE	13.88	0.167	0.14386
2-PHENYL NAPHTHALENE	21.47	0.045	1.96057
1,4-DIPHENYL NAPHTHALENE	22.55	0.039	2.62345
1,5-DIPHENYL NAPHTHALENE	20.92	0.049	1.67926
1,7-DIPHENYL NAPHTHALENE	20.61	0.051	1.53324
1,4,5,8-TETRAPHENYL NAPHTHALENE	32.15	0.013	21.78129
1,1'-BINAPHTHYL	16.68	0.096	0.43211
4,4'-DI(N-BUTOXY)-1,1'-BINAPHTHYL	27.27	0.022	8.19071
2,2'-BINAPHTHYL	22.60	0.039	2.66265
ACRIDINE (A)	27.15	0.022	7.88963
ANTHRACENE (TRACE)	25.41	0.027	5.32374
DEUTERATED ANTHRACENE C14D10 (TRACE)	25.23	0.028	5.09318
9-METHYL ANTHRACENE (TRACE)	24.49	0.030	4.25768
1-AMINO ANTHRACENE	22.41	0.040	2.49815
2-AMINO ANTHRACENE	23.65	0.034	3.44694
9-VINYL ANTHRACENE	25.17	0.028	5.01632
9-PHENYL ANTHRACENE (TRACE)	25.81	0.026	5.82859
9,10-DIPHENYL ANTHRACENE	25.61	0.027	5.56794
9,10-DICHLOROANTHRACENE (TRACE)	24.21	0.032	3.96610
BIANTHRANYL (B)	27.25	0.022	8.07810
9,10-DINAPHTHYLANTHRACENE (B)	25.27	0.028	5.13577
9-(P-VINYLPHENYL)-10-PHENYL ANTHRACENE	25.62	0.027	5.57741
9,10-BIS(PHENYL-ETHYNYL)ANTHRACENE (TRACE)	21.77	0.043	2.09965
RUBICENE (B)	27.24	0.022	8.05581
DIBENZ(A,H)ANTHRACENE (B)	30.30	0.016	15.26008
TETRACENE (TRACE) (B)	18.89	0.066	0.89693
RUBRENE (B)	27.19	0.022	7.96521
TETRAHYDROCANNABINOL ACID METHYL ESTER	9.89	0.463	0.01885
PHENANTHRENE	14.26	0.154	0.16776
3,4-BENZOPHENANTHRENE	21.88	0.043	2.16950
2-PHENYL PHENANTHRENE	16.26	0.104	0.36781
CHRYSENE (TRACE)	22.90	0.037	2.86186
TRIPHENYLENE	15.27	0.126	0.25316
DEUTERATED TRIPHENYLENE-C18D12	15.04	0.132	0.23047
PYRENE	31.37	0.014	18.85154
1,3,6,8,-TETRAPHENYL PYRENE	31.58	0.014	19.55475
BIPYRENYL (B)	36.01	0.010	42.98658
O-PHENYLENE-PYRENE	29.20	0.018	12.22804
ANTHANTHRENE (B)	27.04	0.023	7.70414
FLUORANTHENE	27.21	0.022	7.99642
3-PHENYL FLUORANTHENE	28.86	0.019	11.39093
3-AMINO FLUORANTHENE	26.63	0.024	7.03825
3,3'-BIFLUORANTHENYL (B)	31.72	0.014	20.09168
BENZO(GHI)FLUORANTHENE	26.52	0.024	6.86438
9,12-DI(N-OCTADECOXY)-BENZ(E)ACEPHENANTHRYLENE	28.48	0.019	10.52949
DECACYCLENE (B)	34.69	0.011	34.36230
10-ETHOXY-BENZO(KL)XANTHENE	26.11	0.025	6.24813
PERYLENE	21.87	0.043	2.15614
BENZO(GHI)PERYLENE (B)	28.02	0.020	9.55169
BBOT (TRACE)	36.29	0.009	44.99937
QUININE SULFATE (A)	27.15	0.022	7.98549
ACRIDINE YELLOW (A)	19.28	0.062	1.01291
ACRIDINE ORANGE (A)	18.60	0.070	0.81623
ACRIDINE RED (A)	22.98	0.037	2.90617
URANINE (A)	23.48	0.035	3.30415
RHODAMINE B (A)	26.01	0.025	6.10129
RHODAMINE 6G (A)	26.81	0.023	7.32567
EOSIN B (A)	25.02	0.029	4.84188
EOSIN Y (A)	25.41	0.027	5.30926
ROSE BENGAL (A)	26.78	0.023	7.26452
INDENOINDENE	28.42	0.019	10.53769

8.144 Donor Molecule: 2,6-Dimethylnaphthalene (trace)

N,N-DIMETHYLANILINE 3ML/L	7.73	0.969	0.00361
TRIPHENYLAMINE	22.96	0.037	2.46088
SODIUM SALICYLATE (A)	15.30	0.125	0.21528
ANTHRANILIC ACID (A)	26.11	0.025	5.28717
TMPD	22.37	0.040	2.09415
4,4'-DIHYDROXYBIPHENYL (A)	6.56	1.583	0.00135

ACCEPTOR MOLECULES	RO	CO	OVERLAP
BENZIDINE	19.86	0.057	1.03353
1,3,5-TRIPHENYL BENZENE	7.94	0.894	0.00424
1,3-DIBIPHENYLYLBENZENE (B)	18.10	0.075	0.59388
1,3,5,-TRIBIPHENYLYLBENZENE (B)	21.63	0.044	1.72999
FLUORENE (TRACE)	0.00	0.000	0.00000
1-METHYL FLUORENE	0.00	0.000	0.00000
1,2-BENZOFLUORENE	20.67	0.051	1.30989
9,10-DIHYDROPHENANTHRENE	0.00	0.000	0.00000
CARBAZOLE (TRACE)	20.63	0.051	1.29887
N-METHYL CARBAZOLE (TRACE)	23.85	0.033	3.09490
N-VINYLCARBAZOLE	24.01	0.032	3.21759
N-PHENYL CARBAZOLE	23.68	0.034	2.96629
N-METHYL-2,3-BENZCARBAZOLE	25.53	0.027	4.59423
DIBENZOFURAN	0.00	0.000	0.00000
DIBENZOTHIOPHENE (TRACE)	18.75	0.068	0.73301
P-TERPHENYL	8.10	0.842	0.00479
DEUTERATED P-TERPHENYL-C18D14	5.44	2.776	0.00044
4-METHYL TERPHENYL	10.65	0.370	0.02478
2-METHYL-P-TERPHENYL	0.00	0.000	0.00000
3,3''-DIMETHYL-P-TERPHENYL	11.94	0.263	0.04914
4-(3,3-DIMETHYLBUTOXY)-P-TERPHENYL	18.25	0.074	0.62444
4,4''-DIHEXAHYDROFARNESOXY-P-TERPHENYL	21.71	0.044	1.77055
2,2'-METHYLENE-P-TERPHENYL	19.50	0.060	0.92986
2,2'-ETHYLENE-P-TERPHENYL	21.10	0.048	1.49040
2,2'-METHYLENE-5',6''-METHYLENE-P-TERPHENYL (TRACE)	32.33	0.013	19.22246
3-PHENYL DIBENZOFURAN	19.49	0.060	0.92740
BENZO(1,2-B-4,5-B')BISBENZOFURAN (TRACE) (B)	34.02	0.011	26.05629
SEXIPHENYL BENZENE	0.00	0.000	0.00000
P-QUATERPHENYL	23.32	0.035	2.71077
3,3'''-DIMETHYL-P-QUATERPHENYL	24.90	0.029	4.01610
TETRAMETHYL-P-QUATERPHENYL	0.00	0.000	0.00000
3,3''' -DIMETHYL-3',2''-METHYLENE-P-QUATERPHENYL	33.51	0.012	23.80796
3,3''' -DIMETHYL-3',2''-ETHYLENE-P-QUATERPHENYL	33.00	0.012	21.65251
4-(3-ETHYLBUTOXY)-P-QUATERPHENYL (B)	30.15	0.016	12.60626
4,4'''-DI(2-BUTYLOCTOXY-1)-P-QUATERPHENYL	30.81	0.015	14.35538
DIETHYL-P-QUINQUEPHENYL	26.00	0.025	5.20034
DI(3-ETHYLHEPTYL)-P-QUINQUEPHENYL	32.18	0.013	18.61127
TETRAMETHYL-P-SEXIPHENYL	24.56	0.030	3.70353
TETRAMETHYL-P-OCTAPHENYL (B)	36.56	0.009	39.96673
INDOLE (TRACE)	0.00	0.000	0.00000
2-PHENYL INDOLE	26.10	0.025	5.33538
1-METHYL-2-PHENYL INDOLE	20.61	0.051	1.29175
1-NEOPENTYL-2-PHENYL INDOLE	17.49	0.084	0.48431
1-N-BUTYL-2(BIPHENYLYL-4) INDOLE	29.92	0.017	12.03293
1,2-DIPHENYL INDOLE	21.32	0.046	1.58463
1-BIPHENYLYL-2-PHENYL INDOLE	24.07	0.032	3.26732
1,3-DIMETHYL-2-PHENYL INDOLE	21.98	0.042	1.89845
2,3-DIPHENYL INDOLE	27.63	0.021	7.44626
1-METHYL-2,3-DIPHENYL INDOLE	25.07	0.028	4.16316
2-PHENYL-3,2'-METHYLENE INDOLE (TRACE)	29.54	0.017	11.18079
2-PHENYL-3,2'-ETHYLENE INDOLE (TRACE)	32.08	0.014	18.27026
2-PHENYL-3,2'-TRIMETHYLENE INDOLE	28.13	0.020	8.33202
1-METHYL-3,2'-METHYLENE-2-PHENYL INDOLE (TRACE)	31.65	0.014	16.88009
1-METHYL-3,2'-ETHYLENE-2-PHENYL INDOLE	32.29	0.013	18.98397
1-METHYL-3,2'-TRIMETHYLENE-2-PHENYL INDOLE	27.10	0.022	6.65748
4,5-DIPHENYLIMIDAZOLE (A)	16.46	0.100	0.33488
PPF (TRACE)	33.30	0.012	22.87653
PPO (TRACE)	28.23	0.020	8.52494
2-(4-FLUOROPHENYL)-5-PHENYLOXAZOLE (TRACE)	26.87	0.023	6.34629
αNPO (TRACE)	34.37	0.011	27.49467
BBO (B)	39.17	0.007	60.12711
POPOP	37.76	0.008	48.22394
DIMETHYL POPOP (TRACE)	38.32	0.008	52.60950
PPD (TRACE)	0.00	0.000	0.00000
2,5-DI(2-METHYLPHENYL)OXADIAZOLE	7.64	1.003	0.00338
PBD	26.82	0.023	6.28347
BBD (B)	35.14	0.010	31.58071
αNPD (TRACE)	30.51	0.016	13.50072
βNPD	26.13	0.025	5.36417
αNND (TRACE)	32.64	0.013	20.23164
βNND (TRACE)	32.14	0.013	18.55029
TRANS-1-(4-BIPHENYLYL)-2-PHENYL-ETHYLENE	34.81	0.011	29.77416
TRANS-1-(4-BIPHENYLYL)-2-(1-NAPHTHYL)-ETHYLENE	36.29	0.009	38.00505
DIPHENYLSTILBENE (B)	39.59	0.007	64.11137
1,1,4,4-TETRAPHENYLBUTADIENE	37.01	0.009	42.65561
1,6-DIPHENYLHEXATRIENE	40.61	0.007	74.51778
1,8-DIPHENYL-1,3,5,7-OCTATETRAENE	38.85	0.008	57.14595
BPSB (TRACE)	40.77	0.007	76.30788
BIS-MSB	39.82	0.007	66.28335
ESCULIN (A)	30.91	0.015	14.53082
AZULENE	23.49	0.035	2.79030
NAPHTHALENE	4.53	4.817	0.00015
DEUTERATED NAPHTHALENE-C10D8	9.09	0.595	0.00954
1-METHYLNAPHTHALENE 5ML/L	7.32	1.140	0.00261
2-METHYL NAPHTHALENE	8.74	0.669	0.00758
2,3-DIMETHYL NAPHTHALENE	8.94	0.626	0.00866
2,6-DIMETHYL NAPHTHALENE (TRACE)	15.01	0.132	0.19350
1-NAPHTHOL (TRACE)	14.56	0.145	0.16105
2-NAPHTHOL (TRACE)	18.39	0.072	0.65357
1-NAPHTHYLAMINE	23.50	0.034	2.82805
2-CHLORONAPHTHALENE	10.55	0.381	0.02332
ACENAPHTHENE	11.43	0.300	0.03771
1-PHENYL NAPHTHALENE	12.36	0.237	0.06027
2-PHENYL NAPHTHALENE	20.53	0.052	1.26399
1,4-DIPHENYL NAPHTHALENE	21.70	0.044	1.75729
1,5-DIPHENYL NAPHTHALENE	19.89	0.057	1.04506

ACCEPTOR MOLECULES	RO	CO	OVERLAP
1,7-DIPHENYL NAPHTHALENE	19.60	0.059	0.95618
1,4,5,8-TETRAPHENYL NAPHTHALENE	33.08	0.012	21.78203
1,1'-BINAPHTHYL	15.67	0.116	0.25051
4,4' -DI(N-BUTOXY)-1,1'-BINAPHTHYL	26.99	0.023	6.49069
2,2'-BINAPHTHYL	21.31	0.046	1.57798
ACRIDINE (A)	28.46	0.019	8.82341
ANTHRACENE (TRACE)	26.64	0.024	5.95788
DEUTERATED ANTHRACENE C14D10 (TRACE)	26.32	0.025	5.53734
9-METHYL ANTHRACENE (TRACE)	25.88	0.026	4.98843
1-AMINO ANTHRACENE	23.76	0.033	2.98569
2-AMINO ANTHRACENE	24.41	0.031	3.51193
9-VINYL ANTHRACENE	26.60	0.024	5.88339
9-PHENYL ANTHRACENE (TRACE)	27.28	0.022	6.85085
9,10-DIPHENYL ANTHRACENE	27.15	0.022	6.65610
9,10-DICHLOROANTHRACENE (TRACE)	25.82	0.026	4.92217
BIANTHRANYL (B)	28.74	0.019	9.35361
9,10-DINAPHTHYLANTHRACENE (B)	26.69	0.024	6.00249
9-(P-VINYLPHENYL)-10-PHENYL ANTHRACENE	27.18	0.022	6.69690
9,10-BIS(PHENYL-ETHYNYL)ANTHRACENE (TRACE)	22.73	0.038	2.29004
RUBICENE (B)	27.61	0.021	7.35625
DIBENZ(A,H)ANTHRACENE (B)	30.78	0.015	14.12527
TETRACENE (TRACE) (B)	19.57	0.060	0.93238
RUBRENE (B)	26.37	0.024	5.58323
TETRAHYDROCANNABINOL ACID METHYL ESTER	5.68	2.442	0.00057
PHENANTHRENE	14.45	0.148	0.15324
3,4-BENZOPHENANTHRENE	21.18	0.047	1.50257
2-PHENYL PHENANTHRENE	16.08	0.108	0.28972
CHRYSENE (TRACE)	20.93	0.049	1.40480
TRIPHENYLENE	15.09	0.130	0.19815
DEUTERATED TRIPHENYLENE-C18D12	14.85	0.137	0.18040
PYRENE	30.82	0.015	14.30015
1,3,6,8,-TETRAPHENYL PYRENE	33.63	0.012	24.03538
BIPYRENYL (B)	37.48	0.008	46.02468
O-PHENYLENE-PYRENE	30.57	0.016	13.54688
ANTHANTHRENE (B)	27.82	0.021	7.69267
FLUORANTHENE	28.29	0.020	8.51488
3-PHENYL FLUORANTHENE	30.26	0.016	12.75621
3-AMINO FLUORANTHENE	27.57	0.021	7.29330
3,3'-BIFLUORANTHENYL (B)	33.26	0.012	22.47352
BENZO(GHI)FLUORANTHENE	27.28	0.022	6.84909
9,12-DI(N-OCTADECOXY)-BENZ(E)ACEPHENANTHRYLENE	29.68	0.017	11.36011
DECACYCLENE (B)	36.16	0.009	37.15936
10-ETHOXY-BENZO(KL)XANTHENE	27.72	0.021	7.53349
PERYLENE	23.43	0.035	2.75069
BENZO(GHI)PERYLENE (B)	29.65	0.017	11.27912
BBOT (TRACE)	38.22	0.008	51.80595
QUININE SULFATE (A)	27.42	0.022	7.14421
ACRIDINE YELLOW (A)	20.50	0.052	1.23209
ACRIDINE ORANGE (A)	19.10	0.064	0.80519
ACRIDINE RED (A)	23.83	0.033	3.03971
URANINE (A)	23.87	0.033	3.07341
RHODAMINE B (A)	26.99	0.023	6.41858
RHODAMINE 6G (A)	27.96	0.020	7.93001
EOSIN B (A)	25.37	0.027	4.42656
EOSIN Y (A)	25.75	0.026	4.83974
ROSE BENGAL (A)	27.20	0.022	6.72840
INDENOINDENE	27.87	0.021	7.89256

8.145 Donor Molecule: 1-Naphthol (trace)

N,N-DIMETHYLANILINE 3ML/L	7.78	0.950	0.00806
TRIPHENYLAMINE	20.44	0.052	2.62982
SODIUM SALICYLATE (A)	13.63	0.177	0.23125
ANTHRANILIC ACID (A)	22.83	0.038	5.06198
TMPD	19.58	0.060	2.01968
4,4'-DIHYDROXYBIPHENYL (A)	7.67	0.991	0.00743
4,4'-DIMETHOXY BIPHENYL	3.97	7.174	0.00014
4,4'-DIISOBUTOXYBIPHENYL	0.00	0.000	0.00000
BENZIDINE	17.86	0.079	1.17275
4-VINYL BIPHENYL	5.58	2.573	0.00110
1,3,5-TRIPHENYL BENZENE	7.55	1.039	0.00673
1,3-DIBIPHENYLYLBENZENE (B)	16.59	0.098	0.75580
1,3,5,-TRIBIPHENYLYLBENZENE (B)	19.75	0.058	2.15088
FLUORENE (TRACE)	0.00	0.000	0.00000
1-METHYL FLUORENE	0.00	0.000	0.00000
1,2-BENZOFLUORENE	18.49	0.071	1.44045
9,10-DIHYDROPHENANTHRENE	0.00	0.000	0.00000
CARBAZOLE (TRACE)	18.12	0.075	1.27796
N-METHYL CARBAZOLE (TRACE)	20.74	0.050	2.86870
N-VINYLCARBAZOLE	21.03	0.048	3.11708
N-PHENYL CARBAZOLE	20.70	0.050	2.84006
N-METHYL-2,3-BENZCARBAZOLE	22.42	0.040	4.51557
DIBENZOFURAN	0.00	0.000	0.00000
DIBENZOTHIOPHENE (TRACE)	16.56	0.098	0.74666
P-TERPHENYL	9.18	0.578	0.02179
DEUTERATED P-TERPHENYL-C18D14	6.87	1.378	0.00384
4-METHYL TERPHENYL	11.12	0.325	0.06877
2-METHYL-P-TERPHENYL	0.00	0.000	0.00000
3,3''-DIMETHYL-P-TERPHENYL	11.44	0.299	0.08156
4-(3,3-DIMETHYLBUTOXY)-P-TERPHENYL	16.67	0.097	0.77791

ACCEPTOR MOLECULES	RO	CO	OVERLAP
4,4-DIHEXAHYDROFARNESOXY-P-TERPHENYL	19.66	0.059	2.08944
2,2 -METHYLENE-P-TERPHENYL	18.15	0.075	1.29664
2,2 -ETHYLENE-P-TERPHENYL	19.33	0.062	1.88802
2,2-METHYLENE-5,6-METHYLENE-P-TERPHENYL (TRACE)	28.20	0.020	18.18081
3-PHENYL DIBENZOFURAN	18.13	0.075	1.28735
BENZO(1,2-B-4,5-B')BISBENZOFURAN (TRACE) (B)	29.69	0.017	24.69270
SEXIPHENYL BENZENE	0.00	0.000	0.00000
P-QUATERPHENYL	20.96	0.049	3.06629
3,3'''-DIMETHYL-P-QUATERPHENYL	22.27	0.041	4.40126
TETRAMETHYL-P-QUATERPHENYL	0.00	0.000	0.00000
3,3''' -DIMETHYL-3',2''-METHYLENE-P-QUATERPHENYL	29.46	0.017	23.56296
3,3''' -DIMETHYL-3',2''-ETHYLENE-P-QUATERPHENYL	29.03	0.018	21.55079
4-(3-ETHYLBUTOXY)-P-QUATERPHENYL (B)	26.65	0.024	12.90495
4,4'''-DI(2-BUTYLOCTOXY-1)-P-QUATERPHENYL	27.26	0.022	14.76865
DIETHYL-P-QUINQUEPHENYL	23.25	0.036	5.70634
DI(3-ETHYLHEPTYL)-P-QUINQUEPHENYL	28.45	0.019	19.03413
TETRAMETHYL-P-SEXIPHENYL	22.09	0.041	4.20797
TETRAMETHYL-P-OCTAPHENYL (B)	32.14	0.013	39.58599
INDOLE (TRACE)	0.00	0.000	0.00000
2-PHENYL INDOLE	23.28	0.035	5.75318
1-METHYL-2-PHENYL INDOLE	18.56	0.070	1.47731
1-NEOPENTYL-2-PHENYL INDOLE	16.07	0.108	0.62353
1-N-BUTYL-2(BIPHENYLYL-4) INDOLE	26.32	0.025	11.96155
1,2-DIPHENYL INDOLE	19.25	0.063	1.84180
1-BIPHENYLYL-2-PHENYL INDOLE	21.57	0.045	3.62698
1,3-DIMETHYL-2-PHENYL INDOLE	19.57	0.060	2.03039
2,3-DIPHENYL INDOLE	24.33	0.031	7.44037
1-METHYL-2,3-DIPHENYL INDOLE	22.09	0.041	4.17677
2-PHENYL-3,2'-METHYLENE INDOLE (TRACE)	26.16	0.025	11.56798
2-PHENYL-3,2'-ETHYLENE INDOLE (TRACE)	28.01	0.020	17.36769
2-PHENYL-3,2' -TRIMETHYLENE INDOLE	24.89	0.029	8.57494
1-METHYL-3,2'-METHYLENE-2-PHENYL INDOLE (TRACE)	27.71	0.021	16.32056
1-METHYL-3,2'-ETHYLENE-2-PHENYL INDOLE	28.31	0.020	18.48569
1-METHYL-3,2'-TRIMETHYLENE-2-PHENYL INDOLE	23.94	0.033	6.80142
4,5-DIPHENYLIMIDAZOLE (A)	14.78	0.138	0.37757
PPF (TRACE)	29.17	0.018	22.17661
PPO (TRACE)	25.02	0.029	8.87007
2-(4-FLUOROPHENYL)-5-PHENYLOXAZOLE (TRACE)	23.81	0.033	6.58654
αNPO (TRACE)	30.06	0.016	26.39430
BBO (B)	34.29	0.011	58.07999
POPOP	33.13	0.012	47.18730
DIMETHYL POPOP (TRACE)	33.68	0.012	51.96723
PPD (TRACE)	5.94	2.136	0.00160
2,5-DI(2-METHYLPHENYL)OXADIAZOLE	9.26	0.563	0.02299
PBD	23.95	0.033	6.82665
BBD (B)	30.84	0.015	30.98929
αNPD (TRACE)	26.72	0.023	13.07778
βNPD	23.15	0.036	5.56486
αNND (TRACE)	28.54	0.019	19.40842
βNND (TRACE)	28.27	0.020	18.40100
TRANS-1-(4-BIPHENYLYL)-2-PHENYL-ETHYLENE	30.51	0.016	28.97586
TRANS-1-(4-BIPHENYLYL)-2-(1-NAPHTHYL)-ETHYLENE	31.77	0.014	36.74966
DIPHENYLSTILBENE (B)	34.65	0.011	61.78852
1,1,4,4-TETRAPHENYLBUTADIENE	32.41	0.013	41.30005
1,6-DIPHENYLHEXATRIENE	35.66	0.010	73.30382
1,8-DIPHENYL-1,3,5,7-OCTATETRAENE	34.41	0.011	58.96433
BPSB (TRACE)	35.78	0.010	74.78244
BIS-MSB	34.91	0.011	64.53323
ESCULIN (A)	27.04	0.023	13.97340
AZULENE	20.56	0.051	2.68214
NAPHTHALENE	5.08	3.404	0.00063
DEUTERATED NAPHTHALENE-C10D8	8.27	0.790	0.01162
1-METHYLNAPHTHALENE 5ML/L	6.99	1.309	0.00424
2-METHYL NAPHTHALENE	8.93	0.628	0.01846
2,3-DIMETHYL NAPHTHALENE	9.01	0.612	0.01941
2,6-DIMETHYL NAPHTHALENE (TRACE)	13.19	0.195	0.19104
1-NAPHTHOL (TRACE)	14.00	0.163	0.27311
2-NAPHTHOL (TRACE)	16.16	0.106	0.64589
1-NAPHTHYLAMINE	20.63	0.051	2.78132
2-CHLORONAPHTHALENE	10.37	0.402	0.04505
ACENAPHTHENE	11.79	0.273	0.09754
1-PHENYL NAPHTHALENE	11.53	0.292	0.08510
2-PHENYL NAPHTHALENE	18.40	0.072	1.40450
1,4-DIPHENYL NAPHTHALENE	19.42	0.061	1.93829
1,5-DIPHENYL NAPHTHALENE	17.94	0.078	1.20499
1,7-DIPHENYL NAPHTHALENE	17.65	0.081	1.09333
1,4,5,8-TETRAPHENYL NAPHTHALENE	28.96	0.018	21.01102
1,1'-BINAPHTHYL	14.20	0.156	0.29763
4,4' -DI(N-BUTOXY)-1,1'-BINAPHTHYL	23.79	0.033	6.52728
2,2'-BINAPHTHYL	19.27	0.063	1.85256
ACRIDINE (A)	24.96	0.029	8.62130
ANTHRACENE (TRACE)	23.46	0.035	5.95127
DEUTERATED ANTHRACENE C14D10 (TRACE)	23.20	0.036	5.58073
9-METHYL ANTHRACENE (TRACE)	22.72	0.038	4.90407
1-AMINO ANTHRACENE	21.09	0.048	3.12466
2-AMINO ANTHRACENE	21.62	0.044	3.62789
9-VINYL ANTHRACENE	23.46	0.035	5.93539
9-PHENYL ANTHRACENE (TRACE)	23.97	0.032	6.75259
9,10-DIPHENYL ANTHRACENE	24.13	0.032	7.01514
9,10-DICHLOROANTHRACENE (TRACE)	22.95	0.037	5.18723
BIANTHRANYL (B)	25.49	0.027	9.74169
9,10-DINAPHTHYLANTHRACENE (B)	23.74	0.033	6.36621
9-(P-VINYLPHENYL)-10-PHENYL ANTHRACENE	24.17	0.032	7.08395
9,10-BIS(PHENYL-ETHYNYL)ANTHRACENE (TRACE)	20.59	0.051	2.71132
RUBICENE (B)	24.58	0.030	7.83660
DIBENZ(A,H)ANTHRACENE (B)	27.10	0.022	14.10608

ACCEPTOR MOLECULES	RO	CO	OVERLAP
TETRACENE (TRACE) (B)	17.44	0.084	0.99904
RUBRENE (B)	23.56	0.034	6.07468
TETRAHYDROCANNABINOL ACID METHYL ESTER	6.80	1.422	0.00361
PHENANTHRENE	12.68	0.219	0.14977
3,4-BENZOPHENANTHRENE	18.91	0.066	1.63106
2-PHENYL PHENANTHRENE	14.18	0.157	0.29233
CHRYSENE (TRACE)	19.41	0.061	1.91786
TRIPHENYLENE	13.35	0.188	0.20441
DEUTERATED TRIPHENYLENE-C18D12	13.05	0.201	0.17796
PYRENE	27.33	0.022	14.90308
1,3,6,8,-TETRAPHENYL PYRENE	29.82	0.017	24.97940
BIPYRENYL (B)	32.80	0.013	[illegible]
O-PHENYLENE-PYRENE	27.01	0.023	13.78330
ANTHANTHRENE (B)	24.76	0.029	8.18636
FLUORANTHENE	24.80	0.029	8.25898
3-PHENYL FLUORANTHENE	26.63	0.024	12.67211
3-AMINO FLUORANTHENE	24.35	0.031	7.41062
3,3'-BIFLUORANTHENYL (B)	29.39	0.018	22.91228
BENZO(GHI)FLUORANTHENE	23.83	0.033	6.51438
9,12-DI(N-OCTADECOXY)-BENZ(E)ACEPHENANTHRYLENE	26.20	0.025	11.49911
DECACYCLENE (B)	32.00	0.014	38.13894
10-ETHOXY-BENZO(KL)XANTHENE	24.58	0.030	7.84864
PERYLENE	21.23	0.047	3.25276
BENZO(GHI)PERYLENE (B)	26.18	0.025	11.45091
BBOT (TRACE)	33.71	0.012	52.19800
QUININE SULFATE (A)	24.01	0.032	6.90245
ACRIDINE YELLOW (A)	18.47	0.071	1.41139
ACRIDINE ORANGE (A)	17.09	0.090	0.88645
ACRIDINE RED (A)	20.98	0.048	3.02655
URANINE (A)	21.09	0.048	3.12533
RHODAMINE B (A)	23.77	0.033	6.41652
RHODAMINE 6G (A)	24.44	0.031	7.57165
EOSIN B (A)	22.36	0.040	4.43683
EOSIN Y (A)	22.72	0.038	4.88920
ROSE BENGAL (A)	23.92	0.033	6.65302
INDENOINDENE	24.55	0.030	7.90491

8.146 Donor Molecule: 2-Naphthol (trace)

N,N-DIMETHYLANILINE 3ML/L	3.97	7.168	0.00009
TRIPHENYLAMINE	19.26	0.063	1.20277
SODIUM SALICYLATE (A)	13.25	0.192	0.12728
ANTHRANILIC ACID (A)	24.22	0.031	4.73279
TMPD	20.23	0.054	1.61248
BENZIDINE	15.91	0.111	0.38320
1,3,5-TRIPHENYL BENZENE	4.64	4.476	0.00024
1,3-DIBIPHENYLYLBENZENE (B)	13.88	0.167	0.16958
1,3,5,-TRIBIPHENYLYLBENZENE (B)	16.42	0.101	0.46516
FLUORENE (TRACE)	0.00	0.000	0.00000
1-METHYL FLUORENE	0.00	0.000	0.00000
1,2-BENZOFLUORENE	17.85	0.079	0.76332
9,10-DIHYDROPHENANTHRENE	0.00	0.000	0.00000
CARBAZOLE (TRACE)	17.48	0.084	0.67392
N-METHYL CARBAZOLE (TRACE)	21.31	0.046	2.20783
N-VINYLCARBAZOLE	21.03	0.048	2.04301
N-PHENYL CARBAZOLE	20.77	0.050	1.89719
N-METHYL-2,3-BENZCARBAZOLE	23.55	0.034	3.96817
DIBENZOFURAN	0.00	0.000	0.00000
DIBENZOTHIOPHENE (TRACE)	13.41	0.185	0.13774
P-TERPHENYL	0.00	0.000	0.00000
DEUTERATED P-TERPHENYL-C18D14	0.00	0.000	0.00000
4-METHYL TERPHENYL	4.72	4.257	0.00026
2-METHYL-P-TERPHENYL	0.00	0.000	0.00000
3,3''-DIMETHYL-P-TERPHENYL	6.38	1.726	0.00160
4-(3,3-DIMETHYLBUTOXY)-P-TERPHENYL	13.43	0.185	0.13905
4,4''-DIHEXAHYDROFARNESOXY-P-TERPHENYL	16.66	0.097	0.50691
2,2' -METHYLENE-P-TERPHENYL	13.16	0.196	0.12350
2,2' -ETHYLENE-P-TERPHENYL	15.27	0.126	0.30080
2,2'-METHYLENE-5',6''-METHYLENE-P-TERPHENYL (TRACE)	28.48	0.019	12.61653
3-PHENYL DIBENZOFURAN	13.09	0.199	0.11966
BENZO(1,2-B-4,5-B')BISBENZOFURAN (TRACE) (B)	29.42	0.018	15.31576
SEXIPHENYL BENZENE	0.00	0.000	0.00000
P-QUATERPHENYL	18.53	0.070	0.95793
3,3'''-DIMETHYL-P-QUATERPHENYL	20.22	0.054	1.61485
TETRAMETHYL-P-QUATERPHENYL	0.00	0.000	0.00000
3,3''' -DIMETHYL-3',2''-METHYLENE-P-QUATERPHENYL	28.83	0.019	13.54759
3,3''' -DIMETHYL-3',2''-ETHYLENE-P-QUATERPHENYL	28.59	0.019	12.86088
4-(3-ETHYLBUTOXY)-P-QUATERPHENYL (B)	25.75	0.026	6.85965
4,4'''-DI(2-BUTYLOCTOXY-1)-P-QUATERPHENYL	26.21	0.025	7.63387
DIETHYL-P-QUINQUEPHENYL	21.07	0.048	2.07074
DI(3-ETHYLHEPTYL)-P-QUINQUEPHENYL	27.57	0.021	10.32367
TETRAMETHYL-P-SEXIPHENYL	19.43	0.061	1.27562
TETRAMETHYL-P-OCTAPHENYL (B)	32.21	0.013	26.22714
INDOLE (TRACE)	0.00	0.000	0.00000
2-PHENYL INDOLE	21.19	0.047	2.14477
1-METHYL-2-PHENYL INDOLE	16.15	0.106	0.41979
1-NEOPENTYL-2-PHENYL INDOLE	12.65	0.221	0.09721
1-N-BUTYL-2(BIPHENYLYL-4) INDOLE	26.29	0.025	7.77512
1,2-DIPHENYL INDOLE	16.59	0.098	0.49235
1-BIPHENYLYL-2-PHENYL INDOLE	19.51	0.060	1.30127

ACCEPTOR MOLECULES	RO	CO	OVERLAP
1,3-DIMETHYL-2-PHENYL INDOLE	17.92	0.078	0.78438
2,3-DIPHENYL INDOLE	24.22	0.031	4.73738
1-METHYL-2,3-DIPHENYL INDOLE	21.87	0.043	2.57624
2-PHENYL-3,2'-METHYLENE INDOLE (TRACE)	25.24	0.028	6.11288
2-PHENYL-3,2'-ETHYLENE INDOLE (TRACE)	28.70	0.019	13.14192
2-PHENYL-3,2'-TRIMETHYLENE INDOLE	23.66	0.034	4.14760
1-METHYL-3,2'-METHYLENE-2-PHENYL INDOLE (TRACE)	27.96	0.020	11.26161
1-METHYL-3,2'-ETHYLENE-2-PHENYL INDOLE	29.45	0.018	15.35158
1-METHYL-3,2'-TRIMETHYLENE-2-PHENYL INDOLE	23.22	0.036	3.70132
4,5-DIPHENYLIMIDAZOLE (A)	13.39	0.186	0.13671
PPF (TRACE)	29.24	0.018	14.72071
PPO (TRACE)	23.79	0.033	4.28323
2-(4-FLUOROPHENYL)-5-PHENYLOXAZOLE (TRACE)	22.47	0.039	3.04442
αNPO (TRACE)	31.95	0.014	24.89459
BBO (B)	36.94	0.009	59.40833
POPOP	36.30	0.009	53.42701
DIMETHYL POPOP (TRACE)	37.17	0.009	61.50333
PPD (TRACE)	0.00	0.000	0.00000
2,5-DI(2-METHYLPHENYL)OXADIAZOLE	0.00	0.000	0.00000
PBD	21.46	0.045	2.31139
BBD (B)	30.99	0.015	20.85146
αNPD (TRACE)	27.04	0.023	9.19532
βNPD	20.95	0.049	1.99706
αNND (TRACE)	29.97	0.017	17.01007
βNND (TRACE)	27.81	0.021	10.91665
TRANS-1-(4-BIPHENYLYL)-2-PHENYL-ETHYLENE	31.00	0.015	20.84603
TRANS-1-(4-BIPHENYLYL)-2-(1-NAPHTHYL)-ETHYLENE	33.86	0.012	35.24582
DIPHENYLSTILBENE (B)	37.31	0.009	63.04464
1,1,4,4-TETRAPHENYLBUTADIENE	35.13	0.010	43.86170
1,6-DIPHENYLHEXATRIENE	39.09	0.007	83.19903
1,8-DIPHENYL-1,3,5,7-OCTATETRAENE	38.59	0.008	76.75292
BPSB (TRACE)	39.25	0.007	85.37508
BIS-MSB	37.98	0.008	69.98648
ESCULIN (A)	28.80	0.019	13.36893
AZULENE	21.10	0.048	2.05011
NAPHTHALENE	0.00	0.000	0.00000
DEUTERATED NAPHTHALENE-C10D8	6.99	1.309	0.00277
1-METHYLNAPHTHALENE 5ML/L	4.11	6.432	0.00011
2-METHYL NAPHTHALENE	4.78	4.105	0.00028
2,3-DIMETHYL NAPHTHALENE	4.89	3.833	0.00032
2,6-DIMETHYL NAPHTHALENE (TRACE)	9.08	0.597	0.01333
1-NAPHTHOL (TRACE)	8.85	0.644	0.01144
2-NAPHTHOL (TRACE)	14.88	0.136	0.25689
1-NAPHTHYLAMINE	20.43	0.052	1.71601
2-CHLORONAPHTHALENE	6.19	1.889	0.00133
ACENAPHTHENE	5.97	2.104	0.00107
1-PHENYL NAPHTHALENE	8.33	0.774	0.00793
2-PHENYL NAPHTHALENE	16.21	0.105	0.43047
1,4-DIPHENYL NAPHTHALENE	17.35	0.086	0.64571
1,5-DIPHENYL NAPHTHALENE	15.30	0.125	0.30366
1,7-DIPHENYL NAPHTHALENE	15.45	0.121	0.32149
1,4,5,8-TETRAPHENYL NAPHTHALENE	30.82	0.015	19.99187
1,1'-BINAPHTHYL	11.47	0.296	0.05415
4,4'-DI(N-BUTOXY)-1,1'-BINAPHTHYL	23.06	0.036	3.54572
2,2'-BINAPHTHYL	16.82	0.094	0.53595
ACRIDINE (A)	27.42	0.022	9.90932
ANTHRACENE (TRACE)	25.63	0.027	6.62832
DEUTERATED ANTHRACENE C14D10 (TRACE)	25.42	0.027	6.31149
9-METHYL ANTHRACENE (TRACE)	25.47	0.027	6.36485
1-AMINO ANTHRACENE	23.73	0.033	4.15194
2-AMINO ANTHRACENE	23.65	0.034	4.07438
9-VINYL ANTHRACENE	26.31	0.025	7.71814
9-PHENYL ANTHRACENE (TRACE)	26.76	0.023	8.55966
9,10-DIPHENYL ANTHRACENE	27.19	0.022	9.39823
9,10-DICHLOROANTHRACENE (TRACE)	25.82	0.026	6.89506
BIANTHRANYL (B)	28.63	0.019	12.79894
9,10-DINAPHTHYLANTHRACENE (B)	26.54	0.024	8.12610
9-(P-VINYLPHENYL)-10-PHENYL ANTHRACENE	27.24	0.022	9.50134
9,10-BIS(PHENYL-ETHYNYL)ANTHRACENE (TRACE)	23.19	0.036	3.61574
RUBICENE (B)	26.17	0.025	7.46708
DIBENZ(A,H)ANTHRACENE (B)	28.17	0.020	11.64477
TETRACENE (TRACE) (B)	19.13	0.064	1.14100
RUBRENE (B)	22.69	0.038	3.17378
TETRAHYDROCANNABINOL ACID METHYL ESTER	0.00	0.000	0.00000
PHENANTHRENE	12.86	0.210	0.10665
3,4-BENZOPHENANTHRENE	17.30	0.086	0.62619
2-PHENYL PHENANTHRENE	14.66	0.142	0.23337
CHRYSENE (TRACE)	17.06	0.090	0.57772
TRIPHENYLENE	13.26	0.192	0.12799
DEUTERATED TRIPHENYLENE-C18D12	12.93	0.207	0.11015
PYRENE	27.97	0.020	11.21690
1,3,6,8,-TETRAPHENYL PYRENE	33.71	0.012	34.10526
BIPYRENYL (B)	35.79	0.010	48.90581
O-PHENYLENE-PYRENE	29.97	0.017	16.85529
ANTHANTHRENE (B)	27.13	0.022	9.28496
FLUORANTHENE	26.74	0.023	8.49976
3-PHENYL FLUORANTHENE	29.43	0.018	15.12347
3-AMINO FLUORANTHENE	26.42	0.024	7.90168
3,3'-BIFLUORANTHENYL (B)	32.60	0.013	27.91309
BENZO(GHI)FLUORANTHENE	25.63	0.027	6.59491
9,12-DI(N-OCTADECOXY)-BENZ(E)ACEPHENANTHRYLENE	28.72	0.019	13.06067
DECACYCLENE (B)	35.12	0.010	43.64666
10-ETHOXY-BENZO(KL)XANTHENE	27.69	0.021	10.49035
PERYLENE	24.30	0.031	4.79224
BENZO(GHI)PERYLENE (B)	29.42	0.018	15.08664
BBOT (TRACE)	37.55	0.008	65.30269

ACCEPTOR MOLECULES	RO	CO	OVERLAP
QUININE SULFATE (A)	24.31	0.031	4.86529
ACRIDINE YELLOW (A)	21.00	0.048	1.99709
ACRIDINE ORANGE (A)	18.59	0.070	0.96061
ACRIDINE RED (A)	22.91	0.037	3.36233
URANINE (A)	22.03	0.042	2.66217
RHODAMINE B (A)	26.05	0.025	7.26672
RHODAMINE 6G (A)	26.74	0.023	8.50395
EOSIN B (A)	23.34	0.035	3.76139
EOSIN Y (A)	23.77	0.033	4.19158
ROSE BENGAL (A)	24.94	0.029	5.60020
INDENOINDENE	23.35	0.035	3.83500

8.147 Donor Molecule: 1-Naphthylamine

ACCEPTOR MOLECULES	RO	CO	OVERLAP
TRIPHENYLAMINE	11.25	0.314	0.03319
P-XYLENE (TRACE)	0.00	0.000	0.00000
SODIUM SALICYLATE (A)	9.33	0.550	0.01080
ANTHRANILIC ACID (A)	20.28	0.054	1.12863
TMPD	15.27	0.126	0.20650
BENZIDINE	7.15	1.225	0.00218
1,3-DIBIPHENYLYLBENZENE (B)	5.03	3.519	0.00026
1,3,5,-TRIBIPHENYLYLBENZENE (B)	0.00	0.000	0.00000
FLUORENE (TRACE)	0.00	0.000	0.00000
1-METHYL FLUORENE	0.00	0.000	0.00000
1,2-BENZOFLUORENE	11.35	0.306	0.03496
9,10-DIHYDROPHENANTHRENE	0.00	0.000	0.00000
CARBAZOLE (TRACE)	6.41	1.697	0.00114
N-METHYL CARBAZOLE (TRACE)	13.24	0.193	0.08803
N-VINYLCARBAZOLE	11.41	0.301	0.03601
N-PHENYL CARBAZOLE	11.73	0.277	0.04268
N-METHYL-2,3-BENZCARBAZOLE	25.88	0.026	4.84601
DIBENZOFURAN	0.00	0.000	0.00000
DIBENZOTHIOPHENE (TRACE)	3.78	8.309	0.00005
2,2'-METHYLENE-5',6'-METHYLENE-P-TERPHENYL (TRACE)	11.38	0.303	0.03562
3-PHENYL DIBENZOFURAN	0.00	0.000	0.00000
BENZO(1,2-B-4,5-B')BISBENZOFURAN (TRACE) (B)	15.84	0.113	0.25775
SEXIPHENYL BENZENE	0.00	0.000	0.00000
P-QUATERPHENYL	7.30	1.148	0.00249
3,3'''-DIMETHYL-P-QUATERPHENYL	9.72	0.487	0.01382
TETRAMETHYL-P-QUATERPHENYL	0.00	0.000	0.00000
3,3''' -DIMETHYL-3',2''-METHYLENE-P-QUATERPHENYL	16.07	0.108	0.28142
3,3''' -DIMETHYL-3',2''-ETHYLENE-P-QUATERPHENYL	17.91	0.078	0.53800
4-(3-ETHYLBUTOXY)-P-QUATERPHENYL (B)	15.82	0.113	0.25594
4,4'''-DI(2-BUTYLOCTOXY-1)-P-QUATERPHENYL	15.87	0.112	0.26068
DIETHYL-P-QUINQUEPHENYL	9.66	0.497	0.01329
DI(3-ETHYLHEPTYL)-P-QUINQUEPHENYL	17.66	0.081	0.49312
TETRAMETHYL-P-SEXIPHENYL	7.47	1.074	0.00284
TETRAMETHYL-P-OCTAPHENYL (B)	22.88	0.037	2.33684
INDOLE (TRACE)	0.00	0.000	0.00000
2-PHENYL INDOLE	6.89	1.368	0.00175
1-METHYL-2-PHENYL INDOLE	6.33	1.760	0.00106
1-NEOPENTYL-2-PHENYL INDOLE	0.00	0.000	0.00000
1-N-BUTYL-2(BIPHENYLYL-4) INDOLE	18.26	0.073	0.60471
1,2-DIPHENYL INDOLE	6.57	1.578	0.00132
1-BIPHENYLYL-2-PHENYL INDOLE	11.53	0.292	0.03838
1,3-DIMETHYL-2-PHENYL INDOLE	7.29	1.153	0.00247
2,3-DIPHENYL INDOLE	16.93	0.092	0.38261
1-METHYL-2,3-DIPHENYL INDOLE	14.89	0.135	0.17769
2-PHENYL-3,2'-METHYLENE INDOLE (TRACE)	11.29	0.311	0.03389
2-PHENYL-3,2'-ETHYLENE INDOLE (TRACE)	19.25	0.063	0.82916
2-PHENYL-3,2' -TRIMETHYLENE INDOLE	10.68	0.367	0.02429
1-METHYL-3,2'-METHYLENE-2-PHENYL INDOLE (TRACE)	17.43	0.085	0.45706
1-METHYL-3,2'-ETHYLENE-2-PHENYL INDOLE	22.69	0.038	2.22382
1-METHYL-3,2'-TRIMETHYLENE-2-PHENYL INDOLE	11.84	0.270	0.04502
4,5-DIPHENYLIMIDAZOLE (A)	5.91	2.162	0.00070
PPF (TRACE)	18.23	0.074	0.59949
PPO (TRACE)	9.44	0.532	0.01157
2-(4-FLUOROPHENYL)-5-PHENYLOXAZOLE (TRACE)	8.55	0.715	0.00641
αNPO (TRACE)	27.26	0.022	6.65937
BBO (B)	33.71	0.012	23.77072
POPOP	35.73	0.010	33.66034
DIMETHYL POPOP (TRACE)	38.19	0.008	50.06468
PPD (TRACE)	0.00	0.000	0.00000
2,5-DI(2-METHYLPHENYL)OXADIAZOLE	0.00	0.000	0.00000
PBD	7.50	1.059	0.00293
BBD (B)	20.65	0.051	1.26577
αNPD (TRACE)	18.58	0.070	0.67066
βNPD	10.25	0.415	0.01901
αNND (TRACE)	23.84	0.033	2.98907
βNND (TRACE)	16.00	0.109	0.27383
TRANS-1-(4-BIPHENYLYL)-2-PHENYL-ETHYLENE	21.74	0.044	1.71887
TRANS-1-(4-BIPHENYLYL)-2-(1-NAPHTHYL)-ETHYLENE	30.19	0.016	12.25584
DIPHENYLSTILBENE (B)	33.75	0.012	23.94121
1,1,4,4-TETRAPHENYLBUTADIENE	32.96	0.012	20.70945
1,6-DIPHENYLHEXATRIENE	38.08	0.008	49.28443
1,8-DIPHENYL-1,3,5,7-OCTATETRAENE	43.27	0.006	105.47093
BPSB (TRACE)	38.92	0.008	56.14559
BIS-MSB	36.54	0.009	38.46087
ESCULIN (A)	25.06	0.028	4.02036
AZULENE	14.53	0.146	0.15040

ACCEPTOR MOLECULES	RO	CO	OVERLAP
2-NAPHTHOL (TRACE)	3.66	9.107	0.00004
1-NAPHTHYLAMINE	11.76	0.275	0.04320
2-CHLORONAPHTHALENE	0.00	0.000	0.00000
ACENAPHTHENE	0.00	0.000	0.00000
1-PHENYL NAPHTHALENE	0.00	0.000	0.00000
2-PHENYL NAPHTHALENE	0.00	0.000	0.00000
1,4-DIPHENYL NAPHTHALENE	8.06	0.855	0.00448
1,5-DIPHENYL NAPHTHALENE	5.57	2.585	0.00049
1,7-DIPHENYL NAPHTHALENE	6.33	1.766	0.00105
1,4,5,8-TETRAPHENYL NAPHTHALENE	27.07	0.023	6.36279
1,1'-BINAPHTHYL	0.00	0.000	0.00000
4,4'-DI(N-BUTOXY)-1,1'-BINAPHTHYL	13.84	0.169	0.11495
2,2'-BINAPHTHYL	6.67	1.509	0.00144
ACRIDINE (A)	26.66	0.024	5.80441
ANTHRACENE (TRACE)	25.65	0.027	4.60888
DEUTERATED ANTHRACENE C14D10 (TRACE)	25.18	0.028	4.13423
9-METHYL ANTHRACENE (TRACE)	27.22	0.022	6.57666
1-AMINO ANTHRACENE	27.97	0.020	7.67891
2-AMINO ANTHRACENE	26.67	0.024	5.77146
9-VINYL ANTHRACENE	28.77	0.019	9.15144
9-PHENYL ANTHRACENE (TRACE)	28.55	0.019	8.73977
9,10-DIPHENYL ANTHRACENE	31.05	0.015	14.45126
9,10-DICHLOROANTHRACENE (TRACE)	30.24	0.016	12.31270
BIANTHRANYL (B)	32.57	0.013	19.22247
9,10-DINAPHTHYLANTHRACENE (B)	30.73	0.015	13.55556
9-(P-VINYLPHENYL)-10-PHENYL ANTHRACENE	31.38	0.014	15.35907
9,10-BIS(PHENYL-ETHYNYL)ANTHRACENE (TRACE)	34.15	0.011	25.30093
RUBICENE (B)	30.25	0.016	12.23173
DIBENZ(A,H)ANTHRACENE (B)	23.11	0.036	2.45672
TETRACENE (TRACE) (B)	25.33	0.028	4.21747
RUBRENE (B)	22.66	0.038	2.16006
TETRAHYDROCANNABINOL ACID METHYL ESTER	0.00	0.000	0.00000
PHENANTHRENE	8.20	0.812	0.00496
3,4-BENZOPHENANTHRENE	14.41	0.149	0.14498
2-PHENYL PHENANTHRENE	11.47	0.297	0.03706
CHRYSENE (TRACE)	14.23	0.155	0.13475
TRIPHENYLENE	7.09	1.258	0.00207
DEUTERATED TRIPHENYLENE-C18D12	7.05	1.277	0.00201
PYRENE	15.28	0.125	0.20649
1,3,6,8,-TETRAPHENYL PYRENE	38.19	0.008	49.87787
BIPYRENYL (B)	34.53	0.011	27.31397
O-PHENYLENE-PYRENE	33.05	0.012	20.79102
ANTHANTHRENE (B)	33.46	0.012	22.47959
FLUORANTHENE	24.74	0.030	3.69876
3-PHENYL FLUORANTHENE	30.43	0.016	12.76653
3-AMINO FLUORANTHENE	30.22	0.016	12.14735
3,3'-BIFLUORANTHENYL (B)	36.02	0.010	35.04297
BENZO(GHI)FLUORANTHENE	22.65	0.038	2.17232
9,12-DI(N-OCTADECOXY)-BENZ(E)ACEPHENANTHRYLENE	32.09	0.014	17.53166
DECACYCLENE (B)	39.55	0.007	61.06889
10-ETHOXY-BENZO(KL)XANTHENE	31.60	0.014	16.06221
PERYLENE	34.02	0.011	24.84724
BENZO(GHI)PERYLENE (B)	32.83	0.013	20.19208
BBOT (TRACE)	40.43	0.007	70.39884
QUININE SULFATE (A)	14.47	0.148	0.14995
ACRIDINE YELLOW (A)	31.60	0.014	15.88563
ACRIDINE ORANGE (A)	26.22	0.025	5.18293
ACRIDINE RED (A)	23.96	0.033	3.01952
URANINE (A)	26.50	0.024	5.53024
RHODAMINE B (A)	27.64	0.021	7.12192
RHODAMINE 6G (A)	25.76	0.026	4.66903
EOSIN B (A)	23.99	0.032	3.04086
EOSIN Y (A)	25.23	0.028	4.11713
ROSE BENGAL (A)	24.78	0.029	3.69703
INDENOINDENE	9.25	0.564	0.01029

8.149 Donor Molecule: Acenaphthene

N,N-DIMETHYLANILINE 3ML/L	10.77	0.358	0.01987
TRIPHENYLAMINE	25.46	0.027	3.44136
SODIUM SALICYLATE (A)	16.97	0.092	0.30146
ANTHRANILIC ACID (A)	27.38	0.022	5.28051
TMPD	23.72	0.034	2.23781
4,4'-DIHYDROXYBIPHENYL (A)	10.81	0.354	0.02041
4,4'-DIMETHOXY BIPHENYL	6.62	1.540	0.00108
4,4'-DIISOBUTOXYBIPHENYL	0.00	0.000	0.00000
BENZIDINE	22.48	0.039	1.63394
4-VINYL BIPHENYL	9.00	0.613	0.00680
M-TERPHENYL	5.35	2.926	0.00030
1,3,5-TRIPHENYL BENZENE	9.89	0.462	0.01194
1,3-DIBIPHENYLYLBENZENE (B)	21.26	0.047	1.17152
1,3,5,-TRIBIPHENYLYLBENZENE (B)	25.22	0.028	3.26834
FLUORENE (TRACE)	0.00	0.000	0.00000
1-METHYL FLUORENE	0.00	0.000	0.00000
1,2-BENZOFLUORENE	23.28	0.035	2.01044
9,10-DIHYDROPHENANTHRENE	4.04	6.766	0.00006
CARBAZOLE (TRACE)	22.50	0.039	1.64445
N-METHYL CARBAZOLE (TRACE)	24.99	0.029	3.07260
N-VINYLCARBAZOLE	25.53	0.027	3.50206
N-PHENYL CARBAZOLE	25.10	0.028	3.16166
N-METHYL-2,3-BENZCARBAZOLE	26.86	0.023	4.68680

ACCEPTOR MOLECULES	RO	CO	OVERLAP
DIBENZOTHIOPHENE (TRACE)	20.33	0.053	0.89414
P-TERPHENYL	13.00	0.204	0.06158
DEUTERATED P-TERPHENYL-C18D14	10.61	0.375	0.01819
4-METHYL TERPHENYL	15.21	0.127	0.15784
2-METHYL-P-TERPHENYL	0.00	0.000	0.00000
3,3''-DIMETHYL-P-TERPHENYL	15.14	0.129	0.15328
4-(3,3-DIMETHYLBUTOXY)-P-TERPHENYL	21.27	0.046	1.17573
4,4''-DIHEXAHYDROFARNESOXY-P-TERPHENYL	24.88	0.029	3.01321
2,2'-METHYLENE-P-TERPHENYL	23.45	0.035	2.11374
2,2'-ETHYLENE-P-TERPHENYL	24.64	0.030	2.84113
2,2'-METHYLENE-5',6''-METHYLENE-P-TERPHENYL (TRACE)	34.79	0.011	22.43289
3-PHENYL DIBENZOFURAN	23.43	0.035	2.10163
BENZO(1,2-B-4,5-B')BISBENZOFURAN (TRACE) (B)	36.13	0.009	28.08146
SEXIPHENYL BENZENE	0.00	0.000	0.00000
P-QUATERPHENYL	26.36	0.024	4.24641
3,3'''-DIMETHYL-P-QUATERPHENYL	27.86	0.021	5.91766
TETRAMETHYL-P-QUATERPHENYL	5.62	2.516	0.00040
3,3''' -DIMETHYL-3',2''-METHYLENE-P-QUATERPHENYL	36.19	0.009	28.34129
3,3''' -DIMETHYL-3',2''-ETHYLENE-P-QUATERPHENYL	35.64	0.010	25.82443
4-(3-ETHYLBUTOXY)-P-QUATERPHENYL (B)	32.91	0.013	16.00412
4,4'''-DI(2-BUTYLOCTOXY-1)-P-QUATERPHENYL	33.68	0.012	18.42763
DIETHYL-P-QUINQUEPHENYL	29.09	0.018	7.67626
DI(3-ETHYLHEPTYL)-P-QUINQUEPHENYL	35.09	0.010	23.50152
TETRAMETHYL-P-SEXIPHENYL	27.81	0.021	5.86403
TETRAMETHYL-P-OCTAPHENYL (B)	39.31	0.007	46.38286
INDOLE (TRACE)	0.00	0.000	0.00000
2-PHENYL INDOLE	29.10	0.018	7.70186
1-METHYL-2-PHENYL INDOLE	23.38	0.035	2.06701
1-NEOPENTYL-2-PHENYL INDOLE	20.58	0.051	0.96609
1-N-BUTYL-2(BIPHENYLYL-4) INDOLE	32.23	0.013	14.12227
1,2-DIPHENYL INDOLE	24.30	0.031	2.60907
1-BIPHENYLYL-2-PHENYL INDOLE	27.01	0.023	4.89653
1,3-DIMETHYL-2-PHENYL INDOLE	24.41	0.031	2.68213
2,3-DIPHENYL INDOLE	29.83	0.017	8.85361
1-METHYL-2,3-DIPHENYL INDOLE	27.12	0.022	5.00792
2-PHENYL-3,2'-METHYLENE INDOLE (TRACE)	32.44	0.013	14.73360
2-PHENYL-3,2'-ETHYLENE INDOLE (TRACE)	34.00	0.011	19.46030
2-PHENYL-3,2'-TRIMETHYLENE INDOLE	30.90	0.015	10.99653
1-METHYL-3,2'-METHYLENE-2-PHENYL INDOLE (TRACE)	33.77	0.012	18.72842
1-METHYL-3,2'-ETHYLENE-2-PHENYL INDOLE	34.26	0.011	20.35984
1-METHYL-3,2'-TRIMETHYLENE-2-PHENYL INDOLE	29.59	0.017	8.48332
4,5-DIPHENYLIMIDAZOLE (A)	18.59	0.070	0.52336
PPF (TRACE)	35.52	0.010	25.35325
PPO (TRACE)	31.17	0.015	11.61780
2-(4-FLUOROPHENYL)-5-PHENYLOXAZOLE (TRACE)	29.66	0.017	8.61458
αNPO (TRACE)	36.02	0.010	27.36714
BBO (B)	40.83	0.007	57.98907
POPOP	39.02	0.008	44.09153
DIMETHYL POPOP (TRACE)	39.46	0.007	47.14581
PPD (TRACE)	10.16	0.426	0.01408
2,5-DI(2-METHYLPHENYL)OXADIAZOLE	13.68	0.175	0.08370
PBD	29.96	0.017	9.15455
BBD (B)	37.69	0.008	36.14702
αNPD (TRACE)	32.51	0.013	14.84071
βNPD	28.65	0.019	6.99214
αNND (TRACE)	34.37	0.011	20.73422
βNND (TRACE)	34.59	0.011	21.64836
TRANS-1-(4-BIPHENYLYL)-2-PHENYL-ETHYLENE	37.15	0.009	33.05925
TRANS-1-(4-BIPHENYLYL)-2-(1-NAPHTHYL)-ETHYLENE	38.02	0.008	37.80164
DIPHENYLSTILBENE (B)	41.27	0.006	61.80678
1,1,4,4-TETRAPHENYLBUTADIENE	38.45	0.008	40.36789
1,6-DIPHENYLHEXATRIENE	42.10	0.006	69.55192
1,8-DIPHENYL-1,3,5,7-OCTATETRAENE	39.79	0.007	49.52266
BPSB (TRACE)	42.12	0.006	69.70633
BIS-MSB	41.33	0.006	62.23109
ESCULIN (A)	32.37	0.013	14.40417
AZULENE	24.85	0.029	2.94262
NAPHTHALENE	7.21	1.195	0.00179
DEUTERATED NAPHTHALENE-C10D8	10.55	0.381	0.01751
1-METHYLNAPHTHALENE 5ML/L	9.29	0.559	0.00816
2-METHYL NAPHTHALENE	12.29	0.241	0.04390
2,3-DIMETHYL NAPHTHALENE	12.13	0.250	0.04064
2,6-DIMETHYL NAPHTHALENE (TRACE)	15.87	0.112	0.20314
1-NAPHTHOL (TRACE)	18.02	0.076	0.43513
2-NAPHTHOL (TRACE)	19.91	0.057	0.79084
1-NAPHTHYLAMINE	25.35	0.027	3.34548
2-CHLORONAPHTHALENE	13.48	0.183	0.07622
ACENAPHTHENE	15.90	0.111	0.20599
1-PHENYL NAPHTHALENE	14.97	0.133	0.14302
2-PHENYL NAPHTHALENE	23.13	0.036	1.94136
1,4-DIPHENYL NAPHTHALENE	24.29	0.031	2.60179
1,5-DIPHENYL NAPHTHALENE	22.58	0.039	1.67765
1,7-DIPHENYL NAPHTHALENE	22.23	0.041	1.52695
1,4,5,8-TETRAPHENYL NAPHTHALENE	34.67	0.011	21.70233
1,1'-BINAPHTHYL	17.97	0.077	0.42831
4,4'-DI(N-BUTOXY)-1,1'-BINAPHTHYL	29.31	0.018	8.00564
2,2'-BINAPHTHYL	24.38	0.031	2.66285
ACRIDINE (A)	29.36	0.018	7.99597
ANTHRACENE (TRACE)	27.59	0.021	5.52107
DEUTERATED ANTHRACENE C14D10 (TRACE)	27.40	0.022	5.30077
9-METHYL ANTHRACENE (TRACE)	26.33	0.025	4.16120
1-AMINO ANTHRACENE	24.25	0.031	2.53864
2-AMINO ANTHRACENE	25.64	0.027	3.54851
9-VINYL ANTHRACENE	27.17	0.022	5.01916
9-PHENYL ANTHRACENE (TRACE)	27.77	0.021	5.73152
9,10-DIPHENYL ANTHRACENE	27.80	0.021	5.76658
9,10-DICHLOROANTHRACENE (TRACE)	26.25	0.025	4.08132

ACCEPTOR MOLECULES	RO	CO	OVERLAP
BIANTHRANYL (B)	29.49	0.017	8.20902
9,10-DINAPHTHYLANTHRACENE (B)	27.43	0.022	5.31421
9-(P-VINYLPHENYL)-10-PHENYL ANTHRACENE	27.81	0.021	5.77306
9,10-BIS(PHENYL-ETHYNYL)ANTHRACENE (TRACE)	23.61	0.034	2.16242
RUBICENE (B)	29.44	0.018	8.12617
DIBENZ(A,H)ANTHRACENE (B)	32.84	0.013	15.64843
TETRACENE (TRACE) (B)	20.44	0.052	0.91078
RUBRENE (B)	29.31	0.018	7.91148
TETRAHYDROCANNABINOL ACID METHYL ESTER	10.26	0.414	0.01490
PHENANTHRENE	15.40	0.123	0.16817
3,4-BENZOPHENANTHRENE	23.51	0.034	2.11401
2-PHENYL PHENANTHRENE	17.48	0.084	0.35912
CHRYSENE (TRACE)	24.86	0.029	2.96786
TRIPHENYLENE	16.54	0.099	0.25896
DEUTERATED TRIPHENYLENE-C18D12	16.13	0.107	0.22263
PYRENE	34.10	0.011	19.70382
1,3,6,8,-TETRAPHENYL PYRENE	34.17	0.011	19.88292
BIPYRENYL (B)	38.77	0.008	42.38001
O-PHENYLENE-PYRENE	31.56	0.014	12.34648
ANTHANTHRENE (B)	29.24	0.018	7.80653
FLUORANTHENE	29.37	0.018	8.01393
3-PHENYL FLUORANTHENE	31.16	0.015	11.42449
3-AMINO FLUORANTHENE	28.71	0.019	6.99687
3,3'-BIFLUORANTHENYL (B)	34.27	0.011	20.22048
BENZO(GHI)FLUORANTHENE	28.55	0.019	6.75702
9,12-DI(N-OCTADECOXY)-BENZ(E)ACEPHENANTHRYLENE	30.81	0.015	10.67319
DECACYCLENE (B)	37.56	0.008	35.04978
10-ETHOXY-BENZO(KL)XANTHENE	28.28	0.020	6.38761
PERYLENE	23.74	0.033	2.23403
BENZO(GHI)PERYLENE (B)	30.16	0.016	9.40640
BBOT (TRACE)	39.23	0.007	45.49226
QUININE SULFATE (A)	29.29	0.018	7.97592
ACRIDINE YELLOW (A)	20.92	0.049	1.04673
ACRIDINE ORANGE (A)	20.11	0.055	0.82476
ACRIDINE RED (A)	24.83	0.029	2.92689
URANINE (A)	25.31	0.028	3.27872
RHODAMINE B (A)	28.12	0.020	6.17315
RHODAMINE 6G (A)	28.91	0.019	7.29728
EOSIN B (A)	27.00	0.023	4.83474
EOSIN Y (A)	27.43	0.022	5.31696
ROSE BENGAL (A)	28.87	0.019	7.22429
INDENOINDENE	30.44	0.016	10.07494

8.150 Donor Molecule: 1-Phenylnaphthalene

N,N-DIMETHYLANILINE 3ML/L	8.28	0.788	0.00666
TRIPHENYLAMINE	20.53	0.052	1.53340
SODIUM SALICYLATE (A)	13.90	0.167	0.14731
ANTHRANILIC ACID (A)	24.18	0.032	4.06444
TMPD	20.34	0.053	1.44543
4,4'-DIHYDROXYBIPHENYL (A)	7.95	0.891	0.00522
4,4'-DIMETHOXY BIPHENYL	5.58	2.571	0.00063
4,4'-DIISOBUTOXYBIPHENYL	0.00	0.000	0.00000
BENZIDINE	17.72	0.080	0.63562
4-VINYL BIPHENYL	7.62	1.010	0.00407
M-TERPHENYL	4.82	3.986	0.00026
1,3,5-TRIPHENYL BENZENE	7.44	1.086	0.00350
1,3-DIBIPHENYLYLBENZENE (B)	16.44	0.101	0.40608
1,3,5,-TRIBIPHENYLYLBENZENE (B)	19.50	0.060	1.13355
FLUORENE (TRACE)	0.00	0.000	0.00000
1-METHYL FLUORENE	0.00	0.000	0.00000
1,2-BENZOFLUORENE	18.97	0.066	0.95356
9,10-DIHYDROPHENANTHRENE	3.77	8.351	0.00006
CARBAZOLE (TRACE)	18.31	0.073	0.77328
N-METHYL CARBAZOLE (TRACE)	21.42	0.046	1.97677
N-VINYLCARBAZOLE	21.28	0.046	1.90243
N-PHENYL CARBAZOLE	21.03	0.048	1.77479
N-METHYL-2,3-BENZCARBAZOLE	24.58	0.030	4.46163
DIBENZOFURAN	0.00	0.000	0.00000
DIBENZOTHIOPHENE (TRACE)	15.96	0.110	0.33982
P-TERPHENYL	9.78	0.479	0.01805
DEUTERATED P-TERPHENYL-C18D14	8.84	0.648	0.00987
4-METHYL TERPHENYL	11.32	0.309	0.04340
2-METHYL-P-TERPHENYL	0.00	0.000	0.00000
3,3''-DIMETHYL-P-TERPHENYL	11.37	0.304	0.04467
4-(3,3-DIMETHYLBUTOXY)-P-TERPHENYL	16.39	0.102	0.39877
4,4''-DIHEXAHYDROFARNESOXY-P-TERPHENYL	19.35	0.062	1.08152
2,2'-METHYLENE-P-TERPHENYL	17.78	0.080	0.65158
2,2'-ETHYLENE-P-TERPHENYL	18.88	0.066	0.93302
2,2'-METHYLENE-5',6''-METHYLENE-P-TERPHENYL (TRACE)	28.79	0.019	11.68042
3-PHENYL DIBENZOFURAN	17.79	0.079	0.65300
BENZO(1,2-B-4,5-B')BISBENZOFURAN (TRACE) (B)	30.13	0.016	15.34305
SEXIPHENYL BENZENE	0.00	0.000	0.00000
P-QUATERPHENYL	20.73	0.050	1.62998
3,3'''-DIMETHYL-P-QUATERPHENYL	22.13	0.041	2.41002
TETRAMETHYL-P-QUATERPHENYL	5.11	3.359	0.00037
3,3''' -DIMETHYL-3',2''-METHYLENE-P-QUATERPHENYL	29.76	0.017	14.22623
3,3''' -DIMETHYL-3',2''-ETHYLENE-P-QUATERPHENYL	29.45	0.018	13.32254
4-(3-ETHYLBUTOXY)-P-QUATERPHENYL (B)	26.89	0.023	7.72725
4,4'''-DI(2-BUTYLOCTOXY-1)-P-QUATERPHENYL	27.46	0.022	8.76636
DIETHYL-P-QUINQUEPHENYL	23.09	0.036	3.10875

ACCEPTOR MOLECULES	RO	CO	OVERLAP
DI(3-ETHYLHEPTYL)-P-QUINQUEPHENYL	28.74	0.019	11.50655
TETRAMETHYL-P-SEXIPHENYL	21.84	0.043	2.23088
TETRAMETHYL-P-OCTAPHENYL (B)	32.91	0.013	25.91566
INDOLE (TRACE)	0.00	0.000	0.00000
2-PHENYL INDOLE	23.10	0.036	3.12153
1-METHYL-2-PHENYL INDOLE	18.32	0.073	0.77659
1-NEOPENTYL-2-PHENYL INDOLE	15.78	0.114	0.31811
1-N-BUTYL-2(BIPHENYLYL-4) INDOLE	26.91	0.023	7.75502
1,2-DIPHENYL INDOLE	18.98	0.065	0.96043
1-BIPHENYLYL-2-PHENYL INDOLE	21.44	0.045	1.98832
1,3-DIMETHYL-2-PHENYL INDOLE	19.45	0.061	1.11350
2,3-DIPHENYL INDOLE	24.84	0.029	4.79446
1-METHYL-2,3-DIPHENYL INDOLE	22.50	0.039	2.65329
2-PHENYL-3,2'-METHYLENE INDOLE (TRACE)	26.25	0.025	6.71399
2-PHENYL-3,2'-ETHYLENE INDOLE (TRACE)	28.93	0.018	11.97547
2-PHENYL-3,2'-TRIMETHYLENE INDOLE	24.95	0.029	4.94701
1-METHYL-3,2'-METHYLENE-2-PHENYL INDOLE (TRACE)	28.40	0.020	10.73073
1-METHYL-3,2'-ETHYLENE-2-PHENYL INDOLE	29.53	0.017	13.52457
1-METHYL-3,2'-TRIMETHYLENE-2-PHENYL INDOLE	24.11	0.032	4.02355
4,5-DIPHENYLIMIDAZOLE (A)	14.71	0.141	0.20800
PPF (TRACE)	29.77	0.017	14.25887
PPO (TRACE)	25.10	0.028	5.13908
2-(4-FLUOROPHENYL)-5-PHENYLOXAZOLE (TRACE)	23.85	0.033	3.78099
αNPO (TRACE)	31.93	0.014	21.55585
BBO (B)	36.95	0.009	51.61448
POPOP	36.46	0.009	47.63106
DIMETHYL POPOP (TRACE)	37.54	0.008	56.59867
PPD (TRACE)	8.71	0.677	0.00904
2,5-DI(2-METHYLPHENYL)OXADIAZOLE	10.75	0.360	0.03197
PBD	23.71	0.034	3.65334
BBD (B)	31.56	0.014	20.21387
αNPD (TRACE)	27.44	0.022	8.70866
βNPD	22.95	0.037	2.99593
αNND (TRACE)	29.97	0.017	14.77471
βNND (TRACE)	28.61	0.019	11.24925
TRANS-1-(4-BIPHENYLYL)-2-PHENYL-ETHYLENE	31.40	0.014	19.55057
TRANS-1-(4-BIPHENYLYL)-2-(1-NAPHTHYL)-ETHYLENE	33.93	0.011	30.96476
DIPHENYLSTILBENE (B)	37.29	0.009	54.51534
1,1,4,4-TETRAPHENYLBUTADIENE	35.17	0.010	38.31278
1,6-DIPHENYLHEXATRIENE	39.18	0.007	73.22836
1,8-DIPHENYL-1,3,5,7-OCTATETRAENE	39.59	0.007	77.44957
BPSB (TRACE)	39.46	0.007	76.44105
BIS-MSB	38.10	0.008	61.96086
ESCULIN (A)	28.80	0.019	11.58871
AZULENE	21.20	0.047	1.81557
NAPHTHALENE	5.36	2.897	0.00049
DEUTERATED NAPHTHALENE-C10D8	8.18	0.817	0.00617
1-METHYLNAPHTHALENE 5ML/L	7.07	1.266	0.00258
2-METHYL NAPHTHALENE	8.93	0.628	0.01049
2,3-DIMETHYL NAPHTHALENE	8.80	0.657	0.00958
2,6-DIMETHYL NAPHTHALENE (TRACE)	12.34	0.238	0.07293
1-NAPHTHOL (TRACE)	13.12	0.198	0.10501
2-NAPHTHOL (TRACE)	16.10	0.107	0.35871
1-NAPHTHYLAMINE	20.95	0.049	1.73090
2-CHLORONAPHTHALENE	9.64	0.500	0.01652
ACENAPHTHENE	11.22	0.317	0.04119
1-PHENYL NAPHTHALENE	11.40	0.302	0.04537
2-PHENYL NAPHTHALENE	18.22	0.074	0.75266
1,4-DIPHENYL NAPHTHALENE	19.21	0.063	1.03062
1,5-DIPHENYL NAPHTHALENE	17.60	0.082	0.61154
1,7-DIPHENYL NAPHTHALENE	17.43	0.084	0.57609
1,4,5,8-TETRAPHENYL NAPHTHALENE	30.85	0.015	17.46524
1,1'-BINAPHTHYL	13.91	0.166	0.14920
4,4'-DI(N-BUTOXY)-1,1'-BINAPHTHYL	24.01	0.032	3.92299
2,2'-BINAPHTHYL	19.07	0.065	0.98765
ACRIDINE (A)	27.45	0.022	8.65689
ANTHRACENE (TRACE)	25.81	0.026	6.00256
DEUTERATED ANTHRACENE C14D10 (TRACE)	25.52	0.027	5.61243
9-METHYL ANTHRACENE (TRACE)	25.85	0.026	6.05024
1-AMINO ANTHRACENE	24.68	0.030	4.53722
2-AMINO ANTHRACENE	24.38	0.031	4.20888
9-VINYL ANTHRACENE	26.83	0.023	7.53568
9-PHENYL ANTHRACENE (TRACE)	27.18	0.022	8.16731
9,10-DIPHENYL ANTHRACENE	27.99	0.020	9.71670
9,10-DICHLOROANTHRACENE (TRACE)	26.81	0.023	7.50221
BIANTHRANYL (B)	29.46	0.017	13.20778
9,10-DINAPHTHYLANTHRACENE (B)	27.51	0.021	8.73218
9-(P-VINYLPHENYL)-10-PHENYL ANTHRACENE	28.12	0.020	9.97873
9,10-BIS(PHENYL-ETHYNYL)ANTHRACENE (TRACE)	26.74	0.023	7.30306
RUBICENE (B)	27.52	0.021	8.69089
DIBENZ(A,H)ANTHRACENE (B)	28.32	0.020	10.42881
TETRACENE (TRACE) (B)	20.89	0.049	1.66010
RUBRENE (B)	24.36	0.031	4.17781
TETRAHYDROCANNABINOL ACID METHYL ESTER	8.07	0.852	0.00571
PHENANTHRENE	12.98	0.205	0.09787
3,4-BENZOPHENANTHRENE	19.03	0.065	0.96504
2-PHENYL PHENANTHRENE	14.97	0.133	0.22978
CHRYSENE (TRACE)	19.35	0.062	1.06969
TRIPHENYLENE	13.55	0.180	0.12692
DEUTERATED TRIPHENYLENE-C18D12	13.30	0.190	0.11341
PYRENE	28.00	0.020	9.80400
1,3,6,8,-TETRAPHENYL PYRENE	34.66	0.011	34.95881
BIPYRENYL (B)	35.88	0.010	43.11701
O-PHENYLENE-PYRENE	30.74	0.015	16.87201
ANTHANTHRENE (B)	28.74	0.019	11.31155
FLUORANTHENE	26.79	0.023	7.46526

ACCEPTOR MOLECULES	RO	CO	OVERLAP
3-PHENYL FLUORANTHENE	29.76	0.017	14.00891
3-AMINO FLUORANTHENE	27.47	0.022	8.59156
3,3'-BIFLUORANTHENYL (B)	33.43	0.012	28.05540
BENZO(GHI)FLUORANTHENE	25.54	0.027	5.59676
9,12-DI(N-OCTADECOXY)-BENZ(E)ACEPHENANTHRYLENE	29.53	0.017	13.34248
DECACYCLENE (B)	36.25	0.009	45.40715
10-ETHOXY-BENZO(KL)XANTHENE	28.52	0.019	10.88134
PERYLENE	27.08	0.023	7.91238
BENZO(GHI)PERYLENE (B)	30.20	0.016	15.31862
BBOT (TRACE)	38.19	0.008	62.65654
QUININE SULFATE (A)	24.54	0.030	4.47142
ACRIDINE YELLOW (A)	24.54	0.030	4.36596
ACRIDINE ORANGE (A)	21.10	0.048	1.76395
ACRIDINE RED (A)	23.27	0.035	3.17768
URANINE (A)	23.54	0.034	3.39865
RHODAMINE B (A)	26.50	0.024	6.91945
RHODAMINE 6G (A)	26.76	0.023	7.33782
EOSIN B (A)	24.02	0.032	3.84503
EOSIN Y (A)	24.58	0.030	4.40950
ROSE BENGAL (A)	25.54	0.027	5.54726
INDENOINDENE	24.65	0.030	4.60864

8.151 Donor Molecule: 2-Phenylnaphthalene

N,N-DIMETHYLANILINE 3ML/L	3.33	12.118	0.00004
TRIPHENYLAMINE	15.76	0.114	0.44504
SODIUM SALICYLATE (A)	11.36	0.305	0.06247
ANTHRANILIC ACID (A)	22.01	0.042	3.27941
TMPD	17.77	0.080	0.91212
BENZIDINE	12.42	0.233	0.10691
1,3,5-TRIPHENYL BENZENE	3.64	9.239	0.00007
1,3-DIBIPHENYLYLBENZENE (B)	10.62	0.373	0.04194
1,3,5,-TRIBIPHENYLYLBENZENE (B)	12.27	0.242	0.09960
FLUORENE (TRACE)	0.00	0.000	0.00000
1-METHYL FLUORENE	0.00	0.000	0.00000
1,2-BENZOFLUORENE	15.19	0.128	0.35711
9,10-DIHYDROPHENANTHRENE	0.00	0.000	0.00000
CARBAZOLE (TRACE)	13.01	0.203	0.14119
N-METHYL CARBAZOLE (TRACE)	18.32	0.073	1.09715
N-VINYLCARBAZOLE	17.53	0.083	0.84347
N-PHENYL CARBAZOLE	17.44	0.084	0.81861
N-METHYL-2,3-BENZCARBAZOLE	22.79	0.038	4.01284
DIBENZOFURAN	0.00	0.000	0.00000
DIBENZOTHIOPHENE (TRACE)	9.97	0.451	0.02867
P-TERPHENYL	3.08	15.364	0.00002
DEUTERATED P-TERPHENYL-C18D14	0.00	0.000	0.00000
4-METHYL TERPHENYL	4.41	5.204	0.00022
2-METHYL-P-TERPHENYL	0.00	0.000	0.00000
3,3''-DIMETHYL-P-TERPHENYL	5.27	3.064	0.00062
4-(3,3-DIMETHYLBUTOXY)-P-TERPHENYL	9.87	0.465	0.02699
4,4''-DIHEXAHYDROFARNESOXY-P-TERPHENYL	12.39	0.235	0.10558
2,2'-METHYLENE-P-TERPHENYL	9.76	0.481	0.02532
2,2'-ETHYLENE-P-TERPHENYL	11.34	0.307	0.06203
2,2'-METHYLENE-5',6''-METHYLENE-P-TERPHENYL (TRACE)	21.98	0.042	3.27877
3-PHENYL DIBENZOFURAN	9.72	0.487	0.02468
BENZO(1,2-B-4,5-B)BISBENZOFURAN (TRACE) (B)	24.06	0.032	5.65993
SEXIPHENYL BENZENE	0.00	0.000	0.00000
P-QUATERPHENYL	14.22	0.156	0.24094
3,3'''-DIMETHYL-P-QUATERPHENYL	15.91	0.111	0.47129
TETRAMETHYL-P-QUATERPHENYL	0.00	0.000	0.00000
3,3''' -DIMETHYL-3',2''-METHYLENE-P-QUATERPHENYL	23.45	0.035	4.82952
3,3''' -DIMETHYL-3',2''-ETHYLENE-P-QUATERPHENYL	23.78	0.033	5.23746
4-(3-ETHYLBUTOXY)-P-QUATERPHENYL (B)	21.26	0.047	2.67819
4,4'''-DI(2-BUTYLOCTOXY-1)-P-QUATERPHENYL	21.57	0.045	2.92008
DIETHYL-P-QUINQUEPHENYL	16.48	0.100	0.58403
DI(3-ETHYLHEPTYL)-P-QUINQUEPHENYL	22.93	0.037	4.20390
TETRAMETHYL-P-SEXIPHENYL	14.85	0.137	0.31234
TETRAMETHYL-P-OCTAPHENYL (B)	27.61	0.021	12.82480
INDOLE (TRACE)	0.00	0.000	0.00000
2-PHENYL INDOLE	15.92	0.111	0.47390
1-METHYL-2-PHENYL INDOLE	12.31	0.240	0.10139
1-NEOPENTYL-2-PHENYL INDOLE	9.32	0.552	0.01919
1-N-BUTYL-2(BIPHENYLYL-4) INDOLE	22.45	0.040	3.70429
1,2-DIPHENYL INDOLE	12.65	0.221	0.11961
1-BIPHENYLYL-2-PHENYL INDOLE	15.71	0.115	0.43555
1,3-DIMETHYL-2-PHENYL INDOLE	13.80	0.170	0.20106
2,3-DIPHENYL INDOLE	20.64	0.051	2.23441
1-METHYL-2,3-DIPHENYL INDOLE	18.54	0.070	1.17532
2-PHENYL-3,2'-METHYLENE INDOLE (TRACE)	19.88	0.057	1.79332
2-PHENYL-3,2'-ETHYLENE INDOLE (TRACE)	24.67	0.030	6.52700
2-PHENYL-3,2'-TRIMETHYLENE INDOLE	18.42	0.072	1.13675
1-METHYL-3,2'-METHYLENE-2-PHENYL INDOLE (TRACE)	23.68	0.034	5.11735
1-METHYL-3,2'-ETHYLENE-2-PHENYL INDOLE	26.13	0.025	9.20675
1-METHYL-3,2'-TRIMETHYLENE-2-PHENYL INDOLE	18.69	0.068	1.24010
4,5-DIPHENYLIMIDAZOLE (A)	10.55	0.381	0.04021
PPF (TRACE)	24.81	0.029	6.76433
PPO (TRACE)	18.35	0.072	1.11321
2-(4-FLUOROPHENYL)-5-PHENYLOXAZOLE (TRACE)	17.18	0.088	0.74822
αNPO (TRACE)	29.23	0.018	17.96591
BBO (B)	34.51	0.011	48.63313
POPOP	34.90	0.011	51.90982

ACCEPTOR MOLECULES	RO	CO	OVERLAP
DIMETHYL POPOP (TRACE)	36.24	0.009	64.95120
PPD (TRACE)	0.00	0.000	0.00000
2,5-DI(2-METHYLPHENYL)OXADIAZOLE	0.00	0.000	0.00000
PBD	16.19	0.105	0.52496
BBD (B)	26.35	0.024	9.69698
αNPD (TRACE)	23.23	0.036	4.54774
βNPD	16.57	0.098	0.60302
αNND (TRACE)	26.86	0.023	10.86838
βNND (TRACE)	23.26	0.036	4.59725
TRANS-1-(4-BIPHENYLYL)-2-PHENYL-ETHYLENE	26.73	0.023	10.55544
TRANS-1-(4-BIPHENYLYL)-2-(1-NAPHTHYL)-ETHYLENE	31.29	0.015	27.01723
DIPHENYLSTILBENE (B)	34.78	0.011	50.88695
1,1,4,4-TETRAPHENYLBUTADIENE	33.14	0.012	38.01036
1,6-DIPHENYLHEXATRIENE	37.42	0.009	78.89982
1,8-DIPHENYL-1,3,5,7-OCTATETRAENE	38.87	0.008	98.38214
BPSB (TRACE)	37.81	0.008	83.95281
BIS-MSB	36.12	0.009	63.76214
ESCULIN (A)	26.45	0.024	9.87656
AZULENE	18.22	0.074	1.04043
NAPHTHALENE	1.74	84.193	0.00000
DEUTERATED NAPHTHALENE-C10D8	5.21	3.169	0.00058
1-METHYLNAPHTHALENE 5ML/L	3.30	12.475	0.00004
2-METHYL NAPHTHALENE	3.82	8.038	0.00009
2,3-DIMETHYL NAPHTHALENE	3.94	7.291	0.00011
2,6-DIMETHYL NAPHTHALENE (TRACE)	7.05	1.279	0.00358
1-NAPHTHOL (TRACE)	6.82	1.407	0.00295
2-NAPHTHOL (TRACE)	10.84	0.351	0.04738
1-NAPHTHYLAMINE	16.70	0.096	0.63055
2-CHLORONAPHTHALENE	4.84	3.942	0.00038
ACENAPHTHENE	4.95	3.699	0.00043
1-PHENYL NAPHTHALENE	6.16	1.910	0.00160
2-PHENYL NAPHTHALENE	12.16	0.249	0.09436
1,4-DIPHENYL NAPHTHALENE	13.51	0.181	0.17729
1,5-DIPHENYL NAPHTHALENE	11.60	0.287	0.07089
1,7-DIPHENYL NAPHTHALENE	11.87	0.268	0.08135
1,4,5,8-TETRAPHENYL NAPHTHALENE	28.35	0.020	14.93149
1,1'-BINAPHTHYL	8.38	0.759	0.01015
4,4'-DI(N-BUTOXY)-1,1'-BINAPHTHYL	18.82	0.067	1.28891
2,2'-BINAPHTHYL	12.95	0.206	0.13766
ACRIDINE (A)	26.28	0.025	9.46368
ANTHRACENE (TRACE)	24.82	0.029	6.73057
DEUTERATED ANTHRACENE C14D10 (TRACE)	24.47	0.031	6.18532
9-METHYL ANTHRACENE (TRACE)	25.24	0.028	7.42537
1-AMINO ANTHRACENE	24.29	0.031	5.84129
2-AMINO ANTHRACENE	23.34	0.035	4.60082
9-VINYL ANTHRACENE	26.26	0.025	9.40640
9-PHENYL ANTHRACENE (TRACE)	26.53	0.024	10.00908
9,10-DIPHENYL ANTHRACENE	27.59	0.021	12.63044
9,10-DICHLOROANTHRACENE (TRACE)	26.53	0.024	9.98693
BIANTHRANYL (B)	28.91	0.019	16.70866
9,10-DINAPHTHYLANTHRACENE (B)	27.09	0.022	11.30657
9-(P-VINYLPHENYL)-10-PHENYL ANTHRACENE	27.75	0.021	13.05528
9,10-BIS(PHENYL-ETHYNYL)ANTHRACENE (TRACE)	25.83	0.026	8.44685
RUBICENE (B)	26.15	0.025	9.08488
DIBENZ(A,H)ANTHRACENE (B)	25.31	0.028	7.54547
TETRACENE (TRACE) (B)	20.05	0.056	1.84561
RUBRENE (B)	20.25	0.054	1.96273
TETRAHYDROCANNABINOL ACID METHYL ESTER	0.00	0.000	0.00000
PHENANTHRENE	10.96	0.339	0.05047
3,4-BENZOPHENANTHRENE	15.10	0.130	0.34084
2-PHENYL PHENANTHRENE	13.09	0.199	0.14563
CHRYSENE (TRACE)	15.40	0.122	0.38533
TRIPHENYLENE	10.79	0.356	0.04587
DEUTERATED TRIPHENYLENE-C18D12	10.49	0.387	0.03879
PYRENE	22.42	0.040	3.65878
1,3,6,8,-TETRAPHENYL PYRENE	34.27	0.011	46.26347
BIPYRENYL (B)	33.99	0.011	44.16480
O-PHENYLENE-PYRENE	29.88	0.017	20.23505
ANTHANTHRENE (B)	27.69	0.021	12.82656
FLUORANTHENE	25.26	0.028	7.43446
3-PHENYL FLUORANTHENE	28.80	0.019	16.31184
3-AMINO FLUORANTHENE	26.32	0.025	9.44706
3,3'-BIFLUORANTHENYL (B)	32.56	0.013	33.99866
BENZO(GHI)FLUORANTHENE	23.45	0.035	4.75541
9,12-DI(N-OCTADECOXY)-BENZ(E)ACEPHENANTHRYLENE	28.42	0.019	15.02229
DECACYCLENE (B)	35.06	0.010	52.76483
10-ETHOXY-BENZO(KL)XANTHENE	28.13	0.020	14.19544
PERYLENE	26.68	0.024	10.26425
BENZO(GHI)PERYLENE (B)	29.69	0.017	19.61346
BBOT (TRACE)	37.22	0.009	76.18248
QUININE SULFATE (A)	20.28	0.054	2.01877
ACRIDINE YELLOW (A)	23.42	0.035	4.69338
ACRIDINE ORANGE (A)	19.89	0.057	1.76022
ACRIDINE RED (A)	22.15	0.041	3.36032
URANINE (A)	21.64	0.044	2.92211
RHODAMINE B (A)	25.28	0.028	7.41717
RHODAMINE 6G (A)	25.17	0.028	7.23237
EOSIN B (A)	21.86	0.043	3.10154
EOSIN Y (A)	22.48	0.039	3.67229
ROSE BENGAL (A)	23.15	0.036	4.37165
INDENOINDENE	17.76	0.080	0.91273

8.152 Donor Molecule: 1,4-Diphenylnaphthalene

ACCEPTOR MOLECULES	RO	CO	OVERLAP
TRIPHENYLAMINE	12.36	0.237	0.06725
P-XYLENE (TRACE)	0.00	0.000	0.00000
SODIUM SALICYLATE (A)	9.57	0.511	0.01441
ANTHRANILIC ACID (A)	20.07	0.055	1.22143
TMPD	15.43	0.122	0.25279
BENZIDINE	8.99	0.616	0.00996
1,3-DIBIPHENYLYLBENZENE (B)	7.30	1.152	0.00285
1,3,5,-TRIBIPHENYLYLBENZENE (B)	7.97	0.883	0.00486
FLUORENE (TRACE)	0.00	0.000	0.00000
1-METHYL FLUORENE	0.00	0.000	0.00000
1,2-BENZOFLUORENE	12.27	0.242	0.06427
9,10-DIHYDROPHENANTHRENE	0.00	0.000	0.00000
CARBAZOLE (TRACE)	8.95	0.624	0.00971
N-METHYL CARBAZOLE (TRACE)	14.70	0.141	0.19040
N-VINYLCARBAZOLE	13.51	0.182	0.11440
N-PHENYL CARBAZOLE	13.59	0.178	0.11869
N-METHYL-2,3-BENZCARBAZOLE	25.15	0.028	4.69962
DIBENZOFURAN	0.00	0.000	0.00000
DIBENZOTHIOPHENE (TRACE)	5.88	2.197	0.00078
P-TERPHENYL	0.00	0.000	0.00000
DEUTERATED P-TERPHENYL-C18D14	0.00	0.000	0.00000
4-METHYL TERPHENYL	0.00	0.000	0.00000
2-METHYL-P-TERPHENYL	0.00	0.000	0.00000
3,3''-DIMETHYL-P-TERPHENYL	0.00	0.000	0.00000
4-(3,3-DIMETHYLBUTOXY)-P-TERPHENYL	5.92	2.152	0.00082
4,4''-DIHEXAHYDROFARNESOXY-P-TERPHENYL	8.01	0.870	0.00500
2,2' -METHYLENE-P-TERPHENYL	4.99	3.600	0.00029
2,2' -ETHYLENE-P-TERPHENYL	6.90	1.364	0.00203
2,2'-METHYLENE-5',6''-METHYLENE-P-TERPHENYL (TRACE)	15.84	0.113	0.29786
3-PHENYL DIBENZOFURAN	4.83	3.958	0.00024
BENZO(1,2-B-4,5-B')BISBENZOFURAN (TRACE) (B)	18.47	0.071	0.74883
SEXIPHENYL BENZENE	0.00	0.000	0.00000
P-QUATERPHENYL	9.95	0.454	0.01834
3,3'''-DIMETHYL-P-QUATERPHENYL	11.72	0.278	0.04896
TETRAMETHYL-P-QUATERPHENYL	0.00	0.000	0.00000
3,3''' -DIMETHYL-3',2''-METHYLENE-P-QUATERPHENYL	18.16	0.075	0.67444
3,3''' -DIMETHYL-3',2''-ETHYLENE-P-QUATERPHENYL	19.15	0.064	0.92518
4-(3-ETHYLBUTOXY)-P-QUATERPHENYL (B)	16.99	0.091	0.45212
4,4'''-DI(2-BUTYLOCTOXY-1)-P-QUATERPHENYL	17.14	0.089	0.47693
DIETHYL-P-QUINQUEPHENYL	12.01	0.258	0.05653
DI(3-ETHYLHEPTYL)-P-QUINQUEPHENYL	18.63	0.069	0.78333
TETRAMETHYL-P-SEXIPHENYL	10.31	0.408	0.02269
TETRAMETHYL-P-OCTAPHENYL (B)	23.41	0.035	3.08460
INDOLE (TRACE)	0.00	0.000	0.00000
2-PHENYL INDOLE	10.83	0.352	0.03049
1-METHYL-2-PHENYL INDOLE	8.52	0.722	0.00724
1-NEOPENTYL-2-PHENYL INDOLE	5.52	2.667	0.00053
1-N-BUTYL-2(BIPHENYLYL-4) INDOLE	18.82	0.067	0.83491
1,2-DIPHENYL INDOLE	8.76	0.666	0.00853
1-BIPHENYLYL-2-PHENYL INDOLE	12.33	0.239	0.06597
1,3-DIMETHYL-2-PHENYL INDOLE	9.78	0.478	0.01654
2,3-DIPHENYL INDOLE	17.38	0.085	0.51749
1-METHYL-2,3-DIPHENYL INDOLE	15.43	0.122	0.25288
2-PHENYL-3,2'-METHYLENE INDOLE (TRACE)	14.57	0.145	0.18052
2-PHENYL-3,2'-ETHYLENE INDOLE (TRACE)	20.33	0.053	1.32510
2-PHENYL-3,2' -TRIMETHYLENE INDOLE	13.42	0.185	0.11028
1-METHYL-3,2'-METHYLENE-2-PHENYL INDOLE (TRACE)	19.04	0.065	0.89572
1-METHYL-3,2'-ETHYLENE-2-PHENYL INDOLE	22.80	0.038	2.63770
1-METHYL-3,2'-TRIMETHYLENE-2-PHENYL INDOLE	14.13	0.158	0.15016
4,5-DIPHENYLIMIDAZOLE (A)	7.66	0.994	0.00383
PPF (TRACE)	19.92	0.057	1.17697
PPO (TRACE)	13.10	0.199	0.09558
2-(4-FLUOROPHENYL)-5-PHENYLOXAZOLE (TRACE)	12.12	0.251	0.05979
αNPO (TRACE)	26.90	0.023	7.07698
BBO (B)	32.99	0.012	24.03715
POPOP	34.78	0.011	33.01249
DIMETHYL POPOP (TRACE)	37.15	0.009	48.88952
PPD (TRACE)	0.00	0.000	0.00000
2,5-DI(2-METHYLPHENYL)OXADIAZOLE	0.00	0.000	0.00000
PBD	11.04	0.333	0.03414
BBD (B)	21.68	0.044	1.95243
αNPD (TRACE)	19.37	0.062	0.99208
βNPD	12.24	0.244	0.06334
αNND (TRACE)	23.82	0.033	3.42478
βNND (TRACE)	18.22	0.074	0.68915
TRANS-1-(4-BIPHENYLYL)-2-PHENYL-ETHYLENE	22.44	0.040	2.39816
TRANS-1-(4-BIPHENYLYL)-2-(1-NAPHTHYL)-ETHYLENE	29.62	0.017	12.59274
DIPHENYLSTILBENE (B)	33.06	0.012	24.35054
1,1,4,4-TETRAPHENYLBUTADIENE	32.19	0.013	20.69882
1,6-DIPHENYLHEXATRIENE	37.05	0.009	48.17437
1,8-DIPHENYL-1,3,5,7-OCTATETRAENE	42.04	0.006	102.20005
BPSB (TRACE)	37.88	0.008	55.06031
BIS-MSB	35.63	0.010	38.11265
ESCULIN (A)	24.66	0.030	4.20571
AZULENE	15.22	0.127	0.22825
NAPHTHALENE	0.00	0.000	0.00000
DEUTERATED NAPHTHALENE-C10D8	3.39	11.516	0.00003
1-METHYLNAPHTHALENE 5ML/L	0.00	0.000	0.00000
2-METHYL NAPHTHALENE	0.00	0.000	0.00000
2,3-DIMETHYL NAPHTHALENE	0.00	0.000	0.00000
2,6-DIMETHYL NAPHTHALENE (TRACE)	0.00	0.000	0.00000
1-NAPHTHOL (TRACE)	3.20	13.645	0.00002
2-NAPHTHOL (TRACE)	6.81	1.414	0.00189
1-NAPHTHYLAMINE	13.07	0.200	0.09380

ACCEPTOR MOLECULES	RO	CO	OVERLAP
2-PHENYL NAPHTHALENE	8.12	0.834	0.00544
1,4-DIPHENYL NAPHTHALENE	9.81	0.473	0.01685
1,5-DIPHENYL NAPHTHALENE	7.86	0.923	0.00444
1,7-DIPHENYL NAPHTHALENE	8.34	0.772	0.00633
1,4,5,8-TETRAPHENYL NAPHTHALENE	26.61	0.024	6.61605
1,1'-BINAPHTHYL	4.84	3.937	0.00024
4,4'-DI(N-BUTOXY)-1,1'-BINAPHTHYL	14.92	0.135	0.20713
2,2'-BINAPHTHYL	9.10	0.594	0.01070
ACRIDINE (A)	25.99	0.025	5.73404
ANTHRACENE (TRACE)	24.90	0.029	4.45206
DEUTERATED ANTHRACENE C14D10 (TRACE)	24.45	0.031	3.99131
9-METHYL ANTHRACENE (TRACE)	26.48	0.024	6.42505
1-AMINO ANTHRACENE	27.22	0.022	7.50420
2-AMINO ANTHRACENE	26.00	0.025	5.70264
9-VINYL ANTHRACENE	27.97	0.020	8.89455
9-PHENYL ANTHRACENE (TRACE)	27.77	0.021	8.53263
9,10-DIPHENYL ANTHRACENE	30.16	0.016	13.97786
9,10-DICHLOROANTHRACENE (TRACE)	29.38	0.018	11.94657
BIANTHRANYL (B)	31.65	0.014	18.65506
9,10-DINAPHTHYLANTHRACENE (B)	29.85	0.017	13.12522
9-(P-VINYLPHENYL)-10-PHENYL ANTHRACENE	30.48	0.016	14.86754
9,10-BIS(PHENYL-ETHYNYL)ANTHRACENE (TRACE)	33.48	0.012	25.85884
RUBICENE (B)	29.63	0.017	12.40262
DIBENZ(A,H)ANTHRACENE (B)	22.92	0.037	2.69684
TETRACENE (TRACE) (B)	24.87	0.029	4.33804
RUBRENE (B)	22.66	0.038	2.48549
TETRAHYDROCANNABINOL ACID METHYL ESTER	0.00	0.000	0.00000
PHENANTHRENE	8.87	0.641	0.00916
3,4-BENZOPHENANTHRENE	14.10	0.160	0.14665
2-PHENYL PHENANTHRENE	11.50	0.294	0.04341
CHRYSENE (TRACE)	14.03	0.162	0.14274
TRIPHENYLENE	8.27	0.792	0.00601
DEUTERATED TRIPHENYLENE-C18D12	8.08	0.848	0.00524
PYRENE	17.32	0.086	0.50423
1,3,6,8,-TETRAPHENYL PYRENE	37.09	0.009	48.22054
BIPYRENYL (B)	33.69	0.012	27.18827
O-PHENYLENE-PYRENE	32.16	0.013	20.34635
ANTHANTHRENE (B)	32.66	0.013	22.38455
FLUORANTHENE	24.19	0.032	3.72088
3-PHENYL FLUORANTHENE	29.57	0.017	12.38610
3-AMINO FLUORANTHENE	29.44	0.018	11.97762
3,3'-BIFLUORANTHENYL (B)	35.01	0.010	34.03950
BENZO(GHI)FLUORANTHENE	22.35	0.040	2.30740
9,12-DI(N-OCTADECOXY)-BENZ(E)ACEPHENANTHRYLENE	31.28	0.015	17.31780
DECACYCLENE (B)	38.51	0.008	59.91320
10-ETHOXY-BENZO(KL)XANTHENE	30.70	0.015	15.56435
PERYLENE	33.20	0.012	24.72886
BENZO(GHI)PERYLENE (B)	31.91	0.014	19.61559
BBOT (TRACE)	39.29	0.007	68.36361
QUININE SULFATE (A)	16.05	0.108	0.32171
ACRIDINE YELLOW (A)	31.15	0.015	16.77009
ACRIDINE ORANGE (A)	26.27	0.025	6.03579
ACRIDINE RED (A)	23.59	0.034	3.15905
URANINE (A)	26.65	0.024	6.56966
RHODAMINE B (A)	27.17	0.022	7.37774
RHODAMINE 6G (A)	25.77	0.026	5.36784
EOSIN B (A)	24.06	0.032	3.55657
EOSIN Y (A)	25.22	0.028	4.72366
ROSE BENGAL (A)	24.51	0.030	3.97770
INDENOINDENE	12.53	0.227	0.07314

8.153 Donor Molecule: 1,5-Diphenylnaphthalene

TRIPHENYLAMINE	14.66	0.142	0.14107
P-XYLENE (TRACE)	0.00	0.000	0.00000
SODIUM SALICYLATE (A)	10.90	0.345	0.02388
ANTHRANILIC ACID (A)	22.08	0.042	1.63785
TMPD	17.33	0.086	0.38418
BENZIDINE	11.23	0.316	0.02856
1,3,5-TRIPHENYL BENZENE	2.23	40.518	0.00000
1,3-DIBIPHENYLYLBENZENE (B)	9.42	0.534	0.01000
1,3,5,-TRIBIPHENYLYLBENZENE (B)	10.72	0.363	0.02164
FLUORENE (TRACE)	0.00	0.000	0.00000
1-METHYL FLUORENE	0.00	0.000	0.00000
1,2-BENZOFLUORENE	14.29	0.153	0.12120
9,10-DIHYDROPHENANTHRENE	0.00	0.000	0.00000
CARBAZOLE (TRACE)	11.74	0.277	0.03725
N-METHYL CARBAZOLE (TRACE)	17.19	0.088	0.36755
N-VINYLCARBAZOLE	16.24	0.104	0.26083
N-PHENYL CARBAZOLE	16.21	0.105	0.25796
N-METHYL-2,3-BENZCARBAZOLE	26.26	0.025	4.60445
DIBENZOFURAN	0.00	0.000	0.00000
DIBENZOTHIOPHENE (TRACE)	8.27	0.790	0.00458
P-TERPHENYL	0.00	0.000	0.00000
DEUTERATED P-TERPHENYL-C18D14	0.00	0.000	0.00000
4-METHYL TERPHENYL	0.00	0.000	0.00000
2-METHYL-P-TERPHENYL	0.00	0.000	0.00000
3,3''-DIMETHYL-P-TERPHENYL	0.00	0.000	0.00000
4-(3,3-DIMETHYLBUTOXY)-P-TERPHENYL	8.42	0.750	0.00509
4,4''-DIHEXAHYDROFARNESOXY-P-TERPHENYL	10.84	0.352	0.02313
2,2'-METHYLENE-P-TERPHENYL	7.78	0.950	0.00317
2,2'-ETHYLENE-P-TERPHENYL	9.63	0.501	0.01140

ACCEPTOR MOLECULES	RO	CO	OVERLAP
2,2-METHYLENE-5,6-METHYLENE-P-TERPHENYL (TRACE)	20.04	0.056	0.92291
3-PHENYL DIBENZOFURAN	7.68	0.986	0.00295
BENZO(1,2-B-4,5-B')BISBENZOFURAN (TRACE) (B)	22.26	0.041	1.73235
SEXIPHENYL BENZENE	0.00	0.000	0.00000
P-QUATERPHENYL	12.73	0.217	0.06087
3,3'''-DIMETHYL-P-QUATERPHENYL	14.47	0.148	0.13086
TETRAMETHYL-P-QUATERPHENYL	0.00	0.000	0.00000
3,3''' -DIMETHYL-3',2''-METHYLENE-P-QUATERPHENYL	21.74	0.044	1.50240
3,3''' -DIMETHYL-3',2''-ETHYLENE-P-QUATERPHENYL	22.37	0.040	1.78010
4-(3-ETHYLBUTOXY)-P-QUATERPHENYL (B)	19.93	0.056	0.89120
4,4'''-DI(2-BUTYLOCTOXY-1)-P-QUATERPHENYL	20.17	0.054	0.95803
DIETHYL-P-QUINQUEPHENYL	14.94	0.134	0.15882
DI(3-ETHYLHEPTYL)-P-QUINQUEPHENYL	21.65	0.044	1.45870
TETRAMETHYL-P-SEXIPHENYL	13.27	0.192	0.07783
TETRAMETHYL-P-OCTAPHENYL (B)	26.60	0.024	5.01963
INDOLE (TRACE)	0.00	0.000	0.00000
2-PHENYL INDOLE	14.22	0.155	0.11822
1-METHYL-2-PHENYL INDOLE	10.97	0.339	0.02487
1-NEOPENTYL-2-PHENYL INDOLE	7.89	0.912	0.00344
1-N-BUTYL-2(BIPHENYLYL-4) INDOLE	21.51	0.045	1.40353
1,2-DIPHENYL INDOLE	11.26	0.313	0.02911
1-BIPHENYLYL-2-PHENYL INDOLE	14.59	0.144	0.13696
1,3-DIMETHYL-2-PHENYL INDOLE	12.44	0.232	0.05283
2,3-DIPHENYL INDOLE	19.82	0.057	0.85845
1-METHYL-2,3-DIPHENYL INDOLE	17.70	0.081	0.43573
2-PHENYL-3,2'-METHYLENE INDOLE (TRACE)	18.14	0.075	0.50785
2-PHENYL-3,2'-ETHYLENE INDOLE (TRACE)	23.42	0.035	2.34336
2-PHENYL-3,2' -TRIMETHYLENE INDOLE	16.77	0.095	0.31725
1-METHYL-3,2'-METHYLENE-2-PHENYL INDOLE (TRACE)	22.25	0.041	1.72576
1-METHYL-3,2'-ETHYLENE-2-PHENYL INDOLE	25.56	0.027	3.94961
1-METHYL-3,2'-TRIMETHYLENE-2-PHENYL INDOLE	17.20	0.088	0.36893
4,5-DIPHENYLIMIDAZOLE (A)	9.57	0.510	0.01096
PPF (TRACE)	23.29	0.035	2.27034
PPO (TRACE)	16.64	0.097	0.30217
2-(4-FLUOROPHENYL)-5-PHENYLOXAZOLE (TRACE)	15.52	0.120	0.19917
αNPO (TRACE)	29.49	0.017	9.28797
BBO (B)	35.70	0.010	29.17665
POPOP	37.17	0.009	37.16759
DIMETHYL POPOP (TRACE)	39.41	0.007	52.63294
PPD (TRACE)	0.00	0.000	0.00000
2,5-DI(2-METHYLPHENYL)OXADIAZOLE	0.00	0.000	0.00000
PBD	14.43	0.149	0.12891
BBD (B)	25.02	0.029	3.48049
αNPD (TRACE)	22.20	0.041	1.69502
βNPD	15.05	0.131	0.16556
αNND (TRACE)	26.51	0.024	4.91355
βNND (TRACE)	21.63	0.044	1.45851
TRANS-1-(4-BIPHENYLYL)-2-PHENYL-ETHYLENE	25.62	0.027	4.00890
TRANS-1-(4-BIPHENYLYL)-2-(1-NAPHTHYL)-ETHYLENE	32.16	0.013	15.57826
DIPHENYLSTILBENE (B)	35.84	0.010	29.86693
1,1,4,4-TETRAPHENYLBUTADIENE	34.66	0.011	24.36378
1,6-DIPHENYLHEXATRIENE	39.68	0.007	54.92390
1,8-DIPHENYL-1,3,5,7-OCTATETRAENE	44.06	0.005	102.33333
BPSB (TRACE)	40.43	0.007	61.44493
BIS-MSB	38.19	0.008	43.63579
ESCULIN (A)	26.92	0.023	5.36679
AZULENE	17.44	0.084	0.39054
NAPHTHALENE	0.00	0.000	0.00000
DEUTERATED NAPHTHALENE-C10D8	4.56	4.710	0.00013
1-METHYLNAPHTHALENE 5ML/L	1.94	61.093	0.00000
2-METHYL NAPHTHALENE	2.18	42.992	0.00000
2,3-DIMETHYL NAPHTHALENE	2.32	36.051	0.00000
2,6-DIMETHYL NAPHTHALENE (TRACE)	4.66	4.427	0.00015
1-NAPHTHOL (TRACE)	5.06	3.453	0.00024
2-NAPHTHOL (TRACE)	9.54	0.516	0.01075
1-NAPHTHYLAMINE	15.55	0.119	0.20114
2-CHLORONAPHTHALENE	3.41	11.272	0.00002
1 PHENYL NAPHTHALENE	4.88	3.856	0.00019
2-PHENYL NAPHTHALENE	10.76	0.359	0.02220
1,4-DIPHENYL NAPHTHALENE	12.22	0.245	0.04749
1,5-DIPHENYL NAPHTHALENE	10.24	0.416	0.01650
1,7-DIPHENYL NAPHTHALENE	10.62	0.373	0.02048
1,4,5,8-TETRAPHENYL NAPHTHALENE	28.97	0.018	8.32459
1,1'-BINAPHTHYL	7.08	1.260	0.00180
4,4'-DI(N-BUTOXY)-1,1'-BINAPHTHYL	17.60	0.082	0.42190
2,2'-BINAPHTHYL	11.61	0.286	0.03487
ACRIDINE (A)	27.83	0.021	6.53756
ANTHRACENE (TRACE)	26.58	0.024	4.96918
DEUTERATED ANTHRACENE C14D10 (TRACE)	26.13	0.025	4.49145
9-METHYL ANTHRACENE (TRACE)	27.94	0.021	6.68985
1-AMINO ANTHRACENE	28.32	0.020	7.19917
2-AMINO ANTHRACENE	27.07	0.023	5.49120
9-VINYL ANTHRACENE	29.41	0.018	9.08824
9-PHENYL ANTHRACENE (TRACE)	29.31	0.018	8.91178
9,10-DIPHENYL ANTHRACENE	31.55	0.014	13.83817
9,10-DICHLOROANTHRACENE (TRACE)	30.65	0.016	11.62186
BIANTHRANYL (B)	33.10	0.012	18.44826
9,10-DINAPHTHYLANTHRACENE (B)	31.17	0.015	12.85596
9-(P-VINYLPHENYL)-10-PHENYL ANTHRACENE	31.85	0.014	14.63075
9,10-BIS(PHENYL-ETHYNYL)ANTHRACENE (TRACE)	34.11	0.011	21.84281
RUBICENE (B)	30.73	0.015	11.65988
DIBENZ(A,H)ANTHRACENE (B)	25.29	0.028	3.67575
TETRACENE (TRACE) (B)	25.45	0.027	3.76456
RUBRENE (B)	23.47	0.035	2.31594
TETRAHYDROCANNABINOL ACID METHYL ESTER	0.00	0.000	0.00000
PHENANTHRENE	10.33	0.405	0.01732
3,4-BENZOPHENANTHRENE	15.36	0.123	0.18511

ACCEPTOR MOLECULES	RO	CO	OVERLAP
2-PHENYL PHENANTHRENE	12.85	0.211	0.06372
CHRYSENE (TRACE)	15.41	0.122	0.18982
TRIPHENYLENE	9.98	0.450	0.01408
DEUTERATED TRIPHENYLENE-C18D12	9.72	0.488	0.01197
PYRENE	20.87	0.049	1.16597
1,3,6,8,-TETRAPHENYL PYRENE	38.87	0.008	48.31755
BIPYRENYL (B)	36.05	0.010	30.81554
O-PHENYLENE-PYRENE	33.72	0.012	20.41464
ANTHANTHRENE (B)	33.66	0.012	20.27649
FLUORANTHENE	26.16	0.025	4.49246
3-PHENYL FLUORANTHENE	31.36	0.015	13.30999
3-AMINO FLUORANTHENE	30.63	0.016	11.46865
3,3'-BIFLUORANTHENYL (B)	38.73	0.009	34.31732
BENZO(GHI)FLUORANTHENE	24.18	0.032	2.79699
9,12-DI(N-OCTADECOXY)-BENZ(E)ACEPHENANTHRYLENE	32.64	0.013	16.89550
DECACYCLENE (B)	40.22	0.007	58.71228
10-ETHOXY-BENZO(KL)XANTHENE	32.12	0.013	15.41843
PERYLENE	34.01	0.011	21.56756
BENZO(GHI)PERYLENE (B)	33.49	0.012	19.81779
BBOT (TRACE)	41.39	0.006	70.61610
QUININE SULFATE (A)	18.96	0.066	0.66070
ACRIDINE YELLOW (A)	31.63	0.014	13.87381
ACRIDINE ORANGE (A)	26.67	0.024	4.98403
ACRIDINE RED (A)	24.75	0.030	3.18171
URANINE (A)	27.22	0.022	5.64207
RHODAMINE B (A)	28.44	0.019	7.33956
RHODAMINE 6G (A)	27.24	0.022	5.66386
EOSIN B (A)	25.01	0.029	3.39528
EOSIN Y (A)	26.14	0.025	4.42218
ROSE BENGAL (A)	25.72	0.026	4.01532
INDENOINDENE	16.05	0.108	0.24369

8.154 Donor Molecule: 1,7-Diphenylnaphthalene

ACCEPTOR MOLECULES	RO	CO	OVERLAP
TRIPHENYLAMINE	15.58	0.118	0.22944
P-XYLENE (TRACE)	0.00	0.000	0.00000
SODIUM SALICYLATE (A)	11.58	0.288	0.03872
ANTHRANILIC ACID (A)	23.16	0.036	2.45602
TMPD	18.35	0.072	0.61016
BENZIDINE	11.79	0.273	0.04331
1,3,5-TRIPHENYL BENZENE	2.51	28.266	0.00000
1,3-DIBIPHENYLYLBENZENE (B)	9.79	0.477	0.01419
1,3,5,-TRIBIPHENYLYLBENZENE (B)	11.04	0.332	0.02926
FLUORENE (TRACE)	0.00	0.000	0.00000
1-METHYL FLUORENE	0.00	0.000	0.00000
1,2-BENZOFLUORENE	15.32	0.124	0.20745
9,10-DIHYDROPHENANTHRENE	0.00	0.000	0.00000
CARBAZOLE (TRACE)	12.12	0.251	0.05102
N-METHYL CARBAZOLE (TRACE)	18.49	0.071	0.64109
N-VINYLCARBAZOLE	17.25	0.087	0.42289
N-PHENYL CARBAZOLE	17.28	0.087	0.42720
N-METHYL-2,3-BENZCARBAZOLE	25.52	0.027	4.38035
DIBENZOFURAN	0.00	0.000	0.00000
DIBENZOTHIOPHENE (TRACE)	8.57	0.712	0.00636
P-TERPHENYL	0.00	0.000	0.00000
DEUTERATED P-TERPHENYL-C18D14	0.00	0.000	0.00000
4-METHYL TERPHENYL	0.00	0.000	0.00000
2-METHYL-P-TERPHENYL	0.00	0.000	0.00000
3,3''-DIMETHYL-P-TERPHENYL	0.00	0.000	0.00000
4-(3,3-DIMETHYLBUTOXY)-P-TERPHENYL	8.62	0.699	0.00661
4,4''-DIHEXAHYDROFARNESOXY-P-TERPHENYL	11.15	0.323	0.03098
2,2'-METHYLENE-P-TERPHENYL	8.00	0.873	0.00424
2,2'-ETHYLENE-P-TERPHENYL	9.90	0.461	0.01520
2,2'-METHYLENE-5',6''-METHYLENE-P-TERPHENYL (TRACE)	20.87	0.049	1.32901
3-PHENYL DIBENZOFURAN	7.91	0.903	0.00396
BENZO(1,2-B-4,5-B')BISBENZOFURAN (TRACE) (B)	23.66	0.034	2.81394
SEXIPHENYL BENZENE	0.00	0.000	0.00000
P-QUATERPHENYL	13.27	0.192	0.08783
3,3'''-DIMETHYL-P-QUATERPHENYL	15.25	0.126	0.20226
TETRAMETHYL-P-QUATERPHENYL	0.00	0.000	0.00000
3,3''' -DIMETHYL-3',2''-METHYLENE-P-QUATERPHENYL	23.07	0.036	2.42025
3,3''' -DIMETHYL-3',2''-ETHYLENE-P-QUATERPHENYL	23.77	0.033	2.88773
4-(3-ETHYLBUTOXY)-P-QUATERPHENYL (B)	21.17	0.047	1.44283
4,4'''-DI(2-BUTYLOCTOXY-1)-P-QUATERPHENYL	21.43	0.045	1.55244
DIETHYL-P-QUINQUEPHENYL	15.71	0.115	0.24203
DI(3-ETHYLHEPTYL)-P-QUINQUEPHENYL	22.97	0.037	2.34643
TETRAMETHYL-P-SEXIPHENYL	13.80	0.170	0.11122
TETRAMETHYL-P-OCTAPHENYL (B)	28.18	0.020	8.00471
INDOLE (TRACE)	0.00	0.000	0.00000
2-PHENYL INDOLE	14.69	0.141	0.16185
1-METHYL-2-PHENYL INDOLE	11.42	0.300	0.03573
1-NEOPENTYL-2-PHENYL INDOLE	8.08	0.849	0.00448
1-N-BUTYL-2(BIPHENYLYL-4) INDOLE	22.82	0.038	2.26205
1,2-DIPHENYL INDOLE	11.73	0.277	0.04198
1-BIPHENYLYL-2-PHENYL INDOLE	15.42	0.122	0.21527
1,3-DIMETHYL-2-PHENYL INDOLE	12.98	0.205	0.07684
2,3-DIPHENYL INDOLE	21.00	0.048	1.36922
1-METHYL-2,3-DIPHENYL INDOLE	18.78	0.068	0.70211
2-PHENYL-3,2'-METHYLENE INDOLE (TRACE)	19.02	0.065	0.75947
2-PHENYL-3,2'-ETHYLENE INDOLE (TRACE)	25.07	0.028	3.97829
2-PHENYL-3,2'-TRIMETHYLENE INDOLE	17.58	0.082	0.47351

ACCEPTOR MOLECULES	RO	CO	OVERLAP
1-METHYL-3,2'-METHYLENE-2-PHENYL INDOLE (TRACE)	23.84	0.033	2.94521
1-METHYL-3,2'-ETHYLENE-2-PHENYL INDOLE	27.06	0.023	6.28343
1-METHYL-3,2'-TRIMETHYLENE-2-PHENYL INDOLE	18.19	0.074	0.58118
4,5-DIPHENYLIMIDAZOLE (A)	10.04	0.442	0.01648
PPF (TRACE)	24.95	0.029	3.86813
PPO (TRACE)	17.33	0.086	0.43635
2-(4-FLUOROPHENYL)-5-PHENYLOXAZOLE (TRACE)	16.13	0.107	0.28346
αNPO (TRACE)	30.88	0.015	13.81111
BBO (B)	36.93	0.009	40.32291
POPOP	37.89	0.008	47.04869
DIMETHYL POPOP (TRACE)	39.73	0.007	62.38466
PPD (TRACE)	0.00	0.000	0.00000
2,5-DI(2-METHYLPHENYL)OXADIAZOLE	0.00	0.000	0.00000
PBD	14.94	0.134	0.17915
BBD (B)	26.66	0.024	5.75148
αNPD (TRACE)	23.62	0.034	2.77691
βNPD	15.91	0.111	0.26035
αNND (TRACE)	27.99	0.020	7.68141
βNND (TRACE)	23.11	0.036	2.44834
TRANS-1-(4-BIPHENYLYL)-2-PHENYL-ETHYLENE	27.27	0.022	6.57382
TRANS-1-(4-BIPHENYLYL)-2-(1-NAPHTHYL)-ETHYLENE	33.33	0.012	21.77599
DIPHENYLSTILBENE (B)	37.15	0.009	41.78346
1,1,4,4-TETRAPHENYLBUTADIENE	35.65	0.010	32.57251
1,6-DIPHENYLHEXATRIENE	40.55	0.007	70.54238
1,8-DIPHENYL-1,3,5,7-OCTATETRAENE	43.36	0.005	104.87862
BPSB (TRACE)	41.13	0.006	76.87535
BIS-MSB	39.06	0.008	56.34794
ESCULIN (A)	28.04	0.020	7.73382
AZULENE	18.48	0.071	0.62487
NAPHTHALENE	0.00	0.000	0.00000
DEUTERATED NAPHTHALENE-C10D8	4.70	4.316	0.00017
1-METHYLNAPHTHALENE 5ML/L	2.16	44.457	0.00000
2-METHYL NAPHTHALENE	2.51	28.282	0.00000
2,3-DIMETHYL NAPHTHALENE	2.59	25.742	0.00000
2,6-DIMETHYL NAPHTHALENE (TRACE)	4.99	3.606	0.00025
1-NAPHTHOL (TRACE)	5.22	3.143	0.00033
2-NAPHTHOL (TRACE)	9.76	0.481	0.01395
1-NAPHTHYLAMINE	16.51	0.099	0.32512
2-CHLORONAPHTHALENE	3.55	9.979	0.00003
ACENAPHTHENE	3.04	15.908	0.00001
1-PHENYL NAPHTHALENE	4.99	3.600	0.00025
2-PHENYL NAPHTHALENE	11.10	0.327	0.03023
1,4-DIPHENYL NAPHTHALENE	12.85	0.211	0.07225
1,5-DIPHENYL NAPHTHALENE	10.63	0.372	0.02327
1,7-DIPHENYL NAPHTHALENE	11.09	0.328	0.02991
1,4,5,8-TETRAPHENYL NAPHTHALENE	30.11	0.016	11.82275
1,1'-BINAPHTHYL	7.24	1.181	0.00232
4,4'-DI(N-BUTOXY)-1,1'-BINAPHTHYL	18.66	0.069	0.67616
2,2'-BINAPHTHYL	12.10	0.253	0.05043
ACRIDINE (A)	28.46	0.019	8.42822
ANTHRACENE (TRACE)	26.97	0.023	6.12599
DEUTERATED ANTHRACENE C14D10 (TRACE)	26.58	0.024	5.61438
9-METHYL ANTHRACENE (TRACE)	27.98	0.020	7.61685
1-AMINO ANTHRACENE	27.41	0.022	6.66556
2-AMINO ANTHRACENE	26.14	0.025	5.01350
9-VINYL ANTHRACENE	29.21	0.018	9.83688
9-PHENYL ANTHRACENE (TRACE)	29.36	0.018	10.15726
9,10-DIPHENYL ANTHRACENE	30.91	0.015	13.80097
9,10-DICHLOROANTHRACENE (TRACE)	29.82	0.017	11.11505
BIANTHRANYL (B)	32.42	0.013	18.36366
9,10-DINAPHTHYLANTHRACENE (B)	30.39	0.016	12.45959
9-(P-VINYLPHENYL)-10-PHENYL ANTHRACENE	31.13	0.015	14.39425
9,10-BIS(PHENYL-ETHYNYL)ANTHRACENE (TRACE)	30.56	0.016	12.77081
RUBICENE (B)	29.36	0.018	10.04613
DIBENZ(A,H)ANTHRACENE (B)	26.48	0.024	5.46086
TETRACENE (TRACE) (B)	23.22	0.036	2.45429
RUBRENE (B)	21.99	0.042	1.77207
TETRAHYDROCANNABINOL ACID METHYL ESTER	0.00	0.000	0.00000
PHENANTHRENE	11.07	0.329	0.02957
3,4-BENZOPHENANTHRENE	15.87	0.112	0.25362
2-PHENYL PHENANTHRENE	13.60	0.178	0.10131
CHRYSENE (TRACE)	16.16	0.106	0.28400
TRIPHENYLENE	10.57	0.379	0.02232
DEUTERATED TRIPHENYLENE-C18D12	10.30	0.409	0.01917
PYRENE	21.83	0.043	1.72346
1,3,6,8,-TETRAPHENYL PYRENE	38.34	0.008	50.17792
BIPYRENYL (B)	36.78	0.009	39.16854
O-PHENYLENE-PYRENE	33.21	0.012	21.01666
ANTHANTHRENE (B)	31.60	0.014	15.66769
FLUORANTHENE	27.08	0.023	6.23577
3-PHENYL FLUORANTHENE	31.61	0.014	15.74196
3-AMINO FLUORANTHENE	29.55	0.017	10.43428
3,3'-BIFLUORANTHENYL (B)	36.25	0.009	35.78786
BENZO(GHI)FLUORANTHENE	24.89	0.029	3.75762
9,12-DI(N-OCTADECOXY)-BENZ(E)ACEPHENANTHRYLENE	31.71	0.014	16.03078
DECACYCLENE (B)	39.19	0.007	56.81512
10-ETHOXY-BENZO(KL)XANTHENE	31.50	0.014	15.45760
PERYLENE	31.25	0.015	14.64107
BENZO(GHI)PERYLENE (B)	33.19	0.012	21.16683
BBOT (TRACE)	41.20	0.006	77.43796
QUININE SULFATE (A)	20.20	0.054	1.08997
ACRIDINE YELLOW (A)	27.86	0.021	7.33082
ACRIDINE ORANGE (A)	23.39	0.035	2.56410
ACRIDINE RED (A)	24.30	0.031	3.23130
URANINE (A)	24.59	0.030	3.46644
RHODAMINE B (A)	27.81	0.021	7.25055

ACCEPTOR MOLECULES	RO	CO	OVERLAP
RHODAMINE 6G (A)	27.10	0.022	6.20569
EOSIN B (A)	23.87	0.033	2.90287
EOSIN Y (A)	24.75	0.029	3.60635
ROSE BENGAL (A)	25.13	0.028	3.94752
INDENOINDENE	16.69	0.096	0.34711

8.155 Donor Molecule: 1,4,5,8-Tetraphenylnaphthalene

ACCEPTOR MOLECULES	RO	CO	OVERLAP
ANTHRANILIC ACID (A)	6.91	1.356	0.00244
TMPD	0.00	0.000	0.00000
N-METHYL-2,3-BENZCARBAZOLE	18.40	0.072	0.86645
αNPO (TRACE)	8.15	0.827	0.00656
BBO (B)	14.15	0.158	0.17985
POPOP	17.76	0.080	0.70101
DIMETHYL POPOP (TRACE)	22.21	0.041	2.67638
αNND (TRACE)	3.93	7.344	0.00008
βNND (TRACE)	0.00	0.000	0.00000
TRANS-1-(4-BIPHENYLYL)-2-PHENYL-ETHYLENE	0.00	0.000	0.00000
TRANS-1-(4-BIPHENYLYL)-2-(1-NAPHTHYL)-ETHYLENE	13.15	0.197	0.11588
DIPHENYLSTILBENE (B)	13.86	0.168	0.15892
1,1,4,4-TETRAPHENYLBUTADIENE	15.93	0.111	0.36448
1,6-DIPHENYLHEXATRIENE	17.90	0.078	0.73593
1,8-DIPHENYL-1,3,5,7-OCTATETRAENE	30.48	0.016	17.83627
BPSB (TRACE)	20.30	0.053	1.56542
BIS-MSB	18.72	0.068	0.96172
ESCULIN (A)	10.03	0.444	0.02277
AZULENE	14.46	0.148	0.20014
1,4,5,8-TETRAPHENYL NAPHTHALENE	12.09	0.253	0.06978
1,1'-BINAPHTHYL	0.00	0.000	0.00000
4,4'-DI(N-BUTOXY)-1,1'-BINAPHTHYL	0.00	0.000	0.00000
2,2'-BINAPHTHYL	0.00	0.000	0.00000
ACRIDINE (A)	13.99	0.163	0.16758
ANTHRACENE (TRACE)	12.09	0.253	0.06981
DEUTERATED ANTHRACENE C14D10 (TRACE)	11.44	0.298	0.05033
9-METHYL ANTHRACENE (TRACE)	16.06	0.108	0.38394
1-AMINO ANTHRACENE	23.10	0.036	3.36688
2-AMINO ANTHRACENE	22.86	0.037	3.15978
9-VINYL ANTHRACENE	18.49	0.071	0.89121
9-PHENYL ANTHRACENE (TRACE)	16.80	0.094	0.50263
9,10-DIPHENYL ANTHRACENE	21.87	0.043	2.43562
9,10-DICHLOROANTHRACENE (TRACE)	22.35	0.040	2.77459
BIANTHRANYL (B)	22.66	0.038	3.01674
9,10-DINAPHTHYLANTHRACENE (B)	22.32	0.040	2.75516
9-(P-VINYLPHENYL)-10-PHENYL ANTHRACENE	22.50	0.039	2.88977
9,10-BIS(PHENYL-ETHYNYL)ANTHRACENE (TRACE)	38.84	0.008	75.69317
RUBICENE (B)	33.24	0.012	29.55351
DIBENZ(A,H)ANTHRACENE (B)	12.41	0.234	0.08156
TETRACENE (TRACE) (B)	28.93	0.018	12.91258
RUBRENE (B)	30.50	0.016	17.63754
TETRAHYDROCANNABINOL ACID METHYL ESTER	0.00	0.000	0.00000
PHENANTHRENE	0.00	0.000	0.00000
3,4-BENZOPHENANTHRENE	5.78	2.318	0.00083
2-PHENYL PHENANTHRENE	0.00	0.000	0.00000
CHRYSENE (TRACE)	3.66	9.154	0.00005
TRIPHENYLENE	0.00	0.000	0.00000
DEUTERATED TRIPHENYLENE-C18D12	0.00	0.000	0.00000
PYRENE	4.92	3.744	0.00032
1,3,6,8,-TETRAPHENYL PYRENE	25.36	0.027	5.91589
BIPYRENYL (B)	20.20	0.054	1.51535
O-PHENYLENE-PYRENE	27.74	0.021	10.06830
ANTHANTHRENE (B)	32.80	0.013	27.59746
FLUORANTHENE	11.72	0.278	0.05768
3-PHENYL FLUORANTHENE	18.69	0.069	0.94808
3-AMINO FLUORANTHENE	27.47	0.022	9.48995
3,3'-BIFLUORANTHENYL (B)	26.17	0.025	7.13851
BENZO(GHI)FLUORANTHENE	15.68	0.116	0.33086
9,12-DI(N-OCTADECOXY)-BENZ(E)ACEPHENANTHRYLENE	26.88	0.023	8.38725
DECACYCLENE (B)	33.71	0.012	32.39439
10-ETHOXY-BENZO(KL)XANTHENE	22.09	0.042	2.58923
PERYLENE	34.25	0.011	35.74950
BENZO(GHI)PERYLENE (B)	20.58	0.051	1.69417
BBOT (TRACE)	26.45	0.024	7.64195
QUININE SULFATE (A)	0.00	0.000	0.00000
ACRIDINE YELLOW (A)	38.66	0.008	73.41494
ACRIDINE ORANGE (A)	35.65	0.010	45.07779
ACRIDINE RED (A)	30.68	0.015	18.25900
URANINE (A)	37.45	0.009	60.52218
RHODAMINE B (A)	34.93	0.010	39.71964
RHODAMINE 6G (A)	36.71	0.009	53.58406
EOSIN B (A)	34.98	0.010	40.06215
EOSIN Y (A)	36.07	0.010	48.20741
ROSE BENGAL (A)	33.29	0.012	29.77084

8.156 Donor Molecule: 1,1′-Binaphthyl

ACCEPTOR MOLECULES	RO	CO	OVERLAP
N,N-DIMETHYLANILINE 3ML/L	4.31	5.593	0.00006
TRIPHENYLAMINE	18.84	0.067	0.43896
SODIUM SALICYLATE (A)	13.42	0.185	0.05725
ANTHRANILIC ACID (A)	25.84	0.026	2.89680
TMPD	20.91	0.049	0.81671
BENZIDINE	15.11	0.130	0.11698
1,3,5-TRIPHENYL BENZENE	4.67	4.389	0.00010
1,3-DIBIPHENYLYLBENZENE (B)	13.06	0.201	0.04889
1,3,5,-TRIBIPHENYLYLBENZENE (B)	15.21	0.127	0.12221
FLUORENE (TRACE)	0.00	0.000	0.00000
1-METHYL FLUORENE	0.00	0.000	0.00000
1,2-BENZOFLUORENE	17.95	0.077	0.32782
9,10-DIHYDROPHENANTHRENE	0.00	0.000	0.00000
CARBAZOLE (TRACE)	15.97	0.110	0.15318
N-METHYL CARBAZOLE (TRACE)	21.62	0.044	1.00212
N-VINYLCARBAZOLE	20.91	0.049	0.81986
N-PHENYL CARBAZOLE	20.75	0.050	0.78388
N-METHYL-2,3-BENZCARBAZOLE	27.54	0.021	4.22862
DIBENZOFURAN	0.00	0.000	0.00000
DIBENZOTHIOPHENE (TRACE)	12.47	0.231	0.03708
P-TERPHENYL	4.20	6.053	0.00005
DEUTERATED P-TERPHENYL-C18D14	0.00	0.000	0.00000
4-METHYL TERPHENYL	5.83	2.263	0.00039
2-METHYL-P-TERPHENYL	0.00	0.000	0.00000
3,3''-DIMETHYL-P-TERPHENYL	6.78	1.436	0.00096
4-(3,3-DIMETHYLBUTOXY)-P-TERPHENYL	12.34	0.238	0.03480
4,4''-DIHEXAHYDROFARNESOXY-P-TERPHENYL	15.36	0.123	0.12968
2,2'-METHYLENE-P-TERPHENYL	12.34	0.238	0.03484
2,2'-ETHYLENE-P-TERPHENYL	14.17	0.157	0.07962
2,2'-METHYLENE-5',6''-METHYLENE-P-TERPHENYL (TRACE)	26.73	0.023	3.58301
3-PHENYL DIBENZOFURAN	12.29	0.241	0.03409
BENZO(1,2-B-4,5-B')BISBENZOFURAN (TRACE) (B)	28.79	0.019	5.59121
SEXIPHENYL BENZENE	0.00	0.000	0.00000
P-QUATERPHENYL	17.41	0.085	0.27373
3,3'''-DIMETHYL-P-QUATERPHENYL	19.28	0.062	0.50440
TETRAMETHYL-P-QUATERPHENYL	0.00	0.000	0.00000
3,3''' -DIMETHYL-3',2''-METHYLENE-P-QUATERPHENYL	28.07	0.020	4.80158
3,3''' -DIMETHYL-3',2''-ETHYLENE-P-QUATERPHENYL	28.29	0.020	5.01982
4-(3-ETHYLBUTOXY)-P-QUATERPHENYL (B)	25.35	0.027	2.59735
4,4'''-DI(2-BUTYLOCTOXY-1)-P-QUATERPHENYL	25.74	0.026	2.84693
DIETHYL-P-QUINQUEPHENYL	20.02	0.056	0.63372
DI(3-ETHYLHEPTYL)-P-QUINQUEPHENYL	27.29	0.022	4.03915
TETRAMETHYL-P-SEXIPHENYL	18.20	0.074	0.35808
TETRAMETHYL-P-OCTAPHENYL (B)	32.65	0.013	11.82166
INDOLE (TRACE)	0.00	0.000	0.00000
2-PHENYL INDOLE	19.58	0.060	0.55449
1-METHYL-2-PHENYL INDOLE	15.11	0.130	0.11705
1-NEOPENTYL-2-PHENYL INDOLE	11.67	0.281	0.02495
1-N-BUTYL-2(BIPHENYLYL-4) INDOLE	26.55	0.024	3.42280
1,2-DIPHENYL INDOLE	15.54	0.119	0.13830
1-BIPHENYLYL-2-PHENYL INDOLE	18.89	0.066	0.44514
1,3-DIMETHYL-2-PHENYL INDOLE	16.84	0.094	0.22455
2,3-DIPHENYL INDOLE	24.43	0.031	2.07614
1-METHYL-2,3-DIPHENYL INDOLE	21.96	0.042	1.09781
2-PHENYL-3,2'-METHYLENE INDOLE (TRACE)	24.08	0.032	1.91483
2-PHENYL-3,2'-ETHYLENE INDOLE (TRACE)	29.05	0.018	5.87963
2-PHENYL-3,2' -TRIMETHYLENE INDOLE	22.37	0.040	1.22907
1-METHYL-3,2'-METHYLENE-2-PHENYL INDOLE (TRACE)	28.02	0.020	4.74716
1-METHYL-3,2'-ETHYLENE-2-PHENYL INDOLE	30.67	0.016	8.12870
1-METHYL-3,2'-TRIMETHYLENE-2-PHENYL INDOLE	22.47	0.039	1.26518
4,5-DIPHENYLIMIDAZOLE (A)	12.80	0.213	0.04334
PPF (TRACE)	29.35	0.018	6.26464
PPO (TRACE)	22.37	0.040	1.23064
2-(4-FLUOROPHENYL)-5-PHENYLOXAZOLE (TRACE)	20.99	0.048	0.84059
αNPO (TRACE)	34.31	0.011	15.88996
BBO (B)	40.60	0.007	43.51802
POPOP	41.20	0.006	47.49309
DIMETHYL POPOP (TRACE)	42.97	0.006	61.02507
PPD (TRACE)	0.00	0.000	0.00000
2,5-DI(2-METHYLPHENYL)OXADIAZOLE	0.00	0.000	0.00000
PBD	19.91	0.057	0.61264
BBD (B)	31.14	0.015	8.93569
αNPD (TRACE)	27.45	0.022	4.17785
βNPD	20.06	0.055	0.64114
αNND (TRACE)	31.53	0.014	9.58399
βNND (TRACE)	27.67	0.021	4.40222
TRANS-1-(4-BIPHENYLYL)-2-PHENYL-ETHYLENE	31.51	0.014	9.56370
TRANS-1-(4-BIPHENYLYL)-2-(1-NAPHTHYL)-ETHYLENE	36.82	0.009	24.21464
DIPHENYLSTILBENE (B)	40.89	0.007	45.40904
1,1,4,4-TETRAPHENYLBUTADIENE	39.03	0.008	34.28107
1,6-DIPHENYLHEXATRIENE	44.14	0.005	71.69990
1,8-DIPHENYL-1,3,5,7-OCTATETRAENE	46.55	0.004	98.07526
BPSB (TRACE)	44.68	0.005	77.14662
BIS-MSB	42.62	0.006	58.09603
ESCULIN (A)	31.09	0.015	8.78560
AZULENE	21.55	0.045	0.96029
NAPHTHALENE	2.36	34.014	0.00000
DEUTERATED NAPHTHALENE-C10D8	6.45	1.665	0.00071
1-METHYLNAPHTHALENE 5ML/L	4.24	5.866	0.00006
2-METHYL NAPHTHALENE	4.91	3.778	0.00014
2,3-DIMETHYL NAPHTHALENE	5.10	3.380	0.00017
2,6-DIMETHYL NAPHTHALENE (TRACE)	8.97	0.620	0.00514
1-NAPHTHOL (TRACE)	8.73	0.673	0.00436

ACCEPTOR MOLECULES	RO	CO	OVERLAP
2-NAPHTHOL (TRACE)	13.46	0.184	0.05848
1-NAPHTHYLAMINE	19.96	0.056	0.61955
2-CHLORONAPHTHALENE	6.23	1.849	0.00058
ACENAPHTHENE	6.43	1.685	0.00070
1-PHENYL NAPHTHALENE	7.80	0.943	0.00222
2-PHENYL NAPHTHALENE	15.01	0.132	0.11264
1,4-DIPHENYL NAPHTHALENE	16.44	0.101	0.19415
1,5-DIPHENYL NAPHTHALENE	14.29	0.153	0.08366
1,7-DIPHENYL NAPHTHALENE	14.53	0.146	0.09240
1,4,5,8-TETRAPHENYL NAPHTHALENE	33.35	0.012	13.33623
1,1'-BINAPHTHYL	10.51	0.386	0.01326
4,4'-DI(N-BUTOXY)-1,1'-BINAPHTHYL	22.49	0.039	1.26907
2,2'-BINAPHTHYL	15.85	0.112	0.15599
ACRIDINE (A)	30.98	0.015	8.58090
ANTHRACENE (TRACE)	29.31	0.018	6.16825
DEUTERATED ANTHRACENE C14D10 (TRACE)	28.90	0.019	5.66461
9-METHYL ANTHRACENE (TRACE)	30.06	0.016	7.15217
1-AMINO ANTHRACENE	29.35	0.018	6.15118
2-AMINO ANTHRACENE	28.22	0.020	4.85431
9-VINYL ANTHRACENE	31.37	0.014	9.22132
9-PHENYL ANTHRACENE (TRACE)	31.57	0.014	9.60530
9,10-DIPHENYL ANTHRACENE	33.13	0.012	12.78405
9,10-DICHLOROANTHRACENE (TRACE)	31.93	0.014	10.23503
BIANTHRANYL (B)	34.75	0.011	17.03641
9,10-DINAPHTHYLANTHRACENE (B)	32.58	0.013	11.53887
9-(P-VINYLPHENYL)-10-PHENYL ANTHRACENE	33.35	0.012	13.29425
9,10-BIS(PHENYL-ETHYNYL)ANTHRACENE (TRACE)	32.69	0.013	11.70756
RUBICENE (B)	31.62	0.014	9.59536
DIBENZ(A,H)ANTHRACENE (B)	29.73	0.017	6.68940
TETRACENE (TRACE) (B)	24.90	0.029	2.28901
RUBRENE (B)	24.56	0.030	2.10569
TETRAHYDROCANNABINOL ACID METHYL ESTER	0.00	0.000	0.00000
PHENANTHRENE	12.94	0.207	0.04600
3,4-BENZOPHENANTHRENE	17.97	0.077	0.32726
2-PHENYL PHENANTHRENE	15.35	0.124	0.12774
CHRYSENE (TRACE)	18.15	0.075	0.34926
TRIPHENYLENE	12.94	0.206	0.04602
DEUTERATED TRIPHENYLENE-C18D12	12.58	0.225	0.03882
PYRENE	27.02	0.023	3.78985
1,3,6,8,-TETRAPHENYL PYRENE	41.05	0.006	46.15402
BIPYRENYL (B)	40.11	0.007	40.30516
O-PHENYLENE-PYRENE	35.75	0.010	20.03511
ANTHANTHRENE (B)	33.98	0.011	14.78505
FLUORANTHENE	29.68	0.017	6.61451
3-PHENYL FLUORANTHENE	34.15	0.011	15.30079
3-AMINO FLUORANTHENE	31.83	0.014	9.97834
3,3'-BIFLUORANTHENYL (B)	38.97	0.008	33.74615
BENZO(GHI)FLUORANTHENE	27.59	0.021	4.25356
9,12-DI(N-OCTADECOXY)-BENZ(E)ACEPHENANTHRYLENE	34.23	0.011	15.50013
DECACYCLENE (B)	42.19	0.006	54.09694
10-ETHOXY-BENZO(KL)XANTHENE	33.75	0.012	14.30178
PERYLENE	33.38	0.012	13.29007
BENZO(GHI)PERYLENE (B)	35.54	0.010	19.48387
BBOT (TRACE)	44.36	0.005	73.77685
QUININE SULFATE (A)	24.11	0.032	1.92600
ACRIDINE YELLOW (A)	29.80	0.017	6.71419
ACRIDINE ORANGE (A)	24.97	0.029	2.32353
ACRIDINE RED (A)	26.39	0.024	3.24269
URANINE (A)	26.53	0.024	3.34721
RHODAMINE B (A)	30.17	0.016	7.24078
RHODAMINE 6G (A)	29.70	0.017	6.58666
EOSIN B (A)	26.20	0.025	3.10268
EOSIN Y (A)	27.06	0.023	3.76968
ROSE BENGAL (A)	27.67	0.021	4.30770
INDENOINDENE	21.71	0.044	1.02795

8.157 Donor Molecule: 4,4'-Di (*n*-butoxy)-1,1'-binaphthyl

ACCEPTOR MOLECULES	RO	CO	OVERLAP
TRIPHENYLAMINE	12.02	0.258	0.02332
P-XYLENE (TRACE)	0.00	0.000	0.00000
SODIUM SALICYLATE (A)	9.95	0.454	0.00751
ANTHRANILIC ACID (A)	22.13	0.041	0.90415
TMPD	16.50	0.100	0.15559
BENZIDINE	7.86	0.921	0.00183
1,3-DIBIPHENYLYLBENZENE (B)	5.70	2.409	0.00027
1,3,5,-TRIBIPHENYLYLBENZENE (B)	0.00	0.000	0.00000
FLUORENE (TRACE)	0.00	0.000	0.00000
1-METHYL FLUORENE	0.00	0.000	0.00000
1,2-BENZOFLUORENE	12.09	0.253	0.02422
9,10-DIHYDROPHENANTHRENE	0.00	0.000	0.00000
CARBAZOLE (TRACE)	7.05	1.277	0.00095
N-METHYL CARBAZOLE (TRACE)	14.22	0.156	0.06395
N-VINYLCARBAZOLE	12.48	0.230	0.02926
N-PHENYL CARBAZOLE	12.75	0.216	0.03326
N-METHYL-2,3-BENZCARBAZOLE	29.24	0.018	4.78035
DIBENZOFURAN	0.00	0.000	0.00000
DIBENZOTHIOPHENE (TRACE)	4.29	5.682	0.00005
2,2'-METHYLENE-5',6''-METHYLENE-P-TERPHENYL (TRACE)	12.56	0.226	0.03053
3-PHENYL DIBENZOFURAN	0.00	0.000	0.00000
BENZO(1,2-B-4,5-B')BISBENZOFURAN (TRACE) (B)	17.23	0.087	0.20262
SEXIPHENYL BENZENE	0.00	0.000	0.00000

ACCEPTOR MOLECULES	RO	CO	OVERLAP
P-QUATERPHENYL	8.17	0.821	0.00231
3,3'''-DIMETHYL-P-QUATERPHENYL	10.55	0.381	0.01070
TETRAMETHYL-P-QUATERPHENYL	0.00	0.000	0.00000
3,3''' -DIMETHYL-3',2''-METHYLENE-P-QUATERPHENYL	17.24	0.087	0.20311
3,3''' -DIMETHYL-3',2''-ETHYLENE-P-QUATERPHENYL	19.20	0.063	0.38702
4-(3-ETHYLBUTOXY)-P-QUATERPHENYL (B)	16.92	0.092	0.18126
4,4'''-DI(2-BUTYLOCTOXY-1)-P-QUATERPHENYL	16.94	0.092	0.18302
DIETHYL-P-QUINQUEPHENYL	10.57	0.379	0.01084
DI(3-ETHYLHEPTYL)-P-QUINQUEPHENYL	18.96	0.066	0.35903
TETRAMETHYL-P-SEXIPHENYL	8.37	0.764	0.00267
TETRAMETHYL-P-OCTAPHENYL (B)	24.69	0.030	1.74771
INDOLE (TRACE)	0.00	0.000	0.00000
2-PHENYL INDOLE	7.78	0.951	0.00172
1-METHYL-2-PHENYL INDOLE	7.02	1.293	0.00093
1-NEOPENTYL-2-PHENYL INDOLE	0.00	0.000	0.00000
1-N-BUTYL-2(BIPHENYLYL-4) INDOLE	19.63	0.059	0.44268
1,2-DIPHENYL INDOLE	7.27	1.164	0.00115
1-BIPHENYLYL-2-PHENYL INDOLE	12.31	0.240	0.02694
1,3-DIMETHYL-2-PHENYL INDOLE	8.11	0.838	0.00222
2,3-DIPHENYL INDOLE	18.26	0.073	0.28615
1-METHYL-2,3-DIPHENYL INDOLE	16.00	0.109	0.12959
2-PHENYL-3,2'-METHYLENE INDOLE (TRACE)	12.36	0.237	0.02763
2-PHENYL-3,2'-ETHYLENE INDOLE (TRACE)	20.49	0.052	0.57160
2-PHENYL-3,2'-TRIMETHYLENE INDOLE	11.65	0.283	0.01941
1-METHYL-3,2'-METHYLENE-2-PHENYL INDOLE (TRACE)	18.62	0.069	0.32307
1-METHYL-3,2'-ETHYLENE-2-PHENYL INDOLE	24.36	0.031	1.61534
1-METHYL-3,2'-TRIMETHYLENE-2-PHENYL INDOLE	12.85	0.211	0.03499
4,5-DIPHENYLIMIDAZOLE (A)	6.59	1.560	0.00064
PPF (TRACE)	19.47	0.061	0.42159
PPO (TRACE)	10.51	0.385	0.01047
2-(4-FLUOROPHENYL)-5-PHENYLOXAZOLE (TRACE)	9.53	0.517	0.00581
αNPO (TRACE)	29.84	0.017	5.42915
BBO (B)	37.15	0.009	20.19788
POPOP	39.68	0.007	29.93834
DIMETHYL POPOP (TRACE)	42.66	0.006	46.12320
PPD (TRACE)	0.00	0.000	0.00000
2,5-DI(2-METHYLPHENYL)OXADIAZOLE	0.00	0.000	0.00000
PBD	8.44	0.743	0.00282
BBD (B)	22.01	0.042	0.87869
αNPD (TRACE)	20.07	0.055	0.50498
βNPD	11.18	0.320	0.01519
αNND (TRACE)	25.85	0.026	2.29885
βNND (TRACE)	17.26	0.087	0.20506
TRANS-1-(4-BIPHENYLYL)-2-PHENYL-ETHYLENE	23.26	0.036	1.22245
TRANS-1-(4-BIPHENYLYL)-2-(1-NAPHTHYL)-ETHYLENE	33.24	0.012	10.36540
DIPHENYLSTILBENE (B)	37.17	0.009	20.23337
1,1,4,4-TETRAPHENYLBUTADIENE	36.45	0.009	17.96587
1,6-DIPHENYLHEXATRIENE	42.16	0.006	43.04423
1,8-DIPHENYL-1,3,5,7-OCTATETRAENE	48.71	0.004	101.77302
BPSB (TRACE)	43.27	0.006	50.33061
BIS-MSB	40.54	0.007	34.00267
ESCULIN (A)	27.51	0.021	3.33300
AZULENE	15.75	0.114	0.11536
NAPHTHALENE	0.00	0.000	0.00000
DEUTERATED NAPHTHALENE-C10D8	0.00	0.000	0.00000
1-METHYLNAPHTHALENE 5ML/L	0.00	0.000	0.00000
2-METHYL NAPHTHALENE	0.00	0.000	0.00000
2,3-DIMETHYL NAPHTHALENE	0.00	0.000	0.00000
2,6-DIMETHYL NAPHTHALENE (TRACE)	0.00	0.000	0.00000
1-NAPHTHOL (TRACE)	0.00	0.000	0.00000
2-NAPHTHOL (TRACE)	4.19	6.095	0.00004
1-NAPHTHYLAMINE	12.57	0.225	0.03060
2-CHLORONAPHTHALENE	0.00	0.000	0.00000
ACENAPHTHENE	0.00	0.000	0.00000
1-PHENYL NAPHTHALENE	0.00	0.000	0.00000
2-PHENYL NAPHTHALENE	0.00	0.000	0.00000
1,4-DIPHENYL NAPHTHALENE	8.75	0.668	0.00348
1,5-DIPHENYL NAPHTHALENE	6.24	1.837	0.00046
1,7-DIPHENYL NAPHTHALENE	7.00	1.304	0.00091
1,4,5,8-TETRAPHENYL NAPHTHALENE	29.77	0.017	5.33301
1,1'-BINAPHTHYL	0.00	0.000	0.00000
4,4'-DI(N-BUTOXY)-1,1'-BINAPHTHYL	14.77	0.139	0.08049
2,2'-BINAPHTHYL	7.46	1.078	0.00134
ACRIDINE (A)	29.56	0.017	5.11922
ANTHRACENE (TRACE)	28.43	0.019	4.05431
DEUTERATED ANTHRACENE C14D10 (TRACE)	27.87	0.021	3.60297
9-METHYL ANTHRACENE (TRACE)	30.56	0.016	6.24766
1-AMINO ANTHRACENE	31.71	0.014	7.72193
2-AMINO ANTHRACENE	30.27	0.016	5.84192
9-VINYL ANTHRACENE	32.33	0.013	8.74298
9-PHENYL ANTHRACENE (TRACE)	32.03	0.014	8.27273
9,10-DIPHENYL ANTHRACENE	34.99	0.010	14.02167
9,10-DICHLOROANTHRACENE (TRACE)	34.16	0.011	12.14626
BIANTHRANYL (B)	36.72	0.009	18.72843
9,10-DINAPHTHYLANTHRACENE (B)	34.68	0.011	13.26871
9-(P-VINYLPHENYL)-10-PHENYL ANTHRACENE	35.39	0.010	14.98377
9,10-BIS(PHENYL-ETHYNYL)ANTHRACENE (TRACE)	39.53	0.007	28.81529
RUBICENE (B)	34.61	0.011	12.97180
DIBENZ(A,H)ANTHRACENE (B)	25.07	0.028	1.90104
TETRACENE (TRACE) (B)	29.29	0.018	4.76849
RUBRENE (B)	26.55	0.024	2.64352
TETRAHYDROCANNABINOL ACID METHYL ESTER	0.00	0.000	0.00000
PHENANTHRENE	8.72	0.675	0.00341
3,4-BENZOPHENANTHRENE	15.88	0.112	0.12321
2-PHENYL PHENANTHRENE	12.36	0.237	0.02755
CHRYSENE (TRACE)	15.56	0.119	0.10958
TRIPHENYLENE	7.64	1.002	0.00154

ACCEPTOR MOLECULES	RO	CO	OVERLAP
DEUTERATED TRIPHENYLENE-C18D12	7.58	1.028	0.00147
PYRENE	16.76	0.095	0.17056
1,3,6,8,-TETRAPHENYL PYRENE	42.98	0.006	48.11086
BIPYRENYL (B)	38.36	0.008	24.35160
O-PHENYLENE-PYRENE	37.28	0.009	20.30414
ANTHANTHRENE (B)	38.27	0.008	23.88669
FLUORANTHENE	27.29	0.022	3.15959
3-PHENYL FLUORANTHENE	33.96	0.011	11.70802
3-AMINO FLUORANTHENE	34.32	0.011	12.36860
3,3'-BIFLUORANTHENYL (B)	40.54	0.007	33.82464
BENZO(GHI)FLUORANTHENE	25.05	0.028	1.88679
9,12 DI(N-OCTADECOXY)-BENZ(E)ACEPHENANTHRYLENE	36.36	0.009	17.59134
DECACYCLENE (B)	44.76	0.005	60.81181
10-ETHOXY-BENZO(KL)XANTHENE	35.62	0.010	15.61235
PERYLENE	39.08	0.007	27.07545
BENZO(GHI)PERYLENE (B)	36.96	0.009	19.50036
BBOT (TRACE)	45.35	0.005	66.56485
QUININE SULFATE (A)	15.49	0.120	0.10709
ACRIDINE YELLOW (A)	36.84	0.009	18.87864
ACRIDINE ORANGE (A)	31.11	0.015	6.84748
ACRIDINE RED (A)	27.31	0.022	3.13277
URANINE (A)	31.46	0.014	7.32421
RHODAMINE B (A)	31.50	0.014	7.37140
RHODAMINE 6G (A)	29.63	0.017	5.10599
EOSIN B (A)	28.07	0.020	3.69298
EOSIN Y (A)	29.49	0.017	4.96902
ROSE BENGAL (A)	28.40	0.020	3.96621
INDENOINDENE	10.17	0.425	0.00859

8.158 Donor Molecule: 2,2′-Binaphthyl

ACCEPTOR MOLECULES	RO	CO	OVERLAP
TRIPHENYLAMINE	14.70	0.141	0.18541
P-XYLENE (TRACE)	0.00	0.000	0.00000
SODIUM SALICYLATE (A)	11.23	0.316	0.03674
ANTHRANILIC ACID (A)	22.67	0.038	2.47322
TMPD	17.85	0.079	0.59148
BENZIDINE	10.60	0.375	0.02620
1,3-DIBIPHENYLYLBENZENE (B)	8.65	0.692	0.00771
1,3,5,-TRIBIPHENYLYLBENZENE (B)	9.44	0.532	0.01306
FLUORENE (TRACE)	0.00	0.000	0.00000
1-METHYL FLUORENE	0.00	0.000	0.00000
1,2-BENZOFLUORENE	14.68	0.141	0.18354
9,10-DIHYDROPHENANTHRENE	0.00	0.000	0.00000
CARBAZOLE (TRACE)	10.57	0.379	0.02565
N-METHYL CARBAZOLE (TRACE)	17.42	0.085	0.51276
N-VINYLCARBAZOLE	15.99	0.109	0.30706
N-PHENYL CARBAZOLE	16.09	0.107	0.31860
N-METHYL-2,3-BENZCARBAZOLE	24.99	0.029	4.41346
DIBENZOFURAN	0.00	0.000	0.00000
DIBENZOTHIOPHENE (TRACE)	6.96	1.324	0.00210
P-TERPHENYL	0.00	0.000	0.00000
DEUTERATED P-TERPHENYL-C18D14	0.00	0.000	0.00000
4-METHYL TERPHENYL	0.00	0.000	0.00000
2-METHYL-P-TERPHENYL	0.00	0.000	0.00000
3,3''-DIMETHYL-P-TERPHENYL	0.00	0.000	0.00000
4-(3,3-DIMETHYLBUTOXY)-P-TERPHENYL	6.99	1.308	0.00216
4,4''-DIHEXAHYDROFARNESOXY-P-TERPHENYL	9.48	0.525	0.01341
2,2'-METHYLENE-P-TERPHENYL	5.89	2.191	0.00077
2,2'-ETHYLENE-P-TERPHENYL	8.16	0.823	0.00545
2,2'-METHYLENE-5',6''-METHYLENE-P-TERPHENYL (TRACE)	18.75	0.068	0.80036
3-PHENYL DIBENZOFURAN	5.70	2.411	0.00064
BENZO(1,2-B-4,5-B')BISBENZOFURAN (TRACE) (B)	21.86	0.043	2.00400
SEXIPHENYL BENZENE	0.00	0.000	0.00000
P-QUATERPHENYL	11.77	0.274	0.04899
3,3'''-DIMETHYL-P-QUATERPHENYL	13.89	0.167	0.13213
TETRAMETHYL-P-QUATERPHENYL	0.00	0.000	0.00000
3,3''' -DIMETHYL-3',2''-METHYLENE-P-QUATERPHENYL	21.59	0.044	1.86185
3,3''' -DIMETHYL-3',2''-ETHYLENE-P-QUATERPHENYL	22.62	0.039	2.45224
4-(3-ETHYLBUTOXY)-P-QUATERPHENYL (B)	20.11	0.055	1.21257
4,4'''-DI(2-BUTYLOCTOXY-1)-P-QUATERPHENYL	20.33	0.053	1.29459
DIETHYL-P-QUINQUEPHENYL	14.18	0.157	0.14962
DI(3-ETHYLHEPTYL)-P-QUINQUEPHENYL	21.91	0.043	2.02391
TETRAMETHYL-P-SEXIPHENYL	12.20	0.246	0.06071
TETRAMETHYL-P-OCTAPHENYL (B)	27.22	0.022	7.44010
INDOLE (TRACE)	0.00	0.000	0.00000
2-PHENYL INDOLE	12.82	0.212	0.08175
1-METHYL-2-PHENYL INDOLE	10.07	0.438	0.01925
1-NEOPENTYL-2-PHENYL INDOLE	6.51	1.621	0.00141
1-N-BUTYL-2(BIPHENYLYL-4) INDOLE	22.01	0.042	2.08226
1,2-DIPHENYL INDOLE	10.35	0.404	0.02262
1-BIPHENYLYL-2-PHENYL INDOLE	14.56	0.145	0.17452
1,3-DIMETHYL-2-PHENYL INDOLE	11.56	0.289	0.04400
2,3-DIPHENYL INDOLE	20.24	0.054	1.25927
1-METHYL-2,3-DIPHENYL INDOLE	18.06	0.076	0.63654
2-PHENYL-3,2'-METHYLENE INDOLE (TRACE)	17.29	0.087	0.49119
2-PHENYL-3,2'-ETHYLENE INDOLE (TRACE)	24.22	0.031	3.70422
2-PHENYL-3,2'-TRIMETHYLENE INDOLE	15.91	0.111	0.29855
1-METHYL-3,2'-METHYLENE-2-PHENYL INDOLE (TRACE)	22.62	0.039	2.45599
1-METHYL-3,2'-ETHYLENE-2-PHENYL INDOLE	26.44	0.024	6.24998
1-METHYL-3,2'-TRIMETHYLENE-2-PHENYL INDOLE	16.76	0.095	0.40798
4,5-DIPHENYLIMIDAZOLE (A)	9.06	0.602	0.01018

ACCEPTOR MOLECULES	RO	CO	OVERLAP
PPF (TRACE)	23.75	0.033	3.29741
PPO (TRACE)	15.51	0.120	0.25612
2-(4-FLUOROPHENYL)-5-PHENYLOXAZOLE (TRACE)	14.34	0.152	0.15985
αNPO (TRACE)	30.24	0.016	13.95271
BBO (B)	36.25	0.009	41.28677
POPOP	37.17	0.009	47.99882
DIMETHYL POPOP (TRACE)	38.99	0.008	63.72428
PPD (TRACE)	0.00	0.000	0.00000
2,5-DI(2-METHYLPHENYL)OXADIAZOLE	0.00	0.000	0.00000
PBD	13.06	0.201	0.09158
BBD (B)	25.69	0.026	5.27560
αNPD (TRACE)	22.63	0.039	2.45720
βNPD	14.47	0.148	0.16889
αNND (TRACE)	27.30	0.022	7.56473
βNND (TRACE)	21.59	0.044	1.86236
TRANS-1-(4-BIPHENYLYL)-2-PHENYL-ETHYLENE	26.41	0.024	6.21163
TRANS-1-(4-BIPHENYLYL)-2-(1-NAPHTHYL)-ETHYLENE	32.66	0.013	22.09677
DIPHENYLSTILBENE (B)	36.46	0.009	42.74468
1,1,4,4-TETRAPHENYLBUTADIENE	35.00	0.010	33.37900
1,6-DIPHENYLHEXATRIENE	39.88	0.007	73.11988
1,8-DIPHENYL-1,3,5,7-OCTATETRAENE	42.57	0.006	107.43862
BPSB (TRACE)	40.36	0.007	78.56909
BIS-MSB	38.33	0.008	57.64341
ESCULIN (A)	27.46	0.022	7.82020
AZULENE	17.60	0.082	0.53553
NAPHTHALENE	0.00	0.000	0.00000
DEUTERATED NAPHTHALENE-C10D8	4.01	6.943	0.00008
1-METHYLNAPHTHALENE 5ML/L	0.00	0.000	0.00000
2-METHYL NAPHTHALENE	0.00	0.000	0.00000
2,3-DIMETHYL NAPHTHALENE	0.00	0.000	0.00000
2,6-DIMETHYL NAPHTHALENE (TRACE)	0.00	0.000	0.00000
1-NAPHTHOL (TRACE)	3.78	8.306	0.00005
2-NAPHTHOL (TRACE)	8.06	0.855	0.00505
1-NAPHTHYLAMINE	15.55	0.119	0.25943
2-PHENYL NAPHTHALENE	9.62	0.502	0.01464
1,4-DIPHENYL NAPHTHALENE	11.63	0.285	0.04547
1,5-DIPHENYL NAPHTHALENE	9.29	0.557	0.01188
1,7-DIPHENYL NAPHTHALENE	9.85	0.468	0.01680
1,4,5,8-TETRAPHENYL NAPHTHALENE	29.50	0.017	11.97047
1,1'-BINAPHTHYL	5.72	2.391	0.00065
4,4'-DI(N-BUTOXY)-1,1'-BINAPHTHYL	17.68	0.081	0.56012
2,2'-BINAPHTHYL	10.76	0.359	0.02859
ACRIDINE (A)	27.92	0.021	8.60243
ANTHRACENE (TRACE)	26.47	0.024	6.26457
DEUTERATED ANTHRACENE C14D10 (TRACE)	26.16	0.025	5.84038
9-METHYL ANTHRACENE (TRACE)	27.41	0.022	7.71072
1-AMINO ANTHRACENE	26.82	0.023	6.70138
2-AMINO ANTHRACENE	25.55	0.027	5.01107
9-VINYL ANTHRACENE	28.67	0.019	10.06255
9-PHENYL ANTHRACENE (TRACE)	28.73	0.019	10.21241
9,10-DIPHENYL ANTHRACENE	30.37	0.016	14.21817
9,10-DICHLOROANTHRACENE (TRACE)	29.14	0.018	11.08425
BIANTHRANYL (B)	31.89	0.014	19.04902
9,10-DINAPHTHYLANTHRACENE (B)	29.78	0.017	12.62241
9-(P-VINYLPHENYL)-10-PHENYL ANTHRACENE	30.56	0.016	14.73082
9,10-BIS(PHENYL-ETHYNYL)ANTHRACENE (TRACE)	29.52	0.017	11.90489
RUBICENE (B)	28.60	0.019	9.85216
DIBENZ(A,H)ANTHRACENE (B)	26.00	0.025	5.61053
TETRACENE (TRACE) (B)	22.56	0.039	2.37137
RUBRENE (B)	21.16	0.047	1.61461
TETRAHYDROCANNABINOL ACID METHYL ESTER	0.00	0.000	0.00000
PHENANTHRENE	10.64	0.372	0.02658
3,4-BENZOPHENANTHRENE	15.48	0.120	0.25089
2-PHENYL PHENANTHRENE	13.30	0.190	0.10129
CHRYSENE (TRACE)	15.76	0.114	0.27974
TRIPHENYLENE	9.80	0.475	0.01626
DEUTERATED TRIPHENYLENE-C18D12	9.58	0.509	0.01421
PYRENE	20.18	0.054	1.23128
1,3,6,8,-TETRAPHENYL PYRENE	37.59	0.008	50.97094
BIPYRENYL (B)	36.14	0.009	40.38629
O-PHENYLENE-PYRENE	32.49	0.013	21.17203
ANTHANTHRENE (B)	30.80	0.015	15.37007
FLUORANTHENE	26.53	0.024	6.31077
3-PHENYL FLUORANTHENE	31.05	0.015	16.20680
3-AMINO FLUORANTHENE	28.88	0.019	10.43694
3,3'-BIFLUORANTHENYL (B)	35.54	0.010	36.37234
BENZO(GHI)FLUORANTHENE	24.53	0.030	3.94274
9,12-DI(N-OCTADECOXY)-BENZ(E)ACEPHENANTHRYLENE	31.02	0.015	16.09100
DECACYCLENE (B)	38.31	0.008	56.85665
10-ETHOXY-BENZO(KL)XANTHENE	30.88	0.015	15.72935
PERYLENE	30.32	0.016	13.99317
BENZO(GHI)PERYLENE (B)	32.49	0.013	21.31772
BBOT (TRACE)	40.46	0.007	79.49944
QUININE SULFATE (A)	19.08	0.064	0.88705
ACRIDINE YELLOW (A)	26.74	0.023	6.58596
ACRIDINE ORANGE (A)	22.43	0.040	2.29021
ACRIDINE RED (A)	23.79	0.033	3.26461
URANINE (A)	23.71	0.034	3.19675
RHODAMINE B (A)	27.23	0.022	7.33643
RHODAMINE 6G (A)	26.57	0.024	6.33191
EOSIN B (A)	23.20	0.036	2.80885
EOSIN Y (A)	24.05	0.032	3.48189
ROSE BENGAL (A)	24.47	0.031	3.86571
INDENOINDENE	14.83	0.137	0.19561

8.160 Donor Molecule: Anthracene (trace)

ACCEPTOR MOLECULES	RO	CO	OVERLAP
ANTHRANILIC ACID (A)	12.35	0.238	0.07288
TMPD	4.65	4.461	0.00021
N-METHYL-2,3-BENZCARBAZOLE	24.72	0.030	4.66959
DI(3-ETHYLHEPTYL)-P-QUINQUEPHENYL	3.50	10.399	0.00004
TETRAMETHYL-P-SEXIPHENYL	0.00	0.000	0.00000
TETRAMETHYL-P-OCTAPHENYL (B)	6.84	1.397	0.00112
2,3-DIPHENYL INDOLE	5.39	2.851	0.00051
1 METHYL-3,2-ETHYLENE-2-PHENYL INDOLE	4.34	5.465	0.00014
αNPO (TRACE)	14.33	0.152	0.17800
BBO (B)	24.40	0.031	4.33447
POPOP	29.27	0.018	12.91740
DIMETHYL POPOP (TRACE)	32.21	0.013	22.87073
PPD (TRACE)	0.00	0.000	0.00000
2,5-DI(2-METHYLPHENYL)OXADIAZOLE	0.00	0.000	0.00000
PBD	0.00	0.000	0.00000
BBD (B)	0.00	0.000	0.00000
αNPD (TRACE)	4.28	5.711	0.00013
βNPD	0.00	0.000	0.00000
αNND (TRACE)	7.00	1.303	0.00243
βNND (TRACE)	0.00	0.000	0.00000
TRANS-1-(4-BIPHENYLYL)-2-PHENYL-ETHYLENE	3.95	7.256	0.00008
TRANS-1-(4-BIPHENYLYL)-2-(1-NAPHTHYL)-ETHYLENE	21.98	0.042	2.31626
DIPHENYLSTILBENE (B)	23.63	0.034	3.57067
1,1,4,4-TETRAPHENYLBUTADIENE	25.13	0.028	5.16055
1,6-DIPHENYLHEXATRIENE	30.30	0.016	15.88369
1,8-DIPHENYL-1,3,5,7-OCTATETRAENE	40.29	0.007	87.25602
BPSB (TRACE)	31.85	0.014	21.39597
BIS-MSB	29.05	0.018	12.32236
ESCULIN (A)	17.31	0.086	0.55292
AZULENE	10.39	0.399	0.02539
1,4,5,8-TETRAPHENYL NAPHTHALENE	19.20	0.063	1.02597
1,1'-BINAPHTHYL	0.00	0.000	0.00000
4,4'-DI(N-BUTOXY)-1,1'-BINAPHTHYL	0.00	0.000	0.00000
2,2'-BINAPHTHYL	0.00	0.000	0.00000
ACRIDINE (A)	21.27	0.046	1.89914
ANTHRACENE (TRACE)	21.81	0.043	2.21370
DEUTERATED ANTHRACENE C14D10 (TRACE)	20.41	0.053	1.48734
9-METHYL ANTHRACENE (TRACE)	23.08	0.036	3.09842
1-AMINO ANTHRACENE	27.04	0.023	7.95135
2-AMINO ANTHRACENE	26.22	0.025	6.60398
9-VINYL ANTHRACENE	25.37	0.027	5.46165
9-PHENYL ANTHRACENE (TRACE)	24.59	0.030	4.53404
9,10-DIPHENYL ANTHRACENE	28.92	0.018	11.96885
9,10-DICHLOROANTHRACENE (TRACE)	29.70	0.017	14.02633
BIANTHRANYL (B)	29.70	0.017	14.04143
9,10-DINAPHTHYLANTHRACENE (B)	29.69	0.017	13.99251
9-(P-VINYLPHENYL)-10-PHENYL ANTHRACENE	29.58	0.017	13.68328
9,10-BIS(PHENYL-ETHYNYL)ANTHRACENE (TRACE)	36.45	0.009	47.44682
RUBICENE (B)	30.32	0.016	15.71903
DIBENZ(A,H)ANTHRACENE (B)	16.94	0.092	0.48333
TETRACENE (TRACE) (B)	26.62	0.024	7.18781
RUBRENE (B)	24.04	0.032	3.90072
TETRAHYDROCANNABINOL ACID METHYL ESTER	0.00	0.000	0.00000
PHENANTHRENE	0.00	0.000	0.00000
3,4-BENZOPHENANTHRENE	9.85	0.468	0.01879
2-PHENYL PHENANTHRENE	0.00	0.000	0.00000
CHRYSENE (TRACE)	6.04	2.032	0.00100
TRIPHENYLENE	0.00	0.000	0.00000
DEUTERATED TRIPHENYLENE-C18D12	0.00	0.000	0.00000
PYRENE	8.53	0.720	0.00794
1,3,6,8,-TETRAPHENYL PYRENE	34.65	0.011	35.33969
BIPYRENYL (B)	28.43	0.019	10.80112
O-PHENYLENE-PYRENE	30.99	0.015	17.93558
ANTHANTHRENE (B)	34.31	0.011	33.18008
FLUORANTHENE	17.41	0.085	0.57023
3-PHENYL FLUORANTHENE	25.89	0.026	6.14844
3-AMINO FLUORANTHENE	29.62	0.017	13.68990
3,3'-BIFLUORANTHENYL (B)	33.10	0.012	26.79579
BENZO(GHI)FLUORANTHENE	19.79	0.058	1.22487
9,12-DI(N-OCTADECOXY)-BENZ(E)ACEPHENANTHRYLENE	31.05	0.015	18.26483
DECACYCLENE (B)	37.87	0.008	59.69035
10-ETHOXY-BENZO(KL)XANTHENE	29.85	0.017	14.48371
PERYLENE	35.25	0.010	39.03272
BENZO(GHI)PERYLENE (B)	28.81	0.019	11.69556
BBOT (TRACE)	36.15	0.009	45.64244
QUININE SULFATE (A)	0.00	0.000	0.00000
ACRIDINE YELLOW (A)	34.15	0.011	32.06474
ACRIDINE ORANGE (A)	28.62	0.019	11.12433
ACRIDINE RED (A)	23.25	0.036	3.19789
URANINE (A)	28.53	0.019	10.89645
RHODAMINE B (A)	27.12	0.022	8.05172
RHODAMINE 6G (A)	25.13	0.028	5.09635
EOSIN B (A)	24.91	0.029	4.83648
EOSIN Y (A)	26.29	0.025	6.67391
ROSE BENGAL (A)	24.66	0.030	4.54487

8.161 Donor Molecule: Deuterated Anthracene ($C_{14}D_{10}$) (trace)

ACCEPTOR MOLECULES	RO	CO	OVERLAP
ANTHRANILIC ACID (A)	12.63	0.222	0.09376
TMPD	4.98	3.619	0.00035
N-METHYL-2,3-BENZCARBAZOLE	24.45	0.031	4.91519
TETRAMETHYL-P-OCTAPHENYL (B)	7.35	1.125	0.00367
2,3-DIPHENYL INDOLE	5.83	2.263	0.00091
αNPO (TRACE)	14.98	0.133	0.26141
BBO (B)	24.72	0.030	5.27684
POPOP	29.11	0.018	14.03871
DIMETHYL POPOP (TRACE)	31.95	0.014	24.52079
PPD (TRACE)	0.00	0.000	0.00000
2,5-DI(2-METHYLPHENYL)OXADIAZOLE	0.00	0.000	0.00000
PBD	0.00	0.000	0.00000
BBD (B)	0.00	0.000	0.00000
αNPD (TRACE)	0.00	0.000	0.00000
βNPD	0.00	0.000	0.00000
αNND (TRACE)	7.54	1.042	0.00428
βNND (TRACE)	0.00	0.000	0.00000
TRANS-1-(4-BIPHENYLYL)-2-PHENYL-ETHYLENE	0.00	0.000	0.00000
TRANS-1-(4-BIPHENYLYL)-2-(1-NAPHTHYL)-ETHYLENE	22.15	0.041	2.72823
DIPHENYLSTILBENE (B)	24.01	0.032	4.42704
1,1,4,4-TETRAPHENYLBUTADIENE	25.23	0.028	5.95049
1,6-DIPHENYLHEXATRIENE	30.64	0.016	19.10011
1,8-DIPHENYL-1,3,5,7-OCTATETRAENE	39.85	0.007	91.84852
BPSB (TRACE)	31.68	0.014	23.32673
BIS-MSB	29.00	0.018	13.72893
ESCULIN (A)	17.41	0.085	0.64457
AZULENE	10.09	0.436	0.02397
1,4,5,8-TETRAPHENYL NAPHTHALENE	19.32	0.062	1.19992
1,1'-BINAPHTHYL	0.00	0.000	0.00000
4,4'-DI(N-BUTOXY)-1,1'-BINAPHTHYL	0.00	0.000	0.00000
2,2'-BINAPHTHYL	0.00	0.000	0.00000
ACRIDINE (A)	21.09	0.048	2.03062
ANTHRACENE (TRACE)	22.03	0.042	2.63972
DEUTERATED ANTHRACENE C14D10 (TRACE)	20.98	0.048	1.97001
9-METHYL ANTHRACENE (TRACE)	22.57	0.039	3.05299
1-AMINO ANTHRACENE	26.54	0.024	7.98853
2-AMINO ANTHRACENE	25.69	0.026	6.57830
9-VINYL ANTHRACENE	25.09	0.028	5.74795
9-PHENYL ANTHRACENE (TRACE)	24.09	0.032	4.50832
9,10-DIPHENYL ANTHRACENE	28.66	0.019	12.75863
9,10-DICHLOROANTHRACENE (TRACE)	29.09	0.018	13.93891
BIANTHRANYL (B)	29.54	0.017	15.27708
9,10-DINAPHTHYLANTHRACENE (B)	29.27	0.018	14.45884
9-(P-VINYLPHENYL)-10-PHENYL ANTHRACENE	29.24	0.018	14.34908
9,10-BIS(PHENYL-ETHYNYL)ANTHRACENE (TRACE)	35.55	0.010	45.95752
RUBICENE (B)	29.68	0.017	15.56054
DIBENZ(A,H)ANTHRACENE (B)	16.92	0.092	0.53965
TETRACENE (TRACE) (B)	26.08	0.025	7.15286
RUBRENE (B)	23.39	0.035	3.72202
TETRAHYDROCANNABINOL ACID METHYL ESTER	0.00	0.000	0.00000
PHENANTHRENE	0.00	0.000	0.00000
3,4-BENZOPHENANTHRENE	10.17	0.426	0.02552
2-PHENYL PHENANTHRENE	0.00	0.000	0.00000
CHRYSENE (TRACE)	6.65	1.524	0.00200
TRIPHENYLENE	0.00	0.000	0.00000
DEUTERATED TRIPHENYLENE-C18D12	0.00	0.000	0.00000
PYRENE	8.81	0.654	0.01082
1,3,6,8,-TETRAPHENYL PYRENE	34.21	0.011	36.80844
BIPYRENYL (B)	28.22	0.020	11.62598
O-PHENYLENE-PYRENE	30.44	0.016	18.10976
ANTHANTHRENE (B)	33.43	0.012	31.95210
FLUORANTHENE	17.43	0.085	0.64431
3-PHENYL FLUORANTHENE	25.75	0.026	6.69759
3-AMINO FLUORANTHENE	29.03	0.018	13.63328
3,3'-BIFLUORANTHENYL (B)	32.63	0.013	27.68384
BENZO(GHI)FLUORANTHENE	19.46	0.061	1.24680
9,12-DI(N-OCTADECOXY)-BENZ(E)ACEPHENANTHRYLENE	30.47	0.016	18.35867
DECACYCLENE (B)	37.18	0.009	60.14424
10-ETHOXY-BENZO(KL)XANTHENE	29.45	0.018	15.00794
PERYLENE	34.35	0.011	37.55524
BENZO(GHI)PERYLENE (B)	28.32	0.020	11.86526
BBOT (TRACE)	35.83	0.010	48.71516
QUININE SULFATE (A)	0.00	0.000	0.00000
ACRIDINE YELLOW (A)	33.29	0.012	30.94977
ACRIDINE ORANGE (A)	27.85	0.021	10.60792
ACRIDINE RED (A)	22.71	0.038	3.12166
URANINE (A)	27.73	0.021	10.33673
RHODAMINE B (A)	26.49	0.024	7.87004
RHODAMINE 6G (A)	24.43	0.031	4.83908
EOSIN B (A)	24.24	0.031	4.61528
EOSIN Y (A)	25.59	0.027	6.38683
ROSE BENGAL (A)	24.06	0.032	4.41216

8.162 Donor Molecule: 9-Methylanthracene (trace)

ACCEPTOR MOLECULES	RO	CO	OVERLAP
ANTHRANILIC ACID (A)	5.66	2.472	0.00069
TMPD	0.00	0.000	0.00000
N-METHYL-2,3-BENZCARBAZOLE	22.60	0.039	2.79602

ACCEPTOR MOLECULES	RO	CO	OVERLAP
αNPO (TRACE)	6.65	1.522	0.00182
BBO (B)	15.68	0.116	0.31267
POPOP	21.81	0.043	2.26759
DIMETHYL POPOP (TRACE)	29.12	0.018	12.80288
TRANS-1-(4-BIPHENYLYL)-2-(1-NAPHTHYL)-ETHYLENE	15.91	0.111	0.34187
DIPHENYLSTILBENE (B)	15.87	0.112	0.33693
1,1,4,4-TETRAPHENYLBUTADIENE	19.91	0.057	1.30931
1,6-DIPHENYLHEXATRIENE	20.70	0.050	1.65413
1,8-DIPHENYL-1,3,5,7-OCTATETRAENE	36.71	0.009	51.17372
BPSB (TRACE)	26.22	0.025	6.83573
BIS-MSB	24.14	0.032	4.16290
ESCULIN (A)	10.45	0.392	0.02742
AZULENE	11.49	0.295	0.04771
1,4,5,8-TETRAPHENYL NAPHTHALENE	15.27	0.126	0.26623
1,1'-BINAPHTHYL	0.00	0.000	0.00000
4,4'-DI(N-BUTOXY)-1,1'-BINAPHTHYL	0.00	0.000	0.00000
2,2'-BINAPHTHYL	0.00	0.000	0.00000
ACRIDINE (A)	18.40	0.072	0.81777
ANTHRACENE (TRACE)	11.83	0.270	0.05791
DEUTERATED ANTHRACENE C14D10 (TRACE)	10.73	0.362	0.03222
9-METHYL ANTHRACENE (TRACE)	22.02	0.042	2.39626
1-AMINO ANTHRACENE	26.22	0.025	6.78788
2-AMINO ANTHRACENE	25.69	0.026	5.99682
9-VINYL ANTHRACENE	24.02	0.032	4.02968
9-PHENYL ANTHRACENE (TRACE)	22.47	0.039	2.70560
9,10-DIPHENYL ANTHRACENE	26.81	0.023	7.79081
9,10-DICHLOROANTHRACENE (TRACE)	25.59	0.027	5.88305
BIANTHRANYL (B)	28.79	0.019	11.94841
9,10-DINAPHTHYLANTHRACENE (B)	26.29	0.025	6.92053
9-(P-VINYLPHENYL)-10-PHENYL ANTHRACENE	27.19	0.022	8.46069
9,10-BIS(PHENYL-ETHYNYL)ANTHRACENE (TRACE)	37.90	0.008	61.51722
RUBICENE (B)	30.78	0.015	17.59148
DIBENZ(A,H)ANTHRACENE (B)	15.46	0.121	0.28697
TETRACENE (TRACE) (B)	27.95	0.020	9.89216
RUBRENE (B)	25.91	0.026	6.25691
TETRAHYDROCANNABINOL ACID METHYL ESTER	0.00	0.000	0.00000
PHENANTHRENE	0.00	0.000	0.00000
3,4-BENZOPHENANTHRENE	6.34	1.759	0.00136
2-PHENYL PHENANTHRENE	0.00	0.000	0.00000
CHRYSENE (TRACE)	0.00	0.000	0.00000
TRIPHENYLENE	0.00	0.000	0.00000
DEUTERATED TRIPHENYLENE-C18D12	0.00	0.000	0.00000
PYRENE	4.36	5.401	0.00014
1,3,6,8,-TETRAPHENYL PYRENE	31.98	0.014	22.38956
BIPYRENYL (B)	25.43	0.027	5.67017
O-PHENYLENE-PYRENE	29.97	0.017	15.04931
ANTHANTHRENE (B)	34.55	0.011	35.48219
FLUORANTHENE	14.56	0.145	0.19980
3-PHENYL FLUORANTHENE	22.69	0.038	2.85864
3-AMINO FLUORANTHENE	29.45	0.018	13.55420
3,3'-BIFLUORANTHENYL (B)	30.93	0.015	18.30562
BENZO(GHI)FLUORANTHENE	18.43	0.071	0.82016
9,12-DI(N-OCTADECOXY)-BENZ(E)ACEPHENANTHRYLENE	30.79	0.015	17.81603
DECACYCLENE (B)	36.68	0.009	50.57487
10-ETHOXY-BENZO(KL)XANTHENE	26.75	0.023	7.68616
PERYLENE	36.14	0.009	46.45825
BENZO(GHI)PERYLENE (B)	27.62	0.021	9.31864
BBOT (TRACE)	32.95	0.013	26.85236
QUININE SULFATE (A)	0.00	0.000	0.00000
ACRIDINE YELLOW (A)	36.20	0.009	46.62789
ACRIDINE ORANGE (A)	30.92	0.015	18.07931
ACRIDINE RED (A)	24.48	0.031	4.44958
URANINE (A)	31.33	0.015	19.55626
RHODAMINE B (A)	28.31	0.020	10.64782
RHODAMINE 6G (A)	27.97	0.020	9.91176
EOSIN B (A)	27.38	0.022	8.71059
EOSIN Y (A)	28.67	0.019	11.49604
ROSE BENGAL (A)	26.31	0.025	6.85885

8.163 Donor Molecule: 1-Aminoanthracene

1,8-DIPHENYL-1,3,5,7-OCTATETRAENE	15.77	0.114	0.18337
BPSB (TRACE)	0.00	0.000	0.00000
BIS-MSB	0.00	0.000	0.00000
ESCULIN (A)	0.00	0.000	0.00000
AZULENE	19.03	0.065	0.55731
1-AMINO ANTHRACENE	17.30	0.086	0.31912
2-AMINO ANTHRACENE	16.88	0.093	0.27570
9-VINYL ANTHRACENE	0.00	0.000	0.00000
9-PHENYL ANTHRACENE (TRACE)	0.00	0.000	0.00000
9,10-DIPHENYL ANTHRACENE	4.85	3.913	0.00016
9,10-DICHLOROANTHRACENE (TRACE)	5.55	2.613	0.00035
BIANTHRANYL (B)	4.98	3.623	0.00018
9,10-DINAPHTHYLANTHRACENE (B)	7.33	1.134	0.00186
9-(P-VINYLPHENYL)-10-PHENYL ANTHRACENE	7.54	1.042	0.00220
9,10-BIS(PHENYL-ETHYNYL)ANTHRACENE (TRACE)	41.66	0.006	61.90900
RUBICENE (B)	39.23	0.007	42.90034
DIBENZ(A,H)ANTHRACENE (B)	0.00	0.000	0.00000
TETRACENE (TRACE) (B)	32.20	0.013	13.19660
RUBRENE (B)	37.86	0.008	34.65468

ACCEPTOR MOLECULES	RO	CO	OVERLAP
1,3,6,8,-TETRAPHENYL PYRENE	8.81	0.654	0.00559
BIPYRENYL (B)	0.00	0.000	0.00000
O-PHENYLENE-PYRENE	25.99	0.025	3.65771
ANTHANTHRENE (B)	29.24	0.018	7.44322
FLUORANTHENE	2.32	35.906	0.00000
3-PHENYL FLUORANTHENE	9.53	0.516	0.00897
3-AMINO FLUORANTHENE	25.45	0.027	3.22500
3,3'-BIFLUORANTHENYL (B)	16.40	0.101	0.23211
BENZO(GHI)FLUORANTHENE	8.21	0.809	0.00365
9,12-DI(N-OCTADECOXY)-BENZ(E)ACEPHENANTHRYLENE	16.65	0.097	0.25354
DECACYCLENE (B)	31.47	0.014	11.51971
10-ETHOXY-BENZO(KL)XANTHENE	0.00	0.000	0.00000
PERYLENE	31.36	0.014	11.32273
BENZO(GHI)PERYLENE (B)	0.00	0.000	0.00000
BBOT (TRACE)	0.00	0.000	0.00000
QUININE SULFATE (A)	0.00	0.000	0.00000
ACRIDINE YELLOW (A)	43.52	0.005	80.21056
ACRIDINE ORANGE (A)	43.19	0.006	76.58160
ACRIDINE RED (A)	40.62	0.007	52.70315
URANINE (A)	46.55	0.004	119.97481
RHODAMINE B (A)	46.37	0.004	116.72274
RHODAMINE 6G (A)	48.19	0.004	147.37572
EOSIN B (A)	45.55	0.005	105.04198
EOSIN Y (A)	46.77	0.004	123.12602
ROSE BENGAL (A)	44.36	0.005	89.41515

8.164 Donor Molecule: 2-Aminoanthracene

ACCEPTOR MOLECULES	RO	CO	OVERLAP
1,8-DIPHENYL-1,3,5,7-OCTATETRAENE	19.16	0.064	0.53798
BPSB (TRACE)	0.00	0.000	0.00000
BIS-MSB	0.00	0.000	0.00000
ESCULIN (A)	0.00	0.000	0.00000
AZULENE	18.21	0.074	0.38946
1-AMINO ANTHRACENE	19.95	0.056	0.68392
2-AMINO ANTHRACENE	19.77	0.058	0.64689
9-VINYL ANTHRACENE	0.00	0.000	0.00000
9-PHENYL ANTHRACENE (TRACE)	0.00	0.000	0.00000
9,10-DIPHENYL ANTHRACENE	7.96	0.888	0.00276
9,10-DICHLOROANTHRACENE (TRACE)	8.12	0.835	0.00312
BIANTHRANYL (B)	7.70	0.978	0.00228
9,10-DINAPHTHYLANTHRACENE (B)	9.75	0.483	0.00933
9-(P-VINYLPHENYL)-10-PHENYL ANTHRACENE	10.13	0.431	0.01173
9,10-BIS(PHENYL-ETHYNYL)ANTHRACENE (TRACE)	43.86	0.005	76.81969
RUBICENE (B)	39.41	0.007	40.16121
DIBENZ(A,H)ANTHRACENE (B)	0.00	0.000	0.00000
TETRACENE (TRACE) (B)	33.15	0.012	14.31298
RUBRENE (B)	37.35	0.009	29.08186
1,3,6,8,-TETRAPHENYL PYRENE	11.49	0.295	0.02497
BIPYRENYL (B)	6.96	1.329	0.00124
O-PHENYLENE-PYRENE	28.23	0.020	5.47232
ANTHANTHRENE (B)	33.39	0.012	15.03411
FLUORANTHENE	3.99	7.057	0.00004
3-PHENYL FLUORANTHENE	12.08	0.254	0.03376
3-AMINO FLUORANTHENE	27.77	0.021	4.96023
3,3'-BIFLUORANTHENYL (B)	19.77	0.058	0.64840
BENZO(GHI)FLUORANTHENE	10.49	0.388	0.01447
9,12-DI(N-OCTADECOXY)-BENZ(E)ACEPHENANTHRYLENE	20.62	0.051	0.83588
DECACYCLENE (B)	34.04	0.011	16.78658
10-ETHOXY-BENZO(KL)XANTHENE	5.98	2.096	0.00050
PERYLENE	35.30	0.010	20.97336
BENZO(GHI)PERYLENE (B)	5.04	3.495	0.00018
BBOT (TRACE)	6.04	2.034	0.00050
QUININE SULFATE (A)	0.00	0.000	0.00000
ACRIDINE YELLOW (A)	45.07	0.005	90.18630
ACRIDINE ORANGE (A)	43.22	0.006	70.08260
ACRIDINE RED (A)	38.69	0.008	35.85465
URANINE (A)	46.05	0.005	102.41852
RHODAMINE B (A)	44.01	0.005	77.72829
RHODAMINE 6G (A)	46.28	0.005	105.34584
EOSIN B (A)	43.89	0.005	76.51768
EOSIN Y (A)	45.13	0.005	90.55164
ROSE BENGAL (A)	42.08	0.006	59.40974

8.165 Donor Molecule: 9-Vinylanthracene

ACCEPTOR MOLECULES	RO	CO	OVERLAP
N-METHYL-2,3-BENZCARBAZOLE	21.25	0.047	0.89071
DIBENZOFURAN	0.00	0.000	0.00000
BBO (B)	9.29	0.558	0.00622
POPOP	13.75	0.172	0.06539
DIMETHYL POPOP (TRACE)	22.73	0.038	1.33150
TRANS-1-(4-BIPHENYLYL)-2-(1-NAPHTHYL)-ETHYLENE	9.93	0.457	0.00930
DIPHENYLSTILBENE (B)	10.11	0.433	0.01034
1,1,4,4-TETRAPHENYLBUTADIENE	15.35	0.124	0.12631
1,6-DIPHENYLHEXATRIENE	15.11	0.130	0.11540
1,8-DIPHENYL-1,3,5,7-OCTATETRAENE	35.55	0.010	19.40889
BPSB (TRACE)	18.14	0.075	0.34468

ACCEPTOR MOLECULES	RO	CO	OVERLAP
BIS-MSB	17.51	0.083	0.27918
ESCULIN (A)	0.00	0.000	0.00000
AZULENE	15.05	0.131	0.11031
1,4,5,8-TETRAPHENYL NAPHTHALENE	11.09	0.328	0.01802
1,1'-BINAPHTHYL	0.00	0.000	0.00000
4,4'-DI(N-BUTOXY)-1,1'-BINAPHTHYL	0.00	0.000	0.00000
2,2'-BINAPHTHYL	0.00	0.000	0.00000
ACRIDINE (A)	13.03	0.202	0.04726
ANTHRACENE (TRACE)	6.76	1.445	0.00093
DEUTERATED ANTHRACENE C14D10 (TRACE)	6.01	2.064	0.00046
9-METHYL ANTHRACENE (TRACE)	14.87	0.142	0.09633
1-AMINO ANTHRACENE	27.52	0.021	4.16792
2-AMINO ANTHRACENE	27.33	0.022	4.00338
9-VINYL ANTHRACENE	20.10	0.055	0.63710
9-PHENYL ANTHRACENE (TRACE)	15.83	0.113	0.15204
9,10-DIPHENYL ANTHRACENE	25.34	0.027	2.55467
9,10-DICHLOROANTHRACENE (TRACE)	26.73	0.023	3.52172
BIANTHRANYL (B)	25.61	0.027	2.72428
9,10-DINAPHTHYLANTHRACENE (B)	26.40	0.024	3.26446
9-(P-VINYLPHENYL)-10-PHENYL ANTHRACENE	26.31	0.025	3.19895
9,10-BIS(PHENYL-ETHYNYL)ANTHRACENE (TRACE)	45.15	0.005	80.85866
RUBICENE (B)	37.33	0.009	25.68769
DIBENZ(A,H)ANTHRACENE (B)	13.96	0.164	0.07159
TETRACENE (TRACE) (B)	33.22	0.012	12.83337
RUBRENE (B)	33.25	0.012	12.84540
TETRAHYDROCANNABINOL ACID METHYL ESTER	0.00	0.000	0.00000
PHENANTHRENE	0.00	0.000	0.00000
3,4-BENZOPHENANTHRENE	3.83	7.978	0.00003
2-PHENYL PHENANTHRENE	0.00	0.000	0.00000
CHRYSENE (TRACE)	0.00	0.000	0.00000
TRIPHENYLENE	0.00	0.000	0.00000
DEUTERATED TRIPHENYLENE-C18D12	0.00	0.000	0.00000
PYRENE	0.00	0.000	0.00000
1,3,6,8,-TETRAPHENYL PYRENE	28.42	0.019	5.07148
BIPYRENYL (B)	22.34	0.040	1.20091
O-PHENYLENE-PYRENE	32.70	0.013	11.68745
ANTHANTHRENE (B)	39.17	0.007	34.67087
FLUORANTHENE	12.55	0.226	0.03772
3-PHENYL FLUORANTHENE	21.16	0.047	0.86435
3-AMINO FLUORANTHENE	32.57	0.013	11.40999
3,3'-BIFLUORANTHENYL (B)	30.72	0.015	8.08963
BENZO(GHI)FLUORANTHENE	18.55	0.070	0.39234
9,12-DI(N-OCTADECOXY)-BENZ(E)ACEPHENANTHRYLENE	32.16	0.013	10.63212
DECACYCLENE (B)	39.72	0.007	37.50725
10-ETHOXY-BENZO(KL)XANTHENE	25.70	0.026	2.78262
PERYLENE	40.79	0.007	44.18784
BENZO(GHI)PERYLENE (B)	20.94	0.049	0.81460
BBOT (TRACE)	30.03	0.017	7.08721
QUININE SULFATE (A)	0.00	0.000	0.00000
ACRIDINE YELLOW (A)	44.27	0.005	71.70556
ACRIDINE ORANGE (A)	39.45	0.007	35.83571
ACRIDINE RED (A)	31.17	0.015	8.71770
URANINE (A)	40.74	0.007	43.44932
RHODAMINE B (A)	35.41	0.010	18.73484
RHODAMINE 6G (A)	37.77	0.008	27.55583
EOSIN B (A)	36.49	0.009	22.41570
EOSIN Y (A)	37.84	0.008	27.89990
ROSE BENGAL (A)	33.67	0.012	13.82742

8.166 Donor Molecule: 9-Phenylanthracene (trace)

ACCEPTOR MOLECULES	RO	CO	OVERLAP
ANTHRANILIC ACID (A)	6.32	1.775	0.00096
TMPD	0.00	0.000	0.00000
N-METHYL-2,3-BENZCARBAZOLE	23.70	0.034	2.66109
αNPO (TRACE)	7.21	1.191	0.00213
BBO (B)	15.77	0.114	0.23142
POPOP	21.63	0.044	1.53989
DIMETHYL POPOP (TRACE)	29.00	0.018	8.92512
TRANS-1-(4-BIPHENYLYL)-2-(1-NAPHTHYL)-ETHYLENE	15.67	0.116	0.22321
DIPHENYLSTILBENE (B)	15.83	0.113	0.23694
1,1,4,4-TETRAPHENYLBUTADIENE	19.91	0.057	0.93527
1,6-DIPHENYLHEXATRIENE	20.89	0.049	1.24873
1,8-DIPHENYL-1,3,5,7-OCTATETRAENE	38.39	0.008	47.82389
BPSB (TRACE)	25.83	0.026	4.46671
BIS-MSB	23.90	0.033	2.80150
ESCULIN (A)	10.95	0.340	0.02605
AZULENE	12.41	0.234	0.05411
1,4,5,8-TETRAPHENYL NAPHTHALENE	15.16	0.129	0.18213
1,1'-BINAPHTHYL	0.00	0.000	0.00000
4,4'-DI(N-BUTOXY)-1,1'-BINAPHTHYL	0.00	0.000	0.00000
2,2'-BINAPHTHYL	0.00	0.000	0.00000
ACRIDINE (A)	18.11	0.075	0.53077
ANTHRACENE (TRACE)	12.34	0.238	0.05323
DEUTERATED ANTHRACENE C14D10 (TRACE)	11.25	0.314	0.03060
9-METHYL ANTHRACENE (TRACE)	21.36	0.046	1.42910
1-AMINO ANTHRACENE	27.39	0.022	6.29179
2-AMINO ANTHRACENE	26.94	0.023	5.70137
9-VINYL ANTHRACENE	24.22	0.031	3.02579
9-PHENYL ANTHRACENE (TRACE)	22.09	0.042	1.74391
9,10-DIPHENYL ANTHRACENE	27.91	0.021	7.08393
9,10-DICHLOROANTHRACENE (TRACE)	27.49	0.022	6.46813

ACCEPTOR MOLECULES	RO	CO	OVERLAP
BIANTHRANYL (B)	29.60	0.017	10.07781
9,10-DINAPHTHYLANTHRACENE (B)	27.93	0.021	7.10112
9-(P-VINYLPHENYL)-10-PHENYL ANTHRACENE	28.41	0.020	7.87247
9,10-BIS(PHENYL-ETHYNYL)ANTHRACENE (TRACE)	40.64	0.007	66.78185
RUBICENE (B)	32.96	0.012	18.96842
DIBENZ(A,H)ANTHRACENE (B)	16.17	0.106	0.26820
TETRACENE (TRACE) (B)	30.04	0.016	10.90365
RUBRENE (B)	28.04	0.020	7.18972
TETRAHYDROCANNABINOL ACID METHYL ESTER	0.00	0.000	0.00000
PHENANTHRENE	0.00	0.000	0.00000
3,4-BENZOPHENANTHRENE	6.30	1.790	0.00094
2-PHENYL PHENANTHRENE	0.00	0.000	0.00000
CHRYSENE (TRACE)	0.00	0.000	0.00000
TRIPHENYLENE	0.00	0.000	0.00000
DEUTERATED TRIPHENYLENE-C18D12	0.00	0.000	0.00000
PYRENE	4.66	4.407	0.00016
1,3,6,8,-TETRAPHENYL PYRENE	32.53	0.013	17.74043
BIPYRENYL (B)	25.77	0.026	4.38918
O-PHENYLENE-PYRENE	31.40	0.014	14.23296
ANTHANTHRENE (B)	36.44	0.009	34.92933
FLUORANTHENE	14.69	0.141	0.15079
3-PHENYL FLUORANTHENE	23.17	0.036	2.31459
3-AMINO FLUORANTHENE	31.05	0.015	13.31267
3,3'-BIFLUORANTHENYL (B)	31.95	0.014	15.90408
BENZO(GHI)FLUORANTHENE	19.23	0.063	0.75670
9,12-DI(N-OCTADECOXY)-BENZ(E)ACEPHENANTHRYLENE	32.11	0.014	16.36686
DECACYCLENE (B)	38.45	0.008	47.99188
10-ETHOXY-BENZO(KL)XANTHENE	28.05	0.020	7.31139
PERYLENE	37.90	0.008	44.16889
BENZO(GHI)PERYLENE (B)	27.20	0.022	6.06962
BBOT (TRACE)	33.99	0.011	23.14290
QUININE SULFATE (A)	0.00	0.000	0.00000
ACRIDINE YELLOW (A)	39.04	0.008	52.35481
ACRIDINE ORANGE (A)	33.60	0.012	21.28313
ACRIDINE RED (A)	25.97	0.026	4.53683
URANINE (A)	33.95	0.011	22.65892
RHODAMINE B (A)	30.05	0.016	10.90330
RHODAMINE 6G (A)	29.88	0.017	10.53935
EOSIN B (A)	29.29	0.018	9.33983
EOSIN Y (A)	30.68	0.015	12.33083
ROSE BENGAL (A)	28.01	0.020	7.14906

8.167 Donor Molecule: 9,10-Diphenylanthracene

ACCEPTOR MOLECULES	RO	CO	OVERLAP
N-METHYL-2,3-BENZCARBAZOLE	22.68	0.038	1.00050
DIBENZOFURAN	0.00	0.000	0.00000
BBO (B)	8.36	0.767	0.00251
POPOP	13.20	0.194	0.03896
DIMETHYL POPOP (TRACE)	23.82	0.033	1.34039
TRANS-1-(4-BIPHENYLYL)-2-(1-NAPHTHYL)-ETHYLENE	9.50	0.522	0.00540
DIPHENYLSTILBENE (B)	9.88	0.463	0.00686
1,1,4,4-TETRAPHENYLBUTADIENE	16.05	0.108	0.12560
1,6-DIPHENYLHEXATRIENE	15.62	0.117	0.10673
1,8-DIPHENYL-1,3,5,7-OCTATETRAENE	38.02	0.008	22.08535
BPSB (TRACE)	18.37	0.072	0.28288
BIS-MSB	18.02	0.076	0.25184
ESCULIN (A)	0.00	0.000	0.00000
AZULENE	15.15	0.129	0.08732
1,4,5,8-TETRAPHENYL NAPHTHALENE	11.42	0.300	0.01631
1,1'-BINAPHTHYL	0.00	0.000	0.00000
4,4'-DI(N-BUTOXY)-1,1'-BINAPHTHYL	0.00	0.000	0.00000
2,2'-BINAPHTHYL	0.00	0.000	0.00000
ACRIDINE (A)	13.29	0.191	0.04046
ANTHRACENE (TRACE)	6.12	1.953	0.00039
DEUTERATED ANTHRACENE C14D10 (TRACE)	5.04	3.491	0.00012
9-METHYL ANTHRACENE (TRACE)	14.46	0.148	0.06715
1-AMINO ANTHRACENE	29.36	0.018	4.67230
2-AMINO ANTHRACENE	29.14	0.018	4.46252
9-VINYL ANTHRACENE	21.28	0.046	0.68188
9-PHENYL ANTHRACENE (TRACE)	16.13	0.107	0.12941
9,10-DIPHENYL ANTHRACENE	27.15	0.022	2.93485
9,10-DICHLOROANTHRACENE (TRACE)	28.84	0.019	4.22121
BIANTHRANYL (B)	27.26	0.022	3.01119
9,10-DINAPHTHYLANTHRACENE (B)	28.37	0.020	3.81785
9-(P-VINYLPHENYL)-10-PHENYL ANTHRACENE	28.24	0.020	3.71481
9,10-BIS(PHENYL-ETHYNYL)ANTHRACENE (TRACE)	47.22	0.004	80.43703
RUBICENE (B)	38.52	0.008	23.61685
DIBENZ(A,H)ANTHRACENE (B)	14.81	0.138	0.07767
TETRACENE (TRACE) (B)	34.46	0.011	12.15461
RUBRENE (B)	33.81	0.012	10.78726
TETRAHYDROCANNABINOL ACID METHYL ESTER	0.00	0.000	0.00000
PHENANTHRENE	0.00	0.000	0.00000
3,4-BENZOPHENANTHRENE	3.51	10.303	0.00001
2-PHENYL PHENANTHRENE	0.00	0.000	0.00000
CHRYSENE (TRACE)	0.00	0.000	0.00000
TRIPHENYLENE	0.00	0.000	0.00000
DEUTERATED TRIPHENYLENE-C18D12	0.00	0.000	0.00000
PYRENE	0.00	0.000	0.00000
1,3,6,8,-TETRAPHENYL PYRENE	30.32	0.016	5.69297
BIPYRENYL (B)	23.87	0.033	1.35609
O-PHENYLENE-PYRENE	34.61	0.011	12.50075

ACCEPTOR MOLECULES	RO	CO	OVERLAP
ANTHANTHRENE (B)	41.69	0.006	38.30999
FLUORANTHENE	13.40	0.186	0.04246
3-PHENYL FLUORANTHENE	22.57	0.039	0.96823
3-AMINO FLUORANTHENE	34.49	0.011	12.22979
3,3'-BIFLUORANTHENYL (B)	32.85	0.013	9.18909
BENZO(GHI)FLUORANTHENE	19.83	0.057	0.44559
9,12-DI(N-OCTADECOXY)-BENZ(E)ACEPHENANTHRYLENE	34.30	0.011	11.91290
DECACYCLENE (B)	42.04	0.006	40.09562
10-ETHOXY-BENZO(KL)XANTHENE	27.54	0.021	3.20198
PERYLENE	43.37	0.005	40.54310
BENZO(GHI)PERYLENE (B)	21.83	0.043	0.79773
BBOT (TRACE)	32.07	0.014	7.98706
QUININE SULFATE (A)	0.00	0.000	0.00000
ACRIDINE YELLOW (A)	45.96	0.005	68.22585
ACRIDINE ORANGE (A)	40.27	0.007	30.83708
ACRIDINE RED (A)	31.18	0.015	6.64450
URANINE (A)	41.22	0.006	35.44374
RHODAMINE B (A)	35.69	0.010	14.94677
RHODAMINE 6G (A)	37.24	0.009	19.27419
EOSIN B (A)	36.14	0.009	16.10584
EOSIN Y (A)	37.62	0.008	20.47304
ROSE BENGAL (A)	33.78	0.012	10.72939

8.168 Donor Molecule: 9,10-Dichloroanthracene (trace)

ACCEPTOR MOLECULES	RO	CO	OVERLAP
N-METHYL-2,3-BENZCARBAZOLE	19.38	0.061	0.70815
DIBENZOFURAN	0.00	0.000	0.00000
BBO (B)	7.61	1.017	0.00259
POPOP	11.20	0.318	0.02644
DIMETHYL POPOP (TRACE)	20.02	0.056	0.85994
TRANS-1-(4-BIPHENYLYL)-2-(1-NAPHTHYL)-ETHYLENE	8.17	0.821	0.00398
DIPHENYLSTILBENE (B)	8.21	0.808	0.00410
1,1,4,4-TETRAPHENYLBUTADIENE	13.50	0.182	0.08101
1,6-DIPHENYLHEXATRIENE	12.45	0.232	0.04970
1,8-DIPHENYL-1,3,5,7-OCTATETRAENE	33.41	0.012	18.49774
BPSB (TRACE)	14.86	0.136	0.14366
BIS-MSB	14.51	0.146	0.12494
ESCULIN (A)	0.00	0.000	0.00000
AZULENE	13.99	0.163	0.09870
1,4,5,8-TETRAPHENYL NAPHTHALENE	9.12	0.589	0.00772
1,1'-BINAPHTHYL	0.00	0.000	0.00000
4,4'-DI(N-BUTOXY)-1,1'-BINAPHTHYL	0.00	0.000	0.00000
2,2'-BINAPHTHYL	0.00	0.000	0.00000
ACRIDINE (A)	10.71	0.364	0.02019
ANTHRACENE (TRACE)	5.50	2.685	0.00037
DEUTERATED ANTHRACENE C14D10 (TRACE)	4.79	4.061	0.00016
9-METHYL ANTHRACENE (TRACE)	12.09	0.253	0.04169
1-AMINO ANTHRACENE	26.41	0.024	4.49613
2-AMINO ANTHRACENE	26.18	0.025	4.26474
9-VINYL ANTHRACENE	18.16	0.075	0.47870
9-PHENYL ANTHRACENE (TRACE)	13.26	0.192	0.07256
9,10-DIPHENYL ANTHRACENE	23.83	0.033	2.44216
9,10-DICHLOROANTHRACENE (TRACE)	26.12	0.025	4.23193
BIANTHRANYL (B)	23.24	0.036	2.10264
9,10-DINAPHTHYLANTHRACENE (B)	25.17	0.028	3.38393
9-(P-VINYLPHENYL)-10-PHENYL ANTHRACENE	25.03	0.029	3.27313
9,10-BIS(PHENYL-ETHYNYL)ANTHRACENE (TRACE)	42.81	0.006	81.19471
RUBICENE (B)	35.05	0.010	24.34688
DIBENZ(A,H)ANTHRACENE (B)	12.18	0.248	0.04359
TETRACENE (TRACE) (B)	30.71	0.015	11.06120
RUBRENE (B)	31.06	0.015	11.79667
TETRAHYDROCANNABINOL ACID METHYL ESTER	0.00	0.000	0.00000
PHENANTHRENE	0.00	0.000	0.00000
3,4-BENZOPHENANTHRENE	3.16	14.133	0.00001
2-PHENYL PHENANTHRENE	0.00	0.000	0.00000
CHRYSENE (TRACE)	0.00	0.000	0.00000
TRIPHENYLENE	0.00	0.000	0.00000
DEUTERATED TRIPHENYLENE-C18D12	0.00	0.000	0.00000
PYRENE	0.00	0.000	0.00000
1,3,6,8,-TETRAPHENYL PYRENE	26.65	0.024	4.76902
BIPYRENYL (B)	21.12	0.047	1.18320
O-PHENYLENE-PYRENE	31.30	0.015	12.43236
ANTHANTHRENE (B)	38.25	0.008	41.55372
FLUORANTHENE	11.95	0.262	0.03891
3-PHENYL FLUORANTHENE	20.03	0.056	0.86030
3-AMINO FLUORANTHENE	31.12	0.015	11.99821
3,3'-BIFLUORANTHENYL (B)	29.45	0.018	8.67098
BENZO(GHI)FLUORANTHENE	17.62	0.082	0.39853
9,12-DI(N-OCTADECOXY)-BENZ(E)ACEPHENANTHRYLENE	30.57	0.016	10.83503
DECACYCLENE (B)	37.90	0.008	39.14749
10-ETHOXY-BENZO(KL)XANTHENE	24.20	0.032	2.67769
PERYLENE	39.70	0.007	51.92861
BENZO(GHI)PERYLENE (B)	18.54	0.070	0.54111
BBOT (TRACE)	27.90	0.021	6.29310
QUININE SULFATE (A)	0.00	0.000	0.00000
ACRIDINE YELLOW (A)	41.68	0.006	68.96798
ACRIDINE ORANGE (A)	36.71	0.009	32.14266
ACRIDINE RED (A)	29.08	0.018	7.94276
URANINE (A)	37.74	0.008	37.94983
RHODAMINE B (A)	33.11	0.012	17.31172
RHODAMINE 6G (A)	35.07	0.010	24.43376

ACCEPTOR MOLECULES	RO	CO	OVERLAP
EOSIN B (A)	33.84	0.012	19.71831
EOSIN Y (A)	35.11	0.010	24.63396
ROSE BENGAL (A)	31.36	0.015	12.49156

8.171 Donor Molecule: 9-(*p*-Vinylphenyl)-10-phenylanthracene

ACCEPTOR MOLECULES	RO	CO	OVERLAP
N-METHYL-2,3-BENZCARBAZOLE	18.62	0.069	0.30526
DIBENZOFURAN	0.00	0.000	0.00000
BBO (B)	5.19	3.198	0.00014
POPOP	9.36	0.545	0.00496
DIMETHYL POPOP (TRACE)	19.04	0.065	0.35019
TRANS-1-(4-BIPHENYLYL)-2-(1-NAPHTHYL)-ETHYLENE	6.59	1.566	0.00060
DIPHENYLSTILBENE (B)	7.11	1.242	0.00095
1,1,4,4-TETRAPHENYLBUTADIENE	12.78	0.214	0.03198
1,6-DIPHENYLHEXATRIENE	12.07	0.255	0.02269
1,8-DIPHENYL-1,3,5,7-OCTATETRAENE	33.28	0.012	9.92460
BPSB (TRACE)	13.86	0.168	0.05200
BIS-MSB	13.87	0.168	0.05233
ESCULIN (A)	0.00	0.000	0.00000
AZULENE	16.02	0.109	0.12182
1,4,5,8-TETRAPHENYL NAPHTHALENE	8.77	0.664	0.00334
1,1'-BINAPHTHYL	0.00	0.000	0.00000
4,4'-DI(N-BUTOXY)-1,1'-BINAPHTHYL	0.00	0.000	0.00000
2,2'-BINAPHTHYL	0.00	0.000	0.00000
ACRIDINE (A)	10.11	0.433	0.00785
ANTHRACENE (TRACE)	3.90	7.538	0.00003
DEUTERATED ANTHRACENE C14D10 (TRACE)	0.00	0.000	0.00000
9-METHYL ANTHRACENE (TRACE)	10.55	0.381	0.01011
1-AMINO ANTHRACENE	28.09	0.020	3.57822
2-AMINO ANTHRACENE	28.02	0.020	3.52525
9-VINYL ANTHRACENE	17.38	0.085	0.20247
9-PHENYL ANTHRACENE (TRACE)	12.36	0.237	0.02620
9,10-DIPHENYL ANTHRACENE	23.16	0.036	1.13228
9,10-DICHLOROANTHRACENE (TRACE)	24.86	0.029	1.73030
BIANTHRANYL (B)	22.71	0.038	1.00628
9,10-DINAPHTHYLANTHRACENE (B)	24.38	0.031	1.53742
9-(P-VINYLPHENYL)-10-PHENYL ANTHRACENE	24.50	0.030	1.58251
9,10-BIS(PHENYL-ETHYNYL)ANTHRACENE (TRACE)	47.70	0.004	85.38127
RUBICENE (B)	39.42	0.007	27.06141
DIBENZ(A,H)ANTHRACENE (B)	11.92	0.264	0.02103
TETRACENE (TRACE) (B)	35.19	0.010	13.76407
RUBRENE (B)	35.20	0.010	13.70111
TETRAHYDROCANNABINOL ACID METHYL ESTER	0.00	0.000	0.00000
PHENANTHRENE	0.00	0.000	0.00000
3,4-BENZOPHENANTHRENE	2.26	38.856	0.00000
2-PHENYL PHENANTHRENE	0.00	0.000	0.00000
CHRYSENE (TRACE)	0.00	0.000	0.00000
TRIPHENYLENE	0.00	0.000	0.00000
DEUTERATED TRIPHENYLENE-C18D12	0.00	0.000	0.00000
PYRENE	0.00	0.000	0.00000
1,3,6,8,-TETRAPHENYL PYRENE	26.11	0.025	2.31848
BIPYRENYL (B)	20.55	0.052	0.55188
O-PHENYLENE-PYRENE	34.02	0.011	11.26746
ANTHANTHRENE (B)	40.99	0.006	34.55005
FLUORANTHENE	11.58	0.288	0.01765
3-PHENYL FLUORANTHENE	20.77	0.050	0.58669
3-AMINO FLUORANTHENE	33.90	0.011	11.02370
3,3'-BIFLUORANTHENYL (B)	30.69	0.015	6.10365
BENZO(GHI)FLUORANTHENE	18.33	0.073	0.27754
9,12-DI(N-OCTADECOXY)-BENZ(E)ACEPHENANTHRYLENE	32.95	0.012	9.35398
DECACYCLENE (B)	41.17	0.006	35.37032
10-ETHOXY-BENZO(KL)XANTHENE	23.06	0.036	1.10280
PERYLENE	42.84	0.006	45.05428
BENZO(GHI)PERYLENE (B)	17.84	0.079	0.23598
BBOT (TRACE)	26.70	0.024	2.65483
QUININE SULFATE (A)	0.00	0.000	0.00000
ACRIDINE YELLOW (A)	46.91	0.004	77.12969
ACRIDINE ORANGE (A)	41.74	0.006	38.17998
ACRIDINE RED (A)	33.22	0.012	9.68796
URANINE (A)	43.05	0.006	45.96960
RHODAMINE B (A)	37.81	0.008	21.07423
RHODAMINE 6G (A)	40.03	0.007	29.65413
EOSIN B (A)	38.53	0.008	23.57672
EOSIN Y (A)	39.94	0.007	29.25312
ROSE BENGAL (A)	35.97	0.010	15.60245

8.172 Donor Molecule: 9,10-Bis(phenylethynyl)anthracene (trace)

ACCEPTOR MOLECULES	RO	CO	OVERLAP
AZULENE	21.82	0.043	0.76901
NAPHTHALENE	0.00	0.000	0.00000
1-AMINO ANTHRACENE	7.15	1.223	0.00097
2-AMINO ANTHRACENE	7.29	1.156	0.00108
9,10-BIS(PHENYL-ETHYNYL)ANTHRACENE (TRACE)	41.20	0.006	35.19488
RUBICENE (B)	43.49	0.005	48.39755
DIBENZ(A,H)ANTHRACENE (B)	0.00	0.000	0.00000
TETRACENE (TRACE) (B)	34.36	0.011	11.83325

ACCEPTOR MOLECULES	RO	CO	OVERLAP
RUBRENE (B)	42.71	0.006	43.38267
O-PHENYLENE-PYRENE	22.33	0.040	0.89358
ANTHANTHRENE (B)	9.03	0.608	0.00391
FLUORANTHENE	0.00	0.000	0.00000
3-PHENYL FLUORANTHENE	0.00	0.000	0.00000
3-AMINO FLUORANTHENE	20.33	0.053	0.50916
3,3'-BIFLUORANTHENYL (B)	0.00	0.000	0.00000
BENZO(GHI)FLUORANTHENE	0.00	0.000	0.00000
9,12-DI(N-OCTADECOXY)-BENZ(E)ACEPHENANTHRYLENE	0.00	0.000	0.00000
DECACYCLENE (B)	27.30	0.022	3.11320
10-ETHOXY BENZO(KL)XANTHENE	0.00	0.000	0.00000
PERYLENE	11.02	0.334	0.01294
BENZO(GHI)PERYLENE (B)	0.00	0.000	0.00000
BBOT (TRACE)	0.00	0.000	0.00000
QUININE SULFATE (A)	0.00	0.000	0.00000
ACRIDINE YELLOW (A)	45.43	0.005	63.11648
ACRIDINE ORANGE (A)	48.21	0.004	89.97247
ACRIDINE RED (A)	46.99	0.004	76.80904
URANINE (A)	52.95	0.003	157.89849
RHODAMINE B (A)	53.59	0.003	168.97707
RHODAMINE 6G (A)	55.62	0.003	211.58259
EOSIN B (A)	52.46	0.003	148.89537
EOSIN Y (A)	53.82	0.003	173.84104
ROSE BENGAL (A)	51.16	0.003	127.86368

8.178 Donor Molecule: Phenanthrene

TRIPHENYLAMINE	11.44	0.299	0.12933
P-XYLENE (TRACE)	0.00	0.000	0.00000
SODIUM SALICYLATE (A)	9.00	0.614	0.03070
ANTHRANILIC ACID (A)	18.53	0.070	2.32505
TMPD	14.44	0.149	0.52214
BENZIDINE	7.18	1.211	0.00791
1,3-DIBIPHENYLYLBENZENE (B)	5.28	3.038	0.00126
1,3,5,-TRIBIPHENYLYLBENZENE (B)	4.91	3.776	0.00082
FLUORENE (TRACE)	0.00	0.000	0.00000
1-METHYL FLUORENE	0.00	0.000	0.00000
1,2-BENZOFLUORENE	11.85	0.269	0.16018
9,10-DIHYDROPHENANTHRENE	0.00	0.000	0.00000
CARBAZOLE (TRACE)	6.52	1.616	0.00444
N-METHYL CARBAZOLE (TRACE)	13.74	0.172	0.38904
N-VINYLCARBAZOLE	11.47	0.296	0.13199
N-PHENYL CARBAZOLE	11.89	0.266	0.16346
N-METHYL-2,3-BENZCARBAZOLE	20.65	0.051	4.42687
DIBENZOFURAN	0.00	0.000	0.00000
DIBENZOTHIOPHENE (TRACE)	3.98	7.079	0.00023
P-TERPHENYL	0.00	0.000	0.00000
DEUTERATED P-TERPHENYL-C18D14	0.00	0.000	0.00000
4-METHYL TERPHENYL	0.00	0.000	0.00000
2-METHYL-P-TERPHENYL	0.00	0.000	0.00000
3,3''-DIMETHYL-P-TERPHENYL	0.00	0.000	0.00000
4-(3,3-DIMETHYLBUTOXY)-P-TERPHENYL	0.00	0.000	0.00000
4,4''-DIHEXAHYDROFARNESOXY-P-TERPHENYL	4.77	4.128	0.00068
2,2'-METHYLENE-P-TERPHENYL	0.00	0.000	0.00000
2,2'-ETHYLENE-P-TERPHENYL	4.06	6.689	0.00026
2,2'-METHYLENE-5',6''-METHYLENE-P-TERPHENYL (TRACE)	11.98	0.260	0.17106
3-PHENYL DIBENZOFURAN	0.00	0.000	0.00000
BENZO(1,2-B-4,5-B')BISBENZOFURAN (TRACE) (B)	16.06	0.108	0.99372
SEXIPHENYL BENZENE	0.00	0.000	0.00000
P-QUATERPHENYL	7.26	1.171	0.00847
3,3'''-DIMETHYL-P-QUATERPHENYL	10.26	0.414	0.06776
TETRAMETHYL-P-QUATERPHENYL	0.00	0.000	0.00000
3,3''' -DIMETHYL-3',2''-METHYLENE-P-QUATERPHENYL	16.51	0.099	1.17091
3,3''' -DIMETHYL-3',2''-ETHYLENE-P-QUATERPHENYL	17.76	0.080	1.80922
4-(3-ETHYLBUTOXY)-P-QUATERPHENYL (B)	15.76	0.114	0.88330
4,4'''-DI(2-BUTYLOCTOXY-1)-P-QUATERPHENYL	15.90	0.111	0.93491
DIETHYL-P-QUINQUEPHENYL	9.95	0.454	0.05626
DI(3-ETHYLHEPTYL)-P-QUINQUEPHENYL	17.30	0.086	1.54466
TETRAMETHYL-P-SEXIPHENYL	7.47	1.074	0.01007
TETRAMETHYL-P-OCTAPHENYL (B)	21.84	0.043	6.24792
INDOLE (TRACE)	0.00	0.000	0.00000
2-PHENYL INDOLE	7.21	1.194	0.00814
1-METHYL-2-PHENYL INDOLE	6.38	1.720	0.00392
1-NEOPENTYL-2-PHENYL INDOLE	0.00	0.000	0.00000
1-N-BUTYL-2(BIPHENYLYL-4) INDOLE	17.60	0.082	1.71444
1,2-DIPHENYL INDOLE	6.61	1.551	0.00482
1-BIPHENYLYL-2-PHENYL INDOLE	11.33	0.308	0.12218
1,3-DIMETHYL-2-PHENYL INDOLE	7.26	1.170	0.00848
2,3-DIPHENYL INDOLE	16.21	0.105	1.04439
1-METHYL-2,3-DIPHENYL INDOLE	14.40	0.150	0.51378
2-PHENYL-3,2'-METHYLENE INDOLE (TRACE)	11.79	0.273	0.15557
2-PHENYL-3,2'-ETHYLENE INDOLE (TRACE)	19.60	0.059	3.27315
2-PHENYL-3,2'-TRIMETHYLENE INDOLE	11.08	0.329	0.10727
1-METHYL-3,2'-METHYLENE-2-PHENYL INDOLE (TRACE)	18.10	0.075	2.02950
1-METHYL-3,2'-ETHYLENE-2-PHENYL INDOLE	21.35	0.046	5.45670
1-METHYL-3,2'-TRIMETHYLENE-2-PHENYL INDOLE	12.21	0.246	0.19145
4,5-DIPHENYLIMIDAZOLE (A)	5.78	2.314	0.00217
PPF (TRACE)	18.96	0.066	2.68804
PPO (TRACE)	9.74	0.484	0.04959
2-(4-FLUOROPHENYL)-5-PHENYLOXAZOLE (TRACE)	8.80	0.655	0.02701
αNPO (TRACE)	24.86	0.029	13.53503

ACCEPTOR MOLECULES	RO	CO	OVERLAP
BBO (B)	29.85	0.017	40.52296
POPOP	30.75	0.015	48.39139
DIMETHYL POPOP (TRACE)	32.37	0.013	65.70175
PPD (TRACE)	0.00	0.000	0.00000
2,5-DI(2-METHYLPHENYL)OXADIAZOLE	0.00	0.000	0.00000
PBD	7.70	0.981	0.01207
BBD (B)	20.50	0.052	4.28740
αNPD (TRACE)	18.13	0.075	2.04897
βNPD	10.60	0.376	0.08206
αNND (TRACE)	22.19	0.041	6.87082
βNND (TRACE)	16.55	0.099	1.18701
TRANS-1-(4-BIPHENYLYL)-2-PHENYL-ETHYLENE	21.20	0.047	5.23859
TRANS-1-(4-BIPHENYLYL)-2-(1-NAPHTHYL)-ETHYLENE	26.84	0.023	21.43846
DIPHENYLSTILBENE (B)	30.03	0.017	41.97735
1,1,4,4-TETRAPHENYLBUTADIENE	28.88	0.019	33.17751
1,6-DIPHENYLHEXATRIENE	32.86	0.013	71.97714
1,8-DIPHENYL-1,3,5,7-OCTATETRAENE	35.22	0.010	108.51547
BPSB (TRACE)	33.42	0.012	79.74349
BIS-MSB	31.68	0.014	57.85973
ESCULIN (A)	22.54	0.039	7.52016
AZULENE	13.71	0.174	0.37626
NAPHTHALENE	0.00	0.000	0.00000
DEUTERATED NAPHTHALENE-C10D8	2.02	54.015	0.00000
1-METHYLNAPHTHALENE 5ML/L	0.00	0.000	0.00000
2-METHYL NAPHTHALENE	0.00	0.000	0.00000
2,3-DIMETHYL NAPHTHALENE	0.00	0.000	0.00000
2,6-DIMETHYL NAPHTHALENE (TRACE)	0.00	0.000	0.00000
1-NAPHTHOL (TRACE)	0.00	0.000	0.00000
2-NAPHTHOL (TRACE)	4.07	6.651	0.00026
1-NAPHTHYLAMINE	12.00	0.259	0.17262
2-PHENYL NAPHTHALENE	5.06	3.457	0.00097
1,4-DIPHENYL NAPHTHALENE	8.51	0.725	0.02206
1,5-DIPHENYL NAPHTHALENE	5.60	2.550	0.00179
1,7-DIPHENYL NAPHTHALENE	6.37	1.734	0.00386
1,4,5,8-TETRAPHENYL NAPHTHALENE	24.22	0.031	11.54647
1,1'-BINAPHTHYL	0.00	0.000	0.00000
4,4'-DI(N-BUTOXY)-1,1'-BINAPHTHYL	13.81	0.170	0.40119
2,2'-BINAPHTHYL	6.64	1.525	0.00499
ACRIDINE (A)	23.10	0.036	8.69493
ANTHRACENE (TRACE)	21.72	0.044	6.01668
DEUTERATED ANTHRACENE C14D10 (TRACE)	21.49	0.045	5.64266
9-METHYL ANTHRACENE (TRACE)	23.05	0.037	8.57884
1-AMINO ANTHRACENE	22.26	0.041	6.91011
2-AMINO ANTHRACENE	21.08	0.048	4.97863
9-VINYL ANTHRACENE	23.89	0.033	10.60648
9-PHENYL ANTHRACENE (TRACE)	24.10	0.032	11.19949
9,10-DIPHENYL ANTHRACENE	25.08	0.028	14.20537
9,10-DICHLOROANTHRACENE (TRACE)	24.11	0.032	11.19871
BIANTHRANYL (B)	26.36	0.024	19.13013
9,10-DINAPHTHYLANTHRACENE (B)	24.59	0.030	12.59225
9-(P-VINYLPHENYL)-10-PHENYL ANTHRACENE	25.26	0.028	14.79244
9,10-BIS(PHENYL-ETHYNYL)ANTHRACENE (TRACE)	24.20	0.032	11.37921
RUBICENE (B)	23.63	0.034	9.86428
DIBENZ(A,H)ANTHRACENE (B)	20.99	0.048	4.89369
TETRACENE (TRACE) (B)	18.46	0.071	2.24640
RUBRENE (B)	17.22	0.088	1.47891
TETRAHYDROCANNABINOL ACID METHYL ESTER	0.00	0.000	0.00000
PHENANTHRENE	8.77	0.662	0.02637
3,4-BENZOPHENANTHRENE	12.72	0.218	0.24232
2-PHENYL PHENANTHRENE	10.81	0.354	0.09190
CHRYSENE (TRACE)	12.86	0.210	0.26002
TRIPHENYLENE	7.24	1.177	0.00835
DEUTERATED TRIPHENYLENE-C18D12	7.17	1.211	0.00789
PYRENE	14.43	0.149	0.51888
1,3,6,8,-TETRAPHENYL PYRENE	31.29	0.015	53.40978
BIPYRENYL (B)	29.86	0.017	40.43010
O-PHENYLENE-PYRENE	26.96	0.023	21.75108
ANTHANTHRENE (B)	25.46	0.027	15.46255
FLUORANTHENE	21.89	0.043	6.27330
3-PHENYL FLUORANTHENE	25.74	0.026	16.55003
3-AMINO FLUORANTHENE	23.89	0.033	10.54144
3,3'-BIFLUORANTHENYL (B)	29.50	0.017	37.46171
BENZO(GHI)FLUORANTHENE	20.06	0.055	3.70609
9,12-DI(N-OCTADECOXY)-BENZ(E)ACEPHENANTHRYLENE	25.62	0.027	16.07242
DECACYCLENE (B)	31.74	0.014	57.98951
10-ETHOXY-BENZO(KL)XANTHENE	25.59	0.027	16.00924
PERYLENE	25.13	0.028	14.26369
BENZO(GHI)PERYLENE (B)	27.25	0.022	23.36429
BBOT (TRACE)	33.53	0.012	81.08156
QUININE SULFATE (A)	14.79	0.138	0.60549
ACRIDINE YELLOW (A)	21.86	0.043	6.18716
ACRIDINE ORANGE (A)	18.40	0.072	2.20340
ACRIDINE RED (A)	19.61	0.059	3.22547
URANINE (A)	19.50	0.060	3.11260
RHODAMINE B (A)	22.42	0.040	7.19408
RHODAMINE 6G (A)	21.82	0.043	6.11352
EOSIN B (A)	19.03	0.065	2.69487
EOSIN Y (A)	19.74	0.058	3.35562
ROSE BENGAL (A)	20.07	0.055	3.70097
INDENOINDENE	9.65	0.497	0.04684

8.179 Donor Molecule: 3,4-Benzophenanthrene

ACCEPTOR MOLECULES	RO	CO	OVERLAP
ANTHRANILIC ACID (A)	10.46	0.391	0.08083
TMPD	5.67	2.455	0.00206
N-METHYL-2,3-BENZCARBAZOLE	20.51	0.052	4.57483
DI(3-ETHYLHEPTYL)-P-QUINQUEPHENYL	5.15	3.268	0.00116
TETRAMETHYL-P-SEXIPHENYL	0.00	0.000	0.00000
TETRAMETHYL-P-OCTAPHENYL (B)	8.29	0.785	0.02012
INDOLE (TRACE)	0.00	0.000	0.00000
2-PHENYL INDOLE	0.00	0.000	0.00000
1-METHYL-2-PHENYL INDOLE	0.00	0.000	0.00000
1-NEOPENTYL-2-PHENYL INDOLE	0.00	0.000	0.00000
1-N-BUTYL-2(BIPHENYLYL-4) INDOLE	5.01	3.563	0.00098
1,2-DIPHENYL INDOLE	0.00	0.000	0.00000
1-BIPHENYLYL-2-PHENYL INDOLE	0.00	0.000	0.00000
1,3-DIMETHYL-2-PHENYL INDOLE	0.00	0.000	0.00000
2,3-DIPHENYL INDOLE	6.40	1.708	0.00425
1-METHYL-2,3-DIPHENYL INDOLE	3.91	7.495	0.00022
2-PHENYL-3,2'-METHYLENE INDOLE (TRACE)	0.00	0.000	0.00000
2-PHENYL-3,2'-ETHYLENE INDOLE (TRACE)	0.00	0.000	0.00000
2-PHENYL-3,2'-TRIMETHYLENE INDOLE	0.00	0.000	0.00000
1-METHYL-3,2'-METHYLENE-2-PHENYL INDOLE (TRACE)	0.00	0.000	0.00000
1-METHYL-3,2'-ETHYLENE-2-PHENYL INDOLE	6.30	1.791	0.00387
αNPO (TRACE)	13.46	0.183	0.36994
BBO (B)	20.27	0.054	4.27403
POPOP	23.89	0.033	11.45828
DIMETHYL POPOP (TRACE)	27.29	0.022	25.42155
PPD (TRACE)	0.00	0.000	0.00000
BBD (B)	0.00	0.000	0.00000
αNPD (TRACE)	6.06	2.013	0.00306
βNPD	0.00	0.000	0.00000
αNND (TRACE)	8.54	0.718	0.02403
βNND (TRACE)	0.00	0.000	0.00000
TRANS-1-(4-BIPHENYLYL)-2-PHENYL-ETHYLENE	5.78	2.319	0.00231
TRANS-1-(4-BIPHENYLYL)-2-(1-NAPHTHYL)-ETHYLENE	18.28	0.073	2.30298
DIPHENYLSTILBENE (B)	19.87	0.057	3.79859
1,1,4,4-TETRAPHENYLBUTADIENE	21.00	0.048	5.27911
1,6-DIPHENYLHEXATRIENE	24.73	0.030	14.07421
1,8-DIPHENYL-1,3,5,7-OCTATETRAENE	33.45	0.012	85.68428
BPSB (TRACE)	26.47	0.024	21.18431
BIS-MSB	24.30	0.031	12.66758
ESCULIN (A)	14.36	0.151	0.54267
AZULENE	8.49	0.732	0.02269
1,4,5,8-TETRAPHENYL NAPHTHALENE	16.19	0.105	1.11023
ACRIDINE (A)	17.88	0.078	2.01133
ANTHRACENE (TRACE)	17.04	0.090	1.51299
DEUTERATED ANTHRACENE C14D10 (TRACE)	16.50	0.100	1.24403
9-METHYL ANTHRACENE (TRACE)	20.05	0.055	4.00619
1-AMINO ANTHRACENE	22.62	0.039	8.18336
2-AMINO ANTHRACENE	21.90	0.043	6.74454
9-VINYL ANTHRACENE	21.71	0.044	6.44220
9-PHENYL ANTHRACENE (TRACE)	20.97	0.049	5.23140
9,10-DIPHENYL ANTHRACENE	24.22	0.031	12.38361
9,10-DICHLOROANTHRACENE (TRACE)	23.99	0.032	11.71174
BIANTHRANYL (B)	25.48	0.027	16.81793
9,10-DINAPHTHYLANTHRACENE (B)	24.23	0.031	12.41727
9-(P-VINYLPHENYL)-10-PHENYL ANTHRACENE	24.62	0.030	13.64608
9,10-BIS(PHENYL-ETHYNYL)ANTHRACENE (TRACE)	30.02	0.017	44.46213
RUBICENE (B)	24.91	0.029	14.49226
DIBENZ(A,H)ANTHRACENE (B)	14.24	0.155	0.51334
TETRACENE (TRACE) (B)	22.02	0.042	6.91541
RUBRENE (B)	19.68	0.059	3.52900
TETRAHYDROCANNABINOL ACID METHYL ESTER	0.00	0.000	0.00000
PHENANTHRENE	0.00	0.000	0.00000
3,4-BENZOPHENANTHRENE	8.53	0.720	0.02384
2-PHENYL PHENANTHRENE	3.39	11.449	0.00009
CHRYSENE (TRACE)	6.23	1.847	0.00363
TRIPHENYLENE	0.00	0.000	0.00000
DEUTERATED TRIPHENYLENE-C18D12	0.00	0.000	0.00000
PYRENE	7.57	1.032	0.01160
1,3,6,8,-TETRAPHENYL PYRENE	29.18	0.018	37.83572
BIPYRENYL (B)	23.93	0.033	11.54261
O-PHENYLENE-PYRENE	25.79	0.026	17.89613
ANTHANTHRENE (B)	28.52	0.019	32.86259
FLUORANTHENE	14.75	0.139	0.63155
3-PHENYL FLUORANTHENE	21.60	0.044	6.22161
3-AMINO FLUORANTHENE	24.71	0.030	13.85028
3,3'-BIFLUORANTHENYL (B)	27.70	0.021	27.66324
BENZO(GHI)FLUORANTHENE	16.43	0.101	1.20732
9,12-DI(N-OCTADECOXY)-BENZ(E)ACEPHENANTHRYLENE	26.12	0.025	19.41665
DECACYCLENE (B)	31.55	0.014	59.99719
10-ETHOXY-BENZO(KL)XANTHENE	24.61	0.030	13.66127
PERYLENE	29.43	0.018	39.64599
BENZO(GHI)PERYLENE (B)	24.87	0.029	14.53563
BBOT (TRACE)	30.31	0.016	47.67767
QUININE SULFATE (A)	0.00	0.000	0.00000
ACRIDINE YELLOW (A)	28.02	0.020	29.38875
ACRIDINE ORANGE (A)	23.44	0.035	10.06095
ACRIDINE RED (A)	19.33	0.062	3.16771
URANINE (A)	23.35	0.035	9.85259
RHODAMINE B (A)	22.52	0.039	7.92503
RHODAMINE 6G (A)	20.72	0.050	4.80247
EOSIN B (A)	20.51	0.052	4.52267
EOSIN Y (A)	21.67	0.044	6.27915
ROSE BENGAL (A)	20.41	0.053	4.39063

8.180 Donor Molecule: 2-Phenylphenanthrene

ACCEPTOR MOLECULES	RO	CO	OVERLAP
TRIPHENYLAMINE	6.71	1.478	0.00404
P-XYLENE (TRACE)	0.00	0.000	0.00000
SODIUM SALICYLATE (A)	6.32	1.773	0.00281
ANTHRANILIC ACID (A)	16.20	0.105	0.79298
TMPD	11.54	0.291	0.10397
1,2-BENZOFLUORENE	6.60	1.553	0.00366
9,10-DIHYDROPHENANTHRENE	0.00	0.000	0.00000
CARBAZOLE (TRACE)	0.00	0.000	0.00000
N-METHYL CARBAZOLE (TRACE)	7.09	1.255	0.00561
N-VINYLCARBAZOLE	5.60	2.543	0.00137
N-PHENYL CARBAZOLE	5.96	2.110	0.00198
N-METHYL-2,3-BENZCARBAZOLE	22.27	0.041	5.31411
2,2'-METHYLENE-5',6''-METHYLENE-P-TERPHENYL (TRACE)	5.19	3.207	0.00086
3-PHENYL DIBENZOFURAN	0.00	0.000	0.00000
BENZO(1,2-B-4,5-B')BISBENZOFURAN (TRACE) (B)	7.82	0.937	0.01006
SEXIPHENYL BENZENE	0.00	0.000	0.00000
P-QUATERPHENYL	0.00	0.000	0.00000
3,3'''-DIMETHYL-P-QUATERPHENYL	4.89	3.833	0.00060
TETRAMETHYL-P-QUATERPHENYL	0.00	0.000	0.00000
3,3''' -DIMETHYL-3',2''-METHYLENE-P-QUATERPHENYL	9.18	0.578	0.02646
3,3''' -DIMETHYL-3',2''-ETHYLENE-P-QUATERPHENYL	12.07	0.254	0.13617
4-(3-ETHYLBUTOXY)-P-QUATERPHENYL (B)	10.27	0.413	0.05161
4,4''-DI(2-BUTYLOCTOXY-1)-P-QUATERPHENYL	9.99	0.449	0.04378
DIETHYL-P-QUINQUEPHENYL	4.30	5.640	0.00028
DI(3-ETHYLHEPTYL)-P-QUINQUEPHENYL	12.40	0.234	0.16004
TETRAMETHYL-P-SEXIPHENYL	0.00	0.000	0.00000
TETRAMETHYL-P-OCTAPHENYL (B)	17.03	0.090	1.07355
INDOLE (TRACE)	0.00	0.000	0.00000
2-PHENYL INDOLE	0.00	0.000	0.00000
1-METHYL-2-PHENYL INDOLE	0.00	0.000	0.00000
1-NEOPENTYL-2-PHENYL INDOLE	0.00	0.000	0.00000
1-N-BUTYL-2(BIPHENYLYL-4) INDOLE	13.24	0.193	0.23662
1,2-DIPHENYL INDOLE	0.00	0.000	0.00000
1-BIPHENYLYL-2-PHENYL INDOLE	7.55	1.038	0.00819
1,3-DIMETHYL-2-PHENYL INDOLE	0.00	0.000	0.00000
2,3-DIPHENYL INDOLE	12.56	0.226	0.17254
1-METHYL-2,3-DIPHENYL INDOLE	10.69	0.366	0.06557
2-PHENYL-3,2'-METHYLENE INDOLE (TRACE)	5.62	2.522	0.00139
2-PHENYL-3,2'-ETHYLENE INDOLE (TRACE)	12.22	0.245	0.14677
2-PHENYL-3,2' -TRIMETHYLENE INDOLE	5.31	2.989	0.00099
1-METHYL-3,2'-METHYLENE-2-PHENYL INDOLE (TRACE)	10.19	0.423	0.04922
1-METHYL-3,2'-ETHYLENE-2-PHENYL INDOLE	16.60	0.098	0.91839
1-METHYL-3,2'-TRIMETHYLENE-2-PHENYL INDOLE	6.06	2.008	0.00219
4,5-DIPHENYLIMIDAZOLE (A)	0.00	0.000	0.00000
PPF (TRACE)	10.56	0.380	0.06117
PPO (TRACE)	3.52	10.238	0.00008
2-(4-FLUOROPHENYL)-5-PHENYLOXAZOLE (TRACE)	0.00	0.000	0.00000
αNPO (TRACE)	22.08	0.042	5.07544
BBO (B)	27.94	0.021	20.81638
POPOP	30.25	0.016	33.49681
DIMETHYL POPOP (TRACE)	32.58	0.013	52.14873
PPD (TRACE)	0.00	0.000	0.00000
2,5-DI(2-METHYLPHENYL)OXADIAZOLE	0.00	0.000	0.00000
PBD	0.00	0.000	0.00000
BBD (B)	13.47	0.183	0.26326
αNPD (TRACE)	13.63	0.177	0.28192
βNPD	4.86	3.901	0.00058
αNND (TRACE)	18.46	0.071	1.74075
βNND (TRACE)	8.33	0.774	0.01476
TRANS-1-(4-BIPHENYLYL)-2-PHENYL-ETHYLENE	15.13	0.129	0.52818
TRANS-1-(4-BIPHENYLYL)-2-(1-NAPHTHYL)-ETHYLENE	24.92	0.029	10.46765
DIPHENYLSTILBENE (B)	27.90	0.021	20.61566
1,1,4,4-TETRAPHENYLBUTADIENE	27.55	0.021	19.07539
1,6-DIPHENYLHEXATRIENE	31.97	0.014	46.60959
1,8-DIPHENYL-1,3,5,7-OCTATETRAENE	37.07	0.009	112.68724
BPSB (TRACE)	32.99	0.012	56.30186
BIS-MSB	30.77	0.015	37.06704
ESCULIN (A)	20.48	0.052	3.22720
AZULENE	10.08	0.437	0.04514
NAPHTHALENE	0.00	0.000	0.00000
DEUTERATED NAPHTHALENE-C10D8	0.00	0.000	0.00000
1-NAPHTHYLAMINE	6.95	1.334	0.00496
2-CHLORONAPHTHALENE	0.00	0.000	0.00000
ACENAPHTHENE	0.00	0.000	0.00000
1-PHENYL NAPHTHALENE	0.00	0.000	0.00000
2-PHENYL NAPHTHALENE	0.00	0.000	0.00000
1,4-DIPHENYL NAPHTHALENE	4.05	6.712	0.00020
1,5-DIPHENYL NAPHTHALENE	0.00	0.000	0.00000
1,7-DIPHENYL NAPHTHALENE	0.00	0.000	0.00000
1,4,5,8-TETRAPHENYL NAPHTHALENE	22.21	0.041	5.23438
1,1'-BINAPHTHYL	0.00	0.000	0.00000
4,4' -DI(N-BUTOXY)-1,1' -BINAPHTHYL	8.80	0.656	0.02052
2,2'-BINAPHTHYL	0.00	0.000	0.00000
ACRIDINE (A)	22.38	0.040	5.49140
ANTHRACENE (TRACE)	21.59	0.044	4.43419
DEUTERATED ANTHRACENE C14D10 (TRACE)	21.11	0.048	3.87275
9-METHYL ANTHRACENE (TRACE)	23.49	0.034	7.34184
1-AMINO ANTHRACENE	23.97	0.032	8.21111
2-AMINO ANTHRACENE	22.70	0.038	5.92536
9-VINYL ANTHRACENE	24.72	0.030	9.94548
9-PHENYL ANTHRACENE (TRACE)	24.65	0.030	9.79641
9,10-DIPHENYL ANTHRACENE	26.60	0.024	15.41278
9,10-DICHLOROANTHRACENE (TRACE)	26.06	0.025	13.64121

ACCEPTOR MOLECULES	RO	CO	OVERLAP
BIANTHRANYL (B)	27.89	0.021	20.48320
9,10-DINAPHTHYLANTHRACENE (B)	26.38	0.024	14.66585
9-(P-VINYLPHENYL)-10-PHENYL ANTHRACENE	26.91	0.023	16.51842
9,10-BIS(PHENYL-ETHYNYL)ANTHRACENE (TRACE)	28.50	0.019	23.04520
RUBICENE (B)	25.70	0.026	12.39626
DIBENZ(A,H)ANTHRACENE (B)	17.70	0.081	1.33964
TETRACENE (TRACE) (B)	21.15	0.047	3.84893
RUBRENE (B)	18.85	0.067	1.92978
TETRAHYDROCANNABINOL ACID METHYL ESTER	0.00	0.000	0.00000
PHENANTHRENE	4.80	4.042	0.00054
3,4-BENZOPHENANTHRENE	11.95	0.263	0.12746
2-PHENYL PHENANTHRENE	8.62	0.699	0.01799
CHRYSENE (TRACE)	11.57	0.289	0.10519
TRIPHENYLENE	4.00	7.000	0.00018
DEUTERATED TRIPHENYLENE-C18D12	3.95	7.235	0.00017
PYRENE	11.95	0.262	0.12765
1,3,6,8,-TETRAPHENYL PYRENE	32.91	0.013	55.21878
BIPYRENYL (B)	29.00	0.018	25.91434
O-PHENYLENE-PYRENE	28.19	0.020	21.61372
ANTHANTHRENE (B)	28.32	0.020	22.32668
FLUORANTHENE	20.56	0.051	3.28943
3-PHENYL FLUORANTHENE	25.91	0.026	13.14955
3-AMINO FLUORANTHENE	25.71	0.026	12.43875
3,3'-BIFLUORANTHENYL (B)	30.89	0.015	37.67466
BENZO(GHI)FLUORANTHENE	18.42	0.072	1.69554
9,12-DI(N-OCTADECOXY)-BENZ(E)ACEPHENANTHRYLENE	27.28	0.022	17.89069
DECACYCLENE (B)	33.84	0.012	64.68193
10-ETHOXY-BENZO(KL)XANTHENE	27.16	0.022	17.48197
PERYLENE	28.85	0.019	24.95075
BENZO(GHI)PERYLENE (B)	28.41	0.020	22.89588
BBOT (TRACE)	34.57	0.011	74.35984
QUININE SULFATE (A)	8.44	0.744	0.01596
ACRIDINE YELLOW (A)	26.12	0.025	13.64951
ACRIDINE ORANGE (A)	21.90	0.043	4.74658
ACRIDINE RED (A)	20.13	0.055	2.86645
URANINE (A)	22.26	0.041	5.24008
RHODAMINE B (A)	23.17	0.036	6.66150
RHODAMINE 6G (A)	21.25	0.047	3.95892
EOSIN B (A)	20.17	0.054	2.89780
EOSIN Y (A)	21.24	0.047	3.95468
ROSE BENGAL (A)	20.73	0.050	3.41866
INDENOINDENE	4.24	5.870	0.00026

8.181 Donor Molecule: Chrysene (trace)

ACCEPTOR MOLECULES	RO	CO	OVERLAP
TRIPHENYLAMINE	5.89	2.189	0.00223
P-XYLENE (TRACE)	0.00	0.000	0.00000
SODIUM SALICYLATE (A)	5.44	2.785	0.00138
ANTHRANILIC ACID (A)	15.79	0.114	0.82488
TMPD	11.22	0.317	0.10618
1,2-BENZOFLUORENE	5.73	2.372	0.00190
9,10-DIHYDROPHENANTHRENE	0.00	0.000	0.00000
CARBAZOLE (TRACE)	0.00	0.000	0.00000
N-METHYL CARBAZOLE (TRACE)	5.93	2.142	0.00233
N-VINYLCARBAZOLE	4.88	3.858	0.00072
N-PHENYL CARBAZOLE	5.15	3.281	0.00100
N-METHYL-2,3-BENZCARBAZOLE	21.46	0.045	5.16989
BENZO(1,2-B-4,5-B')BISBENZOFURAN (TRACE) (B)	6.65	1.518	0.00465
SEXIPHENYL BENZENE	0.00	0.000	0.00000
P-QUATERPHENYL	0.00	0.000	0.00000
3,3'''-DIMETHYL-P-QUATERPHENYL	0.00	0.000	0.00000
TETRAMETHYL-P-QUATERPHENYL	0.00	0.000	0.00000
3,3''' -DIMETHYL-3',2''-METHYLENE-P-QUATERPHENYL	8.03	0.863	0.01438
3,3''' -DIMETHYL-3',2''-ETHYLENE-P-QUATERPHENYL	11.51	0.294	0.12400
4-(3-ETHYLBUTOXY)-P-QUATERPHENYL (B)	9.09	0.596	0.03009
4,4'''-DI(2-BUTYLOCTOXY-1)-P-QUATERPHENYL	8.67	0.686	0.02276
DIETHYL-P-QUINQUEPHENYL	0.00	0.000	0.00000
DI(3-ETHYLHEPTYL)-P-QUINQUEPHENYL	11.89	0.266	0.15047
TETRAMETHYL-P-SEXIPHENYL	0.00	0.000	0.00000
TETRAMETHYL-P-OCTAPHENYL (B)	16.51	0.099	1.08049
INDOLE (TRACE)	0.00	0.000	0.00000
2-PHENYL INDOLE	0.00	0.000	0.00000
1-METHYL-2-PHENYL INDOLE	0.00	0.000	0.00000
1-NEOPENTYL-2-PHENYL INDOLE	0.00	0.000	0.00000
1-N-BUTYL-2(BIPHENYLYL-4) INDOLE	12.84	0.211	0.23949
1,2-DIPHENYL INDOLE	0.00	0.000	0.00000
1-BIPHENYLYL-2-PHENYL INDOLE	6.53	1.610	0.00413
1,3-DIMETHYL-2-PHENYL INDOLE	0.00	0.000	0.00000
2,3-DIPHENYL INDOLE	12.14	0.250	0.17087
1-METHYL-2,3-DIPHENYL INDOLE	10.31	0.409	0.06399
2-PHENYL-3,2'-METHYLENE INDOLE (TRACE)	4.81	4.012	0.00067
2-PHENYL-3,2'-ETHYLENE INDOLE (TRACE)	11.20	0.318	0.10541
2-PHENYL-3,2'-TRIMETHYLENE INDOLE	4.48	4.971	0.00043
1-METHYL-3,2'-METHYLENE-2-PHENYL INDOLE (TRACE)	8.67	0.686	0.02275
1-METHYL-3,2'-ETHYLENE-2-PHENYL INDOLE	15.76	0.114	0.81611
1-METHYL-3,2'-TRIMETHYLENE-2-PHENYL INDOLE	5.19	3.195	0.00105
4,5-DIPHENYLIMIDAZOLE (A)	0.00	0.000	0.00000
PPF (TRACE)	9.12	0.590	0.03078
αNPO (TRACE)	21.60	0.044	5.40233
BBO (B)	27.10	0.022	21.04925
POPOP	29.49	0.017	34.89924

ACCEPTOR MOLECULES	RO	CO	OVERLAP
DIMETHYL POPOP (TRACE)	31.61	0.014	52.82160
PPD (TRACE)	0.00	0.000	0.00000
2,5-DI(2-METHYLPHENYL)OXADIAZOLE	0.00	0.000	0.00000
PBD	0.00	0.000	0.00000
BBD (B)	11.90	0.265	0.15183
αNPD (TRACE)	13.36	0.188	0.30323
βNPD	0.00	0.000	0.00000
αNND (TRACE)	18.24	0.074	1.96295
βNND (TRACE)	6.99	1.309	0.00626
TRANS-1-(4-BIPHENYLYL)-2-PHENYL-ETHYLENE	14.25	0.155	0.44664
TRANS-1-(4-BIPHENYLYL)-2-(1-NAPHTHYL)-ETHYLENE	24.21	0.032	10.67585
DIPHENYLSTILBENE (B)	27.10	0.022	20.99784
1,1,4,4-TETRAPHENYLBUTADIENE	26.74	0.023	19.35683
1,6-DIPHENYLHEXATRIENE	30.77	0.015	44.95230
1,8-DIPHENYL-1,3,5,7-OCTATETRAENE	35.79	0.010	110.64314
BPSB (TRACE)	32.09	0.014	57.84542
BIS-MSB	29.88	0.017	37.71989
ESCULIN (A)	19.97	0.056	3.37512
AZULENE	9.12	0.589	0.03024
1-METHYLNAPHTHALENE 5ML/L	0.00	0.000	0.00000
2-METHYL NAPHTHALENE	0.00	0.000	0.00000
2,3-DIMETHYL NAPHTHALENE	0.00	0.000	0.00000
2,6-DIMETHYL NAPHTHALENE (TRACE)	0.00	0.000	0.00000
1-NAPHTHOL (TRACE)	0.00	0.000	0.00000
2-NAPHTHOL (TRACE)	0.00	0.000	0.00000
1-NAPHTHYLAMINE	6.09	1.979	0.00273
2-CHLORONAPHTHALENE	0.00	0.000	0.00000
ACENAPHTHENE	0.00	0.000	0.00000
1-PHENYL NAPHTHALENE	0.00	0.000	0.00000
2-PHENYL NAPHTHALENE	0.00	0.000	0.00000
1,4-DIPHENYL NAPHTHALENE	0.00	0.000	0.00000
1,5-DIPHENYL NAPHTHALENE	0.00	0.000	0.00000
1,7-DIPHENYL NAPHTHALENE	0.00	0.000	0.00000
1,4,5,8-TETRAPHENYL NAPHTHALENE	21.59	0.044	5.37105
1,1'-BINAPHTHYL	0.00	0.000	0.00000
4,4'-DI(N-BUTOXY)-1,1'-BINAPHTHYL	7.64	1.003	0.01064
2,2'-BINAPHTHYL	0.00	0.000	0.00000
ACRIDINE (A)	21.76	0.043	5.62007
ANTHRACENE (TRACE)	20.63	0.051	4.08711
DEUTERATED ANTHRACENE C14D10 (TRACE)	19.96	0.056	3.35942
9-METHYL ANTHRACENE (TRACE)	22.86	0.037	7.59945
1-AMINO ANTHRACENE	23.23	0.036	8.25128
2-AMINO ANTHRACENE	21.98	0.042	5.92358
9-VINYL ANTHRACENE	23.89	0.033	9.82020
9-PHENYL ANTHRACENE (TRACE)	24.10	0.032	10.35992
9,10-DIPHENYL ANTHRACENE	25.56	0.027	14.71438
9,10-DICHLOROANTHRACENE (TRACE)	25.57	0.027	14.75257
BIANTHRANYL (B)	26.68	0.024	19.03175
9,10-DINAPHTHYLANTHRACENE (B)	25.56	0.027	14.71087
9-(P-VINYLPHENYL)-10-PHENYL ANTHRACENE	25.95	0.026	16.10506
9,10-BIS(PHENYL-ETHYNYL)ANTHRACENE (TRACE)	27.45	0.022	22.45659
RUBICENE (B)	24.81	0.029	12.24019
DIBENZ(A,H)ANTHRACENE (B)	16.61	0.098	1.11109
TETRACENE (TRACE) (B)	20.23	0.054	3.59775
RUBRENE (B)	17.87	0.078	1.70828
TETRAHYDROCANNABINOL ACID METHYL ESTER	0.00	0.000	0.00000
PHENANTHRENE	4.03	6.810	0.00023
3,4-BENZOPHENANTHRENE	11.53	0.291	0.12498
2-PHENYL PHENANTHRENE	8.40	0.756	0.01870
CHRYSENE (TRACE)	11.81	0.271	0.14460
TRIPHENYLENE	3.45	10.870	0.00009
DEUTERATED TRIPHENYLENE-C18D12	3.45	10.859	0.00009
PYRENE	11.74	0.277	0.13904
1,3,6,8,-TETRAPHENYL PYRENE	31.90	0.014	55.50777
BIPYRENYL (B)	28.12	0.020	26.14860
O-PHENYLENE-PYRENE	27.36	0.022	22.00875
ANTHANTHRENE (B)	27.47	0.022	22.57303
FLUORANTHENE	20.25	0.054	3.64121
3-PHENYL FLUORANTHENE	25.08	0.028	13.13243
3-AMINO FLUORANTHENE	24.89	0.029	12.45542
3,3'-BIFLUORANTHENYL (B)	29.92	0.017	37.77889
BENZO(GHI)FLUORANTHENE	17.79	0.079	1.67337
9,12-DI(N-OCTADECOXY)-BENZ(E)ACEPHENANTHRYLENE	26.41	0.024	17.87028
DECACYCLENE (B)	32.80	0.013	65.30996
10-ETHOXY-BENZO(KL)XANTHENE	26.31	0.025	17.54421
PERYLENE	28.00	0.020	25.30716
BENZO(GHI)PERYLENE (B)	27.62	0.021	23.47641
BBOT (TRACE)	33.39	0.012	73.25699
QUININE SULFATE (A)	7.38	1.114	0.00863
ACRIDINE YELLOW (A)	24.95	0.029	12.64063
ACRIDINE ORANGE (A)	20.61	0.051	4.01492
ACRIDINE RED (A)	19.41	0.061	2.80106
URANINE (A)	20.93	0.049	4.40304
RHODAMINE B (A)	22.35	0.040	6.52882
RHODAMINE 6G (A)	20.18	0.054	3.54430
EOSIN B (A)	19.13	0.064	2.56724
EOSIN Y (A)	20.19	0.054	3.55197
ROSE BENGAL (A)	19.95	0.056	3.30590

8.182 Donor Molecule: Triphenylene

ACCEPTOR MOLECULES	RO	CO	OVERLAP
TRIPHENYLAMINE	10.02	0.445	0.09492
P-XYLENE (TRACE)	0.00	0.000	0.00000
SODIUM SALICYLATE (A)	8.25	0.796	0.02965
ANTHRANILIC ACID (A)	17.07	0.090	2.30843
TMPD	13.15	0.197	0.48316
BENZIDINE	6.48	1.643	0.00698
1,3-DIBIPHENYLYLBENZENE (B)	4.04	6.784	0.00061
1,3,5,-TRIBIPHENYLYLBENZENE (B)	0.00	0.000	0.00000
FLUORENE (TRACE)	0.00	0.000	0.00000
1-METHYL FLUORENE	0.00	0.000	0.00000
1,2-BENZOFLUORENE	10.12	0.432	0.10091
9,10-DIHYDROPHENANTHRENE	0.00	0.000	0.00000
CARBAZOLE (TRACE)	5.81	2.282	0.00362
N-METHYL CARBAZOLE (TRACE)	12.06	0.255	0.28967
N-VINYLCARBAZOLE	10.10	0.434	0.10004
N-PHENYL CARBAZOLE	10.51	0.385	0.12702
N-METHYL-2,3-BENZCARBAZOLE	19.26	0.063	4.73717
DIBENZOFURAN	0.00	0.000	0.00000
DIBENZOTHIOPHENE (TRACE)	3.03	16.039	0.00007
2,2'-METHYLENE-5',6''-METHYLENE-P-TERPHENYL (TRACE)	10.05	0.441	0.09678
3-PHENYL DIBENZOFURAN	0.00	0.000	0.00000
BENZO(1,2-B-4,5-B')BISBENZOFURAN (TRACE) (B)	14.21	0.156	0.77548
SEXIPHENYL BENZENE	0.00	0.000	0.00000
P-QUATERPHENYL	6.32	1.774	0.00599
3,3'''-DIMETHYL-P-QUATERPHENYL	8.79	0.658	0.04345
TETRAMETHYL-P-QUATERPHENYL	0.00	0.000	0.00000
3,3''' -DIMETHYL-3',2''-METHYLENE-P-QUATERPHENYL	14.31	0.153	0.80529
3,3''' -DIMETHYL-3',2''-ETHYLENE-P-QUATERPHENYL	15.75	0.115	1.43005
4-(3-ETHYLBUTOXY)-P-QUATERPHENYL (B)	13.98	0.164	0.69925
4,4'''-DI(2-BUTYLOCTOXY-1)-P-QUATERPHENYL	14.07	0.160	0.72967
DIETHYL-P-QUINQUEPHENYL	8.75	0.668	0.04216
DI(3-ETHYLHEPTYL)-P-QUINQUEPHENYL	15.45	0.121	1.27265
TETRAMETHYL-P-SEXIPHENYL	6.38	1.719	0.00638
TETRAMETHYL-P-OCTAPHENYL (B)	19.78	0.058	5.60029
INDOLE (TRACE)	0.00	0.000	0.00000
2-PHENYL INDOLE	5.73	2.381	0.00333
1-METHYL-2-PHENYL INDOLE	5.63	2.502	0.00301
1-NEOPENTYL-2-PHENYL INDOLE	0.00	0.000	0.00000
1-N-BUTYL-2(BIPHENYLYL-4) INDOLE	15.90	0.111	1.51210
1,2-DIPHENYL INDOLE	5.87	2.212	0.00385
1-BIPHENYLYL-2-PHENYL INDOLE	10.20	0.422	0.10542
1,3-DIMETHYL-2-PHENYL INDOLE	6.44	1.673	0.00673
2,3-DIPHENYL INDOLE	14.63	0.143	0.91820
1-METHYL-2,3-DIPHENYL INDOLE	12.99	0.204	0.44949
2-PHENYL-3,2'-METHYLENE INDOLE (TRACE)	9.95	0.454	0.09116
2-PHENYL-3,2'-ETHYLENE INDOLE (TRACE)	17.01	0.091	2.27254
2-PHENYL-3,2' -TRIMETHYLENE INDOLE	9.50	0.522	0.06897
1-METHYL-3,2'-METHYLENE-2-PHENYL INDOLE (TRACE)	15.68	0.116	1.39698
1-METHYL-3,2'-ETHYLENE-2-PHENYL INDOLE	19.76	0.058	5.57932
1-METHYL-3,2'-TRIMETHYLENE-2-PHENYL INDOLE	10.56	0.380	0.13032
4,5-DIPHENYLIMIDAZOLE (A)	5.16	3.247	0.00179
PPF (TRACE)	16.30	0.103	1.75925
PPO (TRACE)	8.24	0.801	0.02940
2-(4-FLUOROPHENYL)-5-PHENYLOXAZOLE (TRACE)	7.50	1.061	0.01674
αNPO (TRACE)	22.79	0.038	13.06394
BBO (B)	27.65	0.021	41.62726
POPOP	28.60	0.019	50.96235
DIMETHYL POPOP (TRACE)	30.00	0.017	67.69828
PPD (TRACE)	0.00	0.000	0.00000
2,5-DI(2-METHYLPHENYL)OXADIAZOLE	0.00	0.000	0.00000
PBD	6.30	1.786	0.00591
BBD (B)	18.21	0.074	3.42507
αNPD (TRACE)	16.18	0.106	1.67745
βNPD	9.26	0.564	0.05930
αNND (TRACE)	20.39	0.053	6.72744
βNND (TRACE)	14.46	0.148	0.85982
TRANS-1-(4-BIPHENYLYL)-2-PHENYL-ETHYLENE	19.04	0.065	4.45810
TRANS-1-(4-BIPHENYLYL)-2-(1-NAPHTHYL)-ETHYLENE	24.83	0.029	21.79564
DIPHENYLSTILBENE (B)	27.76	0.021	42.63350
1,1,4,4-TETRAPHENYLBUTADIENE	26.77	0.023	34.18989
1,6-DIPHENYLHEXATRIENE	30.78	0.015	79.02894
1,8-DIPHENYL-1,3,5,7-OCTATETRAENE	32.85	0.013	116.14774
BPSB (TRACE)	31.04	0.015	83.20950
BIS-MSB	29.39	0.018	59.84895
ESCULIN (A)	20.80	0.050	7.55831
AZULENE	12.85	0.211	0.41538
NAPHTHALENE	0.00	0.000	0.00000
DEUTERATED NAPHTHALENE-C10D8	0.00	0.000	0.00000
1-METHYLNAPHTHALENE 5ML/L	0.00	0.000	0.00000
2-METHYL NAPHTHALENE	0.00	0.000	0.00000
2,3-DIMETHYL NAPHTHALENE	0.00	0.000	0.00000
2,6-DIMETHYL NAPHTHALENE (TRACE)	0.00	0.000	0.00000
1-NAPHTHOL (TRACE)	0.00	0.000	0.00000
2-NAPHTHOL (TRACE)	2.91	18.238	0.00006
1-NAPHTHYLAMINE	10.45	0.392	0.12215
2-CHLORONAPHTHALENE	0.00	0.000	0.00000
ACENAPHTHENE	0.00	0.000	0.00000
1-PHENYL NAPHTHALENE	0.00	0.000	0.00000
2-PHENYL NAPHTHALENE	0.00	0.000	0.00000
1,4-DIPHENYL NAPHTHALENE	7.28	1.162	0.01396
1,5-DIPHENYL NAPHTHALENE	4.77	4.111	0.00112
1,7-DIPHENYL NAPHTHALENE	5.64	2.492	0.00304
1,4,5,8-TETRAPHENYL NAPHTHALENE	22.36	0.040	11.63061

ACCEPTOR MOLECULES	RO	CO	OVERLAP
1,1'-BINAPHTHYL	0.00	0.000	0.00000
4,4'-DI(N-BUTOXY)-1,1'-BINAPHTHYL	12.27	0.242	0.31955
2,2'-BINAPHTHYL	5.75	2.347	0.00342
ACRIDINE (A)	21.50	0.045	9.17682
ANTHRACENE (TRACE)	20.56	0.051	7.04139
DEUTERATED ANTHRACENE C14D10 (TRACE)	20.37	0.053	6.65758
9-METHYL ANTHRACENE (TRACE)	20.95	0.049	7.85937
1-AMINO ANTHRACENE	20.55	0.052	6.94944
2-AMINO ANTHRACENE	19.53	0.060	5.11615
9-VINYL ANTHRACENE	22.07	0.042	10.70714
9-PHENYL ANTHRACENE (TRACE)	22.04	0.042	10.64434
9,10-DIPHENYL ANTHRACENE	23.50	0.034	15.59811
9,10-DICHLOROANTHRACENE (TRACE)	22.45	0.040	11.85817
BIANTHRANYL (B)	24.63	0.030	20.66523
9,10-DINAPHTHYLANTHRACENE (B)	22.99	0.037	13.65914
9-(P-VINYLPHENYL)-10-PHENYL ANTHRACENE	23.62	0.034	16.07555
9,10-BIS(PHENYL-ETHYNYL)ANTHRACENE (TRACE)	21.74	0.044	9.74475
RUBICENE (B)	21.82	0.043	9.97247
DIBENZ(A,H)ANTHRACENE (B)	19.71	0.058	5.43671
TETRACENE (TRACE) (B)	16.81	0.094	2.07977
RUBRENE (B)	15.62	0.117	1.33984
TETRAHYDROCANNABINOL ACID METHYL ESTER	0.00	0.000	0.00000
PHENANTHRENE	7.30	1.150	0.01421
3,4-BENZOPHENANTHRENE	11.85	0.269	0.25743
2-PHENYL PHENANTHRENE	9.90	0.461	0.08821
CHRYSENE (TRACE)	11.94	0.263	0.27101
TRIPHENYLENE	6.33	1.764	0.00604
DEUTERATED TRIPHENYLENE-C18D12	6.29	1.795	0.00584
PYRENE	12.90	0.208	0.43007
1,3,6,8,-TETRAPHENYL PYRENE	29.03	0.018	55.35717
BIPYRENYL (B)	27.63	0.021	41.28931
O-PHENYLENE-PYRENE	24.94	0.029	22.20585
ANTHANTHRENE (B)	23.20	0.036	14.40732
FLUORANTHENE	20.28	0.054	6.43849
3-PHENYL FLUORANTHENE	23.95	0.033	17.43810
3-AMINO FLUORANTHENE	22.01	0.042	10.48395
3,3'-BIFLUORANTHENYL (B)	27.38	0.022	38.88235
BENZO(GHI)FLUORANTHENE	18.31	0.073	3.48709
9,12-DI(N-OCTADECOXY)-BENZ(E)ACEPHENANTHRYLENE	23.69	0.034	16.30708
DECACYCLENE (B)	29.35	0.018	58.97320
10-ETHOXY-BENZO(KL)XANTHENE	23.82	0.033	16.92652
PERYLENE	22.70	0.038	12.61136
BENZO(GHI)PERYLENE (B)	24.99	0.029	22.57824
BBOT (TRACE)	31.20	0.015	85.52333
QUININE SULFATE (A)	12.90	0.209	0.43187
ACRIDINE YELLOW (A)	19.51	0.060	5.09099
ACRIDINE ORANGE (A)	16.60	0.098	1.92868
ACRIDINE RED (A)	18.11	0.075	3.25704
URANINE (A)	17.77	0.080	2.91067
RHODAMINE B (A)	20.72	0.050	7.29344
RHODAMINE 6G (A)	20.05	0.055	6.00182
EOSIN B (A)	17.33	0.086	2.50396
EOSIN Y (A)	17.98	0.077	3.11846
ROSE BENGAL (A)	18.42	0.072	3.60921
INDENOINDENE	8.26	0.794	0.02987

8.183 Donor Molecule: Deuterated Triphenylene ($C_{18}D_{12}$)

ACCEPTOR MOLECULES	RO	CO	OVERLAP
TRIPHENYLAMINE	12.03	0.257	0.20754
P-XYLENE (TRACE)	0.00	0.000	0.00000
SODIUM SALICYLATE (A)	9.22	0.571	0.04198
ANTHRANILIC ACID (A)	18.57	0.070	2.78685
TMPD	14.64	0.143	0.67151
BENZIDINE	8.69	0.681	0.02961
1,3-DIBIPHENYLYLBENZENE (B)	6.95	1.334	0.00773
1,3,5,-TRIBIPHENYLYLBENZENE (B)	7.60	1.021	0.01321
FLUORENE (TRACE)	0.00	0.000	0.00000
1-METHYL FLUORENE	0.00	0.000	0.00000
1,2-BENZOFLUORENE	11.98	0.260	0.20229
9,10-DIHYDROPHENANTHRENE	0.00	0.000	0.00000
CARBAZOLE (TRACE)	8.69	0.682	0.02949
N-METHYL CARBAZOLE (TRACE)	14.38	0.151	0.60438
N-VINYLCARBAZOLE	13.08	0.200	0.34328
N-PHENYL CARBAZOLE	13.21	0.194	0.36459
N-METHYL-2,3-BENZCARBAZOLE	19.93	0.057	4.23493
DIBENZOFURAN	0.00	0.000	0.00000
DIBENZOTHIOPHENE (TRACE)	5.66	2.467	0.00226
P-TERPHENYL	0.00	0.000	0.00000
DEUTERATED P-TERPHENYL-C18D14	0.00	0.000	0.00000
4-METHYL TERPHENYL	0.00	0.000	0.00000
2-METHYL-P-TERPHENYL	0.00	0.000	0.00000
3,3''-DIMETHYL-P-TERPHENYL	0.00	0.000	0.00000
4-(3,3-DIMETHYLBUTOXY)-P-TERPHENYL	5.79	2.301	0.00260
4,4''-DIHEXAHYDROFARNESOXY-P-TERPHENYL	7.66	0.994	0.01392
2,2'-METHYLENE-P-TERPHENYL	4.96	3.664	0.00103
2,2'-ETHYLENE-P-TERPHENYL	6.62	1.540	0.00580
2,2'-METHYLENE-5',6''-METHYLENE-P-TERPHENYL (TRACE)	15.15	0.129	0.82810
3-PHENYL DIBENZOFURAN	4.80	4.039	0.00084
BENZO(1,2-B-4,5-B')BISBENZOFURAN (TRACE) (B)	17.98	0.077	2.31137
SEXIPHENYL BENZENE	0.00	0.000	0.00000
P-QUATERPHENYL	9.55	0.514	0.05202

ACCEPTOR MOLECULES	RO	CO	OVERLAP
3,3'''-DIMETHYL-P-QUATERPHENYL	11.31	0.309	0.14377
TETRAMETHYL-P-QUATERPHENYL	0.00	0.000	0.00000
3,3''' -DIMETHYL-3',2''-METHYLENE-P-QUATERPHENYL	17.64	0.081	2.06454
3,3''' -DIMETHYL-3',2''-ETHYLENE-P-QUATERPHENYL	18.51	0.070	2.75184
4-(3-ETHYLBUTOXY)-P-QUATERPHENYL (B)	16.46	0.100	1.36125
4,4'''-DI(2-BUTYLOCTOXY-1)-P-QUATERPHENYL	16.63	0.097	1.44887
DIETHYL-P-QUINQUEPHENYL	11.59	0.287	0.16634
DI(3-ETHYLHEPTYL)-P-QUINQUEPHENYL	17.94	0.077	2.27289
TETRAMETHYL-P-SEXIPHENYL	9.89	0.463	0.06424
TETRAMETHYL-P-OCTAPHENYL (B)	22.32	0.0[illegible]8	8.42223
INDOLE (TRACE)	0.00	0.000	0.00000
2 PHENYL INDOLE	10.38	0.400	0.08572
1-METHYL-2-PHENYL INDOLE	8.20	0.811	0.02090
1-NEOPENTYL-2-PHENYL INDOLE	5.39	2.860	0.00168
1-N-BUTYL-2(BIPHENYLYL-4) INDOLE	18.04	0.076	2.35592
1,2-DIPHENYL INDOLE	8.44	0.745	0.02473
1-BIPHENYLYL-2-PHENYL INDOLE	11.92	0.264	0.19604
1,3-DIMETHYL-2-PHENYL INDOLE	9.40	0.538	0.04741
2,3-DIPHENYL INDOLE	16.59	0.098	1.42178
1-METHYL-2,3-DIPHENYL INDOLE	14.81	0.138	0.71940
2-PHENYL-3,2'-METHYLENE INDOLE (TRACE)	13.94	0.165	0.50195
2-PHENYL-3,2'-ETHYLENE INDOLE (TRACE)	19.81	0.057	4.13487
2-PHENYL-3,2'-TRIMETHYLENE INDOLE	12.93	0.207	0.32049
1-METHYL-3,2'-METHYLENE-2-PHENYL INDOLE (TRACE)	18.54	0.070	2.78066
1-METHYL-3,2'-ETHYLENE-2-PHENYL INDOLE	21.72	0.044	7.16632
1-METHYL-3,2'-TRIMETHYLENE-2-PHENYL INDOLE	13.67	0.175	0.44740
4,5-DIPHENYLIMIDAZOLE (A)	7.37	1.117	0.01102
PPF (TRACE)	19.43	0.061	3.68533
PPO (TRACE)	12.52	0.228	0.26461
2-(4-FLUOROPHENYL)-5-PHENYLOXAZOLE (TRACE)	11.60	0.286	0.16734
αNPO (TRACE)	24.77	0.029	15.67620
BBO (B)	29.59	0.017	45.51211
POPOP	30.22	0.016	51.68591
DIMETHYL POPOP (TRACE)	31.58	0.014	67.09248
PPD (TRACE)	0.00	0.000	0.00000
2,5-DI(2-METHYLPHENYL)OXADIAZOLE	0.00	0.000	0.00000
PBD	10.57	0.379	0.09580
BBD (B)	21.05	0.048	5.95224
αNPD (TRACE)	18.55	0.070	2.78347
βNPD	11.85	0.269	0.19037
αNND (TRACE)	22.40	0.040	8.61606
βNND (TRACE)	17.75	0.080	2.14028
TRANS-1-(4-BIPHENYLYL)-2-PHENYL-ETHYLENE	21.66	0.044	7.03687
TRANS-1-(4-BIPHENYLYL)-2-(1-NAPHTHYL)-ETHYLENE	26.67	0.024	24.42873
DIPHENYLSTILBENE (B)	29.78	0.017	47.27265
1,1,4,4-TETRAPHENYLBUTADIENE	28.53	0.019	36.46655
1,6-DIPHENYLHEXATRIENE	32.45	0.013	79.01874
1,8-DIPHENYL-1,3,5,7-OCTATETRAENE	34.16	0.011	106.99986
BPSB (TRACE)	32.79	0.013	84.25532
BIS-MSB	31.19	0.015	62.36026
ESCULIN (A)	22.46	0.039	8.71742
AZULENE	14.47	0.148	0.61676
NAPHTHALENE	0.00	0.000	0.00000
DEUTERATED NAPHTHALENE-C10D8	3.24	13.205	0.00008
1-METHYLNAPHTHALENE 5ML/L	0.00	0.000	0.00000
2-METHYL NAPHTHALENE	0.00	0.000	0.00000
2,3-DIMETHYL NAPHTHALENE	0.00	0.000	0.00000
2,6-DIMETHYL NAPHTHALENE (TRACE)	0.00	0.000	0.00000
1-NAPHTHOL (TRACE)	3.18	13.923	0.00007
2-NAPHTHOL (TRACE)	6.60	1.556	0.00568
1-NAPHTHYLAMINE	12.71	0.218	0.28814
2-PHENYL NAPHTHALENE	7.72	0.972	0.01457
1,4-DIPHENYL NAPHTHALENE	9.47	0.527	0.04937
1,5-DIPHENYL NAPHTHALENE	7.54	1.044	0.01260
1,7-DIPHENYL NAPHTHALENE	8.03	0.865	0.01835
1,4,5,8-TETRAPHENYL NAPHTHALENE	24.11	0.032	13.29641
1,1'-BINAPHTHYL	4.80	4.056	0.00084
4,4'-DI(N-BUTOXY)-1,1'-BINAPHTHYL	14.47	0.148	0.62711
2,2'-BINAPHTHYL	8.72	0.674	0.03021
ACRIDINE (A)	22.74	0.038	9.36650
ANTHRACENE (TRACE)	21.53	0.045	6.75477
DEUTERATED ANTHRACENE C14D10 (TRACE)	21.26	0.047	6.26874
9-METHYL ANTHRACENE (TRACE)	22.17	0.041	8.05208
1-AMINO ANTHRACENE	21.39	0.046	6.44098
2-AMINO ANTHRACENE	20.37	0.053	4.80626
9-VINYL ANTHRACENE	23.09	0.036	10.24157
9-PHENYL ANTHRACENE (TRACE)	23.21	0.036	10.57850
9,10-DIPHENYL ANTHRACENE	24.32	0.031	13.97402
9,10-DICHLOROANTHRACENE (TRACE)	23.28	0.035	10.74979
BIANTHRANYL (B)	25.51	0.027	18.61588
9,10-DINAPHTHYLANTHRACENE (B)	23.81	0.033	12.30083
9-(P-VINYLPHENYL)-10-PHENYL ANTHRACENE	24.45	0.031	14.41622
9,10-BIS(PHENYL-ETHYNYL)ANTHRACENE (TRACE)	22.66	0.038	9.09572
RUBICENE (B)	22.78	0.038	9.39362
DIBENZ(A,H)ANTHRACENE (B)	21.33	0.046	6.37067
TETRACENE (TRACE) (B)	17.52	0.083	1.94552
RUBRENE (B)	16.84	0.094	1.53161
TETRAHYDROCANNABINOL ACID METHYL ESTER	0.00	0.000	0.00000
PHENANTHRENE	8.59	0.705	0.02756
3,4-BENZOPHENANTHRENE	12.63	0.222	0.27556
2-PHENYL PHENANTHRENE	10.91	0.344	0.11531
CHRYSENE (TRACE)	12.92	0.207	0.31701
TRIPHENYLENE	7.97	0.884	0.01754
DEUTERATED TRIPHENYLENE-C18D12	7.85	0.924	0.01604
PYRENE	16.26	0.104	1.25757
1,3,6,8,-TETRAPHENYL PYRENE	30.20	0.016	51.09081
BIPYRENYL (B)	29.39	0.018	43.55033

ACCEPTOR MOLECULES	RO	CO	OVERLAP
O-PHENYLENE-PYRENE	26.12	0.025	21.37515
ANTHANTHRENE (B)	24.30	0.031	13.83713
FLUORANTHENE	21.68	0.044	7.01051
3-PHENYL FLUORANTHENE	25.15	0.028	17.04598
3-AMINO FLUORANTHENE	23.02	0.037	10.01030
3,3'-BIFLUORANTHENYL (B)	28.57	0.019	36.59133
BENZO(GHI)FLUORANTHENE	19.98	0.056	4.29241
9,12-DI(N-OCTADECOXY)-BENZ(E)ACEPHENANTHRYLENE	24.81	0.029	15.69496
DECACYCLENE (B)	30.65	0.016	55.71394
10-ETHOXY-BENZO(KL)XANTHENE	24.75	0.029	15.54154
PERYLENE	23.60	0.034	11.60942
BENZO(GHI)PERYLENE (B)	26.16	0.025	21.65692
BBOT (TRACE)	32.61	0.013	81.28611
QUININE SULFATE (A)	15.61	0.118	0.98979
ACRIDINE YELLOW (A)	20.39	0.053	4.83033
ACRIDINE ORANGE (A)	17.32	0.086	1.81548
ACRIDINE RED (A)	19.19	0.063	3.35816
URANINE (A)	18.67	0.069	2.85156
RHODAMINE B (A)	21.93	0.042	7.47553
RHODAMINE 6G (A)	21.59	0.044	6.80351
EOSIN B (A)	18.60	0.070	2.78517
EOSIN Y (A)	19.21	0.063	3.37706
ROSE BENGAL (A)	19.70	0.059	3.93078
INDENOINDENE	12.06	0.255	0.21134

8.184 Donor Molecule: Pyrene

ANTHRANILIC ACID (A)	13.55	0.180	0.14338
TMPD	8.03	0.864	0.00623
N-METHYL-2,3-BENZCARBAZOLE	24.46	0.031	4.93832
DIBENZOFURAN	0.00	0.000	0.00000
3,3''' -DIMETHYL-3',2''-ETHYLENE-P-QUATERPHENYL	6.83	1.404	0.00237
4-(3-ETHYLBUTOXY)-P-QUATERPHENYL (B)	4.28	5.694	0.00014
4,4'''-DI(2-BUTYLOCTOXY-1)-P-QUATERPHENYL	0.00	0.000	0.00000
DIETHYL-P-QUINQUEPHENYL	0.00	0.000	0.00000
DI(3-ETHYLHEPTYL)-P-QUINQUEPHENYL	7.92	0.899	0.00576
TETRAMETHYL-P-SEXIPHENYL	0.00	0.000	0.00000
TETRAMETHYL-P-OCTAPHENYL (B)	11.76	0.275	0.06140
INDOLE (TRACE)	0.00	0.000	0.00000
2-PHENYL INDOLE	0.00	0.000	0.00000
1-METHYL-2-PHENYL INDOLE	0.00	0.000	0.00000
1-NEOPENTYL-2-PHENYL INDOLE	0.00	0.000	0.00000
1-N-BUTYL-2(BIPHENYLYL-4) INDOLE	8.38	0.761	0.00804
1,2-DIPHENYL INDOLE	0.00	0.000	0.00000
1-BIPHENYLYL-2-PHENYL INDOLE	0.00	0.000	0.00000
1,3-DIMETHYL-2-PHENYL INDOLE	0.00	0.000	0.00000
2,3-DIPHENYL INDOLE	8.86	0.642	0.01129
1-METHYL-2,3-DIPHENYL INDOLE	6.63	1.537	0.00197
2-PHENYL-3,2'-METHYLENE INDOLE (TRACE)	0.00	0.000	0.00000
2-PHENYL-3,2'-ETHYLENE INDOLE (TRACE)	6.10	1.968	0.00120
2-PHENYL-3,2'-TRIMETHYLENE INDOLE	0.00	0.000	0.00000
1-METHYL-3,2'-METHYLENE-2-PHENYL INDOLE (TRACE)	0.00	0.000	0.00000
1-METHYL-3,2'-ETHYLENE-2-PHENYL INDOLE	10.07	0.438	0.02432
αNPO (TRACE)	17.86	0.079	0.75166
BBO (B)	25.69	0.026	6.66470
POPOP	29.76	0.017	16.07550
DIMETHYL POPOP (TRACE)	33.41	0.012	32.12560
PPD (TRACE)	0.00	0.000	0.00000
2,5-DI(2-METHYLPHENYL)OXADIAZOLE	0.00	0.000	0.00000
PBD	0.00	0.000	0.00000
BBD (B)	5.35	2.922	0.00055
αNPD (TRACE)	9.19	0.576	0.01404
βNPD	0.00	0.000	0.00000
αNND (TRACE)	12.71	0.218	0.09791
βNND (TRACE)	0.00	0.000	0.00000
TRANS-1-(4-BIPHENYLYL)-2-PHENYL-ETHYLENE	9.07	0.600	0.01293
TRANS-1-(4-BIPHENYLYL)-2-(1-NAPHTHYL)-ETHYLENE	23.06	0.036	3.48025
DIPHENYLSTILBENE (B)	25.28	0.028	6.04392
1,1,4,4-TETRAPHENYLBUTADIENE	26.25	0.025	7.55223
1,6-DIPHENYLHEXATRIENE	30.97	0.015	20.40642
1,8-DIPHENYL-1,3,5,7-OCTATETRAENE	39.98	0.007	93.84724
BPSB (TRACE)	32.79	0.013	28.74681
BIS-MSB	30.16	0.016	17.37739
ESCULIN (A)	18.33	0.073	0.87916
AZULENE	13.07	0.200	0.11244
1,4,5,8-TETRAPHENYL NAPHTHALENE	20.38	0.053	1.65444
1,1'-BINAPHTHYL	0.00	0.000	0.00000
4,4'-DI(N-BUTOXY)-1,1'-BINAPHTHYL	0.00	0.000	0.00000
2,2'-BINAPHTHYL	0.00	0.000	0.00000
ACRIDINE (A)	22.07	0.042	2.67098
ANTHRACENE (TRACE)	21.30	0.046	2.16193
DEUTERATED ANTHRACENE C14D10 (TRACE)	20.59	0.051	1.76319
9-METHYL ANTHRACENE (TRACE)	24.40	0.031	4.87559
1-AMINO ANTHRACENE	26.52	0.024	7.96680
2-AMINO ANTHRACENE	25.48	0.027	6.27163
9-VINYL ANTHRACENE	26.27	0.025	7.58437
9-PHENYL ANTHRACENE (TRACE)	25.67	0.026	6.61283
9,10-DIPHENYL ANTHRACENE	28.94	0.018	13.55062
9,10-DICHLOROANTHRACENE (TRACE)	28.44	0.019	12.18065
BIANTHRANYL (B)	30.55	0.016	18.72325
9,10-DINAPHTHYLANTHRACENE (B)	28.78	0.019	13.09030

ACCEPTOR MOLECULES	RO	CO	OVERLAP
9-(P-VINYLPHENYL)-10-PHENYL ANTHRACENE	29.35	0.018	14.69926
9,10-BIS(PHENYL-ETHYNYL)ANTHRACENE (TRACE)	33.90	0.011	34.58536
RUBICENE (B)	29.71	0.017	15.57056
DIBENZ(A,H)ANTHRACENE (B)	17.17	0.088	0.59115
TETRACENE (TRACE) (B)	25.22	0.028	5.86002
RUBRENE (B)	24.91	0.029	5.41350
TETRAHYDROCANNABINOL ACID METHYL ESTER	0.00	0.000	0.00000
PHENANTHRENE	0.00	0.000	0.00000
3,4-BENZOPHENANTHRENE	10.91	0.345	0.03895
2-PHENYL PHENANTHRENE	5.55	[illegible]	0.00068
CHRYSENE (TRACE)	8.75	0.668	0.01042
TRIPHENYLENE	0.00	0.000	0.00000
DEUTERATED TRIPHENYLENE-C18D12	0.00	0.000	0.00000
PYRENE	10.03	0.444	0.02358
1,3,6,8,-TETRAPHENYL PYRENE	35.21	0.010	43.86280
BIPYRENYL (B)	29.22	0.018	14.36119
O-PHENYLENE-PYRENE	30.35	0.016	17.84598
ANTHANTHRENE (B)	32.37	0.013	26.37657
FLUORANTHENE	18.51	0.071	0.92756
3-PHENYL FLUORANTHENE	26.38	0.024	7.75211
3-AMINO FLUORANTHENE	28.62	0.019	12.55349
3,3'-BIFLUORANTHENYL (B)	33.10	0.012	30.19750
BENZO(GHI)FLUORANTHENE	19.52	0.060	1.27194
9,12-DI(N-OCTADECOXY)-BENZ(E)ACEPHENANTHRYLENE	30.45	0.016	18.29367
DECACYCLENE (B)	37.09	0.009	59.43741
10-ETHOXY-BENZO(KL)XANTHENE	29.42	0.018	14.95374
PERYLENE	33.32	0.012	31.34318
BENZO(GHI)PERYLENE (B)	30.17	0.016	17.38035
BBOT (TRACE)	36.56	0.009	55.10325
QUININE SULFATE (A)	0.00	0.000	0.00000
ACRIDINE YELLOW (A)	31.86	0.014	23.79964
ACRIDINE ORANGE (A)	28.54	0.019	12.25219
ACRIDINE RED (A)	27.17	0.022	9.08775
URANINE (A)	30.02	0.017	16.59238
RHODAMINE B (A)	31.41	0.014	21.69720
RHODAMINE 6G (A)	30.80	0.015	19.31557
EOSIN B (A)	29.27	0.018	14.22924
EOSIN Y (A)	30.22	0.016	17.23778
ROSE BENGAL (A)	29.62	0.017	15.25839

8.185 Donor Molecule: 1,3,6,8-Tetraphenylpyrene

ACCEPTOR MOLECULES	RO	CO	OVERLAP
ANTHRANILIC ACID (A)	8.29	0.786	0.00267
TMPD	0.00	0.000	0.00000
N-METHYL-2,3-BENZCARBAZOLE	26.74	0.023	2.98905
DIBENZOFURAN	0.00	0.000	0.00000
αNPO (TRACE)	9.33	0.550	0.00542
BBO (B)	18.31	0.073	0.30912
POPOP	24.25	0.031	1.66764
DIMETHYL POPOP (TRACE)	31.97	0.014	8.74062
TRANS-1-(4-BIPHENYLYL)-2-(1-NAPHTHYL)-ETHYLENE	17.66	0.081	0.24874
DIPHENYLSTILBENE (B)	18.13	0.075	0.29166
1,1,4,4-TETRAPHENYLBUTADIENE	22.22	0.041	0.98633
1,6-DIPHENYLHEXATRIENE	23.89	0.033	1.52293
1,8-DIPHENYL-1,3,5,7-OCTATETRAENE	43.39	0.005	54.31266
BPSB (TRACE)	28.53	0.019	4.41439
BIS-MSB	26.42	0.024	2.78206
ESCULIN (A)	12.95	0.206	0.03882
AZULENE	14.69	0.141	0.08081
1,4,5,8-TETRAPHENYL NAPHTHALENE	16.84	0.094	0.18649
1,1'-BINAPHTHYL	0.00	0.000	0.00000
4,4'-DI(N-BUTOXY)-1,1'-BINAPHTHYL	0.00	0.000	0.00000
2,2'-BINAPHTHYL	0.00	0.000	0.00000
ACRIDINE (A)	19.90	0.057	0.50930
ANTHRACENE (TRACE)	15.21	0.127	0.10163
DEUTERATED ANTHRACENE C14D10 (TRACE)	14.03	0.162	0.06269
9-METHYL ANTHRACENE (TRACE)	23.17	0.036	1.26810
1-AMINO ANTHRACENE	30.82	0.015	6.96765
2-AMINO ANTHRACENE	30.30	0.016	6.28332
9-VINYL ANTHRACENE	26.83	0.023	3.05134
9-PHENYL ANTHRACENE (TRACE)	24.15	0.032	1.62488
9,10-DIPHENYL ANTHRACENE	31.48	0.014	7.95399
9,10-DICHLOROANTHRACENE (TRACE)	31.74	0.014	8.35512
BIANTHRANYL (B)	32.99	0.012	10.52020
9,10-DINAPHTHYLANTHRACENE (B)	31.90	0.014	8.60346
9-(P-VINYLPHENYL)-10-PHENYL ANTHRACENE	32.18	0.013	9.05736
9,10-BIS(PHENYL-ETHYNYL)ANTHRACENE (TRACE)	44.53	0.005	63.01260
RUBICENE (B)	37.08	0.009	20.91908
DIBENZ(A,H)ANTHRACENE (B)	18.09	0.075	0.28718
TETRACENE (TRACE) (B)	32.50	0.013	9.51286
RUBRENE (B)	32.62	0.013	9.68877
TETRAHYDROCANNABINOL ACID METHYL ESTER	0.00	0.000	0.00000
PHENANTHRENE	0.00	0.000	0.00000
3,4-BENZOPHENANTHRENE	7.29	1.153	0.00123
2-PHENYL PHENANTHRENE	0.00	0.000	0.00000
CHRYSENE (TRACE)	3.35	11.854	0.00001
TRIPHENYLENE	0.00	0.000	0.00000
DEUTERATED TRIPHENYLENE-C18D12	0.00	0.000	0.00000
PYRENE	5.80	2.290	0.00031
1,3,6,8,-TETRAPHENYL PYRENE	36.36	0.009	18.84903
BIPYRENYL (B)	28.83	0.019	4.68935

ACCEPTOR MOLECULES	RO	CO	OVERLAP
O-PHENYLENE-PYRENE	35.04	0.010	14.97838
ANTHANTHRENE (B)	41.02	0.006	38.72503
FLUORANTHENE	16.48	0.100	0.16375
3-PHENYL FLUORANTHENE	25.97	0.026	2.50243
3-AMINO FLUORANTHENE	34.67	0.011	14.05572
3,3'-BIFLUORANTHENYL (B)	35.96	0.010	17.60934
BENZO(GHI)FLUORANTHENE	21.71	0.044	0.85460
9,12-DI(N-OCTADECOXY)-BENZ(E)ACEPHENANTHRYLENE	36.03	0.010	17.81216
DECACYCLENE (B)	42.92	0.006	50.57436
10-ETHOXY-BENZO(KL)XANTHENE	31.83	0.014	8.49401
PERYLENE	42.51	0.006	47.95805
BENZO(GHI)PERYLENE (B)	29.79	0.017	5.71615
BBOT (TRACE)	38.20	0.008	25.38636
QUININE SULFATE (A)	0.00	0.000	0.00000
ACRIDINE YELLOW (A)	44.02	0.005	58.69071
ACRIDINE ORANGE (A)	38.79	0.008	27.42427
ACRIDINE RED (A)	30.21	0.016	6.11157
URANINE (A)	39.66	0.007	31.28698
RHODAMINE B (A)	34.52	0.011	13.62410
RHODAMINE 6G (A)	36.25	0.009	18.24871
EOSIN B (A)	35.08	0.010	14.98911
EOSIN Y (A)	36.47	0.009	18.91981
ROSE BENGAL (A)	32.75	0.013	9.92161

8.187 Donor Molecule: *o*-Phenylenepyrene

ACCEPTOR MOLECULES	RO	CO	OVERLAP
AZULENE	18.83	0.067	1.58113
NAPHTHALENE	0.00	0.000	0.00000
1-AMINO ANTHRACENE	5.99	2.077	0.00168
2-AMINO ANTHRACENE	5.83	2.253	0.00142
9,10-BIS(PHENYL-ETHYNYL)ANTHRACENE (TRACE)	27.16	0.022	14.44382
RUBICENE (B)	31.69	0.014	36.21877
DIBENZ(A,H)ANTHRACENE (B)	0.00	0.000	0.00000
TETRACENE (TRACE) (B)	22.27	0.041	4.38370
RUBRENE (B)	32.10	0.014	39.14752
O-PHENYLENE-PYRENE	15.16	0.128	0.43672
ANTHANTHRENE (B)	7.87	0.918	0.00859
FLUORANTHENE	0.00	0.000	0.00000
3-PHENYL FLUORANTHENE	0.00	0.000	0.00000
3-AMINO FLUORANTHENE	13.92	0.166	0.26200
3,3'-BIFLUORANTHENYL (B)	0.00	0.000	0.00000
BENZO(GHI)FLUORANTHENE	0.00	0.000	0.00000
9,12-DI(N-OCTADECOXY)-BENZ(E)ACEPHENANTHRYLENE	0.00	0.000	0.00000
DECACYCLENE (B)	18.45	0.071	1.42268
10-ETHOXY-BENZO(KL)XANTHENE	0.00	0.000	0.00000
PERYLENE	9.54	0.515	0.02730
BENZO(GHI)PERYLENE (B)	0.00	0.000	0.00000
BBOT (TRACE)	0.00	0.000	0.00000
QUININE SULFATE (A)	0.00	0.000	0.00000
ACRIDINE YELLOW (A)	30.10	0.016	26.72587
ACRIDINE ORANGE (A)	34.03	0.011	55.66638
ACRIDINE RED (A)	38.05	0.008	108.39066
URANINE (A)	38.29	0.008	112.89386
RHODAMINE B (A)	44.04	0.005	260.55603
RHODAMINE 6G (A)	43.56	0.005	244.45623
EOSIN B (A)	40.63	0.007	160.79772
EOSIN Y (A)	41.54	0.006	183.74720
ROSE BENGAL (A)	42.11	0.006	199.03999

8.189 Donor Molecule: Fluoranthene

ACCEPTOR MOLECULES	RO	CO	OVERLAP
N-METHYL-2,3-BENZCARBAZOLE	11.19	0.319	0.04806
DIBENZOFURAN	0.00	0.000	0.00000
POPOP	4.24	5.885	0.00014
DIMETHYL POPOP (TRACE)	11.16	0.322	0.04725
1,1,4,4-TETRAPHENYLBUTADIENE	7.49	1.064	0.00433
1,6-DIPHENYLHEXATRIENE	6.86	1.387	0.00254
1,8-DIPHENYL-1,3,5,7-OCTATETRAENE	20.56	0.051	1.84099
BPSB (TRACE)	7.46	1.077	0.00423
BIS-MSB	7.77	0.953	0.00539
ESCULIN (A)	0.00	0.000	0.00000
AZULENE	17.66	0.081	0.72314
1,4,5,8-TETRAPHENYL NAPHTHALENE	4.90	3.791	0.00034
1,1'-BINAPHTHYL	0.00	0.000	0.00000
4,4'-DI(N-BUTOXY)-1,1'-BINAPHTHYL	0.00	0.000	0.00000
2,2'-BINAPHTHYL	0.00	0.000	0.00000
ACRIDINE (A)	5.57	2.583	0.00073
ANTHRACENE (TRACE)	0.00	0.000	0.00000
DEUTERATED ANTHRACENE C14D10 (TRACE)	0.00	0.000	0.00000
9-METHYL ANTHRACENE (TRACE)	5.37	2.892	0.00059
1-AMINO ANTHRACENE	18.17	0.075	0.87309
2-AMINO ANTHRACENE	18.08	0.076	0.84720
9-VINYL ANTHRACENE	10.35	0.403	0.03003
9-PHENYL ANTHRACENE (TRACE)	6.86	1.384	0.00256
9,10-DIPHENYL ANTHRACENE	14.01	0.163	0.18444

ACCEPTOR MOLECULES	RO	CO	OVERLAP
9,10-DICHLOROANTHRACENE (TRACE)	15.16	0.128	0.29570
BIANTHRANYL (B)	13.58	0.179	0.15314
9,10-DINAPHTHYLANTHRACENE (B)	14.84	0.137	0.26010
9-(P-VINYLPHENYL)-10-PHENYL ANTHRACENE	14.89	0.136	0.26532
9,10-BIS(PHENYL-ETHYNYL)ANTHRACENE (TRACE)	36.37	0.009	55.88846
RUBICENE (B)	34.14	0.011	37.99661
DIBENZ(A,H)ANTHRACENE (B)	7.09	1.257	0.00309
TETRACENE (TRACE) (B)	27.69	0.021	10.88545
RUBRENE (B)	33.10	0.012	31.55768
1,3,6,8,-TETRAPHENYL PYRENE	15.83	0.113	0.38275
BIPYRENYL (B)	12.41	0.234	0.08916
O-PHENYLENE-PYRENE	23.86	0.033	4.46851
ANTHANTHRENE (B)	28.02	0.020	11.75538
FLUORANTHENE	7.01	1.296	0.00291
3-PHENYL FLUORANTHENE	12.86	0.210	0.11018
3-AMINO FLUORANTHENE	23.53	0.034	4.11096
3,3'-BIFLUORANTHENYL (B)	19.33	0.062	1.27073
BENZO(GHI)FLUORANTHENE	11.35	0.306	0.05200
9,12-DI(N-OCTADECOXY)-BENZ(E)ACEPHENANTHRYLENE	20.66	0.051	1.89234
DECACYCLENE (B)	28.85	0.019	13.94090
10-ETHOXY-BENZO(KL)XANTHENE	13.95	0.165	0.18014
PERYLENE	29.45	0.018	15.83566
BENZO(GHI)PERYLENE (B)	10.48	0.388	0.03241
BBOT (TRACE)	16.08	0.107	0.42288
QUININE SULFATE (A)	0.00	0.000	0.00000
ACRIDINE YELLOW (A)	37.49	0.008	66.90656
ACRIDINE ORANGE (A)	37.18	0.009	63.50471
ACRIDINE RED (A)	36.58	0.009	57.39609
URANINE (A)	40.23	0.007	101.98549
RHODAMINE B (A)	42.02	0.006	131.80273
RHODAMINE 6G (A)	42.80	0.006	147.42650
EOSIN B (A)	40.26	0.007	102.09144
EOSIN Y (A)	41.28	0.006	118.77318
ROSE BENGAL (A)	40.17	0.007	100.58082

8.190 Donor Molecule: 3-Phenylfluoranthene

ACCEPTOR MOLECULES	RO	CO	OVERLAP
N-METHYL-2,3-BENZCARBAZOLE	9.30	0.556	0.00730
DIBENZOFURAN	0.00	0.000	0.00000
DIMETHYL POPOP (TRACE)	9.06	0.602	0.00622
PPD (TRACE)	0.00	0.000	0.00000
1,1,4,4-TETRAPHENYLBUTADIENE	6.03	2.042	0.00054
1,6-DIPHENYLHEXATRIENE	4.74	4.190	0.00013
1,8-DIPHENYL-1,3,5,7-OCTATETRAENE	21.24	0.047	1.03066
BPSB (TRACE)	4.97	3.651	0.00017
BIS-MSB	5.63	2.507	0.00036
ESCULIN (A)	0.00	0.000	0.00000
AZULENE	19.81	0.058	0.66482
1,4,5,8-TETRAPHENYL NAPHTHALENE	3.39	11.520	0.00002
1,1'-BINAPHTHYL	0.00	0.000	0.00000
4,4'-DI(N-BUTOXY)-1,1'-BINAPHTHYL	0.00	0.000	0.00000
2,2'-BINAPHTHYL	0.00	0.000	0.00000
ACRIDINE (A)	3.85	7.828	0.00004
ANTHRACENE (TRACE)	0.00	0.000	0.00000
DEUTERATED ANTHRACENE C14D10 (TRACE)	0.00	0.000	0.00000
9-METHYL ANTHRACENE (TRACE)	3.21	13.535	0.00001
1-AMINO ANTHRACENE	20.39	0.053	0.80509
2-AMINO ANTHRACENE	20.38	0.053	0.80286
9-VINYL ANTHRACENE	8.70	0.679	0.00490
9-PHENYL ANTHRACENE (TRACE)	5.19	3.206	0.00022
9,10-DIPHENYL ANTHRACENE	12.92	0.207	0.05246
9,10-DICHLOROANTHRACENE (TRACE)	13.81	0.170	0.07816
BIANTHRANYL (B)	12.29	0.241	0.03887
9,10-DINAPHTHYLANTHRACENE (B)	13.94	0.165	0.08260
9-(P-VINYLPHENYL)-10-PHENYL ANTHRACENE	14.33	0.152	0.09728
9,10-BIS(PHENYL-ETHYNYL)ANTHRACENE (TRACE)	41.70	0.006	58.58054
RUBICENE (B)	38.95	0.008	38.65989
DIBENZ(A,H)ANTHRACENE (B)	5.51	2.674	0.00032
TETRACENE (TRACE) (B)	31.97	0.014	11.89317
RUBRENE (B)	37.65	0.008	31.52136
1,3,6,8,-TETRAPHENYL PYRENE	15.35	0.124	0.14669
BIPYRENYL (B)	11.57	0.289	0.02704
O-PHENYLENE-PYRENE	27.20	0.022	4.51476
ANTHANTHRENE (B)	31.49	0.014	10.91041
FLUORANTHENE	6.56	1.585	0.00090
3-PHENYL FLUORANTHENE	14.02	0.162	0.08551
3-AMINO FLUORANTHENE	26.86	0.023	4.19207
3,3'-BIFLUORANTHENYL (B)	21.25	0.047	1.03351
BENZO(GHI)FLUORANTHENE	12.45	0.232	0.04179
9,12-DI(N-OCTADECOXY)-BENZ(E)ACEPHENANTHRYLENE	23.28	0.035	1.78616
DECACYCLENE (B)	32.92	0.013	14.19990
10-ETHOXY-BENZO(KL)XANTHENE	12.12	0.251	0.03576
PERYLENE	33.23	0.012	15.07549
BENZO(GHI)PERYLENE (B)	9.31	0.555	0.00731
BBOT (TRACE)	13.84	0.169	0.07921
QUININE SULFATE (A)	0.00	0.000	0.00000
ACRIDINE YELLOW (A)	43.01	0.006	70.38056
ACRIDINE ORANGE (A)	42.48	0.006	65.19697
ACRIDINE RED (A)	41.30	0.006	54.78821
URANINE (A)	45.87	0.005	103.22987
RHODAMINE B (A)	47.38	0.004	124.91496

ACCEPTOR MOLECULES	RO	CO	OVERLAP
RHODAMINE 6G (A)	48.46	0.004	143.27346
EOSIN B (A)	45.65	0.005	100.01839
EOSIN Y (A)	46.82	0.004	116.65819
ROSE BENGAL (A)	45.29	0.005	95.31924

8.191 Donor Molecule: 3-Aminofluoranthene

AZULENE	21.73	0.044	1.41252
NAPHTHALENE	0.00	0.000	0.00000
1-AMINO ANTHRACENE	7.85	0.926	0.00318
2-AMINO ANTHRACENE	7.61	1.013	0.00266
9,10-BIS(PHENYL-ETHYNYL)ANTHRACENE (TRACE)	32.50	0.013	16.01784
RUBICENE (B)	37.77	0.008	39.20156
DIBENZ(A,H)ANTHRACENE (B)	0.00	0.000	0.00000
TETRACENE (TRACE) (B)	26.94	0.023	5.18911
RUBRENE (B)	38.16	0.008	41.71977
O-PHENYLENE-PYRENE	18.25	0.074	0.50182
ANTHANTHRENE (B)	10.56	0.380	0.01888
FLUORANTHENE	0.00	0.000	0.00000
3-PHENYL FLUORANTHENE	0.00	0.000	0.00000
3-AMINO FLUORANTHENE	16.93	0.092	0.32051
3,3'-BIFLUORANTHENYL (B)	0.00	0.000	0.00000
BENZO(GHI)FLUORANTHENE	0.00	0.000	0.00000
9,12-DI(N-OCTADECOXY)-BENZ(E)ACEPHENANTHRYLENE	0.00	0.000	0.00000
DECACYCLENE (B)	22.35	0.040	1.69418
10-ETHOXY-BENZO(KL)XANTHENE	0.00	0.000	0.00000
PERYLENE	12.76	0.215	0.05894
BENZO(GHI)PERYLENE (B)	0.00	0.000	0.00000
BBOT (TRACE)	0.00	0.000	0.00000
QUININE SULFATE (A)	0.00	0.000	0.00000
ACRIDINE YELLOW (A)	36.14	0.009	30.18218
ACRIDINE ORANGE (A)	40.81	0.007	62.47458
ACRIDINE RED (A)	44.70	0.005	107.44515
URANINE (A)	45.69	0.005	123.02910
RHODAMINE B (A)	51.60	0.003	254.34912
RHODAMINE 6G (A)	51.55	0.003	253.41191
EOSIN B (A)	48.21	0.004	169.39436
EOSIN Y (A)	49.32	0.004	194.33144
ROSE BENGAL (A)	49.33	0.004	194.11040

8.193 Donor Molecule: Benzo(*ghi*)fluoranthene

N-METHYL-2,3-BENZCARBAZOLE	8.84	0.648	0.01165
DIBENZOFURAN	0.00	0.000	0.00000
DIMETHYL POPOP (TRACE)	8.63	0.695	0.01012
PPD (TRACE)	0.00	0.000	0.00000
1,1,4,4-TETRAPHENYLBUTADIENE	5.87	2.216	0.00100
1,6-DIPHENYLHEXATRIENE	5.10	3.376	0.00043
1,8-DIPHENYL-1,3,5,7-OCTATETRAENE	19.92	0.057	1.51967
BPSB (TRACE)	5.39	2.859	0.00060
BIS-MSB	5.78	2.316	0.00091
ESCULIN (A)	0.00	0.000	0.00000
AZULENE	16.61	0.098	0.50336
1,4,5,8-TETRAPHENYL NAPHTHALENE	3.64	9.283	0.00006
1,1'-BINAPHTHYL	0.00	0.000	0.00000
4,4'-DI(N-BUTOXY)-1,1'-BINAPHTHYL	0.00	0.000	0.00000
2,2'-BINAPHTHYL	0.00	0.000	0.00000
ACRIDINE (A)	4.09	6.514	0.00012
ANTHRACENE (TRACE)	0.00	0.000	0.00000
DEUTERATED ANTHRACENE C14D10 (TRACE)	0.00	0.000	0.00000
9-METHYL ANTHRACENE (TRACE)	3.66	9.120	0.00006
1-AMINO ANTHRACENE	19.00	0.065	1.14293
2-AMINO ANTHRACENE	19.12	0.064	1.18504
9-VINYL ANTHRACENE	8.16	0.823	0.00721
9-PHENYL ANTHRACENE (TRACE)	5.16	3.254	0.00046
9,10-DIPHENYL ANTHRACENE	11.68	0.281	0.06181
9,10-DICHLOROANTHRACENE (TRACE)	12.55	0.226	0.09519
BIANTHRANYL (B)	11.19	0.319	0.04797
9,10-DINAPHTHYLANTHRACENE (B)	12.92	0.208	0.11317
9-(P-VINYLPHENYL)-10-PHENYL ANTHRACENE	13.17	0.196	0.12718
9,10-BIS(PHENYL-ETHYNYL)ANTHRACENE (TRACE)	37.25	0.009	64.51716
RUBICENE (B)	34.15	0.011	38.02203
DIBENZ(A,H)ANTHRACENE (B)	5.37	2.882	0.00059
TETRACENE (TRACE) (B)	28.78	0.019	13.70752
RUBRENE (B)	32.68	0.013	29.21854
1,3,6,8,-TETRAPHENYL PYRENE	14.16	0.157	0.19645
BIPYRENYL (B)	10.39	0.398	0.03076
O-PHENYLENE-PYRENE	24.70	0.030	5.48367
ANTHANTHRENE (B)	28.39	0.020	12.69247
FLUORANTHENE	5.88	2.206	0.00100
3-PHENYL FLUORANTHENE	13.30	0.190	0.13477
3-AMINO FLUORANTHENE	24.52	0.030	5.26054
3,3'-BIFLUORANTHENYL (B)	19.90	0.057	1.50937
BENZO(GHI)FLUORANTHENE	12.10	0.252	0.07655

ACCEPTOR MOLECULES	RO	CO	OVERLAP
9,12-DI(N-OCTADECOXY)-BENZ(E)ACEPHENANTHRYLENE	22.21	0.041	2.91950
DECACYCLENE (B)	29.99	0.017	17.59135
10-ETHOXY-BENZO(KL)XANTHENE	11.25	0.314	0.04945
PERYLENE	29.72	0.017	16.71408
BENZO(GHI)PERYLENE (B)	8.48	0.733	0.00908
BBOT (TRACE)	12.85	0.211	0.10967
QUININE SULFATE (A)	0.00	0.000	0.00000
ACRIDINE YELLOW (A)	38.19	0.008	74.67958
ACRIDINE ORANGE (A)	37.22	0.009	63.96169
ACRIDINE RED (A)	35.33	0.010	46.57322
URANINE (A)	39.93	0.007	97.36615
RHODAMINE B (A)	40.50	0.007	105.60483
RHODAMINE 6G (A)	41.59	0.006	124.18815
EOSIN B (A)	39.31	0.007	88.38029
EOSIN Y (A)	40.36	0.007	103.74960
ROSE BENGAL (A)	38.72	0.008	80.72262

8.194 Donor Molecule: 9,12-Di(*n*-octadecoxy)-benz(*e*)acephenanthrylene

ACCEPTOR MOLECULES	RO	CO	OVERLAP
N-METHYL-2,3-BENZCARBAZOLE	8.38	0.760	0.00403
DIBENZOFURAN	0.00	0.000	0.00000
DIMETHYL POPOP (TRACE)	8.09	0.845	0.00326
PPD (TRACE)	0.00	0.000	0.00000
1,8-DIPHENYL-1,3,5,7-OCTATETRAENE	24.24	0.031	2.34912
BPSB (TRACE)	0.00	0.000	0.00000
BIS-MSB	0.00	0.000	0.00000
ESCULIN (A)	0.00	0.000	0.00000
AZULENE	16.84	0.094	0.25985
1-AMINO ANTHRACENE	23.45	0.035	1.92105
2-AMINO ANTHRACENE	23.60	0.034	1.99226
9-VINYL ANTHRACENE	8.20	0.811	0.00353
9-PHENYL ANTHRACENE (TRACE)	0.00	0.000	0.00000
9,10-DIPHENYL ANTHRACENE	13.59	0.178	0.07324
9,10-DICHLOROANTHRACENE (TRACE)	14.20	0.156	0.09516
BIANTHRANYL (B)	12.94	0.206	0.05447
9,10-DINAPHTHYLANTHRACENE (B)	15.05	0.131	0.13464
9-(P-VINYLPHENYL)-10-PHENYL ANTHRACENE	15.71	0.115	0.17416
9,10-BIS(PHENYL-ETHYNYL)ANTHRACENE (TRACE)	44.14	0.005	84.94250
RUBICENE (B)	37.77	0.008	33.14195
DIBENZ(A,H)ANTHRACENE (B)	4.94	3.708	0.00017
TETRACENE (TRACE) (B)	33.15	0.012	15.23086
RUBRENE (B)	34.83	0.011	20.35754
1,3,6,8,-TETRAPHENYL PYRENE	16.97	0.092	0.27637
BIPYRENYL (B)	12.18	0.248	0.03788
O-PHENYLENE-PYRENE	30.01	0.017	8.39381
ANTHANTHRENE (B)	35.45	0.010	22.92681
FLUORANTHENE	6.94	1.337	0.00130
3-PHENYL FLUORANTHENE	16.34	0.103	0.22007
3-AMINO FLUORANTHENE	29.85	0.017	8.13833
3,3'-BIFLUORANTHENYL (B)	24.60	0.030	2.56260
BENZO(GHI)FLUORANTHENE	14.59	0.144	0.11174
9,12-DI(N-OCTADECOXY)-BENZ(E)ACEPHENANTHRYLENE	27.31	0.022	4.80042
DECACYCLENE (B)	36.30	0.009	26.29328
10-ETHOXY-BENZO(KL)XANTHENE	11.66	0.282	0.02919
PERYLENE	37.14	0.009	30.31080
BENZO(GHI)PERYLENE (B)	9.58	0.508	0.00899
BBOT (TRACE)	13.36	0.188	0.06607
QUININE SULFATE (A)	0.00	0.000	0.00000
ACRIDINE YELLOW (A)	44.30	0.005	86.65752
ACRIDINE ORANGE (A)	40.81	0.007	52.87263
ACRIDINE RED (A)	34.98	0.010	20.86795
URANINE (A)	42.71	0.006	69.39886
RHODAMINE B (A)	39.99	0.007	46.54349
RHODAMINE 6G (A)	41.70	0.006	59.96547
EOSIN B (A)	39.74	0.007	44.91604
EOSIN Y (A)	40.96	0.007	53.92575
ROSE BENGAL (A)	38.30	0.008	35.93434

8.196 Donor Molecule: 10-Ethoxybenzo(*kl*)xanthene

ACCEPTOR MOLECULES	RO	CO	OVERLAP
N-METHYL-2,3-BENZCARBAZOLE	23.28	0.035	1.29874
DIBENZOFURAN	0.00	0.000	0.00000
BBO (B)	8.96	0.621	0.00425
POPOP	13.43	0.185	0.04799
DIMETHYL POPOP (TRACE)	24.08	0.032	1.59001
TRANS-1-(4-BIPHENYLYL)-2-(1-NAPHTHYL)-ETHYLENE	9.61	0.504	0.00644
DIPHENYLSTILBENE (B)	9.87	0.465	0.00757
1,1,4,4-TETRAPHENYLBUTADIENE	16.40	0.101	0.15865
1,6-DIPHENYLHEXATRIENE	15.80	0.113	0.12720
1,8-DIPHENYL-1,3,5,7-OCTATETRAENE	38.55	0.008	26.66031
BPSB (TRACE)	18.41	0.072	0.31836
BIS-MSB	18.14	0.075	0.29143
ESCULIN (A)	0.00	0.000	0.00000

ACCEPTOR MOLECULES	RO	CO	OVERLAP
AZULENE	15.44	0.121	0.10860
1,4,5,8-TETRAPHENYL NAPHTHALENE	11.50	0.294	0.01888
1,1'-BINAPHTHYL	0.00	0.000	0.00000
4,4'-DI(N-BUTOXY)-1,1'-BINAPHTHYL	0.00	0.000	0.00000
2,2'-BINAPHTHYL	0.00	0.000	0.00000
ACRIDINE (A)	13.35	0.188	0.04637
ANTHRACENE (TRACE)	6.52	1.612	0.00063
DEUTERATED ANTHRACENE C14D10 (TRACE)	5.81	2.277	0.00032
9-METHYL ANTHRACENE (TRACE)	14.56	0.145	0.07771
1-AMINO ANTHRACENE	28.93	0.018	4.75448
2-AMINO ANTHRACENE	28.70	0.019	4.52546
9-VINYL ANTHRACENE	21.63	0.044	0.83540
9-PHENYL ANTHRACENE (TRACE)	16.24	0.105	0.14961
9,10-DIPHENYL ANTHRACENE	27.52	0.021	3.54023
9,10-DICHLOROANTHRACENE (TRACE)	29.42	0.018	5.28197
BIANTHRANYL (B)	27.56	0.021	3.57471
9,10-DINAPHTHYLANTHRACENE (B)	28.91	0.019	4.74796
9-(P-VINYLPHENYL)-10-PHENYL ANTHRACENE	28.56	0.019	4.42097
9,10-BIS(PHENYL-ETHYNYL)ANTHRACENE (TRACE)	46.06	0.005	77.04613
RUBICENE (B)	37.95	0.008	23.93328
DIBENZ(A,H)ANTHRACENE (B)	15.06	0.131	0.09528
TETRACENE (TRACE) (B)	33.55	0.012	11.48684
RUBRENE (B)	33.75	0.012	11.84290
TETRAHYDROCANNABINOL ACID METHYL ESTER	0.00	0.000	0.00000
PHENANTHRENE	0.00	0.000	0.00000
3,4-BENZOPHENANTHRENE	3.69	8.877	0.00002
2-PHENYL PHENANTHRENE	0.00	0.000	0.00000
CHRYSENE (TRACE)	0.00	0.000	0.00000
TRIPHENYLENE	0.00	0.000	0.00000
DEUTERATED TRIPHENYLENE-C18D12	0.00	0.000	0.00000
PYRENE	0.00	0.000	0.00000
1,3,6,8,-TETRAPHENYL PYRENE	30.61	0.016	6.69627
BIPYRENYL (B)	24.06	0.032	1.58232
O-PHENYLENE-PYRENE	33.91	0.011	12.26788
ANTHANTHRENE (B)	40.70	0.007	36.84886
FLUORANTHENE	13.50	0.182	0.04936
3-PHENYL FLUORANTHENE	22.45	0.040	1.04126
3-AMINO FLUORANTHENE	33.78	0.012	11.99216
3,3'-BIFLUORANTHENYL (B)	32.57	0.013	9.69664
BENZO(GHI)FLUORANTHENE	19.79	0.058	0.48918
9,12-DI(N-OCTADECOXY)-BENZ(E)ACEPHENANTHRYLENE	33.65	0.012	11.79033
DECACYCLENE (B)	41.25	0.006	39.75203
10-ETHOXY-BENZO(KL)XANTHENE	28.25	0.020	4.14064
PERYLENE	42.18	0.006	45.65787
BENZO(GHI)PERYLENE (B)	21.98	0.042	0.91853
BBOT (TRACE)	32.73	0.013	10.02489
QUININE SULFATE (A)	0.00	0.000	0.00000
ACRIDINE YELLOW (A)	44.84	0.005	65.35075
ACRIDINE ORANGE (A)	39.78	0.007	31.83549
ACRIDINE RED (A)	32.52	0.013	9.47377
URANINE (A)	41.02	0.006	38.24675
RHODAMINE B (A)	37.13	0.009	20.99000
RHODAMINE 6G (A)	38.82	0.008	27.40635
EOSIN B (A)	37.30	0.009	21.58508
EOSIN Y (A)	38.66	0.008	26.72824
ROSE BENGAL (A)	35.13	0.010	15.06627

8.197 Donor Molecule: Perylene

ACCEPTOR MOLECULES	RO	CO	OVERLAP
1,8-DIPHENYL-1,3,5,7-OCTATETRAENE	16.00	0.109	0.12986
BPSB (TRACE)	0.00	0.000	0.00000
BIS-MSB	0.00	0.000	0.00000
ESCULIN (A)	0.00	0.000	0.00000
AZULENE	18.36	0.072	0.29221
1-AMINO ANTHRACENE	19.79	0.058	0.46374
2-AMINO ANTHRACENE	18.79	0.067	0.33976
9-VINYL ANTHRACENE	0.00	0.000	0.00000
9-PHENYL ANTHRACENE (TRACE)	0.00	0.000	0.00000
9,10-DIPHENYL ANTHRACENE	0.00	0.000	0.00000
9,10-DICHLOROANTHRACENE (TRACE)	0.00	0.000	0.00000
BIANTHRANYL (B)	0.00	0.000	0.00000
9,10-DINAPHTHYLANTHRACENE (B)	4.50	4.919	0.00006
9-(P-VINYLPHENYL)-10-PHENYL ANTHRACENE	4.41	5.209	0.00006
9,10-BIS(PHENYL-ETHYNYL)ANTHRACENE (TRACE)	47.39	0.004	86.93497
RUBICENE (B)	41.78	0.006	40.58867
DIBENZ(A,H)ANTHRACENE (B)	0.00	0.000	0.00000
TETRACENE (TRACE) (B)	36.33	0.009	17.63370
RUBRENE (B)	38.94	0.008	26.58207
1,3,6,8,-TETRAPHENYL PYRENE	6.66	1.514	0.00068
BIPYRENYL (B)	0.00	0.000	0.00000
O-PHENYLENE-PYRENE	29.88	0.017	5.46849
ANTHANTHRENE (B)	34.49	0.011	12.99421
FLUORANTHENE	0.00	0.000	0.00000
3-PHENYL FLUORANTHENE	7.60	1.020	0.00149
3-AMINO FLUORANTHENE	29.42	0.018	4.98179
3,3'-BIFLUORANTHENYL (B)	17.57	0.083	0.22712
BENZO(GHI)FLUORANTHENE	5.69	2.426	0.00026
9,12-DI(N-OCTADECOXY)-BENZ(E)ACEPHENANTHRYLENE	16.10	0.107	0.13437
DECACYCLENE (B)	36.08	0.010	16.95799
10-ETHOXY-BENZO(KL)XANTHENE	0.00	0.000	0.00000
PERYLENE	37.85	0.008	22.68577

ACCEPTOR MOLECULES	RO	CO	OVERLAP
ACRIDINE YELLOW (A)	48.85	0.004	104.07910
ACRIDINE ORANGE (A)	45.91	0.005	71.62998
ACRIDINE RED (A)	38.63	0.008	25.32012
URANINE (A)	48.56	0.004	100.19823
RHODAMINE B (A)	43.68	0.005	52.89856
RHODAMINE 6G (A)	46.75	0.004	79.59464
EOSIN B (A)	44.50	0.005	59.19419
EOSIN Y (A)	45.86	0.005	70.93031
ROSE BENGAL (A)	41.79	0.006	40.61750

8.199 Donor Molecule: BBOT (trace)

ACCEPTOR MOLECULES	RO	CO	OVERLAP
N-METHYL-2,3-BENZCARBAZOLE	19.73	0.058	0.58525
DIBENZOFURAN	0.00	0.000	0.00000
BBO (B)	6.79	1.430	0.00097
POPOP	10.81	0.354	0.01588
DIMETHYL POPOP (TRACE)	20.39	0.053	0.71334
TRANS-1-(4-BIPHENYLYL)-2-(1-NAPHTHYL)-ETHYLENE	7.81	0.939	0.00226
DIPHENYLSTILBENE (B)	8.10	0.843	0.00280
1,1,4,4-TETRAPHENYLBUTADIENE	13.81	0.170	0.06892
1,6-DIPHENYLHEXATRIENE	13.16	0.196	0.05150
1,8-DIPHENYL-1,3,5,7-OCTATETRAENE	33.40	0.012	13.70976
BPSB (TRACE)	15.32	0.124	0.12832
BIS-MSB	15.17	0.128	0.12093
ESCULIN (A)	0.00	0.000	0.00000
AZULENE	16.37	0.102	0.18720
1,4,5,8-TETRAPHENYL NAPHTHALENE	9.59	0.507	0.00774
1,1'-BINAPHTHYL	0.00	0.000	0.00000
4,4' -DI(N-BUTOXY)-1,1' -BINAPHTHYL	0.00	0.000	0.00000
2,2'-BINAPHTHYL	0.00	0.000	0.00000
ACRIDINE (A)	11.13	0.324	0.01890
ANTHRACENE (TRACE)	5.01	3.563	0.00016
DEUTERATED ANTHRACENE C14D10 (TRACE)	0.00	0.000	0.00000
9-METHYL ANTHRACENE (TRACE)	11.97	0.261	0.02920
1-AMINO ANTHRACENE	26.32	0.025	3.27556
2-AMINO ANTHRACENE	26.18	0.025	3.17332
9-VINYL ANTHRACENE	18.38	0.072	0.38221
9-PHENYL ANTHRACENE (TRACE)	13.57	0.179	0.06194
9,10-DIPHENYL ANTHRACENE	23.65	0.034	1.73341
9,10-DICHLOROANTHRACENE (TRACE)	25.45	0.027	2.69224
BIANTHRANYL (B)	23.52	0.034	1.67805
9,10-DINAPHTHYLANTHRACENE (B)	24.88	0.029	2.34922
9-(P-VINYLPHENYL)-10-PHENYL ANTHRACENE	24.69	0.030	2.23924
9,10-BIS(PHENYL-ETHYNYL)ANTHRACENE (TRACE)	44.82	0.005	79.46409
RUBICENE (B)	38.03	0.008	29.48337
DIBENZ(A,H)ANTHRACENE (B)	12.70	0.219	0.04154
TETRACENE (TRACE) (B)	33.05	0.012	12.76821
RUBRENE (B)	34.83	0.011	17.39784
TETRAHYDROCANNABINOL ACID METHYL ESTER	0.00	0.000	0.00000
PHENANTHRENE	0.00	0.000	0.00000
3,4-BENZOPHENANTHRENE	2.89	18.553	0.00001
1,3,6,8,-TETRAPHENYL PYRENE	26.41	0.024	3.35330
BIPYRENYL (B)	20.76	0.050	0.79245
O-PHENYLENE-PYRENE	31.78	0.014	10.11098
ANTHANTHRENE (B)	38.12	0.008	30.22309
FLUORANTHENE	11.68	0.281	0.02510
3-PHENYL FLUORANTHENE	19.89	0.057	0.61208
3-AMINO FLUORANTHENE	31.60	0.014	9.78192
3,3' -BIFLUORANTHENYL (B)	29.14	0.018	6.04884
BENZO(GHI)FLUORANTHENE	17.55	0.083	0.28891
9,12-DI(N-OCTADECOXY)-BENZ(E)ACEPHENANTHRYLENE	30.56	0.016	8.03575
DECACYCLENE (B)	38.53	0.008	32.12277
10-ETHOXY-BENZO(KL)XANTHENE	24.13	0.032	1.95765
PERYLENE	39.66	0.007	38.33869
BENZO(GHI)PERYLENE (B)	18.66	0.069	0.41872
BBOT (TRACE)	27.93	0.021	4.70867
QUININE SULFATE (A)	0.00	0.000	0.00000
ACRIDINE YELLOW (A)	44.45	0.005	75.38760
ACRIDINE ORANGE (A)	40.73	0.007	44.57022
ACRIDINE RED (A)	34.78	0.011	17.24075
URANINE (A)	42.69	0.006	59.03885
RHODAMINE B (A)	39.61	0.007	37.60343
RHODAMINE 6G (A)	41.65	0.006	50.85127
EOSIN B (A)	39.72	0.007	38.26894
EOSIN Y (A)	40.99	0.006	46.17252
ROSE BENGAL (A)	37.74	0.008	28.13385

8.210 Donor Molecule: Indenoindene

ACCEPTOR MOLECULES	RO	CO	OVERLAP
N,N-DIMETHYLANILINE 3ML/L	2.59	25.848	0.00000
TRIPHENYLAMINE	18.41	0.072	0.31972
SODIUM SALICYLATE (A)	13.34	0.188	0.04616
ANTHRANILIC ACID (A)	26.06	0.025	2.55248
TMPD	20.93	0.049	0.68611
BENZIDINE	14.45	0.148	0.07488

ACCEPTOR MOLECULES	RO	CO	OVERLAP
1,3,5-TRIPHENYL BENZENE	3.21	13.536	0.00001
1,3-DIBIPHENYLYLBENZENE (B)	12.29	0.241	0.02837
1,3,5,-TRIBIPHENYLYLBENZENE (B)	13.98	0.164	0.06165
FLUORENE (TRACE)	0.00	0.000	0.00000
1-METHYL FLUORENE	0.00	0.000	0.00000
1,2-BENZOFLUORENE	17.67	0.081	0.24933
9,10-DIHYDROPHENANTHRENE	0.00	0.000	0.00000
CARBAZOLE (TRACE)	15.16	0.128	0.09999
N-METHYL CARBAZOLE (TRACE)	21.36	0.046	0.78029
N-VINYLCARBAZOLE	20.70	0.050	0.64547
N-PHENYL CARBAZOLE	20.52	0.052	0.61298
N-METHYL-2,3-BENZCARBAZOLE	28.42	0.019	4.27020
DIBENZOFURAN	0.00	0.000	0.00000
DIBENZOTHIOPHENE (TRACE)	10.70	0.365	0.01239
P-TERPHENYL	0.00	0.000	0.00000
DEUTERATED P-TERPHENYL-C18D14	0.00	0.000	0.00000
4-METHYL TERPHENYL	0.00	0.000	0.00000
2-METHYL-P-TERPHENYL	0.00	0.000	0.00000
3,3''-DIMETHYL-P-TERPHENYL	4.22	5.969	0.00005
4-(3,3-DIMETHYLBUTOXY)-P-TERPHENYL	10.87	0.348	0.01361
4,4''-DIHEXAHYDROFARNESOXY-P-TERPHENYL	14.12	0.159	0.06528
2,2'-METHYLENE-P-TERPHENYL	9.96	0.453	0.00805
2,2'-ETHYLENE-P-TERPHENYL	12.47	0.231	0.03095
2,2'-METHYLENE-5',6''-METHYLENE-P-TERPHENYL (TRACE)	26.23	0.025	2.67392
3-PHENYL DIBENZOFURAN	9.83	0.471	0.00745
BENZO(1,2-B-4,5-B')BISBENZOFURAN (TRACE) (B)	28.18	0.020	4.11245
SEXIPHENYL BENZENE	0.00	0.000	0.00000
P-QUATERPHENYL	16.52	0.099	0.16718
3,3'''-DIMETHYL-P-QUATERPHENYL	18.53	0.070	0.33225
TETRAMETHYL-P-QUATERPHENYL	0.00	0.000	0.00000
3,3''' -DIMETHYL-3',2''-METHYLENE-P-QUATERPHENYL	27.52	0.021	3.56566
3,3''' -DIMETHYL-3',2''-ETHYLENE-P-QUATERPHENYL	27.90	0.021	3.85732
4-(3-ETHYLBUTOXY)-P-QUATERPHENYL (B)	24.91	0.029	1.95597
4,4'''-DI(2-BUTYLOCTOXY-1)-P-QUATERPHENYL	25.25	0.028	2.12564
DIETHYL-P-QUINQUEPHENYL	19.21	0.063	0.41354
DI(3-ETHYLHEPTYL)-P-QUINQUEPHENYL	26.88	0.023	3.08209
TETRAMETHYL-P-SEXIPHENYL	17.21	0.088	0.21413
TETRAMETHYL-P-OCTAPHENYL (B)	32.46	0.013	9.56143
INDOLE (TRACE)	0.00	0.000	0.00000
2-PHENYL INDOLE	18.56	0.070	0.33601
1-METHYL-2-PHENYL INDOLE	14.20	0.156	0.06737
1-NEOPENTYL-2-PHENYL INDOLE	10.18	0.424	0.00918
1-N-BUTYL-2(BIPHENYLYL-4) INDOLE	26.36	0.024	2.74546
1,2-DIPHENYL INDOLE	14.56	0.145	0.07841
1-BIPHENYLYL-2-PHENYL INDOLE	18.28	0.073	0.30617
1,3-DIMETHYL-2-PHENYL INDOLE	16.12	0.107	0.14463
2,3-DIPHENYL INDOLE	24.25	0.031	1.66304
1-METHYL-2,3-DIPHENYL INDOLE	21.77	0.043	0.86993
2-PHENYL-3,2'-METHYLENE INDOLE (TRACE)	23.59	0.034	1.41530
2-PHENYL-3,2'-ETHYLENE INDOLE (TRACE)	28.79	0.019	4.65618
2-PHENYL-3,2'-TRIMETHYLENE INDOLE	21.66	0.044	0.84812
1-METHYL-3,2'-METHYLENE-2-PHENYL INDOLE (TRACE)	27.65	0.021	3.66317
1-METHYL-3,2'-ETHYLENE-2-PHENYL INDOLE	30.81	0.015	6.98925
1-METHYL-3,2'-TRIMETHYLENE-2-PHENYL INDOLE	22.03	0.042	0.93891
4,5-DIPHENYLIMIDAZOLE (A)	12.38	0.236	0.02957
PPF (TRACE)	28.97	0.018	4.84816
PPO (TRACE)	21.70	0.044	0.85854
2-(4-FLUOROPHENYL)-5-PHENYLOXAZOLE (TRACE)	20.24	0.054	0.56554
αNPO (TRACE)	34.63	0.011	14.03652
BBO (B)	41.21	0.006	39.83597
POPOP	42.09	0.006	45.15753
DIMETHYL POPOP (TRACE)	44.04	0.005	59.12180
PPD (TRACE)	0.00	0.000	0.00000
2,5-DI(2-METHYLPHENYL)OXADIAZOLE	0.00	0.000	0.00000
PBD	18.76	0.068	0.35862
BBD (B)	30.87	0.015	7.09172
αNPD (TRACE)	27.24	0.022	3.34031
βNPD	19.15	0.064	0.40567
αNND (TRACE)	31.68	0.014	8.26101
βNND (TRACE)	27.23	0.022	3.34689
TRANS-1-(4-BIPHENYLYL)-2-PHENYL-ETHYLENE	31.35	0.015	7.76405
TRANS-1-(4-BIPHENYLYL)-2-(1-NAPHTHYL)-ETHYLENE	37.30	0.009	21.90460
DIPHENYLSTILBENE (B)	41.48	0.006	41.35612
1,1,4,4-TETRAPHENYLBUTADIENE	39.71	0.007	31.79214
1,6-DIPHENYLHEXATRIENE	45.10	0.005	68.28272
1,8-DIPHENYL-1,3,5,7-OCTATETRAENE	48.12	0.004	100.13802
BPSB (TRACE)	45.66	0.005	73.54595
BIS-MSB	43.45	0.005	54.55122
ESCULIN (A)	31.44	0.014	7.86342
AZULENE	21.53	0.045	0.79620
NAPHTHALENE	0.00	0.000	0.00000
DEUTERATED NAPHTHALENE-C10D8	5.95	2.122	0.00037
1-METHYLNAPHTHALENE 5ML/L	2.80	20.306	0.00000
2-METHYL NAPHTHALENE	3.27	12.791	0.00001
2,3-DIMETHYL NAPHTHALENE	3.34	12.002	0.00001
2,6-DIMETHYL NAPHTHALENE (TRACE)	6.34	1.752	0.00054
1-NAPHTHOL (TRACE)	6.54	1.596	0.00065
2-NAPHTHOL (TRACE)	12.25	0.243	0.02783
1-NAPHTHYLAMINE	19.62	0.059	0.46837
2-CHLORONAPHTHALENE	4.38	5.316	0.00006
ACENAPHTHENE	4.03	6.825	0.00004
1-PHENYL NAPHTHALENE	6.08	1.991	0.00042
2-PHENYL NAPHTHALENE	14.10	0.160	0.05474
1,4-DIPHENYL NAPHTHALENE	15.67	0.116	0.12163
1,5-DIPHENYL NAPHTHALENE	13.29	0.191	0.04531
1,7-DIPHENYL NAPHTHALENE	13.74	0.173	0.05527
1,4,5,8-TETRAPHENYL NAPHTHALENE	33.75	0.012	11.98210

ACCEPTOR MOLECULES	RO	CO	OVERLAP
1,1'-BINAPHTHYL	9.08	0.598	0.00462
4,4'-DI(N-BUTOXY)-1,1'-BINAPHTHYL	22.04	0.042	0.93847
2,2'-BINAPHTHYL	15.07	0.131	0.09649
ACRIDINE (A)	31.63	0.014	8.13154
ANTHRACENE (TRACE)	30.06	0.016	6.00081
DEUTERATED ANTHRACENE C14D10 (TRACE)	29.63	0.017	5.50373
9-METHYL ANTHRACENE (TRACE)	30.84	0.015	6.98459
1-AMINO ANTHRACENE	30.47	0.016	6.44088
2-AMINO ANTHRACENE	29.26	0.018	5.05003
9-VINYL ANTHRACENE	32.31	0.013	9.20876
9-PHENYL ANTHRACENE (TRACE)	32.40	0.013	9.38333
9,10-DIPHENYL ANTHRACENE	34.32	0.011	13.21624
9,10-DICHLOROANTHRACENE (TRACE)	33.13	0.012	10.68805
BIANTHRANYL (B)	35.97	0.010	17.53007
9,10-DINAPHTHYLANTHRACENE (B)	33.79	0.012	12.01613
9-(P-VINYLPHENYL)-10-PHENYL ANTHRACENE	34.57	0.011	13.79457
9,10-BIS(PHENYL-ETHYNYL)ANTHRACENE (TRACE)	34.90	0.011	14.44245
RUBICENE (B)	32.95	0.013	10.22862
DIBENZ(A,H)ANTHRACENE (B)	30.05	0.016	5.97084
TETRACENE (TRACE) (B)	26.41	0.024	2.70893
RUBRENE (B)	25.37	0.027	2.13298
TETRAHYDROCANNABINOL ACID METHYL ESTER	0.00	0.000	0.00000
PHENANTHRENE	12.79	0.214	0.03583
3,4-BENZOPHENANTHRENE	17.86	0.079	0.26391
2-PHENYL PHENANTHRENE	15.41	0.122	0.10941
CHRYSENE (TRACE)	18.13	0.075	0.29019
TRIPHENYLENE	12.78	0.214	0.03570
DEUTERATED TRIPHENYLENE-C18D12	12.37	0.236	0.02938
PYRENE	26.93	0.023	3.10444
1,3,6,8,-TETRAPHENYL PYRENE	42.41	0.006	47.00038
BIPYRENYL (B)	40.91	0.007	37.93174
O-PHENYLENE-PYRENE	36.91	0.009	20.24426
ANTHANTHRENE (B)	35.54	0.010	16.19653
FLUORANTHENE	30.14	0.016	6.06245
3-PHENYL FLUORANTHENE	35.02	0.010	14.89728
3-AMINO FLUORANTHENE	33.00	0.012	10.34747
3,3'-BIFLUORANTHENYL (B)	40.21	0.007	34.07055
BENZO(GHI)FLUORANTHENE	27.96	0.020	3.85721
9,12-DI(N-OCTADECOXY)-BENZ(E)ACEPHENANTHRYLENE	35.42	0.010	15.90483
DECACYCLENE (B)	43.67	0.005	55.49643
10-ETHOXY-BENZO(KL)XANTHENE	34.94	0.010	14.71708
PERYLENE	35.23	0.010	15.35799
BENZO(GHI)PERYLENE (B)	36.62	0.009	19.50249
BBOT (TRACE)	45.68	0.005	73.51137
QUININE SULFATE (A)	23.85	0.033	1.50970
ACRIDINE YELLOW (A)	32.12	0.013	8.77985
ACRIDINE ORANGE (A)	27.13	0.022	3.18310
ACRIDINE RED (A)	27.24	0.022	3.26718
URANINE (A)	28.32	0.020	4.11985
RHODAMINE B (A)	31.19	0.015	7.36289
RHODAMINE 6G (A)	30.53	0.016	6.46888
EOSIN B (A)	27.18	0.022	3.22091
EOSIN Y (A)	28.17	0.020	3.99597
ROSE BENGAL (A)	28.44	0.019	4.22778
INDENOINDENE	20.85	0.049	0.67445

AUTHOR INDEX

Numbers in italics refer to the pages on which the complete references are listed.

G

H

I

J

K

SUBJECT INDEX

The numbers in italics refer to the page of the donor in the data tables (Chapter 8).

P

Q

R

S